1986年10月29日从青岛出发前的“极地”船

1987年1月18日南大洋考察（从左至右依次为孙洪亮、王述功、钱志宏、陈皓文、李金洪、张遴梁）

1988年9月27日夜我与二哥陈卓文（经济师）探讨论文写作

1990年3月26日在威海盐场1号虾培养池取水样

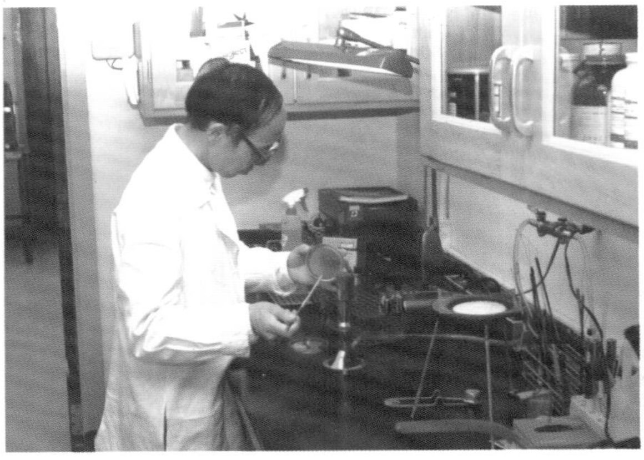

1992年3月5日在美国TA&MU Aquatic Lab作微生物学试验

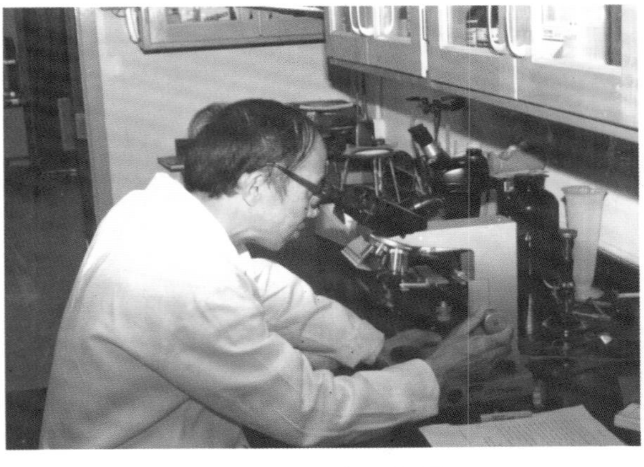

1992年3月19日在美国国家环保局Gulf-Breeze Research Lab 做显微观察

1992年3月6日Prof. D. H. Lewis 为我戴上TA&MU校徽

1992年4月在美国The College of William&Mary
的海洋科学研究所（IMS）作学术报告

1994年1月在南极半岛采集阿德莱岛地衣样品

1994年2月11日在南极乌拉圭站与俄罗斯队员合影

1994年2月11日我与乌拉圭考察工作人员合影

2010年5月9日在威斯康星大学植物园（身后为牛顿
苹果树碑及Neuton带来的苹果种长出的苹果树）
我与夫人及女儿全家合影

谨以此拙作献给我亲爱的母校

——南京大学

海洋极地空气微生物学研究

——陈皓文论文选集

Marine，Polar Regional and Air-borne Microbiological Studies：A Selection of Papers By Chen Haowen

主　编　陈皓文　孙丕喜　战　闰

中国海洋大学出版社
·青岛·

本书收辑陈皓文先生等1978～2012年间34年来在刊物、书报上发表的论著约160篇。另有新作13篇。文章内容涉及普通海洋微生物学、应用和环境微生物学、极地科学、生物活性物质、空气微生物学、海洋地微生物学、海洋微生物学等述评七个方面。书后附有陈先生等迄今所有文论编年史、参与编写/著作的书目等等。

本书可作为生命科学、海洋微生物学、水产科学、极地科学及环境科学等专业的师生及科研工作者参考。

图书在版编目(CIP)数据

海洋极地空气微生物学研究：陈皓文论文选集／陈皓文，孙丕喜，战闰主编．—青岛：中国海洋大学出版社，2012.12

ISBN 978-7-5670-0164-0

Ⅰ．①海…　Ⅱ．①陈…　②孙…　③战…　Ⅲ．①极地—海洋微生物—空气细菌—文集　Ⅳ．①Q939—53

中国版本图书馆 CIP 数据核字(2012)第282564号

出版发行　中国海洋大学出版社
社　　址　青岛市香港东路23号　　**邮政编码**　266071
出 版 人　杨立敏
网　　址　http://www.ouc-press.com
电子信箱　dengzhike@sohu.com
订购电话　0532—82032573(传真)
责任编辑　邓志科　　**电　　话**　0532—85901040
印　　制　青岛海蓝印刷有限责任公司
版　　次　2013年7月第1版
印　　次　2013年7月第1次印刷
成品尺寸　185 mm×260 mm
印　　张　61.25
字　　数　1453千字
定　　价　168.00元

序一

微生物广泛生存于地球气圈、水圈、土圈和岩石圈之中，尽管个体微小，但与人类生产生活、其他生物和非生物之间的关系十分密切，长期以来吸引了众多学者的关注和研究。尤其海洋微生物，人类至今对于它的认识还很肤浅，有学者估计迄今采用常规方法培养出来的海洋细菌，尚不到实际种群数量的1%，所以被学界聚焦为当代生命科学与产业开发研究的重点和热点之一。

陈皓文教授是一位热爱海洋微生物科学而又长期坚持不懈开展学术研究的学者。他曾与约60位同行从事合作研究达47年之久，其涉足的学术领域相当广泛，内容涵盖了普通微生物学、应用海洋微生物学、水产医学微生物学、空气微生物学、微生物生态学、极地生物学和生物活性物质等方面。现已发表的171篇论文中，对于地球三极乃至全球性微生物学研究方面，除了紧跟国际前沿外，有些方面还取得了开创性的成果。陈教授及其合作者在研究中获得的系列成果，无疑给中国和世界微生物学界留下了一笔宝贵的财富。今日他们能够以文集方式付梓，提供给产学研各界参考，将是令学界和业界同仁感到非常高兴的福音。

陈教授不仅数十年如一日全身心投入微生物学研究工作，而且在做人和做学问方面始终远离名利，表现出博学、严谨、专注、勤奋的品格。真诚扶持后学成才是他公益心的突出表现，所以深受学生们的欢迎和敬仰。更加难能可贵的是他在深入开展科研工作的同时，仍不忘抒发自己对祖国、对大自然、对友人和师长的感恩之情，并在文论中作了比较深刻的表达，足令读者肃然起敬。通过拜读《海洋极地空气微生物学研究：陈皓文论文选集》，使我有幸结识一位醇厚、执著的微生物学家，他的学者风范令我深表钦佩。在此，首先祝贺《海洋极地空气微生物学研究：陈皓文论文选集》的顺利出版，并向广大青年学者隆重推荐此书，期望以此为契机，不断扩大和传承老一辈科学家的学术精髓。

中国工程院院士 雷霁霖

2012.3.28

序二

我与陈皓文虽然不是关系密切的同行同事，虽然学术有界，但是学者和学界是互通的。我们相知相识有年数了。在迄今长达47年的科研生涯中，他主要从事海洋微生物学研究。他有强烈的环境忧患和保护意识，紧迫感强。在我国较早地开展海洋环境微生物学方面的工作，比如石油污染和耐重金属菌方面。他一直关注海洋微生物学进展，认为要使应用研究足有成效，必须深入并以坚实的基础性理论为本，为此他既应用传统方法，更重视现代的、先进的微生物学手段。陈教授的工作范围或从事领域相当广泛，足迹遍及祖国各省市区、各领海及大洋、极地；既有微生物的，也有初级生产力的；既有海洋的，也有陆地的，还有生物活性物质等等。他关于极地科学研究所取得的丰硕成果，就全国范围而言，也是不多见的。他最早地在我国开展海洋乃至极地空气微生物调研，其系列成果为阐明微生物在海—气作用中的活动及极地各种空气环境（包括人为的）中微生物状况提供了丰富的资料。而他关于全国、国外乃至环球几十个城市的空气微生物学数据及其分析为掌握特定空气环境状况及其改善提供了大量翔实的信息，迄今尚无他人可及其项背，充分体现出他行远自迩的踏实作风。他关注环境/材料的腐蚀的基础性研究，就腐蚀问题的主要因子之一——硫酸盐还原菌(SRB)发表了多达十余篇论文。这于学者对理解不同地化地质海洋环境中SRB生存状态、腐蚀成因、作用提供了重要的基础性资料，是一组真正的原创性成果，有些很有见地、具有开创性。因此他及其与合作者成果颇丰，对海洋微生物学及相关学科的发展无疑做出了不小的贡献。文如其人，陈皓文教授的大量成果表露出他胸无城府、学风正派、研究踏实，可谓为人师表。值此他的论文选集即将付梓，我高兴地祝贺他事业有成。相信他的业绩将对后来者有所启迪。他的论文选集是相关学科继往开来、提高水平的重要文献，将永载史册。

中国工程院院士 侯保荣

2012.4.6

自序一

1964年春夏之交，我跟随刘志礼老师，偕同我低等植物专门组同学，一起从富饶的长三角来到青岛，以海藻为材料做毕业实习，算是我初识海洋，实际是初识海滨。1965年夏末，我从著名学府南京大学毕业，其灵魂——诚朴雄伟，励学敦行，始终激励并敦促我努力实践。我有机会开始乘海轮从长江口北上，经历36小时的黄海之行，真正见识海洋壮阔，领略海洋魅力。从此在中国科学院海洋研究所工作了20年。从青岛海滨进入胶州湾，南下沿海、港口、码头，以至于中国各领海，期间于1978年发表了第一篇译著。1985年年底以来，一直在国家海洋局第一海洋研究所工作。迄今的47年科研活动，使我不断接触各种生境的微生物，进行包括生态、海洋(水、沉积物、空气)、水产医学、极地、环境污染/腐蚀、地质、地球化学的微生物学和生物活性物质等的研究，足迹遍及全国各地、各领海、大洋、极地和环球，航程逾4万海里。所取样本最深达海底5千米，最高海上空气样本达10千米以上。对人类最大的多维空间——海洋及其微生物有了一个初步的、粗浅的认识，获得了一些成果，即普通海洋微生物学、应用基础海洋微生物学、水产医学微生物学、极地生物学、空气微生物学、环境腐蚀的微生物学、海洋环境微生物学、海洋微生物生态学及生物活性物质研究等。本书收集了我和一些同行发表的多数文章，算是对以往的一个交待。

在工作过程中受到以孙丕喜、高爱国、徐家声、袁俊峰研究员、教授等为首的约60位老中青同事、朋友的帮助和参与，得以完成并发表相关论文，这其中凝聚了他们的辛劳、心血和智慧，我十分欣慰、十分感激、不能忘怀。

在迄今走过的人生历程和科研生涯中(从小学到大学、从中科院海洋所到海洋局一所、从国内到国外)，我得到了许多师长的教导、指点和帮助，其中铭诸肺腑的有常州化机小学的陈宗元校长夫妇，常州五中的颜衷涵校长和奚金芳老师，南京大学的郭影秋和匡亚明校长、朱浩然和曾昭琪教授，海洋所的曾呈奎所长和赵汝英处长，海洋局一所的吴宝铃副所长、蓝乾文和王文兴研究员，美国得克萨斯农工大学(TA & MU)的D・H・Lewis教授等。他们对我的人生、求学、世界观的形成、科研活动等都产生过重大作用和影响，在此再表深深的感谢。本书能顺利出版，特别要感谢海洋局一所环境影响评价中心副主任战闰博士的无私奉献和生态中心孙丕喜研究员的鼎力支持。

海洋浩瀚，学海无涯，我愿与天下的海洋微生物学工作者研究不止。

书中文章，基本保持原貌，改动不多。因篇幅有限，参考文献未被列入，但文中仍保留其标号。

本人才学疏浅，书中会出现一些谬误，恳望读者不吝指正。

2012.1

自序二

作为一个海洋学家，陈皓文教授最擅长的是海洋微生物学研究。47年来，他一直致力于此。当今微生物学发展迅猛，是许多生命科学分支与相关学科同它互为交叉和渗透十分深刻的科学。陈教授不仅关注和研究他本身执著的学科，也十分关注和研究相关的学科。他的研究范围广阔，内容深刻，有的相当领先，具有开创性。这一点从他的论文选集中可一览无余。本书论及微生物生态学、极地科学、空气微生物学、生物活性物、环境污染/腐蚀及保护、水产医学等。像陈教授这样研究之深广，据我所知，并不多见。几十年来，我等60多位人士与陈教授相识相知，共同探讨和不断研究，应该说所得成果颇丰。陈教授高瞻远瞩的眼光、真诚待人的品格、朴实无华的作风、低调谦卑的为人和矢志不渝的求实，深深地影响着我，我从他身上学到了许多优点和长处，终生受益。他经常不计个人得失而潜心扶助年轻人成才出成果，为的是海洋科学健康发展、赶超先进。在与他长期的接触中，我体会到他追求的不是名、利、权，而是真、善、美。他对科学的认真、严谨、专注和勤奋，充分显示出作为一位学者的高风亮节、真才实学和献身精神。因此他是我永远学习的榜样和敬仰的科学家。我相信这本书将是我国海洋科学的一大成果，它的出版将会进一步推动我国的相关研究。我将遂其愿、促圆满。

研究员 孙玉善

2012年2月6日

自序三

认识陈皓文教授是15年前。当时准备做一些海洋物质循环的研究。通过文献知道，在物质的循环过程中，比如磷的优先降解、各种营养物质形态的转换，微生物都起着关键的作用。而我对于微生物知之甚少，就去请教陈教授，由此认识了陈教授并感受到了陈教授严谨的治学精神和勤恳的工作态度。陈教授给我的印象是严谨、勤奋、谦虚、专注。

严谨。陈教授做学问极其认真。对于一个字、一个专业术语，也都是翻找资料细心查证。他写《来克丁学导论》，为了几个词，到我这里查阅七八本专著，仅仅为了查实几个单词译法及确切含义。

勤奋。陈教授及与他人写出了约171篇论文，从海洋到大气、从内陆到极地的微生物学研究都留下了他的印迹。记得1997年一起到深圳出海，陈教授在完成海水微生物分析工作之后，抓住机会，采集大气样品，进行深圳大气微生物的研究。

谦虚。陈教授关于地球三极乃至全球性微生物学领域的研究，有些是具有开创性的。他却没有因此自我标榜，只是说我们现在的工作主要是跟踪，也就是紧跟国际先进水平罢了。

专注。陈教授专注于学问，参与/领衔国家重大/重点项目，包括国家自然科学基金。退休后不是去"走穴"，而仍是专注研究、出版专著、整理论文，不能忘怀的还是学术研究。专注于科学研究事业是陈教授等老一辈科学家留给我们的宝贵财富，是我们年轻一代应该学习、继承和发扬的。

陈教授一生致力于微生物学的研究，执著专注、不浮不躁的治学精神，值得我学习。

敬佩陈教授学术的严谨专注、工作的孜孜不倦、处世的谦虚朴实，能赞助和编著陈教授选集及其出版，为其研究工作尽一点绵薄之力，非常高兴。

博士 战闽

2012年1月26日

目 录

第一篇　普通海洋微生物学

第二篇　应用海洋微生物学

第三篇　极地科学

第四篇 生物活性物质

第五篇 空气微生物学

第六篇　海洋地微生物学

第七篇 海洋微生物学等的研究述评

第一篇

普通海洋微生物学

GENERAL MARINE MICROBIOLOGY

海洋细菌计数和生物量测定方法述评*

摘　要　本文从生态学观点简述了海洋细菌计数和生物量测定方法近十多年来的研究与应用进展。着重述评上荧光显微镜术的应用及其与其他研究技术结合应用于细菌计数上的动态。从细菌细胞主成分—ATP、胞壁酸和脂多糖等的测定分析来估价细菌生物量生产的现状等等。

文章鼓励开发新技术研究，以适应海洋细菌生态学基础研究和应用研究的发展之需。

1. 引言

细菌计数和生物量测定是普通海洋细菌学的基本内容，也是海洋分子生物学发展的基础之一。

显微计数沿用好多年数了。但用以认识可比大小碎屑和无机颗粒中很小细菌的细胞，还相当不精确。总的微生物生物量参数如 ATP，至少反映藻类、细菌和浮游动物这三个不同营养阶的生物量，因而在研究它们之间的关系上意义有限，正在不断改进、完善的方法——上荧光显微术——高度专一的计数技术之一则能对生物容积在有别的生物时对细菌群系的生物量作可靠估计。一些基于化学测定细胞组分的间接方法正受到人们越来越大的注意。

2. 细菌计数

多数海生远洋细菌小（＜0.1 mm），而总数大（10^7～10^9/L）。由此带来两大困难，其一是难于精确估计总数，因为细菌易于逃离注意；其二是当菌数经生物容积换算为生物量时产生较大谬误，因为在由每一细胞大小所致的生物量计算中产生较大差错。因此，改变方法是势在必行的，表 1 列举最普遍使用的直接和间接方法。

表 1　天然水中几个细菌参数的测定法

参数	方法	单位	引申出的单位	摘要/限度	引　证
总数	光学显微镜	N/mL		显微镜学	Sorokin 和 Kadota，1972
总数	上荧光显微镜（EFM）	N/mL		显微镜学	Zimmermann 和 Meyer-Reil，1974；Hobble 等，1977
总数	扫描电镜	N/mL		显微镜学	Zimmermann，1977；Bowden，1977；Larsson 等，1978
总数	EFM	μm^3/mL	μg/mL	显微镜/换算系数	Meyer-Reil，1977

* 原文发表于 1985 年在昆明召开的《全国第一届微生物生态学术会议论文集》。

（续表）

参数	方法	单位	引申出的单位	摘要/限度	引　证
总数	ATP	μg/L	μgC/L	换算系数	Holm-Hansen 和 Booth，1966；Holm-Hansen 和 Parel，1972；Parel 和 Williams，1976
特定群落的活化部分	胞壁酸	μg/L	μgC/L	换算系数	Moriarty，1979
	LPS	μg/L	μgC/L	换算系数	Watson 等，1977
	生活菌数＋MPN	N/mL			Sorokin 和 Kadata，1972
	MPN，带有 ^{14}C 底物	N/mL			Ishida 和 Kadata，1979；Lehmicke 等，1979
	放射自显术	N/mL		显微镜学	Brock 和 Brark，1968；Hoppe，1976
	放射自显术＋EFM	N/mL	总数的％	显微镜学	Meyer-Reil，1978
	电子传递系统—活化细胞＋EFM	N/mL	总数的％	显微镜学	Zimmermann 等，1978
	奈啶酮酸培育＋EFM	N/mL	总数的％	显微镜学	Kogure 等，1979

2.1　上荧光显微术(epifluorescence microscopy，也可译作为表面荧光显微镜术)

自 1676 年以来，显微镜已成为微生物学家不可缺少的一种工具，后来不断有了定量步骤。如 1934 年，有了世界上第一台电镜。1940 年相差镜与电影摄影结合，可研究活体及其差异。1963 年开始，用放射自显术探讨细菌结构，等等。但在技术上和计数上仍有问题。主要是样品经染色或相差镜观察，不易辨认，还普遍有某种主观性。

用吖啶橙荧光染料染细菌，大大有助于天然样品中的细菌技术，甚至在有颗粒时(Erancisco 等，1973)，用聚碳酸盐(Polycarbonate)滤膜可计数小细胞(＜0.5 mm)，并可按细胞大小的级别区分、估计生物容积。这比纤维素酯滤膜好。因为聚碳酸盐滤膜有平坦的表面，孔径一致，再加上荧光染料的滞留力，能保留所有细菌在滤膜上(Zimmermann 和 Meyer-Reil，1974；Hobbie 等，1977)。对荧光下的色泽有新的解释。用前，以 irgalan 黑染色以消除自身荧光，所得菌数大约是纤维素膜所得的两倍。

对计数的条件较严；细菌形态轮廓明确。发荧光亮橙或绿色。以前错认发荧光绿者为活菌，发荧光橙者为死菌(Pugsley 和 Evison，1974)。实际是胞内 RNA 相关量和双股、单股 DNA 的反映。生长快的细胞含较多 RNA 和单股 DNA，它吸收更多的染料，因而发荧光橙色(Dalay 和 Hobbie，1975；Hobbie，1976)。而且颜色大大决定于样品的处理和制备。由荧光橙变绿，可用乙醇基或海水彻底洗涤染色细胞而获得。

天然样品中的微生物细胞，其形态和丰度，即便在短的时间尺度上也因地点和季节而变。用上荧光显微术可研究它们。也容易测定自由漂浮菌与附着菌的比率。用今天的显微计数术研究表明。以前的研究漏看了许多很小的自由生活细胞。许多细胞并不附于颗粒(Ferguson 和 Palumbo，1979；Hobbibaugh 等，1980)。上荧光显微术也用来测天然条件下的细菌生长速率。分裂着的细胞，其百分率必定用显微镜来确定。用于测河口和沿岸水细菌面、体积比 SEM 更精确，但 SEM 比荧光镜和透射镜可提供颗粒等表面的微生物真切所在。

上荧光显微术已与其他计数结合应用了，如呼吸细胞数的测定。这是基于以下观察：好氧细胞能还原人工电子受体，如四唑盐为不溶态的甲瓒，并以颗粒态存于细胞内，这可用光学镜看到。但这在含有许多小细胞的样品，如远岸水样品中，做起来是困难的。

本法的优点是不必培养，用上荧光制备样品直接计算总细胞数及呼吸着的细胞数。

Kogure 等(1979)用酵母膏(0.025%)和奈啶酮酸(0.002%)培养样品 6 小时，所用奈啶酮酸是 DNA 合成的专性抑制者，并阻止 G-细菌细胞分裂，因而活化拉长细胞。在显微镜下易辨。细胞总数和拉长的细胞总数用上荧光显微镜计。得出的生活细胞只是开阔大洋水中总数的 5%～10%，但比洋菜平板上得出的高 1 000 倍。而且表明其多数细胞是自由生活者。杆形的约占 70%。上述方法被广泛应用着(Clavier-Rault，1981；Peele 和 Colwell，1981)。

上荧光显微术与放射自显术相结合。用来测定细菌群系中活化部分对^{3}H或^{14}C标记的有机底物的摄取(Hoppe，1974，1976～1977)。将样品的放射自显影照片(ARG)与传统染料、荧光抗体或荧光染料相结合，使得了解天然生境中一个或几个种的活动和群体动力学成为可能。

免疫荧光显微、吖啶橙直接计数，再与培养法结合可调查、分辨一些不可培养的细菌。

双重过滤术用于区分氚和其他一发射体。

微放射自显术的潜在应用，在生态学研究中尚需大力探讨。

2.2　活菌数技术

测细菌数的经典方法是在洋菜平板上测活细胞的菌落形成单位(CFU)。该法只测到总数的 0.0001%～10%，仅能代表需高营养浓度以生长的腐生细菌，但该法仍广泛应用着(如 Gunn 等，1982)。好处是可测水体卫生状况，能计算经适当稀释的样品的各类菌落数(Horie 等，1981)。此外还有 MPN 法也广为人们所用。

造成总菌数和活菌数间差异达几个量级的主要因素有八条：

(1)天然生境中某些细胞是不活化的。

(2)细胞在所用底物及底物的结合上不生长。

(3)从原则上讲，所用底物合适，但浓度过高，反而加速细菌死亡或过低(常不如此)；由于在固一气相界面的有效氧太多或表面张力太大；细菌不可能生长。

(4)在培养期间，细菌生长成可见菌落太慢，因而难以观察到。

(5)受近亲(近邻)及子代等影响而使某些细胞不活化，因而只有少数细胞形成菌落；其余，不易见到。

(6)采样、溶化洋菜等条件、温度、压力等的影响，使细胞不活化。

(7)细菌的聚集倾向使得一个可见菌落内可能有多于一个的细胞。

(8)细胞还有附壁倾向,使它们从样品中去掉。

Sieburth 等(1979)和 Väätänen(1977)等都对此作过评论,提出过最佳方案,如 Lehmicke 等(1979)在试管液体培养基中加微量标记有机底物,试验产出$^{14}CO_2$,以计算天然水中活跃降解浓度为 g/L 的体外活性物质的那些细菌。

3. 生物量测定

生物量的微观测定需要在细胞数换算为生物容积乃至生物量时用上换算因数。扫描电镜(SEM)可正确测细胞面积。而上荧光显微镜(EFM)比 SEM 则更精确测细胞面积和体积。在 SEM 的样品制备中,细胞缩小了。据报道,河口和沿岸水平均生物容积相当符合:0.05~0.09 μm^3。但 Watson 等(1977)从伍兹霍尔沿岸水中观察到一周内有从 0.28 到0.090 μm^3的减少。对最小的类型,在单群落中的平均生物容积变幅是 0.01 μm^3 或更小。而长杆和丝状细菌细胞。则超过 0.5 μm^3。天然环境中细菌比培养的细菌小不少。从生物容积到生物量的换算中所推荐的值是 $82\times10^{-15}\sim121\times10^{-15}$ g C/μm^3。Romaneko 和 Dobrynim(1978)和 Van Veen 及 Paul(1979)在生物容积到生物量的换算因数中讨论了所用的细菌细胞相对密度值,得到的平均值分别为 1.6×10^{-12}和 0.8×10^{-12} g/μm^3。以连续过滤和电镜观察,发现了迄今最小的海洋原核生物其容积为 0.011 μm^3(Mishustina 等 1981)。

测定细胞的某一组份以换算出生物量。

对收藏样品作显微镜学的研究,其问题是:需较多时间,对结果的观察及解释常带主观性,在将细菌数变为生物量时需测生物容积,数据由 μm^3 转为 μg 时要有一个因数。改进的方法包括基于测定细胞某一组份的浓度及以往测定该组分和生物量间的关系上。Holm-Hamsen(1973)规定了符合充要条件的这种细胞组分,即所测化合物必须:

(1)存在于所有活细胞中,而死细胞无;

(2)不与非生命的、有害物质相关联;

(3)十分均匀地存于细胞中,又不受环境任何压抑;

(4)所用分析技术能精确到亚微克量,材料要较快和易于在有各种样品的大尺度计划中使用。

对所测组分应无一切死细胞的要求是研究群体动力学所需,但食物网研究所需的是不甚完整的死细胞,这是活细胞的有效养料。至于上述第 4 条,应尽可能达到。

3.1 ATP

ATP 满足上述所有各条要求,Karl(1980)对此作了评论。由 ATP 到细胞碳的换算因数因生理条件差异和生物差异而变化相当大,但就整个群落的 C/ATP 言,250,这一因数是可接受的近似值。尽管胞外的 ATP 量有时甚大,腺嘌呤核苷酸(AMP+ADP+ATP)的总量仍是相当稳定的生物量参数。

当有浮游植物时,测细菌群系的 ATP 往往有甚大困难,因而 ATP 浓度(或腺嘌呤核苷酸)在细菌生物量上只给出了一些信息,但对总的微生物生物量(藻类+细菌+微浮游动物)来说却仍有用。

3.2 细菌细胞壁成分

细菌生物量测定上另有两个化学参数引人注目:细菌所特有的胞壁酸和脂多糖(lipo

Poly Saccharide,缩写为 LPS)。胞壁酸是 G-菌细胞壁的一个主成分,也存在于 G-细菌的肽聚糖层中,LPS 在 G-菌中,与外层膜广泛相连,两者都在细胞外层,在将它们换算为生物量前,必须假设或微生物学的测出细胞的平均大小,因为这是多变的。如沉积物中的细胞一般比其上覆水中的大,沿岸水中的比远岸水中的大。Matin 和 Veldkamp(1978)等报导过表面一体积比因细菌生理状态而变。Novitsky 和 Morita(1976)发现一株嗜冷海洋弧菌在饥饿时,其形态由杆状变为球形的变化。另外,天然的沿岸海水样品,其细胞的平均大小有季节变化(Meyer-Reil,1977),这些意味着从 LPS 或胞壁酸到生物量的换算因数将需改变。

海水中多数原核生物是革兰氏阴性的(G^-),因而 LPS 的浓度几乎可反映总细菌群系状况。Sullivan 等(1976)、Ulitzem 等(1979)、Maeda 和 Taga(1979)等对此曾提出几个方法。Watson 等(1977)以马蹄蟹变形细胞溶胞产物(LAL)作试验,对海水样品 LPS 定量,G^-菌(含 LPS)的生物量与样品的 LPS 量相关,再将 LPS 换算为细菌碳,因而,LPS 试验是海水细菌生物量的一个准绳。他们还发现,在 188 个海水(菌)样品中,LPS 和上荧光的细菌数间有甚好的相关性。大肠埃希氏菌的 LPS 对细胞碳的比率是 6.35,海水样品是 5.8～7.8[碳的估计是基于吖啶橙荧光直接计数测定(AODC)]。

Moriarty(1977)由 G^+ 菌与 G^-菌数的相对比例提出了一个换算因数,即胞壁酸量对生物量碳的因数为 100。显然,海洋细菌的 LPS 量大约是胞壁酸量的 15 倍。测 LPS 容易,快速而灵敏,因而更适于迅速测生物量。特别是在将生物量的相对值换算为绝对值,其换算因数及其常数需单独测定时。近年来,Moriarty(1983)在胞壁酸测定上有进展。

4. 结语和展望

如上所述,各国学者用吖啶橙染色,再用上荧光显微术对不同海区的细菌做过大量估计工作。

潮水、昼夜变化及季节变化虽有规律,但并不是不变的。其中一个主要趋势是:细菌数从河口内($>50\times10^8$ 细胞/L)到潮水入口处和沿岸区[$(10\sim50)\times10^8$/L],大约降了两个量级,远至透光的远岸和大洋水[$(0.5\sim10)\times10^8$/L]、深水环境有一个大致随深度而变的梯度:表层$(0.5\sim10)\times10^8$、无光区$<0.1\times10^8$。中间层和近底层时有高数量,甚而在最贫瘠水团,每毫升水仍含几千个细菌。

上荧光和其他光学显微计数术间的差别可用相差镜结果来说明。Wiebe 和 Pomeroy(1972)发现大洋透光层的水,其 40 个样品中不超过 10^6/L 的密度,而佐治亚沿岸陆架中,其 42 个样品中是 $0.01\times10^8\sim1\times10^8$/L。Sorokin(1978)总结:藻红染色应用亮视野观察的大量资料表明,污染河口为 $20\times10^9\sim100\times10^8$/L,寡营养海水和淡水环境表水为 $0.4\times10^8\sim2\times10^8$/L,这与一般估计相近。

上荧光显微术从天然环境样品一般产生最可信细菌数和生物容积数据,但本法测生物容积花时间多,而细菌生物量方面的数据较少。生物量值,远岸表水是 1～5 μg C/L,富营养河口内测达 200 μg C/L。Sorokim(1978)资料是:污染河口和环礁表水是 10～100 μg C/L,寡营养表水是 1～5 μg C/L。从河口到沿海乃至远洋环境间存在生物量梯度与细菌细胞大小有关的状况。

LPS 试验是海水细菌生物量的一个很好测试法。

将直接计数细菌法包括显微术、流动自动分析与间接测试技术（如平板计数、MPN 法等）结合，再加上分子技术，可获得更多正确（包括细菌数量、大小、生理状态、生物量、基因型结构等）的资料，从而使我们进一步理解世界大洋中细菌所起的作用。

建立在工艺技术和哲学假说密切相关支柱上的微生物学尚需技术的发展和结合，以适应研究海洋细菌生态之需。

5. 后记

海洋细菌是海洋环境矿化者和生物量的部分生产者，但有关海洋微生物生态系的全新概念，当时资料不多，主要受制于仪器。当理化和地质方面的海洋仪器进入太空时代时，抓紧研究和提高生态学方法是至关重要的。野外测试的时空尺度要控制，注意力多集中于整个细菌群落，以解决海洋微生物生态学最紧迫问题。要重视引进新技术。扩大研究领域，提高研究水平。

本文写就于 20 世纪 80 年代前期，世界海洋国家相关学者一直不遗余力地进行着这方面的研究、开发。新方法、新技术不断深广，使人们的观点、视野不断更新，海洋微生物学的发展将步入更高阶段。历史难以复制，历史不应忘却。

参考文献 52 篇（略）

鉴定海洋细菌的分类图解*

提出一份新编的鉴定海洋细菌的分类图解。该图解是以有关海洋细菌大量的评论及专著以及伯吉氏的细菌学鉴定手册为基础的，涉及种类较广，并试图以较少的试验来鉴定海洋细菌。

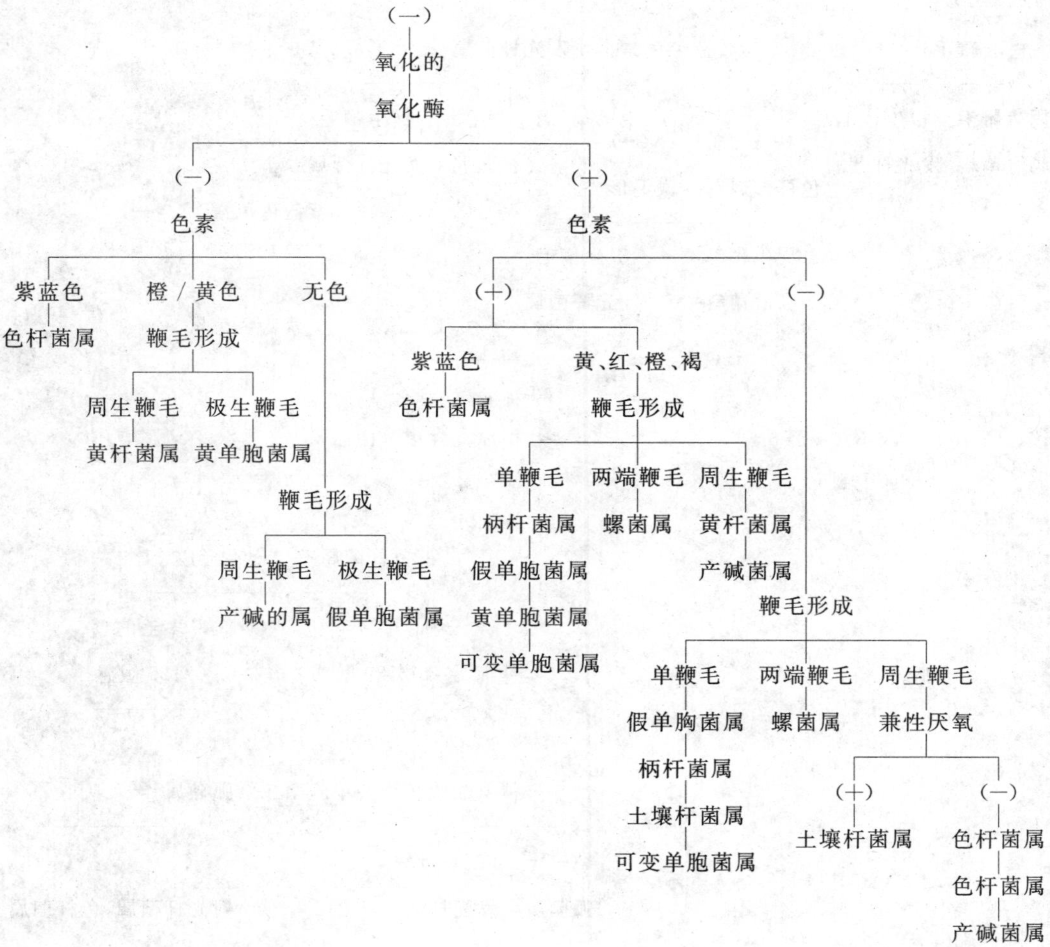

* 原文刊于《海洋译丛》，1985，No. 1：177-179.

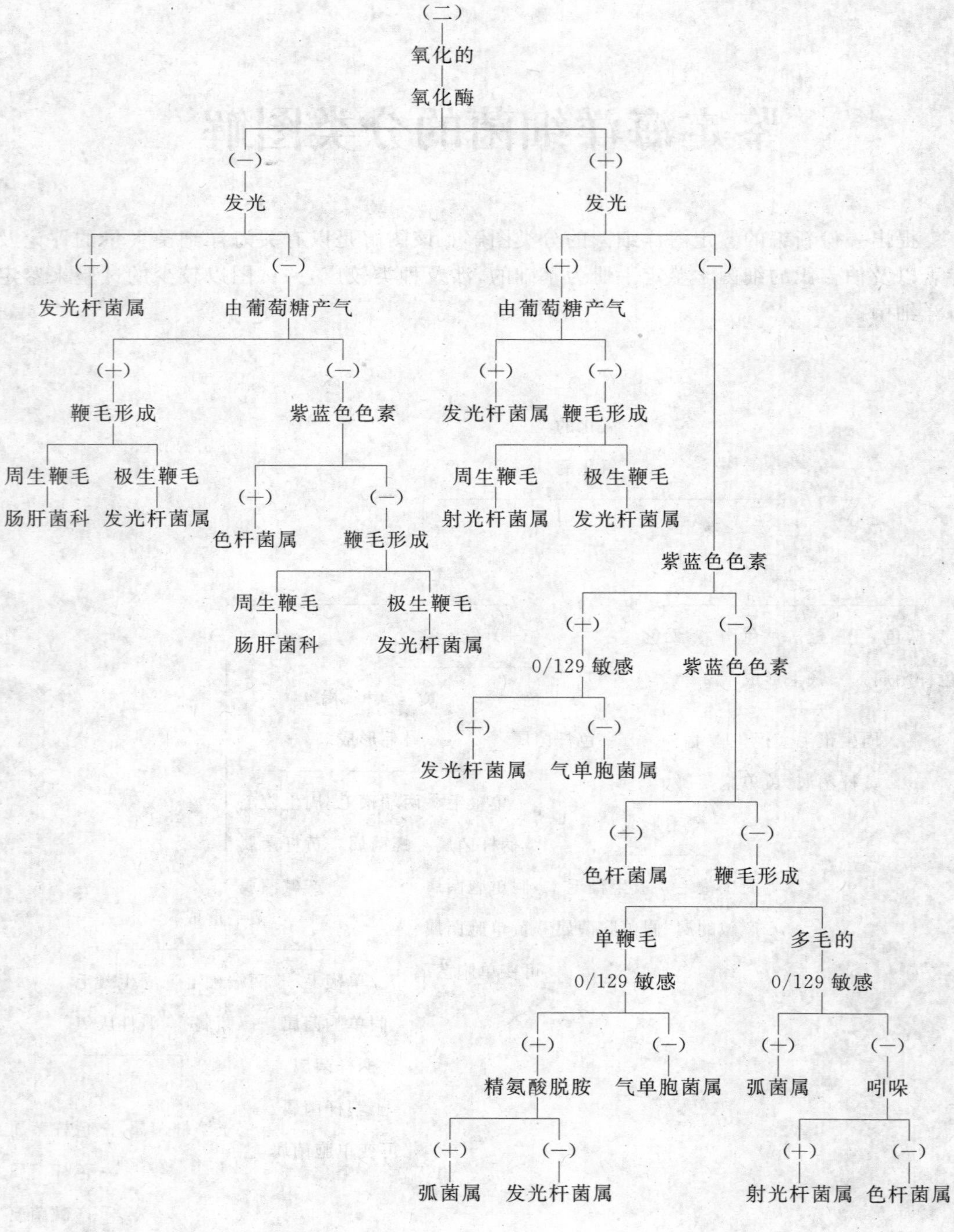

图 1　鉴定能动的革兰氏阴性杆菌的检索表(异化葡萄糖下分四个平行组)

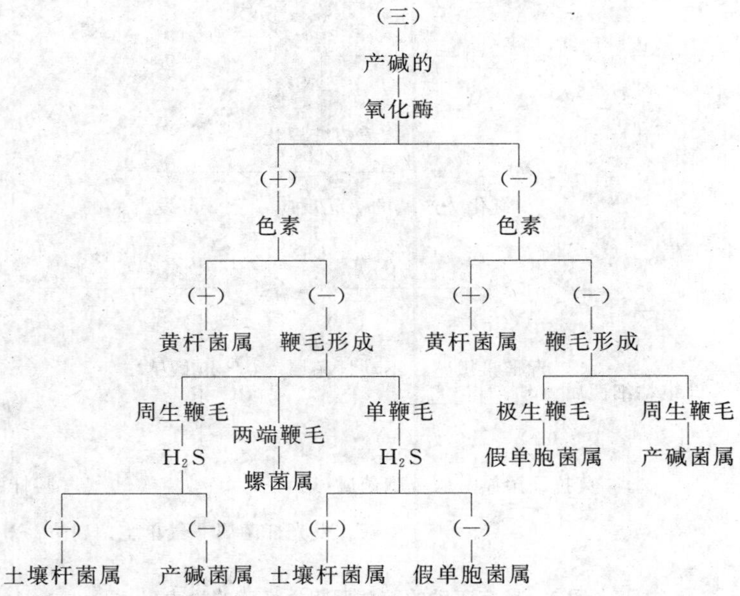
（三）
产碱的
氧化酶
（＋）
（－）
色素
色素
（＋）
（－）
（＋）
（－）
黄杆菌属
鞭毛形成
黄杆菌属
鞭毛形成
周生鞭毛
两端鞭毛
单鞭毛
极生鞭毛
周生鞭毛
H_2S
螺菌属
H_2S
假单胞菌属
产碱菌属
（＋）
（－）
（＋）
（－）
土壤杆菌属
产碱菌属
土壤杆菌属
假单胞菌属

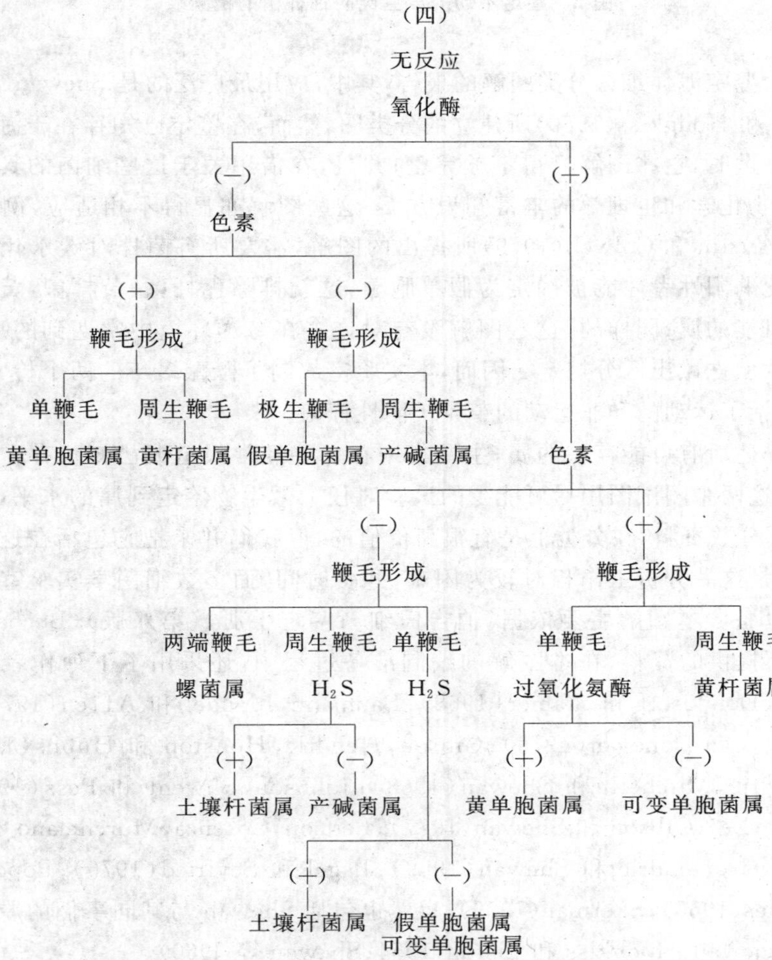
（四）
无反应
氧化酶
（－）
（＋）
色素
（＋）
（－）
鞭毛形成
鞭毛形成
单鞭毛
周生鞭毛
极生鞭毛
周生鞭毛
黄单胞菌属
黄杆菌属
假单胞菌属
产碱菌属
色素
（－）
（＋）
鞭毛形成
鞭毛形成
两端鞭毛
周生鞭毛
单鞭毛
单鞭毛
周生鞭毛
螺菌属
H_2S
H_2S
过氧化氨酶
黄杆菌属
（＋）
（－）
（＋）
（－）
土壤杆菌属
产碱菌属
黄单胞菌属
可变单胞菌属
（＋）
（－）
土壤杆菌属
假单胞菌属
可变单胞菌属

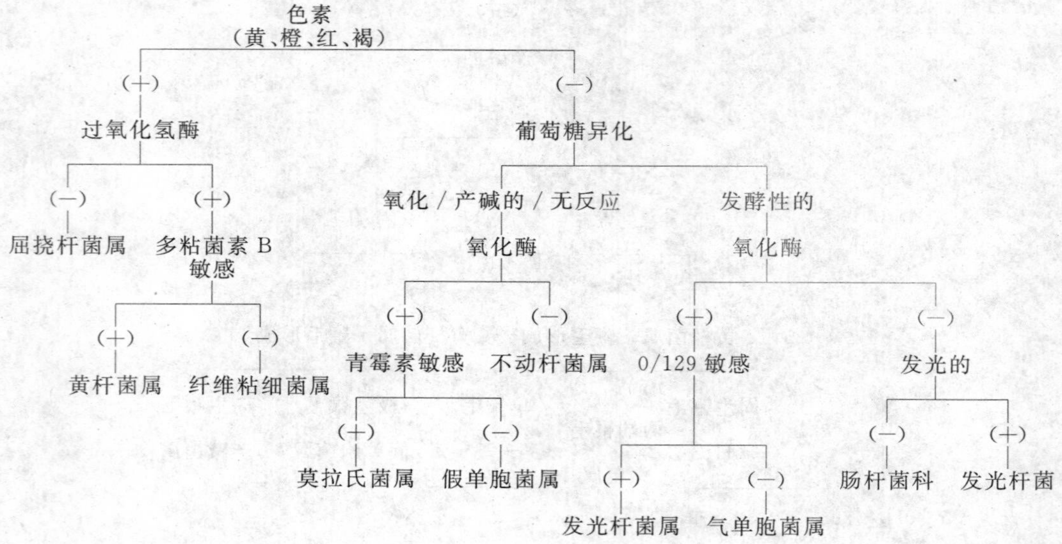

图 2　鉴定不动的革兰氏阴性杆菌的检索表

已有几个鉴定海洋细菌分类图解的报告，其中，应用最广泛的是 Shewan(1963)及 Shewan，Hobbs 和 Hodgkiss(1960)所建立的分类图，然而，在鉴定已知存在于海洋环境中的种类繁多的细菌时，这些图解显得不够完整；并且，在借用关于这些细菌的具有特殊意义的形态学、生物化学和生理学的丰富知识方面，这些图解亦显得不相适应，例如，按 Sochard，Wilson，Austin 和 Colwell(1979)所提出的图解，需要所有菌株对碳水化合物的氧化反应都是氧化酶阳性者才能被确定为假单胞菌，这就排除色杆菌、黄杆菌，黄单胞菌和产碱杆菌一些可能的属；同样的，这些图解没有对许多在海洋环境中常遇到的那些不动的，有发酵作用的菌株做出区分规定。因而，本实验室编制了酝酿多年的两个检索流程图(图 1、2)，其中包括了运动的和非运动的革兰氏阴性杆菌。

图中未描述所有可能存在的属，主要集中在那些经常报道或已知仅存在于海洋环境中的菌属上，这两张图试图用尽可能少的试验对任一微生物鉴定到属的水平，至于属的 G+C 克分子百分数资料，在分类上是有很高价值的，但我们并未把它包括在这两个流程图解中，其原因是这一分析工作相对较为困难，并费时间，而多数细菌学实验室缺乏这类实验所必需的设备。本图解主要依据《伯吉氏细菌鉴定手册》(第八版)(Buchanan 和 Gibbons，1974)所提供的资料，在本图解的编制准备过程中，还采用了下列作者所报道的资料：Baumann，Doudoroff 和 Stanier(1968)；Baumann，Mandel 和 A11en(1972)；Co1well(1969)；Furniss 和 Donovan(1976)；Gibson，Hendrie，Houston 和 Hubbs(1978)；Hayes(1963)；Hendrie，Mitchetll 和 Shewan(1968)；Johnson，Shwent 和 Pess(1968)；Lee 和 Pfeifer(1975)；Lee，Gibson 和 Shewan(1977)；Leifson，Cosenza，Murchelano 和 Cleverdon(1964)；Mitchell，Hendrie 和 Shewan(1969)；Pagel 和 Seyfried(1976)；Popoff 和 Veron(1976)；Rhodes(1965)；Skerman(1959)，自然也包括 Shewan 与其同事们的原始图解(Shewan，1963；Shewan，Hodgkiss 和 Liston，1954；Shewan 等，l960)。

读者可随同参考运用这两个图解的应用实例，即相应的论文(Oliver 和 Smith，1982)。
参考文献 22 篇(略)。

译自《Deep-Sea Research》1982. 6，Part A，Vol. 29，No. 6 A，P. 795-798

MICROBES IN THE SURFACE WATER OF THE SHOALING AREAS OF THE ANTARCTIC KRILL AND THEIR ECOLOGICAL ANALYSIS *

Abstract Analyses and studies were made of the surface water microbes in the shoaling areas of Antarctic Krill and their correlative factors at 25 stations in the South Shetland Islands of Antarctica in January and February, 1987.

Plate culture results showed that the quantity of the microbes was about 21.15.10^3 CFU · L^{-1}; the variations among the stations were relatively large; the bacteria constituted the majority, among which *Pseudomonas* and *Acinetobacter* were abundant. These strains of microbes were characterized by their being psychrophilic and psychrotolerant.

The quantity of microbes has a close relation to water temperature, nutrition of various forms of nitrogen and the Antarctic Krill, and the relationship to the water density, and to the saltmity are inconsistent. The variations in the quantitative distribution of microbes may mean the variations in hydrology, chemistry and biology.

The occurrence of small quantities of Enterobacteriaceae and Bacillus may hint that human activities and foreign intrusion have made an impact on the cold Antarctic waters.

Ⅰ. INTRODUCTION

Up to now different methods have been used to estimate the quantity and biomass of the microbes in the Southern Ocean[1-4]. They are of significance in understanding the true features of the microbes in the Southern Ocean and they are still far from being sufficient. The mineralization of organic matter, and the main process of regeneration and cycling in the Southern Ocean depend to a great extent on the metabolic activities of bacteria and other microbes[5,6], and the microbial colonies affect other members of the ecosystems (in varying degrees). Thus microbes constitute an important component of the food chain of the Southern Ocean[5]. By means of microbial activities, Antarctic Krill which live mainly on diatoms and microbes can then survive and multiply there fore, it is of sig-

* The original text was published in *Collected Oceanic Works*, 1990, 13(2):85-99.
原中文稿载于《中国第三次南大洋考察暨环球重力调查报告》1993,93-103.

nificance to study the relationship between microbes and Krill. In the various activities of Antarctic Krill, including those in the course of shoaling, fundamentally we know nothing about the quantity and population of the microbes and the interrelationship between eco-environmental factors, Krill and microbes. This paper deals with the microbial status and their relation to the eco-environmental factors, in the hope that this will be of help to elucidate the shoaling of Kill with more scientific data and to direct catching activities.

In the period of January to February, 1987, the Chinese Research Vessel "Polar" made three cruises and in-situ analyses were made of the surface water microbes in 25 stations. The results are summarized as follow.

Ⅱ. MATERIALS AND METHODS

1. Culture media

(1)Peptone, Yeast culture medium

Peptone, 5 g; Yeast extract, 1 g; $FePO_4$, 0.01 g; Agar, 18 g; aged Southern Ocean seawater, 1 000 mL; pH, 8.0; used for marine bacteria[7].

(2)PGY culture medium: peptone, 5 g; glucose, 2 g; yeast extract, 1 g; agar, 18 g; aged Southern Ocean seawater, 100 mL; pH 5.5. After sterilization and before apportioning add 30 mg chloromycetin and streptomycin respectively to each 1 000 mL culture medium. Used for marine fungi or yeast.

2. Sampling

Whenever the research vessel reached the established sampling station for microbes, surface water samples were collected immediately with pre-sterilized J-Z strike-open water samplers. The water samples were quickly sent back from the deck to the microbiological lab. The chemical, hydrological and other biological data of the various stations were collected and analyzed by the Southern Ocean Expedition. The localities of the sampling stations and routes are shown in Fig. 1.

3. Counting of microbes and experimentation

(1)Owing to the relatively low quantity of microbes in the Southern Ocean samples were not diluted generally but 0.1 mL of original samples were pipetted directly, emptied into plate culture media and smeared evenly. Then invert them in thermostats at 2℃ and 14℃ and culture for 14 days. The conditions of the microbial growth in the plates were frequently observed. Finally, the CFU (i. e. the colony forming units per liter of water samples, the followings are the same) in unit volume of water were calculated and analyzed according to the length time of growth and the differences between the forms of microbe forms and colour.

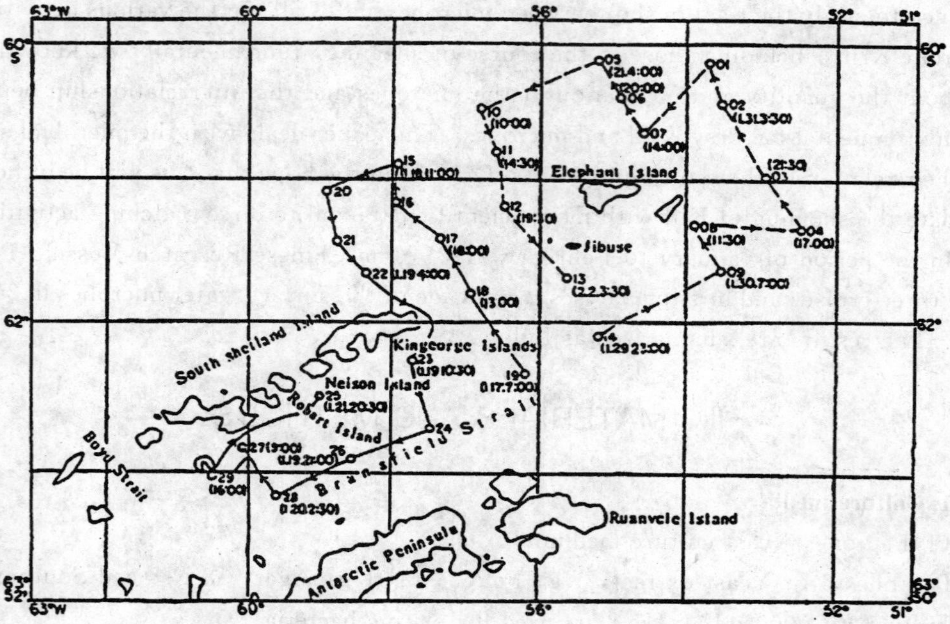

Fig. 1 Location of stations for sampling surface water microbes and routes in the Southern Ocean (the South Shetland Islands areas)

Single representative colonies were picked out, purified, preserved in slant culture medium and brought back home for temperature experiment and classification.

(2) Classification of microbes: For marine bacteria, fungi and yeasts reference is made to the documents[8-12] for identification to the level of genus.

Ⅲ. RESULTS AND DISCUSSION

1. Quantity of microbes and their distributional status

Table 1 lists the quantity of the microbes in the various stations. From Table 1, it can be seen that microbes were detected at every station, i. e., the detection rate of microbes reached 100%. With regard to bacteria, its detection rate at the station was 76%. Bacteria were detected only at stations 2, 3, 10, 11 and 27, but not at stations 4, 14, 17, 18, 19 and 23. The largest quantity of bacteria was detected at station 11 (station 29 will be discussed elsewhere). The detection rate of yeasts was quite low, only 16%. They occurred only at four stations and often appeared together with bacteria and fungi at the same stations (see Fig. 2).

Table 1　Statistics of the Quantity of Microbes in the Surface Water of the South Shetland Islands in the Southern Ocean*

Station	Bacteria	Fungi	Yeasts	Total quantity of microbes
1	16.0	0	28.0	44.0
2	3.0	0	0	3.0
3	125.0	0	0	125.0
4	0	3.0	0	3.0
5	71.5	2.0	0	73.5
6	3.0	3.5	8.0	14.5
7	5.0	1.0	1.0	7.0
8	6.5	1.0	0	7.5
9	1.0	3.0	0	4.0
10	6.5	0	0	6.5
11	293.0	1.0	0	294.0
12	94.5	4.0	0	98.5
13	150.0	1.0	0	151.0
14	0	10.0	0	10.0
15	102.5	1.0	0	103.5
17	0	3.0	0	3.0
18	0	2.5	0	2.5
19	0	20.0	0	20.0
22	1.0	72.5	0	73.5
23	0	15.25	0	15.25
25	1.0	5.0	1.0	7.0
26	35.0	3.0	0	38.0
27	6.2	0	0	6.0
28	2.0	2.0	0	4.0

* The quantity of microbes of the various microbes indicates the number of colony forming units in unit volume, i. e. , $CFU10^3 \cdot L^{-1}$.

Table 2　Statistics on Four Parameters of the Surface Water Microbes in the Waters of the South Shetland Islands in the Southern Ocean

Items	Bacteria	Fungi	Yeasts	Total quantity of microbes
Geometric mean	7.83	2.53	1.24	21.15
Frequency distribution	0～35.0	0～6	1～8	2.5－20
Cumulative frequency	0.72	0.84	0.75	0.60
Range	550.5	71.5	27	550

* The unit for each of the items except cumulative frequency is $CFU10^3 L^{-1}$, and statistics was on the number of bacteria from 25 stations.

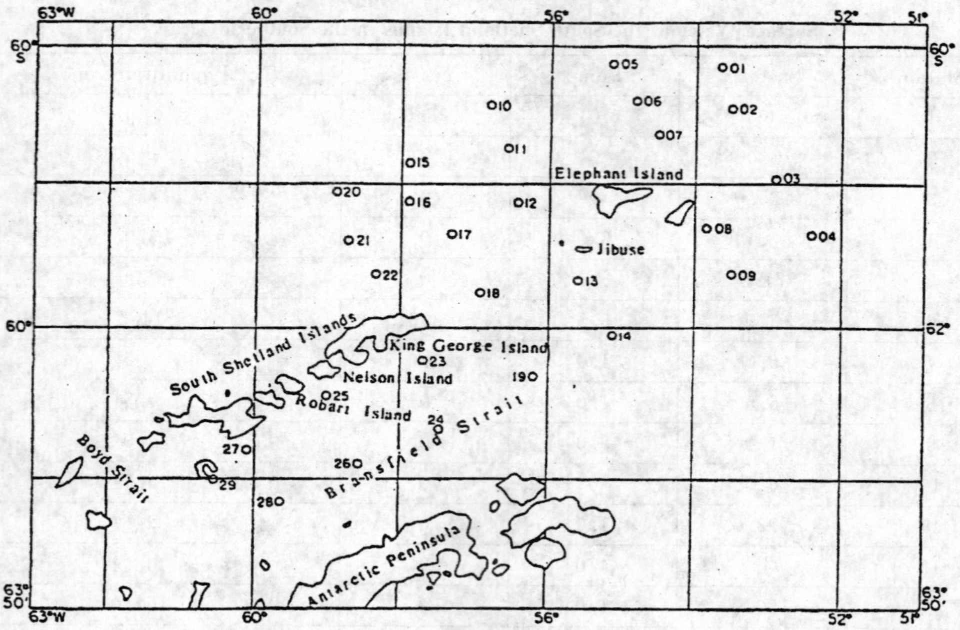

Fig. 2　Diagram showing the detection of various kinds of surface water microbes at the stations in the Southern Ocean (the South Shetland Island area)

Table 2 lists the frequency distribution and cumulative frequency of the various kinds of microbes.

From Table 2, it can be seen that the geometric mean of the number of bacteria is 7.83×10^3 CFU · L^{-1}, The cumulative frequency of the frequency distribution at the bacteria stations where the bacteria number was $(0\sim35)\times10^3$ CFU · L^{-1} reached 0.72. The variation of the quantity of bacteria between the stations was quite large. The geometric mean of the number of fungi is 2.52×10^3 CFU · L^{-1}, the cumulative frequency of the frequency distribution at the station having $(0\sim6)\times10^3$ CFU · L^{-1} reached 0.84, the variations of the quantity of fungi between the stations were the greatest. The geometric mean of the quantity of fungi for four stations reached 1.24×10^3 CFU · L^{-1}, with the frequency distribution mostly being at the value of $(1\sim8)\times10^3$ CFU · L^{-1}, the cumulative frequency reached 0.75 and the range was the smallest. The yeasts may be distributed in low quantities and with small variations in the Southern Ocean.

To sum up, there were quite a few stations where the number of microbes in the surface water was about 21.25×10^3 CFU · L^{-1}, its cumulative frequency reached 0.60, and the variation between stations was quite large (see Table 2). The number of bacteria which occupied a significant proportion of the total number of the microbes was mostly at stations 11 and 13 (except station 29). A high quantity of the microbes occurring in all the stations was mainly caused by bacteria. Hence the action and position of the bacteria in the water bodies of the Southern Ocean are known.

The fact that plate smear culture can culture such a large number of various kinds of microbes proves that the Southern Ocean abounds in microbial biomass[2,14], and which still plays an important role in the food web (and chain) and its productivity in the southern Ocean[5,15,16].

(2) Analysis and comparison between the quantity of the microbes and eco-environmental factors.

Table 3 lists the physical, chemical and biological parameters related to the quantity of the microbes (made by the Southern Ocean Exploration Group of the Third Chinese National Antarctic Research Expedition).

Table 4 is made by correlation analysis of certain parameters in Table 3 with respect to the quantity of the microbes.

Table 4 shows CFU is positively correlated with temperature (T/℃) as well as also with different forms of nitrogen, but is negatively correlated with σ_t, S, Si, Cl, pH respectively.

Even though the period of this investigation was in the summer of the South Pole, all the temperatures of the surface water fluctuated within a narrow range of 2℃ (i. e., −0.371～1.577℃). In this environment, a group of microbes adaptable to this temperature was brought up, in other wards, the microbes here have each an adaptability to low temperature and when nutrient is not a limiting factor or in low-nutrient waters suitable for the microbes[3], a slight increase in temperature will probably speed up the multiplication of microbes. Conversely, a decrease in temperature will probably lower the metabolic rate of microbes. As mentioned above, the quantity of the microbes was the largest at station 29 which is located in Deception Island. According to records and actual observation, there is the heat of the terrestrial heat formed by the volcanic island here, the water temperature is higher than that of the surrounding waters, the sea is calm, and there has been the highest record of primary production. From this it can be seen that an increase of temperature will speed up the multiplication of microbes. It may also be supposed that a sudden excess of water temperature will inhibit the activities or even the death of the microbes of the Southern Ocean[6,10,15].

Table 3　The Physical, Chemical, and Biological Parameters of the Surface Water of the South Shetland Islands Waters of the Southern Ocean

Station No.	T (℃)	S	Cl (×10^{-2})	O_2 (%)	pH	SiO_3−Si (μmol·L^{-1})	σ_t	Total N	NO_2−N (μmol·L^{-1})	NO_3^-−N (μmol·L^{-1})	N/P	E. superba (indiveduals·m^{-3})
1	0.652	34.059	18.84	64.41	8.10	35.91	27.31	19.91	0.17	19.27	10.76	17
2	0.662	34.112	18.88	97.79	8.10	32.35	27.35	12.42	0.14	11.92	6.57	212
3	0.023	34.088	18.87	76.63	8.05	27.99	27.37	14.98	0.21	14.51	7.60	293
4	−0.371	34.332	19.01	76.05	8.06	34.33	27.58	15.42	0.13	14.63	7.87	902

(续表)

Station No.	T (℃)	S	Cl (×10⁻²)	O_2 (%)	pH	SiO_3-Si (μmol·L⁻¹)	σ_t	Total N	NO_2-N (μmol·L⁻¹)	NO_3^--N (μmol·L⁻¹)	N/P	E. superba (indivéduals·m⁻³)
5	0.613	33.471	18.53	68.18	8.17	22.83	26.77	18.00	0.18	16.64	0.29	38
6	0.054	33.450	18.52	83.63	8.13	24.42	26.79	14.18	0.18	13.43	8.20	6536
7	0.750	33.779	18.70	98.31	8.14	28.78	27.17	13.17	0.17	12.26	7.28	293
8	−0.341	34.146	18.90	83.88	8.10	34.33	27.47	13.81	0.16	13.39	7.46	29
9	0.036	34.090	18.87	88.88	8.07	26.40	27.49	13.67	0.13	11.43	7.50	695
10	1.468	33.388	18.49	101.78	8.11	31.95	26.91	16.01	0.22	15.37	9.88	11
11	1.571	33.451	18.52	89.34	8.12	16.09	26.72	16.60	0.27	15.47	11.10	70
12	0.760	33.461	18.52	78.83	8.13	40.67	26.80	22.41	0.17	21.90	16.36	/
13	0.741	34.058	18.86	61.82	8.21	34.33	27.37	15.86	0.17	14.83	10.86	980
14	0.977	34.138	18.90	74.41	8.27	38.29	27.33	10.53	0.15	9.64	7.21	4
15	0.868	33.470	18.53	106.60	8.12	41.46	26.74	19.44	0.19	18.88	10.34	17
17	0.495	33.749	18.97	106.72	8.20	52.17	27.09	15.39	0.19	14.16	7.29	86
18	0.406	34.021	18.85	118.63	8.19	91.25	27.34	17.54	0.19	16.79	8.27	/
19	0.229	34.260	18.97	117.49	8.22	13.95	27.52	15.61	0.21	14.44	5.26	71
22	0.691	34.122	18.89	98.07	8.20	37.50	27.32	21.07	0.17	18.92	12.77	18
23	0.819	33.568	18.58	97.14	8.30	26.00	26.97	8.72	0.15	7.85	10.38	1
25	0.603	33.710	18.66	14.80	8.25	18.47	27.05	2.04	0.09	1.02	1.89	6
26	0.150	34.227	18.95	69.75	8.18	23.23	27.53	18.39	0.14	16.51	11.94	1
27	0.062	33.634	18.62	97.00	8.16	17.68	27.14	6.25	0.13	6.12	4.43	12
28	0.750	34.175	18.92	82.03	8.00	22.44	27.34	15.73	0.12	14.61	10.63	1

Table 4　Correlation Coefficients of the Quantity of Microbes and Some Factors

Pair	*R*	*N*
CFU−T	0.4352	24
CFU−total N	0.3616	24
CFU−(NO_2^-−N)	0.6243	24
CFU−(NO_3^-−N)	0.3591	24
CFU−(NH_4^+−N)	0.0339	24
CFU−N/P	0.4498	24
CFU−S	−0.2802	24
CFU−Si	−0.0969	24
CFU−Cl	−0.0216	24
CFU−pH	−0.1557	24
CFU−σ_t	−0.3512	24

The positive correlation between the microbes and the different forms of nitrogen (and phosphorus) may mean the participation of the microbes in the Southern Ocean in the nitrogen (including phosphorus) cycle.

The negative correlation between the microbes and Salinity (*S*) probably hints that quite a number of microbes are related to the melted ice and snow and the nearby islands. During the period of investigation, floating ice and icebergs of various sizes could be seen on the sea surface, which can force the S and *t* of the water samples to drop, thus making the incoordination between the quantity of microbes and salinity. The surface current in the investigated area was mainly controlled by the west wind drift. The current direction was from west to east and at the same time also affected by east wind drift, and a northward movement vector was also produced. The upwelling of fairly warm deep water-brought replenishment of nutrients to the surface water. Moreover, the various eddies mand drifting ice and snow made the changes is physico-chemical elements still more complicated. These changes can be explained by means of the isoline diagram of dynamic height of surface water (Fig. 3).

On these grounds, the surface of the investigated area may be divided into three regions, namely Region Ⅰ-where the current between the south and the north of Elephant Island flows from the west and the south to the east and north region, and part of it forms anticyclonic gyration in the north west of Elephant Island. The microbes in 12 stations were determined in this region.

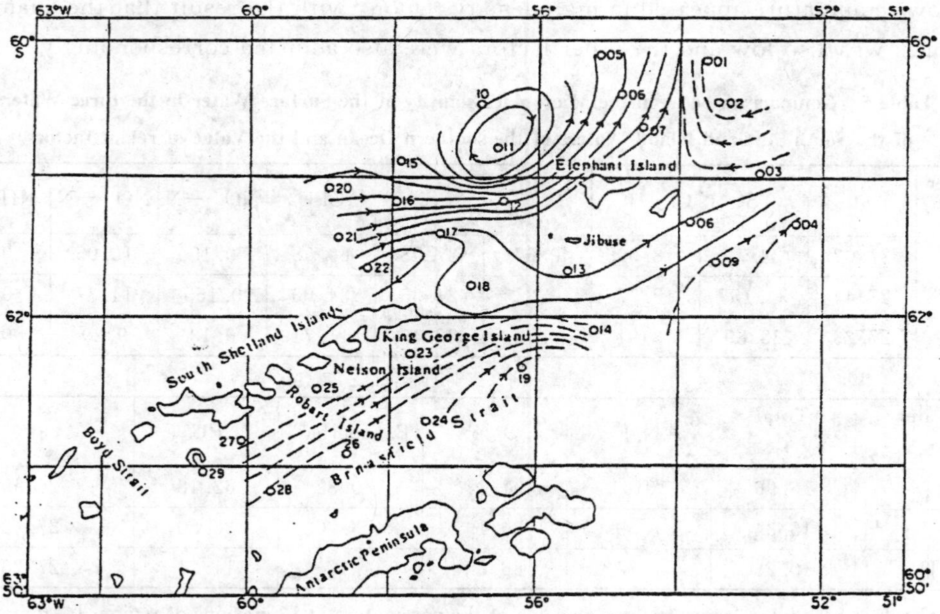

Fig. 3　Diagram showing dynamic heights and isolines of the surface water of the Southern Ocean (the South Shetlands area)

Region Ⅱ-the area in the northeast and southeast of Elephant Island and the southeastern side of Bransfield Strait. The microbes in 6 stations were determined in the region.

Region Ⅲ-the central and north part of Bransfield Strait. The microbes in 6 stations were determined in this region.

As stated above, these three regions have something in common, but the water density (t), S and T vary from region to region, perhaps the variations of the microbes in the various regions are related to these facts. In Table 5, the data of each relevant factor were calculated according to the average value of each region.

From Table 5, it can be seen that the CFU of the three regions can be divided into high, middle and low regions. NO_2^-—N, NO_3^-—N, total N and O_2% and Antarctic Krill (Euphausia superba) are relatively consistent with CFU, but density and salinity are inconsistent with CFU, which indicates that in the waters of low density (t), the S is low. In this region, the temperature being slightly high but O_2% still higher, aerobic microbes multiply relatively fast and the lxuriance of microbes is related to rich organic matter. Therefore there will be a rich supply of the various forms of nitrogen produced from microbial metabolism and decomposition. In addition, the increase of available nitrogen brought about by upwelling current can conduce to the primary production, so the phytoplankton can flourish[5]. Phytoplankton can also provide organic substances of low molecular weight for microbes[4]. It is contrary to the low density region that high salinity and low temperature appeared in high-density region, with the result that the quantity of microbes was also low and the other factors were also adjusted correspondingly.

Table 5 Comparison between the Microbial Quantity of the Surface Water in the three Waters of the South Shetland Island Waters of the southern Ocean and the Value of relant factors)

Water No.	σt	S	Cl·10^{-3}	T(℃)	Si	CFU	NO_2^-—N	NO_3^-—N	NH_3^+—N
Ⅰ	27.02	33.71	18.69	0.67	37.98	69.58	0.19	16.05	0.72
Ⅱ	27.44	34.157	18.77	0.21	24.35	34.00	0.165	14.37	0.86
Ⅲ	27.23	33.89	18.91	0.56	28.40	11.71	0.13	9.29	0.80

Water No.	Total N	N/P	O_2%	E. superba b)	CFU b)	σt b)
Ⅰ	16.96	10.01	91.34	740.82	62.18	26.99
Ⅱ	15.34	7.59	89.19	423.80	36.80	27.42
Ⅲ	10.28	7.75	86.88	4.17	11.71	27.23

a) Each is calculated from the average value of the parameters of 12 stations of region Ⅰ. 6 stations of region Ⅱ and 6 stations of region Ⅲ and from Table 3. b) the average values of the 11 stations of region Ⅰ. 5 stations of region Ⅱ and 6 stations of region Ⅲ. Respectively. The unit for counting E. superba is individuals·m^{-3}.

From Table 5 it can be seen that the variations in the relations between Krill and CFU are relatively concordant, But contrary to *t*, this seems to show that in the surface water environment, both microbes and Krill have common requirements of temperature, salinity and density. It is only natural that the interrelations between microbes, phytoplankton, Krill and other animals are intricate and cannot be summed up in a word.

The results mentioned above are only preliminary but the quantitative variation of microbes may possibly hint some hydrological, chemical and biological phenomena[2], namely, in low-density waters, the presence of a large amount of microbes and phytoplankton may possibly signify the swarming of Krill.

(3) Temperature adapation test of microbes

All the bacterial strains used in this experiment could maintain their vitality at 0℃. After a period of more than one year's transporting of these strains from many places to the home laboratory, tests were made to see their adaptation to temperature. The results are shown in Table 6.

Table 6 shows that most of the tested bacterial strains can grow at 4℃, and that the bacterial strains of the Southern Ocean have stable adaptability to low temperature. For a long time they have possessed adaptive reaction to low temperature inhibition and some of them were estimated to be psychrophiles. Some strains grew faster at 4℃ than at 2℃, which showed that an appropriate increase of temperature was conducive to the growth of part of the Southern Ocean bacteria. This kind of bacteria probably belongs to psychrotolerant ones. The number of bacteria growing at 14℃ and 25℃ decreased greatly; and some were estimated to be psychrotolerant bacteria and the other mesophiles. This indicated that they could survive in a fairly wide range of temperature. Some strains grow rapidly at 25℃, thus revealing that an increase of temperature may accelerate their metabolic rates. It was found that only one strain of bacterium could grow at 37℃, which may show that it had come from outside the Southern Ocean.

Table 6 Susceptibility of the Surface Water Microbes of the Southern Ocean to Temperature

Temperature	No. of bacterial strains tested	The % of the No. of growing bacterial strains in the No. of bacterial strains tested under different temperatures
4	62	75.58(45/62)
14	62	14.52(9/62)
25	62	12.90(8/62)
37	62	1.62(1/62)

The preliminary results of the above temperature experiment showed that the obtained bacterial strains including obligate psychrophilic bacteria and a vast amount of psychrotolerantant bacteria existed in the surface water of the Southern Ocean. Of course, our sampling, culture and preservation of the low temperature microbes and their applications remain to be further studied.

(4) Classification of microbes

75 strains of microbes were randomly taken for classification study, of which 51 were bacteria, 18 were fungi and 6 yeasts.

The classification results show that excepting 4 strains to be identified, 47 of the 51 bactera stains belong to 10 genera and 1 family, which are: *Pseudomonas*, *Moraxella*, *Acinetobacter*, *Flexibacter*, *Flavobacterium*, *Xanthomonas*, *Cytophaga*, *Micrococcus*, *Staphlococcus*, *Bacillus*, *and Enterobacteriaceae*.

Table 7 lists the percentages of each genus of the bacteria detected at 19 stations.

Table 7 Detection Rates of the various Genera and family of bacteria in 19 Stations*

Genus	Detection Rate (%)
Pseudomonas	78.95
Acinetobacter	26.32
Micrococcus	15.79
Flexibacter	21.05
Flavobacterium	10.53
Xanthomonas	10.53
Staphlococcus	10.53
Cytophaga	10.53
Moraxella	5.26
Bacillus	5.26
Enterobacteriaceae	5.26

* Stations where bacteria were detected.

From Table 7, it can be seen that the percentages of the detection of *Pseudomonas* and *Acinetobacter* are the highest. The occurrence of a few Enterobacteriaceae and *Bacillus* suggests that foreign bacteria intrusion and human activities have made impacts on the waters of the Southern Ocean[18].

Table 8 lists the percentages of each genus of bacteria in the number of bacteria identified.

Table 8 Percentages of the Various Genera and Family of Bacteria in the Number of Bacteria Identified

Genus and Family	Percentage (%)
Pseudomonas	46.81
Acinetobacter	10.34
Micrococcus	10.64

(续表)

Genus and Family	Percentage (%)
Flexibacter	8.51
Xanthomonas	4.26
Flavobacterium	4.26
Cytophaga	4.26
Staphlococcus	4.26
Moraxella	2.13
Bacillus	2.13
Enterobacteriaceae	2.13

From Table 8, it can be seen that the major member of the identified bacteria are *Pseudomonas*, *Acinetobacter* and *Micrococcus*. The three members amount for 68.09% of the total. They are the common bacteria found in this investigation, and have a wide distribution and high detection rates. Therefore they are important genera of the bacteria in the surface water of the Southern Ocean.

The 18 strains of fungi belonging to 4 genera are: *Penicillium*, *Penicilliopsis*, *Aspergillus*, *Ramularia*.

Table 9　Detection Rates of 4 Genera of Fungi in 11 Stations

Station NO.	enus of Fungus (%)			
	Penicillum	*Aspergillus*	*Penicilliopsis*	*Ramularia*
3	0	25.0	0	0
5	11.11	0	0	0
6	0	25.0	0	33.3
9	0	25.0	0	0
11	11.11	0	0	0
14	22.22	0	0	33.3
15	11.11	0	0	0
17	0	0	33.3	33.4
23	33.34	0	33.3	0
26	11.11	0	0	0
28	0	25.0	33.4	0

From Table 9 it can be seen that *Penicillium* is in the first place but compared with the results obtained by the First Expedition*, its proportion has dropped, and *Aspergil-*

* Zhang Jianzhong, 1988. Investigation Report on the Southern Ocean (Microbiology, Biology group).

lus has increased. In this investigation, *Cephalosporium* was not found and *Penicilliopsis appeared instead.*

The detection rates of the various genera of fungi at the 11 stations which are mentioned above are shown in Table 10. The results in Table 10 further shows that *Penicillium* is the most widely distributed fungus in the surface water of the Southern Ocean.

A total of 6 strains of yeasts were isolated, one of which was identified to be Debaryomyces, the others remain to be identified.

Table 10 Detection Rates of the Various Genera of Fungi at Stations

Fungus names	Detection Rates (%)
Penicillium	54.55
Aspergillum	36.36
Penicilliopsis	27.27
Ramularia	27.27

Ⅳ. CONCLUSIONS

(1) Although the number of microbes obtained by plate culture is much fewer than those obtained by AODC, the results of this investigation can indicate that there exist a considerable number of aerobic heterotrophic microbes in the surface water of the Southern Ocean, which shows that considerably violent microbial activities still exist in the surface waters of Antarctica.

(2) Within the scope of investigation, 10 genera and 1 family of bacteria, 4 genera of fungi and 1 genus of yeast were preliminary identified, which reflects (from one aspect) the diversity of the species of the microbes in the Southern Ocean. These bacterial strains are characterized by their being psychrophilic and psychrotolerant.

(3) The change in temperature in the Southern Ocean is very small and the characteristics of density and salinity are basically concordant. But the quantitative distribution of microbes is considerably susceptible to slight changes in the hydrological, chemical and biological properties of the surface water. Microbes have a relatively close relation with temperature, nutrition of the various forms nitrogen and Krill, but are inconsistent with water density and salinity. Therefore the variations in the quantitative distribution of microbes may mean the variations in hydrological properties. This expresses the relation of "contradiction-unity" between microbes and environment. Microbes have been making continuous and indelible contribution to the primary production, secondary production and cycling of matter in the Southern Ocean. There exists a definite relation between krill and microbes.

(4) The results also show that there are small quantities of Enterobacteriaceae and *Bacillus* in the water bodies of the Southern Ocean. Does this suggest that human activi-

ties and foreign bacterial intrusion have made impacts on the cold waters of the Southern Ocean? Therefore we must pay attention to the environmental protection of the Southern Ocean.

Acknowledgement：We are indebted to Professor Lan Qianwen for his help.

References 18(abr.)

(Cooperator:Sun Xiuqin　Song Qingyun　Zhang Jinxing　Lu Ying)

南设得兰群岛区域的微生物①

摘　要　1986年12月～1987年3月对南设得兰群岛海区和麦克斯威尔湾的空气、长城湾远岸表层海水、欺骗岛潮间带表层海水、底质和火山灰的微生物作了考察研究。结果表明，两个测区的空气中陆源性微生物的检出率在90%以上，每立方米空气中一般含几个CFU。海洋性微生物的检出率在50%以上，数量较小，每立方米空气仅1个左右的CFU。麦克斯威尔湾比南设得兰群岛海区的空气微生物数稍高些。测区的空气微生物受岛屿的影响较大。空气微生物数呈现出某种昼夜变化。空气微生物数受气温、相对湿度的影响不大明显。每立方厘米的长城湾远岸表层海水含几百个CFU的微生物。其数量随水温的逐月降低减少，并显示出某种昼夜变化趋势。该湾海水比较洁净。空气中的海洋性微生物与海水中的微生物呈现出某种关系。长城湾海水细菌有13属以上。以假单胞菌为常见者。欺骗岛潮间带每立方厘米海水中只有几个CFU，火山灰中的含菌量达每克数千个CFU。

以上结果基本反映了南设得兰群岛一带空气、海洋和陆上的微生物一般状况，并为进一步考察和开发该区域的微生物资源提供了资料。

关键词　南设得兰群岛　麦克斯威尔湾　长城湾　欺骗岛　微生物

一、引言

在1986/1987年的中国第三次南极考察中，对南设得兰群岛一带的微生物作了考察。其主要内容包括南设得兰群岛海区和麦克斯威尔湾的空气微生物、长城湾的海水微生物及欺骗岛的微生物。由此，对该区域空、海、陆的微生物状况有了一个初步的认识，为进一步考察和开发南设得兰群岛区域的微生物资源奠定了初步基础。

二、材料和方法

1. 培养基

海水培养基，成分如下：酵母膏，1 g；蛋白胨，2 g；$FePO_4$，0.01 g；琼脂，18 g；陈南大洋水，1 000 cm^3，pH，8。用于海洋性微生物。淡水培养基，成分如下：牛肉膏，3～5 g；蛋白胨，10 g；NaCl，5 g；琼脂，18 g；蒸馏水，1 000 cm^3，pH，7.2～7.4。用于陆源性微生物。

2. 采样方法

空气微生物样，用JWL－4型空气微生物监测仪测得。采样时间：15或30 min。采样时空气流量为37 dm^3/min。海水微生物样用击开式采水器采集。在上述的海水培养基平皿上涂布。海滨沉积物的微生物也用平皿涂布法。操作步骤见另文(陈皓文等，1990)。

3. 培养和分析

上述样品经10℃温度培养一周后，计算所产菌落数，空气微生物的计算公式如下：

①　原文刊于《南极研究(中文版)》，1992，4(4)：1-6。

$$空气含菌量(CFU/m^3)=\frac{每皿平均菌落数}{采样时间(min)\times采样流量(dm^3\ min)}\times1\,000$$

海水微生物菌落数计算公式如下：

$$海水含菌量(CFU/m^3)=\frac{平均菌落数\ cm^2\times每个平皿的面积(cm^2)}{所分析水样体积(cm^3)}\times1\,000$$

沉积物含菌量测定与此类似。单位是 CFU/g(新鲜、湿重)。挑取和纯化菌落，保存菌株，带回国内作分类研究(欧勒弗，1985；斯克尔曼，1978)。

三、结果和讨论

1. 南设得兰群岛海区的空气微生物

这一海区的空气微生物检测是在"极地"船上进行的。采样结果如表 1 所示。

表 1　南设得兰群岛海区的空气微生物数量表

Table 1　Airborne microbial numbers over area of the South Shetland Islands

采样时间	采样位置	微生物数量(CFU/m^3)		气温	相对湿度
		海洋性	陆源性	(℃)	(%)
1987.1.17	61°51′S,56°53′W	0	4	7.5	91
1987.1.18	60°54′S,58°04′W	0	0.67	4.2	87
1987.1.19	62°22′S,57°40′W	0	2	5.8	88
1987.1.20	62°59′S,60°40′W	0.67	6.7	8.5	92
1987.1.21	63°03′S,60°31′W	3.3	6.7	8.8	94
1987.1.30	61°21′S,53°45′W	0.67	13.3	1.2	96
1987.1.31	60°24′S,54°03′W	0	4.7	4.4	93
1987.2.1	60°50′S,56°34′W	0	10	5.8	86
1987.2.11	61°51′S,56°52′W	0.67	12	4.2	93
1987.1.12	61°11′S,56°15′W	4	2.7	5.0	97
1987.1.13	61°39′S,56°32′W	0	4	2.5	95
1987.1.14	61°45′S,56°15′W	0	0	9.8	91

由表 1 可计算出，调查范围内空气中海洋性微生物的检出率达 58.3%。微生物数量平均为 0.776 CFU/m^3，最高达 4 CFU/m^3。陆源性微生物的检出率为 91.7%。微生物数量平均为 5.56 CFU/m^3。这些结果表明该空气中陆源性微生物的数量和检出率均高于海洋性微生物的，这可能意味着南设得兰群岛海区的许多微生物与附近岛屿有比较密切的联系。

将以上两类空气微生物数量与有关气温、相对湿度等分别作相关分析，结果表明，海洋性微生物与相应气温间的相关系数并不一致大，只是陆源性微生物数与相应气温间的负相关关系达显著水平($-0.687, n=12$)。这表明该海区空气的陆源性微生物对低温比较适应。

2. 麦克斯威尔湾海区的空气微生物

表 2 列出每次采样的结果，这基本上反映了该湾整个夏天空气的微生物数量状况。即海洋性微生物的检出率为 53.7%。其数量平均为 1.53 CFU/m^3。最高数为 1987 年 1 月 10 日的 10.7 CFU/m^3。陆源性微生物的检出率为 94.3%，其数量平均为 6.45 CFU/m^3。最高数为 1987 年 1 月 11 日的 100 CFU/m^3。与上述南设得兰群岛海区的空气微生物比，仅海洋性微生物检出率稍低。其余均是本湾空气的微生物各指标高。这表明本湾空气中的微生物多，数量也大。这可能与本湾离乔治王岛近，人员活动多，其他生物较丰富有关。

该湾空气微生物数在月份间有差别。陆源性微生物数与气温间的关系类似于南设得兰群岛海区的。海洋性微生物数则显出随着气温的逐月降低而逐月减少的趋势。

表 2 麦克斯威尔湾空气微生物数量一览表*

Table 2 List of airborne microbial numbers of Maxwell Bay

采样时间	微生物数量(CFU/m^3)		气温(℃)	采样时间	微生物数量(CFU/m^3)		气温(℃)
	海洋性	陆源性			海洋性	陆源性	
1986.12.27	1.3	1.3	6	1987.1.25	2.7	0.67	2.5
1986.12.28	2.7	1.3	5	1987.1.26	2	0	2.0
1986.12.29	9.3	14.7	3	1987.1.27	0	0	2.7
1986.12.30 15:00	0	6	10	1987.1.28	3.5	4.7	0.9
1986.12.30 20:00	3.3	—	2.0	1987.1.29	0.67	0.67	−0.2
1987.12.31	9.3	13.3	4.8	月平均	1.68	9.24	7.25
月平均	4.32	7.32	5.13	1987.2.2	0	24.7	6.7
1987.1.1	6.7	5	6.5	1987.2.3	1.3	3.3	0.6
1987.1.2	0.67	0.67	6.2	1987.2.4	0	3.3	2.8
1987.1.3 15:30	0	0.67	8.0	1987.2.5	0.67	4	3.5
1987.1.3 22.15	1.3	2	3.5	1987.2.6	0	2	3.0
1987.1.4 03.15	0	0	3.1	1987.2.7	0	0.67	1.2
1987.1.4 10:00	4.7	0.67	4.1	1987.2.8	0	4.7	3.0
1987.1.4 15:30	1.3	2.7	7.1	1987.2.9	0.67	0.67	1.5
1987.1.5	1.3	1.3	7.5	1987.2.10	0.67	2	5.5
1987.1.6	3.3	8.3	2.5	1987.2.15	0	1.3	3.5
1987.1.7	0	0.67	6.8	1987.2.16	0	3.3	5.2
1987.1.8	2	1.3	8.8	1987.2.17	0.5	1.3	4.3
1987.1.9	0	9.3	5.2	1987.2.18	0	0.67	3.5
1987.1.10	10.7	10	0.2	1987.2.19	0.67	0.67	2.8

(续表)

采样时间	微生物数量(CFU/m³)		气温(℃)	采样时间	微生物数量(CFU/m³)		气温(℃)
	海洋性	陆源性			海洋性	陆源性	
1987.1.11	0	100	5.6	1987.2.27 02:00	2.7	2.7	0
1987.1.12	0	20.7	6.5	1987.2.27 20:00	4	1.3	4
1987.1.13	0	5.3	10	月平均	1.12	1.61	3.33
1987.1.14	0.67	2	5.5	1987.3.11	0	2	3.5
1987.1.15	3.3	2.7	3.8	1987.3.12	0	35.3	5.4
1987.1.16	0.67	7.3	3.0	1987.3.13	0	5.3	1.8
1987.1.22	0	2	3.4	1987.3.14	0	2.7	0
1987.1.23	0	7.3	4.0	1987.3.15	0	8	3.8
1987.1.24	0	4.7	2.5	月平均	0	10.66	2.90

* 1986.12.27～1987.1.29 在 621806S,582309W 处采样;1987.2.2～3.15 在 621408S,583504W 处采样

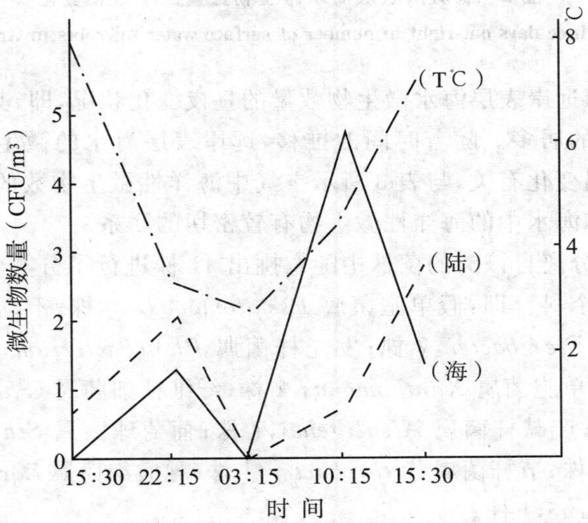

图 1　麦克斯威尔湾空气微生物含量昼夜变化

Fig. 2　Diurnal variation of airborne microbial numbers of Maxwell Bay

图 1 示出麦克斯威尔湾空气微生物数量的昼夜变化状态,该图是根据表 2 所列的 1987 年 1 月 3～4 日之间所作的连续 5 次观察所得数据画出的。该图表明该湾空气中海洋性微生物数量高峰在 10:00,陆源性微生物数量高峰在 15:30。两者在 22:15 也呈现出高数量,两者的低谷均在凌晨 3:15 出现,但是以上这些变化与气温间似无重大相关性。

3. 长城湾远岸表层海水微生物

数量及其变化:对长城湾远岸表层海水中的细菌共作了 12 次测定,主要在两个定点

站上测得,即1987年1月27日～1月28日在621806S,582109W处和1987年2月2日～2月3日在621408S,585504W处。结果表明,其数量范围是0～515 CFU/cm³。平均为376 CFU/cm³左右。这一结果比长城湾近岸水低一个量级。比长城湾海滨水(含海藻)低两个量级。比南大洋水高一个量级(陈皓文等,1990),比纬度较低的中国胶州湾较洁净的海水低三个量级(陈皓文,1981)。表明当今的长城湾水尚比较洁净。

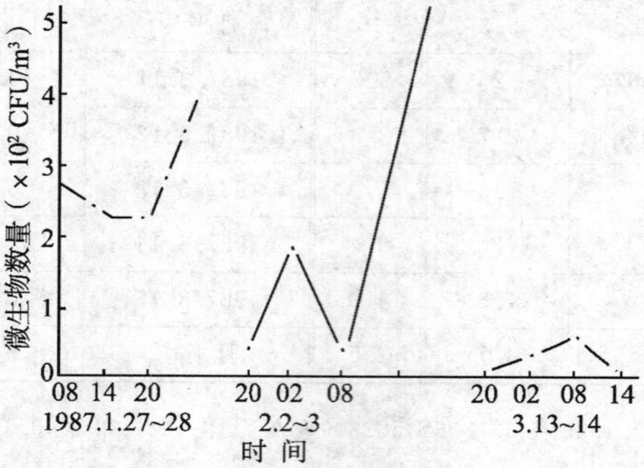

图2 长城湾表层海水微生物数量三个昼夜变化

Fig. 2 Variations for three days out right in number of surface water microbes in Great Wall Bay, Antarctica

图2表明了该湾远岸表层海水微生物数量的昼夜变化状况,即,其数量高峰以2:00的居多,低谷以20:00的居多。随着时间的推移,远岸表层海水的微生物数量在逐月下降,估计这一现象与水温变化有关,与表3所示空气中海洋性微生物数的月间差异相一致,因而暗示了空气中的和海水中的海洋性微生物有较密切的联系。

属群概况:从所分离自该湾的菌株中随意挑出41株进行了分类学研究,结果表明,它们属于13个属和1个科。即:假单胞菌属 *Pseudomonas*,10株;不动杆菌属 *Acinetobacter*,7株;屈挠杆菌属 *Flexibacter*,7株;发光杆菌属 *Photobacterium*,3株,黄杆菌属 *Flavobacterium*,2株;黄单胞菌属 *Xanthomonas*,2株;纤维粘细菌属 *Cytophaga*,2株;萘瑟氏菌属 *Neisseria*,2株;产碱杆菌属 *Alcaligenes*,1株;葡萄球菌属 *Staphylococcus*,1株;八叠球菌属 *Sarcina*,1株;节杆菌属 *Arthrobacter*,1株;乳酸细菌属 *Lactobacillus*,1株;肠杆菌科 *Enterobacteriaceae*,1株。

由上可见,假单胞菌居多,其次是不动杆菌和屈挠杆菌,也较多,这一情况与长城站近岸海滨的细菌组成不很一致(陈皓文等,1992,本期)。

4. 欺骗岛的微生物

欺骗岛在利文斯敦岛南侧13千米处。具体位置是62°59′04″S,60°40′34″W处。是呈马蹄形的岛弧。该岛为活火山,当我们到达该岛水域时,湾内水面平静;气温在10℃左右,由于1970年8月又有火山喷发,所以棕褐色的火山熔岩从山上倾泻入海时留下的情景至今仍历历在目。图3示出了这次微生物考察的站位略图。表3列出了微生物取样的分析结果。

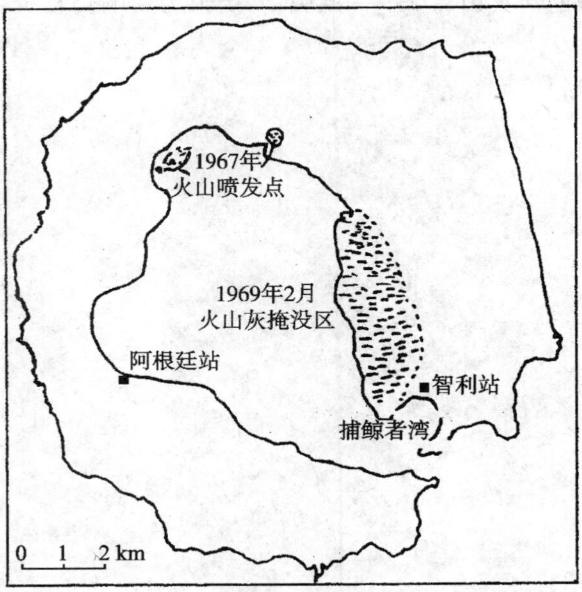

图 3　欺骗岛采样站位图

Fig. 3　The map of sampling stations in Deception Island

表 3　欺骗岛的微生物数量*

Table 3　Microbial numbers in the intertidal zone of Deception Island

取样站址	微生物数量	样品描述
智利 Skur 站	4258.3 CFU/g	火山石、岩浆灰沙，细中砂为主，棕褐色、轻、干，10℃
阿根廷 Marina De Gunea 站	2.5 CFU/cm^3	潮间带表层水，pH 7.2，10℃
阿根廷 Marina De Gunea 站	4753 CFU/g	潮间带表层底质样，10℃

* 采样时间：1987.1.21。微生物数由平皿培养法得到，底质样系湿重

由表 3 可见其冷却的火山岩灰浆沙中存在一些微生物，但比长城站原土的微生物数低了一个量级（陈皓文等，1992，本期），比本表层水又高出三个量级。与长城站潮间带水相比，本岛低了三个量级，尽管据报本岛潮间带以外的附近水体含有丰富的生物量。

四、结语

对南设得兰群岛海区空气微生物与麦克斯威尔湾空气微生物的考察结果反映了这一带海域空气中微生物的一般状况，两区的空气微生物检出率较高，空气中陆源性微生物数量与气温间呈现出相悖的关系。南设得兰群岛海区空气的海洋性微生物数量与气温的变化也不一致。长城湾远岸海水菌数在 3.76×10^2 CFU/cm^3 左右，比该湾海滨含藻海水的菌量低。比南大洋水的含菌量高。结果表明该湾水比较洁净。

空气和海水中的微生物数量均显出一定的昼夜变化趋势。

长城湾水含有的细菌以假单胞菌为多见。此外还鉴定出 12 属以上的细菌。

地理景观较特异的活火山岛弧——欺骗岛，受人为影响仍较少。其陆上和海滨的微生物数量不大。

参考文献 4 篇(略)

(合作者:宋庆云)

MICROBES IN THE AREA OF SOUTH SHETLAND ISLANDS

(ABSTRACT)

Abstract Microbial exploration of the air, sediment and seawater in the area of the South Shetland Islands was performed during December, 1986-March, 1987.

The results obtained show that: for terrigenous microbes over the South Shetland Islands and Maxwell Bay, their detection rates were more than 90% and their amounts were generally a few CFU/m^3; for marine microbes, their detection rates were more than 50% and their amount only 1 CFU/m^3. The amounts of airborne microbes over the area surveyed presented somewhat diurnal variations, They were influenced to some extent by nearby islands. Air temperature and relative humidity, etc. were factors affecting their quantitative variations.

The bacterial amount of the outshore surface water in Great Wall Bay, Antarctica was about $n \cdot 10$ CFU/m^3 ($0 \leqslant n < 10$, the same below), the amount of microbes decreased with the monthly decreases of the temperature of seawater and shows some states of diurnal variations. The seawater in the Bay has been fairly clean. Community contained at least thirteen genera, in which *Pseudomonas* was common.

In the intertidal zone of Deception Island, there were only a few CFU of bacteria per cubic meter in seawater; and $n \cdot 10$ CFU/g (w. w) in surface sediment; and $n \cdot 10$ CFU/g (w. w) in the volcano ash.

The results reflected the general situation about microbes of air, sea and land in the area of the South Shetland Islands. It provides data for further exploration and utilization of microbial resources in the area of the South Shetland Islands, Antarctica.

Key words South Shetland Islands, Maxwell Bay, Great Wall Bay, Deception Lsland, Microbes.

南极长城站微生物考察研究①

摘　要　1986～1987年中国第三次南极考察中，对南极长城站空气、陆上和海滨的微生物作了考察。长城站野外空气中的微生物含量较低，海洋性微生物为148.7 CFU/m^3，陆源性微生物为17.9 CFU/m^3，表明野外空气相当洁净。室内空气微生物数比室外及野外均高得多，霉菌的检出率和含量都不高。微生物因人群活动和环境的不同而异。陆上的微生物含量次序是：冰雪＜湖水＜原土等。海滨的微生物含量相当多，＞10^3 CFU/cm^3（或g）。长城站的微生物组成有11属以上，包括偶见的芽孢杆菌。温度试验结果表明，野外的大部分菌株适于低温生活。没有发现能在≥25℃的温度中存活者。低温是控制长城站微生物生态的一个重要因子。在长城站这类环境中，低温微生物对物质转化和循环发挥着持久的重要作用。

关键词　南极长城站陆、海、空的微生物　属群分析　温度感应

一、引 言

在中国第三次南极考察中，于1986年12月28日-1987年3月11日间对南极长城站作了初步的微生物考察，内容包括空气、陆上和海滨的微生物研究，通过这一考察活动，对长城站的微生物状况有了一个初步的认识。

二、长城站基本概况

南极长城站位于南设得兰群岛乔治王岛南端的菲尔德斯半岛上（62°12′59″S，58°57′52″W）。东临麦克斯威尔湾，南面为菲尔德斯海峡，西距乔治王岛西海岸2.5 km。长城站近岸水深不超过5 m。海底以沙砾石为主，该站为高度不超过155 m的无冰丘陵区，有明显的冻融作用。风化壳呈碳酸盐层状，该站冰雪样品中Na、Cl含量较高，大气中有残留盐分。长城站是南极洲最暖最湿区之一，具极地海洋性气候特点。年平均气温为－2.6℃，极端最高气温在1月。平均相对湿度为85.9%，年降水414.9 mm，2～4月份多雨。陆上植被以苔藓、地衣为主，盛夏时节，湖冰融化，水温可达3℃左右。海水流动，水温在0℃左右。长城站的这些基本自然环境状况为众多的亚极地一极地微生物的繁衍创造了优越的条件。图1为长城站地理环境和采样点概况。

---

① 原文刊于《南极研究》，1992，4(4)：7-13。

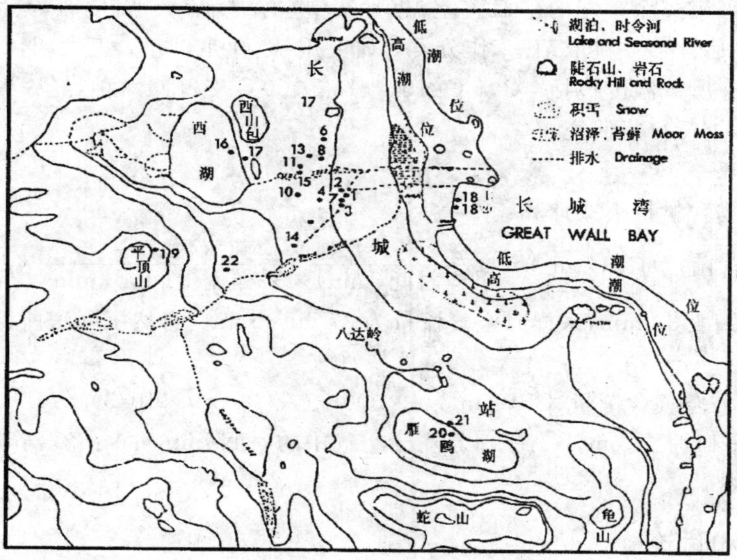

图 1　中国南极长城站采样位置示意图

Fig. 1　Sampling stations in Great Wall Station in Antarctica

1. 餐厅；2. 厨房；3. 医务室；4. 电离层室；5. 通讯室；6. 气象室；7. 1 号楼卫生间；8. 宿舍 213 室；9. 1 号楼门厅；10. 食品库；11. 冷藏房；12. 1 号楼外；13. 2 号楼外；14. 科研楼；15. 冷藏房外；16. 西湖水；17. 西湖湖滨冰雪；18(1). 长城湾海水；18(2). 长城湾海藻；19. 平顶山峰；20. 雁鸥湖水；21. 雁鸥湖滨雪；22. 平顶山傍

1. Dining Room; 2. Kitchen; 3. Clinic; 4. Ionospheric Survey Room; 5. Communication Room; 6. Weather study Room; 7. Washroom of Building No. 1; 8. Living Room No. 213; 9. Hall Way of Building No. 1; 10. Food Store; 11. Refrigeratory store; 12. Out of the Building No. 1; 13. Out of the Building No. 2; 14. Scientific Research Building; 15. Out of the refrigeratory store; 16. Xihu Lake Water; 17. Ice and snow nearby Xihu Lake; 18(1). Sea Water of Great Wall Bay; 18(2). Marine Algae of Great Wall Bay; 19. Pingding Shan Peak; 20. Yanou Lake Water; 21. Snow nearby Yanou Lake; 22. Near the Pingding Shan.

三、材料和方法

1. 培养基

海水培养基：酵母膏，1 g；蛋白胨，2 g；$FePO_4$，0.01 g；琼脂，18 g；陈南大洋水，1 000 cm^3，pH，8。淡水培养基：牛肉膏，3～5 g；蛋白胨，10 g；NaCl，5 g；琼脂，18 g；蒸馏水，1 000 cm^3；7.2～7.4。霉菌培养基：蛋白胨，5 g；葡萄糖，10 g；KH_2PO_4，1 g；$MgSO_4 \cdot 7H_2O$，0.5 g；琼脂，18 g；蒸馏水，1 000 cm^3；pH，7.2～7.4。

2. 取样和试验

空气微生物　空气微生物样品包括离站区较远的野外空气、建筑物附近的室外空气、各种功能性房间的室内空气及室内外衔接部的空气这四类。野外空气微生物样品一般用 JWL-4 型空气微生物监测仪采集，测定是根据固体孔隙撞击可计量空气的原理进行，其余三类空气的微生物用自然沉降法测得。

陆上和海滨的微生物　采集原土、冰雪、湖水、海水和海藻等，定量操作，视情况作必要的稀释，涂布平皿培养基上。

分类　经采集或涂布好的平皿置室温(10℃)左右中培养一周左右，其间经常检查菌落生长情况，最后计数并挑取菌落，作分类等研究用(欧勒弗，1985；王大耜，1977)。

温度试验　所试温度包括4℃、5℃、10℃、15℃、20℃及25℃六种，经一周时间的培养，观察生长情况，进行分析研究。

3. 数据处理

空气微生物含量按下式计算：

$$空气含菌量(CFU/m^3)=\frac{每皿平均菌落数}{采样时间(min)\times 采样流量(dm^3/min)}\times 1\,000$$

采样时间一般30 min，采样时空气流量为37 din³/min。自然沉降法所得菌落数，按下式求出：

$$空气含菌量(CFU/m^3)=1\,000\div(A/100\times t\times 10/5)\times N$$

其中，A：所用平皿面积(cm²)；t：平皿暴露于空气中的时间(min)；N：经培养后平皿培养基上长出的菌落数。

四、结果和讨论

1. 野外空气微生物

长城站野外空气微生物的采样地点在科研栋外西侧平顶山旁雪地上(图1，No. 22)，它代表着长城站自然状态中夏季空气微生物大体状况，表1列出了这些微生物数据及有关参数。

表1　长城站野外空气微生物含量表

Table1　List of field airborne microbial contents over the Great Wall Station

开始测定时间*	微生物含量(CFU/m³)**		天气概况		备注
	海洋性	陆源性	气温(℃)	相对湿度(%)***	
1987.2.20.06:30	0	—	3.7	87	
1987.2.21.21:30	1.3	0.67	2.2	77	
1987.2.22.07:10	0	0.67	2.0	88	
1987.2.23.20:00	—	5.3	2.0	84	
1987.2.24.14:00	2.7	0	6.0	75	
1987.2.25.05:30	0.67	0.67	2.0	78	
1987.2.28.20:40	0.67	1.3	6.0	79	
1987.3.01.04:30	2	0	4.0	89	
1987.3.02.12:00	0.67	2	5.0	97	
1987.3.03.19:45	0.67	2.7	1.0	75	
1987.3.04.22:15	1.3	2	2.0	82	
1987.3.05.21:20	92.6	46.3	2.0	98	自然沉降30 min
1987.3.06.19:30	2870.6	0	1.5	96	自然沉降30 min

(续表)

开始测定时间*	微生物含量(CFU/m³)**		天气概况		备注
	海洋性	陆源性	气温(℃)	相对湿度(%)***	
1987.3.07.14:00	0	185.2	5.1	69	自然沉降 30 min
1987.3.08.20:30	0	0	1.3	87	自然沉降 30 min
1987.3.09.06:30	0	92.6	1.8	95	自然沉降 30 min
1987.3.10.11:30	0	0	3.5	98	自然沉降 30 min
1987.3.10.11:30	0	0	3.5	95	自然沉降 120 min
1987.3.10.11:30	0	0	3.5	95	自然沉降 120 min
1987.3.10.11:30	0	0	3.5	87	自然沉降 180 min
1987.3.10.11:30	0	满皿	3.5	87	自然沉降 240 min

*南极地方时；**由10℃培养中得出；***据长城站1987年地面气象资料

由表1得出表2，表2说明野外空气中陆源性微生物(以淡水培养基上长出的菌落看作陆源性微生物)的检出率高于海洋性微生物(以海水培养基上长出的菌落看作海洋性微生物)，其数量变异小于海洋性的，这可能与测定期间的主要风向有关(常遇西、北风)西、北风时站区冰雪、尘埃，再降落于测点。陆源性微生物含量的平均值也小于海洋性，这一结果与一些城市海滨空气的相似。长城站野外空气微生物仍然较少，与南极南维多利亚干谷一带空气相比，则较大(Cameron, *et al.*, 1970)。

经不同时间长度(30、60、180和240 min)的自然沉降所得的海洋性微生物数均为0，陆源性微生物除一次异常外也较少。这进一步证实了长城站野外的空气质量较好。此外；经相关分析，野外空气微生物含量与相应的气温间正相关性不高，与相对湿度间的相关关系也不大。

表2 长城站野外空气微生物含量测定结果统计表

Table 2 Statistic table of results of field airborne microbial contents over the Great Wall Station

项目	检出率(%)	微生物含量范围(CFU/m³)	微生物含量平均值(CFU/m³)	微生物含量中位数*	测定次数
海洋性	50	0－2870.6	148.7	1.3	20
陆源性	60	0－185.2	17.9	2.0	20

*微生物含量为0时，不计入内

2. 室内外空气微生物

表3列出长城站每次测得的室内外空气微生物的数据。由表3可见室内外空气微生物的测定共分三天进行，15个测区中，室外(即12,12,15)测了7次，室内外衔接部(即9)测了2次，其余室内测区共测了29次。

表4列出了空气微生物含量按不同环境、不同功能类别的测区作比较的结果(参见表3部分数据)。由该表可见，微生物数据在不同环境的空气中是不一样的，按由低到高的数

量排列，次序如下：野外＜建筑物附近的室外＜室内—外衔接部＜室内，这表明长城站空气微生物含量的多寡与人类活动密切相关，受人类活动影响少的自然环境中，空气微生物含量明显地少，人群活动频繁的室内空间，微生物数明显地多，后者的平均值是前者的约45倍，这与"极地"船的情况有类似之处。

就不同功能的室内空间言，微生物含量并不均衡分布，如表4所示，不同功能类型的四种室内空间，其微生物含量大小次序是冰库、食品库＜餐厅＜厨房、医卫＜工作间＜居室。还需指出的是，此处冰库、食品库的空气微生物含量实际上是偏高的，因为尽管其室温较低，但1987年3月11日测量时，已有两只活企鹅在，它们的活动带来了适于低温的微生物繁衍，这证明一些较高级的生物在人为空间存活可能增加微生物含量，除非坚持采取必要的消毒措施。

表3　长城站空气微生物含量

Table 3　Airborne microbial contents over the Great Wall Station

测区代号*	1986.12.28			1987.2.19			1987.3.11		
	a	b	c	a	b	c	a	b	c
1	550.2	353.7	64.3	416.6	0	0	165.1	0	0
2	805.7	452.0	56.1	—	—	—	196.5	0	0
3	255.5	137.6	53.9	212.2	0	0	821.4	0	0
4	255.5	157.2	61.5	1320.5	39.3	3.0	—	—	—
5	884.3	117.9	13.3	487.3	0	0	84.1	—	—
6	845.0	569.9	67.4	338.0	39.3	11.6	90.4	0	0
7	216.2	117.9	54.5	1603.4	0	0	393.0	0	0
8	3654.9	2790.3	76.3	393.0	39.3	10.0	455.9	0	0
9	340.3	0	0	—	—	—	510.9	0	0
10	0	0	0	—	—	—	220.1	0	0
11	78.6	0	0	275.1	0	0	687.8	0	0
12	314.4	0	0	157.2	0	0	117.9	0	0
13	117.9	39.3	33.3	180.8	0	0	0	0	0
14	—	—	—	78.6	0	0	2393.4	0	0
15	—	—	—	589.5	0	0	—	—	—

* a. 微生物总数 CFU/m^3；b. 霉菌数 CFU/m^3；c. b/a%；"—"未测；所有 CFU 均用自然沉降法测得

霉菌是影响空气环境质量的重要因子之一。一些霉菌是致癌因素，因而掌握环境中这一指标早就引人注意(孟昭赫等，1979)。表5列出了长城站空气中霉菌的一些测定结果。

表 4　长城站不同环境中的空气微生物含量比较*

Table 4　Comparison of airborne microbial contents in different environments over the Great Wall Station

环境类别	微生物含量(CUF/m³)	与野外空气微生物含量相比的倍数	备注
建筑物附近室外	211.1	11.8	7 次测定的均值
室内外衔接部	425.6	23.8	2 次测定的均值
室内			
医、卫、餐、厨	480.2	26.8	11 次测定的均值
工作间	538.1	30.1	8 次测定的均值
居室	1395.2	77.9	5 次测定的均值
食品冷库	252.3	14.1	5 次测定的均值
野外	17.9	1	20 次测定的均值

* 本表中的微生物数量指陆源性微生物

表 5　长城站空气中霉菌测定结果

Table 5　Results of determination of airborne mildew over the Great Wall Station

测区类别	1986.12.28	1987.2.19	1987.3.11	平均
室内				
居室(8、14)	100.00(76.30)	50.00(8.33)	0(0)	50.00(28.21)
公用(1、2、3、7)	100.00(58.07)	0(0)	0(0)	33.33(19.36)
工作间(4、5、6)	100.00(42.57)	66.67(3.66)	0(0)	55.56(15.41)
冷库、食品库(10、11)	0(0)	0(0)	0(0)	0(0)
室外	0	0	0	0
内外衔接部(9)	(0)	(0)	(0)	(0)
建筑物外(12、13、15)	50.00(9.09)	0(0)	0(0)	16.67(3.03)

注:括弧内数据均系霉菌占总菌数的百分率

由表 5 可以看到,总的来讲霉菌的检出率是室内>室外,霉菌占微生物的百分率也是室内>室外。就室内言,是居室>公用室>工作间>室内外衔接部、冷库和食品库(检出率为 0)。3 个月中霉菌检出率和数量呈减少趋势,这可能与温度的降低、环境卫生状况的不断改善有关,长城站的霉菌与“极地”船的相比,检出率低,菌量也小。这是因为长城站环境比“极地”船的大而且稳定,气温偏低,不利于一些菌类生长①。

以上微生物含量的测定结果为了解长城站空气微生长的本底,保护和增进长城站空气的洁净水平提供了基本数据和参考资料。

① 待发表

表 6　长城站陆上和海滨微生物采集记录

Table 6　Record of collection of microbes from land and seaside over the Great Wall Station

采样日期	采样点号	样品来源	微生物含量(CFU/cm^3 或 g)	备注
1987.2.4	16	水	280	3℃
	17	雪	63	<2℃
	18(1)	海水	2200	未结冰
	18(2)	海藻	103130	
1987.2.25	16	水	3200	3℃
	17	冰、雪	385	
	19	原土、鸟粪		
		苔藓、地衣	44600	
	18(1)	水	1400	3℃
	18(2)	带水海藻	126400	3℃
1987.3.7	16	水	50	
	17	雪	0	
	19	原土	16100	
1987.3.13	20	水	250	湖一侧有垃圾场,不远处有下水道出口
	21	雪	530	
	18(1)	海水	1400	
	18(2)	带水海藻	21 000	
1986.12.31	18(1)	海水	1100	
	18(2)	带水海藻	2 000	
	17	冰雪	0	
	16	化妆水	1 000	

注:除水外,样品重均系湿重

3. 长城站陆上和海滨的微生物

陆上和海滨微生物样品分 4 天采集,共 20 份,代表了长城站短暂的夏季微生物状况(表 6)。将表 6 的结果统计归入表 7,由表 7 可见,长城站湖水和冰雪中的微生物含量均在 10^2 CFU/cm^3 左右,虽然湖水中要比冰雪中多些。原土等样品中的微生物含量比湖水中高两个量级。海藻的微生物含量平均比海滨水大两个量级,在整个夏季,随时间的推移,水温低、高至低,海藻的微生物含量由少增到多后再降,而海水中的则显得变化较小。长城站海水的微生物含量比淡水的高一个量级，比南设得兰群岛海域也高（陈皓文，1990），这些与 Lizuka 等的报道有类似之处（Lizuka，et al.，1974）。

4. 长城站微生物的属群分析

对所分离的部分微生物作了分类学研究（欧勒弗，1985；Buchaman，et al.，1974），结果列入表 8。

表 7　长城站陆、海微生物含量

Table 7　Statistics of terrestrial and marine microbial contents over the Great Wall Station

样品来源	微生物含量范围	均值
冰雪	$5\times10^1\sim3\times10^3$	9×10^2
湖水	$0\sim1\times10^3$	3.3×10^2
原土、鸟粪、苔藓、地衣	$1.6\times10^4\sim4\times10^4$	2.8×10^4
海水	$1\times10^3\sim2\times10^3$	1.5×10^3
海藻	$2\times10^3\sim1.3\times10^6$	3.5×10^5

注：微生物含量单位，水：CFU/cm^3；其余为 CFU/g(湿)

由表 8 可见，长城站至少有 11 属以上的菌类。分析还表明，海洋性微生物中以假单胞菌、黄杆菌、不动杆菌居多，陆源性微生物中以动胶菌、葡萄球菌居多，偶见芽孢杆菌，已显露出人为影响。长城站的细菌组成与维多利亚干谷的并不完全一致（Nichunzhi，*et al.*，1987），与 Morita（1975）所报道的也有差异。这说明长城站有其自身的微生物特征。

表 8　长城站所鉴定的微生物属名表

Table 8　List of generic names of identified microbes of the Great Wall Station.

微生物属名	分类来源	微生物属名	分类来源
黄杆菌 *Flavobacterium*	海水、海藻	芽孢杆菌 *Bacillus*	湖水、室外空气
纤维黏细菌 *Cytophaga*	室外空气	动胶菌 *Zoogloea*	水、室内外空气
莫拉氏菌 *Moraxella*	室外空气	葡萄球菌 *Staphylococcus*	水、室外空气
不动杆菌 *Acinetobacter*	海水、湖水、室内空气	足球菌 *Pediococcus*	室外空气
库特氏菌 *Kurthia*	海水	酵母 *Saccharomyces*	室外空气
假单胞菌 *Pseudomonas*	海水		

5. 长城站微生物对温度的感应

将分离到的部分野外细菌菌株，计 64 株，作了温度试验，试验结果如表 9 所示。

表 9 说明长城站的微生物，大部分适于 4～15℃的温度。15℃以上的温度中，大部分菌株均不能生存，在 25℃温度时，尚未见到生存者，表明低温对于长城站的主要微生物是适宜的。较高的温度不利于其生存（Gessey，el al.，1975）。这意味着低温微生物在长城站乃至南极的物质转化及其循环中起着重要作用，有关研究尚待继续。

表 9　部分微生物菌株对温度的适应

Table 9　Adaptation of a part microbial strains to cultured temperature

参试菌株数	试验温度℃	生长菌株数%
64	4	32.81
64	5	37.50
64	10	15.63
64	15	10.94
64	20	3.13
64	25	0

五、结语

通过夏季微生物考察，初步获得长城站陆、海、空微生物状况的认识，野外的空气相当洁净。偶尔遇到了较大微生物数量，还有芽孢杆菌，这一异常现象可能与大气尘埃及溶胶沉降、人群活动等有关。室内空气微生物含量较高，其中以居室中的为最。冰雪、原土和生物等样品的微生物测定结果意味着长城站这类亚南极—南极环境孕育着大量嗜冷—耐冷的微生物。

参考文献 11 篇(略)

（合作者：宋庆云　卢　颖）

MICROBIOLOGICAL RESEARCH EXPLORATION IN THE GREAT WALL STATION, ANTARCTICA

Abstract During the third CHARE in 1986/1987, microbes in outdoor and indoor air, ice, snow, lake water, virgin soil, offshore seawater and seaweeds were investigated.

The results obtained indicate: there were relatively few microbes in the open air of the Great Wall Station, Antarctica, i. e. the quantity of marine microbes was about 148.7 CFU/m^3 and that of terrigenous microbes about 17.9 CFU/m^3. Which indicated that the open air was quite clean. Contents of air—borne microbes indoors were much higher than those outdoors and in the field air. The detection rates and amount of mildew were not high. The amounts of airborne microbes in different areas varied with human activities and environmental parameters. The results presented here suggested to us that we should continuously improve the purity of indoor air in the Great Wall Station.

The amounts of microbes from different sources in the Great Wall Station were arranged in order of magnitude, i. e. ice and snow $<$ lake water $<$ virgin soil. The content of microbes in the sea shore was considerable, i. e. about 10^3 CFU/cm^3 (or g). The microbes analysed consisted of more than 11 genera in which Bacillus was occasionally found. The result obtained from temperature test indicated that most of the bacterial strains from the open air were adapted to lower temperature. There were no strains capable of living at or over 25℃ as yet. Low temperature is one of the important factors controll ing microbial ecology in the Great Wall Station. In such an environment as the Great Wall Station, cryophiles play a persistent and important role in the transformation and circulation of materials.

Key words Great Wall Station, Microbes In Land, Sea And Air, Analysis Of Genera And Groups Of Microbes, Microbial Response To Temperature

南极磷虾的微生物研究*

一、引言

南大洋生态系统是一个各成员互为影响、互为作用的复杂动态变化系统，其重要成员——南极磷虾的生长繁殖、集散活动等受种种环境、生态因子的制约，无疑地也与环境的微生物相关联。南大洋水环境生存着为数不少的微生物[1~5]，磷虾体表乃至体内毫无例外地经受着微生物的作用[4]。磷虾滤食硅藻等微小生物，其中浮游植物赖以生存的氨、磷酸盐、二氧化碳及其他营养物，很大部分得靠细菌分解有机体产生。细菌等微生物将南大洋中溶解的有机质转变为自身细胞的物质或其他形式的有机质，为磷虾等所利用。此外，细菌等微生物还帮助磷虾消化食物。另一些微生物对磷虾的健康、磷虾制品的保鲜具有直接的经济意义。

随着南极磷虾的开发利用，相关的微生物学课题必然会提到研究日程上来。至今所见有关研究文献实在太少[6,7]。本文着重研究新鲜南极磷虾体内外的微生物状况，目的旨在为捕捞、贮存、运输和加工南极磷虾诸过程提供一些微生物学方面的资料。

1987 年 1～2 月，中国“极地”号考察船在南设得兰群岛海域考察中，曾在南极磷虾集群区根据船上配备的高频鱼探仪监测磷虾集群，进而试捕磷虾，并分五批次对新鲜磷虾作了微生物分析。这 5 个站的大体范围为 60°25′～61°43′S 和 54°50′～58°29′W，面积约 17 000 n $mile^2$（图 4-2-1）。

二、材料和方法

1. 培养基

（1）蛋白胨酵母膏培养基。蛋白胨 2 g；酵母膏 1 g；$FePO_4$ 0.01 g；琼脂 18 g；陈南大洋水 1 000 mL；pH＝8.0。用于海洋细菌。

（2）TCBS 培养基。蛋白胨 10 g；酵母膏 5 g；枸橼酸钠 10 g；氯化钠 15 g；牛胆汁粉 5 g；枸橼酸铁铵 1 g；硫代硫酸钠 10 g；蔗糖 10 g；去氧胆酸钠 3 g；0.2％麝香草酚蓝 20 mL；蒸馏水和陈南大洋水各 500 mL；pH＝8.4。用于弧菌。

（3）蛋白胨葡萄糖、酵母膏培养基。蛋白胨 5 g；葡萄糖 2 g；酵母膏 1 g；琼脂 18 g；陈南大洋水 1 000 mL；pH＝5.5。消毒后分装前每升培养基各加氯霉素和链霉素 30 mg。用于海洋真菌或酵母。

以上培养基作成平面或斜面，用于微生物的培养、分离或保种。

2. 采样

当捕到南极磷虾的网拉至后甲板后，立即用洁净镊子挑取若干只活磷虾，放入洁净培

* 原文载于《中国第三次南大洋考察暨环球重力调查报告》1993，103-108

养皿中，送至船上试验室待试。

采样站位参见图 4-2-1。

3. 微生物分析

(1)定量取磷虾体表部分，于玻璃研磨器中研成匀浆，视情况加稀释用盐水摇匀后待用。

(2)对所取磷虾消化道作同上处理。

(3)将上述经适度稀释并摇匀的样品立即在各自的平皿培养基上涂布。在 2 或 14℃条件下培养 2 周。待长出菌落后计算 CFU[1]，进行分析比较，并挑取单菌落，作纯化处理，于斜面培养基上培养保存，直至运回国内实验室中作进一步研究。

以上各项，均需严格按无菌操作步骤进行。

(4)微生物分类至属的水平。细菌、真菌和酵母分别以文献[8~12]的规定进行。

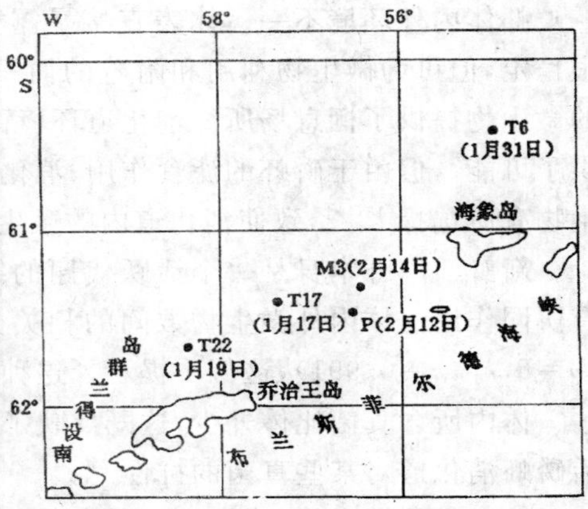

图 4-2-1　磷虾微生物分析站位示意图

对磷虾做微生物分析的磷虾采集站

时间：1987 年，日期系南极地方时

海区：南大洋乔治王岛至海象岛海城

站置：T6 站——60°25′02″S，54°50′0″W

T17 站——61°27′33″S，57°20′50″W

T22 站——61°42′13″S，58°28′22″W

P 站——61°28′21″，56°25′19″W

M3 站——61°21′0″S，56°26′0″W

三、结果和讨论

(一)南极磷虾微生物数量的估计

表 4-2-1 列出了磷虾消化道和体表的微生物数量。

表 4-2-1　不同站位磷虾体内外微生物数量的比较*

站号	体表	消化道内	备注
T6	(200375) (细菌)	30737.5 (细菌)	网内被囊类动物多
T17	493 (细菌 488，真菌 10，酵母 0)	164.5 (细菌 147，真菌 2.5，酵母 15)	未测
T22	T51 细菌 750，酵母 1	825 (细菌)	网内有不少被囊类动物混入
P	64 000 (细菌)	7037.5 (细菌)	网内磷虾很纯净
M_2	14825 (细菌 3750，弧菌 10075，真菌 1 000)	5075 (细菌 5000，弧菌 75)	网内磷虾纯，杂物较少

* 体表以 $CFU \cdot cm^{-2}$ 计；消化道以 $CFU \cdot g^{-1}$ 计。

由表 4-2-1 可见，南极磷虾虾体常聚集环境中的微生物。体表的微生物数量为 493～200 375CFU · cm^{-2}，几何均值为 10 249 CFU · cm^{-2}。真菌和酵母数量小。消化道的微生物数量为 151～30 737.5 CFU · g^{-1}，其几何均数为 2 674 CFU · g^{-1}。真菌和酵母仍少。虾体内外的微生物数量均比水中的多[1]。

以上结果表明，磷虾本身就是微生物的附着基质，加之其体小，水多，故能吸引环境中部分微生物的到来。

虾体内外环境不一。体表直接暴露于水环境中。比起消化道来,作为环境的体表不很稳定,但可为微生物利用和附着的面积却大了。营养物较稀,但种类较多,因而为众多的微生物提供了栖息场所。消化道环境稳定,营养丰富,从而给滞留为数不少的微生物提供了可能。但由于磷虾的滤食作用、消化道的抗菌活性和较少的氧气,对进入消化道的微生物有所选择[13,14],致使消化道内的微生物数一般少于体表的微生物数。

新鲜活磷虾本身是一个活跃代谢的统一有机体,故体内外控制微生物群落的机制应有协同作用。体内外微生物数间的相关分析表明,两者间的正相关系数高达 0.992 以上($n=5, r_{0.005}=0.991$),这似乎提示了这种共同点。

体内所含真菌比体外少,这既反映了环境中真菌和酵母菌较少这一事实[1],也表明了活磷虾消化道对某些真菌的抗性[7]。

消化道的微生物数量比水环境中的多这一事实也表明磷虾吞食水中的细菌等微生物,显示了微生物在磷虾摄取的食物方面占有一定地位。

表 4-2-1 的结果还表明,在不同批次磷虾的微生物数量之间,存在着比较明显的差距。这与不同站位间的水文、化学、生态状况有关,也与不同水环境中的微生物状况有关。

(二)磷虾微生物分类

1. 细菌

从磷虾上挑出细菌共 42 株,其中 17 株来自体表,25 株来自体内。对其中的 29 株分至属的水平,它们是:

假单胞菌 *Pseudomonas*

不动杆菌 *Acinetobacter*

黄杆菌 *Flavobacterium*

莫拉氏菌 *Moraxella*

葡萄球菌 *Staphylcoccus*

微球菌 *Micrococcus*

弧菌 *Vibrio*

表 4-2-2 列出了磷虾身上各属菌在各站位的分布情况。

表 4-2-2 不同属的磷虾细菌在站位上的分布

站号	假单胞菌	不动杆菌	莫拉氏菌	黄杆菌	弧菌	葡萄球菌	微球菌	总计
T6	1	1	1	0	0	0	0	3
T17	3	1	0	0	1	1	0	6
T22	0	0	1	0	0	0	1	2
P	6	0	3	1	0	0	0	10
M_3	4	4	0	0	0	0	0	8
总	14	6	5	1	1	1	1	29

从表 4-2-2 可见,假单胞菌最多,假单胞菌、不动杆菌和莫拉氏菌分布最广。黄杆菌、葡萄球菌、微球菌和弧菌较少。T17 站磷虾的菌属最多,T6 和 P 站次之,T22 和 M3 站最少。磷虾微生物的这种站间差异不仅表现在微生物数量上,而且表现在属群组成上,可能

与环境一生态因子的复合作用有关。表 4-2-2 仅反映了此次对有限磷虾所做的分析结果,只能从一个侧面说明不同站位磷虾微生物的差别。由于水体的不同,船只的冲击,也许还有人为的因素均可能对磷虾的微生物组成产生种种影响。表 4-2-3 所列资料,可能有助于分析和理解这些差异。

表 4-2-3　磷虾微生物数量及有关因子 *

站号	磷虾微生物数		N/P	NO_2-N (μmol/L)	NO_3-N (μmol/L)	T℃	S	南极大磷虾 ($\times m^{-3}$)
T6	200375(表)	30737.5(内)	8.20	0.18	13.43	0.054	33.450	6536
T17	493(表)	151(内)	7.29	0.19	14.66	0.495	33.749	86
T22	751(表)	825(内)	12.77	0.17	18.92	0.691	34.122	18

* NO_2-N,NO_3-N,N/P 由王方国等测定;T,S 由李金洪等测定,大磷虾由陈时华等测定。“表”指大磷虾体表,“内”指大磷虾消化道,“表”的微生物数单位为 CFU·cm^{-2}。

磷虾体内外的微生物属群比较如表 4-2-4 所示。

表 4-2-4　磷虾体内外细菌组成(%)

部位＼菌种	假单胞菌	不动细菌	莫拉氏菌	黄杆菌	弧菌	葡萄球菌	微球菌
消化道	45.45	27.27	13.64	4.55	4.55	4.54	0
体　表	57.14	0	28.57	0	0	0	14.29

表 4-2-4 表明,磷虾体内细菌以假单胞菌为主,次为不动杆菌;体表细菌则以假单胞菌为主,次为莫拉氏菌。由此初步结果可见,与磷虾相关的细菌主要是假单胞菌、不动杆菌和莫拉氏菌。

磷虾消化道细菌属的组成表明,其细菌群系不是单一的,这与其他海洋动物有类似之处[15]。磷虾的微生物以革兰氏阳性菌为主,这与其水环境的微生物属群有某些致性,但其菌谱比水环境的窄[1],表明磷虾消化道中的部分细菌可能是消化道中的共生菌,因而于食物消化有益,也显示了磷虾对一些革兰氏阳性菌有抗性[4,7,13]。

从磷虾身上分离出来的一些细菌,对磷虾的保存可能具有某些积极的作用,因而将影响磷虾的经济效益。与此有关的问题,尚待深入研究。

2. 真菌和酵母

只从 T17 站的磷虾外壳上挑出一株枝孢菌(*Ramulispora*)。

共挑出 6 株酵母,其中 1 株来自 T17 站磷虾消化道,4 株来自体表,还有 1 株来自 T22 站磷虾的体表。这 6 株酵母中,有两株是红酵母(*Rhodotorula*),其他 4 株待定。这些结果暗示酵母可能是磷虾食饵之一。

由于一些真菌和酵母具有寄生性和致病性,因而它们对磷虾的影响值得研究。

四、小结

(1)南极磷虾虾体常集聚环境中海洋好气异养微生物。体表的微生物数量常大于体内消化道。其中，细菌占优势。

(2)经初步鉴定，与磷虾相关的细菌有 7 属，以假单胞菌、不动杆菌和莫拉氏菌为主。细菌属的组成在磷虾体内有差别。消化道内少见革兰氏阳性菌，推测可能与磷虾消化道对一些革兰氏阳性菌有抗性有关。磷虾的菌谱比水中的窄。

与南极磷虾相关的真菌和酵母数量不大，群属也少。

(3)初步分析表明，磷虾微生物的数量、属群受环境一生态因子的制约。

参考文献 15 篇(略)

(合作者：孙修勤　宋庆云)

MICROBES IN THE AREA OF THE SOUTH SHETLAND ISLANDS*

Abstract Investigations were made on the microbes in the air, bottom sediments, surface seawater in the sea area of the South Shetland Islands during December, 1986～March, 1987.

The results obtained show that: from the air over the South Shetland Islands and the Maxwell Bay, the detection rate of terrigenous microbes was more than 90% and their amount was generally a few CFU/m^3; the detection rate of marine microbes was more than 50% and their amount only 1 CFU/m^3. The amounts of airborne microbes over the surveyed area were the witness to their diurnal variations. They were influenced to some extent by nearby islands. Air temperature and relative humidity, etc. were factors affecting their quantitative variations.

The bacterial amount of the outshore surface sea water in the Great Wall Bay, Antarctica was about $n \times 10$ CFU/cm^3 ($0 \leqslant n < 10$, the same below), the amount of microbes decreased with the monthly dropping of seawater temperature and showed a state of diurnal variations. The seawater in the Bay has been fairly clean. The microbial community consisted of at least thirteen genera, in which *Pseudornonas* was common.

In the intertidal zone of the Deception Island, there were only a few CFU of bacteria per-cubic meter in seawater; and $n \times 10$ CFU/g(w. w) in the surface sediment; and $n \times 10$ CFU/g(w. w) in the volcano ash.

The results reveal the general features of the microbes from the air, sea and land in the South Shetland Islands, Antarctica. It provides data for further study of microbial resources in that area.

Key words South Shetland Lslands, Maxwell Bay, Great Wall Bay, Deception Lsland, Microbe.

1. Introduction

An investigation was made on microbes in the area of the South Shetland Islands, Antarctica from December, 1986 to March, 1987. Its main contents included analysis of the airborne microbes over the sea area of the South Shetland Islands and Maxwell Bay, those from the waters of the Great Wall Bay and the Deception Island. A preliminary understanding of the microbial characteristics of the air, sea and land of these areas has

* The original text was published in *Antarctic Research*, 1994, 5(2): 1～8.

been arrived at, which has laid a preliminary basis for further investigation of the microbial resources in the area of the South Shetland Islands, Antarctica.

2. Material and method

2.1 *Culture media*

Seawater culture medium: its ingredients: yeast extract, 1 g; peptone, 2 g; $FePO_4$, 0.01 g; refined agar 18 g; aged seawater from the Southern Ocean, 1 000 cm^3; pH, 8. It is used for culturing marine microbes.

Fresh water culture medium: its ingredients: beef extract, 3～5 g; peptone, 10 g; Nad, 5 g; refined agar, 18 g; distilled water, 1 000 cm^3; pH, 7.2～7.4. It is used for culturing terrigenous microbes.

2.2 *Methods for sampling*

A JWL-4 type airborne microbe inspecting instrument is used for collecting airborne microbes. Collecting time is 15 or 30 min. Air flow rate during sampling is 37 dm^3/min.

An impacting type water sampler is used for collecting sea water microbes.

The spread suspension method for microbial samples is performed on the plate of the above-listed seawater culture medium in culture Petri dishes. This procedure is used for the microbes from beach sediments, too.

The detailed method has been described in another paper (Chen et al., 1990).

The above-mentioned samples are cultured at 10℃ for about 10 days, then, the colonies forming units (CFU) on the culture plates in Petri dishes are counted. The calculating formulas for airborne microbes are as follows:

Counts of airborne microbes (CFU/m^3)

$$=\frac{\text{Average CFU/dish}}{\text{Sampling time(min.)} \times \text{Air flew rate during sampling}(dm^3/\text{min})} \times 1\,000$$ and

Counts of seawater microbes (CFU/cm^3)

$$=\frac{\text{Average CFU}/cm^2 \times \text{Area of each dish}(cm^2)}{\text{Volume of seawater sample}(cm^3)}$$

The procedure for determination of the microbial count of sediments is similar to what has been mentioned above, and its unit is CFU/g (fresh, wet). Some colonies were selected, purified, kept in the refrigerator (0℃), and brought back home for taxonomic study (Oliver, 1982; Skerman, 1967).

3. Result and discussion

3.1 *airborne microbes over the South Shetland Islands*

The inspection of the airborne microbes over the South Shetland Islands was performed on the Vessel "Ji Di".

The results are shown in Table 1. From Table 1, it can be calculated that the detection rate of airborne microbes in the investigated area was 58.3%, and the number of the microbes averaged 0.776 CFU/m^3 with a maximum of 4 CFU/m^3. The detection rate of the airborne terrigenous microbes was 91.7%, and their number averaged 5.56 CFU/

m^3. These results show that the number and detection rate of the terrigenous microbes were both higher than those of the marine ones, which probably suggests that many airborne microbes of the sea area of the South Shetland Islands are closely related with those of the nearby islands.

The results from the correlation analysis of number of the above two airborne microbes and the relevant air temperature or relative humidity showed that the number of the marine microbes has little correlation with the air temperature but remarkably the number of terrigenous microbes has a negative correlation with the corresponding air temperature ($R=-0.687$, $n=12$), indicating that the airborne terrigenous microbes over that sea area are adaptable to low air temperature.

Table 1 airborne microbial numbers over the sea of the South Shetland Islands

Sampling date	Sampling location	Microbial numbers (CFU/m^3)		Air condition	
		Marine	Terrigenous	Temperature (0℃)	Relative humidity (%)
Jan. 17, 1987	6151S, 5653W	0	4	7.5	91
Jan. 18, 1987	6154S, 5804W	0	0.67	4.2	87
Jan. 19, 1987	6222S, 5740W	0	2	5.8	88
Jan. 20, 1987	6259S, 6040W	0.67	6.7	8.5	92
Jan. 21, 1987	6303S, 6031W	3.3	6.7	8.8	94
Jan. 30, 1987	6121S, 5345W	0.67	13.3	1.2	96
Jan. 31, 1987	6024S, 5403W	0	4.7	4.4	93
Feb. 1, 1987	6050S, 5634W	0	10	5.8	86
Feb. 11, 1987	6051S, 5652W	0.67	12	4.2	93
Feb. 12, 1987	6111S, 5615W	4	2.7	5.0	97
Feb. 13, 1987	6139S, 5632W	0	4	2.5	95
Feb. 14, 1987	6145S, 5615W	0	0	9.8	91

3.2 *The airborne microbes over the sea area of the Maxwell Bay*

The sampling records of the airborne microbes over the area of the Maxwell Bay as listed in Table 2 show the quantity of the airborne microbes over that area throughout the austral summer, i. e. the detection rate of the airborne marine microbes was 53.7%, its amount averaged 1.53 CFU/m^3, the highest one was 10.7 CFU/m^3, on 10th Jan., 1987. The detection rate of the airborne terrigenous microbes was 94.3%, their quantity averaged 6.45 CFU/m^3 and the highest one was 100 CFU/m^3 on the 11th Jan., 1987. It seems that only the detection rate of the marine microbes in the Maxwell Bay was slightly lower as compared with that of the airborne terrigenous ones over the South Shetland Islands. But both the kind and amount of the microbes in the bay were more than the latter, which may be ascribed to its geographic location not far from the King George Island, where plenty of people live and other organisms were more abundant.

Table 2 List of airborne microbial numbers of the Maxwell Bay *

Sampling	airborne microbial Number (CFU/m³)		T** (℃)	Sampling	airborne microbial Number (CFU/m³)		T** (℃)
Date and Time	Marine	Terrigenous		Date and Time	Marine	Terrigenous	
Dec. 27, 1986	1.3	1.3	6	Jan. 25, 1987	2.7	0.67	2.5
Dec. 28, 1986	2.7	1.3	5	Jan. 26, 1987	2	0	2.0
Dec. 29, 1986	9.3	14.7	3	Jan. 27, 1987	0	0	2.7
Dec. 30, 1986. 15:00	0	6	10	Jan. 28, 1987	3.5	4.7	0.9
Dec. 30, 1986. 20:00	3.3	—	2.0	Jan. 29, 1987	0.67	0.67	−0.2
Dec. 31, 1986	9.3	13.3	4.8	monthly mean	1.68	9.24	7.25
monthly mean	4.32	7.32	5.13	Feb. 2, 1987	0	24.7	6.7
Jan. 1, 1987	6.7	5	6.5	Feb. 3, 1987	1.3	3.3	0.6
Jan. 2, 1987	0.67	0.67	6.2	Feb. 4, 1987	0	3.3	2.8
Jan. 3, 1987. 15:30	0	0.67	8.0	Feb. 5, 1987	0.67	4	3.5
Jan. 3, 1987. 22:15	1.3	2	3.5	Feb. 6, 1987	0	2	3.0
Jan. 4, 1987. 03:15	0	0	3.1	Feb. 7, 1987	0	0.67	1.2
Jan. 4, 1987. 10:00	4.7	0.67	4.1	Feb. 8, 1987	0	4.7	3.0
Jan. 4, 1987. 15:30	1.3	2.7	7.1	Feb. 9, 1987	0.67	0.67	1.5
Jan. 5, 1987	1.3	1.3	7.5	Feb. 10, 1987	0.67	2	5.5
Jan. 6, 1987	3.3	8.3	2.5	Feb. 15, 1987	0	1.3	3.5
Jan. 7, 1987	0	0.67	6.8	Feb. 16, 1987	0	3.3	5.2
Jan. 8, 1987	2	1.3	8.8	Feb. 17, 1987	0.5	1.3	4.3
Jan. 9, 1987	0	9.3	5.2	Feb. 18, 1987	0	0.67	3.5
Jan. 10, 1987	10.7	10	0.2	Feb. 19, 1987	0.67	0.67	2.8
Jan. 11, 1987	0	100	5.6	Feb. 27, 1987. 02:00	2.7	2.7	0
Jan. 12, 1987	0	20.7	6.5	Feb. 27, 1987. 20:00	4	1.3	4
Jan. 13, 1987	0	5.3	10	monthly mean	1.12	1.61	3.33
Jan. 14, 1987	0.67	2	5.5	Mar. 11, 1987	0	2	3.5
Jan. 15, 1987	3.3	2.7	3.8	Mar. 12, 1987	0	35.3	5.4

(continue)

Sampling	airborne microbial Number (CFU/m³)		T** (℃)	Sampling	airborne microbial Number (CFU/m³)		T** (℃)
Date and Time	Marine	Terrigenous		Date and Time	Marine	Terrigenous	
Jan. 16, 1987	0.67	7.3	3.0	Mar. 13, 1987	0	5.3	1.8
Jan. 22, 1987	0	2	3.4	Mar. 14, 1987	0	2.7	0
Jan. 23, 1987	0	7.3	4.0	Mar. 15, 1987	0	8	3.8
Jan. 24, 1987	0	4.7	2.5	monthly mean	0	10.66	2.90

* Two sampling positions: one located at 621806S, 582309W from Dec. 27, 1986～Jan. 29, 1987; the other at 621408S, 583504W from Feb. 2～Mar. 15, 1987. ** T: Air temperature.

The airborne microbes over the Maxwell Bay were different in quantity in different months. The relationship between the airborne terrigenous microbial number and air temperature was similar to that of the South Shetland Islands. The airborne marine microbial number showed the trend of decreasing month by month with the air temperature decreasing month by month.

Fig. 1 was drawn by using the data of the observation for five times running from 3rd to 4th Jan., 1987. It shows that quantatively the peak of marine airborne microbes in the air over the bay was at 10:00, and that of the terrigenous microbes at 15:30. Large numbers of them also appeared at 22:15 but they counted least at 3:15 before dawn. It seems that there is little correlation between their appearance and air temperature.

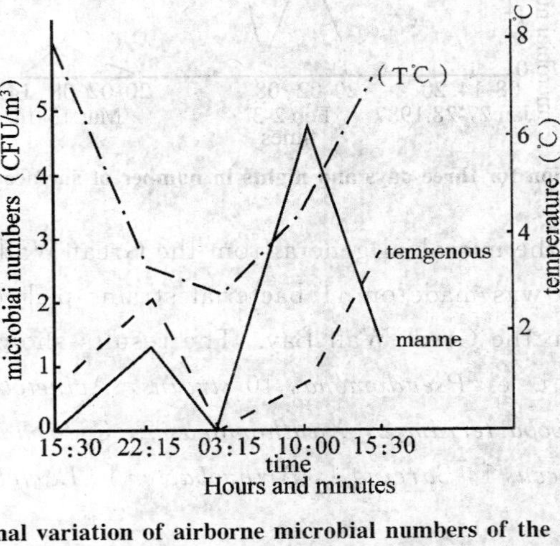

Fig. 1　Diurnal variation of airborne microbial numbers of the Maxwell Bay

3.3　*The outshore surface water microbes of the Great Wall Bay*

3.3.1　*Quantity and its variation*

Investigations were made on the outshore surface seawater bacteria in the Great Wall Bay for 12 times and the data were obtained mainly at two fixed stations, i. e., one located at 621806S, 582109W from the 27th to the 28th Jan., 1987; and the other at 621468S, 555504W from 2nd to 3rd Feb., 1987. The results showed that their number ranged from 0 to 515 CFU/cm^3 with an average of about 376 CFU/cm^3. The number was obviously less than that of the inshore surface-sea water of the Great Wall Bay by one order of magnitude, or less than that of beach seawater of the Bay by two orders of magnitude; and more than that of the water of the Southern Ocean by one order of magnitude (Chen et al., 1990), but less than that of the clean seawater of the Jiaozhou Bay, China by three orders of magnitude (Chen, 1981). All of this showed that the seawater of the Great Wall Bay is pretty clean at present.

Fig. 2 shows the diurnal variations of the microbial number of the outshore surface seawater of that Bay, i. e. its numerical peak appeared mostly at 2:00, and its low level at 20:00. The microbial numbers in the outshore surface water decreased day by day as time went on, which was inferred to be related to the seawater temperature changes and was consistent with the monthly variation of marine microbial number in the air as shown in Table 3. So it is suggested that there is a some what close relationship between marine microbes in the air and in seawater.

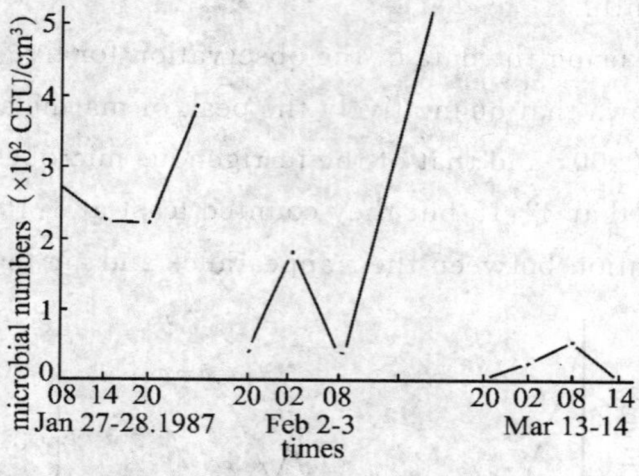

Fig. 2 Variation for three days and nights in number of surface water microbes

3.3.2 Outline of the microbial genera from the Great Wall Bay

A taxonomic study was made on 41 bacterial strains picked out randomly from among isolated ones from the Great Wall Bay. The results show that they belong to 13 genera and one family, i. e. *Pseudomonas* 10 *strains*, *Acinetobacter* 7, *Flexibacter* 7, *Photobacterium* 3, *Flavobacterium* 2, *Xanthomonas* 2, *Cytophaga* 2, *Neisseria* 2, *Alcaligenes* 1, *Staphylococcus* 1, *Sarcina* 1, *Arthrobacter* 1, *Lactobacilus* 1 and *Enterobacteriaceae* 1 *strain*.

From the above, it can be seen that *Pseudomonas.* is in the majority, next come *Acinetobacter* and *Flexibacter*. This bacterial composition is not in common with that from the beach nearby Great Wall Station, Antarctica (Chen *et aL.*, 1992).

3.4 *The microbes of the Deception Island*

Deception Island (62°59′04″S, 60°40′34″W) is located 13 km from the southern side of the Livingston Island. It is an active volcano island, and the island arc takes the shape of a horse's hoof. The surface water within the bay was calm when we arrived at the island, and the temperature was about 10℃. This volcano erupted once again in August, 1970. The remaining traces of the dark brown volcanic lava flowning into sea still came clearly into view.

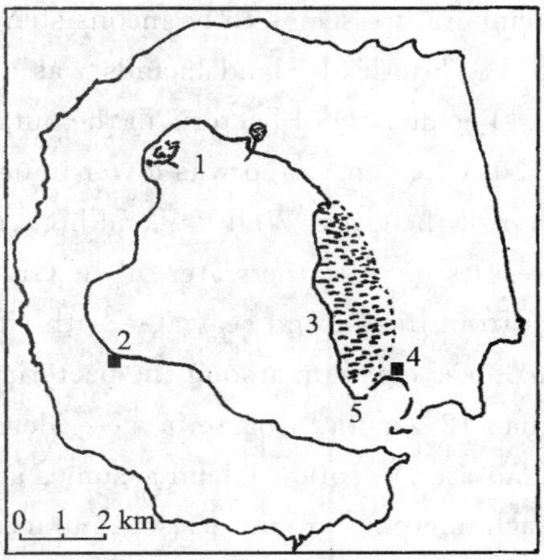

Fig. 3 The map of sampling stations in the Deception Island

1. Volcano's ejecting point in 1967. 2. Marina De Gunea Station of Argentina.

3. Area covered with volcanic ashes, Feb., 1969. 4. Skur Station of Chile. 5. Whaler's Bay

Table 3 Microbial numbers in the intertidal zone of the Deception Island

Sampling location	Microbial number	Description of samples
Skur Station of Chile	4258.3 CFU/g	Volcanic rocks, magn and ashes, fine-medium sands mainly, brown colour, light, dry, 10℃
Marina De Gunea Station of Argentina	2.5 CFU/cm^3	Surface water in tidal zone, pH 7.2, 10℃
Marina De Gunea Station of Argentina	4753 CFU/g	Surface sediment samples were fresh and wet 10℃

* Sampling date: Jan. 21, 1987. Microbial numbers were determined according to culture Petri dish method, the sediment samples were fresh and wet.

Fig. 3 shows a map of sampling stations in the Deception Island and Table 3 lists the analytic results of the samples of microbes. From the table, it can be seen that some microbes exist in the cooling volcanic ashes, but their quantities were less than those in the virgin soil of Great Wall Station by one order of magnitude (Chen et al., 1992), and more than those of surface seawater of the island by three orders of magnitude, and less than those of the intertidal waters of Great Wall Station by three orders, even though it was said that the waters nearby the tidal zone of the island contained a high biomass.

4. Conclusion

The results of the investigations of the airborne microbes over the sea areas of the South Shetland Islands and the Maxwell Bay indicate that detection rates of the airborne microbes of the two areas were fairly high. The relationship between the number of terrigenous microbes and air temperature seems to be inconsistent. The change in the number of microbes in the air of the South Shetland Islands was also not conformable to the variation of air temperature. The density of bacteria in the outshore seawater of the Great Wall Bay was about 3.76×10^2 CFU/cm^3, and was lower in quantity than that in the micro-algae-bearing beach water of the Great Wall Bay and higher than that of the waters of the Southern Ocean. The results indicate the water of the Great Wall Bay is quite clean.

The microbial numbers from the air and seawater both showed a certain trend diurnal variations. *Pseudomonas.* was common among the bacteria from the waters the Great Wall Bay. Besides, more than 12 genera of bacteria were identified.

The active volcanic island arc-Deception Island assumes a specific geographical landscape, its terrestrial and beach microbes are not present in large numbers. The results indicate that it is still less subjected to the effects of human activities.

References 5(abr.)

(Cooperator: Song Qingyun)

桑沟湾细菌的研究*

摘　要　于1989/1990年度，用MPN法在三种培养基上据桑沟湾生态环境和增养殖状况，选择7个站作其表层水和表层沉积物的海洋好气异养细菌、几丁质降解菌和弧菌的时空分布研究，分析了各介质中细菌学指标间的相关性，估计了该水域中细菌含碳量，认为该湾系富生产力的浅水生态系统，但也面临着一定的致病条件。

关键词　桑沟湾　细菌　生态环境　海水增养殖

桑沟湾是山东半岛东端由丘陵环抱的一个半封闭海域，受海水增养殖、水产捕捞和陆源污染的影响，环境压力处于不断增强之中，其生态系统已发生很大变化[1]，预计还将向不利生态平衡方向发展。曾对该水域环境、动植物作过一些研究[1~4]。但有关微生物，如总好气异养细菌(general aerobic hetertrophic bacteria 简写作 H，下同)，条件致病菌——几丁质降解菌(Chitin—degrading bacteria，简写为 C，下同)和弧菌(Vibrio，简写为 V，下同)及其变动状况和分析，只是陈世阳等(1987)、陈碧鹃等(1999)对山东乳山湾有相关研究发表过，对本湾尚未见诸报道。而海洋细菌恰恰是特定水域物质循环、能流和生产力的一个要素，C，V，C/H 和 V/H 等又是衡量环境状况、养殖业健康水平的重要参数和指标[5,6]，为此本文论述该湾表层水和底质中该3类细菌的时空状况，初步估计细菌生物量，为将来评价该湾细菌的作用提供相应基础资料。

1. 材料和方法

1.1　站位及相关参数描述

图1为该湾测站分布状态。共设7个站，各站的一些环境—生态因子数据大体范围如文献[1]所列。按离岸远近划分，4个站为近岸站，第3站为中部区，另2站(第2站和第4站)为远岸站。

1.2　培养基

所用培养基共3种，即海洋好气异养菌培养基[5]、几丁质降解菌培养基和弧菌培养基[7]。

1.3　取样和计数

按国家技术监督局发布的《中华人民共和国国家标准·海洋调查规范(试行本)》中有关微生物调查的要求取样。用MPN法测试。样品置室温中培养1～2周后，记数并换算成单位体积(cm^3)或单位重量(g，新鲜湿底质，下同)样品中的细胞数，分别计算 H，C，V 及 C，V 各占 H 量的比率，即以 C，V，C/H 或 V/H 等细菌学参数说明在调查区中 H 及这2类条件致病菌含量、所占比率，作出对养殖生态环境影响的评价。据异养菌(H)总量换算出相应细菌含碳量(生物量)，单位为 $mg\ C \cdot m^{-3}$ 或 $mg\ C \cdot t^{-1}$(下同)，估计细菌在该湾生产力中的作用。

* 原文刊于《海岸工程》，2001，20(1)：72-80。

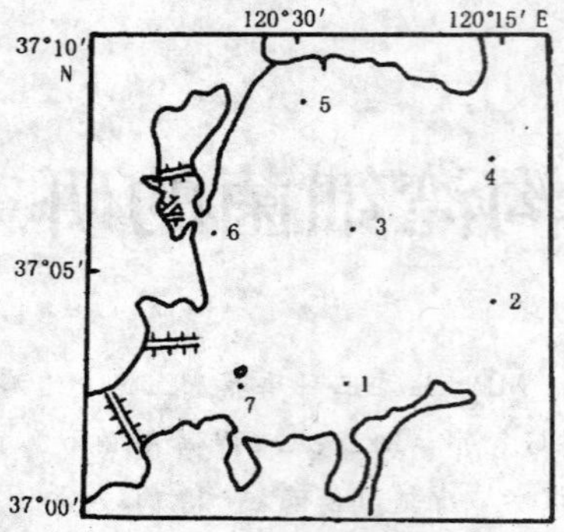

图 1　桑沟湾表层水和底质细菌学测站示意图

2. 结果和讨论

2.1　H,C,V,C/H 和 V/H 等的一般状况

表 1、2 分别列出了所测站表层水和底质中的 H,C,V,C/H 和 V/H 等各个数据，由该两表可见水或底质中 H 的大体范围及各自的平均值。底质中的 H 约比水中的高 1 个量级。水或底质中的 C 和 V 的范围及各自的平均值也是底质中高于水中。底质中 C/H 高，约是水中的 3.5 倍。底质 V/H 也高，是水质中的 4.5 倍多，证实底质更适于 C 和 V 生长。

2.2　H,C 和 V 等的季节变化状态

图 2 中的 4 张图表达了表层水或底质中 H,C 和 V 等的季节变化势态，由图 2 可见，水及底质中 H,C,V 的大小排序及 C/H,V/H 的季节走势。

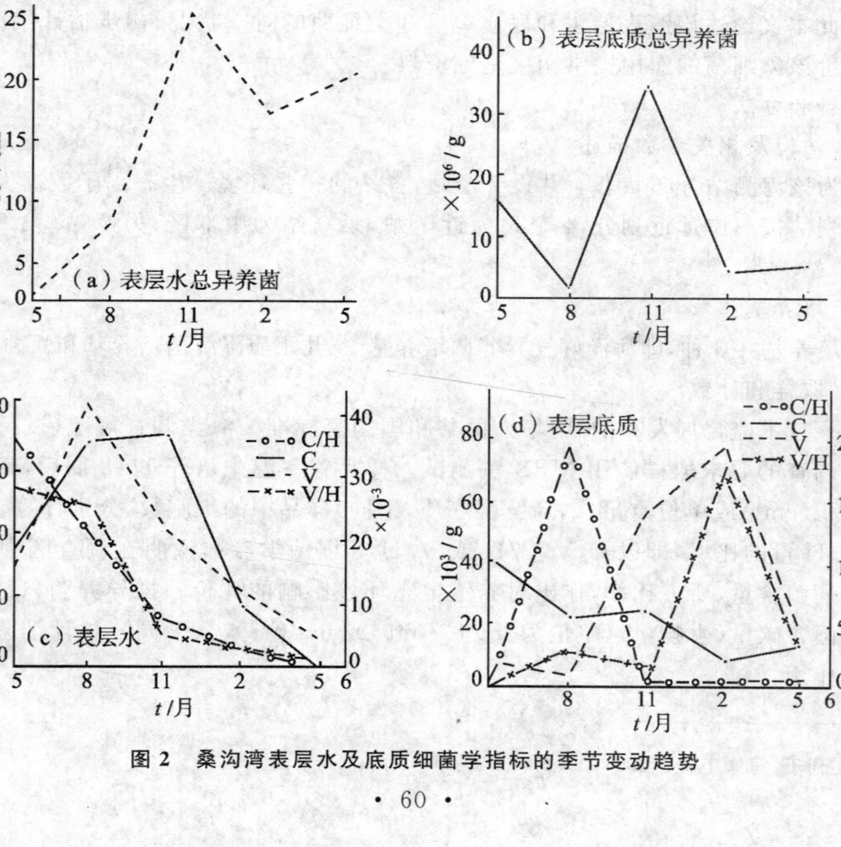

图 2　桑沟湾表层水及底质细菌学指标的季节变动趋势

表 1　桑沟湾表层水细菌含量状况表

站号	细菌简称	1989－05			1989－08			1989－11			1990－02			1990－05			平　均		
		含量	C/H	V/H	含量	C/H	V/H	含量	C/H	V/H	含量	C/H	V/H	含量	C/H	V/H	含量	C/H	V/H
1	H	1.5×10^{6}			2.1×10^{6}			3.9×10^{6}			1.1×10^{6}			1.1×10^{7}			2.57×10^{6}		
	C	1.1×10^{3}	7.3×10^{-3}		3.0×10^{0}	1.4×10^{-5}		1.2×10^{3}	30.8×10^{-4}		2.7×10^{2}	24.5×10^{-5}		3.0×10^{1}	2.7×10^{-6}		5.21×10^{2}	2.0×10^{-4}	
	V	4.0×10^{1}		2.8×10^{-4}	4.3×10^{2}		20.5×10^{-4}	9.3×10^{2}		23.8×10^{-4}	2.0×10^{2}		18.2×10^{-5}	6.0×10^{1}		5.4×10^{-6}	3.32×10^{2}		1.3×10^{-4}
2	H	4.3×10^{4}			1.1×10^{6}			7.2×10^{4}			1.1×10^{6}			3.4×10^{6}			3.33×10^{6}		
	C	3.0×10^{1}	7.0×10^{-4}		1.1×10^{4}	1.0×10^{-2}		9.1×10^{1}	12.6×10^{-4}		2.1×10^{3}	19.1×10^{-3}		9.0×10^{1}	26.5×10^{-5}		2.66×10^{3}	8.0×10^{-3}	
	V	1.9×10^{1}		4.4×10^{-4}	9.0×10^{0}		8.0×10^{-6}	1.5×10^{3}		20.8×10^{-3}	1.5×10^{3}		136.4×10^{-4}	/		/	7.57×10^{2}		2.3×10^{-3}
3	H	1.5×10^{4}			1.1×10^{6}			2.0×10^{6}			2.0×10^{4}			9.3×10^{3}			2.69×10^{6}		
	C	1.1×10^{3}	7.3×10^{-2}		4.0×10^{0}	3.6×10^{-6}		3.0×10^{1}	1.5×10^{-4}		2.0×10^{2}	1.0×10^{-2}		3.0×10^{1}	3.2×10^{-3}		2.73×10^{2}	1.0×10^{-3}	
	V	3.0×10^{1}		2.0×10^{-3}	9.0×10^{1}		8.2×10^{-5}	4.0×10^{1}		2.0×10^{-4}	9.0×10^{1}		4.5×10^{-3}	1.6×10^{2}		17.2×10^{-3}	8.2×10^{1}		3.0×10^{-4}
4	H	2.0×10^{4}			1.1×10^{6}			1.5×10^{6}			2.0×10^{4}			2.4×10^{4}			5.33×10^{6}		
	C	1.1×10^{3}	5.5×10^{-2}		2.0×10^{1}	1.8×10^{-5}		2.0×10^{2}	1.3×10^{-4}		1.5×10^{2}	7.5×10^{-3}		/	/		3.68×10^{2}	6.9×10^{-4}	
	V	4.0×10^{1}		2.0×10^{-3}	4.0×10^{1}		3.6×10^{-5}	3.9×10^{2}		2.6×10^{-4}	9.0×10^{1}		4.5×10^{-3}	1.6×10^{2}		6.7×10^{-3}	1.44×10^{2}		2.7×10^{-4}
5	H	2.9×10^{4}			1.1×10^{6}			7.0×10^{4}			3.0×10^{4}			1.1×10^{6}			4.66×10^{6}		
	C	1.1×10^{3}	3.8×10^{-2}		2.9×10^{1}	2.6×10^{-5}		9.0×10^{1}	12.9×10^{-4}		4.0×10^{1}	1.3×10^{-3}		3.0×10^{1}	2.7×10^{-5}		2.58×10^{1}	5.5×10^{-4}	
	V	9.0×10^{1}		3.1×10^{-3}	2.1×10^{3}		19.1×10^{-4}	9.3×10^{2}		132.9×10^{-4}	3.0×10^{1}		1.0×10^{-3}	3.0×10^{1}		2.7×10^{-5}	6.36×10^{2}		13.6×10^{-4}
6	H	1.1×10^{6}			1.1×10^{6}			4.6×10^{6}			3.0×10^{4}			1.1×10^{6}			1.59×10^{6}		
	C	1.1×10^{3}	1.0×10^{-3}		2.4×10^{2}	21.8×10^{-4}		1.1×10^{4}	23.9×10^{-4}		6.0×10^{2}	2.0×10^{-2}		3.0×10^{1}	2.7×10^{-5}		2.59×10^{3}	1.6×10^{-3}	
	V	2.3×10^{2}		2.1×10^{-4}	6.4×10^{2}		5.8×10^{-4}	2.9×10^{3}		6.3×10^{-4}	3.0×10^{2}		1.0×10^{-2}	2.9×10^{2}		26.4×10^{-5}	8.72×10^{2}		5.5×10^{-4}
7	H	4.6×10^{6}			1.1×10^{4}			1.1×10^{7}			1.1×10^{7}			1.1×10^{6}			4.71×10^{6}		
	C	1.1×10^{3}	2.4×10^{-3}		1.1×10^{3}	1.0×10^{-1}		1.1×10^{2}	1.0×10^{-5}		2.1×10^{2}	1.9×10^{-5}		3.0×10^{1}	2.7×10^{-5}		5.1×10^{2}	10.8×10^{-5}	
	V	4.6×10^{3}		1.0×10^{-2}	1.1×10^{4}		1.0×10^{0}	9.3×10^{2}		84.5×10^{-6}	1.9×10^{3}		17.3×10^{-5}	9.0×10^{2}		81.8×10^{-5}	3.87×10^{3}		8.2×10^{-4}
平均	H	1.04×10^{6}			1.09×10^{6}			3.31×10^{6}			1.9×10^{6}			2.53×10^{6}			3.55×10^{6}		
	C	9.47×10^{2}	3.6×10^{-3}		1.77×10^{3}	2.2×10^{-3}		1.82×10^{3}	7.1×10^{-4}		5.1×10^{2}	2.9×10^{-4}		4.0×10^{1}	1.9×10^{-5}		9.93×10^{2}	6.9×10^{-4}	
	V	7.2×10^{2}		2.7×10^{-3}	2.04×10^{3}		2.3×10^{-3}	1.09×10^{3}		42.7×10^{-5}	5.87×10^{2}		3.3×10^{-4}	2.67×10^{2}		12.7×10^{-5}	5.56×10^{2}		6.4×10^{-4}

注：含量单位为个/立方厘米；“/”——未取样

表 2　桑沟湾表层底质细菌含量状况表

站号	细菌简称	1989—05			1989—08			1989—11			1990—02			1990—05			平　均		
		含量	C/H	V/H	含量	C/H	V/H	含量	C/H	V/H	含量	C/H	V/H	含量	C/H	V/H	含量	C/H	V/H
1	H	1.1×10^{6}			3.0×10^{6}			1.5×10^{7}			1.5×10^{6}			2.9×10^{7}			9.38×10^{6}		
	C	7.5×10^{3}	6.8×10^{-3}		1.1×10^{6}	366.7×10^{-3}		3.0×10^{2}	2.0×10^{-5}		2.0×10^{3}	1.3×10^{-3}		3.0×10^{2}	10.3×10^{-6}		2.4×10^{2}	2.6×10^{-3}	
	V	3.0×10^{3}		2.7×10^{-3}	2.3×10^{3}		76.7×10^{-4}	4.6×10^{4}		306.7×10^{-5}	2.4×10^{4}		1.6×10^{-2}	1.1×10^{6}		37.9×10^{-4}	3.71×10^{4}		4.0×10^{-3}
2	H	9.0×10^{6}			/			/			/			2.1×10^{6}			1.5×10^{6}		
	C	3.0×10^{3}	3.3×10^{-3}		/	/		/	/		/	/		1.1×10^{6}	523.8×10^{-4}		5.65×10^{4}	3.8×10^{-2}	
	V	2.4×10^{4}		26.7×10^{-3}	/		/	/		/	/		/	3.0×10^{2}		14.3×10^{-5}	1.215×10^{4}		8.1×10^{-2}
3	H	3.0×10^{6}			/			1.5×10^{6}			2.1×10^{6}			/			1.3×10^{6}		
	C	2.4×10^{4}	8.0×10^{-2}		/	/		3.0×10^{2}	2.0×10^{-4}		4.3×10^{3}	20.5×10^{-4}		/	/		9.53×10^{3}	7.3×10^{-3}	
	V	4.0×10^{3}		13.3×10^{-3}	/		/	1.1×10^{6}		73.3×10^{-3}	1.9×10^{3}		9.0×10^{-4}	/		/	3.86×10^{4}		29.7×10^{-2}
4	H	6.0×10^{6}			3.0×10^{6}			/			1.1×10^{6}			1.2×10^{6}			5.3×10^{6}		
	C	4.3×10^{3}	7.2×10^{-3}		3.0×10^{2}	1.0×10^{-3}		/	/		4.0×10^{2}	3.6×10^{-4}		3.0×10^{2}	2.5×10^{-3}		1.325×10^{3}	2.5×10^{-3}	
	V	3.0×10^{3}		5.0×10^{-3}	3.0×10^{3}		1.0×10^{-2}	/		/	9.3×10^{3}		84.5×10^{-4}	2.0×10^{3}		16.7×10^{-3}	4.325×10^{3}		8.2×10^{-3}
5	H	3.0×10^{6}			6.0×10^{6}			4.6×10^{7}			3.0×10^{4}			4.0×10^{6}			9.47×10^{6}		
	C	1.1×10^{6}	366.7×10^{-3}		9.0×10^{2}	1.5×10^{-3}		9.1×10^{2}	2.0×10^{-5}		2.0×10^{3}	666.7×10^{-4}		3.5×10^{3}	87.5×10^{-4}		2.346×10^{4}	24.8×10^{-4}	
	V	9.0×10^{3}		3.0×10^{-2}	4.0×10^{3}		6.7×10^{-3}	2.7×10^{2}		5.9×10^{-5}	4.3×10^{3}		143.3×10^{-3}	2.4×10^{3}		6.0×10^{-3}	4.48×10^{2}		4.7×10^{-4}
6	H	7.7×10^{6}			3.0×10^{6}			4.3×10^{6}			1.1×10^{7}			1.1×10^{6}			3.48×10^{6}		
	C	2.4×10^{4}	34.3×10^{-3}		1.5×10^{3}	5.0×10^{-3}		1.5×10^{4}	34.9×10^{-4}		4.0×10^{4}	36.4×10^{-4}		3.0×10^{2}	27.3×10^{-5}		1.616×10^{4}	4.6×10^{-3}	
	V	3.0×10^{3}		4.3×10^{-3}	3.0×10^{3}		1.0×10^{-2}	1.5×10^{4}		34.9×10^{-4}	4.3×10^{6}		390.9×10^{-4}	9.0×10^{2}		81.8×10^{-5}	9.04×10^{4}		2.6×10^{-2}
7	H	1.1×10^{6}			4.3×10^{6}			1.1×10^{8}			1.1×10^{7}			1.5×10^{6}			4.736×10^{7}		
	C	1.1×10^{6}	1.0×10^{-3}		2.0×10^{3}	46.5×10^{-5}		1.1×10^{5}	1.0×10^{-3}		2.1×10^{2}	1.9×10^{-5}		3.0×10^{2}	2.0×10^{-4}		4.45×10^{4}	9.4×10^{-4}	
	V	3.0×10^{3}		2.7×10^{-5}	3.0×10^{3}		69.8×10^{-5}	9.3×10^{5}		1.0×10^{-3}	1.9×10^{3}		17.3×10^{-5}	1.9×10^{3}		12.7×10^{-4}	2.396×10^{3}		5.1×10^{-4}
平均	H	4.41×10^{6}			3.86×10^{6}			3.54×10^{7}			4.46×10^{6}			6.48×10^{6}			1.11×10^{7}		
	C	3.23×10^{5}	24.8×10^{-4}		2.21×10^{5}	19.7×10^{-3}		2.53×10^{4}	7.1×10^{-4}		8.25×10^{2}	18.3×10^{-4}		1.84×10^{4}	33.5×10^{-4}		2.17×10^{4}	2.4×10^{-3}	
	V	7.0×10^{3}		4.3×10^{-4}	3.06×10^{3}		2.6×10^{-3}	4.18×10^{4}		1.6×10^{-3}	7.24×10^{5}		17.6×10^{-3}	1.85×10^{5}		34.4×10^{-4}	2.65×10^{4}		2.9×10^{-3}

注:含量单位为个/克;“/”——未取样

将表1、2与图2结合分析，并以1989年5月和1990年5月的数据平均值代表春季，其他3个月分别代表夏、秋、冬季，则其结果表达出的季节变化势态为：H在水中大小排序为秋＞冬＞春＞夏，底质中为秋＞春＞冬＞夏。C在水中为秋＞夏＞春＞冬，底质中为春＞秋＞夏＞冬。V在水中为夏＞秋＞冬＞春，底质中为冬＞秋＞春＞夏。C/H在水中为夏＞春＞秋＞冬，底质中为夏＞春＞冬＞秋。V/H在水中也为夏＞春＞冬＞秋，底质中为冬＞夏＞春＞秋。

2.3　H,C和V等的站区差别

表3列出的是H,C和V等参数的站区差异。

这是将所测7站按离岸近远划分为3个站区而得出的比较。由此表可见，H和C在水中或底质中均为近岸多，C及H在中部少。离陆越远，H越显出减少之势。这暗示底质中H和V受陆地的影响高于水中，其环境的稳定性大于水。测区范围虽不大，但除陆岸对海洋中细菌分布有影响外，还有其他因素影响，因而也会出现非同一趋势之处。

表3　桑沟湾五个细菌学参数在不同类别站区间的比较

站区分类	样品类别	H	C	V	C/H	V/H
近岸	水	2.334×10^6	9.70×10^4	1.43×10^3	0.0006	0.000715
	底质	1.74×10^7	2.703×10^4	3.90×10^4	0.0027	0.0077
中部	水	2.69×10^5	2.73×10^2	8.2×10^1	0.0010	0.0003
	底质	1.3×10^6	9.53×10^3	3.86×10^4	0.0073	0.0297
远岸	水	4.33×10^5	1.514×10^3	4.48×10^2	0.0043	0.0013
	底质	1.015×10^6	2.89×10^4	8.24×10^3	0.020	0.0081

注：近岸站区由4站平均，中部区仅1站，远岸站区由2站平均得出

将各菌的站区差与所测相应浮游动物生产力站区差对照[2]，可显示下列信息：浮游动物生产力大区，H和V有相应增高现象出现，C/H和V/H的站区差与相应生产力站区差似吻合得更多一些。这暗示了C和V等异养细菌含量与动物间的密切关系[8]。动物的排泌物、残骸等可促进相关微生物繁殖。

2.4　H,C和V等细菌参数间的相关分析

对表水、表层底质及水——底质间的H,C,V,C/H及V/H分别作相关分析，结果列于表4之中。

表4　桑沟湾表层水、底质及水——底质H,C,V,C/H,V/H的相关系数表＊

样品类别	变量对子	相关系数(R)	
水	H－C	－0.5552	(＞r0.20)
	C－C/H	0.7404	(＞r0.10)
	H－V	－0.3241	(＞r0.50)
	V－V/H	0.1445	(＞r0.50)

(续表)

样品类别	变量对子	相关系数(R)	
底质	H—C	0.4521	(>r0.50)
	C—C/H	0.6452	(>r0.20)
	H—V	−0.0932	(>r0.50)
	V—V/H	0.6865	(>r0.10)
水—底质	H—H	0.3528	(>r0.50)
	C—C	0.4337	(>r0.50)
	V—V	0.7545	(>r0.05)
	C/H—C/H	0.9900	(>r0.001)
	V/H—V/H	−0.3121	(>r0.50)

* :$n=7$

由表4可见H—C,H—V在水中均为负相关。H—V这一负相关,在底质中则变得很小,而H—C的相关性在底质中却是正的。说明这两对相关性,尤其是H—C在不同介质中显出不同的含量构成。水中H的消长与C的增减明显是同向的,底质中则是相向的。H—V的负相关性说明HV的消长在水或底质中均是逆向的,而以水中的稍明显些。H—C在2种介质中的含量关系不仅表达出2类菌的含量变化趋势,更意味着2种介质(环境)对它们的影响。

水或底质的C—C/H,V—V/H相关性均为正的。比较常规地反映出C和V含量及其与H的比率关系。

水与底质间细菌参数相关分析结果也可从表4中得知。

H—H,C—C,V—V及C/H—C/H在水——底质间各自的正相关性表明此3类菌在两介质间有一定的相辅相成的依存关系,尤其是V,最为明显。底质的微生物一般情况下均比水中多。水微生物含量的多寡一般意味着相关底质微生物含量的多寡。V/H在水——底质间的轻微负相关表达出V与H的比率关系不完全与V的消长相一致。水中V所占比率的增减可能意味着与底质中V比率的逆向增减,也反映了介质的不同影响着V的存在和消涨,虽然V含量在两介质间有很好的一致。C和V均是条件致病菌,它们各自在两介质间的正相关性意味着特定底质是水中此类菌的一个重要来源和藏匿之处。水既是它们的传媒,也是它们的重要生存基质。两介质的相互作用使细菌在环境中发展并决定和指示经济生物的兴衰。

2.5 细菌生物量初步估计

Watson(1978)对上升流生态系中细菌的作用做过估计[8],指出1个海洋细菌相当于含1.04×10^{-14}g的碳,并借此代表细菌生物量。据此,本湾水或沉积物中总异养菌含碳量状况可从表5中看出。

表 5　桑沟湾细菌含碳量状况表

采样时间	表层水细菌含碳量/mg C·m^{-3}	表层底质细菌含碳量(湿)/mg C·m^{-3}
1989-05-23	2.704	169.208
1989-08-26	8.528	12.064
1989-11-26	26.520	368.160
1990-02-27	18.304	46.384
1990-05-25	21.840	59.280
采样站号	表层水细菌含碳量/mg C·m^{-3}	表层底质细菌含碳量(湿)/mg C·m^{-3}
1	26.728	97.532
2	3.463	15.600
3	2.798	13.520
4	5.543	5.512
5	4.846	98.488
6	16.536	36.192
7	48.984	492.544

表 5 数据是将所有水或底质样中的 H 按同一采样时间各站平均或同站位各采样时平均而得出的，由该表计算，表层水细菌含碳量在采样时间上的均值约为 15.579 mg C·m^{-3}。其所代表的季节按大小排序为：秋＞冬＞春＞夏，其态势秋为峰，冬春次之，夏为谷。表层底质细菌含碳量在采样时间上的均值约为 131.019 mg C·t^{-1}。在季节上的大小排序为秋＞春＞冬＞夏。说明无论是水还是沉积物中，细菌含碳量即细菌生物量均倾向于秋高夏低。由该表还可算出细菌含碳量的站间差异，即各站按水中大小排序为第 7 站＞第 1 站＞第 6 站＞第 4 站＞第 5 站＞第 2 站＞第 3 站。底质中为第 7 站＞第 5 站＞第 1 站＞第 6 站＞第 2 站＞第 3 站＞第 4 站。即第 7 站细菌含碳量最高，第 3 站的低。若按离岸近、中、远来分，3 类站区细菌含碳量差异的计算结果为：

水：近岸区(24.274 mg C·m^{-3})＞远(4.503 mg C·m^{-3})＞中(2.798 mg C·m^{-3})。底质：近岸区(181.194 mg C·t^{-1})＞中(13.520 mg C·t^{-1})＞远(10.556 mg C·t^{-1})，同样表达出近岸区细菌含碳量即细菌生物量高，远、中部区低之势。

3. 结语

本文论述了桑沟湾表层水和底质中细菌学几个指标的时空变化。水中总异养菌，几丁质降解菌或弧菌的含量分别为 1.50×10^{6} cm^{-3}，1.03×10^{3} cm^{-3} 或 9.56×10^{2} cm^{-3}，底质中则分别为 1.04×10^{7} g^{-1}，2.51×10^{4} g^{-1} 或 3.01×10^{4} g^{-1}。给出了水和底质中几丁质降解菌和弧菌含量占总菌量比率的状态及变化。水、底质中总菌量之季节变化势态分别为秋＞冬＞春＞夏、秋＞春＞冬＞夏。几丁质降解菌分别为秋＞夏＞春＞冬、春＞秋＞夏＞冬。弧菌则分别为夏＞秋＞春＞冬、冬＞秋＞春＞夏。水、底质中各类菌量的空间变化大体显出近岸区大于中部或远岸区之势。分析了各细菌学指标，即 H，C，V，C/H 和 V/H 在水、底质中或水——底质间的相关性，总体看水中 H—C 和 H—V 为负相关，底质中

H—C 却为正相关，水或底质中 C—C/H 和 V—V/H 均为正相关，表达出水中 C、V 与 H 在含量上的不一致性。H—C 在两种介质中的不一致暗示了 H 和 C(V)对生态环境因子的不同反应。多数相关性表明 H、C、V 在水和底质间有较好的交流，尤以 C/H 及 V 为明显。

由总菌含量估计的水域细菌含碳量(生物量)，水中约 15.57 mg Cm^{-3}，显出秋＞冬＞春＞夏之势，底质中约 131.019 mg C・t^{-1}(湿重)，显出秋＞春＞冬＞夏之势。依据所得细菌含量、生物量及营养盐等，认为该湾细菌等代表的微生物生物量不低，可以推测该湾是富生产力的浅水生态系统[9]。温热时节较高的弧菌含量及弧菌比率，意味着养殖生物面临着较多的致病条件，因而加强该湾水产医学和环境保护等的研究仍是维持该生态系统良性可持续发展所必需的前提工作之一。

致谢：我所车敦彩参加过部分工作。海上工作得到项目组成员大力帮助。

参考文献 9 篇(略)

RESEARCH ON BACTERIA IN SANGGOU BAY

Abstract　The space and time distributions and variation of the numbers of general aerobic heterotrophic bacteria, chitin-degrading bacteria and vibrio and their proportions in the total amount of bacteria in the surface water and bottom sediment of the Sanggou Bay are analysed, and the correlation analysis between bacteriological parameters is also made.

The bacterial biomass is estimated, the marine multiplication and cultivation, ecological and environmental conditions in the Bay are assessed, and it is shown from above analysis and study that a further microbiological study is needed for the successful marine multiplication and cultivation and the ecological and environmental protection in the Sanggou Bay.

Key words　Sanggou Bay; Bacteria; Ecological Environment; Marine Multiplication And Cultivation

桑沟湾表层水细菌与生态环境因子的关系*

摘　要　就桑沟湾表层水总异养细菌(H)、几丁质降解菌(C)和弧菌(V)含量及C/H、V/H与N、P、水温等9个生态环境因子间分别作了相关分析,结果表明:在不断增强的环境负荷下,H显出与营养盐(如-N、-P等)、COD等大体一致的增减趋势,与水温间有一定的负相关。C、V(及C/H、V/H)的消长基本与水温相平行,而与溶解氧(DO)相悖。结果意味着该湾水中适冷菌较多、生存时间较长,同时尚有外来中温性细菌,包括部分C、V的加入。

关键词　桑沟湾　细菌学参数　致病菌　生态环境因子

桑沟湾是位于山东半岛东端的半封闭海域,因受海水增养殖、水产捕捞和陆源污染的影响而致的生态环境负荷正处于不断增强之中,表现在营养盐的增多,溶氧的减少等方面[1]。这迫使其中的细菌等生物不得不被严格选择,并与之相适应。结果是几丁质降解菌、弧菌等条件致病菌等的滋生,生态环境趋于恶化。可是对桑沟湾这类水体的生态学研究成果迄今仅限于环境因子和动植物方面的[1~5]。有关总异养菌、几丁质降解菌,弧菌等的状况及与生态环境因子间的分析研究一直未见报道。本文仅对此作相关的分析研究,以期有助于了解该湾细菌变动规律和评估增养殖条件。

1.材料和方法

1.1　站位设置

图1示出该测站布设状况。共设7站。按离岸远近划分,4个站为近岸站。3号站代表中部区,2和4号站为远岸区。

测定时间:1989-05～1990-05。每季测一次,共5次。

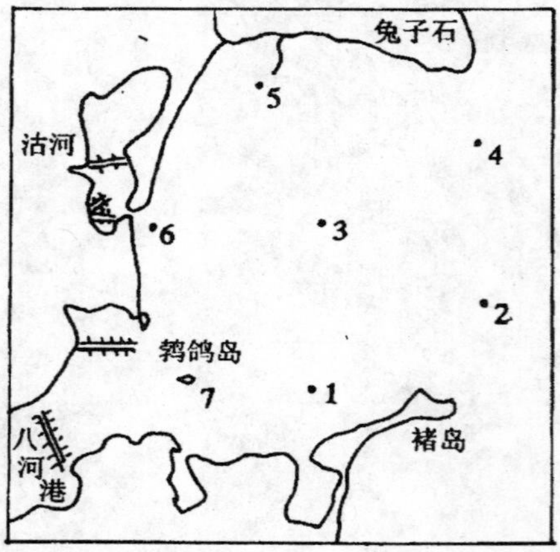

图1　桑沟湾细菌采样站位图

Fig. 1　Sampling stations for bacteriology and eco-environmental factors in Sanggou Bay

1.2　培养基

所用培养基共3种,即海洋好气异养菌培养基[6]、几丁质降解菌培养基和弧菌培养基[7]。

1.3　取样、分析和计数

按《海洋调查规范(试行本)》取样、分析和计数。表层海水三类细菌即总异养菌量(简写为H,下同,培养基:葡萄糖2 g,蛋白胨5 g,酵母膏1 g,陈海水1 000 cm^3,pH 7.7)、几丁质降解菌量(简写为C,下同),弧菌量(简写为

*　基金项目:国家科学技术委员会重点支持项目(K88-0211)
原文刊于《海洋环境科学》,2001,20(3):29-33.

V,下同,培养基为 TCBs),分别用 MPN 法测试,即在试管定量培养基中,按需将样品作 10 倍浓度的连续梯度稀释,经培养后,对照菌数检索表换算为单位体积(cm^3)水中的细胞数,得出 H、C、V 及 C、V 各占 H 量的比率(即 C/H、V/H,下同)[6]。

同时,按《海洋调查规范》取得表层海水的生态环境参数。即水温(T)、溶解氧(DO)、pH、—P、—Si、—N、—N、总无机氮(TIN)、COD 等。

1.4　各类细菌参数与生态环境因子关系分析

将水样 H、C、V、C/H、V/H 与水温、—N、TIN、—Si、—P 或 DO 等分别作相关分析,可知各类细菌及其参数与相应生态环境因子间的关系。依此分析和评价该湾生态环境状况。

2. 结果和讨论

2.1　H、C、V、C/H、V/H 等的一般状况

表 1 列出了所测站位表层水中每季代表月的 H、C、VC/H、V/H 等数据。由该表可见水中各参数平均值,并分析其时空分布变化[8]。

2.2　主要生态环境因子数据

表 1 中同时列出所测站位表层水水温、DO、pH、—P、—Si、—N、—N、—N、TIN、COD 等各自的均值和波动范围。表 2 则是 1983～1984 和 1989～1990 两个 7a 间该湾表层水一些环境生态参数的比较。与 1983～1984 年度相比[1],本次所测值(除水温外)大都有不同程度提高,而 DO 则下降。

表 1　桑沟湾表层水三类细菌及相关生态环境因子

Tab. 1　Statistics of parameters of three indexes of bacteria and relative eco-environmental factors of surface water

取样时间	T/(℃)	DO	PO_4^{3-}—P	SiO_3^{2-}	NO_2^-—N	NO_3^-—N	NO_4^+—N	TIN	COD	H/(×10^5)	C/(×10^2)	V/(×10^2)	C/H	V/H
1989—05	14.78	5.838	0.466	4.522	0.343	0.388	4.177	4.908	3.690	2.60	9.47	7.20	0.0253	0.00252
1989—08	25.23	4.722	0.141	2.809	1.559	0.008	0.761	0.925	1.551	8.20	17.70	20.40	0.0022	0.0025
1989—11	12.14	5.838	0.873	7.957	1.501	4.070	3.437	9.007	0.970	25.50	18.20	10.90	0.00071	0.000427
1990—02	2.54	7.934	0.859	5.017	0.100	1.199	2.365	3.663	0.987	17.60	5.10	5.87	0.00029	0.00033
1990—05	14.56	7.002	0.841	6.071	0.390	1.365	1.085	2.838	1.071	21.00	0.40	2.67	0.000019	0.000127

注:H、C、V 均为细胞数·cm^{-3}、C/H、V/H 为 C、V 各与 H 之比;

DO 为 mL·dm^{-3};COD 为 mg·dm^{-3},余为 mol·dm^{-3}。

表 2　桑沟湾表层水中一些生态环境参数 7 年前后的比较

Tab. 2　Comparison on eco-environmental parameters in surface water seven years late

年度	T/(℃)	DO	PO_4^{3-}—P	SiO_3^{2-}	NO_2^-—N	NO_3^-—N	NO_4^+—N	TIN	COD
1983—06～1984—05	13.9	6.440	0.360	4.250	0.100	0.460	0.490	1.050	0.860
	1.2	4.380	0.000	1.100	0.000	0.000	0.00	3.190	0.450
	～27.5	～8.680	～1.480	～13.250	～1.000	～5.320	～3.190	～9.510	～1.320

（续表）

年度	$T/(℃)$	DO	$PO_4^{3-}-P$	SiO_3^{2-}	$NO_2^{-}-N$	$NO_3^{-}-N$	$NH_4^{+}-N$	TIN	COD
1989－05～1990－05	13.8	6.267	0.636	5.275	0.498	1.200	2.365	4.063	1.65
1989－05～1990－05	3.2	4.507	0	0.600	0	0	0	0.341	0.77
1989－05～1990－05	～26.8	～8.197	～1.200	～13.388	～2.704	～5.422	～9.620	～17.743	～5.97
1989－05～1990－05 1983－06～1984－05	0.993	0.973	1.767	1.241	4.98	3.609	4.827	3.870	4.291

a）单位：DO 为 $mL \cdot dm^{-3}$，余均为 $mol \cdot dm^{-3}$

2.3 H、C、V 等与生态环境因子的关系

表 3 列出了 H、C、V 等细菌学参数与 NO_3^-－N、DO、水温等生态环境因子间的相关关系分析结果，列出了 24 个具有一定意义的相关系数。

表 3　桑沟湾表层水细菌参数与一些生态环境因子的相关分析

Tab. 3　Correlation analysis between bacteriological parameters and eco-environmental factors in surface water of Sanggou Bay

分析因子对	相关系数	备注	分析因子对	相关系数	备注
T－H	－0.4350	$>r_{0.50}$	$(SiO_3^{2-}-Si)-H$	0.8147	$>r_{0.10}$
T－C	0.4794	$>r_{0.50}$	$(SiO_3^{2-}-Si)-C$	－0.0501	$<r_{0.50}$
T－V	0.7029	$>r_{0.20}$	$(SiO_3^{2-}-Si)-V$	－0.3750	$<r_{0.50}$
T－C/H	0.4853	$>r_{0.50}$	$(PO_4^{3-}-P)-H$	0.8008	$\approx r_{0.10}$
T－V/H	0.6385	$>r_{0.50}$	$(PO_4^{3-}-P)-C$	－0.4780	$>r_{0.10}$
$(NO_3^{-}-NN)-V/H$	0.3621	$<r_{0.50}$	$(PO_4^{3-}-P)-V$	－0.7821	$>r_{0.20}$
$(NO_3^{-}-N)-H$	0.8246	$>r_{0.10}$	DO－H	0.4333	$>r_{0.50}$
TIN－H	0.4796	$>r_{0.50}$	DO－C	－0.7918	$\approx r_{0.10}$
TIN－V	－0.2369	$<r_{0.50}$	DO－V	－0.8310	$>r_{0.10}$
TIN－C	0.3132	$<r_{0.50}$	COD－H	0.3967	$>r_{0.50}$
COD－C	－0.2563	$<r_{0.50}$	COD－C	0.3429	$>r_{0.50}$
COD－C/H	－0.2301	$<r_{0.50}$	COD－V	0.2574	$<r_{0.50}$

注：除 COD－H 等 5 对变量是 7 外余均为 5。

2.3.1　细菌学参数与生态环境因子间的主要正相关性

与 H 有最好正相关关系的因子包括 NO_3^-－N、SiO_3^{2-}－Si、PO_4^{3-}－P。与 H 有一定正相关关系的因子包括 TIN 或 DO。与 V 有较好正相关关系的因子包括水温。分别与 C、C/H 或 V/H 有不大正相关关系的是水温。分别与 H、V 有不大正相关的是 COD。

结果表明：H 与 N、Si、P 等主要营养盐各有较好的依存关系。水中 N、－N 较多。与相近条件的湾比，其 TIN 已达富营养化程度（表 2）[9]。H 在 11 月份前后最多，春季和初

夏最少，反映出该水域秋冬季高氮盐和春夏之交低氮盐量之势。这印证了该湾养殖盛期(6～9 月份)的特征和所在海区经受渤海沿岸流等地理学条件的影响[1]。

所测 P 含量比以往明显增加(表 2)[1]。

H—P 的较好正相关性表明 H 对 P 有一定依存关系。在一定范围内，P 增加促使浮游植物和细菌生长。后者的代谢活动又使 P 在环境中再循环。P 在 11 月份前后的增多以及 H-P 的正相关表明，此时含 P 量即处于回升阶段并维持高值，这既可推测当时较高 P 含量的外海水流入，又说明浅底沉积物颗粒上浮对水中 P 和 H 起并行的作用。同理，夏天水生物大量生长耗 P 多，P 量降，H 也少。

N 与 P 分别对 H 有较好的正相关性，说明 N 与 P 间也应主要存在正相关关系。

1983～1984 年间，水的 Si 含量已超米氏常数，但仍不丰富[1]。本测定表明 Si 量有较大增长(表 2)。Si 的增加促进 H 增多，暗示水 H 中有许多成员以硅盐构架的颗粒物为依附对象(也是营养成分之一)，有助于其沉浮。H—P、H—Si 两个正相关可能预示 P-Si 间也有正相关关系。

H—DO 的一定正相关性表明水中 DO 多，有利于 H 生长，证实 H 的好气性。但本测定中 DO 平均值小于 1983～1984 年间的(那时多数测点呈过饱和状态)(表 2)。这意味着此时 DO 可能成为 H 的限制因子。DO 减少也可指示浮游植物减少。溶氧一般是夏低冬高，即水温高，DO 少[1]，因而不利 H。但夏季浮游植物的繁殖可增水中 DO，这可部分抵消升温的不利效果。

V—T 的正相关性表明一定水温范围内弧菌消长依赖于水温降升。V 表现为水温的函数。V 常是病原菌。水温上升，可促使养殖生物繁殖。污染严重的环境却使 V 量升高。结果表明 V 喜生于高水温、低 DO 和少 P 条件中，表现在 V/H—T 上则显示为正相关。表明那时 V 对 H 的比率增大。

C—T、C/H—T 各有不大正相关关系。表明 C 与水温有一些并行不悖关系，但不大明显，因而推测此关系也不稳固。说明 C(C/H)的波动尚有其他重要因子影响。也说明当时，水温不成为 C 的限制因子，即 C 在 20～37℃时，条件合适，均可最适地生长[7]。

以上各正相关关系说明 N 等盐类及溶氧主要影响 H。水温主要影响 V、C、C/H、V/H。测区增多的营养盐和一定的 DO 促使普通好气异养菌增多。水温升高在一定程度上加速弧菌、几丁质降解菌的繁殖。

COD 与 H 或 V 间的不大正相关显示 H、V 在一定程度上随水体有机(污染)物含量的增减而增减之势，本测定表明 COD 是往年的 4 倍多(3.69/0.86，表 2)，预示 H、V 都可能大幅增多，尤其是 V(V/H)的增多，意味着水产病害的发生机率将加大。

2.3.2 细菌学参数与生态环境因子间主要的负相关性

与 V 最好的负相关性存在于 V—DO，较好的是 V—P，较差的是 V—TIN。C—DO 负相关较好，C—P 次之，较差的是 C—Si、C—COD。

以上结果说明，V 最不受限于 DO，暗示许多 V 可能是厌氧(包括兼性厌氧)者。DO 越多，V 越少。表明通气或流动的水体增加 DO 而使 V 减少，这由底质中 V 多，水中 V 少得以证实(如表 1 及文献[8])，故水中 V 可能很多是来自潜伏于底质的。测区 TIN 对 V 大概有较小的抑制作用，或者不构成对其生长的限制(表 2)。

DO 对 V 如此，对 C 也有类似关系，即一些 C 也许对 DO 不感“兴趣”，甚至不好氧。

C—P 的负相关类似于 V—P。P 多,C 也会减少。C—Si 的相斥性不大,即 Si 的增减对 C 的消长并不构成重大负相关。水中 Si 无疑比底质中少,但水 C/H 却小于底质 C/H,水 C 绝对值小于底质 C。表明水 C 对 Si 少依存性,意味着

水 C 可能多是浮游性的。同样 C—COD、C/H—COD 的不大相斥性,意味着有机(污染)物不限制水体 C(C/H)发展,从反面印证有机(污染)物含量的增大。

H—T 的负相关性。一般认为海洋细菌是适冷/嗜冷菌。可是近岸浅湾水细菌种群构成受陆源、大气和人为的重大影响,水中必有一些中温菌。测区测时最低水温是 2.2℃。此时适冷菌仍可缓慢生长,而非不生长。嗜冷菌已处于最低生长温度之上,应比适冷菌生长得更好。中温菌可能很少,或处于潜伏休眠态。春秋季的水体,上述三类菌可并存,以适冷和中温菌为多。到了夏季,最高水温达 26.5℃。25℃左右的水温可经常并持续出现,中温菌大量繁衍。适冷菌可最适生长的时间比中温菌的多。由此可见测区水中大部分时间应以适冷菌为主,中温菌次之,嗜冷菌最少。H—T 负相关性反映出测区水中 H 构成的多元性。主体的适冷菌对水温提高的反应显出它取逆向发展之势,适当降低水温反而有利于和促进它大量增加。

至于前述的 C 或 V 与水温的两个正相关恰好说明 C、V 中可能以中温菌为主,这或许包括陆源、大气和人为带入的污染菌。水温的提高为经常滞留沿岸的 C、V 而不是所有 H 的繁殖创造了一个重要条件。它们可能对升温有更热烈的响应。合适的营养(包括养殖生物的排泌物、残骸和溶解物等)为它们提供了营造和繁衍机体的物质基础。这从一个侧面暗示 C、V 来源的复杂性。也特别提示海产增养殖者在温热季节须关注 C、V 繁衍泛滥致病的严重性。而 V、C 与 DO 及与营养盐的相关性启示防治水产病害的思路。

3. 结语

与 7 年前(即 1983 年度)比,该湾水体营养盐和有机污染物等含量均普遍提高,而溶解氧降低。这可能为加快富营养化进程创造了条件,意味着环境污染的加重和生态环境的恶化。一些迹象已证实该湾良性的水生态系统已遭破坏,且愈益不平衡。在此条件中适应细菌,包括条件致病菌增多。由于普通异养菌、几丁质降解菌、弧菌的群系组成和来源有所不同,它们对水环境生态因子就表现出不完全一致的感应。异养菌含量在总体上较并行地反映出营养盐状况,即它的消长是营养盐(如—N、—P 等)及 COD(在一定程度上)的一个函数。异养菌含量与水温间一定程度上的逆相关意味着水细菌仍有许多是适冷者。几丁质降解菌、弧菌均是条件致病菌,其含量及与异养菌总量之比的增减将直接反映出养殖水质、生态环境的优劣和发生病害的条件。它们(包括 C/H、V/H)与水温的基本相平衡、与溶氧的相悖暗示外来及中温的弧菌和几丁质降解菌的加入,从而增加致病菌繁殖机会。同时本结果也揭示只要不断改善生态环境、控制营养盐浓度及其合适比率,采取科学防治病害措施,仍有望促进该湾细菌等微生物生态及增养殖正常化进程。

致谢:车敦彩副研究员参加部分微生物取样和分析工作,项目组成员在野外和生态环境因子测试工作中提供重要支持,李瑞香副研究员对本文提出宝贵意见,在此一并致谢!

参考文献 9 篇(略)

CORRELATION ANALYSIS ON BACTERIOLOGICAL INDEXES AND ENVIRONMENTAL PARAMETERS FOR SURFACE WATER OF SANGGOU BAY

Abstract The results of correlation analysis among some bacteriological indexes and environmental parameters for surface water of Sanggou Bay were discussed. The results show that: Under the increasing environmental load, heterotrophic aerobic bacterial counts(abbr. H) show similar trends as nutrients such as NO_3-N、-P etc, and COD, a negative correlationship exists between H and T(water temperature). The variation of Chitin-degrading bacterial counts(abbr, C), vibrio counts(abbr. V.), C/H or V/H were parallel to T, and opposite to dissolved oxygen content. This condition implicates that there were more psychrotrophic in the water of the Bay besides to some mesophilic bacteria(possibly *eurytherrnis* ones), including chitin-degrading bacteria and vibrio entering the Bay.

Key words Sanggou Bay; Bacteriological Indexes; Pathogen; Environmental Factor

东海、黄海及渤海测区维生素 B_{12} 产生菌生态学的研究*

摘　要　对取自中国东海、黄海、渤海测区水和表层沉积物中的 2206 株细菌用大肠埃希氏菌（*Escherichia coli* 113-3）作维生素 B_{12} 产生菌的初筛及其 B_{12} 产力测定。发现 B_{12} 菌 443 株，阳性率高达 20.08%。对 B_{12} 菌的生态学分布研究表明，沉积物中、近岸和低纬度区的 B_{12} 菌阳性率一般分别高于海水、远岸和高纬度区。海水 B_{12} 菌阳性率呈现的垂直分布趋势是：上层高、中层次、下层低。季节性变化有秋季高于初夏、夏季高于冬季之特点。分析了不同海区 B_{12} 菌阳性率分布与一些生态环境因子间的关系。

关键词　东海、黄海及渤海　维生素 B_{12} 产生菌　B_{12} 菌分布的生态学

1. 引言

许多生物自身不能合成维生素 B_{12}（vitamin B_{12}，简称 B_{12}，下同），必须由其他生物提供。海洋中溶解状态的 B_{12} 虽然微量但积极参与生命活动，是海洋初级生产和环境质量的一个不可忽视的因子。

40 多年来，人们虽对海洋细菌产 B_{12} 的现象引起了注意（Burkholder，1959；仓田等，1968，1969，1970。），也认识到海洋中 B_{12} 的合成主要由细菌等微生物所进行（Burkholder，1958；Provasoli，1963）。但仅就 B_{12} 产生菌（简称 B_{12} 菌，下同）的分离和分布做过一些初步研究。近些年人们更注重海洋 B_{12} 和生理、生态等意义。世界 B_{12} 生产迄今尚未见到用化学手段，仍靠微生物法。从海洋中提取活性物质正是当今引人注目的热门课题之一，人们尝试着从海洋中筛选并改造出高效优质的 B_{12} 产生菌。较全面地掌握海洋 B_{12} 菌分布的生态规律，将进一步有助于理解 B_{12} 菌及 B_{12} 在海洋生态、生理和环境中的作用和地位，是我国海洋微生物学家的一项任务。本文论述我国东海、黄海、渤海 B_{12} 菌分布的生态学。

2. 材料和方法

2.1　样品的采集

水样和沉积物样分别以 HQM_1 型采水器和表层沉积物采集器采集。测区的大体范围是：28～29°N、123°30′E 以西海区，测定的大体时期和海区分别是：1972 年 5、6 月、1973 年 9 月于东海，1972 年 12 月黄海，1972 年 9 月渤海，1974 年 6 月胶州湾。共得水样 105 份，沉积物样 106 份。每个站从表层、10、25、40、55 m 等直到海底上 5 m 处各收集 100 cm^3 水样进无菌水样瓶中，用蔡氏滤器经孔径为 0.45 μ 的滤膜抽滤 40 cm。

将 1 g 新鲜湿沉积物样入 9 cm^3 稀释用盐水中，摇匀，以 MPN 法取 0.1 cm^3 的稀释沉积物样上清液。

2.2　菌株的分离和纯化

---

*　原文载于《中国当代优秀领导理论文选》，2004，P575-578 和《中国改革发展理论文集》，2004，P. 416-420.

抽滤过水样的滤膜贴于 Zobell 2216E 固体培养基上。将 0.1 cm^3 的沉积物液滴于另一平皿的同样培养基上涂布均匀。于室温照常规培养后，计算异养菌总数，挑取单菌落至斜面上储存备用。

2.3　菌株的 B_{12} 生产

纯化的菌株分别接种进液体或固体的 Zobell 2216E 培养基中培养 18～24 h。挑取一环单菌落接种到改进的 Zobell 2216E 液体培养基上（仓田，1969），于恒温 25℃回旋振荡培养（150 转/分）72～120 h。

2.4　B_{12} 产力的测定

2.4.1　活化测定菌 *Escherichia coli* 13-3。取 18～24 h 菌龄的该菌悬液同 45℃的鉴定培养基混合一起倾注成固体平面，于 37℃培养 18～24 h。

2.4.2　取出培养液，观察终浊度等生长情况，pH 值调至 5，将培养液 151b 15 min 消毒。

2.4.3　上述处理好的培养液滴于牛津杯中，滴满为度，以未接菌的培养基和标准 B_{12} 稀释液作对照，重复两分，37℃培养。24 h 后，牛津杯周围有感应且与纸层析结果一致者初判为产 B_{12} 阳性菌。（详见华北制药厂检验科《维生素 B_{12} 含量测定。微生物杯碟法，1959》）。

2.5　数据整理

按菌种采集的不同海区、季节、特定地理环境和不同样品来源，将 B_{12} 菌数和好气异养菌数进行比较，求出 B_{12} 菌数占好气异养细菌（简称异养菌，下同）数之百分率，以此分析 B_{12} 菌分布的生态学。

3. 结果

从我国东海、黄海、渤海及青岛近海区获异养菌 2206 株，其中 B_{12} 菌 443 株。菌数在异养菌总数中的百分率见表 1。它们在异养菌数中占有相当大的比例，达 20.08%（443/2206）。

3.1　B_{12} 菌%空间变化

经统计，出现 B_{12} 菌的水样站计 64 站，占水样站总数（105）的 61%；出现 B_{12} 菌的沉积物样站计 80 站，占总数（106）的 75.5%，因此，无论在外海还是在沿岸水和沉积物中，B_{12} 菌是普遍存在的。

表 1　东、黄、渤海 B_{12} 菌与异养菌数量比较

Table 1　Comparision between Vitamin B_{12}-Producing bacterial numbers and the number of aerobic heterotrophic bacteria in the East China Sea, Yellow Sea and Bohai Sea

海区	采样	B_{12} 菌数	异养菌数	B_{12} 菌（%）	采样年月
东、黄、渤海	沉积物	154	662	23.26	1972 年
	水	85	720	11.81	1972 年
东海	沉积物	89	267	33.33	1973.9
	水	28	162	17.28	1973.9
胶州湾	沉积物	87	395	22.03	1974.6

表 1 列出了具体的三个海区水和沉积物中 B_{12} 菌%的比较。这说明沉积物的 B_{12} 菌数

及其占异养菌总数%均高于海水中的。沉积物和水中 B_{12} 菌的比较，均表明沉积物 B_{12} 菌出现率高于水样。

3.1.1　B_{12} 菌%在沉积物和水中的比较

表 2　海水及表层沉积物中 B_{12} 菌百分率的比较

Table 2　Comparision between the percentages of Vitamin B_{12}-Producing bacterial numbers from the sea water and those from the surface Sediment of the sea areas investigated

海区	采样	B_{12} 菌数	异养菌数	B_{12} 菌(%)
渤海湾	沉积物	24	81	29.63
	水	3	58	5.17
苏北海区	沉积物	19	82	23.17
	水	16	175	9.14

采样时间：渤海湾—1972 年 9、10 月，苏北区—1972 年 12 月

表 2 是渤海及苏北区(32°30′～35°N，119°30′～123°30′E)沉积物和水中 B_{12} 菌的比较。结果也表明沉积物 B_{12} 菌出现率高于水样。

3.1.2　距岸远近和纬度高低对 B_{12} 菌%的影响

表 3　距岸远近对 B_{12} 菌分布的影响

Table 3　Influence of the distance from the shores surveyed on Vitamin B_{12}-producing bacterial distribution

距岸距离	采样	B_{12} 菌数	异养菌数	B_{12} 菌/异养菌(%)	海区	采样年月
近	水	5	23	21.74	黄海	1972 年
远	水	8	49	16.33	黄海	1972 年
近	沉积物	18	61	29.51	东海	1972 年 5～6 月
远	沉积物	7	41	17.07	东海	1972 年 5～6 月

距岸远近对 B_{12} 菌数百分率的影响是明显的，表 3 给出了这一影响结果，如黄海(34°～36°N)1972 年水样，在 120°30′E(距陆近)上百分率是 21.74%，在 123°30′ E(距岸远)只 16.33%。该差异在某些沉积物样中也呈现出来了，而以长江口区的更为显著。

表 4　不同纬度上 B_{12} 菌百分率比较

Table 4　Comparison between Vitamin B_{12} －Producing bacterial strains percentages on different latitudes

纬度	采样	B_{12} 菌数	异养菌数	B_{12} 菌(%)	采样季节
29°N	水	8	39	20.51	秋
38°30′N	水	3	81	3.70	秋

不同纬度的 B_{12} 菌百分率差异比较列于表 4。表 4 说明在测区内，纬度低的水中 B_{12} 菌数及其所占%高于纬度高区。这种差异在由南往北较小的具体海区中有一个逐步变化的过程。表 5 反映了瓯江口、长江口、山东半岛沿海及辽东半岛沿海 B_{12} 菌%的变化。

表 5　不同地理位置海区的样品 B_{12} 菌百分率比较

Table 5　Comparison of Vitamin B_{12}-Producing bacterial percentages in the tested Samples from different geographic local sea areas

采样区	采样	B_{12} 菌数	异养菌数	B_{12} 菌(%)
瓯江口	水	5	36	13.89
	沉积物	22	88	25.00
长江口	水	18	103	17.48
	沉积物	54	134	40.30
山东半岛沿海	水	26	209	12.44
	沉积物	22	130	16.92
辽东半岛沿海	水	3	66	4.55
	沉积物	19	98	19.39

纬度高的长江口区 B_{12} 菌%高于纬度低的瓯江口区，意味着长江给海洋 B_{12} 菌造成了更合适的环境，与纬度高低不矛盾。同理，所测辽东半岛沿海沉积物 B_{12} 菌%高于山东半岛沿海的。

3.1.3　B_{12} 菌在水层中的分布

表 6　B_{12} 菌数在海水层中的垂直分布

Table 6　Perpendicular distribution variation of Vitamin B_{12}-producing bacterial strains percentages

采样水层	各层中的 B_{12} 菌数	水中 B_{12} 菌总数	B_{12} 菌数(%)	采样海区年月
0 和 20 m	26	31	83.87	东海 1972 年 5、6 月
25 和 40 m	4		12.90	
55 m 及以深	1		3.23	
0 和 20 m	13	28	46.43	东海 1972 年 9、10 月
25 和 40 m	9		32.14	
55 m 及以深	6		21.43	

B_{12} 菌数及其所占比率在海水中的变化呈现明确的垂直分布趋势：上层水中高，中层水次之，下层水最低。无论初夏还是秋季都如此。在初夏的表水中最高。中下层锐减，而秋季则变化较为缓和，表 6 说明了这一点。

3.2　B_{12} 菌数及%时间变化

表 7　B_{12} 菌百分率之季节变化

Table 7　Seasonal Variation distribution of the number of Vitamin B_{12}-producing bacteria in the different layers of Sea water determind

采样季节	采样	B_{12} 菌数	异养菌数	B_{12} 菌(%)	采样年月
夏(1972 年)	水	10	34	29.41	苏北海区
冬(1972 年)	水	16	1775	0.90	

(续表)

采样季节	采样	B_{12}菌数	异养菌数	B_{12}菌(%)	采样年月
初夏(1972 年)	沉积物	9	9	28.13	舟山渔场
秋(1973 年)	沉积物	17	17	41.46	

表 7 列出了两个海区水域沉积物中 B_{12} 菌数及%在时间(季节)上的变化。B_{12} 菌%在季节上的变化,因海区而异,如在舟山渔场海区,秋季高于初夏,苏北海区则夏高于冬,水样尤为明显(表 7)。

4.讨论

调查结果表明,沉积物中 B_{12} 菌百分率高于海水,这是因为底质中有机碳、氨基酸和无机离子等比海水丰富。沉积较大的分置作用,使大量营养颗粒和固体微粒下沉,增加繁殖和存活机会。底质,尤其是泥一水界面,生态环境相对稳定,营养极为丰富,B_{12} 菌大量滋生(Gillespie 等,1972)、仓田等(1968)。

离岸远近对 B_{12} 菌率的影响之关键一点是陆地排水多少。长江口海区受巨量温暖淡水与较冷的海水相遇,造成了生物栖息理想界面,其泥沙所含有机、无机物,给 B_{12} 菌造成了巨大的粘附表面和丰富的营养物质。类似现象在瓯江口也可看到。而离陆地排水越远,大部分情况下是离岸越远,盐分越大,营养越少,B_{12} 菌少,加之由陆地水、气带来的细菌不适大洋环境,可能死亡。远岸水,海区开阔,B_{12} 菌所依附的载体及营养等生态条件变差,B_{12} 菌必然少。

B_{12} 菌数的纬度分布差异,水中明显于沉积物。纬度低,在一定范围,温度增高,B_{12} 菌的生长繁殖活跃,其代谢率也加快,由此影响 B_{12} 菌种群数量变化。低纬度海区、适宜的水温同丰富的营养,较大的沾附表面和其他生物的大量繁殖,一起使特定海区的合适的采样时期内 B_{12} 菌率提高。反之,纬度较高的测区 B_{12} 菌可能相应减少。

在海湾内,B_{12} 菌百分率相对较高,丰富的营养和较稳定的环境是两个原因。如冈市(1971,1972)所说,内湾富营养和水域维生素浓度较高。渤海湾、胶州湾受城市排污影响较大污水带来的营养不断下沉,尤使海底 B_{12} 菌数增加成为可能。另外石油产品等污染物的流入,对一些利用烃类物质的 B_{12} 菌可能有利。

舟山渔场,B_{12} 菌百分率显著大于别的海区,一个有利因子仍是较合适的营养。春夏之交和秋季正是鱼产卵期或渔汛,各种生物活动频繁,B_{12} 菌率会增高。这证实了 Lebedeva 等(1971)所指出的 B_{12} 及其产生菌与鱼类的密切关系。

山东半岛周围海区,水 B_{12} 菌分布状态受近岸环境影响也大。除与营养、地理位置和污染物的影响有关外,若将山东半岛海区周围近岸一站的水样,以 37°N 划分为两部可看出,B_{12} 菌率是南高于北。这是因为 9 月开始,较冷的黄海沿岸流入北侧,不利于 B_{12} 菌的繁殖,甚而不出现 B_{12} 菌。这与该海区某些浮游植物数量显著减少相一致。南侧水相反,黄海沿岸流淌弱,环境稳定,一些冷流还被折回(南黄海北部石油污染调查组,1977),水温适中,B_{12} 菌增多。

B_{12} 菌数季节性变化,正如湖泊层理期末尾下层滞水带的水充满温跃层那样(Cavacari 等,1977),在舟山渔场和东海测区,秋季 B_{12} 菌较高,可能原因是冷暖海流的交汇。测时,河水带来了大量有机质及相当多的秋雨，加上一些浮游植物的衰落而致。冬季苏北海区B_{12}菌数低，寒冷是主因。春、夏B_{12}菌数相当高，且有上升之势。但B_{12}及B_{12}菌数之季节变化，尤其在不同海域海水中，统一又一成不变的规律尚难找到。

在水中，B_{12}菌的垂直分体如同 Cassedebat 在阿尔及利亚奥兰港不远海区发现的细菌数与深度之关系一样呈反比关系。研究表明近岸浅海水细菌层叠现象比大洋中的显著。表水与其中的一些颗粒形成薄膜相连一起，是B_{12}菌等微生物的机械支持物，表层浮游植物常多于别的层次，其大量繁殖和随之而来的大量死亡，将伴之以较多的B_{12}菌，尽管类群不一。一些浮游生物与B_{12}菌等微生物的密切关系反映了B_{12}及其产生菌在藻类、仔虾等生长发育上的重要作用。海流的上升经表水等层次带来的营养也是春夏之交和秋季明显出现B_{12}菌数垂直分布的原因。但秋季B_{12}菌数在特定海区水层中的差异可小于春、夏之际（表 6）。

大体说来，高数量B_{12}菌常产生高浓度的B_{12}。但在特定海区，不能由此简单得出上述结论。B_{12}虽然主要由B_{12}菌产生，但其他合成者，如某些海藻等也产生。海洋B_{12}浓度受种种因子影响，海水比底质明显。海水的高 pH 值，直射光，尤其紫外线辐射、变化的水温，需B_{12}的各种生物及化学要素都与B_{12}的浓度变化有关。本结果表明上层水B_{12}菌多于中、下层。一般言B_{12}浓度在光照层少，下层高于中、上层。两种矛盾的关系表明光照层B_{12}菌多，利用B_{12}者也多，尤其是硅藻、绿藻，再加上不稳定的理化因子，使B_{12}菌易分解破坏。在中、下层，B_{12}菌减少，但利用B_{12}者也少，较稳定环境促使B_{12}积累，越下则略增，直至沉积物—底水界面处。但B_{12}的异常增加，对特定生态可产生不良效果。

综上所述，海洋B_{12}菌数及其百分率变化，反映了某些生态学现象。它与海洋各种生态——环境因子密切相关。B_{12}菌及所产生B_{12}作为一种环境生态因子反过来又影响甚而规定其他生物种群和数量变化。据B_{12}菌和B_{12}浓度变化在一定范围和程度上可推测相关营养、生态因子的差异，并为寻找合适的B_{12}菌提供了重要线索。B_{12}及其产生菌所具有生态意义是重要的。

（合作者：刘秀云　高月华）

生命的三界学说以及生命的第三形式*

摘　要　1997 年，Woese 等人用比较 16 S rRNA 序列的方法将生物分成三大类群，即真核类群，细菌类群和古生物类群。古细菌是既不同于原核生物也不同于真核生物的第三类生物。本文对生物三界学说的由来和发展、古细菌的分类、研究现状以及展望作一概述。

关键词　三界学说　古细菌　16 S rRNA

1. 古细菌的发现与三界学说的创立

长期以来，两界学说认为生物界划分为原核、真核两大界，真核生物由原始的原核生物进化而来。细胞与分子生物学的研究也表明，现在的原核生物在极早的时候就已分化为两大类：真细菌(eubacteria)与古细菌(archaebacteria)。真细菌包括几乎所有的细菌、蓝细菌、放线菌及支原体等；而古细菌是一些生长在特殊环境中的细菌，后来发现的其他古细菌中，盐细菌生长在高浓度的盐水如沿海的盐碱地、大盐湖或死海之类的内陆水域中。Woese(1990)对各种类型生物的 rRNA 核苷酸序列同源性进行测定，选择高度保守的 SS rRNA(small subunit rRNA、原核 16 S、真核 18 S)作为绘制系统进化树的标尺，将生命体系划分为三界：第一界是真核生物、第二界是原核生物、第三界是古核生物。在生物进化过程的早期，由生物的共同祖先分 3 条路径进化，即形成了生命的 3 种形式：真细菌(Eubacteria)，包括蓝细菌和各种原核生物(没有细胞核)；真核生物(Eucaryotes)，包括真菌、动物和植物(有细胞核)；古细菌(Archaea)，包括产甲烷菌、极端嗜盐菌和嗜热嗜酸菌。古细菌通常称为生命的第三形式[1,2]，对古细菌 RNA 的研究发现，它们与真核生物的关系比与真细菌的关系更为密切。

2. 生命第三形式的主要特点

Woese 的三界学说将古核生物分为两大类：Euryarchaeota(包括产甲烷类及表型相关种属)和 Crenarchaeota(在表型上接近原始的 Archaea)。古细菌体积小，直径 0.5～5 μm，形状有杆状、球形、螺旋形、丝状，基因组大小为 2～4 Mbp，多数为嗜热、嗜酸、自养型。古细菌有丰富的组蛋白，DNA 以核小体的形式存在[3,4]。

古细菌与一般细菌的氧化磷酸化产能不同。如产甲烷菌(Methanogen)和硫细菌(Sulfolobus)用氢还原 CO_2、S 生成 CH_4、H_2S 来获得能量，有些古核生物利用钨代替高温下不稳定的 NADH 参与电子传递链。

古细菌具有比细菌更多样化的质膜结构。古细菌质膜由特殊脂类组成，甘油和脂肪酸链之间的结合是通过醚键而不是酯键来实现的。

* 原文刊于《生物学通报》，2005，40(9)：19-20，第二作者

古细菌没有以胞壁酸为基质的粘肽类细胞壁，而真细菌具有与胞壁酸相结合的细胞壁。古细菌对作用于真细菌细胞壁的抗生素如青霉素、头孢霉素以及抑制真细菌翻译过程的氯霉素和环丝氨酸不敏感，而对抑制真核生物翻译过程的白喉毒素却十分敏感。

与真细菌相比，不同类型古细菌的多种蛋白质一级结构中疏水残基的平均数量增多，增强了蛋白质的高温稳定性。例如，*S. solfataricus* S 腺苷半胱氨酸水解酶多肽链中完全没有半胱氨酸，与高温生活环境相适应；*Pyrococuus* sp. KOD I 天冬氨酸—氨酰 tRNA 合成酶的 N 端与真细菌同源，C 端与真核生物同源。因此古核生物酶可能是真核和原核生物酶的原型[2]。

古核生物的遗传信息传递机制与真核生物相似：①DNA 及基因结构具有重复序列和内含子；②古核生物中依赖 DNA 的 RNA 聚合酶亚基和延长因子，与真核生物的同源性高于与原核生物的同源性；③具有与真核生物组蛋白 H_2A、H_2B、H_3、H_4 对应的组蛋白，但种属之间序列差异较大，具体包装过程及核小体的周期规律也不清楚；④古核生物氨酰—tRNA 合成酶基因序列与真核生物质的同源性较高；⑤对 5S rRNA 的分子分析及二级结构的研究表明古细菌与真核生物相似，与原核生物差距甚远；⑥蛋白质合成起始以甲硫氨酸作为新生肽链的 N 端氨基酸，全部以 AUG 为起始密码子。以上分析说明古细菌与真核生物在进化上的关系较原核生物更为密切，真细菌很可能首先从古细菌和真核生物的共同干线中分支出来[2]。

3. 生命第三形式的分类特征

产甲烷菌（*Methanococcus*）能利用氢和二氧化碳生成甲烷，有些还能利用甲酸或甲醇、乙酸或甲胺产生甲烷和二氧化碳。根据 16 S rRNA 序列的不同又可将其分为 3 种类型：Ⅰ型包括甲烷杆菌和甲烷短杆菌，细胞壁主要成分为拟肽聚糖，膜的磷脂有 C_{20} 二醚和 C_{40} 四醚；Ⅱ型只有甲烷球菌，细胞壁由蛋白质组成，膜脂含 C_{20} 二醚；Ⅲ型，包括甲烷螺菌和甲烷八迭球菌，前者细胞壁由蛋白质组成，膜脂含有 C_{20} 二醚和 C_{40} 四醚；后者细胞壁由酸性杂多糖组成，膜脂含有 C_{20} 二醚。

嗜盐菌（Extreme halophiles）。现已发现 2 个属：盐球菌，不运动，细胞壁由酸性杂多糖组成，膜含 C_{20} 二醚；嗜盐杆菌，末端生鞭毛，能运动，细胞壁由糖蛋白组成，膜含 C_{20} 二醚。

嗜盐菌从细胞结构到组成成分上，都有一些耐盐的因素：①盐细菌有电子传递链，能进行有氧呼吸，而且在氧含量低且有光照时，也能利用光能产生 ATP；②细胞壁成分特殊，不含肽聚糖而以脂蛋白为主；③极端嗜盐菌还积累或产生一些维持细胞内外渗透压平衡，又有助于细胞代谢活动的相溶性物质，如积累 K^+、合成糖、氨基酸等；④菌体内的酶是嗜盐性的，酶中所含的酸性氨基酸比率较高，这有助于对高盐环境的适应。

嗜热嗜酸菌。已鉴定出 3 个类群：硫化细菌、热支原体和热变形杆菌。一般最适生长温度在 90℃以上的微生物称为极端嗜热菌，极端嗜热菌大多是古细菌，生活在深海火山喷口附近或其周围区域。嗜热菌多为异养菌，许多能将硫氧化以取得能量，其耐热机制一般是多种因子共同作用的结果：①细胞膜不是真正的双层，单层的脂类末端具有极性基团，降低了高温环境中的膜流动性；膜的化学成分随环境温度的升高发生变化，增加了膜稳定性；②重要代谢产物能迅速合成，tRNA 周转率提高；DNA 中的 G、C 的含量较高，使生物体中的遗传物质更加稳定；一些组蛋白也增加 DNA 的耐热性；③蛋白质一级结构中个别氨基酸的改变导致其热稳定性的改变，二级结构中包括稍长的螺旋结构，三股链组成的B折叠结构，C末端和N末端氨基酸残基间的离子作用以及较小的表面环等使其形成紧密而有韧性的结构，利于热稳定[5,6]；④斯坦福大学发现古细菌中一种含钨酶(一般生物中的钨没有此作用)，认为钨在耐高温的古细菌的代谢中起关键性作用。

4. 生命第三形式研究进展以及展望

以前生命第三形式的研究只是局部的，针对一种或几种蛋白或基因，DNA图谱的工作尚处在物理图谱阶段[2]。1996年8月，*SCIENCE* 发表了美国基因组研究中心L. Bult领导的研究小组的最新学术成果——产甲烷球菌的全基因组序列，同源性搜索和GENE MARK的基因定位方法的结果表明，在产甲烷球菌基因组中，只有38%的基因对应于数据库中已知功能的基因，所以证明产甲烷球菌存在着大量的新基因序列，它同时具有类似细菌和真核生物的基因序列。Bult的研究证明archaea既不是典型的细菌也不是典型的真核生物，而是介乎两者之间的生命体。对它的研究无疑丰富了三界学说的研究内容，是三界学说发展的重要进展。*Nature* 称这个成果是三界学说发展史上的"里程碑"，为古核生物的研究提供了充分的序列材料。使在基因组整体水平上研究古核生物成为可能。古核生物新基因序列的测定为分子生物学家提供了丰富的序列资料，例如Bateman在产甲烷菌的DNA编码的蛋白质中发现一个新的域，有33个同源拷贝，并且有这种拷贝的蛋白质功能差异很大；有研究表明用从古细菌中提取的醚甘油脂配制成的脂质体疫苗佐剂可诱导强烈的体液、细胞以及记忆应答，有强烈的佐剂作用。

美国基因组研究中心下一个将要测定的古核生物全基因组序列是 *Crenarchaeota*，因为从系统发生树来说，它更接近根部，其基因组序列具有更多祖先细胞的特征。Margulis认为三界并非平行起源，最早的真核生物是由古核与原核生物的某些种属发生永久性的细胞融合而来。Baldauf用延伸因子EF-Tu、EF-G的基因作为分子系谱学的标尺来划分生命，将生命划分为原核生物与古核生物两大类，真核生物则落入古核生物中的一个亚类。因此，三界学说还有待于进一步从不同层次深入研究和认识。

参考文献6篇(略)

(合作者：刘洪展　郑凤荣)

第二篇

应用海洋微生物学

APPLIED MARINE MICROBIOLOGY

B$_1$ 环境污染与保护
ENVIRONMENTAL POLLUTION AND PROTECTION

海洋环境中石油的微生物降解*

随着石油残留在水中时间的延长，生物处理便开始见效并迅速达到有意义的水平(Templeton，1971；Miget，1973；pilpel，1968)。有90多种微生物(细菌和真菌)能生活在石油中，通过生物氧化作用降解石油。海洋中这种降解作用的程度决定于水温及漏失石油的某些挥发性成分(萘、煤油等)，其中一些能杀菌或抑菌。当除去这些，原油更易被降解，不过挥发性成分在溢油几个月后才能开始消迹。

其实，所有的石油对微生物降解都是敏感的，但影响烃的生物降解力之最重要因子似乎是分子本身的结构。烷烃比芳烃或萘化合物更易被较多微生物迅速降解。烷烃系列中直链的化合物比支链化合物对微生物氧化作用更敏感，链长的增加似乎增加了微生物氧化作用的难度，这可能主要归因于它们在水中的低溶度，其次是缺乏酶功能发挥作用。

石油烃成分其质与量的差异影响降解的敏感性。调查了巴尔的摩港油污区水、沉积物细菌对两种原油、两种精制油的降解程度。南路易斯安娜原油(简称SLCO——译者注，下同)含饱和烃最高，达56%。科威特原油(简称KCO)含34%饱和烃和18%树脂(吡啶、喹啉、咔唑、亚砜和酰胺)。2号燃油含61%饱和烃、39%芳烃。6号燃油(Bunker C)含60%沥青质(酚类、脂肪酸、酮、酯和卟啉)及发泡物(Walker，Colwell和Petrakis，1975)。上述每种油在接菌7天后，留下的总量用柱层析和数字化质谱仪分析，发现数量大致一样，表明其降解量相近。然而对四种油成分的选择降解却说明每种油的成分都有不同的降解敏感性。因此生物降解充其量是个不平衡的过程；石油的某些成分被选择地降解了，而另一些则留下了。

Mulkins-phillips和Stewart(1974；1974a)比较了诺卡氏菌属(*Nocardia*)一些种对委内瑞拉原油(VZCO)和阿拉伯原油(ABCO)的降解，发现前者77%、后者94%的*n*-链烷及前者13%、后者35%的不溶组分被降解了。Atlas和Bartha(1973年)证实两种石蜡质原油一瑞典原油(SCO)和LCO(路易斯安娜原油)被海水的混合细菌培养物降解得相近

* 原文刊于《应用微生物》，1978，No. 4：70-75.

(70%)。Jobson 等(1974)比较了所谓的“低级”和“高级”原油上微生物生长情况。Westlake 等(1974)报道了原油组分特别是饱和直链组分、树脂和沥青质对生物降解的影响。每种油实际上都有影响微生物降解作用的特殊组分,某些油比别的油降解得更完全。石油的某些组分容易且较快地降解(n-链烷),而另一些则存在下来,要是能全被降解的话,则相当缓慢。

油污区的微生物区系不同于原始状态环境的微生物区系,石油降解速率也不一样。如果一个环境中原来就有油存在,那么解油者可能有所选择,要是这样,降解可能较快,但不一定进行得彻底。比如两个这样的环境,一个是油污区即巴尔的摩港,一个是未油污区即东湾,它们都位于美国东岸一大河口同一地理位置即切萨皮克湾中,以该两海区的沉积物、水样作降解接种体,发现巴尔的摩港细菌(早就暴露在水中)降解得较有效。风蚀样品的资料表明蒸发作用会失去一些成分,但微生物生物降解更为重要,换言之,代谢作用,使SLCO组分消失。比较了不同地理位置两个油污港口沉积物试样接种体对SLCO的降解,发现巴尔的摩港比圣胡安港的更有效,在实验室最适条件下培养几周后,巴尔的摩港培养物中树脂和沥青比圣胡安港增多了,因此两个港口的微生物群系实际是不同的。

缓解或不解的成分以“难解”物积累或是微生物合成的结果之一(Walker 和 Colwell,1976a)。Atlas Bartha 和 Van Leeuwenhoek(1974)指出,烧瓶培养中脂肪酸等化合物的累积可阻止油的完全生物降解。Walker 和 Colwell(1976)报道了 *Acinetobacter iwoffi*、*Pseudomonas* sp. *Nocardia* sp. *Corynebacterium* sp. 及一种未定名菌株对SLCO、Brass河尼日利亚原油、VZCO Anaco 和 Altamont 原油等四种原油的降解。也调查了切萨皮克湾巴尔的摩港 Colgate Creek 的水样混合培养物。在海洋航行中对从迈阿米海滩、佛罗里达和马尾藻海采得的柏油球(Tarballs)进行分析研究。(Brunmek. Duckworth 和 Stephen,1968,Ramsdale 和 Wilkinson,1968)。在AMCO中,有一种诺卡氏菌生长时直链烷烃 $C_{18}-C_{36}$ 增加了。经 *Acinetobacter iwoffi*、*Pseudomonas* sp.,一个未定名的 *Nocardia* 属的一种菌株和AMCO的混合培养物的生物降解后,留下的蜡质残留物不含低沸点直链烷烃,而含相当多的高沸点直链烷烃 $C_{27}\sim C_{36}$ 及 $C_{38}\sim C_{43}$。迈阿米远海和佛罗里达的柏油球也含长的直链烷烃,马尾藻海深海站柏油球却仅含微量长直链烷烃。因此柏油球的组分实际上标志着暴露在外界的原油被微生物降解的程度。

地理位置分开的未污染区,各有独特的微生物区系,对输入的油感应也不同。这从两个未污染环境(一个河口区,一个深海站)的沉积物的种菌对SLCO的降解能力比较中可看到。深海站沉积物试样比切萨皮克湾东湾河口沉积物试样有更大的解油潜力。当这两站沉积物与污染港沉积物比较时,污染港口沉积物清楚显示有最大降解潜力,分析上述河口和深海站水、沉积物情况,发现浅海和深水环境特别是温度和压力这些参数的差异对烃的降解有明显的作用。

Schwarz、Walker 和 Colwell(1974,1974a)以 ^{14}C 标记烃和混合烃基质作利用烃的实验。十六烷上的培养物在1气压下、4℃时,于4周内达最大生长,即静止期,在500大气压下,最大生长只能在培养32周后才达到。在1大气压下,于8周后,94%的十六烷被利用了,而500大气压下,却要40周才达到。因此,在低温、高压时,十六烷的利用是相当有限的(Schwarz,Waller 和 Colwell,1974,1974,1975a)。这些资料有力地表明进入深海的油,被那里的微生物群系降解得很慢。

文献中大多数资料，由于需要，都是讨论石油（或其馏分）的分批或烧瓶培养以及细菌的混合或纯培养。这些文献提供了有用的情报和自然界这类问题的模式系统，但为便于确定海洋环境石油的微生物降解各个参数。必需暂不考虑自然生态系统的复杂性。这样分批培养的解油过程，如同多数实验室中进行的那样，是有限的。经烧瓶或分批培养降解后，将会发生无机营养的消耗、毒物和难解有机物的产生。实验室研究证实生物降解可为脂肪酸限制（Atlas，Bartha 和 Van Leewenhoek，1974）。这些成分最可能因代谢活动的积累和抑制所干扰。沥青质部分即有机复合物和难以进一步降解的不溶物之增加也可因油的微生物作用而致。然而，石油降解的这个方面却被忽视了。因此应发展分析方法，以了解降解出来的更多难溶物。

营养尤其是硝酸盐和磷酸盐的限制，是影响海洋环境石油降解的一个重要因子。有迹象表明，许多实验室内的研究在模拟现场条件时都很难产生有一定意义的石油降解（Atlas 和 Bartha，1972）。当加硝酸盐和磷酸盐到有油的水沉积物混合的流动海水池中时，观察到生物降解增加了 8～9 倍（Atlas 和 Bartha，1972；1973b）。用 SCO 作唯一碳源丰富人造海水培养基（含 SCO 1%）时，辛磷酸盐和缓慢释放的石腊化脲相结合，效果如用硝酸盐和磷酸盐一样。分批培养的切萨皮克湾水的细菌对混合烃基质的利用增加了。该湾水的盐度为 14～18，而普通海水是 35，该湾无机氮含量也高（Walker 和 Colwell，1975a）。类似的效果在海水培养的细菌中也可看到（Walker 等，1975）。因此，石油降解的 N：P：C 比值必须接近于现场研究中石油降解的增长（Walker 和 Colwell，1975）。

气候条件影响石油的微生物降解。Raritan 湾的两个站在 1971 年 7 月～1972 年 5 月期间分 5 个阶段进行试验，发现石油降解者最高数出现在 7 月（Atlas 和 Bartha，1973a）。水温在夏季最高，温度与其他条件相结合，强烈地影响所选择的微生物群系之升降（Kaneko 和 Colwell，1972）。

然而，解油者数量的增加可在两个不同的时期内：12～2 月和 6～7 月出现（切萨皮克湾）（Colwell 等，1974）。冬季无嗜冷解油菌的富集，因为当比较 1 月和 9 月该湾油污区沉积细菌在 0℃时解油情况时，发现解油量无重大差别。这可能表明广温性解油微生物存在着。一种选择即因油的进入而引来解油菌，是存在的。上述实验还表明温度并非解油微生物群系的限制因子。总的降解速率仍是冬季低于夏季。

除了特殊条件多少由该区给定的地理区域确定外，影响温带和北极区微生物降解作用的因子仍是季节条件的那些因子。例如，进入永久性寒冷环境的油将为早就存在或适于该环境的嗜冷微生物群系所降解。低温时，必须考虑的基本影响是物化作用即挥发作用的减少和挥发性石油烃溶水性的增加（Atlas 和 Bartha，1972a；Walker 和 Colwell，1974）。世界各大洋各特定水域都有其本地的微生物群系，其数量是沿岸区、河口水域比深海丰富（Colwell，Walker 和 Nelson，1973；Colwell 等，1974，1976）。其群系直接与环境相适应。石油的输入将改变环境，引起对解油微生物的选择。降解的快慢决定于营养（N、P）、温度、油—水基质的混合和其他参数。北极环境的微生物降解包括有本地区的嗜冷微生物降解，降解虽发生，但比温带暖和季节慢。因此必须协力发展一种包括微生物降解作用在内的先进方法，以对抗油漏。

需要阐明现场细菌、酵母和丝状真菌在石油降解中个别的和相互的作用，因为纯培养物在自然环境中总较少，还是混合群系起作用。必须确定每个微生物群是拮抗的还是协同的。资料表明了油区解油细菌、酵母和真菌的总数，该数与解油活力相关，但总的生活群系（数量）等生物量不清楚。因为降解类型不一，必须研究各种类型的微生物群系（Walker，Austin 和 Colwell，1974）。最近已发现解油的无叶绿素藻类及其解油样式（Walker 和 Colwell 正在出版中，Walker，Colwell 和 Petrakis，1975a）。因此连藻类也必须考虑在所测解油群系之中。

Miget 等（1969）、Atlas 等（1972b），Reisfeld 等（1972）和 Cerriglia 等（1973）报道过细菌和真菌的纯培养物对石油的利用，报导 *Flavobacterium*，*Brevibacterium* 和 *Arthrobacter* spp. 一些细菌利用 35%～70%的石蜡质原油，而真菌 *Penicillium* 和 *Cunnungamella* spp. 利用 85%～92%的石蜡质原油。Walker 和 Colwell 证明酵母 *Canaida* 和真菌 *Penicillium* 比细菌 *Pseudomonas* 和 *Coryneiorms* 降解 SLCO 更广，但没有一篇所引文献报导过石油全被降解的。这是必须考虑的一个重要事实。对细菌、酵母和真菌合起来用的解油范围及其后情况已作过研究（Conrad，Walker 和 Colwell，1976），初步结果表明降解顺序大致由细菌开始，主要利用烷烃和选用环烷烃，其后酵母和真菌在群系数量上获得优势，并有降解环烷烃和芳烃的能力。

用混合烃基质进行了试验（Walker 和 Colwell，1974，1974a）。观察到在含有混合烃基质的烧瓶中细菌的生长超过酵母和真菌的生长。此外，注意到生长在混合烃基质上的混合培养中几属细菌有摆动。

混合烃基质上细菌、酵母和丝状真菌的结合比它们各自使用时，降解作用大二倍（Walker 和 Colwell，1974）。令人奇怪的是当讨论用撒布微生物降解式控制溢油时却往往不强调此点，况且酵母和真菌群系在开阔的大洋水中总是稀少的（Zobell，1946），因此深海石油微生物降解的机会和范围比沿岸或河口区少得多，亦即残留物在更长的时期内持续下来。

微生物播种进石油（Sebba，1971），已有一些研究者介绍过。实验室播种（Liu 和 Dutka，1972；Robichaux 和 Myriek，1972；Lakock 和 Severence，1973）和模拟现场条件中播种（Atlas 和 Bartha，1973b；Miget，1973）都试过。两者都表明细菌能促进石油降解。但播种必须只在某些情况中进行，而且应十分谨慎（Cobet，Guard 和 Chatigny，1973）。单一微生物种对溢油的许多化合物代谢并非都有酶活性。混合的微生物包括真菌、酵母和细菌，是可取的。无论是纯培养物或是混合培养物，其加富培养都可用作接种体。

有几个海区，曾运用微生物制剂除油。但存在着几个问题：应用方法的选择，即湿剂、干粉、丸剂还是气溶胶；潜伏的致病性或不利于人类的反应，即微生物制剂中人类病原体或偶发性病原体；对制剂的过敏反应；环境中微生物过度的增加影响自然状态的生物，乳化油的残物有更大的潜毒；生物降解形成了胞外代谢产物等等，问题不少。Cook，Massey 和 Ahearn（1973）的研究暗示石油的微生物降解会增加油的毒性和油的成分。酵母分解期代谢产物及酵母对高沥青质原油的乳化作用有害鱼类。

不仅作了一些模拟海洋环境设计的现场实验，而且还做了一些具体的现场实验，这就是 Jobson 等（1974）对油污土地作过播种细菌的试验。与 N、P 源相比发现加入细菌对除油没有多少效果。而最大的危险却呈现出来了—新菌种的导入影响自然群落，例如鱼被杀死或其他生物的消失。

石油的生物降解速率取决于温度、盐度、无机营养的浓度、水中油分散范围、微生物的种类和数量、油的化学构成、开放现场，还是封闭的(实验室)系统及其他各种因子。Zobell (1969)在总结69年前完成的工作时报道原油、润滑油、切削油和废油的降解速率约0.02～2.0 g/(m^2·d)(24～30℃)。

油的生物降解速率是按几天或几周之后所降解的数量计算的，一般采用的单位是g/(m^2·d)、g/(m^3·d)、mg·天·细菌细胞或百分比。表1摘录了几个研究者给出的解油速率。对较少的研究资料进行比较是困难的。当比较所测石油生物降解速率的研究时，实验条件的差异是另一问题(Johnson，1970；Kator，Oppenheimer 和 Miget，1971；Kinney，Button 和 Schell，1967；Robertson 等，1973；Bridie 和 Bos，1971)。总的说来，测定包括实验始末的油量，而且假设速率是线性的。

表1 不同条件下石油降解速率(海水温度2～16℃，无机氮浓度：<1～500 gN/L)

所用细菌	石油种类	实验条件	结果摘要	引证
菜园土的几属好氧菌	通用的混合烃	分批培养(无机盐培养等)20℃和37℃之间	28℃时测定的某些物质0.4～0.75 g/(m^2·d)	Sohngen，N. L. (1913)，Zontralbl. Bakt.，Parasitkde. Ab. Ⅱ:37，595
土壤好氧菌	Emba原油和润滑油	分批培养、无机盐培养基中加硝酸盐或氨，23℃	1.2 g/(m^2·d)原油(加入量是45%)0.4 g/(m^2·d)天润滑油	Tausson. V. O. 和 Shapiro，S. I. (1934) Mikrobiologiva 3:79
以海洋细菌为优势的富集培养	洁净的精制矿油	可能是分批培养25℃老化海水中加0.5%的KNO_3	矿油的氧化作用以O_2的吸收，CO_2的量和细菌生长为指准。0℃和4℃时的Q10是3.0，25℃时油降解的平均值是1.2×10^{-10}(mg/天/细菌细胞数)若油一致地分布在水里，数量恒定保持在8×10^5个生物/mL的话，解油速率将是350 g/(m^3·d)(25℃)和36.5 g/(m^3·d)(5℃)	Zobell，C. E. (1964) Advances in Water Poll. Res. 3:86. 本报告也见于 Airst water Poll. (1960) 7:173.
油氧化菌的混合培养	美国原油	分批培养，海水培养基[添有0.01% $(NH_4)_2HPO_4$ 1g/100 mL]振荡通气，油分散在烧结石棉上	30天内，除去重量的17.8%～9.8%平均45%	Zobell，C. E. 和 Prokop，J. V. (1966) Z. allg. M. krbiol. 6:143
自然海水群系	原油和几种精制品	分批培养，18℃海水培养基中添有磷酸盐等	说明了各种物理、化学因子对石油降解的干扰：氮磷的存在明显增加柴油的溃解(8周内)，易降解物质的存在"宽恕了"油，温度的影响也列出了	Gunkel，W. (1967)，Helgolander. wiss. Meeresunters. 15:210

（续表）

所用细菌	石油种类	实验条件	结果摘要	引证
自然海洋群系	科威特油的气化残物	有自然 meio 和微生物区系的海沙，粒度 250 μm 浸透海水的海沙柱，海沙或轻或重地浸过油。10℃	氧气的吸收以"B. O. D"值为 5.0 作解油的指标，作者估计油的消失是 0.09 g/(m²·d)－0.04 g/(m²·d)，决定于给油量，这些速率用于几个月和计算 10%的油，初步的气相色谱指明主要是烷烃的消失，留下的 90%"无限慢地"衰变。	Johnson，R.（1970），J. mar. biol. Ass. Uk 50：925
一些油氧化生物的混合培养	路易斯安娜原油	以无机 N、P，和酵母膏丰富海水，200 mL 海水加 70 mg 油，20℃和 30℃振荡培养及模拟野外研究的大容器（900 升）海水富含（NH_4），50～100 mL 原油	开初的氧化作用归之于小于 C 的直链烷烃之溃解，速率由减少起，其后，有一个增加，原油的 50%失去了。没发现芳烃的利用，在大容器中，细菌是由油的消失及其物理性质的改变	Kator H.，Oppenhoimer，C. H. he Miget，R. J.（1971）Prevention and Controlofoil spills. Amer. Cau petroleum Institute Confereuce 1971，pp. 287

摘自 Ploodgate（1972a）

关于河口沉积物细菌降解饱和的、芳香的和其他石油组分的研究表明不同成分间的生物降解范围和速率有明显的差别（Walker，Colwell 和 Petrakis，1976）。SLCO 总残留物降解的最高值发生在对数生长期（1.42 mg/d），如所预料的，在静止生长期平稳地下降。对数生长期之后，沥青质和芳香烃增加，但饱和烃在整整 7 周中一直被降解。

以油污过的河口沉积物对 SLCO 作生物降解，其比率通过适于获得第 1、2 或第 3 级多项图内经计算的曲线测定（$P \geqslant 0.05$）。饱和烃是唯一显示持续减少的组分。芳香烃于起初减少末期增加，如 3～7 周。而吡啶、喹啉、咔唑、亚砜基和酰胺也是初减后增。沥青质最类似风蚀和生物降解的产物。显然，Walker，Colwell 和 Patrakis（1976a）所得资料表明石油虽同时被降解，但各组分的降解速率却不同。

细菌氧化产物在某种意义上说是个生态因子，微生物完全的生物氧化作用主要产生 CO_2、水、硫酸盐和硝酸盐，但许多生物学反应并未充分进行，中间产物如酸、醛、酮、醇类，过氧化物等将溶于水，从流动性样品中去掉。

微生物侵蚀水上油膜或薄油层，将导致颜色、成分、荧光性，乳化和其他可见性质的变化。油将变得黏稠，渐渐分解成缓慢下沉的小滴，一部分将与微生物群系的极大增加一起乳化在水里，于是培养基变得模糊、混浊，有时则带有色泽（Kator，1973）。油的这种可见消失未必指示它完全毁灭。而微生物的侵蚀可变油为 CO_2，一部分将变为水溶性的其他物质，乳化在水里或沉没到容器底部。

参考文献（略）

译自 *Impact of oil on the marine environment*（Reports and Studies），1977，No6，6：29-37，路成铭校。

胶州湾潮间带和沿岸区的耐铬菌*

摘要　本文首次报道了胶州湾潮间带和沿岸区的耐铬菌数量分布。讨论了耐铬菌的若干生态意义。发现沉积物表层的耐铬菌数量大。但水中的耐铬菌出现频率高于沉积物的。找到了CT/AA与样品含铬量间的正相关。指出CT/AA可作为环境铬污染的一个重要指数。试验了耐铬菌耐 Cr^{3+}、Cr^{6+}、Hg^{2+} 和降解原油的能力。认为耐铬菌对环境污染有适应、忍耐、抵抗乃至降解的能力。耐铬菌在环境污染物(首先是铬污染)的迁移、转化活动中起重要作用。它们是水生微生物区系的一个重要组成部分。耐铬菌的耐铬能力为它能参与铬污染的消除活动赋予强大活力。

海洋环境重金属污染由于宏观难以监测,在一段时期内未受重视。随着问题严重化和检测手段日臻完善,人们对此已高度关注。人体及生物(包括海洋生物)不适当地摄入铬,其不利影响乃至致癌作用是毋庸置疑的。

胶州湾大量收纳含铬废物,长期遭受铬废物污染。实测结果表明该湾河口水和沉积物中的含铬量大都超标,而由此造成的环境污染问题尚未解决。该湾潮间带等的微生物首当其冲遭受和迁移着铬的污染,在铬的持续影响下,已发生重大变化,形成了一类独特的微生物群——耐铬菌。目前已有一些涉及铬对海洋微生物生态影响的研究报告,但有关它与潮间带等海洋环境的微生物学关系研究报告则较少。本文试图报道该区域耐铬菌数量分布,讨论耐铬菌生态意义。

关键词　胶州湾　潮间带　沿岸区　耐铬菌

调查和实验

一、站址选择和样品采集

1. 站位描述

五个定点站:Ⅰ. 石油化工厂后海滩,宽广而坡度小的滩涂,海边堆放大量冶炼钢铁废渣及排放石油化工等废物,滩涂呈绿灰色,岸边沥青渣、焦油多,表泥下面呈灰黑色,硫化氢味重;Ⅱ. 板桥坊河国棉八厂入海口,沙泥底质,有泡花碱、拆船、纺织等废物排放;Ⅲ. 李村河出海口,黑泥底质,河床较宽,有印染、造纸、化工、肉类加工等废物排放;Ⅳ. 海泊河出海口,黑泥底质,有染料、农药、电镀、造纸、车辆制造等废物排放,水质差;Ⅴ. 太平角山海关路海滨,沙质底,礁石上有些淤泥、藻类附着,水质较洁净。

不定期站:Ⅵ. 化纤材料厂出海口,是红星化工厂废物入海必经之路,还收受油漆、皮革、化工等废物,离铬渣山约1500 m,泥色乌黑,硫化氢味重,水质差;Ⅶ. 红星化工厂东门

* 全文在山东省微生物学会1981年年会上宣读过。

原文刊于《环境科学学报》,1981,1(4):313-323.

桥埝排水沟，离铬渣山 100 m，是最大的铬污染源，海潮上涨不到；Ⅷ. 大港矿砂码头，污染重。站位详见图 1。

2. 采样

定点站从 1979 年 5 月到 1980 年 8 月基本上每月取样一次。不定期站按需采集。无菌采取表膜下的表层水 50 mL 入消毒过的细口瓶。用酒精擦拭且火焰消毒的采集刀刮除表面沉积物，无菌地采取沉积样 50 g 到预先消毒好的小广口瓶中，样品瓶置于冰瓶，在 4 h 内回实验室分析。

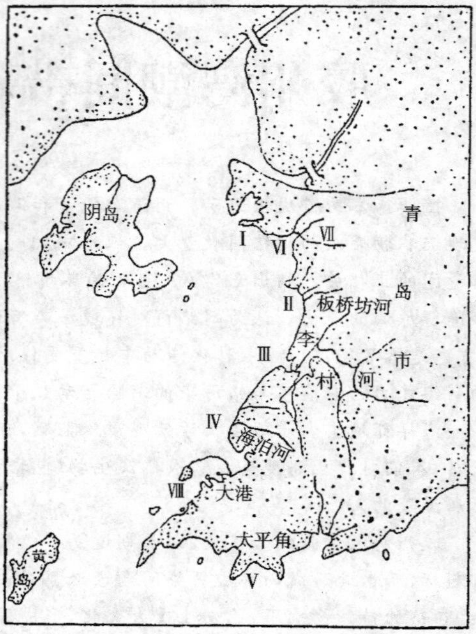

图 1　采样站位图

Fig. 1　Location of sampling station in the zone surveyed

二、微生物学实验

1. 菌落计数

(1)海水培养基上耐铬菌 CFU 计数

在海水肉汁胨平板上涂布稀释样品作海洋好气异养菌落形成单位(Colony Forming Unit，即 CFU)计数用。以此为基本培养基(下同)，加入 168×10^{-6} 的 Cr^{3+} ($CrCl_3\cdot6H_2O$ 形式，下文不作特别说明者均为此)作耐铬菌 CFU 计数用(考虑到本调查区内含铬量高，故用铬量比他人的高)24℃培养 14 日，三次检查 CFU 数，从含铬培养基上长出的菌落，就是耐铬菌(Chromium-Tolerant Bacteria，缩写作 CTB，下同)，纯化后保存备用。上述各数均换算成每毫升水样或每克湿沉积物样中含有的 CFU 数，分别以 AA 代表总异养菌 CFU 数，以 CT 代表耐铬菌 CFU 数，以 CT/AA 代表两者之比。

(2)淡水培养基上耐铬菌 CFU 数计数

培养基成分：蛋白胨，10 g；牛肉膏，3 g；氯化钠，5 g；蒸馏水，1 000 mL，琼脂固化，pH 7.4，加或不加 168×10^{-6} 的 Cr^{3+} (Ⅶ站用)。培养和计数法同上。

2. 耐铬力试验

(1)以海水肉汁胨液培作对照，加入不同浓度 Cr^{3+} 为试验耐铬菌耐铬力的培养基，振荡培养 14 日，以有菌生长者为耐铬力阳性。

(2)耐铬菌耐 Cr^{6+} 试验。加不同浓度 Cr^{6+} ($K_2Cr_2O_7$ 形式)至海水肉汁胨液培中试验耐铬菌耐 Cr^{6+} 能力。

(3)耐铬菌耐汞及解油试验

a. 在上述含 Cr^{3+} 的液培中加入不同浓度的 Hg^{2+} ($HgCl_2$ 形式)试验耐铬菌耐汞能力。

b. 在下列培养基中试验耐铬菌降解原油的能力：氯化铵，2 g；磷酸氢二钾，0.7 g；磷酸二氢钾，0.3 g；陈海水，1 000 mL，胜利原油 5%(W/W)，pH 7.4。

培养和检查法同上。

三、样品中铬含量的测定

在取微生物分析用样的同时，采取化学分析用样，采用二苯酰二肼比色法测含铬量*。

四、数据处理

对部分数据，根据中国科学院数学研究所编《常用数理统计方法》作方差分析和一元回归分析，说明站位（即地理学特异性）对 CT 等月分布的影响及铬含量与 CT/AA 等的相关关系。

结　果

调查和实验结果列于图 2 和表 1-6 中。

表 1　沉积物中耐铬菌数量比较**（CFU/g）

Table 1　Comparison of numbers of CTB in sediments at five stations

站位	CT	AA	CT/AA
Ⅰ	9.69×10^{6}	3.34×10^{8}	0.0290
Ⅱ	1.21×10^{7}	5.40×10^{9}	0.0022
Ⅲ	4.13×10^{7}	4.34×10^{9}	0.0095
Ⅳ	1.08×10^{8}	9.87×10^{8}	0.1084
Ⅴ	4.69×10^{6}	2.27×10^{9}	0.0021
平均	3.52×10^{7}	2.67×10^{9}	0.0304

**本表以 1979 年 7 月 20 日后的数据统计。

表 2　水样中耐铬菌数量比较***（CFU/g）

Table 2　Comparison of numbers of CTB in waters at five stations

站位	CT	AA	CT/AA
Ⅰ	4.45×10^{5}	6.36×10^{7}	0.0070
Ⅱ	1.03×10^{5}	7.08×10^{6}	0.0145
Ⅲ	3.51×10^{6}	1.13×10^{7}	0.3106
Ⅳ	3.10×10^{6}	4.28×10^{7}	0.0724
Ⅴ	5.69×10^{4}	6.76×10^{5}	0.0842
平均	1.44×10^{6}	2.51×10^{7}	0.0977

***本表以 1979 年 7 月 20 日后的数据统计。

* 此项由青岛红星化工厂铬盐研究所协助测定。

表 3　五站耐铬菌数量的方差分析

Table 3　Analysis of variance of CT at five stations surveyed

样品	F	显著性
水	4.19	**
沉积物	3.26	*

* 表示差异显著；** 表示差异高度显著

表 4　耐铬菌对递增的铬浓度的耐力

Table 4　Tolerance of tested strains of CTB to successive concentrations of Cr^{3+} ion

铬浓度(ppm)	参试菌株数	生存数	存活率(%)
0	24	24	100
84	20	20	100
168	22	22	100
336	22	15	68.18
840	22	12	54.55
1176	24	4	16.67
1680	24	0	0

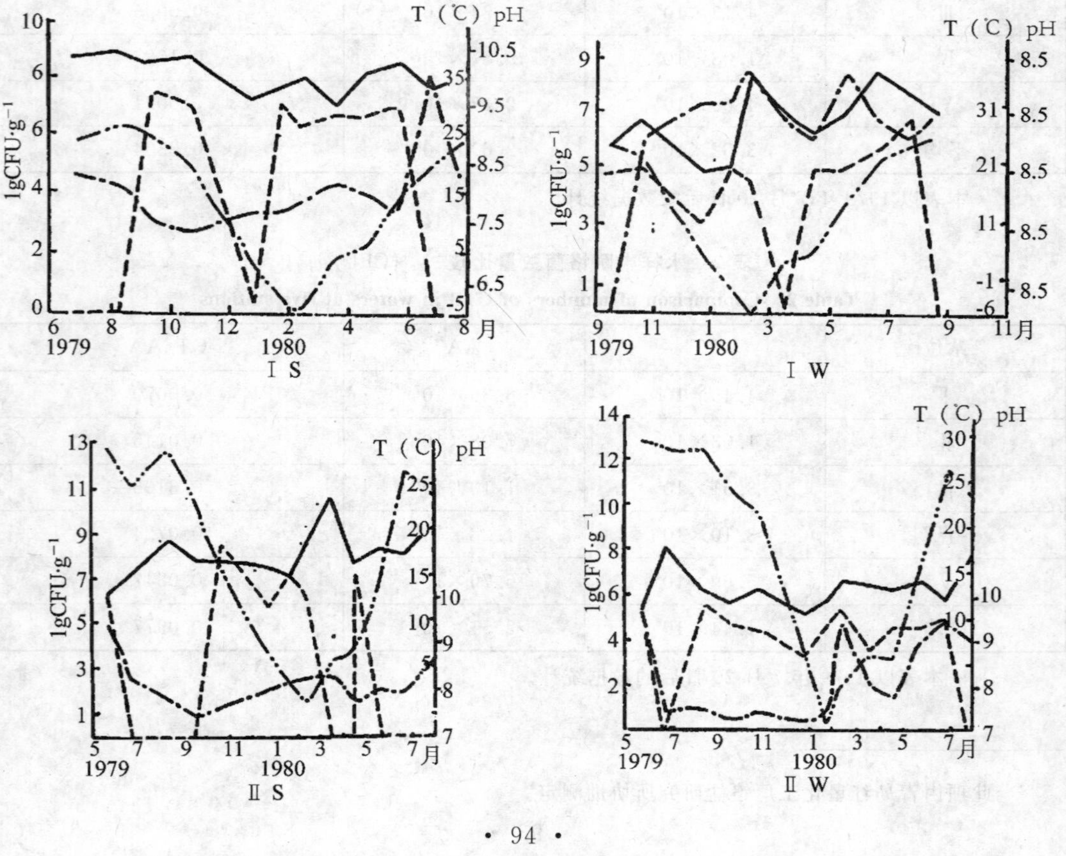

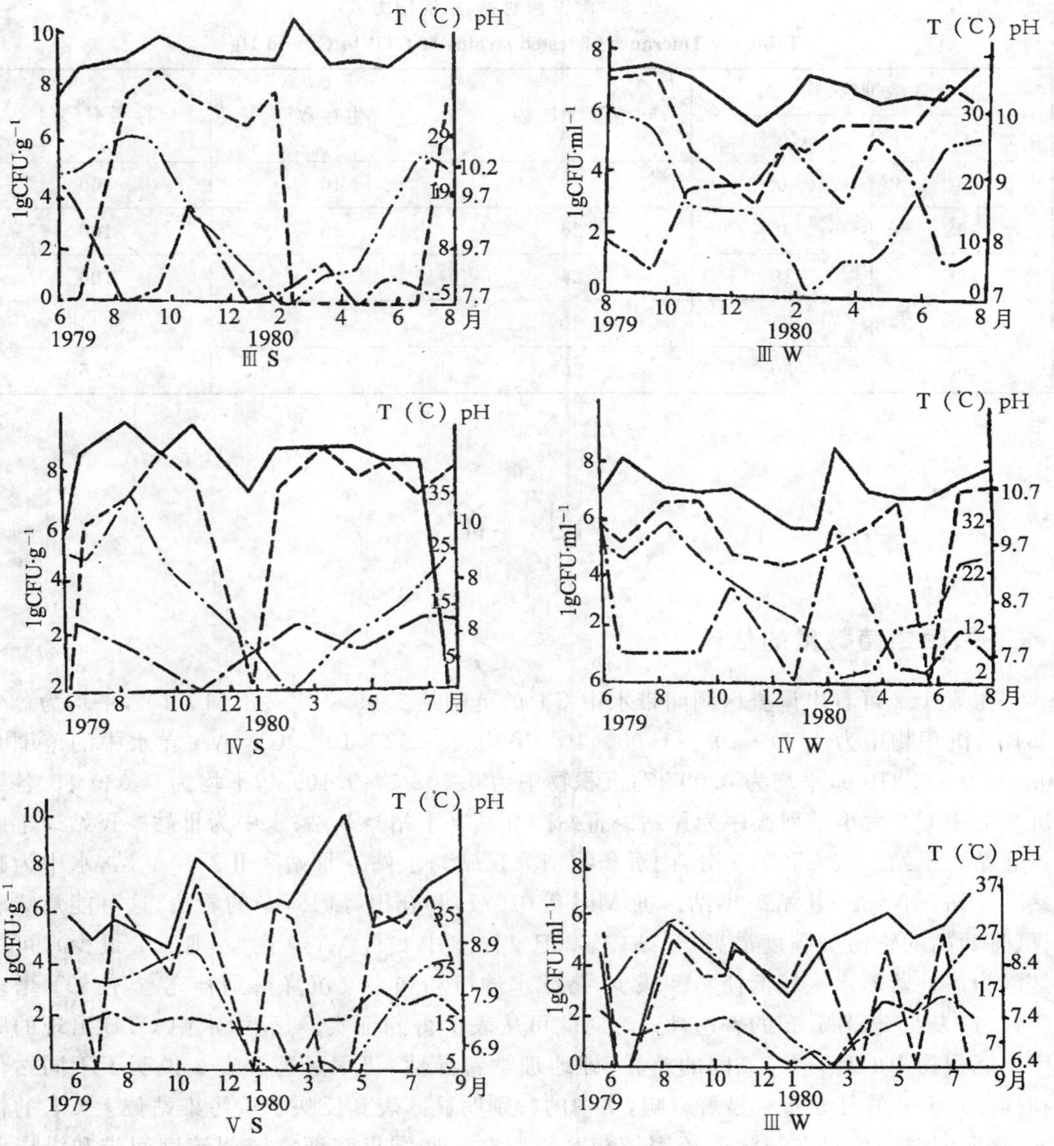

图 2　各站 CT、AA、T、pH 的逐月变化

当 CFU 为 0 时，纵坐标均以 lg1 处理 ······ CT；—— AA；—·— pH；—··— 温度(T) S—沉积物样；W—水样

Fig. 2　Monthly changes of CT(1 ······ 1) AA(1 —— 1) T(1 —·— 1) pH(1 —··— 1) at five stations

表 5　耐铬菌对递增的六价铬浓度的耐力

Table 5　Tolerance of tested strains of CTB to successive concentrations of Cr^{6+} ion

Cr^{6+} 浓度($\times 10^{-6}$)	参试菌株数	生存数	存活率(%)
10	24	24	100
100	24	22	91.7
168	24	19	79.2
500	24	4	16.7

表 6 耐铬菌对铬、汞的耐力

Table 6 Tolerance of tested strains of CTB to Cr and Hg

重金属浓度($\times10^{-6}$)		参试菌株数	生存数	存活率(%)
Cr^{3+}	Hg^{2+}			
0	0	10	10	100
168	0	26	26	100
168	10	26	26	100
168	20	26	23	88.46
168	30	26	15	57.69

讨 论

一、耐铬菌数量的估计

由表 1、2 可看出调查区潮间带水中 CT 的范围是 $5.69\times10^4\sim3.51\times10^6$,平均为 1.44×10^6;沉积物中为 $4.69\times10^6\sim1.08\times10^8$,平均为 3.52×10^7。CT/AA 在水中的范围是 0.007 0～0.310 6,平均为 0.097 7;沉积物中为 0.002 1～0.109 4,平均为 0.030 4。各站沉积物中 CT 大小排列次序为Ⅳ站>Ⅲ站>Ⅱ站>Ⅰ站>Ⅴ站;水中为Ⅲ站>Ⅳ站>Ⅰ站>Ⅱ站>Ⅴ站。按 CT/AA 分,则沉积物中为Ⅳ站>Ⅰ站>Ⅲ站>Ⅱ站>Ⅴ站,水中为Ⅲ站>Ⅴ站>Ⅳ站>Ⅱ站>Ⅰ站。据 Mill 等(1977)和仓田等(1978)的看法,这可能意味着Ⅳ、Ⅲ两站的铬污染程度最高,Ⅴ站较轻,但Ⅴ站水中 CT/AA 甚高(尽管 CT 最少),可能是因为该处营养水平较低使 AA 减少,铬靠水的流动和大气沉降得到补充[7],使 CT 增多所致,这表明 CTB 分布的普遍性[6]。对此可从表 7 得到证实。显然,水中 CTB 出现的频率比沉积物中的高,但各站间的差异(即地理学特异性)仍很显著。表 3 说明了不同站位和样品对 CT 的月分布有显著影响,水中的特别明显。表 8 反映了重污染站位上 CT 的情况,表明耐铬菌的存在与污染源密切相关。所有这些结果都证实胶州湾潮间带和沿岸区普遍存在着高数量的耐铬菌;相对稳定的沉积物聚居着大量耐铬菌。本结果与某些报道相吻合[3,6,15]。

表 7 耐铬菌出现频率比较

Table 7 Comparison of occurrence frequency of CTB samples in water and sediment

项 目	沉积物	水
采样次数	65	63
耐铬菌出现次数	41	53
耐铬菌出现次数/采样次数的%	63.1	84.1

表 8　不定期站上耐铬菌数量比较

Table 8　Comparison of numbers of CTB at irregular stations

站位	样品	CT	AA	CT/AA	采样年月
Ⅵ	沉积物	5.7×10^{7}	7.7×10^{7}	0.7403	1980.7
Ⅶ	沉积物	6×10^{7}	5×10^{6}	12	1980.7
Ⅷ	沉积物	6×10^{7}	3×10^{7}	2	1979.8
Ⅵ	水	1.3×10^{7}	3.69×10^{7}	0.352	1980.7
Ⅶ	水	5×10^{4}	1×10^{5}	0.5	1980.7
Ⅷ	水	2.5×10^{5}	5×10^{4}	5	1979.10

二、耐铬菌数与铬浓度的关系

耐重金属微生物与环境重金属污染物含量间的关系是密切的[5,9,12,16,19]，Kurata 等以耐镍菌数与总异养菌数之比作环境镍污染的指数[11]。表 9 列出了一些站位上 CT/AA 等与样品含铬量的情况。由相关分析的结果表明，CT/AA 与 Cr 浓度间存在着较好的正相关，尤其水中，为 0.925，而 AA 与 Cr 浓度间有负相关。可以推测，在合适条件下，随着环境铬含量的提高，异养菌有所抑制，AA 有所降低，而许多耐铬菌却有所适应，提高了 CT，导致 CT/AA 增高。CTB 的大量增加，可能表明某区域内微生物群体的构造和功能失稳，自然生态失调，这在Ⅶ站相当明显。该站离铬污染源最近，铬浓度高（水样中 27×10^{-6}～192.6×10^{-6}，沉积物中 133.3×10^{-6}），pH 偏碱性，温度适宜，营养丰富，于是耐铬菌大量繁殖，CT/AA 则提高（水中高达 13.25、沉积物中为 0.89）。因此，CT/AA 的提高在一定程度上反映了环境铬污染的加重，以此为铬污染的一个重要指数，再结合其他参数，能更全面地反映铬污染生态环境的内在质量。

表 9　耐铬菌与样品铬含量的相关分析

Table 9　Relationship of CT/AA, AA and Cr ($\times10^{-6}$) content in samples

样品	铬含量（$\times10^{-6}$）	AA	γ	CT/AA	γ	采样站	采样年月
水	0.0207	4×10^{4}	−0.193	0.1	0.925	Ⅴ	1979.9
	0.0438	1.23×10^{7}		0.0033		Ⅲ	1979.10
	0.045	1.65×10^{6}		0.0606		Ⅰ	1980.4
	0.05	1.15×10^{6}		0.0435		Ⅴ	1980.4
	0.053	2×10^{6}		0.25		Ⅲ	1980.4
	0.0636	1.72×10^{6}		0.0174		Ⅱ	1979.10
	0.0888	2.88×10^{7}		0.5521		Ⅳ	1979.9
	0.6	1.5×10^{5}		4.6667		Ⅵ	1980.7
	27	3.68×10^{6}		0.0011		Ⅶ	1980.5
	192.6	2×10^{5}		13.25		Ⅷ	1980.7

(续表)

样品	铬含量($\times10^{-6}$)	AA	γ	CT/AA	γ	采样站	采样年月
沉积物	0	1.6×10^{8}	−0.378	0	0.201	Ⅴ	1980.4
	0.28	4.5×10^{7}		0.2222		Ⅰ	1980.4
	0.6	5.5×10^{8}		0.0909		Ⅵ	1980.4
	3	7.7×10^{7}		0.4584		Ⅵ	1980.7
	133.3	6.1×10^{6}		0.8852		Ⅶ	1980.7
	150	1.11×10^{8}		0.009		Ⅵ	1980.5

三、耐铬菌数量的季节变化

把图 2 中的 10 张图结合 CT/AA 年变化作比较,以 CT、AA 和 CT/AA 出现峰谷的季节变化来分析*,可看到沉积物中 C T 和 CT/AA 的峰多在夏秋,AA 的峰多在冬、夏,三者的谷多在秋、冬、春;水中的 CT 峰多在夏季,CT/AA 峰多在春、冬,AA 峰多在春、夏,而三者的谷较集中在秋季。整个调查区中的情况大致是 CT 峰主要在夏季,AA 的季节变化较为平缓,CT/AA 峰夏季较多。三者的谷主要在秋、春。这些情况与切萨皮克湾的有相似处[15]。值得注意的是,沉积物中 CT/AA 与 AA 的季节变化恰恰相反:CT/AA 是秋>夏>春>冬,AA 是冬>夏>春>秋;水中的 CT/AA 峰多在春末下降,AA 峰却在春季上升,CT/AA 与 AA 季节变化的这一相反关系同 CT/AA-Cr 和 AA-Cr 含量的相关情况相一致。

CT/AA、CT、样品温度(T)、样品 pH 值年变化的初步相关分析表明:CT/AA 与 T、与 pH 间的正负相关不相上下,CT 与 T、与 pH 间的正相关多于负相关;污染较轻的站位(如Ⅴ站)其 CT 与 T、与 pH 的相关性好于污染较重站位;沉积物中的相关性好于水中的,但 CTB 等与环境间的关系是复杂的,对此问题有待探讨。

四、耐铬菌对污染物的忍耐、降解

1. 耐铬菌耐 Cr^{3+}

由表 4 可看到耐铬菌能忍耐递增的铬浓度,最高者可达 1 176$\times10^{-6}$,没有一株能耐 1680$\times10^{-6}$的。将此与 Kurata 等的结果比较,证实胶州湾潮间带耐铬菌耐铬力比日本燧滩海域中的还高[12]。

2. 耐铬菌耐 Cr^{6+}

许多耐铬菌不仅能耐高浓度的 Cr^{3+},表 5 还表明它们能忍耐 Cr^{6+} 的毒性。这意味着它们本身就是耐 Cr^{6+} 菌,耐 Cr^{6+} 的能力比 Petrillic 等所报道的高[17,20]。这一特性还可从下列小实验得到证实:将 4 株耐铬菌在含 833$\times10^{-6}$ Cr^{6+} 的海水—自来水中培养 7 天后,

* 青岛地区四季划分:11/4-4/7 为春季,5/7-12/9 为夏季,13/9-11/11 为秋季,12/11-10/4 为冬季,实测 1979.9.14 水温是 23-28℃、气温是 21-28℃,故将此日划入夏季计算。

其色泽由鲜橙色转变为浑黄蓝色并有沉淀析出，说明经耐铬菌作用已发生了 Cr^{6+} 转变为 Cr^{3+} 的还原过程。环境中的耐铬微生物依靠其忍耐、抵抗等代谢机制，可改变铬离子的价态或浓缩、排除铬污染物。

3. 耐铬菌耐汞和降解原油

表 6 说明了许多耐铬菌同时又是耐汞菌。约 88% 的菌株能耐 20×10^{-6} 汞，约 58% 能耐 30×10^{-6} 汞。解油试验表明所试菌株均能使培养基发生明显变化，如油珠破裂、变形，培养基乳化，pH 下降等。这些结果的实际意义是：在遭受包括铬在内的复合污染环境中，微生物对污染物可获得选择、适应、忍耐、抵抗乃至降解的能力。一些微生物能同时对多种污染物发生转化、迁移或降解作用，它们代表着水生微生物区系的一个重要部分，深入研究它们在环境自净作用中的活动机制是重要的。

小 结

(1)本文初步报道青岛胶州湾潮间带和沿岸区耐铬菌的数量分布。结果表明调查区中普遍存在高数量耐铬菌。CT/AA 以近最大铬污染源的红星化工厂东门桥塆（即Ⅶ站）最高，沿岸区以大港矿砂码头（即Ⅷ站）的最高。潮间带以海泊河（Ⅳ站）和李村河出海口处（Ⅲ站）为最高。站位间差异显著。耐铬菌出现的频率在水中高于沉积物中，其数量则是沉积物中的多于水中的。

(2)在合适条件下，单位水体积和/或沉积物重量的耐铬菌数与总异养菌数之比常跟铬污染物密切相关。CT/AA 与 Cr 浓度间常有正相关。AA 与 Cr 浓度间常有负相关，因此 CT/AA 的提高可作为环境铬污染加重的一个重要指数。

耐铬菌数量有一定的季节变化，高峰主要在夏季，低谷主要在秋、春。

(3)一些耐铬菌有强大的耐铬力，它们对环境铬污染的迁移和转化起重要作用。

(4)一些耐铬菌能忍耐、抵抗汞污染，能降解石油，表明耐铬菌在复合污染环境中有适应、迁移或降解多种污染物的能力。

耐重金属的微生物在污染环境中的生态活动是有意义的。

本文写作得到吴宝铃教授的热心指导；陈弱、孙国玉、陈世阳、沈玉如、丁美丽、刘长安和海洋所微生物组其他同志提出了宝贵意见；特此致谢。

参考文献 21 篇（略）

Cr-TOLERANT BACTERIA IN THE INTERTIDAL AND COASTAL ZONE OF JIAOZHOU BAY

Abstract The work was conducted during the period of 1979. 5～1980. 8. The primary results obtained are as follows:

1. There are a great number of Cr-tolerant microorganisms, particularly Cr-tolerant bacteria (abb. CTB) and a higher ratio of CT/AA in every station of the surveyed zone. The amount of Colony-Forming Units (abb. CUF) of CTB(abb. CT) is $4.69\times10^6-1.08\times10^8$ CFU/g with an average of 3.52×10^7 CFU/g in sediment, and $5.69\times10^4-3.51\times10^6$ CFU/mL, with an average of 1.44×10^6 CFU/mL in water, respectively. The amount of (ratio of) CT/AA is 0.0021～0.1094, with an average of 0.0304 in sediment and 0.007～0.3106, with an average of 0.0977 in water, respectively.

2. Most CTB are often accompanied by a higher concentration of chromium, varying with stations and samples. Some results have demonstrated a close relationship between CTB and Cr-concentration of sample. The correlation between CFU number of total heterotrophic bacteria (abb. AA) and Cr-concentration is negative, while, positive correlation is found only between CT/AA and Cr-concentration. Therefore, increase of CT/AA may be regarded as an important index of severity of Cr-contamination in environment.

CT shows certain kind of seasonal fluctuations.

3. 16.67% of the strains selected can tolerate a Cr^{3+} concentration of 1176×10^{-6}; 16.7% can tolerate 500×10^{-6} Cr^{6+}. It is found that many of CTB have a steady Cr-tolerant ability. So CTB may be thought of as an important remover of Cr-pollutant in environment.

4. 57.69% of the strains selected can tolerate 30×10^{-6} Hg^{2+} (in the presence of 168×10^{-6} Cr^{3+}); All of the strains can be utilized to degrade petroleum.

Based on these data, it may be concluded that CTB has a powerful ability to adapt, to remove and/or degrade environmental complex pollutants. They may play an important role in purifying action in the environment so-called self-purification.

CTB therefore in contaminated environment takes up important ecological significance.

胶州湾潮间带和沿岸区的耐汞菌*

Ⅰ.耐汞菌的数量分布

在受污染的特定海洋环境有较大数量的耐汞菌(Mercury-Tolerant Bacteria,本文简称MTB)。耐汞菌大量出现是它们对汞污染适应、选择、忍耐以至对抗的群体生态反应。耐汞菌数量变化与汞污染关系密切,这方面的研究国外多有报道[5,6,10,11]。但关于潮间带和沿岸区的报道尚不多。本文试图就胶州湾潮间带和沿岸区耐汞菌的数量分布作调查研究,以估计其生态意义。

一、调查和实验

(一)采样站位

5个定点调查站址如前文所述[1]**。其余10站概况列于表1中,采样站位详见图1。

表1　十个不定期站位概况

站位号	地名	概况描述
2	粮油制桶厂后海滩	污脏、沉积物表层滋生大量蓝绿藻、工厂排污
4	沧口海水浴场	比13站脏、海砂有些发黑、磷肥厂排污等
6	水清沟入海口段	收纳化工、发电、造纸、轻工等排污和生活废弃物
8	昌乐路沟下游	收纳印染、车辆制造、电镀、电器仪表、金属切削、塑料制品和生活污水。COD高、悬浮物多、混浊、有高汞史
9	大港五区	海轮停泊、货物装卸、浮油多
10	大港三区	大船较少
11	养殖所码头	尚干净
12	二中后海滩	工业、民用生活污物和肥料压力,水质肥、沉积物发黑
14	感光材料厂出水口附近	尚干净
15	湛山湾	水较干净,中细砂底质为主

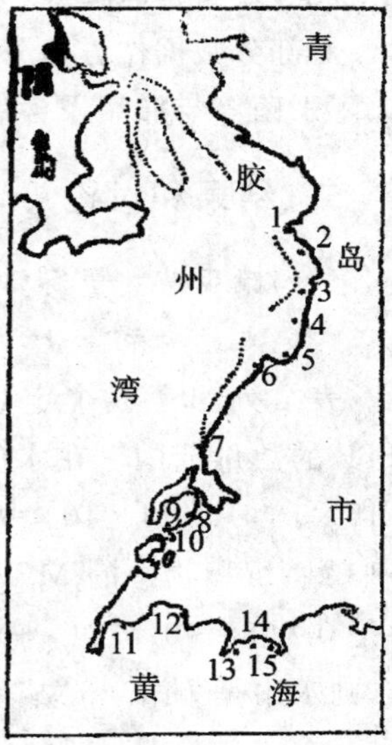

图1　耐汞菌测站示意图

* 原文刊于《环境科学》,1983,4(1):15-20.

** 本文的站址序号由湾内到湾外,自北至南排列。本文中1、3、5、7、13号为文献[1]中Ⅰ、Ⅱ、Ⅲ、Ⅳ、Ⅴ号站,其他站位也有变化,10站前为湾内,11-15站为湾外。

（二）样品采集

方法详见前文[1]。在微生物采样时，平行采集化学分析用样品，现场测定温度和 pH 值等。

（三）好气异养菌和耐汞菌数量的计数

1. 海洋好气异养菌：在海水肉汁胨琼脂平皿上计算海洋好气异养 菌落形成单位数(CFU)（简称 AA，下同）[1]。

2. 海洋好气异养耐汞菌：在上述培养基中添加不同定量的 Hg^{2+}（$HgCl_2$ 形式）收集耐汞菌，一般均以含 10×10^{-6} Hg^{2+} 的培养基上的 MTB 的 CFU（即 MT）作调查区内耐汞菌数量估计的依据。

3. 淡水和半咸水琼脂培养基上的耐汞菌：淡水培养基本成分如前述[1]。半咸水培养基成分如下：A 蛋白胨 5 g；氯化铵 2 g；磷酸铁 0.1 g；磷酸二氢钾 0.3 g；磷酸氢二钾 2 g；陈海水 1 000 mL。B 蒸馏水 1 000 mL。两份 A 液和一份 B 液相混，琼脂 2%。pH 7.4。以上两培养基按需添加 Hg^{2+}。

上述培养基上各加稀释样品 0.1 mL 涂布，于 24℃培养 14 日，三次检查菌量，换算成每毫升表水或每克湿沉积物表层样中的 MTB 数（即 MT）。从含汞平皿上挑取耐汞菌菌落，纯化，保存备用。

（四）样品中汞含量等参数的测定*

汞含量用 YYG-77 冷原子荧光测汞仪测定。COD 用碱性高锰酸钾法测定[3]。

（五）数据处理

对部分数据作方差分析和一元回归分析[2]，以说明耐汞菌数的地理学差异和季节分布、耐汞菌等与样品汞含量等参数的关系。

二、结果和讨论

（一）耐汞菌数量及其变化

1. 定点站耐汞菌数

表 2 列出了五个定点站上的 MT。MT/AA 的年平均数，它们基本上代表着该区 MTB 的数量范围。在水中是 $3.37\times10^{3}\sim6.16\times10^{6}$，平均为 1.77×10^{6}；在沉积物中是 $2.17\times10^{5}\sim3.14\times10^{7}$，平均为 1.45×10^{7}，按大小排列，则水中是 7 站＞3 站＞5 站＞1 站＞13 站；沉积物中的 MT 比水中的高。这吻合于 Nelson 等的估计[7]，MT/AA 范围在水中是 0.0055～0.1381，平均 0.0602；沉积物中是 0.0001～0.0242，平均 0.0099。按大小排，则水中是 7 站＞3 站＞5 站＞1 站＞13 站；沉积物中是 1 站＞7 站＞5 站＞3 站＞13 站。

---

* 汞主要由本所郑舜琴、张淑美、庞学忠测定；COD 等由青岛化工厂赵敦瑾等测定，在此一并致谢。

表 2 五个定点站上 MT、MT/AA 的比较

站号	沉积物		水	
	MT	MT/AA	MT	MT/AA
1	9.17×10^{4}	0.0242	5.74×10^{5}	0.0090
3	2.24×10^{7}	0.0049	1.45×10^{6}	0.0755
5	3.14×10^{7}	0.0098	6.42×10^{5}	0.0728
7	9.15×10^{6}	0.0106	6.16×10^{6}	0.1382
13	2.17×10^{5}	0.0001	3.37×10^{3}	0.0055
平均	1.45×10^{7}	0.0099	1.77×10^{6}	0.0602

从三种培养基上得出的培养情况列于表 3，证明潮间带大量的微生物和 MTB 也适于淡水和半咸水环境。含 10×10^{-6} Hg^{2+} 培养基上的 MT 在淡水培养基上最少，海水肉胨培养基上最多。只有半咸水培养基上 MTB 能耐 50×10^{-6} Hg^{2+}，说明汞在半咸水和海水里比在淡水里可能形成更多的 $HgCl^{-1}/HgCl$ 络合物，其毒性低于 Hg^{2+}[4]。这还表明潮间带的 MTB 主要来自陆水，它们有广泛的盐度适应性。

MT、MT/AA 等逐月和季节变异画在图 2-3 中，图 2-3 描绘出 MT 的主峰月在春末和夏季月份，这可能与大量生源物质及温度增高有关。这在沉积物 5 站和水的 7 站较为明显。图 2 还表明沉积物 MT 有明显的季节变化，其主峰在夏，次峰在春，低谷常在冬。水中的次序则为春＞夏＞秋＞冬（图 3）。

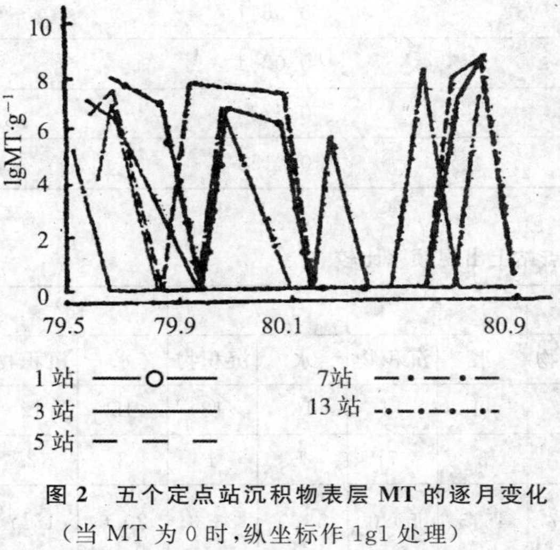

图 2 五个定点站沉积物表层 MT 的逐月变化
（当 MT 为 0 时，纵坐标作 lg1 处理）

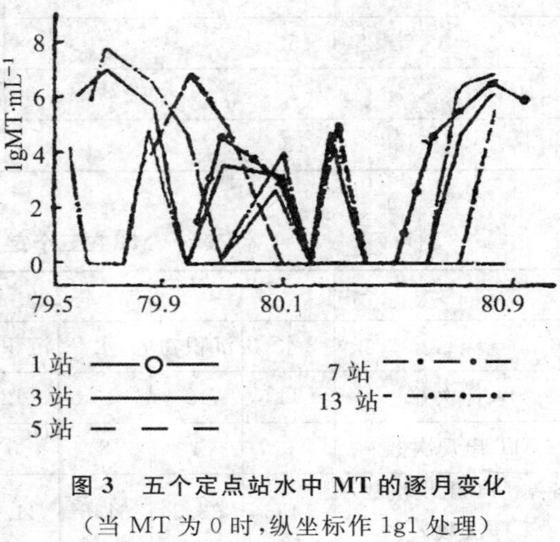

图 3 五个定点站水中 MT 的逐月变化
（当 MT 为 0 时，纵坐标作 lg1 处理）

表 3　7 号站在不同培养基上 MT、MT/AA、AA 比较 *

培养基中汞($\times 10^{-6}$) / CFU 数等 / 培养基种类	0	5		10		50	
	AA	MT	MT/AA	MT	MT/AA	MT	MT/AA
海水肉胨	3.93×10^{7}	2.35×10^{6}	0.0598	3.75×10^{6}	0.0954	0	0
半咸水	1.69×10^{7}	1.45×10^{6}	0.0855	7.5×10^{6}	0.0444	5×10^{4}	0.0029
淡水	1.82×10^{7}	1.35×10^{6}	0.0742	3×10^{6}	0.0165	0	0

* 水样，1981.6.11

表 4　五个定点站上 MT/AA、AA 季节分布方差分析

样品	项目	F 值	显著性
水	MT/AA	1.58	*（$\alpha=0.25$）
沉积物		2.94	*（$\alpha=0.25$）
水	AA	1.96	*（$\alpha=0.25$）
沉积物		0.61	不显著

* 表示差异显著，MT/AA、AA 的自由度各为(3.16)、(4.15)

表 5　五个定点站上水 MT/沉积物 MT 的比较

站号	水 MT/沉积物 MT
1	0.0625
3	0.0647
5	0.0204
7	0.6732
13	0.0155

表 6　MT 在五个定点站上出现频率比较

站号	1		3		5		7		13	
样品	水	沉积物	水	沉积物	水	沉积物	水	沉积物	水	沉积物
采样次数	11	12	13	12	11	14	13	12	15	12
MT 出现次数	6	4	8	5	5	7	7	4	3	2
各站 MT 出现次数/采样次数×%	54.6	33.3	61.5	41.7	45.5	50.0	53.9	33.3	20.0	16.7
五站 MT 出现次数/采样平均次数×%	46.0(水)　35.5(沉积物)									

不同站位与样品影响到 MT/AA 及其季节变化分布上的差异。表 4 表明的站位对

MT/AA 的影响，即地理学差异是显著的。这种地理学差异还表现在水 MT/沉积物 MT 的比率差异上(表 5)。这一比率的增大，反映了两种介质中 MTB 数量差异的缩小及环境汞污染的加重[6]，表 6 则说明水和沉积物这两种介质对 MTB 出现频率有不同的影响：在水中的 MTB 出现频率大于沉积物。这与汞污染物随水的流动和陆地排污有关。

表 7　不定期站上 MTB 的比较

站号	样品	MT	MT/AA	采样日
2	沉积物	1.72×10^{8}	0.2147	79.6.22
4	沉积物	4×10^{6}	0.1429	79.6.22
6	沉积物	1.83×10^{7}	0.0764	81.3.21
12	沉积物	$<1\times10^{6}$	<0.0025	79.6.4
6	水	9.92×10^{3}	0.2332	81.3.21
6	水	9×10^{3}	0.2606	81.5.20
8	水	1.38×10^{3}	0.0157	81.5.20
8	水	6.98×10^{3}	0.0191	81.5.29
9	水	2.19×10^{8}	4380	79.10.10
10	水	0	0	80.3.16
11	水	5×10^{4}	0.0039	79.10.10
12	水	7×10^{5}	0.0368	79.6.4
14	水	1×10^{5}	0.0035	79.10.10
15	水	0	0	81.3.22

2. 不定期站上的耐汞菌数

表 7 列举了 9 个不定期站上 MT 等数据，它们呈现出湾内多于湾外、污染区的高于清洁区、沉积物高于水的趋势，其中以湾内 2 站最高(沉积物)。水中以 9 站最多，水中以湾外的 15 站、沉积物以湾外的 12 站最少(参见表 8)。

(二)MT、MT/AA 与环境理化参数的关系

1. MT、MT/AA 与样品汞含量的关系

表 8 列出了 18 个样品 MT、MT/AA 与 Hg 含量间的关系，由表 8 绘出的图 4 表达了 MT 与 Hg 的直线回归，若以 W 代表水样，则 MT(W)—Hg(W)间的回归方程为 $MT=8.838\times10^{4}+1.5141\times10^{5}\times Hg(W)$，回归截距 $a=8.838\times10^{4}$，其物理意义可理解为相当于汞污染消除后水体的水中耐汞菌的一般数。湾外的 11、13、14、15 等站的 MT 均低于此值。MT/AA(W)—Hg(W)间也有正相关存在。耐汞菌与汞的相关性说明在一定条件下，MT、MT/AA 的增高部分地反映了特定环境汞污染的加重[6]。

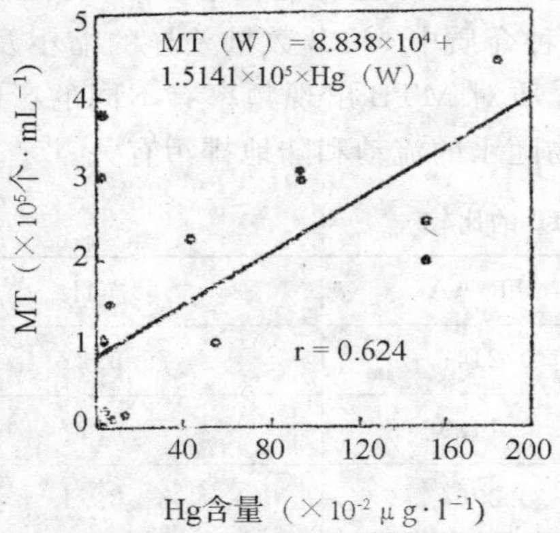

图 4　18 个水样 Hg 含量与 MT 的关系

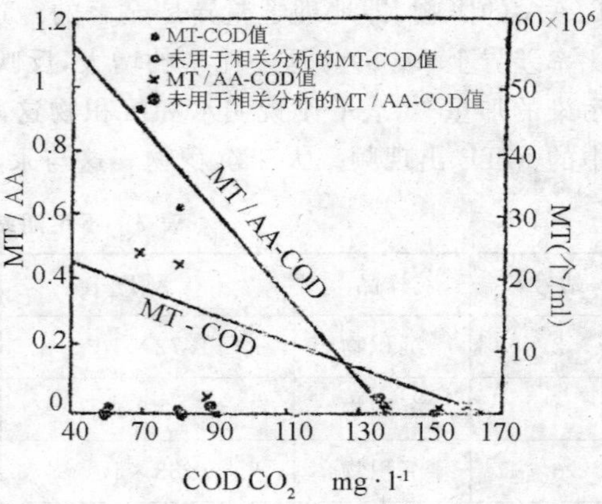

图 5　水样 COD 与 MT、MT/AA 的关系

表 8　水样含汞量与 MT、MT/AA 的关系

Hg(μg/L)	MT	r(Hg— MT)	MT/AA	r(Hg—MT/AA)	站号	采样日
0.014	0	0.624	0	0.081	13	82.8.27
0.016	5×10^3		0.0222		9	82.8.27
0.029	5×10^3		0.0172		15	82.8.27
0.032	0		0		11	82.8.27
0.033	3.05×10^5		0.1419		1	82.8.27
0.033	3.8×10^5		0.0638		2	82.8.27
0.043	2×10^4		0.0290		14	82.8.27
0.052	1.45×10^5		0.2148		4	82.8.27
0.061	1×10^4		0.1053		10	82.8.27
0.152	1.5×10^4		0.0278		3	82.8.27
0.435	2.3×10^5		0.0208		8	82.8.27
0.53	1×10^5		0.3333		7	81.3.21
0.923	3.1×10^5		0.1538		6	82.8.27
0.93	3×10^5		0.3158		7	81.3.21
1.5	2×10^5		0.0071		8	81.5.20
1.5	2.5×10^5		0.0450		8	81.5.20
1.83	4.45×10^5		0.0434		12	82.8.27
0.037	1.05×10^5		0.0598		5	82.8.27

表9 COD与MTB的关系*

COD(mg/L)	MT	MT/AA	站号	采样日
68.6	2.5×10^6	0.9259	6	81.5.20
80.8	2.35×10^6	0.6267	6	81.5.20
88.9	3.5×10^9	0.9100	6	81.5.20
136.0	2.5×10^5	0.0450	8	81.5.20
152.0	1×10^5	0.0100	8	81.5.20
160.0	2×10^9	0.0077	8	81.5.20
160.0	0	0	8	81.5.20

*水样

2.耐汞菌与COD的关系

表9列举MT、MT/AA与样品COD的关系，根据表9的MT-COD、MT/AA-COD关系画出了图5。由此可知水中MT、MT/AA与COD间有良好的负相关，r值分别为-0.83和-0.75(α均为0.05)。其相应的回归方程分别为MT(W)$=32.7362\times10^5-0.2281\times10^5\times$COD(W)和MT/AA(W)$=1.1371-0.0075\times$COD(W)。截距$a=32.7362\times10^5$可理解为随着COD降低及其他条件适于MTB时的最高MT水平。而截距$a=1.1371$则意味着在上述情况中MT/AA的可增之值。这两种相关性从反面暗示MTB与O_2的密切关系[8]。COD消耗O_2从而阻抑好气MTB生长繁殖，减缓汞的好气性甲基化进程，因而影响到环境中汞的迁移[9]。

3.耐汞菌等与样品温度的关系

MT、MT/AA、AA与样品温度(T)的相关分析结果列于表10中。MT-T的r值(=0.05)在水中只有5站的接近于有意义的正相关。MT/AA与T的r值(=0.05)仅3站的沉积物表达出较好的线性正相关，5站的水仍接近于线性关系。

AA-T的10个r值有一半为正的，其中的两个表达出良好的相关性。总的看来，三对相关分析表明水中比沉积物中相关性好。MT受温度的影响较大，它对温度的反应比普通微生物要敏感些。同时也说明包括MTB在内的微生物，其数量分布除受温度影响外，还受其他环境因子的制约。

三、小结

1.本工作表明耐汞菌的出现频率水中比沉积物中高，其数量是沉积物中多于水中：大体范围是$2.17\times10^5\sim3.14\times10^7$个/克(沉积物)和$3.37\times10^3\sim6.16\times10^6$个/毫升(水)。沉积物以李村河出海口(5站)的最多，水以海泊河出海口(即7站)最多。MT/A.A在水中以海泊河出海口最高，沉积物以石油化工厂后海滩(即1站)最高。13站上无疑总是最低、最少，湾内高于并多于湾外的。

2.MT、MT/AA与Hg量为正相关。MT、MT/AA的提高至少部分地意味着特定环境汞污染的加重。把MT这个参数结合进汞污染的环境监测中去是有意义的。

3.MT、MT/AA与COD间有负相关。MT、MT/AA比AA受温度影响大，MT最高数季节在春末和夏季。MT、MT/AA 表现出的季节分布趋势受地理学特异性影响大。

4.虽然大量出现的耐汞菌能参与汞污染的迁移与净化活动，但样品中不时出现的超标高汞量等表明陆地仍有不适当排污，这对微生物正常生态活动仍有抑制的可能。

曾呈奎所长和吴超元副所长对本文提出过宝贵意见，在此深表感谢。

参考文献 11 篇(略)

胶州湾潮间带和沿岸区的耐汞菌

Ⅱ.耐汞菌的耐污力*

胶州湾潮间带和沿岸区遭受程度不等的汞污染[2]**，生活在此类环境中的普通好气异养细菌(本文称作普通菌)对汞污染有多大的耐受力？在长期的汞污染应力下，有可能孕育驯化出大量耐汞菌(简称 MTB)[2]。这两类菌耐汞污染的能力有何差别和联系？环境污染是复杂的，MTB 对多种污染物适应并产生多大忍耐、对抗甚或降解能力？本文试图探讨上述问题，以进一步了解 MTB 等的活动规律并为环境污染的生物治理提供依据。

一、材料和方法

1.试验菌种来源

所用菌种是 1978～1982 年多次采自胶州湾潮间带和沿岸区表层水与表层沉积物的细菌，在含汞或不含汞的海水肉汁胨琼脂平皿上分离纯化而来。

2.培养基和培养方法

(1)海水肉汁胨培养基(下称海肉胨，固体或液体的)　用于估计单位体积(毫升)或单位重量(克)样品中普通菌菌落形成单位(即 CFU)的总数，即 AA。

(2)含汞培养基　在上述培养基中按需添加不同量的汞($HgCl_2$)作成平面，斜面或液体的培养基，用于试验普通菌或 MTB 的耐污力。

(3)含油培养基　在无碳源液体培养基中按需添加原油(取自胜利油田)或原油—汞混合物，用以试验 MTB 忍耐或降解石油的能力。

(4)含多种重金属的培养基　在海肉胨液体培养基中按需添加不同组合的 Hg^{2+}、Pb^{2+}[$Pb(NO_3)_2$]、Cr^{6+}($K_2Cr_2O_7$)或原油等，用以试验所收藏的 MTB 忍耐或降解复合污染物的能力。

(5)将接入菌种的平面或斜面培养基置温箱培育，液体培养基振荡培养(24℃，7～14 天)。

二、结果和讨论

耐汞菌的数量与培养基中汞含量间在一定条件下的正相关性表明了它们与汞的关系

---

* 本文曾在 1983 年举行的全国污染生态学术会议上宣读过。

** 尹相淳等，海洋研究，1981，1，64.

原文刊于《海洋通报》，1984，3(5)：71-75.

十分密切[2]。普通菌与汞的关系常是不密切或负相关。随着汞污染加重，普通菌数量相应地减少，结果列于表1。当汞浓度较低时（如5×10^{-6}），普通菌存活百分数减少得已相当明显；随着汞含量的增加，AA继续下降，其中部分菌株可孕育对汞污染的选择适应；汞含量为50×10^{-6}时，水中的AA小于沉积物中的AA，表明水中菌耐汞力小于沉积物中的细菌；当汞浓度为100×10^{-6}时，平皿上很难见到任何菌数（只在一个沉积物样中分离到MTB菌落），这表明从所调查站位上直接筛选出耐力大于100×10^{-6}汞的菌株的频率很小。

表1　普通菌在递增汞浓度培养基中的存活＊

样品 \ AA等 \ Hg^{2+} ($\times10^{-6}$)	0	5	10	50	100
水	9.23×10^{6}	2.13×10^{6}	1.18×10^{6}	4.17×10^{4}	0
沉积物	1.57×10^{8}	5.34×10^{7}	1.34×10^{7}	3.33×10^{6}	0

＊1981年6月11日和1982年3月20日从6、7、13站共6次采样的平均值。AA的单位是每毫升水或每克沉积物中的菌落数。

随机取自7站和15站的普通菌株（站位详见文献[1],[2]），试验在不同汞污染下存活百分数，结果列于表2。这些菌的耐汞菌株百分数随汞浓度增加而减少，减少的趋势在水和沉积物中是一致的。经相关分析发现这两站普通菌中耐汞菌株百分数在水和沉积物中都与汞含量有良好的正相关性。Nelson等也曾发现过此类现象[6]。结果还表明7站菌的耐汞菌株百分数大于15站，这与两站的汞污染状况相一致[2]。沉积物中耐汞菌的百分数一般高于水中，因而，在沉积物中汞的迁移过程更活跃些。

表2　普通菌对递增汞浓度的耐菌株百分数

样品	站号	Hg^{2+} ($\times10^{-6}$)						参试菌株数
		0	5	10	50	100	200	
水	7	100	26.3	15.8	5.3	5.3	0	19
	15	100	16.7	12.5	0	0	0	24
沉积物	7	100	29.4	17.6	5.9	5.9	0	17
	15	100	23.5	11.8	5.9	0	0	17

7站代表污染区，15站代表清洁区。将表2的数据按水和沉积物分别作两因素无重复数据的方差分析后发现，耐汞菌株百分数在站位间的差异是显著的（水中$F>F_{0.05}$，沉积物是$F>F_{0.1}$）；在所试汞浓度间的差异亦特别显著，这与汞污染程度的差异是一致的[2]。这些差异可能与相应群体组成的差异有关。例如，15站常可分离到琼脂分解菌，初试其耐汞高限为20×10^{-6}；7站则未分离到这种细菌。该菌似乎对污染较敏感，这似乎表明汞污

染的地理学差异是造成各自微生物生态群的一个重要因素。详情有待于进一步研究。

表 3　MTB 对递增汞浓度的耐力

Hg^{2+} ($\times 10^{-6}$)	5	10	100	200	304	350	400	450	500
参试菌株数	46	46	26	41	46	46	46	46	46
$\frac{忍耐株数}{参试株数}$%	100	100	65.4	46.3	15.2	6.5	4.3	2.2	0

表 3 列出了 MTB 对递增汞浓度忍耐菌株百分数的情况。将表 3 与表 2 比较可见，MTB 比普通菌有更大的耐汞力。如，耐 100×10^{-6} 的 MTB 有 65.4%，普通菌平均只占 2.8%(0～5.9%)；耐 200×10^{-6} 的 MTB 约 46%，普通菌则为 0%；尚有 2.2%的 MTB 能耐 450×10^{-6} 汞(另有一些真菌耐汞力更大)。这说明这两类生态群的细菌对汞污染的生理反应不一样。MTB 耐汞是对汞污染长期影响的一种选择适应[7]。普通菌和耐汞菌对汞的生理反应在特定条件下可能向相反方向转化。

耐 300×10^{-6} 汞的 MTB 革兰氏染色反应结果表明，它们都是革氏阴性杆菌，这与陆生、淡水菌不同[3]* 1)。普通菌的染色反应不一。

以污染区 6 个站(即 1、3、5、6、7 和 10 站)海水和沉积物的 MTB 对 $(100\sim 600)\times 10^{-6}$ Hg^{2+} 的耐力分别作方差分析，并未发现它们在站位间、汞浓度间存在明显差异，即其 MTB 耐汞力基本相近。这表明，这些站位所在区域汞污染状况基本相同。

持续的汞污染下出现的 MTB 可能忍耐和对抗多种含汞化合物[4,7,8]它们将积极参与汞污染物的迁移和净化活动。

表 4　MTB 对复合污染的忍耐或降解

污染物编号 *	参试菌株数	忍耐或降解株数/参试株数(%)
1	44	40.9
2	39	46.2
3	44	34.1
4	44	11.4
5	55	0
6	44	90.9
7	44	2.3

* 编号 1. Hg^{2+}、Cr^{6+} 各为 100×10^{-6}；2. Hg^{2+}、Pb^{2+} 各为 100×10^{-6}；3. Hg^{2+}、Pb^{2+}、Cr^{6+} 各为 100×10^{-6}；4. Hg^{2+}、Pb^{2+}、Cr^{6+} 各为 200×10^{-6}；5. Hg^{2+}、Pb^{2+}、Cr^{6+} 各为 300×10^{-6}；6. 石油 5%(V/V)；7. 石油 5%(V/V)、Hg^{2+} 400×10^{-6}。

表 4 的实验结果说明，绝大多数 MTB 能在含 5%的石油培养基中生存，这是因为它们利用石油烃为唯一的碳源[4]。当它们被接入含油—汞培养基中时，存活数大减，这是因为

* 1)杨惠芳等，京津渤区域 6 环境研究报告第二辑(科研报告部分)，中国科学院环境科学委员会，1981.

油浓缩汞，两者毒性增大，抑菌或杀菌力增强的缘故[9]。

表 4 还列举 MTB 对重金属等复合污染物的耐力。在 Hg^{2+}、Cr^{6+} 和 Pb^{2+} 各为 200×10^{-6} 的复合污染物中，有 11.4% 的 MTB 存活。这表明尽管 MTB 对不同重金属的忍耐有相对的独立性[4]，但还不能同时忍耐多种重金属的协同作用[5]，以适应环境的复合污染。MTB 有可能成为这种环境复合污染沿食物链迁移的一个重要生物因子[9]。这说明 MTB 在工业区附近的潮间带沿岸区及河口区的生态作用是十分重要的。

表 5 比较了沉积物和海水中 MTB 耐复合污染物的情况。表 5 表明受污染的沉积物环境比受污染的海水环境更适于驯化各自 MTB 的耐污力，Hg^{2+}、Pb^{2+}、Cr^{6+} 和石油等污染物在环境中经微生物的迁移形式不尽相同。

表 5　MTB 耐复合污染在海水和沉积物间的比较

污染物编号*	海水中忍耐株数/参试株数(%)	沉积物中忍耐株数/参试株数(%)
1	34.6(26**)	50(18)
2	40.9(22)	52.9(17)
3	26.9(26)	44.4(18)
4	7.7(26)	16.7(18)

* 污染物浓度：$\times10^{-6}$；编号 1. Hg^{2+}、Cr^{6+} 各为 100×10^{-6}；2. Hg^{2+}、Pb^{2+} 各为 100×10^{-6}；3. Hg^{2+}、Cr^{6+}、Pb^{2+} 各为 100×10^{-6}；4. Hg^{2+}、Cr^{6+}、Pb^{2+} 各为 200×10^{-6}。

** 括号中数字为参试菌株数。

三、小结

(1)胶州湾潮间带和沿岸区普通好气异养细菌对汞污染有一定耐力，其数量随所试汞浓度增加而减少，减少的程度在站位间和所试汞浓度间是不同的。

(2)在持续的汞污染下，普通菌可能转变为耐汞菌。在汞污染环境中，耐汞菌的耐汞力不因地理学差异而有明显不同。

(3)绝大多数耐汞菌能降解石油，有些能忍耐不同浓度组合的汞、铬、铅和石油等的复合污染，沉积物耐汞菌耐污能力较大。

耐汞菌不仅是环境汞污染的重要指标，而且是污染物迁移过程的一个重要参与者。从耐汞菌中培育出净化环境的微生物是治理污染的一条重要途径。

参考文献 9 篇(略)

MERCURY-TOLERANT BACTERIA IN THE INTERTIDAL AND COASTAL ZONE OF THE JIAOZHOU BAY: TOLERANCE OF MERCURY-TOLERANT BACTERIA

(ABSTRACT)

Abstract The primary results obtained from this research work are presented as follows:

1. Many common aerobic heterotrophic bacteria in the surveyed area have shown some ability to tolerate mercury pollutants, and this tolerant ability will decrease as the concentration of mercury pollutants increases. The differences in tolerance are significant among the stations as well as the concentrations. Between water and sediment samples there exists a good positive correlation.

2. Under the stress of chronic mercury pollution, some of the selected common aerobic heterotrophic bacteria would transform into mercury tolerant bacteria(abb. MTB).

3. The selected strains of MTB about 2%, can tolerate Hg^{2+}—concentration up to 450×10^{-6}. The geographical difference in tolerance of MTB against Hg^{2+} is not obvious. Most MTBs can degrade petroleum, and some can tolerate a few composite pollutions with varied concentrations and compositions such as Hg^{2+}, Cr^{6+}, Pb^{2+} and petroleum etc.

The MTB in sediments has a more powerful tolerance.

MTB is not only a vital indicator of environmental mercury pollution, but also a necessary remover of environmental pollutants. Using this kind of microbe to treat pollutants would be one of the important ways for eliminating environmental pollution.

渤海湾耐重金属菌数量分布的初步分析*

渤海湾如同其他许多海域一样,也经受铅、汞、铬、镍、镉等重金属污染[6]。一些研究表明,相应耐重金属菌数量等指数的变化可反映某一环境污染状况[2,3,9-11],它将为监测、评价和防治海洋污染提供依据,故了解本湾耐重金属菌等数量变动规律很有必要。

1979年,我们对渤海湾耐重金属菌等数量分布作了两航次调查,本文就所得资料作一分析。

一、调查站位

1979年7月调查了38°18′－39°04′N,117°37′－118°30′E海区内4个站的表层水和7个站的表层沉积物样,11月调查了其中1个站的水和10个站的沉积物样,以了解近海陆源污染、海上作业和海流等因子对耐重金属菌等数量分布的影响。站位详见图1。

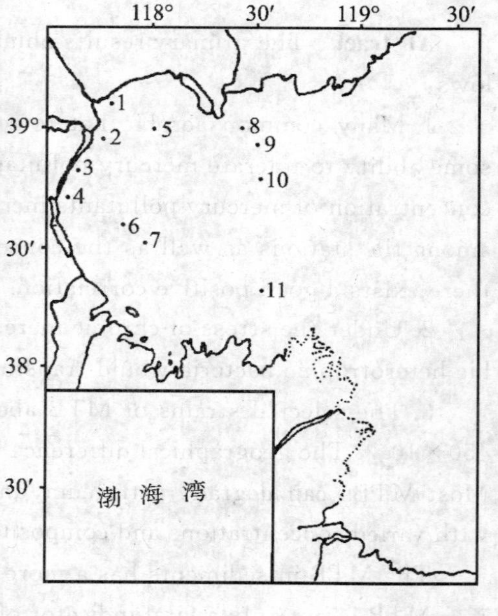

图1　渤海湾耐重金属菌调查站位(1979)

二、材料与方法

1. 微生物学分析用水和沉积物样均按常规操作采集,同时辅以收集有关水文参数。

2. 在海水肉胨琼脂平皿上做普通海洋好气异养菌(下称普通菌)菌落计数。每毫升水或每克湿沉积物中的菌落数(AA)代表全海区单位体积水或单位重量沉积物中平均总活菌数。

3. 在上述溶化的培养基中分别无菌地添加Pb^{2+}($Pb(NO_3)_2$)、Cr^{3+}($CrCl_3 \cdot 6H_2O$)、Cd^{2+}($CdCl_2 \cdot 2.5H_2O$)、Hg^{2+}($HgCl_2$)或Ni^{2+}($NiCl_2 \cdot 6H_2O$)液制成含铅、铬、镉、汞或镍的培养基平皿,其浓度是Pb^{2+}、Cd^{2+}、Ni^{2+}各为100×10^{-6},Cr^{3+}为168×10^{-6},Hg为100×10^{-6},以此培养耐铅菌(LTB)、耐铬菌(CTB)、耐镉菌(CdTB)、耐汞菌(MTB)或耐镍菌(NTB)。每毫升水或每克湿沉积物中相应菌落数分别以LT、CT、CdT、MT、及NT表示,并与相应AA作比较或计算出有关百分比。

4. 样品在船上直接涂布至平皿,条件不够时则保存于冰瓶回岸上实验室立即分析。24℃培养14日。计算每皿菌落数。挑取并纯化菌株,备用。

---

* 原文刊于《海洋通报》,1985,4(6):27-33.

本文摘要以《渤海湾耐重金属菌的数量分布》发表于中国海洋湖沼学会第四届全国海洋代表大会暨学术年会的论文摘要汇编中。

三、结果与讨论

1. 普通海洋好气异养菌数量分布及其趋势

表 1～2 列出有关微生物的数据，说明沉积物的 AA 高于水中，夏季（以 7 月代表）的多于冬季的（以 11 月代表）。将此与蓟运河下游河段、胶州湾的比较，可见本湾 AA 介于这两者间[2-4]，表明本湾营养程度甚高[5]。关于 AA 季节变化和站位分布差异，以 1，3、4、6、7 和 11 六个站的沉积物数据作两因素无重复数据的方差分析（表 3），可见其季节变化差异明显，站位分布差异无明显特异性，后者表明普通菌数量在全海区有均衡分布趋势。

表 1　渤海湾 7、11 月表层沉积物中的耐重金属菌数量分布　单位：个菌落 · g^{-1}（湿）

培养基中所加重金属 / 站号	Pb^{2+}	Cr^{3+}	Cd^{3+}	Hg^{2+}	Ni^{2+}	—*a*)
1	6.84×10^8 (0)	9×10^6 (0)	0 (0)	0 (5×10^5)	—*b*) (1×10^6)	7.52×10^8 (8.5×10^6)
2	(5×10^5)	(5×10^5)	(0)	(0)	(5×10^5)	(5.5×10^6)
3	1.61×10^8 (5×10^5)	3×10^6 (2×10^6)	0 (0)	0 (5×10^5)	—*b*) (1×10^6)	1×10^8 (6.75×10^7)
4	1.09×10^8 (5×10^5)	5.6×10^7 (2.5×10^6)	0 (0)	0 (0)	*b*) (2.5×10^6)	4.05×10^8 (9×10^6)
5	(5×10^5)	(4.5×10^6)	(0)	(0)	(5×10^5)	(5.96×10^8)
6	3.8×10^7 (5×10^5)	8×10^6 (3×10^6)	0 (0)	7×10^6 (5×10^5)	—*b*) (5×10^6)	6.1×10^7 (2.38×10^8)
7	7.2×10^7 (1.5×10^6)	2×10^6 (1×10^6)	0 (0)	0 (4×10^6)	—*b*) (3×10^6)	2.25×10^8 (5×10^6)
8	5.7×10^7	6×10^6	0	7×10^6	—*b*)	1×10^6
9	(5.0×10^3)	(1×10^4)	(0)	(—*b*)*b*)	(1×10^4)	(2.12×10^7)
10	(5×10^3)	(1×10^4)	(0)	(—*b*)*b*)	(5×10^3)	(4.9×10^5)
11	3.79×10^8 (0)	1×10^7 2×10^4	0 (0)	0 (—*b*)*b*)	—*b*) (1×10^4)	3×10^7 (2.19×10^7)

注：括号中数字系 11 月的。*a*）未加任何重金属；*b*）未测。

2. 据耐铅菌出现频率、数量分布等，推断铅污染较重

耐重金属菌在所测站位上出现频率列于表 4，结果表明：它们在水中的频率高于沉积物中的，表示污染物由水体转入沉积物，情况与胶州湾的类似[2,3]。耐镉菌出现频率低，意味着镉污染轻；耐铅菌出现频率很高，暗示它与铅污染密切。

据表 1～2 得出表 5，它给出了渤海湾耐重金属菌的平均数量范围，其大小排列在水中

是 LT>MT>CT>CdT，在沉积物中是 LT>CT>MT>NT>CdT。各类菌数/AA 的排列次序同上。

表 2　渤海湾 7、11 月表层水中耐重金属菌数量分布(单位：个菌落数·mL^{-1})

培养基中所加重金属 / 站号	Pb^{2+}	Cr^{3+}	Cd^{2+}	Hg^{2+}	—a)
1	—b)	—b)	0	0	—b)
7	8.9×10^5	1.6×10^5	0	3.8×10^5	1.06×10^6
8	3.9×10^5	6.5×10^4	0	5×10^5	6.3×10^5
11	1×10^5 (4×10^2)	—b) (—b)b)	0 (3×10^1)	6×10^4 (—b)b)	4.4×10^5 (1.3×10^3)

括号(　)中数系 11 月的。

a)未加任何重金属；*b*)未测。

表 3　渤海湾表层沉积物各类菌数方差分析

菌类名	站位间的显著性	季节间的显著性
LTB	0.99($<F_{0.25}$)	5.58($>F_{0.1}$)**
CTB	1.09($<F_{0.25}$)	2.61($>F_{0.25}$)*
MTB	0.60($<F_{0.25}$)	0.0021($<F_{0.25}$)
AA	0.57($<F_{0.25}$)	2.30($>F_{0.25}$)*

1)据 1979 年 7、11 月六个站的数据。MTB 站位间自由度为 5 和 1，季节间自由度为 1 和 9；其余菌站位间自由度为 5 和 1，季节间自由度均为 1 和 5。MTB 用单因素分析，余为两因素分析。

2)*—差异显著：**—差异特别显著。

表 4　渤海湾各类菌出现频率统计

样品	事项	菌类名					
		LTB	CTB	CdTB	MTB	NTB	AA
沉积物	检出数	16	16	0	6	10	17
	样品数	17	17	17	17	10	17
	检出数、样品数(%)	94.1	94.1	0	42.9	100	100
水	检出数	4	2	1	3	—a)	4
	样品数	4	2	5	3	—a)	4
	检出数、样品数(%)	100	100	20	100	—a)	100

a)未测。

表 5　渤海湾各类菌在水和沉积物间数量比较

样品＼菌类名	LTB	CTB	CdTB	MTB	NTB	AA
沉积物	8.85×10^{7}	6.33×10^{6}	0	1.39×10^{6}	1.35×10^{6}	1.5×10^{8}
水	3.45×10^{5}	1.13×10^{5}	6×100	2.35×10^{5}	—	5.33×10^{5}

单位:水中为个菌落·mL^{-1},沉积物中为个菌落·g^{-1}(湿)。

由上所述,据耐重金属菌与相应污染物的相关关系[2,3,9-11],推测本湾铅污染重于其他金属污染。

将本湾耐重金属菌数与某些海区及有关河流的相比[2-4,9,10],可知本湾耐重金属菌数所占比例不大,表明本湾耐重金属菌数与不甚严重的重金属污染状况吻合[6](见表 7)。

沉积物中各菌数一般高于水中,说明水底沉积物环境更有利于它们生存。以 7、8 和 11 站水和沉积物中各菌数作相关分析后发现的正相关暗示,可以从某站沉积物细菌数推测相应上复水的相关菌数[11]。

3. 耐重金属菌季节变化与陆源排污关系及在全海湾的分布趋势

表 6 代表沉积物各自平均菌数夏冬两季变化。可见夏季各菌数大大超过冬季,尤以耐铅菌显著。

表 6　渤海湾表层沉积物各类菌数季节间的比较

季节＼菌类名	LTB	CTB	MTB	CdTB	AA
夏	2.14×10^{8}	1.34×10^{7}	2×10^{6}	0	2.25×10^{8}
冬	4.01×10^{6}	1.35×10^{6}	7.86×10^{5}	0	9.78×10^{7}
夏/冬	533.67	9.93	2.54	—	2.31

夏、冬分别由 7、11 月代表,单位:个菌落·g^{-1}(湿)

前述 6 个站沉积物各菌数方差分析结果(表 3)表明,LT、CT 的季节变化差异显著,站位分布差异不大。可见它们与铅、铬等陆源污物排放的明显季节性是一致的[6]。MTB 的情况与胶州湾潮间带和沿岸区的则不同[3]。

各类菌数站位差异不明显说明相应污物在全海区有自河口向口外海区扩散之势。入海污物经水文理化和生物作用相当迅速地稀释、扩散或沉降入底,因而相应微生物就顺应此方向发展,出现菌数时高时低、此起彼伏的现象,表明重金属污染不仅仅是点源性的。

4. 耐重金属菌与普通菌的相关性

将上述六站沉积物各菌数与相应 AA 作相关分析后发现,LT、CT、NT 与 AA 均为正相关,但只 LT-AA 间为显著相关,表明在目前渤海湾污染状况下(表 7),各类菌尚能共存,即普通菌尚能容忍并适应这种轻度污染。

表7 渤海湾不同介质中重金属元素含量[6]

元素名	海水($mg \cdot L^{-1}$a)		底质($mg \cdot kg^{-1}$)		间隙水($mg \cdot L^{-1}$)	
	范围	平均值	范围	平均值	范围	平均值
Hg	痕量—0.049b)	0.017	0.005～0.56	0.065	—	—
Pb	0.02～0.18	0.068	11.6～41.2	22.4	0.67～1.2	0.85
Cd	0.02～0.40	0.18	0.04～0.54	0.15	0.067～0.17	0.11
Cr	痕量－1.63	0.40	26.7～66.7	49c)	—	—
Ni	—	—	16.8～45.2	34.4	—	—

a)溶解态；b)总汞含量；c)样品用 HNO_{3+} H_2O_2 消化，测得结果偏低，约相当于全 Cr 量的 60%～70%

MT-AA 间、MT-MT/AA 间的负相关或不好的正相关性说明耐汞菌与普通菌有某种相斥关系，当一定汞污染压力下耐汞菌大量增加，普通菌相当快地减少时，MT/AA 提高，这意味着汞害加重[3,8,11-14]。所测站位不时出现的 MT 较大幅度起落反映汞污染的多途径性(如 6 号站—四号采油平台的生产和运输活动造成的影响)。

图 2 是 LT 与 AA 的相关图，方程式中的 6.69×10^7 是回归截距，其物理意义是沉积物中铅污不重时(如 22.4×10^{-6})AA 的一般数量。

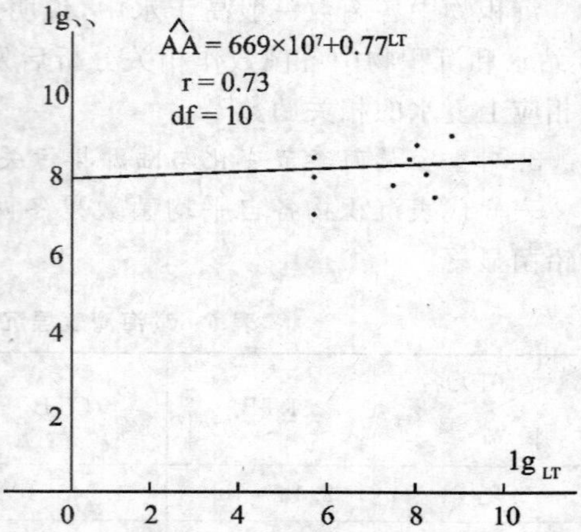

图 2 LT-AA 的相关图

5.耐重金属菌数与海湾水文动力学过程关系初析

渤海湾三面环陆，其微生物注定受陆地和水文气象影响[7]。若以 38°30′～38°45′N，118°E 带为中部海区(见图 1)，将所有测站作东、中、西和南、中、北分，其各菌数以大小排(表 8)，可知菌数分布状况大致是西高东低；中部是 7 月的低于 11 月的；7 月，北部的高于南部，11 月则相反。各菌数/AA 的状况也大体类似。

表8 渤海湾沉积物各菌量与离陆远近关系比较

月份	离陆	LT	CT	CdT	MT	NT	AA
7	西	西＞东＞中	西＞东＞中	中＝东＝西	西＝中＝东	—	西＞东＞中
	南	北＞中＞南	南＞北＞中	北＝南＝中	北＝中＝南		北＞南＞中
11	西	中＞西＞东	中＞西＞东	中＞西＞东	中＞西	中＞西＞东	中＞西＞东
	南	南＞中＞北	南＞北＞中	南＝北＝中	南＞中＞北	中＞南＞北	北＞南＞中

将各站菌数按海区不同部位作相关分析后发现：相关系数在北—南间由 7 月的正值

转为11月的负值*，反映了陆地径流，包括黄河水的影响自夏向冬减弱之势。西—东间由7月的负值转为11月的正值，暗示沿岸流的影响由夏向冬增强，并且也有深水流的影响。

夏季北部海区，暖流带来并沉积较多高盐污物；冬季，海流对近岸区产生较大影响，西部岸流又给南部带来较多营养和污物[5]，从而促进有关海区的微生物繁衍。

中部站位某些沉积物的菌数在7、11月的逆转现象与沉积物间隙水中重金属等含量变化是吻合的[1,5,6]，反映了陆源排水、不同温盐度水团交锋及海洋采油等因子对各类菌的短周期多变量作用。大量微生物的活动反过来又促进有关沉积物的pH、Eh、Es和氧张力等环境参数的改变[1,7]。

综上所述，陆源排污是造成近岸海区耐重金属菌等数量变动的一个重要因素。入海污物靠海湾水文动力学的强大动力而迁移，进而决定各类微生物分布。但因素是多种的，环境是复杂的。有关诸问题，尚待大力研究。

四、小结

1. 渤海湾许多好气异养细菌能忍耐、对抗一定浓度的重金属污染，普遍存在耐铅菌，耐汞菌、耐铬菌和耐镍菌，偶见耐镉菌。各类菌出现频率水中高于沉积物中，数量则一般低于沉积物中，但各类菌数与普通菌数之比值一般是水中高。渤海湾人为富营养化程度甚高。

2. 各类菌大部分与普通菌在数量上有正相关关系，表明重金属等污染不重。各类微生物尚能共处，说明其生态系统尚属正常。

3. 大部分耐重金属菌数量呈现出西部近岸高于远岸海区之势。北—南间、西—东间有季节性逆转，全区为夏高冬低。渤海湾各类菌受陆源和非点源排污的影响较大。海湾水文动力学过程是左右微生物分布的重要因素之一。

4. 各污染物不适当地持续排放，将对海湾微生物种群和数量结构产生不利影响，从而降低其在净化活动中的能力。控制和防治重金属等的污染是至关重要的。

参考文献14篇(略)

PRELIMINARY ANALYSIS OF THE NUMERICAL DISTRIBUTION OF HEAVY-METAL-TOLERANT BACTERIA IN THE BOHAI BAY

(ABSTRACT)

Abstract This paper is a preliminary analysis of and discussion on the numerical distribution of heavy-metal-tolerant bacteria in the Bohai Bay from the point view of environmental ecology.

A lot of common aerobic heterotrophic bacteria have been found in this bay. They have the ability to tolerate and resist heavy metal contamination under certain concentrations. Moreover, many lead-, mercury-, chromium-, and nickel-tolerant bacteria have been found, but cadmium-tolerant bacteria are rated last in terms of both number and occurrence frequencies.

From the results obtained, one can see that the occurence frequencies of each group of bacteria are higher on the surface water than on the surface sediments. Generally speaking, the ratio of the heavy-metal-tolerant bacteria number to that of the common bacteria (AA) is higher in water. The results suggest the artificial eutrophic level in the Bohai bay, especially in its estuary and coastal zone, is rather high.

Between the numbers of most groups of tolerant bacteria and AA there exist some positive correlations. It shows that heavy metal pollution is not serious in this bay. Therefore these bacteria are able to coexist in the present environment of this bay. The microbial ecosystem of the bay is still normal.

Generally speaking, the number of tolerant bacteria in the nearshore sea-water is higher than that in the offshore area. In the investigated sea area, there are some seasonal changes between the south and north sections, as well as between the west and east Sections.

Each group of bacteria found in this bay is subject to pollution from land and other sources. The hydrological kinematics of sea may be one of the important factors affecting microbial distribution.

If a large amount of different pollutants were discharged into the sea improperly and continually, some adverse effects would be brought upon the structure of microbial population and their purificatory activity would be reduced. Therefore it is extremely important to control and regulate heavy metal pollution in this bay.

潮间带耐铅菌时空变化*

摘要　铅是环境中最丰富的毒性污染物之一,海洋铅污染主要来自海洋以外的环境[1]。青岛有关工厂每年排出数百万吨的含铅废水入海,从而使胶州湾及其潮间带发生铅的污染。据调查,95%的水样含铅量都曾超过高限**,入湾河口铅量曾超标几倍到几十倍[2]。大量铅入海,第一个受害者便是潮间带和沿岸区。铅等工业污染物持续影响潮间带是铅等毒性金属并入海洋食物链的一个重要场所,其中微生物已发生着相应的深远变化,国外已有一些报告述及该论题[3~5],但未见国内有关报导。本文就胶州湾潮间带耐铅菌(Lead-Tolerant Bacteria 缩写为 LTB)的数量变化作一报导。

关键词　胶州湾　潮间带　耐铅菌　时空变化

一、调查和实验

(一)站位概况、样品采集、微生物分析

站位描述、采样方法和微生物学操作大体如文献[6]所述。站位详见图1,2-16号站分别为文献[6]中的Ⅰ-Ⅴ号站。

(二)耐铅菌

在海水肉汁胨固体培养基内按需添加 $Pb(NO_3)_2$,使其最终铅(Pb^{-2})达 100×10^{-6},在此培养基上长出的菌(14日)为 LTB。用每毫升水或每克湿沉积物样中的 LTB 菌落形成单位数即 CFU 为 LT,与相应的不加铅培养基上长出的普通菌 CFU(为 AA)之比——LT · AA^{-1}(%)来评价 LTB 的变化及其意义。

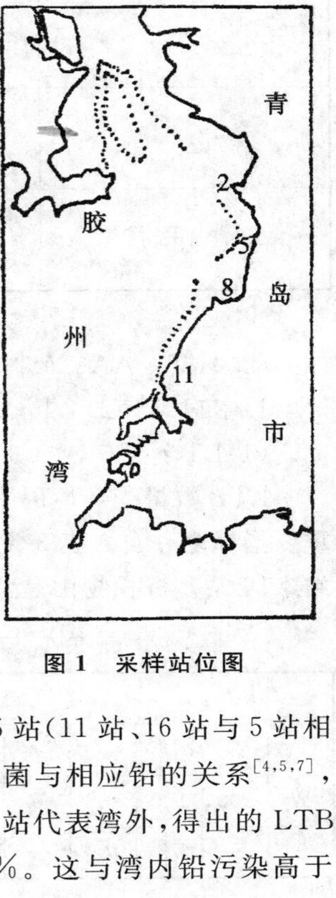

图1　采样站位图

二、结果和讨论

(一)耐铅菌的出现率

1.站位、水和沉积物间 LTB 出现率的比较

表1列出五个站位水和沉积物中 LTB 的出现率(W-水样,S-沉积物样,下同)。由该表可见,水中 LTB 出现率次序是:5站,11站,2站(8站与2站相等),16站。沉积物中是2站,5站(11站、16站与5站相等),8站。结果表明,5站 LTB 出现率较高,16站较低。据耐铅菌与相应铅的关系[4,5,7],这意味着5站比16站有更高的铅污率。将2-11站代表湾内,16站代表湾外,得出的 LTB 出现率分别是湾内:W78.7%;S78.7%。湾外:W71.4%;S76.9%。这与湾内铅污染高于

* 原文刊于《中国环境科学》,1986,6(6):31-35.

** 山东海洋学院化学系关于胶州湾污染的调查。

湾外的趋势是吻合的[8]。

2.不同月份及季节间LTB出现率的比较

由表2可见,水中LTB共检测17次,有10次为100%检出(即每份检样都有LTB,下同),占总数的58.8%。沉积物中LTB检测14次,有6次为100%检出,占42.9%。从每月份看,水中出现率一般高于沉积物中的。按各季LTB出现率平均值看*,水中为:秋(100%)>冬(85%)>春(83.3%)>夏(≤4%);沉积物中为:秋(100%)>冬(80%)>春(75%)>夏(65%)。这表明LTB出现率在两种介质中趋势一致,均显出秋冬季节较高,这可能与大量有机质分解有关。

表1　耐铅菌出现率的比较

站号	样品	出现率(%)	检样数
2	W	72.7	11
	S	91.7	12
5	W	92.3	13
	S	76.9	13
8	W	72.7	11
	S	69.2	13
11	W	76.9	13
	S	76.9	13
16	W	71.4	14
	S	76.9	13

(二)LT·AA^{-1}的时空变化

1.不同站位上LT·AA^{-1}的比较

(1)LT·AA^{-1}<1时的站际变化。

LTB数量受环境因子制约,它们在不同地理位置站位间会显出差异,LT·AA^{-1}是其重要指标之一[4]。表3说明,当LT·AA^{-1}<1时,其水中次序是8站>2站>5站>11站>16站。沉积物中是2站>8站>5站>11站>16站。这意味着16站铅污染轻些。

表2　不同月份间LTB出现率(检出数·检样数$^{-1}$%)的比较

采样	1979年									1980年							
	5	6(4)	6	7	8	9	10(0)	10	12	1	2	3	4	5	6	7	8
W	100	100	100	0	100	100	100	100	60	100	100	80	60	40	100	20	0
S	—	75	100	—	100	100	—	100	60	100	80	80	60	40	100	60	0

* 1979年5~6月,1980年4月~6月作春季;1979年7~9月,1980年7~8月为夏季;1979年10月为秋季;1979年12月~1980年3月为冬季,下同,参见文献[6]。

表 3　LT · AA^{-1}(<1)的站位比较

站号	2		5		8		11		16	
样品	W	S	W	S	W	S	W	S	W	S
LT · AA^{-1}	0.174	0.264	0.161	0.223	0.383	0.240	0.118	0.203	0.116	0.150

表 4　LT · AA^{-1}(≥1)出现率(%)的站位比较

站号	2		5		8		11		16	
样品	W	S	W	S	W	S	W	S	W	S
LT · AA^{-1} (≥1%)	0	16.7	23.1	23.1	0	0	30.8	23.1	30.8	23.1

(2)LT · AA^{-1}≥1 时的站际变化。

在一些站位上不时出现 LT · AA^{-1}≥1 的突跃现象，这是大量普通菌转化为对铅有适应能力的适应菌之结果。表 4 说明：水中次序是 11 站，16 站，5 站，2 站(8 站与 2 站相等(0))；沉积物中是 11 站(5 站、16 站与 11 站相等)，2 站，8 站(0)。这意味着 11 站时有大量有机质及铅污染发生，使 LT 突增(有关资料，另文发表)。

2. 不同月份(季节)中 LT · AA^{-1} 的比较

(1)LT · AA^{-1}在时间序列上的变化。

LTB 数量在时间序列上显出差异，表 5 列出 LT · AA^{-1}的周年变化。

当 LT · AA^{-1}<1 时，五个站位的水和沉积物 LT · AA^{-1}平均值如表 6 所示，水中最高 LT · AA^{-1}发生在 1979 年 6、9 和 10 月，沉积物中最高 LT · AA^{-1}发生在 6 月和 10 月。按各季 LT · AA^{-1}(<1)的平均值分，则水中是：秋(0.392)>春(0.222)>夏(0.205)>冬(0.122)；沉积物中是：秋(0.459)>夏(0.201)>春(0.195)>冬(0.159)。总的看，是秋高冬低：秋 0.426；春 0.207；夏 0.203；冬 0.140(见表 7)。这种势态可能与浮游植物的盛衰[9]及铅在环境中的积累有关。表 8 列出≥1 的 LT · AA^{-1}出现率情况，由该表看到，LT · AA^{-1}(≥1)在水中，1979 年 5 月、6 月(4 日)和 10 月(10 日)的检样均是 100%(即检样中 LT 全部≥AA，下同)，占 LT · AA^{-1}(≥1)检样的 37.5%。LT · AA^{-1}有 9 次(表中为 0 的，下同)。沉积物中 LT · AA^{-1}≥l 主要发生在 5～6 月间，但没有一次 LT 全部≥AA 的。LT≤AA 有 8 次。故水中 LT 易突跃，沉积物中较稳。

按各季 LT · AA^{-1}(≥1)出现率平均值比较，结果如表 9。即水中是春>夏>秋>冬；沉积物中也如此。说明 LT 在春夏比秋冬易突跃，这可能与温度、营养及铅量增高有关(资料另文发表)。

表 5　各站 LT·AA^{-1}的周年变化一览表

站号	采样期 / LT·AA^{-1} / 样品	1979 年								
		5	6(4)	6	7	8	9	10(10)	10	12
2	W	—	—	—	—	—	0.25	—	0.7454	0
	S	—	—	0.631	—	0.945	0.25	—	0.4675	0
5	W	>1	—	>1	—	0.388	0.333	—	0.186	0.01899
	S	>1	—	>1	—	0.06	1	—	0.2727	0
8	W	—	—	—	—	0.5	0.9931	—	0.8917	0
	S	0.5873	—	0.5370	—	0.5922	0.1587	—	0.7740	0.0223
11	W	—	>1	0.768	—	>1	>1	—	0.0175	0.0221
	S	0.5	—	>1	—	0.01974	1	—	0.5625	0.125
16	W	1	—	—	0	0.1884	>1	>1	0.1176	0.66
	S	0	—	1	—	—	0.5	—	0.22	0.5

站号	采样期 / LT·AA^{-1} / 样品	1980 年							
		1	2	3	4	5	6	7	8
2	W	0.5	0.000195	0.04795	0.3030	0.0458	0.0168	0	0
	S	0.1148	0.1052	>1	0.03125	0.0595	>1	0.0323	—
5	W	0.2679	0.03509	0.04545	0.3125	0.0189	>1	0	—
	S	0.85	0.0311	0.000357	0	0.1111	0.9091	0	—
8	W	0.1813	0.0088	0.222	0.75	0	0.6667	0	—
	S	0.2733	0	0.1	0	0	0.01	0.0308	—
11	W	0.1852	0.0098	0.056	0	0	>1	0	—
	S	0.2876	0.2	0	0.3333	0	1	0	—
16	W	>1	0.0625	0	0	0	>1	0.0185	—
	S	0.25	>1	0.00862	0.0002	0	>1	0.0196	0

注:(　)中数字为日期。

本结果与美国切萨皮克湾海滨及小湾的耐铅菌指数比较,一般都较低[4]。

(2)LT·AA^{-1}月际差异的方差分析

对 1979 年 9 月～1980 年 7 月间 10 对 LT·AA^{-1}数据作单因素分组数据的方差分析(重复数不等,固定模型),结果表明水中 LT·AA^{-1}的月际差异显著($F=7.54$,$>F_{0.005}$),暗示水中 LTB 与水温、降水等有关。沉积物中则不甚显著($F=1.81$,$>F_{0.25}$)。

表 6　水和沉积物中 LT · AA^{-1}(<1)的月际变动(五站统算)

样品＼采样期 LT · AA^{-1}(<1)	1979 年								
	5	6(4)	6	7	8	9	10(10)	10	12
W	—	0	0.768	0	0.359	0.525	—	0.392	0.110
S	0.362	—	0.602	—	0.4045	0.303	0	0.459	0.110

样品＼采样期 LT · AA^{-1}(<1)	1980 年							
	1	2	3	4	5	6	7	8
W	0.0284	0.0233	0.074	0.273	0.0129	0.342	0.0037	0
S	0.355	0.084	0.027	0.073	0.034	0.460	0.017	0

注:(　)中数字为日期。

表 7　LT · AA^{-1}(<1)的季节变化

样品	季节			
	春	夏	秋	冬
W	0.222	0.205	0.392	0.122
S	0.195	0.201	0.459	0.159

表 8　水和沉积物中 LT · AA^{-1}(≥1)出现率(检出数 · 检样数$^{-1}$%)的周年变动

样品＼采样期 LT · AA^{-1}(≥1)	1979 年									1980 年							
	5	6(4)	6	7	8	9	10(10)	10	12	1	2	3	4	5	6	7	8
W	100	100	50	0	25	40	100	0	0	20	0	0	0	0	60	0	0
S	66.7	—	60	—	0	40	—	0	0	0	20	20	0	0	60	0	0

注:(　)中数字为日期。

表 9　LT · AA^{-1}(≥1)的出现率%的季节变化

样品	季节			
	春	夏	秋	冬
W	35	18.8	16.7	5
S	33.3	13.3	0	10

(3)LT · AA^{-1}－T 的相关分析

表 10 列出 LT · AA^{-1}与 T(样品温度)的相关系数,该表证实了 LT · AA^{-1}－T 有一定正相关。耐铅菌对温度的感应类似于耐汞菌的[10],(一些普通菌与温度间的关系以不显

著的负相关为主)。

(4)LT · AA^{-1}的年度差异

LTB 在不同年度间的变化可能反映铅污染动态。将 1979～1980 年度与 1984～1985 年度取自 5 站和 16 站四季五个月沉积物 LT · AA^{-1} 作单因素分组数据(重复数相等,固定模型)的方差分析结果表明,5 站 LT · AA^{-1} 年度差异显著($F=3.68, >F_{0.1}$),16 站则差异甚微($F<1$),说明 5 站铅污染状况变动甚大,16 站较稳(铅污染轻)。

表 10　LT · AA^{-1}－T 的相关系数表

站位	2	5	8	11	16
W	0.151 (n=11,<0.5)	0.608 (n=13,>0.05)	0.439 (n=11,>0.2)	0.741 (n=13,>0.01)	0.173 (n=11,<0.5)
S	0.622 (n=12,>0.05)	0.306 (n=13,>0.5)	0.310 (n=12,>0.5)	0.349 (n=13,>0.5)	0.134 (n=11,<0.5)

(三)LT · AA^{-1}在水与沉积物间的关系

将前述有关数据(即 W LT · AA^{-1}-S LT · AA^{-1},$n=33$)作相关分析(图 2),得出的 $r=0.349(>\alpha_{0.05})$表明水中与沉积物中的 LTB 有甚明确的正相关,水中 有向相应沉积物富集的可能[3]。

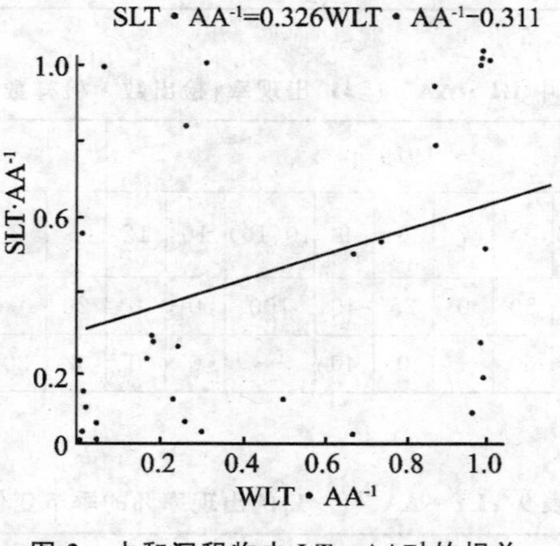

图 2　水和沉积物中 LT · AA^{-1}的相关

三、结语

铅污染物在环境中的长期积累可能产生的一个效应即是某些适应菌——耐铅菌等的大量繁殖,这些微生物借其生命活动在一定程度上改变铅等物质在环境中的状态,参与铅等的定位、形态选择和迁移活动。一些迹象表明耐铅菌与铅间存在一定联系。耐铅菌数量的一些规律初步总结于下:

1.湾内潮间带耐铅菌出现率和 LT 均高于湾外的,这可能意味着耐铅菌分布与铅分布有一致之处,尚待深入探讨。

2.大体说,LT·AA^{-1}显出秋高冬低之势。在水中,于春夏变动大,易突跃,因而有一定的月际—季节差异。LT·AA^{-1}与温度间有正相关,也可能与营养及铅量等有关。

3.湾内的LT·AA^{-1}变动大于湾外,这一迹象可能与湾的地理特征、陆上排污变化有关。

4.耐铅菌与铅污染有关,因此推测本调查范围内铅污染状况是春夏高于冬、湾内重于湾外、湾内和温热季节变动大并进而影响整个潮间带的铅分布状态。

参考文献10篇(略)

THE CHANGES OF TIME AND SPACE OF LEAD-TOLERANT BACTERIA IN INTERTIDE ZONE

(ABSTRACT)

Abstract In this paper, the investigations and experiments on changes of time and space of lead-tolerant bacteria (LTB) numbers (LT) in the intertide of Jiao Zhou Bay over one year are reported.

The result indicates that the occurance frequency, number and LT · AA^{-1}(%) of LTB in the intertide are larger inside the bay than in the outside. There are some positive correlations between LT · AA^{-1} and temperature of the samples. LT · AA^{-1} generally tends to be higher in autumn and lower in winter, and its change occurred in water and warm season appears to be obvious and greater suddenly. LT · AA^{-1} is connected to Pb-pollution. Therefore, it is inferred that Pb-pollution in spring and summer is higher than that in water and more serious in the inside than in the outside; also its change is greater in the inside than in the outside.

铅污染与微生物*

一、铅的污染概况

铅是亲硫元素，多产于硫化矿氧化带，它是地球中储藏量较多的元素之一，铅又是亲气元素，有大气污染和参与全球循环的特点，因而它也是环境中最丰富的有毒污染物之一[4]，长期来，铅矿的开采、铅的应用和扩散，使许多区域受到铅污染，主要来自海洋之外的海洋铅污染使河口、沿岸区和海岸带等污染较重。

河口等海岸带积聚的大量有机质和颗粒物，是铅等污染物、也是细菌等微生物的良好基物，海洋细菌可从基本海水培养基或海洋本身浓缩少量离子[12]，这在浅海区域更为突出，因而海洋微生物干扰、影响铅等的迁移活动。各级生物也参与铅等的迁移和积累活动。人们食用海鲜，使铅等沿食物链进入人体，使人受害。

二、铅对生物影响的概述

铅对动物的影响主要是它在体内的蓄积抵制正常酶功能，紊乱一些生命元素在体内正常运行，造成各种铅中毒症状。

一些植物以铅作为一种必需微量元素。有些植物有耐铅力，能在体内积累、转移铅。这种性能可能与遗传有关。人们利用这一特性将一些植物作铅污染的指示者和净化器。

对微生物而言，铅常是有毒的。如无机 Pb^{2+} 抑制海藻光合作用和真菌生长，抑制蓝细菌固氮，抑制真菌孢子萌发和菌丝生长。但许多证据也表明微生物能耐铅毒，有些与其固有特性相关[6]。一些细菌，如节杆菌和丝细菌，能在铅矿上生长。有些菌能在不损害其生命活动时固定铅[26]。

三、微生物与铅的相互作用

（一）微生物对铅的反应

多数对铅敏感的细菌，当培养基中含低量铅时就被抑制，大肠埃希氏菌（NCTC/0418）对 0.001 mol 铅离子很敏感。随铅浓度递增，大量铅敏感菌受抑，只少数能形成菌落，当平皿中铅浓度为 200×10^{-6} 时，尚可见褐—赭色菌落。此色泽似与铅浓度有关，铅浓度低，该色菌落色浅。无铅则不见此色素，表明褐—赭色菌落是铅污染的结果[24]。在铅浓度为 400×10^{-6} 时，耐铅菌生存下来，经 120 h 铅培养的好气异养菌数比空白对照时的大[23]。Jensen 培养铅浓度为 $5\,000\times10^{-6}$ 的土样（含 2%油）一月后，细菌数达 $2\,425\times10^{6}\cdot g^{-1}$（干土），占对照土样菌数的 85.4%，4 月后占 74.2%。表明耐铅菌对铅有强大耐力[11]。

---

* 原文刊于《环境科学》，1986，7(6)：87-89 转 81。

耐铅力又是有条件的。如用铅培养的海水菌适于铅污染。当耐铅的海水菌培养在无铅培养基后,便逐渐再次失去耐铅力。海洋沉积物菌也有类似现象。

(二)微生物积累铅的能力及部位

1.积累能力。在铅影响下生长起来的耐性菌株,其生理生化活性将发生相应变化。例如,它们可能产生出质粒以抗铅,金黄色葡萄球菌、一些肠菌即是[14,22]。另一重要变化是铅的积累。Aickin 等试验的枯草芽孢杆菌、食爬虫假单胞菌、弱氧化葡萄糖酸杆菌、荧光假单胞菌、肺炎克雷伯氏菌等能在醋酸铅培养基上摄取 0.1%～36%(占生物干重)的铅、黑曲霉、炭黑曲霉等九株真菌的最高摄铅量达 19%[1]。

来自海洋、河口和陆地的细菌,包括海生产碱菌(*Alcaligenes marinus*)、海水浮游可变单胞菌(*Altermonas haloplanktis*)、大肠埃希氏菌、荧光单胞菌、卵状假单胞菌、恶臭假单胞菌及其他假单胞菌、巨大芽孢杆菌、金黄色葡萄球菌、费氏弧菌等在海水培养基上生长时,原子吸收分光光度计表明其细胞含铅量达 $10^3\times10^{-6}$ · 干重$^{-1}$。这些 Mn、Ni 或 Cd 的积累量多,但比其他金属的少[12],藤黄微球菌和固氮细菌分别从含 $PbBr_2$ 2.5 mg · mL^{-1} 的肉汤培养基中固铅达 4.9 和 3.1×10^2 mg · g^{-1}(细胞干重)。柠檬酸细杆菌的休眠细胞铅达其干重的 10%。当把它与荧光假单胞菌培养在以柠檬酸盐缓冲、磷酸甘油为唯一磷源的醋酸铅培养基上培养时,积铅达干重的 1/3,浮游球衣菌用 100×10^{-6} Pb[$Pb(NO_3)_2$]培养 10 天,最高积铅量为 9700×10^{-6},是开始加入培养基中铅量的 2.3%,0.1、1.0 和 100 $\times10^{-6}$的铅刺激该菌,积铅量随铅浓度增加而增加,随该菌数减少而减少[20]。除细菌外,藻类如斜生栅列藻积铅达干重的 137962×10^{-6}。最大积累发生在 10×10^{-6}的铅培 5 天时。在这之前,铅进入细胞的速度随铅浓度增加而加快[13]。

2.积铅部位及积铅物质。铅的积累主要与胞壁/胞膜相关。X 射线分析、扫描电镜和扫描透射电镜检查都证实此点。藤黄微球菌积累的铅有 99.3%在胞壁+胞膜的片段上,固氮细菌是 99.1%,柠檬酸细杆的休眠细胞表面有铅的集中化现象。浮游球衣菌的铅很大部分与黏液荚膜物质有关。表明在决定多少金属进入细胞上,细胞外层很重要[2]。将棕黄色小球菌用 100 mg · L^{-1} 的 $PbBr_2$ 或 $Pb(NO_3)_2$ 培养基培养,吸取的铅,有 75%～82%在细胞类脂萃取物中。这种萃取物系积铅的天然混合物。类脂组分不为铅的专性稳定结合提供环境可适性,但天然类脂混合物却为铅提供成核条件[27]。

一些结果证实,细胞表面沉积的铅是 $PbHPO_4$ 的形式。

有些菌的细胞质也积铅,浮游球衣菌即是淡水的斜生栅列藻将部分铅均匀地分布在胞质和内质网上。棕黄色小球菌也有类似现象。这说明微生物积铅部位因细胞结构不同而异。此外,还受一些条件制约。

四、铅的微生物毒理效应

(一)破坏细胞

铅对质膜结构造成不利影响,使一些原生质产生胞溶现象,或使一些菌发生质壁分离,并造成其间体结构膜破裂征象的变异。用三(羟甲基)氨基甲烷和 EDTA 处理萃取的脂类部分,或用 p一氯汞基苯磺酸还原硫基,对铅的存留影响较小[21]。

(二)抵制胞内酶活动

铅影响淀粉酶产生菌的蛋白质合成,使淀粉酶合成受抑。淀粉酶合成的减少同对铅敏感的淀粉酶产生菌数减少相平行。淀粉酶合成的恢复与该菌数增加相关[5]。Babich 等发现[3]2 mg·g^{-1}的铅(醋酸铅)可减少一葡萄糖苷酶、液化酶和转化酶的微生物合成,铅也抵制纤维素酶、脲酶等。

(三)影响细菌生长、能量代谢和离子运输

三乙基铅和三丙基铅阳离子对大肠埃希氏菌有这些影响,还程度不等地抵制细胞生长和完整、细胞及膜颗粒的摄氧。较高浓度的三乙基铅明显抵制整个细胞的呼吸速率[7]。

(四)改变生长规律

当浮游球衣菌生长在含 $Pb(NO_3)_2$ 的培养基中时,迟滞期消失,生物量增加。但因铅刺激而迅速生长的细胞后来却很快死亡了,大肠埃希氏菌在临界 Pb^{2+} 浓度(45 m)中生长时,活力减了 89%,迟滞期延长了。

(五)影响铅毒的若干因素

1. pH 和培养期。用气单胞菌、假单胞菌、产碱菌、不动杆菌、黄杆菌在 0.5%营养肉汤和 0.1%酵母膏中培养,其转变 Me_3PbOAC 为 Me_4Pb 的过程决定于 pH。pH 5 时,Me_4Pb 最多;pH 9 时减少。在化学上限定的培养基中,上述菌无一株能将无机铅转化。

在特定铅浓度下,摄铅量一般随培养期的延长而增加。Pb^{2+} 对酶的抑制常出现于培养末期,表明铅效慢,随时间延长而铅毒增加。耐性菌群系也不断发展起来。

2. 培养基中含铅量。一般是铅浓度高,入胞的渗透作用快,最高积累发生得早。如斜生栅列藻、浮游球衣菌等即是。但积铅量增减,与铅浓度不始终成比例。

3. 铅的化学形式及其与细胞接触的时间。铅的化学形式不同,对酶合成的抵制程度也不同。四甲基铅比四乙基铅毒性小。烷基铅毒性比无机铅毒高几个量级,PbS、氧化铅类、特别是 PbO_2 比 $PbBr_2$、$Pb(NO_3)_2$ 和金属 Pb 不易被微生物攻击。三丙基铅比三乙基铅易脂溶,抵制作用大,但三乙基铅浓度大时,抵制作用也十分明显。

Greszta 等(1979)列出不同金属联合对土壤微生物的毒性次序是:Cd-Pb-Zn＞Pn-Cu、Pb-Zn＞Pb-Cu。

不同形式的铅接触细胞的方式和条件不同。同一形式的铅盐随与细胞接触时间加长而增毒。

4. 胞壁成分。不同的胞壁成分不同地选择吸收金属盐,这是胞壁的屏障作用。含多糖多,选择结合的金属也多。胞壁结构和组成限制入胞金属化学形式和数量。因而能减低某些金属的毒害作用[20]。

5. 培养基成分。许多培养基能大量结合重金属离子。酪蛋白、氨基酸结合所有金属。一些阳离子对许多配位体有亲和力,它们在 Irving-Williams 系列中的顺序为:$Hg^{2+} > Pb^{2+} \gg Cu^{2+} \gg Cd^{2+}$[18]。非发酵性碳源比葡萄糖更能使三乙基醋酸铅(5 m)发挥抵制力[7]。

6. 其他。铅毒还决定于介质的理化性质和生物性状。铅的氯化物、硫酸盐和醋酸盐的毒性在土壤间有差异。这归之于它们在溶液中留存的量。但黏土的螯合作用减低铅毒。若土壤金属浓度已饱和,少加金属即会显著增毒,因为溶液中的浓度直接随所加金属量增而增[19]。此外,水的硬度、腐殖酸等也有影响。

铅污染土中的微生物活动有时还决定于地面植物状况,其根际效应比铅污染状况还重要。

五、铅的微生物生态效应

铅的性质使它与环境颗粒物、有机质密切相关。一些结果表明它们间有正相关。许多微生物也有类似性质。如 Humber 河口(英)的附着细菌、附着细菌/悬浮固体与颗粒铅间均有正相关,细菌数含量与有机质含量有关[9,10],因而耐铅菌与铅污染必定相互作用。当铅污染导入境内后,一个明显的生态效应便是耐铅菌的选择效应。其淘汰作用使铅敏感菌减少乃至消失,耐铅菌适应并增多。微生物群体这种朝耐铅方向移动所需时间甚至不会太长。适应下来的活菌数有时甚至比对照的更多[23]。

铅污染进入环境的另一生态效应是原来群落组成的变化,种的多样性将要减少,严重时甚至导致群落结构失稳。如荷兰的一些土壤便有此情况。随着培养基中铅的增加,革兰氏阴性杆菌比例增高,棒形菌数比例减少[6]。这表明革兰氏阴性杆菌是耐性菌数增加的主要贡献者,它占耐性株数的 74%。棒形菌占 19%,其他杆形菌占 5%。敏感株中相应比例分别为 58%、38%和 3%。所以,耐铅菌有生物化学和形态学上的特点[25]。

在含铅量正常或较低的环境中,仍然有相当比例的耐铅菌。这与其分类特点及遗传学性质有关。但一般而言,含铅多的环境,耐铅菌比例高,耐重金属性最终也是获得性的[19,23]。这种筛选作用在一定程度上带有间接性。反之,在后一环境中,仍可发现为数不少的铅敏感菌,这意味着铅在环境中分布的不均匀性。如土壤团粒结构在一些方面是内外有别的。相应含铅量及其有关微生物也是内外不一的[6]。

对铅的感应也决定于微生物的地理学起源,北大西洋 Iberia/NW 非洲的菌比 Spitzbergen 的菌更耐铅[25]。

在某一环境测得相当严重的铅污染,也有相当数量和活力的微生物,这还意味着铅毒仍不至于危及有关微生物。因为铅与控制细菌的参数有关。这些参数可能减毒,表现出细菌与铅在一定程度上互不影响,因只对共同因子发生感应。尽管如此,耐铅菌比例仍不失为特定条件下反映铅污染水平的重要指标之一[8,10,15,16]。

六、耐铅菌的生态意义及应用

耐铅菌在江、河、湖、海等环境中参与铅等元素的地球化学过程,对铅在环境中的定位和富集起重要作用。它们的活动促使铅等重金属在河口等环境中滞留,反过来又增加金属溶解度,促进金属的流动化[17]。一些耐铅菌积累铅,并经食物链传递,其中包括铅的甲基化作用。因而耐铅菌的活动具有一定的生态意义。人们利用这些特性,赋予耐铅菌以一些应用价值:

1. 检测和净化铅污染等。耐铅菌积累大量铅,甚至耐、抗其他污染。显出多重抗性,因而可用于净化环境。此外铅敏感菌和耐铅菌都以一定意义对铅污染起指示作用。

2. 一些微生物与铅关系密切。已发现一些方铅矿结晶中的许多古细菌可用于研究成岩作用早期晶体生长上。已用于贫矿浸出。已提出了商业规模应用微生物浸出技术处理复杂 Pb-Zn 硫化矿的流程图和工艺学。

3. 一些含质粒的耐铅菌是当今兴起的生物工程材料之一。

参考文献 27 篇(略)

潮间带耐铅菌与铅等重金属有机质关系初析*

胶州湾潮间带有较多铅、铜、镉、锌及有机碳、氢、氮等污染物，因而也有较多耐铅菌(Lead-Tolerant Bacteria 缩写作 LTB，下同)等相应的微生物。LTB 等微生物参与这些污染物的价态，类型变化和迁移活动。潮间带 LTB 量的时空变化曾有过报道，但对 LTB 与环境因子间的关系未探讨过，而 LTB 的利用则有赖于环境因子，因此探索 LTB 与铅等重金属、有机质的关系，对阐明污染物归宿、生态变化和污染治理都有重要意义，本文就此作一尝试。

一、采样、分析和实验

1. 采样及站位概况　潮间带表层海水样取自 1983 年 7 月和 10 月份，其沉积物样取自 1983 年 4 月，1984 年 3、6、8、10 月和 1985 年 2、5 月份，共 19 站(见图 1)。

2. 化学分析　用 Davis A1660 阴极射线示差极谱仪测样品 Pb、Cd、Cu、Zn 含量，用 Perkin-Elmer 240C 型元素分析仪测总有机碳 TOC(%)、总有机氢 TOH(%)和总有机氮 TON(%)。

3. 细菌含量测定　单位体积(重量)mL(g)样品中 LTB 的 CFU 作 LT，不添加重金属的培养基中得出的细菌 CFU(以下称普通菌)作 AA，将 LT・AA^{-1}等与 Pb 等重金属、有机质含量统计分析。

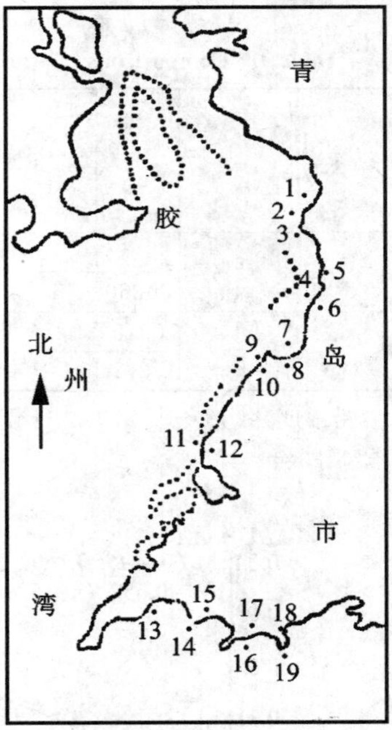

图 1　潮间带采样站位图

Fig. 1　Location of sample collected stations over intertidal zone

1-3 站石化石厂后海滩；4-6 站板桥坊河口；7-8 站李村河出海口；9-10 站水清沟入海口段；11-12 站海泊河口；13 站栈桥浴场海滩；14-15 站二中后海滩；16-17 站太平角；18-19 站湛山湾；1-12 站为湾内站，13-19 站为湾外站。

二、结果和讨论

1. LT・AA^{-1}和 Pb 的方差分析　表 1 列出 Pb、LT・AA^{-1}两月间的变化，表明铅已达污染程度。对表 1 的 Pb、LT・AA^{-1}各作单因素分组重复数相等的方差分析表明，两者各自的站际差异不大($F<1$)，湾内铅污染和 LT・AA^{-1}仍高于湾外。同法分析，LT・AA^{-1}月间差异明显($F=8.901, >F_{0.01}$)，符号检验分析表明 Pb 量的月间差异显著($P<0.05$)，表明夏季陆源排污大于秋季，故 Pb 和 LT・AA^{-1}的变化并行不悖，暗示 LTB 参与环境铅的迁移活动。

2. LT・AA^{-1}与 Pb 等的相关性　图 2 示出水中 LT

* 原文刊于《生态学报》，1989，9(3)：280-282。

· AA^{-1}与 Pb 等 4 对相关分析结果。表明当培养基中 Pb^{2+} 浓度为 100×10^{-6}时，LT · AA^{-1}不仅与样品含 Pb 量的正相关性达显著水平，而且 LT · AA^{-1}-Cu，LT · AA^{-1}-Cd、LT · AA^{-1}-Zn 的 r 值也达显著正相关。进一步说明 LTB 与 Pb 等关系密切，这为 LTB 产生耐铅力乃至参与环境净化创造了条件，即 LTB 参与环境中 Pb 等重金属的降解、定位和迁移活动。沉积物中情况与此相似，LT · AA^{-1}-Pb 的相关性高于 Doelman 等所报道的。当培养基含 200×10^{-6} Pb^{2+} 时，LT · AA^{-1}-Pb 的 r 值仍为正值，说明 Pb 对 LTB 有较大、较持久的影响。特定条件下的 LT · AA^{-1}变动反映环境铅污染状况，随培养基 Pb^{2+}浓度递增，LTB 出现时间推迟、菌落变小，其褐一赭色加深，表明铅污染压力深刻影响 LTB 细胞生理生化及形态变化。

与 LTB 不同，AA-Pb 的 r 值是-0.131($n=16$)、-0.048($n=16$)和 0.13($n=20$)(取样同前)，表明普通菌对铅敏感，重金属对普通菌有害。

表 1　11 个水样中铅及 LT · AA^{-1}在两个月间的比较

Table 1　Comparisons of lead concentrations and LT · AA^{-1} of 11 water samples in two months

月份 \ Pb ($LTAA^{-1}$) \ 站号	1	2	3	4	5	6	7	8	10	11	16	平均
7	0.155	0.155	0.075	0.038	0.036	0.031	0.093	0.065	0.104	0.78	0.031	0.142
	0.260	1	0.366	0.279	0.342	0.143	0.236	0.179	0.250	0.327	0	0.306
10	0.19	0.15	0.05	0.11	0.17	0.16	0.04	0.04	0.07	0.07	0.13	0.107
	0.2727	0.1875	0.0028	0.0833	0.0788	0.0263	0.0157	0.0248	0.0046	0.0307	0	

铅浓度：微克 · 升$^{-1}$；第二行数字为 LT · AA^{-1}，其 LT 系培养基中 Pb^{2+} 为 200×10^{-6}时得出，样品取自 1983 年。

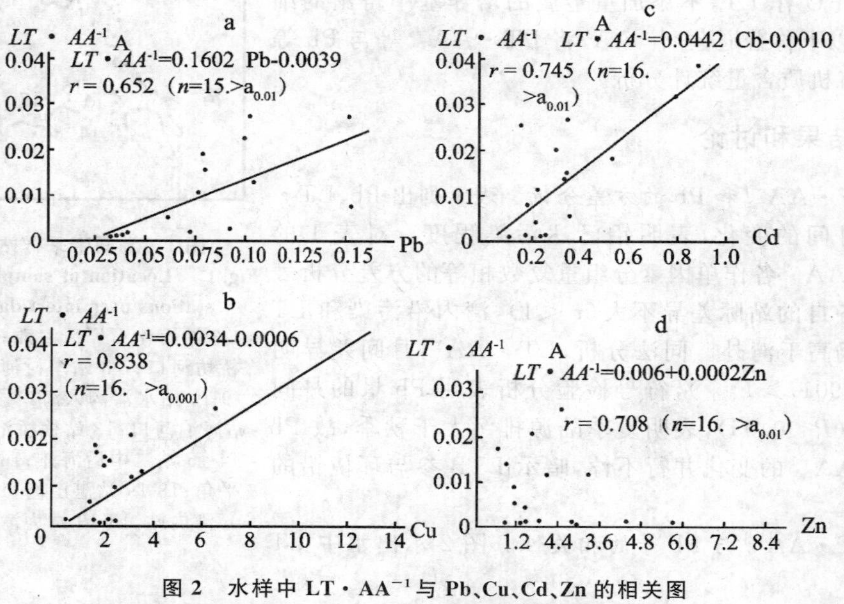

图 2　水样中 LT · AA^{-1}与 Pb、Cu、Cd、Zn 的相关图

金属浓度：g · L^{-1}　采样时间：1983 年 7 月。

Fig. 2　Correlations of LT · AA^{-1}-Cd, LT · AA^{-1}-Pb, LT · AA^{-1}-Cu, LT · AA^{-1}-Zn, The concentrations of heavy metals: g · L^{-1}, Time of sample collection: 1983.7

3. TOC、TOH、TON 与 $LT \cdot AA^{-1}$ 关系的分析　表 2 列出 12 个沉积物样 TOC、TOH、TON 与 $LT \cdot AA^{-1}$ 数据，表明本潮间带已有有机质污染，湾内 TOC 普遍高于湾外，自北向南减，中部又起峰的状态，TOH、TON 也有类似分布趋势。$LT \cdot AA^{-1}$ 在湾内平均为 0.159，湾外为 0.60，这与 TOC 等的分布吻合，而且也印证了湾内铅含量比湾外高，这些与青岛工厂布局、港口设施和生活排污有关。

表 2　有机质含量(%)及 $LT \cdot AA^{-1}$ 在十二个沉积物样中的比较

Table 2　Comparisons of organic matter concent(%) as well as $LT \cdot AA^{-1}$ in 12 sediment samples in different stations

站　号	4	6	7	8	11	13
有机碳 TOC	10.4399	30.575	10.7824	16.5946	20.7729	4.71142
有机氢 TOH	0.410004	0.557395	1.27948	2.15441	2.46529	0.0958692
有机氮 TON	0.240407	0.597972	0.319602	1.32841	1.29904	0.0737404
$LT \cdot AA^{-1}$ *	0.2979	0.1333	0.0334	0.0612	0.267	0.20
站　号	14	15	16	17	18	19
有机碳 TOC	2.74333	1.29951	0.090794	0.458834	0.490771	0.866911
有机氢 TOH	0.254632	0.0917354	0.0440447	0.0628017	0.082925	0.084283
有机氮 TON	0.137368	0.034286	0.0137196	0.025893	0.0239926	0.0320214
$LT \cdot AA^{-1}$ *	0	0.0634	0	0.0277	0.0182	0.1111

* 其 LT 系培养基含 200×10^{-6} Pb^{++} 得出，1985 年 5 月取样。

表 3 列出 $LT \cdot AA^{-1}$-TOC 等的相关分析结果，表明 $LT \cdot AA^{-1}$ 与三种有机质间均为正相关，说明 LTB 还受有机质的制约。当 TOC 等有机物含量不很高时，LT 随 TOC 浓度升高而增长。而 AA 与 TOC 等的 3 个正相关系数均小，表明 LTB 与普通菌对有机质的依赖性不同，而且有机质种类、形态及生物有效性等均影响微生物在环境中的作用(Tan 等，1981)。

表 3　沉积物中 $LT \cdot AA^{-1}$ 与有机质含量(%)的相关性

Table 3　Correlation between $LT \cdot AA^{-1}$ and organic matter content (%) in the sediments

对子　pairs	相关系数　r
$LT \cdot AA^{-1}$－TOC	0.462(n=12)
$LT \cdot AA^{-1}$－TOH	0.309(n=12)
$LT \cdot AA^{-1}$－TON	0.355(n=12)
$LT \cdot AA^{-1}$－TO(C+H+N)	0.416(n=12)

三、结语

本研究表明，胶州湾潮间带有铅等重金属及有机质污染，存在较多耐铅菌。LT・

AA^{-1}－Pb、LT·AA^{-1}－有机质等呈正相关，它们在时空变化上一致。LTB 不同于普通菌，它在铅污染环境中，将参与微生物及其与相应污染物间的适应变化导致本环境中微生物的生态分布特征。本研究为控制和治理铅等的污染提供了一些重要参数。

胶州湾表层海水石油烃含量状况与评价*

摘　要　对采自多至17个测点的胶州湾表层水石油烃含量逐月用荧光法做了测定，作出了石油烃浓度的时空分布变化并结合水温、盐度、悬浮物、无机氮等相关因子作了分析。结果表明其全年的平均浓度为12.15 g/L^{-1}，以2006年9月的为最高，达193.00 g/L^{-1}，以2007年2月的为最低，仅1.42 g/L^{-1}，季节变化显示为夏＞秋＞冬＞春之势。测点间显示为东北部和西南部出现高浓度烃的几率大致相似，而湾口部易出现低浓度烃。湾中部其平面分布趋势不明确。与4个参数间的相关分析中仅烃与水温间略呈显著正相关关系。认为该湾污油含量分布状态多变，乃受制于多因素。

关键词　胶州湾　表层水　石油烃　时空分布　评价

1. 引言

世界海洋经受石油烃污染已有相当长的历史了，港湾、浅海及河口等遭受其害尤为明显[1-3]。胶州湾是中国乃至世界的知名中小型港湾之一，但它是个综合性的、半封闭浅水式、水交换并不很畅通却集港口、旅游、生产、养殖和接纳/消化污物于一身的水体，负担很重。对胶州湾性状、结构与功能等的状况与评价一直引人注意，并做了大量调查与研究，包括对其石油烃污染的调研[3-6]。但海湾污染是个动态的过程，人们必须经常关注之。石油烃类一般轻于水，表层水的石油烃污染状况受环境生态因子的影响大，并相互作用，因而变化也大。本文论述于2006年3月～2007年5月间，对该湾表层水石油烃污染状况所作的调查研究结果，以期有助于进一步认识、评价和推动治污工作。

2. 材料与方法

2.1　调查测点

于2006年3月至2007年5月按月在东经120°750～120°3403和北纬36°0150～36°1746范围内的海区，根据历史上所掌握的地质、地理、水文和污染状况，布设了11测点及6个参考测点，布点总体呈现为湾内沿岸和东北部密集于湾外、中、南和湾口部，详见图1。每月出海调查一次。用预先洁净了的有机玻璃采水器采集表层水样。采样和分析严格按国家《海洋监测规范——海水分析》(GB17378.4-1998)和《荧光分光光度法》要求执行。同时同步观测相关的水化学因子和水文数据资料。

* 本人为第二作者。

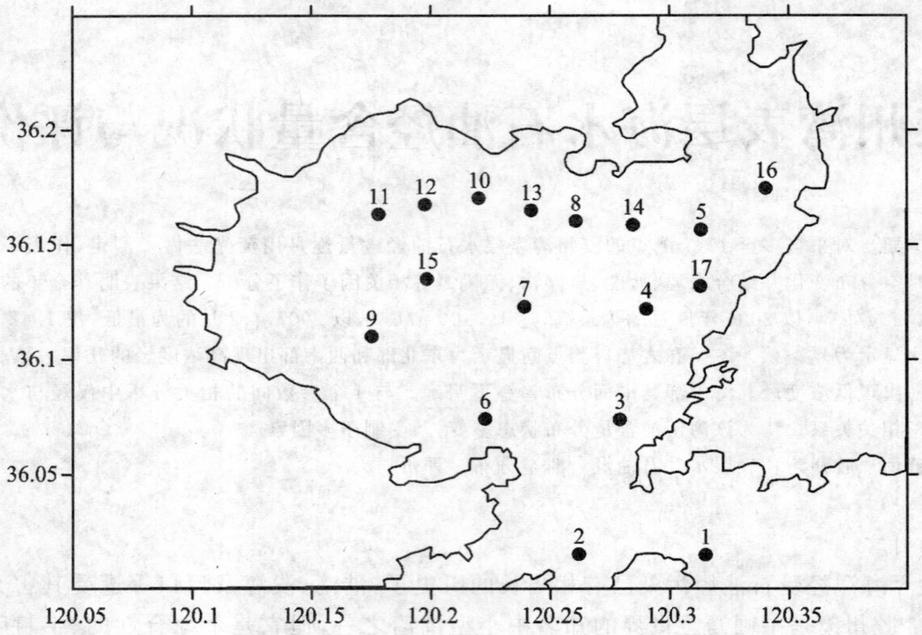

图 1 胶州湾表层水石油污染状况调查测点图(2006.3～2007.5)

Fig. 1 Map of Sampling stations for condition of petroleum pollution in surface layer seawater in the Jiao zhou Bay (2006.3～2007.5)

2.2 实验室分析

将严格保存于调查船上的水样当日取回陆上实验室,即刻用荧光法测定,其检出限为 2 g/L^{-1}。统计有关数据,分析研究和评价之。

3. 结果与讨论

3.1 胶州湾表层石油烃浓度概况

表 1 列出的是调查海区内一周年逐月对 11 个测点表层海水石油烃浓度的所有测定的统计结果。

表 1 胶州湾表层海水石油烃含量(g/L^{-1})及相关指标表

Table 1 Petroleum hydrocarbon contents and relative indexes in surface layer of seawater in the Jiaozhou Bay

项目	2006 年										2007 年	
	3 月	4 月	5 月	6 月	7 月	8 月	9 月	10 月	11 月	12 月	1 月	2 月
石油烃	4.55	6.93	12.04	13.92	19.47	20.50	21.08	2.06	18.01	7.59	18.27	2.00
水温(℃)	6.82	9.64	15.12	22.85	23.93	24.02	21.10	22.02	14.75	8.25	3.74	4.34
盐度	30.93	30.94	30.85	30.74	28.04	30.68	27.40	30.58	29.25	28.94	29.85	29.45
悬浮物	8.44	8.27	17.02	8.27	8.22	8.22	18.31	6.88	15.25	31.78	9.30	15.81
无机氮	275.47	128.49	217.95	172.63	128.52	85.96	284.53	290.19	287.84	131.97	749.51	173.35

由该表可知,一共在 11 个测点上于 2006 年 3 月～2007 年 2 月间至少做了 12 个月的

每月一次测定，发现所有测点测次均有油污物。对该表初步统计得出胶州湾全年表层水污油浓度平均值为 12.15 g/L^{-1}，其波幅是 2006 年 9 月为最高浓度 21.08 g/L^{-1}，最小浓度为 2007 年 2 月的 1.42 g/L^{-1}，两者相差 14.8 倍。测点间浓度波幅为 2007 年 1 月份的 No.1 测点的 1.00 g/L^{-1}～2007 年 2 月份 No.9 测点的 193.00 g/L^{-1}，极值达 192.00 g/L^{-1}，两者相差达 193 倍。这表明该湾污油浓度比 1997～2004 年的高，其波动较大[3-4]。比渤海湾的也高和大[7]。但比湄洲湾的低和小[8]。按近岸水域石油污染标准看，本结果表明胶州湾水体油污浓度处于较高状态。

3.2　胶州湾表层水石油烃浓度的时间差异/变化

图 2 画出的是胶州湾表水各测点每月平均石油烃浓度的周年变化状态，由该图可见，其波动并不规律。在全年间，该浓度基本上在月际间上下跳动。从 2006 年 3 月起一直趋高，于 2006 年 9 月现顶峰，很快次低谷立即出现于 2006 年 10 月，最低谷出现于 2007 年 2 月，峰值是低谷的 63.3 倍。

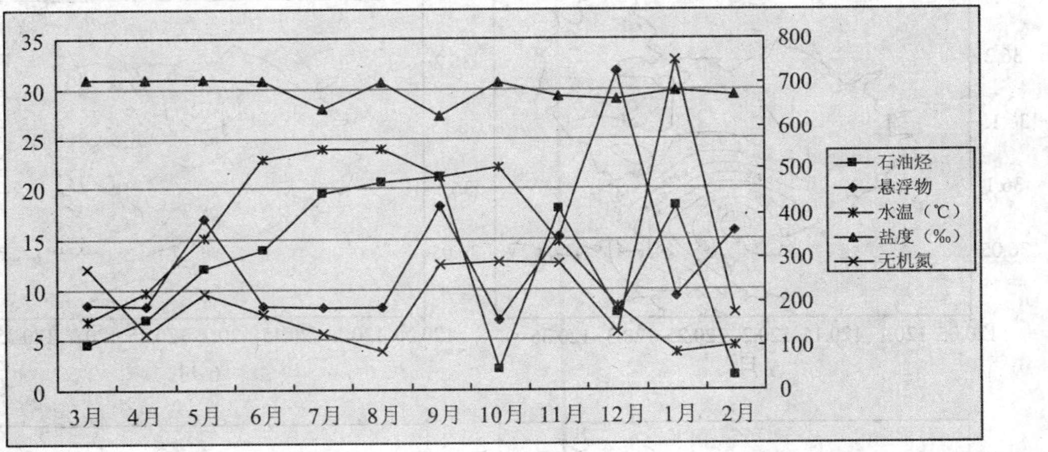

图 2　胶州湾石油烃浓度月际变化(2006.3～2007.2)

Fig. 2　The monthly changes of concentrations of Petroleum hydrocarbons in surface layer seawater in the Jiaozhou Bay (2006.3～2007.2)

青岛因地理环境而致气象(候)状况不同于内陆，其四季中以 12、1、2 月份为冬季，3、4、5 月份为春季，6、7、8 月份为夏季，9、10、11 月份为秋季，而胶州湾海水则比陆上气温变化慢且迟，因此仍可以陆上气温的四季变化借鉴来分别海水温度的四季变化。由此得出，该湾污油浓度的四季变化高低排序如下：夏(17.96 g/L^{-1})＞秋(13.72 g/L^{-1})＞冬(9.21 g/L^{-1})＞春(7.84 g/L^{-1})。此变化状态与表层水温四季变化不完全一致，和悬浮物含量四季变化趋势呈逆向变化(下述)。

由上可见，胶州湾表层水中污油浓度以夏季最高，春季最低，两者相差约 2.29 倍。这不完全相同于 1997～2004 年间所呈现出的季节变化势态[4]，与 1987 年的结果比则大相径庭[3]。

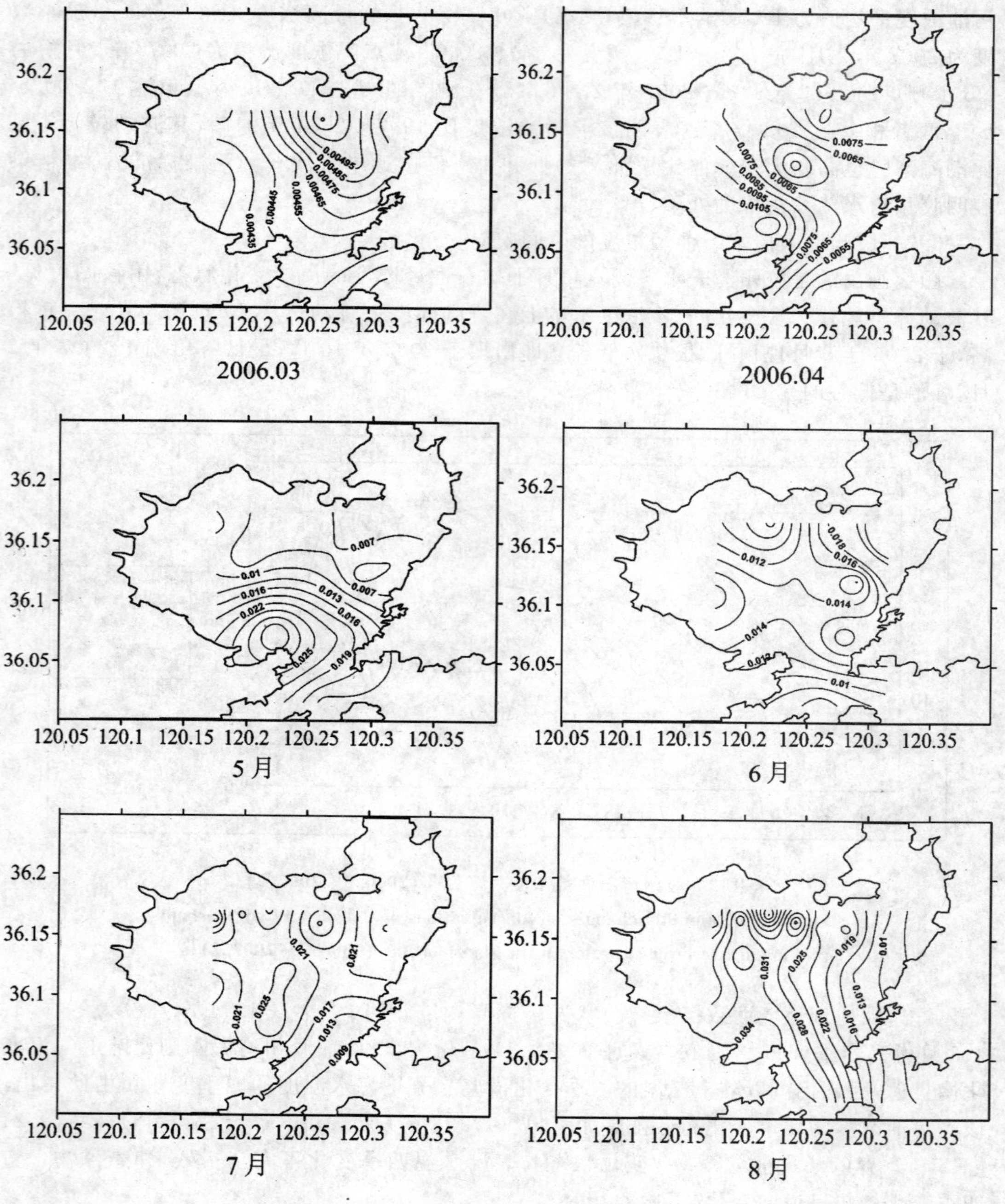

图 3 胶州湾表层水石油烃含量逐月波动态势图

Fig. 3 Movement forms of contents of petroleum hydrocarbon month by month in surface layer seawater in the Jiaozhou Bay

(续表 3)

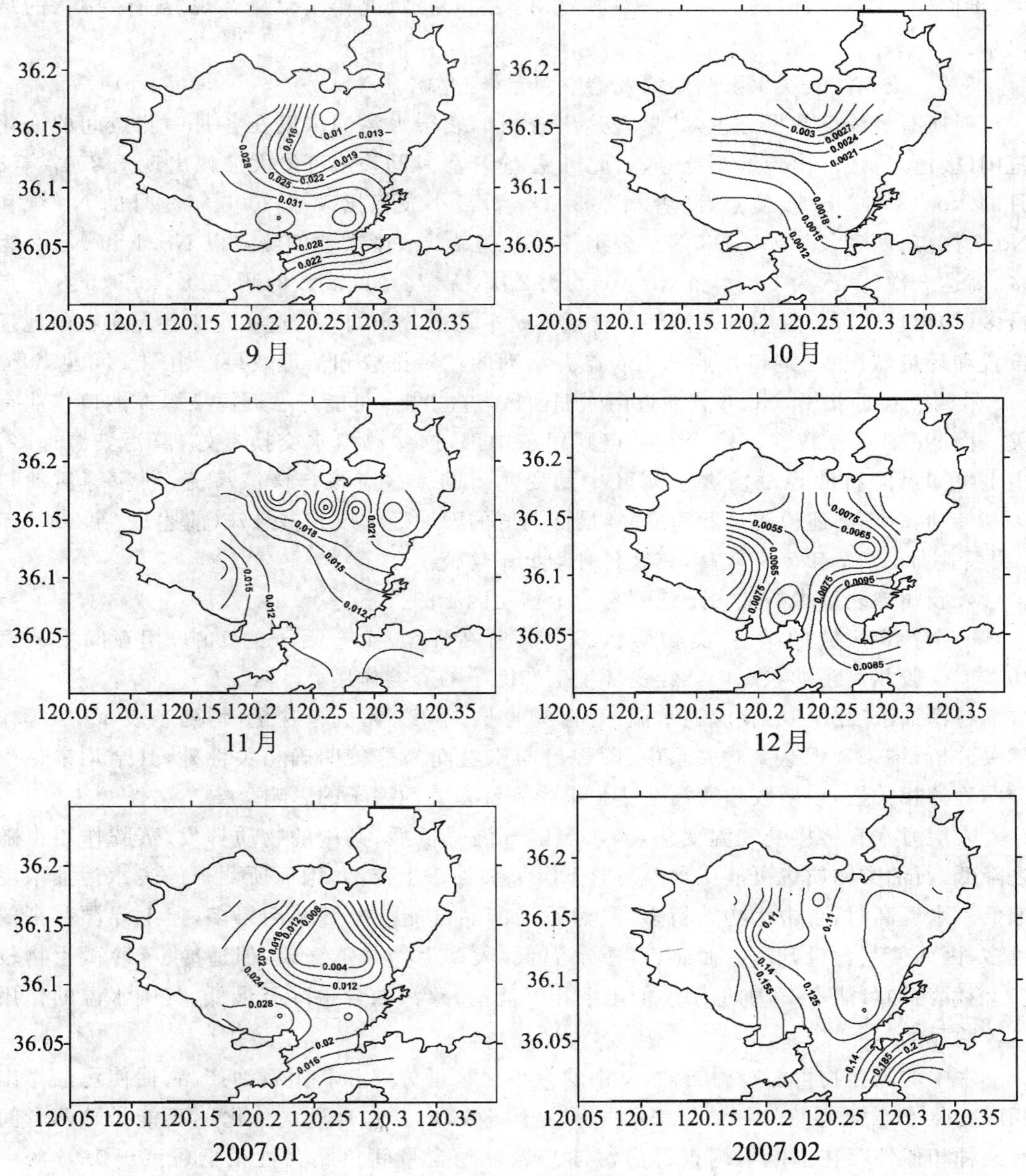

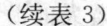

图 3 的 12 张图表达的是湾表层水石油烃浓度各月的运动分布状态。由此大体可见，各月状态相似者不多，仅 2006 年 3 月与 11 月其高低浓度出现站位相似，即高者均在东北端，低者均在西南部。但流向却相反，3 月份浓度较规律地由西向东递增，而 11 月却基本上是由南向北增。再看 2006 年 9 月与 10 月，两者正好发生了逆转。总体看，东北向或东北部出现高浓度机率与西南向或西南部出现高浓度机率相似，而湾口部低浓度机率较高。这些现象表明东北和西南部均可类似出现高浓度，口部则易出现低浓度区。湾中石油烃

浓度平面分布基本上看不出规则的月际分布状态。由此可推测，对一个特定的偌大的海区，难以凭一期有限时段的观测来对特定对象的分布状况作出十分客观、规律、中肯的结论来。

3.3　胶州湾表层水石油烃浓度的空间差异/变化

对所有烃浓度数据分析，结果表达出胶州湾表层水污油浓度在各月中测点间的变化范围（极值）及极值出现的测点号，由此可见，全年各月间该浓度最大波幅出现于2007年5月的No.13与No.15测点，两者相差5.34倍，最小波幅出现于2006年3月的No.1和No.8测点，两者仅相差1.28倍。全年各月出现最低浓度次数的测点以No.1和No.2为多，高达3次，其次为No.4和No.16，均为2次。最高浓度次数的测点以No.5、6、8三点最多，均出现3次。说明胶州湾污油含量不是平均地分布的[3-4]，这决定于各点所在的地理位置和环境状况、受港口作业和工农旅游活动的油污排放和治理效果。由图1可见，No.5、8号测点在近北岸，No.6点在近南岸偏西区，这些测点可能与靠近污染源或港口作业有关。而No.1、2点在湾口外，No.4点则偏东中部，这些测点水交换良好，离污染源远些。由上可知该湾石油烃浓度在全测区中是变化着的。这应该与该湾海流、潮汐等活动相关[5-6]。同时表2还说明月际间最高/最低均值的石油烃浓度相差悬殊（前述）。除2007年2月特高外，2007年5月较高，其余各月均低于均值。

4. 胶州湾表层水石油污染浓度与若干理化因子的关系

一个特定区域的某种/类物质状态受其所在环境/生态因子的影响，相互间关系密切[1-7,9]。胶州湾污油状态同样经受环境生态因子的影响作用。

将污油浓度分别与相应的水温、悬浮物含量、盐度、无机氮含量（均列于表1中）等作相关分析，结果发现这4对关系中，除烃量与水温间尚显较明确相关性外，其余均未见十分明确的相关性，说明胶州湾污油变化更受到许多其他因子的影响。

表层海水直接接受气温变化，当水温随气温降低后，其中的物质挥发/蒸腾作用也随之降低，石油类物质也可能比水温高时下沉势头强于上浮力，因而所测得的表水污油浓度偏低。水温高时，水中生物繁衍多，虽然石油产品可能被分解利用得多，但上浮漂移来的也多，因而使其含量升高。而冬、春季水温低，大型生物并不繁荣，但适冷、嗜冷微生物较多，消解污油可能多[1]，加上水层的上下相互流动，致污油含量降至最低，其中水温的作用是较明显的（$r=0.480, n=12, >R_{0.10}, <R_{0.05}$）。

水中的悬浮物可能吸附石油，其中的微小生物可分解和利用石油产品，促使污油作出相应变化[10]。悬浮物和污油虽来源并不完全一致，成分和含量也变化*，不能排除两者间有某种相似性存在。但本调查得出污油与悬浮物含量间负相关性很低（$r=-0.07, n=12$），缘由当待探讨。

海水盐度会影响污油的存在状态，进而影响其被分解利用和归宿。该湾表层水盐度的季节分布*，结果显示如下状态，秋(31.07)＞冬(31.00)＞春(30.92)＞夏(30.32)，这一状态与污油量的状态证实两者间有正相关关系，但不甚密切（$r=0.387, n=12, >R_{0.25}, <R_{0.10}$）。

*　待发表资料

石油及其制品在海水中经过一段时间的物理、化学和生物作用后，将不断降解、分解为低分子物，其中一部分将变成无机形式。石油烃含量与无机氮含量间将在海水中可能形成某种消长关系[1,4,6]。本调查对所得亚硝酸铵、硝酸铵和其他铵氮三者的总量进行统计，得出它们的季节变化状态*。无机氮总量的季节变化大小排序如下：冬(504.00 g/L^{-1})>秋(305.00 g/L^{-1})>夏(176.00 g/L^{-1})>春(170.00 g/L^{-1})，这一次序与污油的并不一致，这意味着两者间正相关性不很强($r=0.183$，$n=12$)，消涨关系并不十分对应。冬季水中两者浓度均偏高，都可能与水温低相关。此时生物已过茂盛期，并大量死亡，微生物等分解释放出的氮被利用的也少了，但污油仍被降解不少。夏季水生物茂盛，直至秋季开始衰落，加上陆上径流于秋季递减，均使油污减少。夏季水中无机氮减至最低，与植物利用氮的机率和能力增大有关，也与阳光的光化学和海水的蒸发/蒸腾作用增强相关。

5. 结语

本文于2006年3月～2007年5月对胶州湾表层水污油浓度作了测定，结果表明全测区污油检测率达100%，污油平均浓度约为12.15 g/L^{-1}，波幅为1.00～193.00 g/L^{-1}，与以往比显得浓度高些，波幅大些。污油浓度表达出一定的时空分布变化：季节上表达出夏秋高冬春低，与陆源污染、港口作业较近之区，浓度偏高。另一空间变化则表达为湾中部、湾外及附近海区浓度偏低之势。分析了与水温、悬浮物、盐度及无机氮四者间的关系，总体看污油浓度受水温、盐度的影响稍大，而受无机氮含量等的影响小。由此推测，其含量首先应决定于工农娱乐旅游活动和港口作业的油污排放和治理状态。污油流动还决定于湾流潮汐日光和风力动向。胶州湾的生态系统对油的消解和转化也一直起着作用[1,4]。

胶州湾污油浓度的变化是个动态的过程，它随机地受所在环境—生态因子，包括海洋流体动力的重大影响和作用，此一时的状态并不一定与彼一时的结果相吻合，但长时持久广泛地对它作出精确的检测将对石油污染的动向和控制产生重要和确切的指导作用。

参考文献11篇(略)

(合作者：孙丕喜　战闰等)

PETROLEUM HYDROCARBON CONTENT IN SURFACIAL SAMPLES OF SEAWATER FROM THE JIAOZHOU BAY AND APPRAISAL ON IT

(ABSTRACT)

Abstract Petroleum hydrocarbon content in surfacial seawater sampled from 17 stations in the Jiao zhou Bay was determined using fluorescent method month by month (2006. 3～2007. 5).

Their change condition of distribution of time and space was given, and the relationships between it and temperature, salinity, suspended solides, and inorganic nitrogen etc. in the water were analyzed respectively. The result obtained indicated that: the average concentraction of petroleum hydrocarbon was 12. 15 g/L^{-1}, its highest one, i. e. 193. 00 g/L^{-1} was happened in September, 2006, the lowest one was happened in February, 2007. Their season change showed that: summer＞autumn＞winter＞spring. The distribution among sampling stations indicated that: occurrence chance of high conceutration in east-north part was similar to part of west-south in the Bay, and their lower concentrations were occurenced in the mouth area of the Bay, and their trend of plane distribution was not very clear in central part of the Bay. There was a rather obvious positive relationship only between it and water temperature in the correlation analysis between 4 parameters and petroleum hydrocarbon concentration respectively. The paper considers that the content and its distributional condition in the Bay are changing and controlled by many factors.

Key words Jiaozhou Bay, Surficial Seawater, Petroleum Hydrocarbon, Time-Space Distribution, Appraisal

B₂ 海水养殖病害防治

(PREVENTION AND TREATMENT ON DISEASES AND EVILS OF MARICULTULRES)

越冬亲虾及其环境的细菌学研究*

摘　要　通过对亲虾越冬系统的外蓄水池水、预热池水、虾池水、亲虾用饵料、虾体及其病患作细菌分析,发现从外蓄水至虾体这五个环节存有细菌等微生物富集过程。稳定的环境、过多的营养、残饵、污物、病虾等是亲虾池细菌富集的要因;不健康亲虾富集并扩散病原微生物。对虾病病因、种类、亲虾成活率等的分析,发现至少31%的疾病与细菌等微生物有关。亲虾成活率与环境微生物一些参数,如几丁质降解菌含量、C/H%和V/H%间有很好的负相关。结果表明,监测亲虾越冬期各类细菌消涨势态,防治病害等是提高亲虾质量数量乃至育苗成功不可少的工作。

关键词　越冬　亲虾　环境　细菌　富集

一、引言

亲虾是人工育苗最基本条件之一。亲虾安然越冬的三大因子是健壮的体质、优良的饵料、正常的环境。后两因子则是保障健康的要素,而且任何时候都经受着微生物的作用。因此做好亲虾及其环境的微生物监测十分重要。

中国开展对虾人工越冬已多年,但就越冬亲虾及其环境微生物作研究,迄今未见全面报道。

笔者于1988/1989年冬春季节,对寿光县水产养殖公司越冬亲虾及其环境的细菌状况作了调研,旨在掌握越冬期细菌等的一些活动规律。本文提供这一研究的有关资料。

二、材料和方法

(一)培养基

1.海洋异养细菌培养基:葡萄糖2 g;蛋白胨5 g;酵母膏1 g;蒸馏水、陈海水各500 cm^3;pH 7.7。

2.弧菌培养基(TCBS)[5]。

---

*　原文刊于《黄渤海海洋》,1993,11(2):38-46。

3. 几丁质降解菌培养基[4]。

(二)取样和实验

每隔 10 d 取亲虾越冬池、室内沉淀预热池或室外蓄水池等的水样、虾样或饵料样,按 MPN 法接种液体培养基。于 25℃恒温培养 5 d 后,计数分析。

以上各步均严格按微生物学操作规程进行。

(三)计数和分析

经培养的水样、对虾血液等均以单位体积(cm^3),对虾甲壳以单位面积(cm^2),内脏和饵料以单位湿重(g)计算其细菌数量(即 Colony forming Unit,缩写为 CFU)。总异养菌、弧菌、几丁质降解菌的含量分别以 H,V,C 表示。V/H%和 C/H%则表示 V 和 C 各占 H 的比率(下同)。比较分析不同功能、类别、时间和状态样品中的细菌数量状况,寻找细菌与环境因子、亲虾成活率间等的某些关系。

三、结果和讨论

(一)外蓄水池水的细菌

表 1 示出了亲虾越冬期间外蓄水池中三类细菌含量、水温等的变化数据。

由表 1 可见,H 变化较平稳。冬末,水温始回升,细菌数量变得较多。水温与 H 间经分析有较明确正相关($r=0.646$,$>r_{0.05}$)。这表明水温是控制室外蓄水池微生物含量变动的要因之一。

表 1　冬季室外蓄水池水细菌含量变化表

H	V	V/H(%)	C	C/H(%)	T(℃)	d
4.6×10^3	2.3×10^1	0.5	6.0×10^0	0.13	1.8	1988.12.19
6.0×10^2	9.3×10^1	15.5	2.3×10^1	3.83	2.0	1988.12.29
4.3×10^3	2.3×10^0	0.05	6.0×10^0	0.14	2.0	1989.1.9
6.0×10^4	6.0×10^2	1.0	6.0×10^0	0.01	0.0	1989.1.25
9.1×10^3	6.0×10^2	6.59	7.3×10^0	0.08	3.7	1989.2.15
9.3×10^3	6.0×10^1	0.645	6.0×10^0	0.645	3.7	1989.2.25
3.6×10^2	6.0×10^1	16.67	/	/	3.0	1989.3.8
1.6×10^5	6.0×2	0.375	6.0×10^0	0.0038	7.5	1989.3.18

注:H 为总异养菌细胞数/cm^3;V 为弧菌细胞数/cm^3;C 为几丁质降解菌细胞数/cm^3;d 为取样日期(下同)。

V/H%平均为 5.17%,最高者是最低者的 333 倍多;C/H%平均为 0.608%,最高者是最低者的 1 000 多倍;V 比 C 大两个量级。

(二)室内沉淀预热池水的细菌

亲虾分别饲养在东、北、太阳能三个单元。每单元各有相应室内沉淀预热池以澄清水体,增高水温。北单元和东单元预热池水的细菌测定结果列在表 2 之中。由该表可见,对东单元两个池子,北单元 3 个池共作 10 次测定。平均看,东单元的 H 是 4.82×10^6,V 是

2.21×10^3,C 是 2.25×10^1;$V/H\%$是 0.046,$C/H\%$是 0.000 47。北单元的 H,V,C 分别是 6.97×10^4,2.56×10^2,6×10^0;$V/H\%$是 0.368,C/H 是 0.000 086。表明北单元三类细菌含量均比东单元的低,如 H 低两个量级,V 和 C 都低一个量级,表明北单元预热池水中细菌含量小。相反 $V/H\%$和 $C/H\%$均以北单元的高,表明北单元比东单元含有更高比率的 V 和 C,这将会影响到亲虾池的细菌状况。

将表 1 和表 2 比较、计算,可见北单元预热池水三类细菌含量与外蓄水池的处于同一量级;东单元的 H 比外蓄水池的高两个量级;V 与 C 则均比外蓄水池的高一个量级。

表 2　室内沉淀预热池水细菌含量状况表

单元	池号	H	V	C	d
东	15	3.6×10^4	9.1×10^2	6.0×10^0	1989.1.25
	22	2.4×10^7	9.1×10^3	7.5×10^1	1989.2.15
	22	3.3×10^3	9.3×10^2	3.0×10^0	1989.2.25
	22	2.1×10^4	7.3×10^1	/	1989.3.8
	22	4.26×10^4	6.0×10^1	6.0×10^0	1989.3.18
	平均	4.82×10^6	2.21×10^3	2.25×10^1	/
北	1	4.4×10^4	9.1×10^0	6.0×10^0	1988.12.19
	8	7.5×10^4	2.9×10^1	6.0×10^0	1988.12.29
	5	6.0×10^4	6.0×10^2	6.0×10^0	1989.1.25
	8	9.3×10^3	4.3×10^1	/	1989.2.15
	5	1.6×10^5	6.0×10^2	6.0×10^0	1989.3.15
	平均	6.97×10^4	2.56×10^2	6.0×10^0	/
东+北	总平均	2.445×10^5	1.235×10^3	1.425×10^1	/

从外蓄水池进入室内预热池,是细菌的一级富集,这主要与增温有关。预热池水温比外蓄水池的一般高 4℃～13℃,加上相对稳定的水体,为细菌等的繁衍提供了可能。

(三)亲虾池水中细菌

对北单元 17 个亲虾池、东单元 8 个池、太阳能池单元的 3 个池子水的细菌含量分别作的 23 次、18 次和 4 次测定的结果列于表 3。

三单元相比,东单元的 H 和 V 均最高;北单元的最低;三单元的 C 基本持平。$V/H\%$是太阳池单元的最低;$C/H\%$则是东单元的最低。北单元有最小总异养菌含量,但 $V/H\%$却高;东单元细菌含量大,但 $V/H\%$却较小;太阳池的基本处于中间状态。由上可见,亲虾池与室内预热池相比,前者的细菌含量更进一步富集。

表 3 还说明,东 23 池的表、底层水细菌含量差别不大,暗示池内充氧尚可。唯底层水的 V 稍大,可能与亲虾栖息底部者多、有一些排泄物和残饵等有关。经分析,H 不仅与溶氧间相关性不大,与 pH 值、水温间的相关性也不大(仅据所测虾池数据,未发表资料)。这表明这些因子在亲虾池中相当稳定。不难看出,促使细菌等微生物在环境稳定的亲虾池中富集的原因可能还与其营养、虾只健康状况有关。

表 3　越冬亲虾池水的细菌含量状况表

单元池号		*H*	*V*	*C*	*d*
东	7	3.6×10^{4}	2.0×10^{2}	2.4×10^{4}	1988.12.29
	4	2.4×10^{7}	2.1×10^{4}	6.0×10^{2}	1988.12.9
	2	2.4×10^{7}	4.6×10^{4}	9.4×10^{2}	1988.12.9
	7	2.4×10^{7}	2.4×10^{5}	6.0×10^{0}	1988.12.16
	3	1.5×10^{7}	1.6×10^{3}	3.6×10^{0}	1988.12
	16	2.4×10^{8}	1.1×10^{5}	6.0×10^{0}	1988.12
	23	3.3×10^{6}	1.1×10^{4}	6.0×10^{0}	1988.12
	24	2.4×10^{9}	2.1×10^{4}	6.0×10^{0}	1988.12
	26	1.1×10^{7}	9.3×10^{4}	6.0×10^{0}	1988.12
	1	2.7×10^{6}	2.4×10^{5}	6.0×10^{0}	1989.1.25
	5	3.3×10^{6}	2.4×10^{4}	6.0×10^{0}	1989.1.25
	9	4.4×10^{6}	1.1×10^{5}	6.0×10^{0}	1989.1.25
	10	3.5×10^{7}	1.5×10^{4}	6.0×10^{0}	1989.1.25
	11	2.1×10^{8}	4.3×10^{4}	6.0×10^{0}	1989.1.25
	12	6.0×10^{6}	1.5×10^{4}	6.0×10^{0}	1989.1.25
	14	6.0×10^{5}	3.5×10^{3}		1989.2.15
	18	1.1×10^{5}	4.3×10^{3}		1989.2.15
	21	1.5×10^{5}	3.6×10^{3}	2.0×10^{1}	1989.2.15
	23	2.4×10^{8}	3.6×10^{3}		1989.3.8
	24	3.3×10^{6}	3.0×10^{2}		1989.3.8
	5	4.8×10^{5}	6.0×10^{0}	6.0×10^{3}	1989.3.18
	10		5.0×10^{2}	5.0×10^{2}	1989.3.18
	23	4.0×10^{7}	6.0×10^{3}	6.0×10^{0}	1989.3.18
	平均	1.595×10^{8}	4.4×10^{4}	1.691×10^{3}	
北	1	1.1×10^{7}	0.91×10^{0}	0.6×10^{0}	1988.12.19
	4	2.4×10^{7}	2.0×10^{3}	2.4×10^{4}	1988.12.29
	8	3.6×10^{4}	1.5×10^{3}	2.4×10^{4}	1988.12.29
	2	3.6×10^{5}	6.0×10^{2}	6.0×10^{2}	1989.1.9
	5	6.4×10^{5}	6.0×10^{2}	6.0×10^{2}	1989.1.9
	7	2.4×10^{5}	3.6×10^{2}	6.0×10^{2}	1989.1.9
	3	6.0×10^{5}	6.0×10^{2}	6.0×10^{0}	1989.1.16
	6	2.0×10^{6}	9.6×10^{2}	6.0×10^{0}	1989.1.16
	5	6.0×10^{5}	3.6×10^{2}	6.0×10^{0}	1989.1.25
	5	2.4×10^{5}	2.4×10^{3}		1989.2.15
	6	1.1×10^{7}	1.5×10^{4}	1.1×10^{1}	1989.2.15
	7	4.6×10^{6}	4.3×10^{3}	2.8×10^{1}	1989.2.15
	8	1.5×10^{6}	3.5×10^{3}	4.3×10^{1}	1989.3.9
	3	9.3×10^{4}	3.6×10^{2}		1989.3.9
	4	1.5×10^{5}	3.6×10^{2}		1989.3.9
	5	1.6×10^{5}	6.0×10^{2}	6.0×10^{0}	1989.3.18
	7	2.7×10^{5}	9.1×10^{2}	6.0×10^{0}	1989.3.18
	8	1.3×10^{6}	6.0×10^{2}		1989.3.18
	平均	4.78×10^{6}	1.95×10^{3}	3.57×10^{3}	

（续表）

单元池号		H	V	C	d
太	3	1.1×10^{7}	1.5×10^{3}	2.4×10^{4}	1988.12.29
	4	1.1×10^{7}	2.9×10^{3}	3.0×10^{2}	1989.1.9
	1	2.4×10^{8}	2.4×10^{4}	6.0×10^{0}	1989.1.16
	1	1.6×10^{5}	3.3×10^{4}	$2.7\times10^{1}1$	1989.3.18
	平均	6.55×10^{7}	1.54×10^{4}	6.08×10^{3}	
东＋北＋太	总平均	8.77×10^{7}	2.46×10^{4}	2.87×10^{3}	
东	23 表	2.4×10^{7}	4.6×10^{4}	2.1×10^{1}	1989.2.25
	底	2.4×10^{7}	1.1×10^{5}	2.0×10^{1}	

（四）饵料的细菌状况

仅以越冬期亲虾的主要饵料——沙蚕的两次细菌含量测定为例说明，结果如下：冰沙蚕及其所带化冰水：H 为 6.0×10^{5}/g，V 为 4.3×10^{2}/g，C 为 6.0×10^{3}/g；存放鲜沙蚕缸中的液体：H 为 $2.4\times10^{11}/cm^{3}$，V 为 $2.4\times10^{6}/cm^{3}$，C 为 $6\times10^{1}/cm^{3}$。

由此可见，活沙蚕存放室外缸中，因时间过长，沙蚕及其余水经历结冰—化冻过程而死亡、变质，致使大量微生物繁殖。投用这种沙蚕，不仅使虾池水质很快污染，而且成为繁衍微生物的良好场所，包括条件致病菌的得逞，因此必须很好清洁沙蚕。

（五）亲虾的细菌状况

1. 表 4 列举虾体不同部位、器官等的细菌含量

由该表可见，一般说，死虾甲壳、肠和血液的细菌含量、V/H%和 C/H%均比濒死虾和活虾的高，这说明较高比率的 V 和 C 等促成亲虾病变、死亡。富集此类细菌最多的亲虾池及变质饵料可能成为亲虾死亡率提高的重要起因。

表 4　越冬亲虾体不同部位细菌含量比较表

来源	部位	H	V	V/H(%)	C	C/H(%)
死虾	甲壳			0.1027		0.0102
	肠	2.9×10^{5}	1.21×10^{5}	4.17		
	血液	1.68×10^{7}	7.42×10^{4}	0.442	7.52×10^{2}	0.0045
濒死虾	肠	1.1×10^{7}	4.6×10^{3}	0.0418		
活虾	甲壳			0.085		
	肠	4.615×10^{6}				
	血液	1.255×10^{5}	2.765×10^{3}	0.220	2.1×10^{2}	0.0167

2. 不同方法处理的甲壳细菌含量比较

甲壳是亲虾机体不可分割的一部分，它与环境微生物接触最多及最早，因而其附着的细菌状况不仅与体内的病理状况有关，且表达出环境质量信息。

表 5 列举出不同处理的甲壳细菌数量状况，由该表可见，死虾甲壳的 H 是：未消毒＞消毒；活虾甲壳的 V 和 C 比死虾的各小一个量级。C 是未消毒（2.4×10^{2}）＞消毒（7.5×

10^1)>余水(3.6×10^1);濒死虾附有较多 H 和 V。这与甲壳质量较差等原因有关;余水中菌数多寡与甲壳附菌及被洗涤能力有关。

表 5　不同处理的甲壳细菌含量等的比较

虾体状况	未消毒			消毒			稀释甲壳余水		
	H	V	C	H	V	C	H	V	C
死	2.4×10^6	2.4×10^3	6.0×10^0						
	2.4×10^7	9.1×10^2	1.5×10^2	2.4×10^5	2.1×10^2	1.1×10^3	2.4×10^5	2.4×10^4	2.4×10^3
	2.4×10^5	6.0×10^1	1.5×10^2						
	3.6×10^4	7.5×10^1	2.4×10^3						
	4.4×10^4	2.4×10^4	6.0×10^0	2.4×10^6	2.4×10^4	2.7×10^2	1.5×10^4	2.4×10^4	2.0×10^1
平均	5.3×10^6	5.3×10^4	5.4×10^2	1.3×10^6	1.2×10^4	6.9×10^2	1.3×10^5	2.4×10^4	1.2×10^3
濒	2.4×10^3	9.3×10^0	2.3×10^0						
死	2.4×10^7	3.0×10^1	6.0×10^1						
	2.4×10^7	2.4×10^5	6.0×10^2						
平均	1.6×10^7	8.0×10^4	2.21×10^2						
活	2.4×10^7	1.1×10^4	9.1×10^1						
	4.6×10^2	4.3×10^2	3.6×10^1	2.4×10^5	2.0×10^1	7.5×10^1	1.1×10^5	9.1×10^1	3.6×10^1
	2.4×10^6	1.1×10^4	2.4×10^2						
平均	8.8×10^6	7.5×10^3	1.2×10^1						

注:未消毒甲壳,指取样后立即做细菌计数。消毒甲壳,指用蘸有消毒海水的药棉三遍擦拭的甲壳。稀释甲壳余水,指消毒海水洗涤甲壳三遍后的混合水。

由表 5 可得出下列数据,即死虾、濒死虾、活虾的未消毒甲壳上 $V/H\%$ 分别是 1,0.5 和 0.085;$C/H\%$ 分别是 0.01,0.001 3 和 0.000 1。这表明 $V/H\%$ 和 $C/H\%$ 均是死虾>濒死虾>活虾。

不同健康水平亲虾的甲壳细菌状况暗示其所在水体的环境质量,表明了水的质量直接影响甲壳微生物组成结构。洁净的水对甲壳微生物具稀释作用。相反,大量死虾及其甲壳又将恶化水质,进而对活虾带来后患。

(六)不同健康状况亲虾及其器官与外蓄水池等越冬设施的细菌数量比较

表 6 列举了亲虾及其相关的外蓄水池水的细菌含量等有关数据。由该表可见,虾的甲壳、肠、血液比外蓄水含有较多的 C,V,H;死虾对这三类菌有更高的富集能力。

综上所述,越冬设施直至亲虾本身,包括投入饵料,存在着细菌的富集现象。越接近虾体,尤其是体质越不佳的亲虾,对细菌等微生物数量的富集力越大,这启示我们,越冬系统每环节的微生物监测都不能忽视。

表 6 虾体富集细菌含量的比较表

虾体状况部位	虾的三类细菌含量			外蓄水池细菌含量			虾的细菌含量/外蓄水池细菌含量 %	取样时间
	H	V	C	H	V	C		
死虾甲壳			2.4×10^{3}			7.3×10^{0}	≈328.8	1989.2.15
活虾甲壳		4.3×10^{2}			6.0×10^{1}		≈7.2	1989.2.25
死虾肠		2.4×10^{5}			6.0×1		4.000	1989.2.25
死虾血液	2.4×10^{7}			1.6×10^{5}			150.0	1989.3.18
活虾血液		8.4×10^{3}			6×10^{2}		13.5	1989.3.18

注:甲壳单位以 cm^2 计,肠的单位以 g 计,外蓄水池水以 cm^3 计,因而该两数字只近似反映有关比数。

(七)细菌与亲虾病害、成活率的关系分析

共检查越冬病虾 90 尾。由症状、镜检及微生物培养等,大体将疾病归纳于表 7。该表说明,常见亲虾病约 13 种。从病原和病因学角度分,虾病的 31.1%直接与细菌相关;32.2%的疾病由创伤及其后继病菌感染引起。因此亲虾疾患受细菌等微生物影响很大[1];环境及饵料的不适,有效营养的缺乏,使虾体质减弱;其他生物在病菌的先导和协同作用下,使伤残虾体诱发多种疾病[2,3],一尾病虾往往染有多种疾病。因而亲虾病害的防治,要在兼顾多种病因的同时,重点考虑微生物病因。

细菌等微生物影响的重要结果之一是亲虾成活率,表 8 列举这一情况。将表 8 的数据分别作相关分析,发现亲虾成活率与 C,$C/H\%$,$V/H\%$间各有明确的负相关关系。r 值分别为$-0.997\ 5$,$-0.997\ 0$ 和 $0.892\ 0$。亲虾成活率最大的东 23 池有最少的 C,同时 $V/H\%$和 $C/H\%$也最低;H 与亲虾成活率间的正相关性不大($r=0.452\ 3$)。这些结果似乎表明虾池中正常的异养菌一般不至于成为亲虾成活率的制约因素,而弧菌和几丁质降解菌占的比率(即 $V/H\%$和 $C/H\%$)越大,对亲虾成活率的提高越不利;几丁质降解菌的含量与亲虾成活率高低特别相悖。因而,控制几丁质降解菌等条件致病菌病原的状况是提高亲虾成活率的必要措施。

表 7 越冬亲虾疾病种类分析表

编号	病死、病因	疾病名称	病虾数量(尾)	名类病虾数/病虾总数(%)
1	不良环境因素	肌肉坏死病	1	1.1
2	细菌等	弧菌病	16	17.8
3	细菌等	丝状菌病	12	13.3
4	微生物	黑斑病	6	6.7
5		黑鳃病	5	5.6
6	其他生物	褐斑病	16	17.8
7		白斑病	1	1.1
8	创伤	头胸甲鳃区坏死病	4	4.4
9		头胸甲内水疱病	2	2.2

(续表)

编号	病死、病因	疾病名称	病虾数量(尾)	名类病虾数/病虾总数(%)
10	等	寄生性纤毛虫病	8	8.9
11		固着性纤毛虫病	9	10.0
12		线虫病	9	10.0
13	营养缺乏	黑死病	1	1.1

表 8　四个新虾池亲虾成活率与一些细菌参数间的比较

池号	C	$C/H\%$	$V/H\%$	亲虾成活率%*
东 23	6×10^{0}(二次平均)	0.000 006	0.007	84.01(247/294)
北 5	2.04×10^{2}(三次平均)	0.02	0.104	82.99(517/623)
北 7	2.1×10^{2}(三次平均)	0.001 8	0.016	82.91(485/585)
北 8	1.2×10^{4}(二次平均)	1.27	0.198	71.81(433/603)

*计算时间:从 1988 年 12 月 1 日算起,截止于 1989 年 3 月 1 日。

四、结　语

亲虾越冬系统是有关各部密切联系的整体。该系统中,特别是亲虾池,环境参数稳定,营养丰富杂质多,生物密度高生物量大。这些不仅不对几丁质降解菌、弧菌等条件致病菌构成威胁,反而有利于其富集,因而明显影响到亲虾成活率。为了提高亲虾成活率,必须调控好亲虾越冬各环节的细菌等组成结构,使细菌数量不致过大,条件致病菌所占比率不致过高。因而要在沉淀过的海水进入预热池前,抓住滤水这一环。投喂优质饵料。常换水、常充氧,及时吸污,以减少残毒含量和中毒可能。精心护理亲虾,减少机械创伤。必要时适当施药。保持环境、工具和操作人员清洁卫生,杜绝染病机会。这样,亲虾成活率有望进一步提高。

参考文献 5 篇(略)

(合作者:袁春艳　王　伟)

BACTERIOLOGICAL RESEARCH ON PARENT PENAEID SHRIMP (*PENAEUS SINENSIS*) AND ITS WATER ENVIRONMENT DURING OVERWINTERING PERIOD

(ABSTRACT)

Abstract In the system of wintering for parent penaeid shrimp (*Penaeus sinensis*), the bacteriological analyses on water stored in an outdoor pond, indoor-pre-warm pond water and culturing pond water, diet for parent shrimps, shrimp bodies and diseased shrimp were performed. It is considered that there is an enrichment process of bacteria in the five links from outdoor-stored water to shrimp bodies. Stable environment, plenty of nutrients, residal food, pollutants, diseased shrimps are important factors of bacterial enrichment in culturing ponds for parent shrimps. Parent shrimps, especially, unhealthy shrimps enrich and diffuse microbes further, including those pathogenic organisms. The relationships of the causes and kinds of disease survival rate of parent shrimps were analysed, It is found at least 31% of shrimp diseases was related to bacteria and other biopathogens. These diseases affected level of health of parent shrimp directly. There were better negative correlations between the survival rate of parent shrimps and some parameters of environmental microbes, for example, the content of chitin-degrading bacteria, $C/H\%$, $V/H\%$.

The results obtained indicate: monitoring the declining and growing situation of each kind of microbes, disease control and cure for parent shrimps are necessary work for raising quality and quantity of parent shrimps and even for successful seedling rearing of penaeid shrimps.

Key words Wintering Period Parent Shrimps Environment Bacteria Enrichment

中国海产养殖生物病原微生物研究进展*

摘要 此文论述近11年(1992～2002)来中国海产养殖生物的微生物学研究进展,除了病毒外,它记述中国主要海产养殖生物,即扇贝等软体动物,对虾等甲壳动物,鱼等脊索动物和海藻等的病原微生物研究状况、病害防治方法、光合细菌的应用及病原微生物学研究方法等。文章最后提出几条建议,以期推进中国海产养殖生物病原微生物学研究水平更上一层楼。

关键词 中国 海产养殖 病原微生物 光合细菌 养殖环境 弧菌 真菌

近十一年来(1992～2002),中国海产养殖生物的微生物学研究取得了长足进步。它作为微生物学家和水产学家共同关注的课题而得到了交叉发展和共同增长。对海洋微生物的研究除了海洋病毒外,主要的成果之一体现在海产细菌学、真菌学及养殖环境微生物生态学等的研究上。所主要研究的海产养殖生物包括扇贝等软体动物、虾蟹等甲壳动物、鱼等脊索动物及海带等藻类。本文就我国一些海产养殖生物的微生物学研究进展作一回顾,以期进一步总结过去,展望未来。

1 研究方法的进步和研究范围的深广化

1.1 研究方法和手段的进步

电子显微镜的应用逐渐普遍,这包括对皱纹盘鲍肠上皮组织立克次体、珍珠贝和大珠母贝类立克次体超微结构及组织病理学、对虾支原体、褐斑病及丝状细菌的观察、褐藻酸降解菌侵染海带过程的研究等等[1-9]。

相关手段包括PCR用于副溶血弧菌的检测、LPS抗血清作溶藻弧菌间接ELISA检测及双抗夹心ELISA研究皱纹盘鲍病、酶联免疫吸附试验、间接荧光抗体诊断、弧菌同工酶分析、16S rRNA基因片段克隆测定病原菌株、重组酵母对牙鲆非特异性免疫能力的研究及创伤弧菌疫苗对青石斑鱼的免疫学等等[10-18]。

1.2 研究范围的深广化

病原微生物的接触与入侵,必然引起养殖生物体内生理生化反应。如海带褐藻酸降解菌产酶条件及褐藻酸酶对褐藻细胞的解毒作用、海带超氧化物歧化酶(SOD)活性及其作用、栉孔扇贝血细胞抗菌力、弧菌病在被入侵者体内的病变部位,由此必然加深对病原菌的认识,如分离出真鲷弧菌病病原的外毒素、对副溶血弧菌胞外产物致病性分析及气单胞菌基因的研究,有助于更实质理解宿主体防病抗病能力及病原菌入侵机制[19-26]。

2 养殖对象的病原微生物及其防治

2.1 扇贝等软体动物的病原微生物

* 原文刊于《中国水产》2003年专刊——2003青岛·全国水产养殖研讨会论文集,60-66转45

表 1　软体动物病原微生物及其防治

动物名称	疾病名称	病原微生物	防治	备注（参考文献号）
栉孔扇贝	类支原体病、类衣原体病	类支原体、类衣原体	侵染部位广，数量多	27、28
海湾扇贝	消化盲囊、性腺，其他器官病变病	弧菌、飘浮弧菌	新霉素 B、磺胺药＋TMP、多黏菌素、辛洛西林	29、30
	幼虫面盘解体病	鳗弧菌	红霉素、SMZ	*
	飘浮细菌病	飘浮细菌	新霉素 B、磺胺药等	*
	立克次体样生物病	立克次体样生物		30
	衣原体样生物感染	衣原体样生物		31
	弧菌病	黑美人弧菌		32
皱纹盘鲍	肿瘤病	细菌		
	稚鲍脓疱病	河流弧菌－Ⅱ、创伤弧菌	免疫、噬菌体、复方新诺明	34～37
	幼鲍溃烂病	荧光假单胞杆菌Ⅳ型等细菌	土霉素、盐酸环丙沙星、ELISA 检测	38
	脓毒败血症	坎氏弧菌		39
杂色鲍	溃疡病	亮弧菌Ⅱ（*Vibro splendidus*-Ⅱ）、荧光假单胞菌		40
	肌肉萎缩症	副溶血弧菌		
鲍＊＊	暴发性流行病	溶藻弧菌、副溶血弧菌		39
马粪海胆	黑嘴病	坚强芽孢杆菌	对磺胺甲基异恶唑、复方新诺明、头孢噻肟	41
虾夷球海胆	斑点病	噬纤维菌科的 G^- 长杆菌		日本报导
珍珠贝　合浦珠母贝	弧菌病	溶藻弧菌		42
珍珠贝　大珠母贝	类立克次体病	类立克次体		43～45
青蛤	弧菌病	弧菌		46

＊见管华诗主编，1999，海水养殖动物的免疫、细胞培养和病害研究（M），山东科学技术出版社，济南；P130-149。

＊＊鲍还有 6 种其他疾病。

表 1 列举九种软体动物相关病原微生物引起的约 19 种疾病。由该表可见，软体动物类病原之一类主要是细菌，并以弧菌为主。此外对蛏子也作过病害研究[47]。

2.2 虾等甲壳动物的病原微生物

表 2 主要列举了中国对虾细菌病及其防治措施。由该表可见,中国对虾约二十九种疾病中以细菌引起的为主,涉及的细菌以弧菌为主,还有微球菌、普通变形杆菌、肠杆菌、真菌镰刀菌、类支原体等 17 属/类微生物。

表 2 中国对虾细菌病等种类及防治措施

疾病名称	病原	防治措施	备注 (参考文献号)
败血症	海弧菌、溶藻弧菌、产气弧菌	齐复康Ⅱ号	48
褐斑病 黑鳃病 } 败血病	甲壳腐蚀细菌 a) 丝状细菌引起,镰刀菌、其他细菌参与	土霉素、磺胺异恶唑、甲醛保持良好水质,合理投饵	49～50
烂鳃病	可能是弧菌、假单胞菌或气单胞菌	每天换水一次,沸石粉、光合细菌	b)
白斑病	杆状病毒与弧菌、白斑综合征病毒	土霉毒	5l～58
暴发性流行病	弧菌,微球菌、病毒	弧菌制剂	59
仔虾糠虾肠道堵塞肠道杆菌病	肠杆菌	清洁海水、非源蜕皮激素	60
肠结节病	支原体	/	61
红腿/红肢病	镰刀菌、尖孢镰孢菌,溶藻弧菌、副溶血弧菌、坎贝氏弧菌、哈氏弧菌、需钠弧菌、弗尼斯弧菌、鳗弧菌、非 01 群霍乱弧菌、普通变形杆菌	FTL、IPS、口服免疫型药,乌梅、石榴、五味子、5,7 一二-8-羟基喹啉蛭弧菌	62～70
黑斑病	白斑病后期	土霉素、	69
幼体、仔虾弧菌病	麦氏弧菌、哈维弧菌、弯曲弧菌、霍利斯弧菌、产气弧菌	四环素、呋喃唑酮、磺胺(SMZ)	70～72
幼体弧菌病	鳗弧菌、副溶血弧菌、溶藻弧菌、气单胞菌	土霉素	b)
幼体肠道细菌病	不动细菌	土霉素、链霉素	b)
荧光病	弧菌		
甲壳附肢溃疡病	细菌		
肠炎	细菌		
弧菌病	产气弧菌	季铵盐类(BKC)	74
红体病		降密度、迟放苗	

（续表）

疾病名称	病原	防治措施	备注（参考文献号）
肝胰腺虾死病	弧菌、病毒、类支原体	恶喹酸	73
烂眼病	弧菌 903、非 01 群霍乱弧菌	4×10^{-6}漂白粉、虾康一、二号	75～76
肌肉白浊病（肌肉坏死）	由烂眼病等引起或高温低溶氧	4×10^{-6}漂白粉、虾康一、二号，改善水质	77～78
黄鳃病	黑鳃病初期	清除残物，呋喃唑酮、甲醛	
丝状细菌病	发状白丝菌＋发硫菌＋聚缩虫	保持良好水质、合理投饵	6
幼体病	哈维氏弧菌	齐复康Ⅱ号，有机碘，喹诺酮	79～80
细菌病	细菌	次氯酸钠，二氯异氰尿酸钠	78
气单胞菌病	嗜水气单胞菌、豚鼠气单胞菌、索布雷气单胞菌	氯霉素、诺氟沙星、呋喃唑酮、复方新诺明	b)
霉菌病	链壶菌	氟乐灵、制霉菌素	
鳃组织病	细菌、真菌及藻类太多	及时控水	81～82
镰刀菌病	镰刀菌 c)	三氯异氰尿酸钠、洁藻灵、制霉菌	83～84
呼吸破坏	黑鳃病引起		84

a)包括弧菌、假单胞菌、气单胞菌、粘细菌和贝内克菌等；

b)见管华诗主编，1999，海水养殖动物的免疫、细胞培养和病害研究，山东科学技术出版社，P92-102；

c)指腐皮镰刀菌、尖孢镰刀菌、禾谷镰刀菌、三线镰刀菌、茄类镰刀菌（由环境、水质、饵料、动植物引起的疾病在此不一一列举）。

表 3 则列举了斑节对虾等五种对虾的细菌病及防治对策。表 3 表明斑节对虾等五种对虾的约二十二种疾病病原主要是弧菌、气单胞菌、极毛杆菌和假单胞菌等。

此外，锯缘青蟹有非液化摩拉氏菌引起的黄体病[95]及细菌性感染病[96]和远海梭子蟹[97]及三疣梭子蟹的疾患。对贻贝、毛蚶、菲律宾蛤仔的抗菌能力也作过研究[98]。

表 3　斑节对虾等 5 种对虾的细菌性疾病及防治法

动物名称	疾病名称	病原体	防治法	备注（参考文献号）
斑节对虾	受精卵，无节幼体发育受阻	8 种弧菌，2 种未定弧菌		85
	发光病	亮弧菌、荧光假单胞菌、哈维氏弧菌、发光杆菌	Lg/m^3 ClO_2，30 mL/m^3 甲醛	86～87
	红体病	嗜水气单胞菌、副溶血弧菌	多西环素、诺氟沙星、红霉素	88
	弧菌病	副溶血弧菌，溶藻弧菌等弧菌等	/	89

(续表)

动物名称	疾病名称	病原体	防治法	备注（参考文献号）
南美白对虾	发光病	同斑节对虾		86
	幼体真菌病	链壶菌	制霉菌素、氟乐灵，	90
	亲虾甲壳溃疡病	弧菌	福尔马林、抗生素、维生素	91
	弧菌病	副溶血弧菌、鳗弧菌		92
	红体病	未定		92
	黄腿、黄鳃、肌肉白浊病			92
	黑鳃、红须、红腿、红尾、断须、烂鳃、活力减、游动慢、摄食少			92
	红腿	副溶血细菌。鳗弧菌、气单胞菌		92
	肿鳃	细菌/支原体		92
	肠胃病	细菌		92
	坏死性肝胰脏炎	Protebacteria 的一种杆菌		92
	分枝杆菌病	瘰疬分枝杆菌、偶发分枝杆菌		92
	立克次体病	未定		92
	黄鳃病	细菌		蔡强等
	褐斑病	机械损伤致弧菌、气单胞菌、粘细菌入侵		蔡强等
日本对虾	荧光极毛杆菌病	荧光极毛杆菌		
长毛对虾	病毒病诱发细菌并发症	弧菌 PR-9301 和 9303	弧菌制剂	93
近缘新对虾	细菌病	假单胞菌		94

2.3　鱼等脊索动物的病原微生物

迄今国内沿海养殖的鱼有十余种，其病害也引人注目，表 4 列举的是养殖鱼等脊索动物主要的细菌病。周宗澄就鲻鱼、黄鳍鲷消化道细菌及其作用做了研究[131～132]。

此外，美国青蛙可患膨气病，其病原是嗜水气单胞菌和假单胞菌，庆大霉素、氯霉素及黄连、黄芩有杀抑作用[132]。

以上 2.1～2.3 说明许多海产动物的疾病由弧菌引起，证实弧菌是海洋养殖环境中的优势菌，具有高度多样性和适应性，可以成为条件致病菌，适当条件时，即成为流行病、暴发病一类元凶。

2.4　海藻的病原微生物

此处主要指海带、紫菜等的病害，就海带言，是病害研究得最多的一种海藻。海带病大体分夏苗及栽培期病两类。由细菌引起的病主要发生在苗期，包括绿烂病、脱苗烂苗病、幼体畸形病等。栽培期的细菌病主要有柄粗叶卷病、斑点腐烂病等。主要由褐藻酸降解菌及产 H_2S 细菌引起；主要的病原菌为溶胶弧菌[113]。对褐藻酸降解菌培养条件[19]、侵染藻体的过程[9]、褐藻酸酶形成条件[20]、褐藻酸裂解酶[134]及对藻体的影响[135]作了较深的观察或研究。藻体对该类菌的抗性着重于超氧化物歧化酶的研究[135]。紫菜微生物病有赤腐病（赤腐病菌等腐霉菌引起）、壶状菌病（壶状菌病寄生所致）、绿斑病（丝状细菌等为病原）等。

海带、紫菜病迄今大都从养殖环境及管理的改善上着手防治。周丽、黄健、王丽丽、严正凛等综述了海带、紫菜和江蓠等的病害研究概况[136～139]。

3　光合细菌的应用

对光合细菌的重视是顺应养殖业疾病发生的产物。有关研究涉及光合细菌的培养、分离、鉴定、生长条件及其影响因子和应用[140～144]。所应用的光合菌主要有荚膜红假单胞菌（*Rhodopseudomonas capsulata*）、球形红假单胞菌（*R. spheroides*）、沼泽红假单胞菌（*R. palustris*）、胶质红假单胞菌（*R. gelatinosa*）、嗜酸红假单胞菌（*R. acidophila*）、细着色菌（*Chromatium gracile*）、最小着色菌（*C. minutissimum*），它们的具体应用状况如表 5 示。

此外，朱励华等作了相关的综述报告[156]。

表 4　养殖鱼等的细菌性疾病及防治措施

鱼的名称	疾病	病原体	防治措施	备注（参考文献号）
大黄鱼	弧菌病	弧菌、溶藻弧菌、副溶血弧菌	一些放线菌可拮抗、庆大霉素	99～101
	门多萨假单胞菌病	门多萨假单胞菌（*Pseudomonas mendocina*）	对多西环素、四环素、环丙沙星等 7 种药物敏感	102
	肠炎病	嗜水气单胞菌	禁用药浴，先停饵后少饵，并适当抗生素	103
	滑走细菌病（烂尾烂嘴病）	滑走细菌等	加强管理，鲜饵，适当密度，抗生素，ClO_2 浴	103
	急性型开鳃白身病	一种长杆菌	福尔马林、恩诺沙星、ClO_2	103
	苗慢性型开鳃白身病	一种短杆菌		103

（续表）

<table>
<tr><th colspan="2">鱼的名称</th><th colspan="2">疾病</th><th>病原体</th><th>防治措施</th><th>备注
（参考文献号）</th></tr>
<tr><td colspan="2" rowspan="7">鳗鲡</td><td colspan="2">烂尾病</td><td>鳗鲡气单胞菌</td><td></td><td>107</td></tr>
<tr><td colspan="2">鳃病</td><td>嗜水气单胞菌，柱状屈挠杆菌</td><td></td><td>104</td></tr>
<tr><td colspan="2">烂鳃</td><td>恶臭假单胞菌</td><td></td><td>104～106</td></tr>
<tr><td colspan="2">爱德华氏病</td><td>爱德华氏菌</td><td>二氧化氯（ClO_2）</td><td>108</td></tr>
<tr><td colspan="2">黏液细菌性鳃病</td><td>细菌</td><td>换水排污，呋喃类或磺胺</td><td>104</td></tr>
<tr><td colspan="2">鳃霉病</td><td>霉菌</td><td>ClO_2、硫酸铜、生石灰</td><td>104</td></tr>
<tr><td colspan="2">鳃、肾炎</td><td></td><td>ClO_2、防水质坏，5%～10%食盐水浸浴</td><td>104</td></tr>
<tr><td colspan="2" rowspan="3">鲈鱼</td><td colspan="2">鳗弧菌病</td><td>鳗弧菌</td><td>疫苗</td><td>109</td></tr>
<tr><td colspan="2">花鲈鳗弧菌病</td><td>鳗弧菌</td><td>疫苗</td><td>109</td></tr>
<tr><td colspan="2">鱼苗出血烂尾病</td><td>鳗弧菌</td><td>弧菌疫苗</td><td>109</td></tr>
<tr><td rowspan="13">鲷</td><td rowspan="6">真鲷</td><td colspan="2">弧菌病</td><td>最小弧菌(Vbrio mimicus)</td><td>/</td><td>24</td></tr>
<tr><td>烂尾</td><td rowspan="5">松鳞</td><td></td><td></td><td rowspan="5">110</td></tr>
<tr><td>结节</td><td></td><td></td></tr>
<tr><td>腹胀</td><td></td><td></td></tr>
<tr><td>白云</td><td></td><td></td></tr>
<tr><td>鳃、肾结节</td><td></td><td></td></tr>
<tr><td>平鲷</td><td colspan="2">弧菌病</td><td>溶藻弧菌</td><td></td><td>110</td></tr>
<tr><td rowspan="6">黑鲷[a)]</td><td colspan="2">肠道白浊病</td><td>弧菌</td><td>保洁、合理养殖密度、活饵、呋喃唑酮</td><td>111</td></tr>
<tr><td colspan="2">腹水病</td><td>爱德华氏病</td><td>保洁、换水、土霉素</td><td>111～112</td></tr>
<tr><td colspan="2">腹胀满病</td><td>溶藻弧菌</td><td>保洁、喂卤虫＋人工饵</td><td>112</td></tr>
<tr><td colspan="2">皮肤溃疡病</td><td>假单胞菌、鳗弧菌</td><td>保洁、漂白粉、土霉素、免受伤、呋喃唑酮、磺胺</td><td>112</td></tr>
<tr><td colspan="2">眼病</td><td>假单胞菌、弧菌</td><td>鱼肝油、免受伤、复合维生素</td><td>112</td></tr>
<tr><td colspan="2">石斑鱼</td><td colspan="2">弧菌病</td><td>河流弧菌、创伤弧菌</td><td>创伤弧菌、强毒株福尔马林灭活免疫（FV）、热灭活免疫苗（HV）</td><td>113～114</td></tr>
</table>

（续表）

鱼的名称	疾病	病原体	防治措施	备注（参考文献号）
点带石斑鱼	溶藻弧菌病	溶藻弧菌	复方新诺明、环丙沙星	115
斑点叉尾鮰	链球菌病、组织损伤	链球菌		116
尼罗罗非鱼	尾鳍腐烂病	嗜水气单胞菌无气亚种、链球菌		116～117
	脑膜炎	细菌		116
	心肝脾卵巢肾多发性浆膜炎	细菌		116
	卵巢炎	细菌		116
	心肌炎	细菌		116
杜氏鰤	细菌病	弧菌（Y-1 菌）	呋喃，喹诺酮，大环内脂	118
牙鲆[b)]	爱德华氏菌病	迟缓爱德华氏菌（*Edwardsiella tarda*）		119
	细菌病	兔莫拉氏菌	喂重组酵母	120
	病毒、原虫吸虫病	病毒、原虫、吸虫、鳗弧菌	鳗弧菌疫苗	121
	盾纤虫病、链球菌并发症	盾纤虫、杀鲑弧菌	H_2O_2 药浴，适当淡水浸浴，适度 $CuSO_4$、土霉素	122
大菱鲆[c)]	疖疮病	嗜水气单胞菌、杀鲑弧菌		123
	腹水病	爱德华氏菌		123
	滑走细菌病	海岸屈挠杆菌		123
	链球菌病	链球菌		123
	细菌性败血症	嗜水气单胞菌	123	
红鳍东方鲀	卵子丝状细菌病	毛霉亮发菌	洁净水淘洗、孵化、冲洗	124
	链球菌病	链球菌	防过密，鲜饵，清洁，H_2O_2 浴，切记抗生素	125
鲟 d)	出血病	气单胞菌	ClO_2、抗生素	126
	肠炎	细菌	土霉素	126
	幼鱼肿嘴	细菌	加强管理，改善水质，用不变质饲料	126
	稚、幼鱼疥疮	疥疮型点状产气单胞杆菌	漂白粉、土霉素	126
	卵霉病	丝水霉、鞭毛绵霉	改善水质，苏打食盐水浸	126
	水霉病	水霉	食盐水浸泡，提高水温	126

(续表)

鱼的名称	疾病	病原体	防治措施	备注（参考文献号）
鱼	大批死亡	嗜水气单胞菌		123
	大批死亡	运动性气单胞菌、杀鲑气单胞菌、鳗弧菌、鱼害巴斯德菌、鲁克耶尔森氏菌、嗜水气单胞菌、溶藻弧菌	API-20E 系统鉴定诊断	128-130
鲢鳙	出血病			
海马	肠胃炎病	细菌	保洁、洁饵、土霉素、青霉素	127

a)黑鲷除细菌病，还有原生动物、甲壳动物及其他病原引起的 9 种疾病；

b)有各种疾病 20 种，其中细菌病 8 种，但多为 20 世纪 90 年代前或国外报道；

c)尚有其他原因致病 13 种；

d)还有其他病 13 种。

表 5　光合细菌在海水养殖中的应用

动物名称		所用光合菌	效果	备注(参考文献号)
海湾扇贝	幼虫	光合细菌(PSB)	优于仅喂单胞藻	145
	人工育苗	光合细菌(PSB)	效果好	146
	育苗	紫色无硫细菌	西瓜红	147
蟹	暂养苗	5×10^{-6} 光合细菌	降换水成本，促生长	148
栉孔扇贝	幼虫	紫色无硫细菌	促生长发育效果好	149
对虾	幼体	光合细菌	开口饵料，除有机质、有害物，净化改善水质	150～151
	育苗	光合细菌	饲料添加剂，提高成活率，抗病，促生长发育	152
		沼泽、球形、荚膜红假单胞菌	饵料、增重明显	153
轮虫		光合细菌	最有利繁殖	154
褶皱臂尾轮虫		球形红假单胞菌	与单胞藻异样，效果好	155

4　海产养殖生物的生态——环境微生物

高密度、高投饵及循环差的养殖方式加上池塘外环境的恶化等多种因素促进了病害的发生。反思之后的人们重视养殖生态环境微生物学的研究，养殖生物体内外环境的改善有望使养殖水平更上一层楼。

对养殖环境除了摸清池塘外的环境状况外，大量人力也投入到池塘内水质、底泥理化生物状况的调研。其中一项便是相关的微生物学调查，这包括不同生长发育期养殖生物的环境调查。如对虾的亲虾越冬期[157]：投苗前[158]、育苗期[159~161]、养成期[162~167]、及池塘水质与底质状况调研[158]。还有池塘水质改善[168~172]等。一定程度上某些方面摸清了不同阶段的环境微生物种群数量及其变动状况和规律、病原微生物的活动，包括条件致病体及预防治疗措施等，为评价、制定、改进养殖生态-环境条件和标准，提高养殖生物的防病抗病能力迈出了一大步。新技术的采用将为研究养殖生物及其环境的病原微生物学提供后劲[173]。

5　展望

在1992～2002年间(包括少量2003年成果)，中国学者就海产养殖的微生物学问题进行了大量研究。1992年前的42年，中国的有关研究，请参见即将出版的 *History of marine microbiological research in China from* 1949 *to* 1991。涉及细菌、真菌的论文占了不少(200余篇)。确认了一些病原生物和防治病害方法，其中最重要的病原体是几种弧菌。人们对病原微生物本身的研究，加深了对它们的认识，对于疾病的三因子即宿主、病原和环境内在机制、内外环境及条件的研究已达到相当的深度和广度，从而为疾病的防治打下了较雄厚的基础，挽回了许多损失。但这些比起先进水平和人们的期望还有较大的差距。因此提议在下列方面作适当调整，以更快、更扎实、更全面地解决现实的病害问题。从而进一步提高海产微生物学和养殖的研究水平。

5.1　高瞻远瞩统筹兼顾，重点突出学科，安排好全国的相关科研工作，使科研计划项目合理分配落实到单位和人，减少不必要低层次的分散重复和混乱。

5.2　增加经费的投入，稳定科研人员队伍。水产病害的研究范围广泛，人员众多，而且有一定难度，企图短平快，一蹴而就，一劳永逸，急功近利地解决问题是不现实的，也是不可能的。保证具有良好科研素质人员的稳定工作，重要的是基础性研究工作，就能使相应研究持续稳步健康地开展。

5.3　加强课题间国内外的联系协作和交流，不断开拓和创新，提高有效劳动率和成果转化率。

5.4　采用和发展先进手段，加深和加强对探讨问题的认识和预见，使病害发现得较早，解决得更好。

5.5　强化养殖环境治理，更好更广地认识养殖生物本身及其内外环境条件的改善。认识和调动养殖生物的防病、抗病物质、因子和能力，变害为利，综合利用，造福人类，使自然条件和人为环境完美结合，并良性永续地发展。

参考文献66篇(略)

养殖鲍的疾病与敌害*

摘　要　鲍是一种重要的海珍品。世界上至少有7个国家开展鲍鱼养殖，我国年产养殖鲍约2.4千吨以上。养殖鲍面临14种疾病，包括病毒、细菌、真菌和动物等引起的和管理不当及环境不适造成的疾病，另有一些敌害生物危害。本文分别描述以上疾患及相关防治措施。

关键词　鲍　疾病　灾害

鲍是最名贵的海珍品之一，由于其营养价值高和味道鲜美，深受中国和世界许多国家的欢迎。近年来世界鲍的产量(野生和养殖)平均在1.5万吨左右，而需求量却在2.0万吨左右(图1)，有大约5 000吨的需求缺口。目前鲍的市场主要在亚洲，中国和日本的需求为最大，且中国对鲍的需求呈逐年上升趋势。目前我国的鲍产量大约在2 400吨(高绪生2 000)，所以只能依靠大量进口。

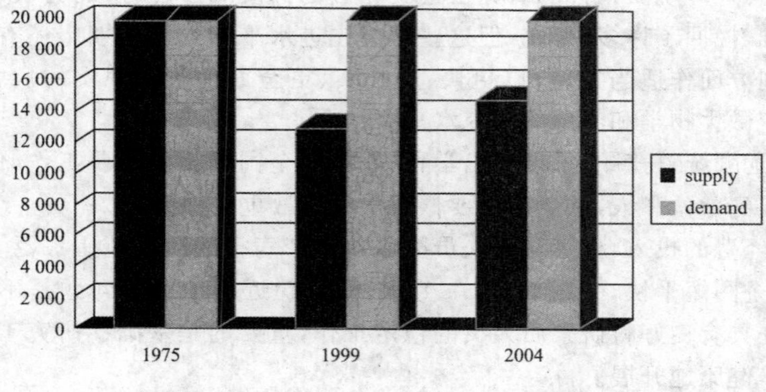

图1　**1977—1999—2004**世界鲍供求关系

1. 养殖鲍及其疾病概述

自古以来，鲍就被列为海产"八珍"之冠，其食用、药用及装饰工艺用途都很广泛。我国鲍的人工养殖起步于20世纪50年代，80年代初即开始工厂化养殖，主养区在辽、鲁、闽、粤。世界上，日本、朝鲜、美国、墨西哥、新西兰及澳大利亚等国均有生产。

养殖的鲍属腹足纲、原始腹足目、鲍科、鲍属(*Haliotis*)。世界上主要养殖的鲍有(1)桃红鲍(*H. corrugata*)、(2)黑鲍(*H. cracheroidii*)、(3)皱纹盘鲍(*H. discus hannai*)(4)杂色鲍(九孔鲍)(*H. diversicolor*)、(5)大鲍(*H. gigantea*)、(6)绿鲍(*H. fulgens*)、(7)黑唇鲍(*H. tuber*)、(8)红鲍(*H. rufescens*)、(9)西氏鲍(*H. siebodii*)等16种以上。我国主要养殖皱纹盘鲍和杂色鲍两种。鲍的养殖同样面临着亟待解决的病害问题。

---

* 原文刊于《中国水产》2004年专刊，青岛.全国水产养殖研讨会论文集。221-226。

鲍的疾病主要有病毒病、原核生物样病、幼虫面盘纤毛脱落病、细菌病、真菌病、单孢子虫病、才女虫病、缨鳃虫病、其他寄生虫病、气泡病、缺氧暴死、肌肉萎缩症、肿嘴病、缺钙碎壳症等14类，此外还遭遇敌害和灾害。

2. 鲍病分述

2.1　病毒病

裂壳病(Crack shell disease)(病毒病之一种)。病原是球状病毒(*Spherical virus*)。该病毒大小90～40 nm。被膜双面体。囊膜内有一层内膜包围的核衣壳。超薄切片见发生基质中有该病毒粒子。它由两部分组成，最外由双层囊膜全封闭，厚8～10 nm，核衣壳直径70～100 nm，发生基质由中等电子密度细颗粒聚集成，外有双层膜包被，形成封入体。封入体破裂即逸散出病毒粒子。病毒侵染足、外套膜、肝、嗉囊、鳃，但仅在上述器官、松散的组织间隙、血窦中的血细胞内和肝脏细胞质见到病毒粒子。张明等综述过裂壳病等3种病况及防治，王勇江等对该病毒作过专门研究[1,3]。

此病发生于皱纹盘鲍、九孔鲍及日本盘鲍中。福建还发现低温病毒病。

2.2 原核生物样病(WS萎缩并发症)

黑鲍的WS病，即萎缩并发症(withering syndrome)，也称萎缩症，又名消瘦病，还叫末梢病(Terminal disease)。该病致黑鲍大批死亡，结果是存活率大大降低，残存者体重大大减轻。其病原是一种胞内原核生物，吸盘原虫、肾球虫、簇生原虫或立克次体样原核生物。该病发生区域包括广为传播的美国加州海峡群岛和diablo小湾。1986年在加州SNI(San Nicolas islands)发现。1992年4月份在SNI最西端首见，之后以不同的死亡率发现于一些点上。而SNI东北向范围无WS。

该病病因：海水温度的抬高，可能对黑鲍数量的回收有强大负面撞击作用，如1990～1993年El Nio对加州沿岸影响即是。密度过大也使该病难以消除。曾有人对WS的鲍体组织作化学分析，并将黑鲍暴露到有多氯联苯、含氯农药或多环芳烃的环境中，并未发现WS与污染物间有什么明确的关系。其他影响代谢系统的诱因不清。

该病症状：染病鲍体质衰弱，变疲、昏睡不动，组织退色，其足肌萎缩。病鲍比正常鲍消耗海带等大型藻类少了4倍，而耗氧少了3倍，但排氨量又高出3.8倍。

杆形、富含核糖体、革兰氏阴性的、立克次体样的原核生物感染衬在胃肠里的上皮细胞和消化窦酶的分泌细胞中，之后充满于宿主细胞质。结果使消化酶颗粒消失，酶分泌细胞及其组织变形。消化窦细胞广泛感染。肌肝糖不足，肌纤维空虚。染病鲍不得已只好用足肌蛋白为能源。

该病防治：控制水温至合适程度，将健鲍移至潮下带去或深水、开阔、有波浪处。减低鲍的密度。

黑鲍(*Haliotis cracherodii*)、皱纹盘鲍的消化道，后者的鳃、外套膜、足等的上皮组织、固有结缔组织内均有立克次体样生物感染，引起萎缩综合征。黑鲍的此病见于美国的加州，皱纹盘鲍的此症发生于山东沿海。

2.3 幼虫面盘纤毛脱落病

估计其症状与海湾扇贝的相似。当人工培育的单轮幼体向面盘幼体转化时的皱纹盘鲍，有此病发生。面盘与软体分离，在水中浮游。发病率可高达10%～36%。1980年流行于大连金州区一些苗场。1993年在蓬莱苗场也发生过。原因是亲鲍积温不够。雌鲍促熟时，7.6℃以上的有效积温低于800℃所得幼虫易患此病。孵化时水质不好，换水量小于2倍水体的容积，易患此病。为此要提高有效积温达1 000℃左右，并改善水质[14]。

2.4 细菌病

聂宗庆等(2000)研究认为相关病原有(1)荧光假单胞菌Ⅳ(*Pseudomonas fluorescorts* Ⅳ)、弧菌属的(2)溶藻酸弧菌(*Vibrio alginolyticus*)、(3)坎氏弧菌(*V. campbellii*)、(4)河流弧菌(*V. fluvialis II*)、(5)梅氏弧菌(*V. metschinikovii*)、(6)创伤弧菌(*V. vulnificus*)等。

细菌性的鲍病因病原和症状的不同而有几种称呼，如脓疱病(Pustule disease)，它因荧光假单胞菌Ⅳ所致，也可由病原菌河流弧菌引起，还可因一种G^-的具鞭短杆形细菌致病。而坎氏弧菌则引发出脓毒败血症(Septicopyemia)。溶藻弧菌使红鲍致病，叫做弧菌病(Vibriosis)。创伤弧菌也使皱纹盘鲍致病。

2.4.1 皱纹盘鲍脓疱病1

病原是一极生单鞭毛的革兰氏阴性短杆形细菌。它使足肌感染率达100%。染病后的槽养鲍足肌微隆起成白色疱。疱破流出白色浓汁，含少量组织成分和能动的杆形细菌，留下2～5 mm深孔在足肌上。此时病鲍附着力弱，体翻转至死。该病可使死亡率达50%～60%，侵害地区：大连。侵害时间：1993年夏天发现，当年死去60吨，损失2 000万元。该菌对氟本尼考、复方新诺明敏感，对诺氟沙星中度敏感。6.3×10^{-6}氟本尼考处理，可使死亡率降低73%。噬菌体已用于该病防治[5]。

2.4.2 皱纹盘鲍脓疱病(Pustule disease)

病原是河流弧菌-Ⅱ。

染此病的鲍，早期可见损害从足下表面始，扩展至足的深部。纤维坏死、瓦解或破碎，足肌白丘脓疱，有溢出液，致溃烂坏死，食欲退，活力降直至死。病灶中只剩下血淋巴细胞、弧菌及细胞碎片。组织切片中可见到鲍足的肌原纤维排列混乱、疏松。横纹肌的横纹模糊。细胞核肿大，核膜破裂，核质均质化，呈无结构状态。核仁消失。肌糖原也消失。线粒体内嵴模糊及至破裂。晚期病灶中有该菌。外壁变薄。核区明显。胞质内含有颗粒糖原。细菌周围的组织被侵蚀成空斑。

该病原侵害龄鲍至成鲍，流行区：大连(1993.9～1995.12)。

河流弧菌-Ⅱ的D、N和T株对18种抗生素有不同的敏感性。该菌对一些抗生素有抗药性。抗药性的不同可能与基因突变有关。抗生素为细菌提供了耐药突变株的选择环境，从而使耐药菌株大量繁殖。该病原菌可与鲍血清发生凝集反应。菌苗效果明显，可提高成活率达50%以上。鲍的血淋巴细胞具吞噬细菌能力。在菌苗作用下，血淋巴细胞增殖并活跃，吞噬异源致病菌。免疫因子也被激活，参与免疫过程，提高鲍成活率。用同工酶法诊断此病。适当用氟本尼考、复方新诺明、诺氟沙星等可治[2,6,7]。

2.4.3 幼鲍溃疡病(Fester disease)

此病有时称作一种脓疱病。工厂养成中的皱纹盘鲍得此病，病原是荧光假单胞菌Ⅳ(*Pseudomonas fluorescens* Ⅳ)。可用ELISA(酶联免疫吸附检测法)诊断，发病之处在山东俚岛，常年均见，死亡率高。陈志胜等(2000)研究过相关的病原菌。

2.4.4 由溶藻酸弧菌引起的弧菌病(Vibriosis)

1993年，Elston等首先发现了美国幼红鲍(*Haliotis rufescens*)的这种弧菌病，同时发现其他相似弧菌。溶藻弧菌为革兰氏阴性(G^-)、兼性厌氧,不产气、氧化酶阳性、单极鞭毛杆菌。免疫荧光研究表明它具优势抗血清性。染病鲍主要是壳长小于 1 cm 的幼体。组织学上显出它们有典型的细菌感染。细菌沿血窦和神经鞘生长。鲍的上皮脱落或破裂,外围神经迅速退变。感应性的寄主细胞不能有效阻滞感染推进。病鲍不活泼,体色蜕,无刺激反应,萎缩。之后,幼鲍死亡,鲍场的生产就萧条下来。

Sindermann 早在 1988 年就有类似发现,但没有确证为溶藻酸弧菌,只是症状相似而已。

该病发生于美国康涅狄克州兽医学院的养鲍室中。外界因素是水温的提高(由 18 升至 20℃)和过饱和的充氧。蓄养的野生鲍因受伤而感染,当水温达 20℃以上就生病,当水温在 25～27℃时,染病更多。

预防此病发生,要确保幼鲍有合适环境条件,对受伤鲍,要及时用药物洗伤口。药物用 25～50 g/m^3 的氟本尼考海水溶液。用此浸洗鲍 0.5～1 h,或 1%的复方新诺明海水液浸洗 5 min。用药后将鲍在干净空气中晾 10～15 min,药液可渗入病灶发挥作用。之后放鲍入海水中养。但溶藻弧菌也多发生于低温,其感染速度快,处理难。

2.4.5 皱纹盘鲍脓毒败血症(Septicopyaemia)

病原是坎氏弧菌(*Vibrio campbellii*)。染病鲍围心腔因充菌而过分膨大。外观见鲍足僵硬呈收缩状,壳上有大小不一脓疱。腹足底脓疱尤其多且大,有的鲍之腹足呈松弛舒展状。壳上偶见脓疱,肌肉疏松,腹足可轻易捏穿。血涂片、脓涂片可见细菌及血淋巴碎片。肌肉组织切片也见菌。肌纤维结构破坏。病/死鲍之血淋巴、脓汁中的细菌检出率达 95%以上。马健民等(1996)指出该病发生于辽宁大连沿海,如长海、大连的鲍鱼场的鲍及天然鲍身上。流行时间已持续 3 年,溶藻弧菌使台湾的九孔鲍幼苗严重脱落死亡,该菌致鲍所需硅藻饵料缺乏而死亡。当然也可能是病毒的原因,故称为渔业"口蹄疫"[8]。

育苗池中幼鲍用福尔马林(5～20 g/m^3)浸浴 6～12 h,可杀死其附着板上的原生动物、线虫和部分桡足类动物,还可杀菌,对幼鲍及饵料硅藻则无不良影响。

对上足分化幼虫和围口壳幼虫期鲍,可用 0.5～5 g/m^3 浓度的氟本尼考浸浴 48 h、1～10 g/m^3 的土霉素浸浴 48 h 或 0.25～0.5 g/m^3 的盐酸环丙沙星浸浴 24 h,均可有效抑杀细菌,而对幼鲍及饵料硅藻无害,但也不杀伤原生动物、线虫和桡足类动物[9]。此外 1 种弧菌使九孔鲍得破腹病,外伤感染使盘鲍得化脓病。

2.5 真菌病

病原是密尔福海壶菌(*Haliphthoros milfordens*)。该菌菌丝直径为 11～29 μm,分枝不多,以整体繁殖产孢。菌丝任一处均可由排放管放出生成的游动孢子。游动孢子具 2 条倒生鞭毛。休眠孢子球形,直径 6～10 μm,它们寄生在鲍的外套膜上足或足背面。

该菌侵染西氏鲍(*Haliotis siebodii*)及其他鲍。染病鲍病患处隆起,显微镜检查其水浸片就可见菌丝。夏季捕得的鲍即使饲养在 15℃的循环海水中,不过 10 天就得病,之后不多几天即死。病原何处来?未见报道。用 10×10^{-6} 的次氯酸钠未杀死其游动孢子,但可起预防作用。

火田井于 1982 年在日本养殖鲍中就发现过此病,1995 年还报道盘鲍因艾特金菌(*Atkimsieus* sp.)而得卵菌病。

2.6 单孢子虫病(派金虫病)

病原是奥氏派金虫(*Perkinsus olseni*)。其孢子球形,直径 16.6 μm 左右,具壁。胞质内含一大液泡,直径 10 μm。液泡含 50～100 个直径约 1 μm 的小颗粒。营养体直径约 13.8 μm,具一球形核,偏于一侧。另有嗜曙红液泡一个。常见裂殖体,可分裂出多个子细胞,其营养体可在巯基醋酸酯培养基上发育。

染病黑唇鲍(*Haliotis ruber*)血淋巴中有其聚集的褐色细胞团,并游离于循环系统内。鲍的足、外套膜、闭壳肌内部或表面有淡黄/褐色柔软的脓疱,直径 1～8 mm 不等,球形或半球形,内含脓汁由大量营养体、裂殖体、白细胞组成。脓疱有壁,与其周围组织隔开。可据症状和脓疱,由压片再镜检确诊此病。

流行于澳大利亚南部 20℃温度、30 盐度的海水中。

另一种派金虫即 *P. atlanticus*(大西洋派金虫)也可使黑唇鲍染病。此外 *Labyrintho* 的原生动物对堪察加鲍幼体感染性很强。

2.7 才女虫病

病原是五种才女虫,它们是刺才女虫(*Polydora armata*)、凿贝才女虫(*P. ciliata*)、东方才女虫(*P. websteri*)、贾氏才女虫(*P. giardi*)和韦氏才女虫(*P. websteri*)。其中以刺才女虫为主,其次是韦氏才女虫等,贾氏才女虫很少。

受侵害鲍是杂色鲍(*Haliotis diversicolor*),多是壳长＞29 mm 的成鲍。症状是鲍壳被穿孔,且随鲍长而增多。一些染病鲍因寄生大量才女虫,健康状况下降,鲍肉重量减轻,使业户遭受经济损失。

此病由小岛在 1982 年发现于日本。穿贝海绵(*Cliona*)也是壳上常见寄生虫。

2.8 缨鳃虫病

病原是缨鳃虫(*Sabelid polychaete*),可能是帚毛虫属(*sabellaria*)的成员。寄生于美国加州红鲍(*Halictis rufescens*)身上。受其寄生,鲍贝壳前缘被感染,生长受到严重影响[10]。

2.9 其他寄生虫病

有关的寄生虫是假钩棘头线虫(*Echinocephalu pseudouncinatus*),偏顶蛤类的 *Panaietis haliotis* 和一种复殖吸虫。

假钩棘头线虫的第二期幼虫寄生于美国南加州的桃红鲍(*Haliotis corrugata*)中。该幼虫平均大小是 $20\times0.65\ mm^2$,头球围绕有 40～50 个钩子,排成 6 行。它的寄生部位是鲍的鳃及足腹部。形成的包囊使所在组织出现疱状突起,鲍肌肉明显变瘦,附着力降低。

Panaietis haliotis 寄生于日本沿岸的大鲍(*H. gigantea*)口腔中,未见有致病力。

未定名的一种属于孔肠科的复殖吸虫之包蚴发现于大不列颠水体的鲍中。包蚴呈黄色。

人工养幼鲍有 16 目 48 属 68 种纤毛虫,以槽纤虫、钟虫、双眉虫为优势种[11]。

贝类动物鳃毒及外套腔中寄生/共栖几种纤毛虫,如纤细盾纤毛虫、双壳吸虫、原鱼钩虫等,都可能致病[12]。球虫拟克洛西虫(*Pseudoklossia*)感染美国红鲍等,引起毁灭性综合征(WS)。

2.10 气泡病(消化不良症)

病因有两个:

一是稚(幼)鲍的集约化养殖系统中的大量海藻。海藻靠强光和不畅通水体进行光合作用，使水中含有过饱和氧，导致鲍生此病。

患者的上皮组织下、肌肉、结缔组织、消化道和血管中都有气泡者，鲍可浮于水面。鲍的口部褪色，齿舌扩张，触角不伸，器官肿胀，上足呈鳞茎状，不动。纤维性神经鞘与神经细胞及神经鞘周围组织分离，大血细胞之液泡扩大，受伤之小血细胞的细胞核附近有液泡出现。吸附力降甚至无，摄食减甚至无。

病因之二是饲喂不合适的合成饵料，有的合成饵料中的蛋白消化异常，鱼粉或乳制品中酪蛋白变性。不合适的饵料使鲍消化不良，消化道充满气泡，甚至使鲍不得已而上浮至水表层。此病也称消化不良症。气泡病患者可继发溶藻酸弧菌病。

诊断方法：观察并检查养殖系统及其条件，即海藻、光照和水交换及饵料状况。再查鲍生长情况，如体内外器官组织，发现气泡，即可证实为本病。

防治方法：消除不良养殖条件，适当降低水温和提高换水率等，改进饵料成分，包括多喂生鲜海藻等，适当推迟剥离时间[2]。

2.11 缺氧暴死

已发现工厂化养殖的九孔鲍得此病。病鲍爬至门边，足底朝上仰死，足肌软白，病因是密度过大，其他水生物过多，卫生差，水、气不足。气候不利时易发生。防治方法是勤换水，清池，加大气量，减低密度，及时清除死者[13]。

2.12 肌肉萎缩症

气泡病—肌肉萎缩症。

气泡病若处于水温上升时期，又受到 200 μm 大小以下的细菌菌落或更小的病毒感染可能转变成肌肉萎缩症[2]。如前 2.2 中所述，萎缩综合征病原还可能包括立克次体样生物。

以上疾病见于日本东北至北海道，中国山东长岛的皱纹盘鲍也见此病，中津川俊雄(1995)、松桥聪(1996)均对此病有研究。

该病病因是 *Amyotrophia*，它侵害 3 种鲍，即 *H. discus discus*，*H. discus hannei*，*H. madaka* 的幼贝。濒死的幼贝显出肌萎缩，贝的前缘有少量缺痕，附着基质的能力减低。切片可见源于神经组织异常细胞团块，具有典型的特征性组织病理变化，广见于各种器官中，而以在神经干中和足肌末端神经中更常见些。但光镜和电镜及实验性感染均未查出真正的病原。患此症后，可使幼贝大量死亡。桃山和夫等在 1999 年报道该病。另外原因可能与海水中重金属离子超标及近亲繁殖有关。

发病时间：(1988～1996)春至早夏，发病地区：日本西部。

2.13 缺钙碎壳症

病因是大量使用回水、饵料中钙含量过低或寄生虫钻凿贝壳等。

病鲍主要是处于中间育成升温阶段的中国鲍(皱纹盘鲍)。

2.14 肿嘴病

病鲍吻部肿胀。病因是残饵变质、水交换差。因此要及时清理残饵，保持海藻鲜度。采用 5 倍的新诺明药浴可以治愈。

3. 鲍的敌害

鲍的敌害生物包括肉食性鱼类、贝类、棘皮动物和甲壳动物等。肉食性鱼如鲷、鲨鱼、鲀、鲈、鲽和鳗、海鲋鱼等，它们可大量吞食幼鲍等个体，使鲍受损。

软体动物如章鱼、荔枝螺等是鲍的食用者。

棘皮动物主要是海星、海胆等，伤害鲍并与鲍争饵料。

甲壳动物主要是各种蟹类，包括日本蟳、黄道蟹等，它们都可捕食鲍，尤其是幼鲍。

寄生虫性敌害病，引起外伤感染者，可用磺胺、抗生素类药涂擦、洗涤或浸泡、注射。

前述的才女虫既是病原生物又是敌害生物。

参考文献 8 篇（略）

（合作者：王　波）

珠母贝的疾害*

摘　要　珠母贝是热带、亚热带地区的重要养殖对象之一，经济价值很高。养殖珠母贝面临病原微生物和低等动物所致的一些疾病，同时又遭遇敌害及灾害。本文论述珠母贝养殖过程中的上述疾病、敌害、灾害及防治措施。

关键词　珠母贝　疾病　敌害　灾害　防治

1　养殖珠母贝及其疾病、敌、灾害概述

珠母贝属于瓣鳃纲珍珠贝目珍珠贝科(Pteriidae)。至今作为养殖对象的珠母贝有2属6种，它们是珠母贝属(*Pinctada*)的长耳珠母贝(解氏珠母贝，*P. chemnitzi*)、涂抹珠母贝(*P. fucata martensii*)、珠母贝(黑蝶贝，*P. margaritifera*)、合浦珠母贝(马氏珠母贝，*P. martensii*)、大珠母贝(白蝶贝，*P. penguin*)、珍珠贝属(*Pteria*)的企鹅珍珠贝(*P. penguin* 或 *Magnavicula penguin*)(或叫珍珠蚌，*Margaritana margaritifera*)。

养殖珠母贝的最终产品是珍珠。由于珍珠是贵重装饰品，又是很好的药材，因而世界上许多国家，主要是中、日、缅、澳、泰、马、印尼和菲律宾等亚洲国家大规模养殖，其中以日本的养殖最为发达。珠母贝疾病及敌、灾害带来的经济损失很值得重视。

珠母贝的疾病主要有3类，一是微生物病；二是簇虫病；三是蠕虫病。此外尚有1种未明种类的单孢子虫，球形体，大小5～7 μm，它寄生于黑蝶贝，但对它的致病性不明(Nasr，1982)。珠母贝的敌害主要来自6类生物。珠母贝的灾害来自恶劣天气、海况和环境恶化[1]。

2　珠母贝疾病分述

2.1　微生物病

2.1.1　细菌病：病原是革兰氏阴性短杆菌(*Brevibacterium*)和双球菌(*Diplococcus*)。对珠母贝施术是珍珠生产的关键技术。施术时期的水温过高和过低，水的相对密度过低(≤1.015)、不合适的施术工具及施术人员的较低水平等都有可能导致病原菌入侵。入侵处一般是手术贝伤口处，其症状是肌肉系统先发生腐败分解，然后液化以致死亡。该病主要发生在合浦珠母贝上，其他贝上也有。预防法：一是注意卫生，二是经常监测水质，确保不受污染[2,3]。

马氏珠母贝亲体在不适当的投饵时，其幼体可能患面盘纤毛脱落病。

2.1.2　类立克次体病：病原是类立克次体生物(RLO)。该病发现于中国海南岛一带的大珠母贝、合浦珠母贝中。RLO使它们强烈致病。

RLO包涵体寄生于宿主多处内脏组织的细胞质里，具明确的嗜曙红细胞特性，感染外套膜、鳃、消化道上皮/表皮细胞、上皮下结缔组织物血管内皮细胞。它们多形态(Pleio-

* 原文刊于《水产养殖》，2005，26(1)：34-37.

morphic)、大小平均967 nm×551 nm。透射电镜切片清楚表达出原核的细菌样细胞，它包括两个trilaminar膜，一个是增多着的电子致密的周质核糖体区，另一个具似线的DNA类核结构。这些生物以两种方式存在于寄主中，一是自由地生活于胞质，另一是位于Phagolysosomes膜内。细胞有两种变异体，一是小细胞变异体(small cell variant，SCV)，另一是大细胞变异体(large cell variant，LCV)。SCV的周质中富含核糖体，在胞质内显示电子致密，在中央类核区比LCV窄。增殖模式为横向二分裂和出芽(budding)，该生物的包涵体发育期分作颗粒前期和颗粒期(颗粒Ⅰ-Ⅲ期)，根据两个propagative形式、两个细胞类型和胞内所在位置，该生物的发育周期包括营养分化期和出芽分化期。证据表明该生物不同于立克次氏小体。该生物使染病贝多种器官变性坏死，完整性被破坏，破坏程度与RLO包涵体数量密切相关，而此破坏又与RLO的生长繁殖密切相关。幼贝对RLO极易感染，即对RLO抵抗力极弱。受其感染，贝体呈急性变质性炎症时，细胞呈崩解性、溶解性或退变性坏死。呈慢性增生性炎症病理时，显出实质细胞增生，成纤维细胞增生为纤维细胞并纤维化。此病以此分为急性和慢性两期。死亡高峰期为急性期，高峰后，幸存者转向溶解性坏死、退变性坏死，慢性期呈现出增生、修复。此疾病后期愈后显著良好，尤以肝胰腺和胃肠壁修复较快、较好。

该病由吴信忠等在1999年首报。他们于2002进一步确证大珠母贝幼虫群体死亡与RLO显著相关[4-5]。

2.1.3　类病毒病

Munday等(1998)在西澳大利亚的大珠母贝之消化腺中普遍发现其细胞内有病毒样颗粒包涵体，但无明显的病症。

2.1.4　病毒病

病原：肿疡病毒(*papovavirus*)。染病对象及部位：大珠母贝唇须。染病率：4%(2/50)，发现时间和地区：1993年澳大利亚(Nonon，等，1993)。

宫崎照雄2000年发现西日本(1994)海区的稚、亲贝及插核用贝软体部、闭壳肌萎缩，变黄色后大量死亡，认为是病毒所致。中国珍珠贝已分离出此病毒。外套膜注射基因重组体干扰素(IFN)可延长寿命至治愈。

2.2　动物病

2.2.1 簇虫病

寄生物为原生动物簇虫*Gregararinea*。该寄生物以包蚴(囊)单细胞形式出现于珍珠贝的消化道组织、细胞中。受害生物是法国的黑蝶贝(*Pinctada margaritifera*)。1985～1986年，法国Tuamotu Archipelago的环礁中，该贝群有高死亡率(约50%)。它在消化道中产生，以后肠为主，可散布至中、上部，有的可抵达消化腺、上皮细胞中。营细胞内寄生生活，其攻击的主要目标是消化道表皮组织。遭强烈寄生的组织，可在次表皮的结缔组织中见其踪迹。染病者病灶包括局部破坏、受寄生的表皮及其细胞变薄，甚至直肠上皮遭完全破坏。6月龄以上的标本，均有不同程度的寄生，自然的比养殖的更严重些(Acad，1993)。

2.2.2　蠕虫病

(1)复殖吸虫病：病原是①珍珠蚌盾腹虫(*Aspidogaster margaritiferae*)，②马氏珠母贝牛头吸虫(*Bucephalas magcritifera*)的胞蚴和尾蚴，③变异牛头吸虫(*B. varicus*)的幼

虫，④另一种复殖吸虫（*Musalia herdmanl*）的囊蚴，⑤一种复殖吸虫（*Muttua margaritiferae*）的囊蚴，⑥牡蛎居肛吸虫（*Proctoeces ostrea*）的囊蚴等。

珍珠蚌盾腹虫寄生在珍珠蚌的围心腔中。

马氏珠母贝牛头吸虫的胞蚴和尾蚴通常寄生于日本的马氏珠母贝中，此外与该吸虫相近的牛头吸虫也寄生于土拉牡蛎中。

变异牛头吸虫的幼虫也寄生于日本马氏珠母贝中，这些牛头吸虫的幼虫感染珠母贝，待插核后易大批死亡，即使活下来，其产出的珍珠质量也低劣。

因寄生有牛头吸虫和肛吸虫幼虫的贝，其贝壳和软体部症状不易识别，要用注射器抽取生殖腺组织作镜检，由是否见其包蚴来确定。发病率可达 30%～60%，发病环境是泥底质、不大畅通的内湾。也无适合方法可防除。

Musalia herdmani 的囊蚴寄生于珠母贝的肌肉、外套膜和足内。

Muttua margaritiferue 则发现于斯里兰卡珍珠蚌的鳃中。

牡蛎居肛吸虫的囊蚴属非包囊型，它除寄生于日本的长牡蛎外，还寄生在日本的合浦珠母贝中。寄生部位在生殖腺、肝、肾、围心腔、心房壁等器官、组织中。数量多时使贝体体质明显衰弱。

(2)绦虫病：该病原包括 *Acrabathrium* 等绦虫的幼虫。它寄生于斯里兰卡珠母贝（*P. vulagaris*）中。

(3)线虫病：寄生珠母贝的线虫是锥吻目（*Trypanorhyncha*）的线虫。其幼虫的寄主是珍珠蚌，主要寄生部位在消化腺和鳃内。大量寄生时可见明显的纤维质包囊。另外还发现其他几种（未定名）线虫，其幼虫有包囊，可寄生于生殖腺、口腔壁和闭壳肌内。*Tetrarhychus* 则寄生在斯里兰卡珠母贝（*Pinctada vulagaris*）中。

2.2.3　其他

(1)才女虫病：病原组成较为复杂，其原发病原主要是凿贝才女虫（*Polydora ciliata*）等，继发病原是细菌。才女虫病被许多人称作是黑心肝病或黑壳病，主要是因为该虫分泌物能腐蚀贝壳，使贝壳内面窝心部或中心部一带呈现为黑痂皮质。黑心肝病实际是多毛类虫寄生虫病继发性脓肿。除凿贝才女虫外的寄生虫还有①阿曼吉娅虫（*Armandia lanceolata*）、②周氏突凿沙蚕（*Leonnates ulusseaumei*）、③松襟虫（*Lysidice uinetta*）、④岩虫（*Marpysa sanguinea*）、⑤短须沙蚕（*Nereis costae*）、⑥镰毛沙蚕（*N. falcaria*）、⑦独齿围沙蚕（*Perinereis cultrifera*）、⑧细裂虫（*Syllis gracilis*），以及可能包括矶沙蚕科欧努菲虫亚科和海女虫科的各一种，此外尚有一种纽虫（*Nenenini*），共 11 种以上。继发病原有未定种的小杆菌、双球菌（*Diplococcus*）或短杆菌（*Brevibacterium*）等。

该病主要侵染 2～4 龄的合浦珠母贝。2 龄以上的珠母贝，其多于 60% 的个体得此病，有病母贝被插核，则 1 周年内的死亡率高达 89%。低龄贝虽最初感染轻，但原地再养 1 年，终死亡率也达 24%。

每一贝体有寄生虫 7 条，以凿贝才女虫为多，平均约 6 条，染病贝生长缓慢甚至变小。1 年中贝生长停滞，壳高仅 0.7 cm。

病贝壳内面可见虫管，壳表面有虫管开口，外观呈相连的假双孔型，以闭壳肌痕上的开口最多。闭壳肌痕区因厚度不够，与发病力强的外套膜接触少，易被穿透。有的群体贝壳内寄生平均 10 条虫子。

染病珠母贝的壳内面开口周围有炎症，局部脓肿，溃烂直至死亡。

该病发现于20世纪60年代的日本及70年代的广东沿岸珍珠养殖场。发病率：5%～71.5%，平均35%。

防治方法：必须采取杀死壳内穿孔寄生虫。防治新虫体附贝，方法的基本之点是用饱和盐水杀虫防治[6-7]。

b.贝壳穿孔病：病原动物种类有6种以上，主要是石蛏（*Lithophaga* sp.）、短壳肠蛤（*Botula silicula*）、楔形开腹蛤（*Gastrothaena cuneiformis*）、小沟海笋（*Zirfaea minor*）、海螂（*Mya* sp.）、翘鳞蛤（*Irus* sp.）。才女虫不多。有些贝壳上有穿孔海绵或贝类动物，其中仍以石蛏占优势。

该病发生在大珠母贝（*Pinctada maxima*）上。受害的贝壳有寄生洞穴，大小不一，深浅不同，乃至穿透。从洞穴可见穿贝者足面紧贴穴底。洞穴因穿孔者物种而异。贝之左右壳被穿孔情况不一，左壳洞穴数多于右壳的，但左壳被穿透的洞穴数又比右壳的少。

这些穿孔者使不同贝龄的大珠母贝均可受害，使壳皮和鳞片剥落，使珍珠层被侵蚀，有的遭严重破坏。甚至发生继发感染脓疡而死。

防治此病，谢玉坎等（1984）已提出了水泥砂浆涂覆法，即在养成期间，对患贝涂覆砂浆，并逐渐牢固与贝壳的结合，使病灶于1d内死去，而不对贝体本身产生任何损害[8]。

此病主要发生于日本、澳大利亚和华南地区。

3　珠母贝的敌害及其防除

珠母贝的敌害生物可以分对珠母贝生产产生直接或间接危害作用的敌害生物和妨碍珠母贝生存生长的生物2大类。

3.1　对生命产生直接或间接危害作用的生物大致可分为4类，即鱼、蟹、软体动物和涡虫等。

3.1.1　鱼类，肉食性鱼类如鲷、鳗、鲀、鹦嘴鱼（*Collyodon* sp.）等侵害幼贝和较大贝。它们以利牙咬破贝笼、嚼食贝肉。食量大、如1条15 cm长的中华独角鲀消化道中有几百万只幼贝，使采苗工作大受影响。黑点裸胸鳝可捕食大珠母贝的育珠贝，中华独角鲍食幼贝。

3.1.2　蟹类，一些青蟹、梭子蟹、石蟹危害幼贝或成贝，以＜5 cm的小贝受害最重。防治方法：在4～5月份尤要加强管理和捕捉蟹。

3.1.3　软体动物，贝类、青螺、嵌线螺和岩棘螺危害大，一旦它们（几个就足够）窜入笼内，不要1个月，珠母贝就可全笼覆灭。但织锦芋螺、方斑东风螺能捕食嵌线螺，对贝有良好作用。

羽状石蛏（*Lithophaga malaccarta*）、肉桂贻贝（*Botula silicula*）、楔形开腹蛤（*Gastrochaena cuneiformis*）和小沟海笋（*Zirgaea minor*）等都属穿孔者，其中以羽状石蛏最多，次为肉桂贻贝，危害程度也最严重。

海南岛的一些养贝场，发现合浦珠母贝笼内有10种嵌线螺，各地优势种不一，但多以毛嵌线螺（*Cymatium pileare*）、环沟嵌线螺（*C. aingulatum*）为主。幼体危害小于60 mm的成体螺。它们均捕食贝，致珠母贝死亡，最严重时间是有大台风的8月份，而死亡高峰期是5～9月，占全年死亡数的30.3%～82.8%，嵌线螺危害加上船舶油污和船舶往来造成的干扰，使当地养贝业大受损失。

3.1.4　章鱼，它用腕直接撬开珠母贝贝壳或分泌酸汁穿孔贝壳，再靠毒液麻醉开壳取肉食之。

3.1.5　涡虫，是指平角涡虫(*Planocera*)。它们可对附着不久的幼贝小贝致严重危害。在其繁殖生长旺季(6月底～10月份)，可大量吃食幼贝。防治方法：饱和盐水浸泡后用淡水冲洗贝笼或清洗后晾干再淡水浸泡一定时间可起到杀除涡虫作用。

3.2　妨碍珠母贝生存生长的生物

主要是附着生物，包括藤壶、海鞘、牡蛎、苔藓虫、海绵、海藻等。在其合适海区，它们可大量附于贝笼及珠母贝上，堵塞水流，减少珠母贝食源，或与其争食。附于贝壳顶部者(如牡蛎、海鞘、藤壶等)使珠母贝难以开壳而死。防除方法：加强管理和清洁珠母贝，适时调节贝笼所处水层。附着生物多发区，要适当换笼。

4　珠母贝的灾害

有关的自然灾害来自对天气、海况和环境状况的变化缺乏准确信息和应变手段。这包括掌握台风、热带风暴、寒潮、降水等天然条件的变化上是否清楚，否则会使珍珠养殖业蒙受大难。养殖场地的选择不充分考虑陆上径流、水库的排放及控制设施、防洪抢险准备工作不充分，会使养贝业处于难堪境地。防寒升温条件不具备也会使贝生活失常，难以适应：此外，不确当的臭氧处理海水对鳃换水不利。海洋开发工程，如石油钻探用的一些消油剂及产生的钻井泥浆均有不同程度的急性毒性效应(邓岳文等，2001)。

近几年来的赤潮频发，大多与水体的富营养化有关，对珠母贝等养殖业危害很大。如日本三重县养殖的马氏珠母贝1996年因炎夏水温高、水质差，发生赤潮致贝大量死亡。日本西部的涂抹珠母贝因太多不可食用的菱形藻致饥饿变弱、再感染传染性病而死(Tomarll等，2001)。为此必须对之有预警措施，及时将笼贝移入潮流畅通、营养程度正常场地加以防避。

参考文献8篇(略)

DISEASES, ENEMIES AND DISASTERS OF MARICULTURED PEARL OYSTERS

Abstract Pearl oyster is one of important mariculture objects in tropical and subtropical areas. It is of very high commercial value. Mariculturing pearl oysters are confronted with some diseases caused by pathogenic microbes and lower animals while they suffer from organisms of enemies and disasters. This paper discusses diseases, enemies disasters above mentioned and some measures aiming at them for prevention and treatment.

Key words Pearl Oyster Mariculure; Disease; Enemy; Disaster; Prevention; Treatment

扇贝的疾病*

摘　要　主要论述中国乃至世界养殖的8种扇贝的疾病、敌害、病害，其疾病有18种以上，分别讨论了这些疾病的流行状况和防治技术。

关键词　扇贝　疾病　敌害　病害　流行　防治

扇贝养殖是世界上一些临海国家的主要养殖业之一，我国养殖的扇贝类动物主要是海湾扇贝、美丽日本日月贝即日本月贝、长肋日月贝、栉孔扇贝、华贵栉孔扇贝、齿舌栉孔扇贝和虾夷扇贝等8种。

世界养殖扇贝产量有40万吨/年左右，在养殖贝类动物中仅次于牡蛎而居第二位。大规模人工养殖发端于20世纪70年代前后。因而迄今在它的育苗和养成期中都存在着一些病害问题，有的还相当严重。主要可归为病毒病、立克次体样生物病、幼虫面盘纤毛脱落、原虫病、海绵动物病、腔肠动物病、复殖吸虫病、线虫病、才女虫病、齿口螺病、扇贝蚤病、蟹奴病、豆蟹病、黑壳病等18种，另有敌害及环境和管理不适引起的病害。以下择其要分述之：

1　病毒病

病原是传染性(肝)胰腺坏死病毒。该病毒除在肝胰腺可见到外，其他组织如性腺、肾、外套膜、直肠、血淋巴中均有。被它感染11个月后，在肝胰腺中仍可测出最高滴度，但未见病毒复制。该病毒可从受感染鱼粪便和腺分泌物传播，鱼因此而衰竭，经环境中的过滤和积累，病毒颗粒传入贝类动物。受IPNV严重感染的贝，其鳃丝整个的表皮因鳃褶的继发解体而有进行性的细胞增生且分散各处。细胞逐步衰落。上皮层内侧显出分明的海绵状病原(spongiosis)。严重感染的鳃缺乏黏液细胞。该病发现于 *Pecten maximas*。1988年7月，较重的残废率影响到挪威西部Bergen附近的养贝场之幼体和成体贝。同时又发现了另一种病毒—*Birnavrus*。由于IPNV在贝体内持续时间长，对它的病理学意义和危害性至今了解不多。

在染病栉孔扇贝消化腺、肾、肠中发现具囊膜的球形病毒，完整者直径130～170 mm，核衣壳直径90～140 mm，无包涵体囊膜表面有囊膜纤突，而健贝中无。

2　原核生物感染

RLO，CLO，MLO，即三体病。

3　衣原体样生物病

Argopecten irradians养殖场其孵化卵和幼体中发现CLOD。该扇贝消化囊中有CLO。原生小体借细胞吞噬容易进入。用H&E(染色)，在光镜下，该包涵体被染成紫色颗粒，出现在消化囊末管上皮细胞的胞内液泡中。包涵体中的颗粒大小和分布均匀。

*　原文载于《学习和实践“三个代表”的重要思想》，P325～326，2005

CLO使加拿大的海湾扇贝、大西洋深水扇贝等贝类动物得病，感染率有的可高达40%。

CLO也在中国山东沿岸养殖的海湾扇贝、虾夷扇贝、贻贝、栉孔扇贝及葛劳贻贝中发现。染病贝消化盲囊小管上皮细胞浆中、小管腔和组织间隙有大量淡褐色圆形颗粒，形成空泡样病变。淋巴细胞、淋巴液、坏死上皮细胞和变性固缩细胞核构成了大片坏死病灶，该CLO系寄主细胞内专性寄生，其检出的感染率可达91%。

4　扇贝的幼虫面盘纤毛脱落病

4.1　症状

扇贝幼虫面盘纤毛脱落，乃至全部脱落，使幼虫失去浮游能力而下沉死亡。

4.2　病因

(1)亲贝性腺尚未成熟，受外界环境变化刺激而产卵，经受精而孵化的幼虫常出现面盘纤毛脱落。

(2)海湾扇贝促熟时间25～29天，积温在351℃～389.4℃所产的卵，孵化出的幼虫培育到5～7天会出现面盘纤毛脱落，经倒池可恢复正常。

(3)促熟时间23天左右的海湾扇贝，积温仅285.6℃时产卵孵化的幼虫育至第6天(壳长约50 μm)时，面盘纤毛脱落，身体下沉而死。

(4)孵化率高的海湾扇贝幼虫可出现大量纤毛脱落。

(5)有效积温67℃的栉孔扇贝之幼虫培育到第7天(壳长60 μm)时有少量纤毛脱落。有效积温61℃时所产幼虫培育第11天时见眼点(壳长192～200 μm)后有纤毛脱落，不变态。有效积温36℃～49℃时所产生的幼虫，眼点出现前即发生大量纤毛脱落而死亡。

(6)促熟中的亲贝所得饵料种类及数量不合适，其所产卵孵化出的幼贝一般发生纤毛脱落。如单喂小球藻、异胶藻或扁藻，效果不好。刚扩种的饵料若包含在未耗尽肥料或老化的饵料中，分泌过多毒素，均会致此病。

(7)投饵量、藻液密度既影响亲贝性腺发育，也与幼贝面盘纤毛脱落有关。性腺发育不正常，其后孵化出的幼贝易患纤毛脱落病。

(8)pH≥8.3～8.5的池水，亲贝性腺不发育、甚至退化。换水后，亲贝虽产卵和有较高的孵化率，但所得幼虫第8天时(及之后)大量发生纤毛脱落致死。

(9)幼虫的密度过大，其面盘纤毛脱落者多。

(10)光照过强(如9 600～10 200 lx)、水温过高(如28.7℃)，使幼虫患此病者增多(如青蛤和紫石房蛤幼体)。

(11)贻贝幼体若在pH5.9的水中，也易发生此病。

4.3　流行情况

该病发生于海湾扇贝、栉孔扇贝的幼体中，如1989—1995年间每年的3～4月份间，山东荣成、蓬莱育苗场的海湾扇贝幼虫，该病发生率有的很高，可致大量死亡。1988年—1996年间每年5月份在山东蓬莱等地的一些育苗场中，有栉孔扇贝幼虫全发生此病，有的甚至导致大量死亡。

日本虾夷贝人工苗，壳长150 μm时，也可患此病，但日本的金井丈夫认为此是一种正常的自然淘汰现象。

4.4　防治技术

(1)亲贝性腺促熟中，必须达到所要积温，投喂合适种类及用量的饵料。海湾扇贝于2

～3 月份入池促熟，经 30～32 天，积温 410～443℃；栉孔扇贝 24～26 天，4℃以上有效积温为 191.6～218.2℃时，幼虫不生此病。

(2)促熟亲贝以 40～50 万 c/mL 的量投喟小新月菱形藻、三角指藻、牟氏角毛藻等硅藻，亲贝产幼体好。

(3)良好的生态条件、幼虫密度：8～10 个/mL，光照：200～400lx，水：pH7.9～8.2，COD≤mg/L，NH_3—N≤100×10^{-9}，DO≥4 mg/L，水清、常换，池底：见清或倒池。

(4)幼贝应投喂等鞭金藻，叉鞭金藻、单鞭金藻等，不投老化饵料和刚扩种的饵料。

(5)培育池中施用青霉素、土霉素等，浓度(1～2)×10^{-6}。防杂菌生长，净化水质。

吕豪等(2003)认为海湾扇贝面盘幼虫解体病的主要因素是亲贝性腺发育不成熟，在亲贝促熟上多下工夫，可以有好的结果。太平洋牡蛎体也可能患盘纤毛脱落病。

4.5　扇贝的敌害

扇贝的敌害主要是肉食动物、附着生物及其他。

肉食动物包括鱼，如鰈、杜父鱼。它们捕食小扉贝、海星、海胆，尤其是蚕食稚贝、幼贝，危害大。蟹则是用螯足钳碎贝壳而食之，壳蛞蝓舔食 500～600 的稚贝；附着生物包括藤壶、牡蛎、海绵、贻贝、金蛤、石灰虫及一些藻类。它们一是附于贝壳，二是附于养殖网笼，影响扇贝活动和摄食。

4.6　监测和诊断法

(1)应用免疫学手段识别有毒藻。

(2)遗传探针技术识别害藻体特异的 DNA 或 RNA 片段。

4.7　海湾扇贝死因探讨

(1)环境条件不适：蓝淑芳(1990)发现如长岛海区 1989 年夏季有强温跃层形成，受太阳辐射风和黄海冷水团影响，温跃层深度常变剧变，扇贝不适而死。赤潮，1989 年中国有 12 次赤潮席卷河北、天津、山东，持续 72 天。迄今中国海发生赤潮的机率有升高之势。

化学污染物，郭新堂(1992)指出，1989 年黄岛油库爆炸，几百万吨原油入海。对突发事故应加强污染监测和防治。

(2)盲目追求早苗：育苗提前到 3 月，幼体先天不足，生长期加长，使闭壳肌指数和重量下降，闭壳肌糖原降低，体质变弱。应移向外海(>10 m 深处)。早期苗易死。分苗、成熟期应避开 9 月份。3 月 15 日后才可促熟亲贝，4 月 15 日采卵，5 月 15 日出池(大于 400 μm)，6 月 10 日达 0.5～1 cm，虾池或海上轮养至 8 月中旬，使扇贝平均高达 0.5 左右分苗才可死亡。

(3)过度密殖：滤水率过低致死，应加大距(5 m)，区距(8 m)。浮筏有效长度≥60 m，筏间距 6～8 m，4 行合一亩。疏散，吊养密度：400 笼×10 层×(300～350)粒卵＝12 万～14 万卵/亩，就行。

(4)单养效果差：可采用对虾、贝类混养或与海带等轮养。

(5)海湾扇贝原产于美国浅水区，成体足丝小或无，附着力差，可用垂下式网笼。到了中国，仍用此法，经不起风浪和大流的袭击，易致死。试验证明，表层下 2 m 水层效果好于 1 m 层和 3 m 层处，垂吊式的好于海底的。对于栉孔扇贝也应选择合适网眼的笼，谨防高温期摔打笼再挂养等操作。

(6)采用新技术：培养三培体，既不育又可避免繁殖季节。体壮、适应力强。

改善饵料质量。紫色无硫细菌、光合细菌用于育苗。

牡蛎的微生物疾病*

摘　要　牡蛎养殖是海水养殖的主要种类之一。悠久的养殖史同时发生发展出因微生物源产生的一些疾病，从而制约养蛎业的发展。本文试图立足我国，放眼世界，论述养殖牡蛎的微生物病原病害及其防治，以利于推动牡蛎养殖业的健康持续发展。

关键词　牡蛎养殖　疾病　病毒　细菌　真菌　病害防治

牡蛎养殖有着悠久的历史，在我国有2 000年以上的养殖史。它是当今世界上贝类养殖的支柱产业之一，各临海国家均有生产，其产量居养殖贝类之首。

在我国主要养殖的种类主要有：长牡蛎（太平洋牡蛎，*Crassostrea gigas*）、褶牡蛎（*Ostrea cucullata*）、近江牡蛎（*O. rivularis*）、大连湾牡蛎（*O. talienwhanensis*）、密鳞牡蛎（*O. denseanellosa*）和僧帽牡蛎（*Saccostrea cucullata*）等6种。其他国家养殖的品种有：欧洲巨蛎（*C. angulata*）、商业巨蛎（澳大利亚岩牡蛎）（*C. commercialis*）、*C. gryphoides*、*C. rhizophorae*、美洲巨蛎（*C. virginica*）、（*Ostrea equestris*）、叶片牡蛎（*O. folium*）、叶牡蛎（*O. frons*）、青牡蛎（即灰黄牡蛎、扁牡蛎 *O. lurida*）、土拉牡蛎（*O. tularia*）、*O. vitrifacta*、*Saccostrea aommercialis*（*sydney oyster*）等。

根据病原，牡蛎的微生物疾患大致可分为病毒病、细菌病、真菌病、三体样病等。每类疾病根据病原微生物又可分出许多种疾病。现就各种疾病的病原生物、起因、症状和病理变化、流行情况和防治方法分别叙述如下[1-3]。

1　病毒性疾病

已从美洲牡蛎、太平洋牡蛎、欧洲扁牡蛎、扁牡蛎、葡萄牙牡蛎、商业巨蛎中发现了至少属于7科不同的病毒[4]。

1.1　牡蛎面盘病毒（OVVD，缘膜病）

OVVD（Oyster velar virus disease）由属于虹彩病毒（*Iridovirus*）的病毒引起，为20面体的DNA病毒。此病可使壳高＞150 μm的长牡蛎致病。据Elston等[5]研究有38％的样品的盖膜上皮有此病毒包涵体（圆球形细胞质的），13％的样品之口、11％的食管上皮有此包涵体，相连的病灶包括细胞肿胀、脱落，细胞内的损伤包括核的膨胀。受病毒质（Viroplasm）连累的线粒体也鼓胀。所有这些病理变化包括细胞的剥离均在病毒粒子形成之前发生[5]。

该病主要发生在太平洋牡蛎（*C. gigas*）中，每年3～6月幼牡蛎损失＞50％，波及牡蛎体的边膜及卵的孵化，使死亡率增高。感染OVVD的牡蛎缺乏活力，内脏团缩入瓣膜内。当边膜从瓣膜扩展时，则不能正常活动和移动。从边膜分离出来的已感染的盖膜上皮细胞呈现出沿边膜周边的疱疹。受感染的牡蛎之盖膜、口、表皮组织有Viroplastic包涵体。

* 原文刊于《水产科学》，2007，26(9)：531-534。

Viroplasm 呈粘粒状电子致密。完整的病毒颗粒中有一个中等稠密区所包围的致密的核心，它从病毒壳体的核心分出，病毒粒子平均长为 228 nm。该壳体含两个双层膜结构，消化道胃部内有海洋真菌(*Hyalochorella marina*)，可能与此病有联系。

OVVD 流行的地理范围：1985 年曾在美国华盛顿州的 Willepa 湾和 Puget 海峡发现此病。该种牡蛎 1902 年自日本传出后已扩展到亚、美、澳等地。该病传播途径可能是由携带此病的亲体向幼体作纵向感染，亲体也许是转移宿主或储存宿主，而主要患病者常是育苗场中的幼体。

防治方法：目前主要是加强管理，建立正确的诊断法。一旦发现，彻底消除。彻底消毒幼体牡蛎群及相关设施。建立无病的孵化储备并保证不被污染。

欧洲巨蛎(*C. angualata*)也有类似 OVVD 的病毒病发生，在病体的鳃和血细胞中发现有此病毒。

1.2　疱疹型病毒

由疱疹病毒(Herpes-type virus)引起。主要患病对象为长牡蛎(*C. gigas*)和美洲牡蛎(*C. virginica*)。该病毒颗粒呈六角形，也有圆形或多角形，直径 70～90 nm，有的单层外膜内具浓密的类核体(Nueleoid)[1]。病毒主要发现于牡蛎卵及幼体中。幼体的面盘、外套膜、鳃组织及围绕消化管的结缔组织均见病灶。感染者消化腺苍灰色。群体出现散发性死亡。不正常的细胞核电镜可见胞内和胞质内有病毒样颗粒，核含球形或多边形颗粒。肌细胞的胞质内可看到裸露的胞质核蛋白。在幼体和年轻牡蛎的其他细胞内可测出胞质泡囊有包被的病毒粒子。在溶胞了的细胞中的胞外空间也见包被的病毒。这些颗粒组成了一个电子致密的病毒核心的衣壳。结缔组织的肌胞(Myoeytes)胞质中含无包被的颗粒。牡蛎卵中可见到不正常颗粒内质网积累，线粒体肿胀，且两者相连，再与电子透明中心的浓密核相连。常见已退化和溶化的被感染的核。

当 *C. gigas* 幼虫进入有该病毒的悬液后，仅 2 d 幼虫即停止游泳并沉于底部。镜检其盖膜全部已受损感染，至第 6 d 积累死亡率达 100%。有时也见高死亡率后的幸存者。虽镜检不出病毒，但仍有不正常核及压缩的染色质。Renault 等由健康的野生幼牡蛎和健康或受中等死亡率影响的养殖幼牡蛎中证实无此病毒存在[6]。

牡蛎疱疹型病毒 1991 年首见于新西兰，发病记录始于 1992 年(Farley 等)。1992—1993 年 2 个夏季在法国西岸进一步爆发。染病牡蛎群体死亡率达 80%～90%。爆发后 1～4 个月内，法国两处养殖场所剩无几。

此病毒储存液存于－20℃中，几个月仍具有感染性。该病可能与水温高有密切关系，病蛎出现在电站排出的热海水中(28～30℃)，降温则可消除此病。将染病的牡蛎转移到邻近温度较低(12～18℃)的天然海水中，可降低死亡率。

美洲巨蛎外膜具纤毛的表皮细胞存 Cowdry 包涵体(Intranudeu incluiiou bodies)，该包涵体似该种牡蛎细胞中所含的疱疹样病毒[7]，幼体中出现过。在食用牡蛎中分出 birnavirus[8]。

1.3　卵巢囊肿病

病原为类似于肿疡病毒(*Papovaviruses*)的病毒，电镜下观察呈 20 面体。该病主要发生于美洲巨蛎身上。染病病灶为生殖腺，可见到生殖腺上皮细胞肥大、异常，细胞核中含膨大颗粒团呈孚尔根阳性反应。美洲牡蛎已有染病记录。近年在佛罗利达的彭萨科拉湾、缅因州、长岛一带也有类似病蛎出现。感染高峰约在盐度14时出现。

该病毒还导致韩国长牡蛎配子体病毒性肥大;东方牡蛎也可染有该病毒样包涵体[9]。

1.4 其他病毒病

C. angulata 体内有3种病毒,一种感染血细胞,另一种感染鳃区组织细胞,第3种感染长牡蛎血细胞。3种病毒生活史类似于OVVD病毒,均属虹彩病毒属,均呈20面对称体,比OVVD病毒大(约380 nm),均可使牡蛎成体发病。

珍珠牡蛎(*Pinctada maxima*)和欧洲扁牡蛎体内有Papova病毒样和Herpes病毒样感染。前者病毒颗粒无被膜,20面体,直径60 nm,属*Papovaviridae*(科)。后者在胀大的细胞核内有球形或多角形病毒颗粒,直径80 nm,似由具核心或不具核心的衣壳组成;细胞质内有被膜,有不对称形颗粒,直径160~180 nm,由具电子致密的核衣壳组成。病毒核心偏位。周围充填电子致密基质,并包以单位膜。形态一致的病毒颗粒有一长250~350 nm突出伸展的尾样结构,横切核衣壳为环形,直径140~150 nm。

从*C. virginica*中发现了呼肠弧病毒。此外,葡萄牙牡蛎感染虹彩病毒。罗文新等也从牡蛎肝胰腺中发现病毒样颗粒[10]。

1.5 嗜碱胞质包涵体

美洲巨蛎的一些成体其肠道上皮细胞中有最大直径为50 μm的嗜碱胞质内包涵体,它颗粒细致。此外一些牡蛎消化道细胞的胞质中含有的这种包涵体则较小,直径为10 μm,有晕轮。体内含杆形体,上述的包涵体数量少。幼体中可见此包涵体,也见有Cowdry A型核内包涵体(Intranuclear inclusion bodies)。

此外,对长牡蛎死因之一进行分析,见其外套膜、肝脏、中肠组织有大量直径为0.1~1.10 μm的球状有机化合物,但无病原体感染,认为是严重有机污染中毒而死,病区:1993年广西北海[11]。

2 细菌病

2.1 弧菌病(Vibriosis)

2.1.1 幼体牡蛎溃疡病

相关病原可能为鳗弧菌(*Vibrio anguillarum*)、溶藻弧菌(*V. alginolyticus*)以及气单胞菌(*Aeromonas* sp.)和假单胞菌(*Pseudomonas* sp.)。1970年Brown等即证实病原是弧菌,因为在幼牡蛎的全身组织中均有分布。用它们做人工感染试验,4~5 h幼体就出现组织消散或溃疡性变化,随之幼体大量死亡。

症状:全身性组织的溃疡,养殖水体中见幼体缺乏活力,下沉以至死亡。光镜观察表明,病体内有大量上述细菌,异常的面盘、溃疡甚而崩解的组织。1981年Elston等提出染色排除法,以0.5%的台盼蓝区分出病患组织,达到早期诊断的目的,当然有条件好者,可作荧光抗体试验,效果更好更准。

流行情况:主要为美洲牡蛎幼虫,也不排除其他种类的牡蛎(长牡蛎、欧洲扁牡蛎、褶牡蛎等)。当条件合适时,弧菌可迅速繁殖传播而致病[12]。

预防措施:①加强养殖水体的卫生管理,使弧菌数不超过10^2/L。对菌含量过多者用50~100 mg/L的复合链霉素(Combostrep)浸浴幼牡蛎,有望治愈此病。此外多黏杆菌素B、红霉素、新霉素等也有疗效,但应注意抗生素的副作用,谨慎使用。②检测出感染者应丢弃。③确保饵料(如单胞藻)无病菌污染。④有效地采取消毒方法,包括过滤、臭氧或紫外线。

2.1.2　幼体弧菌病

病原为类似溶藻弧菌的弧菌等，主要发生在食用牡蛎（*O. edulis*）、欧洲巨蛎（*C. angulata*）和美洲巨蛎身上，能使50％幼体得病、致畸。

另一种弧菌为亮弧菌（*V. plendiduss*），使三倍体长牡蛎稚贝患杆菌坏死病，试验证明氯霉素可完全抑制该菌，红霉素、新生霉素、庆大霉素和链霉素也能有效减轻患病稚贝死亡。

2.1.3　表面弧菌病

病原为弧菌，但并未指出是何种弧菌。主要发生在壳高约0.8～3.0 nm的幼体身上。幼体染病后，贝壳表面首先感染以至不能完全钙化，周边有未钙化的几丁质区，使两壳不正常，壳体缺乏强度、硬度、易碎，与壳瓣分离，由此深入壳内，影响韧带发挥正常功能。体内没有正常的消化活动，无食物，而有脱落的吸收细胞，幼体渐渐生长不良，乃至死亡。

该病主要发生对象为食用牡蛎和美洲巨蛎，也有欧洲巨蛎，此病导致的死亡率为25％～70％。

弧菌病有3种致病类型，①各期均可受影响，幼体牡蛎弧菌病。染病者静止，细菌均附壳内面，沿外套膜生长，入侵内脏腔，打败吞噬细胞，且大肆繁殖。②早期面盘幼体受害。幼体活动异常、缘膜损伤，处于伸张态，牵缩肌剥离。后期，细菌侵袭整个缘膜。③后期面盘幼体受害。幼体处静止态。内脏萎缩呈进行性、广泛性。消化腺吸收，细胞脱离。后期，消化器官呈病灶性损伤[4]。

此外，沈晓盛等[13]曾报道，坎氏弧菌（*V. campbellii*）可能引起太平洋牡蛎生病。Gooch等[12]也指出副溶血弧菌（*V. parahaemolyticus*）使牡蛎致病。

2.1.4　土壤丝菌病（*Nocardiosis*）（PON），也称夏日之死

曾与一种多发性脓肿或局灶性坏死的细菌病相符，1990年Friedman和Hedrick重新定名。该菌为主要病原，该菌G^+或不定，丝状和分枝状，中度抗酸染色。PAS阳性。染病者外观变化不明显。有时闭壳不全或无力。外套膜正常或有黄、绿、褐色小结节。组织切片见生殖腺滤泡、消化道周围的囊样结缔组织呈受该菌侵袭样，有菌落可见。其他部位有少量感染，感染处大量血细胞浸润。病重者见组织坏死。

温暖富营养海区港湾的太平洋牡蛎多发此病，流行于北美西海岸，日、朝沿海。各龄期均染此病，但以2龄以上为重，该病不以水为媒介传播。

2.1.5　牡蛎的足病

为欧洲记录的最古老的牡蛎病，发生在法国Arcachon流域。病症有外收肌的附着部下及邻近的肌组织变得粗糙，水泡状和退化。以后于1919、1923年分别在意大利和英国及欧洲其他国家中发现。于1924年，据牡蛎外膜退缩、消化腺苍白、肌肉退化、肠炎、贝壳生小隆起脓疱等症状判断此病为细菌性疾病（Ortou）。

传染途径：当幼体附于有弧菌的基物后，壳被感染，再侵入韧带、外套膜、鳃，最后致全身感染而死。

预防措施：养殖用水、器械和容器的彻底消毒。所用药品有：漂白剂溶液，可将5.25％的次氯酸钠以1∶100稀释于水中，进行环境消毒，并用干净海水冲洗干净。盛放稚贝的器皿中水应常换并做细菌学检查，以便采取适当措施。染轻病的稚贝，尚可用稀次氯酸钠

溶液(100 mg/L)浸泡 1 min,再用干净海水冲洗净。此法的有效性以处理后贝壳是否完好,闭合是否充分为准[4]。

2.1.6　点状坏死(多发性脓肿)

病原可能为一种无色杆菌(*Achromobacter*)。

细菌学的初步研究表明此菌系 G^-,具鞭毛,能主动运动,个体长 1～3 μm。

此病已获得人工感染结果,濒死于此病的病牡蛎之组织学检查发现其细腻浸润已扩散,组织坏死,呈点状,并伴有大量此菌,最后即是大批死亡。

该菌主要感染各种规格的太平洋牡蛎、长牡蛎[4]。

流行情况:日本广岛湾。可引起大批死亡。之后,1960、1965、1970 年等均有许多报告可见。

2.1.7　多发性脓肿病

病原为一种 G^+ 杆菌,由 20%的牡蛎体上分出。症状似点状坏死症,包括消化腺变苍白、壳张开、散发性死亡等。主要发生在日本松岛湾的牡蛎(*C. nippona*)上及引自美国西岸的该牡蛎之种苗和成体上。

2.1.8　放线菌样病

病原为一种放线菌,该生物有分开的小型菌丝体,约 10.6 μm×0.7 μm,但无树栖形分枝,无星状体,直径为 1.5～2.0 μm。菌丝体尚有直径为 0.8 μm 的小末端孢子,GMS 阳性,不耐酸,当用 GD 和 β&B 染色时不见,H&CE 染色切片可见。

发生于美洲巨蛎体。症状:该菌感染使牡蛎体普遍水肿而消瘦,消化管上皮变薄。检查病体可见菌丝病(Mycelial disease)状的分布及寄主病理学变化。该病记录自美国纽约长岛,夏季最为流行。

放线菌(*Actinomyces*)[和原生质型单孢子虫(*Haplospridium plasmodid forms*)]可引起牡蛎病变继而爆发流行性病。

2.2　其他细菌病

长牡蛎幼体中有时有一类似嗜洋红(细)胞菌、弯曲菌、产黄菌的细菌、G^-、胞壁柔韧可曲,无鞭毛,长杆形,>3.0 μm,大小均一。该菌使幼贝的绞合部韧带发生腐蚀性糜烂,外套膜组织病损。93%的弹力纤维、20%的张力纤维受损。

太平洋巨蛎有致命的发炎 Bacteomia 症状出现,另 Boragjuni 等[14]灿烂弧菌研究证实,*Vibrio splendidus biovar* Ⅱ致幼体病,该菌成分及体外产物均显毒性,但毒素产力与病原性间无明显相关。

国外已有报道,诺卡氏菌也使牡蛎得病。美洲巨蛎因创伤弧菌(*V. vunificus*)得病,Birkenhauer 等研究过二乙酰对此的减毒作用[15],Borazjtmi 等也有相当的研究[14]。美国学者 Noga 等已发现该种牡蛎体内产生新型抗菌蛋白,即美国牡蛎防卫素(AOD)可以抵御疾病。

3　真菌病

3.1　幼体离壶菌病

病原为海洋动腐离壶菌(*Sirolpidium zoophthorum*)。该菌感染幼体,在幼虫体内弯曲生长出菌丝。菌丝有不多的分枝,动孢子囊生出排放管至幼体体外,并释放游动孢子。该孢子梨形,鞭毛 2 根,5 μm×3 μm。孢子游动水中短时后再感染其他幼体。

症状和诊断：显微镜可见患体内有此菌丝。尤其稚贝染病后，即停止生长和活动。用中性红海水液也可区分出真菌和幼体，肉眼则可见死去的多数患病幼体。少数幸存者病愈而得免疫力。

此病主要流行于美洲巨蛎幼体的不同阶段，严重时可导致大批死亡。

迄今尚未有治疗方法，只能及时采取预防措施，如：严格过滤或紫外线消毒育苗用水。一旦检出病体，即刻销毁，所用器具立即彻底消毒。

3.2 壳病

病原为绞扭伤壳菌(*Ostracoblabe implexa*)，系藻菌纲真菌，菌丝上有呈卵形或球形膨大体，球形者膨大为厚壁孢子。低温时(5℃)该菌的厚壁孢子增高增大。15℃下消毒海水培养染病碎片，3～4周可获该菌菌落。酵母蛋白培养基可作纯化培养时用。

症状与诊断法：牡蛎受感染后，壳即被穿孔，尤其闭壳肌处最重。内壁显白色雾状，并显疣状突起，后者变黑或微棕，或淡绿色，甚至有大片壳基质沉淀。但发生的病理变化因地而异，产生的伤害程度也有差异。诊断的主要依据是视培养病灶能否得出相应菌丝来定。

该病有广泛的流行区，如欧洲的荷兰、法国、英国，北美的加拿大和亚洲的印度等地。总感染率10%，<1%的总病例可达赘疣期。发病高峰以水温高于22℃的秋季时节为主，长达2周，病程可能加重。

染病者为欧洲巨蛎和另一种巨蛎(*C. gryphoiodes*)。

迄今尚未找到该病的防治方法。

念珠霉菌(*Monilia* sp.)，后被Alderman等定为藻状菌*Ostracobale implexa*。它可使荷兰*Ostrea edulis*群体死亡。

3.3 肤囊菌病

病原是藻菌纲的海水肤囊菌(*Dermocystidium marinus*＝*Larynthomyx marinus*)，它主要侵染美洲牡蛎、叶牡蛎(*O. frons*)和等纹牡蛎(*O. equestsis*)的成体。

该菌孢子近球形，直径3～10 μm，以5～7 μm居多。胞质内的大液泡偏于一侧，内含液泡体(Vacuoplast)，周围有一层泡沫状细胞质，卵圆形胞核位于胞质较厚一侧。

寄生于牡蛎的该菌，待寄主死后可变成动孢子，释放并游动于周围水中，伺机再附着，并变成变形虫状，借助寄主的鳃、外套膜、消化道而进入上皮组织，至身体各部，并在寄主细胞内或细胞间繁殖，使细胞溶解、组织肿胀、患体消瘦停长。外界条件不利则加速死亡。

该病流行于美国弗吉尼亚、古巴、委内瑞拉、墨西哥和巴西，以夏秋季(8～9月)为死亡率高时(水温30℃，盐度30)，有时可高达95%的死亡率。反之，牡蛎在温度>15℃、<25℃时，也不被该菌感染。

迄今仅有预防法，即彻底清理病灶，消毒用于牡蛎的附着器件，合理密植。有条件的可注入淡水降低盐度，避开25～33℃的水温时间。

4 支原体、衣原体、立克次体样生物病(三体样病)[4]

病原为立克次体样生物(Riehettsia-like Organisms)(RLO)。该病原由Harshbarger等从美国巨蛎、太平洋牡蛎等中发现。1994年Renault等才从长牡蛎鳃上皮细胞质中发现，也可从消化管上皮细胞发现，很像*Wolbachic*属的生物。超微结构研究表达出它们被包围在Parasitophorous空泡中。该空泡有一单位型的膜。有些表面具纤丝。该生物长

0.4～4.0 μm,直径 0.2～2.0 μm,多形态。一些美洲巨蛎的消化腺之嗜碱胞质包涵体(较小者)含的立克次体样生物,外层有 3 层,大者为 1.19 μm×0.41 μm,较大的盲管(Divertleularc)包涵体(100 μm)也含立克次体生物[15]。

支原体样生物(Mycoplasma-like organisms mLO)和衣原体样生物(Chlamydia-like organisms CLO)。

对美国切萨皮克湾的一些美洲巨蛎做 TEM 观察,发现肠小片细胞(杯状细胞)(Goblet cells)含支原体样生物,它们体内类似的兰体(Bluebodies)中也含支原体样生物。

在美州巨蛎中已发现 CLO,其致病性不明。

5 结语

据上所述,可见牡蛎的微生物源疾病主要来自病毒(7 种)、细菌(9 种)、真菌(3 种)和三体样微生物。总体看,相关研究国外做得多且深刻。国外相关的水产医学,包括牡蛎在内,已达到快速鉴定,对病害机理有较深理解,有些已达到基因层次,包括遗传因子等。对耐药机理也有较深广的认识。我们与国外的差距主要表现在:(1)对多数种类的疾病缺乏深刻系统的研究;(2)对疾病的认识多处于不甚了解状态;(3)缺乏统筹、长远和深层次的考虑,使之头痛医头、脚痛医脚;少远见性,常处于被动挨打的局面,对疾病难以做到综合治理和持久效果;(4)这很大方面归因于理念、方法的陈旧,仪器设施更新慢,难以有大面积、高水平成果出现。为此必须加强基础设施建设,提高科技和管理水平,包括疾病测报、防治、药检等水平,提倡和应用微生物调控、修复环境等,使牡蛎养殖业整体素质和水平大提高。随着时间推移,关注程度和经济投入的增多,国内牡蛎的病害研究必将深广。

参考文献 15 篇(略)

(合作者:陈 阳)

牡蛎的寄生虫病*

摘　要　据病原动物类别，牡蛎寄生虫病分作5类，对寄生虫病的病原、流行状况、检测、诊断和防治分别作了讨论。

关键词　养殖牡蛎　寄生虫病　疾病防治

据不完全统计，约有15种牡蛎遭到由6大类动物（主指寄生虫）引起的各种病害，这些动物大约可概括为原虫、蠕虫、寄生性甲壳类动物、附着生物、掘穴生物及海绵。本文对这些疾病的病原、流行及状况、检测和诊断、预防和治疗作一论述以供参考。

1　原虫病

1.1　六鞭毛虫病

1.1.1　病原　肉鞭动物门双滴虫目的尼氏六鞭毛虫（*Hexamita nelsoni*）。该虫梨形身体，8条鞭毛中的6根伸前为前鞭毛，此外身体两侧各有1根向后。

1.1.2　与寄主关系　寄主是*C. gigas*、商业巨蛎（*Crassostrea commercialis*）、青牡蛎（*Ostrea lurida*）和欧洲牡蛎（*O. angulata*）。该虫据说是食用牡蛎或青牡蛎的死因。但有些报告认为两者（六鞭毛虫与牡蛎）间有共栖关系。这要取决于水温的适宜和牡蛎的代谢机能活动强健与否。若是则此时的牡蛎可排除其体内过多的六鞭毛虫。反之，低水温、低代谢机能的牡蛎将促使共栖者六鞭毛虫变为病原或寄生物而致病。

1.1.3　流行　该病主要发生在荷兰和美国，无严重流行病和防治方法报道。

另一种六鞭毛虫*H. inflate*可使美洲牡蛎受损。

1.2　扇变形虫病

该病的病原有两种，一种是帕特扇变形虫（*Flabellula*＝*Vahlkampfia*）（*F. patuxent*），另一种是同属的卡氏扇变形虫（*F. calkinsi*）。均是大约75 μm长的小型变形虫，具宽广的扇状体形，但常有变化。行动时宽度大于长度，宽边向前。体内的原生质前部有时呈峭状或具深裂，也迅速平缩进。有时形成尾状丝，该丝有时细长，有时具瘤。颗粒状内质。泡状胞核，球形核仁。伪足类圆锥形或扇形，是体长的2～3倍。帕特扇变形虫虫体直径可达140 μm，吃食细菌，以复分裂或出芽繁殖。偶尔形成具单核的球形包囊。而卡氏扇变形虫的背腹扁，其包囊虽大小一致，但外形仍不规则。

扇变形虫主要与牡蛎营共栖生活，一般发现于消化道内，而未从其他组织和细胞内发现过。借助染色可从形态上将它们与牡蛎本身的血细胞相区分。

该病主要发现于美洲巨蛎体（Sprague，1970）。

1.3　线簇虫病

* 原文刊于《河北渔业》，2007，10：6-10转22，本人为第二作者。

1.3.1　病原　可能致病的线簇虫有 3 种，即牡蛎线簇虫(*Nematopsis ostrearum*)、普氏线簇虫(*N. prytherchi*)和未定种的线簇虫(*N. prytherchi*)，此外在一些成体牡蛎中尚检测到线簇虫样生物。

牡蛎线簇虫是一种常见的寄生虫，而且在牡蛎的外套膜中寄生最多。寄生在牡蛎等软体动物中时是其孢子，该孢子大小为 16 μm×(11～12)μm。

普氏线簇虫以孢子形式常寄生于牡蛎线簇虫的同一牡蛎寄主上，但寄生部位仅限于鳃，该孢子大小为 19 μm×16 μm。

另一个未设定种的线簇虫的孢子较小，仅 11 μm×7 μm，寄生于鳃部。此外，一个线簇虫样生物寄生于幼体牡蛎上。

1.3.2　症状　线簇虫对牡蛎寄生，一般状态下不威胁到牡蛎的生存，原因可能是双方之间可建立起动态平衡，即牡蛎能排出线簇虫孢子，而不致在体内长期过多积累。一旦线簇虫孢子大量积累，其主要作用是机械障碍，鳃血管的被堵塞等而致病。

1.3.3　诊断　以发现牡蛎外套膜、鳃等器官上有此类孢子可确诊。

1.3.4　流行情况　这些线簇虫孢子和簇虫样生物主要寄生对象是美洲巨蛎、等纹牡蛎等的成体或幼体上。该病主要流行于美国大西洋沿岸，包括纽约长岛海区，我国的台湾所养的巨蛎也有报道。

1.4　单孢子虫病(Haplosporidiosis)

1.4.1　海产派金虫病　海产派金虫的学名为 *Perkinsus marinus*。属原生动物顶复体门派金虫目的单孢子虫类。它可使牡蛎得重病，而且牡蛎年龄越大，越易致病。

1988 年，Andrews 发表了它的生活史报告。它寄生并传播于牡蛎养殖区的不远范围内(病牡蛎周围 15 m)。

海产派金虫孢子近球形，其直径以 5～7 μm 居多。细胞质内有一个偏位的大液泡。液泡含一个大液泡体，形状不定、折光。一层泡沫状的细胞质包围液泡体。偏于泡内一侧的卵形胞核则处于细胞质较厚部，核膜靠一圈无染色带包围。

孢子在宿主体内发育成为游动孢子(具 2 根鞭毛)，将游动于水中，伺机再附于牡蛎身上，脱鞭毛成变形虫状，经鳃、外套膜或消化道入侵上皮组织。借寄主的细胞吞噬作用进入软体部组织再寄生于细胞内或细胞间，作二分裂或复分裂繁殖。

患此病的牡蛎，其结缔组织、闭壳肌、消化道上皮组织和血管是主要受害部位。早期症状是组织发生炎症，并有纤维变性，直到组织的广泛溶解、肿胀或水肿。也有得慢性感染者，其身体渐消瘦，生长停滞，包括生殖腺发育的受阻。重感染者壳口张开而死，一旦条件不利死得更快。

该病诊断方法的原理是借染色培养的营养体或动孢子囊或染色组织切片来鉴别。培养活体组织的培养基是葡糖巯基醋酸盐琼脂。检查活体组织有无海产派金虫的培养基成分：酵母粉 50 g；胰蛋白胨 15.0 g；葡萄糖 5.5 g；巯基乙酸钠 0.5 g；刃天青 0.001 g；氯化钠 2.5 g；*L*-胱氨酸 0.5 g；琼脂 0.5 g；蒸馏水 1 000 mL。培养温度：25～30℃。培养时间：1 d。鲁哥氏碘液可将营养体或动孢囊期扩大形成的壁染黑即可确诊。

该病使美洲巨蛎、叶牡蛎(*Ostrea frons*)和等纹牡蛎及长牡蛎染病。发病地区主要有美国、古巴、委内瑞拉、墨西哥和巴西及中国的台湾等。以夏季和初秋(8～9 月)为发病高峰期，有时染病率可达 90%～99%。水温降，死亡减，冬季不发生该病。

流行条件：较大牡蛎、一年期以上的较大幼蛎，水温：30℃，高密度养殖，流行于病体周围 15 m 以内环境。

也曾有类似于海产派金虫的派金虫寄生者，如奥氏派金虫、大西洋派金虫和 *P. karlssoni* sp. 也可能来寄生[1]。

预防方法：选好低盐度养殖区（<15），彻底清刷牡蛎附着基物及其他任何杂物，安排蛎床的密度要确当，可免传播疾病。免用已染病者作亲蛎，适时收获。该病进程快而短（1 d 左右），一旦发现，欲治难行，只好提早收获。至今尚无有效治疗法。将病蛎移至低盐度区（小于 5 的盐度）仍可忍耐。夏季虽高温，若其他条件不满足仍可迫使其衰落。

1.4.2　单孢子虫病　牡蛎的这类单孢子虫病原是单孢子虫，属于单孢子虫目。使牡蛎致病的单孢子虫至少有三种，它们分别是尼氏单孢子虫（*Haplosporidium nelsoni*）、沿岸单孢子虫（*H. costale*）和装甲单孢子虫（*H. armoricana*）。

① 尼氏单孢子虫个体卵形，6～10 μm 长，一端具盖，盖缘延伸至孢子壁外。寄生在牡蛎的该虫变成为具多个核的多核质体（Plasmodia）。核内体偏于一侧。多核质体大小不一，大者可有 50 μm 长，小者仅为 4 μm 大。它寄生于牡蛎全身。

② 沿岸单孢子虫大小约是 3.1 μm ×2.6 μm。球形质体也小（>5 μm）。它主要寄生于病蛎全身的结缔组织内。

③ 装甲单孢子虫长 5.0～5.5 μm，宽 4.0～4.5 μm，有盖及其长宽突起伸出。孢子有外膜，也叫孢外胞质（extra. spore cytoplasm）。多核质体直径为 17～25 μm，介于上述两种单孢子虫大小之间。

三种单孢子虫各自可引发出以其种名命名的单孢子虫病，该类病的共同症状是使患蛎变得消瘦、生长停滞。病理切片则可检查到多核质体。严重时可致牡蛎大批死亡。

染尼氏单孢子虫病的牡蛎成贝、稚贝组织细胞浸润呈白细胞状，组织水肿。感染早期，多核质体仅发现于鳃上皮组织或肝管、消化管的上皮组织中。重感染者的细胞萎缩，组织坏死并可看到大量单孢子虫孢子，肝水管因充塞成熟单孢子虫而呈微白色，之后色素细胞增多。有些病蛎的壳基质上形成大小不一的褐色疱状被囊。病体内的单孢子虫孢子借寄主变形细胞的吞食而进入全身各处，甚而所有组织均可见多核质体，称作海滨病（seaside disease）。此病也可由沿岸单孢子虫引起。

亚甲蓝染色全身严重感染该单孢子虫时的血涂片，可见各组织血窦中均有其踪迹。

该病流行区在美国东岸德拉华湾和切萨皮克湾、中国台湾和朝鲜等。被害动物是美洲巨蛎和太平洋牡蛎。流行季节为 5 月中旬至 9 月，6～7 月为高发月，8～9 月份死亡多，9 月达高峰。其他时间为潜伏期，但体弱者也会死去。

发病条件：15～35 的盐度，以 20～25 盐度时更多发病。低盐度区，死亡率在 50%～70%，高盐度区的死亡率高达 90%～95%。

至今对该病仍缺乏治疗方法，只好加强管理和预防，这包括将染病牡蛎移至低盐度（<15）海区以控制病势恶化。另一措施是从病区选择仍为强壮的幸存者作亲体以繁殖具抗病力的后代。

沿岸单孢子虫病主要感染结缔组织。该虫在牡蛎中形成多核质体、孢子囊和孢子。镜检染色的结缔组织切片则可发现多核质体，最早出现于 3 月份的检样中，5 月份居多，至 6 月份可见孢子，7 月及以后月份的样品检不出病原，症状不明显。

该病感染对象主要是老年(2～3 龄)的美洲巨蛎,也流行于美国弗吉尼亚湾沿岸,那儿海水盐度范围是 25～30,表明该病的发生条件是高盐度。此外发病季节在春季 5 月至夏季 7 月,发病高峰在 5 月中旬～6 月初。时间短、季节性强、来势猛,退却快而突然。死亡率为 20%～50%。也能使幼蛎染病,采取过滤颗粒和紫外线适当照射有效[2]。

预防措施主要是促进牡蛎加速生长,提早至该病流行前收获,实在不能提早收时,则也应将病蛎转移至低于盐度 25 的海区中。

装甲单孢子虫病则发生在欧洲牡蛎身上,流行区域在荷兰和法国等地。诊断、症状大体类似于前两种病。另外,原生质型单孢子虫(*Haplospridium plasmodic forms*)可引起牡蛎病变甚而爆发流行性病。

尼氏单孢子虫发育及感染阶段:

原质团期:4～30 μm,最大 50 μm,含 1～60 个左右的核。核两型,小者 1.5～1.6 μm,有致密染色帽,新感染的病蛎上鳃室的上皮细胞中多见。大者 22.5～30 μm,有核内棒,感染 10 d 后的牡蛎血中多见并持续长达 1 年,胞质中央一核。核内体包围之。

孢子形成期和孢子期:孢子囊内含 24～48 个孢子,成熟孢子具有盖孔一个,盖下有小球体或高尔基复合体,孔周是孢子外壁细胞质,核周是单孢子小体形成区。

上皮期:感染初期,病症限于鳃上皮细胞中。虫体以原生质团形式在上皮细胞内大量繁殖。上皮细胞层与基底膜分离。浸润的血细胞和细胞碎片及虫体广泛脱落。

皮下期:虫体突破基底膜进入上皮下结缔组织。

全身感染期:虫体经血循环迅速扩散至全身引发全身感染期,最终使病体死亡。

该病患者的血清成分变化明显,血清中磷酸己糖异构酶、溶菌酶、血清蛋白等浓度降低。早期感染时磷酸己糖异构酶、天冬氨酸转氨酶等活性明显增加。能量代谢受到干扰。生殖力和贝体条件指数下降。这是由于糖原储存降低,摄食、竞争等能量代谢降低所致。近年,已可用 DNA 探针和 PCR 法测定尼氏孢子虫的寄生了[3]。

该病受低盐度(10)影响。5℃的水温,可使其减低活力。大于 20℃时,牡蛎可抑制或排出该单孢子虫。

1.4.3　包纳米虫病(Bonamisis,血细胞寄生虫病 Haemocytic parasitosis)

该病原即是包纳米虫属的牡蛎包纳米虫(*Bonamia ostreae*)等,与上述单孢子虫同属怪孢门。该原生动物是 1979 年才被发现的。1980 年则被定名为此新属新种。但 1960 年,美国加州的牡蛎及欧洲牡蛎(*O. edulis*)得过小细胞病(microcell disease),后来证实它也属包纳米虫病[4]。

该虫主要寄生于牡蛎颗粒性血细胞和其他两种血细胞的细胞质内,每个血细胞有时可含 10 个虫体。在牡蛎的坏死结缔组织中也易见此游离的虫体,而在胃或鳃的上皮细胞之间或间质细胞之间偶见其游离者。

该球形虫之直径 2～3 μm,细胞核嗜碱,一层细胞膜内的细胞质含线粒体和一个单孢子体(haplosporosomes)及一个浓密体(dense body)。线粒体含具少数冠嵴的球形颗粒。虫体在寄主血细胞内生长分裂,感染周期可长达一年。可被卵细胞吸入再传染新一代。

病蛎鳃丝或外套见白色小溃疡。有的溃疡呈较深的穿孔。病体早期的病理切片可在鳃、结缔组织、胃、外套膜等处见颗粒性血细胞增殖病灶。晚期病理切片则有浓密血细胞浸润。患者循环粒细胞及一些其他细胞酶活性减少。该病主要发生在欧洲牡蛎、食用牡蛎的 3～4 龄成体上、*Tiostrea chilensis*、也许还有爬网牡蛎（*Tistrea lutaria*）身上。据说不感染长牡蛎（*Crassostrea gigas*）。

主要发病地区：首先发现在法国布列塔尼，很快传染到苏格兰、西班牙、荷兰及爱尔兰、北美。另一种类牡蛎包纳米虫的传染病则发生在新西兰。发病的季节一般是夏季，死亡率一般在 40%～60%，严重时可达 80%，也能累及年幼牡蛎。

用心室印痕和组织学方法及原位杂交（FISH）和 PCR 扩增等分子诊断法可诊断该感染[4]。

该病原的感染力较强，含有该病原的悬液具传染性，借助其在寄主细胞内定位接触 30 min 即可感染健康牡蛎，其中以颗粒血细胞中的病原感染力最强。

防治措施迄今并不完善，用细胞松弛素 B 处理该寄生虫，可减少寄主的血细胞感染率，另外该寄生虫受热、含水甲醛或紫外线处理后，均失活。合成抑制剂也可用来处理它。国外已考虑从来克丁（Lectin）-糖的相互作用角度来解决此病原[5]。

1.4.4　马太尔虫病　病原之一是马太尔虫属的折光马太尔虫（*Marteilia refringens*），也属怪孢门。该虫寄生于牡蛎的胃上皮细胞和消化腺细胞内，它的多核质体幼小时的直径 7～15 μm。扁圆或圆形的孢子之形成期时的直径大至 15～30 μm，而且其包含体具特征性的折光，强烈嗜曙红。它主要侵害欧洲牡蛎、长牡蛎、食用牡蛎。发病时期 2～10 月份，感染率 0～100%，地区间也有差异。

另一种病原是悉尼马太尔虫（*M. sydneyi*），该病首先见于法国的布列塔尼（1967），之后又传入荷兰和西班牙等国。它引起商业巨蛎的大批死亡，但仅发现于澳大利亚东岸水体。它也使团聚牡蛎（*O. glomerata*）致 QX 病，在夏季爆发，死亡率达 98%。QX 病据说是酸性和金属过量等环境导致，病因复杂，俗称昆士兰复杂症，也侵袭岩石牡蛎（*Saccostrea glomerata*）。

迄今，有关养殖区水温、盐度、水深与牡蛎得此病有何关系尚不清楚，也无防治方法的报告。

近年日本学者报告另一种是 *Marteilioides chungmuensis*，其孢子体寄生于长牡蛎和美洲牡蛎卵巢，引发卵巢肥大症（Marteilioides 感染症），多发生于夏、秋。牡蛎产卵后本应退缩卵巢，但却致卵巢软体部显出异常卵块，使其失去商业价值。

此病也发生于韩国南部沿海及美洲。至 1995 年并未找到治疗法，只得将未染病的牡蛎移至无感染区[1]。但 2003 年以来虫体分离及基因测序等有了长足进步[6]。

1.5　纤毛虫病

纤毛虫病根据病原分可包括 3 类疾病，为纤毛虫病、钩毛虫病和海星精巢病，每类病又可据具体的病原种名而分得更细。

纤毛虫病病原虽均属纤毛虫纲，但各自所属的目不一致。

1.5.1　纤毛虫病　共两个病原，一种是楔形虫（*Sphenophrya* sp.），另一个是触毛目（*Thigmotricha*）的鳃纤毛虫（*Thigmatriche*）。楔形虫虫体较长，一条触手生于前端。幼虫有纤毛几行，成虫则蜕去。靠出芽生殖。主要寄生在牡蛎鳃上，可形成大的包囊。发病区在美国。

鳃纤毛虫寄生于美国长岛的小美洲牡蛎（*American oysters*），可能是在其消化道中。发现时间：1975 年 2 月至 1976 年 10 月。

1.5.2 钩毛虫病 病原系转吻目的钩毛虫属(*Ancistrocoma*)的派塞尼钩毛虫(*A. pelsemeeri*)。此病原行寄生或共栖生活。宿主多为海产双壳贝类,包括巨蛎在内。也寄生于美国大西洋沿岸的牡蛎之消化道内,数量或多或少。可能致病。该虫体较尖长(62 μm 左右),背腹略扁(12.5 μm 左右),宽仅 16 μm 左右。有一触手在外,以附着宿主上皮细胞。触手在体内部分,于体长的 2/3 处。变曲弧形。体内包含 1 长形大核和 1 小核。体背、腹部共 14 条纵行纤毛线,由前延伸向后,腹面中部的 5 条较短。

1.5.3 无口目的病原即海星精巢虫(*Ochitophrya stellarum*)。该生物的生活史未见报道。

对一些美洲牡蛎的病害作研究后发现该虫长 35～65 μm。全身有均匀斜向平等的纤毛。体内具一球形大核。消化道,包括其肠道的上皮组织中均有此不多的病原入侵,但可严重感染致病。病原因是海星生殖腺而得此病名。

1.6 涡虫病

病原属于涡虫纲(*Turbellaria*)*Dalyellieida* 亚目的原生动物一蠕虫。表面有纤毛。发现在美国长岛蠓湾中的年幼美洲牡蛎上(1975 年 2 月～1976 年 10 月)。牡蛎肠腔或体腔是其生存之处,春季居多,大多是共生者。

涡虫(*Stylochus frontalis*)可杀死美国佛罗里达海岸 50%～75%的美洲牡蛎。涡虫栖息于牡蛎的隐蔽和孵化处。它分泌大量黏液使牡蛎窒息而死,同时还直接掠食牡蛎。牡蛎的壳变得易打开。涡虫还侵害另一种牡蛎(*C. rhizohporae*)。

此外,扁虫(*Sininmicus*, *S. ellipiticus*)的大量增繁殖可使澎湖、新竹的牡蛎在高温高盐季节死亡率上升。

1.7 丹门岛(Denman island)病

此病病原为原虫 *Mikrocytes mackini*。它能使生长 2 年以上(以 5～7 年为多)的美洲牡蛎得病,病蛎软体,体表见黄或黄绿色脓疱,以后使疱状结缔组织也见病灶,组织坏死、闭壳肌感染。此病见于加拿大,以 4～8 月份(10～15)℃为多发期,死亡率高者达 53%。此病与牡蛎所在潮区有关,高潮区少病[6]。

2 蠕虫病

蠕虫作为一大类病原,主要包括一些复殖吸虫和绦虫,而且主要是它们的幼虫阶段,约有 6 种吸虫、1 种绦虫。

2.1 复殖吸虫

指腹口类牛头科的牛头吸虫属,包括海门牛头吸虫(*Bucephalus haimeanus*)、巾带牛头吸虫(*B. cuculus*)。居肛吸虫属(*Proctoeces*)的牡蛎居肛吸虫(*P. ostrea*)、斑点居肛吸虫(*P. maculatus*)、多刺棘缘虫(*Acanthoparyphium spinulosum*)[吴信忠文中:棘缘(吸虫)属是 *Echinopharyphium*,棘头(虫)属是 *Acanthocephalus*][7]、东京拟裸茎吸虫(*Gymnophalloides tokiensis*)。另外还有超寄生在牛头吸虫包蚴中的道佛微粒子虫(*Ameson dollfusi*)[7]。

海门牛头吸虫的包囊以地中海欧洲牡蛎(寄生组织是生殖腺和消化腺)为第一中间宿主,以后再寄生其他动物。巾带牛头吸虫则是以包囊先寄生于美国美洲巨蛎生殖腺和消化腺中。其包蚴有时又被一种单孢子虫超寄生,有时道佛微粒子虫也超寄生其中,使包蚴及牡蛎外套膜和内脏均发黑。单孢子在包蚴破裂后进入牡蛎组织,使原本被吸虫感染甚而无繁殖力的体弱牡蛎加速死亡。吸虫的感染率并不都很高，有的牡蛎经吸虫感染后反而风味更优良、蛎体更丰满。土拉牡蛎（*Ostrea tularia*）中也有牛头虫的幼虫寄生，受此感染牡蛎死亡率不低。此病发现于新西兰。

2.2　牡蛎居肛吸虫

牡蛎居肛吸虫的囊蚴无包囊，长椭圆形（1.5 μm×0.4 μm），其中间寄主是日本的长牡蛎。寄生在生殖腺中。同时寄主的鳃和外套膜中尚可发现东京拟裸茎吸虫囊蚴。斑点居肛吸虫的幼虫在欧洲牡蛎中可见。多刺棘缘虫囊蚴则寄生于美国美洲巨蛎中。

2.3　绦虫

在此指疣头绦虫（*Tylocephalum*）（即圆头属）的钩球蚴及另一种未确定的疣头绦虫幼虫。疣头绦虫的钩球蚴寄生于胃和鳃，并使寄主上皮组织有明显的细胞反应。它的寄主是美洲巨蛎，后一种未确定的绦虫则寄生于我国台湾省和日本的牡蛎中。

3　寄生甲壳类动物引起的疾病

使牡蛎生病的甲壳类动物可分为两大类，即桡足类动物和蟹类动物。

3.1　桡足类动物病

病原有两属三种，两种是居贻贝蚤属（*Mytilicola*）的，其一种是肠居贻贝蚤（*M. intestinalis*），另一种是东方贻贝蚤（*M. orientalis*），通称扇贝蚤，总起来称为居贻贝蚤病（或红虫病、贝肠蚤病）。另一属是近似鱼蚤的动物（*Ergasilus*），但未定到种。

肠贻贝蚤主要侵害贻贝，其次也寄生于欧洲产牡蛎中。东方贻贝蚤呈蠕虫状，由各体节愈合而成。雌雄虫大小不同，雌体长 6～11 mm，雄体不超过 3.55 mm 长，多数为橘红色，淡黄或黄褐色也有。横断面观呈扁平而腹略圆。身体两侧各有 5 对突起，自背侧伸出。单眼长在头部背面。触角两对。大小颚各一，均退化。小颚竟退化至消失状。虽有两颚足，但雌体的第二颚足也已消失了。上下唇形状不同，前者三角状，后者椭圆形，4 对胸肢很短小，尾叉的小刚毛数不一，少则没有，多至 4 根。

该虫寄生于牡蛎消化道内后，使消化道组织损伤，黏膜组织破坏，肌肉消瘦，生长不良，以至造成散发性死亡，幸存者也失去食用价值。

受害者还包括长牡蛎和青牡蛎（*Ostrea lurida*）。流行区首先是在日本，后又因美国从日本引进蛎苗而传染于美国。中国尚未见此报道。对居贻贝蚤病，尚无防治法可借鉴。似鱼蚤的鳃桡足类动物已在美洲牡蛎中发现（Quick，1971）。

3.2　豆蟹病

引起此病的豆蟹有 5 种。它们常以寄生危害牡蛎，并引发微生物病。表 1 列举这五种豆蟹的主要特征及其危害性。

表 1　5 种豆蟹的主要特征比较

中文名	学名	体态描述	寄主	流行地区	对宿主产生的病害状况
近缘豆蟹	*P. affinis*	头胸甲近圆形，大小 12.7 mm×13.5 mm	密鳞牡蛎等	山东、日本、菲律宾、泰国等地	夺取食物，妨碍摄食，伤害鳃体，身体瘦弱，♀→♂重者死亡。

(续表)

中文名	学名	体态描述	寄主	流行地区	对宿主产生的病害状况
弋氏豆蟹	*P. gordanae*	头胸甲近圆形,大小3.3 mm×3.5 mm	牡蛎等	山东、辽东、日本等地	
牡蛎豆蟹	*P. ostreum*		牡蛎等		损鳃、吸食营养、衰弱,有时感染率达90%,甚而死亡。
中华豆蟹	*P. sinensis*	头胸甲近圆形,大小8.0 mm×11.2 mm(♀),3.4 mm×3.7 mm♂(表面光滑、稍隆。前后侧角弧形,侧缘拱、后缘中凹。额窄下弯。圆眼窝子,眼柄短,腹大。♂体额向前突出,腹窄长	褶牡蛎等	辽东、朝鲜、日本等地	夺取食物
另一种豆蟹	*Pinnotheses* sp		牡蛎等		因它吸引桧叶螅而使寄主有特殊刺激风味

它们可因行寄生生活而使牡蛎等瓣鳃类动物生病以致死亡。三种豆蟹(即近缘豆蟹、弋氏豆蟹、中华豆蟹)寄生于外套腔中,体形与它们未寄生营自由生活者相似,但寄生后体色变白或淡黄,壳变薄而软,眼睛及螯均退化。

3.3 掘穴生物

包括凿贝才女虫(*Polydora ciliata*)、凿穴蛤(*Adesmace*)。它们穿破牡蛎壳(常是壳顶部),穴居之。同时也为细菌性疾病的发生创造了温床,结果是使牡蛎从附着器脱落入泥。

4 海绵感染引发疾病

有两种穿贝海绵(*Cliona*),即 *C. celata* 和 *C. vastifica*。它们在悉尼牡蛎(*Saccostea aommercialis*)的贝壳上挖穴,使牡蛎壳的强度降低,闭壳肌的附着受到妨碍。与此同时射线照相法观察到壳上还有致病性很强的寄生虫悉尼马太尔虫(*Marteilia sydreyi*),它也引起团聚牡蛎 QX 病。海绵的致病性不明,穿贝海绵既可寄附于活牡蛎,也可在死牡蛎上。两者比例为 28∶195。感染率为 49.9%。它们对牡蛎的感染与牡蛎的养殖期沉浸期的延长有关。预防方法是将牡蛎养放在潮间带中潮线区的贝盘或柱子上。本病记录自澳大利亚昆士兰东南部的匹姆帕玛河[8]。

5 结语

综上所述,动物引起的牡蛎病害也是多种多样的,对牡蛎养殖产生种种负面影响。它们对牡蛎产生的危害大小、范围、程度不一，最起码在不同程度上降低风味品质、营养价值和商业利润。有的可导致牡蛎很高的死亡率，甚至全军覆没。由于动物病原个体较大，普通光镜也能用来观察。相应的研究国外较为深入，但其病害仍难以根治。有待深入的研究。

参考文献 8 篇(略)

(合作者:孙丕喜)

养殖牡蛎的敌害*

摘　要　论述了养殖牡蛎面临的敌害问题，所涉及的敌害包括绿蛎病、瘤、敌害生物、赤潮、环境压力和综合性因素的恶化等，相信中国的相关研究将走向深广。

关键词　牡蛎　病害　敌害　环境

牡蛎肉嫩、汤鲜、营养价值高，有些还具药效，为世人所推崇。但它的养殖面临着多种问题，除来自微生物和动物引起的疾病以外，它还遭遇到人为环境的压力、来自本身的恶变及各种敌害生物的侵袭。

1　绿蛎病

病源有两种，一是舟形藻(*Navicula* sp.)，另一是过多的含锌物所致。这一般是牡蛎吸收了过多的环境污染锌造成。

2　瘤

4 种牡蛎有产生瘤的条件，如美洲巨蛎，就有血细胞生瘤条件(neoplastic)，可染血瘤病(neoplasia)，其组织、细胞病理、病因及其发病机理也有研究。该生瘤条件与病毒有关。

美国 Yaquina 湾的青牡蛎生瘤有季节性和传染性。

2.1　增生无序(Prolifesative disorder)

少数成年美洲牡蛎，其泡沫状结缔组织中有异常增生变化，其特征是出现过量的未分化的海蕾细胞(blastoid cells)。细胞大核偏心、嗜碱，由较浅的染色质围绕，有丝分裂异常。

2.2　配子母细胞过度生长(Hypertrophiod gametocytes)

除牡蛎的围心区生长出长达 3 cm 的间质瘤外，一些美洲牡蛎生殖腺体中有过度生长的配子母细胞和退化的配子母细胞，称作为卵(细胞)囊肿(Ovacystis)，这很容易地联想起肿疡病毒。配子母细胞在雌体中呈现为卵形，但受影响的美洲牡蛎是雄体的，而外表上则这些细胞及相关病毒在雄体中很少，因此表明该病可影响未分化的任一性别的配子母细胞。换言之，受影响的细胞可能代表随性别转回雄体的雌体期残卵，因牡蛎在一定程度上是间性的(intersexual)。典型精巢可能有少数卵受此病毒感染，染此病者不多，病理学意义也不大。

3　牡蛎的敌害

3.1　鱼类

以肉食性鱼为主，包括河豚、黑鲷、海鲫及鳐鱼(Raja)等。鳐鱼用其强硬的齿咬破近江牡蛎的贝壳再吃之。对付鳐鱼的办法是在水中插有可移动的竹木等物，以达到恐吓或预防其游近之目的。害鱼对蛎苗的危害更为严重，防范方法只有在苗区围网或诱捕。

* 原文刊于《河北渔业》，2008，No. 1(总 No. 169)：31-32 转 42.

3.2　贝类(残食生物)

以肉食性腹足类动物为主,包括红螺(*Rapana*)、荔枝螺(*Purpura gradata*)、蚵螺(*P. clavigera*)、香螺(*Cymatium pileare*)、玉螺(*Natiea* sp.)、纺锤管蛾螺(*Siphonalia fusoides*)及土嘴瓜(*Mytilopsis sallei*)等。

蛎敌荔枝螺的螺壳面具结节或棘状突起。隐居于潮间带及低潮线下之岩石下。春末至5月产卵,5 000～10 000个/只。酸液穿孔贝壳,麻痹养殖如牡蛎等贝类,并使之开壳再食其内脏团。对幼蛎,其穿孔本领很大,危害更重,严重时可使一半遭殃。有报道的受害区如闽粤一带沿海养蛎场。

玉螺,侵害方法是先以外套膜包围被害对象,再从穿孔腺分泌液汁溶贝壳后即可伸吻入贝壳内食肉之。危害时间是4～11月份。

属壳菜蛤科的土嘴瓜营底上或水中固着生活,在蛎场大量生长后即将蛎串包满,使蛎难以摄食和呼吸而至死去,这在台湾西海岸的内湾或潮沟处时有发生[1]。

此外一些贝类动物产生麻痹性贝毒致牡蛎死亡,这在日本也常见。

对付方法:找准出现时机和现象以捕捉之。每年10～11月份左右玉螺排出形似无顶笠帽的卵,这样可以拾除之。0.2～0.3 mg/L的石灰水液喷洒养殖池,可迫使玉螺出土而捕之。以10 g/m^{2}的用量在涨潮前泼洒五氯酸钠溶液,可使大部分玉螺死去。

红螺栖息地在砂泥或沙砾底质的低潮线下。它夏季繁殖,其卵附于蛎壳或石块上。它对近江牡蛎的危害很大。防治方法基本同上。甲壳类*Mytilicola orientalis*可使牡蛎生长不良而死去。

3.3　棘皮动物

包括海燕(*Asterina*)、海盘车(*Asterias*)、海星等。危害大连湾牡蛎和密鳞牡蛎。一个海盘车吃坏20只牡蛎。不过有的人认为,海盘车可蚕食藤壶及螺,从而为牡蛎除害。这类棘皮动物生活在高盐海区。每个海星每天损坏牡蛎20只。

3.4　甲壳动物

危害牡蛎的甲壳类动物有锯缘青蟹等蟹类。它们以其强大的螯足钳破蛎壳而食之。也藏身于牡蛎苗附着器空隙,伺机嚼食蛎苗。

3.5　附着生物(竞争生物)

包括藤壶(*Balanus*)、紫贻贝、海鞘(*Ascidiacea*)、苔藓虫(*Bryozoa*)、石灰虫,薮枝虫(*Obelia*、薮枝螅水母)、金蛤等。侵害主要以与牡蛎争夺固着基、氧气和食料为手段,因而影响牡蛎固着和生长。其中以藤壶为最重。藤壶繁殖季节早于牡蛎。掌握好采苗季节,适时投放固着基,以有效避免藤壶固着。真枝螅附生于太平洋牡蛎鳃上可致牡蛎大量死亡。褶牡蛎则是太平洋牡蛎的竞争者[2]。聂宗庆,曾用油喷灯或火把燎烤养殖绳可烧死贻贝、海鞘、藤壶等污损生物,还可促使牡蛎生长。

3.6　鸟类

鸟类也是牡蛎养殖的一害,如蛎鹬(*Haematopus*),它们啄食牡蛎等贝壳类动物。此外包括河豚等肉食性鱼类,也是养殖牡蛎之一害。

以上关于敌害生物的研究,在中国主要是针对近江牡蛎的[3]。

4　赤潮

其危害是赤潮生物的毒素所致。毒素一是毒化水质,二是毒害养殖生物。赤潮生物包括原甲藻、亚历山大藻、米氏裸甲藻、棕囊藻、多环旋沟藻等。严重赤潮可使50%以上牡蛎死亡。养殖密度过大也可发生赤潮[4]。此外，夜光虫、角藻等的大量繁殖也参与进来。广东沿海牡蛎也曾遭麻痹性毒素的影响[5]。

消灭方法：$CuSO_4$ 结晶泡成的溶液，浓度为0.5～1.0 mg/L，即可有效杀灭。

5 重金属、石油等

养殖水体过量的汞、镉、铅、锌、铜、铬等重金属及硫化物、砷、有机氯、多氯联苯、総丁基锡、河鲀毒等可使牡蛎的生态环境及免疫功能大受影响[6-7]。石油污染已使厦门一带的僧帽牡蛎抗氧化酶不能正常发挥作用。

6 综合性因素的恶化

养殖环境污染加重及水体质量（包括过高水温和盐度）的恶化，加上单一品种密度太大，超负荷运转，病、虫、灾害性生物大肆活动，使整个水体的生态平衡遭到破坏。牡蛎在繁殖季节里生理活性降低，繁殖生理受到障碍，其生命周期中止或病态变异。加上养殖工艺落后，作业方式不完善等急功近利的措施，最终导致养殖大面积亏损。经济效益总体下滑，后劲不足[4,6]。为了抵御病害和环境恶化，寻找和发掘牡蛎本身的御敌机制也就十分重要，这包括防卫素(defensins)的应用，抗感染、抗逆分子、热休克蛋白基因的提取与克隆等[3、8-9]。

7 结语

牡蛎病患的发生、发展，深受各种因子制约，既有自然的，但更多的是来自环境和外界的压力。因此人们在牡蛎的养殖上必须作周密、长远的思考和安排，既要获取一定的经济效益，更要使牡蛎的养殖业能健康持续发展。综观全球，对养殖牡蛎病患的研究，迄今国外的应比国内的广泛深入一些。国内尚需努力[10]。随着牡蛎病患传统诊断防治技术的有效加强，更应引、进和创新诊治理念、方法，包括分子生物学技术的器材、方法和手段。另一个不可忽视的大问题，便是减轻污染，改善环境，提高从业人员的素质，使牡蛎健康生长、正常繁育并永续利用。

参考文献10篇（略）

（合作者：王　波）

ENEMIES AND EVILS OF CULTURED OYSTERS

(ABSTRACT)

Abstract This paper expounds that the problem of enemies and evils which has been confronting with cultured oysters. Some problems involved in this content include green disease of oysters, tumour, hostile and evil organisms, red tide, environmental pollution and disturbance of synthetical factors, etc. It believes that the researches relating to this subjeet will be deepening and widening in China.

Key words Cultured Oysters　Enemies & Evils　Tumours　Ambient Pressures

B_3 应用及应用基础
(APPLY AND APPLIED BASES)

海洋中维生素 B_{12} 供求变化及其生态意义*

摘　要　从微生物角度论述海洋中 Vit B_{12} 的分布变化及测定法、需求与生产及与相关因子间的关系。

关键词　海洋　维生素 B_{12}　生产与需求　分布及其变化　影响因子

维生素 B_{12} 是一类积极参与生理活动的物质，在细胞营养和机体代谢中起着重要作用。人们长期缺乏 B_{12}，会导致多种疾病。

B_{12} 作为海洋中的一种生态因子，活跃在海洋生物生产力和食物链中，影响着海洋生物的盛衰和种群的连续。研究海洋中的 B_{12} 也能加深理解海洋生物、物理、化学的某些现象及其相互关系。

海洋中的 B_{12} 是由生物合成的，其中又以细菌等微生物为主。海洋中有着丰富的各类 B_{12} 产生菌。

本文主要从微生物角度论述海洋维生素 B_{12} 的分布变化、B_{12} 的供求以及某些环境因子对 B_{12} 的影响。

一、海洋中 B_{12} 的分布、变化及测定法

近几年来，人们调查了太平洋北部、日本近海和欧美沿岸的大西洋及其内海等水的 B_{12}，发现不同海区的 B_{12} 浓度差异很大，其范围是 0～26.0 ng/L(如太平洋西北部)。在北海是 0.13～2.00 ng/L(Cowey,1956)，北极圈附近太平洋海域是 0～3.39 ng/L，日本福山湾是 0.42～6.43 ng/L[2]，而不列颠哥伦比亚沿岸水域则是 0～18.7 ng/L(Cattell,1973)。

近海 B_{12} 浓度常高于外海。加利福尼亚沿岸水，其 B_{12} 浓度高达 0.4～6.5 ng/L，而大西洋中部表水 B_{12} 浓度平均只有 0.1 ng/L[3]。不同水域的悬浮颗粒和碎屑的 B_{12} 浓度，一

* 原文刊于《海洋科学》,1980,3,51-54.
本文承吴超元、纪明侯、孙国玉等同志热情指导，谨致谢忱。

般也是沿岸的高于外海的，如美国佐治亚州 Sapelo 岛沿岸的大西洋和 Duplin 河一带，B_{12}的最高值出现在该河上游水中，而此处正位于肥沃的 Spartira 沼泽地带。Doboy 海峡则次之。离岸较远的大洋水，其悬浮颗粒和碎屑中的 B_{12} 往往最低(Starr，1956)。

海底沉积物中 B_{12} 含量常比海水的含量高。日本相模湾和骏河湾底泥，有的样品含 B_{12} 高达 1.14 ng/g(湿泥)，日本舞鹤湾沉积物及其表层的 B_{12} 含量比海水则高得多[1]。

B_{12} 在海水中的垂直分布趋势是上层少下层多。在透光层，B_{12} 浓度非常低。如马尾藻海，100 米以上的海水，B_{12} 一般只 0.01～0.1 ng/L；200 米和 200 米以下，则高于表层；近 500 米处高达 1.5 ng/L。

B_{12} 浓度的分布还有一定的季节变化。日本相模湾表水中，B_{12} 在初春最低，夏季最高，秋末冬初接近最低(Cohwada 等，1972)。这种状况因不同海区而异，且与浮游植物密切相关。美国长岛海峡水，B_{12} 浓度在冬季最高，达 16.0 ng/L；随着晚冬硅藻藻花的出现，B_{12} 明显下降；到初春时，硅藻的繁殖激减，B_{12} 最少，仅 3.4 ng/L；从春至夏再增(Vishniac 等，1961)。

B_{12} 浓度及其分布的变化首先决定于 B_{12} 产生菌的种类和数量，而这两者又受磷酸盐、温度、叶绿素和浮游生物等环境因子的制约。B_{12} 菌产生的 B_{12} 溶至海洋又受季节、河流、潮汐、底质等因子的作用，互为影响。因此，B_{12} 浓度及分布的问题是复杂的。

对 B_{12} 的测定，除对非生物材料可用同位素法外，一般常用微生物法。其原理是借专性需 B_{12} 的微生物对微量 B_{12} 灵敏的生长感应来推断 B_{12} 浓度。针对海洋特点，须排除高盐度的干扰作用。不少学者采用了各种脱盐法，但回收率低且操作繁难。近几年开始寻找耐盐微生物，国际间常用 *Cyclotella nanal* 3－1[3]、*Amphidinium arteri* (Cattell，1973)，还有采用赖氏乳杆菌[2]、大肠杆菌突变株(Propp，1970)等。它们在对 B_{12} 的灵敏度、耐盐性及操作上都有其优越性。Heed(1972)的方法则在测定 B_{12} 的速度和精度上更有提高。

二、海洋生物对 B_{12} 的需要

1953 年，Darken 所研究的 534 株细菌中有 8%依赖 B_{12} 生活。海洋细菌纯培养的研究揭示，它们常需一种或多种 B 族维生素[4]。Burkholder(1963)从海水、海泥分离的 1,272 株菌中有 12%需 B_{12}；1968 年，他与同事又证实有 20%以上的菌株需 B_{12}[4]。1976 年，Berland 等分离的海水细菌有 3.2%需 B_{12}。一些海洋酵母、自营菌也需 B_{12}。一些低等真菌如 *Thraustochytrium globosum* 只是在有 B_{12} 时才能生长，当浓度高达 100 ng/L 时，其生长则显著加快。

从五十年代起，一些学者研究了 B_{12} 活性化合物在许多藻类的生理上的作用，指出 B_{12} 是其生长因子[5]。1963 年，Provasoli 试验了属于不同门类的海藻，发现 57%需 B_{12}，其中有扁藻 *Platymonas tetrathele*、中肋骨条藻、塔形冠盖藻、金藻 *Monoehysis lutheri*、隐藻 *Hemiselmis virescens*、蓝藻的伊朗席藻、甲藻的多边膝沟藻等。1971 年，Rheinheimer 指出很多海藻不能合成维生素，需由其他生物提供。试验表明，有 50%的海洋蓝藻需 B_{12} 等多种维生素，25%的海洋硅藻需 B_{12}。浮游植物的培养研究表明，它们之中有的必须有 B_{12} 才能生存，有的则需 B_{12} 刺激生长，另一些则在营养上起补助作用。B_{12} 对保障某些大洋植物如颗石藻类(*Coccolithophorids*)的外壳(coccolith)正常结构也是不可缺少的。此外，B_{12} 对一些不需它的浮游植物之生长还会有某种抑制作用(Carlucci，1974)。

在大尺度生态系统中，B_{12}如同别的活性物一样，也是周围生物重要的一种环境因子。在某些水域，当下层带释放B_{12}补充给透光层时，常伴随有双鞭藻、硅藻和绿藻生物量的增加，而当透光层B_{12}浓度最低时，蓝藻生物量增加了[6]。一些海洋硅藻的单位细胞数及马尾藻海的浮游植物消长、种群组成、移栖现象均与B_{12}密切相关。

有些较高等的海洋动植物虽能贮存一些B_{12}，但本身并无合成能力，要由B_{12}生产者供给，其生长和分布都受B_{12}制约。于是海洋维生素B_{12}生产、贮存、消费三个环节包括的如共生和寄生等关系就要建立起来。许多海洋经济动物从饵料、消化道内的B_{12}微生物区系或从两者兼得B_{12}[7]，微生物和鱼、特别和幼鱼是密切相连的。一些渔场(如舟山渔场等)的B_{12}浓度和B_{12}产生菌数量、种类较为丰富。用B_{12}等维生素或其产生菌同其他饵料饲养经济动物已越来越被人们所重视。可见，B_{12}是各类海洋生物一种重要的生态因子。

三、海洋中B_{12}的微生物生产

自1948年分离到B_{12}以来，人们就其生物合成进行了广泛的研究，很快于五十年代初从土壤、有机肥料、池塘等环境中筛选过大量B_{12}产生菌(Smith，1960)，指出放线菌和细菌在B_{12}生产上较有希望(Durton等，1951；Lochhead，1957)。简单棒杆菌在每升正链烷中产B_{12}0.6 mg(Koichi，1966)。反硝化假单胞杆菌在厌氧条件下产20 mg/L，而有名的薛氏丙酸杆菌则产高达58 mg/L以上的B_{12}。此外，一些光合细菌、浮游藻类也是B_{12}的生产者(依姆歇涅斯基，1978；Nishijma等，1979)。

近年来，人们已把注意力放到海洋。海洋也是产维生素B_{12}微生物的一个天然场所。如前所述，海洋中的B_{12}主要由细菌等微生物活动所产生。这在大洋中尤其如此，B_{12}主要由异养细菌合成和排泌[8]。

运用由普通氮、碳源和某些无机盐类组成的培养基能筛选出许多B_{12}菌[1,4]，但应尽量减少培养基中的干扰成分，以无机盐代替复杂有机物或根据测定生物感应范围来选择。但以这些技术分离纯培养较麻烦，从高浓度营养基质上长出的细菌转至较低浓度培养基上时，微生物的生长和合成能力可能降低。因此，有人提出新法并成功地从入海口、沿岸区和大洋中分离到B_{12}菌[9]。

凭借不同方法，人们从海洋的不同样品中分离到大量B_{12}菌。1953年，Ericson等发现海水和海藻表面的34株细菌中有24株是B_{12}菌，即11株假单胞杆菌，8株无色细菌，2株黄杆菌，2株芽孢杆菌和1株欧氏植病杆菌。Starr等[7]发现他们所研究的海洋细菌有30～70%是B_{12}菌，对地中海几种鱼的消化道之异养菌调查后，确证68株细菌中有28株能合成B_{12}，它们是杆菌、假杆菌、假单胞杆菌、弧菌、芽孢杆菌、微球菌、八叠球菌等属中的16个种。最大B_{12}活力的菌是嗜盐杆菌和微嗜氮芽孢杆菌[10]。可能是微嗜氮芽孢杆菌(*B. oligoritrophilus*)。1976年，Berland等从普通表层海水中分离的232株细菌中得到26%的B_{12}产生菌及其分布，指出海洋底质的B_{12}菌比海水中的多。对我国大陆架、沿岸区和海湾的调查也表明，B_{12}菌种类多、数量大。

研究工作表明，微生物产生的B_{12}大致分两种方式进入培养基或周围的环境，主要是在对数生长期后大量细胞代谢并排泌B_{12}；其次是在衰亡期，由死亡细胞解体而游离出B_{12}。这样海洋B_{12}需求者就源源不断地取得了B_{12}。

四、若干环境因子与 B_{12} 生产的关系

海洋微生物合成 B_{12} 的能力首先决定于本身，有些仅能满足自身所需，有些则合成过剩的 B_{12} 或自身非必需的 B_{12}。Ericson 等测定的菌种，B_{12} 产率是 $5\sim1.40\times10^5$ ng/L。Starr 等所研究的 366 号菌产 B_{12} 约 1.84×10^5 ng/L。这说明海洋中有些 B_{12} 菌的生产能力可与陆生菌相匹敌。

细菌给予环境的 B_{12} 水平是有变化的，这与它们的区系，特别是活力指数如细菌群体的生长速度变化有关(Vacelet，1975)，而后者又与环境条件息息相关。从细菌合成 B_{12} 的能力看，一个突出的方面就是与钴离子的关系。大量实验工作证明，钴离子是合成 B_{12} 的要素之一，能明显影响 B_{12} 生产(Hodge 等，1952；Berry 等，1966；Wagmen 等，1969)。1975 年，Nishio 等以克氏杆菌 101 做试验，发现无钴时，不合成 B_{12}；当钴浓度从 0.02 ng/L 增至 2. ng/L 时，B_{12} 产量从 23 μg/L 增至 75 μg/L。我们的工作也证实了这点。研究还表明，无论是混合培养还是纯培养，适当添加钴离子对 B_{12} 生产确有益。

B_{12} 浓度与钴离子的这种密切关系，在 Benoit 等(1957)对 Linsley 池塘的调查也得到证实。B_{12} 浓度在表水中低，最明显的原因就是所提供的钴很有限。通常，钴，特别是在溶解状态时，下层滞水带常多于温跃层。深水中钴的含量常较丰富。Gillespie 等(1972)指出，Kinnert 湖 40 米深处的水中 B_{12} 浓度最大，该处为下层滞水带[6]。一些近海沉积物表层有较多 B_{12} 菌，可能反映了它们与钴离子的密切关系。

水—沉积物界面 B_{12} 浓度高，还可能与丰富的营养、较高的静水压和厌氧的环境有关。在那里，B_{12} 菌的种类不同于一般表水，B_{12} 菌数量和合成能力较大，而耗用量相对较少，于是 B_{12} 有可能积聚起来。Szontagh 等在 1972 年指出，污泥的混合细菌厌氧发酵 B_{12} 试验表明，加压和控制通气是获得高产 B_{12} 的手段。这与从某些高等动物粪便和肠含物分离出 B_{12} 高产的厌氧微生物情况相吻合(Uphill 等，1977)。由此启示我们，实际应用海底 B_{12} 菌时必须考虑到这些条件。

在许多自然生态系统中，本地的微生物区系有极高的生产速率，这是经混合培养得出的估计。培养一周，将生产高达 23 ng/mL 的 B_{12}，比纯培养高出 30 倍(Gillespie 等，1972)。显然 B_{12} 菌的适当混合培养既能较客观地反映现场 B_{12} 生产，又有实际的应用价值。

实验表明，B_{12} 菌在不同培养基上有不同的合成能力。Lebedeva 等[10]证实，经特殊射线辐射的无 B_{12} 培养基，其上的 9 种 17 株 B_{12} 菌，包括蜡状芽孢杆菌、嗜盐杆菌等比在未经辐射的培养基上有较高的维生素合成能力。一些烃化物是细菌合成 B_{12} 的基质，如 C_{11}—C_{18} 的直链烷、乙醇、α-酮戊二酸酯、苹果酸、苯甲酸酯和 Fumarate 等是棕色芽孢杆菌积累 B_{12} 所要同化的基质，尤其是 C_{16}—C_{18} 的直链烷是该菌 B_{12} 生产的最好基质。一些石油产品的微生物降解过程伴随有 B_{12} 的产生：经受石油等污染的一些海、湾，B_{12} 菌的数量仍不低，可能说明其本身就是石油等的降解者或是它们对环境适应的结果。以廉价而普通的烃化物作微生物生产 B_{12} 等的唯一碳源是近年来制取维生素的新动向，这在环境微生物领域中也具实际意义。

产 B_{12} 的嗜冷细菌是高纬度或低温水域 B_{12} 的主要供应者，也是它们长期适应低温的结果。

由微生物等产生的B_{12}与海洋初级生产力有密切关系。Menzel 等(1962)证明，海水光亮带B_{12}浓度与浮游植物碳吸收基本上是平行的，这是B_{12}积极影响浮游植物组成和群体碳吸收的反映。B_{12}浓度的变化是海洋初级生产力的一个有用参数。

由于浮游生物与B_{12}的关系密切，B_{12}也是水体营养程度的一个指标。近几年来，人们研究了沿岸区、海湾等水域的赤潮问题，发现引起赤潮的原因复杂，并指出其中的一些赤潮浮游植物常常是B_{12}需求者。因此，人们可以根据调查海区的B_{12}分布状况和需B_{12}浮游植物的组成情况，估计水体富营养化过程和发生赤潮的可能性。

综上所述，多年来，各国学者对海洋中的B_{12}作了广泛研究，基本搞清了世界主要海区的B_{12}状况。在我国，这一课题也于 1972—1977 年间初步开展，对东、黄、渤海三海区及相关海湾的B_{12}产生菌数量分布、B_{12}产生菌培养基和培养条件作了调查研究，取得了一定成果，并在 1978 年 11 月的全国海洋湖沼学会学术讨论中以摘要形式发表。要把这一工作继续深入下去，必须对B_{12}测定方法作改进，现今仍沿用陆上方法测海洋样品有一定缺陷，要采用灵敏简便的方法。B_{12}产生菌的筛选和培养方法也需改进，以适应进一步开展海洋B_{12}菌普查的需要。目前国内生产的B_{12}主要从生产抗生素发酵液中提取，质和量都不能满足日益增长的需要。开展B_{12}产生菌的海洋调查，可能发掘出高效B_{12}菌种。应用先进技术设备对B_{12}等在海洋环境及各级营养水平中的循环和作用的调查研究将会推动海洋科学的深入发展。

参考文献 10 篇(略)

微生物富集铀*

摘　要　论述微生物与铀的关系、富集铀的微生物和藻及其集铀活动、应用前景展望。

关键词　铀　集铀微生物及其活力　应用前景

一、微生物与铀

有些微生物与铀关系密切,这些微生物天然地生存在含铀的矿物或基质中。其中一些具有强大耐铀力,如氧化亚铁硫杆菌(*Thiobacillus ferrooxidans*)、氧化硫硫杆菌(*Thiobacillus thiooxidans*),前者耐铀浓度是1～10 g/L,后者达12.5 g/L(Torma,1977)。微生物的这一耐铀特性启示我们:从特定环境中能分离出适应高铀浓度的、富集铀的菌株。这类环境,可能优势地生长着耐铀菌。用扫描电镜研究某风化铀矿样品后发现,其沥青上大量生长有球状和杆状细菌。该黑页岩含铀量高出其他页岩的10%。

一些非硫杆菌属的细菌在含合适有机基质和铀的花岗岩中不仅耐铀且能活跃地溶解铀化物,在花岗岩、半花岗岩、沥青片麻岩上生长的曲霉(*Aspergillus*)和青霉(*Penicillium*)能从基质中将大量的铀吸附至菌丝体。曾发现一些青霉能从氯化铀酰溶液中积铀,另一些丝状真菌能吸附其他重金属和锕系元素。

鉴于铀在国民经济中的重要性及其在微生物学研究上的特殊用途,人们加强了这方面的研究,如耐铀微生物细胞学、铀的生物效应、微生物富集铀的机制等。

Bahadur 等从土壤中分离到三种不运动的微生物,即大球形、杆形和小球形生物。它们在25～50 μm的醋酸铀酰上生长、固氮,而耗碳很少。

固氮菌属(*Azotobacter*)的一些固氮种,当有醋酸铀酰时,加钼酸盐能使固氮作用增大(Jyotishmati 等,1978)。

关于耐铀—抗铀微生物的细胞结构,以氧化亚铁硫杆菌言,Dispirito 等(1982)发现有些氧化亚铁硫杆菌的细胞生有鞭毛,鞭毛有极生和周生之分;有的还有几根伞毛(pili)。鞭毛的维数、细微结构和定向作用不尽相同,无鞭毛者仍含运动细胞。

就细胞的吸铀部位而言,青霉和曲霉将铀吸至菌丝表面;酿酒酵母(*Saccharomyces cerevisiae*)积累并沉积铀在胞外;细胞吸铀的基本物质,在少根根霉(*Rhizopus arrhizus*)中,是细胞壁结构中的几丁质或壳聚糖组分,铜绿假单胞菌将铀积在胞内。

耐铀菌质粒的最新研究进展表明,耐铀菌研究已进入分子水平,人们可以凭借遗传工程手段筛选、培育出优异的富集铀的生物。对其生长动力学的研究,指出了富集铀的最佳条件和实用途径。

---

* 原文刊于《工业微生物》,1984,No:44-47.

二、富集铀的微生物

现将迄今发现的与铀的富集关系较密切的微生物列举如下；

细菌：*Acinetobacter*（不动杆菌）、*Bacillus*（芽孢杆菌）、*Desulfotomaculum ruminis*（瘤胃脱硫肠状菌）、*Desulfovibrio*（脱硫弧菌）、大肠埃希氏菌（*Escherichia coli*）、*Ferrobacillus ferrooxidans*（氧化亚铁亚铁杆菌）、*Ferrobacteria*（亚铁杆菌）、*Thiobacteria*（硫杆菌）：*Thiobacillus ferrooxidans*（氧化亚铁硫杆菌）、*Th. denitrificans*（脱氮硫杆菌）、*Th. acidophilus*（嗜酸硫杆菌）、*Th. thioparus*（排硫硫杆菌）、*Th. thiooxidans*（氧化硫硫杆菌）、*Pseudomonas aeruginosa*（铜绿色假单胞杆菌）、*Zoogloea ramigera* 115（生丝动胶菌（115）以及一些其他球状、杆状菌和乳酸细菌 *Lactic acid bacteria*）。

放线菌：*Streptomyces viridochromogenes*（绿色产色链霉菌）；

酵母：*Candida*（假丝酵母）、*Hansenula*（汉逊酵母）、*Rhodotorula*（红酵母）、*Saccharomyces cerevisiae*（酿酒酵母）；

真菌：*Aspergillus*（曲霉）、*A. niger*（黑曲霉）、*Penicillium*（青霉）、*P. digitatum*（指状青霉）、*Rhizopus*（根霉）、*Trichoderma*（木霉）。

藻类：*Chara*（轮藻）、*Chlamydomonas*（球衣藻）、*Chlorella regularis*（整齐小球藻）、*Oscillatoria*（颤藻）、*Platymonas*（扁藻）、*Spirogyra*（水绵）、*Synechococcus elongatus*（细长聚球藻）。

三、铀富集菌的集铀活动

（一）氧化亚铁硫杆菌、氧化硫硫杆菌的浸铀

这类菌有强大耐铀力，人们曾对它们作过广泛研究。氧化亚铁硫杆菌是浸铀的佼佼者，如在 Forstan（澳）沉积铀矿中，该菌能回收 5.4%的铀。它们特别适于从低品位矿中浸铀。有人用它来集铀，9 天就将 68%的铀浸出；将余渣再研磨，浸铀率又提高到 87%。

为了提高该菌的耐力和浸铀率，*Bruymesteyen* 等（1980）用一系列转移技术，使其耐力从<50 mg/L 提高到 500 mg/L。Barbic 等（1976）指出，铀的富集与相应菌的组成、数量直接成比例。

对影响氧化亚铁硫杆菌浸铀活动的因子做了大量研究，包括温度、pH、铁离子浓度、有效氧、细菌适应性、有机化合物、颗粒大小、矿浆浓度等。该菌的细胞悬浮物促进与氧化作用相关的摄氧；在稀硫酸中，其化学氧化速率不为高压消毒过的细胞所影响。

不同药品对该菌生长发生不同作用，如 2 mM 硫酸铀酰抑其生长，但若加 200 mmol/L K^+、Na^+、Li^+ 或 NH_4^+ 的硫酸盐到培养基中，这种抑制作用便得以缓解。提高硫酸浓度，铀毒增加，但不影响其正常生长速度（Tuovienen 等，1974）。不过 Barbic 等仍得出相反结论，认为适当加大硫酸浓度，将大大刺激浸矿速率，使浸铀增加 20%的原因很可能是他们的试验条件不一。

用该菌的适应株浸铀，其 Fe^{2+} 氧化作用活化是-13.9 ± 0.1 kcal/mol，失活能是53.3 ± 0.2 kcal/mol（细菌热死），温度系数 Q_{10} 为 1.8，由此得出的浸铀率达 80%～100%。

影响氧化亚铁硫杆菌浸铀活动的因子主要有下述两种：①金属硫化物被细菌氧化的直接可溶性；②黄铁矿的金属硫化物或氧化物被细菌氧化出高铁离子而产生的溶化作用

(Kelly,1977)。

该菌集铀分一步连续浸出和两步浸出法,后者系从黄铁矿或 Fe^{2+} 来源的细菌产 Fe^{3+},然后在酸液中借 Fe^{3+} 化学富集。前法则用连续培养装置,使细菌在硫酸亚铁上最适 pH、温度和通气量分别是 2.3～2.7、29～34℃和 5.0×10^{-7} g·mL O_2/mL·min·atm。当硫酸亚铁培养基中矿重占 10%时,浸铀以 30℃、pH2 和 8.5×10^{-7} g·mLO_2/mL·min·atm 为最适。单阶段连续培养中,铀矿物的最适稀释速率高达 0.10·小时$^{-1}$,UO_2 的连续浸提,还决定于矿物中高铁离子的反应状态。

氧化亚铁硫杆菌和氧化硫硫杆菌混合浸铀能使耗酸减少、出铀增加。氧化亚铁硫杆菌和嗜酸硫杆菌的混合洗涤悬液可直接氧化 U-IV,其摄氧率决定于细胞浓度、生物的既往史、U-IV 及其抑制剂的量、pH 和硫酸铁。氧化亚铁硫杆菌与红酵母混用时,可不被 10^{-4} m 的丙酮酸所抑制。

自 20 世纪 60 年代以来,人们发展了氧化亚铁硫杆菌等的常规池浸铀工艺,并逐步进入实用化生产,如地下集铀、工业堆摊或废石堆浸铀。氧化亚铁硫杆菌在炼铀工业中的作用日益引人注目。

(二)真菌富集铀

1. 青霉和曲霉富集铀

这两种霉有一定耐铀力,它们不为 5×10^{-4} m/dm^3 的铀所抑制。铀与培养基中的磷酸盐离子复合而附于菌丝表面,菌丝起多功能离子交换器作用,铀可借洗脱而获得。干菌丝,是三维结构的吸附剂,用增强树脂处理后,可防止以菌丝组成的吸附器被泄漏。

当培养基中营养不足时,它们的集铀与其培养基中的生物溶解铀不相关,与菌丝生产也不相关,而与所在的岩石性质、含铀量相关。此时,因营养匮乏而菌丝生长迟缓,但积铀却最高,达 2.8 mg/g(生物干重)。在营养正常时,只有 0.18 mg/g。

青霉菌富集铀的条件:4 克新鲜青霉在 20～30℃生长 1～4 小时后可移去 75%～90%的 UO_2^{2+}[铀溶液的浓度为$(1\sim10)\times10^{-6}$],pH5.5～7.5,以减少铀的富集。pH2.5 时,可保持上述(75%～95%)移去率的三分之二以上。用吐温 80 预培养,可改进铀的富集率。煮沸、巴斯德灭菌和三氯乙酸预处理也增加铀的富集率。

指状青霉从氯化铀酰水溶液中富集铀时,叠氮化物不抑制该真菌本身及与壁相关的生物聚合物的富集铀过程。经水煮沸,用酒精、二甲基亚砜、KOH 处理,其摄铀力增至 $10\,000\times10^{-6}$,但甲醛无助于增高铀的富集率。

黑曲霉和乳酸细菌在亚硫酸盐液中(含铀)经 7 天培养,富集了 12%的铀,使含铀量达 153×10^{-6}。

2. 少根根霉富集铀

少根根霉(*Rhizous arrhizus*)摄铀量占其粗干重的 20%(Santa,1982)。pH、Cu^{2+}、Zn^{2+} 和 Fe^{2+} 都能影响其铀的富集活动,这些金属离子是其集铀活动的竞争性离子,致使铀的富集量不稳定(Tsezes,1983)。当 pH4 时,该霉菌富集铀最多,大于 180 mg/g,分别是离子交换树脂和活性炭的 20 和 3.3 倍。

3. 酿酒酵母富集铀

酿酒酵母能从稀铀水溶液中摄铀。Weideman 等(1981)对其大量回收铀时的转移系数及生产曲线作过预测。实验表明,该酵母有 32%的细胞将铀吸附在胞外。积铀量达细胞干重的10%～15%。用化学法预处理和后处理,可分别增加吸铀和从胞外移去铀。去铀后的细胞尚可反复用于吸铀。

4.其他菌对铀的富集

铜绿假单胞菌、生枝动胶菌也表达出富集铀的能力(Norberg等,1984)。它们常与环境的污染物在一起,因而在净化铀的污染及其他污染中发挥作用。

四、用微生物从海水回收铀

人们早已开始探索从海水提铀,据知海水中含溶存的铀达40亿吨,可是,铀在海水中的浓度既均匀又稀少,一般仅3 μg/L。靠化学法回收十分困难,于是,不得不促使人们从生物学特别是微生物学角度去探求。近年来的事实证明,这种探索是富有意义的,迄今已找到了大量微生物,如细菌、放线菌、酵母、丝状真菌及藻类有富集海水中铀的能力,还有一些突变菌株可用于集铀。

已使用绿色产色链霉菌和整齐小球藻从海水和淡水中提铀,它们的固定化细胞吸铀率向人们展示了明朗的工业应用前景。

可以预料,随着富集铀微生物的开发和改造,从海水中提铀的效率将会逐步提高。

参考文献12篇(略)

SUMMARY BRIEF ON CYTOBIOLOGICAL STRUCTURE AND CLASSIFICATION OF SHRIMP HEMOCYTES*

Introduction

Farming of economically important shrimps (e. g. Penaeus vannamei and *P. Chensis*) is becoming more and more popular in the world.

The shrimp in intensive or semi-intensive culture resulted in the development of epizootics that is often explosive and same times leads to the loss of the whole stock.

The health condition monitoring of farmed populations is not very easy, especially in developing countries because of the lacking of techniques to control.

The situation of hemocytes connected to the defensive and physiological functions. Therefore study on the hemocytes and hemograms is important for showing the blood status of infected shrimp. However, prevention and treatment for shrimp remain unsolved.

We developed a method for the hemocytes formation of two species along with the differentiation and function with cytomorphological and cytochemical techniques. This paper describes preliminarily some aspects of procedures of separation and analysis of shrimp (*P. vannamei*) hemocytes.

Sampling and experiment

Animal

The shrimp used in experiment was matured *Penaeus vannamei* and provided by Marine Sciences Institute, Port Aransas, USA, and Kept in tank with circulating seawater at 15℃±2℃.

Preparation of gradients

Continuous gradients of 5 mL 30% percoll (Sigma Chemical Co, USA) in 3. 2% NaCl were prepared in a round bottomed opon-top ultra-clean™ tubes (Spinco. Division of Beckman Instruments, INC, USA) (by centrifugation in an angle-head rotor at 25, 000 g for 45 min at 7±3℃) (LS-80 mol/L Ultracentrifuge, Spinoco Division of Beckman Instruments, ING. , Palo Alto California, USA).

---

* 本文是 1991/1992 年间在美国 TA&MU 做高级访问学者时在 D·H·Lewis 教授指导下写成,并于 1992 年春在香港举办的国际水产学大会上由 Lewis 教授宣读。

To minimize cell attachment to the wall of centrifuge tube, all tubes were washed in 6 M curea, pH 2.0, rcinsed thoroughly in distilled water and dryed in the air.

Blood sampling and separation

A volume of 0.9 mL of shrimp haemolymph was withdrawn from the ostium of heart using a 25 g needle into a 1.0 mL tuberculin syringe containing 0.2 mL heparin (5,000 unit/mL, Elrins Siam, Inc. Cherry Hill, N. J. USA), Precooled to 7℃ as an anticoagulant.

A volume of 1.0 mL of hemolymph was mixed in 1∶1% formalin (formaldeayde solution 37% *W/W*, 10 mL, Na_2HPO_4 (anhyd.) 0.65 g, $NaH_2PO_4H_2O$, 0.40 g, 32 salinity into a vacutainer containing EDTA buffer (Becton, Dickinson and Company, USA).

One millilitre of the diluted hemolymph was then added immediately to the top of pre-prepared percoll gradients and the tubes were centrifuged (Beckmou Model JZ-21 centrifuge) in 2900 g for 10 min (7℃).

The cell bands in the whole density gradient were collected with FS101-gradient tube fractionator (Hoefer Scientific Instruments) from the bottom of the tubes in 0.5 mL fractions into clean tubes (Tendo tronics).

The resulted cell bands were withdrawn separately into a sample chamber, filtered and placed on microscopic slide, and centrifuged again a cytospin (Shandon Cytospin 2, Scimeatrics, INC.) at 2 000 r/min for 15 min.

Analyses of hemocytes fractions from *P. vannami*

a) The smear of hemocytes on to microscopic slides and stain them.

The slides were dried in the air naturally and stained with the Diff Quik stain set.

The procedure is as followed as:

Fixative solution, 1 min→Solution Ⅰ, 2 min→solution Ⅱ, 10 sec. →washed with tap water→dried in air.

b) Microscopic observation of hemocytes and stored from refrigerator after collecting and storing.

The hemocytes in different periods of time were withdrawn from vacutainer into a chamber assembly of the cytospin and centrifuged at 2 000 r/min for 5 min (15℃).

Stain the smears of hemocytes on slides.

Examine those samples with light microscopy.

Test some immunological phenomenon of shrimp haemLymph.

Better methods of withdrawing, separating and staining of shrimp hemocytes were established.

The cytomorphological changes of hemocytes that collected from shrimp and stored in diffent periods of time were compared each other, we found some phenomenon: coagulation, cytolysis plasmoptysis, cytokinesis, etc.) and three types of the shrimp hemocytes, i. e. agranular hemocytes, small granule hemocytes, large granule hemocytes were recognized.

Immunological phenomenon of the shrimp hemocytes were tested preliminarily.

Acknowledgement

I am grateful to Doctor Ziye Yu for his skillful help.

海洋维生素 B_{12} 产生菌培养条件研究*

摘　要　用大肠埃希氏菌(*Escherichia coli*)113-3 测定 11 株分离自我国海域的海洋细菌维生素 B_{12}(VB_{12})产力。试验了它们产生维生素 B_{12} 的培养基及培养条件。结果表明,以海水浸提豆饼并添加一些营养组成的培养基适于进一步确定产 B_{12} 的菌种。培养基中适当添加二氯化钴等盐类有利于 B_{12} 合成。培养基酸碱度在所试菌种发酵过程中随时间延长而提高。保持偏碱性约 96 h,可获得较好的 B_{12} 生产效益。所测菌株中,No. 2627 菌株表现良好。No. 2627 菌株的混合培养,B_{12} 产力常常有潜力和优势可取。

关键词　海洋细菌　维生素 B_{12} 生产　B_{12} 菌培养条件

维生素 B_{12}(本文简称为 B_{12},下同)是自然界已知结构中较复杂的非聚合有机化合物。在医药、人和生物营养上具重要应用价值,如可促进、恢复造血功能,对恶性贫血症有疗效、可防止脂肪在肝脏中沉着,还可治疗神经痛等疾患、在各种水生态系统中发挥不可替代的重要作用。迄今尚无法借助化学手段来实际生产 B_{12}。长期以来人们只从陆源或淡水微生物中筛选有限的 B_{12} 生产菌株(简称 B_{12} 菌,下同)以发酵法生产 B_{12}。随着人口膨胀和资源短缺的压力,探索 B_{12} 菌的眼光已更多地投向海洋。我们对取自我国海域的一些初筛为 B_{12} 菌的菌作培养条件等研究。以期为确定和构建出优秀工程 B_{12} 菌提供依据。

1. 材料和方法

1.1　培养基:共计有 7 种,见表 1。

表 1　海洋维生素 B_{12} 菌试用培养基成分一览表

编号	培养基名称	培养基成分及制作
1	改进的 Zobell 2216	蛋白胨 5 g,牛肉膏 3 g,葡萄糖 1 g,陈海水 1 000 cm^3。
2	玉米粒海水	玉米粒 200 g+陈海水 1 000 cm^3,3 d 后,2 次热压,8 层纱布滤得清液,加 10%:KNO_3、10%KH_2PO_4、10%NH_4NO_3 各 10 cm^3。
3	豆饼海水	豆饼 40 g+陈海水 300 cm^3,煮沸后浸泡半日,热压后取清液,加陈海水至 1 000 cm^3,10%KH_2PO_4 5 cm^3。
4	花生饼海水	花生饼 40 g+陈海水 300 cm^3,煮沸,浸泡半日,热压后取清液,加陈海水至 1 000 cm^3,10%KH_2PO_4 5 cm^3。

* 原文刊于《中国海洋药物》,2000,6(总第 78 期),9-13。

(续表)

编号	培养基名称	培养基成分及制作
5	玉米浆海水	玉米浆 5 g＋陈海水 500 cm^3；10％KNO_3、10％ NH_4NO_3、10％ KH_2PO_4 各 5 cm^3
6	豆饼、玉米浆海水	No.3 培养基：No.5 培养基＝1：3 混合
7	花生饼＋玉米浆海水	No.4 培养基：No.6 培养基＝3：10 混合

注：以上各种培养基都加有 0.01％的 $FePO_4$，30 cm^3 分装前，调 pH 至 7.4，10lb 30 消毒；玉米粒、豆饼和花生饼为市售：济南第二制药厂产玉米浆。

1.2　参试菌株　从收藏的初筛自中国海的部分 B_{12} 菌菌株中随机抽取 14 株，编号分别为 No.37、54、79、172、209A、279、822、933、2000、2036、2494、2627、2749 及 3595。用前在 No.1 液体培养基中活化 24 h 后接一环入表 1 的相应培养基中 25℃振荡培养（120r・min^{-1}）数日，待测 B_{12} 产力。

1.3 B_{12} 产力的鉴定　用来鉴定参试菌株 B_{12} 产力的菌种为大肠埃希氏菌（*Escherichia coli*）113-3。*E coli* 113-3 培养和 B_{12} 的鉴定和差别方法见华北制药厂的“维生素 B_{12} 含量测量（微生物杯碟法）”。根据该 *E coli* 113-3 在培养基上显示的生长圈大小，并与所用标准 Vit. B_{12} 浓度作对照，以比较和分析所试菌 B_{12} 产力的大小。

以下 1.4～1.7 除特指外，所用培养基为 No.3（下述），培养法同 1.2。B_{12} 产力按 1.3 测。

1.4　Co^{2+} 盐对 B_{12} 产力的影响试验

在培养基中添加 $CoCl_2$，其浓度从 0～25 g・m^{-3}（W/V）。参试单株菌分别为 No.2036、2627 或 3595。混合株为 No.2036＋No.2627，连续培养 5 d 后，分别测 B_{12} 产力。

1.5　不同培养时段 B_{12} 产力的比较试验

比较 4 株菌（即 No.54、2627、2749、3595）及不同培养时间段（从 22～216 h 间）分 10 次，即 22、46、67、81、96、120、144、168、190 及 216 h 的 B_{12} 产力。

1.6　B_{12} 生产过程中 pH 值变化的测定

分别试验两单株细菌（即 No.2627 和 No.35 95）B_{12} 生产过程中 pH 值变化，以掌握 pH 值与 B_{12} 产力间的关系，培养时间最长持续 216 h。

1.7　单株或混株培养的 B_{12} 产力比较

分别选取 No.54、2036、2627、2749 和 3539 为单株，选取 No.54＋No.2627＋No.3595、No.2036＋No.2627 或 No.2627＋No.2749 为 3 个混合菌株培养，比较 72 h 和 120 h 时不同菌株及其组合的 B_{12} 产力。

2　结果和讨论

2.1　培养基的比较

表 2　海洋维生素 B_{12} 菌在下同培养基上 B_{12} 产力比较

菌株编号 No.		37	79	172	209A	279	822	933	2 000	2494	2627	空白对照	生长圈平均值(mm)
培养基编号	1	15.7	15.7	16.2	13.1	晕圈	晕圈	15.0	太小	晕圈浓	13.7	晕圈	14.9
	2	12.0	14.6	14.9	15.9	0	0	14.4	太小	15.3	17.1	0	14.9
	3	太小	14.8	16.2	15.6	0	0	15.2	太小	17.5	19.2	0	16.4
	4	太小	13.5	13.1	14.2	0	0	17.63	太小	14.6	19.3	0	16.3
	5	16.3	9.5	15.8	14.4	0	0	14.6	太小	11.8	18.1	晕圈	14.4
	6	18.3	晕圈	18.1	15.3	0	0	太小	13.5	9.1	9.4	—	14.8
	7	0	太小	17.9	13.6	—	—	—	—	—	—	晕圈	15.8
	平均	15.6	13.6	16.6	14.6	—	—	15.3	$<$13.5	13.7	16.1	≈0	—

注：* 培养时间：振荡 58 h＋静止 38 h；Vit B_{12} 产力以 *E. coli* 113—3 生长圈直径(mm)表示(下同)；标准 Vit B_{12}(0.005 $mg \cdot m^{-3}$)的 *E. coli* 生长圈均为 13.5 mm，未参试："晕圈" "太小"为不计入者。

＊＊ 仅计算"0"以外的有效数。

表 2 比较了表 1 中各培养基中 B_{12} 产力的状况。由表 2 可见，10 株菌在 7 种培养基中平均的 B_{12} 产力大小(即 *E. coli* 生长圈直径 mm，下同)排序按培养基号分别为 No. 3(16.4)＞No. 4(16.0)＞No. 7(15.8)＞No. 1(14.9)＝No. 2(14.9)＞No. 5(14.4)＞No. 6(14.0)。这说明豆饼(No. 3)、花生饼(No. 4)或花生饼加玉米浆(No. 7)的培养基效果好于改进的 Zobell 2216(即 No. 1 培养基)。No. 1 培养基主成分含牛肉膏、蛋白胨等，使其上生长出的细菌可携带多种维生素(包括 B_{12})，因而将干扰真正 B_{12} 菌的阳性率，抑或降低其 B_{12} 产力，而且该培养基选择性不强。它不仅可作为从海洋中也可在室内初筛 B_{12} 菌用。统计该表数据还说明 10 株菌在 7 种培养基中给出的 B_{12} 产力，各自平均的大小排序为 No. 172＞2627＞37＞933＞209A＞2494＞79＞2000＞279＝No. 822。由此初步判断 No. 2627 菌株可能有较好 B_{12} 产力。方差分析结果显示，不同培养基间、不同菌株间均各有显著差异($P<0.01$)。

以下各试验一般均在 No. 3 培养基基础上进行。

2.2　钴盐对 B_{12} 产力的作用

钴盐对 B_{12} 菌产力的作用可见表 3。由该表可见，培养基中添加有二氯化钴($CoCl_2$)，其菌株经 5 d 培养，无论是单株还是混合培养，其 B_{12} 产力均比未加钴盐的大。平均地说，B_{12} 产力大小排序为：加 25 $g \cdot m^{-1}$(16.97 mm)＞5 $g \cdot m^{-3}$(16.60 mm)＞无钴的(12.30 mm)。但 No. 2627 单株培养，其 5 $g \cdot m^{-3}$ 的 B_{12} 产力却大于 25 $g \cdot m^{-3}$ 时的，这暗示特定菌株的不同培养方式，钴盐的作用是不一样的。还表达出过高钴离子浓度不利于 B_{12} 生产。正如 Yongsmith 等(1982)指出的那样，12 $g \cdot m^{-3}$ 浓度的 $CoSO_4 \cdot 7H_2O$ 最适于丙酸杆菌(*Propioribacteriwn arl AKU*1251)在基本培养基中积累 B_{12}，而处于固定化状态的该菌在最适培养中，需要 20 $g \cdot m^{-3}$ 浓度的该盐才可使其产生和积累更多 B_{12}[7]。

表 3　添加氯化钴对海洋维生素 B_{12} 产生菌产力的影响

菌株编号	培养方式	加钴量(g·m^{-3})	B_{12}产力(*E. coli* 生长圈直径 mm)
2036	单株	0	10.6
		0	13.2
2627	单株	5	19.8
		25	17.4
		0	12.6
3595	单株	5	16.0
		25	19.3
		0	12.8
2036	混合	5	14.0
2627		25	14.2

注：所用培养基为表 1 中的 No. 3、另加有 10％$MgSO_4$ 液 10 cm^3，5％N-苯基甘氨酸 1 cm^3，加 Co^{2-} 或不加 Co^{2-}。培养时间：连续 5 d。

2.3　B_{12}生产过程中 pH 值的变化

图 1 可见 No. 2627 和 3595 两株细菌在各自 B_{12}生产过程中培养液 pH 值的变化状态(培养基同表 3，但未加 Co^{2+})。由该图可见，pH 值于发酵初期(50 h 以前)下降很快。B_{12}产量也不大。到了中期(50～100 h)，B_{12}产力处于上升、稳定阶段，而 pH 也逐渐碱化。到了后期(100 h 以后)，pH 值更大 B_{12}，生产和积累走向衰落。随着生产进程加深。pH 值碱化、B_{12}积累下降、说明过高 pH 值的培养液不利于所试菌株的 B_{12}生产。当然这一情况尚需具体分析，比如在另一个实验中，一直处于弱酸环境中的 N0. 2634 菌株之 B_{12}产力水平一直不高、变幅下大。这样生产过程势必拉长、成本增加。总之在一定限度内，pH 值由小逐步变大至适当的碱性时，可能有利于 B_{12}菌生长和 B_{12}积累。

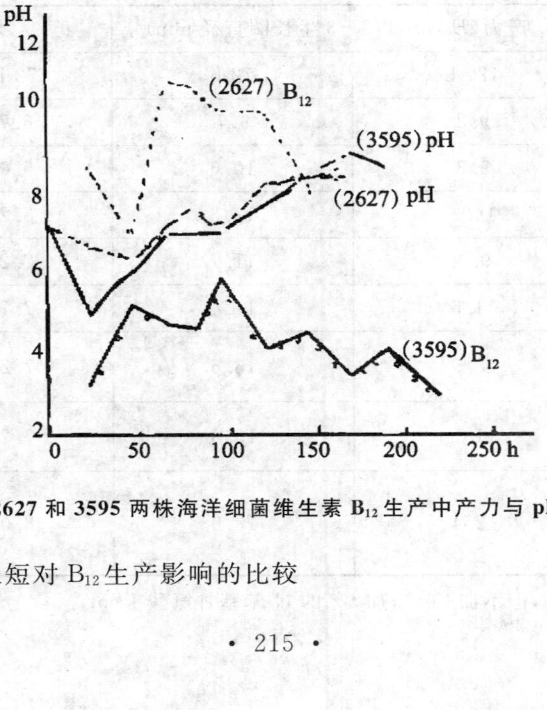

图 1　No. 2627 和 3595 两株海洋细菌维生素 B_{12}生产中产力与 pH 值变化图

2.4　培养时间长短对 B_{12}生产影响的比较

表 4　不同培养时段海洋维生素 B_{12} 产生菌产力比较

培养时间 h		22	45	57	81	96	120	111	158	190	216
菌株编号	51	0	9.3	10.4	10.1	11.5	太小	10.3	1.5	—	—
	2637	12.6	6.9	19.9	19.8	19.0	19.0	17.0	17.7	—	—
	2713	12.5	12.6	15.7	12.7	14.0	5.7	21.2	11.8	—	—
	3535	12	14.3	13.7	13.6	14.9	13.2	13.5	12.5	13.1	12.6

注：所用培养基同表 3。但不加 Co^{2+}、Vit. B_{12} 产力以 *E. coli* 给出的生长圈直径(mm)表示。"—"未参试。

表 4 是 4 株菌分别在 22～216 h 间的 10 个培养时段上得出的 B_{12} 产力比较，统计表 4 的数据得出：4 株菌 B_{12} 产力的各个平均值大小依各培养时段排序如下：67 h(14.9)＞96 h(11.9)＞81 h(14.0)＞120 h(14.0)＞168 h(13.7)＞190 h(13.1)＞46 h(13.0)＞216 h(12.5)＞22 h(14.0)。

B_{12} 产量出现高峰时间。两株菌(No. 2627 和 No. 2719)在 67 h，另两株菌 No. 54 和 No. 3595 则在 96 h 时。No. 51 在 168 h 时也见 B_{12} 产量高峰。在 67～96 h 间，No. 54 菌株 B_{12} 的平均产力为 10.7 mm，No. 2627、2749 和 3595 分别为 19.4、14.1 和 11.1 mm。这期间 B_{12} 产力均比各自的全时期平均 B_{12} 产力大(相应地、No. 51、2627、2749 和 3595 产力分别为 9.0、18.2、12.5 和 13 mm)，表明 67～96 h 为这 4 株菌 B_{12} 生产高峰期。各株菌平均的 B_{12} 产力大小排序为：No. 3539、No. 2749，No. 54。方差分析结果也表明该 4 株菌 B_{12} 产力间差异十分明显($P<0.005$)。总体讲，更短或更长的收获期很可能得不到好的 B_{12} 生产效益。

2.5　单株或混合培养时 B_{12} 产力的比较

表 5　单株或混合培养时海洋维生素 B_{12} 产生菌产力的比较

菌株号(No.)	培养方式	B_{12} 产力(*E. coli* 113－3 生长圈直径 mm)		培养液终 pH 值	
		120 h	72 h	72 h	120 h
54	单株	9.8	9.7	6.7	7.2
2036	单株	9.8	10.6	5.0	5.7
2627	单株	14.3	13.3	8.2	7.7
2749	单株	9.6	11.7	7.2	7.2
3595	单株	11.5	11.4	7.7	7.7
2036 2627	混合	18.9	13.6	7.2	8.1
2627 2749	混合	11.7	13.6	7.2	7.2
54 2627 3595	混合	12.4	13.9	8.0	8.0

注：所用培养基同表 3，但不加 Co^{2+}，加 5% 的 N-苯基甘氨酸 1 cm^3。

表5列出了5个单株培养和3种不同混合培养菌的 B_{12} 产力比较。据该表作的统计可见，72 h时5个单株培养的 B_{12} 产力平均为11.1 mm，3种混养平均 B_{12} 产力为12.7 mm，120 h时单养的 B_{12} 产力平均为11.3 mm，混养平均产力为13.5 mm，最差的单株菌是No. 2749，最好株No. 2627。最差混养是No. 2749＋ No. 2627。总的来看，混养 B_{12} 产力好于单养。72 h和120 h的单株和混株培养，其 B_{12} 的各自平均产力分别为12.1和11.6 mm，表明培养时间短，B_{12} 产量显尚低之势（与2.4中所述一致）。试验还表达出保持7.2～8.2的pH值至少96 h的合适培养液对 B_{12} 生产有利。

3　结语

按微生物学方法用 *E. coli* 113-3测试了11株海洋细菌的 B_{12} 产力。结果表明，改进的Zobell 2216培养基可作为海上或室内初筛 B_{12} 菌用培养基。在以豆饼为基质的海水培养基中，参试菌将可显示较明确的 B_{12} 产力。在此基础上选择菌株作添加 Co^{2+}、不同发酵时间、单养或混养及发酵期pH值变化与 B_{12} 产力的关系等的试验表明，适当添加 Co^{2+} 至培养基可促使 B_{12} 的合成和生产。合适调节和保持发酵液偏碱性（如pH值为7.2～8.4）达96 h有利于 B_{12} 生产和积累。一定菌株的混养优于单养，结果表明菌株No. 2627单养或与No. 2036菌株混养于豆饼等浸提的海水培养基中，添加钴离子等适当盐类，经4 d以上的发酵可能得到较好的 B_{12} 效益。对海洋维生素 B_{12} 产生菌的性状及其 B_{12} 生产条件值得进一步探讨。

致谢：本实验张学雷同志对部分数据进行了统计学处理，在此表示感谢。

参考文献7篇（略）

（合作者：刘秀云　高月华）

STUDY ON CULTURE CONDITION OF MARINE VITAMIN B_{12}-PRODUCING BACTERIA

(ABSTRACT)

Abstract The vitamin B_{12} productivity of marine bacterial strains isolated from the China Sea was determined microbiologically using *Escherichia coli* 113-3.

Tests were made on the culturing media and culture conditions for Vitamin B_{12} production. The results showed that the culturing media in which the extract from soybean cake with seawater and some nutriments were added was suitable to recognize further some marine Vitamin B_{12}-producing bacterial strains. The medium with appropriate cobalt chloride and some salts added may be useful for Vitamin B_{12} biosynthesis. The pH value of the fermentation for the testing strains was raised with the lapse of time for fermentation. A better Vitamin B_{12} production may be obtained by keeping the fermentation liquid slightly basic till about 96 h. No. 2627 strain in all the strains tested showed well in the productivity and superiority for Vitamin B_{12} production.

Key words Marine Bacteria, Vitamin B_{12} Production, Culturing Conditions of Vitamin B_{12}-Production Bacteria.

第三篇

极地科学

POLAR SCIENCES

C$_1$ 微生物学
(MICROBIOLOGY)

THE AMOUNT OF HYDROCARBON BACTERIA IN THE GREAT WALL BAY AND ITS ADJACENT AREA*

Abstract During the summer of Antarctica in 1993/1994, the species and amount of hydrocarbon bacteria in the Great Wall Bay and its adjacent sea area have been studied. *Flavobacterium*, *Pseudomonas*, *Kurthia* and *Actinetobacter* have been identified. The number of them varied from 3 cell/L to 1100 cell/L. The number in the inner bay is larger than that out of it. The petroleum dispersing is a very important way of the changing of hydrocarbon bacteria.

Key words Great Wall Bay Antarctica Hydrocarbon Bacteria

1. Introduction

The pollution of the petroleum hydrocarbon is extraordinary in the Antarctic sea area. On Jan. 19th, 1989, the luxurious pleasure boat "Bahia Paraso" of Argentina dashed on the rocks at the Aner Island of the Antarctic Peninsula (Li et al. 1992; Kenni-cutt et al. 1991) at least leaked 1.5×10^5 gallon fuel oil and lube oil. Not long after it, a Peru oil tanker sunk near the King George Island and the adjacent area was polluted; in addition, investigative ships sometimes leaked oil, and the waste oil which was used to warm persons inputted into the sea inevitably. So people has paid more attention to the oil pollution of these areas. Ocean has the ability to self-purify the oil pollution. During the removing period of ocean petroleum hydrocarbon, hydrocarbon bacteria play an important role in catabolizing the petroleum hydrocarbon. And it is also the essential indicate or-

* The original text was published in *C. J. P. S.*, 1998, 9(1): 66-70. I was the third writer.
原中文稿刊于《极地研究》,1998,10(4):26-30,本人为第三作者。

ganism of the pollution of the ocean petroleum hydrocarbon. There are many research reports on microbes in the Antarctic sea area (Chen and Song 1992; Chen et al. 1992). But the hydrocarbon bacteria have not attracted our much attention. Thus, we studied the dispersing and changing of the hydrocarbon bacteria monthly during the summer of Antarctica from Dec. 1993 to Mar. 1994, and want to get some information about the hydrocarbon bacteria in Antarctica.

2. The general situation of the research area

The studied area includes the whole Great Wall Bay, a part of the Maxwell Bay and the Ardely Bay (Fig. 1 shows the fixed research stations and investigative stations.)

The Files Peninsula is of Sub-Antarctic ocean climatically and the weather is warm and wet as compared with other Antarctic area. It has much cloud in the whole year, high humidity, lots of rain and more snowstorm. At the Great Wall Station, the average temperature is −4.5℃, and the average precipitation is 414.9 mm yearly and the maximum is in March, the minimum is from June to August. The highest air temperature is 11.7℃ in summer, and the lowest is −26.6℃ in winter. The temperature changes very quickly. During several hours, it can change more than ten centigrade degree. The air temperature and humidity are both suitable for the organisms to live in summer. The wind speed is fast in the area and average wind velocity is 17～18 m/s. The average atmospheric pressure is 987.0hPa yearly. When it is influenced by the cyclone, it can be lower than 940hPa which is usual to the atmospheric pressure of the center of the hurricane.

The inner Great Wall Bay is long-shape and its area about 3 km^2. The mouth of the bay is divided into two waterways by the Gulangyu Island. There is a stone dike between the Ardely Island and the Files Peninsula, where the rubber ships can go through smoothly during the ebb tide. When the tide is coming in, sea water of the Ardely Bay can go into the Great Wall Bay through the waterways. Besides this, more of the sea water of the Great Wall Bay goes through its mouth to exchange with the water of the Maxwell Bay. The outside sea water moves from the bottom of bay to the inner. While the inside water moves to the outside from the surface layer. By

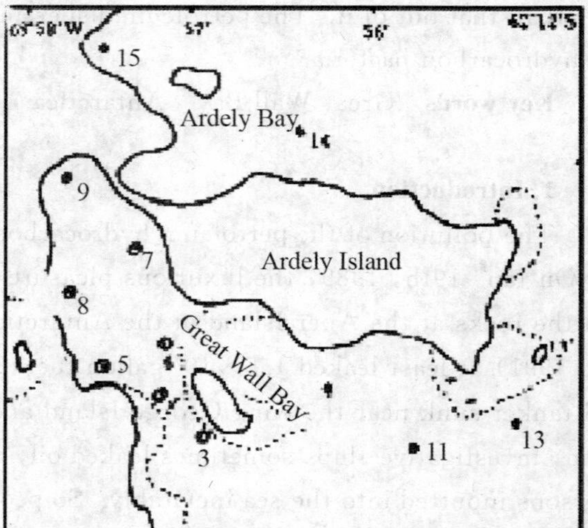

Fig. 1 The research area and the sampling stations

this way, the inner Great Wall Bay can eliminate some pollution.

3. Material and method

On Dec. 13th, 1993, Jan. 7th, 1994 and Feb. 3th 1994, the measurement was carried out at the 9 fixed research stations and water samples were used to analyze the hydrocarbon bacteria. On Dec. 13th, 1993, samples were collected from Nos. 10, 11, 12, 13, 14 and 15 stations. For the continuous storm and high wave, samples were collected only from Nos. 1～9 fixed stations and ships returned for the high wave.

The sea water of surface layer (0.5 m) was collected by bacterial sampler. Then samples were analyzed about the amount of hydrocarbon bacteria by Try-Tube Method in the laboratory. After being purificated in disc, the hydrocarbon bacteria were brought to China to identify.

The culture medium of hydrocarbon bacteria is followed: NH_4Cl 2 g; $K_2 hPO_4$ 0.7 g; KH_2PO_4 0.3 g; filtered sea water 1 L; diesel oil 0.5 m; pH 7.8; 8P, 15 min.

4. Results and analysis

4.1　The species of the hydrocarbon bacteria

By primary identity, four genera of hydrocarbon bacteria occur in the sea area researched. They are *Flavobacterium*, *Pseudomonas*, *Kurthia and Acinetobacter*. Table 1 shows their main physiological characters.

Table 1　The main physiological characters of the hydrocarbon bacteria

Genus	Gram stain	H-L culture	Indole	Mythyl red test	Voges proskauer	Gelatin hydrolysis	Catalase	Oxidase	Cysts formed	Flagellar arrangement
Flavobacterium	−	−	−	−	/	−	+	−	−	Peritrichous
Acinetobacter	−	+	−	/	/	−	+	−	/	Pertrichous
Pseudomonas	−	−	−	−	+		+	+	/	Monotrichous
Kurthia	+	+	−	−	/	−	+	−	/	Peritrichous

4.2　The amount dispersion of the hydrocarbon bacteria

Table 2 and Table 3 show the results.

Table 2　The number of hydrocarbon bacteria in the Great Wall Bay and its adjacent area

Number	Environment situation	Result/(cell · L^{-1})		
		Dec. 13, 1993	Han. 7, 1994	Feb. 3, 1994
1	Near the Two Peaks Island	210	6	1 100
2	Near the Jiuquan River, 400 m	1100	6	1 100
3	The mouth of inner bay		11	>1 100
4	The mouth of inner bay	53	16	>1 100
5	Near oil treasury	290	11	>1 100

(continue)

Number	Environment situation	Result/(cell・L^{-1})		
		Dec. 13, 1993	Han. 7, 1994	Feb. 3, 1994
6	Near the gulangyu Island	1 100	16	>1 100
7	Near the Ardely Island	53	20	>1 100
8	Near Great Wall Station	>11 100	15	>1 100
9	Habitat of seals	>1 100	20	>1 100
10	The mouth of inner bay	1 100	3	
11	Outside of the Ardely Island	36	3	
12	Outside of the Ardely Island	42		
13	Outside of the Ardely Island	1 100		
14	Outside of the Ardely Island	210		
15	Outside the dock of Chile station, about 300 m	460		

Table 3 The change of numbers of hydrocarbon bacteria in the research area (cell/L)

Sea area	Number of station	Dec. 1993	Jan. 1994	Feb. 1994	Means
Inner bay	5,6,7,8,9	>728.6	16.4	>1 100	>615
Mouth of inner bay	3,4	>578.5	13.5	>1 100	>564
Outside sea area	1,2,10,11,12,13,14	532.25	3	1 100	545
Means		574.65	11	>1100	546

* The inner bay means the sea area in the Gulangyu Island.

(1) The number of the hydrocarbon bacteria of each station varied from 3 cell/L to 1 100 cell/L, and the degree of the average changes is 3 order of magnitude. The total average number is 546 cell/L. There are 10 stations whose number exceeds the inspection range (1 100 cell/L). Among them, there are 3 times in Dec. 1993 and 7 times in Feb. 1994. There are 15 stations whose number is lower than 100 cell/L. There are 11 times in Jan. 1994 and 4 times in Dec. 1993(Table 2).

(2) The number in the inner bay is higher than that out of it. This tendency can be learned from these 3 investigations. It is not obvious when viewed at an angle of the average value of each sea area (Table 3).

(3) The number of the hydrocarbon bacteria, which is in the inner Great Wall Bay, the mouth of the inner bay, varied obviously in a month. The average value of each station is the maximum in Feb. 1994, and it is higher than 1 100 cell/L. The minimum is in Jan. 1994 and its value is only 11 cell/L The number of Dec. 1993 is between them.

5. Discussion

The researching history of the hydrocarbon bacteria dated more than one hundred years. There are more than 70 generic microbes, which can catabolized hydrocarbon and exist in ocean widely (Kennicutt et al. 1991). However, there is less report on Antarctic hydrocarbon bacteria in the documentations.

Four genera of microbes, which catabolize hydrocarbon, have been checked in the Great Wall station's adjacent area. They belong to *Favobacterium*, *Pseudomonas*, *Acineto bacteria* and *Kurthia*. Their number varies from 3 cell/L to 1 100 cell/L. The total average value is 546 cell/L every time. The distinction of the results of the different days is obvious. The minimum is 3 cell/L and the maximum is 1 100 cell/L.

By the general analysis clearly, the number of hydrocarbon bacteria in the inner bay, which is higher than that out of the bay. This result is obtained according to the rich hydrocarbon resource of inner bay, slight wave, little exchange with other sea area and stable environment.

Because of the several days' storms, the exchanging quantity of sea water between the inside and outside is very big on the Great Wall Bay in Jan. 1994. It should be the main reason why the amount of hydrocarbon bacteria in Jan. 1994 is much lower than that of other two research times.

In the present research, there is no obvious relation between the density change of hydrocarbon bacteria and the content of hydrocarbon. The hydrocarbon content is relatively stable. While the hydrocarbon bacterial density varies violently (we should report the prouted condition of the sea area in another paper).

As compared with the hydrocarbon bacteria density of near-shore sea area in our country, the value of the Antarctic Great Wall Bay and adjacent sea area is lower. For instance, in the Bohai Sea (Wang 1983), the hydrocarbon bacteria density varies from 1×10^2 cell/L to 1.536×10^6 cell/L from 1980 to 1981. There are 14 genera. *Pseudomonas*, *Acinetobacter*, *Bacillus*, *Corynebacterium* are the dominant species. In the Jiaozhou Bay (Ding et al. 1979) the hydrocarbon bacteria density varies from 2×10^4 cell/mL to 1.4×10^5 cell/mL. The amount of July is higher than that of March. *Pseudomonas*, *Flavobacterium*, *Acinetobacter* are frequent species. In the Dalian bay (Li et al. 1987), the hydrocarbon bacteria density varies from 1.102×10^3 cell/L to 1.5×10^5 cell/L during May of 1982-May of 1983. There are 12 genera. In the area of Helogo land of the North Sea (Gunkel 1973), there are a large number of oil-oxidizing bacteria in the whole year. It varies from 930 cell/L to 4.6×10^5 cell/L, and the minimum is in winter.

At the same time of the present researching of the Antarctic oil hydrocarbon, the hydrocarbon bacteria of the Jiaozhou Bay in winter were used for comparing study. In the corresponding period, the hydrocarbon bacteria of the Jiaozhou Bay varies from 10 cell/mL to 900cell/mL. There are 10 genera. Amony them, *Pseudomonas*, *Acinetobacter*, *Flavobacterium* were checked out from Antarctica and the Jiao zhou Bay samples. In addition, *Coryebacterum*, *Arthrobacter*, *Alealzenes*, *Micrococcus*, *Staphylococcus*, etc. were checked out from the Jiaozhou Bay samples. This shows that the hydrocarbon bacterial amount of the great Wall Bay and adjacent sea area is lower than that of temperate zone and the species are also little in number than that of temperate zone.

The temperature of Antarctic sea water is low throughout the year. In the most days of a year, sea water was covered by ice. So the oil hydrocarbon vsvolatilized showly. Once being polluted, it is difficult to remove it. In the studied area, the activity of the persons if frequent. Much fuel oil and lube oil flowed into the sea water for the all kinds of the reasons. These made oil pollution and the oil hydrocarbon study become the important task of environmental conservation. Hydrocarbon bacteria was the index of oil hydrocarbon polluting degree. It also plays important role in eliminating the pollution (ZoBell 1973). It is very important to enhance the stud on this aspect.

Acknowledgments We are very grateful to Mr. Lu Hua and Mr. Hong Xuguang for the helping of sampling in Great Wall Station, Antarctica. This study was supported by the State Antarctic Committee of China (No. 85-905-02).

References9(abr.)

(Cooperators: Junfeng Yuan Yongqi Li Baoling Wu)

南极菲尔德斯半岛生物体的微生物含量分析*

摘　要　于1993/1994年以平板培养法用5种基本培养基对南极菲尔德斯半岛地区的生物体作了微生物含量分析。所研究的海藻、苔藓、地衣、沙蚕、小红蛤、帽贝、鱼、企鹅和鲸等的体表和体内样本计33个。分析和比较了各类生物体内外所含的陆源性细菌、海洋性细菌、肠细菌、真菌及耐酸细菌等的含量。在一定程度上培养温度的提高可促进样本中原有的微生物迅速大量繁殖，表明低温仍是制约南极微生物的重要因子。

关键词　南极　菲尔德斯半岛　生物体　微生物含量

1. 前言

一些现象已证实人类活动的影响已对南极原始的生态产生作用。有些甚至是不可逆转的。西南极菲尔德斯半岛上考察站分布最为密集，人类活动的影响相对较大，研究该区微生物的来源和数量变化不仅有助于追踪人类对南极生态环境影响，同时有助于进一步认识南极生物间的相互关系。

就目前该地区的生态研究而言.有关微生物研究的报道并不多见(Cameron and Morelli，1972；陈皓文等，1992；肖昌松等，1994)。作者于1993～1994年对于取自该地区的33份生物样本的微生物含量做了研究。因而，对生物样本所含的各类微生物之状况有了进一步的认识。本文对生物体内和体表的微生物测定结果进行分析报道。

2. 材料和方法

2.1　培养基制作

海水培养基、营养琼脂培养基、真菌培养基、乳糖蛋白胨培养液和品红亚硫酸钠培养基分别按文献(陈皓文等，1992；陈皓文等，1994；全国主要湖泊水库富营养化调查研究课题组，1987)所述制作。酸性培养基是指pH为6.0的营养琼脂。

2.2　样本采集

在营养琼脂培养基上长出的菌落视为陆源性的微生物，在海水培养基上长出的视为海洋性的微生物(陈皓文等，1994)。

苔藓、地衣的植体研磨碎，作不同稀释后，取悬浮液涂布在相应平皿培养基上。海藻藻体、动物体不同部位的样品，均定量称取，按需稀释后作平皿涂布计数。企鹅胃含物样品系人为迫使呕吐得来，鱼消化道含物样品是解剖鱼体后分段取出的。鲸已死去，采取其各部分肌肉样品作平皿涂布。

以上操作均严格无菌。采样后尽快返回实验室，培养温度一般在15℃，培养时间在7～14 d。

3. 结果和讨论

在菲尔德斯半岛共采集生物各种样本计33份。每份样本的微生物采集记录及相应

---

*　原文刊于《极地研究》，1998，10(4)：290-293。

的微生物含量如表1所示。

表1　菲尔德斯半岛地区生物样本采集记录及其微生物含量[细菌或真菌含量单位 $\times 10^4$ 个/克(鲜重)]*

Table 1　collecting record of living materials and their microbial contents of fields Peninsula, Antarctica

样品	时间	地点	一般描述	陆源性细菌	海洋性细菌	肠系细菌	真菌	耐酸菌
苔藓(a)	93.12.21	长城站后山海关山坡	黑色或深褐色,水分占24.8%,研磨液混黄色	400				
	93.12.22	长城站后山海关山坡	研磨液较清,水分占71.1%,黄褐色	675				
地衣(b)	93.12.22	长城站后山海关山坡	干,孢囊、深褐色,水分占3.3%,研磨液乳白	0				
海藻(c)	93.12.21	长城站海滨	藻体,水分占97.1%	0.1	0.4			
	93.12.21	长城站海滨	体内液汁,pH6.7	0	0			
沙蚕(d)	93.12.21	长城站海滨	体外附液	8.87	13.6		2.92	
	93.12.21	长城站海滨	体内混合液	6.34			1.17	
	93.12.21	长城站海滨	体内混合液	1.33	2.67		0.11	
	93.12.21	长城站海滨	整体研磨液				0.56	
	93.12.21	长城站海滨	整体研磨液				0.222	
小红蛤(e)	94.1.14	长城站海滨	蛤体体壳洗涤水		152			
	94.1.14	长城站海滨	蛤体研磨液		1910			
	94.1.14	长城站海滨	蛤体		23.7			
帽贝(f)	94.1.14	长城站海滨	贝体体壳洗涤水		1.89			
	94.1.14	长城站海滨	贝体体壳洗涤水	0.75	2.05			
	94.1.14	长城站海滨	贝肉		12			
	94.1.14	长城站海滨	贝黏液	0.799	41.7			
	94.1.14	长城站海滨	整贝体		4.2			
南极鳕鱼(g)	94.2.15		消化道液黄褐色,pH4.5	0	0			0
	94.2.15		大肠内含物,黑褐色,含未消化藻片		2.64		1.02	1.02
	94.2.15		直肠内含物,淡黄色粪便	0.9	0.4			0.4

(续表)

样品	时间	地点	一般描述	陆源性细菌	海洋性细菌	肠系细菌	真菌	耐酸菌
企鹅(h)	93.12.24		胃含物,以磷虾为主,水分占84.3%	1.9	2.9			
	93.12.24		活体上的羽毛(表层以下)	111	633	1.11		
	93.12.31		鲜粪,稀,水分占97.9%,磷虾碎屑,杂有碎砂,虾红色	54	21.3 (15℃) 54.4 (37℃)	15		
鲸(i)	93.12.25	生物湾海滨	身体前部,较深处肉,海水不大碰着	60				
	93.12.25	生物湾海滨	身体前部肉,海水可浸处	43.5			大片	
	93.12.25	生物湾海滨	尾脊部肉已明显腐烂	39600				
	93.12.25	生物湾海滨	身体后部肉已腐烂,海水不大碰着	5550				
	94.1.18	生物湾海滨	头部肉海水浸着	583		0	0.557	
	94.1.18	生物湾海滨	颈部肉水气交界面	193		0		
	94.1.18	生物湾海滨	胸腹部肉,暴气	0		0		
	94.1.18	生物湾海滨	背部肉,暴气	0.513				
	94.1.18	生物湾海滨	背部肉,较深层	5.56 0.167 (24℃)				

* 微生物含量以每克(鲜重)样品中含有的菌落数个数计,少数生物以个体计。培养温度一般为15℃;(a)主要是镰刀藓、卷毛藓和裂叶苔等;(b)石萝等;(c)倒卵银杏藻、小腺串藻;(d)沙蚕属;(e)小红蛤;(f)帽贝;(g)南极鳕鱼;(h)阿德雷企鹅;(i)逆戟鲸;大片:连片,无法分辨菌落。

由表1可见,各生物样本的微生物含量大小不能一概而论,就植物而言,苔藓＞海藻＞地衣。就动物而言,鲸＞企鹅＞蛤＞沙蚕＞贝＞鱼。所测的地衣植体,生长在山坡迎风面,十分干燥,其微生物含量很少。沙蚕体内外的陆源性细菌含量同处一个数量级,体外附有较多真菌。企鹅的粪便含大量细菌和真菌,而胃含物所含细菌最少。小红蛤研磨液含较多的海洋性细菌。帽贝贝体外附有较多的海洋细菌。鱼肠含物微生物含量,大肠中的高于直肠中的,其滴液中未见菌落。胃含物偏酸,可能抑制细菌生长。而肠道中则见不少耐酸细菌,细菌菌落偏细小。所研究的逆戟鲸,尸体已搁浅海滨岩石旁多日,鲸体正遭破坏分解。浅表肌肉已腐败,并吸引一些燕鸥啄食。因而其肌体伴生较多的微生物,包括真菌，它们不在37℃生长，只在低于20℃中生长。当最初的样本培养于0℃左右达32 d时，大部分未见菌落长出。待移至15℃中，则很快就生长出大量菌落，小而无色的菌落占多数。其中不少是球状菌。也有一些真菌长出。未见大肠菌。推测这一夏季期间，鲸体腐败可能是周围环境（包括空气）污染的来源之一。在长城站区，顺风时就能嗅到其尸体发出的腐臭味。

4. 结语

菲尔德斯半岛33个生物样本的微生物考察结果表达了该区域一些生物体内外微生物含量的一般信息。文章记录并比较了所测定动植物样本所含的陆源性细菌、海洋性细菌、肠系细菌、真菌及耐酸细菌等的含量及其差异。结果还表明样本在低温条件保存一定量的微生物。搁浅的外来生物的微生物大量繁殖，有可能带来附近的污染。培养温度的提高使相应生物所含的微生物含量的检出数增加，表明低温是制约南极微生物的重要因子。

参考文献5篇（略）

（合作者：袁俊峰　张树荣）

ANALYSIS OF MICROBIAL CONTENTS OF LIVING MATERIALS IN FILDES PENINSULA, ANTARCTICA

(ABSTRACT)

Abstract The analysis of microbial contents of living materials in Fildes Peninsula, Antarctica was made by using the plating method with five basic culture media during the austral summers from 1993 to 1994.

The 33 samples of the surface and inner materials of their bodies were collected from marine algae, mosses, lichen, sea worm, little red clam, cap shaped shell, fish, penguin and whale. It was analyzed and compared to the contents of terrigenous. marine, intestinal, acid-fast bacteria and fungi of each category of organisms. The appropriate rise of incubation temperature promoted the multiplication of the microbes rapidly and a large number at some degree. It suggests that the low temperature is still an important factor for restricting the microorganisms in Antarctica.

Key words Antarctica, Fildes Peninsula, Living Materials, Microbial Contents.

南极菲尔德斯半岛环境微生物含量估计*

摘　要　对1993～1994年南极夏季取自菲尔德斯半岛地区的空气、雾、雨、冰、雪、海水、淡水、陆上、湖泊和海滨的土、砂、石、不同深度湖泊沉积物的206份样品的微生物，包括海洋性细菌、陆源性细菌、菌物(即真菌)、大肠埃希氏菌、耐铬、镉、汞、银、铁、阴离子洗涤剂(LAS)菌、耐酸菌及烃氧化菌等的含量做了分析，对该区域生态环境中的微生物状况有了更深的了解。

关键词　南极　菲尔德斯半岛　环境微生物

南极乔治王岛的菲尔德斯半岛同南极其他地区一样，本应千万年来人类未曾涉足而保持原始状态。但一些现象证实，人类活动已对其原始的生态状况产生影响，其中有些影响甚至是不可逆转的[1]。因此，人类必须去考察、比较、分析它的现状，才有可能推断南极地区的过去并预测它的将来。

菲尔德斯半岛是进行科学考察的独特场所，很多国家相继在此建起了科考站进行多学科考察[1-3]，但就微生物学而言，所发表的研究成果并不多。

我们在对该区生物体作微生物分析的同时[4]，对1993～1994年南极夏季取自该地区海、陆、空环境的206份样本的微生物含量作了分析研究，并由此对样本所代表的环境的各类微生物之状况有了更进一步的估计和认识。而这将为认识和开发利用南极微生物资源提供更多基础资料。

1. 材料和方法

1.1　培养基制作

海水培养基、营养琼脂培养基、菌物培养基、乳糖蛋白胨培养液和品红亚硫酸钠培养基分别按文献[1,4-6]所述制作。含铬($K_2Cr_2O_7$)、镉($CdCl_2 25H_2O$)、汞($HgSO_4$)、铁($FeCl_3 \cdot 6H_2O$)、银(Ag_2SO_4)的培养基和含阴离子洗涤剂(LAS)或石油烃的培养基则是在相应培养基中加入有关的金属盐类等而成，其离子浓度一般为50 g/dm³。LAS的浓度分低(100 μg/dm³)和高(233.4 μg/dm³)2种。酸性培养基是指pH为6.0的营养琼脂。

1.2　样本采集与实验室操作

在菲尔德斯半岛及近海共采集环境微生物样本8类206份。按空、陆、海三大环境类别分，则空中95份、海水28份和陆上83份，均严格无菌操作。采样后尽快返回实验室。培养温度一般为15℃。各微生物样品类别详见表1。

* 原文刊于《黄渤海海洋》，2001，19(1)：49-54.
基金项目：国家“八五”重点攻关项目资助(905-02-02)

表 1　南极菲尔德斯半岛环境微生物样品类别

Table 1　Sample Classification of environmental microorganisms in the Fildes Peninsula, Antarctica

样品类别		样本数
1 空气	1-1 室内	25
	1-2 室外	66
2 雾		2
3 雨		2
4 雪与冰	4-1 正在降着的雪	2
	4-2 积雪(或冰)	7
5 水	5-1 海水	28
	5-2 湖水(淡水)	12
	5-3 饮用水	7
6 陆上土、砂、碎石		31
7 海滨砂		11
8 湖泊沉积物	8-1 湖滨沉积物	1
	8-2 湖表层沉积物	4
	8-3 湖中表层下分层沉积	8

1.2.1　气样　将倾碟好了的有关培养基，按需置于室内外空气中，酌情确定曝皿时间(自然沉降法)，然后取回实验室培养计数菌落形成单位(CFU，下同)，并换算为单位体积(m^3)空气中的 CFU 数。以在营养琼脂培养基上长出的菌落视作陆源性的，在海水培养基上长出的视作海洋性的。在酸性培养基上长出的为嗜酸或耐酸菌[5]。

1.2.2　雨样　计时收集降雨一段时间并仍在降着的雨水入无菌空皿中。定量吸取所收集的雨水涂布至上述有关平皿培养基上，培养计数；以单位体积(m^3)水中的 CFU 计。

1.2.3　雾样　选大雾时间，置有关培养基平皿于雾中，计时曝皿收集雾微生物。以单位体积(m^3)雾气计 CFU。

1.2.4　雪样分 2 种：(1)收集降雪一段时间并仍在降着的雪样入无菌空皿后，移入室内，适温融化，定量涂布于相应平皿培养基上；(2)选取 7 个时段的积雪样本(期间或有新雪加入)融化后作平皿涂布。其中 2 份样品取自阿德雷岛，1 份中明显混有单胞藻，另 1 份明显混有原生动物。以化雪水单位体积(cm^3)计 CFU。

1.2.5　水样　定量取海滨水、潮间带水、表层海湾水、砂间水、海洋生物体内外水、湖水和饮用水(经管道并泵入室内由水龙头取得的湖水，下同)。按需稀释作平皿涂布培养，以所用水单位体积(cm^3)计 CFU 数。此外，对饮用水做滤膜萌发法测定：过滤(0.3 孔径的滤膜)100 cm。的水样，并贴好该滤膜，在品红亚硫酸钠培养基上培养(37℃)，步骤同文献[1]，以确定样品中的大肠埃希氏菌(简称大肠菌，下同)含量。

耐重金属菌、耐 LAS 菌则在相关培养基上培养并计数[7]。烃氧化菌培养在含石油烃的液体培养基中后计数(MPN 法)。

1.2.6　土、砂、碎石样本　均在采集完之后作定量称取并按需稀释。涂布至相应平皿培养基上。以单位重量(g)的鲜样本来计数 CFU。必要时换算为干重计。

上述采样的主要代表性站点的设置见图 1。图中各数字分别代表:1-1 考察站室内;1-2 室外;3. 降雨;4-1 降雪;4-2 积雪或冰;5-1 海水;5-2 湖水;5-3 饮用水;6. 陆上土、砂、碎石;7. 海滨;8. 湖泊沉积物。所采集样本的类别统计见表 1。

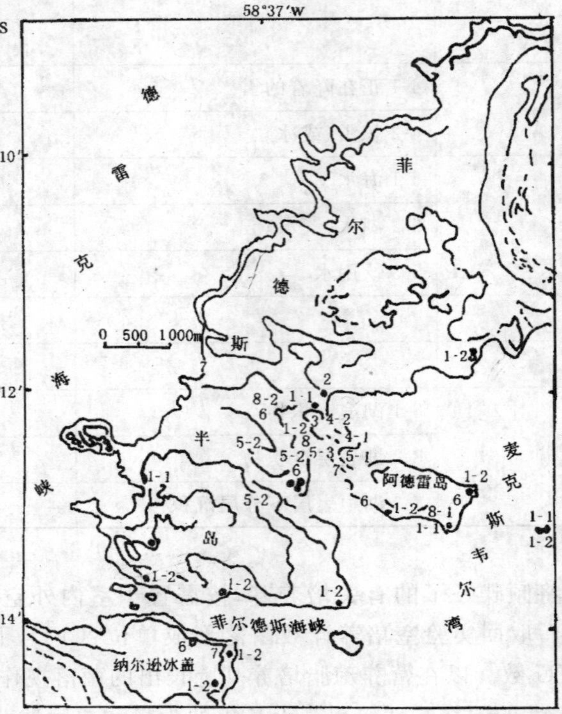

图 1　南极菲尔德斯半岛环境微生物代表性采样点示意图

Fig. 1　The representative sampling sites for environmental microorganisms in the Fildes Peninsula, Antarctica

2. 结果和讨论

表 2 为所有样本的微生物含量分析结果。现将各主要类群样本的平均微生物含量分别叙述。(1)陆源菌,室内空气:3.81×10^{2} CFU/m^{3};室外空气:3.03×10^{3} CFU/cm^{3};陆上土砂、碎石等:陆源细菌 8.57×10^{5} CFU/g,海洋细菌 4.23×10^{3} CFU/g;湖水:2.58×10^{3} CFU/cm^{3};海水:1.82×10^{3} CFU/cm^{3};饮用水:1.11×10^{2} CFU/cm^{3};湖表层沉积物:1.75×10^{5} CFU/g;(2)海水海洋菌:5.51×10^{3} CFU/cm^{3};(3)其他各类微生物含量如下:1)菌物,海滨砂:1.20×10^{3} CFU/g(陆源性的为 2.50×10^{2} CFU/g);海水:5.50×10^{1} CFU/cm^{3}[9];2)大肠菌群,湖水:1.33×10^{3} CFU/dm^{3}(由 cm^{3} 换算为 dm^{3},下同);海水:5.00×10^{-2} CFU/dm^{3};饮用水:6.70×10^{-1} CFU/dm^{3};3)海滨砂耐银菌:6.00×10^{1} CFU/g;4)耐铬菌:8.93×10^{3} CFU/g;海水:8.61×10^{2} CFU/cm^{3};5)耐镉菌,海滨砂是:8.00×10^{1} CFU/g;海水中:未测出;6)耐汞菌,湖水:4.13×10^{1} CFU/cm^{3};7)耐低浓度 LAS 菌,海滨砂:7.57×10^{3} CFU/g,海水:7.52×10^{2} CFU/cm^{3};8)海水中的石油烃氧化菌含量平均是:

5.45×10² 个细胞/cm³，范围 3.00×10⁰～10×10³ 个细胞/cm³。湾内＞湾外，2月＞12月＞1月。

结果表明，室内空气微生物含量比起 7 a 前，总体上看呈上升趋势，室外空气微生物含量比 7 a 前增长更多[8]，尽管仍比中、低纬度地区一些数据低得多[10]。雾气中的微生物与一般空气中的含量相差不大。正在下的雨水其微生物含量比积存于湖中的水的低得多，前者不足后者的 0.002 倍[5/(2.58×10³)]，仅是海水中的 0.0027 倍[5/(1.82×10³)]。

表 2　南极菲尔德斯半岛海、陆、空 206 份环境样本中的微生物含量统计表

Table 2　Statistics of microbial content from the 206 environmental samples of sea, land and air in the Fildes Peninsula, Antarctica

样本类别	陆源性细菌	海洋性细菌	肠系细菌	菌物	耐重金属菌					耐酸菌	耐 LAS 菌		石油烃降解菌
					耐 Cr^{6+}	耐 Cd^{2+}	耐 Hg^{2+}	耐 Ag^{2+}	耐 Fe^{3+}		低	高	
1-1	3.81×10^2												
1-2	3.03×10^3												
2	1.89×10^3	2.15×10^3											
3	5.00×10^0	1.00×10^1											
4-1	5.00×10^0	0		满皿						0			
4-2	1.85×10^4	5.00×10^3								0			
5-1	1.82×10^3	5.51×10^3	5.00×10^{-2}	5.50×10^1	8.61×10^2	0	0	0			7.52×10^2	1.00×10^2	5.45×10^2
5-2	2.83×10^3	1.50×10^1	1.33×10^3		0	0	4.13×10^1	0	0	0	3.00×10^3		
5-3	1.11×10^2		0.67×10^0										
6	8.57×10^5	4.27×10^3											
7	3.49×10^4	4.25×10^4		2.50×10^2	8.93×10^3	8.00×10^1	0	6.00×10^1			7.57×10^3		
				1.20×10^3									
8-1	1.90×10^4			5.00×10^2									
8-2	1.75×10^5												
8-3	5.08×10^2												

注：样本类别名称参见表 1 含量（CFU 或细胞数）中，空气单位以 m³，水样以 cm³ 或 dm³ 计，其他以湿重 g 计；菌物栏中 1.20×10³ 系海洋性的

降雪中的微生物含量比积雪中的少得多。而积雪中的一般呈现出雪积存时间越长，微生物含量越大之势，并且也比 1986～1987 年的高了 2 个量级[1]，类似于 Wynn-Williams(1991)的报道[11]。淡水中的微生物含量比以前测定的高出 1 个量级[1,3]。

浅海近岸水中所测的微生物含量仍比远岸水的高[12]。潮间带或海滨水呈现出比海湾水含较多的陆源性细菌数。

饮用水中的微生物含量比一般室外淡水的低。

就大肠菌群而言，3 种类型的水中均有不同程度的检出率，湖水、海水和饮用水中分别为 33.3%，2.0%和 57.1%。平均含量依大小分，次序排列为湖水＞饮用水＞海水。长城湾海水中大肠菌数与以往的持平[13]。尽管 1993～1994 年夏季考察站队员人数与历年各次考察相比为最少，但外来考察观光者人数依然不减。饮用水中的大肠菌数虽然并不大，但仍然必须引起注意。

空气、雪的微生物、大肠菌的数量可能意味着外来影响正在加强。陆土等的微生物含

量并不很低，与 1991～1992 年夏季的基本持平[2]，而比 7 a 前的又高出一个量级[1]。陆上的 3 个类型环境，即阿德雷岛、鼓浪屿和长城站的样本相比，长城站土样的微生物含量最低。这可能与土壤的营养成分相关。从外观可见，所采长城站的样本比前二者的贫瘠。若将 3 个样本所在环境的测点相连一线，从所测结果可发现，测点离海越近，样本中海洋性微生物含量越大。这可能与那儿的生存环境对它们较为优越有关。总的讲，海滨砂中的陆源性微生物含量比离海相对远的陆土中的要少。

所得的资料表明，湖泊表层沉积物的微生物检出率为 71.4%，含量范围是 1.00×10^2～5.80×10^5 CFU/g，平均含量为 1.75×10^5 CFU/g。湖滨沉积物菌物为 5.00×10^2 CFU/g。湖底质表层以下的沉积物，其细菌检出率较低，仅为 40%，平均含量范围是 0～2.5×10^3 CFU/g，平均值为 5.08×10^2 CFU/g，且因沉积物样本所在深度和组分不同而异，变化较大。湖滨表层砂土样的微生物含量较高，其量级达 10^4 CFU/g。

从所采底质和水样中，发现耐铬菌的机会较多，其次是耐镉菌（包括一些耐镉菌物），耐银菌亦有所发现。这 3 类菌的含量基本呈依次递减。少见耐汞菌，未见耐铁菌。这与中、低纬度一些区域所见的耐重金属含量及所耐重金属浓度相比均低得多[14]。这一结果显示出所调查区域含有一定量的金属本底及其相应的微生物。海滨砂和海水中普遍存在着耐 LAS 的微生物，表明合成的洗涤剂可能对海滨和近岸水的相应微生物之生长繁育有一定刺激作用，也反映了所测环境受到了某种程度的污染。烃氧化菌的存在和发展意味着相关海域存在着一定的石油污染，同时它们也担负起净化油污的重要责任[15]。

3. 结语

菲尔德斯半岛 206 份环境样本的微生物考察结果表达了该区域立体的生态环境微生物含量的一般信息。这一结果受制于复杂的环境一生态因子，因而，在一定程度上表达了该区域的生态学状况和环境信息。与以往结果相比，室内外空气微生物含量，从总体看呈现出上升趋势，尽管局部样本的微生物含量有所下降[12]。陆地样本的微生物含量也显出增多现象，虽然长城站样本的微生物含量变化不大。海水中普通微生物的含量与以往相比变化不大。存在着一定数量的大肠菌、耐重金属菌、耐 LAS 菌和烃氧化菌。表明沿岸水域可能存在着相应的污染，尽管水体的交换能力和自净能力尚健。结果启示，淡水湖泊的检测和饮用水设施的完善尚待加强。有迹象显示，菲尔德斯半岛的土著微生物和生态环境已受到人类及其他入侵者活动的影响，且日益明显。因此，对该区域的微生物研究仍要强化。

参考文献 15 篇（略）

（合作者：袁俊峰　曹俊杰　张树荣）

ESTIMATION ON ENVIRONMENTAL MICROBIAL CONTENT IN THE AREA OF FILDES PENINSULA, ANTARCTICA

(ABSTRACT)

Abstract A further exploration was made of the microbial content in the area of Fildes Peninsula, Antarctica during the austral summers of 1993/1994. 206 samples of non-living materials collected from marine, land and air environments were tested.

This paper recorded the content of different microbes including marine, terrigenous, intestinal bacteria; Cr-, Cd-, Hg-, Ag-, Fe-, negative ion detergent (LAS)-tolerant bacteria; acid-fast bacteria; petroleum hydrocarbon oxidizing bacteria; panomycetes, etc. These microbes were isolated from air, fog, rain, ice and snow, sea water, freshwater, including drinking water from water supply system; soil, sand, crashed stones from lakesides and beaches; lake sediments from different depths and so on.

A further understanding has been obtained of the microbiological condition in the ecological environments of the surveyed area.

Key words Antarctica; Fildes Peninsula; Environmental Microbes

ABUNDANCE OF GENERAL AEROBIC HETEROTROPHIC BACTERIA IN THE BERING SEA AND CHUKCHI SEA AND THEIR ADAPTATION TO TEMPERATURE *

Abstract The abundance of general aerobic hetertrophic bacteria (GAB) from the water and sediment in the Bering Sea and the Chukchi Sea was determined by using petri dish cultivation and counting method. The abundance of GAB among the different sea areas, sampling sites, layers of sediments surveyed and adaptability to differential temperatures was studied. The result obtained showed that: the occurrence percentage of GAB in the surface water was higher than that in sediment, but the abundance was only 0.17% of sediment. The occurrence percentage of GAB in surfacial layer of sediment as higher than that in the other layers. The occurrence percentage, abundance and its variation of GAB in the Being Sea were higher than that in the Chukchi Sea respectively. The average value of the abundance in the all sediment surveyed was 3 166.3×10^{2} CFU·g^{-1} (wet.). The abundance of GAB in sediment showed a trend: roughly higher in the lower latitudinal area than higher latitude. The results from temperature test mean that: the majority of bacteria tested were cold-adapted ones, minority might be mesophilic bacteria. The results indicated that: Arctic ocean bacteria had a stronger adaptability to environmental temperature.

Key words Arctic, Marine General Aerobic Heterotrophic Bacteria. Abundance, Adaptability To Temperature.

1. Introduction

In the deep sea with a water temperature lower than 5℃ and the sea bottom, the organic matters are oxidized mainly by the aerobic bacteria (Bender and Heggie 1984; Herbert 1986), Which means that the aerobic bacteria in this environment are rather abundant. The arctic sea area is an area where environment is permanently cold, but some research results (Sahm et al. 1998) indicate that there are dominant bacteria-eubacteria in the arctic seawater and sediment, and their adaptation to temperature has no substantial difference in comparison with that of the bacterial strains from the temperate zone and

---

* The original text was published in *Chin. Pol. Sci*, 2004, 15(1): 39-41.
原中文稿刊于《极地研究》,2003,15(3):171-176.

warm-temperate zone, which means that many microbes in cold area persistently live in very low temperature, but their respiration and growth activities can quickly achieve an optimum state in the event that the microbes come in a temperatures higher than the on-site temperature. For example, the environmental temperature in the Antarctic area is permanent lower than 0℃, but it is shown that the optimum temperature for microbial metabolism is at least 10℃ higher than the local temperature. The real psychrophilic bacteria have been found in the Antarctic area, but the main isolated aerobic bacteria are rather psychrotrophic bacteria than psychrophiles (Delille and Perret 1989; Delille 1993), which indicates that the low temperature is an important factor inhibiting the on-site microbial physiologic and biochemical metabolism.

Some studies on the general aerobic heterotrophic bacteria (GAB) in some arctic sea areas have been made (Sahm et al. 1998; Forrari and Hollibargh 1999; Bano and Hallibaugh 2000; Ravenschlag et al. 2001), but the study on GAB in the Bering Sea and Chukchi Sea has not been reported yet. In this paper, the abundance of GAB in sediments and seawater in before mentioned sea areas is studied, which will help analyze and study the microbial condition and function of GAB.

2. Sample collection and analysis

The samples were collected during the first Chinese Arctic Research Expedition from July to September, 1999. Sediment samples were collected at 10 stations in the Bering Sea and at 24 stations in the Chukchi Sea, the sediments from the Bering Sea shelf are fine sand or silt, and the sediments from the Chukchi Sea shelf are mainly silly clay or clayey silt. Two surface water samples were collected at Station W_1 (74°50′N, 160°33′W) and Station W_2 (77°4′N, 160°33′W) in the Chukchi Sea, respectively, and the surface water temperature was about −1.4℃. The locations of sediment sampling sites and their environmental conditions were described by Chen and Gao (2000).

After being collected onto the ship deck by box corer, the sediment samples were immediately sealed up and stored in clean plastic bags for later analysis in the domestic laboratory, and the sediment samples were divided into surface layer sample (upper 0～5 cm), subsurface layer sample (upper 5～10 cm) and mixed layer sample for microbiological as say.

The helicopter was used to collect the surface water samples at stations S_1 and W_2. and the collected water samples were sealed up and stored in clean water sample bottles for microbiological assay in the domestic laboratory.

The different sediment samples were diluted to specified concentrations, respectively, 0.1 cm^3 diluted sediment samples were then taken to be smeared on Zoball 2216E plates, and 0.1 cm^3 raw water samples were taken to be smeared on above-mentioned solid culture medium plates. The above sample plates were cultured for over 2 weeks at 4℃, 15℃, 25℃ and 28℃ temperatures, respectively. The colony forming units (here after referred to as CFU) were then observed and counted, and the CFU per g (wet sediment) or per cm³ water were calculated. The abundances of GAB in different sea areas, at different survey sites and in different layers were analyzed and compared with each other.

3. Results and discussions

3.1 Analysis if GAB content condition in the survey area

The data for GAB in the water samples and sediment samples at different survey sites are listed in Table 1.

Table 1 Status of abundance of GAB from surface water and sediment in the Being See and Chukchi Sea

Kinds of samples	No. of samples	Names of stations	Longitude (W)	Latitude (N)	GAB CFU×10^2 cm^{-3} or [g^{-1} (wet)]		
					Water or surface sediment (upper 0~5 cm)	Subsurface layer (5~10 cm)	Mixed layer (non-divided)
	1	W1	160°33′	74°50′	0	/	/
warer	2	W2	160°33′	77°04′	4.4	/	/
	average	/			2.2	/	/
sediment	1	C1	170°06′	67°30′	150.0	/	/
	2	C2	169°59′	68°00′	/	0	/
	3	C3	170°00′	68°30′	0		/
	4	C4	169°59′	69°00′	/	/	30.0
	5	C5	170°00′	69°20′	0	/	/
	6	C7	170°01′	70°00′	/	/	0
	7	C8	172°15′	69°59′	/	0	/
	8	C9	175°00′	71°01′	0	/	/
	9	C10	175°02′	70°30′	/	0	/
	10	C11	173°54′	71°00′	182.0	/	/
	11	C12	172°30′	71°01′	/	200.0	/
	12	C14	167°31′	70°00′	/	/	0
	average				66.4	50.0	10.0
	13	C16	169°53′	66°30′	/	/	1,100.0
	14	C17	169°59′	67°00′	/	/	800.0
	15	C24	170°00′	70°59′	/	/	0
	16	C25	168°53′	71°42′	/	/	0
	17	C26	168°40′	72°00′	/	0	/
	18	C27	168°38′	72°30′	0	/	/
	19	C28	165°03′	73°00′	0	/	/
	20	C28	165°03′	73°00′	/	0	/
	21	C29	164°57′	73°26′	0	/	/
	22	C30	160°01′	71°15′	/	/	0

(continue)

Kinds of samples	No. of samples	Names of stations	Longitude (W)	Latitude (N)	GAB CFU×10^2 cm^{-3} or [g^{-1}(wet)]		
					Water or surface sediment (upper 0～5 cm)	Subsurface layer (5～10 cm)	Mixed layer (non-divided)
	23	C31	159°34′	71°041′	/	0	/
	24	C32	159°09′	72°06′	0	/	/
	25	C33	158°56′	72°22′	/	0	/
	average				0	0	380. 0
	26	B1-9	179°26′	60°15′	/	/	0
	27	B1-10	179°05′	60°25′	/	/	18,000. 0
	28	B1-11	178°45′	60°32′	/	/	0
	29	B1-12	178°14′	60°40′	0	/	/
	30	B1-13	177°45′	60°55′	/	0	/
	31	B2-10	178°26′	59°30′	0	/	/
	32	B2-11	178°10′	59°33′	/	70,000. 0	/
	33	B2-12	177°51′	59°43′	360. 0	/	/
	34	B5-9	175°59′	58°34′	/	/	20,000. 0
	35	B5-10	175°54′	58°40′	/	/	
	average				120. 0	35,000. 0	7,600. 0

* count after 12 days culture under 4℃

It can be calculated by using the data in Table 1 that the overall detection rate of GAB in the sediment samples at all survey sites is 28. 6%(10/35), and the overall average abundance of GAB is 3,166. 3×10^2 CFU · g^{-1}(wet) with a variational range of 0 to 70,000. 0×10^2 CFU · g^{-1}. Described by sea area, the detection rate of GAB in sediment from the bering Sea is 40. 0%, and the average abundance is 10,836. 0×10^2 CFU · g^{-1} with a variational range of 0 to 70,000. 0×10^2 CFU · g^{-1}; the detection rate of GAB in sediment from the Chukchi Sea is 24. 0%, and the average abundance is 98. 5×10^2 CFU · g^{-1} with a variational range of 0 to 1,100. 0×10^2 CFU · g^{-1}, which indicates that the detection rate, average abundance and abundance range of GAB in sediment from the Bering Sea are all higher than those from the Chukchi Sea, and the bottom water temperature and salinity in the Bering Sea during that period were higher than those in the Chukchi Sea (Chen and Gao 2000). The above-mentioned detection rates of GAB are all detection results for the 4℃ culture temperature, and the detection results for other 3 culture temperatures show much greater detection rates of GAB, being up to 88. 7% (as described below). The detection rate of GAB in the surface water sample is 50. 0%, and the average abundance is 2. 2×10^2 CFU · g^{-3}, which is much less than that in the sediment, the former is only 0. 07% of the latter, namely, the abundance of GAB in the sediment is 1,439. 2 times great as that in the surface water, which is 6 orders of magnitude lower than that determined by using DAPI (4, 6-diamidino 2-phenylindol dyeing method

(Sahm et al. 1998).

The GAB abundance distributions in the Bering Sea and the Chukchi Sea are listed in Table 2 and Table 3, respectively.

Table 2 Regional contrast of GAB abundances in Bering Sea

Region		GAB(CFU10^2 g^{-1})		
		Surface	Subsurface layer	Mixed layer
Latitude (N)	60°-60°	0	0	6,000.0
	60°-59°	180.0	70,000.0	/
	59°-58°	/	/	15,000.0
	average	120.0	350,000.0	10,200.0
Longitude (W)	180°-179°	/	/	6,000.0
	179°-177°	120.0	35,000.0	/
	177°-175°	/	/	15,000.0
	average	120.0	35,000.0	10,200.0

It is shown in Table 2 that for the Bering Sea the GAB abundance in sediment collected in the lower latitudes is greater than that in the higher latitudes, and the GAB abundance in the mixed layer sediment is greater in the eastern Bering Sea than in the western Bering Sea.

It is shown in Table 3 that for the Chukchi Sea the GAB abundance in the surface and mixed layer sediments collected in the lower latitudes is greater than that in the higher latitudes, but that in the subsurface layer sediments is greater in the higher latitudes than in the lower latitudes; and the GAB abundance in the sediments from the Chukchi Sea has no significant difference in east-west direction.

Table 3 Regional contrast of GAB abundances in Chukchi Sea

Region		GAB(CFU10^2 g^{-1})		
		Surface	Subsurface layer	Mixed layer
Latitude (N)	75°-70°	30.0	33.3	0
	70°-66.30°	50.0	0	613.3
	average	36.9	28.6	275.7
Longitude (W)	175°	0	0	/
	170°	30.0	100.0	482.5
	165°	0	0	/
	160°	0	0	0
	average	5.0	40.0	386.0

As shown in Table 2 and 3, the GAB contents in the surface and subsurface layer sediments from the Bering Sea 61°—60°W and the Chukchi Sea 175°W and 165°W are all undetected, and those in the sediments from the Chukchi Sea 160°W are undetected, which means that the sediment environment in the Bering Sea might be more suitable for GAB to survive than that in the Chukchi Sea. The cause for the above phenomena remains to be explored.

3.2　Sensitiveness of GAB to temperature

The environmental temperature is one of important factors to control the microbial survival. With other conditions being the same, and appropriate increasing temperature can facilitate and promote the microbial survival and reproduction to result in the increase in microbial abundance. The survey sea area is permanently cold sea area and the microbes in the area live permanently in cold environment, so the microbes can be both psychrophilic bacteria and psychrotrophic bacteria. Once an appropriate rise in temperature occurs, the psychrotrophic bacteria in the moderate temperature (25℃ to 45℃) conditions become rather active and begin to reproduce. Four replicas of 35 bacterial strains were cultured at 4℃, 15℃, 25℃ and 28℃ temperatures separately in order to show their sensitiveness to temperature, and the culture results are listed in Table 4.

Table 4　Response to differential culture temperatures of GAB from Bering Sea and Chukchi Sea

Culture temperature(℃)	Occurence % of GAB
4	28.6(10/35)
15	28.6(10/35)
25	22.9(8/35)
28	8.6(3/35)

* Number of testing samples: 354, time of cultivation: 15 d

It is shown in Table 4 that GAB all occurred at the 4 culture temperatures. The oecurrence frequencies of GAB at 4℃ and 15℃ temperatures account for 64.5% (57.2/88.7), and those at 25℃ and 28℃ temperatures account for 35.5% (31.5/88.7), which indicates that most of GAB are favorable for living at lower temperature and about one third of GAB are favorable for living at lower temperature and about one third of GAB are favorable for living at higher temperature (25℃ to 28℃). If the experiment can be viewed as an optimum temperature experiment, according to the opinion of Isaksen and Jrgensen (1996), the experiment results indicate that most of the detected bacteria in sediment are psychrophilic bacteria or psychrotrophic bacteria, and the rest might be mostly mesophilic bacteria with a small amount of psychrotrophic bacteria (Delille and Perret 1989). The results also show the interaction between bacteria and temperature, namely, temperature screens bacteria and bacteria select the adaptable temperature. It is shown in the experiment results that the bacteria in the Arctic sea area have considerable adaptability to the environmental temperature.

4. Conclusions

The GAB abundances in 2 surface water samples from the Chukchi Sea and in 35 sediment samples from the Chukchi Sea and Bering Sea were determined, and their differences in sea area, survey site and sediment depth and their sensitiveness to the culture temperature were analyzed. Generally speaking, the detection rate of GAB in water sample is higher than that in sediment sample, but the GAB abundance in water sample is low, being only 2.2×10^2 CFU · g^{-3}, and the average abundance of GAB in sediment is $3{,}166.3\times10^2$ CFU · g^{-1}. The detection rate of GAB in the surface layer sediment is higher than that in the subsurface and mixed layer sediments, but the GAB abundance in the surface layer sediment is not necessarily great. The detection rate, abundance and abundance of GAB in sediment from the Bering Sea are all greater than those from the Chukchi Sea. The differences of GAB abundance in varied the latitude show a trend that GAB abundance in lower latitudes higher than that in higher latitudes.

It is shown in the temperature experiment results that most of the determined GAB samples are psychrotrophic bacteria, a part of them are inferred to be psychrophilic bacteria, and a small amount of them are mesophilic bacteria; the bacteria in the Arctic sea area have adaptability to the environmental temperature with some variational range, so an appropriate rise in temperature can make their growth, reproduction and metabolic activities be accelerated or enhanced. However, the sustained global warming caused by different factors including the man-made factor may be unfavorable for conserving the biodiversity of psychrophilic bacteria and even psychrotrophic bacteria. The marine bacteria in the Arctic area remain to be further studied.

Acknowledgements The authors would like to thank the Polar Expedition Office of State Oceanic Administration of China and the crew on "Xuelong" expeditionary ship for their support and assistance.

References10(abr.)

Cooperators: Gao Aiguo　Sun Haiqing　Jiao Yutian

C_2 叶绿素-a 与初级生产 (CHLOROPHYLL-A AND PRIMARY PRODUCTION)

中国南极长城湾 1994 年 1～2 月间海水初级生产初析*

摘　要　对中国南极长城湾内水体夏季的初级生产力于 1994 年 1～2 月间作了估计。发现其初级生产碳量约为 0.799 $g \cdot m^{-3} \cdot d^{-1}$。2 月份高于 1 月份。表层、中层和下层相比，下层的日产量仍高。意味着该湾水体处于真光层范围内。水柱上下均有相应的初级生产。水温提高似有利于初级生产。初级生产的粗生产量与叶绿素 a(Chl-a)、NH_4^+-N、总 N、$\sum$N/P 的变化较为一致。营养盐相当充足。而与 NO_2^--N、PO_4^- x-P 较不一致。与异养菌数也相悖。推测异养菌有较强的分解有机质能力和矿化作用活动。

关键词　中国南极长城湾　海水初级生产　生态因子

海洋生态系统的初级生产主要靠海藻的光合作用固定太阳能而完成。海洋初级生产是海洋生态系统最初也是最基础的能量积累过程。有了该过程，无机碳才得以转化为有机碳。浮游植物等自养生物是海洋初级生产力的重要成员，在其生产过程中必然地释放氧气。根据其产氧量推算出固定的碳量而可以了解特定海区的初级生产状况。

张坤诚等(1994)曾对中国南极长城站近岸海域的初级生产作过估计。本文就长城湾海水 1994 年 1～2 月间(夏季)的初级生产再作一次估测。

1　材料与方法

于 1994 年 1～2 月间，对中国南极长城湾水体的初级生产作了初步测定。所用方法是黑白瓶测氧法(全国主要湖泊、水库富营养化调查课题组，1987)。选择的测点位于长城湾内偏顶部中间，即 7、8 两站间(图 1)。测定分两次进行。第一次是 1994 年 1 月 18 日。第二次是 1994 年 2 月 15 日。在测点上，首先测好透明度(约 8.5 m)。按透明度情况将水柱

---

*　系“八五”国家重点项目——南极重点地区生态系统研究。

原文刊于《植物生态学报》，1997，21(3)：285-289.

分作表层(0～0.5 m)、中层(0.5～4 m)及下层(4～8.5 m)。将有机玻璃采水器悬入各层采集水样,每层4瓶。所取各层的初始氧瓶立即带回实验室作固定氧用。其余各瓶按所在层次透明度的位置悬入长城站地区的西湖相应水层中24小时。取回,测定各自的溶解氧。换算得出固定的碳量。各项操作均按文献所列"调查规范"严格执行。对照有关环境一生态参数作初级生产分析(全国主要湖泊、水库富营养化调查研究课题组,1987)。

图1 中国南极长城湾1994年1～2月间海水初级生产测定站位示意图(第七、八站间的＊号处)

Fig. 1 Sketch of the location for determing primary production of sea water in Chinese Great Wall Bay, Antarctica during Jan. ～Feb., 1994(Located the star between the seventh and the eighth site)

2 结果和讨论

所得结果列于表1中。由表1可见1994年1月测定的粗生产量在水柱中的差异由高到低的顺序是下层、中层、表层。2月份的是表层、中层、下层。正好由1月份的转了过来。呼吸作用量由高到低的顺序在1月份是中层、表层、下层。2月份是表层、下层、中层。1月份的正好是12月份的转向2月份的过渡状态。净生产量由大到小的顺序在1月份的是下层、中层、表层,2月份的是下层,表层,中层。经计算,日产量也基本如此趋势,即每平方米水面下各水层的日产量在两次测定中由大到小的次序均是下层、中层、表层。结果意味着中、下层此类较暗的环境中光合作用仍相当强烈,生产力尚高。中下层仍在真光层深度范围内,仍有相应光照强度满足水下浮游植物、底栖植物之需,这可能是该湾浅水特征的一种反映。

表1　中国南极长城湾海水初级生产力估计表

Table 1　Estimination of primary production of sea water in Chinese Great Wall Bay, Antarctica

测定日期 Date of determination	层次 Layers determined	净生产量 Net primary production（$mgO_2 \cdot dm^{-3}$）	呼吸作用量 Consumption from respiration（$mgO_2 \cdot dm^{-3}$）	粗生产量 Gross primary production（$mgO_2 \cdot dm^{-3}$）	每平方米水面下各层水日产量 The production per day in each layer under a square meter water
1994-01-18	表层 Surface layer（0～0.5 m）	－0.100	0.110	0.010	0.005
	中层 Intermediate Layer（0.5～4.0 m）	－0.070	0.130	0.060	0.210
	下层 Lower layer（4.0～8.5 m）	0.100	－0.030	0.070	0.315
	全水柱 Whole water column	—	—	—	0.530
1994-02-15	表层 Surface layer（0～0.5 m）	0.059	0	0.059	0.295
	中层 Intermediate Layer（0.5～4.0 m）	0.490	－1.570	0.490	1.715
	下层 Lower layer（4.0～8.5 m）	1.090	－0.640	0.450	2.025
	全水柱 Whole water column	—	—	—	4.035

1)计算方法参见文献(全国主要湖泊、水库富营养化调查研究课题组，1987)For calculating method refer to references (A national investigative group on eutrophication of main lakes and reservoirs in China, 1987)

两次测定的结果均取自同处，但相差甚远。2月份的净生产、粗生产及日产量均高于1月份的，相反，1月份的呼吸作用量却高于2月份的。这可能与水温等环境—生态参数的变化有关。表2列举了有关参数，由该表可见，1月份表层水温低于2月份的(相应地，中、下层水温均呈此状态，略低)。可表明水温的提高有利于初级生产。初级生产，无疑主要由浮游植物等来完成。因而结果意味着2月份的浮游植物的繁荣。表2中的Chl-a浓度在两次测定中与浮游植物的粗生产量变化较一致①。此外，初级生产量与环境中的

① 陈皓文等，中国南极长城湾及其邻近海区表水1993/1994夏季叶绿素a浓度及分布的初析。

NH_4^+-N、NO_3-N、总氮及∑N/P 间均呈现出相似的增减势，而与 NO_2^--N、PO_4^--P 间呈现出相逆的倾向。这可暗示，所调查区的海水中的营养成分较充足①，而不明显限制浮游植物的光合作用和初级生产。初级生产力与异养菌数间也呈现为相悖倾向②，这表明多量的异养菌可能较多地破坏和利用浮游植物排泌物及浮游植物等的有机体(Watson，1981；Helm-Hansen et al.，1977)，最终将使浮游植物等初级生产者减少而得以循环。

表 2　中国南极长城湾海水初级生产力的有关参数统计表

Table 2　Statistical table of relative parameters of primary production in sea water of Chinese Great Wall Bay，Antarctica

日期 Date of determination	1994-01-18	1994-02-15	备注 Remark
层次 Layers	表层 Surface layer	表层 Surface layer	有关数据参见营养盐、叶绿素 a、异养菌等 3 文
叶绿素 a(Chl-a)	0.61	2.93	For the parameters concerned，refer to three papers on nutrient salts，Chl-a，heterotrophic bacteria of Chinese Great Wall Bay，Antarctica
水温℃ Water temperature(℃)	2.00	2.20	
NH_4-N	1.21	1.45	
NO_3-N	2.08	6.78	
NO_2-N	0.25	0.20	
PO_4-P	1.68	1.41	
N	3.54	8.40	
N/P	2.11	5.96	
CFU	16.35×10^2	7.35×10^2	
毛生产 Gross primary production	0.01	0.59	

Chl-a 浓度($mg \cdot m^{-3}$)、$NH_4{}^+$-N 等 4 种营养盐浓度($\mu g \cdot dm^{-3}$)、异养菌菌落形成个数($CFU \cdot cm^{-3}$)、Concentration of Chl-a($mg \cdot m^{-3}$)，Concentrations of four salts($g \cdot dm^{-3}$)，Colony forming units of heterotrophic bacteria($CFU \cdot cm^{-3}$)。

假定浮游植物产生 1 g 氧气相当于固定了 0.35 g 碳的话(全国主要湖泊、水库富营养化调查课题组，1987)，那么两次测定的结果表明，1 月份和 2 月份水柱中的日产碳量各为 $0.186\ g \cdot m^{-3} \cdot d^{-1}$ 和 $1.412\ g \cdot m^{-3} \cdot d^{-1}$。后者是前者的 7.6 倍。平均为 $0.799\ g \cdot m^{-3} \cdot d^{-1}$。与 1982/1983 年度南极戴维斯站沿岸海水的测定结果相比，本湾这次所测得的初级生产力较高(吕培顶等，1986)。与 1987/1988 年度南极之夏本湾的测定结果持平(张坤城等，1994)。

3　结语

本文报道了 1994 年 1～2 月间中国南极长城湾水体的初级生产状况。结果表明，所测

① 陈皓文等，中国南极长城湾及其邻近海区表水营养盐浓度及与相关因子关系初析，海洋湖沼通报。

② 陈皓文等，中国南极长城湾及其邻近海区表水异养菌数估测及相关因子关系初析。

水柱中的初级生产日产碳量平均为 0.799 $g \cdot m^{-3} \cdot d^{-1}$，与以往结果持平。初级生产与有关生态因子间的分析表明，水温的提高似有利于初级生产。初级生产与 NH_4^+-N、NO_3^--N、总 N 的浓度及 N/P 间似各有类似的增减趋势，而与 NO_2^--N、PO_4^--P 浓度间则呈相逆状态，无机营养充足。与异养菌含量间呈相悖关系。故本结果也意味着本湾水体中多量异养菌积极参与有机质的降解、物质转化和能量传递活动。

参考文献 5 篇（略）

（合作者：朱明远　洪旭光　袁俊峰）

A RUDIMENTARY ANALYSIS ON PRIMARY PRODUCTION OF SEAWATER IN CHINESE GREAT WALL BAY, ANTARCTICA DURING JAN. ~FEB. 1994

(ABSTRACT)

Abstract Summer primary production of seawater in Chinese Great Wall Bay, Antarctic during Jan. ~Feb. 1 994 was estimated. The results revealed that the yield of the carbon primary production was about 0. 799 $g \cdot m^{-3} \cdot d^{-1}$ in the waters. We found that the yield in the deeper water column was fairly high as compared to the yield in the upper and middle columns. It showed that the deeper water surveyed lies still in the range of the euphotic zone. This meant that there was a quite high primary production in the waters. The yield of the gross production was rather consistent with the variates, such as. chlorophyll a, NH_4^+-N, NO_3^--N, ΣN and $\Sigma N/p$. This seemed to indicate that there were sufficient nutrient salts there, which, however, was not consistent with the concentrations of NO_2^--N, PO_4^- x-P. This was not in coordination with the abundance of heterotrophic bacteria and suggested that these bacteria had a capability of degrading organic materials as well as strong mineralization.

Key words Chinese Great Wall Bay, Antarctica, Primary Production Of Seawater, Ecological Factors

A PRIMARY ANALYSIS ON CONCENTRATION AND DISTRIBUTION OF CHLOROPHYLL-A IN SURFACE WATER OF GREAT WALL BAY AND ITS ADJACENT AREA DURING AUSTRAL SUMMER 1993/1994*

Abstract The concentration and distribution of chlorophyll-a in surface seawater of Chinese Great Wall Bay and its adjacent waters, Antarctica were determined once monthly, altogether three times during an austral summer period from Dec. 1993 to Feb. 1994.

The results obtained show:

The concentration of Chl-a is about 1.29 mg/m^3, ranging from 0.18～6.75 mg/m^3. The concentrations are higher inside the Bay than that outside and in Ardely Bay, This situation is consistent with that respect of the water temperature of surface layer in the area surveyed. However, the time for the occurrences of high and low values of Chl-a concentration is not the same as those for water temperature.

The results of analysis correlation between each of the 4 concentrations of nutrient salts and Chl-a show that there is an apparent positive correlation only between(NO_2^--N) and Chl-a in the sea water surveyed.

It is inferred from the results obtained that the distribution and changes of Chl-a in the Bay and its adjacent waters are still interfered with by other ecological factors including activities of the heterotrophic microbes.

Keywords Chinese Great Wall Bay in Antarctica, surface sea water in austral summer, concentration and distribution of Chlorophyll-a

1 INTRODUCTION

The main cellular organell for carrying on photosynthesis by plants is autoplast and the center of photosynthesis is Chlorophyll-a(abbr. Chl-a the same below).

Since the main plant-phytoplankton in the sea/ocean accomplishes its primary production and performs energy transformation mainly upon Chl-a, the concentration of Chl-a and its content are closely related to the quantity and species of its photosynthesis rate.

---

* The original text was published in *J. Glaci. Geocr.*, 1998, 20(4): 312-315.

Besides the distribution of the Chl-a concentration in the sea and environmental parameters confirm each other. The standing crop or stocking crop of phytoplankton can be found from Chl-a. Hence Chl-a becomes an important indicator of marine ecosystems.

The Chinese Great Wall Bay and its adjacent waters are inshore shallow water bodies. The understanding of the situation about the concentration and distribution of Chl-a in the bay waters has aroused people's attention and some research results have been published (Wu and Zhu, 1992; Lu and Huang, 1989).

This paper presents a preliminary analysis of Chl-a concentration monitored during December 1993～February 1994, in the surface water of the Bay and its adjacent sea area.

2 MATERIALS AND METHODS

Samples totaling up to 29 were taken from the surface water of the Chinese Great Wall Bay respectively from stations No. 4, 5, 6, 7, 8 and 9 (zone Ⅰ), 5 samples from sites No. 1, 2, 3, 10 and 11 from outside the Bay respectively (i. e. zone Ⅱ), and three samples (i. e. No. 12, 13, 14) respectively from zone Ⅲ near by the side of Ardely Bay. Samples were taken three times respectively in December 1993, January 1994 and February 1994.

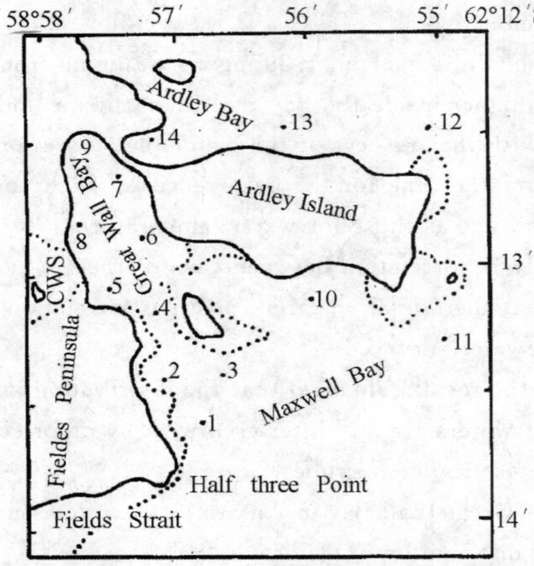

Fig. 1 Sampling stations for Chlorophyll-a from surface water in Chinese Great Wall Bay and its adjacent sea area, Antarctica, during austral summer of 1993/1994

Details about the sampling stations, etc. were described in Fig. 1 and another paper (Chen et al., 1998).

Turner fluorimeter was used and determination was made according to the methods described in the references (State Technological Superintendency Administration, 1991). Micropore (0.65 μ) filters were used as filter membranes, except otherwise noted. The standard Chl-a was purchased from Shanghai.

When taking the water samples for analyzing the Chl-a data on water temperature, transparency and nutrient salt concentration were monitored together with the primary production and some relevant microbial conditions. Samples were determined one by one so as to illustrate the relationship between Chl-a and these parameters.

3 RESULTS AND DISCUSSION

Table 1 listing the Chl-a concentrations are determined during Dec. 1993～Feb. 1994. it can be seen that the Chl-a concentrations averaged 1. 29 mg/m^3, the extreme difference is 6. 57 mg/m^3. The highest value is 37. 50 times higher than the lowest one and also higher than that one of Prydz Bay during 1989～1990, East Antarctica.

Table 1 Table of Chl-a concentrations in surface water of the Chinese Great Bay and its adjacent sea area, Antarctica

No. of sampling station	1993-12	1994-01	1994-02	Average value
1	0. 53	0. 29	0. 33	0. 38
2	0. 53	～	0. 22	0. 38
3	1. 00	～	0. 26	0. 63
4	0. 34	0. 38	0. 49	0. 40
5	1. 16	～	0. 32	0. 74
6	1. 02	1. 46	0. 61	0. 70
7	0. 63	0. 75	3. 10	1. 28
8	1. 12	～	2. 76	1. 94
9	6. 75	2. 30	3. 84	4. 30
10	0. 31	—	—	0. 31
11	5. 37	0. 81	—	2. 78
12	1. 53	—	—	1. 53
13	0. 39	—	—	0. 39
14	0. 49	—	—	0. 49
average	1. 51	0. 73	1. 33	1. 16

Note: Unit of Chl-a concentration: mg/m^3; "～": No determined; "—": Undetected.

Table 2 Comparison of Chl-a concentrations from surface water in three surveyed zones respectively in the Chinese Great Wall Bay and its adjacent sea area, Antarctica

Surveyed zone	1993-12	1994-01	1994-02	Average value
Ⅰ	1. 84	0. 97	1. 85	1. 55
Ⅱ	1. 55	0. 24	0. 27	0. 69
Ⅲ	0. 80	—	—	0. 80
average	1. 40	0. 61	1. 06	1. 01

Note: Unit of Chl-a concentration: mg/m^3; "—": Undetected.

Table 2 was obtained by dividing these data into three groups corresponding to the 3 zones in the area surveyed.

The data in Table 2 showed that the average Chl-a concentration for zone Ⅰ is greater than that for zone Ⅱ (for short, zone Ⅰ >zone Ⅱ) (for zone Ⅲ, only one datum available, discussion follows). This state of regional distribution of the surface water Chl-a concentration i. e. the distribution in zone Ⅰ and Ⅱ is the corresponding surface water temperature (T), i. e. where there is lower water temperature, there is also lower Chl-a concentration.

A comparison of Table 2 and Table 3 showed that the variation in the Chl-a concentration in the surface water determined in each zone was consistent with that of the corresponding surface water temperatures measured on the same date, e. g., the data in December showed that for both Chl-a and T, zone Ⅰ >zone Ⅱ >zone Ⅲ. The other two determinations also resulted in the same way. But the variation in the occurrence time for the average Chl-a concentration and that of water temperature (T) were not consistent in terms of the occurrence time. For example, in zone Ⅰ, the descending order of the occurrence date in terms of the magnitude of Chl-a concentration was Feb. 3, 1994, Dec. 13, 1993, Jan. 17, 1994, but for that of water temperature the order was Jan. 17, 1994, Feb. 3, 1994, Dec. 13, 1993, For Chl-a, in zone Ⅱ, the order was: Dec. 13, 1993, Feb. 3, 1994, Jan. 17, 1994, but for that of water temperature, Jan. 17, 1994, Feb. 3, 1994, Dec. 13, 1993. Looking at the three zones together, we can find that: in the whole area investigated the time order of the occurrence of the dates Chl-a concentration was: Dec. 13, 1993～Feb. 3, 1994～Jan. 17, 1994, but for the water temperature, the order was Jan. 17, 1994, Feb. 3, 1994, Dec. 13, 1993.

Table 3 Comparison of average values of surface water temperature in three surveyed zones respectively in the Chinese Great Wall Bay and its adjacent sea area, Antarctica

Surveyed	1993-12	1994-01	1994-02	Average value
Ⅰ	1.29	1.63	1.48	1.47
Ⅱ	1.00	1.30	1.20	1.17
Ⅲ	−0.73	—	—	−0.73
Average values	0.52	1.47	1.34	0.64

Note: Unit of water temperature, ℃; "—": No determined.

The analytical results of the correlation between Chl-a concentration and water temperature (in Table 4) show that most of the correlation coefficients were not significant and moreover they were mainly negative. Among them, the negative one in December, 1993 was significant. Besides, the correlation in February, 1994 in zone Ⅰ was significant but also positive. This result might suggest the acceleration of the course of Chl-a degradation to phaeophytin at a certain temperature in the shallow sea area such as this surveyed area. Thus it can be seen that the variation pattern of the relationship between water temperature and Chl-a concentration in the surveyed area is not fixed at a certain period. This may suggest that the Chl-a concentration is subjected to the interference from other factors to a fairly large extent. It is possible that if only one method is used to analyze the relationship between the factors and trends for the change in Chl-a concentration, it is liable to come to a one-sided.

Table 4 Comparison correlation coefficient of water temperature and Chl-a content of surface water in the Chinese Great Wall hy and its adjacent sea area, Antarctica

Surveyed zones	1993-12	1994-01	1994-02	1993-12～1994-02
Ⅰ	−0.54(n=5)	−0.81(n=6)	0.92(n=6)	—
Ⅱ	0.92(n=5)	−0.34(n=4)	—	—
Ⅰ Ⅱ Ⅲ	—	—	—	−0.56(n=26)

Note: "—": Indicate them were not computed respectively

Nutrient salts, such as ammonium salts, nitrates, nitrites, phosphates and a ratio, i. e. N/P are the needs of phytoplankton and other organisms for their existence and the environmental indications. The concentrations of nutrient salts should affect obviously their photosynthesis and growth. The analytical results of the correlation between Chl-a concentration and concentration of nutrient salts listed in Table 5 show that their relationships are not completely clear.

Only the correlation coefficient between Chl-a and NO_2-N is more significant positive. In some sites surveyed, a contrary condition occurred between Chl-a and nutrient salts, for example, at station No. 2 (in the December), the Chl-a concentration was higher, but the concentration of NH_4-N was lower (Table1) which indicated that the propagation of phytoplankton consumes nutriments in the specific environment. Under certain conditions, their correlation appeared to be negative, such as the relationship between (Chl-a) ～(PO_4-P). In this case, it is possible that bacteria might take part in a mechanism of the recycling of nutrients (Smith, 1992).

Table 5 Table of comparison correlation coefficient among Chl-a concentration and four nutrient salts concentrations of surface water in the Chinese Great Wall Bay and its adjacent sea area, Antarctica

Pairs of parameters	*R*	*n*
Chl-a-(NH_4-N)	0.084	23
Chl-a-(NO_3-N)	0.069	23
Chl-a-(NO_2-N)	0.357	22
Chl-a-(PO_4-P)	−0.126	19
Chl-a-(N/P)	0.025	33

The results obtained in the paper might suggest that there are sufficient nutrients, such as N, P etc. in the sea area surveyed and that situation does not become a limit to photosynthesis-growth and reproduction of the relevant phytoplankton and autotrophic organisms.

4 CONCLUSIONS

The highest concentration of Chl-a in surface water in the area surveyed was 6.75 mg/m^3 and the lowest one only 0.18 mg/m^3, the range of the respective average concentration of the sites surveyed was 0.31～4.30 mg/m^3. The mean concentration for the whole area investigated was about 1.99 mg/m^3, slightly higher than that during the summer of 1988～1989 and lower than that one during the summer of 1985～1986 in Great Wall Bay. Antarctica. The highest and lowest Chl-a concentrations occurred in December 1993 in station No. 9 and on January, 1994 in station No. 11. The distribution of Chl-a has a more obvious regionalism, i. e. the concentration of Chl-a within Great Wall Bay is higher than outside. It is inferred that this situation may be affected by the land environment and melting snow entering into the sea, beside its relation to the supply of nutrients and heterotrophic organisms in the area surveyed.

The analytical results of the correlation between Chl-a concentration and some environmental-ecological factors indicated that there is not a fixed pattern of correlation among them, i. e. to a certain extent there are some relationships of supplementing each other as well as some phenomenon of opposing each other and yet also complementing each other indicating that there occur a variety of interaction among the Chl-a concentration and even phytoplanktonic distribution against various environmental-ecological factors.

References5(abr.)

(Cooperators: Zhu Ming-yuan, Hong Xu-guang, Wu Bao-ling)

南极长城湾夏季叶绿素a变化的研究*

摘 要 1992年12月～1993年3月、1993年12月～1994年2月、1994年12月～1995年3月连续3个夏季期间在南极长城湾进行了海洋生态系统调查，利用获得的浮游植物现存量（叶绿素a）资料进行分析研究发现1992/1993年夏季叶绿素a含量最高，平均值为3.79mg/m³，变化幅度也最大，为10.10 mg/m³。连续3个夏季的年际变化趋势是3.79 mg/m³(1992/1993)＞1.80 mg/m³(1993/1994)＞1.20 mg/m³(1994/1995)。而变化的幅度也同样依次为10.10 mg/m³＞6.57 mg/m³＞2.41 mg/m³。所获得的结果与南极地区相同季节其他近海海水中的叶绿素a含量进行比较得知：除Gerlache海峡（1959年1月）11.60 mg/m³和McMurdo海峡（1961年12月）56.10 mg/m³外，本夏季连续调查结果属中上水平。本文还对该海域叶绿素a的垂直分布状况进行了分析。

关键词 叶绿素a含量　变化幅度　垂直分布　长城湾　南极

1 前言

自20世纪80年代以来，我国对南极海洋生态系统进行过多次较全面的调查研究，获得了大量有价值的研究资料，为了解、研究南极海洋提供了有用的参考资料。据调查南大洋及南极沿岸海域中浮游植物现存量（Chl-a）的分布，存在着明显的地理变化和季节变化。一般说来，南极大陆以及亚南极岛屿近岸水域，Chl-a的蕴藏量要比外海大得多，例如南极半岛以西的斯科特湾、罗斯海、威德尔西南海区的浮游植物生物量很丰富（El-Sayed，1990）。所以南极菲尔德斯半岛及其邻近地区生态系统的调查研究被列为中国"八五"南极科学考察重大攻关项目七个专题之一。Chl-a的含量是计算海洋初级生产力的重要参数。人们常用水域中Chl-a作为海洋浮游植物蕴藏量的指标（吕培顶，1986a）。也可以用作表征水域中生产力状况的指标（费尊乐，李宝华，1990）。所以在对海洋生态系统的调查研究中，Chl-a的研究一直是海洋生态研究中的主要内容之一。

自1981年以来.对南极Chl-a的调查研究比较多（吕培顶，1986a；1986b；张坤城，吕培顶，1986；吴宝铃等，1992；吕培顶，渡边研太郎，1994；国家海洋局极地考察办公室，1998），但南极同一海区连续几个夏季Chl-d的调查研究未见报道。本文利用南极长城湾Chl-a连续3个夏季的调查资料进行分析研究，旨在为研究该海区海洋生态系统变化趋势提供参考数据。

2 材料和方法

南极长城湾3个夏季的调查时间分别为1992年12月～1993年3月，1993年12月～1994年2月和1994年12月～1995年3月，调查海区及观测站位置由图1示出：现场作业主要是在7个站位上进行的，其中4个站（B_1，B_2，B_5，B_9）在长城湾内湾，2个站（B_{11}和B_{12}）在长城湾外湾，一个站（B_{17}）在长城湾外。3个夏季每月各站采样1次，采样及测定方法按

* 原文刊于《极地研究》，1999，11(2)：113-121.本人为第四作者。

《海洋调查规范》(国家技术监督局,1992)中的海洋生物调查规范进行,一定体积水样经 25 mm 的玻璃纤维 GF/F 滤膜过滤,用 10 mL 90%的冷丙酮溶液暗中低温(<4℃)萃取 14~24h,采用 Turner Desings 10 型荧光计测定 Chl-a 的相对荧光值,并计算其浓度。

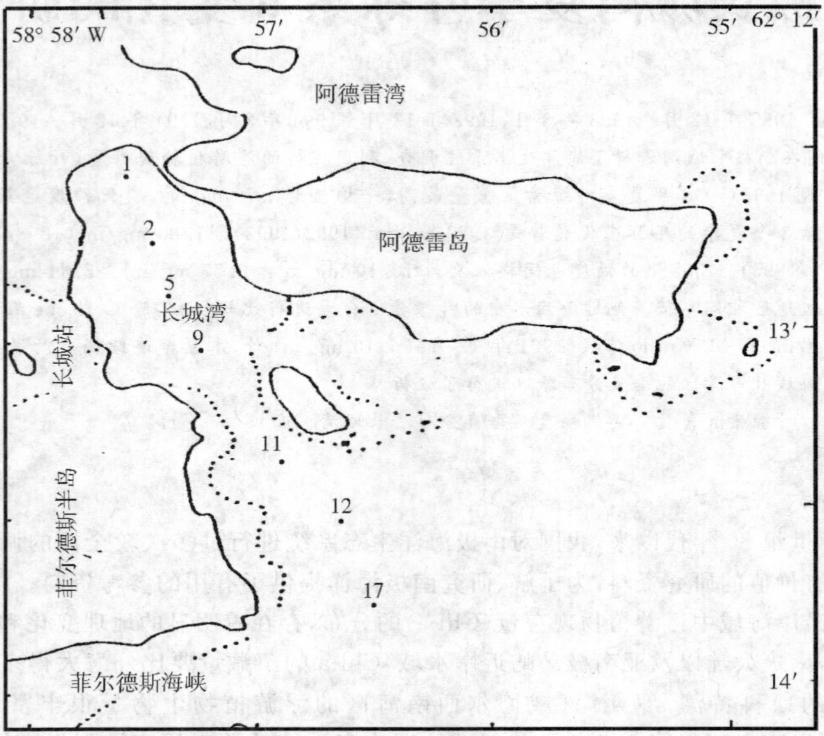

图 1 调查海区及观测站位置

Fig. 1 Investigated region and sampling stations

3 结果与讨论

3.1 Chl-a 的含量及变化范围

长城湾及其毗邻水域夏季 Chl-a 含量的变化范围:1992 年 12 月~1993 年 3 月为 1.41~11.51 mg/m^3,平均值为 3.79 mg/m^3。最高值出现在 1992 年 12 月的 B_1 号站(11.51 mg/m^3),最低值在 1993 年 3 月的 B_9 号站(1.41 mg/m^3)。1993 年 12 月~1994 年 2 月为 0.18~6.75 mg/m^3,平均值为 1.80 mg/m^3,最高值出现在 1993 年 12 月的 B_9 号站,最低值在 1994 年 1 月 B_{11} 号站。1994 年 12 月~1995 年 3 月为 0.82~3.23 mg/m^3,平均值为 1.20 mg/m^3,最高值出现在 1995 年 1 月的 B_9 号站,最低值在 1995 年 3 月的 B_{11} 号站。后 2 个夏季最高值和最低值出现的站位均相同,只是最低值站出现的月份不同。Chl-a 的变化幅度是第一个夏季(10.10 mg/m^3)>第二个夏季(6.57 mg/m^3)>第三个夏季(2.41 mg/m^3)。3 个连续夏季 Chl-a 含量的变化范围及平均值均高于 1988 年 11 月~1989 年 2 月,第三个夏季 Chl-a 含量的变化范围、平均值及变化幅度均低于 1985 年 11 月~1986 年 2 月(表 3)。在南极这种海洋环境污染不甚严重的特殊海洋环境中,光和温度是影响南极长城湾 Chl-a 的主要因素，从南极各夏季温度的变化也可以看出 Chl-a 这种变化的原因。1985 年 11 月～1986 年 2 月南极长城湾水温的变化范围为 1.4～2.2℃，平均值是 0.4℃；1988 年 11 月～1989 年 2 月，变化范围为－0.8～2.9℃平均值为 1.05℃；1992 年 12 月～1993 年 3 月，其变化范围为 0.7～3.8℃，平均值为 2.54℃；1993 年 12 月～1994 年 2 月，其变化范围为－0.73～1.63℃，平均值为 0.89℃。由此看出，在南极这种特殊的海洋环境中，海水温度在－0.8～3.8℃范围内，Chl-a 的含量变化和幅度与温度呈正相关。

3.1.1　3 个夏季各月份 Chl-a 的平均值的比较

调查海区夏季各月份 Chl-a 的平均值列入表 1。从表 1 可以看出，第一个夏季和第二个夏季 Chl-a 以 12 月份为最高，其含量分别为 6.33 mg/m³ 和 2.65 mg/m³，第三个夏季是 1 月份为最高。第一、三夏季都是 3 月份最低。引起 Chl-a 这种变化的因素很多，除营养盐以外，温度、光照是影响 Chl-a 含量和变化的直接控制因素。

表 1　长城湾夏季 Chl-a 的含量及平均值

Table 1　The average concentration of Chlorophyll-a in summer in the Great Wall Bay

		第一个夏季				第二个夏季			第三个夏季			
		92.12	93.1	93.2	93.3	93.12	94.1	94.2	94.12	95.1	95.2	95.3
Chl-a (mg/m³)	月平均值	6.33	4.34	2.45	2.05	2.65	0.92	1.18	1.06	1.72	1.05	0.97
	总平均值	3.79				1.80			1.20			
表层水温 (℃)	月平均值	1.54	2.25	2.83	3.55	0.52	1.47	1.34	—	—	—	—
	总平均值	2.54				1.11			—			

3.1.2　不同年份不同站位 Chl-a 的平均值

不同年份不同站位 Chl-a 的平均值由表 2 示出。不同年份夏季 Chl-a 含量在各站上的变化非常明显。1992 年 124～1 993 年 3 月 Chl-a 含量的最高值在 B_1 号站(湾内)，依次向外海递减，Chl-a 含量最高值与最低值的递减幅度为 3.00 mg/m³，这个夏季 Chl-a 含量均较高，其变化幅度大。1993 年 12 月～1994 年 2 月 Chl-a 含量除 B_9 号站外(最高值站，4.30mg/m³)，其他各站 Chl-a 含量均低于第一个夏季。而且变低的幅度较大。1994 年 12 月～1995 年 3 月各站 Chl-a 含量均高于 1.00 mg/m³，而且分布均匀，均低于第一个夏季。这是因为后两个夏季的水温均较低引起的。这三个夏季 Chl-a 平均含量的总趋势是第一个夏季(3.79 mg/m³)＞第二个夏季(1.80 mg/m³)＞第三个夏季(1.20 mg/m³)。

表 2　不同站位各夏季 Chl-a 的平均值

Table 2. The average concentration of Chlorophyll-a in the different stations in different summer

站 位	B_1	B_2	B_5	B_9	B_{11}	B_{12}	B_{17}	平均值
1992 年 12 月～1993 年 3 月	5.23	4.19	4.22	4.01	—	2.86	2.23	3.79
1993 年 12 月～1994 年 2 月	0.38	0.38	0.74	4.30	2.78	—	—	1.80
1994 年 12 月～1995 年 3 月	—	1.06	1.13	1.53	1.06	1.04	1.23	1.02

3.2　不同年份各站 Chl-a 的月变化

不同年份各站 Chl-a 的月变化由图 2 示出。图中看到,第一个夏季(图 2a)各站各月 Chl-a 含量的分布是:12 月和 1 月份含量高且变化幅度大,而 2 月和 3 月份 Chl-a 的含量相对较低,而分布均匀,前两个月(12 月和 1 月)与后两个月(2 月和 3 月)Chl-a 含量的分布状况差异很大。而第二个夏季(图 2)各站的 Chl-a 含量分布十分相似。都是在出号站为最高 B9 号站各月 Chl-a 含量依次为 12 月($6.75\ mg/m^3$)>2 月($3.84\ mg/m^3$)>1 月($2.3\ mg/m^3$)。第三个夏季(图 2c)Chl-a 的含量在各月各站上分布(除 1 月份以外)较均匀,Chl-a 的含量均在 $0.82 \sim 1.20\ mg/m^3$ 之间,变化幅度较小,为 $0.38\ mg/m^3$。1 月份 Chl-a 含量在各站上的分布与第二个夏季的非常一致,也是在出号站为最高。

3.3 长城湾 Chl-a 的垂直分布

由于南极考察条件的限制,使得 Chl-a 含量的垂直分布报道较少(吕培顶,1986b;吴宝铃等,1992)。本文因第二个夏季未能获得 Chl-a 的垂直分布资料,只能利用在第一、三个夏季获得部分站位垂直分布资料的平均值显示南极长城湾不同年份夏季 Chl-a 的垂直分布特征,并以第三个夏季为例探讨各月 Chl-a 的垂直分布状况。

3.3.1 不同年份夏季 Chl-a 的垂直分布

不同年份夏季 Chl-a 的垂直分布由图 3 示出。1992 年 12 月(第一个夏季)Chl-a 的垂直分布是:表层水中 Chl-a 的含量低于表层水以下各层,并随水深的加大而增加,高值在深水层($30\ m, 8.52\ mg/m^3$)。1994 年 12 月(第三个夏季)Chl-a 的垂直分布状况与第一个夏季的一致,也是随深度的增加 Chl-a 的含量呈递增趋势。长城湾海水向下辐射慢,衰减系数 Kd 数值 1992 年 12 月~1993 年 3 月在 $0.15\ m^{-1} \sim 2.35\ m^{-1}$ 之间,透光层厚度在 18.4~33.0 m 之间。1994 年 Kd 值相对偏低. 在 $0.06\ m^{-1} \sim 0.19\ m^{-1}$ 之间,透光层厚度在 23.1~76.8 m 之间。所以水深也不会太大地影响光对浮游植物的作用。在南极,营养盐不是浮游植物生长繁殖的限制因子。Eppley(1970)曾经报道,可能会限制海洋浮游植物生长的营养盐米氏常数(Ks)为硝酸盐 1.49~91 μg/L 磷酸盐 3.7~17 μg/L。长城湾海水中营养盐浓度较高。1992 年 12 月~1993 年 3 月夏季。硝酸盐浓度平均值是 187 μg/L,是无机氮的主要形式,磷酸盐 53 μg/L,而且各月营养盐浓度的波动范围不大,整个调查水域营养盐分布均匀,这两种营养盐浓度均比影响海洋浮游植物生长的营养盐米氏常数高出 2 倍以上,所以无论是表层水中还是深层水中营养盐不会是限制长城湾浮游植物生长的因素。

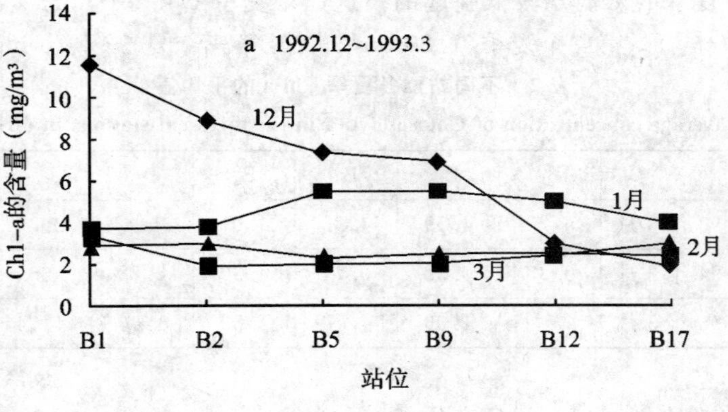

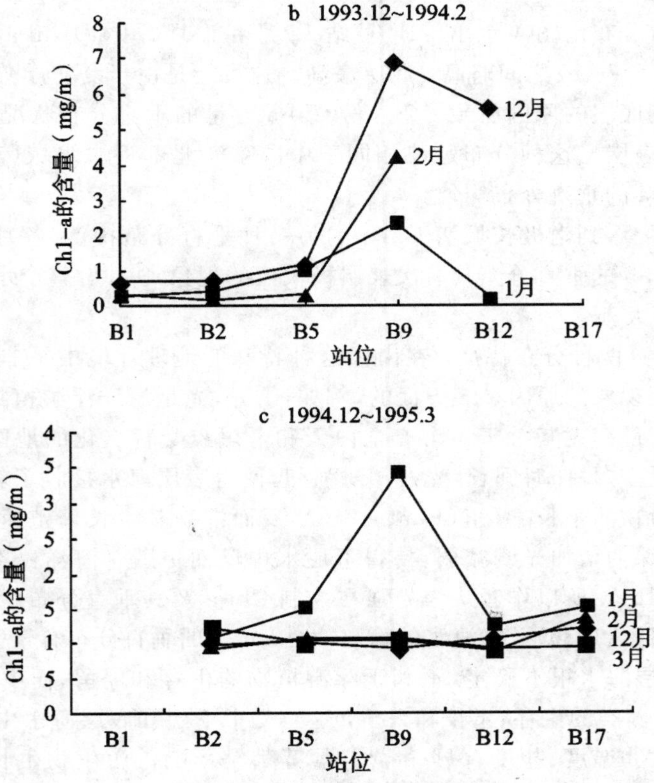

图 2 不同年份各站 Chl-a 的月变化

Fig. 2　The monthly variations of Chlorophyll-a in different stations in different year

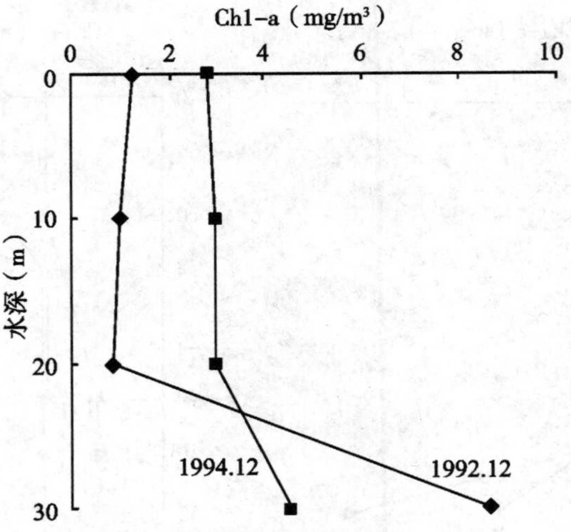

图 3　不同年份夏季 Chl-a 的垂直分布

Fig. 3　The vertical distributions of Chlorophyll-a in summer in different year

3.3.2　Chl-a 垂直分布的月变化

本文以 B_2 和 B_{17} 号站为例显示南极长城湾夏季各月份 Chl-a 的垂直分布状况。Chl-a 在不同月份的垂直分布由图 4 示出。B_2 号站 12 月和 1 月(图 4a)Chl-a 含量的垂直分布 15 m 以上均匀,15～20 m 之间略高,底层最高。这可能是由于夏初浮游植物快速繁殖后一部分细胞沉降到底层的原因所致。2 月份 Chl-a 含量的垂直分布状况在南极是较少出现的,最高值在次表层。这种分布趋势相同于中国东海北部近岸水域(李宝华,费尊乐,1992),3 月份 Chl-a 的垂直分布均匀。

由图 4b 中我们看到南极长城湾 Chl-a 的另一种垂直分布特点,各月 Chl-a 的垂直分布均匀,2 月份 Chl-a 的垂直分布具有次表层极大值的现象,3 月份整个水柱 Chl-a 含量均有变化,但波动不大。

以上 Chl-a 的这两种分布特征,在南极这种特殊的地理环境中是非常典型的。造成这种垂直分布特征现象的原因是多方面的,例如:①光,光是海洋浮游植物赖以生存、生长以及繁殖的重要条件。尤其对于一个有全白天和全黑夜交替变化的地区,光就显得更为重要。夏季或称暖季,日照时间长,光线相对强烈,使得表层海水接收了较强光线的照射,不利于海洋浮游植物的生长、繁殖(Bunt,1968),反而抑制其生长繁殖速度。而在表层以下水层中,随着水深的增加光线减弱,使得下层水中反而更适合海洋藻类的生存、生长和繁殖(吕培顶,1986b)。所以作者认为光强在控制 Chl-a 及其垂直分布中起着非常重要的作用(张坤城,吕培顶,1986)。②海水的稳定性对 Chl-a 的垂直分布有较大的影响,夏季由于风浪的作用,表层海水很不稳定,不利于浮游植物的生存和停留,而中下层和底层海水相对稳定,有利于藻类的生存、生长和繁殖,这也是中下层和底层海水中 Chl-a 含量高于表层的原因之一。③温度、盐度、pH 和营养盐等都是影响浮游生物生长的较主要因素。它们的变化与 Chl-a 的变化及垂直分布密切相关。反之,如光线不强烈、风浪不大等,表层海水中诸要素均比中下和底层海水中更适合藻类的生存和繁殖,就会出现 B_{17} 号站的垂直分布状况。

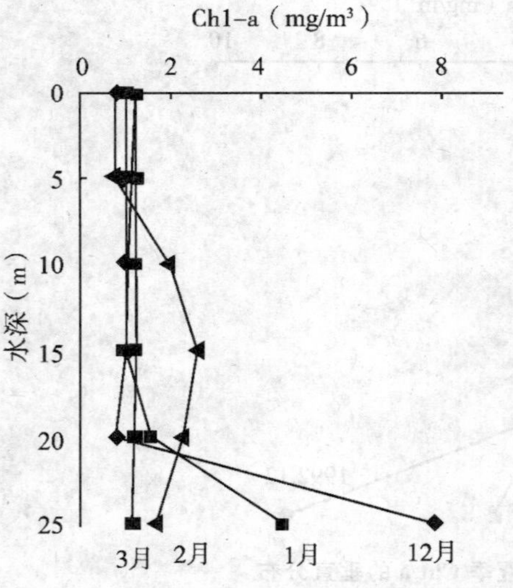

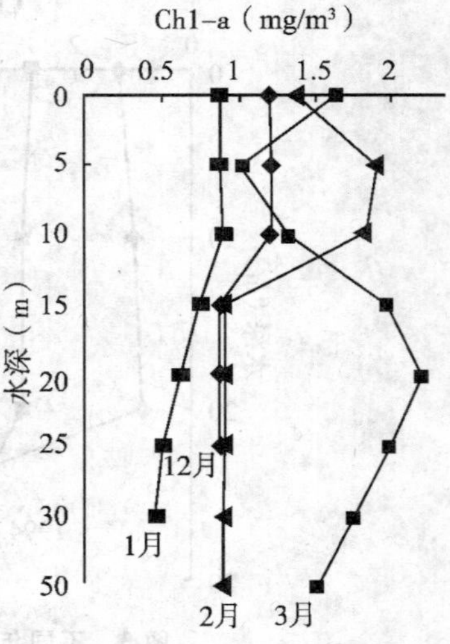

图 4　B_2 和 B_{17} 号站 Chl-a 的垂直分布

Fig. 4　The vertical distributions of Chlorophyll-a at the stations B_2 and B_{17}

3.4　与南极其他地区夏季 Chl-a 含量的比较

由历年来夏季 Chl-a 变化范围及平均值(表 3,以长城湾为主)看到,20 世纪 50 年代末、60 年代初,Chl-a 含量的变化范围及平均值差异较大,最大值与最小值之间的变化幅度高于50.00mg/m^3。自 1982 年以来 Chl-a 含量的变化范围及平均值变幅不大。

表 3　南极不同地区夏季 Chl -a 含量的变化范围及平均值(mg/m^3)

Table 3　The comparison of Chlorophyll-a in summers in different regions of Antarctica

地区	年份	变化范围(mg/m^3)	平均值(mg/m^3)	资料来源
长城湾	1992 年 12 月～1993 年 2 月	1.41～11.51	3.79	本文
长城湾	1988 年 11 月～1989 年 2 月	0.16～1.33	0.44	吴宝铃等(1992)
长城湾	1985 年 11 月～1986 年 2 月	0.23～4.91	1.66	吕培顶等(1994)
长城湾	1991 年 1 月～2 月	0.96～2.61	1.72	陈波,何剑锋(1994)
长城湾	1993 年 12 月～1994 年 2 月	0.18～6.75	14.8	本文
长城湾	1994 年 12 月～1995 年 3 月	0.82～3.23(表)		
0.10～8.17(深)	1.2			
1.33	本文			
戴维斯站	1982 年 11 月～1983 年 1 月	0.14～3.52	1.16	吕培顶(1986)
Admiralt y Bay	1978 年 12 月～1979 年 12 月	0.26～4.43		Rakusa-Suszczewski (1980)
Weddell 海	1963 年 12 月～1964 年 1 月	0.10～4.30	0.73	El-Sayed etal. (1965)
Gelrache 海峡	1959 年 1 月	0.30～26.80	11.6	Burkhoder and Sieburth (1961)
McMurdo 海峡	1961 年 12 月		56.1	Bunt (1968)
McMurdo 海峡	1962 年 12 月		2.5	Bunt (1968)
McMurdo 海峡	1955 年 12 月		0.14～3.52	Bunt (1963)

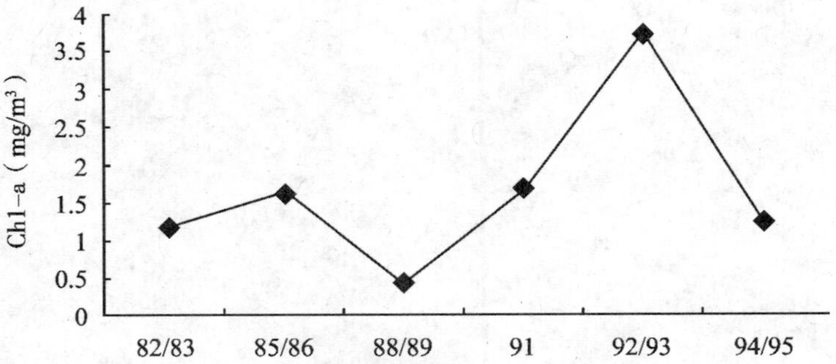

图 5　长城湾 Chl-a 的年际变化

Fig. 5　The year to year variations of Chlorophyll-a in Great Wall Bay

3.5 长城湾 Chl-a 的年际变化

长城湾及邻近海域自 1985 年以来夏季 Chl-a 含量的年际变化非常显著(图 5),最高值在 1992 年 12 月～1993 年 3 月(3.79 mg/m^3),变化幅度也最大,为 10.10 mg/m^3,最低值出现在 1988 年 11 月～1989 年 2 月(0.44 mg/m^3),其变化幅度也最小,为 1.17 mg/m^3。南极长城湾仍然是一个受人干扰甚微的自然本底环境,由此可以认为 Chl-a 的年际变化主要是受年际自然环境状况的控制,对于这种变化的深入研究将有助于了解长城湾环境的变化,特别是气候变化对生态系统的影响及其反馈作用即海洋生态系统对环境变化的响应。

参考文献 17 篇(略)

(合作者:朱明远　李宝华　黄凤鹏)

STUDY ON CHOLOROPHYLL-A CONTENT AND ITS VARIATION IN GREAT WALL BAY, ANTARCTICA DURING THE AUSTRAL SUMMERS FROM 1992 TO 1995

(ABSTRACT)

Abstract　During three successive austral summers from Dec. to March of 1992/1993, 1993/1994 and 1994/1995, investigation was carried out on phytoplanktonic Chlorophyll-a content in Great Wall Bay, Antarctica. The results showed that among the three summers the concentration of Chlorophyll-a in the summer of 1992/1993 was the highest with an average value of 3.79 mg/m^3, the range of Chlorophyll-a variation in this year was the largest, too. The year-to-year variation of average value of Chlorophyll-a was 3.79 mg/m^3(1992/1993) > 1.80 mg/m^3(1993/1994) >1.20 mg/m^3(1994/ 1995). The variation range of Chlorophyll-a was 10.10 mg/m^3(1992/1993)>6.57 mg/m^3(1993/1994)>2.41 mg/m^3(1994/1995). Compared with Chlorophyll-a content in austral summer in other area of Antarctica, the Chlorophyll-a in Great Wall Bay in the three year was relatively high though it was lower than that of 11.60 mg/m^3 in Jan. 1959 in Gerlache Strait and 56.10 mg/m^3 in Dec. 1961 in McMurdo Sound. The vertical distribution of Chlorophyll-a in water column was also presented.

Key words　Chlorophyll-A Content, Range Of Variation, Vertical Distribution, Great Wall Bay

C_3 环境化学与污染
(ENVIRONMENTAL CHEMISTRY AND POLLUTION)

南极特定区域污染及其影响*

摘　要　文章论述南极特定区域石油、生活污水(粪便、洗涤剂等)对环境和生物的污染及空气微生物状态。证实南极,包括它的生态系统,已受到人类活动的影响。提醒人们对此引起高度重视。强调在认识、开发和利用南极的同时,应遵守《南极条约》,以保护和促进南极及其生态系统良性发展。

关键词　南极　污染　环境　生态系统

1.引言

南极洲是神秘的白色大陆。围绕它及相应的广阔水域——南大洋是浩瀚的蓝色世界。南极远离人类。迄今它仍是地球上人类涉足最晚、人群活动最少的区域。然而随着南极科考活动的展开,近20年来,南极正逐步被人类所认识。

科技的进步使一个国家或地区的重大活动常常带有全球性,因而南极不可避免地受到人类的干扰。据不完全资料,本文就有关区域的石油、生活污水对环境和生物的污染及空气微生物状况作一概述。

2.1　石油污染

我国学者于20世纪80年代初就开始关注南极的污染问题。实际上,南极半岛海域的油污染已有一段历史了。1989年前,阿根廷"天堂湾"号油轮曾因触礁而泄漏了17万加仑的成品油至海中。之后,一秘鲁船因遭飓风袭击而触礁在佐治亚岛附近,致使周围染油。各国科考站油库和船只航行的泄漏或冒滴现象时有发生,如美国一考察站曾因一容器破裂而泄油5.2万加仑。由于南极的低温及大量海冰雪而使进入该水域的油污不易消除,对环境和生物造成的污染、损害在某种意义上重于地球的其他区域。

在菲尔德斯半岛南部海域,通过对20份表层水样和4份生物样的分析表明[1],水的油污染浓度范围达512.25 μg/L～4169.72 μg/L。油污染程度最重处是海豹湾,其次是智利马什基地前海区。最轻的是在长城湾(污染程度均高于我国第三类水质标准)。潮间带生

* 原文刊于《环境与发展》,1996,11(4):36-37.

物比鱼类受油污的程度要重。

1993/1994年度南极之夏，对取自有关水域的33份水样作了油污分析。结果总的说来比5年前的情况有所好转，海水油浓度大约降了一个量级。这归之于各考察站的努力和石油烃的自然净化。但进一步的检查，发现所检样品中符合一类标准的仅占2份。符合二类标准的也仅占2份。而超过三类标准的却有8份(即>0.5 mg/L)。而且均取自长城湾。油污染的地域性分布呈现为越近海岸边污染越重[2]。这表明南极一些海区迄今存在着明显的石油污染。

石油污染对环境与生物产生了损害作用。对南极的鲸(尾部表面，腰部，头部表面及其深层肌肉)、皮毛海狮(肝、脂肪)、金图企鹅(胸大肌)、贼鸥(胸大肌)、帽贝(全组织)等作烃含量测定[2]，证实了它们有石油烃的存在。其中的帽贝，随着环境中石油烃含量升高而其浓缩系数降低，表明它对高污染的石油烃的抵抗力不可能无限增加。人所共知，石油烃对生物的毒性与环境因子密切相关，其中温度是个重要因子。所幸南极常年低温。低温中的毒物对生物的损害较小，但影响却是深远的。

令人担忧的是石油对生态系统、食物链的影响。菲尔德斯半岛潮间带生态系的食物链中不同层次的消费者有黑背海鸥、鱼、软体动物(主要是帽贝，一些腹足类动物及石鳖)、棘皮动物、甲壳类动物等等[3]。像帽贝等活动能力弱的动物，其避污能力也低，却又是许多鸟只的首要食源。油污一旦严重威胁到它们，油污物将传递至更高级营养阶中。处于食物网、链不同层次和许多环节中的微生物同样遭受到油的影响，从而对整个生态系统造成危害。

油污物在南极水域同样也被置于自然净化过程之中。其重要的作用者便是石油烃氧化菌(*Hydrocarbon oxiding microbes*)。对有关海湾作的调查表明[2]，那儿夏季时广布着此类微生物，检出率达100%，唯其数量不大，范围在3～1 100个/升，大多数是1立升水中只几个此类菌，属群不多，仅发现4属，即黄杆菌(*Flavobacterium*)、假单胞菌(*Pseudomonas*)、不动杆菌(*Acinetobacter*)和库特氏菌(*Kurthia*)。这意味着南极某些海域的这类特点，即那儿虽有石油污染，但仍有一定的净化作用。关键是油污净化少而慢。因而提醒人们在南极使用和分散石油及其制品时必须慎之又慎。

2.2　生活污水等对环境的污染

对1994年某站的供水及入海污水状况作了调查。发现样品中大肠菌类的检出率不低，如潮水中为33.3%，饮用水为2.0%，海水中为57.1%。大肠菌含量分别为$1.33\times10^{-3}/cm^3$、$5.00\times10^{-2}/cm^3$和$6.71\times10^{-1}/cm^3$。1990年1～3月份，对菲尔德斯半岛南部海水样中粪大肠菌的检测表明，其检出率较高，含量基本与这一数据持平。但是1993/1994年度夏时科考人员大大减少，生活污水排放量理应减少许多。结果还表明，大肠菌的分布趋势呈现出离岸越近，数量越高之状。这证实了它与人类活动的密切关系[4]。

另一个生活污水成分是阴离子洗涤剂[活性成分为十二烷基磺酸钠(LAS)]。据统计，某站全夏将有4.5×10^6 mg的洗涤剂排放入海。实测结果已表明它在特定海区的存在。LAS对鳕鱼、真鲷等鱼的卵的发育会产生不利影响。但所调查水域中的LAS浓度已远远高于鳕鱼的承受力。从而对它们及近亲产生危害[2]。

还对几百个海水、淡水、冰雪、原土、沙砾和生物样品作过包括耐重金属菌类在内的测定工作。这些重金属盐包括Cd^{2+}、Cr^{6+}、Hg^{2+}、Ag^{2+}、Fe^{3+}及酸等。结果反映出这些样品中确实不同程度地存在着有关的细菌、真菌。这意味着南极蕴藏着(或携入)一定量的金属盐及其相应的微生物,它将参与或影响当地的物质循环和能量传递,也为未来可能出现的污染物之降解活动提供了某种可能性。此外对DDT、六六六、狄氏剂的测定也曾进行过,均证实南极曾被它们污染并受到伤害[5]。

2.3 空气微生物

Cameron等在20世纪70年代曾指出过麦克默多站钻探处有污染的空气微生物。测定结果证实人类活动已对南极空气微生物产生了一定的影响[6]。将菲尔德斯半岛的部分地区分为洪荒区、大型动物出没区和科考站及其附近三个区。其室外的空气微生物含量间存在较大差别,尤其是在洪荒区和科考站区间每立方米空气中所含微生物量(简写为CFU)分别为16和4 994个。差距是明显的。就科考站及其附近区而言,室外空气微生物含量是建筑物附近的高于站区稍远地方的。室内空气微生物含量一般均高于室外的,有的还高出许多倍。就是说,空气微生物含量与种群组成状况,可能反映和预测人们所在环境的空气质量。测定还表明,各国科考站相比,空气微生物含量间存在着一定差异。这差异与相关站的人员素质、生活、工作条件和习惯密切相关。

3. 结语

南极特定区域确实存在环境污染问题。对南极的被污染,人类应该站在历史的、前瞻全球的高度上来认识,由于南极在地球三极中举足轻重,它实实在在地规定和影响着全球变化,因而对它的污染应该引起高度重视。

参考文献6篇(略)

(合作者:战 闰)

南极麦克斯韦尔海湾阴离子洗涤剂污染研究*

摘　要　首次报道了南极麦克斯韦尔海湾海域阴离子洗涤剂污染的变化和分布状况。研究海域水体阴离子洗涤剂含量在0.148～0.997 mg/L之间，若按中国国内地表水水质标准0.3 mg/L评价，全部20个站次中有6站次超过标准，超标率达30%，表明该地区已有一定程度的阴离子洗涤剂污染。主要污染源是各考察站的生活污水。

关键词　南极麦克斯韦尔湾　阴离子洗涤剂污染

南极麦克斯韦尔海湾是南极考察最活跃的区域，仅乔治王岛无冰区就有智利、阿根廷、波兰、前苏联、巴西、乌拉圭、韩国、秘鲁、中国等国家的考察站及众多的避难所。每年有大量的考察队员和后勤保障人员在该地区工作；有远多于南极工作人员的旅游者乘船乘飞机到此访问。即使是极地的严冬，各考察站还有很多越冬队员。这些人员在工作和生活期间，不可避免地对这一海域带来一定程度的污染，尤其是洗涤剂，在生活、工作中用量较大，而且这种污染物完全是人工合成的，对南极生物究竟有怎样的影响尚难预料，是南极环境中不可忽略的污染因素，应引起足够的重视。

阴离子洗涤剂中十二烷基磺酸钠(LAS)是最常用的成分，本文是"南极长城站重点地区生态系统研究"课题组在1993年12月～1994年3月南极夏季对中国长城站及邻近海域中阴离子洗涤剂的分布及变化研究结果，对该海域的污染问题进行了探讨。

1　研究海域概况

本研究区域包括了麦克斯韦尔湾及其附属海湾(见图1)。

2　材料与方法

测站站位见图1。

用采样器采取表层0.5 m处海水，带回实验室进行分析。

1994年2月4日在15个站位；1994年2月26日，对6个站位采集了水样，回实验室进行了阴离子洗涤剂分析，所有水样采回后按《海洋监测规范》(国家海洋局1991年发布)

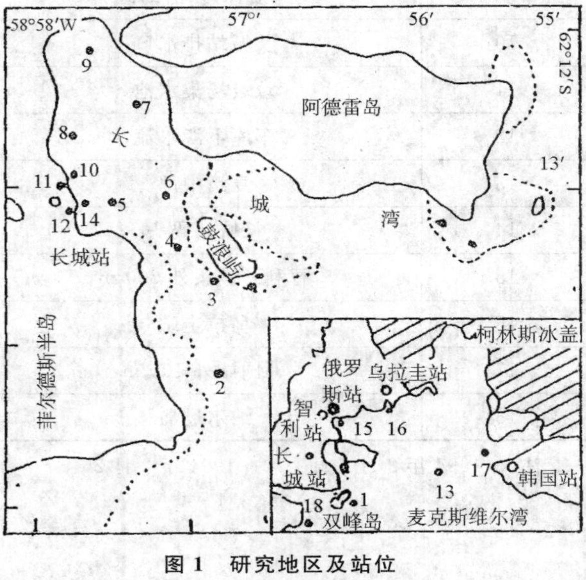

图1　研究地区及站位

Fig. 1　The research area and the sample stations

*　国家"八·五"重大课题资助项目

原文刊于《中国环境科学》1998,18(2):151～153,本人为第二作者。

进行处理,测定分析方法采用亚甲基蓝分光光度法。

研究区域环境和气候状况请参考文献[1]。

3　结果

各测站的十二烷基磺酸钠测值见表1,各海域的平均值见表2。

此次研究的阴离子洗涤剂测值介于0.148～0.997 mg/L之间,最低值是湾内8号站(1994年2月4日),最高是乌拉圭站。高于0.3 mg/L的测值占6个,主要是各考察站附近,长城站入海的淡水流、离排污口近的8号站、智利站码头外、韩国站外、乌拉圭站外,湾外海水仅酒泉河口的测值均大于0.3 mg/L。

表1　南极麦克斯韦尔海湾阴离子洗涤剂污染检测结果(mg/L)

Table 1　The detergent (LAS) content of sample stations (mg/L)

序号	站位	测值*	测值**
1	双峰岛内侧	0.239	
2	酒泉河外400 m	0.401	
3	内湾口,鼓浪屿侧	0.165	
4	内湾口,望龙岩下	0.202	0.239
5	油库下海域	0.310	
6	鼓浪屿上顶	0.266	
7	湾内,阿德雷岛侧	0.229	
8	湾内,长城站侧	0.148	0.229
9	湾顶,海豹活动区	0.199	0.222
10	长城站排水口	0.256	
11	垃圾场集水池	0.212	
12	车库下淡水流	0.337	
13	麦湾中部	0.286	
14	长城站码头	0243	
15	智利站码头外300 m	0.361	
16	乌拉圭站	0.997	
17	韩国站外100 m	0.395	
18	纳尔逊岛	0.243	

注:*1994年2月4日　**1994年2月26日

表 2　麦克斯韦尔海湾的阴离子洗涤剂(LAS)含量(mg/L)

Table 2　The detergent (LAS) content of Maxwell Bay (mg/L)

海域	站号	测值*	测值**	两次测值平均	海域	站号	测值*	测值**	两次测值平均
内湾	5,6,7,8,9	0.230	0.226	0.228	智利站	15	0.361		
内湾口	3,4	0.183	0.239	0.211	韩国站	17	0.395		
入海淡水	12	0.338			乌拉圭站	16	0.997		
排污口	10	0.256			外海	11,2,13,18	0.309	0.243	0.276
垃圾场	11	0.212							

注：*1994 年 2 月 4 日　**1994 年 2 月 26 日

4　讨论

南极是受到人类污染最少的地区，然而，为了研究和保护南极所建立的科学考察站和航行该海域的各类船只却不可避免地成为污染物的来源[1～7]，给南极的环境保护带来一定的问题。

阴离子洗涤剂是海域尤其是人类生活区附近重要的环境污染物质，生活废水中阴离子洗涤剂来源于洗涤剂的使用，据作者统计，长城站各种洗涤剂的月使用总量约在 30 包以上，总计约 15 kg，按阴离子洗涤剂中十二烷基磺酸钠含量 10％计，每月有 1.5×10^6 mg 十二烷基磺酸钠进入长城湾，整个夏季的输入达到 4.5×10^6 mg(45 kg)，海域的阴离子洗涤剂污染是不小的。

除了考察人员生活废水排放外，每年还有大量的旅游人员和旅游船进入该海区，也会带来大量的污染。

阴离子洗涤剂主要的毒害作用是影响鱼卵的正常发育[2]，鳕鱼卵比真鲷卵对十二烷基磺酸钠(LAS)更敏感，尤其在胚胎闭孔前是敏感时期，当 LAS 含量为 0.02 mg/L 时，鳕鱼卵的孵化率下降 40％，且 66％的初孵仔鱼脊椎骨畸形。本次研究的海域大部分测站的测值都低于国内地表水水质标准，但都远高于鳕鱼的耐受能力，南极近海有大量的南极鳕鱼生长，南极鳕鱼和鳕鱼有亲缘关系，有必要进一步开展 LAS 对南极鳕鱼和其他海洋生物危害的研究。

参考文献 7 篇(略)

(合作者：袁峻峰　吴宝玲　李永祺)

THE ANION DETERGENT POLLUTION OF ANTARCTIC MAXWELL BAY AND ITS ADJACENT SEA AREAS

(ABSTRACT)

Abstract The change and distribution of anion detergent pollution in the antarctic Maxwell Bay and its adjacent sea areas were reported for the first time. During the summer of antarctic in 1993～1994, the concentration of the detergent was varied from 0.148 to 0.997 mg/L with 30% super standardized. This pollution comes from these stations.

Key words Antarctic Maxwell Bay; Anion Detergent Pollution

南极麦克斯韦尔湾及邻近海域高锰酸钾指数研究*

摘　要　为了研究南极水域的污染状况，于 1993 年 12 月～1994 年 2 月对南极麦克斯韦尔湾及邻近海域的高锰酸钾指数（COD_{Mn}）的变化和分布状况进行了监测。39 个站次的测值在 0.334～1.959 mg/L之间，符合中国《海水水质标准》(GB 3097-82)第一类水质(3.000 mg/L)要求。各考察站附近的测值高：长城湾内高于湾口及湾外。分析了湾内、湾口和湾外在 1993 年 12 月、1994 年 1 月及 2 月的变化情况和原因：对 COD_{Mn} 的主要来源进行了研究分析并对南极长城站及邻近海域的环境保护和研究提出了建议。

关键词　南极　麦克斯韦尔湾　COD_{Mn}

1　前言

乔治王岛是南极考察最频繁的区域。仅无冰区就有智利、阿根廷、波兰、俄罗斯、巴西、乌拉圭、韩国、秘鲁和中国等国家的考察站以及众多的避难所。每年夏季有大量的考察队员和后勤保障人员在该地区工作：还有很多的旅游者乘船乘机到此访问。智利的空军基地非常繁忙：即使是极地的严冬，各考察站还有很多越冬队员，这些人员在工作和生活期间。不可避免地对这一海域带来一定程度的污染(Mahian et al.，1991；R ailshack. 1992)。除了人类活动以外，该地区有大量的企鹅、海豹和其他生物生活，无疑也造成巨大的环境压力，研究这些污染的强度及其对南极生态系的影响，对南极环境保护、科学考察和区域管理是很重要的。

高锰酸钾指数（COD_{Mn}）是水体有机污染的表观反映，是海洋环境研究中常用的指标。关于南极水域 COD_{Mn} 污染的研究未见文献报道。在南极海域，调查 COD_{Mn} 的本底值，并由此来推断海域的有机污染状况及变化规律，对南极海洋学研究和环境保护具有积极的意义。

本文报道了"南极长城站重点地区生态系统研究"课题组在 1993 年 12 月至 1994 年 2 月南极夏季对麦克斯韦尔湾及邻近海域 COD_{Mn} 的分布及变化研究结果，对该海域的污染问题进行了探讨。

2　研究海域概况

研究区域包括麦克斯韦尔湾及其附属海湾(图 1)。

麦克斯韦尔湾位于南极乔治王岛南部，是南极考察站密集地区。中国南极长城站位于长城湾畔。长城湾为一长形内湾，面积约 3 km^2，鼓浪屿将长城湾分为内外两个部分，内湾比较封闭。本次考察的 3、4、5、6、7、8、9 位于长城内湾；16、17、18 为湾内污染重点监测站。长城湾的海水主要通过长城湾口与麦克斯韦尔湾进行交换，湾外海水主要由湾底向

* 国家"八五"重大项目资助；参加采样的还有陆华、洪旭光等同志，本人为第二作者。

原文刊于《极地研究》，1998，10(3)：212-216。

内,湾内海水主要由表层向外。

图 1　采样站位

Fig. 1　Research area and the sample stations

长城湾隔阿尔德斯岛与菲尔德斯半岛间有一砾石堤与阿德雷湾的爱特莱伊湾相通,退潮时可步行通过,涨潮时橡皮艇可顺利通过;涨潮时的海水可从此水道进入长城湾。阿德雷湾口比较开阔,海水交换比较好,智利、俄罗斯、乌拉圭站位于阿德雷湾畔。本次的15站位于该湾。

韩国站位于巴通半岛,海水交换较好,附近为19号站。1、2、10、11、12、13、14号站位于长城内湾外的近岸海域,20号站在麦克斯韦尔湾中部。

麦克斯韦尔湾的阿德雷岛面积约4 km^2,每年夏季大约有1万只企鹅在上生活,该岛海豹较少,未超过20头,长城湾湾顶也有近10头海豹生活,贼鸥、黑背鸥、巨海燕等鸟类在附近大量存在,湾内海洋生物丰富,海藻、软体动物大量存在,海水初级生产力较高,浮游生物丰富,这些生物能产生大量的有机物质。

流入麦克斯韦尔湾的小溪很多,溪流水来自苔藓地衣丰富的区域,也带来大量的有机物。

3　材料与方法

采样时用采样器采取表层0.5 m处海水,回实验室进行分析。

1993年12月13日,1994年1月7日和1994年2月3日,对9个定位站采集水样进了COD_{Mn}分析;12月13日还采集了10、11、12、13、14及15站的水样:1月7日由于连续几天风暴,涌很大,仅在1～9号定位站及10、11号站采样;2月3日在1～9号定位站采样，亦因涌大而返回，在站区采集了16、17、18三个站的水样，2月6日采集了19号和20号站的水样，该涌亦很大。所有水样采回后按《海洋监测规范》(国家海洋局，1991)进行处理，COD_{Mn}测定分析方法采用碱性高锰酸钾法。

4　结果与分析

本次研究的结果见表1。

表1　南极麦克斯韦尔湾及邻近海域COD_{Mn}检测结果(mg/L)

Table 1　The COD_{Mn} of Maxwell Bay and its adjacent area

海域	站位	测值			平均
		93.12	94.1	94.2	
内湾*	5 油库下海域	1.12	0.437	0.826	
	6 鼓浪屿上顶	0.904	0.555	1.080	
	7 湾内，阿德雷岛侧	0.767	0.555	0.792	
	8 湾内，长城站侧	1.333	0.538	0.995	
	9 湾顶，海豹活动区	1.396	0.674	1.452	
	平均	1.104	0.551	1.029	0.895
内湾口	3 内湾口，鼓浪屿侧	0.527	0.471	1.022	
	4 内湾口，长城站侧	0.826	0.628	0.877	
	平均	0.676	0.554	0.949	0.727
内湾外近海海域	1 双峰岛	0.334	0.572		
	2 长城湾口	0.672			
	10 阿岛与鼓浪屿间	0.566	0.572		
	11 湾外，阿德雷岛侧	0.354	0.893		
	12 湾外，阿德雷岛侧	0.708			
	13 湾外，阿德雷岛侧	0.503			
	14 湾外，阿德雷岛侧	0.511			
	平均	0.526	0.679	0.727	0.609
智利站	15 码头外300米	1.474			
入海淡水	16 长城站				1.418
积水池	17 长城站垃圾场				1.012
排污口	18 长城站				1.959
韩国站	19 码头外300米				0.596
外海	20 麦克斯韦尔湾				0.572

* 内湾指长城湾的鼓浪屿以内的海域

(1)各站位的COD_{Mn}测值介于0.334～1.959 mg/L，最低值是湾外1号站，最高是长城站排污口。高于 1.000 mg/L 的测值仅 10 个。主要是排污口、入海的淡水流、智利站码头外、长城湾顶、离排污口近的 8 号站，距长城站码头和库较近的 5、6 号站；湾外海水的测值都小于 1.000 mg/L。

(2)由表 1 可知，长城站及邻近海域的 COD_{Mn} 测值高于外海，但与国内《海水水质标准》(GB 3097-82)的要求比较，符合规定的第一类水质要求，没有明显的污染迹象。

(3)由表 1 可知，长城湾内 COD_{Mn} 高于湾口，湾口高于湾外海域，表明 COD_{Mn} 污染从湾内到湾外逐渐降低，湾外的 COD_{Mn} 在一定程度上来自湾内，长城湾内外水质交换是长城湾 COD_{Mn} 稀释的主要方式之一。

(4)由表 1 可知，长城湾内外海域 COD_{Mn} 有明显的月变化，1 月份的 COD_{Mn} 值明显降低，这和 1 月初连续几天的风暴明显相关，风暴引起强烈的湾内外水交换，降低了 COD_{Mn} 含量。但在总的变化趋势上，COD_{Mn} 从 12 月到 2 月明显增加。

(5)9 号站的 COD_{Mn} 值较高，应归因于该区域有海豹和企鹅的活动；同时，这里是长城湾的海流上升区湾底的有机质在此上升，从而引起 COD_{Mn} 增高。

(6)1、2 月湾内和湾口的 COD_{Mn} 相差不大，主要是水交换强烈所引起采样期涌浪很大。1 月湾外 COD_{Mn} 含量大于湾内和湾口，可能主要因为 11 号站受风浪影响。将底质搅起而造成高值，使湾外平均值高于湾内和湾口。

(7)虽然测站离智利站的海滩、码头、垃圾场较远．但 COD_{Mn} 的测值仍较高(1.474 mg/L，15 号站)。智、俄两站紧靠在一起，人员、设备很多，建站很早，加上马尔什基地的影响，较高的 COD_{Mn} 值是可以理解的。这一高值在 12 月份测出．说明该海域有机污染比较严重。

(8)韩国站的 COD_{Mn} 测值较低，为 0.586 mg/L，为各考察站站前的最低值。一方面采样站距码头较远、海区开阔交换迅速和采样时风浪大有关，另一方面韩国站管理完善，污染控制较好也是重要的原因。

(9)在麦克斯韦尔湾中部的 COD_{Mn} 测值为 0.572 mg/L，在该次研究中是比较低，也符合其面积大，与外海交换好的基本情况。

5　讨论

COD_{Mn} 是海域环境污染的重要指标，可作为有机污染的评价代表。一般情况下，大洋、外海的 COD_{Mn} 值低于 1.000 mg/L(李永祺，丁美丽，1991)。长城站及邻近海域的 COD_{Mn} 部分测值已超过大洋、外海的数据，已有一定程度的污染，但尚不严重。1984 年，渤海海河口区三个区域(顾宏堪，1991)，COD_{Mn} 测值范围分别是 10.1～238.0 mg/L(大沽排污口区)、7.6～33.2 mg/L(永定新河河口段)和 1.0～14.1 mg/L(河口近岸海域)，本次研究海域的污染是不能与之相提并论的。

从污染来源分析看，长城站排污量较小(＜10 吨/天)，排污口附近的测值较小，为 1.959 mg/L(18 号站)，高出附近入海淡水约 0.5 个单位(1.418 mg/L，16 号站)，但长城湾有三条主要的淡水河流入．流量较大．而流域的植被非常茂盛，所能提供的有机物是比较丰富的，可能是海湾有机质的主要来源(吴宝铃等，1992；杨和福，1992)。

根据本次研究的结果．提出如下建议：

(1)南极是特殊地区．不能用国家和地区的水质标准监测其水质。作为最重要的保护地，我们认为应该进行进一步研究，为制定南极海域水质标准提供依据；

(2) 长城湾水域环境受海流影响很大,要可靠的反应水质状况必须多站位、多次取样;

(3)虽然长城站的污水排放总量较小,污水已经过处理,但排污口的 COD_{Mn} 明显高于其他区域,建议建立陆上沉淀系统,污水在陆地上经沙砾过滤后再渗入海域;

(4)本次研究未能判断长城湾生物自产有机物占总 COD_{Mn} 的比例,也未对入海淡水和沿岸渗入水进行研究,建议在今后的研究计划中列入。

参考文献7篇(略)

(合作者:袁俊峰 李永祺 吴宝铃)

THE CONTENT OF COD_{Mn} OF THE GREAT WALL BAY AND ADJACENT SEA AREAS, ANTARCTICA

(ABSTRACT)

Abstract During 1993. 12～1994. 2, the COD_{Mn} of the Great Wall Bay, Antarctica and its adjacent sea areas was investigated. The COD_{Mn} of 39 stations was from 0. 3342 mg/L to 1. 9590 mg/L; the high value was at the research stations, and the value of the inner Great Wall Bay was higher than that of the outside. The reason of the change of the COD_{Mn} was studied; the main resource of the COD_{Mn} was analysed. Some of the pollution come from the research stations, but most of it may cone from the living beings. The wind and wave is one of the most important factors affect the values of COD_{Mn}. Suggestion were given to protect the environment of Antarctica.

Key words Great Wall Bay, Antarctica, COD_{Mn}

CONCENTRATION AND DISTRIBUTION OF INORGANIC NUTRIENT SALTS OF NITROGEN AND PHOSPHORUS IN SURFACE WATER OF CHINESE GREAT WALL BAY AND ITS ADJACENT SEA AREA, ANTARCTICA, DURING AUSTRAL SUMMER OF 1993/1994*

Abstract The concentrations and distributions of four species of inorganic nutrient salts in surface sea water of Chinese Great Wall Bay and its adjacent sea area, Antarctica were surveyed during austral summers of 1993～1994.

The results obtained are as follows:

The general concentrations of NH_4-N, NO_3-N, NO_2-N and PO_4-P are 2. 13, 7. 07, 0. 74 and 1. 12 g. dm^{-3}, respectively. Generally speaking, the concentrations are higher inside the Bay than those outside and the higher values of those inorganic salts often appear in December. The time or the station where their higher and lower concentrations appear are not completely the same. The descending order of the magnitudes of $\Sigma N/P$ value according to the arrangement monthly is: Dec. Feb. Jan. The variations of water temperature are mostly consist with the concentrations of NH_4-N, PO_4-P and N/P, but not often with NO_2-N. The low concentrations of nutrient salts as shown in some stations may be related with absorption by phytoplankton, perhaps including autotrophic bacteria.

The results suggest that the area surveyed has a typical characteristic of an inner shallow water bay, sub-Antarctica. The nutrients needed for phytoplankton are sufficient, and the Bay surveyed is still in a normal condition.

Keywords Chinese Great Wall Bay and its adjacent sea area in Antarctica, Surface sea water, Inorganic nutrient salts of nitrogen and phosphorus

1 INTRODUCTION

Some inorganic salts, including three nitrogenous salts, i. e. NH_4-N, NO_3-N and

* The original text was published in *J. Glaci. Geocr.*, 1998, 20(4): 306-311.
中文刊于《海洋湖沼通报》,1996,No,4:13-19.

NO_2-N, and phosphates, are important chemical parameters in marine environment. They are produced by the joint efforts of the various kinds and levels of organisms in marine food chains and webs with environmental factors, and are important nutrients upon which marine plants depend for existence. Their growth and decline effect determine the prosperity/decline and distribution of marine plankton and microorganisms. They further determine marine productivity and maintain development of marine ecosystems. So marine chemists, ecologists and environmental scientists have paid attention to investigations on the concentration, distribution, states and changes of nutrients for a long time.

Only a few research reports on nutrient salts of marine environment in Antarctica, including shallow sea ecosystem, have been seen till now (Wang et al., 1986; Lu and Watanabe, 1994).

The Chinese Great Wall Bay (GWB for short the same fellow) is a semi-open type shallow inner bay nearby the Chinese Great Wall Station in Maxwell Bay. It is very necessary to take researches on its nutrient salts and their distribution conditions and its constitution of eco-structure, cycle of materials, and energy metabolism. During 1988～1989, Lu and Watanabe (1994) investigated some nutrient salts in the Bay, but detailed analyses and reports about its salts in the surface seawater have not been made. The paper reports the study results of four kinds of nutrients in the surface water of the Bay and its adjacent sea area during the austral summers of 1993/1994.

2 MATERIAL AND METHODS

2.1 Sampling times

December 13, 1993, January 7 and February13, 1994.

2.2 Sampling stations

14 stations in all. They were divided into three survey zones. There were stations in the inner part of GWB, contiguous to western side of GuLangyu Island. They were No. 4, 5, 6, 7, 8 and 9 stations named zone No. Ⅰ. Zone Ⅱ was located outside the Bay, i. e. Contiguous to East side of the island. They are five sampling stations, i. e. No. 2, 3, 10 and 11 stations respectively. Zone Ⅲ was located outside the Bay at Ardley Bay and nearby northern side of Ardley Island. The detailed sampling condition was drawn in Fig. 1.

2.3 Sampling and Analysis

Each sample in the bay was collected by a rubber craft. Enough water sample at each sampling station was collected into polyethylene bottle, filtered through 0.8 Millipore filter membrane, and kept in < -5℃ temperature till analysis not longer than ten hours. The determination of samples was followed according to the reference No. 3 cited (State Technological Superintendency Administration, 1991). The standard NH_4-N, NO_3-N, NO_2-N and PO_4-P were purchased from the Second Institute of Oceanography, SOA. At the same time, samples were also collected for determining the relative parameters including Chloraphyll-a, heterotrophic bacterial count (CFU · m^{-3}), water temperature, etc.

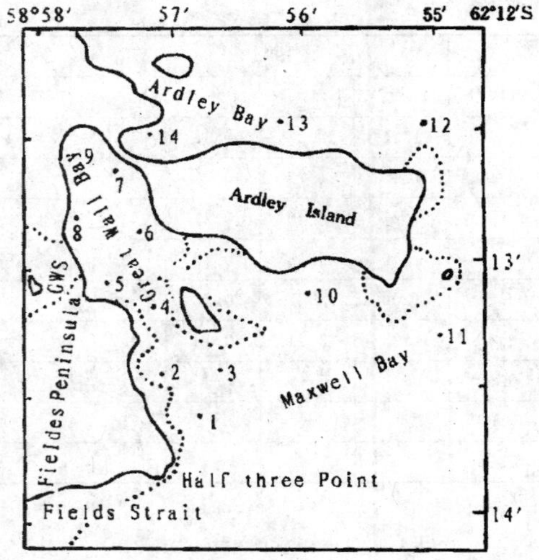

Fig. 1　Sampling Stations for four inorganic nutrient salts from surface water in Chinese Great Wall Bay and its adjacent sea area, Antarctica during austral Summers of 1993/1994

3　RESULTS AND DISCUSSION

The analytical results of the determination of the concentrations of four nutrient salts and the corresponding water temperatures were listed in Table 1.

Table 2 showed the states of the average concentration of the salts in the surface water surveyed. It indicated the general trend of the concentration of the four nutrient salts in the three zones. They were described as follows: the order of the zones arranged in a descending order, among zones surveyed magnitudes of the nutrient concentrations they possess respectively:

NH_4-N: Ⅰ, Ⅱ, Ⅲ;

NO_3-N: Ⅲ, Ⅱ, Ⅰ;

NO_2-N: Ⅲ, Ⅰ, Ⅱ;

PO_4-P: Ⅱ, Ⅰ, Ⅲ.

The results showed that, in many situations, the concentrations of the four nutrient salts in the Bay were higher than those of the other two zones.

Table 1　Table of concentration of the four nutrient salts of surface water in GWB and its adjacent area during the summer of 1993/1994

Stations	1993-12				1994-01				1994-02			
	NH_4^+-N	NO_3^--N	NO_2^--N	PO_4^--P	NH_4^+-N	NO_3^--N	NO_2^--N	PO_4^- x-P	NH_4^+-N	NO_3^+ x-N	NO_2^--N	PO_4^--P
1	1.80	13.88	1.45	0.64	0.92	8.47	0.26	1.68	0.85	7.15	0.18	1.36
2	2.40	13.00	1.70	2.02	1.56	2.85	0.13	1.50	0.98	7.30	0.20	1.28
3	9.05	11.38	0.25	0.47	1.29	2.95	0.05	1.78	1.00	7.48	0.12	1.58
4	8.40	12.00	1.25	0.36	0.84	1.97	0.27	1.93	1.02	7.17	0.81	1.34
5	7.00	8.88	1.40	0.08	1.27	2.24	0.23	2.00	1.01	7.08	0.24	1.36

(continue)

Stations	1993-12				1994-01				1994-02			
	NH_4^+-N	NO_3^--N	NO_2^--N	PO_4^--P	NH_4^+-N	NO_3^--N	NO_2^--N	PO_4^- x-P	NH_4^+-N	NO_3^+ x-N	NO_2^--N	PO_4^--P
6	3.15	8.75	1.65	～	1.76	2.22	0.18	1.54	1.31	7.65	0.20	1.36
7	3.08	7.76	2.55	0.20	1.15	1.76	0.23	1.63	0.97	7.06	0.20	1.22
8	4.64	7.63	2.30	～	1.27	2.40	0.27	1.78	1.92	6.45	0.20	1.60
9	3.64	7.63	2.20	～	1.40	2.38	0.19	1.98	1.35	7.75	0.34	1.80
10	2.35	12.63	1.25	0.25	1.00	2.82	0.23	1.30	—	—	—	—
11	1.60	9.13	～	0.45	1.28	2.91	0.14	1.73	—	—	—	—
12	0.40	10.00	1.75	0.59	—	—	—	—	—	—	—	—
13	0.40	9.30	1.26	0.63	—	—	—	—	—	—	—	—
14	0.40	19.62	2.04	0.56	—	—	—	—	—	—	—	—

Note: Sampling time, unit of concentration: $g \cdot dm^{-3}$;"～"indicates"not detected";"—"indicates"not determined".

Corresponding water temperatrues(T, ℃)recorded at some stations were:

Dec. 1993-Station No. 1: 2.20, 2: 2.20, 3: 2.20, 4: 2.40, 5: 2.45, 7: 0.40, 8: −0.20, 9: −0.10, 10: 1.30, 11: −0.60;

Jan. 1994-No. 1: 1.40, 2: 1.40, 3: 1.60, 4: 1.40, 5: 1.80, 6: 1.70, 7: 1.70, 8: 2.00, 9: 1.70, 11: 1.20;

Feb. 1994-No. 4: 1.20, 5: 1.30, 6: 1.40, 7: 1.60, 8: 1.80, 9: 1.60.

Table 2 Comparison of average values of four nutrient salts from surface water of three zones in GWB and its adjacent sea area($g \cdot dm^{-3}$)

No. surveyed zone	NH_4^+-N	NO_3^--N	NO_2^--N	PO_4^--P
Ⅰ	2.50	6.04	0.78	1.10
Ⅱ	1.95	7.71	0.45	1.20
Ⅲ	0.40	9.97	1.68	0.59

Table 3 listed the monthly differences of the concentrations among the three zones surveyed.

From Table 3, the monthly differences of salts concentration in the three surveyed zones were found. The month arranged in a descending order according to magnitude of nutrient concentration among months were:

No. Ⅰ zone: NH_4-N: Dec. Jan. Feb.; NO_3-N: Dec. Feb. Jan.; NO_2-N: Dec. Jan. Feb.; PO_4-P: Jan. Feb. Dec..

No. Ⅱ zone: NH_4-N: Dec. Feb. Jan.; NO_3-N: Dec. Feb. Jan.; NO_2-N: Dec. Feb. Jan., PO_4-P: Jan. Feb. Dec..

(For No. Ⅲ zone only the data for December available).

The results indicated that: in many situations the higher concentration of the salts often occurred in December, the concentrations of NH_4-N, NO_3-N and NO_2-N, but the lower concentration of PO_4-P often occurred in Dec.

Table 3 Comparison of the average salts concentrations among three months of surface sea water in GWB and its adjacent area

Numbers of Sampling stations	1993-12				1994-01				1994-02			
	NH_4^+-N	NO_3^--N	NO_2^--N	PO_4^--P	NH_4^+-N	NO_3^--N	NO_2^--N	PO_4^--P	NH_4^+-N	NO_3^--N	NO_2^--N	PO_4^--P
Ⅰ	4.96	10.00	1.89	0.11	1.28	2.65	0.23	1.80	1.26	7.19	0.23	1.45
Ⅱ	3.44	9.73	0.93	0.77	1.21	4.00	0.16	1.60	0.94	7.31	0.17	1.41
Ⅲ	1.40	9.97	1.68	0.59	—	—	—	—	—	—	—	—

Note: Unit of nutrient salts: g · dm^{-3}; "—"indicates"not determined."

Table 4 listed the respective average concentration of the four nutrient salts at each station. It showed that the highest average value of NH_4-N occurred at station No. 3, the lowest one at stations No. 12, 13 and No. 14. The highest and lowest concentrations of NO_3-N occurred at stations No. 14 and No. 8 respectively. For NO_2-N, the highest and lowest occurred at stations No. 14 and No. 11 respectively. For NO_2-N, the highest and lowest occurred at stations No. 14 and No. 11 respectively. For PO_4-P, the highest and lowest concentrations occurred at stations No. 2 and No. 14 respectively. It might be inferred that the lower concentrations of nutrient salts in some surveyed stations were associated with the uptake of phytoplankton(Wang, 1989).

Table 4 Comparison of the average concentrations of the four nutrient salts in the respected surveyed from surface water in GWB and its adjacent area(μg · dm^{-3})

Numbers of Sampling stations	NH_4^+-N	NO_3^--N	NO_2^--N	PO_4^--P
1	1.19	9.83	0.63	1.23
2	1.65	7.72	0.68	1.60
3	3.78	7.27	0.14	1.28
4	3.42	7.05	0.57	1.09
5	3.09	6.07	0.62	1.15
6	2.07	6.21	0.68	0.97
7	1.73	5.53	0.99	1.02
8	2.61	5.49	0.92	1.11
9	2.07	5.90	0.91	1.26
10	1.68	7.73	0.74	0.78
11	1.44	6.02	0.07	1.09
12	0.40	10.00	1.75	0.59
13	0.40	9.30	1.26	0.63
14	0.40	12.63	2.04	0.56

One of the important conditions for growth and development of phytoplankton and microbes is suitable N/P. Table 5 listed the ratio of N and PO_4-P concentration. From Table 5, the highest value of N/P occurred at station No. 5 in Dec. 1993, the lowest one was obtained from station No. 4 in January, 1994. The descending order of N/P arranged according to the monthly variation was: Dec. Feb. and Jan. Viewed roughly, the variation in ΣN/P as pretty big, which might be caused by the bigger range of variation in N-content and smaller one in the P-content, but also it was related to the different types of circulation of N and P in the environment. This state of ΣN/P might reflect the characteristic of the shallow inshore waters of the bay and its adjacent sea area, i. e. the effects of the fresh water coming from the land nearby the coastal line was strong during the summer and it also indicated that there was still a better exchange capacity of water between the sea surveyed and the open sea, only the variation being bigger.

Table 5 Table of N/P in surface seawater in GWB and its adjacent sea area

Numbers of Sampling stations	1993-12	1994-01	1994-02
1	26.77	5.74	6.01
2	8.47	3.03	6.63
3	44.00	2.41	5.44
4	60.14	1.60	6.25
5	216.00	1.87	6.13
6	—	2.70	6.74
7	66.95	1.93	6.75
8	—	2.28	5.36
9	—	2.01	5.24
10	64.92	3.12	—
11	23.84	2.50	—
12	20.59	—	—
13	17.40	—	—
14	23.32	—	—
Average	34.36	2.60	6.01

Note: N at each station was total value of NH_4-N, NO_3-N and NO_2-N, unit of concentration: $\mu g \cdot dm^{-3}$

Table 6 listed the results of the correlation analysis among some related parameters in that investigation. In that table, the positive correlative relationships of (NH_4-N)T, (PO_4-P)-T, ΣN/P-T were illustrated. Among them, positive correlative coefficient of (PO_4-P)-T reached a significant level (n=26, r=0.299, $r_{0.1}$=0.260). Left the negative correlative coefficients of ΣN-P, (NO_3-N)-(PO_4-P) were all significant. Those results indicated that the rise/fall of water temperature has a more positive effect on concentration of NH_4-N, PO_4-P and $\Sigma N/P$, but a more negative effect on concentration of NO_2-N. The relationships of (PO_4-P)-T, (PO_4-P) - (NO_3-N) were not consistent with those obtained from the waters of the south Shetland Islands (State Antarctic Research Expedition Committee of People's Republic of China, 1993). The significant negative correlative relationship between $\Sigma N/P$ and P might suggest that phytoplankton has a common need for N, P, but the contents and concentrations of N and P needed are not completely consistent. It seems to be that the content of N salt absorbed by phytoplankton was more than that of phosphorus. In addition, the result might suggest that the relevant resource of nutrients were more sufficient.

Table 6 The table of results of correlative analysis of the four nutrient salts of surface water in GWB and its adjacent sea area

Pairs	*r* values	*N*
$\Sigma N/P$	−0.761	34
(NH_4-N)-T	0.216	27
(NO_3-N)T	−0.049	27
(NO_2-N)-T	−0.234	26
$\Sigma N/P$-T	0.094	31
(PO_4-P) -T	0.299	26
(NO_3-N)-(PO_4-P)	−0.362	26

Table 7 was a summation of the table mentioned above, illustrating a general range of concentrations of the four nutrients, the times and stations for the occurrence of highest or lowest concentrations. Another result illustrated that the general trend of the four nutrient concentrations is that the nutrient concentration inside GWB is higher than that outside. The latter trend was similar to the distribution of the corresponding Chl-a concentration. The concentration of nitrates in surface water of this bay during summer seems to be higher than that from the coastal zone of Davis Station, Antarctica. The range of variation was rather big, though the result was similar to that from Pryolz Bay during 1989/1990, (Wang, 1989). The fluctuating range of PO_4-P concentration was bigger than that in the coastal zone of Mawson Station (Bunt, 1957) and Pryolz Bay. The results confirmed further that the nutrient salts of GWB and its adjacent sea area were still in normal conditions. The relevant situation awaits further research.

Table 7 Record table of five main parameters of four nutrient salts in the surface water in GWB and its adjacent area, Antarctica

Nutrient salts	Average/$\mu g \cdot dm^{-3}$	ranges/$\mu g \cdot dm^{-3}$	Station and month of the highest	Station and month of the lowest
NH_4-N	2.13	0.40～9.05	No. 3 1993-12	No. 12,13,14 1993-12
NO_3-N	7.07	1.76～13.88	No. 1 1993-12	No. 7 1994-01
NO_2-N	0.74	0～2.55	No. 7 1993-12	No. 11 1993-12
PO_4-P	1.12	0～2.02	No. 2 1993-12	No. 6,8,9 1993-12

4 CONCLUDING REMARKS / ONCLUSIONS

From the results obtained. the concentrations of NH_4^+-N, NO_3^--N, NO_2^--N and PO_4^--P of the surface water in the area surveyed were about 2.13, 7.07, 0.74 and 1.12 $\mu g \cdot dm^{-3}$, respectively.

Comparing the concentrations of the four salts inside and outside GWB, it's found that they are generally higher inside than outside, which might be related to the water entering the sea from melted snow on the land.

In many situations, the salt concentrations often appeared higher during December, especially salts of NH_4^+-N and NO_3^--N. The months arranged in regard to the ratio ΣN and P were: Dec, Feb, Jan. There are more or less similarities between the variation of heterotrophic bacterial content and this ratio. Their contents, appeared to descend monthly (Chen et al., 1998).

The stations where higher or lower concentrations of some salts occurred were not definitely identical with those when higher or lower concentrations of other salts occurred. Some salt concentrations which were rather low at some stations, might be related to the absorption of phytoplankton. There were significant negative correlative relationships between (NO_2^--N), and CFU $\cdot m^{-3}$, (PO_4^--P) and CFU $\cdot m^{-3}$. It might hint that the NO_2^--N, PO_4^--P and so on degraded by bacteria were absorbed more rapidly by phytoplankton and some autotrophic bacteria.

The water temperature had more positive effects on change in concentration of NH_4^+-N, PO_4^--P and $\Sigma N/P$, i. e., the higher the water temperature, the higher concentration of the three salts, but it has a more negative effect on NO_2^--N.

The result indicated that the area investigated had richer nutrient salts which were in normal conditions.

The results in this paper displayed basically a shallow-water characteristics of the four nutrient salts in the Chinese Great Wall Bay and its adjacent sea area, Antarctica.

References8(abr.)

Cooperators: Zhu Mingyuan, Hong Xuguang, Lu Hao, Yuan Junfeng

南极石油烃污染的自然消除过程研究*

摘 要 1993/1994 年的南极夏季,在中国南极长城站利用可控生态系统对石油烃在南极海域的自然降解进行了研究,同时对研究地区少量的石油污染进行了同步监测.当实验水体加入 10 mg/L[柴油(-30 号):机油(美孚标准机油)比为 1:1(重量比)混合物]的低污染量时,水体溶解 2.776 mg/L,第二天达到 6.816 mg/L 的最高值,以后降解缓慢,第 21 天与环境背景值接近;加入 100 mg/L 的高污染量石油烃时,水体溶解 11.881 mg/L,第三天达到 23.885 mg/L 的最高值,第 58 天时接近环境背景。在研究海域,溶解污染石油烃含量为 2.477 mg/L 可以迅速消除,第 8 天为 0.204 mg/L,接近调查的本底值(0.292 mg/L,同期长城湾平均含量)。

关键词 石油烃 降解 南极长城站

石油烃污染是南极环境污染中最引人注目的问题.大量的石油制品以各种方式进入南极水域,不论是跑、冒、滴、漏,还是严重的泄漏事故,都对南极环境构成巨大的威胁[1,4]。在南极环境条件下,石油烃污染究竟遵循怎样的规律,有什么特点,国内外都未见系统研究报道。在南极夏季,我们对此问题进行了初步的实验研究,本文报道石油烃进入水体后,自然降解过程中总烃量的变化过程。

1 材料及方法

1.1 实验区及监测站的位置

实验区位置选在长城湾内湾口附近,其上方 40 m 左右是长城站油库,离长城站船码头和车库 100 m 左右:监测站在实验区下方,监测少量石油烃入海后的变化情况(图 1)。

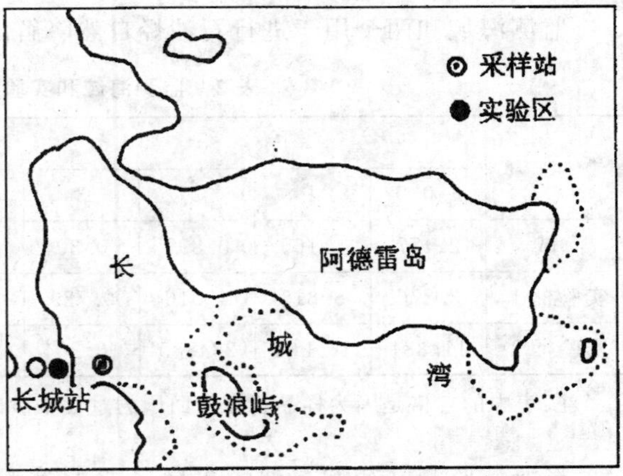

图 1 实验区及采样站位置

1.2 方法

使用未锈蚀的油桶洗涮干净后装海水 150L 备用,水面离桶口 0.5 m,桶口敞开,阳光可射到水面。实验开始时,在实验桶里加入刚配制好的柴油(—30#):机油(美孚标准机油)1:1(重量比)混合物,1 号桶加至 10.00 mg/L,2 号桶加至 100 mg/L;采样前用木棒搅拌样桶中的海水,充分搅匀后避开油膜,均立即采上层水样 1L,并记录温度、风力、风向等指标。所有水样采回后按《海洋监测规范》(国家海洋局 1991 年发布)进行处理,用处理好的正己烷萃取两次,上层萃取液立即用酒精喷灯封入 10 mL 安瓿瓶,冰箱避光保存,带回国内实验室分析。油

* 原文刊于《首都师范大学学报(自然科学版)》,2001,22(3):44-47,本人为第三作者。

含量测定采用紫外分光光度法，油标作参照，石英皿盛样，用255 mL 波长测定。

1993年底准备实验时，实验用桶在洗涤时有少量柴油被逸入海里，从1994年1月1日至2月28日对长城湾洗涤点进行了监测，对柴油进入长城湾后的自然消除进行了观测，取样时取表层无油膜海水1 L，处理方法同上。

2　实验结果

2.1　实验期间主要环境因素变化

实验期间，主要为阴雨天气，平均气温4.7℃，海水平均温度1.4℃，实验桶平均水温3.9℃，平均风速4～5级。情况见表1。

表1　实验期间主要环境因素变化情况

	日　期(日/月)									平均
	1/1	2/1	3/1	4/1	8/1	14/1	21/1	7/2	28/2	
气温(℃)	8	−1	6.6	2.4	8.8	5.9	3.5	5.4	2.8	4.7
气象	晴	阴雪	阴雨	阴	晴	晴	阴雪	阴雨	阴雨	
风速(级)	3～5	5～6	4～5	5～6	2～3	3～4	6～7	4～5	4～5	4～5
槽温(℃)	7.0	0.0	4.7	3.0	7.0	6.0	2.7	3.0	1.5	3.9
海水温(℃)	2.2	0.0	2.0	3.8	3.5	2.0	2.0	−1.0	−1.9	1.4

2.2　石油烃总量的变化过程

监测海域和两个用于进行石油烃自然降解的实验桶的石油烃测定结果见表2.

表2　监测海域和实验桶的石油烃变化　　$mg \cdot L^{-1}$

	天数/d								
	0	1	2	3	4	8	21	37	58
海域	2.477	1.165	0.897	0.30	0.467	0.204	0.292	0.349	0.784
实验桶1	2.276	6.816	4.016	0.729	1.119	0.705	0.425		
实验桶2	11.881	16.264	23.885	16.264	8.569	5.112		0.890	0.607

注：表中的空格是因为样品管在带回国的过程中破碎

将海域监测站初始石油烃含量，1、2号实验桶的加入石油烃量作100%，计算石油烃的残存率，作图2。

从结果分析可见，海域的石油烃消除比较快，第三天即达0.300 mg/L的低值；第四天回升，第八天就达其最低值0.204 mg/L，仅残存8.23%，91.77%已被消除。从图表可见，第四天的回升可能是滩底吸附烃的释放，第21天海域的石油烃含量增加，应归因于湾内总环境的变化，后期石油烃含量增加与本次考察中海域的石油烃污染调查的结果一致(见另文)。

加入实验桶的石油烃首先有蒸发、溶解、吸附的过程，加上油膜的存在，虽然加油后立即搅匀取样，水中的烃含量仍然较低，1号桶按10.00 mg/L加入，烃含量测出值为

2.28 mg/L;2 号桶按 100.00 mg/L 加入,测出值为 11.881 mg/L。

第 1 天,实验桶中的烃含量增加,1 号桶升到其最高值,达 6.816 mg/L;2 号桶的烃含量也增加到 16.264 mg/L。

第 2 天开始,1 号桶的烃含量逐步下降、其间虽有小的回升,但总变化趋势很清楚:第 21 天达最低值 0.425 mg/L,与环境背景值接近。从图表还可见,1 号桶的降解主要在前三天,以后的烃含量变化不大。该桶第 3 天起就未再见油膜。

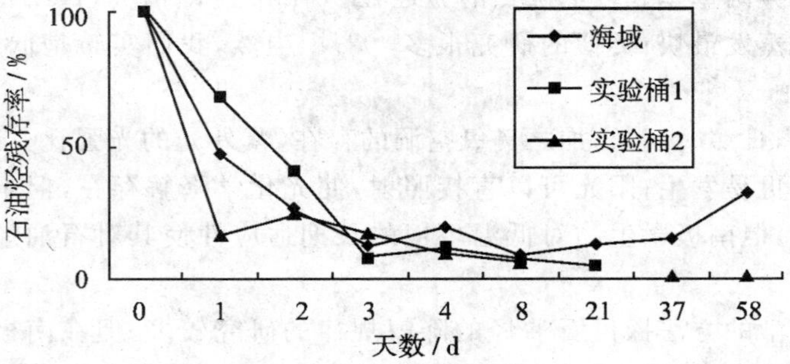

图 2 南极石油烃降解过程分析

2 号桶初始石油烃加入量大,第 2 天水中烃含量达最高值,为 23.885 mg/L,以后稳步下降。与加入量相比,第 1 天仅降解 21.26%,第 4 天残存 8.57 mg/L,即前 4 天降解了 64.12%,以后的降解比较缓慢,到实验结束时,2 号桶内烃残存量为 0.607 mg/L,实验桶中 97.46%的石油烃已被降解。该桶从第 15 天起未见油膜。

3 讨论

石油烃污染是南极环境污染的重要因素,进入南极海域的石油烃在自然条件下如何变化、消除是被普遍关心的问题。石油烃进入水域后,要经历扩散、蒸发、溶解、乳化、海气直接交换、化学更换、沥青块形成和沉积等一系列消散过程,每一步骤都受到环境条件的影响。由于南极具有独特的环境条件和生物组成,石油烃进入南极海域后的降解速度、降解过程也有其特点,我们的实验目的在于研究南极特殊环境和生物条件下石油烃消除的一般规律。

从油污染的消散监测结果可以看到,在长城湾的自然条件下,海水交换、稀释是非常重要的。监测的第 1 天即降低了 52.97%,第 4 天烃含量已降到 0.47 mg/L,降低了 81.13%,第 8 天降至 0.20 mg/L,此时长城湾内的平均烃含量为 0.29 mg/L(袁峻峰,另见文),可以认为污染已基本消除,而此时实验桶中的含烃量仍较高,说明海水的交换在烃的消散中占有重要的地位。

实验开始的前 3 天,两个桶的表面都有油膜,1 号桶第 4 天油膜消失,2 号桶第 6 天消失,实验过程中,未见沥青块形成。

不同浓度的实验桶中的烃消除速率差异很大,浓度越高,消除越慢,尤其是前 4 天、前 8 天,相差 1 倍左右;但最终都能完全消除,使烃含量与环境非常接近。所以,除了海水交换外,理化和生物过程仍能清除污染石油烃。

由于石油烃在海水中的溶解度低,不超过 12～30%[2],与其成分的相对分子质量成反比，在极地低温条件下更难溶解，所以，当加入量大时，水中石油烃的含量较难测准确，用海水中烃含量来判断烃消解的详细过程有一定难度，在研究中应该注意。用烃组分的变化来进行研究，其结果能更好地反映南极海水对石油烃的降解过程。

在我们的实验条件下，由于水面离桶口近 0.5 m，桶的直径近 0.9 m，所以桶内水面的风速很小：影响石油烃蒸发的主要环境因子是温度。低的环境温度和实验水温，使低分子的烃蒸发较慢，第 4 天才完全消除；而在温暖的气候条件下，将 1.04L 的原油加入水池，其中的低分子成分异丙基苯和较低沸点的芳烃在 90 min 内即消失（Harrism，1975，见[2]），一般以前两小时蒸发最快；类似的研究很多[3,4,6]。当然，我们实验是低起始量，低分子的烃也很快消失了。

南极的夏季，日照时间长，加之臭氧空洞的存在，紫外光的光强比其他地区高得多，石油烃的化学更换更易发生；阳光可以直接照射，使光化学降解存在，但本次研究未进行探讨。虽然水温低，但南极微生物对低温环境的长期适应性使其对石油烃的降解作用不容忽视有关研究。

用实验生态学的方法探讨石油烃的降解规律的研究较多，但在南极环境条件下进行的研究却未见报道，本研究是一次新的尝试，希望能对以后的研究提供参考。

在分析结果时应该注意，实验是在受控生态系统中进行的，与海水的实际条件有一定差异，这些差异在多大程度上影响实验的结果，尚难定论，需要进一步研究。

参考文献 6 篇（略）

（合作者：袁俊峰　李永祺　吴宝铃）

THE DEGRADATION OF OIL IN THE ANTARCTIC SEA

(ABSTRACT)

Abstract During the summer of Antarctic in 1993-1994. We have used the controlled ecosystem to study the degradation of oil dispel in Greatwall Station, Antarctic. When added 10 mg/L mixed oil [diesel oil, 30 # : machine oil = 1 : 1, weight], the dissolved parts is 2.78 mg/L and the highest value was reached at the 2nd day, is 6.816 mg/L. Then the content of oil is decline slowly, and reached the background value at the 8th day(0.29 mg/L, the average value of Greatwall Bay, at the same time). When added 100 mg/L mixed oil, the dissolved part is 11.88 mg/L, the highest value was reached at the 3rd day, is 23.89 mg/L, and the background was neared at 58th day. By the research area, a little oil was washed into the sea, the content of oil was 2.48 mg/L, and to 0.204 mg/L at the 8th day, reached the background value.

Key words Petroleum Hydrocarbon, Dispel, Greatwall Bay, Antarctic.

第四篇

生物活性物质

BIOACTIVE SUBSTANCES

D$_1$ 来克丁学(LECTINOLOGY)

外源凝集素——水产动物御敌的有力兵器*

摘　要　外源凝集素是非免疫起源的热敏蛋白质或糖蛋白,由于其多价构型及对特定细胞多糖具结合亲和力,它们选择凝集某些细胞(包括脊椎动物血球及一些病原微生物)和沉淀某些复杂碳水化合物;其凝集活动常被单糖所抑制,有的却需某些二糖、三糖或多糖。外源凝集素分离自广泛的低等和较高等的生物。一些地区水产养殖业受挫,因素较多,但对发病机理认识之缺乏和调动养殖对象本身防病积极性之不足,是其重要原因。外源凝集素可在诸如凝集外来细胞、调理作用等的防御活动中扮演识别因子,协助一些生理活性物发挥作用,参与吞噬、胞囊、凝固、止血及创伤修复等作用。在生物学、细胞学、生物化学、医学和食品科技上有广泛用途。迄今外源凝集素的许多性能并未全清,在自然界中的精细生理作用仍是个谜。本文论述外源凝集素的定义、性能及用途,以提请有关人士重视。

关键词　外源凝集素　凝集活动　水产动物　防御机制　外来入侵者

被称作为外源凝集素的 Lectins 是凝集素中的一大类具独特生理活物的物质。无脊椎动物(如甲壳类动物)血淋巴中 Lectin 的生成,也许与血浆中的凝集素原(agglutinogen)等有关。除血清中有一定 Lectin 外,血细胞的 Lectin 主要在透明细胞(hyaline cell)里或其上。进行凝集活动是甲壳纲、头足纲中许多动物血淋巴透明细胞的主要机能之一[1]。在动物对外来入侵活动进行的识别、防御、凝集、吞噬、胞囊及其随后而来的创伤修复等作用中,Lectin 协同其他凝集素、调理素、内毒素、胞溶素、溶菌酶、凝固酶、抗菌蛋白等发挥作用,其中的凝集活动 agglutination activity 占有十分重要的地位[2,3]。尽管外来入侵(生物的和非生物的)与主体间的作用与反作用十分频繁、复杂,但一旦入侵者采取灵活战略、策略,顽强攻击,冲破警戒,主体的躲避、防御等战线受到严重威胁,而使其难以健康、正常生存之时,便是病害具备了爆发条件,有关生物,包括野生的,尤其是集约、半集约化养殖的,招致大批感染,疾病流行之日就已来临[4],后者恰恰又密切关系着养殖业和国计民生。这迫使人们不得不从根本上去加深对动物的,包括 Lectin 及其凝集作用在内的防御体系和疾病发生机理的认识,进而调动其机体的积极作用,以使养殖业建立在越来越科学化的基础上。

---

* 原文刊于《黄渤海海洋》,1995,12(3):61-71。

国家自然科学基金资助项目(39370539)。

本文论述 Lectins 的来源、性质、功能和应用等方面,以提请有关人士,特别是水产养殖界同仁对 Lectin 的重视,并拓宽它们的应用途径。

1. 定义

凝集素中一大类动植物乃至微生物起源的多价非免疫蛋白或糖蛋白,一般被认作为外源凝集素——Lectins[5]。它是 Boyd 于 1963 年提出,后被大家认可的。在此时前后所发现的血细胞凝集素 hemagglutinin(其中有许多是植物凝集素 phytagglutinin,又名植物血细胞凝集素 phytahemagglutinin),一般均纳入 Lectin 范畴之中。

Lectins 这样的凝集素最先发现于植物中,尤其是豆科植物种子中;其后从植物的根、叶和树皮中均有发现。从许多无脊椎动物、低等脊椎动物的体液、血淋巴及动植物浸汁、病毒、细菌等微生物中都发现了 Lectins。根据性质和功能,可定义 Lectins 是一大类对特定细胞多糖具结合亲和力的、能选择凝集脊椎动物一些血球和某些微生物细胞的、多价构型的热敏蛋白质或糖蛋白复合物[6]。

迄今,以商品形式面世的 Lectin 已有 50 余类,其中大多仍提炼自植物[7]。随着 Lectin 资源的开发、提取技术的进步和应用的日益扩大,Lectins 家族将日益壮大,新的 Lectin 将不断面市。

2. 研究简史

Lectins 的最早研究始于 1888 年,Stimark 发现了蓖麻子白朊 ricin 和与红豆碱相关的 abrin,1919 年,Summer 从刀豆(*Canavalia eusiformis*)中获取,并于 1936 年确认为第一个结晶形的血细胞凝集素-刀豆球朊 A(*Cancanavalin A*)。1964 年,从巨大石房蛤(*Saxidomusgigantens*)的盐浸汁中提取出专性凝集表型为 A_1 和 A_1B 的人红细胞的 Lectin[8]。1965 年,Allen 及其合作者研究了大量植物浸汁对人血球的凝集活动[9]。

从无脊椎动物中提取 Lectins 的努力可上溯到 1903 年。Noguchi 从一种鲎(*Limuluspolyphemus*)的体腔液中证实了血细胞凝集素的存在[6]。自 70 年代以来,有关无脊椎动物 Lectin 的研究有了更长足的进步。Pauley 主要从事腹足纲动物 Lectin 的研究[10];后来,Yeaton(1981)列举出截至那时所研究过发现有 Lectins 的 103 种无脊椎动物[6],这些动物分布在 9 个门、亚门或纲中,并进而推断所有无脊椎动物,在其生活史的不同阶段,可能都有 Lectin[6]。1983 年,Ratclitte 等对来自 7 个目的 25 种昆虫,研究了它们的血细胞产生的凝集活动 Hemagglutinating activity 简写为 HA),认为其凝集素具识别因子 recognition factors 作用[12]。Fisher 等(1991)对六种海产软体动物(牡蛎、蛤、乌贼、章鱼、鱿鱼和海兔)Lectins 作过研究。在所试范围内,发现大约 68.1×10^{-2} 的菌株和 7 种脊椎动物红细胞被凝集[29]。表 1 列举了近十几年来被研究过 Lectin 的动物名录及其凝集特异性等。

表 1　一些动物的外源凝集素及其特效性

动物种名	外源凝集素所取部位名称	特效的糖类等	其他	引证文献号
长红猎蝽 *Rhodnius prolixus*	嗉囊、中肠、血淋巴 Lectin	N-乙酰-D-甘露糖胺 N-乙酰-D-半乳糖胺 一半乳糖，一粒糖		[11]
沙漠蝗 *Schistococca gzeagria*	无细胞血淋巴、蝗虫凝集素		对抗布氏锥体虫及一种利什曼虫	[12,13]
美洲大蠊 *Periplaneta america*	凝集素	土冉糖、海藻糖		[12]
巴氏白岭 *Phlebatomus papatasi*	前、中、后肠；凝集素	土冉糖、海藻糖		[16]
Megabalanus volcano	血淋巴，MVA＋1MVA2	D-半乳糖		[14,15]
M. rosa	体腔液；BRA-2,BRA-3		具调理性明显	[17]
红头丽蝇 *Calliphora erythrocephala*	凝集素	甘露糖		[18]
灰翅夜蛾 *Spodoptera exigua*	幼虫血淋巴，凝集素	真菌细胞表面半乳糖等		[19,22]
刺舌蝇 *Glossina morsitans*	血淋巴，中、后肠；似 Lectin 的凝集素	D(＋)-葡糖胺		[19,20]
须舌蝇 *G. palpalis*	凝集素	对糖有广泛特效性		[19]
一种舌蝇 *G. tuchinoides*	血淋巴，非特效			[23]
另一种舌蝇 *G. austeri*	血淋巴，非特效		具异质性	[23]
一种拟菊海鞘 *Batrylloides leachii*	血淋巴，HA-2	乳糖		[24]
黄蛞蝓 *Limax flaus*	LFA	唾液酸		[25,26]
陆地蜗牛 *IIelix Pomatia*	Lectin			[27]
褐云玛瑙螺 *Achatina fulica*	血淋巴，ATNH	唾液酸		[28,29]
大型白蛤 *Calyptogena magnifica* 一种蛸 *Octopus maya* 莱氏拟乌贼 *Sepitoeuthis lessoniana* 一种乌贼 *Sepia officinalis* 加州海兔 *Aplysia californica*	血清，Lectin 血清，Lectin 血清，Lectin 血清，Lectin 血清，Lectin	对 68.1×10^{-2} 的所试菌株和 7 种脊椎动物血有高度特效性		[30]

(续表)

动物种名	外源凝集素所取部位名称	特效的糖类等	其他	引证文献号
贻贝 *Mytilus edulis*	无细胞血清,NAN	N-乙酰神经氨酸	增强对酵母的吞噬作用	[30~33]
一种沼虾 *Macrobrachium rosenbergii*	血清、复合 Lectins	神经氨酸		[34]
美洲龙虾 *Homarus americanus*	血淋巴;Lag1,Lag2	N-乙酰神经氨酸 N-乙酰半乳糖		[35,36] [37,38,43]
一种鲎 *Limulus polyphemus*	血淋巴;LPA	唾液酸		[38,39]
印度鲎 *Carcinoscorpius rotundacaula*	血淋巴 Carcinoseorpin	唾液酸		[40,43~45]
麻蝇 *Sarcophaga* sp.	血淋巴		在变态期有助于机体清除不必要的细胞和组织碎片	[31,41]
家蚕 *Bombyx mori*	血淋巴			[46]
一种黄道蟹 *Cancer antennarius*	血清	D-乙酰神经氨酸		[24]
弗及尼亚巨蛎 *Crassostra virginica*	血清		凝集霍乱弧菌	[5]
一种鲑 *Oncorhynchus tshawytscha*	卵	凝集人B型血、鱼红细胞、抵制四种鱼病源菌、被D-半乳糖,L-鼠李糖特效抑制		[42]

3. Lectin 的提取和纯化法

用常规的蛋白质化学技术,如盐的分馏、离子交换,或不同吸附剂的层析技术来提取或纯化 Lectin。下面举出 9 个 Lectins 的基本提取实例。

(1)琼脂糖-胎球蛋白和琼脂糖-多聚乙酰神经氨酸柱上的硫酸铵沉淀、制备淀粉成块电泳、凝胶过滤及亲和色谱的序列结合。从美洲龙虾血淋巴中分离出 Lag[34~35,42]。

(2)在不溶性聚亮氨酰 A 型血物质的柱子上吸附粗提物,再用 N-乙酰-D-半乳糖胺洗脱。从一种马草(*Dolichos biflours*)种子和陆地蜗牛(*Helix pomatia*)中提取出对 A 型血特效的 Lectin[9]。

(3)将 A 型血物质偶联到琼脂糖 B 上以吸收植物 *Vicia cracca* 的 Lectin,该 Lectin 对 A 型血也特效[9]。

(4)用 trifunctional fucosyl 染料专门沉淀 *Lotus tetragonolobus* 种子提浸汁，再以离子交换树脂移去染料而得大量对 *L*-墨角藻糖有特效的 Lectin[9]。

(5)交联葡萄糖凝胶吸附、*D*-葡萄糖或低 pH 缓冲液洗脱，纯化出刀豆球朊 A 和扁豆、豌豆 I,ectins[9]。

(6)用偶联牛下颌粘蛋白的琼脂糖做亲和色谱，从一种黄道蟹纯化出对 9-0-7 或 4-0-乙酰唾液酸特效的 Lectin，或用上述琼脂糖装 Econo 柱也纯化出该 Lectin[9]。

(7)聚乙(烯)二醇 8 000(PEG)测定束缚 Lectin[43]。

(8)用有 0.02 mol $CaCl_2$ 的 Affi-Gel 卵白朊柱束缚，T′ris-HCl 缓冲盐水溶液洗涤几遍，再用含 0.002 mol 的 EDTA 的 Tris-HCl 缓冲盐水溶液从柱中移出，得凝集素片段。灰翅花蛾幼虫的血淋巴 Lectin 如此提炼[21]。

(9)制备盘式-电泳纯化出贻贝血清凝集素，再用聚丙烯酰胺凝胶电泳(PAGE)分清[31]。

4. Lectins 的性质

4.1　化学性质

无脊椎动物的 Lectins 是具有生物学功能的主要血淋巴蛋白[6]。至今发现的 Lectins 仅刀豆球朊 A 是无糖的，其余均是糖蛋白。糖以共价键形式结合进 Lectin 中，糖的种类主要包括有甘露糖、氨基葡萄糖、半乳糖，而少见木糖，阿拉伯糖。这些糖的种类使 Lectin 表达出是具有动物糖蛋白的典型结构物。蛋白质部分主要是天冬氨酸、丝氨酸和苏氨酸，它们占其氨基酸总量的 30×10^{-2}，少见含硫氨基酸，因而 Lectins 又显出许多植物蛋白质特性，以上这些成分还有多种构型。Lectins 的金属离子常是 Ca^{2+} 和 Mg^{2+}，这是许多进行糖的结合或凝集活动所需的[9]。

Lectins 的相对分子质量变化范围大，如麦胚凝集素(WGA)的相对分子质量是 26 000，利马豆的是 369 000，鲎的是 400 000。即使同一 Lectin，相对分子质量也并不一定不变。如刀豆球朊 A，其相对分子质量范围是 55 000～100 000。造成相对分子质量差异大的原因可能是因为许多 Lectins 由亚单位构成，且进行着缔合一解离反应。亚单位本身可能还由多肽片段组成，片段之间可互补、组装，此方面的最明显一例是刀豆球朊 A[9]。

此外，还有异外源凝集素 isolectins，如 LAg1 和 LAg2，它们的结构不尽相同，功能也不一，却同时来自美洲龙虾。

血细胞表面的 Lectins 有多价构型，结合糖类的能力不一，它们主要结合单糖、寡糖，特别沉积多糖和糖蛋白。已知有 7 种糖及其变型物被凝集，它们是 *L*-岩藻糖、*D*-半乳糖、*N*-乙酰-*D* 半乳糖胺、*D*-葡萄糖、*N*-乙酰-*D*-葡糖胺、甘露糖和 *N*-乙酰(或羟乙酰)神经氨酸[6]。

一个 Lectin 分子必须至少有两个或两个以上偶数的结合点。如扁豆 Lectin 有两个点用于甲基 α-D-吡喃半乳糖甙；葡萄园蜗牛的 Lectin 有 6 个点用于 N-乙酰-D-吡喃半乳糖等等。结合点位的偶数现象，再加上结构中糖与蛋白两部分的对称性，是许多 Lectins 的两个共性。

Lectins 结合糖类的专一性范围不一。少数 Lectins 的范围相当窄，如植物 *Vicia cracca* 的 Lectin 对 A 型血的凝集特效，而 *Vicia graminea* 的却对 N 型(原文如此)血特效，尽管两种植物同为一属。一些专一性强的 Lectins 优先凝红细胞，并与相应血型可溶物形成沉淀。特效性强而窄的 Lectin 较少，如 Allen 等(1965)筛选的用于人血凝集试验的 2 663 种植物浸汁中，只 90 种是特效性强的，而非特效的 Lectins 却有 711 种。这表明非特效性 Lectin 的资源更为丰富。但大批对人血非特效性的 Lectin 虽然有较广的凝集对象，有时却也发现其中的少数对结合某些糖有高度专一性。刀豆球朊 A，大豆凝集素和 ricin 是 3 例。

Lectins 特效性的类型不断丰富，除了以对红细胞起作用来分外，还有专门凝集肉瘤一150 细胞、白细胞或其他微生物细胞者等等。

多数 Lectins 的凝集活动并非必然地与血型决定簇(determinant)相关，但必须借助结合糖的受体而与细胞反应，而细胞表面上的这些受体是因血型决定簇变化的。

被结合的 Lectins 之数量与细胞的凝集能力之间不一定有直接相关关系，尽管这些细胞可分作正常细胞、癌变细胞和正在解朊着的细胞等。细胞表面上的糖受体相对分布状态对 Lectin 的凝集作用在某种程度上有决定意义，而这些分布状态在上述几类细胞表面上恰恰是不一样的。

用蛋白酶，如胰蛋白酶、链霉蛋白酶等温和处理 Lectin，可使其凝集活动的敏感性得到提高。一些添加剂、金属阳离子也影响 Lectin 活动[6]。

Lectins 进行的凝集反应常被单糖所抑制，被抑制的敏感性差别大。某些典型特效 Lectins 易被相应血型物质中部分糖类所抑制，如 A 型血抗原特效的 Lectin 被 N-乙酰-D-半乳糖所抑制；H(O)型特效的 Lectin 被 *L*-岩藻糖所抑制。

4.2　物理学性质

温度：一般 Lectin 在冰冻、低温状态时，性质较为稳定；多数 Lectins 是热敏的，不耐热。巨大石房蛤的 Lectin 在 70℃时即失活，也不可被透析[8]；美洲巨蛎的血清，其细菌凝集作用(Bacterial agglutination)简写为 BA(下同)，在 80℃时被抑制[5]；沙漠蝗、蟑螂和美洲大蠊的凝集活动用 65℃的温度处理 30 即失活[19]。

盐度：当盐度高过 24～30 时，美洲巨蛎血清的 BA 增强[5]。

4.3　生物学性质

4.3.1　不连续性

主要包括下列四点：

1. 同一个体的不同生长阶段、不同生境，其 Lectins 具高度变异性，如弗吉尼亚巨蛎的血清对霍乱弧菌的凝集活动因巨蛎的生长阶段和生境不同而异。野外的霍乱弧菌生态学分布受制于它们的宿主的血清凝集素活动的干扰[5]。

2. 个体不同，使 Lectins 的 BA 效价水平不同。一个属的不同种，Lectin 不一定相同；有的种有，有的种只在某一阶段有。

3. 同一群体中的个体之间，其 Lectin 浓度变化不一定与年龄、性别相关。如软体动物、节足动物或棘皮动物中就有例。它们的 Lectin 有时或缺，相关的几个因子可能是季节、健康状况、生长阶段(如蜕皮)等。如棘皮动物 *Anthocidaris assipina*(紫海胆)的 Lectin 效价就有季节变化，4 月、5 月间，全有；7 月、8 月间大多数有；11 月则只少数有[6]。

4. 同一研究对象的研究结果不一，并非误差所致，但又查不出何因[6]。

4.3.2　可诱导性

一些昆虫不必经注射死或活的细菌入体内，仅借皮下组织针刺伤体壁就可诱导出

Lectin。如一种麻蝇*Sarcophaffe preagrina* 的 Lectin 即可依此而导出,并能活化小耗子的巨噬细胞[40]。

4.3.3　调理性

许多研究表明,Lectin 的凝集活动有高度的调理作用(Opsonic action),促使血细胞更易吞噬外来颗粒。如贻贝 *Mytilus edulis* 的一种 Lectin 能使其血细胞增长对酵母细胞的吞噬作用,促使入侵者易被主体确认[23]。如灰翅夜蛾幼体具有血凝活性的血清及对半乳糖有特效的纯化凝集素对暴露出半乳糖残基的真菌细胞具有调理作用,使后者更易被主体认出,促使了吞噬、摄入活动[18]。一种拟菊海鞘 *Botrylloides leachii* 的血淋巴中对乳糖有特效的 Lectin 即 HA-2 能调理鼠巨噬细胞,使颗粒物更有效地被摄取[23]。

5. Lectin 的功能

5.1　有助机体 Ca^{2+}、糖输运、维系共生功能

在机体的 Ca^{2+} 和糖输运中起作用。在贝类中,为它们同一些植物共生起维系作用[14]。

5.2　共轭接合功能

与其他糖蛋白或糖发生共轭接合,就如高等动物中抗体与抗原体系中的相互作用[43]。

5.3　清除杂物功能

在甲壳动物变态期间,参与清除机体不必要的细胞、组织片段或残余物[18]。

5.4　识别-防御机制,参与其他活动功能

由于无脊椎动物缺乏脊椎动物那样的抗微生物体系,仅有准免疫(quasi immunity)活动(包括 Lectins 在内的活性物质所起的作用)。这方面的例子很多,如弗及尼亚巨蛎血清明显地凝集霍乱弧菌,Lectin 可充作识别分子(recognition molecuies),识别自身和非自身物质,包括外来入侵的病原菌,将它们拒之于门外,在细胞表面受体上确认,区分它们,进行一系列凝集活动。另外协助其他生理活性物质,参与止血、凝固、胞囊、吞噬直至创伤修复等一系列作用,以保障机体健康,正常生活[1,2,5,6,12]。

6. Lectins 的应用

(1)用于细胞表面研究,为结构分析提供材料,用于分离细胞,包括病原菌和多糖接合子 glycoconjugates 的亚细胞定位,含糖聚合物的分离,纯化等[6,24]。

(2)促有丝分裂剂,用于解释休止细胞转化为活跃生长细胞过程中的生化活动[41]。

(3)一种模式与糖类的反应可充作为蛋白质-碳水化合物反应的一般模式之一[24]。

(4)作为肿瘤细胞的抑制剂和特效药。对那些已沾染了致癌病毒,化学致癌物或自我致癌的哺乳动物的组织培养细胞[9],某些 Lectins 能优先凝集它们[9]。

(5)加深对遗传学的认识,促进细胞遗传学的扩展,增进对染色体异常和人类疾患间关系的认识[9]。

(6)提供一系列有用的探针(包括膜探针)。与 Pneumococcal 细菌上糖类一特效抗体作比较,为一些糖的确认,如多糖及糖蛋白、甘醇类脂提供联系。对 Lectins 一级、二级、三级结构的研究及理化性质的研究,将导致对糖特效性的分子基础和细胞功能机理的深入理解,终将阐明它在自然界中的生理作用[19,33]。

(7)加深对水产动物防病治病认识,发展水产医学。对水产动物 Lectins 生理功能和防御机制的理解,为防病治病可能提供一条有效途径,为水产医学的发展拓展道路。

(8)促进食品构成科学化,来自丰富资源的 Lectins 长期吸引着食品科技人员和营养学家的关注。越来越多的科学化食品势将考虑 Lectins 的用途[9]。

迄今,虽然有关 Lectins 的研究有了较大的发展,但依然存在大量课题有待人们去探讨。如 Lectins 的许多精细生理作用尚未阐明,尤其是在自然界中的地位尚不明确。好在 Lectins 留给人们许多值得研究的途径和吸引力,如有效性、巨大的多样性、易分离性和有趣而奇异的长处,随着时间的推移和人们认识的提高,Lectins 的科技之花,包括在水产业上的应用,将会盛开。

参考文献 46 篇(略)

(合作者:孙丕喜　宋庆云)

LECTINS, A POWERFUL DEFENSIVE WEAPON OF AQUATIC ANIMALS

(ABSTRACT)

Abstract Lectins are heat-sensitive proteins of glycoproteins of non-immune origin. They can selectively agglutinate some cells including vertebrate erythrocytes and pathogenetic microbes because of their polyvalent configuration and binding affinity for specific cell surface polysaccharides.

Lectins are isolated from a wide variety of biological resources such as seaweeds, sponges, molluscs, crustaceans, fish eggs, body fluids of invertebrates and lower vertebrates, etc.

Huge scale intensive and semi-intensive aquaculture including marine culture has suffered setbacks in some places. Although it is true, there are many causative factors for this, yet there are two important aspects, i. e. lack of knowledge of disease mechanism and insufficient mobilization of the positive factors of the cultured species, that should not be neglected.

Lectins and their agglutination activities act by recognizing and differentiation between self and non-self materials, guard against foreign invasion, repair wounds, assist in phagocytosis, encapulation, opsonization, and so on. Besides, lectins are widely useful in biology, cytology, biochemistry, medicine, food science and technology. But their precise physiological role in nature is unknown.

This article discusses definition, brief research history, properties and uses of lectins. It hoped that people, especially those aquaculturists, will pay more attention to the investigation and application of lectins.

Key words Lectins, Agglutination, Aquatic Animals, Defensive Mechanisms, Foreign Invasions

中国的来克丁研究*

摘　要　认为凝集素、选择素或外源凝集素等中文名称不能概括“Lectin”这一名称的原有含意，且可能混淆“凝集素”(agglutinin)与来克丁(Lectin)之间的差别，可能妨碍中国来克丁学(Lectinology)的研究和发展。建议直接从英文名词，即Lectin(或拉丁文名词Legere)译成中文名称《来克丁》。文章评论了20世纪70年代末以来中国来克丁研究的状况，中国学者分离、纯化来克丁的方法.所研究的来克丁的生物学和化学性质，来克丁在中国的应用，也包括国内市场上国产来克丁的一些情况。着重描述了一些研究得较透的来克丁特性、功能和用途。对我国来克丁，包括海洋生物来克丁将来的研究提出了一些建议。

关键词　来克丁 Lectin 译称　凝集素　性状　分离纯化技术　应用　研究方向

1. 提出以“来克丁”一词译称“Lectin”供商榷

凝集素(agglutinin)主要指一些能与体内或体外同种抗原结合后，就与颗粒抗原(如细菌等)凝集的抗体[1]，此外，以前也有将除抗体外的其他能凝集颗粒的物质如Lectin归入凝集素中的。这可能是由于凝集素这类抗体广泛活动于血清、免疫反应的凝集作用中，且被发现得较早的缘故。“Lectin”一词只是到20世纪50年代才被用于血凝素上。至那时，Lectin发现于少数植物中，并被认识到其主要起凝集作用。制成试剂或药品则是后来的事。因此人们沿用凝集素这一名称于Lectin上。随着Lectin研究和应用的扩展，尽管凝集素研究也在发展，但Lectin在种类、数量和用途上大有超越凝集素之势。国际间有关的权威性医学、生物学检索刊物多年来一直把Lectin单独列成一个重要门类。最新资料表明，凝集素不可能将所有Lectin包括在其名下。一般最多将植物来源的Lectin看作与它有关，或者说凝集素与Lectin有交汇之处。而且在Lectin条目之下，常常聚集有凝集素的信息。迄今Lectin系列中的试剂或药品已达51种以上，各种品牌约276种。而单独列出的凝集素却不多，也没对凝集素再作新的解释[2]。综合散见于中外不同资料对这两者的释义，可以看到它们间有些许共通、交汇，但又确实存在许多相异之处，这主要在于以下几点：

表1　凝集素与Lectin的差异

、凝集素	Lectin
由凝集原刺激而生 (英汉辞海)*	非免疫起源

* 原文刊于《广西科学》，1997，4(3)：161-169.
国家自然科学基金项目(项目编号 39370539)

(续表)

凝集素	Lectin
与细胞表面特殊抗原结合产生凝集反应,(英汉生物化学词典、英汉生命科学辞典)	凝集细胞和/或沉淀复杂糖类,有些具糖专一性,或血型特异性;凝集作用常被合适单糖或寡糖抑制,接合作用不改变其专一性[2]。(the Merck Index)
抗体(英汉生物化学词典)	非抗体,但有的具类似抗体的活性(英汉生物化学词典、英汉辞海)
来自正常或免疫血清(英汉生命科学词典)	广泛分布于各级各类各种生物体[2]
主要用于血清学、免疫学[4]	用途广泛,如正常和病理条件的组织化学研究、淋巴细胞亚群研究,细胞和其他颗粒的分级分离、血球分型、红细胞多重凝集反应,刺激淋巴细胞有丝分裂、医疗、卫生诊断等[2,4]

* 小括号、中括号中的内容系所引用文献名称或文后所附文献号,下同。

由上可见,大部分 Lectin 不但已独立于原先意义中的凝集素,而且已发展成崭新的分支学科,即 Lectinology 或者 Palaimmunology(副免疫学),在这个学科之下,已分出 17 项专题研究栏目,比起现代凝集素研究,有了大大的拓展。

国内目前除上述将 Lectin 称作为凝集素外[3,98],还有称之为"选择素"或"外源凝集素"的[4,5]。"选择"两字虽保留了它的拉丁文 Legere 含意,也有别于仅仅"凝集"之意,但未能清楚表达出 Lectin 的特性和功能。近几年来文献中已出现了与汉语巧合的选择素(selectin)或叫选凝素的物质名称[6,7],(恕在此不论及 selectin 这一名称的起源和含意),一些文章将它归入 Lectins 之中。因而仅将 Lectin 称作选择素将越来越限制 Lectinology 的发展。Lectins 在功能上除了可分作外源的和相对于它们的内源的(endogenous)外,更有庞大的 Lectins 家族。鉴于此,可否像将 chitin 译成"几丁质""那样,将"Lectin"译成"来克丁"? 这样做的好处有以下几点:①与英语同音,保持了它的原始发音,国际通用;②避免并廓清了它的不同汉语译意而得以统一称呼;③有别于一般意义的凝集素,减免了为在"凝集素"之名外添加任何修饰性字眼的烦恼。凝集素、来克丁均可自由发展;④为 Lectinology 在中国的发展留下更广阔天地。

暂时困难之处在于不习惯。似有添乱之嫌。因为多年来,Lectins 的中文资料中有许多是借用"凝集素"之名的,而且存在相当多的含混不清,不易改去。不必强求过去的东西。"来克丁"之称一旦为大家认可,读起来也就会朗朗上口,逐步推广。当否,请商榷。

2. 我国来克丁学(Lectinology)研究现状

我国学者对来克丁的重视,始于 20 世纪 70 年代末和 80 年代初。早期的文章主要是传递国际间来克丁的一些研究信息,介绍来克丁学的研究动态和方法。1981 年开始有我国自己的研究报道[8,9]。对有关基础性研究等资料不完全统计显示,80 年代前后至今发表的文章(中文)至少有 110 多篇。文章按年代分布如下:1979 年,1 篇;1980 年至 1989 年,50 篇;1990 年至 1996 年约 60 篇,其中 1983 年、1991 年、1995 年和 1996 年发表的文章最多,各为 11 篇。其次为 1993 年和 1994 年,各为 9 篇。至截稿时,我国来克丁研究的主要成果表现在以下四个方面。

2.1 来克丁的分离、纯化技术研究

我国的来克丁分离纯化技术最早是引用外国的，继而则改进和发展出具有我国风格的一些方法，包括吸附剂、层析柱成分的改进等等。方法大体上有两种，即分部分离蛋白质法和亲和层析法。迄今，纯粹用分部分离蛋白质法的已不多见(如丁邦裕等(1991)用pH值为2的酸性水抽提江苏豌豆种子粉后，用$(NH_4)_2SO_4$分级沉淀，得PHA的一种同功来克丁即PHA-p[10])。除以甲壳素(几丁质)为载体的亲和层析简便分离纯化出芝麻来克丁(SIL)[11]和用葡聚糖凝胶为载体分出兵豆来克丁(LCA)外[112]，不少人用Sepharose 4B为载体，如舍蝇来克丁(MDL)即是这样分出的[13]。受体或配体种类有所变化，如用鸡卵类粘蛋白(OM)-Sephaxose 4B亲和层析分离出番茄来克丁(LEL)[14]、Sepharose 4B-甘露糖等分离纯化出大蒜来克丁(*Allium sativura Lectin*，即ASL)[15]。用Sepharose-胎球蛋白从猪精子和精子浆中分纯出猪精子来克丁BSL(Boar spesm Lectin)，或经DEAE-纤维素柱盐离子浓度梯度洗脱后再将猪胃粘蛋白配上而纯化出一种菜豆来克丁(PLA)[17]和桑寄生来克丁(LPL)[17]。还用Sepharose 4B与猪胃黏蛋白又纯化出野花生豆来克丁CML(*clrotalaria mwronata Lectin*)[19]。此外，Sepharose 6B-鼠李糖柱还分离纯化了蓖麻蚕血淋巴来克丁(*Philosania cynthia Lectin* 即PCL和PCL_1)[20,21,23]。有的以Sephadex-G50固相化亲和吸附为基础作荧光标记来分纯一些来克丁。相当多的来克丁的分离纯化是用分部分离蛋白质法和层析术合用或联用而实现的，如大蒜来克丁(ASL)等[15]和半夏来克丁[9,24]即是。还可将载体Sepharose 4B和Sephadex-G100或Sepharose 4B和Sephadex-G200合用于来克丁提纯。前者如蓖麻来克丁(RCA)[25,26]，后者如CML和PCL[19,20]。也有以葡聚糖凝胶一对氨基苯砜乙基(ABSE)-交联琼脂组成的亲和吸附剂分出三齿草藤来克丁(VBL)[27]；用猪甲状腺球蛋白代替葡聚糖凝胶制成上述亲和柱并以脲解吸而分出多种来克丁[28]。较传统的方法如分子筛层析和离子交换技术的应用为分离纯化来克丁提供了简便易行之路，如常山鲜叶中提取DEL[29]。两种方法的混用或不同载体、配体联用分离纯化来克丁达到了取长补短的作用。

对要求并不十分严格的定性分析，谢继锋[30]发展出一种快速法，即玻璃表面固着法，此法可迅速定性测出来克丁。

3. 所研究的来克丁之一般性状

3.1 陆上生物的来克丁

我国学者所研究的来克丁材料主要取自陆上生物。这一般取决于有关学者所处的工作地区。表2列举出我国学者所研究过的陆上动植物、真菌来克丁来源、名称及一般性状等。由表2可见，所研究过的陆上来克丁已有43种以上。其来源主要在豆科植物，约占37.2%，其他植物的来克丁分布于11个科中。还有10种来克丁取自动物及人。只有一种来克丁取自菌类。对淡水动物来克丁的提取已获得一定成果[13]。

表 2 我国一些陆生菌类、动植物来克丁来源、性状一览表

Table 2 A list of sources, characteristics of lectins produced by some terrestrial microbes, plants and animals in China

来克丁名（缩写）Name of lectins (abbreviated)	来源生物类别 Classification of source organisms		性质、功能 Characteristics, functions	引证文献号 Number of reference cited
	科 Family	属或种名 Genera or species		
SLL	多孔菌科 Polyporaceae	白茯苓 *Smilax lauceoefolia*	相对分子质量 31 000，亚基相对分子质量 8 000，凝兔红细胞，甲状腺球蛋白、胃黏蛋白、黏蛋白抑凝，促有丝分裂原	32
BPA	豆科 lguminosae	羊蹄甲 *Baubinia purpurea*	具九肽、十肽结构，有稳定空间构象的实体，十肽立体结构不紧密，与钙、乳糖结合	33
CML	豆科 lguminosae	野花生豆/*crotalaria mucronata*	亚基相对分子质量 4 9000，专凝人 A 型细胞，N-乙酰半乳糖胺抑凝，促有丝分裂原	19
ConA	豆科 lguminosae	刀豆 *Coneanavalia ensiformis*	分子中铽（Ⅲ）和钴（Ⅱ）间能量转移，与葡萄糖专一结合，与一些细胞及其受体反应	22、34
DIA	豆科 lguminosae	旋扭山绿豆 *Desmodium intortum*	相对分子质量 32 000，强烈凝兔红细胞，专性结合半乳糖，抑凝最强	35
E-PHA	豆科 lguminosae	红腰豆 *Phaseolus vulgaris*	可与其他来克丁组成同相凝集柱	36
LCA	豆科 lguminosae	兵豆 *Lens culinaris*	凝人红细胞，甲基-*D*-甘露糖最抑凝	12
LGA	豆科 lguminosae	小扁豆	可与其他来克丁组成系列来克丁柱层析分析 *N*-连接型糖链结构	28
MDL	豆科 lguminosae	岩豆 *Milletia dielsiana*	由寡糖链糖基组成	38
PCL	豆科 lguminosae	红花菜豆 *Phaseolus coccineus var. rubronanus*	四聚体，四个相同亚基组成，与高甘露糖型寡糖结合	39
PHA-P	豆科 lguminosae	江苏菜豆 *Phaseolus vulgaris P*	相对分子质量 128 000，亚基相对分子质量 31 000、33 000，LA、L_3E_1、L_2E_3 和 E_4 为同功来克丁	10

(续表)

来克丁名(缩写) Name of lectins (abbreviated)	来源生物类别 Classification of source organisms		性质、功能 Characteristics, functions	引证文献号 Number of reference cited
	科 Family	属或种名 Genera or species		
PL	豆科 lguminosae	黑色菜豆 *Phaseolus* sp.	亚基相对分子质量 35 000 和 38 000,凝人 A 型红细胞	37
PLA	豆科 lguminosae	一种菜豆 *Phaseolus limensis*	四个亚基组成 $(17\ 200)_2$ + $(14600)_2$,专凝人 A 型红细胞,葡萄糖、甘露糖专一结合	17
PVL	豆科 lguminosae	沙克沙豆 *Phaseolus vulgaris* sp.	相对分子质量 3 026 000,单亚基,凝兔、小鼠,鸡及人 A、B、O 型红细胞,猪甲状腺球蛋白抑	100
VBL	豆科 lguminosae	三齿草藤/ *vicia bangei*	相对分子质量 24 600,凝红细胞,Man、GlcNAc、GLc 抑凝,促有丝分裂原	27
VCL	豆科 lguminosae	饭豇豆 *Vigna cylindrical*	测量了它的等电点	40
MSL	豆科 lguminosae	常春油麻藤 *Mucana sempervirens*	相对分子质量 131 800,亚基相对分子质量 66 000 和 33 000。专凝人 A 型血	61
AHL	木樨科 Oleaceae	白桂木 *Artocarpus hypargyreus*	相对分子质量 15 000,19 000,与 IgG 的某些亚类结合,凝多种动物细胞及人的 AB 型、ADO 型红细胞	41
DFL	虎耳草科 Saxifragaceae	常山 *Dichroa fabrifuga*	亚基相对分子质量 37 000,凝动物槽子、肿瘤细胞	29
Jaclin	桑科 Moraceae	木菠萝 *Artocarpus integrifulia*	强烈凝淀含 DGaLBL→3DGalNAc 结构的无活性抗冻糖蛋白,最强抑制剂 $DGaLB_1$→3DGalNAcl→ϕNa	42
LEL	茄科 Solanaceae	蕃茄 *Lyocpersicon eseulentum*	凝人及多种动物红细胞及某些培养的细胞,专性结合 N-乙酰葡萄糖胺寡聚糖	14

（续表）

来克丁名（缩写） Name of lectins (abbreviated)	来源生物类别 Classification of source organisms		性质、功能 Characteristics, functions	引证文献号 Number of reference cited
	科 Family	属或种名 Genera or species		
LPL	桑寄生科 Loranthusceae	桑寄生 *Loranthus parasiticus*	相对分子质量 67 500，专凝兔红细胞，GAI、GAINAc、山梨糖、岩藻糖、松三糖抑凝，促有丝分裂原	18
PTL	天南星科 Araceae	半夏 *Pinellia ternata*	相对分子质量 44 000，凝兔红细胞，色氨专一结合低聚糖、聚糖链，结合甘露糖即抑凝，每个亚基含一个酸残基，共四个亚基	24
RCA、RCA5、RCA120	大戟科 Euphorbiaceae	蓖麻 *Ricinus communis*	相对分子质量 120 000，有兰个主要的寡糖部分，RCA_5 不抑制^{125}I-胰岛素与其受体结合，有三种蛋白	25、26
SIL	脂麻科 Sesamumceae	芝麻 *Sesamum indicum*	少数分子中两条肽链是拆开的，N-乙酰葡萄糖胺、单糖不抑制它与红细胞的凝集	11
TKL、TKL1	葫芦科 Cucurbitaceae	瓜蒌 *Trichosanthes kirilowii*	相对分子质量 60 000，等电点 5.5，专一结合半乳糖、乳糖、葡萄糖，TKL_1 与天花粉毒蛋白相似，无金属离子结合部位	43、44、45、46
WGA	禾本科 Gramineae	小麦 *Wheatgerm* (*Tritcum vulgaris*)	分子由两个相同原体组成，WGA Ⅱ 的原体为 171 个氨基酸残基的肽链，分四段。WGA 高级结构有明显对称性。特异结合 N-乙酰葡糖胺和 N-乙酰神经氨酸及其衍生物和寡聚糖	47
ACL	百合科 Liliaceae	洋葱 *Allim cepa*	亚基相对分子质量为 26 500 和 47 500，凝兔、狗红细胞及人 ABO 型红细胞最弱，甘露糖及含甘露糖的复合糖类抑凝	48
ACYL	百合科 Liliaceae	黄洋葱 *Allim cepa yellow*	凝兔、狗红细胞及人的 ABO 型红细胞，半乳糖、岩藻糖抑凝	48

（续表）

来克丁名（缩写）Name of lectins (abbreviated)	来源生物类别 Classification of source organisms		性质、功能 Characteristics, functions	引证文献号 Number of reference cited
	科 Family	属或种名 Genera or species		
AFL	百合科 Liliaceae	青葱 *Allim fistulosum*	凝兔、狗红细胞及人的 ABO 型红细胞，甘露糖及含甘露糖的复合糖类抑凝	48
ALL	百合科 Liliaceae	香葱 *Allim ledebourianum*	凝兔、狗红细胞及人的 ABO 越红细胞，含甘露糖的复合糖类抑凝	48
AOL	百合科 Liliaceae	韭菜 *Allim odorum*	凝兔、狗红细胞及人的 ABO 型红细胞，古甘露糖的复合糖类抑凝	48
ASL	百合科 Liliaceae	大蒜 *Allim sativum*	凝兔、狗红细胞及人的 ABO 越红细胞最强，含甘露糖的复合糖类抑凝	48
PCLⅡ	百合科 Liliaceae	囊丝黄精 *Polygonatum cyrtonema*	相对分子质量 159 000，只凝兔红细胞，可被 D-甘露糖和猪甲状腺球蛋白抑，含 Mg^{2+}、Ca^{2+}	49
AFL	人科 HcIminidae	人羊水 *Anniofic fluid*	凝兔红细胞，含低聚糖甘露糖的糖蛋白与 AFL 结合最强	101
AGL	螺科 Ampullariidae	大瓶螺蛋白腺 *Ampullaria gigas*	相对分子质量 15 000。无血型专一性，但凝人 A 型血最强，被乳糖、半乳糖抑	31
APL	蚕蛾科 Bombycoidea	中国柞蚕蛹血淋巴 *Antheraea pernyi*	相对分子质量 38 000，凝兔红细胞，-甲基-D-葡萄糖苷和 D-半乳糖抑凝，高凝度乳糖、N-乙酰-D-半乳糖胺也抑	50
MAPA	合鳃鳗科 Synphobranchidae	黄鳝血浆 *Monopterus albus*	相对分子质量 46 800，对半乳糖、乳糖专一。对人 B 型血作用大。血凝活性和糖结合专一性不需金属离子	74
BSL	猪科 Suidae	猪精子 *Boar sperm*	相对分子质量 56 000，亚基相对分子质量为 13 600(β)和16 000(α)，与 ZP_3 透明带糖蛋白结合，亦与精子蛋白结合	16

(续表)

来克丁名(缩写) Name of lectins (abbreviated)	来源生物类别 Classification of source organisms		性质、功能 Characteristics, functions	引证文献号 Number of reference cited
	科 Family	属或种名 Genera or species		
MDL	家蝇科 Muscidae	舍蝇 *Musca domertica*	相对分子质量 34 000,凝人 B 型血及小白鼠,兔红细胞,专一结合半乳糖、D-和 L-岩藻糖	13
PCL	蚕蛾科 Bombycoidea	蓖麻蚕蛹 *Philosamia cynthia*	凝固定的人 A、O 型红细胞,不凝新鲜血球。一种 PCL 专一结合鼠李糖和 L-岩藻糖	20、21、30
PCL_1	蚕蛾科 Bombycoidea	蓖麻蚕血淋巴 *Philosamia cynthia*	鼠李糖专一	19
TSL	蝰蛾科 Viperaceae	竹叶青蛇 *Trimeresurus stejnegeri*	相对分子质量 6 000,半乳糖专一,半乳糖系列的糖较强抑凝	53
PAL	猪科 Suidae	猪脂细胞膜 *Porcine adipocyte*	相对分子质量 89 000 和 112 000,促游离脂肪细胞凝集,麦芽糖、甘露糖、甘露聚糖抑凝最强	51,52
HSL	人科 Hominidae	人胃 *Humen stomach*	相对分子质量 12 000,单亚基,乳糖最有效抑制、结合半乳糖苷	97

有关工作大都集中于有限的来克丁生化性质和糖结合专一性上。这些成果为研究来克丁的其他生理生态特性、作用及应用打下了一定的基础。

3.1.1　海洋生物的来克丁

我国海洋生物来克丁的研究起步大大晚于陆上生物来克丁的研究。迄今其发展仍缓慢。仅有吴素兰等[54]、苏拔贤等[55]、丁伟及陆皓文等[5]发表过 4 篇文章。吴、苏等对百拾余种海洋动物的肌肉、腺体、外皮、肝、鳃、某些基部或淋巴等作来克丁类物质的筛选,发现其中 47 种有之。我们初步分离筛选过三疣梭子蟹、日本鲟、毛蚶、泥蚶、蛤蜊、螺、蛏子、扇贝、中国对虾、虾蛄、尼罗罗非鱼等经济海产动物体液的来克丁含物,并初步测定其中一些来克丁含物的糖专一性,也进行了对人、兔血红细胞及一些病原菌的凝集反应试验。结果表明所研究的水产动物有产生来克丁的活力,以此调理体内代谢、清除废物、抵御外来(非自我)入侵和攻击。这是水产动物机体本身御敌和痊愈机制的重要活动之一[5]。

3.1.2　我国研究较深的来克丁之主要性状

我国最早研究的来克丁还数刀豆球朊 A(ConA)。1981 年,朱正美等[22]就 Con A 作过荧光标记研究，发现它与葡萄糖专一结合；在对其有稳定作用的溶液中反应，结合物的质量和得率都可提高。就来克丁与细胞凝集研究而言，大多以 ConA 等作过试验。他们就此发表过 4 篇相关文章[34,56～58]，主要涉及 ConA、PHA 对正常细胞、胚胎及肿瘤细胞的凝集反应及与细胞膜的作用，正常鼠肝及腹水型肝癌的 ConA“受体”及对该“受体”的分析。就 Con A 中金属离子铽(Ⅲ)和钛(Ⅱ)间能量转移作过研究，认识到这可用作测定金属离子结合位置间距离的荧光探针。据此还了解到稀土金属离子钆(Ⅲ)对 PHA 活性的影响；还可研究溶液状态中生物大分子的微区结构[59,90]。周红等[61]用化学修饰、内源荧光和荧光淬灭研究油麻藤来克丁 J-MSLC(*Mucana sempervirens*)溶液构象变化和微环境构象特征。

有关 WGA 分子结构的研究，董素才等[47]曾作过述评，指出它是由两个相同原体组成，其高效结构有明显对称性；它特异结合 N-乙酰葡糖胺和 N-乙酰神经氨酸及其衍生物和寡聚糖。

我国学者对下列来克丁的认识有所加深，即 FKL、RCA、PCL 和 TSL 等。就 TKL 言，除确认它糖专一性和/或糖结合性外，还研究了它对红细胞补体敏感溶血作用的影响、免疫学、同工来克丁双链的拆分和鉴定及与其相似蛋白质、荧光光谱[62～65]，证实了该来克丁分子有两条链、与 TCSCL 天花粉毒蛋白是非常相似的蛋白质[65]。在分离出 PTL 后，深入研究了它的色氨酸残基修饰与活性的关系。并掌握了它在植体中的分布状况[9,24,66]。纯化出 RCA，并对其理化性状作过描述，包括它的 LD_{50} 值。继而进一步纯化出它的寡糖链。分析了它的组成及结构[23,25,26]。对蓖麻蚕来克丁(PCL)，于 1983 年就开始了研究[21]。彭金荣等从该血淋巴中分离出分别对鼠李糖、岩藻糖、半乳糖专一的来克丁；后来进一步对蓖麻蚕的两种来克丁在其体内的分布与生长关系作过研究，指出它们在血淋巴中的活力随蚕的生长期不同而变，两者的变化也非同步进行[20,21,50,67]。对 TSL，曾描述过它的一般性状，还着重研究了它的糖结合性[68]。

关于红花菜豆来克丁(PCL)，也作了一些研究。这包括其一般性状的分析。用气相层析术和外切糖苷酶的专一性，测定其糖组成和寡糖链顺序，它与其寡糖部分分子间的聚合现象及凝集细胞。促淋巴细胞转化、抑制细胞生长作用等[39,69,70]。黄精来克丁Ⅱ(PCLⅡ)(*Polygonatum cyrtonema* Lectin Ⅱ)的研究，是由鲍锦库等专门进行的[49,71]。他们用 $(NH_4)_2SO_4$ 沉淀其根茎物，猪甲状腺球蛋白-Sepharose 4B 亲和层析，CM-Sepharose 柱离子交换层析、Sephadex G-100 凝胶过滤得 PCLⅡ。分析了它的部分性质，并且就其分子稳定性与生物学活性作过探讨，发现了一些修饰分子的巯基和色氨酸等基团对该来克丁分子的生物活性的关系。

此外，就来克丁的疏水性[72]、一些来克丁结合部位糖的特异性[39]、来克丁碎片的糖结合活性[73]、活性肽构象预测、结构功能关系作了探索[33]。对内源来克丁也作过一些研究。指出了它与发育分化有关。肿瘤细胞表面的来克丁关系着肿瘤的转移活动[75～77]。

3.1.3 来克丁的应用研究

对 ConA 的研究，除了认识它本身性状外，还将它运用到有关领域中，如荧光标记的 ConA 亲和吸附固相化，可用于其他来克丁的分离[22]。ConA 与其他来克丁组成的系列固相亲和层析术，已用于分离多种来克丁或某些抗原[28,78]。还可研究鼠肝碱性磷酸酶(ALP)的糖链结构及诱发肝癌 ALP 糖链中的平分型 GlcNAe 残基变化[36]。ConA-FITC

用于细胞、组织切片的荧光染色及荧光显微照相记录，效果很好[22]。用 ConA 等来培养静止软骨细胞，发现它能完全抑制软骨细胞的 DNA 合成。不过，这一抑制过程是可逆的。ConA 还能促进软骨细胞成熟化，且使之产生终末分化[78,80]。研究过 ConA 和 PHA 对急性白血病细胞系细胞生长的影响。ConA 双向免疫电泳可研究人体-抗胰蛋白酶(糖蛋白)分子变异体，为临床诊断提供依据[81]。

WGA 用于降低红细胞膜阴离子通透性，发现 WGA 浓度越大，该通透性降得越快。N-乙酰葡萄糖胺对 WGA 的这一作用有恢复效应[82]。

刘黎等[83,84]在 1984 年、1987 年分别用花生凝集素(PNA)研究了鸡胚视顶盖中视神经纤维发育的时空图象，描绘了视网膜发育过程中它结合糖蛋白的图谱，阐述了与 PNA 结合，即含 D-半乳糖残基的糖蛋白在视网膜发育中的意义。用大豆来克丁(FITC-SBA)或 WGA、SBA 研究细胞表面的来克丁受体亲近程度及林蛙卵裂活动[85~87]。

用几种不同糖专一性的来克丁研究了绵羊精子的射出作用。指出绵羊射出精子的 Con A 受体主要分布在头部质膜，而且 Con A 与其受体的结合是专一的[88]。EL 介导的 7 种 MTX-拟糖蛋白(糖化的白蛋白)-氨甲蝶呤可分析对肝癌细胞的靶间杀伤作用[77]。用来克丁分离纯化出含糖高分子或含糖基化碱基的不同 tRNA，进而推断其所含糖基，也分离了带有不同表面糖分子的细胞，如淋巴细胞等。兵豆来克丁(LCA)可用于鉴别肝癌 AFP 糖基变异体[12]。1992 年，许兵等[89]发现 WGA 有促进中国对虾抗病力的效果，这是我国来克丁应用于海产养殖的首次报道。

美商陆来克丁(PWM)的应用研究也有所报道。如将它用于观察对人单核细胞及白血病细胞的 HLA-DR 分子表达、对人胎儿胸腺细胞表型的影响等。发现 PWM 可用于肿瘤免疫治疗、可加强人 LAK 细胞的增殖活性和杀伤力，诱导人单核细胞表达 IL-2 受体 α 链，介导人单核细胞去诱导白血病细胞凋亡。还发现 PWM 是 T 细胞、B 细胞的多克隆激活剂；可用来诱导单核细胞依赖的外周血单个核细胞(PBMC)活化；可明显诱导胎儿胸腺细胞增殖[90~92]。巨同忠等[93] 1996 年用包括有兵豆来克丁(LCA)的亲和柱，即 LCA-Sepharose CL-4B 柱与非同位素标记底物来测定核心 α-66-岩藻糖转移酶。在对 TKL 性质有了广泛认识的基础上；用它与 Sepharose 4B 组成亲和柱分离一些相关的糖蛋白，并探讨了 TKL 可能参与所在植体中的御敌活动[94]。我国来克丁在临床医学、疾病防治中的应用取得了进展。这在肿瘤、癌症诊断治疗中较明显。所用的来克丁种类甚多，如 PNA、uEA-1、DBA、LCA、Con A、CEA、PHA、IL、PTL、TAT、等。有关在医学上的应用，因篇幅有限，恕不赘述。近几年来人们已注意到选凝素(Selectin)研究和应用的势头正在加强[6,7]。

由上可见，我国来克丁的应用研究已在一些领域中迈出了可喜的一大步。

3.2　我国市场上的国产来克丁

如前所述，国际间商品化来克丁的品牌已有几百种，我国不同程度研究过的来克丁已达 43 种。我国研究过的来克丁有几种是国际间已上市的来克丁，如 BPA、Con A、LCA、L. EA、RCA 及其同工来克丁、PHA 等。豇豆来克丁，国内外虽都取自同一属的，但种不一样。国外的是取自 *Vigna radiata* 的 VRL 等，我们的是另一种，即 V. cylindrica，得的是 VCL。来自菜豆的来克丁，外国做出了 3 种，我国则研究了 5 种(如 E-PHA、PL、PCL、PHA-p、PLA 等)。两者相比，只一种，即 *Phaseolus vulgaris* 的来克丁是一致的。比较国内外植物来克丁,发现番茄、蓖麻,兵豆、羊蹄甲和江苏菜豆都已用来作过来克丁学的研究。此外也许是我们所拥有的特色了。在国内市场上所见的国产来克丁试剂却只有7～8种而已。如取自四川芸豆的PHA。另外,鲎试剂,也可看做是比较接近来克丁的生物制剂。由此可见,国内外的差距是明显的。我们必须急起直追,提高来克丁市场的国产化比率。

4.我国来克丁研究展望

国际间来克丁研究十分活跃,伸展面日益扩大。用于研究来克丁的材料不断增多。加深了对来克丁本身的认识。这是建立在对来克丁性状、特征日益精细的分离、纯化和分析的基础上的,也离不开对来克丁的分类、生物合成和遗传学的理解。人们越来越认识到来克丁在新陈代谢、生理活动中的作用,包括免疫、受体、促细胞分裂作用等。由此推动了来克丁的广泛应用,如在临床诊断、血液学等方面。因而促使人们去了解来克丁的药物学、药效和药物代谢动力学及部分来克丁的毒理(性)学等。因此来克丁研究历经一个世纪至今仍是一个经久不衰的生长点。回头看看我国来克丁学发展过程。它虽仅有20多年历史,但取得了可喜进步。这体现在所研究的材料日益丰富,对一些来克丁性状有所认识,应用途径扩展。与国际上比,差距又十分触目。首先作为一个资源大国,来克丁材料十分丰富。我们不必完全走外国人走过的路子,对所有生物一一筛选,但毕竟我们迄今所用的材料太有限。对种类繁多的动物,我们只做过几种来克丁的较详细研究。至于微生物,可说是少有问津。精细研究并制成试剂的国内来克丁太少,在很大程度上限制了来克丁在许多领域中的应用。我国是一个海洋国家。作为生命摇篮的海洋,以其特殊的生态环境养育许多陆地上没有的生物,是生物天然物质的资源库,是许多新化合物的世界。包括来克丁在内的海洋天然物质的开发是国际三大热点之一。我国还是个海产大国。完全应该结合实际情况,从大量水生物中提取来克丁[31,54,95,96]。来克丁普遍被认为是生命机体一类对外来物的识别因子,并在体内(外)起调理作用。它在认识和解决集约化、半集约化的养殖生物的防病治病中应发挥重要的、不可替代的作用。而这一点,即来克丁在养殖生物的御敌机制和活动,迄今还刚刚引起国人,包括水产界同仁的注目[5]。对来克丁认识的普及和研究的深入必将能为提高我国水产医学、养殖业总体水平作出贡献。对海洋来克丁等生化信号分子的开发、利用和生化工程技术的拓展,将会使传统的海洋产业,如水产养殖业及其病害防治转化为高技术类型。由此改变许多海洋产业结构和形态。

由于我国这些年来来克丁研究的重点在陆上和少数淡水生物。因而造就了此方面相当坚实的科研力量。至此就应该强化结合海洋、陆地和内陆水域的来克丁研究力量,使一部分陆地或淡水的来克丁科研基础向海洋转移和倾斜,以加速海洋来克丁研究的成长。这是培育海洋高技术所需,是建设海产强国、"海上中国"战略目标所需的。

来克丁的广泛应用,在很大程度上还受制于来克丁制品的纯度。这有赖于我国学者日臻的来克丁分离纯化技术的出现和运用。来克丁纯度的不断提高、花色品种的日益增多及在基因分子水平上认识和应用它们,我国来克丁学发展前景将是十分诱人的。

参考文献101篇(略)

(合作者:孙丕喜)

LECTIN RESEARCH IN CHINA

(ABSTRACT)

Abstract　It is considered that the Chinese terms, Ningjisu, Xuanzesu or Waiyuan-ningjisu, etc., can not generalize the original implication of the term "Lectin", and may mix up the difference in meaning between the two terms, agglutinin and lectin, which will probably hinder from the development and research on lectinology in China. It is suggested that the English term Lectin(or a Latin term, Legere) be translated directly into the Chinese term—Laikeding. The paper also reviews the status of Lectin research in the period from the end of the 1970 s to now in China. discusses Chinese scholars' methods of isolation and purification for lectins, the biological, chemical properties and application of lectins in China, with emphasis on a description of some more thorough researches on the characteristics, action and uses of lectins. Also included in the paper is a catalogue of Chinese lectin products in domestic markets. Some suggestions are put forward for 1ectins(including ones from marine organisms) research in the future in China.

Key words　Lectin, Chinese Term for Lectin, Agglutinin, Character, Technique of Isolation and Purification, Application, Direction of Research

从来克丁物质与微生物相互作用看水产无脊椎动物对疾病的防御活动*

摘　要　本文从水产无脊椎动物与环境的关系论述其来克丁物质的功能;来克丁与微生物间的相互作用包括相互作用中有关来克丁、微生物种类,作用部位和作用物质等。从来克丁在水产无脊椎动物识别外来物质,包括抵御病原体的作用看,健康发展的水产业之一项关键任务便是认识和掌握来克丁在预防病害活动中的机理。

关键词　来克丁与微生物的相互作用　水产养殖业　防病机理　病菌

1　水产无脊椎动物的生存和发展及应运而生的来克丁物质

1.1　水产无脊椎动物与环境间相互作用

海洋生物(包括海洋无脊椎动物)与其生存环境共同组成一个互有密切联系的统一体,但相互间的界限依然存在,即有有机体内外、胞内外之分。体外有携带食物颗粒的海水、病原体和捕食者;体内包括各种循环系统、组织器官和细胞。胞外主要指外壳和软组织内的血清、体液、入侵病原体等;胞内主要是具有各种功能的细胞结构物及成分。体内外以动物的外皮层或钙质化的外壳为界,胞内外以各细胞膜(壁)来分。外壳和软组织、血管壁和软结缔组织、循环细胞表面等构成了无脊椎动物(如甲壳类等)体内外、胞内外相互作用中一、二级的隔室(compartments)、界面[1]。

1.2　应运而生的来克丁

海洋无脊椎动物,尤其是当今处于陆岸集约式或半集约式养殖环境中的生物,因其生活在人为环境中,多半仍未能精确计算和控制其环境容量,所养品种单调,密度很高,使各类非养殖对象(病原生物)增多。许多无脊椎动物常以滤食为摄食"手段",难免"泥沙"俱摄,其体液仍未能处于正常脊椎动物那样的"无菌"状态中。在这日益增大的危险中,为了有效摄食,需要有精良的防御措施相匹配,即胞内消化和准确防卫两者的默契配合。为此必须进化、发展和尽可能完善自身的防卫体系。在它漫长的进化过程中发展出了来克丁(lectins)——防卫体系的重要成员。尤其是在含有各种病原体环境中的海洋生物,除非它们被疾病缠身而不能自拔者,都毫无例外地需要来克丁,以便在细胞、分子水平上确认自我,分辨非自我,阻止入侵病原生物在体内繁殖。保持对细胞生长和繁殖的调控,阻止癌细胞无限增生,包括来克丁在内的这套系统的能力对于所有有生命的多分子(multimolecular)物体的内外防御、共生和繁殖的调控等都是基本的。

无脊椎动物来克丁的结构和功能显得相当繁杂,但基本序列都保存着广泛的同源现象:尽管脊椎动物、无脊椎动物的保护体系在本质上是一致的,但水平上明显存在很大差

* 原文刊于《黄渤海海洋》,1998,16(4):55-64.国家自然科学基金资助项目(39370539)。

距，或者说无脊椎动物的保护体系仅呈现为哺乳动物保护体系的原始相。如无脊椎动物的保护体系之化学专一的多样性较少，抗原决定簇的反应并不丰富多彩，无免疫球蛋白，准免疫作用所起的保护性记忆过程短，“记忆力”也弱，但它之所以仍能在当今的生物之林中占有颇为重要的地位，来克丁这个识别因子所起的作用是不可磨灭的。因而来克丁的识别、防御活动对认识防病治病机理和提高淡水、海水养殖的科学性是很有意义的。

最早发现于植物，尤其是豆科植物中的来克丁被认为是植物的识别因子，其识别单位被定义为植物受体蛋白(Plant receptor protein)。迄今市售来克丁大都来自植物，动物来源的来克丁类试剂之所以少，原因不外乎取得样品的不易和纯化工作的难度。所以目前动物来源的来克丁实际上大都只能称为拟来克丁物质(lectin—like substances)，虽然这方面的研究报道越来越多，这意味着最初的 lectin 这一名称并不完全适合于动物[2]，甚至某些生物。据动物来克丁/拟来克丁物质的作用分，所使用的名称颇多，如保护素(protectins)、凝集素(agglutinins)、沉淀素(precipatins)、预防素(defensins)、杀菌素(bactericidins)、杀菌蛋白(bactericidal proteins)等，这反映了来克丁物质的不同纯度和存在状态。已知的绝大多数来克丁物质是糖蛋白，但所发现的和将被发现的来克丁物质会越来越多，形成的家族将日益庞大，而且其纯化迄今仍存在着较大难度，所以不少研究者称这类糖蛋白为多体来克丁(multimeric lectins)。这就意味着多体来克丁是某些单体来克丁的复合、聚合或联合物。它们需具有多相结合、复合或聚合活动的能力，在适当条件下，它又可分散，并各司其职。

2　来克丁与微生物相互作用方面

2.1　水产无脊椎动物来克丁性状、功能

就无脊椎动物言，许多来克丁存在于血清、体液或血细胞上[3]。它与外来物质(赋予生命形体的物质)发生的相互作用/反应依赖于血球表面组分和外来分子结构间的作用[4]。当它与微生物细胞(本文指病原菌)相遇时，接触的一般是糖类[5]。来克丁与特定糖类的结合除在受精作用、藻类配子融合、酵母有性相容、粘霉细胞间交流和附着、八瓣类蛤与藻类共生、组织大分子或多酶复合物参与 Ca^{2+} 或糖的输运和在发出穿过生物膜的信号中起作用外[6]，还在以下六方面发挥功效：①调节幼体附着和变态；②凝集细胞；③充作调理素(调理素是血清中作用于颗粒或细胞以增加吞噬细胞的消化作用的一些物质，而当仅有血球时则不发挥作用)；④吞噬作用[7]；⑤形成特定的识别因子，调节同种细胞间的相互吸附能力，使过多繁育的(proliferative)细胞不被自我受体所认同，不建立联系，反而被捣毁[8]；⑥阻止多精入卵，保护卵的正常发育。

无脊椎动物来克丁的形态和专一性虽不完全一致，但当它与受体相遇时都需要偶数的结合位点，以与配体结合[9]。这其中的一半结合点用于结合抗原，另一半与血球细胞表面反应。来克丁借助结合点促进对外来者的反应，一部分外来物被认同为自我——被血球所消化或被螯合(sequestrate)，有些则被去除(eliminate)以至逐出。这些结合点一般状态下应是隐蔽的、不活泼或不明朗的。必要时来克丁调整与非自我的反应，通过寻求分子的重排得以实现识别活动，此时自身决定簇即暴露在外；也可增加有效的非掩盖的结合点来完成与非自我的识别反应过程；不与外来细胞表面发生反应时，则避免与自身不适反应。这表明它们对本身血细胞具高度亲和性，以阻止外来入侵者的干扰或破坏[5~10]。

2.2　来克丁与微生物的相互作用

对来克丁防御、识别活动机制的认识最初是由它与微生物的相互作用研究中获悉，而且迄今仅对不多的来克丁，如 ConA，WGA，PHA，Limulin，Gigalin 等做了较多的研究[5]，其中以监测淋巴细胞上 α-D-吡喃甘露糖和 α-D-吡喃葡萄糖(D-Gle)残基开始，再扩展至测定细菌、真菌表面上的这些糖类。另外以根瘤一根瘤菌间的共生关系演变为认识寄主一病原体活动中来克丁的作用。经过几十年的努力，终于取得了当今来克丁一微生物相互作用研究的较大进步。

曾被用来研究与微生物相互作用的来克丁名录(表 1)共约 66 种，其中来自植物的有 19 种，占 28.8×10^{-2}；来自动物的是 43 种(包括脊椎动物 6 种，海洋无脊椎动物 19 种，占总数的 28.8×10^{-2})，占 65.2×10^{-2}。此外，尚有 3 种由细菌，1 种由霉菌产生的来克丁。

表 1　与有关微生物相互作用的来克丁和似来克丁物质名录

Table1　Name list of lectins and lectin-like substances interacting on microbes concerned

1. ACL(*Aplysia california lectin*)	加州海兔来克丁
2. ADL(*A. depilans lectin*)	地中海海兔生殖腺来克丁
3. AFL(*A. fasauta lectin*)	一种海兔的来克丁
4. AMH(*Arenicola marina hemocytes*)	海生沙娲蠋血球来克丁
5. APL(*Antheraea Pernyi lectin*)	中国栎蚕蛹血来克丁
6. APP(*Acer psendoplatanus protein*)	小无花果树细胞蛋白质
7. Aprotinin	小牛肺似来克丁物
8. Axinella Ⅳ	小轴海绵 4 号提取物
9. BSL(*Bandeiraca simplicifolia lectin*)	西非单叶豆
10. CAL(*Colonial ascidians lectin*)	一种群落性囊海鞘来克丁
11. CH(*Chieabolus hemolymph*)	
12. ConA(*Concanavalia A*)	刀豆球朊 A
13. CRL(*Carcinoscorpius ratunde cauda lectin*)	马蹄蟹来克丁
14. CSL(*Callinectus sapidus lectin*)	蓝蟹来克丁
15. CSL(*Chinook salmon ora lectin*)	大鳞大麻哈鱼卵来克丁
16. CVL(*Crassostrea virginica lectin*)	弗吉尼亚蠔来克丁
17. DBL(*Dolichos biflorus lectin*)	双花扁豆来克丁
18. DCL(*Didomnum candidum lectin*/$\frac{\text{DCL I}}{\text{DCL II}}$)	一种膜海鞘的两种来克丁
19. DDL(*Dictyostelium discoideum leetin*)(*Slime Faold lectin*)	黏菌来克丁
20. DFL(*D. fulle lectin*)	一种膜海鞘来克丁
21. DML(*D. molle lectin*)	另一种膜海鞘来克丁
22. EFL(*E. coil fimbrial lectin*)	大肠杆菌菌毛来克丁
23. Gigalins(*Gigalin E and Gigalin H*)	太平洋牡蛎血淋巴来克丁

(续表)

24. HAH(*Helix aspersa hemacytes lectin*)	褐色菜园大蜗牛血细胞来克丁
25. HAL(*Homorus americanus lectin*)	美洲巨螯虾来克丁
26. HCL(*Hyalophora cacropialarvae lectin*)	
27. HPL(*Halichondria panacea lectin*)	面包软海绵来克丁
28. HPL(*Halocynthia pysiforuis lectin*)	
29. HPL(*Helix pomatia lectin*)	一种大蜗牛的来克丁
30. JWL(*Jimson weed seeds lectin*)	罗陀罗籽来克丁
31. Lamulin	似昆布氨酸物
32. LPL(*Limulus palyphemus lectin*)	取自鲎的一种来克丁
33. LEP(*Lycopersicun esculentum steam protein*)	番茄茎蛋白质
34. LFH(*Lithobius foricatus hemolymph*)	
35. LPP(*Limulus polyphemus protein*)	一种鲎的蛋白质
36. LTL(*Lumbricus terrestris humor lectin*)	土壤蚯蚓来克丁
37. MML(*Mercenaria mercenaria lectin*)	硬壳蛤来克丁
38. MSP(*Manduca sexta protein*)	
39. OKL(*Oncorhynchus kisutch egg lectin*)	银大麻蛤鱼的来克丁
40. OLL(*Otala lacteal lectin*)	一软体动物来克丁
41. OVL(*Octopus vugaris hemolymph lectin*)	真蛸来克丁
42. PAL(*Pseudomonad aeruginosa lectin*)	铜绿假单胞菌来克丁
43. PCL(*Paraobarius corneus lectin*)	蝲蛄来克丁
44. PCL(*Phaseolus coccineus var. rechronanus lectin*)	淡水蜗牛来克丁
45. PCL(*Planobarius corneus lectin*)	淡水蜗牛来克丁
46. PHA(*Phaseolus vulgaris agglutinin*)	菜豆凝集素
47. PMI(*Petromyzon marinus immunoglobulin*)	海八鳗免疫球蛋白
48. PNA(*Arachis hypogaea lectin*)	花生凝集素
49. PPL(*P. plumose ova lectin*)	一种鱼的来克丁
50. PVP(*Phaseolus vulgaris seedlings protein*)	菜豆苗来克丁
51. RCA(*Ricinus communis agglutinin*)	蓖麻凝集素
52. RVH(*Rhapidostreptus virgator hemolymph*)	
53. SBA(*Glycine max lectin*)	大豆凝集素
54. SCH(*Scolopendre cingulata hemolymph lectin*)	一种蜈蚣血淋巴来克丁
55. SCL(*S. cantharus ova lectin*)	一种海鲷卵来克丁

(续表)

56. SCP(*Samia Cynthia larvae hemo-protectin*)	一种幼虫血蛋白质
57. SFL(*Serratia marcescins fimbrial lectin*)	黏质赛氏杆菌菌毛来克丁
58. STL(*Solanum tuberosum lectin*)	马铃薯来克丁
59. THL(*Tridancnids hemolymph lectin*)	砗磲血来克丁
60. TPL(*Tobacco plant lectin*)	烟草来克丁
61. Trifolium	
62. VCL(*Vicia crecca lectin*)	草藤来克丁
63. VFL(*V. faba lectin*)	蚕豆来克丁
64. VRL(*Vigna radiate lectin*)	豇豆来克丁
65. WFL(*Wisteria floribunda lectin*)	另一种紫藤来克丁
66. WGA(*Jriticum rulgare lectin*)	麦胚凝集素

与上述来克丁发生相互作用的微生物菌种名录(表2)共99个种(属等),来自细菌的27属主要是大肠杆菌、微球菌、萘瑟氏球菌、假单胞菌、沙门氏菌、链球菌和弧菌;真菌13个属主要由各属酵母组成。

表2　与来克丁和似来克丁物质发生反应的微生物名录

Table 2　Name list of microbes interacting on lectins and lectin-like substances

1. *Acetobacter capsulatum*	荚膜醋杆菌	51. *Pseudomonas*	假单胞菌
2. *Acholeplasma*	无胆甾菌形体	52. *P. aeruginosa*	铜绿假单胞菌
3. *A. laidiowii*	莱氏无胆甾菌形体	53. *P. insolita*	稀有假单胞菌
4. *Actinomyces*	放线菌	54. *P. putida*	恶臭假单胞菌
5. *Aeromonas salmonieida*	杀鲑单胞菌	55. *P. solanacearum*	青枯假单胞菌
6. *Aspergillus*	曲霉	56. *Rhizobium*	根瘤菌
7. *A. fumigatus*	烟曲霉	57. *Rhizobium japonicum oryzue*	日本根瘤菌
8. *Bacillus megatherium*	巨大芽孢杆菌	58. *Rhizopus*	米根霉
9. *B. Subtilis*	枯草芽孢杆菌	59. *Poryzuehodosprillum rubrum*	深红螺菌
10. *Blastomyces dermatiddis*	皮尖芽生菌	60. *Rhodotorula*	红酵母
11. *Candida*	假丝酵母	61. *R. glutinis*	粘红酵母
12. *C. albicans*	白假丝酵母	62. *Saccharomyces*	酵母
13. *C. utilis*	产朊假丝酵母	63. *S. cerevimae*	啤酒酵母
14. *Cryptococcus*	隐球酵母	64. *S. H. A. H.*	
15. *C. neoformans*	新隐球酵母	65. *Salmonella*	沙门氏菌

（续表）

16. *Enterobacter aerogenes*	产气肠细菌	66. *S. minnesota*	明尼苏达沙门氏菌
17. *Eschena coli*	大肠杆菌	67. *S. montevideo*	蒙得维的亚沙门氏菌
18. *E. coli HB* 101	大肠杆菌 *HB*101	68. *S. typhimurium chemotype* R_a-R_b	鼠伤寒沙门氏菌 R_a-R_b 型
19. *E. coli K*1	大肠杆菌 *K*1	69. *S. zuerieh*	朱氏沙门氏菌
20. *E. coli K*12	大肠杆菌 *K*12	70. *Salmonella R mutant*	粗糙型沙门氏菌突变株
21. *E. coli. K*92	大肠杆菌 *K*92	71. *Sarcina*	八叠球菌
22. *E. coli R mutant*	大肠杆菌粗糙型突变株	72. *Schizosaccharomyces*	裂殖酵母
23. *Erwinia*	欧文氏菌	73. *S. pombe*	粟酒裂殖酵母
24. *E. cananas*		74. *Serratia marcescens*	粘质赛氏杆菌
25. *Fusarium*	镰刀菌	75. *Shigetla flexnesi*	费氏志贺氏菌
26. *F. roseum*	粉红镰刀菌	76. *Sporobolomyces*	掷殖醋母
27. *F. Solanl*	茄病镰刀菌	77. *Staphylococcal*	葡萄球菌类
28. *Helieobaeter pylofi*	人的伤寒病原菌	78. *Stophylococcus aureus*	金黄色葡萄球菌
29. *Lactobaeillus*	乳杆菌	79. *S. epidermidis*	表皮葡萄球菌
30. *L. plantarum*	植物乳杆菌	80. *Streptococcus*	链球菌
31. *Lichen symbionts*	地衣共生菌	81. *S. C group*	*C* 群链球菌
32. *Micrococcus*	微球菌	82. *S. faecalis*	粪链球菌
33. *M. luteusikticus*	溶壁微球菌	83. *S. lactis*	乳链球菌
34. *M. lysodeikticus*	溶壁微球菌	84. *S. millerl*	微小链球菌
35. *Myeobacterium*	分枝杆菌	85. *S. mutant*	血清型链球菌变种
36. *M. bovis*	牛分枝杆菌	86. *S.* 6 715	链球菌变种 6 715
37. *Mycoplama*	菌形体	87. *S. pneumoniae*	肺炎链球菌
38. *M. mycoides var. copi.*	萝状菌形体复制变种	88. *S. pneumoniae*12	肺炎链球菌 12 型
39. *Neisseria*	萘瑟氏球菌	89. *S. pneumoniae* 14	肺炎链球菌 14 型
40. *N. gonorrhoeae*	淋病萘瑟氏球菌	90. *S. salivarius*	唾液链球菌
41. *N. meningitidis*	脑膜炎淋病萘瑟氏球菌	91. *S. sanguis*	血链球菌
42. *N. meningitidis B group*	*B* 群萘瑟氏球菌	92. *Tilletia controversa*	腥黑粉菌
43. *N. perflava*	深黄萘瑟氏球菌	93. *Treponema*	螺旋体
44. *Nitrobacter*	硝化杆菌	94. *T. pallidum*	苍白密螺旋体
45. *N. winogradskyi*	维氏硝化杆菌	95. *Trichoderma vicide*	绿色木霉

(续表)

46. *Paracoccidiaides brasiliansis*	巴西副球菌	96. *Vibrio anguinarum*	鳗弧菌
47. *Penicillium*	青霉	97. *V cholerae*	霍乱弧菌
48. *Photobacterium leignsthi*	莱氏发光杆菌	98. *V harveyi*	哈氏弧菌
49. *Pneumococcal type* 12	12 型肺炎球菌	99. *V. parahaemolyticus*	副溶血弧菌
50. *P. type* 14	14 型肺炎球菌		

12 种主要来克丁及似来克丁物与有关微生物发生作用的情况见图 1，由图 1 可见，与来克丁 ACI，ADL，AFL，ConA，CVL，HPL，Limulin，PHA，PNA，RCA，SBA 及 WGA 等发生相互作用的微生物种(属)数分别为 5 种以上(5，5，76，6，5，8，5，5，8，12 和 10)，其中用来研究 ConA 发生相互作用的微生物种类最多，表明对 ConA 的认识比较透彻。除了上述 12 种来克丁外，其他来克丁与有关微生物发生相互作用的状况列于表 3。表 3 列出来克丁 42 种以上，与其发生作用的微生物种类有 32 个以上。

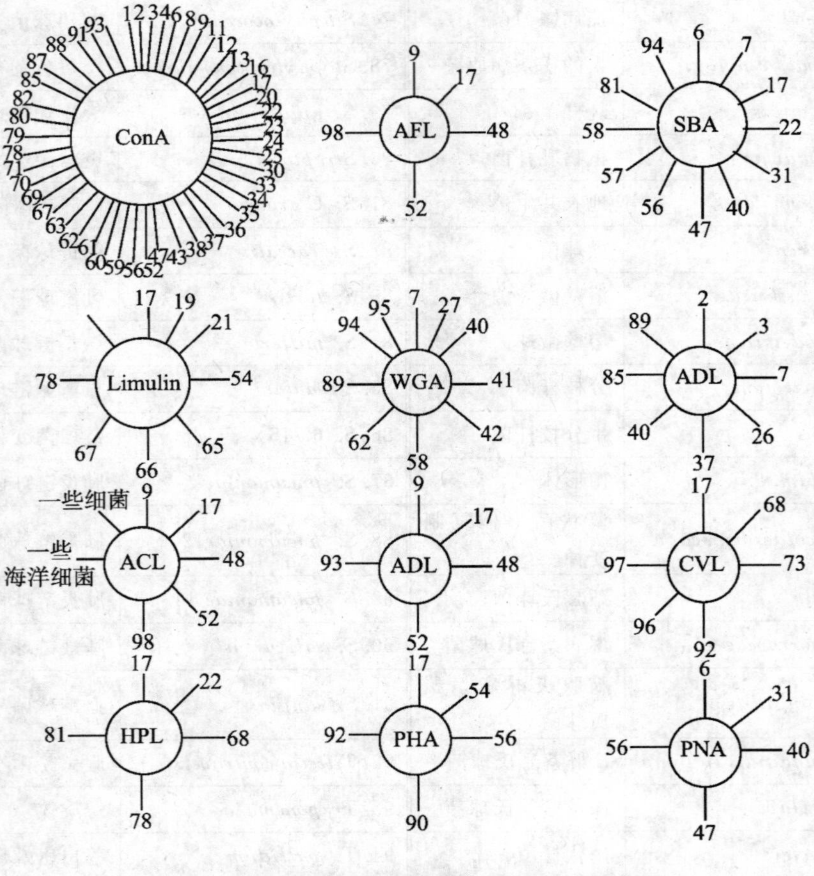

图 1　12 种来克丁与有关微生物发生反应示意图

Fig. 1　Sketch map of 12 kinds of lectins reacting on microbes concerned

虽然不同来克丁与发生作用的微生物细胞上的部位或配位体不尽相同,但运用同种微生物于不同来克丁反应的研究上时有所见[11]。这表明某种微生物在一定条件下可以与不同来克丁发生凝集作用,即有的来克丁可以作用于相同的微生物,如 ConA 与 SBA 各可作用于表 2 中所列的 6,17,22,47 和 56 号菌株。PHA 和 Limulin 均各与 17 和 54 号菌株发生作用。PNA 和 RCA 或 SBA 均各能与 40 号菌株作用。图 2 为上述 12 种来克丁与相关微生物菌株间的交叉作用状况。

表 3　42 种来克丁、似来克丁物与一些微生物发生相互作用的对应表

Table 3　Corresponding table of interactions of 42 kinds of lectins and lectin-like substances concerned with microbes

Lectins 来克丁名称缩写	作用的微生物编号及其他微生物等	Lectins 来克丁名称缩写	作用的微生物编号及其他微生物等
ACL	9.17.48.52.98. 另有其他细菌及一些海洋细菌	LEP	病原真菌
ADL	9.17.48.52.98.	LFH	34.97.
AFL	9.17.48.52.98.	LPP	77.
AMH	46.某些 G^+,G^- 细菌	LTL	Chitin 某些 G^+,G^- 细菌
APL	17	MML	17.68.75.
APP	一些病原真菌	MSP	某些细菌
Aprotinin	10.15.46	PAL	16.17.某些海洋细菌
Axinella Ⅳ	45.	PCL①	78.
BSL	50.73.	PCL②	18.63.
CH	34.97.	PPL	55.
CSL①	5.	PVP	一些病原真菌
CSL②	99.	PVH	34.97.
CVL	17.68.75.92.96.97.	SCH	34.97.
DBL	16.17.81.	SCP	一些细菌
DDL	细菌	STL	Entermoba histermoeba
EFL	Entermosba histemoeba 寄生组织的靶糖	TPL	一些病原体
HAH	64.	VCL	一些寄生组织靶糖
HCP	细菌	VFL	63.一些真菌
HPL	53.	VLL	一些土壤细菌
JWL	Chitin. 一些真菌	WFL	16.17.81.
一些来克丁	98.	另一些来克丁	地衣共生菌

* 来克丁、似来克丁物名称详见表 1,微生物名称详见表 2,表中阿拉伯数字为表 2 中相应微生物编号

来克丁一旦与微生物接触,发生"交战"的"战场"最重要之处就是细胞外层结构[12],如细胞表面、细胞壁上、胞壁内、荚膜、酵母等真菌的母体与芽体的隘痕、芽生孢子、大分生孢子[13],还有胞质外表面、胞质膜、原生质膜外表面、原生质球、泡囊、液泡中等。

微生物与来克丁物质发生相互作用的主要化学物质是糖类[4]。表 4 列出了来克丁与有关微生物发生作用的糖类等化学物质名称共约 40 种,其中主要是聚糖、多糖、磷壁酸质等,尚有不多几种寡糖、有机酸和酶等。

2.3 来克丁与病原菌的反应过程

许多无脊椎动物对病原体的感染作出的反应由循环血细胞调控,这种调控作用主要发生在它们(寄主及其细胞)的表面,其上的多糖一蛋白质复合物,主要成员便是来克丁物质[15]。它们或许是可以多重的(multiple)。同功的来克丁(isolectins)和具有异质结合点的(heterogenic binding sites)。当它们结合到入侵病原体时,将根据需要变成单体,或可能被激活,并释放出多功能低聚物,包括聚多亚基蛋白质(polymeric multisubunit protein)中的糖半部。这些便是一大类重要的识别一确认因子[16]。它们的浓度增大,常具有特定的专一性,游离或匹配至抗原上,也可迁移、活化、聚散或释放,从而对外来的非自我物起凝集作用[17]。

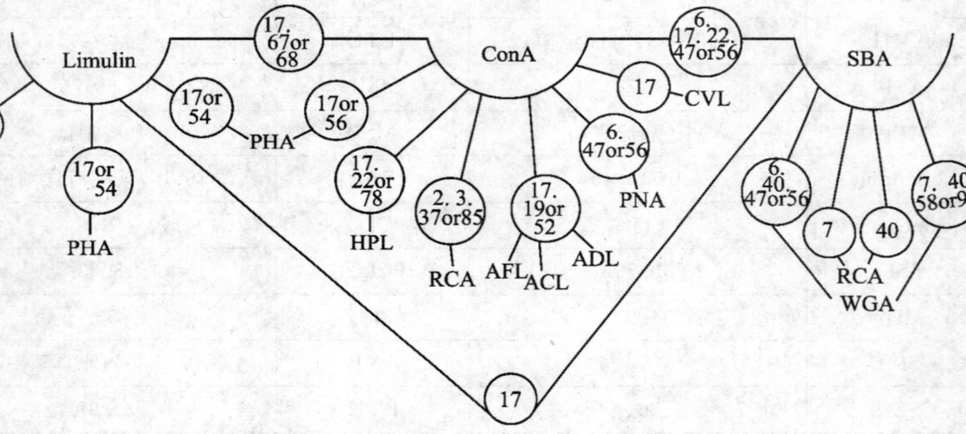

图 2 12 种 Lectins 与有关微生物交叉相互作用示意图

Fig. 2 Sketch map of intercrossing reactions between 12 kinds of lectins and microbes concerned

表 4 来克丁与微生物发生反应的主要化学成分表

Table 4 Main chemical components involved in the interaction between lectins and microbes

1. 阿拉伯半乳聚糖 4	2. 葡聚糖	3. 甘露聚糖
4. α-甘露聚糖	5. β-甘露聚糖	6. β-链甘露糖聚合物
7. 半乳糖	8. 果聚糖	9. 己糖
10. 二-N-乙酰基	11. N-乙酰葡糖胺	12. N-乙酰糖
13. 脂多糖	14. 不含 α 链的 α-D-乙酰胞外多糖	15. 荚膜多糖

(续表)

16. 酸性多糖	17. O-多糖	18. NANA 或 KDO 之类的同功聚合物
19. 垣酸	20. 葡糖基磷壁酸	21. α-葡糖基垣酸
22. α-D-葡糖取代的垣酸	23. 脂垣酸	24. N-酰基氨基酸
25. D-半乳糖醛酸	26. α-D-甘露糖残基	27. 曲二糖残基
28. D-半乳糖胺残基	29. α-脱氧-D-葡糖基	30. 终端 N-乙酰-D-半乳糖胺残基
31. N-乙酰-D-半乳糖胺残基	32. α-甲基糖残基	33. N-乙酰神经氨基残基
34. 曲二糖残基	35. 唾液酸糖醛	36. D-半乳糖苷
37. 鳞酸胆碱	38. 多聚半乳糖醛酸梅	39. 内-β-1,2-葡聚糖酶

一些在发挥作用后去凝,其中的一部分是与寄生物表面糖接合子(glycoconjugates)相连者。此外许多无脊椎动物的识别—确认因子也是糖基转移酶(glycosyltransferases),它既有合成自身糖侧链的能力,又可分辨出非自我,即抗感染。所以来克丁的多体以及它们转变成的低聚物,与高聚物之间不断发生变换。识别因子层出不穷,防御活动推陈出新,整个防卫系统有望永葆青春[18]。

人们已发现寄主对细菌等病原生物的凝集作用及杀菌作用两者间有高度的正相关性[5],这意味着来克丁的凝集活动不是孤立进行的,恰是与抑菌—杀菌作用、调理化作用相辅相成的,其相互协同、促进和完善自身的防卫活动,从而使类似脊椎动物免疫球蛋白早期前体的无脊椎动物来克丁及其类似物有了用武之地[19]。

寄主对病原体识别作用之丧失可能表明两种状态,其一,寄主已处于无力合成识别因子时,或合成了无功能形态的识别因子;其二,病原体等微生物产生了识别因子一时难以识别的细胞表面结构物,或模拟、伪装成寄主的自我物而不被发现[20]。

3　*来克丁对水产无脊椎动物的重要性加深了对来克丁认识的必要*

诚然,水产无脊椎动物具有识别、防御、抵抗、抑制或杀灭外来入侵功能的活性物质、因子及其机制不断有所发现。这意味着人们对这些生物为了保护和发展自身的能力的认识越来越全面和深入了。对整个生物界,就来克丁的性质、作用、功能等多方面的知识,近十年中有了崭新的进展,因而来克丁学(lectinology)的形成就是水到渠成的事实了。许多发展中国家的水产养殖业面临一个又一个棘手的病害问题,严酷事实不得不使我们必须清醒地总结以往的经验教训,摈弃急功近利的短视做法,按自然规律办事。从来克丁学的观点来看,一个重要的任务便是全面增加对包括其他在内的疾病防治作用机理的认识,从内外因素、生态系统上设法创造、帮助和促进养殖对象去健全、完善和稳定发展其自身的防卫体系,并保障其正常功能的持续发挥,那么来克丁类物质的应用将会有广阔的天地。

参考文献 20 篇(略)

(合作者:孙丕喜)

THE DEFENSIVE ACTIVITIES OF AQUACULTURED INVERTEBRATES AGAINST DISEASES AND VIEWED FROM THE INTERACTIONS BETWEEN LECTINS/LECTIN-LIKE SUBSTANCES AND MICROORGANISMS

(ABSTRACT)

Abstract This paper discuss the functions of lectins and lectin-like substances existing in invertebrates as viewed from the relationship between aquacultured invertebrates and their environments. This study of the interaction of microorganisms with lectins and lectin-like substances includes the kinds of lectins and related microorganisms, the sites for their interaction, and the chemical substances acted upon, mainly carbohydrates. The function of lectins to recognize foreign substances including pathogens indicates that the key task of developing aquaculture healthily is to understand and master the mechanism involved in the disease prevention activities of lectins.

Key words Interaction Between Microorganisms And Lectins/Lectin-Like Substances Aquaculture Mechanism Of Disease Prevention Pathogens

毛蚶(*Arca subcrenata*)体液来克丁的凝集作用*

摘　要　用甲状腺球蛋白琼脂糖 4B 亲和层析毛蚶(*Arca subcrenata*)体液来克丁(ASL)。试验了 ASL 对 9 属 33 株来自不同生境的细菌、安哥拉兔血细胞和人的 A、B、O 型血红细胞的凝集作用。结果表明 ASL 对 24.2×10^{-2} 的参试菌株具凝集活性(BA)。被凝者主要是来自南极及毛蚶和对虾体的细菌。对所试红细胞则显示了效价不一的凝集活动(HA)。ASL 的凝血活性受半乳糖、乳糖的一定抑制。文章讨论了 ASL 在毛蚶的防御活动中可能起的作用。

关键词　毛蚶体液来克丁　凝集作用　细菌　血红细胞　御敌机制

各个生命体之所以能自立于浩瀚生物之林，其重要一点是它们自身有一套与环境、与他物相适应的防卫体系，这防卫体系中的一个重要成员便是来克丁(Lectin)[1,2]。迄今已从处于不同分类水平的一些生物，包括海洋经济动物中发现有来克丁的存在[3~5]。大部分来克丁是生物机体产生的糖蛋白或纯蛋白。它们中有许多分布于血淋巴等体液中。来克丁经与特定对象细胞表面糖类等物质结合而使之凝集，达到识别异己的目的，或刺激某些细胞的分裂而具有多种生理生化功能[6]。生物通过来克丁对外来入侵者，包括病原微生物发生作用、抵御或对付它们，同其他机制一道，有效地保障自身正常健康地生活于环境之中[7]。毛蚶便是一例。

毛蚶(*Arca subcrenata*)大量栖息于有一定淡水影响的内湾或较平静的浅海软泥或含沙泥质海底，是一种常见的海水养殖对象，被人们广泛食用。虽然各国学者已对海洋生物来克丁作了一些研究[3~5]，但对毛蚶来克丁状况迄今只进行过少量相关探讨[8]。研究毛蚶来克丁(Arca Subcrenata Lectin，缩写为 ASL，下同)的性状将会对毛蚶养殖的病害防治机制提出新的认识。来克丁作为一类生物活性物质的另一特点是它分布的广泛性，因而很值得开发应用。本文报道了毛蚶来克丁的凝集作用的初步研究成果。

1　材料和方法

1.1　毛蚶体液的抽取

从市场上采购来的成体活毛蚶擦净后经实验室 2d 培养，剖开表面，以一次性使用的无菌注射器抽取其体腔中的体液(抗凝剂：柠檬酸钠，用量 1∶1*V*/*V*)，摇匀后，4 000 r/min 离心 15 min，取上清液，足量后即用或冷藏(4℃)备用。

1.2　ASL 的制备

1.2.1　亲和层析柱的制作

甲状腺球蛋白琼脂糖 4B(Thyroglobulin sepharose 4B)填充柱子，以 1 mol/L 的 NaCl 溶液洗涤，制成亲和层析柱[9]。

* 国家自然科学基金项目(39370539)资助。

原文刊于《黄渤海海洋》，1999，17(4)：60-65.

1.2.2 ASL的收集

上述毛蚶体液与4 mol/L的NaCl溶液以3∶1(V/V)混合,缓缓流经层析柱,以NaCl溶液和蒸馏水分别轻洗柱子,自动收集器获得洗脱液,光栅分光光度计测洗脱液峰值。选取适当峰值范围内的洗脱液待试。以天花粉来克丁(Trichosanthes Keirilowii Lectinc,即TKL)为平行参照物。

1.3 安哥拉兔和中国人血红细胞的制备

常规抽得安哥拉兔血液、中国人的A、B、O三型血液各$2cm^3$分别滴入抗凝剂中,以5倍体积的生理盐水稀释,4 000 r/min离心15 min,去上清液,重复多次至上清液无色,收集积淀来的红细胞,制成2×10^{-2}浓度的红细胞悬液。红细胞(Red Blood Cells,简写为RBCs,下同)浓度为$10^{9+}/cm^3$,直接供试或4℃存放备用。

另有一部分毛蚶血淋巴经离心后直接用于人血红细胞的凝集试验。

1.4 菌悬液的制备

参试菌共有33株,它们分别是分离自南极空气的2株芽孢杆菌(*Bacillus* sp.)、1株微球菌(*Micrococcus* sp.)、南极逆戟鲸尸体的2株库特氏菌(*Kurthia* sp.),养殖的尼罗罗非鱼的3株短杆菌(*Brevibacterimn* sp.)、3株嗜纤维菌(*Cytophaga* sp.)、3株微球菌、2株弧菌(*Vibrio* sp.),日本鳄的2株黄杆菌(*Flavobacterium* sp.)、2株诺卡氏菌(*Nocardia* sp.),菲律宾蛤仔的2株假单胞菌(*Pseudomonas* sp.),毛蚶的1株微球菌,中国对虾的2株藻酸降解弧菌(*Vibrio alginolyticus*)、3株坎普氏弧菌(*V campbellii*)、3株河川弧菌(*V. fluvialis*)和2株副溶血弧菌(*V. parahaemolyticus*)。培养和分离方法如文献[10~11]所述。

从上述刚长好的相应菌株斜面菌苔上以PBS轻轻刮洗生长物,并分别收集至$3cm^3$ PBS之中,以4 000r/min离心得一定体积的菌细胞悬液(浓度为10^7个/cm^3),待试。

1.5 细菌的凝集试验和血凝活性的测定

细菌凝集活性(Bacterial agglutinating activity,简写为BA,下同),指特定来克丁凝集细菌等微生物的活力。血凝活性(Hemagglutinating activity,简写为HA,下同),指来克丁凝集某一动物(多以哺乳动物做试验)红细胞(RBCs)的活力。BA或HA以来克丁的凝集价(titer)表达。它是以经倍比稀释后的来克丁液凝集特定浓度细胞的最小倍比稀释倍数的倒数来表示的。

ASL的BA或HA测定是在V型血凝板(microdilution plates)上完成的[12]。另外辅以玻片表面固着法测定[13]。96孔的血凝板上,各孔在滴入试验物(ASL或其倍比稀释液25 μl)后置于小型混合器上摇匀,经2h的适温培养观察结果。以肉眼见孔中网状或块状物者可初步定为凝集作用阳性,再辅以显微观察确定之[13]。以稀释倍数越大者为凝集价越高者。

1.6 凝集作用的抑制试验

将显示有较强凝血活性的各ASL液或混合物,在血凝板上作15种糖类对它的HA的抑制试验。每孔依次精确加入ASL、一种待试糖、RBCs悬液各25 μL,并作倍比稀释,培育和观察方法同上。RBCs采自安哥拉兔,表1列出的是参试糖类及所用浓度状况。

表 1　毛蚶来克丁对红血细胞凝集作用的抑制试验用糖及其浓度表

Table 1　Name and their concentrations of sugars used in inhibitory test of ASL against RBCs

试剂名称	所试浓度
唾液酸 Neuranainic acid	10.0 mg/dm^3
N-Z 酰唾液酸 N-Acetyl neuramine acid	10.0 mg/dm^3
葡聚糖 Dextran	10.0 mg/dm^3
L-岩藻糖 *L*-fucose	0.2 mol
半乳糖 Galatiose	0.2 mol
N-乙酰氨基半乳糖 N-acetyl-D-galaetos amine	0.2 mol
葡萄糖 Glucose	0.2 mol
N-乙酰-D-氨基葡萄糖 N-acetyi-D-galactos amine	0.2 dm^3
胍 Guanidine	10.0 mg/dm^3
乳糖 Lactose	0.2 mol
甘露聚糖 Mannan	10.0 mg/dm^3
甘露糖 Mannose	0.2 mol
黏蛋白 Mucin	10.0 mg/dm^3
卵类黏蛋白 Ovomucold	10.0 mg/dm^3
甲状腺球蛋白 Thyroglobulm	10.0 mg/dm^3

* 红血细胞采自安哥拉兔

2　结果

2.1　ASL 的 BA 测定

ASL 对 33 株细菌的凝集活性试验结果列于表 2 之中。

试验结果表明，所试菌株中有 8 株被凝集价为 4 的 ASL 所凝聚，这 8 株菌分别为 2 株芽孢杆菌、2 株库特氏菌、2 株微球菌和 2 株溶血弧菌。被凝集的菌株数占参试菌株数的 24×10^{-2}(8/33)。被 ASL 所凝的菌株中有 5 株分离自南极，即芽孢杆菌、库特氏菌各 2 株和 1 株微球菌，占所试南极菌株数的 100×10^{-2}(5/5)。ASL 对分离自毛蚶的微球菌及对虾的副溶血弧菌具有凝集活力。这说明 ASL 对一些微生物细胞有凝集作用。毛蚶 ASL 对地域上与其相去甚远的南极菌株显出特别普遍的凝集活力。

2.2　ASL 的 HA 测定

表 3 列出毛蚶来克丁凝集 4 种 RBCs 的凝集价。

表 2　毛蚶来克丁对一些细菌的凝集价

Table 2　BA titers of ASL againsts some bacterial cells tested

参试菌株名称	菌株数	凝集价
芽孢杆菌 *Bacillus* sp.	2	4
短杆菌 *Brevibacterium*	3	

（续表）

参试菌株名称	菌株数	凝集价
嗜胞菌 *Cytophaga* sp.	3	—
黄杆菌 *Flavobacterium* sp.	2	—
库特氏菌 *Kurthia* sp.	2	4
微球菌 *Micrococcus* sp.	2	4
微球菌 *Micrococcus* sp.	3	—
诺卡氏菌 *Nocardia* sp.	2	—
假单胞菌 *Pseudomonas* sp.	2	—
弧菌 *Vibrio* sp.	2	—
褐藻酸降解菌 *V. alginolyticus*	2	—
坎普氏弧菌 *V. campbellii*	3	—
河川弧菌 *V. fluvrialis*	3	—
副溶血菌 *V. parahaemolyticus*	2	4

表 3　毛蚶来克丁对血红细胞的凝集价

Table 3　HA titers of ASL againsts some RBCs tested

参试血红细胞类别	毛蚶来克丁洗脱液峰值编号等	凝集价
安哥拉兔	No. 33	4
	34	8
	35	16
	36	16
	37	16
	38	16
	39	8
	40	4
	41	4
	42	4
	33～42 混合	8
A 型（人）	经离心的血淋巴上清液	8
B 型（人）	经离心的血淋巴上清液	4
O 型（人）	经离心的血淋巴上清液	<4

在洗脱液峰值 No. 33～42 间的 10 份 ASL 液中，发现有 9 份对兔 RBCs 显出了凝集活

性,其中以 No.34～39 的 ASL 凝集价为高。No.35～38 的 ASL,洗脱液的凝集价为 16。对人的三型 RBCs,ASL 的 HA 水平均低于对兔的,其中对 A 型 RBCs 的 HA 有最高凝集价(达 8)。对 B 型血的居于中间。对 O 型血的为最低,凝集价不超过 4。结果说明在所试范围内,ASL 具有一定的凝血作用,并显出凝血强度的某些差异,但未见该 ASL 的凝血作用在所试动物种属上的专一性(Speciality)。

2.3 ASL 凝集作用的抑制

用表 1 中所列的 15 种糖类试剂进行 ASL 的 HA 被抑试验,结果表明,仅有浓度为0.2 mol 的半乳糖显出较明确的抑制作用。乳糖(所试浓度为 0.2 mol)的抑制作用不大,其余各试剂均未显抑制力。

3 讨论

来克丁性质的主要标志之一是对特定细胞的凝集作用。该作用过程主要靠其分子的亚基与存在于靶生物表面上的来克丁能识别的糖基结合而完成。专一性强的来克丁,其糖结合部位识别和结合特定细胞表面抗原决定簇。

本文用 ASL 所进行的血凝作用初步表明它缺乏这类明确的专一性[14]。可以推测 ASL 识别的糖基具有普遍性,即在许多动物红血细胞中都可能有此类来克丁识别的糖基,也反映出毛蚶在进化历程中的某些遗传保守适应性质[2]。迄今尚无证据可以排除它在其他动物血细胞凝集作用上将产生的专一性。

ASL 之所以只与一些而不是与所有供试微生物菌株的细胞起凝集作用,其中可能包含着 2 个原因:①决定于 ASL 所含的识别分子;②被凝集的菌株细胞中是否有相应的抗原决定簇,即这些微生物细胞表面有无 ASL 的糖可结合性存在[1~2]。ASL 的凝血作用被半乳糖、乳糖所抑可从一个侧面映证此论点。

ASL 对所试南极菌株的凝集作用,表明某些南极微生物可能难以象不被 ASL 凝集的细菌那样,去侵袭在差异大的地理环境中生存的无脊椎动物体进而在其体内繁衍开来,也反映了南极生态系中的微生物的脆弱性[15]。ASL 的 BA 现象还意味着,在环境中与毛蚶相伴的许多微生物,其生态区系性质受制于来克丁之类的生物活性物质[4]。

本试验所用的 ASL 主要是含来克丁浓度不高的体液,其凝集价偏低。对毛蚶进行适当的诱导,有可能提高 ASL 的凝集价。提取、制备纯度高的 ASL 将会使其显出更明确的生物学活性,由此全面认识 ASL 在机体防御活动中的自我—非自我识别能力、对入侵微生物的准免疫和潜在的吞噬作用、调理活动。这些工作尚待深入进行。

致谢 中国科学院生物化学研究所王克夷研究员指导和制备 ASL、提供兔血细胞、TKL 和部分血凝试验器材,青岛市人民医院王正强医师协助制备人血细胞,青岛海洋大学徐怀恕教授提供部分菌株,在此一并致以谢忱。

参考文献 15 篇(略)

(合作者:孙丕喜)

AGGLUTINATION OF HUMORAL LECTIN FROM BLOODY CLAM (*ARCA SUBCRENATA*)

(ABSTRACT)

Abstract This paper discusses the probable function of the humoral lectin from the bloody clam (*Arca subcrenata*), i. e. ASL in its defensive mechanism. The humoral lectin was analysed by affinity chromatography using thyroglobulin sepharose 4B.

Experiments were made of the agglutination of 33 strains of bacteria belonging to 9 genera and coming from different biotic environments, red blood cells from the Angora rabbit and Chinese people.

The results indicated that ASL had 24.2×10^{-2} bacterial agglutinating activity (BA) against tested bacterial strains from Antarctica, bloody clam and Chinese penaeid shrimp. The different titers were presented of hemagglutinating activity (HA) of ASL which was inhibited by galactose and lactose at certain concentrations.

Key words Arca Subcrenata Lectin (ASL) Agglutination Bacteria Red Blood Cells (RBCs) Defensive Mechanism

微生物凝集素的研究*

摘　要　本文首先分析微生物凝集素种类及其来源生物类别、主要性状，如糖类专一性及它们与一些蛋白质、酶、抗生素等间的反应，讨论病毒、细菌、黏霉凝集素等的主要功能，凝集素在海洋细菌、藻和无脊椎动物中的活动。文章着重阐述由细菌凝集素介导的凝集素吞噬作用及其意义和价值。最后文章认为对微生物凝集素(包括海洋微生物凝集素)的研究以及对其有益方面的开发已经并将继续有助于人们去探索水产养殖生物御敌机制、病害防治方法和人类医学的发展。

关键词　微生物凝集素　凝集素吞噬作用　识别　水产养殖生物　医学科学

一、微生物凝集素及其种类分析

生物学意义和用途已探讨得比以前更清楚了的凝集素(lectins)广泛存在于原核至真核的许多生物之中。微生物凝集素(microbial lectins，ML)的最先发现时间是1908年，Gugot从大肠杆菌(*Escherichia coli*)中找到了能凝集血细胞的凝集素。至20世纪80年代初，已发现了188种以上的ML物质。近十几年来的研究，又进一步扩充了ML的范围。其总数约212个。表1列出了被研究得较多的五类ML数[1]。

表1　已发现的微生物凝集素类别及种数

凝集素类别	凝集素种数
病毒凝集素(VL)	10
细菌凝集素(BL)	132
真菌凝集素(FL)	57
细胞黏霉菌凝集素(CSML)	6
原生动物凝集素(PL)	7

分析ML的来源生物状况得知，在1986年以前，它们大都来自高等生物潜在的致病菌或与哺乳动物密切相关的种类。在71个ML中，以分离自动物不同部位的微生物种类的百分比多寡排序为：肠道(36.6%)、呼吸道(23.9%)、口腔(18.3%)、尿殖道(14.1%)、血细胞(2.8%)、其余(4.2%)，此外来自环境的有33个。由表1可见，ML来源生物类别按多寡排序为：细菌＞真菌＞病毒＞原生动物＞细胞黏霉菌。它们分别是4属7种病毒、33属122种突变株的细菌、40属53种真菌(包括细胞黏菌2属6种)和6属6种原生动物。种属多样化增强，尤其是真菌类凝集素的增幅更为明显，即来源真菌属、种和凝集素

* 国家海洋局生物活性物质重点实验室资助项目
原文刊于《自然杂志》，2000，22(2)：95-100。

数量分别是以往的2.38、2.35和2.24倍。ML的总数量则是以往的1.39倍。

分析得知，ML在微生物体的部位以细胞表面为常见。它们可以与细胞联系不那么密切，如特定的表面细胞器上，包括细菌菌毛(firnbria)或纤毛(pili)。这些物质有时被称作为粘附素(adhesins)，它们中具有许多凝血能力。这些部位还可包括病毒的刺突、真菌的子实体、分生孢子、菌丝等。ML不限于生物的某个位点，有的可存在于胞内颗粒，如原生质、周质等。有些则游离于细胞，分泌至培养基中。一种微生物不一定只产生一种凝集素，如嗜水气单胞菌 *Aerorrmnas hydrophila* 即有两种与细胞相连的凝集素(一是血凝素，一是粘附素)，又有游离于细胞的血凝素，还有菌毛凝集素(pili agglutinin)等等。

二、微生物凝集素的一些主要性状

如同多数其他生物来源的凝集素一样，大多数ML的一个重要特性是对某些/某种糖有结合性能，它具有可变换性和非共价性，其中有的结合能力的特异/专一性很强。迄今发现的ML结合的糖及其衍生物之范围虽然有限，但也达60种以上. 表2列出了ML所结合糖的种类数。分析认为，这些糖大多属于单糖、寡糖或者双糖之类，而以甘露糖为多见，其次是半乳糖等。与同病原菌发生反应的动植物凝集素的结合的糖种类相比，有一定差别[2]，而且ML结合的糖类范围正在扩展之中。这与当今从微生物寻找凝集素的努力有关[3]。

表2 抑制微生物凝集素活动的一些糖类、酶、蛋白质和抗生素种数

凝集素抑制物类别	抑制物种数
糖类	61
酶类	7
蛋白质类	19
抗生素类	9

表2还列出了与ML相结合或起抑制作用的一些蛋白质、酶或抗生素种类数，即它们大多是对某种/一些ML的活性起抑制剂作用的，但种类上比糖的少。同时，还必须注意到有些ML并不引起血凝作用，但可以起其他作用。

三、微生物凝集素的功能

ML之所以如此引人注目，一是因为研究微生物本身所需。凝集素是微生物自身进化发展之需。一些微生物，包括各类病原菌，需要与寄主发生各种关系以发展自己。这类微生物如沙门氏菌、链球菌、弧菌、衣原体、拟杆菌等。它们的凝集素是感染动物的毒力决定子，其中以大肠杆菌对甘露糖和galgal——专一的菌毛凝集素为突出例子。二是因为一些有助于其产生菌选择识别寄主的表皮(面)细胞，或者识别高等动物的白细胞。在此，该凝集素起着细胞与细胞间相互作用中的识别分子(recognition molecules)作用。再深入一步看，识别作用本身又是个复杂的过程，有多种活性物质参与。凝集素在此起识别活动的决定子(determinat)作用，如粘球菌(*Myxococcus*)的细菌所显示的那样。这说明ML在介导各种正常和病理过程中起重要作用。这些决定子作用首先由细胞表面凝集素发挥。凝集素吞噬作用(lectinoph agocytosis)则是近几年来议论的主题之一。三是因为 ML 广泛用于生化研究工作之中[4]。

四、微生物凝集素的识别作用

识别作用在所有生物学现象中是一个中心事项,且是细胞间反应过程中的首要步骤.它参与受精、胚胎生成、细胞迁移、器官形成、免疫防御和微生物感染中。细胞识别作用的不适当功能可能引起疾病,故涉及凝集素与细胞糖类结合等在内的细胞与表面编码分子基础的理解关系着生物学和医学的许多领域[5]。

凝集素作为识别分子的理论兴起于 80 年代。几乎所有细胞之表面均携有糖,糖在编码生物信息中具有潜势。凝集素与相关糖间的关系好比钥匙与锁间的互补关系[5]。两者间的反应调节识别作用,同时也意味着该反应具依赖糖的性质。

流感病毒与其靶细胞间的相互作用是凝集素——糖类识别系统的一个很好例证[4]。该病毒的血凝素 IVH(influenza virus hemagglutinin)在其靶细胞表面上与唾液酸化的糖类相结合导致它附于此细胞上,致病毒与细胞膜融合,允许病毒基因组释放入细胞质并复制。借助唾液酸酶从胞膜移去唾液酸,则可去除结合、阻止感染。唾液酸化酶再附作用或含寡糖基的唾液酸插入唾液酸酶处理的细胞膜再储存该细胞结合此病毒的能力而感染之。唾液酸血凝素相互作用的有关知识为获取阻断病毒附于细胞的抗病毒药物之设计提供了可能的理论基础[6]。

细胞表面凝集素在肠系细菌种类如大肠杆菌(*E. coli*)、沙门氏菌(*Salmonellae spp.*)中常以菌毛的亚显微毛发状形式出现。*E. coli*1 型菌毛是甘露糖专一的,它优先结合动物细胞表面糖蛋白的寡甘露糖(oligomannose)和杂化寡糖上。后两者在整合素(integrain) $CD_1 1$/CD18 上。*E. coli* 的 P 菌毛专与含 Gal_{21}—4Gal 的糖脂结合。S 菌毛则对 $NeuAc_{\alpha}$—3Gal 专一。与糖结合的位点一般位于菌毛端上和沿着其长轴的大间隔中。口腔的放线菌(如内氏放线菌 *Actinomyces naeslundi* 和粘放线菌 *A. viscosus*)凝集素对半乳糖专一结合,以此促使细菌在口腔和牙齿表层膜的表面上群集并吸附于牙齿珐琅质上。由此可知,细菌表面凝集素在引起感染中起关键作用。这一作用靠调节细菌附于寄主表皮/层细胞来实现。

粘霉的一种盘基网柄菌(*Diztyostelium discoideum*)之凝集素是半乳糖专一的 discoidin,它集中于细胞表面。该凝集素在此粘霉从单细胞、营养形态到集合体形态的分化发育过程中起调节作用,即起细胞与细胞间的确认作用。借此,细胞相互黏附,其间该凝集素靠糖自主机制经三肽 Arg—Glv—Asp 这一短段分子结合到细胞上[7]。

五、凝集素在海洋细菌、动植物附着和共生活动中的作用

1. 附着

痢疾内变形虫(*Entamoeba histolytica*)有两种凝集素,一是对 N-乙酰葡糖胺的 1—4 连的低聚物专一;另一个对半乳糖和 N—乙酰半乳糖胺专一。该变形虫靠凝集素使人体的结肠粘膜紊乱,入侵体内可诱发赤痢。由于一些糖对此凝集素有抑制作用,从而减少了它附于肠道细胞及其他靶细胞上的能力,说明了凝集素-糖的相互作用调介该变形虫的附着[8]。

在无脊椎动物的变态和附着活动中，微生物组成的表膜，包括其中的ML起着重要作用。微生物表膜的外多糖和蛋白质等调节无脊椎动物不可逆转的附着作用。其中的凝集素构成了生化识别因子。海生假单胞菌（*Pseudomonas marina*）的膜在诱导多毛类动物石灰虫 *Janua Dexiospira brasiliansis* 等的幼体附着和变态的行为上最有控制力。该动物幼虫表面上的凝集素调节这一过程。它能识别 *P. marina* 外聚物中的糖（主要是葡萄糖），并与之结合，而 *P. marina* 产生的凝集素又与该幼虫发生识别和补充结合活动[9]。理解并运用微生物在动物变态中此类凝集素作用的知识，对提高水产养殖生物幼体的附着和成活率是很有用处的。

2. 共生

凝集素在生物的共生活动中起作用，如海绵的一种凝集素，即由 *Halichondra paricea* 产生的 Halichondra lectin 是稀有假单胞菌（*Pseudomonas insolita*）生长所必需的。借此，该菌可寄生于海绵中，该菌产生的多糖推测被海绵凝集素识别为自我。同时（这一）共生活动中的凝集素可能起着“种的专一识别”作用，以确定种水平上特定的寄主——共生物关系[10]。

砗磲（*Tridacna maxima*）的血淋巴凝集素即 tridacnin 专一结合具有一个-半乳糖（苷）构型的糖，而半乳糖苷糖类（galactosyl sugars）恰是双鞭藻 *Synibiodinum*（裸甲藻）（=Gymnodinium）*microadriaticum* 等藻的细胞壁组分。在砗磲与该藻建立的共生关系中凝集素起了选择作用。该凝集素识别出此藻细胞壁上的结合位点，中介其合成者不同生长阶段中共生者在体内的输运。证明了该凝集素在共生的种和株专一性上起识别作用[11]。

六、凝集素与吞噬细胞间建立的凝集素吞噬作用

一些携有凝集素的细菌等微生物可容易地结合到吞噬细胞，如人的多形核白细胞或鼠的腹膜巨噬细胞的糖上，通过ML与吞噬细胞的结合激活了后者的代谢活动，消化并最终消灭相应细菌，此属于非调理的吞噬作用（nonopsonic phagocytosis），1988年被 Ofek 等称作为凝集素吞噬作用（lectinophagocytosis），具有重要的临床价值[1,12]。

当今参与细菌凝集素吞噬作用的一些物质有许多被称作为粘附素（adhesins）。粘附素属于凝集素范畴[13]。能在哺乳动物组织上群集的细菌均借其细胞表面的菌毛、纤毛等发状附器完成此过程。它包括凝集素识别吞噬细胞表面的特定糖，但是这些糖不一定妨碍已调理细菌与噬细胞发生反应，介导吞噬作用，刺激对细菌的杀灭活动。迄今已发现12种细菌表面凝集素与吞噬细胞发生此类相互作用。它们分离自10种细菌（详见表3）。

表3 细菌表面凝集素介导的凝集素吞噬作用

凝集素名称	糖	细菌	受试细胞
1型菌毛	甘露糖	大肠杆菌 鼠伤寒沙门氏菌 肺炎克雷伯氏菌	人粒细胞、巨噬细胞 小鼠腹膜巨噬细胞 大鼠腹膜巨噬细胞
2型菌毛	半乳糖 3N-乙酰 半乳糖胺	粘放线菌 内氏放线菌	人粒细胞 人粒细胞

(续表)

凝集素名称	糖	细菌	受试细胞
PrsCFl6517 菌毛	N-乙酰半乳糖胺 β3N-乙酰半乳糖胺	大肠杆菌	猪中性白细胞
P 型菌毛	半乳糖,1,4 半乳糖	大肠杆菌	半乳糖,a 半乳糖 4S(含糖脂),包被的人粒细胞
菌毛	N-乙酰半乳糖胺	牙龈卟啉单胞菌	小鼠腹膜巨噬细胞
丝状凝血素 百日咳毒素	半乳糖	副百日咳博德特氏菌	人和大鼠齿槽巨噬细胞
外膜	N-乙酰半乳糖胺 半乳糖	啮蚀艾肯氏菌 具核梭杆菌	豚鼠巨噬细胞 人粒细胞
细胞表面	N-乙酰半乳糖胺 N-乙酰神经氨酸	腐生葡萄球菌	人粒细胞
菌毛	配位体	铜绿假单胞菌	巨噬细胞
S 型菌毛	Neu Aca2,3 Gal-专-	大肠杆菌	
ML 蛋白	岩藻糖	酿脓链球菌	土著动物细胞
SSA	唾液酸	血链球菌	

* 资料来源:多引自文献[6]

包括 *E. coli* 等肠杆菌细菌在内的Ⅰ型菌毛与各种动物及人的细胞(如噬细胞)表面上的甘露糖受体结合。多数病原性 *E. coli* 还有 P 和 S 菌毛。它们很少或根本不结合到人的粒性细胞上,但经其他方式凝集一些动物血细胞。与 P 菌毛相关,却与 P 菌毛相反的 Prs,其携带者(*E. coli*)可结合到猪的中性粒细胞上。此外,*E. coli* 尚有非菌毛的粘附素(nanfmbrial adhesins,NFA)。它们是 NFA-1 和 NFA-3,能介调细菌结合到人的粒细胞上。鼠伤寒沙门氏菌(*Salmonlla typhimudum*)等细菌产生 MR(mannose resistant)血凝素(属粘附素之类)参与非调理吞噬作用。

该菌表达出粘附性,并被鼠腹膜巨噬细胞所消化[14]。

淋病奈瑟氏球菌(*Neisseiar gonorrhoeae*)的外膜蛋白(outer membrane proteins,OPa)被认为是凝集素样成分,它可与吞噬细胞相互反应。而该菌表面上的唾液酸化的脂寡糖(lipooligosaccharide)则参与对抗吞噬[15]。一种条件致病菌铜绿假单胞菌(*Psueomonas aeruginosa*)的两种胞内凝集素即 PA-I 和 PA-Ⅱ,在该菌使烧伤或囊性纤维变性病患者感染期间,它们调停细菌与表皮细胞、吞噬细胞间的反应[3]。口腔感染中有 *A. naeslundii* 和 *A. viscosus* 参与,它们有一种被称为 a 型菌毛的粘附素,借此粘附素作用,这些放线菌结合到人粒性细胞上,经凝集素吞噬作用,过氧化物得以刺激,乳铁蛋白(lectofernn)得以释放,相应病菌则被消化和杀死[13]。结核分枝杆菌(*Mycobacterium tuberculosis*)表面上的脂阿拉伯甘露聚糖(lipoarabinomarman)介调结合到人巨噬细胞上的活动。

副百日咳博德特氏杆菌（*Bordetella portussis*）的丝状凝血素（filamentous hemagglutinin）含有 RGD 三联体，它促进细菌粘附至哺乳动物噬细胞（及其他细胞）整合素 CR_3 上。与此相反，另一个含 RGD 蛋白被称为 Pertactin，则介导细菌结合至 Hela 细胞的整合素上。该病原菌由此识别、介调不活化/活化的受体的表达[16]。相类似的是鸟分枝杆菌（*M. avium*）和胞内分枝杆菌（*M. intracellulare*），它们的复合物则靠鼠巨噬细胞或人单核细胞上的整合素 αvB3（victronectin）受体等来介调识别[8]。利斯曼虫的原丝胞（promastiagates）借其多种粘附素可结合到哺乳动物巨噬细胞整合素 CD11b/CD18（CR_3）的两个位点：一个位点结合脂多糖，如脂磷聚糖；另一个是 iC3b，它识别 gp63（纤维结合素样分子）[77]。克鲁氏锥体虫（*Tryparosoma Cruzgi*）和白色假丝酵母（*Candida albicans*）非调理结合噬细胞则是通过整合素样分子来介调的，这些分子存在于此类真核微生物表面。并可结合到噬细胞表面 fibronectin（纤维结合素）的 RGD 序列中[18]。肺炎菌 *Pneumocystis carinii* 和 *C. albicans* 均是特别感染获得性免疫缺陷症病人的条件致病菌。巨噬细胞甘露糖受体（即 Man/GlcNAc 专一凝集素）具有结合和消化上述两微生物的能力。白色葡萄球菌（*Staphlococcus albus*）经 1a 抗原（巨噬细胞的）结合到巨噬细胞则是凝集素吞噬作用的另一个模式，借此模式可导致免疫反应的发展。有关巨噬细胞凝集素的其他内容因本文篇幅所限，在此不多述及。

七、凝集素吞噬作用的意义和价值

Ofek 等（1995）[6] 指出，微生物侵犯非调理的吞噬活动。采用的两个战术之一主要是依赖其表达的粘附素，这包括 *E. coli* 的各种菌毛、淋病萘瑟氏球菌（*N. gonorrhoeae*）的 Opa、*S. pyogenes* 的脂磷壁质等。借此，细菌结合到周围动（植）物表皮和吞噬细胞的受体上，使后者感染。细菌得以在粘质表层中生存并集群. 在深层组织则去冒犯以致消灭吞噬作用。为了生存和继续感染，活体中这些细菌合成荚膜等表面聚合物. 借助改变表面的网状符号、表面疏水性或表面粘附素的形式和态势等去干预粘附素-受体的相互作用。另一战术则是位相变化（phase *vmiation*）。在此变化中，生物从一个变异体以较高频率振荡表达其粘附素以应对无相应（粘附素的）振动的另一个变异体。不产生粘附素的变异体通常在富噬细胞的位点占优势，既结合噬细胞也结合粘细胞，那些表达粘附素的变异体则在噬细胞位点占优势，另有一些微生物产生粘附素结合噬细胞和粘细胞，但不改变其粘附力，它们很像仅群集于粘液表面而不能在深层组织中成功感染。这类微生物可能归入动植物的正常微生物区系中。

因此微生物粘附素的作用至少有两个：一是在寄主粘液表层促进感染；二是在深层组织中增强吞噬和消灭寄主。为了在各种表面上群集，微生物产生复式粘附素。借其多重分子机制和复杂的遗传调节，此类粘附素无须回避调理吞噬即可抵达深处引起感染。为了预防此点，宿主必须加强非调理吞噬作用功能，首先是借此来预警和监察有症状的感染，引发发炎因子释放。

不同模式的研究表明，病原-巨噬细胞相互作用的结果可不依赖于利用摄取病原的受体。而整合素既可在胞内也可在胞外发挥作用。它们可确认“自我”和识破“异己”。而且非调理（如凝集素的）和调理的吞噬作用是相关联的[19]。非调理吞噬活动中巨噬细胞表面受体的分子区与调理吞噬中的受体分子区可以是同一的也可以是区别的，即微生物等的非调理识别中的一些受体可能参与吞噬作用等许多其他生物学活动。这是在世代进化压力下不断筛选优化出的识别系统，目的是增加宿主防御中对抗感染的天赋免疫力以延续和发展自身于生物之林中[20]。因此，对微生物和吞噬细胞间包括凝集素在内的非调理机制的理解也将有助于人们去认清和驾驭水产养殖生物的御敌机制。调整和发挥微生物凝集素对人们生产和生活有利的一面，将有助于发掘和利用微生物这一资源宝库以造福人类。

八、结语

本文论及微生物凝集素的种类、来源生物. 据不完全统计，迄今已从 83 属 188 种微生物中发现了 212 种凝集素(物)。从真菌中寻找凝集素的努力得到了加强。凝集素所结合的糖种类不断增多。试验了蛋白质、酶、抗生素等与凝集素间的作用。

文章论述了微生物凝集素的三个主要功能。分述了病毒、细菌、粘霉凝集素的作用以及微生物凝集素参与的原生动物、藻类、贝类动物附着、变态、共生活动。评述了凝集素吞噬作用的研究状况、价值及意义。对微生物(包括海洋微生物)凝集素及微生物与凝集素关系的研究将有助于促进人体和水产养殖生物的健康。

参考文献 20 篇(略)

(合作者：牟敦彩　李光友)

RESEARCH ON MICROBIAL LECTINS

(ABSTRACT)

Abstract This Paper discusses microbial leatins. It includes that analyses on their kinds, properties, effects, functions, lectinophagocytosis, significances, and so on.

Key words Microbial Lectins, Lectinophagocytosis, Recognition, Aquacultured Bionts, Medicine Sciences

微生物来克丁分析方法*

摘　要　根据微生物来克丁增长的功能和用途，要求建立精细的来克丁分纯和鉴别法。阐述枯草芽孢杆菌、根瘤土壤杆菌、少孢节丛孢、齐整小菌核、草菇等的来克丁之分纯和测定法。指出微生物来克丁先进测试方法将促进微生物，包括海洋微生物来克丁的广泛应用。

关键词　微生物来克丁　分离　纯化　应用　功用

引言

来克丁学(Lectinology)正在兴起。近二三年来国际间每年有400～500篇论文发表，涉及约20个专题。微生物来克丁的研究大约占其10%左右，且显出方兴未艾之势。有微生物参与的来克丁，包括其自身产生的来克丁，涵盖代谢、遗传、免疫、受体(及致裂原)、化学(及纯化、分析)、药理、治疗等几个方面。代谢方面的研究大约占29%。尚有27%左右的文章论及分纯及化学等。这表明微生物来克丁研究已相当侧重认识其性状和功能，反映微生物作为来克丁的重要来源库越来越被重视，由此使微生物来克丁的开发和应用不断增长。本文从微生物来克丁的功用出发指出有关的分纯和分析工作的重要性，并较详细列举5种微生物的来克丁分纯提取和测试方法，以促进我国来克丁学的发展。

1　微生物来克丁的功能与用途

微生物等来源的来克丁(lectins)的功能与用途越来越被认识，如它能使病原微生物识别血红细胞甚至其他细胞上共同的糖类[1]。比如由于枯草芽孢杆菌(*Bacillius subtilis*)316 M等腐生菌的胞外来克丁与相应的细菌毒性紧密相连，而使它对某些糖具选择和可逆结合力。来克丁可作为研究糖接合之分子的试剂，在具细胞表面功能和生物学性质的液体中作测定用。来克丁制备物在某些疾病诊断试验系统中也有应用。已认识到获取罕见、独特的专一性微生物来克丁制备物这一生物工程学研究的巨大实用性[2]。

来克丁广泛用于癌细胞表面糖基化组分的纯化、结构与功能的研究及其检测。在允许癌转移的受体测定中有用[3]。来克丁可作为研究与正常组织和临床分离的细菌株有潜力的工具，如来克丁反应性(lectin reactivity)在区分芽孢杆菌的种类及奈瑟氏球菌科成员上是一个迅速而有效的手段[4]。

酿酒酵母絮凝作用在酿酒工业中有商业重要性。厂家希望在单细胞生产阶段就能迅速稀释糖成酒精，在麦芽糖用尽后，以期絮凝出清澈的啤酒。获得的酵母产物又可适用再调制进日后的酿造之中。啤酒酵母表面来克丁(*Sacchromyces cerevisiae* surface lectin，缩

* 原文刊于《青岛科学》，2000，2(3)：34～38。

写为 SCL)正发挥其不可替代的作用[5]。

一些真菌附于寄主表面这一黏附现象是真菌在人体、动植物体上聚居乃至发病的一项重大事件,但对其附着机制知之不多。现仅对糖～蛋白质(来克丁相互作用介导真菌～寄主,植物病原、真菌～真菌,真菌～藻类,嗜线虫真菌～动物间的粘附/居着)活动有所了解[6,7]。

微生物来克丁在 20 世纪 80～90 年代仅处于振兴阶段,由于其众多而纷繁的培养、合成条件和作用机理等问题研究的滞后,这些又与其来克丁方法息息相关,因此它在许多方面已被动植物来克丁所掩盖或领先。迄今商品化微生物来克丁在国际上也是凤毛麟角。开发海洋微生物资源越来越被人们认识,对其来克丁研究更为紧迫。

经历徘徊和探索之后,人们进一步认识到只有更正确地理解微生物来克丁的作用,才能发挥它们的功能与用途。这些都要求和必须建立在对其性质有明晰的认识和基础上[9]。

人们就微生物来克丁的提取、纯化技术已作了不懈努力[2],为描述其性状特征提供了先决条件。然而迄今中国大陆尚未见微生物来克丁提取方法的报道[10],本文叙述 5 种微生物的来克丁提取、分析方法,以期激励进一步作出更多贡献,促进人们对微生物来克丁的日趋重视。

2 枯草芽孢杆菌来克丁的提取

2.1 枯草芽孢杆菌(*Bacillus subtilus* 316M)来克丁(Bacillus subtilus lectin,缩写为 BSL)的提取[4]

B. *subtilus* 316M 培养液按照 Kudrya 等(1994)的方法可先取得碱性丝氨酸蛋白酶,该酶的阳离子交换剂洗脱液含 BSL。提取方法如下:回旋式蒸发器浓缩该洗脱液,于 5～7℃温度中缓慢加入固体$(NH_4)_2SO_4$。搅拌过夜,得 60%饱和度液。5℃ 200 g 离心 15 min。所得积淀物溶于小体积水,再对蒸馏水透析 12h 后冻干。

此法第一阶段即可获纯度是直接从培养液提取法两倍的来克丁。经$(NH_4)_2SO_4$沉积后,来克丁浓度可提高 12 倍,凝血活性则大大提高。

糖试验表明 BSL 对 N-羟乙酰神经氨酸、N-乙酰神经氨酸、N-乙酰胞壁酸和 D-半乳糖醛酸高度专一,其分段之等电点主要是 PI 3.1～3.75、5.10、6.8～7.15。该法需求较多的培养液。

2.2 枯草芽杆菌结合唾液酸来克丁(Sialic acid binding lectin, SBL)的提取[9]

该法可称作为一步性疏水相互作用色谱法(One-step Hydrophobic interaction chromatography,HIC)。

2.2.1 弱疏水吸着剂的合成

玻璃珠(MPD-2 000VGKh,平均孔径 2 000A,颗粒大小:0.16～0.31 mm)。细硅胶粒(平均孔径 500A,颗粒大小:5 μm)。

丁基-Toyopearl 650C 凝胶,Trp-Trp 二肽、溶菌酶、丁基-PA-硅。

步骤:将 64±0.5 mg 的填料装入 1.0 mL 的微量离心管中。轻搅,分散颗粒入 0.4 mL 的悬浮液中并定量转入校准了的玻管(50×3.5 mm I.D)中,记下经 3～4 d 沉积后的沉降高度,估计出乙醇-水混合物表面张力。

在 1×9 cm 的玻柱中作等度疏水相互作用色谱(Isocranc hydrophobic-interactio

chromatography)。流速:1.0 mL/min,蠕动泵控制。

不同浓度的$(NH_4)_2SO_4$液分别溶入2 mg溶菌酶或0.5 mg的Trp-Trp二肽,再各溶入1 mL pH7.0的0.1 mol/L的磷酸缓冲液(Buffer,下同),待用于上述柱中。LKB Uvicord SⅡ用于色谱时测吸收部分(λ:280 nm)。

该来克丁粗提物之梯度疏水相互作用色谱是在一个填有丁基-PA玻珠的21×1.5 cm I.D的柱子中以流速1.0 mL/min进行的。

提取物Liophilized后,得34 mg样品。将此溶入20 mL 10 mmol/L的磷酸钠(含60%饱和度的$(NH_4)_2SO_4$)中,以同样缓冲液平衡柱子。先以起始缓冲液50 mL(60%饱和度至0%)梯度洗,再以$(NH_4)_2SO_4$ 20 mL洗涤柱子收集2 mL部分,并在280 nm处分析吸收率。收集有凝血活性的部分。在Pm-10膜上超滤膜透析,4%～30%的丙烯酰胺非还原条件电泳,$Ag(NO_3)_2$染色蛋白质。

此法即非还原条件下的梯度凝胶电泳(SDS-PAGE),获80×10^3的纯来克丁。还原条件下的SDS-PAGE则揭示出另外两个来克丁亚单位,其相对分子质量分别为43×10^3和36×10^3。

该法之特点是利用丁基-PA-玻璃(珠)的疏水性(即高表面张力)使疏水来克丁与其他蛋白质分开。此法可得结合唾液酸来克丁(SBL)[9]。

3　根癌土壤杆菌两种来克丁的纯化[11]

3.1　培养液的预处理

经合适预培养后的1 L根癌土壤杆菌(*Agrobacterium tumefaciens*)82.139菌悬液接入100 L的液体甘露醇培养基中,经23℃间断通气搅拌培养至指数生长期共24 h即用连续离心系统(2 000 r/min)收获。按Depierreux等(1991)的方法浸提细菌,得5 g团块。将它悬浮于50 mL缓冲液(它含0.2 mol/L $MgCl_2$和0.05 mol/L Tris-5 ml等,pH9.0)中,于室温中振荡1 h,4℃ 8 000 g离心20 min得上清。0.22 μm滤膜过滤。Tris-丙烯GF05柱上用凝胶过滤,在0.1 mol/L醋酸钠缓冲液pH5.0)中平衡。所获蛋白质部分以空柱体积洗脱。

3.2　亲和层析

亲和载体由6-氨基己糖亚氨基甲叉-琼脂糖-A4匹配σ-异硫氰酸苯-β-半乳糖吡喃糖苷组成。0.1 mol/L的碳酸氢钠缓冲液(pH9.5)充分洗涤填充珠子。在含有1 mLN,N-二甲基甲酰胺的200 mgσ-异硫氰酸苯-糖苷中混合过夜(4℃),在玻璃滤器上以蒸馏水洗涤该珠子(未置换的氨基在0℃时以加入200 mL的乙酸酐酰基化),所得凝胶搅拌4 h(4℃),蒸馏水洗涤后稳定于pH5.0的0.1 mol/L醋酸钠缓冲液中。

3.3　凝胶电泳(SDS-PAGE)

在填充有0.1%SDS(pH8.5)的12.5%丙烯酰胺凝胶块上进行电泳。蛋白质用一个Centricon 10浓缩器浓缩,考马斯亮蓝染色。所用相对分子质量标准是磷酸化酶6(相对分子质量97 400)、牛血清白蛋白(相对分子质量-66 200),卵清蛋白(相对分子质量45 000)、碳酸酐酶(相对分子质量31 000)、大豆胰蛋白酶抑制剂和白卵溶菌酶(相对分子质量14 400)。

所得来克丁之表面相对分子质量是58×10^3,等电点是pH5.5(PI5.5)。它有效凝集人B型红血球。

3.4　凝集马红细胞的来克丁的提取

通过 α-*D*-半乳糖柱子的蛋白质部分再送至 α-*L*-鼠李糖-吡喃-(6-硫氨甲酰基-己基亚氨基)甲叉-琼脂糖-A4 柱上。以 0.1 mol/L 乙酸钠缓冲液(pH5.0)平衡该柱，pH8.0 的 0.25 mol/L NaCl 液洗脱。所得即多肽。SDS-PAGE 显示其相对分子质量为 40×10^3，等电聚焦点是 5.0 PI。

4　少孢节丛孢来克丁的分析[7]

4.1　蛋白质的分纯

经 7 d 培养的少孢节丛孢(*Arthrobotry oligospora*)菌丝用尼龙网过滤，得 500 mg 干物，以含 0.15 mol/L NaCl 的 20 mm-磷酸钠缓冲液(pH7.4)(3×50 mL)洗涤之，分入 4 个 50 mL 管中(5 mL PBS/管、0.1 m mol/L-PMSF/管)。该菌丝在超级 Turrax 匀质器中分解 4×30S，间歇冰浴。匀浆经 13 500 g 20 min，0.22 μm 膜(Millex-GV，微孔)滤上清液，超滤器浓缩，得滤物、菌丝团等。亲和层析用猪胃黏蛋白-CNBr 活化的琼脂糖 4B 纯化。所获浸提物转至含 4 mL 的黏蛋白亲和凝胶试管中，保留含悬浮物的凝胶在长颈摇瓶中 80 min。测上清液凝血活性。PBS 洗涤凝胶至 280 nm 的最低吸收值处(0.25 mL/min · 4 min)。束缚在柱上的蛋白质用现制 0.1 mol/L-NH_4 OH(pH11)淘洗再浓缩。

依 Borrebaeck 等(1984)所述方法用 GalNac 琼脂糖凝胶分离蛋白，280 nm 紫外吸收测定。

4.2　糖蛋白的 PAGE 和测定

用 2 050 侏儒电泳仪，15%(*W*/*V*)的凝胶按 Laemmli(1970)法做间断 SDS-PAGE，50 mmol/L-Tris/HCl 缓冲液稀释样品(1∶1 体积比)。该缓冲液中含 11.6%(*W*/*V*)甘油，0.001%(*W*/*V*)溴酚兰，有/无 5%(*W*/*V*)DDT 的 2%(*W*/*V*)SDS。100℃热 3 min 后凝胶电泳。相对分子质量用卵铁转移蛋白(76-78)$\times10^3$、牛血清蛋白(66.25×10^3)、卵清蛋白(45×10^3)、碳脱水酶(30×10^3)、肌红蛋白(17.2×10^3)和细胞色素(12.3×10^3)等标准来测算。考马斯亮兰染色。糖蛋白用 Dubray & Bezand(1982)法在 SDS-PAGE 凝胶上以高碘酸-希夫试剂(PAS)测定。

4.3　凝胶过滤

以 Pharmacia Hiload 16/60 柱的 Superdex 200 制备级凝胶，测天然形态蛋白质相对分子质量。该柱与 HPLC(Phanrmacia LKB；Pump：2248，检测器：可变波长测器 2141)相连，PBS 平衡(流速：0.5 mL/min)，214 和 280nm 处洗脱，HeliFrac(Phamaacia)收集各部分，比较纯化蛋白质洗脱液体积和标准蛋白质体积(Pharmacia 低相对分子质量标准 Kit 蛋白质和醛缩酶 158×10^3)以计算表观相对分子质量。

4.4　等电聚焦和聚焦色谱

等电点聚焦(Pharmacia 固定干片)，pH4-7，Multiphore Ⅱ系统(Pharmacia)按厂家说明测量(10℃)。PI 用一个宽的 PI 校准 Kit(pH3～10，Pharmacia)测量。考马斯亮蓝染凝胶。

聚焦色谱在连接有 Pharmacia HPLC 的 Mono PHR 5/20(Pharmacia)上进行。柱子用 0.025 mol/L-Dis-Tris(pH7.1)平衡。同一缓冲液溶解待测来克丁(230 g)，聚合缓冲液 74(pH5.0)(Pharmacia)洗脱。280nm 处测定(4℃，10 mL · min^{-1})。

4.5　免疫印迹

真菌培养物各部的蛋白质以 SDS-PAGE 凝胶电印迹，该凝胶则是有 Immobilon 转移膜(微孔 AB，瑞典，即蛋白质印迹用 PVDF 膜)的。

4.6 蛋白质的切割

水解无色杆菌(*Achromobacter lyticus*)蛋白酶在 2 mol/LM-盐酸胍中消化 25 g 纯来克丁(Riviere 等，1991)。另将 20 g 来克丁作羟胺切割试验。Brownlee C4 Aquapore 柱(2.1×30 m)(在 HPLC 上)分离肽。Gold(Beckman)系统泵、二级管系列 990(水)作测器。回复相色谱所用缓冲液：(A)0.1%(*V/V*)的三氟乙酸(TFA)(水中)，(B)0.1%TFA(*V/V*)、10%水(*V/V*)，90%乙腈(*V/V*)。梯度是 2%～62%B(60 min)(羟胺 30 min)、62%～90%B(3 min)，流速为 10 μL/min。214 nm 处洗脱，手工收集。

4.7 肽的序列

用配备有一个 On-line(在线计算的)苯基硫代乙内酰脲(PIH)氨基酸分析仪(120 型应用生物系统)的序列分析仪测。

4.8 氨基酸分析

用 4-乙烯基吡啶(在 6 md/L-盐酸胍中)还原和烷基化样品，冰冻干燥，水稀释至 3 倍量，再加 20 倍体积的冷丙酮/0.2%HCl，沉淀，置－20℃中数小时。低速离心回收样品。酸性丙酮重复沉淀后再溶于水。样品于 110℃ 22 h 6 mol/L-HCl，2%(*W/V*)酚气相水解处理。Beckman 6300 测，检测剂：水合茚三酮。Shimadzu R4A。整合器定量，440 nm 处色谱。手工计算。顺序同源性(Sequence homology)用蛋白质鉴定源(PIR，30)和 Intelti 遗传学成套设备研究。

4.9 凝血测定

V 型血凝板上测定。胰蛋白酶化和脱水戊二酸处理的人 A 型血细胞用 2%PBS 稀释。倍比稀释 20 μL 样品加入每孔，与同体积红血球悬液混合，室温中 1 h，以终稀释液的倒数记录其血凝活性。

4.10 各细胞分部的凝血活动

将菌丝匀质化、超声处理呈团块做分部研究。以超声波移去胞外聚合物，悬浮此团块在 25 mL PBS 中。4℃超声振荡 30 s，13 500 g 10 min PBS(2×25 mL)洗涤，在 25 mL PBS 中再超声处理两次。间隙离心，转样品入含酸洗玻璃珠(0.5 mm 直径)和 0.1 mol/L PMSF 的 PBS(1：2：1 团块/缓冲液/玻璃珠，体积比)的试管中。全混合器冰上分部 3×60 s。洗涤的细胞壁碎片至 280 nm，吸光率低于 0.08(6×75 mL PBS，13 500 g 10 min)细胞壁团块再悬浮，混合摇动 30s。所得上清物全用于测凝血。

4.11 结合胞壁来克丁的测定

将 100 mg(湿重)纯化的细胞壁样品转入 1.5 mL 的微量离心(Eppendorf)管中。管中含有 1 mL 下列酶溶液(1 mg/mL；PBS)，即新酶 234、溶胞酶或几丁质酶。回旋振荡 90 s。30℃水浴锅中培养 22h。10 000 g 离心 10 min，测上清血凝活性。

结果发现主要凝血活性存在于匀质化和超声波处理后的上清中，胞壁中来克丁很少。

所得少孢节丛孢来克丁(Arthrobotry oligospora lectin，AOL)凝集人 A 型血。单糖、双糖均不抑制其凝集作用。但即使是 0.001 mg/mL 的低浓度胎球蛋白、去唾液酸胎球蛋白或粘蛋白均抑其凝血活动。也不显酸性或碱性的磷酸酶、触酶、过氧物酶、蛋白酶、α-甘露糖苷酶、β-葡糖苷酶、几丁质酶或 β-N 乙酰-氨基葡糖苷酶的酶活。分析表明 AOL，是一个二聚糖蛋白，相对分子质量为 36×10^3。等电点是 pH6.5，不含硫氨基酸。

5　齐整小菌核来克丁分析法要点

5.1　离子交换色谱

取 Okon 等(1973)推荐的合成培养基培养 5 d 的齐整小菌核(*Sclerotium rolfsii*)培养滤液 50 mL，对 20 mmol/L Tris-HCl 缓冲液(pH7.0)透析 24 h。此缓冲液 6 mL 洗涤 DEAE-Sepharose 柱得束缚蛋白，逐段 NaCl 梯度(流速：0.5 mL/min)洗脱至 0.2 mol/L，将洗脱出的蛋白质对蒸馏水透析 24 h(5×51)。同上柱子快速阴离子交换色谱(Pharmacia，1.5×2.5 cm)流过，得流出物。

5.2　SDS-PAGE

在 10%(*W/L*)的聚丙烯酰胺直角平板凝胶中用小型 Protean Ⅱ系统(Bio-Rad)作 SDSPAGE，在还原条件下重复 4 次。

结果表明所期望的齐整小菌核来克丁(Sclerotium rolfsii lectin，SRL)不结合在柱子上，流出物含 0.95 μg 蛋白质/mL(占总量的 16%)，凝集 *E. coli* 活性之效价(mg/mL)从 666.7 增加到 4 000。NaCl 洗脱的部分无凝集活性。SRL，的凝集活动不被任一单糖或双糖所抑，但浓度分别达 125 和 62.5 g/mL 的胎球蛋白、去唾液酸胎球蛋白更缺抑凝效力。仅猪胃黏蛋白和去唾液酸粘蛋白，其浓度分别低至 0.48 和 0.96 μg/mL 时仍有抑凝效力。

该来克丁含两个主要带子，相应于 55 和 60×10^3 的相对分子质量。不同于以往的 SRI。

6　草菇来克丁的测定

可从草菇(*Voivariella volvacea*)的子实体和菌丝分别提取到来克丁，它们分别被命名为 VVL-F 和 VVL-Me。

6.1　VVL-F 和 VVL-M 的分离

VVL-F 的分纯：

收获 7 d 30℃振荡培养的 V. *volvacea* 菌丝，以下各步均在 4℃中进行。2 kg 新鲜子实体在 2 L 5%(*V/V*)的乙酸液中匀质化(其中含 0.1%(*V/V*)的巯基乙醇)3h，10 000 g，20 min 离心，加固体硫酸铵至上清中得 95%饱和度。搅拌 1 h，20 000 g 30 min 离心得沉淀物，溶入 10 mmol/L 的磷酸钠缓冲液(pH8.2)中透析。DE-52 纤维素柱(2.5×22 cm)(配有上述 buffer 平衡)分步上清。以 10 mmol/L 醋酸钠 buffer(pH4.8)平衡。10 mmol/L 磷酸钠(pH8.2)的平衡 buffer 洗提血凝活性。用 CM-52 纤维素柱(2.5 cm×22 mn)收集活性部分。pH4.8 的 buffer 充分洗涤，吸附物含凝血活性。以醋酸盐 buffer-线性梯度(0～0.3 mol/L NaCl)洗提。活性部分将结合和用于 Mono S HR 之 5/5(具 FPLC 本系)柱上，再在 pH5.2 的 10 mmol/L 醋酸钠 buffer 柱中洗提。用蒸馏水透析，得纯化来克丁，冷冻干燥。

VVL-M 分纯：

所得培养液经粗棉布过滤得菌丝，再以蒸馏水洗出菌丝，其余各步同 VVL-F。

6.2　凝血试验

在 U 型血凝板中于室温中进行，试验对 2%红细胞(磷酸盐 buffer 的盐水中)的凝集作用。红细胞分别来自 Wistar，Sprague-Dawley 和 Lewis3 种鼠及其他动物。用的糖分别是 D-葡萄糖、D-半乳糖、L-岩藻糖、α-乳糖、D-甘露糖、蔗糖、D-乙酰-D-半乳糖胺、D-葡糖胺、N-乙酰-D-葡糖胺、N-乙酰唾液酸等。糖蛋白是卵清蛋白、甲状膜球蛋白、人绒笔腺促性腺激素和乙酰肝素等。试验糖和糖蛋白对 VVL 凝集作用的抑制活动。

6.3　SDS-PAGE

按 Laemmli 法在 Bio-Rad 微型蛋白凝胶设备中进行。考马斯亮蓝 R-250 或高碘酸-锡夫氏技术染色测糖，罗氏法测蛋白质浓度，牛血清白蛋白为标准。

6.4　凝胶过滤

在 Superdex 75 HR 10/30 柱(FPLC 系统)上测 VVL 相对分子质量。用 4 种标准蛋白为标示物:细胞色素 C(MW12.4×10^3)、胰凝乳蛋白酶 A(25×10^3)、卵清蛋白(45×10^3)和 BSA(67 ×10^3)。

6.5　氨基酸序列测定

VVL 的 N-末端氨基酸用自动埃德曼降解法测。微量序列测定用 Hewllett Packard 1000 A 蛋白序列仪(有 HPIC 系统)。

6.6　促有丝分裂活性的测定

无菌取出 8 周龄的 BAIB/C 鼠的脾，分离得脾细胞，将其悬浮于 RPMI 1640 培养基(辅有 10%胎半血清、10 单位的青霉素/mL、10 μg 链霉素 mL、200 mmol/L-谷氨酸)中，浓度为 5×10^6 细胞/mL。96 孔培养板上的每孔中加入 5×10^5 细胞/50 μL/孔，系列 VVL 和 ConA。再加入 50 μL 培养基中的 ConA，37℃5%∞b 气体培养 42h，再添 10 μL[3H]-胸苷(0.25Ci)，同样再培养 6h，自动细胞收集器收集细胞至一个玻璃滤器上，Beckman 或 LS6000SC 内烁计数器测放射性。

6.7　RNA 的浸提和 RT-PCR

如上分离得的 BALB/C 鼠脾细胞再悬浮于培养碟中(10^7 细胞/60 mm)。同上条件培养以各种浓度的 VVL，和 ConA 处理的细胞 4h，总细胞 RNA 按 Chomczynski 等的方法分离。MMLV 逆转录酶从 1 μg 的总 RNA20 μL 中制备寡(脱氧胸苷)-引物 cDNAs Oligc，(dT)-原型 cD-NAs。扩增产物用琼脂糖凝胶电泳

分析，其强度用溴化乙锭染色，在 UV 下具像。

6.8　结果

测定结果表明 VVL-F 和 VVL-M 是一样的来克丁，相对分子质量为 32×10^3，很像是非一共价联的同二体蛋白，无糖，但所取部位不同，其淋巴细胞激活作用大约是 ConA 的 10 倍。显著刺激白介素-2 和干扰素-γ 基因表述。甲状腺球蛋白可抑 VVL 的凝血性，可结合到寡糖上。VVL 是强力致裂原。总之 VVL 是具有强大免疫调节活性的、结构新型的真菌来克丁。

7　结语

本文从理解微生物来克丁功能和用途的角度出发，认为必须加强对微生物来克丁的分纯、鉴定研究，为此阐述了枯草芽孢杆菌结合唾液酸来克丁、根癌土壤杆菌来克丁、少孢节丛孢来克丁、齐整小菌核来克丁、草菇来克丁的分纯分析法。由此可见，由于微生物及其来克丁差异大，对培养产物的前处理分析方法就需应株而异。不同的亲和层析术应用最广。对不同来克丁，尤应据其糖结合性考虑采取适宜分离手段，如凝胶渗透、离子交换

色谱不适于某些结合唾液酸来克丁的提纯。另外，一种微生物菌株可能产生几种活性物质，包括不同级别的代谢产物。可在采取适当的提取方法时从同一培养物中分离出所需物质。微生物(包括海洋的)来克丁分析方法的日益精细将拓展来克丁学视野，加快和增涨来克丁应用步伐。

(合作者：李光友)

D_2 防卫素(DEFENSINS)

四种海产双壳贝类血淋巴中抗菌物质的诱导及其活性测定*

摘　要　以贻贝(*Mytilus edulis*)、毛蚶(*Scapharca subcrenata*)、海湾扇贝(*Argopecten irrdians*)和菲律宾蛤仔(*Ruditapes philippinesis*)为试验材料,分别用活的和灭活的大肠杆菌为诱导物进行诱导,采用3种革兰氏阳性细菌和3种革兰氏阴性细菌作为指示菌,对4种供试双壳贝类血淋巴进行抗菌活性测定。结果表明:①贻贝在诱导前其血淋巴对金黄色葡萄球菌和四联微球菌有抗菌活性,免疫诱导可使血淋巴对鳗弧菌出现抗菌作用;②毛蚶在诱导前其血淋巴中含有对金黄色葡萄球菌和鳗弧菌有抗菌活性的物质,诱导对这两种菌的抑菌活性有明显的提高,但未诱导出另外4种供试细菌有活性的物质;③海湾扇贝在诱导前其血淋巴对所有供试细菌均无抗菌活性;但经诱导后产生对金黄色葡萄球菌、副溶血弧菌和鳗弧菌有抗菌活性的物质,且活菌诱导比灭活菌诱导效果好;④菲律宾蛤仔在诱导前其血淋巴仅对金黄色葡萄球菌和鳗弧菌表现出抗菌活性。经诱导后其血淋巴不仅对这两种菌的抗抑菌活性增强,而且还诱导出对副溶血弧菌有抗菌活性的物质。

关键词　双壳贝类　血淋巴　抗菌物质　活性测定

在自然条件下,贝类与其他水生养殖动物一样,在其生活的环境中存在大量的不同种类的病原菌。但在一般情况下,并不是所有的这些病原菌都感染贝类并造成危害,主要是因为贝类具有一套比较完善的防卫系统,能抵御外来病原菌的侵袭。使其免受感染。在贝类的防卫反应过程中,除了血细胞本身的作用外,其血淋巴中含有的多种具生物活性物质如溶菌酶、凝集素、来克丁、溶血素和抗菌肽等[5,6]也起到十分重要的作用。在国外,双壳贝类血淋巴中的抗菌物质尤其抗菌肽是近年来研究的一个热点,国内虽对某些贝类动物的多肽(如扇贝多肽)等作过生理活性或药物进行诱导,但尚未见到有关抗菌活性方面研究的报道。本文分别用活的和灭活的大肠杆菌为诱导物进行诱导,采用3种革兰氏阳性细菌和3种革兰氏阴性细菌作为指示菌,对贻贝、毛蚶、海湾扇贝和菲律宾蛤仔4种供试双壳贝类的血淋巴进行抗菌活性测定。

* 原文刊于《海洋科学,Marine Sciences》,2002,26,(8):5-8,本人为第四作者及项目负责人。

1 材料与方法

1.1 试验材料

1.1.1 贝类

贻贝(*Mytilus edulis*)、毛蚶(*Scapharca subcrenata*)、海湾扇贝(*Argopecten irrdians*)和菲律宾蛤仔(*Ruditapes philippinesis*)。分别于2001年冬末和2002年春初自山东胶南海青养殖场采集新鲜健康大小相近的贝类,移入实验室后立即放入盛有新鲜海水并装有通氧设备的玻璃水槽中饲养,海水温度18～20℃。

1.1.2 菌株

革兰氏阳性细菌 枯草杆菌(*Bacillus subtilis*)、金黄色葡萄球菌(*Staphylococus aureus*)、四联微球菌(*Micrococcus tetragenus*);革兰氏阴性细菌;大肠杆菌(*Escherichia coli*)、副溶血弧菌(*Vibrio parahaemolyticus*)、鳗弧菌(*V. anguillarum*)。其中大肠杆菌、枯草杆菌、金黄色葡萄球菌、四联微球菌、副溶血弧菌由青岛海洋大学(现中国海洋大学)食品工程系提供,鳗弧菌由中国科学院海洋研究所提供。

1.1.3 培养基

营养琼脂培养大肠杆菌、枯草杆菌、四联微球菌、金黄色葡萄球菌,ZoBell 2216E[4]培养基培养副溶血弧菌、鳗弧菌。

1.2 试验方法

1.2.1 大肠杆菌诱导物的制备

大肠杆菌菌株接种于营养琼脂斜面上,37℃培养24 h,生理盐水洗下菌苔,2 000 r/min离心10 min,去上清液,沉淀再用生理盐水洗3次,配成浓度为3×10^9个菌体/mL的菌悬液,0.1 MPa灭菌20 min或不灭菌,分别用于灭活菌和活菌免疫诱导。

1.2.2 诱导方法

小心掰开供试动物的贝壳,用微量进样器轻微刺入体内,注射10 μL大肠杆菌灭活杆菌或活菌菌液3×10^7个菌体/只。然后,放回水槽内继续养殖,定期采集血淋巴。

1.2.3 血淋巴的采集

注射活菌或灭活菌后,每隔24 h用无菌注射器从供试贝类体内采集一次血淋巴,每次每种贝类采集10只,血样混合后,10 000 r/min低温(4℃)离心10 min,取上清液,经0.22 m孔径无菌过滤器过滤备用。按同样方法从未经诱导处理的贝类体内采集血淋巴作为对照血淋巴。

1.2.4 抗菌活动测定

参照Bulet等1991年所采用的平板生长抑制法。取制备的血淋巴上清液6 μL,点在含有供试细菌的营养琼脂或Zobell 22216E平板上的一个孔穴(φ2.5 mm)中,每一样品点样5个孔穴,对照孔穴点等量的灭菌海水。点样后,将平板倒置于37℃(大肠杆菌、枯草杆菌、四联微球菌、金黄色葡萄球菌)或26℃(副溶血弧菌、鳗弧菌)温箱中,培养24 h或48 h后观察结果,测量并记录抑菌圈的直径(减去孔穴直径)。

2 结果与讨论

抑菌试验结果见表1～4。

从表1中可以看出,贻贝在诱导前其血淋巴对金黄色葡萄球菌和四联微球菌有抑菌活性,免疫诱导对血淋巴抑制这两种菌的活性无大影响,但诱导后72h的血淋巴对鳗弧菌出现抑菌作用,96～120h 时达到高峰。贻贝在诱导前和诱导后其血淋巴对大肠杆菌、枯草杆菌和副溶血弧菌均无抑菌活性。

由表 2 得知,毛蚶在诱导前其血淋巴中含有对金黄色葡萄球菌和鳗弧菌有抑菌活性的物质,经过诱导后对这两种菌的抑菌活性有明显的提高,诱导后 48～72h 对金黄色葡萄球菌抑菌活性最强;72h 对鳗弧菌抑菌作用明显提高,其后一直保持较高的抑菌活性。活菌的诱导效果比灭活菌的诱导效果略好。诱导前和诱导后的毛蚶血淋巴对大肠杆菌、枯草杆菌、四联球微菌和副溶血弧菌均无抑菌活性。

表 1　贻贝诱导前后其血淋巴的抗菌活性比较

Tab. 1　Antibacterial activity spectrum of hemolymph from unchallenged and challenged *Mytilus edulis*

处理	菌株	诱导后时间(h)					
		0	24	48	72	96	120
A	大肠杆菌	0	0	0	0	0	0
	枯草杆菌	0	0	0	0	0	0
	四联微球菌	0.5	0.6	0.7	0.5	0.7	0.5
	金黄色葡萄球菌	1.5	2.1	1.5	1.6	2.0	1.7
	副溶血弧菌	0	0	0	0	0	0
	鳗弧菌	0	0	0	0.3	1.0	2.3
B	大肠杆菌	0	0	0	0	0	0
	枯草杆菌	0	0	0	0	0	0
	四联微球菌	0.5	1.6	0.5	0.3	0.3	0.2
	金黄色葡萄球菌	1.5	2.1	3.1	1.5	1.7	1.5
	副溶血弧菌	0	0	0	0	0	0
	鳗弧菌	0	0	0	0.3	2.5	2.3
C	大肠杆菌	0	0	0	0	0	0
	枯草杆菌	0	0	0	0	0	0
	四联微球菌	0	0	0	0	0	0
	金黄色葡萄球菌	0	0	0	0	0	0
	副溶血弧菌	0	0	0	0	0	0
	鳗弧菌	0	0	0	0	0	0

注:A 代表用灭活菌诱导的免疫血淋巴;B 代表用活菌诱导的免疫血淋巴;C 代表灭菌海水(后同)。

表 3 表明,海湾扇贝在诱导前其血淋巴对所有供试细菌均无抑菌活性,但经诱导后 48h 产生对金黄色葡萄球菌、副溶血弧菌和鳗弧菌有抑菌活性的物质,而且活菌诱导比灭活菌诱导效果好。诱导后的血淋巴对大肠杆菌、枯草杆菌和四联微球菌仍无活性。

从表 4 可以看出,菲律宾蛤仔在诱导前,其血淋巴仅对金黄色葡萄球菌和鳗弧菌表现出抑菌活性,经诱导后其血淋巴不仅对这两种菌的活性增强,而且还诱导出对副溶血弧菌有抑菌活性的物质，但未诱导出对大肠杆菌、枯草杆菌和四联微球菌有抑菌活性的物质。

本试验结果表明，诱导免疫对各种供试贝类血淋巴的抗菌活性影响不尽一致，如贻贝、海湾扇贝和菲律宾蛤仔经诱导后其血淋巴可产生新的抗菌活性，而诱导后毛蚶血淋巴不产生新的抗菌活性。就海湾扇贝而言，活菌诱导比灭活菌诱导效果好，而另外3种贝类，活菌诱导与灭活菌诱导的效果基本一致。这说明不同种类的贝类动物在防卫系统以及所产生的抗菌物质种类和数量等方面存在一定的差异。因此，在研究贝类抗菌物质时，应根据研究对象、目的和指示菌种类选择是否需要免疫诱导以及何时的诱导方式。

表2　毛蚶诱导前后其血淋巴的抗菌活性比较

Tab. 2　Antibacterial activity spectrum of hemolymph from unchallenged and challenged *Scapharca subcrenata*

处理	菌株	诱导后时间(h)					
		0	24	48	72	96	120
A	大肠杆菌	0	0	0	0	0	0
	枯草杆菌	0	0	0	0	0	0
	四联微球菌	0	0	0	0	0	0
	金黄色葡萄球菌	1.1	2.0	4.5	4.5	1.0	0.9
	副溶血弧菌	0	0	0	0	0	0
	鳗弧菌	0.3	0.2	1.0	2.0	1.5	2.3
B	大肠杆菌	0	0	0	0	0	0
	枯草杆菌	0	0	0	0	0	0
	四联微球菌	0	0	0	0	0	0
	金黄色葡萄球菌	1.1	2.1	4.5	5.0	1.5	1.0
	副溶血弧菌	0	0	0	0	0	0
	鳗弧菌	0.3	0.2	0.9	2.5	2.5	2.4
C	大肠杆菌	0	0	0	0	0	0
	枯草杆菌	0	0	0	0	0	0
	四联微球菌	0	0	0	0	0	0
	金黄色葡萄球菌	0	0	0	0	0	0
	副溶血弧菌	0	0	0	0	0	0
	鳗弧菌	0	0	0	0	0	0

表 3　海湾扇贝诱导前后其血淋巴的抗菌活性比较

Tab. 3　Antibacterial activity spectrum of hemolymph from unchallenged and challenged *Argopecten irrdians*

处理	菌株	诱导后时间(h)					
		0	24	48	72	96	120
A	大肠杆菌	0	0	0	0	0	0
	枯草杆菌	0	0	0	0	0	0
	四联微球菌	0	0	0	0	0	0
	金黄色葡萄球菌	0	0	0.1	0.5	0.5	1.5
	副溶血弧菌	0	0	1.5	0.5	1.0	2.0
	鳗弧菌	0	0	0.1	0.5	1.0	2.0
B	大肠杆菌	0	0	0	0	0	0
	枯草杆菌	0	0	0	0	0	0
	四联微球菌	0	0	0	0	0	0
	金黄色葡萄球菌	0	0	0.1	1.0	1.5	1.5
	副溶血弧菌	0	0	1.7	1.5	2.3	2.3
	鳗弧菌	0	0	0.2	2.0	2.5	2.3
C	大肠杆菌	0	0	0	0	0	0
	枯草杆菌	0	0	0	0	0	0
	四联微球菌	0	0	0	0	0	0
	金黄色葡萄球菌	0	0	0	0	0	0
	副溶血弧菌	0	0	0	0	0	0
	鳗弧菌	0	0	0	0	0	0

表 4　菲律宾蛤仔诱导前后其血淋巴的抗菌活性比较

Tab. 4　Antibacterial activity spectrum of hemolymph from unchallenged and challenged *Ruditapes philippinesis*

处理	菌株	诱导后时间(h)					
		0	24	48	72	96	120
A	大肠杆菌	0	0	0	0	0	0
	枯草杆菌	0	0	0	0	0	0
	四联微球菌	0	0	0	0	0	0
	金黄色葡萄球菌	2.0	10.2	4.0	4.5	5.0	3.7
	副溶血弧菌	0	0	1.5	2.0	2.0	2.3
	鳗弧菌	1.5	1.3	1.5	1.5	2.0	2.3

(续表)

处理	菌株	诱导后时间(h)					
		0	24	48	72	96	120
B	大肠杆菌	0	0	0	0	0	0
	枯草杆菌	0	0	0	0	0	0
	四联微球菌	0	0	0	0	0	0
	金黄色葡萄球菌	2.0	10.8	4.0	3.5	4.0	3.0
	副溶血弧菌	0	0	1.8	2.0	2.0	3.0
	鳗弧菌	1.5	1.5	2.0	1.7	3.0	2.3
C	大肠杆菌	0	0	0	0	0	0
	枯草杆菌	0	0	0	0	0	0
	四联微球菌	0	0	0	0	0	0
	金黄色葡萄球菌	0	0	0	0	0	0
	副溶血弧菌	0	0	0	0	0	0
	鳗弧菌	0	0	0	0	0	0

Mialhe等1995年提出,在软体动物中,由于不存在免疫球蛋白,缺乏抗体介导的免疫反应,因而它不能像脊柱动物那样通过接种抗原来达到产生免疫抗体的自我保护的目的。周永灿等1997年认为,由于贝类的防卫系统具有非特异免疫的性质,适当的诱导可提高贝类细胞的吞噬能力及多种免疫因子的数量和活性。本试验中,通过诱导不仅使有些贝类血淋巴原有的抑菌活性增加,而且还诱导出新的抑菌活性,这说明某些贝类动物的非特异性免疫因子存在一定程度的可诱导性。本试验诱导出的新抑菌活性是由于激活或提高了血淋巴中原有抗菌物质的数量和活性,还是诱导出了新的抑菌物质,有待于进一步研究。

参考文献6篇(略)

(合作者:魏玉西　郭道森　李　丽)

INDUCTION AND ASSAYS OF ANTIBACTERIAL ACTIVITY FROM THE HEMOLYMPH OF FOUR KINDS OF BIVALVE MOLLUSCS

(ABSTRACT)

Abstract　Four kinds of bivalve mollusks, *Mytilus edulis*, *Scapharca*, *Argopecten irrdians* and *Ruditapes philippinesis*, were challenged with a suspension of heat-killed or live *Escherichia coli* and hemolymph was collected periodically. The activities of antibacterial substances in the hemolymph from challenged and unchallenged milluscs were measured using a plate growth inhibition assay with Gram-positive bacterial strains and Gram-negative bacterial strains, repectively. The results showed that: (1) The hemolymph of unchallenged *Mytilus edulis* was active against *Staphylococcis aureus* and *Micrococcus tetragenus*, and *activities* againsts inhibit *Vibrio anguillarum* were present in the hemolymph of challenged *Mytilus edulis*. (2) The hemolymph of unchanllenged *Scapharca* subcrenata presented acticities againsts *Staphlococcus aureus* and *Vibrio anguillarum*, and the activities against the two strains were highly improved after challenge. (3) The hemolymph of unchallenged *Argopecten irrdians* had no antibacterial activities against all of bacterial strains tested. However, the hemolymph of challenged *Argopecten irrdians* was found to contain antibacterial substances againsts inhibit *Staphylococcus aureus*, *Vibrio parahaemolyticus* ang *V. anguillarum*, and the activities in the hemolymph challenged with live *Escherichia coli* were stronger than those in the hemolymph challenged with heat-killed *Escherichia coli*. (4) The antibacterial activities were observed against inhibit *Staphylococcus aureus* and *Vibrio angullarum* in the hemolymph of unchallenged Ruditapes philippinesis. After challenge, the activities against inhibit *Staphylococcus aureus* and *Vibrio anguillarum* increased and a new activity against inhibit *Vibrio parahaemolyticus* was found in the hemolymph.

Key words　Bivalve Molluse, Hemolymph, Antibacterial Substance, Bioassay

贻贝防卫素的研究进展*

摘　要　防卫素是贻贝等海洋软体动物的重要抗微生物肽，迄今发现的贻贝防卫素根据其初级结构、性质和共有的半胱氨酸列阵分成：贻贝防卫素（MDA 和 MDB）、地中海贻贝防卫素（MGD1 和 MGD2）、Myticins（MytieinA 和 Myticin B）、Mytilins（Mytilin A，Mytilin B，Mytilin C，Mytilin D 和 Mytilin G1）和 Mytimycin。阐述它们的化学性质、结构、产生和生物活性等。不同环境中的贻贝不易为重病所折磨，能抵御各类微生物的入侵和保卫自身。对贻贝防卫素的研究有助于人们理解贻贝及其他海洋软体动物的先天免疫，并提高海水养殖的水平。

关键词　贻贝　防卫素　抗微生物肽　先天免疫　海洋软体动物

贻贝属于贻贝目贻贝科贻贝属（*Mytilus*）。在我国贻贝有 30 余种，经济价值较高者有 10 余种，其中的食用贻贝（*M. edulis*）、翡翠贻贝和厚壳贻贝为我国主要养殖种。包括贻贝在内的软体动物的许多种类具有重要的经济价值，但其对抗微生物时惊人的防卫活性却被忽视了，直至 1996 年仍未见报道。实际上，软体动物明显地对一些微生物病原敏感。它们依赖血细胞或淋巴细胞样细胞对入侵微生物作细胞防卫反应。近几年来的研究已发现抗微生物肽是软体动物在漫长进化历程中保存下来的天然免疫作用之一。软体动物抗微生物肽具备一种崭新的无脊椎动物活性作用模式，与哺乳动物抗微生物肽密切相关，如贻贝血细胞与人单核细胞/巨噬细胞系统相似，但又经与昆虫抗微生物肽结构相关的分子中介，显出与节肢动物防卫素有共同的祖先起源。贻贝防卫素等抗微生物肽主要由其血细胞进行生产、贮存、分泌、转运和抗/杀菌等活动，并在循环细胞中丰富。贻贝等双壳类软体动物的血细胞浓度为（2～4）$\times 10^6$ 细胞/毫升。其中的粒细胞是防卫素进行免疫反应的主要细胞效应物，经循环细胞运行全身。防卫素等抗微生物肽可在身体许多部位及器官，包括在体表中表达并形成防病原入侵和保护自身的第一道防线。

目前发现的防卫素大多是一类小相对分子质量、阳离子、富半胱氨酸、-螺旋的、具抗特定微生物活性的肽。已被阐明的贻贝防卫素根据其初级结构中共有的半胱氨酸序列，分为 4 类：（1）贻贝防卫素（MDA 和 MDB）和地中海贻贝防卫素（MGDl 和 MGD2）；（2）Myticins（Myticin A，Myticin B）；（3）Mytilins；（4）Mytimycin。

本文旨在论述贻贝 4 类防卫素化学性质、结构、产生和生物学功能等，以利于加深对贝类动物防御病害机制的认识和理解。

1. 贻贝防卫素和地中海贻贝防卫素

1.1　贻贝防卫素（MDs）

分离自食用贻贝（*M. edulis*）血浆的防卫素，一个被命名为贻贝防卫素 A（Mytilus de-

* 原文刊于《广西科学》2003，10（2）：129-134

国家自然科学基金项目资助（编号：40086001）。

fensin A，缩写为 MDA)，另一个被命名为贻贝防卫素 B(Mytilus defensin B，缩写为 MDB)。MDA 的相对分子质量为 4 314.3，含 37 个氨基酸残基，其中包括 6 个半胱氨酸残基(Cys-)，构成 3 个分子内二硫键。其 21 位和 28 位上未获得乙内酰苯硫脲氨基酸，可能被修饰或携带一个基团取代(或保守性置换)。MDB 相对分子质量为 4 392.4，有 35 个氨基酸残基，也含 6 个半胱氨酸残基。其 28 位上无信号。在其 COOH 端可能有少数几个额外的残基，与 MDA 类似。

已测定出的 MDA 和 MDB 的氨基酸序列如下(为 2 个防卫素共有的半胱氨酸列阵)。

MDA　GFG [C] PNDYP [C] HRH [C] KSIPGRXGGY [C] GGXHRLR [C] T [C] YR

MDB　GFG [C] PNDYP [C] HRH [C] KSIPG. RXGGY [C] GGXHRLR [C] T [C]

分析显示 MDAs 与蜓(*Aeschna fyanea*)和蝎(*Leiurs quinquestriatus*)的防卫素的序列有 70%一致，进化上同源。上述三防卫素排成一列，与伏蝇防卫素相对。MDA 的三级结构在 NH_2-端已导出一个间隙，相应于伏蝇防卫素扩展的环。MDAs 的 NH_2：一端环短于伏蝇防卫素的。MDB 和 MDA 的序列靠得很近，都显出半胱氨酸位置高度保守，与节肢动物防卫素类似。表明这些贻贝肽与节肢动物防卫素有真正的进化同源性，防卫素序列分部的系统发育分析又揭示出它们属于缓慢进化组的代表，相应于优势组成性表达的防卫素。

MDA 和 MDB 存在于含大颗粒的粒细胞的大颗粒及含小颗粒的粒细胞中，或与 Mytilins 一起，部分地分布于相同血细胞亚群、消化道表皮及肠道细胞的颗粒结构中，也在细胞浸润的消化道表皮中表达。该防卫素的基因以一个单拷贝存在于基因组中，其信使浓度在循环血细胞中不增加。MDA 和 MDB 主抗 G^+ 细菌，包括一些海洋无脊椎动物的病原，详见表 1。

1.2　地中海贻贝防卫素(MGDs)

地中海贻贝防卫素是分离自地中海贻贝(*Mytilus galloprovimcialis*)血浆和血细胞的防卫素或防卫素样肽，即 Mytitus galloprovincialis defensin 或 defensin-like peptides，缩写为 MGDs，它分为 MGDl 和 MGD2。

MGDs 是节肢动物防卫素家族的一个原始新成员。MGDl 相对分子质量为 4×10^3，最初由 Hubert 等[1]在 1996 年发现自地中海贻贝的酸性无细胞血淋巴上清液，后来从血细胞中的富细胞器的片段中也分离得到。由 RF-PCR(反转录转座聚合酶链反应)获得由读码框架 ORF 末端和 3′-未翻译区组成的 MGD1 的 cDNA 片段，表示 MGDl 与节肢动物防卫素的分子有类似性。MGD1 分子中除 6 个半胱氨酸外还有另外 2 个半胱氨酸和 1 个修饰了的氨基酸。MGDl 具有规范的似节肢动物防卫素的 CSαβ(Cysteine-Stabilizedamostif)结构，主要包括 1 个螺旋部分(Asn7-Ser16)和 2 个逆平行的链(Arg20-(Ays25 和 Cys33-Arg37)，一起形成蝎毒中常见的由 4 个二硫键稳定的 α-β 模板。除 Cys21-Cys38 二硫键是溶剂暴露的之外，其他 3 个二硫键均为高度束缚的特别疏水的芯。在 Cis 结构中的 C_4-P_5 酰胺键特化出 MGD1 结构。

MGD1 含 39 个氨基酸残基。第 39 位上有甘氨酸残基，C-端酰胺化。

MGD1 对人的红细胞和原生动物无细胞毒性。抗微生物活性如表 1 所示。此生物学活性由分子中 3 个二硫键产生，Trp28 的羟化作用不包含在生物学活性之中，即 Cys21-

Cys38 键对于此活性不是本质的。整个分子结构中显示出位于 3 个疏水簇中的典型阳电荷分布和疏水侧链，它们保证产生杀细菌效能及对 G^+ 细菌的专一性。现已能人工合成出 MGD-1，其抗细菌活性类似天然 MGDl。

MGD1 存在于血细胞的富含细胞器的片段中。经细菌诱导的贻贝，触发血浆中 MGD1 的浓度在 24h 后增加。血淋巴在被注射即时或之后，MGD 免疫反应向胞膜迁移，并释放 MGD1 进入循环系统作全身性的抗微生物感应。循环血细胞中 MGD 信使浓度不在细胞被诱导后立即增加，而是在 48 h 后戏剧性地增加。MGD1 的天然形式纯化自血细胞，说明是在血细胞中从前体加工成 MGD1 活性化合物的。MGD 前体含 39 个氨基酸残基的成熟肽及 1 个额外的阴离子的 H 残基的 C-端序列。而成熟肽可通过常规加工机制产生，并循环酰胺化作用信号"G-R-R"。MGD 的成熟经下列 3 个步骤完成：①经识别该大前体的二元序列(R-R)的加工酶在 41 位和 42 位的 R 和 D 间作剪切；②一个羧肽酶 α-样的酶释放 R-40 和 R-41；③对前体的羧基端 G 残基识别的酶催化氧化的酰胺化作用[1~3]。

MGD2 最初分离自地中海贻贝，后来在食用贻贝中也测到 MGD2 的 mRNA。MGD2 分子含 39 个氨基酸残基及 8 个 Cys-，与 MGD1 的氨基酸序列相比，有 6 个不同的氨基酸残基。MGD2 的 cDNA 序列有 1 个具 21 残基的 N-端原序列，含有 1 个信号序列的疏水芯特征。细菌注入触发转录物，使 MGD2 在 6h 和 24h 后的浓度减少，但与 Mytilin B 相反，MGD2 的 mRNA 水平直至采样期结束仍保持低下(这是受细菌攻击 72h 后的情况)。锉擦贻贝 6h 后 MGD2 的 mRNA 浓度会增加，锉擦 24h 后尤其明显。热休克(33℃)贻贝 90 min，其血细胞 RNA 显出与 MGD2 cDNA 的带杂交，强度明显增加，在热休克后转至 15℃ 24h，此杂交强度逐渐降至对照水平。MGD2 基因表达呈现出为热压力所调整。锉擦和热休克贝壳使 MGD2 信使水平大为提高，意味着压力下该基因表达可能诱导出来。化脓性损伤触发循环血细胞中 MGD2 信使浓度增加[4]。

表 1　贻贝防卫素的抗微生物活性

Table 1　Antimicrobial activities of mussel defensins

贻贝防卫素种类 Mussel defensins	对抗微生物的状况 Antimicrobial conditions	备注 Remarks
MGD1	抗 G^+ 细菌：藤黄傲球菌，大肠埃希氏菌、解藻朊酸弧菌、副溶血弧菌、灿烂弧菌；也抗一些 G^- 细菌，合成的 MGD-1 抗细菌活性类似天然 MGD1。 Anti-G^+ bacteria: *M. lateus*, *E. coli.*, *V. alginolytius*, *V. parahaerrmlyticus.*, *V. splendius*; and anti-some of G-bacteria. Manmade MGI-1 k similar to natural MGD-1 in antibacterial&ctivity.	人 O 型血球 20%溶解 Dissolution of 20% O-type blood ceils of human
Myticins A	明显对抗 G^+ 细菌：藤黄微球菌、巨大芽孢杆菌、绿色气球菌。 Anti-G^+ bacteria markedly: *M. luteus*, *B. megatherium*, *A. viridans*.	

(续表)

贻贝防卫素种类 Mussel defensins	对抗微生物的状况 Antimicrobial conditions	备注 Remarks
B	抗上述三菌力比A更强,还抗大肠埃希氏菌、尖孢镰刀菌。 It is more powerful in anti three species of bacteria mentioned above, and also antigonize *E. coli*, *F. oxysporum*.	
Myticins	总体看对抗 G^- 细菌的活性强于抗 G^+ 细菌。 It is nxore powerful in anti-G-bacteria than in anti-G＋ bacteria overall.	
A	抗 G^+ 细菌:藤黄微球菌、大肠埃希氏菌,抗 G^- 细菌稍弱。 Anti-G^+ bacteria: *M. 1uteus*, *E, cobi*, weaker in anti-G- bacteria.	
B	抗灿烂弧菌、尖孢镰刀菌。 Anti-*V. splendius*, *F. oxysporum*.	
C	抗灿烂弧菌。 Anti-*V. splendoius*.	抗海水派金虫 Anti-*P. marinus*
D	抗 G^+ 细菌,如金黄色葡萄球菌、藤黄微球菌,但杀菌力小于 Mytilin C,也抗尖孢镰刀菌。 Anti-G^+ bacteria, such as *S. aurctus*, *M. luteus*, but its bactericidal effect is weaker than mytilin C. Anti-*F. oxysporurn.*	
Gl	抗 G^+ 细菌,活性强于 Mytilin D。 Stronger than rnytilin D in anti-G^+ bacteria.	具有溶胞作用 induce cytolysis
Mytimycin	延滞一些真菌的生长,如黄色镰孢、Nowakouskiella crasa 等。 Block growth of some fungi, such as *F. culmorum*, Nowakouskiella crassa etc.	
MDs A	主抗 G^+ 细菌,包括一些海洋无脊椎动物病原。 Anti-G^+ bacteria mainly, as well as some pathogens of marine invertebrates.	
B	主抗 G^+ 细菌,包括一些海洋无脊椎动物病原。 Anti-G^+ bacteria mainly, as well as some pathogens of marine invertebrates.	

贻贝在不同发育期，MGD2 是受调控的，但直至幼体附着和变态后未测出基因表达。在卵期、幼体阶段或早期的幼体后期均无 MGD2 基因信号，至幼体后期第 25 天，才有微弱的信号并增至受精后 32d，在血细胞中有高水平的表达，但总的成体组织中未测出转录。成体贻贝在受物理和温度压力时，MGD 的基因可能是过渡表达的，当受细菌挑战时表达则减退[4]。编码 MGD2 的同工型的基因已克隆出来并测完其序列，结果揭示出它有 4 个外显子、3 个内显子。MGD2 基因可能以 1 个单拷贝存在于基因组之中。

MGD2 样物质主要分布在 2 种血细胞亚型中，MGD2 的基因是连续表达的，且经细菌攻击后不被诱导，MGD2 基因表达水平变化发生在幼体变态期及之后。3 个不同 cDNA 共享一个高度同源，即相当于 MGD1 的 1 个 cDNA 片段，相当于 MGD2 的 cDNA 及相当于 MGD2b 的 cDNA 片段。

成熟的 MGDs 之列阵和半胱氨酸排位如下（[C]为 2 个防卫素共有的半胱氨酸列阵）：

MGD1　GFG [C] PNNYQ [C] HRH [C] KSIPGR [C] GGY [C] GGWHRLR [C] T [C] YR [C] G

MGD2　GFG [C] PNNYA [C] HQH [C] KSIRGY [C] GGY [C] AGWFRLR [C] T [C] YR [C] G

MGD2 是 1 个新的 MGD 同工型，是在筛选血细胞 cDNA 文库中描述出 MGD2 的 cDNA 特征的。

MGDs 具有原始前体结构，在血细胞中合成和加工。MGDs 参与后来的全身性防卫过程。可能在抗感染过程中的不同阶段、对抗不同病原中发挥作用[1]。

2. Myticins

1999 年，Mitta 等[5]分离地中海贻贝血细胞和血浆得出一类新的富半胱氨酸抗微生物肽，即 Myticin。它共有 A、B 两个同工型。

血细胞外，Myticin A 还分离自地中海贻贝的血浆（未经细菌诱导的），之后，Myticin B 又从食用贻贝血细胞中分离出来。

Myticin A 相对分子质量为 4.43×10^3（4437.28±2.46），含 40 个残基，包括 8 个半胱氨酸残基或 4 个二硫键。Myticin A 初级结构中的半胱氨酸列阵不同于以前描述过的许多富半胱氨酸抗微生物肽。克隆的 cDNAs 序列分析表明其前体由带有 20 个氨基酸的推

断性信号肽的 96 个氨基酸和 36 个残基的 C-端伸展组成。该肽是在血细胞中作为前蛋白原合成的，然后在该活性肽贮存之前经各种解朊过程加工而成，其前体主要在血细胞中表达[5]。

Myticin A 有 36 个残基的 N-端序列。包括 7 个烷基化半胱氨酸，其序列为：HSHACTSYWCGKFCGTASCTHYLCRVLHPGKMCACV。

从表 1 可看到，Myticin A 有明显杀伤作用，如对藤黄微球菌、巨大芽孢杆菌和绿色气球菌（*A. viridans*）（最小杀菌浓度 MBC 为 2.25～4.5 mol/L），但对原生动物寄生虫（如美洲巨蛎）则没有活力。

Myticin B 相对分子质量为（4 563.45±1.32），有 8 个半胱氨酸残基。部分 N-端序列获得包含 1 个烷基半胱氨酸的 7 个残基。

Myticin B 对 G^+ 细菌的最小杀菌浓度(MBC)为 1～2 μmol/L。对 G^- 细菌大肠埃希氏菌(*E. coli*)D31 则有中等活性(MBC 为 2.0～10 μmol/L)。对丝状真菌尖孢镰刀菌的最小抑菌浓度(MIC)为 5～10 μmol/L。

Myticins 的氨基酸序列如下(C为 2 个防卫素共有的半胱氨酸列阵):

A　HSHA C TSYW C GKF C GTAS C THYL C RVLHPGKM C A C VH C SR

B　HPHV C TSYY C SKF C GTAG C TRYG C RNLHRGKL C F C LH C SR

Myticin 基因以单拷贝存在于基因组中,至少含有 1 个内显子—其剪接部位于衔接点,遵从 GF-AG 规则。这是与规范的外显子-内显子衔接点共有的序列。Myticins 从 96 个残基组成的前体分子加工来,该前体含 1 个有 20 个氨基酸残基的信号肽,1 个有 40 个残基的成熟肽和附加的 36 个残基的 C-端序列。成熟的 Myticin、可经常规加工机制产生,首先释放前一片断,然后内蛋白酶 R 在 40 位和 41 位上的 N-V 间剪切。20 个残基的 N-端片断假设是易位到糙面内质网的管腔的 1 个信号序列。C-端部分的功能意义不明,在 2 个同工型间高度保守且同工型 A 和 B 分别含 5 个和 6 个酸性氨基酸,该酸性区可与肽的阳离子部分作用而稳定前体的结构,以进行解朊过程或阻止碱性肽部分的膜相互作用。Myticin 的 C-端伸展可作为将 1 个信号交付给该肽至 1 个特定的血细胞部位而作用。此蛋白质部分在成熟肽后的这种定位对无脊椎动物抗微生物肽前体而言是不一般的。

贻贝血细胞是 Myticins 的生产和贮存位点,也是 Myticin 前体产生位点。Myticins 在血细胞富细胞器部分的酸性浸汁中很丰富。Myticins 在血细胞中从 mRNA 加工为活性肽。未经细菌诱导的贻贝,也可能已处于刺激的免疫状态,从而导致血浆中该肽的释放和血细胞活化。

Myticins 的三维结构高度束缚。其一级结构中的半胱氨酸位置不同于以往描述的无脊椎动物富半胱氨酸抗微生物肽,如昆虫防卫素、大防卫素,果蝇抗真菌肽、tachyplesine(最先分离自日本鲎细胞的防卫素)、死亡素、buthimine mytilin、MGD1、peraedin(白对虾血细胞抗微生物肽)、哺乳动物防卫素 Protegrins、brevinins(猪白细胞小分子抗菌肽)和植物防卫素 r-thionins、Ib-AMP1-4 等。Myticin 还具有溶菌作用,但比 Mytilin 的慢得多。Myticin 作为其他抗微生物肽的一个合作者参与贻贝免疫。

Myticin 基因在血细胞中充分表达,在外膜、触唇和鳃里,其 RNAs 显出有与此相同迁移率的淡带[3,6]。

3. Mytilins

Mytilins 分子富含半胱氨酸(8 个 Cys-),共 34 个残基,相对分子质量约为 3 877.79。Mytilins 的三维结构高度束缚,二级结构中的二硫键的连接和半胱氨酸列阵不同于迄今已知节肢动物、蛙、哺乳动物或植物的富半胱氨酸的抗微生物肽。Mytilins(同工型 A 和 B)在贻贝血中的浓度约 22 μmol/L,这是所试多数细菌的 MIC 范围。它们抗 G^- 细菌的活性强于抗 G^+ 细菌。

Mytilin 包括 5 个同工型,它们分别是 Mytilin A、Mytilin B、Mytilin C、Mytilin D 和

Mytilin G1。MytilinA 和 Mytitin B 分离自食用贻贝血浆，Mytilin B、Mytilin C、Mytilin D、Mytilin G1 来自地中海贻贝血细胞。

Mytilin A，相对分子质量为 3 773.7，含 34 个氨基酸残基，包括 8 个 Cys-。在其分子的 25 位上是 Gly-(甘氨酸残基)。对 G^+ 细菌的抗性大于对 G^- 细菌。如抗藤黄微球菌、大肠埃希氏菌等。它能使遭遇到的细菌在几秒钟内胞质膜的通透性壁障瓦解，部分去极化，膜质 ATP 减少，呼吸受抑制直至死亡[7,8]。

Mytilin B 含 34 个残基，其中有 8 个 Cys-，相对分子质量约 3 974.3 或(3 973.09±0.95)。在血浆中的浓度约 2 μmol/L。对抗相关微生物如尖孢刀镰菌和灿烂弧菌的 MIC 只需几分钟就可达到。Mytilin B 作为血细胞中加工成活性化合物的前体分子而产生。该前体由 22 个残基前片断，34 个残基的成熟肽和 48 个富酸性氨基酸残基的 C-端序列组成。Mytilin B 的信使和该肽存在于肠细胞(包括潘尼氏细胞)的颗粒结构中，得到细菌诱导的贻贝之 Mytflin B 信使浓度在循环血细胞中暂时减少，其后在血浆中浓度增加，超过 MIC，参与抗微生物的局部防卫，以杀灭细菌。Mytilin B 是发育调节的，在幼体附着和变态之后，其基因不表达。该基因含 4 个外显子，3 个内显子[7]。

Mytilin C 相对分子质量为(4 287.05±0.29)，含 31 个残基，其中有 7 个为 Cys-。对抗(杀死)细菌如灿烂弧菌的 MIC 仅需几分钟即显现出来。另外它也抗原生动物海水派金虫(*Perikinus marinus*)。Mytilin C 同 Mytilin B 的一级结构中均共享 1 个高度同源，两者功能互补。

Mytilin D 分离自地中海贻贝。相对分子质量约 3 877.79，含 34 个残基，包括 8 个 Cys-。分子中只有 1 个氨基酸不同于 Mytilin A，即在 25 位上有 Arg(精氨酸)，而不是 Gly(甘氨酸)。Mytilin D 的主要生物学活性是明显拮抗 sG^+ 和 G^- 细菌。但杀菌力(对藤黄微球菌和金黄色葡萄球菌)小于 Mytilin C 和 MytilinG1 的。Mytilin D 对尖孢镰刀菌的活性也较强[7]。

Mytilin Gailloprovincialis 1(即 Mytilin G1)，分离自地中海贻贝血浆。其相对分子质量为4 118.3，含 36 个氨基酸残基，包括 8 个 Cys-。与其他 Mytilins 共享一致的半胱氨酸列阵。但与它们只有有限的同源。MytilinG1 仅对 G^+ 细菌有活性。其 MIC 需 6h 才显出。其浓度达 5.6 μmol/L 时，对金黄色葡萄球菌和尖孢镰刀菌也不显活性。此外 Mytilin G1 还具溶胞作用[7,8]。

上述 Mytilins 的抗微生物活性详见表 1。

Mytilins 的 5 个同工型的氨基酸序列如下(C 为 2 个防卫素共有的半胱氨酸列阵)：

A　G C ASR C KAK C AGRR C KGWASASFRGR C Y C K C FR C

B　S C ASR C KGH C RARR C GYYVSVLYRGR C Y C K C LR C

C　S C ASR C KSR C RARR C RYYVSVRYGGF C Y C R C C

D　G C ASR C KAK C AGRR C KGWASASFRRR C Y C K C FR C

G1 VVT C GSL C KAH C TFRK C GYFMSVLYHGR C Y C R C LL C

富多样性的 Mytilin 同工型具多样性的生物学意义，其生物学活性互补，共同抑杀入侵者。

37%的循环血细胞含 Mytilins，主要贮存于颗粒细胞，特别是表达大颗粒（即溶酶体样结构）的血细胞中，并浓缩于大颗粒。与防卫素 MDAs 两者似乎部分地分布于同一或不同的血细胞亚群中。含 Mytilin 的粒细胞可包含在抗感染反应的不同阶段及吞噬细胞中。Mytilin 在血浆中的增加是通过脱粒作用实现的，然后再释放入循环系统作全身抗微生物反应。其基因是组成性的，不随细菌挑战而被诱导。Mytilin 广泛分泌（包括经胞泌作用的释放）进入与血细胞直接的环境中向细菌入侵位点转运，在胞内对被卷入的细菌行使杀死作用。含有 Mytilin 的免疫反应性细胞（亚型及多泡囊颗粒）吞噬细菌，首先在吞噬小体样结构中内化细菌，致细菌与 Mytilin 共同位于（相）同一细胞器并融合。Mytilin 表达细胞很少或不在消化管道上皮中表现，但在鳃中则表达得很好。在与外环境相连的表皮中，Mytilins 阳性细胞特别丰富，从而构成阻止病原入侵的第一道防线。

Mytilin cDNA 分离自血细胞 mRNA。血细胞是 Mytilin 及其前体合成位点。Mytilin 在血细胞中加工为活跃化合物[4]。

总之，Mytilins 包含在对感染的反应之中，它们可在几小时内由血细胞转至感染部位，与被卷入的细菌同处，随后即杀灭之。血细胞的脱粒作用促使其浓度在血浆中增加，并作全身性抗微生物反应。Mytilins 的释放使血浆浓度超过必要的 MBC 值以杀死多数入侵细菌。

4. Mytimycin

Mytimycin 分离自未经病原诱导的食用贻贝血浆，相对分子质量为 6.3×10^3，有 32 个残基，包括 12 个 Cys-。Mytimycin 是专抗真菌的贻贝肽，已获得含 32 个残基的 NH_2—端的部分序列，它约相当于全序列的一半。在蛋白质资料库中未搜索到与此有任何同源的肽。该肽对一些真菌的生长有延滞作用。Mytilmycim，有很紧密的结构形式，其部分的 NH_2—端序列如下[7,9]：

DCCRKPFRK ACWDCTAGTPYYGYSTRNIFGCTC

……

5. 结语

迄今的贻贝防卫素研究成果主要在法国，可见法国养殖贻贝史的悠久、经济价值的重要及对其先天免疫研究的重视。我国贻贝的规模养殖至今已有 40 余年，也存在一定的病害问题，可是迄今，未见到任何关于它的防卫素等抗微生物肽及先天免疫的研究报道。尽管养殖中的贻贝能较好适应环境，也不易为重病所折磨，但我国的贻贝是否有防卫素等抗微生物肽在防御敌害，在保卫自身的活动中起作用，是值得研究的课题。人们可以从贻贝防卫素成果获得启发，以遗传选择或与贻贝类软体动物先天免疫为基础，增强养殖贝类的免疫能力，改进品质，提高产量。

参考文献 9 篇（略）

（合作者：魏玉西　郭道森）

PROGRESS OF RESEARCH ON MUSSEL DEFENSINS

(ABSTRACT)

Abstract Defensins contained in mussel and other marine mollusca are important antimicrobial peptides. The mussel defensins known update are divided into four groups according to character, primary structure and mutual cysteine sequence. They are mytilus defensins(MDA and MDB), Mytilus gallovoprovincial defensins(MGDl and MGD2); Myticins A and B; Mytilins A, B, C, D, G1; and Mytimycin. Their chemical character and structure, production and bioactivities are explained. The mussels living in various environments are not subjected to serious diseases luquente, and resist invading of diverse pathogens and protect themselves from enemy microbials. The researches on mussel defensins are contribute to understand of the innate immunity of mussel and other marine mollusea, and improvement of maricultural techniques.

Key words Mussel, Defensin, Antimicrobial Peptide, Innate Immunity, Marine Mollusca

菲律宾蛤仔(*Ruditapes philippinarum*)血淋巴中抑菌活性物质的活性影响因素*

摘　要　为了建立蛤仔血淋巴中抑菌活性物质的快速高效分离、纯化方法，利用正交设计法对影响菲律宾蛤仔血淋巴中抑菌物质抑菌活性的3个因素进行了探讨。结果表明，在本实验条件下，pH对抑菌活性影响最大，温度次之，处理时间的影响最小；而以在pH值为4，温度小于65℃的介质中活性物质能保持较高抑菌活性。提示菲律宾蛤仔血淋巴中抑菌活性物质对热有一定的耐受性，且适于偏酸性的环境。

关键词　菲律宾蛤仔(*Ruditapes philippinarum*)　血淋巴　抑菌物质

许多双壳类动物在我国是种类多、数量大、经济价值高的食用种，有些已成为主要养殖对象，如贻贝(*Mytilus galloprovincialis*)、栉孔扇贝(*Chlamys farreri*)、海湾扇贝(*Argopecten irradiants*)和菲律宾蛤仔(*Ruditapes philippinarum*)等。其中，菲律宾蛤仔仅在胶州湾的年捕获量就达10余万吨[1]。近年来，病害对贝类养殖业造成的危害逐年加剧，但大多数贝类病害一直缺乏理想的防治办法[2]。作为无脊椎动物免疫学的一部分，贝类的免疫生物学研究，无论在物种起源与进化，以及免疫系统发生的探索等方面，还是在生产应用方面均有重要意义[3]。已有实验证明，双壳贝类依靠细胞及体液成分抵御潜在病原的入侵[4]，其中贝类血细胞在机体的防御机制中起主要作用，而淋巴液中因含多种生物活性物质，如能水解脊椎动物红细胞的高相对分子质量细胞毒性蛋白和新发现的抗菌肽等[5~6]，已引起人们的广泛关注。我们在已进行的青岛地区海域中双壳贝类血淋巴中抗菌活性物质的筛选时发现，菲律宾蛤仔血淋巴对金黄色葡萄球菌(*Staphyloccocus aureus*)和鳗弧菌(*Vibrio anguillarum*)具较强的抗菌活性[7]。

本文以金黄色葡萄球菌为指示菌，对菲律宾蛤仔血淋巴中抑菌活性物质的活性影响因素进行探讨，为其中抑菌活性物质的分离、纯化方法的确立奠定必要基础。

1　材料与方法

1.1　试验材料

菲律宾蛤仔(*R. philippinarum*)采自胶南海青，处理前于18～20℃海水中通氧暂养；金黄色葡萄球菌(*Staphyloccocus aureus*)由原青岛海洋大学食品工程系提供，培养基为营养琼脂。

1.2　试验方法

* 国家自然科学基金资助项目(40086001)Supported by the National Natural Science Foundation of China

原文刊于《应用与环境生物学报》，2003，9(3)：271～272，本人为第四作者及项目负责人。

1.2.1　血淋巴的采集　取鲜活且大小均匀的菲律宾蛤仔 500 只，将采集的血淋巴于 10 000r/min 离心 30 min(4℃)，取上清液，经 0.22 μm 无菌过滤器过滤得血淋巴滤液(pH 6.5)置于 4℃冰箱备用。在后续实验中，pH 值小于 6.5 的血淋巴滤液须以 1.0 mol/L 盐酸调节。

1.2.2　抑菌活性测定　参照文献[8]所采用的平板生长抑制法进行。取经处理过的血淋巴上清液 6 μL，滴在含有供试细菌的营养琼脂平板上的一个孔穴(d2.5 mm)中，每一样品滴样 5 个孔穴。滴样后将平板置于 37℃温箱中，培养 24h 后观察结果，测定并记录抑菌圈的直径(减去孔穴直径)。

1.2.3　试验方案设计　选温度、时间及 pH 值 3 个因素，每个因素设计 5 个水平，参照 $L_{25}(5^6)$设计正交实验方案[9]。因子与水平排列见表 1，实验设计方案见表 2。血淋巴滤液不经任何处理作为对照组直接进行抑菌活性测定。每组实验重复 3 次，结果取其平均值。

表 1　因子与水平排列

水平 Levels	因子 Factors		
	pH	θ/℃	t/min
1	2	4	10
2	3	25	30
3	4	45	60
4	5	65	90
5	6.5	80	120

2　结果与讨论

影响血淋巴抑菌活性的因素很多，且相互关系复杂。本实验按表 2 设计方案考察了抑菌活性物质纯化过程中的主要影响因素(pH 值、温度和时间)对其抑菌活性的影响，并将结果填入表 2，同时按文献[9]进行数据处理。由表 2 中的 pH 值、温度和时间 3 因素的分散度可判定，pH 值对活性的影响最大，其次是温度，影响最小的是时间。

表 2 正交设计方案及结果

Tab 2　P1an of orthogonal design and results

实验号 No.	因子 Factors			抑菌圈 Zone diameter (d/mm)
	pH	θ/℃	t/min	
1	2	4	10	3.5
2	2	25	30	3.5
3	2	45	60	3.4
4	2	65	90	3.4
5	2	80	120	2.9
6	3	4	30	3.6
7	3	25	60	3.9

（续表）

实验号 No.	因子 Factors pH	 θ/℃	 t/min	抑菌圈 Zone diameter（d/mm）
8	3	45	90	4.0
9	3	65	120	3.6
10	3	80	10	3.5
11	4	4	60	4.0
12	4	25	90	4.1
13	4	45	120	4.0
14	4	65	10	4.9
15	4	80	30	3.0
16	5	4	90	2.5
17	5	25	120	3.1
18	5	45	10	2.8
19	5	65	30	2.6
20	5	80	60	2.4
21	6.5	4	120	2.5
22	6.5	25	10	3.4
23	6.5	45	30	2.6
24	6.5	65	60	2.6
25	6.5	80	90	1.0
Ⅰ	16.7	16.1	18.1	总和 Total $T=80.8$
Ⅱ	18.6	18.0	15.3	
Ⅲ	20.0	16.7	16.3	
Ⅳ	13.4	17.1	15.0	
Ⅴ	12.1	12.8	16.1	
Ⅰ/5	3.3	3.2	3.6	总平均 Total average $\mu=T/25$ $=3.2$
Ⅱ/5	3.7	3.6	3.1	
Ⅲ/5	4.0	3.3	3.2	
Ⅳ/5	2.7	3.4	3.0	
Ⅴ/5	2.4	2.6	3.2	
分散度 Degree of separation	1.8	0.6	0.2	

(1)对照组直接进行抑菌活性测定，其抑菌圈直径为3.4 mm。pH值为2.0的盐酸溶液无抑菌活性，由此可排除调节血淋巴滤液样品pH值时所用盐酸的影响。另据预备实验证实，pH值大于6.5的血淋巴滤液抑菌活性较对照组有下降趋势，因此，在设计pH值水平时，以pH值6.5作为上限，最低pH值选为2.0。在pH值的5个水平内，pH＝2.0和pH＝6.5时抑菌圈 d 的平均值分别为3.3 mm和2.4 mm，pH＝4.0时抑菌圈的平均值达4 mm。说明抑菌活性物质对介质的pH非常敏感，这是因为pH严重影响其活性基团的解离，从而影响了其活性[10]。

(2)在5个处理温度的水平内，$t<65$℃时，t 对抑菌活性无显著影响，说明菲律宾蛤仔血淋巴中抑菌活性物质对温度有一定的抗性。但当 $t>80$℃时，其活性显著下降，这是由于高温可致活性物质的降解或变性失活。因此，在进行菲律宾蛤仔血淋巴中抑菌活性物质的分离、纯化时，选择在65℃时先将血淋巴预处理一定时间，不仅可使其中的某些水解酶变性失活，还可将无活性的大分子杂蛋白变性沉淀而除去，从而使血淋巴易于保存，便于后续操作，且使之先得到一定程度的纯化。

(3)从处理时间的5个水平看，处理时间以短时为宜，时间延长则活性降低，但从总体来说，时间是非主要的影响因素。

总之，菲律宾蛤仔血淋巴中抑菌活性物质在偏酸性条件(pH＝4.0)、65℃以下及短时处理活性最强。在后续抑菌活性物质的分离纯化时应尽量满足上述条件，尤其应注意介质的酸碱度范围，以最大限度保存活性并使操作简便、快速、高效。

参考文献10篇(略)

(合作者:魏玉西　郭道森　李　丽)

FACTORS AFFECTING INHIBIT-MICROBIAL ACTIVITY OF ANTIMICROBIAL SUBSTANCES IN HEMOLYMPH FROM *RUDITAPES PHILIPPINARUM*

(ABSTRACT)

Abstract Effects of three factors on the inhibit-microbial activity of inhibit-microbial substances in the hemolymph from *Ruditapes philippinarum* were studied by the method of orthogonal design. The results showed that pH value affected it greatly, and temperature the second, and time the last. Furthermore, inhibit-microbial substances kept high inhibit-microbial activity in a medium of pH4 and temperature of lower than 65℃. It was indicated that inhibit-microbial substances in the hemolymph from *Ruditapes philippinarum* had certain resistance to heat and were suitable partially to acidic environment.

Key words *Ruditapes Philippinarum*; Hemolymph; Inhibit-microbial Substances

防卫素——先天免疫的一种中枢组分*

摘　要　从先天免疫角度，简述防卫素研究的发展历程，阐述防卫素的性质、分类、合成、作用和应用前景。特别关注防卫素在先天免疫中的功能，并强调海水养殖对象等海洋无脊椎动物防卫素的研究状况及与先天免疫的关系。指出海洋生物作为包括防卫素在内的抗生肽资源库的巨大开发利用价值和研制抗菌新药的诱人潜力。

关键词　防卫素　先天免疫　海洋无脊椎动物

一、先天免疫——生物的一种防御机制

地球上的各级各类多细胞生物之免疫系统长期经受感染微生物(主要指病原生物)的胁迫而选择性演化进化，发展出各种防御、防卫机制，其一是对感染具有触发能力以保护宿主生物不被入侵微生物所破坏并中和其毒力因子，这便是系统发育上古老的防卫机制，即先天免疫系统(innate immunity system)，或曰原生宿主防卫机制(primary host defence mechanisms)[1]。该系统因识别微生物病原而使用生殖系编码受体(germline-encoded receptor)，它无疑分布于无脊椎动物之中。脊椎动物另有适应性免疫系统(adaptive immune system)，它以克隆选择为基础。无脊椎动物不含免疫球蛋白，也缺乏对最先遭遇到的病原产生抗体，因而需要依赖先天免疫机制对抗微生物攻击。靠对微生物特异的碳水化合物分子的识别机制，先天免疫系统识别潜在的微生物入侵，经常选择地对抗已包含在被识别的靶中的异己。但直至 20 世纪末，对无脊椎动物的免疫防御实际上是知之甚少的。抗微生物肽(antimicrobial peptides)是所有生物天然免疫系统的重要组分[2]。该肽除了 cathlecidins(具强力膜活性的抗微生物肽类前体，N 端具高度保守的 cathelin(来自猪白细胞的一种抗细菌多肽)样节段和高度可变的 C 端杀微生物域的新奇蛋白)外，其中主要的一类即是 defensins(防卫素)等的富含半胱氨酸的肽[1]，(Hoffmann 等将抗(细)菌肽(主指昆虫的)分作四类。其中一个重点就是防卫素。虽然防卫素研究始于 20 世纪 80 年代，但在后来的 20 年中取得了可喜的进展[3]。本文就防卫素的研究现状作一介绍，特别就海洋无脊椎动物防卫素研究作一分析，以期引起重视。

二、防卫素及相似名称

Defensin 的中译名有预防素、防御素和防卫素等，而译作预防素及相近词的除了 de-

* 国家自然科学基金项目(批准号:40086001)

原文刊于《自然杂志》2003，25(3)：129-135。

本文采用下列缩写词：Ala-丙氨酸；Arg-精氨酸；Cys-半胱氨酸；Glu-谷胱氨酸；Pro-脯氨酸；val-缬氨酸；OM-细胞外膜；CM-细胞质膜；MIC-最小抑制浓度。

fensin 外，还有预防剂(prophylactic)。译作防御素的还有 alexin(亦称补体)、sozin(亦称防卫素)、phylaxin 及 mycophylaxin(即霉菌防御素、御菌素、菌防御素)。译作防卫素的还有 sozin(亦称防御素)及相近的防卫蛋白(sozalbumin)。由上可见，这些中文名称间有相互交叉或混用的现象，必须根据英文原意进行区别。Defensin 不应译作预防素，以免与 prophylactic 混淆。称作防御素的多达 5 个(包括 addiment-补体)。其实它们各有其明确的含意，作为 defensin 可不必插入了。sozin 和 sozalbumin 均属蛋白类，而 defensin 多为肽，与前两者有别，功能也比前两者广泛。根据前两者之一 sozin 的来源，可将其称作人体防卫蛋白，而 sozalbumin 则仍可保留“防卫蛋白”之称。再者 defensin 在上述各名称中是发现较晚者，只需称作“防卫素”即可，取“defensin”英文的防御和保卫之意[14]。当否请专家指教。

三、防卫素的性质

1. 防卫素的一般性质

防卫素最早发现并定名于 1980 年代初。1980 年，Leherer 实验室从兔肺巨噬细胞分出两个阳离子性极强的小分子抗菌肽，后又从兔和人的中性粒细胞浆颗粒中分出小肽分子，他们最先从昆虫(*Hyalophora cecropia*)中提取出一类抗细菌蛋白，将其定名为 attacin(侵蚀素)[15]。迄今已从动植物中发现了 2 000 多种抗菌肽。在此提出“抗菌肽”及前述的“抗微生物肽”这两个概念，是因为虽然一些防卫素是蛋白质，但迄今发现的多数防卫素本质上仍是肽(包括多肽)类物质，它们大大丰富了防卫素的涵盖面。这表明防卫素是肽类物质，但又不同于-般的肽。其主要区别在于，它们属内源性抗生素肽(endogenous antibiotic peptides)范畴。尽管它们因生物种类及结构和序列的不同而有明显多样性，但共同之点一是均有二硫键，借此使分子结构稳定；二是均具 β-结构域；三是含高量的碱性残基，如半胱氨酸；四是倾向于采用一个两亲构象(amphipathic conformation)。但是随着防卫素研究的深广化，防卫素的内涵正在不断扩充。

按防卫素的来源生物大类分，可以将防卫素分为哺乳动物防卫素、昆虫防卫素、无脊椎动物防卫素，植物防卫素等。

2. 哺乳动物防卫素的性质

哺乳动物防卫素是可变阳离子的、较富含精氨酸(Arg)的、非糖基化的由 29～34 个氨基酸残基组成的肽，它典型地含形成三个分子内二硫键的 6 个半胱氨酸，其结构以三个二硫键稳定化的 β-片层所支配，不含 α-螺旋。所有防卫素均以一个特征性的半胱氨酸为模板，为动物细胞内源抗生肽。哺乳动物防卫素总数计 100 多种。此外，人、兔吞噬细胞、中性粒细胞、浆颗粒等的防卫素至 1999 年已发现了 20 多个，构成了最大的抗生肽家族。迄今哺乳动物防卫素是脊椎动物防卫素中研究得最好、最多的一类抗微生物肽。哺乳动物防卫素分作两大类。

一类为 α-防卫素。此类防卫素的主要特点是其二硫键以 Cys-1 与 Cys-6 和 Cys-3 与 Cys-5 方式连接。这是一类较早发现的哺乳动物防卫素。包括 HNP-1，2，3，4。Goebel (2000 年)等列举了从人等组织中鉴定出有 8 个防卫素，对其名称、相对分子质量、氨基酸序列、二硫键及组织专一性等都作了论述。这 8 种防卫素中有 human neutrophil defensins 1-6(人体中性粒细胞防卫素 1-6)、兔的 NP-1，2，3A，3B，4，5、Cs-1、鼠的 RatNP1，2，

3,4 及豚鼠的 GPNP 和小鼠小肠的 cryptainl 等。另一类哺乳动物防卫素是β-防卫素，它们被发现得较晚，其分子较大，有 4 个保守的残基。-防卫素有-defensins 1-2[6]、adrenomedullin（肾上腺髓质蛋白）、BNBD12、hBD-1、hBD-2、RK-2（rabbit kidney defensin，兔肾防卫素）、TAP 等 18 余种。

从哺乳动物中发现的另外两个防卫素：一个是从猪白细胞中分离的 proteginin，其中含有 16～18 个氨基酸残基和两个二硫键；另一个是从猕猴中性粒细胞和单核细胞中分离的猕猴 θ 防卫素 1（rhesus theta defensin 1，缩写为 RTD-1）。这是迄今为止发现的唯一而独特的环状抗生肽[7]。

3. 植物防卫素的性质

植物防卫素至少有 13 个以上，其分子大小和二硫键数目与 α-、β-硫素接近。多由 45～54 个残基组成相对分子质量为 5 000 的分子。带一个阳离子。四对二硫键配对，连接方式一致。均有三个保守的 Gly-、芳香族和 Glu-残基。

4. 无脊椎动物防卫素性质

无脊椎动物防卫素比较庞杂，其中以昆虫类防卫素研究得较早较广。迄今已从十目昆虫中认知出防卫素，这包括较古老的目——蜻蜓目（Odonata）、半翅目（Hemiptera）、脉翅目（Nevroptera）、鳞翅目（Lepidoptera）、鞘翅目（Coleoptera）、膜翅目（Hynenoptera）、双翅目（Diptera）、等翅目（Isoptera）、毛翅目（Triehoptera）及蛛形纲的蝎目（Seorpionida）等。至少有 43 种以上的昆虫类防卫素已经纯化出来或被测序过。比如蜓防卫素（aeschna defensin）、麻蝇防卫素（sarcophaga defensin）[7] 和金龟素（holotricins）等。表 1 列出了昆虫类防卫素名录。

表 1　昆虫防卫素名录 *

防卫素名称		来源生物
中文	英文	名称
贻贝防卫素 A B	mytilus defensin A B	食用贻贝（*Mytilus edulis*）
地中海贻贝防卫素	mytilus galloprovinsialis defensin	贻贝（*M. galloprovinsialis*）
菜蜓防卫素	aesehna defensin	蜓（*Aeschna Cyanea*）
Leiurus 防卫素	leiurus defensin	（*Leiurus quinquestriatus*）
麻蝇防卫素 A B C	Sarcophagia defensin(sapecin) A B C	棕尾别麻蝇（*Sarcophara peregrina*）
尾蛆蝇防卫素	eristalis defensin	尾蛆蝇（*Eristalis tenax*）
果蝇防卫素	Drosophila defensin	黑尾果蝇（*Drosophila melanogaster*）
伏蝇防卫素 A B	Phormia defensin A B	新陆原伏蝇（*Phormia terranovae*）
红头丽蝇防卫素	Calliphora vicina defensin	红头丽蝇（*Calliphora vicina*）

（续表）

防卫素名称		来源生物
中文	英文	名称
厩螫蝇防卫素 1, 2	SMD 1, 2	厩螫蝇（*Sstomoxys calcitrans*）
红蝽防卫素	pyrrhocoris defensin	无翅膜红蝽（*Pyrrhocoris apterus*）
仰泳蝽防卫素	Notonectes defensin	褐仰泳蝽（*Notonecta glauca*）
益蝽防卫素 1, 2	podiscus defensin 1, 2	斑腹刺益蝽（*Podisus maculiventris*）
碧蝽防卫素	palomena defensin	红尾碧蝽（*Palomena prasina*）
稻斑稻石蛾防卫素	Limnophilus stigma defensin	稻斑稻石蛾（*Limnophilus stigma*）
夜蛾素	hellaths virescens defensin	烟芽夜蛾（*Heliothis virescens*）
蜜蜂防卫素	apis defensin，royalisin	意大利蜜蜂（*Apis mellifera*）
熊蜂防卫素	bombus pascuorum defensin	一种熊蜂（*Bombus pascuorum*）
草蛉防卫素	chrysopa perla defensin	草蛉（*Chrysopa perla*）
按蚊防卫素（前防卫素原）	anopheles defensin（proprodefensin）	冈比亚按蚊（*Anopheles gambia*）
伊蚊防卫素	Aedes defensin A, B, C	埃及伊蚊（*Aedes aegyti*）
粉虫防卫素	tenicin 1（tenebrio defensin）	黄粉虫（*Tenebrio molitior*）
双叉犀金龟防卫素	Allomgrina dichotome defensin	双叉犀金龟（*Allomyrina dichotoma*）
金龟素 1, 1A, 1B, 1C	Holotricin 1, 1A, 1B, 1C	东北大黑鳃角金龟（*Holotrichia diomphalia*）
红林蚁防卫素	Formica rufa defensin	红林蚁（*Formica rufa*）
白蚁素	termicin spinigerin（缺 cys-）	刺拟棘白蚁（*Pseudacanthotermes spiniger*）
防卫素	reschna defensin	
防卫素 A, B	zophobas defensin A, B	（*Zophobas atratus*）
一种蝎防卫素	androctonus ausnalis defensin（具发夹结构样β-层的结构）	（*Androclonus ausmalis*）

* 表中凡标有 A, B 和 1, 2 等均为相应同种防卫素的编号以示差异。

昆虫防卫素开头是为分离自新陆原伏蝇幼虫血的两个相对分子质量为 4 000 的抗 G^+ 细菌肽(伏蝇防卫素 A、B)创立的,该肽表达出与 Lehrer 等(1993)命名的哺乳动物吞噬细胞的一群杀菌肽序列相似性。昆虫防卫素的特征有三:一是分子链中 Cys-3 与 Cys-4 间有 3 个保守的 Gly 残基;二是 3 对二硫键配对分别为 Cys-1 和 Cys-4,Cys-2 和 Cys-5 及 Cys-3 和 Cys-6;三是具一个 α-螺旋和 2～3 个反平行折叠片。多数昆虫防卫素是阳离子的、含 34～46 个残基,少数如蜜蜂、大黄蜂防卫素是含 51 个残基、富含 Cys-残基的,相对分子质量约 4 000 的肽,即属于具 $CS_{\alpha\beta}$(即 cysteine-stabilized αβ)折叠的肽类。它们多数出现于许多昆虫经细菌挑战或受伤后的血淋巴中,具有抗某些 G^+ 细菌的活性,并参与昆虫有效的抗菌防卫反应。分离自斑腹刺益蝽的死亡素(thanatin)是迄今最广谱的抗菌肽。昆虫防卫素以分类发散种及序列比较分成两大家族。第一家族主要来自半翅目、双翅目、膜翅目和鞘翅目四目的昆虫(Endoptcrygote insects)。这是一群数量最大的昆虫防卫素。第二家族的成员相互靠得近,是古老的昆虫,即蜻蜓目(Odonata)和蝎目(Scarpionida)成员及两种软体动物,即食用贻贝(Mytilusedulis 和 *M. galleprovincialis*)的防卫素等。

线虫—猪蛔虫(*Ascaris suum*)和 *Caenorhabditis elegans* 的抗细菌肽也有昆虫防卫素的序列特征。

多数昆虫防卫素在浓度为 0.1～1.0 μmol 时在革氏阳性细菌暴露其中几秒钟后就会使其细胞质膜之通透性屏障瓦解。但防卫素的抗菌活性可因二价阳离子,尤其是 Ca^{2+} 的存在而改变。

近几年来,人们开始注意海洋无脊椎动物的防卫现象。海洋无脊椎动物如同许多低等动物一样,体内无吞噬细胞和淋巴细胞或仅有吞噬细胞样细胞,但仍无淋巴细胞,无免疫球蛋白和补体成分。它们靠循环细胞作出吞噬细胞样的、细胞毒的或发炎感应以加宽各种准免疫反应,从而对微生物入侵作出有效抵御和防卫反应。适任的细胞从整个身体体腔补充来,诱导并分泌出某些胞内酶和溶解的及细胞毒的分子等体液防卫因子,方式多是非专一的。其中主要的一点是靠内源性抗微生物阳离子肽来抗御感染。这就是它们的先天免疫。其中可能主要以短的阳离子抗细菌和抗真菌肽、多肽来击退入侵微生物。同源性抗生素肽的动用简单、快速,可在感染后短期内阻遏、延缓病原生物生长或全消其感染,易于调控、节能并节约遗传信息。防卫素即是内源性抗生肽之重要成员。已从有鳃亚门甲壳纲(如虾)、有螯亚门肢口纲(如鲎)、软体动物门的腹足纲、瓣鳃纲、棘皮动物门的海参纲等海洋无脊椎动物中分离出防卫素,这包括食用贻贝防卫素(mytilus defensin,MD-A,MD-B)、地中海贻贝防卫素(MGD-1)及其两个同系物、截尾海兔防卫素(dalabellanin A)、*Tachyaleus tridentatus*(及 *Limulus potyphemus*)的大防卫素(big defensin)等。凡纳对虾(*Penaeus vannamei*)的两个抗微生物肽——penaedins 2,3 则是节肢动物防御分子的新成员。这些分子含的氨基酸残基不等,范围是 35～79。相对分子质量范围 4 400～6 200,均有二硫键。此外,与贻贝防卫素相近,也是分离自此贻贝的还有 6 种抗微生物的防卫分子。如 mytilin 和 mytimycin 均来自食用贻贝血液,其半胱氨酸列阵不同于其他动植物富半胱氨酸的抗微生物肽,还有另外两个半胱氨酸残基。Mytilins 的两个同型是一样的,即富 Cys-(8 个),相对分子质量则小,仅 3 700,对抗 G^+ 细菌甚于 G^- 细菌。Mytimysin 则含 12 个 Cys-,相对分子质量为 6 200,选择性地对抗丝状真菌。鲎的大防卫素分子中 C-端区与哺乳动物 β-防卫素有明显同源性,其半胱氨酸连接性与牛中性白细胞的 β-防卫素相似。氨基和羧基半部间存在功能性差异。N端区主要是疏水的，扮演较强的抗 G^+ 菌活性。C-端则活跃对抗 G^- 菌。说明该防卫素是两个功能域的嵌合体。凡纳对虾防卫素的C端域有Cys-，并形成二硫键，N-端域富含Pro-，呈现出原始的半胱氨酸类型，为一个新种类的富含Cys-的肽。尽管一些海洋无脊椎动物防卫素分子结构各有特色，但特别有趣的一点是，上述软体动物的防卫素在种系发生上与昆虫防卫素如蜓防卫素很相近，并被归入昆虫防卫素超级家族之中，但相互间分子距离变化较大。这暗示了一些防卫素分子结构上在动植物间的相似性，反映出防卫素系统发育上的起源现象。表2列出了典型的海洋无脊椎动物防卫素的主要性状。

四、防卫素的合成与作用原理

1. 防卫素的生物合成

已经克隆出了防卫素代表成员的cDNA。多数情况下，昆虫防卫素是作为前体分子合成的。由一个常规的、疏水的单肽、一个大约30个残基的前序列和位于前肽原羧基端的成熟序列组成。前体在位于前序列和成熟防卫素间的衔接点上有一个二元加工。哺乳动物和植物防卫素的合成也类似于此。哺乳动物和植物防卫素的前体内保守了很少的原域，但阴离子残基中有一特别高量的稀有氨基酸分布。在折叠过程中，原域可充作为一个分子内立体陪伴分子，也可能阻止胞内异常转输期间与其他蛋白质或脂膜相反应[7]。

2. 防卫素的作用机制

防卫素主要作用于一些微生物细胞以及生长旺盛的癌细胞。证据表明，防卫素的攻击目标是细胞质膜(cytoplasmic membrance,CM)。在一定浓度(如10～100 μg/mL)时可损害细胞外膜(outer membrance,OM)。阳离子化程度越高的防卫素，越能活跃对抗OM。防卫素分子中氨基端(即疏水端)两亲性(螺旋插入细胞膜内，在细胞质膜的平面脂双分子层膜中形成依赖电势的弱阴离子筛选孔道，即离子可渗透孔道。或作用于膜蛋白，引起蛋白质凝聚失活，膜变性而成离子通道。经上述孔/通道)，胞内外渗透压改变，细胞内含物如 K^+ 大量漏出。内膜发生部分去极化作用，胞质ATP减少。有的防卫素使细菌呼吸受抑或使DNA的生物合成(包括OM蛋白、胞壁等)也受抑。此时细胞质膜栅栏(屏蔽)彻底瓦解，细菌死亡[8]。总之防卫素这个肽与被攻击对象(靶)的细胞膜间的动力学反应是防卫素生物学活动及此专一性活动的关键所在。

一些研究证实，两亲分子α-螺旋的氨基酸组成及螺旋度与抗微生物活性间密切相关。防卫素的这一作用原理指出了研制毒副作用更小的药物之设计策略和研制方向。

表 2 海洋无脊椎动物防卫素的主要性状

中文名称	英文名称	相对分子质量×10³	主要氨基酸酸及残基数	二硫键数	分子式	生物学活性	产生方式	来源动物名称	引文号	备注
大防卫素	Big defensin	8	79. Ala, Arg, Ser-lval. 6 个 Cys—	3	NPLIPAIYIGATVGPSVWAYlLVALVGAAAVITAANIRRASSDNHSCAGNRGWCRSKCFRHEYVDTYYSAVCGRYFCCRSR	抗 LPS 介导的鲎 C 因子酶原活动抑埃希氏菌、沙门菌、葡萄球菌和白假丝酵母生长	在血细胞的大粒细胞(L9)和小粒细胞(S5)中	*Tachypleus tridentatus*	[11]	与鼠、牛嗜中性细胞防卫素相近
贻贝防卫素	mytilin A	3.7	37. 6 个 Cys—	3	GCASRCKAKCAGRRCKGWASASFRGRCYCKCFRC	抗藤黄微球菌、大肠杆菌、抗 G^+ 菌>G^- 菌	血液	食用贻贝	[9]	有两个防卫素样肽的同系物
	mytilinB	4.4	35. 6 个 Cys—		SCASRCKGHCRARRCGYYVSVLYRGRCYCKCLRC		血液	食用贻贝		
	Mytimycin	6.2	12 个 Cys—		DCCRKPFRKHCWDCTAGTPYYGYSTRNIFGCTC	延迟粗糙脉孢菌和黄色镰孢生长、抗丝状真菌	血液	食用贻贝		
地中海贻贝防卫素	MGD—1	4.4	38. 8 个 Cys—	4	GFGCPNNYQCHRHCKSIPGRCGGYCGGXHRLRCTCYRC	抗藤黄微球菌、*E. coli*、解藻朊酸弧菌、副溶血弧菌 1 和灿烂弧菌	血浆	*Mytilus galloprorincialis*	[12]	节肢动物防卫素成员
	mytiIlin G				SCAVRCKGYCRARRCGYYVSHLYRRRCYCKCLRCFV	有溶胞作用	血浆	*Mytilus galloprorincialis*		
凡纳滨对虾抗微生物肽	Penaedin 1		50.6 个 Cys—	3			血淋巴中由血细胞合成经吐胞作用释放入血浆，存于粒细胞胞质颗粒中	凡纳对虾	[16]	
	Penaedin 2	5.5～5.6		3	ZVYKGGYTRPIPRPPPFVRPGGPIGPYNGCPVSCRGISFSQARSCCSRLGRCCHVGKGYS	抗藤黄微球菌、粗糙脉胞菌 penaeidins 主抗 G^+ 菌，也抗丝状真菌				
	Penaedin 3	5.5～5.6	62，含 Pro—6 个 Cys—							
北美鲎抗生肽	Polyphemus in Ⅰ		18 个 Arg—	2	NH2—R—R—W—C—F—R—V—C—Y—R—G—F—C—Y—R—K—C—R—COOH	抑、抗沙门菌、大肠杆菌、葡萄球菌、隐球菌(Cryptococcus neofeormans mIMF4OO4O) 白假丝酵母 M9	细胞膜上	*Limulus polyphenus*	[17]	有一大防卫素
	Polyphemus in Ⅱ		18 个 Arg—		NH_2—R—R—W—C—F—R—V—C—Y—R—G—F—C—Y—R—K—C—R—COOH		细胞膜上	*Limulus polyphenus*		
	Polyphemus inⅢ		17 个 Arg—		NH2—R—R—W—C—F—R—C—Y—K—G—F—C—Y—R—K—C—R—COOH		细胞膜上	*Limulus polyphenus*		
日本鲎抗微生物肽	Tachyploein (Ⅰ)			2		抗细菌、真菌，与细菌脂多糖形成复合物	血细胞膜上	*Tachypleus tridentatus*	[18]	
黑斑海兔素 P	aplysianin P	60				溶肿瘤细胞	紫腺中紫液	*Aplysia kurodai*	[19]	还有 aplysianinA，E

五、海洋无脊椎动物防卫素的功能

1. 对抗相当广谱的细菌

许多防卫素具有对抗、杀灭病原细菌包括那些有不完全外膜的菌体及分枝杆菌的活力。已从对照和经受细菌挑战的贻贝血中检测到抗细菌和抗真菌肽。比如从食用贻贝(*mytilus edulis*)中纯化出的 mytilin A 即具有强大的杀细菌活力[9]。Talylor 等(2000 年)从柄海鞘(*Styela clava*)的血细胞中纯化出的 styelin D(海鞘素 D)表达出对抗 G^+/G^- 细菌的活性,此活性在 200 mmol/L NaCl 时仍可维持。在低 pH 和(或)高盐度时,此天然肽则能活跃对抗 G^+ 细菌。表明 styelin D 在柄海鞘先天免疫中起重要作用。天然的和合成的海鞘素 D 同样地活跃对抗绿脓杆菌[10]。从柄海鞘细胞分离出的另一类抗微生物肽——clavanins 也具类似抗菌性。上述 mytilinA 对抗 G^+ 菌如绿色气球菌、巨大芽孢杆菌、藤黄微球菌、粪肠球菌和金黄色葡萄球菌。显出的抗性范围 MIC 为 0.6～10 μmol。对 G^- 菌如大肠杆菌 D22、D31 也有类似作用。对海洋细菌如食鹿角菜交替单胞菌,嗜酸植物杆菌和噬纤维菌(*Cytophage drobachines*)也有影响,但对大肠杆菌 1106 和鼠伤寒沙门氏菌无效,Mytilin 对藤黄微球菌还有强大的杀菌力。

2. 对抗真菌

从上述食用贻贝分离出的另一化合物,即含 12 个半胱氨酸的 6 200 的 mytimycin,对丝状真菌粗糙脉孢菌和黄色镰孢有抗性,能延迟其生长。这一化合物全部以类似浓度发现在未经病原生物挑战的的贻贝血中[9]。分离自 *Tachypleus tridentatus* 的大防卫素抗白假丝酵母[11]。凡纳对虾的 penaedins 则对粗糙脉孢菌显抗性,而植物防卫素主要表达出抗真菌性。

3. 对抗病毒

一些防卫素显示出对抗有包膜的病毒,对无衣壳的病毒则无效。杀伤病毒的防卫素包括 NP1～2,HNP1～3 等[14]。所针对的病毒如单疱疹病毒(HSV)、感冒病毒 A/WSV 等。无脊椎动物防卫素的对抗病毒活动迄今尚无报道。

4. 细胞毒作用和溶胞作用

地中海贻贝(*Mytilus galloprovinicialis*)的防卫素 MGD-1 使所试的人 O 型红细胞之血浆引起 20%的总溶血作用。Mytilin G 也有溶胞作用。多形核细胞(PMN)的某种细胞侵害因子与 PMN 胞内防卫素含量成正比,有些防卫素也杀死淋巴细胞、中性粒细胞和内皮细胞,或一些原生动物。PMN 释放的防卫素有利于感染或发炎位点中单核细胞的积累。

黑斑海兔素(aplysianin P)具有溶肿瘤细胞(如对 MM46 肿瘤细胞)的作用,而不溶正常的红细胞、白细胞[14]。

5. 其他作用

防卫素可能在动脉粥样硬化患者的脑动脉内皮细胞和内层平滑肌细胞中发挥作用,如 LP(a)在血管内皮细胞中的定位与维持。NP-3A 具有抑制肾上腺皮质激素结合其受体的活性。RTD-1 的杀菌活动可对抗增强的盐浓度,在血液所有盐浓度中都有活力。有些防卫素具抑制细胞呼吸活动能力,或对细胞体积、趋化性和有丝分裂起调节作用,抑制天然杀伤细胞活动,预防口腔感染,维持健康。

六、防卫素的应用前景

防卫素的研究有助于探讨各级各类生物在系统发育上的关系。

无脊椎动物包括大量海水增养殖对象的先天免疫现象的发掘和研究的深入，将会使人们更多更深地认识和开发利用它们的防卫素于病害防治工作。以往被认为是过时的天然免疫这一概念，将会有很大的突破和发展。先天免疫在水产业上的应用也将大有前途。

人防卫素及防卫素样分子广泛存在于表皮细胞，生殖组织等，β-防卫素又可诱导与人皮肤相连，人肺中盐敏感的β-防卫素可被囊性纤维化灭活，意味着此类肽可能有治疗潜力[14]。

人、牲畜和水产生物对抗生素的过度使用依赖导致抵抗多种药物的细菌迅速进化（演变），最可依赖的药剂在对抗一些菌株上再也无效。从天然免疫系统中发现和探索包括防卫素的天然肽抗微生物剂，已从近几年发展起来。人们据天然防卫素的结构、功能分析，已合成出上千种肽。从中有望获得新的克服抗生素抗力的分子策略。如从人皮肤中生产的β-防卫素肾上腺髓质蛋白 adrenomedullin 极具抗细菌性[13]。又如相当于兔防卫素 NP-2 的β-发夹环的合成肽，其β-发夹域有强大的抗微生物活性。

海洋生物是巨大的抗微生物肽资源库，这些肽具有迅速杀死靶细胞等的独特性质，作用专一，所用量低，效果显著的特点。而且与细胞分化，神经激素递质调节，肿瘤病变，免疫，准免疫调节有关，如南美白对虾（*Penaeus schmitti*）的高血糖肽激素、南非刺龙虾的抑制蜕皮激素（MIH）、单环刺螠虫的刺蜢虫速激肽等的速激肽相关的肽及从招潮蟹、黄道蟹、龙虾、螯虾、对虾、海兔等无脊椎动物中提取出的相应肽及海洋涡虫的 FMRF-amide-相关的神经肽 GYIRF-amide 和 YIRF—amide。已发现海洋原索动物血细胞中有杀菌肽样肽。海洋无脊椎动物也富含脯氨酸的抗微生物肽。

Kelly 等（1994 年）提取的阳离子肽已用于养殖鱼感染的治疗之中，也有用于对抗有包膜病毒和癌细胞及促进损伤的上皮细胞修复[15]。

随着分子生物学，如重组 DNA 技术，分子模型、基因组和转基因等工程的深入开展，人们有望发展出新的抗菌药剂。

参考文献 19 篇（略）

（合作者：洪旭光　张进兴）

DEFENSINS—ONE OF PIVORAL ELEMENTS OF INNATE IMMUNITY

(ABSTRACT)

Abstract This paper expounds briefly the developmental history of defensins research, properties, synthesis, functions, and foreground of application of defensins. It concerns especially their effects in innate immunity including marine invertebrates. Finally it indicates that the worth and Potential of antibiotic peptides including defensins.

Key words Defensins, Innate Immunity, Marine Invertebrates

菲律宾蛤仔(*Ruditapes philippinsis*)血浆中防卫素的纯化及其抗菌功能*

摘　要　经硫酸铵沉淀、凝胶排阻层析和离子交换层析,自菲律宾蛤仔(*Ruditapes philippinsis*)血浆中分离纯化得到一种新的蛋白质——防卫素 RPD-1(Ruditapes philippinsis defensin-1)。经 SDS-PAGE 分析,其估计相对分子质量为 24.8×10^3;经 ABI473 蛋白质自动测序仪测定,其 N 端部分氨基酸序列为 AVPDVAFNAYG。在 NCBI 和 EBI/EMBL 两个蛋白质数据库中未发现任何与该蛋白质有同源性的已知蛋白质。该蛋白质对革兰氏阳性菌和革兰氏阴性菌具有广谱抗性。对金黄色葡萄球菌(*Staphylococcus aureus*)、枯草芽孢杆菌(*Bacillus subtilis*)、四联微球菌(*Micrococcus tetragenus*)、大肠杆菌(*Escherichia coli*)、副溶血弧菌(*Vibrio parahaemolyticus*)和鳗弧菌(*Vibrio anguillarum*)的最小抑菌浓度分别为 9.6 mg/L、76.8 mL、38.4 mg/L、76.8 mg/L、19.2 m/L 和 19.2 mg/L。实验结果证明 RPD-1 为一种新的防卫素蛋白质。

关键词　菲律宾蛤仔防卫素　血浆　纯化　抗菌功能

尽管海洋贝类软体动物本身缺乏免疫球蛋白,但由于其长期生活于富含海洋微生物的环境中,逐渐形成了抵抗外来微生物侵袭的有效体系,其防御机制主要依赖于血细胞和体液的抗菌活性[1,2]。从海洋中已发现腹足类动物有抗细菌或抗微生物的蛋白质,相对分子质量在$(56\sim250)\times10^3$ 之间[3]。从近海蟹分离出的抗菌蛋白估计相对分子质量在 70×10^3 以上[4]。而关于海洋双壳贝类动物体液中抗微生物病原组分的分子结构所知相对甚少。目前,国外学者从虾、贻贝等无脊椎动物中发现的抗菌物质均属于防卫素家族的低相对分子质量蛋白质[5~8],国内尚未见关于海洋动物防卫素的研究报道。所有已发现的防卫素均表现出抗菌活性,但不同来源的防卫素在结构方面却有很大不同,作用方式也具多样性。防卫素在医药、农业等方面的应用具有诱人前景[9]。

本实验利用蛋白质分离、纯化技术,以金黄色葡萄球菌为指示菌,从菲律宾蛤仔血浆中纯化到一种新的防卫素 RPD-1(Ruditapes philippinesis defensin-1)。

1　材料和方法(Materials and Methods)

1.1　菲律宾蛤仔与菌株

菲律宾蛤仔(*Ruditapes philippinsis*)采自青岛胶南海域。革兰氏阳性细菌包括金黄色葡萄球菌(*Staphylococcus aureus*)、枯草牙孢杆菌(*Bacillus subtilis*)、四联微球菌(*Micrococcus tetragenus*)和革兰氏阴性细菌包括大肠杆菌(*Escherichia coli*)、副溶血弧菌(*Vibrio parahaemolyticus*)由中国海洋大学提供。革兰氏阴性细菌鳗弧菌(*Vibrio an-*

* 国家自然科学基金资助项目(No. 4008601)

原文刊于《生物化学与生物物理学报》,2003,35(12):1145-1148,本人为第四作者,系项目负责人。

guillarum)由中国科学院海洋研究所提供。

1.2 主要仪器及试剂

仪器:超滤离心管(Pall-Gelman,美国),BioGel(Bio-Rad,美国),CM-Cellulose(进口分装);HiTrap™ sP HP预装柱(Amersham,瑞典),冷冻离心机(上海),ABI437蛋白质自动测序仪(美国ABI公司),核酸蛋白质检测仪(上海嘉鹏科技有限公司)。

试剂:牛胰蛋白酶抑制剂(Amersham,瑞典),超低相对分子质量蛋白质标识物(Sigma,美国),Tricine(Sigma,美国)。其他试剂均为国产市售(分析纯)。

1.3 RPD-1的分离纯化

1.3.1 硫酸铵分级沉淀

自新鲜菲律宾蛤仔抽取血淋巴,加海洋抗凝剂及牛胰蛋白酶抑制剂,于冷冻离心机(10 000 r/min,4℃)中离心1h去除血细胞得血浆。血浆在慢搅下加入固体硫酸铵使之达25%饱和度(4℃),离心收集上清液(10 000r/min,4℃)。重复上述操作使硫酸铵饱和度分别达35%、45%、55%、65%和75%饱和度,分别收集上清液和沉淀,沉淀以20 mmol/L HCl溶解并检测抗菌活性,方法见1.4.3,以确定最佳硫酸铵饱和度。

1.3.2 凝胶排阻层析

使用Bio-Gel P-10凝胶柱,以20 mmol/L HCl平衡后上样,接着以同样溶液洗脱,分部收集,核酸蛋白质检测仪检测A_{280},并检测各峰活性,方法见1.4.3。

1.3.3 离子交换层析Ⅰ

使用CM-Cellulose离子交换柱,预先以10 mmol/L NaAc(pH3.5)平衡,以0～1.5 mol/L NaCl/10 mmol/L NaAc(pH3.5)梯度洗脱,分部收集,核酸蛋白质检测仪检测A_{280},使用相对分子质量截留值为1000的超滤离心管脱盐、浓缩,并检测各峰活性,方法见1.4.3。

1.3.4 离子交换层析Ⅱ

使用HiTrap SP HP预装柱,预先以50 mmol/L NaAc(pH 3.5)平衡,以0～1.0 mol/L NaCl/50 mmol/L NaAc(pH3.5)进行梯度洗脱,分部收集,核酸蛋白抟检测仪检测A_{280},使用相对分子质量截留值1000的超滤离心管脱盐,检测各峰活性,方法见1.4.3。

1.4 RPD-1的部分性质鉴定

1.4.1 相对分子质量测定

采用三层凝胶的不连续体系进行,参照文献[10]并加以适当改进。分离胶:120 g/L T,45 g/kg C(其中T代表1L凝胶溶液中含有的单体和交联剂的总克数,C代表凝胶溶液中总量为1 kg的单体和交联剂中含交联剂的克数。下同);间隙胶:80 g/L T,24 g/kg C;浓缩胶:60 g/LT,30 g/kg C。染色与脱色方法:电泳完毕的凝胶先在蒸馏水中洗5～10 min,在新配制的50 g/L戊二醛中固定1h。以0.25 g/L考马斯亮蓝G加100 g/L醋酸染色1h,然后以100 g/L醋酸脱色10h。超低相对分子质量标识物组成为:兔肌磷酸丙糖异构酶(26.6×10^3)、马心肌红蛋白(17.0×10^3)、牛奶α-乳清蛋白(14.2×10^3)、抑蛋白酶肽(6.5×10^3)、氧化型牛胰岛素B链(3.5×10^3)、缓激肽(1.1×10^3)。

1.4.2 氨基酸序列测定

使用ABI 473蛋白质自动测序仪进行。

1.4.3 生物活性检测

以金黄色葡萄球菌为指示菌,采用平板生长抑制法[3,11],对抗菌活性物质进行追踪检测。最小抑菌浓度(MIC)的测定采用二倍稀释法进行[12]。

2 结果和讨论(Results and Discussion)

2.1 RPD-1 的分离纯化

2.1.1 硫酸铵分级沉淀

以不同饱和度的硫酸铵处理菲律宾蛤仔血淋巴,所得蛋白质沉淀及上清液对金黄色葡萄球菌的抑制结果见表1。由表可见,25%饱和度的硫酸铵所得沉淀活性低,而上清液抑菌活性较强;在25%~55%饱和度之间,随着硫酸铵饱和度的逐渐提高,蛋白质沉淀物抑菌活性逐渐增强,并在55%饱和度下所得蛋白质沉淀活性达最强。因此,最适硫酸铵饱和度即为55%。在该范围内获得的蛋白质沉淀应包括绝大多数抑菌活性蛋白质,且由于选择的饱和度范围较窄,在有效沉淀浓缩活性蛋白质的同时,可大大减少其他非活性蛋白质,有利于提高后续分子筛层析的分离效率。

Table 1 Antibacterial activity of the plasma from *Ruditapes philippinnesis* by treatment with $(NH_4)_2SO_4$

Saturation degree of $(NH_4)_2SO_4$ (%)	Antibacterial activity	
	Supernatant	Precipitate
25	++	+
35	++	+
45	+	+++
55	+	+++
65	−	+
75	−	−
85	−	−

—, no inhibiting effect; +, little inhibiting effect; ++, some inhibiting effect; +++, remarkable inhibiting effect。

2.1.2 凝胶排阻层析

将55%硫酸铵饱和度所得蛋白质沉淀溶解,上 Bio-gel P-10 凝胶柱,洗脱曲线见图1。该步层析有两个大峰,第二峰具较强抑菌活性。将第二峰洗脱液合并,以超滤离心管进行浓缩。

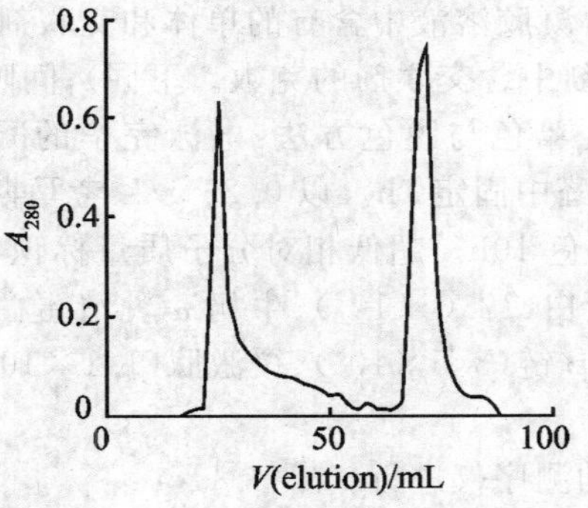

Fig. 1 Chromatography of the active fraction from Bio-Gel P-10

2.1.3　离子交换层析Ⅰ

将2.1.2中所得浓缩液上样于CM-Cellulose离子交换柱，洗脱曲线见图2。该步层析得到3个峰，分别进行抑菌活性测定，结果显示第2、第3峰显示较强抑菌活性，SDS-PAGE检验结果表明这两个峰虽均为不纯组分，但主要蛋白质带一致。因此，将二峰合并，以超滤离心管脱盐、浓缩。

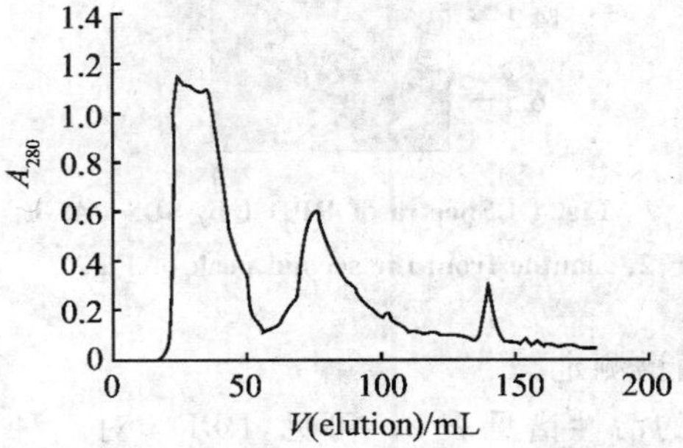

Fig. 2　Chromatography of the active fraction from CM-Cellulose

2.1.4　离子交换层析Ⅱ

将2.1.3中所得浓缩液上样于HiTrap SP HP阳离子交换预装柱，洗脱曲线见图3。该步层析得到两个洗脱峰，经活性测定，结果为第1峰无抑制金黄色葡萄球菌活性，第2峰有较强抑菌活性。

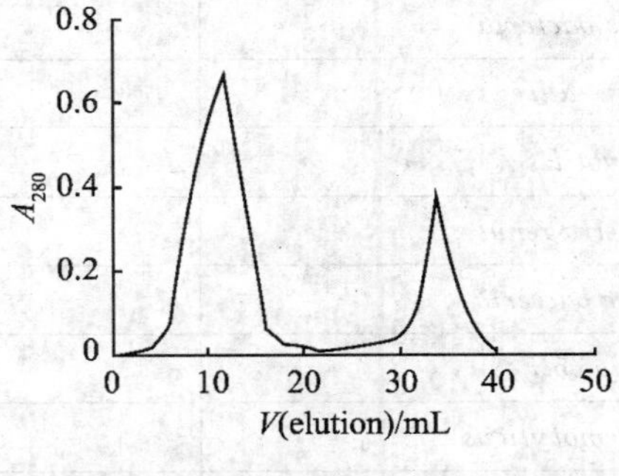

Fig. 3　Chromatography of active fraction from HiTrap SPHP

2.2　RPD-1的部分性质

2.2.1　纯度及相对分子质量测定

还原和非还原条件下的聚丙烯酰胺凝胶电泳(SDS-PAGE)结果证明，离子交换层析Ⅱ所得洗脱峰中第2峰显示一条谱带，为单一组分(RPD-1)，且相对分子质量为24.8×10^3(图4)。

2.2.2　RPD-1 N末端氨基酸残基序列分析

对RPD-1进行N末端氨基酸残基序列分析，共分析了11个氨基酸残基，结果为AVPDVAFNAYG。将该序列与美国国家生物技术信息中心数据库(NCBI)、欧洲分子生物学实验室数据库(EBI/EMBL)中已知的蛋白质序列进行比较，结果显示不存在同源性序列，证明RPD-1是一种新的蛋白质。

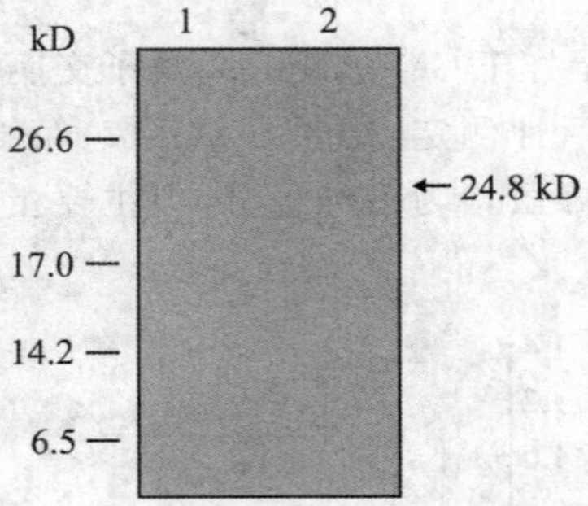

Fig. 4 Spectra of RPD-1 by SDS-PAGE

1. protein marker; 2. sample from the second peak of Fig. 3. Arrow means RPD-1。

2.2.3 RPD-1抗菌谱测定

RPD-1对6种细菌的活性谱见表2。可见，RPD-1对6种受试菌均有较强的抑制作用。其中，RPD-1对金黄色葡萄球菌、枯草杆菌和四联微球菌3种革兰氏阳性菌的最小抑菌浓度(MIC)分别为9.6 mg/L、76.8 mg/L、38.4 mg/L；RPD-1对大肠杆菌、副溶血弧菌和鳗弧菌3种革兰氏阴性菌的最小抑菌浓度(MIC)分别为76.8 mg/L、19.2 mg/L、19.2 mg/L。

Table 2 Antibacterial activity of RPD-1

Bacteria	MIC(mg/L)
Gram-positive bacteria	
Staphylococcus aureus	9.6
Bacillus subtilis	76.8
Micrococcus tetragenus	38.4
Gram-negative bacteria	
Escherichia coli	76.8
Vibrio parahaemolyticus	19.2
Vibrio anguillarum	19.2

综上所述，本文从菲律宾蛤仔血浆中分离出一种新的防卫素，并选用了6种细菌对其进行抗菌谱研究。然而，目前已知的海洋无脊椎动物防卫素大多是从相应的动物血细胞纯化出来的，且具广谱抑菌活性。因此，对菲律宾蛤仔血细胞中防卫素的研究有待进一步展开，以全面了解其在自然条件下的防御机制。同时，防卫素的广谱杀菌活性展现了一条开发新型的、理想的抗生素途径。尤其如今科学发展到了后抗生素时代，由于很多抗生素具有毒副作用、微生物也对很多抗生素产生了抗药性，这加速了人们对防卫素的研究和开发。

参考文献12篇(略)

(合作者：魏玉西　郭道森　李　丽　陈培勋)

PURIFICATION OF A BIG DEFENSIN FROM *RUDITAPES PHILIPPINESIS* AND ITS ANTIBACTERIAL ACTIVITY

(ABSTRACT)*

Abstract A novel big defensin was isolated and characterized from the plasma of *Ruditapes philippinesis*. It was purified to homogeneity by means of precipitation with $(NH_4)_2SO_4$, gel-exclusion chromatography, two kinds of cation-exchange chromatography and named RPD-1. Its relative molecular mass was 24.8 $\times 10^3$ by means of SDS-PAGE. By means of ABI437 amino acid sequence analyser, its 11 NH_2-terminal amino acid sequence is AVPDVAFNAYG. Two databank systems (NCBI and ENI/EMBL) was indexed, no sequence with homology to RPD-1 was found. Furthermore, it exhibited strong inhibition on the growth of Gram-negative and-positive bacteria. Minimal inhibitory concentration (MIC) of RPD-1 against *Staphylococcus aureus*, *Bacillus subtilis*, *Micrococcus tetragenus*, *Escherichia coli*, *Vibrio parahaemolyticus*, *Vibrio anguillarum* was 9.6 mg/L, 76.8 mg/L, 38.4 mg/L, 76.8 mg/L, 19.2 mg/L, 19.2 mg/L respectively.

Key words Ruditapes Philippinesis Defensin; Purification; Amino Acid Sequence; Antibacterial Activity

* This work was supported by a grant from the National Natural Sciences Foundation of China (No. 40086001)

中国的防卫素研究进展状况分析*

摘　要　文章论述中国的防卫素研究状况，主要内容包括世界和中国防卫素类别与种类；中国的防卫素研究之主要成果；中国的防卫素之主要作用/功能及应用前景；中国的防卫素研究论文数量之历年变化及相关的发展趋势。

关键词　中国　防卫素研究　先天免疫

发迹于20世纪80年代的世界防卫素(亦作防御素 defensin)研究经历20多年的努力，已在生命科学中形成了具有自身特色并展现广泛用途的一个崭新领域，新的防卫素不断涌现。我国的防卫素研究则有10余年的历史，但成绩斐然。本文论述中国的防卫素研究现状，并对其未来的研究方向提出一些看法。

一、防卫素的主要类别及种类

防卫素(defensins)是各级各类真核生物产生的一类抗微生物肽(AMPs)。它们大多是带正电荷、含较多精氨酸/半胱氨酸、无糖基，相对分子质量在3 500～4 500，具有一定抗(病原)微生物性能、在免疫反应中起作用，并展现出广泛应用前景的生物活性物[1,2]。根据防卫素的生物来源，可将其分为植物防卫素、昆虫防卫素超家族、哺乳动物防卫素等[3]。以下简述这三类防卫素的分类。

(1)植物防卫素。约有18种，已从马铃薯、萝卜的块茎及叶片、日本梨树、桑树、剑叶龙血树等植体中发现，它们的分子大小与二硫键数类似硫素，主要作用是对抗真菌[4,5]。

(2)昆虫防卫素超级家族，此类防卫素种类涉及许多无脊椎动物，迄今已研究了蚊、蜓、蝇、蝽、蛾、蜂、蛉、粉虫、金龟、蚁、蝎、甲虫等昆虫的40余种防卫素。以其物种的不同分类位置及分子序列可分出两大家族。昆虫防卫素还包括软体动物防卫素，即贻贝、鲎、对虾、蛤仔、毛蚶、黑斑海兔、菲律宾蛤仔等海洋动物防卫素。一些线虫也产生防卫素。它们之所以被归入此类，重要的原因是它们在种系发生及分子序列上与昆虫防卫素有相似性[2,5～9]。实际上将此类防卫素称呼为无脊椎动物防卫素更富包容性。

(3)哺乳动物防卫素。迄今已从人、恒河猴、牛、猪、绵羊、兔、大/小鼠中发现了几十种防卫素。按其分子中氨基酸空间结构、二硫键桥线状分布、整个分子结构等分成三类，即α-防卫素、β-防卫素和θ-防卫素。-防卫素有6种，即hNP1-4和hD5-6；β-防卫素有3种，著名的有hBD1-4。牛舌中的β防卫素不同于-防卫素。另外还发现了至少17种牛β-防卫素，包括BNBD、TAP等。迄今的θ-防卫素发现于猴中性粒细胞，包括3种，即rTD1-3。猪防卫素为β-防卫素，有PD、PBD-I和Protaginin等。兔防卫素有NPI(MCPI)、NP2、

* 原文刊于《自然杂志》，2004，26(5)：264～268.

国家自然科学基金项目批准号：40086001

NP3A、NP3B、NP4-5、Cs-1 及 RK-2 等。绵羊防卫素为 sBD-1。大鼠防卫素有 β-防卫素-2、rBD1-2、RatNP1-4。小鼠防卫素有 β-防卫素-4 及 cryptain。据不完全统计，哺乳动物防卫素迄今至少有 100 余种[10~16]。

表 1 列出了中国学者已研究的防卫素名录。

由表 1 可见，中国学者关注、论述或亲自研究的三类防卫素，涉及约 23 种动植物，约计 46 种之多。

表 1　中国学者研究的防卫素名录

防卫素类别	防卫素名称	引证文献数
植物防卫素	桑树防卫素[11]；Moracims1-4 及 san-1	3
	剑叶龙血树树脂防卫素；7,4'羟基黄烷，7-羟基-4'-甲氧基黄烷，Loureirim	1
	日本梨树花粉类植物防卫素蛋白 PDNI	1
昆虫防卫素超级家族	绿蝇防卫素(phormicin A)	1
	贻贝防卫素(5 类 12 种)	1
	菲律宾蛤仔血浆防卫素-1(RPD-1)	1
	白纹伊蚊防卫素，家蝇防卫素 Mdde	2
	蚊虫防卫素 A[12]	2
	中华按蚊防卫素	1
	埃及伊蚊防卫素	2
	中国对虾抗菌肽	1
	蚯蚓抗菌肽	1
	致乏库蚊防卫素，	1
	对虾一贴 penaeidins(pen－1、－2、－3a、－3b、－3c)[13]	1
哺乳动物防卫素	小鼠肺 β-防卫素－4	1
	大鼠肺 β-防卫素，rBD-1，rBD-2	2
	大鼠肺组织 β-防卫素－2	1
	家兔白细胞防卫素	1
	兔中性粒细胞防卫素(NP1-2)	1
	兔防卫素 NP1-1；兔巨噬细胞阳离子多肽(MCP-1)[14]	2
	绵羊防卫素(sBD-1)	1
	猪 β-防卫素，PD、PBD-1	1
	髓源性防卫素[16]	1
	血浆体液中髓源性防卫素[16]	1
	人中性粒细胞防卫素(HNP)，人 α-防卫素 HNPI	1
	肠组织 β-防卫素	1
	Paneth 细胞防卫素	1
	小肠隐窝潘氏细胞防卫素	1
	β-防卫素-2	1
人 β-防卫素；HBDs(HBD-1，-2，-3)，人类生物防卫素，哺乳动物防卫素[6,7,10,17,18]		8

二、中国的防卫素研究主要成果

表 2 列举了中国的防卫素研究之主要成果。由于中国防卫素研究起步较晚，最早是在 20 世纪 90 年代初，因而所发表的论文中相当一部分是推介国际间相关研究的动态，这对中国学者介入、掌握相关动态、研究方法和开创富有自身特色的研究很有促进作用。由表 2 可见，中国学者的相关研究主要集中在一些防卫素的纯化、生物学特性的分析包括分子的序列测定及作用机制；防卫素的克隆、扩增基因、转录、调控、基因表达、载体构建；防卫素的化学合成，转录入其他生物体后的变化、作用等。将这些研究成果与国际间相比，可见中国的研究发展迅速，有些已达国际水平，但研究缺乏系统性，在深度和广度上也欠缺。对本研究领域尚缺乏深厚的积淀和广泛的基础，相信在不远的将来会有改观。

表 2　中国防卫素研究的主要成果

防卫素名称	主要成果	引证文献数
日本梨树花粉类植物防卫素	编码了它(PDNl)的 cDNA 核苷酸序列	1
桑树 5 种防卫素	其组分的抗/抑 G^+ 菌及支原体、桑叶枯病菌、黄化萎缩病能力	3
剑叶龙血树树脂 3 种防卫素	从被特定真菌感染的该树脂分离得到	1
白纹伊蚊防卫素[12]	克隆基因，扩增出成熟肽基因，分析了分子生物学特性	2
蚊子防卫素[13]	基因克隆及序列、中华按蚊防卫素扩增片段与已往序列比同源性较低，无半胺酸序列	2
昆虫防卫素 A	分子改造后其蛋白保持原有三级结构不变，并在噬菌体表面展示	1
家蝇防卫素(Mdde)	基因的 cDNA 克隆，分析序列	1
贻贝防卫素[5]	报导并综述了 5 类 12 种该防卫素的序列、作用与功能	1
对虾一帖[13]	报导并综述了 3 类 5 种该肽序列、作用、性能	1
菲律宾蛤仔防卫素-1(RPD-1)	分离纯化出 RPD-1 测序及测抑/抗细菌活性	1
中国对虾抗菌肽[19]	它的成熟肽的 cDNA 克隆，即克隆出 chp 基因	1
蚯蚓抗菌肽	分离纯化出 EABP-1，测了抗菌活性	1
小鼠 β-防卫素-4	(克雷伯氏肺炎杆菌内素)诱导基因表达，活化其信号传递通道	1
大鼠肠组织 β-防卫素-2	大肠杆菌感染负相调节其肺组织 rBC-2mRNA 表达	2
大鼠肠组织 β-防卫素-2	基因在不同发育阶段肺组织中的表达	1
大鼠 rBD-2	皮肤中提取、扩增，与 hBD-2 高度同源，基因在大肠杆菌感染损伤皮肤组织中表达明显增强	2
大鼠 β-防卫素 rBD-1	克隆出 cDNA，重组后测序	2
人工合成绿蝇防卫素	抑一些 G^+ 菌、MDA435 乳腺癌细胞及 TSU 前列腺癌细胞体外增殖	1
兔防卫素	以小球藻为载体进行其产业化生产，基因用花粉管导人新疆陆地棉新陆早 7、8 号	2

（续表）

防卫素名称	主要成果	引证文献数
豚鼠中性粒细胞防卫素	杀新型隐球菌，在寄主防御反应中起作用	1
兔中性粒细胞 N P[20]	抗出血热病毒，杀 HL60 细胞，灭活单纯疱症病毒 1 型、流感病毒 2 型	3
NP1	克隆并构建植物表达载体，在百合中的转化，地农杆菌的影响 小麦外源基因瞬间表达词控；大肠杆菌中高效表达设计的合成基因	2
NP2	纯化；抗 5 种念珠海菌	2
兔防卫素(MCP-1)[14]	构建 cDNA 表达质粒并表达克隆测序；基因在转基因番茄中表达	1
绵羊防卫素 sBD-1	克隆、扩增 I cDNA 21bp	2
猪防卫素 PD	基因表达载体构建	1
猪防卫素	基因 PBD-1 在大肠杆菌中表达	1
人中性粒细胞防卫素(HNP)	色谱两步纯化法，杀钩端螺旋体，杀新型隐球菌；抗 5 种念珠菌及之后的超微结构变化	3
HNP1	基因脂质体转导在大气管上皮细胞。黏膜上有效表达；基因转入烟草抗 TMV，在烟草原生质体中表达；基因在人、兔、气管上皮细胞转染表达及在大肠杆菌中表达	3
HNP；NP1，2	抗牙周致病菌，龈下菌斑菌细胞降，球菌增，螺旋体和杆菌减	2
人 β-防卫素(HBDs)[21]	基因在女生殖道及妊娠相关组织中广泛表达	1
人 HBD-2[18]	基因在 CO5-7 细胞转染表达、基因 5′一端转录调控，测序；不同通气方式的机械通气相关性肺炎中的表达	4
人防卫素[18]	基因 cDNA 片段的克隆。可筛选更高效融合杀菌肽基因	2
肺组织 β-防卫素-2	基因在鼠肺组织中有组成型表达，与急性肺损伤(cAll)的发生发展有关	3
人 β-防卫素-3[17]	(cDNA)克隆，基因测序	1
人防卫素	毒杀胃癌细胞及胃癌耐药细胞系	1
防卫素[22]	基因枪法将基因转入玉米再生植株；化学合成，并克隆出完整防卫素	2

三、中国研究的防卫素之主要作用及应用前景

表 3 列举中国学者研究的防卫素之主要作用及应用前景。

由该表可见，中国学者研究的有着明显作用、并展示一定应用前景的防卫素集中在哺乳动物(人、猪、兔、小鼠)及贝类动物和桑树上。所研究的防卫素的主要作用为直接的抗/抑菌等微生物及肿瘤细胞，抵御感染、炎症、创伤及其修复等免疫功能中。防卫素可应用

于消化道、呼吸道及眼科疾患防治，转基因而获得优良性状等。它在超级抗生素等新抗的研究及开发中展示了较好较广阔的前景。

四、中国的防卫素研究论文数量的年际变化

中国的防卫素研究起始于20世纪90年代初，应该说比国际间先进的相关研究晚不了几年。中国的防卫素研究论文所载刊物约64种。所载刊物集中于医学、微生物、免疫、生化、生物工程等生命科学方面，展示了较强的专业性。表4列出了这10余年间中国防卫素研究论文数量统计趋势。据不完全统计，这些年所发表的论文约计95篇。年际间论文数起伏变化，低谷出现于1995～1996年。于1997出现高值，进入21世纪论文数则明显出现趋高之势。预测今后几年防卫素研究论文数将持续于高值范围。

表3　中国研究的防卫素主要作用及应用前景

防卫素名称	主要作用及应用前景	引证文献数
桑树防卫素[11]	抑/抗 G^{+} 细菌，抑支原体	1
剑叶龙血树脂防卫素	中药：血竭，内治跌仆创伤、胸腹瘀痛，外治创伤出血、疮疡不敛	1
贻贝防卫素[5]	海洋软体动物先天免疫组分	1
菲律宾蛤仔防卫素-1(RPD-1)	菲律宾蛤仔的先天免疫，防治病害	1
大鼠 rBD-1	肺组织天然抗微生物感染	2
绿蝇防卫素 A	抑/抗一些细菌或肿瘤细胞	1
大鼠 rBD-2	皮肤组织防御感染性质损伤时重要免疫介质	1
兔 NP	对新陆早7、8号棉作基因转化，抗出血热病毒	1
兔 NP-1[23]	基因转入小球藻，使新抗-兔防卫素产业化；转达基因蕃茄、百合、烟草	5
兔 NP-1,2	杀死 HL60 细胞，灭活一些病毒；抗牙周致病菌、护牙	6
兔 NP-2	抗白念珠菌	1
猪 β-防卫素	用于畜牧业	1
人 α、β-防卫素[18]	口腔医学，抗微生物	4
人防卫素	治肺部炎症，用于损伤修复、特异性免疫反应，克服微生物耐药问题	4
HNP	抗白念珠菌等，抗牙周病菌，护牙(杀伤胃癌细胞及胃癌耐药细胞系)	4
HNP-1	烟草改良	1
β-防卫素	上皮组织天然化学防御屏障，组织烧伤医治、抗生素，杀伤一些细菌和病毒，防治眼表/内感染性疾病病毒	3
β-HBDs 防卫素[21]	维系正常妇女及孕妇生殖道内环境稳定，宿主天然抗感染，光谱示高效抗菌	1
肺Ⅱ型细胞 β-防卫素-2	预防和减缓机械通气相关性肺炎发生发展	1

(续表)

防卫素名称	主要作用及应用前景	引证文献数
潘氏细胞防卫素[15,16]	肠道先天免疫,抑细菌移位,防治肠源性感染,中医药研究	3
血浆体液中髓源性防卫素[9]	诊断炎症性疾患,超级抗生素	1
防卫素[24]	转基因玉米,转基因大肠杆菌,克隆	3

表 4 中国防卫素研究论文数量年际统计

年份	1991	1992	1993	1994	1995	1996	1997	1998	1999	2000	2001	2002	2003
论文数量	2	2	4	3	1	1	8	5	5	8	15	18	22

五、结语

本文论述了自20世纪90年代初以来的13年间中国的防卫素研究状况。包括防卫素研究的内容、论文数量和走势(本文共引证文献74篇)。本分析说明了中国学者已瞄准了国际间学科的发展动向并迅速着手开展了相关研究,而且取得了喜人的成绩。有些成果已跻身于国际先进水平,但发展是不平衡的。按照中国相关研究人员和资源状况,中国学者应该做出更多更高质量的贡献。中国学者要开展防卫素的应用性研究,但更应大力加强防卫素的基础性研究。对具中国特色的物种之防卫素作更深更新的研究,包括相关的基因工程研究。中国又是一个海洋和养殖大国,丰富多样性的海洋野生生物、养殖生物及其严峻的病害防治任务,都要求我们更多关注海洋生物防卫素及其在包括先天免疫在内的免疫功能中的重要作用的研究。同时加强与国际学术界的实质性交往。在以往研究的基础上,中国学者一定会在未来的防卫素研究中取得更丰硕、更多采的高水平成果。

参考文献24篇(略)

(合作者:孙丕喜 战 闰)

ANALYSIS ON STATE OF PROGRESS OF RESEARCHES OF DEFENSINS IN CHINA

(ABSTRACT)

Abstract It analysed on research progress of defensins in china. It included their categories and kinds, functions, effects, foregrounds of applications and dvelopmental trend. It auanged the annual change of quantities of papers related and indiated the direction.

Key words China, Defensins, Researches of Defensins, Innate Immunity

D_3 肽等(PEPTIDES ETC.)

鲎肽素(tachyplesins),tachystatins 和 polyphemusins*

摘　要　本文介绍分离自中国鲎、美洲鲎的三个肽类物质,即鲎肽素(tachyplesin)、tachystatin、polyphemusin 及其氨基酸序列、生物活性、应用前景。

关键词　中国鲎　鲎属　鲎肽素　tachystatin　polyphemusin

TachyplesinI(TPⅠ)是分离自中国鲎(*Tachypleus tridentatus*)血细胞的酸性抽提物,由 Nakamura 等于 1988 年首次发现。以后又发现了它的类似物 TPⅡ。TPⅠ为含有 17 个残基的阳离子肽,其结构如下:

$NH_2KWCFRVCYRGICYRRCRCONH_2$

C 末端为精氨酸 α-酰胺,相对分子质量为 2.263×10^3,内含 2 个二硫键。有三个 X_1CX_2R/K 的重复顺序,其中 X_1 为疏水残基,X_2 为芳香性残基,R 或 K 为碱性残基,从而具有两亲性。两个逆平行的 β 折叠(β—sheet)通过 β—转弯(β-trun)相连,5 个疏水侧链位于一侧,6 个阳离子侧链位于另一侧[1]。TPI 在低浓度时也抑制 G^+/G^- 细菌,如 G^+ 的 *Staphylococcus aureus*(金黄葡萄球菌)209P 和 *S. aureus* ATCC25923、G^- 的 *Salmonella typhimurium*(鼠伤寒沙门氏菌)LT2(S)、*S. typhimurium* SL1102(RE)、*S. minnesota*(明尼苏达沙门氏菌)114W(S)及 *S. minnesota* R595(RE)(3.5 μgTPI/mL)。它能与细菌的脂多糖(LPS)形成一复合物,从而抑制 LPS 对酶原因子 C 的活化,起到抗微生物入侵的作用。在低 pH(如 0.1%三氟酸)和高温时(如 100℃30′)仍高度稳定[2]。

TPⅠ还能抗双壳类动物病原——牡蛎寄生虫 *Bonamia ostreae*(牡蛎包纳米虫)和弗吉尼亚巨蛎的寄生虫海生派金虫(*Perkinsus marinus*)及菲律宾蛤仔的弧菌 PⅠ病原,MIC 是 0.4～0.8 μg/mL。该肽对寄主的细胞无伤害作用,是一类内源寄主防御分子[3]。

* 原文刊于《生命科学》,2003,15(5):304-306.

1989 年，Miyata 等从中国鲎血细胞中分离出另一个 tachyplesin，TPⅠ的同源肽，即 TPⅡ及其同系物 polyphemusinⅠ和Ⅱ。

TPⅡ的序列如下：NH_2-RWCFRVCYRGICYRKCR-$CONH_2$。TPⅠ第一位上的 K 和 15 位上的 R，在 TPⅡ中分别被 R 和 K 代替(K 和 R 的位置正好对调)。TPⅡ抑制微生物的活性几乎与 TPⅠ相同，也抗 G^- 的沙门氏菌、大肠杆菌和 G^+ 的葡萄球菌，能强烈抑制白假丝酵母(*Candida albicans*)M9 和新隐球酵母(*Cryptococcas neoformans*)IMF40040。

tachyplesin 的 cDNAs 序列表明，其前体由 77 个氨基酸残基组成，其前导肽 Presegment 有 23 个残基，成熟肽 TP 的羧基末端与具有酰胺化作用的"Gly-lys-Arg"信号肽相连。在 C 端后另延伸一个 37 残基的肽段，含 14 个酸性残基，其 C 端则是连续 9 个酸性氨基酸簇-D-E-D-E-D-D-D-E-E，这种序列是极罕见的，可能与 TP 的生物合成及体内定位有关。前体蛋白原先转移至细胞器中，随后进行翻译后加工。Northern 杂交分析表明，TP 前体主要在血细胞和心脑组织中表达，而免疫组化表明，TP 主要位于血细胞的小而浓密的颗粒中[4]。

此外，TP 能使鲎的血蓝蛋白转化为有活性的酚氧化酶，这是属于酶原的接触激活，酶原本身并不被裂解或修饰，而是构象起了变化，犹似胶原蛋白能使凝血因子Ⅻ的酶原激活。若 TP 中的 Try 或 Tyr 被修饰，TP 就不再能与血蓝蛋白结合，说明该残基是结合位点。血蓝蛋白本身不能与几丁质结合，但能与覆盖有 TP 的几丁质结合，并通过 TP 被转化为酚氧化酶，起到甲壳动物骨骼愈伤的功能，这也是一种自我保护机制[5]。

Tam 等[6]以 TP 分子为模板，人工合成了各种含 18 残基结构紧密的环肽，它们与 TP 不同的是，TP 中第 5、15 位的两个 Arg 残基用一对二硫键代替，因而合成的环肽含有三对二硫键的结(knot)。分子内的两个反平行 β-折叠，即由此三对相平行的二硫键连接。β-折叠与二硫键成垂直方向。两个二硫键之间均相隔一个残基，整个分子结构像覆盖屋顶的瓦片(β-tile)。瓦面是二硫键，两侧分别是 β 折叠中的疏水及亲水残基，或在同一 β-折叠链上疏水与亲水残基相互间隔。合成的环肽与新发现的猕猴 θ 防卫素(θ defensin)极相似，都具有广谱抗微生物活性，高盐浓度时活性增强，但活力都低于 TP。值得指出的是，它们对红细胞的溶血毒性远低于 TP，仅是后者的 1%左右，因而，有潜在的应用价值，无毒性专一破坏细菌的细胞膜。

除 TPⅠ外，至 1989 年还发现另外 5 个 tachyplesin 的类似物，即 tachystatins。包括中国鲎的 tachystatinA、B、C。它们均能抗 G^+/G^- 细菌及真菌，其中以 C 最为有效。C 与 A、B 间无显著的序列相似，A、B 与漏斗蜘蛛毒液的-agatoxin-ⅣA 序列相似。C 与几个杀昆虫的蜘蛛毒液神经毒素序列相似。

tachystatin A、B、C 明显结合几丁质，此活性与抗真菌活性间有因果关系。tachystatin 能使酵母形态变化。C 还有强大胞溶活性。鲎与蜘蛛进化上靠近，推测 tachystatin 与蜘蛛神经毒在进化上可能来自同一古老的具适应功能的肽[7]。

1989 年，Miyata 等同时从美洲鲎(*Limulus polyphemas*)血细胞中分离出抗微生物肽 polyphemusinⅠ和Ⅱ。这两个肽各含 18 个氨基酸残基，比 TPs 多一个残基，即在 NH_2-末端有一个额外的精氨酸残基，其 C 末端也为精氨酸 α-酰胺化。二硫键结构也与 TPⅠ相同。polyphemusinⅠ与Ⅱ仅在第 10 位上差一个残基，分别为 Arg 和 Lys(图 1)，它们抑制 G^+/G^- 细菌的活性比 TPs 稍弱，也抑制酵母—白假丝酵母 M9。它们都能与细菌 LPS 形成复合物,起到抗微生物的自我防御作用。

Polyphemusin Ⅰ　NH_2-RRWCFRVCYRGFCYRKCR-$CONH_2$

Polyphemusin Ⅱ　NH_2-RRWCFRVCYKGFCYRKCR-$CONH_2$

TP Ⅱ　NH_2-R WCFRVCYRGICYRKCR-$CONH_2$

polyphemusin 和 tachyplesin Ⅱ 的氨基酸序列

综上所述,鲎肽素,tachystatins 和 polyphemusins 等均为甲壳动物体内起免疫作用的肽类物质,因而有望从养殖的海洋生物中提取有用的生物活性多肽。

参考文献 7 篇(略)

TACHYPLESINS, TACHYSTATINS AND POLYPHEMUSINS

(ABSTRACT)

Abstract This paper introduces three peptide substances, i. e., tachyplesins, tachystatins and polyphemusins, isolated from *Tachypleus tridentatus* and *Limulus polyphemas* respectively, and discusses their molecular structure, amino acid sequence, bioactivity and prospects of application.

Key words *Tachypleus Tridentatus*; *Limulus Polyphemas*; *Tachyplesin*; Tachystatin; Polyphemusin

对虾一帖的研究进展*

对虾是全世界海水养殖国家、地区集约化规模养殖对象之一，其经济价值颇高，但连年来对虾养殖面临严重病害，这包括 4 类病毒、7 种细菌、2 属真菌等病原生物引起的疾病。对对虾疾病病理学研究已做了许多工作，但就病原生物的寄主反应方面的研究仍相当有限。最近几年来发现了被命名为 Penaeidins 的对虾抗微生物肽类[1]。

Penaeidins 发现于万氏对虾（*Penaeus vannamei*），迄今尚无相应的中译名称。根据其来源和发音等，建议起名为对虾一帖。因为(1)Penaeidin 来自对虾，其全称的前半部为 Penae-即是对虾拉丁属名称 Penaeus 的前半部 Penae；(2)Penaeidin 的后半部 idin 之发音相似于汉语“一帖”的发音，“帖”有妥适之意，“一帖”连起来为一定妥适，对虾一帖正是对虾借此肽一定妥帖，是对虾吉祥之意。

对虾一帖共分 3 种，它们分别为 Pen-1，Pen-2 和 Pen-3，相对分子质量范围为 5.5～6.6×10^3，其中 Pen-1 为 5.48×10^3，Pen-2 为 5.5×10^3，Pen-3 为 6.6×10^3，是对虾免疫的一类抗微生物效应物，此 3 种对虾一帖最初均纯化自集约化虾场万氏对虾的血浆和血细胞的富细胞器部分中。它们高度同源，均由血细胞生产和贮存于粒细胞胞质颗粒内，借助脱颗粒作用和经细菌挑战后的免疫反应刺激作用（吐胞现象）释放入血液并随之全身循环而存在于不同组织和位置（包括结合至表皮组织）中[2]。在刺激后最初的循环血细胞中对虾一帖 mRNA 水平下降，但血浆中浓度增加[3]。

对虾一帖具有 NH_4 末端富脯氨酸域和由 6 个半胱氨酸组成 3 个分子内二硫键的 COOH 末端域（含 6 个半胱氨酸残基的抗微生物肽，是最初分类出防卫素的要素之一）。两域高度保守，均为翻译后修饰。重组的对虾一帖中的 COOH 末端之非酰胺化作用于抗真菌活动无效，但轻微减退其抗细菌活性。其 NH_4 端域含 4 个精氨酸线基和相对高量的脯氢酸壁基，其中一些包含在保守的 Pro-Arg-Pro 模体中。这种三联体以不同十昆虫富脯氨酸抗微生物肽密切相关成员的方式重复，该 NH_4 端域可能包含在该肽与靶微生物膜相互作用或识别活动过程中。

对虾一帖 COOH 末端域以 6 个半胱氨酸为特征，其中有 4 个以二联体组织起来。

3 种对虾一帖的生物学活性一致，这是由酵母表达的重组肽确立的。它们抗菌和抗真菌性能较高，但主抗 G^+ 菌，其作用模式有不同专一性，这决定于所试菌株。它们抗巨大芽孢杆菌（*Baillus megateriwn*）、抑藤黄微球菌（*Micrococcus duteus*），缓慢杀死浅绿气球菌（*Aerococcus viridans*），能大范围抑制真菌，如尖孢镰刀菌（*Fusariwn cxysporum*）。在低于 MIC（<5 μm）时，对虾一帖减低该菌丝生长和伸长以致形态异常。高浓度（10 μm）时，它可杀该真菌孢子，使其不萌发，以致后来即使用无对虾一帖的培养基培养时，也不萌发。

---

* 原文刊于《海洋科学》，*Marine Sciences*，2004.28，(2)：75-76

基金项目：国家自然科学基金项目(40086001)

3 种对虾一帖虽有以上许多共性，但 3 者间仍有一定差异。Pen-1 和 Pen-2 均含 50 个氨基酸残基，免疫前血清(PI)均为 9.34，但残基在两者组成和排列上不一(图 1)。Pen-1 从 cDNA 文库中克隆出来。3 种对虾一帖虽均有两域组成，但 Pen-3a 的结构在两域中有一个 6 氨基酸的中间序列，它将两个保守域相连在一起。Pen-3 又可分为 3 种，即 Pen-3a，Pen-3b，Pen-3c。Pell-3a 和 Pen-3b 的残基数相同，而 Pen-3c 比它们少一个残基。该三者核苷酸序列的差别还在于第 30 位、33 位和 40 位。Pen-3a 与 Pen-3c 的第 30 位均为 I，而 Pen-3b 的是 V。第 33 位上，Pen -3c 缺如。第 40 位上，Pen-3a 和 Pen-3b 均为 L，Pen-3c 却是 V。而 Pen-3a 的氨基酸残基总数达 62 个，其免疫前血清为 9.84。Pen-3a 翻译后修饰是由一个谷氨酰胺残基(形成焦谷氨酸)的 NH_2 端环化完成的，而 Pen-2 和 Pen-3 则是借助包含在一个甘氨酸残基的消除的 COOH 端酰胺化作用实现修饰的口[3]。

万氏对虾的 3 种对虾一帖序列比较如图 1 所示。

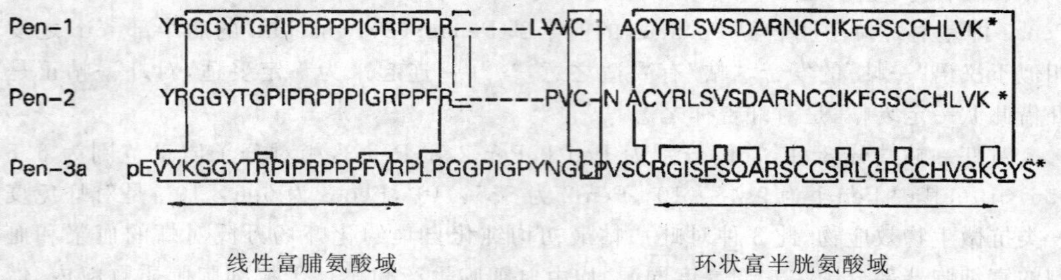

图 1　凡纳对虾 3 种对虾一帖分子结构序列

对虾一帖的完整序列划成一线；间隙(缺口)被导致最适整列线；半胱氨酸(**C**)是黑体的；相同残基和保守替代是在框内部分；星号(*)代表一个 **C** 末端的-酰胺；**pE** 代表焦谷氨酸，引自文献[2]。

对虾一帖的释放过程也与结合几丁质活动相关[4]：它除了抗微生物活性外，还负责几丁质的装配和对虾外骨骼的合成或伤愈以保护虾，特别是在蜕皮周期中暴露至微生物感染时的虾的保护。因此对虾一帖在对虾免疫功能和发育功能间起重要的作用。有关其分子结构的研究进展将促进人们更好理解对虾生理学和它们对病原性创伤反应的能力。

对虾属有 10 余种，分布于世界各海区，它们如何抵御外来入侵，其他种类的对虾是否也有类似作者所述的对虾一帖的抗微生物肽在发挥免疫(包括先天免疫)作用，很值得研究。将来有可能据此人工合成，则于对虾养殖无疑是一大福音。

参考文献 4 篇(略)

无脊椎动物高血糖肽激素和速激肽相关肽*

人们对海洋生物资源的需求越来越迫切。近十几年来，人们从海洋甲壳动物等无脊椎动物中不断发现新的起生理作用等的活性物质，包括高血糖肽激素、速激肽相关肽等等。作者介绍这2种肽类物质的研究进展。

1　高血糖肽激素

高血糖肽激素(hyperglycemic peptide hormone)，即高血糖激素，是窦腺(sinus gland，缩写为SG)中最丰富的一种肽激素，习惯上称作为甲壳动物高血糖激素(Crustacean hyperglycemic hormone，缩写为CHH)。以CHH为首，结合SG中形成的其他激素，如Molt-inhibiting hormone(蜕皮抑制激素)(MIH)、性腺抑制激素(GHI)、卵黄发生抑制激素(Vitellogensis inhibiting hormone)(VIH)和大腭器抑制激素(MOIH)等构成了SG中的肽激素之CHH一族。

已从三叶真蟹(*Carcinus maenas*)、黄道蟹(*Cancer pagarus*)、美洲螯龙虾(*Homarus americanus*)、虾蛄(*Orcanectes limosus*)、原螯虾(*Procambarus bouvieri*)、克氏原螯虾(*Procambarus clarkii*)、陆生等足目种类(*Armadillidium vulgare*)及罗氏沼虾(*Macrobrachium rosenbergii*)、日本对虾(*Penaeus japonicus*)、凡纳滨对虾(*Penaeus vannamei*)、刀额新对虾(*Metapenaeus ensis*)及南方白对虾(*Penaeus schmitti*)等动物中测出过CHH[1]。

1.1　CHH族

CHH主要位于X-器官的窦腺复合物中，它是多功能的神经激素，主要功能是调节血糖水平、肝醣和葡萄糖代谢，刺激卵细胞生长，抑制性腺活动[2]，具有多态性。它由眼柄中髓质末端X器官(MTXO)合成，具72～74个氨基酸，在相同的位置中有6个cys残基。甲壳动物CHH的氨基酸序列高度相似，但其核苷酸和氨基酸序列仅有40%～60%的一致性。单一动物种类中普遍存在多重同工型或不同的CHHs，表明这些神经激素可能由几个基因编码。下面分述之。

1.1.1　刀额新对虾高血糖肽激素，即MeCHH

刀额新对虾(*Metapenaeus ensis*)的前CHH原样(即类似前CHH原，下同)肽由一个信号肽、CHH前体相关肽(CHH-Precursor-related peptide，缩写为CPRP)和CHH样肽组成，该虾肽的信号肽和CPRP是所有CHH中最短者。刀额新对虾高血糖肽激素在眼柄中表达，不在心脏、肝胰腺、肌肉、神经索和孵化前胚胎中表达[3]。其基因组至少含CHH样基因6个拷贝。该基因在染色体上排成一簇。该CHH样基因1.5～2.1 kb，分享98%～100%的氨基酸序列同一性。每个CHH样基因有3个外显子、2个内显子。第1个内

* 基金项目：国家自然科学基金项目(40086001)

原文刊于《海洋科学，Marine Sciences》，2005. 29. 9：83-87

显子分离信号肽，第 2 个内显子在密码区分离成熟肽。该 CHH 样基因的上游 5 区的 150～200bp 含类似于多数真核生物基因特征的启动子。该 CHH 样基因的组织类似于同一种虾和斑纹潽(*Charybdis feriatus*)的 MIH 基因。根据 CHH 基因的结构组织的信息可理解这些基因表达的调节，提供此种群在肽水平上进化的信息。研究 CHH 对生物的分子种系发生(*Molecular phylogeny*)有价值[3]。Perez 等[4]已提议用有关信息进行虾亚属的分类。已分离出 *Metapemaeus ensis* 编码推测性 CHH_5 的一个 cDNA，并描述了其特征，也展示了该 CHH 样基因的结构组织，证实该虾的 CHH 样神经肽是由排列在同一染色体片断上的一簇基因的多重拷贝所编码的。

1.1.2　美洲螯龙虾 CHH

美洲螯龙虾(*Homarus americanus*)窦腺中有 2 个 CHH-免疫反应组。每组都有两个不同样氨基酸序列和相对分子质量的同工型。已分离出两个全长的 cDNA，它们编码两个不同的 CHH 前激素原的 cDNA。此前激素原结构由一个信号肽、一个与 CHH 前体相关的肽和一个高度保守的 CHH 肽组成。X 器官不是 CHH mRNA 的唯一来源[2,5]。

Kleijn 等[2]从美洲螯龙虾窦腺浸汁中提取出两组高血糖神经肽。腹神经系统也表达此 mRNA，而且腹神经系统中的 CHH 可能参与生殖和蜕皮。

此螯龙虾 CHH 分 2 个，一个为 CHH-A，另一个为 CHH-B。CHH-B 不仅具有高血糖活动，对卵母细胞生长还有刺激作用。另一个具有抑制蜕皮和高血糖活性的龙虾神经肽也与 CHH-A 序列高度相关[3]。

1.2　蜕皮抑制激素 MIH

甲壳动物蜕皮甾类化合物(蜕皮类固醇)在软甲亚纲甲壳动物触角或颌节上的一对腺体中，组织学及功能上似昆虫的前胸腺，即 Y 器官中合成。主要作为～蜕皮素发挥作用，后来则转换为 20-OH-蜕皮素(在血淋巴和靶组织中)。窦腺含阻遏循环蜕皮甾类化合物升高的因子，即蜕皮抑制激素，SG 抽提物将阻遏蜕皮甾类化合物的生物合成[6]。

已测定过美洲螯龙虾(*Homarus americanus*)、三叶真蟹(*Carinus maenas*)、凡纳滨对虾(*Penaeus vannamei*)、原螯虾(*Pracambarus bourieri*)和蓝蟹(*Callinectes sapidus*)的 MIH 一级结构。它们显出与几个甲壳动物高血糖激素有高度序列相似性。均属于同一神经肽家族，该类神经肽还包含卵黄生成抑制激素(VIH)[6~8]。

Aguilar 等[9]已揭示出 MIH 的氨基酸序列，它有 72 残基的肽链组成，相对分子质量为8 322，其中有 6 个半胱氨酸形成 3 个二硫键，连接残基 7-43、23-29 和 26-52，已封阻了 N-端和 C-端，缺色氨酸、组氨酸和甲硫氨酸。

1.2.1　斑纹鲟和美洲龙虾 MIH

已克隆出斑纹蟳(*Charybdis feriatus*)MIH 的基因，并分离出几个重叠的基因组无性系和重组了 MIH 基因[10]。该 MIH 基因全长 4.3 kb，由 3 个外显子和 2 个内显子组成。外显子 1 和 2 携带单一肽的编码序列，外显子 2 和 3 组成天然肽的编码序列。该 MIH 基因的外显子—内显子边界也遵循“GFAG 规律”(剪接供体和受体)。该 MIH 的氨基酸序列与蟹 *Callinects sapidus* 和 *Carcinus maenas* 的 MIH 及大红虾的生殖腺抑制激素(GIH)显出高度全面相似性。蜕皮间的后蜕皮前期、孵化卵的雌体虾眼柄可测出 MIH 转录物，脑中也有，但卵巢中无，肝胰腺中无、肌肉和表皮中也无。该蟹的 MIlH 的基因较小较简单，其启动子组成了一些共同的转录因子的推测性元素。该蟹 MH 的基因是在成体眼柄和幼体神经组织中的全部抑皮周期中连续表达的[10]。

Homarus amercanus 的 MIH 的氨基酸序列于 1990 年首报[6]。这是从该龙虾的窦腺浸液中分离出来的。它是 71 个残基的疏水肽，可延长蜕皮间隙期、降低幼体的蜕皮甾类化合物的值。N-端焦谷氨酸残基的移去使前 36 个残基中的 30 个序列化。该 MIH 还有高血糖活性。三叶真蟹（*Carcinus maenas*）的 Cam-CHH 则有 72 个氨基酸残基，相对分子质量为 8524。Hoa-MIH 与 Cam-CHH 有 61% 的序列相似性，6 个 cys 残基位置是一样的，Cam-CHH 有某些 MIH 活性，但与 Cam-MIH 则不同。

1.2.2　日本对虾 MIH

日本对虾（*Penaeus japanicus*）的 MIH_3 有 6 个，均分离于窦腺浸汁，其中之一为 Pej-SGP-IV，它活跃地抑制 ecdysteroid 合成。Pej-SGP-IV 由 7 个氨基酸残基组成，有游离氨基端和羧基端。它的氨基酸序列与三叶真蟹（*Carcinus maenas*）MIH 相当类似，与 Pej-SGP-Ⅲ相似性很少。甲壳动物 X-器官窦腺系统是神经内分泌调节的中心，结构和功能类似于脊椎动物下丘脑神经垂直体后叶系统。甲壳动物许多生理过程如糖原代谢、蜕皮和生殖均为该系统分泌的神经肽激素所调节。眼柄中的 X 器官合成 MIH，故 Y 器官进行的蜕皮类固醇合成被认为是为 X-器官窦腺系统的 MIH 所调节[1]。

CHH、MIH、VIH 组成甲壳动物高血糖激素即 CHH 家族或 CHH/MIH/VIH 家族，它们的共性是：由 72～78 个氨基酸残基组成，在保守的分子中具 6 个 cys 残基的类似氨基酸序列[1～3,5～10]。

在某些 X-器官核周体（Perikaryon）（文中为 Perikarya）中的甲壳动物高血糖激素（CHH）、蜕皮抑制激素（MIH）和卵黄形成抑制激素（VIH）组成结构上同源的神经肽，均在眼柄中经髓质终端Ⅺ器官的神经元合成并沿着轴索传输至神经血（neurohaemal）贮存器官，从窦腺释放入循环中。CHH 在糖代谢，MIH 在生长调节上，VIH 在生殖发育中各起主要作用。三者在血免疫学上的相似性可在免疫组织化学水平上测得[7]。

McGAW 等[12]指出十足类甲壳动物中肽神经激素还包括五肽原肛肽，甲壳动物心脏激活肽（CCAP）和 FMRF 酰胺相关肽 F1 和 F2，它们在功能上类似，与心脏血管调节（蛋白）有关，均位于心脏上游的围心血管中。

不仅从三叶真蟹（*Carcinus maenas*）和美洲螯龙虾（*Homarus americanus*）中已分出 MIHs，还测了序[6,8]。也描述了美洲螯龙虾 VIH[12]。

2　无脊椎动物速激肽相关肽

2.1　速激肽相关肽（TRPs）由来

速激肽（tachykinin）存在于许多动物中，它是一组具有多种生物学活性功能的神经多肽家族，也称 neurokinins（神经激肽），量大且结构各异，其共同之处在于共有保守的 C-末端序列，即苯丙-X-甘-亮-甲硫-NH_2，其中的 x 或为一芳（香）族氨基酸，或为一支链脂肪族氨基酸。它分布在中枢、外周神经系统及许多其他组织中，并因其受体的亚型差异而发挥不同作用，包括从炎症到神经传递的作用。速激肽还包括 P 样物质等。

表 1　无脊椎动物速激肽相关肽来源生物及其氨基酸序列

生物名称	缩写	起源组织	序列	备注
马德拉虫非蠊(*Leucophaea maderae*)	LemTRP-1	脑、中肠	APSGFLGVRa	虫非蠊速激肽相关肽
	LemTRP-2	脑、中肠	APEESPKRAPSGFLGVRa	虫非蠊速激肽相关肽
	LemTRP-3	中肠	NGERAPGSKKAPSGFLGTRa	虫非蠊速激肽相关肽
	LemTRP-4	中肠	APSGFMGMRa	虫非蠊速激肽相关肽
	LemTRP-5	脑、中肠	APAMGFQGVRa	虫非蠊速激肽相关肽
	LemTRP-6	脑	APAAGFFGMRa	虫非蠊速激肽相关肽
	LemTRP-7	脑	VPASGFFGMRa	虫非蠊速激肽相关肽
	LemTRP-8	脑	GPSGFYGVRa	虫非蠊速激肽相关肽
	LemTRP-9	脑	APSMGFQMRa	虫非蠊速激肽相关肽
	LemTRP-10	脑	SGLDSLSGATFGGNRb	虫非蠊速激肽相关肽
非洲蝗虫(*Cocusta migratoria*)	LomTK-Ⅰ	脑集合物	GPSGFYGVRa	蝗虫速激肽
	LomTK-Ⅱ	脑集合物	APLSGFYGVRa	蝗虫速激肽
	LomTK-Ⅲ	脑集合物	APQAGFYFVRa	蝗虫速激肽
	LomTK-Ⅳ	脑集合物	APSLGFHGVRa	蝗虫速激肽
	LomTK-Ⅴ	脑集合物	XPSWFYGVRa	蝗虫速激肽
反吐丽蝇(*Calliphora voonitora*)	CavTK-Ⅰ	全身、头	-APTAFYGVRa	(反吐)丽蝇速激肽
	CavTK-Ⅱ	全身、头	-GLGNNAFVGVRa	(反吐)丽蝇速激肽
草潭库蚊(*Culex salinarius*)	CTK-Ⅰ**	全身、头	APSGFMGMRa	(草潭)库蚊速激肽
	CTK-Ⅱ	全身、头	APYGFTGMRa	(草潭)库蚊速激肽
/	CTK-Ⅲ	全身、头	-APSGFFGMRa	(草潭)库蚊速激肽
埃及伊蚊(*Aedes aegypti*)	Sialokimin Ⅰ[d]	唾液腺	NTGDKFYGLMa	(埃及库蚊)唾液激肽Ⅰ
	ialokimin Ⅱ[d]	唾液腺	DTGDKFYGLMa	(埃及库蚊)唾液激肽Ⅱ
黄道蟹(*Cancel borealis*)	CabTRP-Ⅰa[c]	中枢神经系统	APSGFLGMRa	黄道蟹速激肽相关肽Ⅰa
	CabTRP-Ⅰb	中枢神经系统	SGFLGMRa	黄道蟹速激肽相关肽Ⅰb
刺虫(*Urechis unicinotus*)	UruTK-Ⅰ	中枢神经系统	LAQSQFVGSRa	刺虫速激肽-Ⅰ
	UruTK-Ⅱ	中枢神经系统	AAGMGFFGARa	刺虫速激肽-Ⅱ
无齿蚌(*Anodonta cygnea*)	AncTK	中枢神经系统	PQYGFHAVRa	无齿蚌速激肽
章鱼(*Eledone moschate*)	Eledoisind	唾液腺	pEPS×10^3 AFIGLMa	章鱼涎肽

注：修改自文献[14]；**单一的；a. 分离出肽的组织；b. 该肽非酰胺化；c. 最初自凡纳滨对虾；d. 其序列与脊椎动物速激肽有较大序列同一性

无脊椎动物中与速激肽有某些序列相似性的肽是在1991年被发现的，以后有关研究发展很快[13]。由于它与速激肽相似相关，因而被暂定名为速激肽相关肽(tachykinin-related peptides，缩写为TRPs)，其氨基酸最多的与速激肽有40%的相似性。表1列出了这些TRPs的名称、生物来源和器官、缩写名及氨基酸序列等。

由表1可以查见，至此用于TRPs研究的生物至少有9种，其中来自海洋的仅有2种。

2.2　TRPs所在部位

TRPs在神经系统和肠中的丰度分布与速激肽有某些相似性。它们在所研究的动物中，其所在部位如表2所示。

表2　速激肽相关肽(TRPs)在所研究的动物中的部位表

所研究的动物	TRPs所在部位
鲨、甲壳动物、昆虫	中枢神经系统(CNS)，眼
甲壳动物、昆虫	胃神经系统
昆虫	中肠的内分泌细胞
腔肠动物、扁虫、软体动物	肠神经
马德拉虫非蠊	2个TRPs在中肠，另有4个专一地在脑中，还有3个在脑及肠组织中
昆虫、甲壳动物	脑中有TRPs的神经元
蟑螂	脑、中肠
甲壳动物、蜗牛	中央及外神经系统、内脏肌肉、腺体
双壳类软体动物	成熟的神经系统、一些腺体、内脏肌
小章鱼	唾液腺
大蜗牛	脑、足神经节、原小脑
无齿蚌	脑、足、内脏神经节

注：引自文献[14]

2.3　TRPs性状、作用与功能

TRPs的活动构象和哺乳动物速激肽相似。

鲨脑的P样物质的免疫反应性与昆虫的TRPs类似。在眼里它调节外周视觉加工、光敏的昼夜节律，使小眼结构中的形态变化以增加光敏性。在鲨视觉系统、乌贼巨神经胶质、低等无脊椎动物发育中，TRPs起调控作用。果蝇胚胎发育中有TRPs受体。

黄道蟹的CabTRP-Ia调节和刺激胃神经节Stomatogastric ganglion(STG)中幽门不同类型的运动神经活动。

迄今所分离的无脊椎动物TRPs都对蟑螂后肠自然收缩有一个刺激作用。昆虫和甲壳动物的TRPs在中央和外周神经系统中及内脏肌肉及腺中发挥作用。脊椎动物速激肽作为神经调节子起作用。无脊椎动物中TRPs也起神经调节作用、局部释放的神经调节子和循环激素作用。

Christie等[15]揭示出黄道蟹(*Cancer borealis*)神经系统中的CabTRPIat Ib均含羧基

端序列:-FLG-MR-NH_2。CabTRPIa 在 STG 中有 3.75×10^{-13} mol 调节幽门运动神经类型,对幽门节律有刺激作用,但该作用为脊椎动物 NK-13 受体的一个肽拮抗物-SpantideI 所抑制。

CabTRPI 的量比 CabTRPIb 高出 20 倍,其潜力更比 CabTRPIb 强 500 倍。CabTRPIb 是 CabTRPIa 的降解产物。CabTRPIb 可能在 *C. borealis* 的一些其他系统中起生理作用,包括对 STG 功能在内,对幽门节律无连续干扰。CabTRPIa 与 Ib 的序列也不一,前者为 APSGFLGMR-NH_2,后者为 SGFLGMR-NH_2。

TRPs 以保守的 C-端的五肽 FX_1GX_2 酰胺(X_1 和 X_2 是可变残基)为特征。研究表明 TRPs 在中枢和外周神经系统,在其他组织中是有靶的多功能肽。

TRPs 主要功能归结如下:①对昆虫(如蟑螂)后肠肌肉有肌刺激作用。②诱导蟑螂前肠、输卵管及蛾心肌收缩。③触发蟹胃神经中枢运动物的神经节律。④作为触发蝗虫、蜗牛的去极化或超极化一致的中间神经之节奏。⑤诱导蝗虫心侧体释放脂动激素。⑥在甲壳动物、昆虫胃神经系统及昆虫中肠的内分泌细胞中起 TRP-免疫反应作用。⑦在节肢动物、腔肠动物、扁虫、软体动物中,TRPs 在(肠)神经过程中发挥类似作用。⑧在果蝇胚胎发育过程中 TRPs 也发挥作用。⑨鲨视觉系统、乌贼巨(神经)胶质中起调节作用。⑩其他各种附加作用及至今不明的生理作用。

3 结语

综上所述,高血糖肽激素速激肽相关肽是无脊椎动物体内的重要肽类物质。近年来,中国对海洋的一些无脊椎动物从药理上进行过许多研究,但与先进国家相比,我们的差距还是很明显的摆在面前的一个紧迫任务是逐步应用和发展高新技术,分离、分析肽等活性物质的结构、功能和作用机理,调动它们的免疫功能,包括先天免疫功能。从中筛选出高活性的肽类化合物,有的放矢地进行增养殖及其病害防治,有望人工合成相关的化合物。

参考文献 15 篇(略)

(合作者:刘洪展)

第五篇

空气微生物学

AIR MICROBIOLOGY

E_1 城市(CITIES)

国际五城市空气微生物概况*

摘　要　对瓦尔帕莱索、彭塔阿雷纳斯、坎帕纳、新加坡和青岛的港口等空气微生物作了监测。这五个城市空气中的海洋性微生物及陆源性微生物数量(CFU/m³)按大小次序排列均为:坎帕纳＞青岛＞瓦尔帕莱索＞彭塔阿雷纳斯＞新加坡。对空气微生物数量与气温、风速和相对湿度的关系作了初步比较分析后,认为适宜的气温及较大幅度的相对湿度有利于空气中微生物的增加;大风速、低相对湿度或高气温不利于空气中海洋性微生物的增加。本结果反映有关城市间空气污染的差异,坎帕纳的空气状况较差对所得菌株作了初步分类,其中以不动杆菌较多,次为欧文氏菌等。

关键词　国际五城市　空气微生物　海洋性微生物　陆源性微生物

一、引言

空气微生物是空气洁净度的一个重要标志,也是大气污染状况的重要参数之一,因而监测空气微生物对了解特定区域的空气状况具有重要意义[1,5]。

在1986～1987年中国第三次赴南极考察期间,"极地"船途中停靠了一些港口,包括智利的瓦尔帕莱索(Valparaiso)、彭塔阿雷纳斯(Punta Arenas),阿根廷的坎帕纳(Capana),新加坡(Singapore)和中国的青岛(Qingdao)。利用靠港间隙,监测、比较、分析了以上城市的一些锚地、码头、近海等空气的微生物状况。

二、五市概况

表1列出以上五个城市及其港口的概况。

由表1可见,这五个城市分布于地球的不同地理位置上,纬度间相距最大约89°,经度间相距最大约181°。它们分别有各自的环境状况、气候类型及市政概况,因而具有不同的城市风貌,空气质量上必有差异,有地域代表性。

* 原文刊于《黄渤海海洋》,1993,11(1):50-57。

三、材料和方法

(一)培养基

1. 海洋培养基:蛋白胨 5 g;酵母膏 1 g;磷酸高铁 0.01 g;琼脂 18 g;陈海水 1 000cm^3,pH7.6~7.8。以在此培养基上生长者为海洋性微生物。

制成平皿或斜面培养基。

表 1　五城市概况一览表

市名	地理位置及环境描述		气候类型	市政概况
瓦尔帕莱索	33°01′S,71°38′W	太平洋东岸、瓦尔帕莱索湾南岸	地中海式西风带、信风带交界	25 万人口,南美太平洋沿岸最大港,市内有油漆、制革、造船、金属加工等工业
彭塔阿雷纳斯	53°10′S,70°54′W	南美大陆最南端、麦哲伦海峡东部	温带海洋性南风为主	13 万人口,智利第 12 区首府,有些商业,大工业、农业少,市内建筑物不密集,绿化较好
坎帕纳	34°11′S,58°56′W	南美大西洋西岸拉普拉塔河下游	亚热带季风性湿润气候	钢铁冶金、化工工业,商业不多,离布宜诺斯艾利斯不远
新加坡	1°16′N,103°54′E	马来半岛南端、马六甲海峡口,太平洋、印度洋毗邻处	热带雨林式	242 万人口,东南亚最大港,商业发达,石油炼制、造船、纺织、加工业发达,绿化好
青岛	36°06′N,120°15′E	黄海胶州湾畔崂山脚下	温带季风	市区 100 多万人口,大型港口、轻纺、化工、造船业为主,商业较发达

2. 淡水培养基:牛肉膏 3~5 g;蛋白胨 20 g;氯化钠 5 g;琼脂 18 g;蒸馏水 1 000 cm^3;pH7.2~7.4。捕捉陆源性微生物用。

制成平皿或斜面培养基。

(二)采样

使用 JWI-4 型空气微生物监测仪,按操作规程,将配备好培养基的塑料平皿置入监测仪。估计测区空气状况等以确定采样时间长度,如 5′,15′或 30′不等。采样时,空气流量为 37 dm^3/min;仪器一般置于“极地”船的上甲板开阔空间,高出水面 6~8 m。在抛锚时测,无航行影响,同时记录天气状况等参数。

图 1 示出被采样的五城市地理位置。

(三)结果处理

将采过样的培养皿分别置于恒温箱中培养,海洋性微生物于 25℃以下、陆源性微生物于 37℃温度中培养 1 周以上。其间,经常检查菌落生长情况,最后按公式分别计算微生物数量,公式如下:

$$\text{空气微生物数(CFU/m}^3\text{)}=\frac{\text{每平皿中平均菌数}}{\text{采样时间(min)}\times\text{采样流量(dm}^3\text{/min)}}\times 1000^{*}$$

有时，用自然沉降法收集菌落，计算公式如下：

$$\text{CFU/cm}^3\text{(每立方米大气微生物菌落数)}=1\,000\div\left(\frac{A}{100}\times t\times\frac{10}{5}\right)\times N^{**}$$

A 为所用平皿面积(cm^2)；t 为平皿暴露于空气中的时间(min)；N 为经培养后平板上长出的菌落数。挑取一些菌落，纯化之，备分类研究用[2,4]。

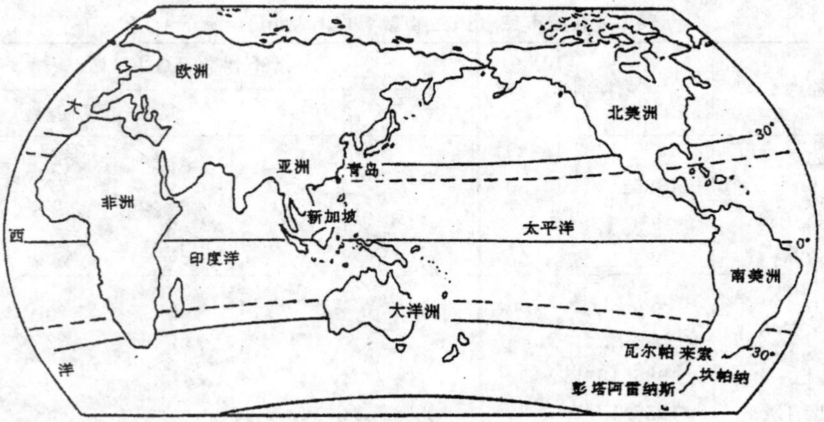

图 1　监测空气微生物的五个城市地理位置示意图

四、结果和讨论

(一)空气微生物的数量分析

表 2 列出了五城市每次监测所得的空气微生物数量。由表 2 可见，在瓦尔帕莱索，测了 4 次，在彭塔阿雷纳斯测了 7 次，在坎帕纳测了 9 次，在新加坡测了 5 次，在青岛测了 5 次，共计测 30 次。从总体上来看，测定时间分布在每日的不同时段中，因而基本上可以反映出各个城市空气微生物的数量的昼夜变化状态。由表 2 还可见其昼夜变化的规律不明确，以坎帕纳为例，从 1987 年 3 月 29 日到 3 月 30 日间，无论是海洋性的还是陆源性的微生物，在昼夜内的不同时间上，其数量均较高，差距较大；新加坡的情况则相反，微生物数量偏低，数量变幅偏小。上述两城市的差异，无疑与它们的空气环境差异有关，彭塔阿雷纳斯的情况类似新加坡。

将表 2 的数据统计，再加上监测微生物的检出率等列成表 3。

表 3 说明了这五个城市空气中的陆源性微生物检出率都达到 100%，表明空气微生物广泛地来自陆上；海洋性微生物检出率平均约 74%，其大小次序是瓦尔帕莱索＝坎帕纳＞新加坡＞彭塔阿雷纳斯＞青岛，这表明瓦尔帕莱索和坎帕纳的空气中海洋性微生物出现

* 军事医科院五所空气微生物学研究组，1985，空气微生物监测仪使用说明书。

** 谢淑敏，1986，大气微生物数量及动态变化规律研究(非正式出版物)。

机会多，不同来源的空气微生物与测定时所在地的地形，包括沿海地形及气象状况等有关。

表3列出了各城市所测大部分空气样品中微生物数的中位数值大小，即海洋性微生物数量的中位数大小次序如下：坎帕纳＞青岛＞瓦尔帕莱索＞彭塔阿雷纳斯＞新加坡；陆源性微生物数量的中位数大小次序是：坎帕纳＞瓦尔帕莱索＞青岛＞彭塔阿雷纳斯＞新加坡。表明这五个城市间空气微生物在海洋和陆地这两个来源上有相似之处。海洋性的和陆源性的微生物均以坎帕纳的为最多，以新加坡的为最少。

表2　五城市空气微生物数量(CFU/m³)一览表

市名	采样时间	微生物数量(CFU/m³)	
		海洋性	陆源性
瓦尔帕莱索	1986.12.2 15:40	448	408
	1986.12.3 10:30	72	72
	1986.12.4 6:45	84	136
	1986.12.5 21:30	24	592
彭塔阿雷纳斯	1986.12.13 15:30	0	22
	1986.12.14 21:30	6	22
	1986.12.15 9:30	0	20
	1986.12.16 3:30	38	68
	1986.12.16 9:30	38	16
	1986.12.18 0:00	8	76
	1986.12.18 6:30	52	24
坎帕纳	1987.3.29 21:00	260	22.7
	1987.3.30 3:00	1816	208
	1987.3.30 9:00	350.7	168
	1987.3.30 15:00	110.7	124
	1987.3.31 21:00	293.3	77.3
	1987.4.1 15:00	276	170
	1987.4.2 21:00	104	60
	1987.4.3 9:00	18	1228
	1987.4.4 7:30	400	78
新加坡	1987.5.4 21:00	65.3	16
	1987.5.5 7:00	56	45.3
	1987.5.6 22:00	1.3	4
	1987.5.7 13:00	1.3	18.7
	1987.5.8 19:00	0	164
青岛	1986.10.31 15:30	586.7	200
	1987.5.15 15:00	0	30.7
	1987.5.16 15:00	0	64
	1987.5.16 17:00	0	24
	1986.5.17 7:00	0	38.7

表 3　五城市空气微生物的一些参数统计表

市名	空气微生物的检出率(%)		空气微生物数的中位数(CFU/m³)		空气微生物数的极差及范围(CFU/m³)			
	海洋性	陆源性	海洋性	陆源性	海洋性		陆源性	
					极差	范围	极差	范围
瓦尔帕莱索	100	100	72	136	424	24～448	520	72～592
彭塔阿雷纳斯	71.4	100	8	22	52	0～52	60	16～76
坎帕纳	100	100	276	168	1798	18～1816	1205.3	22.7～1228
新加坡	80	100	1.3	18.7	65.3	0～65.3	160	4～164
青岛	20	100	117.34*	38.7	586.7	0～586.7	176	24～200

* 系五次测定结果的算术平均值。

由表 3 还可看到空气微生物数量的极差大小，即各地海洋性微生物数量的极差大小次序是：坎帕纳＞青岛＞瓦尔帕莱索＞新加坡＞彭塔阿雷纳斯；陆源性微生物是：坎帕纳＞瓦尔帕莱索＞青岛＞新加坡＞彭塔阿雷纳斯，表明坎帕纳空气微生物数量变化较大，彭塔阿雷纳斯的变化则小。这与该两城市的不同工商业活动，人群集散程度有关。

以上差异无冬季的影响。一般言，空气微生物数量的大小能反映空气污染的重轻程度，因而可以说坎帕纳和瓦尔帕莱索的空气状况较差，新加坡和彭塔阿雷纳斯的空气较洁净，青岛空气状况一般。将该五城市与著名工商业城市天津和中国首都北京的空气微生物数量相比，仍要少得多[3,*]。

上述五城市港口的空气微生物状况的差别还可能决定于所在区域采样时的天气因素差别上，表 4 列出前面所述采样时这五个城市的一些具体天气状况等。

表 4　采集五城市空气微生物样品时的天气状况

市名	采样时节	气温范围(℃)	气温均值(℃)	相对湿度范围(%)	平均值(%)	主要天气过程	平均风速(m/s)	所在区域地形
瓦尔帕莱索	夏初，晴天多，小风为主，1986 年 12 月	14.5～26.5	20.05	58～89	78.5	温带气旋渐强	2.95	安第斯山西侧，坡地
彭塔阿雷纳斯	夏初，雨雪多，大风为主，1986 年 12 月	8.0～17.5	12.43	58～81	70.3	西风带，副极地气压影响	7.86	山丘
坎帕纳	夏末，雨较多，1987 年 3～4 月	16.5～26.0	22.4	73～98	90.7	温带气旋	4.56	丘陵，起伏
新加坡	雨季，1987 年 5 月	28.0～32.5	29.1	81～92	89.2	赤道，低气压	4.0	地势平缓
青岛	秋、春，1986 年 10 月，1987 年 5 月	16.0～17.6	16.92	59～86	78.4	秋季，西北风渐多，春季副热带高气压影响渐强	6.92	丘陵，崂山脚下

* 张宗礼等，1986，城市气挟菌类生态评价是城市规划、管理工作中的重要参数(油印本)。

以钢铁、冶金工业为主，大气尘埃较多的布宜诺斯艾利斯卫星城之一的坎帕纳，采样时，正值夏末秋初，较高的气温，较大的温差，不大的风速和最大的相对湿度，有利于空气微生物的繁殖及其载体的悬浮。

具有地中海式气候的瓦尔帕莱索港，采样时节为初夏，渐升的气温有利于各类微生物生长繁殖，小风速有助于离海平面/地平面不高处附着微生物的颗粒悬浮，因而使空气中微生物增多。

在彭塔阿雷纳斯，气温和相对湿度是五城市中最低者，而风速却居首位，微生物数量大减。因为风速增加，一般将减少空气微生物数量[6]。

新加坡尽管有甚高的相对湿度和最高的气温，但空气微生物数量却最低，这表明高气温不利于空气微生物存活，高相对湿度只利于耐潮湿孢子的补充[6,7]。新加坡洁净空气与其"花园城市"之美名是相一致的。

由上可见，大风速、低相对湿度或高气温不利于空气中海洋性微生物高数量的形成；变幅较大的气温和相对湿度及适宜的气温有利于空气中微生物菌落数的增加。此外，沿海港口受海陆风的影响较大，这可能相应减少空气中的含菌量。人口的密集程度与工农业生产状况则是控制各地空气洁净状况差别的重要人为因素，具体关系尚待进一步探讨。

(二)空气微生物属、科的初步分类

对所得的部分空气微生物菌株作了初步的分类研究，它们是革兰氏阳性菌 3 属、革兰氏阴性菌 4 属及肠杆菌 1 科的细菌。它们分别是：

不动杆菌属　*Acinetobacter*；
黄杆菌属　*Flavobacterium*；
欧文氏菌属　*Erwinia*；
噬纤维菌属　*Cytophaga*；
明串珠菌属　*Leuconostoc*；
葡萄球菌属　*Staphylococcus*；
库特氏菌属　*Kurthia*；
肠杆菌科　*Enterobacteriaceae*。

表 5　被鉴定细菌属科的来源表

细菌属科名称	细菌来源
不动杆菌属	瓦尔帕莱索、彭塔阿雷纳斯、新加坡
欧文氏菌属	瓦尔帕莱索、彭塔阿雷纳斯、新加坡
黄杆菌属	瓦尔帕莱索、彭塔阿雷纳斯、新加坡
噬纤维菌属	坎帕纳
明串珠菌属	瓦尔帕莱索
葡萄球菌属	坎帕纳
库特氏菌属	彭塔阿雷纳斯
肠杆菌科	彭塔阿雷纳斯

表5列出了上述细菌的来源，该表说明彭塔阿雷纳斯的空气中有肠杆菌科细菌，这类细菌数量尚未达到很高程度。

五、结语

微生物在空气中的大量繁殖将会危及环境，因而空气微生物的研究近几十年来在国际间相当活跃。本文初步测定了国际间五城市空气微生物状况，是个尝试。结果表明，这五个城市的空气微生物数量有差别，分布不一。原因是多种多样的，环境的污染程度（很多情况中，主要是人类活动的结果）是个重要因素，反之，空气中的微生物状况又可表达出空气的质量问题。结果还暗示，空气中大量的微生物是世界性分布的，因而其中的病原微生物的扩散和影响值得注意。

参考文献7篇（略）

（合作者：宋庆云）

GENERAL SITUATION OF AIRBORNE MICROBES IN FIVE INTERNATIONAL CITIES

(ABSTRACT)

Abstract The airborne microbes of the harbours of Valpriso, Punta Arenas, Capuna, Singapore and Qingdao were monitored. These harbours are arranged in the order of the amounts of their airborne marine microbes as well as their terrigenous microbes as follows: Capuna > Qingdao > Valpriso > Punta Arenas > Singapore. Through a preliminary comparison and analysis of the relationships between the amounts of air-borne microbes and air temperature, wind speeds or relative humidity, it is concluded that appropriate air temperature and a larger range of relative humidity are advantageous to increase of air-borne microbes, while greater wind speed and lower relative humidity or higher temperature are disadvantageous. The results indicated that there were some differences in levels of pollution in the cities monitored. The air condition of Capuna is rather bad. Preliminary identifications were made of a part of microbial strains obtained among them, quite a few were *Acinetobacter*, next was *Erwinia*.

Key words Five International Cities; Airborne Microbes; Marine Microbes; Terrigenous Microbes

青岛城区室外空气微生物数量的测定*

摘　要　采用平皿沉降法初步测定了青岛城区空气微生物数量分布状况。结果表明，空气微生物平均数量为12 388.2CFU/m³，变幅在1 284～44 053CFU/m³之间。白天、低海拔高程处的空气微生物数量分别比夜间、高海拔高程处高，并随商业、交通的发达程度及居民的密集度而异。新鲜的海洋、海滨空气微生物数量明显比中心城区少。

关键词　室外空气微生物　大气质量　数量分布　测定　青岛市

1　引言

空气微生物的数量状况是反映空气质量的一个重要参数。20世纪70年代以来，其有关研究越来越受到重视，国外已经对城市空气微生物的生态分布与环境污染的关系作了具体的调查和研究[1]，我国也作过类似的调查[2,3]。青岛是一个工商业发达的城市，在这方面的研究还不多[4]，本文就青岛城区室外空气微生物数量状况作一简略介绍。

2　方法

采用平皿沉降法，用普通营养琼脂培养基按规范收集微生物[5]，并计算和分析它的数量状况。

3　结果与讨论

1994年3～4月间收集、统计的青岛市城区室外空气的微生物数量，如表1。

从表1可以看出，青岛城区室外空气微生物数量的算术平均值为12 388.2CFU/m³，F采样点的微生物数量是A采样点微生物数量的34倍。分析结果与北京、天津2个城市相比，微生物数量高于北京天安门广场地区[2]，而低于天津市中心区的平均最高值[3]。青岛是一个海滨城市，此结果意味着它的室外空气并不洁净。

表1　青岛市城区室外空气微生物数量(CFU/m³)

代号	采样点	空气微生物数量	备　注
A	国际礼堂前凉亭	1 284.1	海边
B_1	市府大楼前花园	10 115.8	下午
B_2	市府大楼前花园	6 630.3	夜间
C	市府大楼附近居民院内	5 972.2	夜间
D	青岛山山顶凉亭	6 632.4	海拔130 m，有雾
E	中山路人行天桥	12 028.9	商业中心区段

* 原文刊于《上海环境科学》，1995，14(2)：31-32。

(续表)

代号	采 样 点	空气微生物数量	备 注
F	四方嘉定路花园	44 053.4	树、花、繁杂
G	对照	38.7	文献[4]

空气微生物数量在昼夜间有明显的差异。以下午测的 B_1 点与夜间测的 B_2、C点相比,其微生物数量分别为它们的1.53和1.69倍。这与夜间气温的下降,人群、车辆活动的减少有关。

空气微生物数量受海拔高程影响明显。海拔高程约130 m的D点,测得的空气微生物数量分别是低海拔处(B_1 和E点)的0.55和0.66倍。此结果同海拔高程与空气微生物数量间呈负相关的观点吻合[3]。

空气微生物数量与大气污染程度关系明显。A、B_1、E、F四个采样点分别设在海滨、商业、交通繁忙或居民聚集的不同区域,从统计数据可以看出,F点的微生物数量分别是A、B1及E点的34.3、4.4和3.7倍。说明离海滨地区的距离越远,距商业、交通区及居民区越近,空气微生物数量有增多之势。它表明,作为大气环境质量综合指标之一的空气微生物,受到人群密集程度、商业、交通状况的重要影响。

海洋空气对中心市区的空气微生物数量有调节作用。青岛海滨地区的空气微生物数量少,越接近中心市区、空气微生物数量逐渐增多。海滨地区A点和中心市区E和F点的空气微生物数量分别是近岸海洋G点的33.2、310.8和1138.3倍。这表明海洋空气清新,新鲜的海洋空气随海风飘刮到城市中心区,其调节作用随着离陆地距离的加大,而呈减弱之势。

4 结论

青岛城区室外空气微生物平均数量为12 388.2CFU/m^3 左右,其变动因所处地理位置、采样时间、人群活动、商业交通和离海距离等因素的影响而变。海洋空气对陆地空气微生物有调节作用。离海洋越近,其调节作用越大,随海拔高程的增大,空气微生物数量减少。

参考文献5篇(略)

十三城市空气微生物含量研究*

摘 要 用平皿沉降法对分布于三大洲、五个国家中的十三个城市的空气微生物含量作了估测。结果表明各城市间室内外空气微生物含量存在较大差异。从空气微生物总量、真菌数、真菌数/总菌数%三指标看空气质量,在室外似以彭塔·阿雷纳斯(智)的为佳。室内以温哥华(加)为好。若所测城市以所在国家、洲和半球来分上述三指标,则分别以日本(加拿大)、南美洲、南半球和西半球的空气较洁净些。与七年前相比,世界空气质量有所下降。世人对此须引起足够关注。

关键词 空气质量 空气微生物含量

1. 引言

近20年来,尤其是发展中国家城镇人口的集聚日趋加剧,而城市基础建设却与之脱节。不少人环境保护意识淡薄,追求短期效益,致使城市空气质量恶化,大气中各种颗粒物、气溶胶、微生物颗粒大量增加。据不完全统计,近几年来,世界每年产生的包括微生物颗粒在内的气溶胶粒子等共有 773×10^6 t~2 200×10^6 t。它们已经给空气环境容量造成了巨大的负荷,因而使一些城市的生态环境严重恶化。而为数不少的人并不了解这一状况,包括空气微生物的数量和质量构成状况。实际上不少空气微生物属于坏微生物气溶胶(bad bio aerosols)。它们在空气中大肆繁殖,又与化学污染起协同等作用,从而对人和动植物,对工农牧林等各个领域起到一定程度的破坏作用。为科学治理大气污染和防治流行病而研究空气微生物现已被有识之士所关注。这是保护环境,促进生态系统良性循环的重要环节。城市空气微生物研究有所进步[1-2],但研究区域相当有限。本文就世界十三个城市的空气微生物含量概况作一初步研究,以期能促进空气微生物学尤其是发展中国家空气微生物学研究的发展。

2. 世界十三个城市概况

被研究空气微生物状况的城市有十三个。它们是位于亚洲的北京、青岛、广州、深圳、珠海和日本的东京。位于北美洲的温哥华和多伦多,位于南美洲巴西的圣保罗(sao paulo)、智利的圣地亚哥(Santiago)、瓦尔帕莱索(Valparaiso)、彭塔·阿雷纳斯(Punta Arenas)和阿根廷的比尼亚·德尔马(Vina del Mar)。

十三个城市分布于南、北半球,东、西半球中;有发达国家的,也有发展中国家的;有沿海的,也有内陆的;采样时有夏季的,也有初春冬末的;城市类型,发展状况各不相同,因而有一定的代表性。

十三个城市在地球上所处位置及其地理、环境概况分别如表1所列,如图1所示。

---

* 原文刊于《环境与发展》1995,10(4):5-9 转 18.

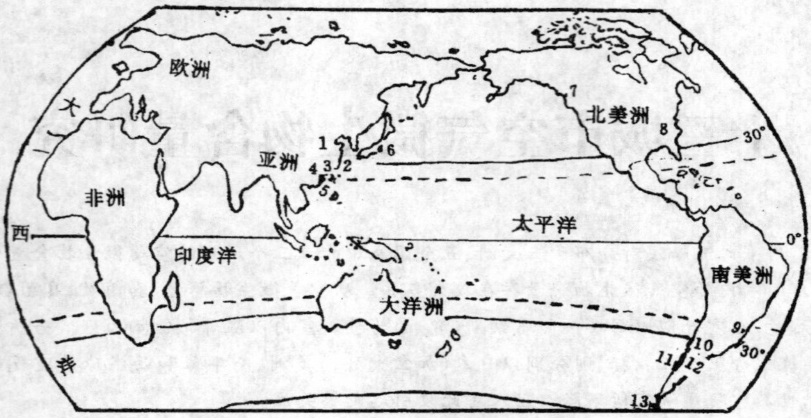

图1　测定空气微生物的世界十三城市地理位置示意图

图例：1. 北京，2. 青岛，3. 广州，4. 深圳，5. 珠海，6. 东京，7. 温哥华，8. 多伦多，9. 圣保罗，10. 圣地亚哥，11. 瓦尔帕莱索，12. 比尼亚德尔马，13. 彭塔·阿雷纳斯

表1　世界十三城市地理、环境概况一览表

城市名称	所在半球和大洲	环境、气候状况	市政概况
北京	北半球、亚洲	温带、大陆性气候	市区较平坦、市中心海拔43.71cm，中国首都
青岛	北半球、亚洲	温带、季风气候	丘陵、滨黄海，工商交通旅游较发达
广州	北半球、亚洲	亚热带气候，夏长无冬，夏季台风，暴雨多	华南中心城市
深圳	北半球、亚洲	夏季台风、暴雨多	广东中心城市，市南交通繁杂
珠海	北半球、亚洲	夏季台风、暴雨多	新兴的广东中心城市之一
东京	北半球、亚洲	冬季较寒冷、干燥	东京湾西北岸，日本首都，政治、军事、工商中心，人口稠密
温哥华	北半球、北美洲	气候良好	濒太平洋伯拉德湾，木材贸易发达，四季花香，交通畅达
多伦多	北半球、北美洲	气候温和	安大略湖北岸，良港，金融、工商、文教中心
圣保罗	南半球、南美洲	高原气候，温和	巴西最大城，各业发达，著名旅游胜地，工业较发达
圣地亚哥	南半球、南美洲	地中海式气候	智利首都，政治、工商、文教中心
瓦尔帕莱索	南半球、南美洲	地中海式气候环境宜人	港口，风景秀丽，旅游胜地，工业发达

（续表）

城市名称	所在半球和大洲	环境、气候状况	市政概况
比尼亚·德尔马	南半球、南美洲	环境宜人	太平洋岸，南美著名游览城
彭塔·阿雷纳斯	南半球，南美洲	温带海洋性气候	南美最南端，市内安静

3. 材料和方法

空气微生物培养基包括收集总菌数的淡水培养基和真菌培养基两种。制作好的培养基平皿分别以平皿沉降法在设定的室内外采样点收集空气总微生物和空气真菌，采样高度均在人群呼吸带内。采样时间长度一般在 5 分钟至 15 分钟之间，有时则根据现场情况决定。经培养后，平皿中的微生物菌落数以单位体积空气中的总微生物和真菌的 CFU·m^{-3}计。作出数据的比较和分析，以估计所在城市中空气微生物含量状况[3]。

4. 结果和讨论

表 2 列出世界十三个城市空气微生物含量状况。该表中空气微生物含量由小到大排列，室外的次序如下：彭塔·阿雷纳斯、瓦尔帕莱索、东京、圣地亚哥、广州、青岛、比尼亚·德尔马、珠海、北京、深圳，室内的次序是：多伦多、温哥华、圣保罗、东京、彭塔·阿雷纳斯、北京、青岛、圣地亚哥。

表 2　世界十三城市空气微生物含量统计

城市名称	空气微生物含量（CFU·m^{-3}）		室内/室外％	采样时间
	室内	室外		
北　京	1 965.0	51 876.0	3.8	1994,3
青　岛	2 445.0	13 457.5	18.2	1994,3-4
广　州		6 740.0	18.2	1994,6-7
深　圳		61 622.4		1994,7
珠　海		19 807.2		1994,7
东　京	471.6	1 519.6	31.0	1994,3
温哥华	56.1			1994,3
多伦多	52.4			1994,3
圣保罗	235.8			1994,3
圣地亚哥	3 320.9	3 930.0	84.5	1994,3
瓦尔帕莱索		818.8		1994,3
比及亚·德尔马		14 510.8		1994,3
彭塔·阿雷纳斯	681.3	659.9	103.2	1994,3

一般而论，空气微生物含量高，意味着所测空气污浊，反之亦然[4]。由此推测，所测城市相比，室外空气质量以彭塔·阿雷纳斯的为最好。而以深圳的最差。室内空气质量以

多伦多的为最佳，以圣地亚哥的为最劣。尽管它尚未超过中国的有关标准[5]。

许多情况下，室内空气要比室外空气干净、稳定些，因而空气微生物含量要低些。表2大体反映出这一情况。但彭塔·阿雷纳斯的却反而是室内的大于室外的。这可能与当时室内气温高于室外，人和杂物较多有关，也可能意味着相应室内空气中颗粒物、CO_2 浓度较高[6]。尽管该市室外空气微生物含量在所测城市中是最低的，但比起七年前的情况，已有较大增长。这在瓦尔帕莱索和青岛也是如此。可能意味着有关城市空气质量有所下降[6]

表3列出的是上述十三个城市空气真菌含量状况，仍以由低到高的 CFU/m³ 排列。可见，室外空气真菌含量次序如下：比尼亚·德尔马、彭塔·阿雷纳斯、青岛、东京、圣地亚哥、瓦尔帕莱索、广州、深圳、北京、珠海。室内的次序是：温哥华、多伦多、圣保罗、彭塔·阿雷纳斯、东京、圣地亚哥、青岛、北京。

表3　世界十三城市空气真菌含量比较表

城市名称	空气真菌含量参数			
	室　外		室　内	
	真菌含量（CFU/m³）	真菌数/总菌数%	真菌含量（CFU/m³）	真菌数/总菌数%
北　京	4 744.4	8.2	628.8	32.0
青　岛	170.3	1.3	387.8	15.9
广　州	1 395.2	20.7		
深　圳	1 729.2	2.8		
珠　海	8 391.6	42.4		
东　京	209.6	13.8	78.6	16.7
温哥华			0	0
多伦多			26.2	50.0
圣保罗			52.4	22.2
圣地亚哥	230.6	5.9	117.9	3.6
瓦尔帕莱索	262.0	32.0		
比及亚·德尔马	0	0		
彭塔·阿雷纳斯	114.6	17.4	65.5	9.6

以真菌数/总菌数%的低高次序分，室外空气城市名次排列如下：比尼亚·德尔马、青岛、深圳、圣地亚哥、北京、东京、彭塔·阿雷纳斯、广州、瓦尔帕莱索、珠海。室内则是温哥华、圣地亚哥、彭塔·阿雷纳斯、青岛、东京、圣保罗、北京、多伦多。

许多真菌是致病因子[7]，且与一些化学污染物一起，可能导致传染病、流行病发生，还常破坏果蔬农牧产品和工业产品质量。其在空气中的含量及在总菌数中所占%的高低均标志着所测空气质量的低高。大量真菌的繁殖与其所在环境较高的相对湿度，温和的气温相关[1,3]。

由表 3 及对它的分析，可知比尼亚·德尔马和温哥华分别是本次测定室外室内空气质量最好的城市，珠海和北京的室内外空气含有最多的真菌。

表 4 是将所测城市的空气微生物含量以国家、洲、半球来分而得出的数据。

由表 4 数据可比较国家间、洲间和半球间空气微生物含量差异，列入表 5。

表 4　世界十三城市所在区域空气微生物含量统计表

政区	空气微生物总量(CFU/m³)		空气真菌(CFU/m³)		空气真菌/总菌数%	
	外	内	外	内	外	内
中国	30 700.6	2 205.0	3 186.1	508.3	15.1	24.4
日本	1 519.6	471.6	209.6	78.6	13.8	16.7
加拿大		54.3		13.1		25.0
巴西		235.8		52.4		22.2
智利	4 980.0	2 001.1	151.8	91.7	13.8	6.6
亚洲	25 837.1	1 627.2	2 690.1	365.1	14.9	21.5
北美洲		54.3		13.1		25.0
南美洲	4 980.0	1 417.2	151.8	78.6	13.8	11.8
北半球	25 837.1	420.4	2 690.1	277.1	29.7	22.4
南半球	4 980.0	1 412.7	151.8	78.6	13.8	11.8
东半球	25 837.1	1 627.2	2 690.1	365.1	14.9	21.5
西半球	4 980.0	1 073.1	151.8	62.2	13.8	15.1
全球	17 494.2	1 153.5	1 674.8	169.7	14.5	18.8

* 据表 3 统计得出

表 4-5 说明室外空气以日本、智利较洁净，室内空气以加拿大和巴西较洁净。各洲间，南美空气较好。四个半球相比，总的看以西半球和南半球的空气较为清洁[7]。

表 5　五国三洲四半球空气微生物含量次序排列表

政区	环境	总菌数(CFU/m³)	真菌数(CFU/m³)	真菌数/总菌数%
国家间	室外	日＜智＜中	智＜日＜中	日＝智＜中
	室内	加＜巴＜日＜智＜中	加＜巴＜日＜智＜中	智＜日＜巴＜中＜加
洲际	室外	南美＜亚	南美＜亚	南美＜亚
	室内	北美＜南美＜亚	北美＜南美＜亚	南美＜亚＜北美
半球间	室外	西＝南＜东＝北	西＝南＜东＝北	西＝南＜东＜北
	室内	北＜西＜南＜东	西＜南＜北＜东	南＜西＜东＜北

若将这十三个分布于世界各地的城市代表全球陆块区域，那么全球陆地上空气微生物含量状况的一瞥便是：城市室外空气微生物含量约 10^4 CFU/m³，真菌约 10^3 CFU/m³，真

菌/总菌数%约14%。室内空气微生物含量约10^3CFU/m^3,真菌约10^2CFU/m^3,真菌/总菌数%约19%左右。这与七年前相比,空气微生物含量明显增多[8],也许意味着人口密集地区空气质量正在下降,这是值得世人引以为戒的。

5.结语

对世界13个城市的测定结果表明,这些城市空气微生物参数间存在较大差别,其含量的极差在室外达60 962.5CFU/m^3,室内达3 268.5CFU/m^3,最小值/最大值的%,室外是1.1,室内是1.6。其原因是错综复杂的。但空气微生物含量等指标低,无疑表明那儿空气质量上乘,而有关指标高,肯定与其空气的、环境的污染状况相关。由于空气微生物参数测定的难度,不可能在广大的时空范围内面面俱到。但本文作出的估计至少从一个侧面、在一定时段中反映出所代表区域的空气微生物和空气质量状况,并进而唤起人们对空气微生物研究的重视。

参考文献8篇(略)

空气微生物对铜陵市环境质量的指示作用*

摘　要　所测铜陵市区空气的平均细菌含量约4 181.5CFU·m^{-3}，总菌含量约4 495.9CFU·m^{-3}。与许多中国的大中城市比，污染不重，结果指示铜陵空气尚处于正常状态之中。

关键词　铜陵，空气微生物含量，环境质量

铜陵市是安徽省中南部位于长江申汉线中段的一个新兴工矿城市。主要市区濒临长江。市内地势平缓。市区山水宜人。人口不足20万。工业以有色金属采掘及较初步的加工业为主。交通、商业和城市设施正在发展之中。现有资料表明，对铜陵市空气环境了解甚少。

空气乃是人们生存的首要条件之一。城市空气与人们的关系尤为密切。空气中一些微生物属于致病病原体。它们的大量繁殖有碍健康，还损害工农林牧副渔产品和商业物品，且常与环境的其他污染物发生协同作用而使环境恶化、生活质量降低。因而，空气微生物含量及其变化状况恰恰是特定区域空气环境质量的重要指标。迄今，各国已加紧研究该重要课题[1-2]，但一些地区，缺乏"环境眼光"，并未认识到它的紧迫性。

本文初步报道铜陵市空气微生物含量状况，讨论与空气质量的关系。唤起有关各方从空气微生物学的角度来认识环境问题。期望铜陵市的环境质量可为其环境资本的持续增长提高到一个更高的水准，以利于铜陵这类城市的健康发展。

1　方法

采用常规平皿自然沉降法测定空气中细菌和真菌含量，两者之和代表空气微生物总量。曝皿时间：5 min，培养温度：真菌25℃，细菌33℃以上。培养时间：48h，以培养皿中的菌落平均数(CFU·皿$^{-1}$)换算为单位体积空气中的微生物菌落形成单位数(CFU·m^{-3})来分析和评论空气污染程度[3]。

2　结果和讨论

在铜陵市区，共布设五个测点以代表不同的功能区域。

测定结果如表1所示。

---

* 原文刊于《长江流域资源与环境》，1996，5(3)：259-261.

表 1　铜陵市区空气微生物含量统计表

测次编号	测点名称	细菌		真菌		总微生物菌量(CFU·m⁻³)	备注
		CFU·m⁻³	占微生物总菌量的%	CFU·m⁻³	占微生物总菌量的%		
1	火车站对面院内	2515.2	88.9	314.4	11.1	2 829.6	5:30 树林中 28℃湿、微风
2	长江商场路边	8 174.4	92.9	628.8	7.1	8 803.2	6:05 车、人不多
3	北京路商场附近居民楼内	2 829.6	94.7	157.2	5.3	2 986.8	0:00 小风、安静
4	居民楼二楼阳台	2 515.2	88.9	314.4	11.1	2 829.6	4:00 小风、安静
5	长途汽车站附近街心铜塑旁	4 873.2	96.9	157.2	3.1	5 030.4	7:00 小风、车、人不多
	平均	4 181.5	93.0	314.4	7.0	4 495.9	/

根据有关文献[4]，城市空气细菌含量(CFU·m^{-3})在 1.0×10^{-3}，-5×10^{3}·m^{-3}时，认为该空气已被轻污染，当该含量<5.0×10^{4}·m^{-3}时的空气属中等污染。由表 1 可见，4/5 的结果表明细菌含量<5×10^{3}·m^{-3}，故所测空气可看作为有轻污染。1/5 的结果>5.0×10^{3}·m^{-3}而<5.0×10^{4}·m^{-3}，表示该处空气达中等污染。全部数据的平均值为 4 181.5·m^{-3}，低于 5×10^{3}·m^{-3}。证实此时的铜陵之空气属轻度污染。与目前许多大中城市比[5,6]，铜陵市的平均空气细菌含量并不算高。

一些真菌产生致癌或腐败物品，因而对铜陵市的空气真菌含量也作了测定。由表 1 可见，该市空气真菌含量不高，平均仅为 314.4CFU·m^{-3}。铜陵空气微生物含量以细菌为主，细菌百分含量占绝对优势。虽然国家并未对空气真菌含量作过明确规定，但若以细菌为参照，那么两者之和作为空气微生物含量的状况，也可从表 1 中得知。其平均值为 4 495.9CFU·m^{-3}，极差为 5 973.6CFU·m^{-3}。指示空气污染仍为较明显的轻污染。

测点间的差距按含量大小排列，次序为 2>5>3>1=4，最高为最低的 3 倍多。表明闹市区的(2)高于居民楼院的(1,4)，街心广场的(5)低于闹市区的(2)。

表 1 还列出了测定空气微生物的时间等参数，说明本文的空气微生物样本采自早晨或夜间。一般说，空气微生物除与化学污染相关外，还与人群聚散、生活及生产活动密切。一些区域白天的空气微生物含量常高于夜间[7]。表 1 也证实了此点。随着人们经济活动和日常生活的节奏变化，铜陵市白天的空气微生物含量将可能升高。因此若有机会，尚需增加对铜陵市空气微生物含量其他时段的观察，以期对该市市区空气质量作出全面的评价。

3 结语

迄今铜陵市空气微生物含量及构成，根据以上论述，较为正常，指示铜陵市区空气质量良好。空气病原微生物常无孔不入地影响空气状况和人们健康。空气微生物状况迄今仍被不少环保行政部门所忽视。这受制于它本身的调查难度及宣传不力，以至于在一定程度上成了“环境盲区”。中国新兴市镇如雨后春笋不断涌现，面临的空气污染实不能等闲视之。本文结果虽表述出调查时铜陵空气微生物状况良好，但查清它的全貌尚需各方付出大量努力。相信将有更多人士关注中小城镇的空气微生物研究并促使经济进步的源泉之一——环境质量不断提高。

参考文献 7 篇(略)

INDICATION ON ENVIRONMENTAL QUALITY OF TONGLING CITY FROM ITS AIRBORNE MICROBIAL CONTENT

(ABSTRACT)

Abstract Airborne microbial content over the urban district of Tongling city was monitored preliminarily. The result shows that the average air borne bacterial and total microbial contents are 4 181. 5CFU · m^{-3} and 4 495. 9CFU · m^{-3} respectively. The contents are not high as compared with those of many other large or middle-sized cities in China. The result indicates that the air environment over Tongling city is in normal condition.

Key words Tongling City, Content of Air Borne Microbes, Quality of Environment

北海市空气微生物含量的时空分布*

摘　要　于 1996/1997 年之交研究了广西北海市空气微生物含量及其分布特征,结果表明北海市室外空气细菌、真菌及总微生物含量分别为 5 709.7CFU/m³、1 847.1CFU/m³ 和 7 556.8CFU/m³。已超过了相关规定的轻度空气污染程度。客房空气质量较好。空气微生物含量显出的测点差意味着北海市闹市区空气质量一般差于非闹市区,随离陆地采样高度的抬升,空气微生物呈减少之势,其昼夜变化特征表现出空气细菌及总量在下午达峰值,上午次之,晨间为谷,而空气真菌在中午峰值之前出现低谷。空气细菌、空气真菌和空气微生物总量与气温间分别有显著正相关关系。北海市空气微生物含量变化显出不全同于我国一些亚热带或内陆城市的势态。

关键词　空气微生物　时空分布空气质量

位于广西南部、濒临北部湾的北海市,既是著名的对外开放口岸,又是重要渔港。受年降水量丰沛的南亚热带季风气候的控制,即使北方已处隆冬时节,这儿依然枝头常绿。近几年经济的蓬勃发展,使得北海市也成了人群越来越密集之地,环境将面临严峻的挑战。

空气是环境的重要成员之一。生存离不开空气,空气微生物与我们关系密切。许多病害的发生发展,物品的霉腐、变质等等都有空气中的坏生物气溶胶(bad bioaerosols)在作怪。它们与其载体——各种颗粒构成了空气污染的重要因子。人们谈论空气污染及其治理,必须对空气微生物状况做到心里有数,才能有的放矢。然而无孔不入又无时无刻不影响我们的微生物恰恰又是常被忽略的。因为它(肉眼)看不见,(触觉)摸不着,无“影”无“踪”。国际上空气微生物学的研究却在大踏步地前进着[1]。它以许多无可争辩的事实告诉人们,对空气微生物再也不能等闲视之了。我国的空气微生物学在经历了近 20 年的研究之后,也取得了可喜的成果[2]。这对人们去了解环境污染和推动其防治、认识应用与环境微生物及正确处理环境与发展的关系等等都起到了并将继续起到重要的作用。

处于发展中的北海,有理由更全面地了解自己及邻近区域的空气微生物状况,以便于更科学地规划好自己的环境。调研北海空气微生物将有利于并促进它经济的健康发展。本文正是基于这一动机在此报道北海市空气微生物含量及其时空变化的初步监测和分析结果。

1　材料和方法

采用平皿沉降法,于 1996/1997 年之交对北海市的 8 个采样点作了 14 次空气微生物含量的监测(另加 1 个室内对照),所用采样介质是普通培养基和真菌培养基。以空气细菌和空气真菌量之和为空气微生物总含量,其单位均是 CFU/m³(即每立方米所测空气中的菌落形成单位数,下同)。采样、培养及结果的计算、分析和评价均用常规法进行[3]。采

* 原文刊于《广西科学》,1998,5(2):83～86。

样点见图1。采样时间等条件见表1。

2　结果和讨论

2.1　北海空气微生物含量概况

表2列出了监测所得出的空气微生物有关数据。

由表2可见，北海空气微生物含量的4个参数，即空气细菌、空气真菌、空气微生物总量、真菌/总菌(%)的一般概况。各个参数的平均值即平均的空气细菌量、空气真菌量、空气微生物总量、空气真菌/空气微生物总量分别为5 709.7CFU/m^3、1 847.1CFU/m^3 和 7 556.8CFU/m^3 和 24.4%。作为对照，测定了一次室内空气微生物含量，结果是未测出细菌，只有含量不大的真菌。说明此时高层空调房间的空气相当洁净，但应减轻湿度，以使真菌量再减。

表1　北海市空气微生物测定点概况表

Table 1　General situation of the sites for sampling airborne microbes above Beihai city

测次编号 No. of Sampling	位置 Site	测定时间 Time for sampling	测处 Position for Measuring	环境参数 Environmental parameter
M_1	码头	1996-12-31T08:30～09:00	室外 Outdoor	22.0℃气温，95%相对湿度，阴天
M_2	码头	1996-12-31T11:35～12:05	室外 Outdoor	24.5℃气温，90%相对湿度，雾
Q	侨港镇	1997-01-01T09:45～09:50	室外 Outdoor	24.0℃气温，小风、小雾、鱼腥臭味
D	地角	1997-01-01T10:45～10:00	室外 Outdoor	24.0℃气温，马路边
R	人民剧场	1997-01-01T11:50～11:55	室外 Outdoor	24.5℃气温，马路边
B_1	北部湾广场	1996-12-31T18:00～18:05	室外 Outdoor	22.0℃气温，静风
B_2	北部湾广场	1997-01-01T06:00～06:05	室外 Outdoor	20.0℃气温，>95%相对湿度，大雾
B_3	北部湾广场	1997-01-01T12:50～12:55	室外 Outdoor	25.0℃气温，95%相对湿度，中雾
B_4	北部湾广场	1997-01-01T22:30～22:35	室外 Outdoor	22.0℃气温，85%相对湿度，静风
X	徐屋小区	1997-01-01T16:00～16:05	室外 Outdoor	26.0℃气温，85%相对湿度，晴，薄雾

（续表）

测次编号 No. of Sampling	位置 Site	测定时间 Time for sampling	测处 Position for Measuring	环境参数 Environmental parameter
L_1	北部湾中路立交桥上	1997-01-03T22：25～22：30	室外 Outdoor	26.0℃气温，85％相对湿度，晴，薄雾
L_2	北部湾中路立交桥上	1997-01-04T11：25～11：30	室外 Outdoor	小风、阴天、薄雾
Y_1	粤海大厦 16 楼上	1997-01-03T22：35～22：40	室外 Outdoor	小风、阴天、薄雾
Y_2	粤海大厦 16 楼上	1997-01-04T11：40～11：45	室外 Outdoor	22.0℃气温，4 级风
K	粤海大厦客房	1997-01-01T16：00～16：05	室内 Indoor	20.0℃气温，二人房间

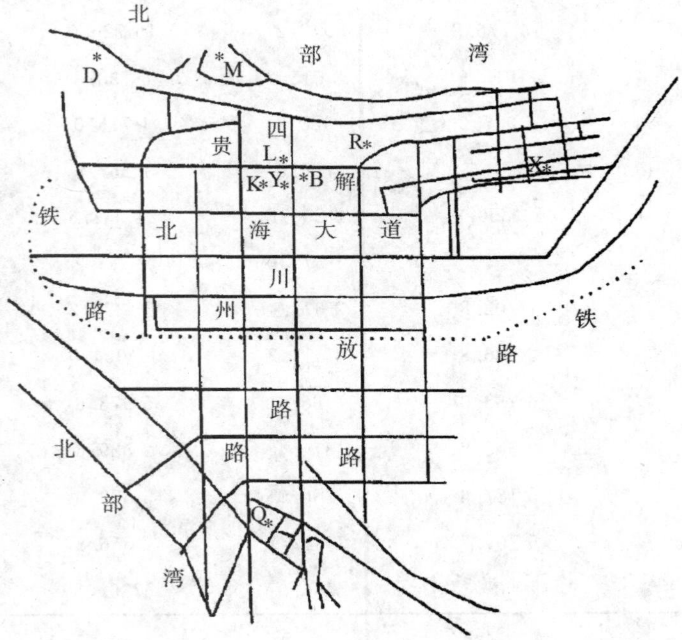

图 1　北海市空气微生物采样点位置示意图

Fig. 1　Sketch of the residential district for sampling air borne microbes above Beihai city

按照有关文献[4]，室外空气细菌含量达到 5 000CFU/m^3 时，空气达到了轻污染上限，当有 50 000CFU/m^3 细菌时，空气有中等污染。因此空气微生物含量越高，意味着空气污染越重、空气质量越差。据此，北海的平均空气细菌含量已是超过轻度污染上限了

(5 709.7/5 000)。若加上空气真菌(国家暂无对此定标准),北海市空气污染大体已是相关限度的1.51倍了(7 556.8/5 000)。其中有57.1%(8/14)超过轻度污染上限,有28.6%(4/14)达到中度空气污染范围。

与我国一些类似规模的城市相比,可以发现北海的空气微生物含量高于内陆的安徽铜陵市(4 495.9CFU/m^3)[5],而大大低于广东珠海(19 807.2CFU/m^3)[6],也比处于亚热带北侧的庐山风景区的低[7]。

表2 北海市空气微生物含量统计表

Table 2 Statistic table of airborne microbes contents above Beihai city

测次编号 No. of Sampling	空气微生物含量 Air-borne Microbial contents (CFU/m^3)			真菌/总菌 Fungi/Total Microbes (%)
	细菌 Bacteria	真菌 Fungi	总菌 Total microbes	
M_1	3 772.8	183.4	3 956.2	4.6
M_2	2 122.2	3 982.4	6 104.6	65.2
Q	3 615.6	471.6	4 087.2	11.5
D	14 776.8	943.2	15 720.0	6.0
R	3 615.6	2 515.2	6 130.8	41.0
B_1	4 087.2	628.8	4 716.0	13.3
B_2	471.6	786.0	1 257.6	62.5
B_3	21 850.8	2 200.8	24 051.6	9.2
B_4	943.2	1 572.0	2 515.2	62.5
X	7 702.8	2 672.4	10 375.2	25.8
L_1	2 986.8	8 017.2	11 004.0	72.9
L_2	6 288.0	786.0	5 973.6	11.1
Y_1	2 515.2	314.4	2 829.6	11.1
Y_2	5 187.6	786.0	5 973.6	13.2
平均 Mean	5 709.7	1 847.1	7 556.8	24.4
K	0	786.0	786.0	100.0

表 3 北海市空气微生物含量在闹市和非闹市区间的比较

Table 3 Comparison of airborne microbes contents between the downtown area and non busy shopping area of Beihai city

测点代表区域 Area	空气微生物含量 Airborne Microbial contents (CFU/m^3)			真菌/总菌 Fungi/Total Microbes (%)
	细菌 Bacteria	真菌 Fungi	总菌 Total microbes	
非闹市区 Non busy shopping	4 303.4	1 827.5	6 130.9	29.8
闹市区 Downtown	7 624.2	1 441.0	9 065.2	15.9

* 非闹市区系 M、Q、X Non busy streets include points M, Q and X; 闹市区系 B、D、R. Downtown areas include points B, D and R; 数据由各区含量平均而得 The numbers are averaged CFU or% contents per area respectively.

2.2 北海空气微生物含量的测点(地区)差

北海空气微生物的 14 次室外测定分布于 8 个测点上。有工商交通或人群活动频繁的闹市区,也有相对安静或较偏的小区、码头等非闹市区,将测点按此分作两个区域的空气微生物含量状况列入表 3。

由表 3 可见,闹市区和非闹市区的空气微生物含量间存在着一定差异,即闹市区空气细菌、总菌含量均比非闹市区的高。这表明许多时候闹市区空气质量比不上非闹市区的好。表 3 还指出非闹市区的空气真菌含量比较高,这主要是测点 M 位于海边,湿度大,耐潮湿孢子的真菌可能更适于湿润空气环境[8]。表 2 的数据还证实像北海这样的城市,闹市区较宽敞的地方,只要保持洁净环境,仍可减少空气微生物。反之处于开发或人群不甚密集之地,只要不保持好环境卫生,就容易使空气变得不清净。

2.3 北海市空气微生物含量测点间的距地高度差

空气微生物的分布不仅在同一距地高度的水平方向上有一定差异,在不同距地高度上也会有某种分布特征。离地面 1.5 m 左右的高度处于人群呼吸带中,其微生物当然与人的呼吸系统疾病关系密切。同时,空气微生物又处于流动状态之中,它也会随气流及其载体的上下波动、生物、物品的上下移运而变化。北海空气微生物含量在垂直方向即不同距地高度上的分布状态列于表 4。

表 4 北海市空气微生物含量距地面不同高度上的统计

Table 4 Statistics of airborne microbes contents at the different altitudes from the earths surface of Beihai city

测点距地面高度 Altitude of sites above the earths surface	空气微生物含量 Airborne Microbial contents (CFU/m^3)			真菌/总菌 Fungi/Total Microbes (%)
	细菌 Bacteria	真菌 Fungi	总菌 Total microbes	
下层 Lower(1.5 m)	11 397.0	1 886.4	13 283.4	14.2
中层 Middle(10.0 m)	4 637.4	4 401.6	9 039.0	48.7
上层 Upper(45.5 m)	3 851.4	550.2	4 401.6	12.5

* 下层点包括 B_1、B_2. Lower levels include sites B_1 and B_2；中层指 L_1、L_2. Middle levels include sites L_1 and L_2；上层系 Y_1、Y_2. Upper levels include sites Y_1 and Y_2；数值由各层平均得出 The numbers are averaged frorm contents per level respectively.

表 4 数据取自相当接近的 3 个层面，因而空气微生物含量高度上的差距可比性强。结果表明空气细菌量大小排序是下、中、上层，空气真菌量的是中、下、上层，总菌量的是下、中、上层，真菌/总菌(%)的是中、下、上层，总的来看，空气微生物表达出下、中层的多于上层的。这一方面反映了空气微生物如同其他生物一样，都得经受地球引力作用的控制，另一方面更多地表明空气微生物与湿润空气的有限上浮、污染物的扩散、人群及其活动的立体移运均相关，它们会导入真菌等微生物至中层甚至上层。北海空气微生物含量在垂直方向上的分布特征与青岛的空气微生物含量在海拔高程上的变化是一致的[9]。

2.4 北海市空气微生物含量在昼夜间的分布特征

一般而言，空气微生物含量在昼夜间的变化应遵循某种规律，这种规律不难想象是与其“生物钟”相关的。在空气受到污染和人为干扰较大的城镇，它的变化将可能异常。

选择 B 测点作特定的空气微生物含量昼夜变化观测。结果见表 5，由此可见，空气细菌量的大小排序是中午，下午，夜间，早晨；真菌的是中午、夜间、早晨，下午；总菌量的是中午，下午，夜间，早晨；真菌/总菌(%)的是夜间一早晨，下午，中午。总的来看，空气微生物含量在该点是中午高，早晨偏低。

表 5 北海市 B 测点上空气微生物含量的昼夜变化

Table 5 Diurnal variation of airborne microbial contents above site B in Beihai city

测定时段 Time interval of sampling	空气微生物含量 Airborne Microbial contents (CFU/m^3)			真菌/总菌 Fungi/Total Microbes (%)
	细菌 Bacteria	真菌 Fungi	总菌 Total microbes	
早晨 Early morning	471.6	786.0	1 257.6	62.5
中午 Noon	21 850.8	2 200.8	24 051.6	9.2
下午 Afternoon	4 087.2	628.8	4 716.0	13.3
夜间 Night	943.2	1 572.0	2 515.2	62.5

再将整个测区有关的10次空气微生物含量以一日的早晨，上午，中午，下午和夜间来分。将各含量以常用对数值作成图2，仍可发现如上的类似昼夜变化。即空气细菌量的大小排序是下午，上午，中午，夜间，早晨；真菌的是中午，下午，夜间，早晨，上午；总菌量的是下午，上午，中午，夜间，早晨。另外，真菌/总菌(%)的大小排宁显出的是晨＝夜，中午，下午，上午。由图2看，下、上午的空气微生物含量偏大，早晨和夜间的偏小，表明北海空气微生物量常在夜晨之间处于低谷，上午下跌至中午后高攀下午峰，再度下降，以上势态与青岛，庐山的有类似之处[7,9]。

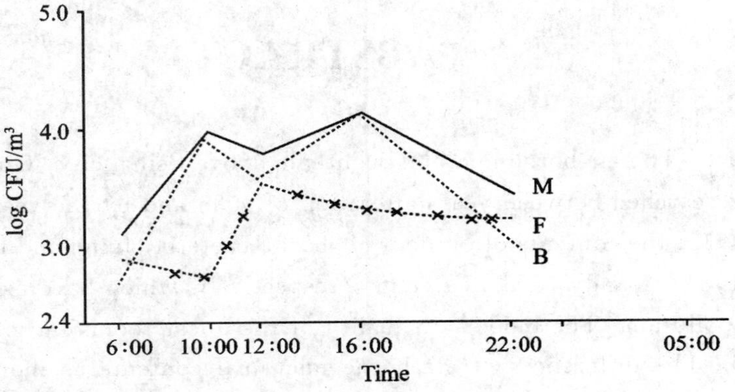

图2　北海市室外空气微生物含量的昼夜变化状态

Fig. 2　Diurnal variation of outdoor airborne microbial content above Beihai city

B：细菌 Bacteria；F：真菌 Fungi；M：总菌 Total microbes

将表1中室外10次采样时的气温(除 Y_2 外)与表2中相应的10次空气细菌、真菌及总菌量分别作相关分析，得知气温与这三者间均有显著正相关关系，即气温—空气细菌量的r(相关系数)为0.723($>r_{0.01}$)，气温—空气真菌量的 $r=0.578$($>r_{0.1}$)，气温—空气微生物总量的 $r=0.822$($>r_{0.005}$)。表明所测北海市的空气微生物受气温的正面影响大。这可从它们上述的昼夜变化中得到印证。当然这些关系不可能排除其他各种因子的制约或干扰，对此应多作研究。

3　结语

空气微生物的时空分布受制于当时的天气、地形、生态和污染条件，反之，它又是上述参数的一种综合表现。空气微生物变化多端，因为其他因子处于变动之中。必须经周密、大量、长期、同步的采样分析，才能得出较全面的结论。即使做到了这一点，也不能以一概全。合适方法的采用十分重要。本文所用之方法虽有粗糙之虞，但简易、方便又很经典，仍广为所用。由本结果可初步认为，北海市空气已超过了轻度污染，但比起我国某些城市，其空气质量不属最差之列。为防患于未然，当加强全市范围的有计划监测。希望有机会进一步采取先进手段，监测更多样本，并与其他因素的测定同步进行，取得更有说服力的信息，尽可能准确掌握空气微生物活动规律，以服务于北海市环境生态资源的永续良性循环。

参考文献9篇(略)

SPATIAL AND TEMPORAL DISTRIBUTIONS OF AIRBORNE MICROBES ABOVE BEIHAI CITY

(ABSTRACT)

Abstract The air borne microbial contents above Beihai city, Guangxi ant distribution were researched between years alteration of 1996 and by the year of 1997. The results show that the contents of outdoor air borne bacteria, fungi and total microbes were 5,709.7, 1,847.1 and 7,556.8CFU/m^3, respectively, which have exceeded the limit of lighter air pollution. The indoor air quality of the guest rooms in hotels monitored was pretty well. The differences of air borne microbial contents as monitored in different sampling sites imply that the downtown air quality is worse than the uptown in Beihai. It appears that the air borne microbial count decreases with the rise in the night. The characteristics of the diurnal variation show that contents of air borne bacteria and total microbes reached their peaks in the afternoon, took second place in the late morning, and lowest in the early morning. But before noon peak level of air borne fungi content is lowest. There were significant positive correlations between the contents of air borne bacteria, fungi and total microbes, and air temperatures respectively. The content changes in air borne microbes above Beihai city show a distinct tendency from those shown in some subtropic or interior cities in China.

Key words Airborne Microbes, Spatial and Temporal Distribution, Air Quality

空气微生物粒子沉降量指示兰州空气质量*

摘 要 用平皿沉降法在兰州市区设立的6个点共10次测定空气中细菌和真菌粒子沉降量。结果表明兰州所测空气细菌、真菌、微生物总量及真菌/总菌百分比分别为18 785.4、458.5、19 243.9CFU/m^3及2.4。细菌含量明显比以往的增多,结果意味着兰州空气质量欠佳并正在趋于变化,必须采取果断、对应措施以改变这一状态。

关键词 兰州 空气微生物 空气质量

1 引言

位于我国西北高原的兰州属温带半干旱气候区,周围有大山,逆温强,风速小。兰州又是以石油、化工为主的工业城市,空气污染明显。烟雾十分重。据报,兰州市即使不烧一吨煤,大气中总悬浮微粒量仍达不到国家二级标准[1]。这些因素使兰州空气中的微生物有了繁盛蔓延的条件。空气污染中的一个活跃分子——微生物,包括污染菌,因其本身微小而常可借助载体沉浮在空气之中,从而与一些化学污染物一起威胁人和动植物的健康、腐蚀败坏许多物件、商品。虽然几年前就有关于兰州室内外空气微生物的调查报告,但不多[2-3]。近年来它的状况有了什么变化、发展趋势如何?都是值得继续探讨的,本文试图就此作一报道。

2 材料和方法

选取兰州市区6个点作10次微生物学的测定,采用平皿沉降法将细菌和真菌培养基曝皿于室外空气中5分钟以收集沉降下来的空气微生物粒子,按常规方法计算出单位体积(m^3)空气中的微生物CFU含量,即CFU/m^3,分析和比较所有采样点的各项空气微生物指标,包括空气细菌、空气真菌的CFU/m^3、空气细菌、真菌两者之和代表空气微生物总量,并计算出真菌占总菌量的百分比率,以评价空气污染程度。采样点及其相关的环境概况列于表1之中。

表1 兰州闹市区空气微生物采样概况表

采样点编号	采样点名称及环境简况	采样起始时间
1	兰州火车站广场,旅客来往多	1996.4.18. 08:30 1996.4.18. 20:30
2	东岗立交桥上,距地面约10米以上,空气灰蒙蒙	1996.4.14. 10:22 1996.4.14. 22:22

* 原文刊于《内蒙古环境保护》,1998,10(2):11-13.

(续表)

采样点编号	采样点名称及环境简况	采样起始时间
3	西什子商业区,此点仍为较开阔之处	1996.4.14. 11:30 1996.4.14. 18:30
4	渭源路航天宾馆外,车人往来,气温:10℃	1996.4.14. 21:55
5	中山桥畔,15℃,无风	1996.4.23. 07:20
6	黄河母亲塑像上,黄河之滨,有人来往走动	1996.4.15. 08:20 1996.4.23. 08:30

3 结果和讨论

表2列出了在所有采样点获得的各空气微生物粒子沉降量的计算结果。

将该表各类数据比较,按大小排序可见各点空气细菌沉降量是4、1、6、5、2、3,真菌的是6、1(=5)、2、3(=4),总菌是4、1、6、5、2、3,真菌%是6、5、2、1、3、4。因此细菌粒子沉降量和总菌粒子沉降量均是以4号点为最高,真菌和真菌所占百分比均以6号点的为最高,反之在3号点的细菌、真菌和总菌量均最低,虽然真菌%在4号点最低。这表明,作为市中心闹区之一的渭源路空气微生物污染最重。相对而言,3号点的空气最为洁净。黄河畔的6号点之空气微生物粒子沉降量数据意味着那儿的较湿润空气似更有利于一些真菌的生长。

由表2的各类数据可得出以下四个平均值,即所测兰州市区的平均空气细菌、真菌、总菌沉降量及真菌/总菌百分比分别为18 785.4、458.5、19 243.9CFU/m^3 和2.4。按有关标准[4],当室外空气中的细菌含量达5 000CFU/m^3 时,该空气即处于中度污染下限,结果表明此时的兰州空气正处于此等污染状态之中,若加上真菌含量,空气污染更重(见表2)。总体看,空气真菌粒子沉降量平均值及其所占百分率并不高。这反映了兰州此间的空气较为干燥,因此推测某些测点空气所含真菌中有一些是具不耐湿孢子者[5]。

表2 兰州闹市区空气微生物粒子沉降量统计表

采样点编号	空气微生物粒子沉降量状况*			
	细菌	真菌	总菌	真菌/总菌%
1	27 431.4	471.6	27 903.0	1.7
2	12 261.6	393.0	12 654.6	3.1
3	11 475.6	157.2	11 632.8	1.4
4	31 125.6	157.2	31 282.8	0.5
5	14 305.2	471.6	14 776.8	3.2
6	16 113.0	1 100.4	17 213.4	6.4

*各类微生物粒子沉降量单位:CFU/m^3;1、2、3、6测点数据是其两次测定之均值

将所得数据与以往结果比较,可以发现兰州市本身的空气微生物含量有了较大的增

长[2]。与一些城市比，兰州的空气微生物含量居高不下[6-7]，这意味着兰州空气状况不佳且趋于恶化。空气微生物状况与其所依附的载体密切相关。一些研究表明，空气微生物含量与 SO_2、NO_x、TSP、降尘量间各有正相关关系存在[8]，这表明空气微生物粒子沉降量状况可以指示空气质量的好坏。根据有关研究[9]，可以认为本结果反映出兰州空气有较多粒径较大的微粒，包括浓度较高、粒径较大的微生物颗粒，发出了兰州空气污染状况亟待改善的信号。改善空气质量的措施应该包括考虑地形、气候等特点而调整工业结构和布局。改造旧城区，科学安排居民楼院和商业网点，不断提高工业和生活采暖效率，尽可能减少人为污染产生的负效应[3]。大量种植花草树木，使城市环境中的自然、生态区域增多、环境状况更趋合理化并可良性持续发展。而这些都必然地要在有较坚实的公众环境忧患意识指导下才能取得。

4 结语

本文用平皿沉降法获得了兰州若干测点的空气微生物粒子沉降量。平皿沉降法测定的空气微生物粒子沉降量既与微生物浓度密切正相关，又与包括载有微生物粒子在内的较大尘埃相关，这意味着兰州空气微生物状况，反映了工业、生活、污染物的种类、形态和含量等条件。尽管测点间存在一定差异，但整体看本结果从一个侧面在一定程度上证实兰州空气质量欠佳，且趋于恶化，为此兰州在科学、果断安排环境检测和防治工作的同时必须加大空气微生物的检测力度，以使兰州空气质量的提高建立在更科学基础上。

参考文献 9 篇（略）

LANZHOU AIR QUALITY TO BE INDICATED BY THE PARTICLE SUBSIDING VOLUME OF AIR MICROORGANISMS

(ABSTRACT)

Abstract At 6 monitoring sites set up in Lanzhou municipal region, the particle subsiding volume of air bacteria and fungi were measured by flat grevitate subsiding methods to ten times, the results showed that the total amount of bacteria, fungus, microorganisms and fungi/all microorganisms percent were 18,785.4、458.5、19,243.9CFU/m^3 and 2.4 respectively. The bacterial content increased, this meant the air quality of Lanzhou city had been deteriorating. Therefore, the decisive counter measures must be taken to change the current status.

Key words Lanzhou Air Microorganisms Air Quality

天水市空气微生物含量分析*

摘　要　用平皿沉降法测定了天水市空气微生物含量,结果证实其空气细菌、真菌及总菌的平均含量分别为 5 312.3、13.1 和 5 325.4 CFU/m^3。空气真菌占总菌量的 0.2%左右,空气微生物以细菌为主十分明显。所测风景区空气细菌含量与城区的平均水平间差异很大。说明天水市空气已受到较轻的中度污染,闹市区的一些测点空气污染已相当严重。

关键词　天水市　空气微生物　空气质量

1　引言

天水市在渭河以南,气候属东南季风区向内陆干旱区过渡类型,较干旱。市区包括秦城和北道两部分,是甘、陕、川交通要道和军事要地,有相当规模的工商活动,这些为天水市的空气微生物提供了特定条件。当空气环境受化学和物理污染时,微生物(以及其他生物)即可与其中的一些起各种作用,包括协同作用,使人和动植物患病,使许多物品腐损变质。为此掌握空气微生物状况至关重要。本文初步报道了天水市的空气微生物调查结果,供天水市作为评价空气质量的参考。

2　材料和方法

用自然沉降法在 15 min 内捕捉空气中的细菌或真菌入细菌培养基或真菌培养基上。采好样的培养皿经适当培养后计数菌落形成单位(CFU),并换算为单位体积空气中各类微生物粒子沉降量(CFU/m^3)及空气真菌/总菌量(%),以此评价测点及所在地区的空气污染程度和质量[1],有关采样点情况及具体采样时间见表 1。

3　结果和讨论

本实验为 6 个采样点,共测定 7 次。天水市空气微生物含量状况见表 2。

表 2 说明 6 个测点空气细菌和总菌含量两个指标间各存在较大差异,空气细菌量和总菌量均以 4 号测点为最高,1 号测点最低。最高细菌含量是最低点的 402 倍,空气真菌含量普遍偏低,只有 2 号测点检测到真菌,但量也不大。1 号测点代表了风景区的状况,其余各点为城内者。将 2～6 号 5 个测点的空气细菌含量平均计算,可得出 24 649.0 CFU/m^3,该值是 1 号点的 104.5 倍,这表明城区与风景区的空气细菌含量差别很大,总菌量在这两者间的差异更大,达 104.6 倍(24 664.7/235.8)。此外城区内各点间空气细菌含量也存在一定差异,按由小到大排序是:3 号、6 号、2 号、5 号、4 号,总菌含量排序也是如此。空气细菌等微生物含量是衡量特定区域空气质量的一个重要指标。按照我国有关部门的推荐,空气细菌含量在 5×10^3 CFU/m^3 以下时的空气为轻度污染,含 $5\times10^3\sim5\times10^4$ CFU/m^3 细菌的空气为中度污染,含$>5\times10^4$ CFU/m^3 细菌的空气为重污染[2]。因此本结果表明所测 1、3、6 号 3 个测点的空气仅为轻污染,2、5 号 2 个测点空气处于中度污染程度,4 号点的

---

* 原文刊于《干旱环境监测》,1998,12(3):152-154.

空气已受到严重污染。

表 1　天水市空气微生物采样记录

测点编号	测点名称	环境概况	采样起始时间
1	麦积山山脚停车场	阴天、薄雾、轻风、10℃，车、人不多，空气有动感	1996-04-12(10:42)
2	天水火车站广场	阴天，附近有建筑施工，车、人来往，但不乱，微风	1996-04-12(14:30)
3	北道花鸟市场公路桥附近	较空旷，河床畔，有小规模建筑施工	1996-04-12(21:00)
4	一马路西、隆昌路口	人行道上，卡车来往，尘土飞扬	1996-04-12(14:45)
5	天水市中心广场文庙前	阴天、薄雾、微风、偶有毛毛雨星，公交车来往，晨练、休闲者不少	1996-04-13(08:10)
6	耤河堤上	河床基本干涸、空旷，路上行人少，附近有医院	1996-04-13(08:47)

表 2　天水市空气微生物含量状况*

测点编号	空气微生物含量指标			
	细菌/CFU·m^{-3}	真菌/CFU·m^{-3}	总菌*/CFU·m^{-3}	(真菌/总菌)/%
1	235.8	0	235.8	0
2**	9 746.4	78.6	9 825.0	0.8
3	2 515.2	0	2 515.2	0
4	94 791.6	0	94 791.6	0
5	13 047.6	0	13 047.6	0
6	3 144.0	0	3 144.0	0

注：* 细菌和真菌之和。 ** 2 次测定平均值。

根据表 1 和表 2 还可看到，凡是尘土飞扬或人群活动集聚处（如 4 号和 5 号测点），空气微生物含量就高。这意味着该处空气质量低劣。

将各点空气细菌含量以几何均数法计算，可以得出其值为 5 312.3 CFU/m^3，真菌含量的算术均数为 13.1CFU/m^3，两者之和为天水市空气微生物含量的均值 5 325.4CFU/m^3。空气真菌与总菌量的百分比为 0.2。

从上述测定结果得知，天水市空气所含的微生物含量已处于较轻的中度污染状态，其中空气真菌含量及在总菌量中所占比值很小，表明天水市空气微生物污染主要由细菌引起。

天水市与国内外一些城市空气微生物含量状况的比较见表 3。

表 3 天水市与其他城市空气微生物含量状况的比较

城市名	空气微生物含量指标				测定日期	引证文献号
	细菌/CFU·m^{-3}	真菌/CFU·m^{-3}	总菌*/CFU·m^{-3}	(真菌/总菌)/%		
天水	5 312.3	13.1	5 325.4	0.2	1996-04	[6]
彭塔阿雷纳斯(智利)	545.3	114.6	659.9	17.4	1994-03	[2]
深圳	59 893.2	1 729.2	61 622.4	2.8	1994-07	[3]
珠海	11 475.6	8 331.6	19 807.2	42.1	1994-07	[3]
庐山	10 092.2	11 900.0	21 992.2	54.1	1995-07	[4]
铜陵	4 181.5	314.4	4 495.9	7.0	1995-07	[5]
北海	5 709.7	1 847.1	7 556.8	24.4	1996-1997 年之交	[6]
洛阳	33 247.8	550.2	33 798.0	1.6	1996-04	[6]
泰安	8 771.8	1 037.5	9 809.3	10.6	1997-09	[6]

*:细菌和真菌之和。

由表 3 可见,参与比较的 9 个城市中的 8 个城市分布于我国各地,有一定代表性。迄今国内外学者用不同方法对大城市空气微生物状况作了不少研究工作,但对中、小城市则研究不多。用经典的平皿沉降法所得的 9 个城市空气微生物含量相比较,可见天水市的空气微生物含量 4 个指标均处于较好状态,反映了天水市的空气质量总体上看是良好的。

参考文献 7 篇(略)

广州市区的空气微生物含量*

摘　要　用平皿沉降法测定了广州市区室内外空气微生物含量。市区室外空气微生物总量、空气细菌量、真菌量分别为 21 744.2 CFU/m³、21 110.6 CFU/m³、633.6 CFU/m³，真菌占总菌量的 2.9%，表明空气已受到中度污染，真菌有时占甚大比率。闹市、交通要道的空气质量一般比近郊、清静处的差；室外空气微生物含量显示出时间变化的某种特点。室内空气微生物监测结果反映出室内空气一般尚可。指出应进一步采取得力的环境保护措施，切实改善空气环境质量。

关键词　空气质量　微生物　数量分布　广州市

1　引言

空气中悬浮颗粒是许多有害微生物的重要载体。气生病菌在适宜条件下可能会诱发多种病害。有些霉菌对人类有致癌作用，有些使物品变质等等[1]。广州处于亚热带海洋季风气候带内，空气温湿，有利于空气微生物的生长繁殖；而地面上人群呼吸带的微生物直接影响人体健康。广州又是华南规模仅次于香港的大都市，人口稠密，工商业发达，交通繁忙。因此，监测广州市区的空气微生物含量，对提高广州市区的大气环境质量、保障人体健康是十分必要的。

2　材料和方法

选择广州市区不同功能区 7 个室外点和两个室内点用平皿沉降法监测空气微生物含量，采样点见图 1。所用采样介质包括普通培养基和真菌培养基。采样、培养和计算方法均按常规进行[2,3]。以空气细菌和真菌含量之和代表空气微生物总量并分析空气微生物状况，作出评价。

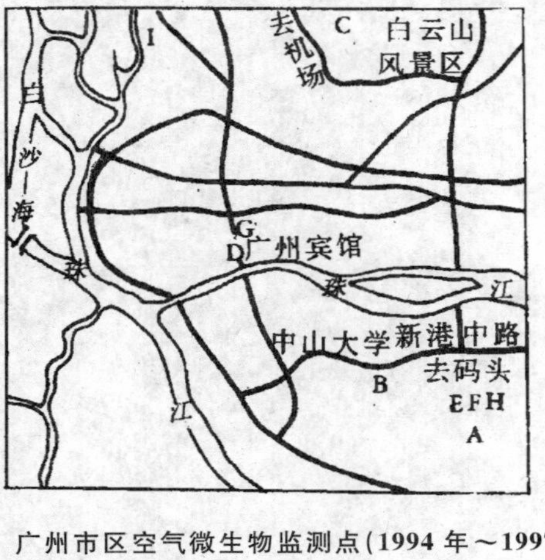

图 1　广州市区空气微生物监测点(1994 年～1997 年)

* 原文刊于《环境监测管理与技术》，1998，10(4)：14-16.

3　结果和讨论

3.1　概况

测定时间：1994 年 6 月～7 月(夏季)、1995 年 5 月(春季)、1996 年 12 月和 1997 年 1 月(冬季)。测定结果见表 1。

由表 1 计算，室外空气微生物总量、空气细菌量、空气真菌量的几何均数分别为 21 744.2CFU/m^3、21 110.6 CFU/m^3、633.6CFU/m^3(即每立方米所测空气中的菌落形成单位数，下同)，真菌/总菌为 2.9%。空气微生物含量状况是衡量特定地区空气质量的一个重要指标，其含量越高，空气质量越差。按照我国有关部门推荐，室外空气细菌浓度达 5 000CFU/m^3 时为轻度污染上限，室内空气细菌浓度为 4 000CFU/m^3 时为轻度污染。由此可见，广州市区室外空气细菌浓度已达中度污染，其超标测次率为 75.0%。室内空气尚可，但有一次超标。若将真菌计入(国家暂无标准)，那么超标率则更高。这表明广州市区空气质量不容乐观。

表 1　广州市区室内外测点空气微生物含量统计(CFU/m^3)

测次代号	测点名称	细菌	真菌	总菌	真菌/总菌%
室　外					
A_1	新洲码头	2 358.0	1 886.4	4 244.4	44.4
B	新港西路某院	8 331.6	903.9	9 235.5	9.8
C	白云山机场候机楼外	7 783.1	235.8	8 018.9	2.9
D	广州宾馆附近	94 811.4	157.2	94 968.6	0.2
E_1	新港中路某立交桥	297 422.4	未测出	297 422.4	—
A_2	新洲码头	3 615.6	1 336.2	4 951.8	27.0
E_2	新港中路某立交桥	52 819.2	1 257.6	54 076.8	2.3
E_3	新港中路某立交桥	121 672.8	314.4	121 987.2	0.3
E_4	新港中路某立交桥	74 984.4	471.6	75 456.0	0.6
H	某招待所	1 886.4	235.8	2 122.2	11.1
I_1	某宾馆	17 763.6	628.8	18 392.4	3.4
I_2	某宾馆	31 125.6	2 829.6	33 955.2	8.3
室　内					
F_1	某招待所	2 279.8	314.5	2 594.3	12.1
F_2	某招待所	1 886.4	235.8	2 122.2	11.1
G	某商店	4 951.8	707.4	5 659.2	12.5

3.2　各功能区室外空气微生物含量的差异

将表 1 所列的 7 个室外测点按功能区划分为 4 个类别，即离市中心较远的码头、机场区(AC 区)、居民区(BHI 区)、交通繁忙路段(E 区)和中心商业区(D 区)，由此得出表 2。

表 2　各功能区划室外空气微生物含量比较(CFU/m³)

功能区划代号	细菌	真菌	总菌	真菌/总菌%
AC	4 048.6	840.8	4 889.4	17.2
BHI	9 653.8	784.7	10 438.5	7.5
D	94 811.4	157.2	94 968.6	0.2
E	109 420.8	571.3	109 992.1	0.5

注:细菌、真菌含量均由其所在区各值之常用对数值的均值换算来。

表 2 显示,不同功能区划的空气微生物含量是不同的。空气细菌量、总菌量 E 区最大,AC 区最小;真菌量及真菌百分比 D 区最小,AC 区最大。它表明,在交通繁杂之处,流动的人、物使尘土、污物颗粒沉浮空中,致微生物沉降量大增(所幸测时的真菌量不是最高);人口密度小、空旷之处空气相对洁净得多,但濒珠江的 A 点测时湿度大,加上温热的天气,空气真菌量及其在总菌量中所占百分比增大。

3.3　室外空气微生物含量的时间变化

3.3.1　交通繁忙路段空气微生物含量的昼夜变化

选择交通繁忙的新港中路某立交桥(E 点)作早晨两次、下午 1 次和夜间 1 次测定,结果见表 3。

表 3　E 测点空气微生物含量昼夜差异(CFU/m³)

所测时段*	细菌	真菌	总菌	真菌/总菌%
早晨	175 120.8	628.8	175 749.6	0.4
下午	121 672.8	314.4	121 987.2	0.3
夜间	74 984.4	471.6	75 456.0	0.6

* 早晨:7:00 左右,下午:14:00～16:00,夜间:21:00 始,下同。

表 3 显示,空气细菌量大小排序为早晨、下午、夜间,真菌量大小排序是早晨、夜间、下午,总菌量大小排序是早晨、下午、夜间,真菌百分比大小排序是夜间、早晨、下午,细菌和真菌百分比在该点昼夜变化似呈相悖倾向。故在一定条件下,该结果反映了广州此类区划环境空气微生物含量昼夜变化特征。

3.3.2　室外空气微生物含量的昼夜变化

对所有室外测点空气微生物含量(总计)以早晨、上午、中午、下午和夜间 5 个时段分,用各含量的常用对数值表达,结果见图 2。

由图 2 可见,细菌量和总菌量的峰值均在早晨,谷值均在中午;真菌量峰值在下午,谷值在早晨。真菌量百分比值的波动状况基本类似于真菌含量本身的波动状况。

3.3.3　室外空气微生物含量的季节差异

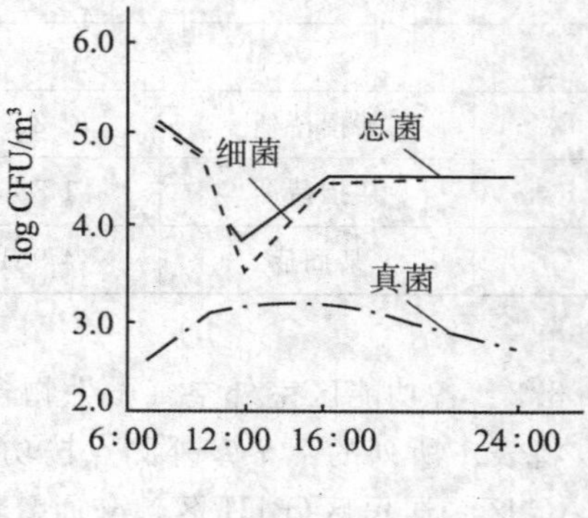

图 2　室外空气微生物含量昼夜变化

上述各次室外测定虽没有连贯入一个年度的四季中，但仍可大致划分为春、夏、冬三季。依此划分的空气微生物含量见表4。

表4　室外空气微生物含量的季节差异＊（CFU/m³）

季节	细菌	真菌	总菌	真菌/总菌%
春	29 847.0	367.3	30 214.3	1.2
夏	4 432.0	1 305.9	5 737.9	22.8
冬	28 196.8	654.0	28 850.8	2.3

＊细菌、真菌含量是各季有关测点的几何均数值，春季采自A、C、D、E点，夏季采自A、B点，冬季采自E、H、I点。

由表4可见，春季采自A、C、D、E点的空气微生物数量最大，这可理解为气温最适宜于微生物的季节最利于它们的繁殖。夏季，采自A和B点的空气微生物中有最大比率的真菌含量，这表明雨季增大的湿度有利于真菌的繁殖，尽管频繁的风雨可起到部分净化空气的作用。冬季，采样结果表明E区和H、I区由于交通繁杂、人口稠密，促使空气污浊，加上闹市区明显的温室效应，为空气微生物的繁殖和扩散创造了重要条件，也使广州此类城市的空气微生物生态的季节差异有趋于模糊之势。

3.4　室内外空气真菌状况

众所周知，空气真菌，尤其是一些霉菌，与疾病关系密切。真菌含量、真菌与总菌的百分数等都反映所测之处和所测之时的空气质量[4-6]。总结以上室内外的真菌状况，可发现居民区和码头（ABHI区）室外空气真菌量大，平均为1 303.5CFU/m³，机场、交通繁杂路段和商业区（CDE区）的则小，平均仅为406.1CFU/m³（见表1）。这暗示两者在空气湿度上存在一定差异。较高的湿度有利于具耐湿孢子的真菌的生存[4]。室内，空气真菌量显示出商店（如G）中的大，招待所（如F）内的小（见表1）。真菌所占百分比，一般是室内的高于室外的（除A外）。空气真菌量在时间上的差异显示出一年之中的夏令时节或一日之中的日出时分起，真菌量增，而冬季或夜间的真菌量减（见表3、表4）。与之相应的真菌百分率在季节差异上也作类似变化（见表4）。真菌量大的测点，常可能有较高的温、湿度和流通性不很大的空气或不洁的环境[4,6]。因此，人们在温热潮湿的时节应采取防霉除霉措施。

4　结语

大范围长期测定空气微生物含量需要雄厚的资金及人力和精力，难度大。以经典的平皿沉降法选取必要的测点调查空气微生物至今仍为各国学者所采用，从相对粗放的结果至少可略知测区空气微生物概况。此次测定结果表明了广州市区空气微生物的一般状况，暴露了广州市区空气质量的一些问题。由此再次证实了在我国某些特定地区环境微生物污染仍十分严重。因此在发展经济的同时，必须正视环境问题，切实采取得力措施改善包括空气质量在内的环境质量，这对造福人类是十分必要的。

参考文献6篇（略）

MICROBIAL CONTENT IN ATMOSPHERE IN GUANGZHOU CITY

(ABSTRACT)

Abstract The indoor and outdoor microbe content in atmosphere in Guangzhou city was estimated by the method of gravity plate. The total microbe, bacteria, fungi contents in the whole city were 21 744. 2, 21 110. 6, 633. 6CFU/m^3 and the percentage of fungi to the total was 2. 9%. The figures indicated the atmosphere had been polluted. The air quality in the areas of downtown or heavy traffic was worse than that in outskirts or less populated areas. The outdoor microbe content showed it varies with time. The indoor figures indicated the quality was acceptable. The paper points out that stronger measures should be adopted to improve the air quality.

Key words Air Quality Microbes Fractionation Guangzhou City

济南、泰安、曲阜空气微生物监测*

摘　要　用平皿沉降法测定了济南、泰安、曲阜的空气微生物含量，包括细菌、真菌、总菌及真菌占总菌量的百分比值。测定结果表明这3个城市的空气已受到中等污染，其中以济南的为重，3市中除一些测点显出有重污染外，尚有少数景点空气污染甚轻，空气质量较好，空气真菌占总菌的百分比不低，既反映初秋测时空气所含的较高湿度，又表达了3市所处的地理位置和环境污染状况，同时分析了空气微生物含量的昼夜变化态势，阐述了空气微生物含量的意义。

关键词　空气质量　微生物监测　济南　泰安　曲阜　大气监测　生物监测　城市

1　引言

同处暖温带半湿润气候带内的济南、泰安和曲阜构成了鲁中重要的旅游地带，其区域内的人口状况，经济活动等必然带来一些环境问题，空气微生物与环境污染密切相关，其状况可以反映空气污染程度[1]。同时三市的自然条件和结构功能又存在一定差异，表现在空气微生物状况上会有所不同，现就三市空气微生物含量及其含意作一初步探讨。

2　材料和方法

采用平皿沉降法于1997年9月间测定三市空气微生物含量，选择济南10个点作13次、泰安5点5次、曲阜6点10次，共计21点28次测定。所用培养介质包括普通营养琼脂培养基和真菌(琼脂)培养基，曝皿时间：5 min。所获样品培养于室温3d以上后，计算各平皿中长出的各类微生物菌落数(CFU/皿)，换算为单位体积空气中微生物含量，即CFU/m^3(简写为个/m^3，下同)[1]。

3　结果和讨论

3.1　三市空气微生物含量概况

表1列出的是三市所有测次中获得的空气微生物含量的4个指标，对这些数据以各市平均，可见曲阜的空气细菌、真菌、总菌及真菌占总菌量的百分比分别为7 687.1，3 081.1，10 768.2 CFU/m^3 及28.6相应地，泰安的是8 771.8，1 037.5，9 809.3 CFU/m^3 及10.6。济南则分别为14 100.8，1 886.4，15 987.2 CFU/m^3 及11.8选4个指标在三市间存在一定差异。按大小排序分，细菌、总菌量是济南最大；真菌及其在总菌量中所占百分比是曲阜＞济南＞泰安。按照我国有关规定(建议)，室外空气细菌含量在5 000～50 000个/m^3时，该空气即处于中度污染状态，由此可见，该三市空气已受到中等污染，其中以济南的为重，三市平均空气细菌含量为10 862.5个/m^3，空气真菌量为2 169.4个/m^3，总菌量为13 031.9个/m^3，空气真菌百分比则为16.6，因此加上真菌量，空气所受的污染将显得更重。

3.2　空气微生物含量在各市测点间的比较

表1中列出曲阜测点6个，No.1的5次测定的细菌、真菌、总菌、真菌占总菌百分比之

* 原文刊于《环境与开发》，1999，14(4)：43-45.

平均值分别为 5 376.2,2 641.0,80 172 个/m^3 及 32.9。将它们与其采 5 个点作比较后可发现曲阜空气细菌污染以 No.4(孔林)为最重,No.3 未测出,其空气微生物全由真菌组成,这反映出当时空气湿度大的特点。No.2、3、4 三个测点代表曲阜的主要景点,其空气细菌、真菌、总菌及真菌百分比各自的平均值分别为 9 164.8,4 087.2,13 252.0/m^3 及 30.8。其中以 No.4(孔林)测点的空气细菌含量为最高,还可能与当时游客较多、林木密度大、空气流动相计小有关;以 No.3(孔府)的空气微生物含量最低(未检出细菌),因而这里空气最为洁净。

泰安 5 个测点中,No.10 的空气微生物含量最高,No.11 最低。由表 l 可见泰安空气微生物含量及其测点间差异主要由细菌引起,而且其真菌含量及其所占百分比也是三市中最低的,测定结果表明泰安市内空气污染重于郊外,街道上重于景点内,山下重于山上。

表 1　曲阜、泰安、济南空气微生物含量测点结果统计

测次编号	测点名称或位置	空气微生物含量指标			
		细菌	真菌	总菌	细菌/总菌%
曲　阜					
1-1	阙里宾馆	6 288.0	4 401.6	10 689.6	41.2
1-2	阙里宾馆	2 200.8	2 672.4	4 873.2	54.8
1-3	阙里宾馆	3 615.6	3 458.4	7 074.0	48.9
1-4	阙里宾馆	5 030.4	1 414.8	6 445.2	22.0
1-5	阙里宾馆	9 746.4	1 257.6	11 004.0	11.4
2	孔庙	8 174.4	5 973.6	14 148.0	42.2
3	孔府	0	3 772.8	3 772.8	100.0
4	孔林	19 335.6	2 515.2	21 850.8	11.5
5	尼山	16 034.4	5 030.4	21 064.8	23.9
6	五巴祠商区马路	6 445.2	314.4	6 759.6	4.7
泰　安					
7	中天门	4 244.4	157.2	4 401.6	3.6
8	御座宾馆门外	10 218.0	1 729.2	11 947.2	14.5
9	岱庙	7 074.0	1 572.0	8 646.0	18.2
10	岱庙前马路	20 750.4	1 414.8	22 165.2	6.4
11	灵岩寺千佛殿前	1 572.0	314.4	1 886.4	16.7
济　南					
12-1	某机关院内	4 087.2	2 672.4	6 759.6	39.5
12-2	某机关院内	5 973.6	1 257.6	7 231.2	17.4
12-3	某机关院内	9 117.6	786.0	9 903.5	7.9

(续表)

测次编号	测点名称或位置	空气微生物含量指标			
		细菌	真菌	总菌	细菌/总菌%
12-4	某机关院内	5 816.4	2 829.6	8 646.0	32.7
13	经十路民主大街口	5 973.6	2 515.2	8 488.8	29.6
14	趵突泉	3 772.8	1 729.2	5 502.0	31.4
15	大明湖湖心	3 772.8	2 358.0	6 130.8	38.5
16	铁公祠门口水上	7 074.0	314.4	7 388.4	4.3
17	北极庙内	4 716.0	1 257.6	5 973.6	21.1
18	历下亭	4 558.8	1 100.4	5 659.2	19.4
19	省摩托车厂门外	59 893.2	2 358.0	62 251.2	3.8
20	市公交王官庄小区站	24 680.4	2 200.8	26 881.2	8.2
21	济南火车站大门外	4 286.8	3 144.0	45 430.8	6.9

* 各类微生物含量单位：个 CFU/m^3，以下 CFU/皿为单位时，以 CFU/m^3 ÷157.2 即是，下同

济南 No. 12 测点空气微生物测定了 4 次，由此，得出的 No. 12 的各指标平均值分别是空气细菌为 6 248.7，真菌为 1 886.4，总菌为 8 135.1 个/m^3，真菌占总菌百分比为 23.2。将该点各指标平均值与其他 9 个点的相应值(表 1)比较得知空气细菌含量，总菌量最高测点是 No. 19，最低测点是 No. 14 和 No. 15，真菌含量最高测点是 No. 21，最低测点是 No. 16，真菌占总菌百分比最大值在 No. 15，最小值在 No. 19。结果表明工业区、马路旁(如 No. 19，20)的空气中微生物含量高，有的测点(如 No. 19)空气细菌含量已很高，这意味着那里空气已受到严重污染。较洁净的测点在有关景区内，如 No. 14，15，其平均细菌含量为3 772.8个/m^3，指示空气只有轻度污染。但该两点空气真菌占总菌的百分比较高，仍然指示那里(水面以上空气)较高湿度利于许多真菌繁殖。总体看济南市某些机关院内空气状况好于工业区或交通繁杂之处。

3.3　空气微生物含量的昼夜变化

空气微生物含量的昼夜变化势态以济南和曲阜为例，即济南以 No. 12 的 4 次测定、曲阜以 No. 1 的 5 次测定来计算并画入图 1 之中说明。图 1-Ⅰ表明济南空气微生物总量的昼夜变化基本上与细菌的势态一致，高峰常在上午(08:00 左右)，下午(12:00 以后)较低，而真菌与此呈相异之势，上午(12:00 以前)不见高峰却见低谷(08:00 左右)、中午(12:00 左右)见高峰。

图 1-Ⅱ表明曲阜空气微生物总量与细菌量的变化态势基本一致，总菌量经凌晨 3 点的较大减少后攀高，各值在夜间(20:00 前后)现减低之势。细菌量的峰值也出现于上午(06:00 以后)真菌的状态则与此相异，低谷出现在晨间前(06:00 前)。上午(09:00 左右)有攀峰值，但比细菌和总菌的峰值来得晚。

由上可见，两地的细菌、真菌和总菌量之变化态势相类似，与庐山相比，则显出不同之处[2]。

4　结语

经测定表明济南、泰安和曲阜的空气已受到中度污染,以济南的为重,各市空气微生物含量指标表现出各测点间的差异。曲阜的孔林(No. 4)污染较重,泰安市内的污染重于郊外,其景点受污染的程度在三市中最低,即尚处于轻度污染状态,济南测定的数据指示该市不洁的工业过程,繁杂的交通状况与所处的地理环境相结合造成了较重的空气污染,有的甚至达到严重污染程度。较安静之处,人口相对稀疏之处或空间开阔之处,空气微生物少,环境污染也轻[3]。所描述的空气微生物昼夜变化状态说明细菌量的高峰在上午,真菌的峰值滞后。总菌量大都是由细菌变动引起,细菌量的谷值在夜间,真菌谷值在晨间出现,这次测定是在一场秋雨过后进行,较高的空气湿度和不大的风力促使不洁空间的真菌滋生增多,也是所处气候带"暖温"和"湿润"特点的反映,由于一些真菌与人体病患、物品变质关系密切[4],因此在潮湿温热天气时应注意防霉除霉。

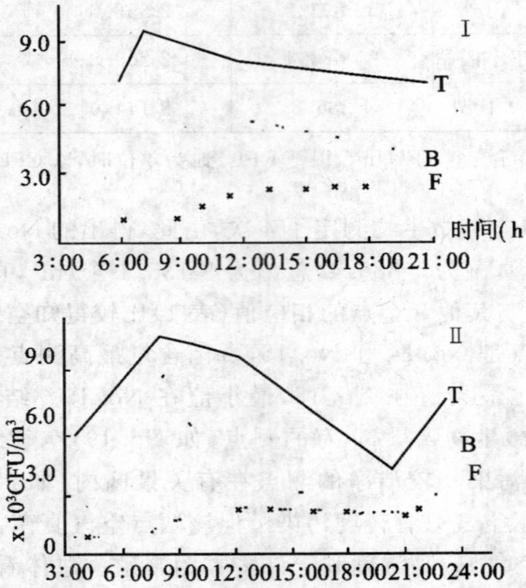

Ⅰ—济南,Ⅱ—曲阜,T—空气微生物总量,(时间 h)

B—空气细菌,F—空气真菌

图 1　济南、曲阜空气微生物含量的昼夜变化状态(1997. 9)

一些研究已表明,空气微生物含量与空气中某些不同形态、性质的化学物质、降尘类别和颗粒大小有一定相关关系[5],也受地形、风源等的影响[6],因此需进一步研究这三市空气微生物与相关因子间的关系,以阐明其空气微生物的时空变化规律。

参考文献 6 篇(略)

SPYING INTO AIR BORNE MICROBES OVER THREE CITIES OF JINAN TAIAN AND QUFU

(ABSTRACT)

Abstract The airborne microbial contents over three cities of Jinan, Taian and Qufu were determined by using the method of gravity plate and their general situation was obtained. It included the contents of airborne bacteria, fungus, total microbes and fungus/total microbes (%) of the three cities. Their air environments have been suffered from the middle pollution. But there was slight pollution over a few sites investigated and their air conditions were better. The result obtained indicates that the percentages of airborne fungus count were not lower from the investigation. It reflects that there was a higher relative humidity on the air determined during the early autumun, expresses the geographic location situated and environmental pollution situation of the three cities as well. At the same time the paper analyses the diurnal change states pf the airborne microbes while it interprets the signficance of the airborne microbial content.

Key words Air Quality; Microbes; Spying Into Jinan, Taian And Qufu

空气微生物粒子沉降量指示太原空气污染状况*

摘　要　用平皿沉降法初步测定了太原的空气微生物粒子沉降量，结果表明太原空气中的微生物含量偏高，其空气细菌、真菌、总菌含量分别为24 819.9、627.4和25 447.3CFU/m^3，真菌在空气微生物总量中所占百分比率为2.47。这意味着太原空气污染相当严重、空气质量不佳，文章分析了造成太原空气微生物含量增高的原因。提出了减轻空气污染、提高空气质量的一些措施。

关键词　空气微生物　太原　空气污染　粒子沉降量　大气监测

1　引言

太原位于我国黄土高原东缘，系三面环山的丘陵盆地，在大陆性气候带内。因能源和重化工工业而使环境污染相当严重，空气中应运而生出许多微生物，包括各种病原细菌、真菌、放线菌、病毒、立克次体等，它们以各种污染颗粒物为载体和媒介而成生物气溶胶沉浮于空气中，散布于生命体内外，致使人和动植物发生病害、物品腐损、文物变质。空气微生物与化学污染物间的协同作用则加重了环境、生态和健康方面的问题。这已引起了各国学者的重视和研究[1-2]，但人们对太原空气微生物状况的了解据知迄今还是不多的.本文就太原空气微生物粒子沉降量状况所作的初步监测结果作一报道，以期太原人民对研究该课题的重视。

2　材料和方法

以细菌培养基和真菌培养基作介质，用平皿沉降法收集太原室外空气中的细菌和真菌。曝皿时间为5分钟，经培养后分别计算出单位体积空气中的细菌、真菌CFU/m^3，细菌和真菌之和为总菌量(CFU/m^3)，并计算出真菌量占总菌量的百分比值，以此作为一种指标来评价太原的空气污染程度[3]，采样点及其概况列于表1之中。

表1　太原空气微生物采样记录

测点编号	测点名称	测点环境概况	采样开始时间
1	晋祠门口	晴、灰蒙蒙、气温23℃，微风，有些游人	1996.4.25 13:07
2	五一广场	晴、灰蒙蒙、气温21℃，和风—清劲风	1996.4.25 15:30
3	火车站对面小杂院	晴、灰蒙蒙、气温21℃，和风	1996.4.25 14:50
4	火车站对面邮政局门口	聚有人群、背阴	1996.4.25 15:00
5	火车站广场	行人来往、空气流动	1996.4.25 15:10

* 原文刊于《内蒙古环境保护》，1999，11(4)：9-10.

3　结果和讨论

表2列出的是太原空气微生物粒子沉降量测定结果，由该表可见，这五个测点空气微生物粒子沉降量间存在着明显的差异。将1号测点与4号测点比，可知后者的空气细菌含量是前者的4.2倍、总菌量的近4倍，空气真菌百分比却反了过来，前者是后者的6.8倍。这意味着4号测点的空气中细菌浓度大大高于1号测点的，即细菌污染重。而1号测点空气中真菌所占比率高，意味着那儿湿度大，易滋生较多耐湿真菌，五个测点各微生物参数按大小排序为：细菌量是4,5,3,2,1，总菌量是4,3,5,2,1，真菌是1(3),2,5(4)，真菌百分比是1,2,3,5,4。

表2　太原空气微生物粒子沉降量统计*

Table 2　Statistics of airborne particles precipitation above Taiyuan

测点编号	空气微生物粒子沉降量			
	细菌	真菌	总菌	真菌/总菌百分比
1	12 104.4	786.0	12 890.4	6.1
2	18 156.6	707.4	18 864.0	3.8
3	28 924.8	786.0	29 710.8	2.6
4	50 933.8	471.6	51 404.4	0.9
5	29 082.0	471.6	29 553.6	1.6

*各类微生物粒子沉降量单位：CFU/m^3，下同 ＊＊两测定平均值

空气微生物粒子沉降量在一定程度上指示所测空气污染程度，空气微生物含量越大，空气污染越重，当空气细菌含量在5×10^3～5×10^4 CFU/m^3时表明空气已处于中度污染状态[3]，由表2可看出，这五个测点的空气都至少遭到中等程度的污染。

五个测点的细菌、真菌粒子沉降量各经几何均数统计后分别得到24 819.9和627.4 CFU/m^3，两者之和为空气微坠物粒子沉降量即25 447.3CFU/m^3，空气真菌量占总菌量的约2.5%，若这五个测点的空气微生物的平均状况有一定代表性的话.那么本结果表明太原空气微生物浓度偏高，总体上暗示出太原空气已受到中等污染，与一些城市相比，太原空气质量不佳[4-5]，这是十分触目的势态。

造成太原空气微生物此种状况是有原因的，它所在的地理条件和人为污染是最基本的，因为这儿位于高原丘陵盆地干旱区，雨水少，尘土易飞扬。工业污染又使大气中颗粒物、气溶胶大增。生产生活燃料结构及其改造滞后。人群的密集，促使气生微生物大增，即使是远离市区的风景区——晋祠的空气也难于幸免污染，这一状况的改变有赖于根据自然条件工业结构和布局的合理化，环保措施的跟上，良好生态环境的恢复和营造以及市民环保意识的增强和行动的落实。可喜的是太原有关方面已经和正在采取有效措施，太原空气恢复清新将指日可待。

4　结语

本文以有限的测点对太原空气微生物的瞬间状态作一论述。虽然不能以点代面，但至少可以从一个角度一个侧面反映出一些问题，即太原空气污染不轻，太原如同国内外的一些发展中城市一样，其空气都经受着污染。人们无法回避它，当代人的重要任务之一便

是处理好下面一对矛盾，即在发展经济的同时，也要抓好精神文明建设，在最早最严重爆发人与自然界间矛盾之处——自然界最脆弱的大气圈表层底部调好相互作用。

参考文献 5 篇（略）

TO INDICATE AIR POLLUTION OF TAIYUAN CITY USING THE PRECIPITABLE AMOUNT OF AIRBORNE MICROBIAL PARTICLES

(ABSTRACT)

Abstract The precpitable amount of airborne microbial particles was primarily examined using the method of gravity plate. The analytic results showed that microbial conbents are 24,819.0, 627.4, 25,447.3CFU/m^3 respectively, which were on the high level in the outdoor air of Taiyuan city. The percentage of airborne fungus content occupied in the total bacterial amount was 2.47. It means that outdoor air pollution of Taiyuan was quite severe and the air quality was not good.

Keywords Airborne Microbes, Taiyuan, Air Pollution

宝鸡市空气微生物粒子沉降量及与铜陵市的比较*

摘　要　选取宝鸡市城区测，再用平皿沉降法测定了空气微生物粒子沉降量。结果表明空气细菌、真菌和总菌量各自的平均值为 7 755.2CFU/m^3、366.8CFU/m^3 和 8 122.0 CFU/m^3。空气真菌占总菌量的比率为 4.5%。从一个侧面反映出宝鸡市的空气环境已受到了中度污染。与铜陵的比较，表达了中国东、西部规模相近城市间空气微生物及由此反映出的空气质量差异。必须提高人们的环境忧患意识，使空气质量更上一层楼。

关键词　宝鸡市　铜陵市　空气微生物　空气质量

宝鸡市位于关中渭河平原西缘，具暖温带季风气候，易出现春旱和伏旱，市区主要位于渭河以北。扼西北——西南交通咽喉的宝鸡，机电、轻纺、食品、冶金、化工、建材工业发达，有大宗物资和人流集散于此，这给宝鸡环境包括周围的空气造成了一定的污染。微生物包括许多病原生物以各种颗粒为载体传播于空气之中，对人和动植物的健康构成了威胁，使许多物品遭受侵袭和腐损，所以空气微生物与人类关系密切。空气微生物状况是环境质量的重要标志[1]。本文论述了宝鸡市空气微生物粒子沉降量状况，由此评价其空气污染程度，并与东部相近规模城市铜陵作一比较，以促进东西部城市空气质量的改进。

1　材料和方法

本次测定在宝鸡市共设立了 3 个点，于 1996 年 4 月 24 日进行，具体的采样点环境概况和采样时间详见表 1。

表 1　宝鸡市空气微生物采样记录

采样点编号	采样点环境概况	采样起始时间
1	宝鸡市工商银行大楼前街心花园。阴天，轻风—软风，气温 15℃，附近有人、车来往	10:20
2	炎帝园内人造小丘亭子外，阴天，和风，灰蒙蒙，游人不多，气温 15℃	10:20
3	火车站广场，天灰蒙蒙的，气温 20℃，轻风	17:30

所用方法是平皿沉降法。空气样本经 5 min 采集至细菌或真菌培养基上，培养 72h 后计数皿中的菌落形成(单位为 CFU)，并换算出单位体积空气中细菌和真菌粒子沉降量(单位为 CFU/m^3)、空气微生物粒子总沉降量以及真菌占总菌量的百分比，由此评价宝鸡空气污染程度[3]。将宝鸡空气微生物状况与东部类似规模城市铜陵作一比较，以了解中国

* 原文刊于《宝鸡文理学院学报(自然科学版)》，2000，20(4)：292-294.

东、西部城市间空气微生物状况的异同之点。

2　结果和讨论

2.1　宝鸡市空气微生物粒子沉降量

表 2　宝鸡市空气微生物粒子沉降量统计

采样站编号	空气微生物粒子沉降量/CFU·m^{-3}			真菌/总菌%
	细菌	真菌	总菌	
1	11 161.2	157.2	11 318.4	1.4
2	7 071.0	314.4	7 388.4	4.3
3	5 030.4	628.8	5 659.2	11.1

由表 2 列出的宝鸡市空气微生物粒子沉降量的统计结果可知，各测点空气微生物粒子沉降量的 4 个参数间有一定差异，按其大小排序如下：空气细菌量是 1 号测点、2 号测点、3 号测点；真菌量是 3 号测点、2 号测点、1 号测点；总菌量是 1 号测点、2 号测点、3 号测点 I 真菌所占比率是 3 号测点、2 号测点、1 号测点。这表明 1 号测点的空气细菌含量大，3 号测点的小，而真菌量的情况恰恰相反，3 号测点大于 1 号测点。空气细菌含量的大小在一定条件下可反映所测空气的被污染程度，其含量越大，空气污染越重。当细菌含量在 5×10^3CFU/m^3 时，该空气即达中等污染程度[2,3]。因此表 2 说明了这 3 个测点的空气已处于中度污染状态，其中 1 号测点的污染最重。空气真菌毒素可以致癌，量多毒害更大[2,3]。3 号测点的空气真菌量及其所占比率最高，当引起注意。各点的细菌加上真菌将使空气的微生物污染更重。

将表 2 所列各参数类各自平均，可得出空气细菌粒子沉降量为 7 755.2CFU/m^3，真菌为 366.8CFU/m^3，总菌量为 8 122.0CFU/m^3，真菌量与总菌量之比为 4.5%。这 3 个测点的情况，可从一个侧面反映出宝鸡市空气微生物一般状态，本结果说明了宝鸡市的空气目前处于中度污染程度，同时在某些地区或范围内的空气则污染较为严重，值得注意。

2.2　宝鸡、铜陵空气微生物粒子参数比较

宝鸡位于中国西部陕西省，铜陵位于中国东部安徽省，两市间直线距离约 1 000 km，其规模相近，但地理环境、工业状况等有一定差异，因而在一定意义上在我国东、西部中、小城市中具有代表性。

于 1995 年 7 月对铜陵市空气微生物用同样的方法进行了测定[4]。表 3 列出了宝鸡和铜陵两市的空气微生物状况及相关条件的比较。

表 3　宝鸡、铜陵空气微生钧和相关条件比较

城市名称	空气微生物状况			真菌/总菌%	相关条件
	细菌	真菌	总菌		
宝鸡	7 755.2	366.8	8 122.0	4.5	渭河平原西端。暖温带，易旱。人、物流集散地，工业结构较复杂的西部城市

(续表)

城市名称	空气微生物状况			真菌/总菌%	相关条件
	细菌	真菌	总菌		
铜陵	4 181.2	314.2	4 495.4	7.0	长江东侧。东南有群山。亚热带湿润季风气候。有色金属工业为主的东部工矿城市
宝鸡/铜陵	1.85	1.17	1.81	0.64	

由表3可知,宝鸡的细菌量、真菌量和总菌量均高于铜陵的,表明前者的空气污染程度高于后者。这可能决定于两地的工业结构、污染程度及气溶胶构成、空气悬浮颗粒总数(TSP)等[5]。铜陵的真菌与总菌比率较高,意味着铜陵空气相对湿度高于宝鸡,它有利于真菌的生存,也反映出铜陵所处纬度较低、经度较高,气温较高,并濒临大河之地理、气候特点。但将本结果与某些城市所测情况相比,又可发现宝鸡市的空气污染不很重、空气质量尚可[6],甚至优于一些风景旅游区[7,8]。

3 结语

像宝鸡这样规模的城市全国有许多座,世界上更多,今天它们都面临着日益严峻的环境问题。但由于原因和条件不同,就环境(包括空气环境)质量而言,东、西部城市如铜陵、宝鸡间存在着一定差别。从宝鸡市的空气微生物粒子沉降量状况说明了它的空气质量已有问题,是不容过于乐观的,因此宝鸡尚须努力,应根据所处位置,安排好生产和生活。我们不能完全排除空气污染,但应尽力将其减轻至最低程度,以确保环境更好。本文结果启示我们,在加速开发西部和城市化健康进程中,应重点地科学地发展中小城镇的建设。发展经济,更应提高环境意识.重视和维护可持续发展的良性生态环境,造福千秋万代。

参考文献9篇(略)

AIRBORNE MICROBIAL PARTICLES PRECIPITATION OVER BAOJI CITY AND COMPARISON WITH TONGLING CITY

(ABSTRACT)

Abstract The airborne microbial particles precipitation from some sites over the urban district in Baoji City was monitored by using the technique of gravity plate. The result obtained was that: the average airborne bacterial, fungus and total microbial particles precipitations were 7 755. 2 CFU/m^3, 366. 8CFU/m^3 and 8 122. 0CFU/m^3 respectively. The percentage of airborne fungus content occupied in total ones was 4. 5%. It reflected from a side that the air environment of Baoji City was suffered immediately from pollution. The result expressed that the difference of air quality on the basic of air-borne microbial conditions between Baoji and Tongling, and the differentiation among the similar scale cities in the Eastern and Western parts in the mainland of China. The people ought to arrange further production and living according to Baoj and Tongling's conditions and heighten public environment concerned consciousness for raising the air quality of Baoji going up to a higher level.

Key words Baoji City; Tongling City; Airborne Microbes; Air Quality

西宁、青海湖、格尔木的空气微生物含量*

摘　要　用平皿沉降法对西宁、青海湖和格尔木的空气微生物含量作测定，分别检测出三地各自平均的空气细菌、真菌、总菌含量、真菌检出率、真菌与总菌量的比率。文章分析了三地空气微生物指标的时空分布状态及其起因，指明必须提高环境保护意识，加强包括空气微生物状况在内的环境监控。根据地理学和生态学特点，安排好生产和生活，有望扭转恶化中的空气环境和提高空气质量。

关键词　西宁　青海湖　格尔木　空气微生物　环境质量

一、引言

凭借沉浮于空气中的不同颗粒物等气溶胶媒体，空气微生物生长、繁殖、散布于周围环境并发生相互作用，因而空气微生物是空气环境状况的重要指标之一，其含量的多寡及其变动指示着特定测区空气中颗粒物的种类、浓度、大小、气象要素及其变化。尤其是离地面 1.5 m 左右高度的人群呼吸带内的空气微生物，包括病原微生物直接关系着人和动植物的健康。空气微生物的测定已日益为国内外环境科学家、生态学家和微生物学家所重视，并对改善空气质量提供了丰实的成果和作出了有益的贡献[1-2]。

二、测区环境条件和生态因子

2001 年 5 月间，作者对青海省西宁、格尔木两市和青海湖区作了空气微生物含量的检测。

西宁、青海湖和格尔木地处青藏高原东北部，属大陆性高原气候，具较低气温、大温差、长日照、少降水之特点。西宁位于群山环抱的湟水谷地中，海拔 2 275 m。西川、北川、南川三条河与湟水穿城而过。是全省政治、经济、交通和文化中心。

青海湖位于青海东北部的大通山、日月山和青海南山之间，海拔 3 195 m，湖岸有广阔的草原，鸟岛位于湖的西北隅。格尔木市则位于柴达木盆地南缘、昆仑山北麓、格尔木河畔，是新兴的盐化工城市。

三、测时的天气状况

表 1 列出了三地在 2001 年 5 月 15-20 日共 26 个测点的天气和环境，以及空气微生物状况。

---

* 原文刊于《国土资源科技管理》2002，19(1)：62-66.

基金项目：国家海洋局生物活性物质重点实验室资助

表 1　西宁、青海湖、格尔木空气微生物测点概况及测点空气微生物状况

测区	测点编号 / 测试日期	测点概况	空气微生物指标			
			细菌	真菌	总菌	F/T%
西宁	1 15 日 13:00[①]	大十字过街桥，风不大，车、人不密集地流动	50 471.9	0	50 471.9	0
	2 15 日 18:00	北杏园附近一公司门口，人不多	6 132.1	0	6 132.1	0
	3 15 日 19:20	西宁火车站正门塑像上，车、人流量大	69 339.9	0	69 339.9	0
	4 15 日 20:30	笑笑公园内，无人	5 660.4	471.7	6 132.1	7.7
	5 16 日 07:30	植物园，无风，嫩松树叶上	2 358.5	471.7	2 830.2	16.7
	6 16 日 08:30	人民公园，无风，白杨树下，有集体晨练活动	6 132.1	0	6 132.1	0
	7 16 日 09:55	中科院盐湖所，微风，太阳直射处 20℃	3 773.6	0	3 773.6	0
	8[②] 16 日 13:30	大十字过街桥，轻风，太阳直射处 25℃	27 358.6	0	27 358.6	0
	9 16 日 18:30	省心血管病院绿篱上，轻风，静，院外有集市	7 075.5	943.4	8 018.9	11.8
	10 16 日 19:30	2 路车终点站，活动中心附近街头，人、车多，有山谷风	43 868.1	0	43 868.1	0
青海湖	11 17 日 17:30	蛋岛，阴云，清劲风，各种鸟孵卵，有些游人	3773.6	0	3773.6	0
	12 17 日 18:00	《鸟是人类的朋友》碑处，风、雪、小雨，无人	471.7	3 773.6	4 245.3	88.9
	13 17 日 18:20	蛋岛门口，风、雪、小雨，有一车来	21 226.4	3 537.6	24 764.0	14.3
	14 17 日 18:30	鸟岛门口窗台，风、雪、小雨	2 358.5	471.7	2 830.2	16.7
	15 18 日 07:00	吉泉乡招待所院内，和风，炊烟，汽油味混杂，晴间多云	9 905.7	471.7	10 337.4	4.5
	16 18 日 10:30[③]	黄玉火车站附近青海湖滨，无人，微风从山坡来，晴间多云，较干燥	11 320.8	0	11 320.8	0
	17 18 日 14:00[③]	离黄玉火车站 1 km，青海湖之滨，无人，湖风，清劲风	2 594.3	0	2 594.3	0
格尔木	18 19 日 09:00	鑫苑宾馆附近（近火车站），有沙、灰尘、晴间多云，微风	20 754.8	471.7	21 226.5	2.2
	19 19 日 09:15	昆仑公园门口，游人少，较干燥	19 811.4	0	19 811.4	0
	20 19 日 09:32	西区大转盘内，阵风清劲风，尘土	74 528.6	1 415.1	75 943.7	1.9

(续表)

测区	测点编号/测试日期	测点概况	空气微生物指标			
			细菌	真菌	总菌	F/T%
格尔木	21 19 日 09:55	青垦市场二楼口,微风,太阳直射	17 452.9	943.4	18 396.3	5.1
	22 19 日 10:30	江源路北端,市郊接合部,沙石路面,开阔	31 603.9	943.4	32 547.3	2.9
	23 19 日 10:50	清真寺院大殿楼梯口,有沙、石,建筑破旧,太阳直射,较干燥	66 038.0	0	66 038.0	0
	24 19 日 11:30	河东集贸市场外购物中心门口碑上	12 264.2	0	12 264.2	0
	25 19 日 22:00	黄河路青藏公路交叉口,软风,车辆来往,开阔	7 547.2	0	7 547.2	0
	26 20 日 14:50	长途汽车站院内,无风,太阳直射处气温 25℃	14 151.0	0	14 151.0	0

注:①检测起始时间,测时一般 5 min;②与①同;③测时 10 min

* 细菌、真菌、总菌含量单位均为 CFU/m^3;F/T%为真菌含量/总菌含量%,下同。

四、材料和方法

培养基:营养琼脂培养基用于测定细菌总数,沙氏培养基用于采集真菌。

方法:根据地势、交通、工商业、建筑物和人群活动等情况设置好测点,将配制好的培养基平皿在测点定时(一般为 5 min),重复两份采集落尘菌,测点概况如表 1 所示。取好样的平皿置 25℃温度中培养 2～5 天,计数各皿的 CFU(即大气微生物菌落数),并以下列公式算出空气微生物(细菌或真菌)浓度(个 CFU/m^3):

$$CFU/m^3 = 1\,000 \div \left(\frac{A}{100} \times t \times \frac{10}{5}\right) \times N$$

式中,A 为平皿面积(cm^2),t 为平皿暴露于空气中的时间(min),N 为经培养后平皿中长出的菌落数。

细菌和真菌合计为(微生物)总菌量,按测点测时等条件比较分析各点的 CFU/m^3(每立方米大气微生物菌落数)等参数及其时空变化,评价其空气环境污染和质量状况。上述培养基制作及结果的处理详见文献[2]、[3]。

五、结果和讨论

(一)西宁空气微生物含量分析

据表 1 西宁一栏,计算出西宁空气中细菌含量的平均值为 22 217.1CFU/m^3。按照有关规定,特定测点空气细菌含量达到 5×10^3～5×10^4 CFU/m^3 时的空气为中度污染,这表明西宁一般的空气中已受细菌中等程度的污染。各测点(次)空气细菌含量大小排序为:No. 3>10>8>9>2(=6)>4>7>No. 5,最大值是最小值的 29.4 倍(69 339.9/2 358.5),表明最大细菌量处的空气污染已超过中度污染的平均水平。N0.3 测点正是人群和车辆高度集中的火车站广场。而 N0.5 测点正是地处较偏僻、人员很少、空气流动小的植物园内。这表明空气细菌含量与地理位置、人群及其活动有密切关系。

将这10次测量按测时划分为上午、下午和晚间三个时段来分析，可知各时段的平均空气细菌含量分别为上午4 088.1CFU/m³、下午32 075.6CFU/m³、晚间24 764.3CFU/m³，大小排序为下午＞晚间＞上午。测定所在的下午正值阴间晴天，气温不高，相对湿度较低，光线和紫外线的消毒作用不大，此时的空气细菌生长繁殖不弱。而上午（如No.7）在太阳光直射下或氧气浓度高时（如No.5和No.6），细菌死亡率高，导致细菌含量降低。

西宁空气真菌量也可从表1中看出。结果表明其检出率为30%（3/10），真菌含量平均值为188.7CFU/m³。真菌含量/总菌量平均为3.6%，最高真菌含量在No.9，最高真菌%在No.5，达16.7%，表明植物多的地方相对湿度可变大，有助于真菌繁殖（如No.9，No.4，No.5）。结果意味着春季的到来、逐渐升高的相对湿度和气温使潜在的真菌趋于活动状态，但与东部城市济南和南部城市北海比，仍较低[6],[7]，这也反映出西宁所处的地理特征。

表1还可得出西宁空气微生物含量状况的总体水平即各测点的平均值为22 405.8 CFU/m³。大部分测点已处于中度污染状态。只有2个测点（20%）空气为轻度污染。细菌加上真菌，使西宁市的空气微生物污染更重。

（二）青海湖空气微生物含量分析

由表1可见青海湖空气微生物含量状况。经计算，青海湖7个测点的空气细菌平均含量为7 378.7CFU/m³，表明青海湖空气总体上处于较低的中度污染水平。其中有4个测点（占57.1%）的空气细菌量较低，说明青海湖空气大多仍较洁净，测点间空气细菌含量有差异，各测点细菌含量大小排序为No.13＞16＞15＞10＞17＞14＞No.12。最高细菌量是最低者的约45倍（21 226.4/471.7）。出现最高细菌量之处在No.3，此处受陆地影响大，风雪雨可携来一些微生物，出现最低细菌含量之处在No.12，这儿离湖近，受人为活动影响小，湿度大，可能降低一些细菌的生存韧性。

将检测时间按上午和下午划分，空气细菌的含量显示出上午＞下午之势，即10 613.3 CFU/m³＞6 603.8 CFU/m³。这表明测定所在的上午时间，空气中的相对湿度比下午更适合于细菌生存。

青海湖7个测点的空气真菌含量平均值为1 179.2 CFU/m³，变幅为0～3 773.6 CFU/m³，检出率高达57.1%（4/7）。各测点真菌含量大小排序为No.12＞13＞14（＝15）＞16＝17＝No.11＝0，出现最高真菌含量之处其细菌含量却最小（No.12）。真菌含量还显出下午＞上午之势，即1 556.6 CFU/m³＞235.9 CFU/m³，说明真菌与细菌对空气条件有不同要求。相对湿度大、较高的气温可有助于真菌繁殖蔓延，以至于有的测点真菌比细菌多得多（如No.12）。

由表1可计算得出青海湖空气微生物含量的平均值为8 557.9 CFU/m³。各测点空气微生物含量大小排序为N0.13＞16＞15＞12＞11＞14＞N0.17，真菌量占总菌量平均为17.5%，真菌加重了青海湖空气污染程度，而且真菌检出率与真菌%这两个指标与南方城市铜陵和海口比也不低。

（三）格尔木空气微生物含量分析

由表1可计算出格尔木9个测点的空气中细菌含量平均值为29 350.2CFU/m³，该值达中度空气污染较重范围，说明该市空气质量较差，极差为66 981.4CFU/m³（即74 528.6

~7 547.2)。有 6 个测点的细菌含量低于此平均值,说明较差的空气质量主要由不多的、污染较重处引起(如 No. 20,23,22),其大小排序为 N0. 20>23>22>18>19>21>26>24>No. 25。No. 20 位于马路中间,空气污物遇较大风而刮入皿中致使菌含量大增,表明高的细菌含量可能与汽车排气、不洁作业或较差路面有关。而 No. 25 虽位于交叉口,但无风又值夜间,气温相对较低,因而细菌少。

将测时划分为上午、中午、下午、晚上,则空气的细菌含量大小排序为上午>下午>中午>晚上,说明逐升的气温(上午)比较冷的夜晚有利于细菌生长繁殖。

格尔木市的空气真菌含量经计算平均值为 419.3CFU/m³。出现真菌的空气,其真菌含量均大于此值。真菌的检出率为 44.4%(4/9),出现真菌的测点之真菌含量排序为 No. 20>21=22>No. 18,其余 5 个测点均未测出真菌。这 5 个测点空气环境似较干燥,人相对也少。按时段分,各测点空气真菌大小排序为上午>中午>晚上=下午=0。

格尔木空气微生物含量的平均值为 29769.5CFU/m³,各测点按大小排序为:No. 20>23>22>21>18>19>26>24>N0. 25,与空气细菌量排序基本一致,说明格尔木空气微生物含量波动主要由细菌引起。真菌占总菌的比率为 1.4%,比同处内陆的天水高,比宝鸡的低。虽然此真菌/总菌比率不高,但不小的真菌检出率说明该市出现真菌有一定机会,所幸格尔木总的来说处于日照长、降水少、气温低的高原区和人口密度较低,抑制了空气真菌的蔓延。

(四)西宁、青海湖和格尔木空气微生物含量的比较

表 2 总结出三地区各自的空气微生物 5 个平均指标,由此可见,细菌含量以格尔木最高,说明其空气细菌污染最重;青海湖的最轻。真菌含量平均值及真菌检出率、真菌占总菌比率以青海湖的最大,说明其空气湿度最适于当时真菌的生长繁殖,但其变幅大,意味着真菌的生存条件波动大。格尔木的总菌量和变幅均最大,说明该市空气微生物污染最重,且主要由细菌引起。

表 2　西宁、青海湖和格尔木各自空气微生物指标的平均值*

测区名	空气细菌		空气真菌		总菌		真菌检出率	F/T%
	平均值	极差	平均值	极差	平均值	极差		
西　宁	22 217.1	66 981.4	188.7	943.4	22 405.8	66 509.7	30.0	3.6
青海湖	7 378.7	20 754.7	1 179.2	3 773.6	8 557.9	22 169.7	57.1	17.8
格尔木	29 350.2	66 981.4	419.3	1 415.1	29 769.5	68 396.5	44.4	1.4

* 微生物含量单位:CFU/m³。

六、结语

上述三地同处西北干旱区,测时又处于升温并波动较大的春季,三地的平均空气细菌含量、真菌含量、总菌量、真菌检出率与真菌占总菌的比率经计算分别为 20 691.3CFU/m³、535.2CFU/m³、21 226.5CFU/m³、42.3%与 2.5%。如果这横贯青海 700 多公里的三地可以代表青海空气微生物污染概貌的话,则结果表明青海省空气细菌污染已达中等程度。空气污染主要由城市细菌引起,空气真菌则加重了这一污染,但各地又有自身功能区

域上的特点。西宁为工商业区，格尔木为新兴工业城市，青海湖为接近原始生态区，这三个不同的环境生态类型，其空气微生物发生发展条件也不尽相同。相对活跃的工业生产和商务活动给处于大陆性高原气候带内的西宁带来了较大的空气微生物含量，空气质量有恶化趋势。应增强环境意识，提高改善环境的使命感。格尔木的状态暗示了新兴城市之初还来得及着力科学规划城市建设，摒弃落后工业设施和流程，加强空气环境的监测和改善。以鸟岛及周围湖区为测点的结果意味着青海湖沿湖草场环境有退化之势，气候变暖，不适当的经济活动等等原因造成的沙漠化现象已开始影响湖区。随着西部开发热潮的到来，西部城市间、东西部地区间空气微生物状况的异同点也将不断地揭示出来，这将有助于全国经济和生态环境的良性可持续发展。

参考文献 7 篇(略)

CONTENT OF AIRBORNE MICROBES ABOVE XINING, QINGHAI LAKE AND GEERMU IN QINGHAI PROVINCE

Abstract The content of airborne microbes above Xining, Qinghai Lake, and Geermu has been determined by using the method of gravity plate, obtaining airborne bacteria, fungi, total content of bacteria, detection rate of airborne fungi, and the rate of fungi and total amount of microbes in those three areas. Based on an analysis of spatial and temporal distribution of airborne microbial indexes and their causes, this paper points out that it is necessary to strengthen awareness of environmental protection and to enhance environmental monitoring including airborne microbial condition.

Key words Xining; Qinghai Lake; Geermu; Airborne Microbes; Environmental Quality

青岛空气微生物状况的测定*

摘 要 本文用自然沉降法于2001和2002年分别对青岛10个测点作19次空气微生物测定。论述青岛空气中陆源和海洋微生物的时空变化状况,平均陆源细菌、真菌、总菌量和(F/T)%分别为8 781.8/3 014.5/11 796.3CFU·m^{-3}及25.6,平均海洋细菌、真菌、总菌量及(F/T)%分别为314.4<95.2/1 209.6CFU·m^{-3}及74.0。表明青岛空气处中度微生物污染状态,陆源污染较重,向滨海区扩张势头较强。结果显示海洋对陆地空气具调节作用。

关键词 空气微生物 空气污染 海洋与内陆 青岛

空气微生物是空气污染的重要因子,它与气溶胶、颗粒物等媒体一起散布并污染环境、左右疾病发生与传播,监测空气微生物状况是掌握其活动和作用的必要前提。青岛作为沿海开放和奥运伙伴城市,其环境状况越来越引人关注[1,2],近几年来,每年又都遭受沙尘暴袭击[3],加上本地尘埃的作祟,对空气微生物结构和组成必产生影响。本文对2001~2002年青岛空气微生物状况进行了测定和分析。

1 测点概况

青岛空气微生物的测点、测时及环境、天气状况见表1。

表1 青岛空气微生物测点概况

测点/时编号(No.)	测点名	环境与天气状况	起始测时
1	青岛高科园仙霞岭路花圃	人、车少、晴间云,9℃,可能有沙尘暴影响,3~4级南风,相对湿度80%。	2002.3.25.0915
2	青岛高科园仙霞岭路花圃	人、车少、多云,10℃,非直射阳光,4级南风	2002.3.25.16:00
3	青岛高科园仙霞岭路花圃	人、车少、晴,25℃,日晒,4级南风	2002.5.23.15:00
4	青岛高科园仙霞岭路花圃	人、车少、晴,21℃,0级风	2002.5.24.09:00
5	前海栈桥	游人多,漫射光,薄雾,23℃,4~5级南风	2001.5.30.15:40
6	前海栈桥	海雾较浓	2001.5.30.17:55

* 原文刊于《山东科学》,2003,16(1):9-13

(续表)

测点/时编号(No.)	测点名	环境与天气状况	起始测时
7	前海栈桥	游人不多,25℃,3 级风	2001.5.30.20:40
8	前海栈桥	晨练者不多,刚日出,3 级东北风	2001.5.31.05:15
9	前海栈桥	游人多,25℃,相对湿度 70%,晴,4 级南风	2001.6.02.10:20
10	前海栈桥	游人如织,晴,25℃,6 级北风	2002.5.25.15:20
11	市场居民楼	晴,18℃,1 级风	2002.5.24.05:55
12	市南居民楼	晴,18℃,1 级风	2002.5.25.06:00
13	李沧广场	人少,晴,27℃,4～5 级风,阳光直射	2002.5.27.12:25
14	胜利桥	车来往多,晴,26℃,4～5 级风,阳光直射花粉飞扬	2002.5.27.13:10
15	中山路过街桥	车、人往来,晴,20℃,5 级东北风	2002.5.27.14:05
16	云南路广州路口	车、人往来,晴,20℃,3 级南风	2002.5.27.16:00
17	市人大前花园	行人少,晴,20℃,0 级风	2002.5.27.20:15
18	延安路老年乐园	休闲者多,22℃,3～4 级西北风	2002.5.28.17:40
19*	国家海洋局北海分局码头	船舶装卸货物,人走动,多云间晴,4～5 级南风,24℃	2002.7.02.13:30

测时持续为 5 min,* 测时 25 min

2　材料和方法

以 Zobell2216E 固体培养基和营养琼脂分别作海洋性和陆源性空气微生物粒子自然沉降用。经 48 和 72h 以上培养后,计数皿中生长出的菌落(CFU),并换算为每单位体积 m^3 空气中的 CFU,包括空气细菌(B)、真菌(F)、总菌(细菌+真菌,T)量及真菌占总菌量的%[(F/T)%]。分析其时空分布,评价空气污染状况和空气质量,方法详见文献[4]。

3　结果和讨论

3.1　青岛空气微生物概况

表 2 列出的是青岛所有测点/时的空气微生物含量指标。

表 2　青岛空气微生物含量状况表

测点编号	空气微生物含量指标							
	陆源性微生物				海洋性微生物			
	B	F	T	(F/T)%	B	F	T	(F/T)%
1	314.4	0	314.4	0	—	—	—	—
2	3 930.0	1 572.0	5 502.0	28.6	—	—	—	—

（续表）

测点编号	空气微生物含量指标							
	陆源性微生物				海洋性微生物			
	B	F	T	(F/T)%	B	F	T	(F/T)%
3	35 212.8	14 462.4	49 675.2	29.1	—	—	—	—
4	17 606.4	2 515.2	20 121.6	12.5	—	—	—	—
5	943.2	1 572.0	2 515.2	62.5	0	314.4	314.4	100.0
6	—	—	—	—	0	2 515.2	2 515.2	100.0
7	157.2	786.0	943.2	83.3	235.8	157.2	393.0	40.0
8	2 043.6	417.6	2 515.2	18.7	786.0	786.0	1 572.0	50.0
9	0	2 672.4	2 672.4	100.0	157.2	1 257.6	1 414.8	88.9
10	6 132.1	10 477.4	16 609.5	63.1	—	—	—	—
11	35 212.8	14 462.4	29 675.2	29.1	—	—	—	—
12	6 710.7	2 020.2	8 730.9	23.1	—	—	—	—
13	10 218.0	471.6	10 689.6	4.4	—	—	—	—
14	13 833.6	471.6	14 305.2	3.3	—	—	—	—
15	3 301.2	628.8	3 930.0	16.0	—	—	—	—
16	14 305.2	943.2	15 248.4	6.2	—	—	—	—
17	943.2	0	943.2	0	—	—	—	—
18	6 288.0	471.6	6 759.6	7.0	—	—	—	—
19	919.4	262.0	1 181.4	22.2	707.4	340.6	1 048.0	32.5

由该表计算出陆源和海洋空气细菌检出率分别为94.4%和66.7%。陆源和海洋空气真菌检出率分别为88.9%和100.0%。表明空气中陆源细菌和海洋真菌检出率分别高于海洋细菌和陆源真菌。空气中陆源细菌、真菌、总菌量及(F/T)%分别为8 781.8、3 014.5、11 796.3 CFU·m^{-3}及25.6。海洋细菌、真菌、总菌量及(F/T)%则分别为314.4、895.2、1 209.6 CFU·m^{-3}及74.0。表明青岛空气陆源微生物多于海洋性的。陆源细菌、真菌、总菌及(F/T)%分别是海洋性的27.9、3.4、9.8及0.35倍。陆源真菌占总菌量的份额少于海洋性的。陆源微生物中细菌是真菌的2.9倍。相反海洋真菌却是细菌的2.8倍。说明温湿海洋空气挟持真菌所占%多于较干燥陆源空气挟持真菌所占%。按文献[4]推荐，说明青岛空气处于中度污染。空气污染以细菌为主，真菌及其作用不可轻估。

本次所测空气微生物含量比1994年3～4月间低，却高于1986～1987年的水平[1~2]。比较海、陆两源，青岛空气微生物仍以陆源为主。陆源微生物含量比海洋性微生物高出不少，这意味着测时有沙尘暴的影响[3]。

陆源细菌、真菌极差分别为35 212.8和14 462.4CFU·m^{-3}，海洋细菌、真菌极差分别为786.0和2 358.0FU·m^{-3}，差距比1986～1987年大而小于1994年[1~2]。

本结果与其他城市比较，发现青岛空气微生物总量比广州、深圳、济南均低，但真菌及其F/T%均比它们高[5~7]。这反映各地天气与空气污染状况差距及青岛空气的潮湿程度。

3.2 青岛空气微生物在时间上差异

图1是青岛空气微生物含量指标昼夜变化状态。由该两图可知陆源微生物三指标的峰与谷值分别在下午和夜晚，海洋细菌与陆源细菌相反，显出晨峰及下午谷。海洋真菌与陆源真菌状态一致。陆源总微生物（即总菌，下同）谷值类似于海洋性的，但却与1986～1987年相悖。海洋性总微生物昼夜变化势相似于1986～1987年和1994年3～4月间[1~2]。结果表明青岛空气微生物污染以夜晚为轻。

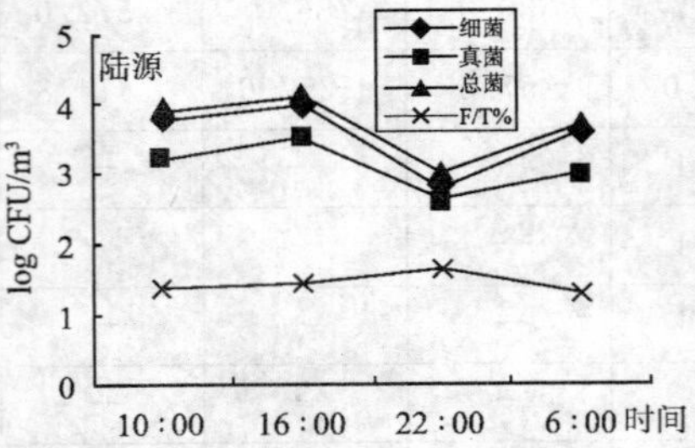

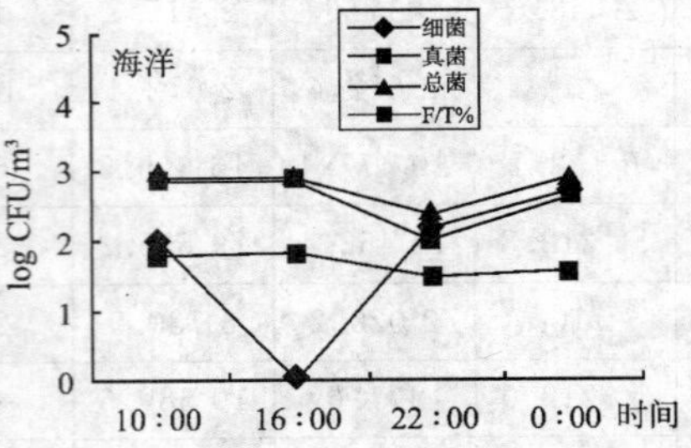

图1 青岛空气微生物的昼夜变化状态

3.3 青岛空气微生物的空间分布状态

将测点/时按5个方位分别得到各方位平均陆源微生物含量指标见表3。

表3说明城市各方位空气微生物含量间存在差异。按各指标大小、高低排序为细菌：东＞北＞西＞中＞南，真菌：东＞南＞西＞北＞中，总菌：东＞北＞西＞中＞南，(F/T)%：南＞东＞西＞中＞北。表明东部空气微生物最多，空气污染最重。这与东部新市民人口密集、有较多正在建筑、装修和排污等造成的颗粒物、气溶胶有关。南部空气微生物污染较轻，其(F/T)%最高。这与近海空气湿度较高及市井成熟相一致。总的看，东部和北部空气已处中度污染，南、中、西轻污染，南部最轻。

表3 青岛不同方位空气微生物含量比较

方位及测点	空气微生物含量指标			
	B	F	T	(F/T)%
东(No.1～4)	14 265.9	4 637.4	18 903.3	24.5
南(No.5～12,15,17)	2 400.5	1 910.0	4 210.5	44.3
西(No.16,19)	7 612.3	602.6	8 214.9	7.3
北(No.13～14)	1 2025.8	550.2	12 576.0	4.4
中(No.18)	6 288.0	471.6	6 759.6	7.0

*括号中数码为各方位所包含的测点，下同。

将各测点/时按城市功能区划归并，即风景游览①、工业②、商业③、居住④和高科园⑤后，各自平均陆源微生物指标排序为细菌：④＞⑤＞②＞③＞①。表明居住区空气污染最重，次为高科园区。风景游览区相对较轻，但(F/T)%最高，仍与较高湿度相关。

按距海远近划分两类测点，即滨海和内陆区各自平均陆源空气微生物指标比较，结果表明，滨海区空气陆源细菌比内陆区少，但真菌较多。两区总菌量不相上下。意味该两区空气微生物污染程度趋同，滨海区空气污染有加重之势。

仅就滨海区两类微生物比较，结果表明陆源微生物即使在沿海仍占据绝对高的优势。陆源细菌、真菌、总菌量分别是相应海洋性指标的 5.9、3.6、4.2 倍，但陆源 F/T 没有海洋性的高。反映海洋真菌及海洋对内陆空气调节作用不可低估。

4　结语

本结果表明青岛空气微生物检出率高，空气正处于微生物中度污染状态。海洋和陆源微生物昼夜变化均较平缓，但海洋细菌变幅稍大。测区空气微生物空间变化表明东部、居住区空气污染较重，南部、西部和风景游览区空气污染较轻。滨海区空气陆源细菌少于内陆区，真菌却较多。滨海与内陆区空气微生物含量相近，滨海区空气陆源微生物增多，意味两区空气污染有趋同现象。陆源污染有趋海势头，也许与沙尘暴有关。新开发东部、高科园区空气微生物含量居高不下，暗示其空气污染已到不容忽视程度。居住区可能存在城市污染错综复杂原因和影响，其后果影响千家万户，更应注意。监测和治理青岛空气及其微生物污染仍是现在面临的重要任务。

参考文献 7 篇(略)

DETERMINATION ON CONDITION OF AIRBORNE MICROBES ABOVE QINGDAO

(ABSTRACT)

Abstract In this paper, the airborne microbes above Qingdao were determined at 10 sites totally 19 times by means of gravity plate respectively during 2001 and 2002. The result obtained expounds the temporal and spatial distributions of terrigenous and marine microbes in air above Qingdao. Average terrigenous bacterial, fungus, total microbial contents and(F/T)% are 8,781. 8, 3,014. 5, 11,796. 3CFU · m^{-3} and 25. 6 respectively. Average marine bacterial, fungus, total microbial contents and (F/T)% are 314. 1, 895. 2, 1 209. 6CFU · m^{-3} and 74. 0 respectively. It indicates that the air above Qingdao is at the middie level of microbial pollution. There are a more serious terrigenous source pollution and stronger trend towards seashore area. The result also shows the regulatory function of sea to land air (atmosphere). The paper analyses the air-borne microbial difference among various functional areas of the city plan, and points out the necessity of strengthening monitoring and administering on air and its microbes of Qingdao.

Key words Qingdao; Airborne Microbes; Air Pollution; Sea, Ocean And Land

乌鲁木齐空气微生物含量的测定*

摘　要　用自然平皿沉降法对乌鲁木齐市13个室内外空气测点作了微生物含量的测定，结果表明平均室外空气的细菌、真菌、总菌含量及F/T%分别为28 583.8、780.5、29 364.3CFU·m^{-3}及2.7，证实了该市空气处于中度微生物污染状态。少量测点显示其空气已受到严重污染，空气质量不佳。室内空气微生物污染也重。空气微生物污染以细菌为主。分析了室外空气微生物含量的时空分布特点、差异，探讨了其成因，并与其他城市作了比较。文章指出增强环境忧患意识，采用切实而有效措施，减少环境污染的必要性。

关键词　乌鲁木齐；空气微生物；空气污染；空气质量

1　引言

乌鲁木齐市位于天山北麓，东临博格达峰，南依天山林带。西北紧靠准噶尔盆地。它是世界上离海洋最远的城市，海拔680～920 m，具大陆性气候、较干燥。城区面积83 km^2，人口150万。它是新疆全区的政治、经济、文化交通的中心。同时也是污染较重的城市之一。

乌鲁木齐市的山谷地形这一地理环境，逆温层造成强起尘因子等气候及复杂的人文条件决定了它的环境，包括空气状况。空气微生物正是受制于它们而生存发展起来。

空气微生物包括对人体、动植物有害的病毒、细菌、支原体和真菌或其孢子，能引起感染、病患等传染病。空气微生物关系到医学、生态、环境等方面。空气污染包括微生物污染对东北亚城市人体健康和生产力造成了多达200亿美元/a的损失[1]，因而研究空气微生物状况显得非常重要和紧迫。当前，空气微生物学的发展取得了长足的进步。研究乌鲁木齐市（简称乌市，下同）的空气污染状况已做了不少工作，但关于空气微生物研究迄今为止仍十分欠缺[2]。本文初步报道该市空气微生物含量状况，以期为进一步开展相应工作提供资料。

2　乌市空气微生物测点状况描述

乌市空气微生物含量状况的测定于2002年8月间进行，选择的测点计12个室外和1个室内。共进行了13次测定，这些测点分布于全市的四个方位，基本可覆盖全市（见表1）。

*　原文刊于《新疆环境保护》2003，25(1)：17～20.

表 1　乌鲁木齐空气微生物测点状况

Fab. 1　The condition of Urumqi airborne microbes

编号	测点名称	环境状况	气温(℃)	采样起始时间
1	乌鲁木齐火车站南端附近商店外	有人来往	26	2002.8.17 22:40
2	红山公园门口内	无人,静	20	8.23 06:45
3	红山公园山顶	海拔 910 m,有晨练者、有烟、雾薄	18	07:00
4	青年路日月星光花园马路	行人少,有出租车来往	18	08:35
5	西山塑料厂住宅院内	近有山丘,城乡接壤,西南风 3 级	21	09:45
6	北京北路过街桥下段	2 级风,偶有雨丝,树旁	26	12:30
	儿童公园内	3 级风	28	13:00
8	新疆大学院内楼荫中	3～4 级风,少人	25	14:15
9	二道桥市场门口	人来往购物,休闲	29	15:40
10	红柳泉	城乡接壤部,飞尘多	27	17:45
11	乌鲁木齐火车南站附近马路边	人来车往	25	18:30
12	乌鲁木齐火车南站客运站台	火车旅客多	24	20:20

注:采样持续时间:5 min,采用北京时间,乌鲁木齐地方时比北京时间约晚 2h。

3　材料和方法

采用自然平皿沉降法。将备有消毒好的营养琼脂和沙氏真菌培养基的平皿分别曝气 5 min,收集距地高 1.5 m 处的空气中的微生物即空气细菌(B)和真菌(F)的菌落(Colony),经至少 2d 的培养(28℃)后,计数各皿中的 CFU、细菌和真菌的 CFU 为总菌含量(T)。换算出单位空气中的 CFU · m^{-3} 和 F/T%。参见有关文献,评价空气微生物污染状况和空气质量[3]。

4　结果和讨论

4.1　乌市室外空气微生物含量的一般状况

表 2 列出了乌市各测点的空气微生物含量状况。

表 2　乌鲁木齐空气微生物测定状况

Tab. 2　The condition of Urumqi airborne microbes

测点编号	空气微生物含量(CFU · m^{-3})			
	B	F	T	F/T(%)
1	10 689.6	0	10 689.6	0
2	19 488.0	406.0	19 894.0	2.0
3	18 549.6	786.0	19 335.6	4.1
4	44 173.2	943.2	45 116.4	2.1
5	13 519.2	1257.6	14 776.8	8.5

（续表）

测点编号	空气微生物含量(CFU・m^{-3})			
	B	F	T	F/T(%)
6	6 759.6	471.6	7 231.2	6.5
7	9 589.2	1 572.0	11 161.2	14.1
8	21 536.4	943.2	22 479.6	4.2
9	57 692.4	1 572.0	59 264.4	2.7
10	87 403.2	314.4	87 717.6	0.4
11	33 955.2	943.2	34 898.4	2.7
12	19 650.0	157.2	19 807.2	0.8
13	13 204.8	0	13 204.8	0

注：No.1～12为室外，No.13为室内。

由该表可见，乌市室外空气中细菌、真菌、总菌及F/T%的各平均值分别为28 583.8、780.5、29 364.3CFU・m^{-3}及2.7。据相关文献推荐[4]。此结果表明其细菌、真菌和总菌含量均意味着空气已处于中度微生物污染状态，说明其空气质量欠佳。真菌含量及在总菌中所占比率与一些城市比不算高[5]，这证实了乌市所处气候较干旱的特点。

4.2　乌市室外空气微生物含量的测点差异

空气细菌含量的变幅为80 643.6CFU・m^{-3}(87 403.2-6 759.6)，最大细菌含量是最小量的12.9倍。空气真菌含量的变幅为1 572.0CFU・m^{-3}(1 572.0-0)，检出率为91.2%。F/T%的变幅为14.1。空气总菌含量的变幅为80 486.4CFU・m^{-3}(87 403.2-6 759.6)，最大含菌量是最小量的12.1倍。以上结果反映出测点即地区间空气质量的差异。

据表2数据，可以得出乌市空气细菌含量测点间的大小排序为No.10＞9＞4＞11＞8＞12＞2＞3＞5＞1＞7＞No.6。空气真菌含量测点间的大小排序为：No.9＝7＞5＞4＝8＝11＞3＞6＞2＞10＞12＞1＝0。空气总菌含量测点间的大小排序为：No.10＞9＞4＞11＞8＞2＞12＞3＞5＞7＞1＞6。F/T%测点间大小排序为：No.7＞5＞6＞8＞3＞9＝11＞4＞2＞12＞10＞No.1。

这表明乌市空气微生物以细菌为主，细菌含量主要地决定了微生物含量，也反映了空气的较干燥性。

若把测点按城市不同功能区划归类，可分作四类(见表3)。

表 3 乌鲁木齐市城市功能区划间空气微生物含量比较

Tab. 3 Comparison of airborne microbial content in Urumqi city function districts

区划编号	空气微生物含量(CFU · m^{-3})			
	B	F	T	F/T(%)
1(No. 6、7)	8 174.4	1 021.8	9 196.2	11.1
2(No. 1、2、3、5、8、12)	17 238.8	591.7	17 830.5	3.3
3(No. 4、11)	39 064.2	943.2	40 007.4	2.4
4(No. 9、10)	72 547.8	943.2	73 491.0	1.3

注:No. 1 区划为人烟较稀处,灰尘少。No. 2 区划为一般市井场所。No. 3 区划为马路旁,飞尘流动较大。No. 4 区划为城乡结合部市场,气温较高,飞尘多。

由表 3 可得出,不同区划间空气微生物含量状况的差异是:空气细菌含量—区划 No. 4>3>2>No. 1。空气真菌含量—区划 No. 1>3=4>No. 2。空气总菌含量—No. 4>3>2>No. 1。F/T%—No. 1>2>3>No. 4。仍然表达出乌市空气微生物含量在功能区划间的差异由细菌为主因子引起。真菌对总菌量只起干扰作用。结果还意味着各城市功能区域环境各因素对各自的空气微生物含量构成发挥着重要作用。

4.3 乌市室外空气微生物含量的昼夜变化

表 4 列出的是乌市室外空气微生物含量的昼夜变化差异。由该表可知,乌市室外空气微生物总含量及细菌含量的昼夜变化差异大小排序均为:下午>上午>清晨>夜晚。这说明空气总菌含量昼夜变化势态与细菌的吻合,而 F 与 F/T%则与此不大一致,即 F 为上午>下午>清晨>夜晚;F/T%则为上午>清晨>下午>夜晚;但四者均显出夜晚是各类微生物含量的最低时段。这是由夜晚的低气温、低人群活动量、低湿度及低空气污染所致。乌市空气微生物含量的昼夜变化势态与广州、泰安的不一致。显出其自身的特点[6,7]。

4.4 乌市室内空气微生物含量

由表 2 可看到作为对照的室内空气微生物含量状况,表明室内空气也处于中度微生物污染水平,但未见空气真菌菌落。说明室内空气相当干燥。意味着该市室内真菌污染不重,即细菌主宰了室内空气微生物质量和数量。室内空气污染大体上轻于大部分室外空间,但大大超过了 GB9663-88 规定的旅店细菌总数(≤4 003 个 · m^{-3})。

表 4 乌鲁木齐市空气微生物含量的昼夜变化差异

Tab. 4 The disparity of airborne microbial content day and night in Urumqi

时段	空气微生物含量(CFU · m^{-3})			
	B	F	T	F/T(%)
夜间	10 689.6	0	10 689.6	0
清晨	19 018.8	596.0	19 614.8	3.0
上午	19 115.5	1 037.5	20 153.0	5.1
下午	49 675.2	746.7	50 421.9	1.5

5　结语

本文初步测定了乌市空气微生物，其室外平均空气微生物含量为 29 364.3CFU·m^{-3}。室内则为 13 204.8CFU·m^{-3}。可以认为乌市大部分空气已处于中度微生物污染水平。少量测点表达出空气已受重污染，即空气质量不佳，室内污染一般轻于室外。空气微生物以细菌为主。室外空气真菌只占总菌含量的 2.7%，证实了空气相当干燥。

空气微生物含量呈现出有别于其他城市的时空分布态势。它反映了乌市特定的地理、生态、环境和其他特征。

本结果启示人们在发展经济、改善物质生活的同时，更要有强烈的环境忧患意识。规划和实施市政建设时要尽可能考虑自身的特点，尽可能避免野蛮作业，减少污染作业以不破坏环境，使之可良性持续发展。

参考文献 7 篇(略)

DETERMINATION ON AIRBORNE MICROBIAL CONTENT IN URUMQI CITY*

Abstract The outdoor/indoor airborne microbial counts indexes over 13 sampling sites were determined using the method of gravity plate in Urumqi city. The result obtained indicates that the average airborne bacterial(B), fungus(F), total microbial counts (T)and F/T% were 28,583.8, 780.5, 29,364.3 CFU · m^{-3} and 2.7 respectively. It was verified that the air of Urumqi had been in all unfavorable situation of middle polluted with microbes, some air environments were in state of serious pollution. The air quality is not good enough. The air microbial pollution was caused mainly by bacteria. The air microbial pollution was also more serious indoors. The spatiotemporal distribution, its character and difference were compared and analysed for outdoor air-borne microbial content. And the causes of formation were probed. The situation was compared to other cities determined. The paper points out that the necessity of strengthening environmental consciousness of suffering and using effective measures to decease environmental pollution.

Key words Urumqi; Air-Borne Microbes; Air Pollution; Air Quality

* Environmental Protection of Xinjiang 2003, 25(1): 17～20

襄樊、荆门、宜昌空气微生物含量状况*

摘　要　于2002年3月间，对位于湖北省的襄樊、荆门和宜昌三市用平皿沉降法测定了各自的空气微生物含量状况。结果表明襄樊空气微生物已处于中度污染状态，且居三市之首。宜昌空气微生物含量最低，空气质量甚佳。空气微生物一般以细菌为主，但宜昌空气中真菌量百分数最高。论述了三市空气微生物含量的昼夜变化。分析了三市空气微生物含量状况及其成因。与其他内陆、沿海城市的空气微生物污染状况作了比较。

关键词　襄樊　荆门　宜昌　空气微生物　空气污染

空气是微生物赖以生存的主要条件之一，其含量的高低和组成关系到生活质量。空气微生物是空气中主要的生命物质，它们借助其中的媒体生繁、悬浮、散布。测定空气微生物的主要成员，即细菌和真菌的状况对估计空气环境的质量是最基本的，掌握好相关的信息才能为改善环境—生态状况做到心中有数。各国学者对此进行了相关研究[1-3]。

湖北位于我国中部，气候类型属北亚热带季风区，空气温湿。人口密集，中小城市多，其中包括襄樊、荆门和宜昌。农业发达，工业生产居中。这些对环境造成了一定的污染压力。已有一些湖北空气环境状况的报告，但就襄樊、荆门、宜昌三市的空气微生物具体状况的调查研究尚未见到。本文对该三市的空气微生物含量状况作一研究，以引起人们对中小城市的空气环境质量及其改善的更多关注。

1　襄樊、荆门、宜昌环境状况概述

表1　襄樊、荆门、宜昌的环境、市政概况

城市名	所属气候区	自然环境及测时天气	人文、经济概况
襄樊	北亚热带季风	鄂西北，濒临汉水。晴间阴雨，气温7～23℃，软风～微风，空气相对湿度60%	交通重镇，经济中心，轻纺、汽车工业发达；73万人口；历史文化名城
荆门	北亚热带季风	鄂中，多种地形，竹皮河、刘家河、漳河水库四干渠相通并穿南至北。阴间小雨，小风，气温12～18℃	交通要隘；炼油、化工、电子、食品加工、建材工业；人口28万
宜昌	北亚热带季风	鄂西，濒临长江、黄柏河。长江三峡东口，川鄂咽喉。软风～微风，气温12～20℃，相对湿度65%	鄂西政治经济文化中心；航运中转站；水电工业为主，有钢铁、机械、化轻工、建材工业；人口39万

表1列出了三市环境及市政概况，三市均为中等城市，其中以襄樊人口最多，是湖北中西部的政治、经济中心。三市地理位置、地形和经济状况有所差异，对各市的空气微生

* 原文刊于《卫生软科学》，2003，17(6B)：149-152.

物必将产生不同影响。

2　襄樊、荆门、宜昌空气微生物采样点环境状况

表2列出了三市空气微生物采样点具体的环境状况及采样时间。采样时间基本上代表了一昼夜的上、中、下午及夜间各个时间段，由此可见空气微生物含量变化状态之一斑。

表 2-1　襄樊空气微生物采样点概况

编号	名称	环境状况	采样起始时间
1	长虹大桥南桥头堡	微风，有薄雾，车来往，人不多	2002.3.11,12:59
2	襄阳公园水上曲桥	轻风，游人少，清净	2002.3.11,12:30
3	古隆中腾龙阁，海拔约 350 m	微风，有雾，气温 16℃，远方电厂可能有漂尘来	2002.3.11,15:40
4	古隆中门口内	轻风，水塘边高树下，树叶少而小	2002.3.11,16:15
5	诸葛亮文化广场，铜塑像围墙上	软风，有些游人	2002.3.11,20:00
6	火车站附近离地面约 6 m	软风，9℃，无人	2002.3.12,04:40
7	襄阳神龙汽车公司一住宅区（东城一区）	轻风，12℃，轻雾	2002.3.12,09:00
8	同 6	软风，有雾，游人不多	2002.3.12,10:35
9	火车站广场	微风—和风，风向不定	2002.3.12,12:40

*曝皿采样　时间持续 5 min，下同。

表 2-2　荆门空气微生物采样点概况

编号	名称	环境状况	采样起始时间
1	龙泉公园	假山上，树下，阴天，软风，18℃	2002.3.12,16:55
2	岚光阁	软风，离山脚约 300 m 高	2002.3.12,17:15
3	掇刀区白石坡	软风，阴天，马路边	2002.3.12,18:15
4	车站路二楼上	软风，无人	2002.3.12,21:00
5	车站路二楼上	无风，小雨，无雨落入皿内，无人，12℃	2002.3.13,02:00
6	车站路二楼上	软风，雨后，无人，12℃	2002.3.13,06:00
7	市政府门口	小雨，软风，雨入皿少，车、人少，13℃	2002.3.13,08:30
8	火车站广场龙泉雕塑台基上	软风，阴天，有雾及明显水蒸气	2002.3.13,12:00

表 2-3 宜昌空气微生物采样点概况

编号	名称	环境状况	采样起始时间
1	火车站附近昭君旅馆三楼阳台	晴,和风,有湿感	2002.3.13,16:50
2	火车站附近昭君旅馆三楼阳台	晴,和风,有湿感	2002.3.13,21:00
3	火车站附近昭君旅馆三楼阳台	雨后,微风	2002.3.14,03:20
4	火车站附近昭君旅馆三楼阳台	阴,微风,有湿感,下面有不少人	2002.3.14,08:00
5	葛洲坝船闸上,离水面约 20 m 高	软风,阴,雾气,游人少	2002.3.14,09:50
6	夷陵长江大桥上	微风,晴,雾气,游人少	2002.3.14,11:00
7	三游洞蹦极挑战极限门口	软风,云层透光,无人	2002.3.14,12:20
8	夷陵广场	晴,有花,游人多	2002.3.14,14:45
9	胜利四路边二楼	软风,晴	2002.3.14,15:15

3 材料及方法

选取代表性站点(表 2)用自然沉降法将琼脂平板曝皿后培养,获取各点的空气微生物菌落(colony forming units,缩写为 CFU,下同)。以普通营养琼脂测定细菌、沙氏培养基测真菌。曝皿时间:5 min,方法详见参考文献[4]。整理所得单位空气体积(m^3)中的 CFU,分析空气细菌、真菌、总菌(细菌+真菌)含量、真菌占总菌量的百分比及它们与相关因子间的关系,估计和评价空气环境质量。

4 结果和讨论

表 3 列出了三市所有测点/测时空气微生物含量的上述四个指标。

表 3-1 襄樊空气微生物含量状况

编号	空气细菌	空气真菌	总空气微生物	空气真菌/总空气微生物(%)
1	9 274.8	1 257.6	10 532.4	11.9
2	6 130.8	2 200.8	8 331.6	26.4
3	2 029.6	1 257.6	3 287.2	38.3
4	2 515.2	628.8	3 144.0	20.0
5	7 860.0	786.0	8 646.0	9.1
6	50 618.4	628.8	51 247.2	1.2
7	11 475.6	471.6	11 947.2	3.9
8	8 017.2	628.8	8 646.0	7.3
9	5 030.4	1 572.0	6 602.4	23.8

表 3-2 荆门空气微生物含量状况

编号	空气细菌	空气真菌	总空气微生物	空气真菌/总空气微生物(%)
1	13 519.2	471.6	13 990.8	3.4
2	4 401.6	3 458.4	7 860.0	44.0
3	19 807.2	786.0	20 593.2	3.8
4	8 646.0	314.4	8 960.4	3.5
5	314.4	471.6	786.0	60.0
6	157.2	157.2	314.4	50.0
7	12 890.4	314.4	13 204.8	2.4
8	786.0	943.2	1 729.2	54.5

表 3-3 宜昌空气微生物含量状况

编号	空气细菌	空气真菌	总空气微生物	空气真菌/总空气微生物(%)
1	10 846.8	471.6	11 318.4	4.2
2	4 558.8	157.2	4 716.0	3.3
3	378.8	0	378.8	0
4	1 515.2	0	1 515.2	0
5	1 136.4	378.8	1 515.2	25.0
6	757.6	378.8	1 136.4	33.3
7	757.6	1 894.0	2 651.6	71.4
8	6 060.8	378.8	6 439.6	5.9
9	3 788.0	1 515.2	5 303.2	28.6

4.1 襄樊空气微生物含量分析

据表 3-1 可知，襄樊市平均的空气细菌、真菌及总菌量分别为 11 439.1、1 048.0、12 487.1CFU/m^3，真菌占总菌量平均为 8.2%，按照文献[4]推荐的评价标准，即某地空气细菌菌落量达到 5 000～50 000 个/m^3 时的空气为中度污染，因此襄樊平均空气细菌含量指示它的空气污染处于中度污染状态。细菌含量是真菌的 10.9 倍，表明襄樊空气污染以细菌为主，真菌则加重了污染强度。

襄樊各测点细菌含量大小排序为 No. 6＞7＞1＞8＞5＞2＞9＞4＞No. 3，极差为 48 588.8，最大是最小的 24.9 倍。真菌含量大小排序为 No. 2＞9＞1＝3＞5＞4＝6＝8＞No. 7，极差为 1 729.2，最大是最小的 4.7 倍。总菌量大小排序为 No. 6＞7＞1＞5＝8＞2＞9＞3＝No. 4，极差为 48 103.2，最大是最小的 16.3 倍。真菌占总菌量的%大小排序为

No. 3＞2＞9＞4＞1 ＞5＞8＞7＞No. 6，极差为 37.1，最大是最小的 31.9 倍。

细菌量之极差反映了测点间空气状况的差异，这是由测点空气环境状态的差异所致，如 No. 6 测时尽管是凌晨时段，而且离地面较高(6 m)，但因它处于闹市人口稠密区，致细菌含量达峰值，最低细菌含量之处在郊外风景区的高山上(No. 3)，这儿人烟稀少，空气洁净度高，细菌量大减，这表明空气细菌与人口集中程度、工商业交通繁华状态相关。海拔的提高在一定条件下可降低细菌含量。No. 6 与 No. 3 的比较反映了空气中污染物浓度、颗粒物粒径随地势升高而变小的态势。

最大空气真菌量出现于 No. 2，此处正是水体上。最低真菌量出现于 No. 7。上午 9 点左右，太阳正处于上升期，光线对真菌的消杀作用强[3]，真菌占总菌百分比偏低。最大总菌量之处即是最大细菌量之处，反之亦然。表明襄樊空气微生物污染主要由细菌引起。空气相对湿度高促进了真菌的生长繁殖。真菌加重了空气污染程度。

4.2　荆门空气微生物含量分析

据表 3.2 可得出荆门市空气中平均的细菌、真菌及总菌量分别为7 565.3、864.6及8 429.9CFU/m³，真菌占总菌量的百分比为 10.3，结果意味着荆门空气处于轻微中度污染状态，空气质量尚佳。

各测点按细菌、真菌、总菌含量大小排序分别为：No. 3＞1＞7＞4＞2＞8＞5＞No. 6，No. 2＞ 8＞3＞1＝5＞4＝7＞No. 6，No. 3＞1＞7＞4＞2＞8＞5＞No. 6；真菌%的大小排序为 No. 5＞8＞6＞ 2＞3＞4＞1＞No. 7，各点细菌、真菌、总菌及真菌百分比的极差分别为 19 650.0、330 162、20 278.8CFU/m³，及 57.6，最大与最小细菌量、真菌量、总菌量及真菌百分比之比值分别为 126.0、22.0、65.5 及 25.0。

出现最大细菌量的 No. 3 测点位于马路旁。车辆、行人来往，使污染物沉浮机会多，又值傍晚时分，使细菌大增。最小细菌量出现于 No. 6 测点次。正值雨后清晨，空气中污物少，也恰是无车无人来往之际，空气细菌大减。总菌量测点的差异同细菌，表明荆门空气微生物污染主要由细菌引起。最大真菌量测点是 No. 2。此处虽系荆门制高点，但树木葱茏，湿度大，助长了它们的生繁，位于市区的 No. 6 测点(次)，雨后空气，污染物颗粒少，空气流动很轻，即使细菌大减，又使真菌量降至最少。

4.3　宜昌空气微生物含量分析

根据表 3 可计算出宜昌市空气细菌、真菌、总菌量及真菌%各自的平均值分别为 3 311.1、574.9、4 106.8CFU/m³ 及 14.8，此结果说明荆门空气微生物污染很轻，空气质量甚佳。

各测点(次)空气细菌、真菌、总菌含量及真菌百分比大小排序分别为：No. 1＞8＞2＞9＞4＞5＞6＝7＞No. 3，No. 7＞9＞1＞5＝6＝8＞2＞3＝No. 4＝0，No. 1＞8＞9＞2＞7＞4＝5＞6＞ No. 3 及 No. 7＞6＞9＞5＞8＞1＞2＞3＝No. 4＝0，它们的极差及最大/最小值则分别为：10 468.0、1 894.0、10 936.6、7.14 及 28.6、29.9。

出现最大空气细菌含量的测点为 No. 1。此处正是闹市区，下午的中段空气晴和温湿，有利于细菌滋生。虽为同处(即 No. 3)，但因测时差异却大大减少了细菌量。因为此时正值雨后，空气中悬浮物必然大为降低，细菌依附物少了，菌量则降。同理此时空气真菌量也降为最低。而出现最大真菌量的测点是 No. 7，这是由湿润天气、浓密树丛及多云的中午时分三者综合造成的。

4.4 三市空气微生物含量的昼夜变化

经统计襄樊和荆门的空气真菌量均以上午为低谷，襄樊和宜昌的空气真菌量均在中午达到高峰，荆门空气真菌量的高峰却显露在下午，而宜昌空气真菌量的低谷出现于夜间。三市空气真菌量昼夜变化有差异，且与各市空气细菌量昼夜变化不一致。

5 结语

比较三市空气微生物含量状况，可见襄樊空气细菌、真菌、总菌含量居三市之首，以宜昌为最少。这说明襄樊空气微生物污染最重。真菌百分比却是宜昌最高，襄樊最低。测时遇雨的荆门和濒临长江的宜昌空气因湿度较大致真菌百分比提高。与同为内陆城市的济南比，襄樊的各项指标均低。荆门与曲阜相比，各项指标也均低。宜昌与泰安比，细菌及总菌量也较低，但真菌量和真菌百分比均高于泰安[5]。与沿海城市比，襄樊空气的细菌含量低于青岛[3]，荆门空气细菌和总菌量均比北海的多，而真菌量及真菌百分比均比北海的少[6]。宜昌与海口相比，除空气细菌量较后者少外，其余各项均显出比海口的高[7]。这说明襄樊等三市空气质量一般好于内陆的济南等三市，它反映了各地工农、商贸、交通等经济状况和所处地理环境位置上的差异，它们一般均差于沿海的海口等三市，这表明海洋空气比陆源空气较洁净。本结果在一定程度上也表达了这三市空气质量及其对所在地域的影响。

参考文献 7 篇(略)

AIRBORNE MICROBIAL CONTENT SITUATION ABOVE THREE CITIES OF XIANGFAN, JINGMEN AND YICHANG

Abstract The air-borne microbial content situation above three cities of Xiangfan, Jingmen and Yichang was determined by using gravity plate during last March, 2002. The result obtained shows that: the air-borne microbial content above Xiangfan indicates that its air condition was at the middle pollution and the most serious level among the three cities determined. The air-borne microbial content was lowest and its air quality was fine in Yichang. In general speaking, the bacteria was the primary content in the air-borne microbial determined. But the fungus content % in the total air-borne microbial content above Yichang city was highest. The situation of air-borne microbial diurnal variation was discussed. This paper compared the air-borne microbial pollution conditions among three cities and some other inland and coastal cities.

Key words Xiangfan, Jingmen, Yichang, Air-Borne Microbial. Air Pollution

南京空气微生物含量状况的研究*

摘　要　主要于2002年5月(及1995年7月)间用自然沉降法22次测定了南京13个测点的空气微生物含量状况。论述其空气、真菌、总菌量(CFU·m^{-3})及真菌占总菌量(F/T)%等指标及其时空分布状态和变化。结果表明,南京平均的空气细菌、真菌、总菌量及F/T%分别为11 969.3、2 323.7、14 293.0CFU·m^{-3}及16.3。说明其空气处中度污染状态,且以细菌为主。真菌加重空气污染。讨论了空气微生物昼夜变化趋势和不同方位间、不同功能区划间、不同取样高度间的差异。南京空气中较重的细菌污染出现于早晨时段,下午则轻。较重的真菌污染发生于下午,中午相对较轻。5个方位中的空气细菌污染状况是西部最重,中部最轻。真菌污染中部重南部轻。在5个不同的城市功能区划中,景区空气细菌、真菌、总菌污染最轻,闹市区空气细菌、总菌污染最重,居民区空气真菌污染最重。就采样点高度言,离地面越高,空气微生物越少。分析了空气微生物状态这些差异的原因。指出其空气质量有所改善,尚需因地制宜采取更有力措施,不断提高。

关键词　南京　空气微生物　空气污染　空气质量

南京位于长江下游,北接江淮平原,东南临长江三角洲,周边丘陵起伏,山水环绕。工商、交通、旅游、科教文发达。气候为亚热带和暖温带过渡区。测时正值梅雨季节,空气温湿。此一时空存在着环境污染问题,因此环境监测和治理十分重要。

空气中的微生物指示环境的污染状况,是环境,尤其是空气质量的指标,因而空气微生物状况是掌握空气污染、保障人们健康十分必要的基础资料[1]。曾对南京的空气微生物做过研究[2],但是空气微生物处于流动和变化之中,又与环境中气溶胶、颗粒物密切相关,它们会产生差异。本文主要就2002年5月间对南京空气微生物状况的调研结果作一报道。

1. 测点概况

本空气微生物调研的测点计13个,它们分布于南京市的东、南、西、北、中五个方位。测点的名称、测时及环境与天气状况详见表1。

表1　南京空气微生物测点环境概况表

测点/时编号(No.)	测点名称	环境概况	测时
1	雨花台烈士纪念碑前	台阶,有游人,阴、雾,20℃,软风	2002.5.18.10:15-1020
2	南京大屠杀纪念墓地广场	多云间雾,轻风(东北向),人少	2002-5.18.11:25-11:30
3	中山陵园入口处	阴软风,游人多,树下	2002.5.18.14:00-14:05
4	金陵饭店门口花园	阴间晴,闹中取静,少人	2002-5.18.15:10-15:15

* 原文载于《中华名人文选大全Ⅰ》,2004,1393-1395.

(续表)

测点/时编号(No.)	测点名称	环境概况	测时
5	南京大学西南大楼门口	阴,软风,有几人来往	2002.5.18.17:40-17:45
6	长江二桥公园	多云间直射光,轻风	2002.5.19.10:10-10:15
7	古鸡鸣寺牌坊上	晴,微风,26℃	2002.5.19.13:30-13:35
8	阅江楼上	晴,清劲风,游人不多	2002.5.19.14:50-14:55
9	南京大学西南大楼门口	有些人来往	2002.5.19.17:15-17:20
10	阳光广场16层楼上	多云	2002.5.19.22:20-22:25
11	阳光广场16层楼上	小雨,阴,强风	2002.5.21.06:00-06:05
12	阳光广场16层楼上	小雨,阴,强风	2002.5.21.09:00-09:05
13	阳光广场16层楼上	小雨,阴,强风,22℃	2002.5.21.12:00-12:05
14	阳光广场16层楼上	毛雨,阴,强风和风	2002.5.21.14:30-14:35
15	阳光广场	小雨,清劲风	2002.5.21.22:05-22:20
16	阳光广场16层楼上	毛雨	2002.5.21.22:20-22:28
17	阳光广场16层楼上	阴,雾,微风	2002.5.22.02:25-02:30
18	阳光广场16层楼上	毛雨	2002.5.22.05:40-05:45
19	阳光广场	小雨,清劲风	2002.5.22.05:55-06:00
20	南京汽车总站停车场	微雨,微风	2005.5.22.09:55-09:60
21	南京火车站广场	小雾和毛雨,轻风,人车来往不多,29℃	1995.7.7.14:10-14:15
22	南京火车站候车室内	人多,30℃左右,有些人走动	1995.7.7.15:10-15:15

2. 材料和方法

暴露营养琼脂培养基于所设测点空气中[1],室温培养48h和72h以上后统计其上形成的菌落数(CFU),换算成每立方米空气中的CFU(CFU·m^{-3}),包括细菌(B)、真菌(F),总菌T(B+F)粒子沉降量,计算出真菌占总菌量的%(F/T%)。采样高度分别设定为离地面1.5 m、35 m和120 m。以上述数据分析、比较空气微生物污染状态时空分布,评价空气质量。方法详见文献[3]。

3. 结果和讨论

3.1　南京空气微生物含量概况

表2列出了所有测点/时的空气微生物含量、数据。由表2可计算出空气细菌的检出率为100%,而真菌为95.2%。说明测时该区空气中均有细菌,真菌略少。

表 2　南京空气微生物含量状况表 *

测点编号	空气微生物含量指标			
	细菌(B)	真菌(F)	总菌(T)	F/T%
1	3 615.6	314.4	3 930.0	80.0
2	15 091.2	1 414.8	16 506.0	8.6
3	2 986.8	1 100.4	4 087.2	26.9
4	3 458.4	1 100.4	4 558.8	24.1
5	4 244.4	2 515.2	6 759.6	37.2
6	3 930.0	786.0	4 716.0	16.7
7	5 816.4	471.6	6 288.0	7.5
8	3 144.0	1 729.2	4 873.2	35.5
9	7 860.0	943.2	8 803.2	10.7
10	7 547.2	5 660.4	13 207.6	42.9
11	25 000.1	0	25 000.1	0
12	3 301.9	3 773.6	7 075.5	53.5
13	19 811.4	471.7	20 283.1	2.3
14	1 415.1	471.7	1 886.8	25.0
15	29 245.4	5 188.7	34 434.1	15.1
16	4 717.0	2 830.2	7 547.2	37.5
17	13 679.3	943.4	14 622.7	6.5
18	14 622.7	1 415.1	16 037.8	8.8
19	31 132.2	11 792.5	42 924.7	27.5
20	21 812.4	5 717.0	27 529.4	20.8
21	28 924.8	157.2	29 082.0	0.5
22	27 352.8	628.8	27981.6	2.2

* 各微生物含量单位：CFU · m^{-3}，下同

经计算平均的空气细菌、真菌含量分别为 11 969.3 和 2 323.7CFU · m^{-3}，细菌约是真菌的 5.2 倍。总菌量为 14 293.0CFU · m^{-3}，F/T%为 16.3。说明空气中细菌比真菌多得多。按照文献[3]的推荐，细菌含量处于 5 000-50 000CFU · m^{-3} 的空气为中度污染，则本结果说明南京此间的空气处于以细菌为主的中度污染之中，而真菌加重了空气污染程度。个别测点(如 No. 19)的空气污染较重。表 2 说明南京空气微生物比 7 年前有所改善，与 4 年前比也有了变化[2]。夏季污染重于春季。室内污染轻于室外，但真菌量大于室外，这可能与当时室内候车者人气较旺有关。

空气细菌、真菌和总菌量及 F/T%的极差分别为 27 830.3、11 792.5、41 037.9CFU ·

m^{-3}及 80.0。

与济南、兰州、广州比，南京空气细菌和总菌量均较低，说明南京空气质量好于它们。但真菌量及 F/T%均较高[4-6]。

3.2 南京空气微生物昼夜变化势态

将所有测时按早晨、上午、中午、下午和夜间 5 个时段划分，进行空气微生物含量指标统计分析后发现细菌量的高峰和低谷分别出现于早晨和下午。而下午正是真菌及其 F/T%的峰值，这是由其中午的谷值后上升来的。总菌量状况类似于细菌。距地面上 35 m 处的空气细菌、总菌量状况类似于此。只是真菌及其 F/T%峰提前至上午，谷值于中午即至。

结果表明南京空气较重的细菌污染出现在早晨时段，下午反而较轻。较重的真菌污染发生在下午，中午时段真菌污染较轻。总体而言，此间南京空气微生物污染主要由细菌引起，这可能与测定时一日的气温、湿度变化状况相关。南京空气细菌、总菌量的峰值与济南、广州类似。真菌峰值仅类似于广州[4]。

3.3 南京空气微生物的空间变化势态

将测点/测时按南京东、南、西、北、中 5 个方位划分，即 No. 3 为东部，No. 1 为南部，No. 2 为西部，No. 6、8、20、21 为北部，No. 4-5、7、9-19 为中部，得出各方位平均空气微生物含量指标于表 3 之中。

表 3 南京各方位微生物含量的比较

方位	空气微生物含量指标			
	B	F	T	F/T%
东	3 458.4	1 100.4	4 558.8	24.1
南	3 615.6	314.4	3 930.0	80.0
西	1 591.2	1 414.8	16 506.0	8.6
北	14 452.6	2 097.4	16 550.0	12.7
中	14 321.0	2 830.2	17 151.2	16.5

由该表可见，空气细菌、真菌、总菌量及 F/T%在各方位上的大小排序分别为西>北>中>南>东，中>北>西>东>南，中>北>西>东>南及南>北>东>中>西。这说明空气细菌污染以西部最重，东部最轻，中部真菌及总菌污染最重，南部最轻，但南部空气真菌量所占份额量大，西部最小。

将各测点按风景游览（No3、6、8）、科教文（No. 5、9）、闹市（No. 4、20-21）和居民（No. 10-19）等四个城市功能区划来分，以比较各区平均的空气微生物含量指标间的差距，结果如表 4 所示。

表4　南京不同功能区划空气微生物含量比较

编号	功能区划分	空气微生物含量指标			
		B	F	T	F/T%
1	风景游览区	5 764.0	969.4	6 733.4	14.4
2	学区	6 052.2	1 729.2	7 781.4	22.2
3	闹市区	18 065.2	2 324.9	20 390.1	11.4
4	居民区	1 547.2	3 251.7	18 301.9	17.8

表4说明景区空气细菌、真菌、总菌污染最轻，闹市区空气细菌、总菌污染最重，居民区空气真菌污染最重。进一步证实人口越密集、工商交通越繁忙，没有及时合适的措施，将使空气微生物污染越重，而人口相对稀疏，环境安稳的区域，空气质量相对易控制，也好。此点已在广州得到了印证[4]。

测点距地面的高低水平也使相应空气微生物存在差异，表5即反映此一状况。

表5　南京不同测点高度空气微生物含量的比较

距地表高度	空气微生物含量指标			
	B	F	T	F/T%
15 m(No. 15、19)	30 188.8	8 490.6	38 679.4	22.0
35 m(No. 10-14、16-18)	11 261.8	1 945.8	13 207.6	14.7
120 m(No. 8)	3 144.0	1 729.2	4 873.2	35.5

表5说明离地面越高，空气微生物越少。越近地面（1 m左右及以上）人群更密，大颗粒越往下沉，环境越污染，自然沉降的微生物粒子越多，空气质量越差。青岛的状况曾经证实过此点[7]。

4. 结语

本文对南京空气微生物状况的调研工作初步揭示出其空气处于以细菌为主的中度污染状态。比较分析了空气微生物的时空分布。表达出城市不同方位、不同功能区划和不同距地面高度间的差异及昼夜变化势态。讨论了空气微生物变化之原因。尽管南京空气污染状况比以前有所好转，但空气质量的提高尚待环境保护意识的进一步增强和因地制宜地采用更为有力的措施。

参考文献7篇（略）

RESEARCH ON CONDITION OF AIR-BORNE MICROBES ABOVE NANJING

(ABSTRACT)

Abstract　This paper mainly discusses the condition of air-borne microbes above Nanjing by using the method of gravity plate during May, 2002. The result obtained describes microbial indices and their temporal and spatial distribution are changes including air-borne bacterial (B), fungous (F), total microbial (T) contents (CFU · m^{-3}) and F/T% etc of Nanjing.

Average airborne bacterial, fungus, total microbial contents and F/T% are 11,969.3, 2 323.7, 14 293.0 CFU · m^{-3} and 16.3 respectively. It illustrates that the air above Nanjing is at the middle level of bacterial pollution. The air-borne microbes are consisted of bacteria mainly, but fungi aggravates the air pollution.

The paper discusses the diurnal variational trend and differences among variant directions city plans or heights of sampling for air-borne microbes, analyses the reasons of their conditions and shows that the air quality of Nanjing has been taken a turn for the better improved, but it is necessary to use stronger ways and means suit measures to local conditions of Nanjing to heighten its air quality further.

Key words　Nanjing, Air-Borne Microbes, Air Pollution, Air Quality.

青岛十大山头空气大肠菌含量分析*

青岛之所以成为著名风景游览胜地，除海滨外，其十大山头公园也是重要支柱。这十大山头包括观海山、观象山、信号山、鱼山、贮水山、浮山、太平山、嘉定山、东南山和楼山等等。

大肠菌群(Coliform bacteria)是包括肠道致病菌在内的主要肠道微生物类别，也是食品/食物污染的主要指标菌[1]。凭借食品的不洁制作和带菌人群的流动扩散，它们进入空气环境、滋生、藏匿和传播。有大量大肠菌的空气必然污浊，进而影响人们的健康。了解、比较和分析不同环境、不同时段的空气环境之大肠菌含量及其变化对大肠菌等的流行病学、环境科学就显得很有意义。对大肠菌的研究，一直引人注意[2,3]，这包括它们的研究方法、它们在食品及其生产环节中的监测等等。但它们在室外空气中的状况，相关研究甚少。本文就青岛十大山头空气中大肠菌的含量分布作一调研，以期从一个侧面来说明青岛空气环境和旅游环境的质量状况并加以改进之。

表1　青岛十大山头环境概况表

测点编号	名称	环境概况	备注
1	观象山	海拔79.0 m，绿色覆盖率92%	市西南部
2	观海山	海拔66.0 m，75%山体为乔灌木覆盖，距海岸线600.0 m	市西南
3	信号山	海拔98.0 m，市区制高点，为榉林等6山围裹，系“红楼暮霞”景点	市西南
4	浮山	主峰海拔384.0 m，石多，绿化尚可	崂山区
5	鱼山	海拔65.0 m，具“海”、“鱼”主题与特色	市西南部，临海
6	太平山	海拔131.5 m以上，有榉林公园，乔木为主要绿化种	市北，近中山公园
7	贮水山	海拔80.0 m，密集居民区	市北，靠中港较近
8	嘉定山	海拔110.4 m，黑松，刺槐多，居民区，离马路近	四方区
9	东南山	山势平缓成居民小区，沧口机场东侧	李沧区西南
10	楼山	海拔98.2 m，烟尘雾霭，位于工业区，但视野开阔	李沧区西部，离海较近

1. 青岛十大山头环境概况

表1列出了青岛十大山头的一般状况。它们所处地理位置不同，环境与生态状况也有差异。这十大山头多已在不同年代被开辟为山头公园，成为青岛的部分旅游景点。由

* 原文刊于《创新与发展》*Innovation and Development* 2005，17：10-12.

该表可见，它们分布于市区的五个主要行政区，地跨由北至南、由西至东的广阔范围内。山头最低海拔为66.0 m，最高海拔384.0 m。除东南山外，平均海拔高123.6 m。

2. 材料和方法

将伊红美蓝琼脂培养基制成平板，按常规用平皿自然沉降法在上述十大山头上采集空气样品。采样时间5′。同时记录现场天气、人文活动等状况，采好样的平皿置37℃中培养48 h。计数各皿菌落数，并换算为单位体积空气中大肠菌含量（CFU·m^{-3}），将相关数据作时空分布分析(4)。

3. 结果与讨论

3.1 青岛十大山头空气中大肠菌含量的空间分布

表2列出的是采集十大山头空气大肠菌样本时的环境状况，由该表可见，测定工作于青岛的春季在三天之中进行，温差不大（14.0℃～21.0℃），平均为15.3℃，风力也不算大，因而测定结果基本可以反映测点春季空气大肠菌状况。

表2 青岛十大山头空气大肠菌采样时环境状况

测点编号	环境概况	采样起始时间
1	市区吹来北偏西风3级，新叶摇晃，可能有尘埃、花粉落下 15.0℃	2005年5月11日 18:00
2	西北阵风5级，有轻烟雾，无人，开阔 14.0℃	2005年5月11日 18:17
3	静风，阳光透过，中雾，有晨练者 14.0℃	2005年5月12日 06:35
4	云雾不大，阳光灿烂，东南阵风5～6级 16.0℃	2005年5月12日 11:17
5	背风背阳偏向海，东南阵风3级 14.0℃	2005年5月12日 13:55
6	电视塔平台上，阵风3级，阳光斜射，有人来往 21.0℃	2005年5月12日 15:40
7	亭侧/树旁，轻风2级，人少 15.0℃	2005年5月13日 08:01
8	近山顶下坡松树下，阵风2级，背风/向风 18.0℃	2005年5月13日 09:15
9	东南风3级，阴，在居民新区 19.0℃	2005年5月13日 11:10
10	小丘灌木旁，阵风4级，阴 21.0℃	2005年5月13日 13:00

表3列出了十大山头空气大肠菌含量的具体数据，由该表可见，十大山头空气中大肠菌的检出率为100%，其平均含量约为424.4CFU·m^{-3}。最高值同样出现于No.1和No.10，即均为707.4 CFU·m^{-3}。最低值出现于No.6。最高值是最低的9.0倍。十大山头中大于该平均值的有5座，低于此平均值的同样也另有5座。

表3 青岛十大山头空气大肠菌含量状况表

测点编号	空气大肠菌含量（CFU·m^3）	取样时段
1	707.4	傍晚
2	393.0	傍晚
3	157.2	晨

(续表)

测点编号	空气大肠菌含量(CFU·m^3)	取样时段
4	471.6	中午
5	314.4	下午
6	78.6	下午
7	471.6	上午
8	314.4	上午
9	628.8	上午
10	707.4	中午

若以 1994 年青岛山山顶空气总微生物含量为基准(5),可见本次测得的大肠菌含量约是它的 6.4%(424.4/6 632.4)。与 2001 年～2002 年青岛全市空气平均细菌含量比(6),则本测定的大肠菌平均含量约是它的 4.8%(424.4/8 781.8)。估计山头以下的市区空气中大肠菌含量将有所增加。

我国至今未有相关标准规定空气大肠菌含量,但本结果可能意味着青岛山头空气有一定程度的大肠菌污染。

十大山头空气大肠菌含量大小排序为:No. 10＝No. 1＞No. 9＞No. 4＝No. 7＞No. 2＞No. 5 ＝No. 8＞ No. 3＞No. 6。大致趋势是北部山头＞南部。南部 No. 2、3、5、6、7 五山头的平均值为 283.0 CFU·m^{-3}。北部 No. 8、9、10 三山头的平均值为 550.2 CFU·m^{-3},这表明北南间的差异,即北部山头空气大肠菌污染状况重于南部。

西部 No. 1、2、3、5、6、7 六座山头的平均值为 353.7 CFU·m^{-3},东部 No. 4 为 471.6 CFU·m^{-3},这表达出西部山头空气以大肠菌为标志的质量好于东部。

这些结果意味着空气大肠菌数量与山头所在地区、工农商学、人群聚散和卫生等状况有关。此外测时的天气等状况也会对大肠菌含量产生别样的干扰,如 No. 1 虽处在南部近海,但测时市北西部吹来了不洁之风致树木摇晃,促使带菌颗粒落入其下部的平皿中,故有较多大肠菌。当然 No. 1 测点海拔也是一因(下述)。No. 6 测点的大肠菌含量最低,表明测时较高山势的该处空气清新,空气质量好。这反映出较高海拔和早晨空气中大肠菌较少这一事实。

十大山头以海拔高度可分为两类:一类为海拔≥98.0 m 的五座,即 No. 3、4、6、8 和 10。另一类海拔高良＜80.0 m 的五座,计算它们的大肠菌含量,前者平均为 353.8CFU·m^{-3},后者平均为 503.0CFU·m^{-3},这表明从总体上看,海拔高,其空气大肠菌含量低,反之海拔低,则大肠菌含量高。这基本上类似于空气微生物总量的分布趋势(5)。但并不尽然,比如 No. 10,海拔也属高列,但其空气大肠菌量最大。这说明海拔高,其空气大肠菌量不一定就低。测点海拔高低一般言是影响大肠菌含量低高的一个重要因素,但大肠菌量尤与空气其他污染、人群集散等状况相关,对具体情况应作具体分析。

3.2 青岛十大山头空气大肠菌含量的时间分布

将十大山头空气大肠菌含量以一日之中的时段来分析,得出的结果列于表 4。

由表4可见，以早晨、上午、中午、下午和傍晚5个时段划分的空气大肠菌含量差异是：中午＞傍晚＞上午＞下午＞早晨，这表明大肠菌含量自早晨至中午一直处于上升势，下午处于较小的低谷，但很快就开始攀升至傍晚的次峰，之后估计夜间即处于低迷状态。这似乎与多数人的活动规律即昼出夜归相同，反映了大肠菌含量的波动与人群聚散的相关性。

表4 青岛十大山头空气大肠菌含量的时段变化

取样时段	空气大肠菌含量平均值(CFU・m^3)
早晨	157.2
上午	471.6
中午	589.5
下午	196.5
傍晚	550.2

总体看青岛十大山头的空气与旅游环境并不十分洁净，尚有较大空间待改进，包括旅游、环境、产业结构、布局的调整、人们环境意识和卫生观念的更新和提高。

4. 结语

本文对青岛十大山头空气中大肠菌含量的时空分布状况作了初步测定分析，结果表明其平均值为424.4CFU・m^{-3}，约是以往所测空气细菌含量的6.4%、总菌量的4.8%。分析证实位于污染重的工业区或海拔较低处的山头，空气大肠菌含量较高，这反映了空气质量的地区差。空气大肠菌含量的时间分布是不平均的，中午高、早晨低，呈不规则曲线状，估计与人群活动及其分布相关。对空气中大肠菌的研究尚待深入。

参考文献6篇(略)

巴塞尔(Basel)空气微生物含量状况*

摘　要　本文用平皿自然沉降法测定了瑞士巴塞尔空气微生物状况，结果表明巴塞尔空气微生物检出率较高，达94.7%，平均的含量为513.0CFU·m^{-3}，其一半以上由细菌产生。分析了它的时空状态和分布。分析确认该市空气微生物含量不高，即便是在闹市也不高。本状况表明巴塞尔空气中微生物污染不重，空气质量上乘，空气微生物处于良好的循环之中：

关键词　巴塞尔　空气微生物　空气污染　空气质量

巴塞尔(Basel)市是瑞士第二大城市，位于瑞士北部，与德国交界。它以化学、制药等工业著称，既是铁路枢纽又是河运大港。该市远处见山。莱茵河由东顺流至西，再向北去而贯穿全市。市容整洁、生态安宁。初夏的巴塞尔最高气温约30.0℃，一般在10.0～25.0℃之间，风力不大，空气温凉而爽。

空气微生物无处不有，它们借助空气为媒而生存、繁衍、传播。包括病原菌在内的空气微生物关系人们、动植物健康。国际间对空气微生物的研究已逾143年，它随空气、环境污染及其防治和医学而发生、发展，但相关研究在世界各国间是不平衡的。欧洲美丽小国瑞士，其优良环境和清新空气一向被人推崇，但对其包括微生物在内的空气状况并未被人们十分关注。本文就瑞士重要城市——巴塞尔的空气微生物含量状况作一研究分析，以期人们对此有所了解并探索其成因，使之更加提高。

1.巴塞尔空气微生物测点概况

表1列出了巴塞尔空气微生物计19个测点/次的基本概况。由该表可见，这19个测点/次分布在巴塞尔市的不同功能区划之中，测定时间为2005.6.1～6.18，分布于昼夜的不同时段中，因而本测定可涵盖整个巴塞尔市区初夏空气微生物含量状况。

2.材料和方法

将Bacto agar(Dickison and company出品)制成平板培养基，按常规去设定的测点用自然平皿沉降法采集空气样本[1]，一般持续时间5～10 min。置采过样的平皿于25℃或37℃分别培养48h，计数各皿中的细菌(B)、霉菌(M)、酵母菌(Y)的CFU，并换算为单位体积空气中的总微生物含量(T)，即CFU·m^{-3}。比较分析它们的时空分布及变化状况，并由此评价巴塞尔市的空气清洁/污染程度和空气质量。

* 原文载于《中国现代化理论成果汇编》，2006，724-728.

表 1　Basel 空气微生物测点/测次概况表

测点/测次编号	地名	环境概况	测定起始时间
1	Bahnhof st Johaiiii 西北	有车人来往，西风 3 级，道路墩上，不远处有 Novatis 烟冒出来，不强阳光直射	2005.6.1　19:30
2	Lachans trasse 的"Sutter"超市旁	住房，上有稀叶树，无阳光直射，无车人来往，阵风 3 级	2005.6.1　19:40
3	Egli seestrasse platz	路边花园草坪小丘，附近有少量汽、火车来往，人少，阵风 3 级，稍凉，阳光透过叶丛来	2005.6.2　08:22
4	Egliseestras se 人行道	行道树浓密，在树、草、住房间，有车来往，人少，无阳光直射	2005.6.2　08:37
5	Towel 园内	高树下，砂砾路上，绿树环抱，路边小丘，马路上有车来往，另一侧在修房，风小，树叶偶动	2005.6.2　10:34
6	Towel 园，近 Gellertstrasse 与 Alban-Anege 交叉	小山丘下平地，静、小风	2005.6.2　10:53
7	Mittlers Brucke(中桥)	时有微小风，夕阳下，车人来往，桥下有行船	2005.6.8　21:00
8	Mittlers Brucke(中桥)	3 级风，行人、车少，气温 8.0℃	2005.6.9　01:29
9	Mittlers Brucke(中桥)	晨曦中，东风 2～3 级，湿气轻，9. 0℃，一辆有轨电车驰过	2005.6.9　04:54
10	Mittlers Brucke(中桥)	无阳光直射，西北风 2～3 级，车人来往，似有 Novatis 白烟飘来	2005.6.9　08:49
11	Mittlers Brucke(中桥)	云层透光来，来往人多，车少，东转西风 3～4 级	2005.6.9　12:57
12	Mittlers Brucke(中桥)	云层透光来，来往人多，车少，东风 3～4 级	2005.6.9　16:55
13	Mittlers Brucke(中桥)	夕阳下，来往人多，车少，风 2～3 级	2005.6.9　21:00
14	City Hall 广场	休闲、市场、人、车、鸽，阵风 3～4 级，23.0℃，桥下无行船	2005.6.10　14:20
15	Hauptstrasse	高出地面 10 m 的路边平台上，1～3 级风，湿气轻，云挡阳光，有花、树，邮局旁，Kiosk 店上平台	2005.6.13　15:57
16	莱茵河畔 Oetlinge rstrasse 附近	2 级风，云遮阳，树林间，来往人不多	2005.6.16 10:59**
17	ST. Matthaus 教堂外	静-3 级风，阳光下，树荫旁砂地上，不远处马路上车人来往	2005.6.16 15:59**

(续表)

测点/测次编号	地名	环境概况	测定起始时间
18	巴塞尔大学医院内花园	叶不很密的树下,直射阳光不多,2～3级风,20.0℃,无旁人	2005.6.17 09:44**
19	闹市邮局门边	时有2～3级风。人车来往,26.0℃	2005.6.17 16:30**

注:*持续时间5秒,**持续时间10秒

3. 结果和讨论

3.1 巴塞尔空气微生物含量的空间状况

表2列出的是巴塞尔空气微生物含量在各测点/测次的状况。

表2 Basel空气微生物含量状况表(CFU·m^{-3})

测点/次编号	细菌(B)	霉菌(M)	酵母菌(Y)	总菌(T)
1	157.2	314.4	0	471.6
2	157.2	157.2	157.2	471.6
3	157.2	157.2	157.2	471.6
4	0	157.2	0	157.2
5	157.2	0	157.2	314.4
6	157.2	314.4	0	471.6
7	157.2	0	0	157.2
8	0	157.2	0	157.2
9	0	157.2	0	157.2
10	157.2	0	157.2	314.4
11	157.2	157.2	0	314.4
12	157.2	471.6	0	628.8
13	157.2	0	0	157.2
14	0	0	0	0
15	314.4	0	314.4	628.8
16	393.0	943.2	0	1 336.2
17	314.4	471.6	78.6	864.6
18	235.8	0	0	235.8
19	2 436.6	0	0	2 436.6

由该表可计算出,巴塞尔空气中细菌、霉菌和酵母菌及总菌的检出率分别为78.9%、57.9%、31.6%及94.7%。这说明巴塞尔空气中细菌污染的机率最大。

由该表可计算出巴塞尔空气微生物含量的平均值为 513.0CFU·m^{-3},这含量比国内所有类似城市的均低得多[2-4]。其中平均的细菌、霉菌和酵母菌含量分别为 277.2、182.0 和 53.8CFU·m^{-3}。这表明巴塞尔空气微生物含量组成以细菌为最多,占总菌含量的 54.0%,霉菌和酵母菌数量则分别为 35.5%和 10.5%。

巴塞尔空气微生物含量在各测点/测次中的分布是不均匀的,表 3 列出了它们在各测点/次间按含量排序的状况。由表 3 结合表 2 可计算出,除未检出的外,总菌、细菌、霉菌和酵母菌在各测点/次间含量的最大差距分别为 2 279.4、2 279.4、786.0 和 235.8 CFU·m^{-3}。这四类菌的最高值分别为最低值的 15.5、15.5、6.0 和 4.0 倍,说明它们在各测点/次间存在一定的差距。

表 3 Basel 空气微生物含量在测点/测次间的大小排序

微生物类别	在各测点/测次间的大小排序
总菌	No. 19>No. 16>No. 17>No. 15=No. 12>No. 1=No. 2=No. 3=No. 6>No. 5=No. 10=No. 11>No. 18>No. 4=No. 7=No. 8
细菌	No. 19>No. 16>No. 17=No. 15>No. 18>No. 13=No. 12=No. 11=No. 10=No. 7 -No. 6 -No. 5-No. 3-No. 2 No. 1>No. 14-No. 9-No. 8-No. 4-0
霉菌	No. 16>No. 17=No. 12>No. 1=No. 6>No. 2=No. 3=No. 4=No. 8=No. 9=No. 11>No. 19=No. 18=No. 15=No. 14=No. 1 3=No. 1 0=No. 7=No. 5=0
酵母菌	No. 15>No. 2=No. 3-No. 5=No. 10>No. 17>No. 1=No. 4=No. 6=No. 7=No. 8=No. 9=No. 11=No. 12=No. 13=No. 14=No. 16=No. 18=No. 19=0

将巴塞尔 19 个测点/测次按城市功能区划的不同分为 A 僻静处(No. 5、6、17、18)、B 河畔(No. 16)、C 桥梁上(No. 7、8、9、10、11、12、13)、D 工厂、街道区(No. 1、2、3、14、15)和 E 市场(No. 14、15)等 5 区,各区空气微生物含量分别列于表 4 中。

表 4 Basel 空气微生物含量在城市不同功能区划中的比较

功能区编号 No.	空气微生物含量(CFU·m^{-3})
A	471.6
B	1 336.2
C	269.5
D	440.2
E	314.4

由表 4 可见各区空气微生物含量大小排序分别为 B>A>D>E>C。这表明开阔、空气流通好的场所,空气微生物含量少;数量小;湿度大、空气流通差的地方,空气微生物尤其霉菌含量多,数量大。但此类差距与国内一些城市比也甚小[5~6]。

表 4 中还显出一有趣现象,即一般城市的闹市,人群聚集处空气常较污浊,其微生物含量偏高[4],而巴塞尔却并不全是如此,如 E 区,空气微生物含量不算高。这表明巴塞尔

所测市场其环境较好,腐烂变质物品少,人们讲究整洁卫生素质高。巴塞尔空气微生物含量的空间分布状态说明其空气质量上乘,空气环境优良。

3.2 巴塞尔空气微生物含量的时间分布状况

据采样时间将巴塞尔空气微生物含量分作夜晚、清晨、上午、中午、下午和傍晚6个时段分析,得出表5-6。表5是一个测点一昼夜间空气微生物含量的变化。表6是所有测点通算出的时间变化。两个表尽管存在一定差异,但共同表达一个信息,即巴塞尔空气微生物含量在下午出现峰值,清晨或夜晚出现低谷,这说明巴塞尔空气微生物含量状况与人们的早出晚归节律及气温的昼夜变化状态类似,并与国内外许多城市基本相似[6~7]。

表5 Basel中桥上空气微生物含量昼夜变化状态

时段	空气微生物含量($CFU \cdot m^{-3}$)
夜晚	157.2
清晨	157.2
上午	314.4
中午	314.4
下午	628.8
傍晚	157.2

表6 Basel空气微生物含量的时间分布状况

时段	空气微生物含量($CFU \cdot m^{-3}$)
夜晚	157.2
清晨	157.2
上午	471.6
中午	314.4
下午	911.8
傍晚	314.4

4.结语

本文用平皿自然沉降法测定了巴塞尔初夏空气微生物含量状况,结果显示巴塞尔空气微生物平均检出率较高,达94.7%,含量达513.0 $CFU \cdot m^{-3}$。其中细菌占一半以上。总体看巴塞尔空气微生物含量不高,证实其空气污染轻,空气质量上乘。测点间存在的差距说明了城市不同功能区划间的差异,但闹市区一些市场的空气微生物含量也不高,说明整个城市空气环境好、生态稳定,其时间分布上的变化符合一般规律。本结果暗示巴塞尔空气微生物生态结构正常,处于良性循环之中。

参考文献7篇(略)

(合作者:李为民)

青岛夏冬空气副溶血性弧菌含量分析*

摘　要　本文用副溶血性弧菌(Vp)培养基在夏冬两季对青岛7个室外1个室内的空气Vp含量作了测定。结果表明,夏冬季空气Vp检出率分别为100和57.1%,其含量分别平均为1 855.0和112.3 CFU·m^{-3},夏季含量明显高于冬季。夏天人群密集之室内Vp含量大增。对室外空气Vp含量的时空分布作了分析。结果指出Vp含量及其分布与人群集散、污染程度关系密切,并受制于气温。

关键词　青岛　空气　副溶血性弧菌

1.引言

副溶血性弧菌(*Vibrio parahemolyticus*,简写为Vp,下同),主要依仗其致病因子-热稳直接溶血素(themostable direct hemolysin,TDH)引起食物中毒和旅游者发生腹泻等胃肠疾病。它的大规模繁衍可能导致流行病的暴发,这种现象尤其在沿海地区更为明显。

环境,包括空气环境中含有Vp,有的可能致病。空气是Vp传播的一个媒介,Vp在空气中的含量、传播、分布及其变化受制于空气中各种生态因子,并指示环境质量状况,但相关研究并不多见。

本文报道Vp在青岛市区一些室内外空气中状况的调查研究。

2.材料与方法

2.1　采样点描述

于2004年7～8月和2005年2月共选择青岛市区7个室外和1个室内采样点。它们分别代表市区室内外区位、环境、污染程度和功能上的差异性。具体状况见表1。

表1　青岛夏冬空气副溶血性弧菌测点状况

测点编号(No.)	测点概况		测定起始时间
1	高科园仙霞岭路	雾、湿、气温>30.0℃、无风	2004.8.4　13:30
	某机关大院	5级风、风向不定,飘尘,人车少,0℃、阳光斜入	2005.2.1　15:45
2	市立医院对面圣保罗教堂围墙上	2～3级东南风,天气状况同上	2004.8.4　18:10
		3～5级西风,1.5℃	2005.2.2　15:08
3	栈桥迴澜阁海堤上	风小,人多,雾浓,湿度大,>28.0℃	2004.8.5　10:1 5
		4～5级西北风,人很少,−1.5℃	2005.2.2　5:50
4	水清沟乐安路口宿舍楼房	5级北风,风口,1.0℃	2005.2.2　10:58

* 原文载于《中国共产党人理论研究选萃》,2006,340-342

(续表)

测点编号 (No.)	测点概况		测定起始时间
5	台东步行街	4 级风,1.0℃,行人来往,阳光斜射。路中央	2005.2.2 9:05
6	齐东路邮局门口	人车来往较多,阴湿,2 级风,气温 27.0℃	2004.8.6 07:15
		3 级风,−2.0℃,车人不多	2005.2.1 17:08
7	市人代会前花园	车人少,阴凉	2005.8.5 21:00
		3 级西风,−3.0℃,放烟爆之后	2005.2.1 21:00
8	青岛火车站售票厅	人气、湿气旺	2005.8.5 13:58
		人多,15.5℃	2005.2.21 13:08

2.2 实验方法

制备好 Vp 平皿,去设定测点按正规方法采样,取样时间 5 min 或 10 min,取回样后于 37℃温度中培养 24h 以上。不断观察和记录皿上出现的 Vp 菌落形成数(CFU,下同)。

2.3 数据处理分析

按照文献中的方法计算出各皿代表的所测单位体积空气(m^3)中的 Vp 数量为 Vp 含量,以此比较分析空气中 Vp 含量的时空分布,并做出相应评价。

3. 结果和讨论

3.1 青岛市区空气 Vp 含量空间状况

表 2 列出了青岛测点夏冬两季室内外空气中 Vp 含量的测定结果。

表 2 青岛夏冬空气副溶血性弧菌(Vp)含量测定结果表

测点	测点编号	夏	冬
室外	1	314.4	314.4
	2	707.4	157.2
	3	707.4	0
	4	/	157.2
	5	/	157.2
	6	7 388.4	0
	7	157.2	0
室内	8	2 122.2	0

由表 2 可统计出青岛室外夏季空气中副溶血性弧菌的检出率为 100%,其含量平均为 1 855.0CFU · m^{-3},波动幅度为 157.2～7 388.4 CFU · m^{-3},最高是最低的 47 倍。按浓度大小排序,其测点编号为 No. 6＞No. 2＝No. 3＞No. 1＞No. 7,这 5 个测点分别代表人口密集区 A(No. 6)、交通要道 B(No. 2)、风景游览区 C(No. 3)、高科园 D(No. 1)和清静开

阔区 E(No. 7)等.城市不同功能区划，其含量大小排序为 A＞B＝C＞D＞E。由此可见人口密集区域此类菌容易聚集，此外交通要道或旅游旺季人口、污染物的聚集也导致此类菌的繁殖。

青岛夏季室内尤其是人群高度密集的室内(No. 8)空气更适宜此类菌的繁衍传播，导致其含量变大，比室外平均值高。因此人们应尽可能不要密集，室内空间应保持新鲜空气流通，外出也应多去开阔清静之处。

表 2 还列举了青岛冬季空气副溶血性弧菌的含量状况，由表 2 统计出冬季该菌检出率为 57.1%，其室外的平均含量为 1 12.3 CFU · m^{-3}，夏季的该值是它的 16.5 倍，意味着夏季室外空气中迷漫着大量副溶血性弧菌。在冬季则出现机率少了许多。冬季该菌含量在测点的大小排序为 No. 1＞No. 2＝No. 4＝No. 5＞ No. 6＝No. 7＝No. 3＝0。除了高科园测点(No. 1)显出高含量外，医院附近的交通要道(No. 2)、人群聚居(No. 4)和商业街(No. 5)，其含量持平且低于 No. 1。No. 1 测点虽人群不密集，但顺风而来的工业区污染和散发的飘尘可能带来副溶血性弧菌。冬季的各点按夏季的城市功能区划分，则 No. 4 和 No. 5 测点可同归 A。菌量大小排序为 D＞B＞A＞C＝E＝0，显出不同于夏季的趋势。

冬季室内副溶血性弧菌检出率为 0，说明其含量很低，甚至为难以检出。

冬天比夏天室内外空气中副溶血性弧菌检出率降低，浓度减淡，说明冬天气温低，此类菌不易繁衍。但不能就此对之掉以轻心。

3.2 青岛市区空气 Vp 含量的时间分布

青岛空气副溶血性弧菌含量昼夜变化分析如下：

夏天的 5 个测点按测定时间分，。则可得出上午(No. 3,6)、中午(No. 1)、下午(No. 2)及夜间(No. 7)等 4 个时段。各个时段测得的副溶血性弧菌含量值显出自上午高峰(4 047.9CFU · m^{-3})起经中午(314.4CFU · m^{-3})和下午(707.4 CFU · m^{-3})的相对平稳后降至夜间的最低值(157.2CFU · m^{-3})。总体看是白天浓度大于夜间。

冬天的 7 个测点按测定时间分，得出上午(No. 3,5)、中午(No. 4)、下午(No. 1、2)及夜间(No. 7)等 4 个时段。其副溶血性弧菌含量大小排序为：下午(235.8CFU · m^{-3})＞中午(157.2CFU · m^{-3})＞上午(78.6CFU · m^{-3})＞夜间＝0，其轨迹显出自早晨至上午、中午处于上升阶段，至下午达高峰，之后则一落千丈，为夜间的 0，也表达为白天浓度高、夜间低之势，这与人群集散、与气温高低也相关。

4. 结语

对青岛夏冬两季室内外空气副溶血性弧菌含量作了瞬时测定，并就时空分布状况作了分析，结果表明空气中副溶血性弧菌含量及分布与人群集散、污染程度关系密切，并受气温制约，夏季更为明显些。该检出率及含量夏天高于冬天，似与气温高低关系大。夏冬室外 Vp 含量的最高测点发生了逆转，但清静、开阔或临海场所，Vp 含量常较低.昼夜变化状态冬夏间有一定差别，但均显出白天高于夜间，夏天人群密集之室内该菌含量大增。副溶血性弧菌与人类健康相关，其在空气中的检出率、含量及分布反映某地空气此类微生物的污染状况，且受制于各种因子。相关研究有待深入。

大同市空气中胃肠炎致病菌含量调查*

摘　要　用平皿自然沉降法对大同市空气中胃肠炎病原菌含量作了初步调查，结果表明大同空气胃肠炎病原菌检出率和含量均居高不下，证实大同空气污染严重，空气质量差，亟待改善。

关键词　大同　空气污染　胃肠病原菌　空气质量

位于华北的大同市虽系历史文化名城，但据 2004 年 7 月 13 日国家环保总局公布的报告可知，大同位居全国十大污染最重城市之第三名，可见其污染之严重性[1]。大同空气污染原因种种，包括地理环境（三面环山成盆地）、气候特征（干寒多风）、工业生产（煤炭挖掘、加工、运输致金属、煤烟等可吸入颗粒多等）及生活/取暖方式落后（多煤炉）等等。这种污染势态必然影响居民生活质量和健康水平。

夏季和初秋气温高、湿度大，是传染病多发季节。人们饮食品种繁杂，不洁食品易出入消化系统等[2]，再加上严重的空气污染，导致空气中散布胃肠炎致病菌，包括急性胃肠病原菌，严重时将会使一些病菌增多，疾病传播流行。

空气是各种微生物的媒介，检测空气中特定微生物的含量及变化有助于人们掌握它们的生存、繁衍和发展状况及规律，从而预警、控制其扩散、蔓延[3]。胃肠炎病原菌是其中一类重要微生物。有关大同空气微生物状况的调研报告比较稀少。检测大同空气中胃肠炎病原菌含量正是本文的内容。由此可从一个侧面反映大同空气环境质量，并唤起人们对这里空气污染的重视。

1. 测点状况

于 2004 年 8 月间选取大同六个测点作空气胃肠炎病原菌调查研究。测点状况见表 1 所示。由该表可见，测点涵盖了大致大同市区不同方位和郊外的一个重要景点，可以代表大同不同城市功能区划中空气的基本状况。

表 1　大同市空气中胃肠炎病原菌测点状况

测点编号	测点名称	测点状况	测空起始时间*
1	大同公园	微风、阴天、气温 18.0℃，小桥下有小河	2004.11.8 08:52
2	上华严寺大雄宝殿门口	阵风 3～4 级，有些香火，气温 18.0℃	2004.11.8 09:19
3	大同火车站广场	阴天，风 3 级，附近马路人车来往，气温 23.0℃	2004.11.8 12:48
4	云冈石窟第 5 窟	中部，释迦坐像旁，气温 22.0℃	2004.11.8 15:30
5	市政府办公大楼外	微风、气温 18.0℃	2004.11.8 18:00
6	东关十字路口	阵风 3 级，气温 15.0℃	2004.11.8 21:00

*采样持续时间 5 min

* 原文载于《永葆共产党人先进本色论文集》，人民日报出版社，P. 730-731，北京，2006.

2.材料和方法

选择胃肠炎病原菌固体培养基。用自然沉降法按常规在测点采集空气样本，同时记录现场天气和人文状况。经 48h 37℃培养，计数平皿上长出的菌落数(CFU)，并换算为单位体积空气中的胃肠炎病原菌含量 CFU・m^{-3}。分析相关数据，评价空气胃肠炎病原菌含量及污染状况[3]。

3.结果和讨论

3.1 大同空气中胃肠炎病原菌含量的空间分布

表 2 大同空气中胃肠炎病原菌含量状况表

测点编号	胃肠炎病原菌含量(CFU・m^{-3})
1	28 406.7
2	23 780.9
3	10 1452.5
4	11 272.5
5	51 402.6
6	62 224.2

表 2 列出了大同市各测点空气中胃肠炎病原菌含量状况，由该表可见大同空气胃肠炎病原菌的检出率达 100%，其平均数为 46 423.2CFU・m^{-3}。这一数值表明大同空气中有大量胃肠炎病原菌，且是青岛的 73.8 倍之多*。其大小排序为 NO.3>NO.6>NO.5>NO.1>NO.2>NO.4，最高值是最低值的 9.0 倍。六个测点中，有三个，即 NO.3、6、5 大于平均值。这六个测点按城市功能区划可分作 A 景区(NO.1、2、4)、B 交通要道(NO.6、3)和 C 离市中心较远的居民聚居区(NO.5)，此三类区划空气中胃肠炎病原菌平均含量各为 21 153.4、81 838.4 和 51 402.6CFU・m^{-3}，说明交通要道空气最坏，传染该病的机率最大，景区相对较好。

3.2 大同空气中胃肠炎病原菌含量的时间分布

将表 2 的数据按一日间的上午、中午和下午和夜间四个时段分，则可看出这四个时段中的空气胃肠炎病原菌含量分别为 26 093.8、101 452.5、31 337.6 和 62 224.2CFU・m^{-3}。它们的大小排序显出中午>夜间>下午>上午，即中午出现峰值，上午见低谷，这一状况不同于青岛夏季的状况 *。这表达出该菌含量的时间分布与人群聚散、胃肠健康状态、污染状况密切相关*。

4.结语

本文初步测定了大同空气中胃肠炎病原菌的含量状况，指出其检出率高，全市平均值为 46 423.2CFU・m^{-3}。空间分布分析表明测点间它们的含量差距大，反映出不同城市功能区划间空气质量上的差异。时间分布分析则表明该含量上午低，中午高。大同空气胃

* 陈皓文，青岛空气胃肠炎病原菌含量的初步分析

肠炎病菌含量表明了大同空气污染较重、空气质量较差，传病可能性大。尚待人们重视和采取有效措施。

参考文献3篇(略)

呼和浩特空气胃肠炎病原菌含量的初步分析*

摘　要　本文用平皿自然沉降法测定和分析了呼和浩特的空气胃肠炎病原菌含量。结果表明，呼市空气胃肠炎病原菌平均含量为22 780.3CFU·m^{-3}，病原菌含量显示出一定的时空分布态势，此态势与空气污染、人群健康水平与流动状况相关。

关键词　呼和浩特　空气　胃肠炎病原菌　污染　健康状况

呼和浩特是内蒙古自治区首府，位于阴山南麓，大黑河以北。全年较干旱。8月份平均气温20.0℃，低温时已有凉意。呼和浩特市有以毛纺、机械工业为中心的综合性工业。这些条件给呼市造成了一定的污染压力。

夏、秋季是人们胃肠病多发季节，这归因于一些人的饮食卫生以及胃肠炎病原菌的传播与蔓延。空气是此类病菌的重要媒介。测定空气中胃肠炎病原菌的状况有助于人们去掌握和监控它们的生存和污染状态、繁殖和传播规律[1]。迄今未见呼市这方面的研究报道。本文试图从一个侧面论述呼市空气中胃肠炎病原菌含量状况，以便分析和评价它们对空气质量的影响。

1. 呼市空气胃肠病原菌测点概况

表1列出了呼市胃肠炎病原菌测点的环境概况。

由该表可见，相关测点共8个，它们分布于全市不同方位的4个功能区划内，其中包括一个远离市区的景点。因而有一定的代表性和普遍意义，可能代表呼市空气这方面的整体状况。

2. 材料和方法

依文献[2]的方法，制作胃肠炎病原菌培养基。按常规去设定的测点采集空气样本，采集时间为5 min，同时记录当时的天气、人文状况。于37℃培养48h后，计数各皿中的CFU，并换算为单位体积空气中的胃肠炎病原菌含量，即CFU·m^{-3}，由此分析此类菌的时空分布状况。

表1　呼和浩特空气胃肠炎病原菌测点状况

测点编号	测点名称	测点状况	测定起始时间*
1	新华广场	人海边缘、小吃摊、烟尘、4级风，干凉	2004.8.13 21:30
2	五塔寺	晴朗、微风、游客很少	2004.8.14 08:55
3	内蒙古农业大学宿舍区	阴凉处、路边、3级风	2004.8.14 11:15

* 原文载于《世界重大学术成果精华》，华人卷Ⅱ，2006.

(续表)

测点编号	测点名称	测点状况	测定起始时间*
4	55 路公交车建材场站(鄂尔多斯大街)	阳光下,车人来往不很多,无遮挡	2004.8.14 14:25
5	昭君墓和亲碑上	有游客,3～4 级风	2004.8.14 16:09
6	鄂尔多斯大街、石羊桥路十字路口	来往车辆多、行人少、4 级风	2004.8.14 17:07
7	旅客列车呼和浩特站	车人来往不多、有闲人、汽车、摩托车、2～3 级风	2004.8.14 20:30
8	市西巴彦淖尔路工人新村	阴天,4 级风	2004.8.15 06:00

*采样持续时间 5′

表 2　呼和浩特空气胃肠炎病原菌含量状况

测点编号	胃肠炎病原菌含量($CFU \cdot m^{-3}$)
1	26 095.6
2	13 047.3
3	37 879.2
4	56 818.8
5	1 683.5
6	8 838.4
7	23 990.2
8	13 889.0

3. 结果与讨论

3.1　呼和浩特空气胃肠炎病原菌含量的空间分布分析

表 2 列出了呼市 8 个测点空气胃肠炎病原菌含量状况。由表 2 可见呼市空气胃肠炎病原菌检出率达 100%,计算得出呼市空气胃肠炎病原菌含量的平均值为 22 780.3$CFU \cdot m^{-3}$,是青岛的 36.0 倍*,但比大同的低,不足大同的一半(0.49)**。低于该平均值的测点计 4 个,即 NO.2、5、6 和 8,占总测点数的 50%。出现于 NO.4 的最高值 56818.8$CFU \cdot m^{-3}$,约是最低值(NO.5)的 33.8 倍,这说明呼市各地的空气胃肠炎病原菌含量存在明显差距。将 8 个测点按城市功能区划可分为 A. 景点区(NO.2、5)、B. 交通要道(NO.4、6)、C. 广场(NO.1、7)和 D 居民区(NO.3 和 8)四种类型。经计算,各区划空气胃肠炎病原菌含量平均分别为 B(32 828.6$CFU \cdot m^{-3}$)＞D(25 884.1$CFU \cdot m^{-3}$)＞C(25 042.9

*　陈皓文,青岛空气胃肠炎病原菌含量分析。

**陈皓文,大同空气胃肠炎病原菌含量状况。

$CFU\cdot m^{-3}$)>A(7 365.4$CFU\cdot m^{-3}$)。这表明呼市各区划间空气胃肠炎病原菌含量间存在差异，以交通要道空气中此类菌污染最重，而景区最轻。前者约是后者的4.5倍。与青岛比，该市交通要道空气胃肠炎病原菌含量是青岛的51.2倍*，呼市空气胃肠炎病原菌含量处于较高水平，其区划间的差别反映了相互间空气质量上的差异。与一般城市类似，即空气中该类菌的含量分布与人群集散、胃肠道健康水平、空气中各种颗粒物、气溶胶，包括人体排泌物的浓度关系大。

3.2 呼和浩特市空气胃肠炎病原菌的含量的时间分布分析

表3 呼和浩特空气胃肠炎病原菌含量的时间分布

时段	$CFU\cdot m^{-3}$
早晨	13 889.0
上午	25 463.3
下午	22 446.9
夜晚	25 042.9

将呼市空气胃肠炎病原菌以采样时间分作早晨、上午、下午、夜晚四个时段来分析，所得结果见表3。计算出胃肠炎病原菌含量在这4个时段上的含量大小排序为上午>夜晚>下午>早晨，显出它们从早晨开始上升，上午达高峰，之后一直处于较高水平。这一状态虽与青岛大多情况不一*，但清晨少则是一致的，高峰值是低谷值的1.8倍以上。

4.结语

本文对呼和浩特市空气胃肠炎病原菌含量作了初步测定和分析，结果表明呼市空气胃肠炎病原菌的检测率达100.0%，平均含量约为22 780.3$CFU\cdot m^{-3}$。胃肠炎病原菌含量在全市空气中分布不平衡，高低差别明显。此含量在不同功能区划中显出交通要道很高，景区最小。空气胃肠炎病原菌含量的时间分布也显出不平衡性，晨间低，上午达最高值。

呼市空气胃肠炎病原菌含量的分布状态决定于空气污染程度，包括带菌者在内的人群生产/生活状态等。这启示人们必须提高环保忧患意识，着力改善环境状况和健康水平，使生存环境和水平日臻完善和提高。

参考文献2篇(略)

(合作者:孙丕喜)

INITIATORY ANALYSIS ON CONTENT OF PATHOGENIC BACTERIA CAUSING GASTROENTERITIS IN AIR OF HUHHOT

(ABSTRACT)

Abstract The content of pathogenic bacteria causing gastroenteritis in air of Huhhot was determined and analyzed initially. The result obtained shows that: the average content of the bacteria was 22 780. 3CFU · m^{-3}. The content appears to have some status of time variation and spacial distribution. The paper infered that this status was relative to air pollution and human healthy level and their more conditions.

Key words Huhhot Air Pathogenic Bacteria Causing Gastroenteritis Pollution Health

延安空气微生物含量状况并与韶山北海比较*

摘　要　用自然沉降法对延安10个测点作了空气微生物含量的测定，结果表明总体看延安已经受到中度微生物污染，少数测点空气已严重污染。对空气微生物含量作了时空分布分析，并与规模类似但地理和气象条件不同的两城市韶山和北海作了比较，表达出延安空气微生物污染较重，但有自身特征。文章最后指出了改善空气质量的必要性。

关键词　延安　空气微生物　空气污染　空气质量

1.引言

延安是中国新民主主义的革命圣地，南距黄帝陵约百十千米。它地处陕北高原，延河流淌，市区东西北向山岭对峙。大陆性气候特征明显，人口不足20万，相当静谧。但电力、机械、化肥、汽车修理、纺织等产业及市政交通等与特定的地理环境造成了不轻的污染，包括对空气的污染，使延安既令人神往又令人担忧。

人们赖以生存的空气质量决定于空气所含污染物之种类和浓度，其中所含的微生物则是重要的指标。空气微生物关系健康，因而对空气微生物学的研究十分重要，并为许多先进国家所重视[1]。中国20世纪80年代以来的相关研究已获可喜成果[2]，但像延安这样的内陆中小城市的空气微生物研究成果并不多见。本文试图报道检测所得延安空气微生物含量状况并与不同地域、气候条件，但有相似之处的韶山、北海两市作比较，以促进延安空气环境的改善和加强我国的生态环境可持续发展之势。

2.材料与方法

将普通营养琼脂和沙氏真菌培养基作为本次空气微生物的培养介质，按常规的自然沉降采样法选取市内不同城市功能区划、方位计10个测点曝皿5 min，培养37℃左右的温度中2～3d后计数各皿中的菌落形成数(CFU，下同)，换算为单位体积空气(m^3)中的空气细菌(B)、空气真菌(F)、空气总微生物含量(T)及真菌占总菌量的%(F/T%)，比较分析各测点空气微生物含量的时空分布变化，判断微生物引起的空气污梁程度以指示所测地区的空气质量[2～3]。

3.测点状况

在延安市区设立10个空气微生物测点，以覆盖和代表延安不同城市功能区划和环境状况。即革命纪念地、宾馆住户和交通要道等，这10个测点的具体状况列于表1之中。

* 原文刊于《创新与发展》，2006，20：86-89

表 1　延安空气微生物测点状况＊

测点编号(No.)	测点名称	测点状况	测定起始时间＊＊
1	石窑宾馆	2 层平台、餐厅前、有些人来往、2 级风、阴天	2004.8.21　23:00
2	杨家岭中央大礼堂院内	阴天、1 级风、少人	2004.8.22　06:47
3	王家坪,八路军总部旧址院内	阴天、1 级风、仅 1 人	2004.8.22　07:30
4	延安革命纪念馆内前广场	2 级风、阴天,有花草树木、晨练者不少	2004.8.22　07:54
5	宝塔山前过桥马路	延河边、交通要道、不宽、阴天	2004.8.22　08:35
6	清凉山旁马路	延河边、山的不通行人阶梯上。车人来往、有风被山挡回	2004.8.22　08:55
7	延安大学宿舍雅园区	延河边、无人、2 级风、阵风 4 级、阴天	2004.8.22　10:00
8	居民窑洞前院高坡	居家杂物、坡下为马路、阴天	2004.8.22　10:30
9	枣园内	树下、2 级风、阴天、附近有些人走动	2004.8.22　12:50
10	延安火车站广场雕塑上	阴天、灰尘飘扬、3 级风	2004.8.22　14:23

4.结果和讨论

4.1　延安空气微生物含量一般状况

所得结果列于表 2 之中,由该表可统计出延安空气微生物的检出率达 100%,其中细菌的检出率达 100%,细菌的平均含量为 23 581.4 CFU・m^{-3},极差为 77 244.0 CFU・m^{-3}。最高的细菌含量是最低者的 62.5 倍.按含量大小排序测点为 No.6>No.10>No.3>No.4>No.5>No.2>No.8>No.7>No.1>No.9,即交通要道处的 No.3 测点空气细菌含量明显高于革命纪念地的 No.9。按照空气细菌污染评价标准[4],可见延安空气已处于中度细菌污染水平。其中有 2 个测点的细菌污染已达严重污染程度(No.10 和 No.6),这是必须引起足够重视的状况。延安平均空气细菌含量高于宝鸡的,意味着延安空气细菌污染重于宝鸡[5]。

表 2　延安空气微生物含量(CFU・m^{-3})等状况

测点编号(No.)	空气微生物			
	细菌(B)	真菌(F)	总菌(T=B+F)	F/T%
1	1 413.0	314.0	1 727.0	18.2
2	9 734.0	0	9 734.0	0
3	34 069.0	628.0	34 697.0	1.8

(续表)

测点编号(No.)	空气微生物			
	细菌(B)	真菌(F)	总菌(T=B+F)	F/T%
4	21 352.0	471.0	21 823.0	2.2
5	18 055.0	0	18 055.0	0
6	78 500.0	314.0	78 814.0	0.2
7	5 024.0	785.0	5 809.0	13.4
8	5 966.0	157.0	6 123.0	2.5
9	1 256.0	628.0	1 884.0	33.1
10	60 445.0	0	60 445.0	0

空气真菌的检出率为70.0%,其平均的真菌含量为329.7CFU·m^{-3},该值比韶山的低[6]。空气真菌含量的极差为785.0CFU·m^{-3},最高真菌量是次高量的5.0倍。按含量大小排序测点为No.7>No.3=No.9>No.4>No.1=No.6>No.8>No.2=No.5=No.10=0,即住户区的N 0.7空气真菌含量明显高于革命纪念地的。

F/T%平均约为1.4,最高F/T%是次高的30.0倍,大小排序为No.9>No.1>No.7> No.8>No.4>No.6>No.3>No.2>No.5=No.10=0。表明绿化较好处的测点空气真菌所占比率大于树少,暴露于日光处的多。

延安平均的空气微生物含量为23 911.1CFU·m^{-3},极差达77 087.0CFU·m^{-3},最高的微生物含量是最低者的45.6倍,按各测点含量大小排序为No6>No.10> No.3>No.4>No.5>No.2>No.8>No.7>No.9>No.1。基本与细菌含量排序一致。即交通要道的No.6测点空气微生物含量明显高于纪念地或居住区的。

总的看,延安空气微生物主要由细菌组成,细菌含量占微生物总量的98.6%。而真菌量只占其不足1.4%(F/T)。表明延安空气中的微生物污染主要由细菌引起,真菌加重了空气污染,这表达了延安所处气候带和地理环境的特点。

表3 延安不同功能区划测点空气微生物含量状况比较

测点编号(No.)	测区名称	空气微生物				备注
		B	F	T	F/T%	测点编号(No.)
A	革命纪念地	16 602.8	431.8	17 034.6	2.4	2、3、4、9
B	宾馆住户	4 134.3	418.7	41 553.0	9.2	1、7、8
C	交通要道	52 313.3	104.7	52 418.0	0.2	5、6、10

微生物含量单位:CFU·m^{-3}

4.2 延安空气微生物含量在城市功能区划间的比较

将延安测点按城市功能区划可分作下列3类,即A-革命纪念地,B-宾馆、住户,C-交通要道。据此可得出表3,由表3可知,各类区划中空气细菌的平均含量大小排序为C>A>

B。真菌的为A>B>C。总菌量为C>A>B,F/T%为B>A>C。表明所测宾馆、住户区总体看空气细菌污染最轻,而交通要道区空气细菌和总量污染最重,虽然后者的真菌量不大。宾馆、住户区真菌量虽小于革命纪念地,但F/T%却最高,表明前者测时的空气湿度大,较大空气湿度助长了具耐湿孢子的真菌的生长繁殖。虽然总体看延安空气真菌量比不上南方地区的。

4.3　延安空气微生物含量的时段变化

表4列出的是将测点按测时的不同归为5个时段后所得空气微生物含量指标的变化状态。

表4 延安空气微生物含量的时段变化状态

空气微生物含量 \ 类别	时段				
	早晨 (No.2-4)	上午 (No.5-8)	中午 (No.9)	下午 (No.10)	夜间 (No.1)
细菌	21 718.3	26 886.3	1 256.0	60 445.0	1 413.0
真菌	366.1	314.0	628.0	0	314.0
总菌	22 084.4	27 200.3	1 884.0	60 445.0	1 727.0
F/T%	1.7	1.2	33.1	0	18.2

* ()中为所取测点号,含量单位:(CFU·m^{-3})

由该表可见,空气细菌含量时段上的大小排序为下午>上午>早晨>夜间>中午,即下午峰,中午谷。空气真菌含量时段上的大小排序为中午>早晨>上午=夜间>下午=0,即中午峰,下午谷。空气总菌含量时段上的大小排序为下午>上午>早晨>中午>夜间,即下午峰,夜间谷。而F/T%含量时段上的大小排序则同于空气真菌的。

结果表明,延安空气微生物含量时段的变化主要由细菌造成,中午开始空气微生物污染重,而夜间趋轻。

这显示出不同于一些城市之态[7]。

4.4　延安、韶山、北海空气微生物含量状况的比较[6,8]

延安位于北方干旱区,韶山在华中的温湿区,而北海处于华南海滨。三地虽有不同功能,相距也有1 850 km左右,但均是中小规模的游览区。三地比较,可见差异。它们的空气微生物含量状况比较如表5所示。

表5　延安和韶山、北海空气微生物含量状况比较

城市名	空气微生物含量状况							
	细菌		真菌		总菌		F/T%检	检测时间
	含量*	检出率(%)	含量*	检出率(%)	含量*	检出率(%)		
延安	23 581.4	100.0	329.7	70.0	23 911.1	100.0	1.2	2004.8
韶山	17 993.0	83.1	757.6	83.1	18 750.6	83.1	4.0	2002.3
北海	5 709.7	100.0	1 847.1	100.0	7 556.8	100.0	24.4	1996/1997

* 含量单位:(CFU·m^{-3})

由该表可见,空气细菌含量大小排序为延安＞韶山＞北海。空气真菌含量大小排序为北海＞韶山＞延安,F/T%也是如此。空气总菌含量大小排序为延安＞韶山＞北海。

这表明,在由北往南的地理走势中,空气微生物污染由重减轻,而其中的真菌污染则由北往南加重。这与南北方降雨量差别,空气温度高低相关。当然也排除不了人为的空气污染。这一差距说明近几年来南方的一些中小城市空气质量好于北方。

5.结语

本文虽然未对延安空气微生物状况作深入调研。但从一个侧面反映了延安的空气污染状况,总体看今天的延安空气尚处于中度微生物污染状态。而少数测点已达到严重污染程度,这是值得引人关注的。空气微生物含量的时空分布表达出延安空气微生物分布的不均匀性,与国内处于不同纬度、气候带的两市相比,延安空气污染重于它们,有些测点空气真菌比率也不低。造成延安空气污染不轻的原因虽然多种多样,但应争取理清。增加对本地地理、气候状况的认识,增强环境忧患意识,借鉴国内外先进的环保经验,根据自身的特点安排好工农业生产和人民生活是十分必要的。差距是环境上的,也是意识中的,差距就是动力。相信延安人民会找出差距,努力奋进。空气质量将会提高,生存环境将会改善。

参考文献8篇(略)

青岛空气中胃肠炎病原菌含量的分析*

摘　要　用胃肠炎病原菌培养基在夏冬两季选择7个室外、2个室内测点测定空气中胃肠炎病原菌含量。结果表明,夏冬两季空气胃肠炎病原菌检出率均为100%,其室外的平均含量分别为628.8和2919.4CFU/m³。夏天的室内空气胃肠炎病原菌含量大于冬天,夏天室内大体上均高于室外值。分析了它们的时空分布状态,说明与人群聚散、污染状况密切相关,而与气温关系不大。

关键词　青岛　空气　胃肠炎病原菌

胃肠炎病原菌可经气溶胶飞沫进入呼吸道、泌尿系统/器官和体液而引起感染或经呼吸道再次感染消化道[1,2]。人们常认为胃肠炎病原菌在夏秋季节数量较大,散布较多、较广。这与食物不洁、人群在户外活动频繁及环境包括空气污染等关系密切。可是迄今相关的专门报道并不多见[3]。因此调查研究特定地域空气中此类菌的含量分布状况就有其必要性。本文论及青岛市区室内外空气中胃肠炎病原菌含量及其时空分布状况。

1　材料与方法

1.1　采样点描述

于2004年7～8月和2005年2月共选择青岛市区7个室外、2个室内采样点,它们分别代表青岛市不同城市功能区划中室内外环境和污染程度等方面的差异性(表1)。

1.2　调研方法

制备好胃肠炎病原菌固体平皿,去设定测点按常规方法采样[2],取样时间5 min或10 min。取回样后于37℃温度中培养24h以上,经常观察和记录皿中出现的菌落形成数(CFU,下同)

1.3　数据处理分析

按照文献[4]中的方法计算出各皿代表的所测单位体积空气中的胃肠炎病原菌数量为其含量(CFU/m³),以此比较分析空气中胃肠炎病原菌含量的时空分布,并做出相应评价。

2　结果和讨论

2.1　青岛空气中胃肠炎病原菌含量一般状况

表2列出了青岛市空气中胃肠炎病原菌含量状况。

由表2可见,青岛夏季空气中胃肠炎病原菌检出率为100%,其室外的平均含量为628.8CFU/m³。波动范围为471.6～786.0 CFU/m³,最高是最低的1.7倍左右。按照各测点的含量大小,它们的排序为No.1＝No.4＞No.2＞No.5＝No.3。这5个测点顺次代表高科园、商业区、交通要道或人口密集区和风景游览区。胃肠炎病菌含量在各点间的差异反映出它们所代表的各功能区划间空气质量的差异。此类菌多,意味着相应空气质量较差,反之亦然。海洋空气对胃肠炎病原菌具有稀释作用。

* 原文刊于《国土与自然资源研究》,2006。1:51-52

表1　青岛夏冬空气胃肠炎病原菌测点状况

测点编号、位置及概况	测定时间
1.高科园仙霞岭路某机关大院 雾，湿，>30.0℃，无风 5级风，有飘尘，0℃，阳光斜入	2004-08-04，13:30 2005-02-01，15:45
2.市立医院对面圣保罗教堂围墙上　2～3级东南风，天气同上 3～5级西风，1.5℃	2004-08-04，18:10 2005-02-02，15:08
3.栈桥迴澜阁海堤上风小，人多，雾浓，湿度大，28.0℃ 4～5级西北风，人很少，−1.5℃	2004-08-05，10:15 2005-02-02，5:50
4.水清沟乐安路口，宿舍楼旁有雾气，有云，3级北风，30.5℃ 5级风，风口，1.0℃	2004-08-01，10:32 2005-02-02，10:58
5.台东步行街3～4级风，30.0℃，人多 4级风，1.0℃，行人来往，阳光斜射	2004-08-01，11:15 2005-02-02，9:05
6.齐东路邮局门口3级风。−2.0℃，车、人不多	2005-02-01，17:08
7.市人代会前花园　放烟爆之后，西风3级，−3.0℃	2005-02-01，21:00
8.中山商城内(地下)人不太多，温度约25.0℃ 人来往多，16.5℃	2004-07-31，11:13 2005-02-02，17:12
9.青岛火车站售票厅人多，15.5℃	2005-02-02，13:08

表2　夏冬季青岛空气胃肠炎病原菌测定情况 CFU/m³

测区区位	测点编号	夏	冬
室外	1	786.0	9 589.2
	2	629.0	1 729.2
	3	471.6	943.2
	4	786.0	3 772.8
	5	471.6	1 257.6
	6	—	2 829.6
	7	—	314.4
室内	8	943.2	471.6
	9	—	314.4

夏季室内空气胃肠炎病原菌含量(943.2CFU/m³)是室外平均值的1.5倍．也比室外最高值还高，反映了该特定场所的空气质量之差，暗示人们尽可能少聚集，尤其是在炎夏。

由表2还可以看出青岛冬季空气中胃肠炎病原菌检出率也为100%，其室外的平均含量为2 919.4CFU/m³，变幅为314.4～9 389.2CFU/m³，最高是最低的30.5倍，按其在各测点的含量大小排序为 No. 1＞No. 4＝No. 6＞No. 2＞No. 5＞No. 3＞No. 7。除上述 5 个测点外，No. 4 和 No. 7 分别代表居民区和清静区，由此可见冬天的青岛空气中呈现出人群聚居之地的胃肠炎病原菌含量高于交通要道、清静区或临海场所，后者该类菌最少. 这表明人群的稀疏和空气的新鲜会降低其含量。No. 1 出现高量菌，而且冬夏一致居高不下。此处虽非热闹之地，也不是交通车辆频繁之处。原因可能是空气污染重，附近有粪便等污染形成的气溶胶飘来[4]。夏冬相比，夏天室外该类菌平均含量仅是冬天的 0.2 倍。这表明夏天室外空气好于冬天。

从表 2 可看出冬天的室内空气中胃肠炎病原菌含量平均值为 393.0CFU/m^3，同期室外该值约是它的 7.4 倍。夏冬相比，夏天室内该类菌含量是冬天的 2.4 倍，表明冬天的室内空气好于夏天。各测点病原菌含量排序在夏冬两季间相比，趋势似有相似之处。这表明 No. 1 测点代表的高科园空气在这方面的污染是相当重的，而测点 No. 3 临海一带代表的则污染较轻。

2.2　青岛室外空气中胃肠炎病原菌含量的时间变化

夏天的 5 个测点按时段可分为上午(No. 3，No. 4)、中午(No. 5)和下午(No. 1，No. 2)3 个时段，这 3 个时段的空气胃肠炎病原菌分别按大小排序为下午＞上午＞中午，表明测区空气中胃肠炎病原菌含量下午达峰值。

冬天的 7 个测点按时段可分作为清晨(No. 3)、上午(No. 5)、中午(No. 4)、下午(No. 1，No. 2，No. 6)和夜晚(No. 7)，各时段的空气胃肠炎病原菌含量按含量大小排序为下午＞中午＞上午＞清晨＞夜晚，也显示为下午达高峰。整个白天的状况冬夏不尽一致，反映处于不同区位和环境状况中的测点有差异。

夏冬相比，室外空气胃肠炎病原菌含量显出前者低于后者。这意味着空气中胃肠炎病原菌含量及其分布与人群聚散和污染状况密切相关，而与气温关系不大。对冬天空气中的胃肠炎病原菌状况应引起警惕。相关研究尚待进行。

参考文献 4 篇(略)

PRIMARY ANALYSIS ON PATHOGENIC BACTERIAL CONTENT CAUSING GASTROETERITIS IN THE AIR ABOVE QINGDAO

Abstract The content of pathogenic bacteria causing gastroeteritis was determined by using the culture medium for this kind of bacteria in the air from 7 outdoor and 2 indoor sampling sites, above Qingdao during summer and winter seasons. The result obtained shows that: the detection percentage of the bacteria was 100% during the two sessons. Its outdoor average contents during summer and winter were 628.8 and 2,919.4 CFU/m^3 respectively. The indoor content in summer than the winter and the indoor content was higher than the outdoor one during the summer, but the outdoor content during the winter was higher than one during the summer. Its time domain distributional state was analyzed and illustrated that it had closed relation to the gathering and dispersion of the multitude and condition of pollution, but had not very close relation with the determining air temperature.

Key words Qingdao; Air; Pathogenic Bacteria of Causing Gastroenteritis

日内瓦空气微生物含量

摘　要　用平皿自然沉降法对日内瓦空气微生物含量作了初步调查。调查发现，日内瓦空气微生物检出率高，25℃和37℃培养所得的空气微生物含量平均值分别为2 515.2和2 719.5 CFU·m^{-3}，指示日内瓦空气经受一定的微生物污染，商场空气微生物含量较高。比较、分析了空气微生物含量、比率等的组成情况及测点，城市功能区划间的差异，证实不同区划的污染物、生态、环境和人文活动等在不同温度的控制下影响了空气微生物类别和含量上的差别。

关键词　日内瓦　空气微生物　空气污染　空气质量

日内瓦(Geneva)市座落于瑞士西部的日内瓦湖畔，与法国接壤，并与阿尔卑斯山相邻。罗纳河从东至西流贯繁华市区。它是著名的旅游胜地和国际性城市，有18.5万人口，是重要的金融、贸易和工业中心。夏天的日内瓦气温较高，但阳光明媚。

空气中有不同类别的微生物存在。过多的微生物包括病原菌含量引起人、动植物病害，因而空气微生物状况是衡量特定区域空气污染程度、空气质量好坏的重要指标，并为各国所重视[1]。有关日内瓦空气微生物的报道不多见，本文记录了日内瓦空气微生物含量状况的初步调研结果，并由此分析、评价空气污染和空气质量。

1. 采样点概况*

日内瓦空气微生物采样点计10个，其中包括一个准室内点，采样点概况描述列于表1之中。这些采样点分布于城市不同功能区域之中，因而可代表或反映全市空气微生物状况。

表1　日内瓦空气微生物采样点环境概况

测点编号	测点名称	测点状况	测定起始时间*
1	中心火车站外大街	晴天，车、人来往	2005.6.10. 14:59
2	日内瓦湖 Mont-Blanc Bridge 边	湖中喷水柱有水雾飘来	2005.6.10. 15:46
3	＋CICR＋5 路公交车 Appia 站附近，联合国红十字会总部，	山丘、绿树下有学生走动，3级风吹来	2005.6.10. 15:29
4	Quai Gustave-Ador 旁，湖滨花园	有游人，离喷泉更近	2005.6.11. 7:09
5	日内瓦湖上	游船在湖上晃动，湖风从门窗进来	2005.6.11. 7:25
6	Bellevalle 码头附近湖面	船只晃动，湖风入窗	2005.6.10. 17:46

* 原文载于《中国学术大百科全书》学术卷Ⅱ(中国文化传媒出版社，北京，P954-956，2006)

（续表）

测点编号	测点名称	测点状况	测定起始时间*
7	8 路公交车 Rued'Italie Riuve 站附近	车人来往	2005.6.10. 18:21
8	16 路公交车 Goulart 站附近	车子不少，路上	2005.6.10. 18:35
9	火车站前南 Rue chabonnieke，饭店前	有车人来往	2005.6.10. 18:51
10	火车站东侧 Migros 店内	有不多的人来往，屋顶开向天空	2005.6.10. 19:01

2. *材料和方法*

选择 Bacto agar(Dickinson and company 出品)培养基用自然沉降法在设定采样点按常规方法采集空气样本[2]。将采过样的平皿置 25℃或 37℃各培养 48h 后计数各皿中长出的菌落数，并换算为单位体积空气中的细菌(B)、霉菌(M)、酵母菌(y)及总菌(T)的含量，即 CFU·m^{-3}，据此作空气微生物含量分析，并评价空气微生物污染状况。

表 2　日内瓦空气微生物不同培养温度所得含量(CFU·m^{-3})比较

采样点编号	B	B/T%	M	M/T%	y	Y/T%	T	B	B/T%	M	M/T%	y	Y/T%	T
	25℃							37℃						
1	157.2	33.3	314.4	66.7	0	0	471.6	157.2	33.3	314.4	66.7	0	0	471.6
2	157.2	0.93	16 820.4	99.1	0	0	16 977.6	943.2	6.7	13 362.0	92.4	157.2	1.1	14 462.4
3	157.2	14.3	943.2	85.7	0	0	1 100.4	314.4	28.6	786.0	71.4	0	0	1 100.4
4	0	0	0	0	0	0	0	314.4	100.0	0	0	0	0	314.4
5	0	0	157.2	100.0	0	0	157.2	786.0	71.4	314.4	28.6	0	0	1 100.4
6	1 100.4	77.8	314.4	22.2	0	0	1 414.8	1 257.6	80.0	157.2	10.0	157.2	10.0	1 572.0
7	628.8	100.0	0	0	0	0	628.8	628.8	100.0	0	0	0	0	628.8
8	0	0	157.2	100.0	0	0	157.2	314.4	100.0	0	0	0	0	314.4
9	2 358.0	88.2	314.4	11.8	0	0	2 672.4	3 772.8	92.3	314.4	7.7	0	0	4 087.2
10	1 572.0	100.0	0	0	0	0	1 572.0	3 144.0	100.00	0	0	0	0	3 144.0

表 3　日内瓦空气微生物不同温度所得含量指标的差异

微生物类别	25℃			37℃		
	检出率(%)	含量	(含量·总量$^{-1}$%)	检出率(%)	含量	(含量·总量$^{-1}$%)
B	70.0	613.1	- 24.4	100.0	1 163.3	42.8
M	70.0	1 902.1	75.6	60.0	1 524.8	56.1
Y	0	0	0	20.0	31.4	1.1
T	90.0	2 515.2	100.0	100.0	2 719.5	100.0

3. *结果和讨论*

表 2 列出了日内瓦 10 个采样点空气微生物含量状况，据该表可计算出 25℃或 37℃培

养出来的各类微生物检出率及平均含量(表 3)。由表 3 可见,总体上看,此时的日内瓦空气微生物检出率较高,但 37℃时的微生物检出率更高于 25℃的。空气微生物总量也是 37℃高于 25℃。这两个平均含量与巴塞尔和伯尔尼的相比,均显出较高水平*。从这一点来讲,日内瓦空气中微生物污染状况较重。但与国内外的其他一些城市比又轻了不少[3-4]。霉菌含量在总菌量中的比率较高,是日内瓦空气微生物污染的主因子,显出与其他城市的不同[5]。

表 4 日内瓦空气微生物不同培养温度所得含量在测点间的排序

微生物类别	25℃	37℃
B	No. 9＞No. 10＞No. 6＞No. 7＞No. 1＝No. 3＞No. 2＞No. 4＝No. 5＝No. 8	No. 9＞No. 10＞No. 6＞No. 2＞No. 5＞No. 7＞No. 4＝No. 8＝No. 3＞No. 1
M	No. 2＞No. 3＞No. 9＝No. 1＞No. 6＞No. 5＝No. 8＞No. 4＝No. 7＝No. 10＝0	No. 2＞No. 3＞No. 1＝No. 5＝No. 9＞No. 6＞No. 4＝No. 7＝No. 8＝No. 10＝0
y	No. 1＝No. 2＝No. 3＝No. 4＝No. 5＝No. 6＝No. 7＝No. 8＝No. 9＝No. 10＝0	No. 2＝No. 6＞No. 1＝No. 3＝No. 4＝No. 5＝No. 7＝No. 8＝No. 9＝No. 10＝0
T	No. 2＞No. 9＞No. 10＞No. 6＞No. 3＞No. 7＞No. 1＞No. 5＝No. 8＞No. 4＝0	No. 2＞No. 9＞No. 10＝No. 6＞No. 5＝No. 3＞No. 7＞No. 1＞No. 4＝No. 8＝0

表 4 列出的是日内瓦空气微生物在不同温度培养时所得含量在测点间的排序比较,这表达了测点间空气质量的差异。室内空气仅测到细菌,其细菌及总菌含量一般均比室外 9 个测点的平均值大(表 5),说明日内瓦室内外空气微生物组成和含量上存在着不同之点。测点各类微生物含量等的极差状况则列于表 6 之中。表 6 说明两种培养温度所得的霉菌含量在测点间的极差均较大,25℃的更大于 37℃时的,因而霉菌是造成微生物含量极差的主因。

表 5 日内瓦室内外空气细菌及总菌含量($CFU \cdot m^{-3}$)

测点类别	25℃		37℃	
	B	T	B	T
室外	506.4	2 620.0	943.2	2 672.4
室内	1 572.0	1 572.0	3 144.4	3 144.4

表 6 日内瓦空气微生物不同温度时的含量等指标的极差比较

微生物指标	25℃	37℃
B 极差	2 200.8	3 615.6
M 极差	16 663.2	13 204.8

* 陈皓文,巴塞尔(Basel)空气微生物含量状况;伯尔尼(Berne)空气微生物含量状况

(续表)

微生物指标	25℃	37℃
y 极差	0	157.2
T 极差	16 820.4	14 148.0
B 极差 · T 极差$^{-1}$%	13.1	25.6
M 极差 · T 极差$^{-1}$%	99.1	93.3
y 极差 · T 极差$^{-1}$%	0	1.1
T 极差 · T 极差$^{-1}$%	100.0	100.0
B/T% 极差	99.6	93.3
M/T% 极差	88.5	84.3
Y/T% 极差	0	8.9
T/T% 极差	100.0	10.0

表 7　日内瓦空气微生物最高/最低含量指标间的差异状况

培养温度	25℃	37℃
B	15.0	24.0
B/T%	250.0	14.9
M	107.0	85.0
M/T%	8.7	11.7
y	/	1.0
Y	/	9.1
T	108.0	46.0

表 7 列出的是最高/最低空气微生物含量指标出现的测点间比率的比较。由该表可知,25℃温度培养时得出的这些指标值有些高于 37℃,尤以 B/T% 和 T 的更为明显,说明较低温时,空气微生物含量指标差异变大的可能性更大些。

表 8　日内瓦城市不同功能区划空气微生物含量等状况

培养温度 / 指标平均含量等 / 区划编号	25℃							37℃						
	B	B/T%	M	M/T%	y	y/T%	T	B	B/T%	M	M/T%	y	y/T%	T
Ⅰ	157.2	0.9	16820.4	99.1	0	0	16977.6	993.2	6.5	1336.2	92.4	157.2	1.1	14462.4
Ⅱ	550.2	70.0	235.8	30.0	0	0	786.0	1021.8	76.5	235.8	17.8	78.6	5.7	1336.2
Ⅲ	78.6	14.3	471.6	85.7	0	0	550.2	314.4	44.4	393.0	55.6	0	0	707.4
Ⅳ	786.0	80.0	196.5	20.0	0	0	982.5	1218.3	88.5	157.2	11.5	0	0	1375.2
Ⅴ	1572.0	100.0	0	0	0	0	1572.0	3144.0	100.0	0	0	0	0	3144.0

表9　日内瓦空气微生物含量等在城市不同功能区划间的排序

培养温度	25℃	37℃
B	Ⅴ＞Ⅳ＞Ⅱ＞Ⅰ＞Ⅲ	Ⅴ＞Ⅳ＞Ⅱ＞Ⅰ＞Ⅲ
B/T%	Ⅴ＞Ⅳ＞Ⅱ＞Ⅲ＞Ⅰ	Ⅴ＞Ⅳ＞Ⅱ＞Ⅲ＞Ⅰ
M	Ⅰ＞Ⅲ＞Ⅱ＞Ⅳ＞Ⅴ＝0	Ⅰ＞Ⅲ＞Ⅱ＞Ⅳ＞Ⅴ＝0
M/T%	Ⅰ＞Ⅲ＞Ⅱ＞Ⅳ＞Ⅴ＝0	Ⅰ＞Ⅲ＞Ⅱ＞Ⅳ＞Ⅴ＝0
y	Ⅰ＝Ⅲ＝Ⅱ＝Ⅳ＝Ⅴ＝0	Ⅰ＞Ⅱ＞Ⅲ＝Ⅳ＝Ⅴ＝0
y/T%	Ⅰ＝Ⅲ＝Ⅱ＝Ⅳ＝Ⅴ＝0	Ⅱ＞Ⅰ＞Ⅲ＝Ⅳ＝Ⅴ＝0
T	Ⅰ＞Ⅴ＞Ⅳ＞Ⅱ＞Ⅲ	Ⅰ＞Ⅴ＞Ⅳ＞Ⅱ＞Ⅲ

将所有采样点按城市功能区划的不同分作五类，即Ⅰ类区为树下，包括No.2测点，Ⅱ类区为湖内，包括No.5、6，Ⅲ类区为湖滨，包括No.3、4，Ⅳ类区为街道，包括No.1、7、8、9，Ⅴ类区为室内，仅No.10测点。五类区划中空气微生物含量指标各自的平均值分别列入表8之中。对表8所列两组五列数据按大小排序的情况列入表9。所以表8、表9两者大致表达出树下空气霉菌(M)、酵母菌(y)最多，而室内霉菌(M)、酵母菌(y)最少，湖滨区(Ⅲ类区)空气中细菌(B)和总菌最少。

4.结语

本文对日内瓦空气微生物作了一次初步的调研，结果表明该市空气微生物的检出率高。两种培养温度所得的空气微生物平均含量分别为2 515.2(25℃)和2 719.5(37℃) CFU·m^{-3}。温度高，促使空气微生物类群更多样化，含量也攀高。本结果意味着日内瓦空气有一定的微生物污染，但并不严重。文章分析了测点间、城市不同功能区划间微生物含量指标的差异。表明城市不同功能区划间空气微生物受制于当时当地气温、污染物、生态环境和人文活动。

参考文献5篇(略)

(合作者：李为民　黄秀兰)

AIRBORNE MICROBIAL CONTENT OVER GENEVA

(ABSTRACT)

Abstract Airborne microbial content was investigated by using the method of gravity plate. The investigation found that the microbial occurrence percentage was higher. The average contents of airborne microbes were 2 515. 2 and 2 719. 5 CFU · m^{-3} when they were cultivated at 25℃ or 37℃ respectively. The result indicated that the air was underwent microbial pollution definitely over Geneva. There was a higher content of airborne microbes in supermarket examined. This paper compared and analyzed the compositional conditions of microbial kinds, their contents and percentages, differences among sampling sites and city plans of functional areas. It was confirmed that the contaminations, eco-environment and activities of human being under controlling of different temperatures impact/effect difference of air-borne microbial kinds and contents.

Key words Geneva, Air-Borne Microbes, Air Pollution, Air Quality

古都洛阳与中国内陆三城市空气微生物的含量比较*

摘　要　用平皿沉降法初步测定洛阳空气微生物含量状况，结果表明该市空气细菌、真菌、总菌量及真菌/总菌百分比大体上是 33 247.8，550.2 CFU/m³，33 798.0 及 1.6，说明空气微生物状况已达中度污染上限，虽然真菌含量%低。分析了造成这一状况的原因，指出了一些改善思路。与中国内陆三个北方城市作了比较，结果表明洛阳空气细菌和总菌量居高不下。分析了四市空气微生物含量状况。

关键词　洛阳　中国内陆北方城市　空气微生物　空气污染　空气质量

一、引言

处于暖温带内的古都洛阳，市区坐落于盆地之中，街道甚为平坦，尘土较多。4～5 月间空气湿度和风力不大，牡丹盛开，游人增多，使本已污染的空气质量呈下降之势，其重要的一个因素是空气微生物。微生物包括一些病原体凭借大气中的颗粒物、气溶胶为载体或营养源沉浮、繁殖于空气，散布至人群和各种生物、物品，从而污染环境，有害健康。这在当今加速的开发西部和促进健康的城市化进程中日益明显，因而研究和比较分析城市空气微生物生态问题就有紧迫性和重要意义。空气微生物学这一课题正受到各国学者的重视[1-2]。据悉，对洛阳等城市空气微生物状况知之甚少。本文试图对洛阳及同处黄河流域的济南、太原和兰州的空气微生物状况作一初步探讨、比较和分析。

二、材料和方法

用平皿沉降法采集洛阳市区室外空气微生物粒子沉降于细菌培养基或真菌培养基上，采样时间为 5 分钟，采得样的培养皿于 28℃培养 2～3 天后分别计数 CFU，即可得出空气中的微生物含量[3-4]。共设测站计 4 个，采样时间和站位概况详见表 1。济南、太原、兰州三市空气微生物测定方法同此，测站见文献[5-7]。

三、结果和讨论

1. 洛阳空气微生物状况

表 2 列出的是洛阳市空气微生物含量状况，由该表可见所测四站中空气微生物含量的四个指标以测站按大小排序如下。空气细菌量：3，1，4，2；空气真菌量：1，3，4（＝2）；空气微生物总量：3，1，4，2。空气真菌/总菌百分比：1，2，4，3。由细菌与总菌量看，两者站际差异次序一致，表明洛阳空气微生物含量及其差异主要由细菌引起。真菌所占比率不高，

* 原文载于《创新中国探索》，2006，P410-413.

反映了该市空气较为干燥。这一点有利于致病真菌的抑制。

据表2数据可以得出该四测站空气微生物含量各指标的算术平均值分别是细菌量—33 247.8 CFU/m³,真菌量—550.2 CFU/m³,总菌量—33798.0 CFU/m³ 和真菌占总菌的百分比—1.6。按国家推荐标准,即室外空气细菌含量在 $10^4/m^3$ 时,空气达中度污染程度。因此,洛阳空气细菌量已处于中度污染上限,若加上真菌含量,则空气微生物含量更高,污染更重,这是令人担忧的。

分析洛阳空气中微生物来源,一是地面尘埃上浮带来的微生物,二是人们生活、工作的不洁活动散布,使处于较干旱的盆地中的洛阳空气质量进一步下降。

要想改变这一状态,必须改进生产、炊事、取暖方式。减少空气中燃煤源颗粒。广种花草树木,提高绿化水平。干旱天气时,街道上适当洒水,减轻尘土飞扬。而这些措施恰恰必须同根据地理条件扬长避短调整和安排工业结构布局同时进行。这样洛阳有可能在发展经济的同时,使城市生态环境逐步走上健康持续发展之路。

表1 洛阳空气微生物测站记录

测站编号	站名	环境描述	采样起始时间
1	牡丹公园	人造山上,有树木、牡丹含苞待放,气温15℃,阴天,软风	1996年4月11日13:30
2	上海市场马路边	车、人往来,尘埃较多,气温15℃,阴天,软风	1996年4月11日13:50
3	洛阳火车站广场	公交车站附近,人车不多,但飞尘较多气温15℃,阴天,软风	1996年4月11日14:30
4	解放路纱厂路口人行道上	车、人来往不多,气温15℃,阴天,软风	1996年4月11日15:30

表2 洛阳空气微生物含量状况表

测站编号	空气微生物含量*			
	细菌	真菌	总菌	真菌含菌%
1	33 169.2	1 100.4	34 269.6	3.2
2	15 562.8	314.4	15 877.2	1.9
3	60 050.4	471.6	60 522.0	0.8
4	24 208.8	314.4	24 523.2	1.3

* 各微生物含量单位:CFU/m³,总菌量是细菌与真菌量之和,下同。

2. 洛阳与中国其他三个内陆城市空气微生物含量状况的比较分析

表 3 洛阳与中国内陆三城市空气微生物含量相关条件比较*

城市名	空气微生物含量*				相关条件	测定时间和引证文献号
	细菌	真菌	总菌	真菌/总菌%		
洛阳	33 247.8	550.2	33 798.0	1.6	暖温带内。盆地中。街道平坦。尘土较多。较干旱。机械、棉纺、建材、炼油等,工业发达。	1996.4 本文
济南	14 100.8	1 886.4	15 987.2	11.8	暖湿带半温润气候。黄河下游南岸。兼有山麓丘陵,平原低地。机械、石化、钢铁、纺织、面粉、造纸业	1997.6 [5]
太原	24 819.9	627.4	25 447.3	2.47	大陆性气候,黄土高原东缘,三面环山丘陵盆地中。能源和重化工工业发达	1996.4 [6]
兰州	18 785.4	458.5	19 243.9	2.4	温带半干旱气候,周围有大山。逆温强,风速小。工业以石油、化工为主,烟雾重。	1996.4 [7]
平均	22 738.5	880.6	23 619.1	3.7		
洛阳/济太平平均值	1.73	0.56	1.67	0.29		

* 济南、太原、兰州三市空气微生物测定法同洛阳的。

表 3 列出的是洛阳与中国三个内陆城市济南、太原、兰州空气微生物含量指标的比较结果。

由此表可见,四市空气微生物均以细菌为主,其中细菌量以洛阳的为最高、最明显。空气微生物总量以洛阳的为最高。真菌所占百分比以洛阳的最低,说明测时的洛阳空气相对湿度低。最高真菌百分比在济南。它既反映了济南所处气候带这一地理条件,也反映了测时和测处空气的相对湿度较高,这有利于真菌的生存[5]。也从一个侧面反映了中国北方东部城市(如济南)与中西部城市(如其他三市)间空气质量的不同。

这四个城市都地处黄河流域,周围都有山岭,市区相对低平,降雨少而集中,干燥,空中尘土和颗粒物使较多空气微生物有了载体和营养源,而各市空气微生物含量四个指标间又都存在一定差异。这可能与各市具体的地理环境、工业状况有关[6-7]。总体看,洛阳空气细菌量、总菌量均大大高于四市相应平均值。仍然表达出洛阳空气微生物污染的不容乐观性。

三、结语

本文重点论述了洛阳市区空气微生物含量状况,指出了空气微生物污染的严重性和改进这一状况的路子。结合比较分析了同处黄河流域的其他三个城市济南、太原和兰州的空气微生物状况。从一个侧面表达了我国北方东、西部城市间空气质量上的差别。

参考文献 7 篇(略)

洛桑空气微生物检测*

摘　要　对洛桑空气微生物含量用平皿自然沉降法作了检测。结果发现洛桑空气微生物检出率高，25℃和37℃培养出的空气微生物总量分别为471.6和655.0 CFU·m^{-3}。表明温度的适当提升可使空气微生物出现机率增多，微生物含量增高。比较分析了测点间城市不同功能区划间各类微生物含量指标的分布、差异和状况。证实该分布、差异和状况与测定当时当地的环境、生态状况、人类活动等密切相关。一般言，结果表明洛桑空气微生物污染很轻，空气质量很好。

关键词　洛桑　空气微生物　空气污染　空气质量

洛桑位于瑞士由阿尔卑斯山陡峭斜坡下至日内瓦湖畔的湖光山色中，人口12.5万，环境优美，是国际奥林匹克之都。闹市新古建筑密集。商业、金融和文教相当发达。同时环境包括空气也存在一定的污染问题。

大气圈下垫层的空气一般都有微生物包括病原菌生存。空气微生物与环境、生态状况息息相关，病原菌尤与人类、动植物健康关系密切，因而检测空气微生物状况十分必要，并为各国所重视[1]。对洛桑空气微生物状况的研究，相关报道甚少。本文报道洛桑空气微生物含量的初步调查结果，以期能引起瑞士环境科学、医务工作者的关注和讨论。

表1　洛桑空气微生物测点概况

测点编号	测点名称	测点状况	测定起始时间*
1	火车站外大街	晴天，车、人来往，有鸽子，最高气温24.0℃	2005.6.15 10:30
2	Sun store 过街道 Passage jumelles	人少，有2～3级风	2005.6.15 11:14
3	5路电车 Vernes 站，近东郊	附近有工地，车辆来往多，阳光直射	2005.6.15 11:51
4	Ouchy 地铁站东	湖边，树荫下，有鸟、湖风2～3级，人车不多，	2005.6.15 12:47
5	Ouchy 地铁站近门口	乘客不多	2005.6.15 13:57
6	火车站铁道边	旅客来往，上下车，静风	2005.6.15 14:07

1.洛桑空气微生物测点概况描述

表1列出了洛桑6个空气微生物测点的大致状况。由该表可见，这6个测点分布于洛桑全市不同城市功能区划的地理环境和方位中，对全市的空气状况有一定的代表性和涵盖度。

* 原文载于《世界重大学术成果精选——华人卷Ⅲ，珍藏版》，2007，73-75。

2.洛桑空气微生物的检测材料和方法

选择 Bacto agar (Dickson company 出品)。用自然沉降法至上述设定测点按常规方法采集气样至培养皿平板上。在25℃和37℃分别培养采得气样的平板达48h。计算平板上长出的菌落数,换算为单位体积空气中的菌落形成数,即 $CFU \cdot m^{-3}$[2]。由此计算空气微生物细菌(B)、霉菌(M)、酵母菌(y)及总菌(T—三者相加的总和)等含量指标,以分析和评价洛桑空气微生物污染状况和空气质量。

表 2 洛桑空气微生物含量状况表 $CFU \cdot m^{-3}$

采样点编号(No.)	25℃						37℃							
	B	B/T%	M	M/T%	y	Y/T%	T	B	B/T%	M	M/T%	y	Y/T%	T
1	0	0	314.4	100.0	0	0	314.4	786.0	71.4	314.4	28.6	0	0	1 100.4
2	0	0	157.2	100.0	0	0	157.2	0		157.2	100.0	0	0	157.2
3	786.0	100.0	0	0	0	0	786.0	628.8	57.1	314.4	28.6	157.2	14.3	1 100.4
4	0	0	0	0	0	0	0	0	0	0	0	0	0	0
5	157.2	100.0	157.2	100.0	0	0	314.4	157.2	25.0	471.6	75.0	0	0	628.8
6	0		1 257.6	100.0	0	0	1 257.6	0		943.2	100.0	0	0	943.2

3.结果和讨论

表2列出的是洛桑6个测点空气微生物含量指标的状况由该表可计算出洛桑各类空气微生物的平均检出率等(表3)。根据表3可知,37℃的培养温度所得的各空气微生物检出率均大于25℃时的,这表明温度的提高将增加微生物的出现机率,微生物污染机率随即增多。据表2数据可得出表4。表4说明了全洛桑空气微生物含量指标的大致状况,25℃和37℃培养出的洛桑空气微生物总量(T)分别为471.6和655.0 $CFU \cdot m^{-3}$。证实温度的适当提高,将助涨空气微生物含量的检出。空气微生物以霉菌为主,即洛桑空气微生物污染以霉菌为主,虽然细菌是不可忽视的因子。与其他城市比,洛桑空气的微生物污染很轻,在瑞士所测城市中则首屈一指,因而表明洛桑空气质量很好[3,4]*

表 3 洛桑空气微生物检出率(%)

培养温度(℃)	B	M	y	T
25	33.3	50.0	0	83.3
31	50.0	83.3	16.7	83.3

表 4 洛桑空气微生物含量指标不同温度培养所得平均比较 $CFU \cdot m^{-3}$

培养温度(℃)	B	B/T%	M	M/T%	y	y/T%	T
25	157.2	35.3	288.2	64.7	0	0	471.6
37	262.0	40.0	366.8	56.0	26.2	4.0	655.0

* 陈皓文,巴塞尔空气微生物含量分析,伯尔尼空气微生物含量。

表 5　洛桑空气微生物含量在测点间的排序

微生物类别	25℃	37℃
B	No. 3＞No. 2＝No. 5＞No. 1＝No. 4＝No. 6＝0	No. 1＞No. 3＞No. 5＞No. 2＝No. 4＝No. 6＝0
M	No. 6＞No. 1＞No. 2＞No. 3＝No. 4＝No. 5＝0	No. 6＞No. 5＞No. 3＝No. 1＞No. 2＞No. 4＝0
y	No. 1＝No. 2＝No. 3＝No. 4＝No. 5＝No. 6＝0	No. 3＞No. 1＝No. 2＝No. 4＝No. 5＝No. 6＝0
T	No. 6＞No. 3＞No. 1＞No. 2＞No. 5＞No. 4＝0	No. 3＞No. 1＞No. 6＞No. 5＞No. 2＞No. 4＝0

表 6　洛桑空气微生物含量等指标测点间极差变化　CFU·m^{-3}

培养温度(℃)	25	37
B 极差	628.8	628.8
B/T%极差	100.0	46.4
M 极差	1 100.4	786.0
M/T%极差	100.0	71.4
y 极差	0	157.2*
y/T%极差	0	14.3*
T 极差	1 100.4	943.2

*与未检出相比,其他均与检出相比较

表 5 则是洛桑所有测点空气微生物含量按大小排序所得的比较结果。该表说明测点间存在一定的差异。25℃时,No. 6 测点的霉菌和总菌含量最高,而细菌和酵母菌量却也是 No. 6 测点最低。37℃时 No. 3 测点的酵母菌和总菌量最高,而 No. 4 测点细菌、霉菌、酵母菌和总菌量则最小。结果表现出测点的空气微生物含量对不同温度有着不同的反应,也意味着 No. 4 和 No. 6 测点空气质量上的较大差异。

表 6 是空气微生物含量指标在测点间的极差之比较。该表表明培养温度低时,各种极差比温度高时的大,证实自然温度低的空气中,其微生物对外界的感应更加强烈些。

将洛桑空气微生物测点按城市不同功能区划分作三类,即Ⅰ类区包括 No. 1、2、3 三个测点为街道,Ⅱ类是湖滨区,有 No. 4 和Ⅲ类区包括 No. 5、6,为铁道区。据此三个功能区各自的平均空气微生物含量指标列入表 7 之中。

表7　洛桑城市不同功能区域间空气微生物平均含量指标的差异　CFU·m^{-3}

功能区划编号	培养温度													
	25℃							37℃						
	B	B/T%	M	M/T%	y	y/T%	T	B	B/T%	M	M/T%	y	y/T%	T
Ⅰ	262.0	62.5	154.2	37.5	0	0	419.2	471.6	60.0	262.0	33.3	52.4	6.7	786.0
Ⅱ	0	0	0	0	0	0	0	0	0	0	0	0	0	0
Ⅲ	78.6	11.1	628.8	88.9	0	0	786.0	78.6	10.0	707.4	90.0	0	0	786.0

由表7可见，三区空气微生物含量指标间存在明显的不同。Ⅱ区即湖滨区未检测出任何空气微生物，意味着此类环境的空气最为洁净，而Ⅲ类区即铁道一带空气最为污浊，且主要由霉菌引起微生物污染。结果表达出功能区划的不同，决定了它们环境条件和状况的不同，由此也产生了不同的微生物质量和数量差别[5]。

4. 结语

本文报道了洛桑空气微生物含量的检测结果。检测表达出洛桑空气微生物的检出率高。25℃和37℃培养所得结果表明洛桑空气微生物总量分别为471.6和655.0 CFU·m^{-3}，证实其空气微生物污染很轻。测点间、城市不同功能区划间空气微生物含量指标均存在一定差距。这些差距意味着测点/区划的空气或环境上的差距。文章对此作了一定的比较和分析，总体看洛桑空气质量优良。

参考文献5篇(略)

（合作者：李为民　陈颖稚）

INSPECTION ON AIR-BORNE MICROBIAL CONTENT OVER LAUSAUNE

(ABSTRACT)

Abstract The air-borne microbial content over Lausanne was inspected by using the method of gravity plate. The result obtained found that the occurrence percentage of air-borne microbes was high over Lausanne. The total air-borne microbial contents over Lausanne were 471.6 and 655.0 CFU • m^{-3} cultivated at 25℃ or 37℃ respectively. The result showed that the proper elevation of temperature would increase of occurrence chance of airborne microbes and raise of the microbial content. It was compared and analysed to the distributance status and their differences of each kind of microbial content indexes among sampling sites or city plan of functional areas. It was confirmed that they were closely related to the environmental-ecological conditions and activity of human being etc at that time and that place of samplings. In general speaking, the result indicated that the microbial pollution in air was not serious over lausanne and its air quality was very good.

Key words Lausanne, Air-Borne Microbes, Air Pollution, Air Quality

伯尔尼(Berne)空气微生物含量状况*

摘 要 本文用平皿自然沉降法,选择两种培养温度调研了瑞士伯尔尼空气微生物含量状况。结果证实伯尔尼空气中微生物检出率高达100.0%,其微生物平均总量大于1 500.0CFU·m^{-3}。培养温度不同出现的各类微生物含量之差异表达在城市不同功能区划空气中。不同温度的空气结合当时当地的污染、生态状况造成了测点、区划间空气微生物污染差异。总体看伯尔尼空气微生物污染不重,意味着它的空气质量尚可。

关键词 伯尔尼 空气微生物 空气污染 空气质量

伯尔尼(Berne)是瑞士首都。它位于瑞士中西部伯尔尼高原。阿勒河(Aare)在市区东部流贯南北。气候温和。人口12.4万。铁路枢纽。有机械、金属制品、化学、纺织工业。老城区虽小,但充满商业气息,经营钟表著称于世。

空气中有包括病原菌在内的各种微生物,并与种种污染物为伴,因此空气微生物含量是衡量特定区域空气质量的一个重要指标,并为世界各国所重视[1]。

伯尔尼环境总体观感优越,其空气感觉也清新,但它的空气微生物具体状况,报道不多。本文就伯尔尼空气微生物含量作一初步调研,以与瑞士空气环境的相关研究者共同探讨之。

1.伯尔尼空气微生物采样点环境概况

表1列出了伯尔尼十个空气微生物采样点的环境概况及采样时间,由该表可见,这些采样点分布于全市不同方位的三种功能区划,即交通要道、静谧区和市场之中,因而于伯尔尼全市言,具有一定的涵盖性和代表性。

表1 伯尔尼空气微生物采样点环境概况

测点编号	测点名称	环境状况	测定起始时间
1	Wankdorf,20路电车终点站附近	晴朗,近地面,时有小风吹动,马路边附近有修路	2005.6 14:30
2	Wankdorf,20路电车终点站附近南	绿荫中有阳光洒下,有火车	2005.6 14:38
3	一教堂外院内 No. 24, Wylerstrasse	树旁台阶上	2005.6 15:09
4	Pfarramt st. Marien	风小	2005.6 15:18

* 原文载于《中国党政干部理论文选》,2008,8,461-465.

(续表)

测点编号	测点名称	环境状况	测定起始时间
5	21路汽车路终点站,Bremgarten附近	路边,车人很少,太阳直射不到,3级风	2005.6 16:46
6	F ährstrasse 的 stauwehrrain 桥下	河畔、树旁、小风、车少,无他人,地势较低	2005.6 17:05
7	Langgasstrasse 附近,小道 Muraltweg 边的院子口	宿舍楼院,12路电车终点站附近小风,绿化好,无阳光直射	2005.5 17:40
8	Adrian voh Bubemberg 雕塑广场	交通要道口,人、车、鸽,无阳光直射,偶有3级风	2005.6 18:02
9	火车站广场 Spitalgasse 附近	无阳光直射,车人来往,偶有4~5级风	2005.6 18:22
10	火车站广场 Spitalgasse 附近	无阳光直射,车人来往背风	2005.6 18:30

* 测定持续时间 5 min

2. 材料和方法

选用 Bacto agar (Dickinson and company 出品)制成平板培养基,在上述设定的测点采用平皿自然沉降法按常规采集空气样本[2]。采样时间持续 5 min。将取得了样本的培养皿倒置入25℃或37℃各培养48 h后分别计数各皿中的细菌(B)、霉菌(M)、酵母菌(Y)和此三者之和——总菌(T)的菌落,并换算为单位空气体积中的菌落数,即空气微生物含量——CFU · m^{-3},以此比较分析空气微生物含量等指标及空间分布状况,评价伯尔尼的空气微生物污染状况和空气质量。

3. 结果和讨论

表2由两表组成,表达伯尔尼十个测点两种培养温度时得出的空气微生物含量状况。

由表2可计算出伯尔尼空气微生物的检出率,状况如表3所示。

由表3可见,37℃时伯尔尼各类空气微生物检出率均高于25℃时的,虽然两温度时的空气微生物总检出率一样高。说明培养温度对不同空气微生物的检出率影响较大。

据表2可得出25℃、37℃培养得出的微生物含量平均值状况,结果列于表4之中。

表 2-1 伯尔尼空气微生物含量状况(25℃培养)

测点编号	空气微生物含量(CFU · m^{-3})						
	细菌(B)		霉菌(M)		酵母菌(y)		总菌(T)
	含量	B/T%	含量	M/T%	含量	y/T%	CFU · m^{-3}
1	0	0	943.2	100.0	0	0	943.2
2	314.4	33.3	628.8	66.7	0	0	943.2
3	471.6	30.0	1 100.4	70.0	0	0	1 572.0

(续表)

测点编号	空气微生物含量(CFU·m⁻³)						
	细菌(B)		霉菌(M)		酵母菌(y)		总菌(T)
	含量	B/T%	含量	M/T%	含量	y/T%	CFU·m⁻³
4	471.6	50.0	471.6	50.0	0	0	943.2
5	1 729.2	68.8	786.0	31.2	0	0	2 515.2
6	1 572.0	83.3	314.4	16.7	0	0	1 886.4
7	471.6	30.0	1 100.4	70.0	0	0	1 572.0
8	3 772.8	82.8	786.0	17.2	0	0	4 558.8
9	0	0	0	0	157.2	100.0	157.2
10	0	0	157.2	100.0	0	0	157.2

表 2-2 伯尔尼空气微生物含量状况(37℃培养)

测点编号	空气微生物含量(CFU·m⁻³)						
	细菌(B)		霉菌(M)		酵母菌(y)		总菌(T)
	含量	B/T%	含量	M/T%	含量	y/T%	CFU·m⁻³
1	0	0	1 886.4	100.0	0	0	1 886.4
2	157.2	16.7	628.8	66.0	157.2	16.7	943.2
3	471.6	21.4	1 572.0	71.5	157.2	7.1	2 200.8
4	314.4	20.0	943.2	60.0	314.4	20.0	1 572.0
5	157.2	5.6	1 100.4	38.8	1572.0	55.6	2 829.6
6	628.8	33.4	628.8	33.3	628.8	33.3	1 886.4
7	314.4	14.3	1 729.2	78.6	157.2	7.1	2 200.8
8	786.0	62.5	471.6	37.5	0	0	1 257.6
9	0	0	314.4	66.7	157.2	33.3	471.6
10	157.2	50.0	157.2	50.0	0	0	314.4

表 3 伯尔尼空气微生物在不同培养温度时的检出率(%)比较

培养温度	B	M	y	T
25℃	70.0	90.0	10.0	100.0
37℃	80.0	100.0	70.0	100.0

表 4　伯尔尼空气微生物含量在不同培养温度中出现的差异

培养温度	各类空气微生物含量平均值			
	细菌(B)	霉菌(M)	酵母菌(Y)	总菌(T)
25℃	880.3	628.8	15.7	1 524.8
37℃	298.7	943.2	314.4	1 556.3

表 4 说明两种温度培养出的各类空气微生物含量不尽一致。总体看，虽然差距不大，但 37℃时得出的含量仍高于 25℃，这主要由霉菌和酵母菌造成。说明伯尔尼的空气霉菌和酵母菌仍适于较高温度生长繁殖。伯尔尼空气微生物含量总体看高于巴塞尔*[3]，但仍低于国内外许多城市[4][5]。

从表 2 的两表也可看出各测点空气微生物含量间存在一定差异。该测点差异按大小排序的结果列于表 5 之中。

表 5　伯尔尼空气微生物不同温度培养所得的含量在测点间的排序

温度培养	微生物类别	空气微生物含量在测点间的大小排序
25℃	B	No. 8＞No. 5＞No. 6＞N0. 3＝No. 7＝No. 4＞No. 2＞No. 1＝N0. 9＝No. 10＝0
	M	No. 7＝No. 3＞No. 1＞No. 5＝N0. 8＞No. 2＞No. 4＞No. 6＞No. 10＞No. 9＝0
	y	No. 9＞No. 1＝No. 2＝No. 3＞＝No. 4＝No. 5＝No. 6＝No. 7＝No. 8＝No. 10＝0
	T	No. 8＞N0. 5＞No. 6＞No. 3＝No. 7＝No. 1＝No. 2＝No. 4＞No. 9＝No. 10
37℃	B	NO. 8＞No. 6＞No. 3＞No. 7＝No. 4＞No. 2＝No. 5＝No. 10＞No. 5＞No. 9＝0
	M	No. 1＞No. 7＞No. 3＞No. 5＞No. 4＞No. 6＝No. 2＞No. 8＞No. 9＞No. 10
	y	No. 5＞No. 6＞No. 4＞No. 2＝No. 3＝No. 7＝No. 9＞No. 1＝No. 8＝No. 10＝0
	T	No. 5＞No. 7＝No. 3＞No. 1＝No. 6＞No. 4＞No. 8＞No. 2＞No. 9＞No. 10

将测点间含量极差等状况列于表 6。由该表可见，空气细菌含量在温度低时测点间差异大，细菌在总菌中的比率反而变小。温度高时空气霉菌、酵母菌含量的测点差异大，说明温度高时它们生长繁殖机会都有可能增多。

表 6　伯尔尼空气微生物含量极差及在总菌量中所占比率的极差比较*

培养温度	B 含量极差	B/T%极差	M 含量极差	M/T%极差	y 含量极差	y/T%极差
25℃	3 458.4	52.8	943.2	83.3	157.2*	100.0*
37℃	628.8	56.9	1 729.2	66.7	1 414.8	48.5

除*外均系检出率为阳性点间的比较

如前所述可将测点按城市功能区划的不同分为三类，即Ⅰ类为交通要道测点，包括 No. 1、2、5；Ⅱ类为较静谧的测点，包括 No. 3、4、6、7 四点，Ⅲ类为城市广场测点，包括 No. 8、9、10 三点。将各类区划的空气微生物含量各自平均，结果列于表 7 之中。表 7 证实两种培养温度时，各自的各类微生物含量均有上下波动，但总体看 37℃时的波动幅度高于

25℃时的。

表7 伯尔尼城市不同功能区划空气微生物含量比较

培养温度	空气微生物含量(CFU·m^{-3})											
	区划编号											
	Ⅰ				Ⅱ				Ⅲ			
	B	M	Y	T	B	M	Y	T	B	M	Y	T
25℃	681.2	786.0	0	1467.2	746.7	746.7	0	1493.4	1257.6	314.4	52.4	1624.4
37℃	104.8	1205.2	576.4	1886.4	432.3	1218.3	314.4	1965.0	314.4	314.4	52.4	681.2

表8 伯尔尼空气微生物含量在城市不同功能区划间的比较

培养温度	B	M	y	T
25℃	Ⅲ＞Ⅱ＞Ⅰ	Ⅰ＞Ⅱ＞Ⅲ	Ⅲ＞Ⅱ＞Ⅰ	Ⅲ＞Ⅱ＞Ⅰ
37℃	Ⅱ＞Ⅲ＞Ⅰ	Ⅱ＞Ⅰ＞Ⅲ	Ⅰ＞Ⅱ＞Ⅲ	Ⅱ＞Ⅰ＞Ⅲ

将表7的数据比较得出的排序列于表8之中，由此可见，不同培养温度时的微生物含量不同，因而影响到它们在城市不同功能区划中的排序。总的看，大致倾向是25℃培养得Ⅲ区空气中细菌、酵母菌和总菌偏多，而37℃培养得Ⅱ区空气中的细菌、霉菌和总菌偏多。这是由不同功能区划的地理生态环境、空气洁净程度，包括各区划人为污染和测时小气候的状况所决定的。

4. 结语

本文对伯尼尔空气微生物含量作了一次粗放的调查，结果表明伯尔尼空气微生物检出率高。两种培养温度所得的空气微生物含量及其在总菌量中所占比率存在一定差异，但总体看37℃时的微生物多，尤其霉菌及酵母量增多，所占比率增高。分析了测点间、城市各功能区划间空气微生物状况的差异。调查证实伯尔尼空气微生物含量不算大，即微生物污染较轻，意味着伯尔尼空气质量尚可，适于人类生存程度较高。

参考文献5篇(略)

（合作者：李为民）

CONDITION OF AIR-BORNE MICROBIAL CONTENT OVER BERNE

(ABSTRACT)

Abstract The distribution of air-borne microbial content over Berne, Switzerland was examined using the method of gravity plate and two kinds of cultivating temperatures. The result obtained verifies that: the occupancy percentage of air-borne microbes is at 100.0% over Berne and its total average content is higher than 1,500.0 CFU · m^{-3}. The differences of each kinds of microbes occurred at different cultivating temperatures expresses that: to combine different temperatures with polluting/ ecological conditions at that time and the place causes the difference of air-borne microbial pollution in the air over different city plans of functional areas. In general speaking, the air-borne microbial pollution was not serious over Berne. The result suggests that the air quality was not still bad in Berne.

Key words Berne, Air-Borne Microbes, Air Pollution, Air Gravity

银川空气微生物含量状况分析*

摘　要　用自然沉降法测定了银川 12 个测点的空气微生物含量，结果表明银川空气受微生物污染已达到中等程度，个别测点污染严重；分析了银川空气微生物含量的时空分布状态，指出了监测银川空气污染、提高空气质量的重要性。

关键词　银川　空气微生物　空气质量

银川市位于宁夏区贺兰山和黄河之间的平原上，其气候属大陆性，冬春时有沙尘暴。工业有机电、钢铁、仪表、化工和毛纺等。市内外有些景点/区，如宁园、西夏王陵等。市区人口约 50 万。这些对环境造成一定的污染压力。

空气是人类赖以生存的要素之一，又是各种微生物，包括病原微生物生存、繁殖和散布的重要媒介。空气微生物种类、含量、分布与其所在环境状况密切相关，反映环境质量，并对人畜禽等健康有重大影响。当今空气微生物学研究已成为世界各先进国家环境科学、微生物学和医学的交叉学科[1]。但据知对银川空气微生物的研究成果甚少。本文论述银川空气微生物含量及其变动，以供有关人员参考。

1　材料与方法

将普通营养琼脂和沙氏真菌培养基作为本次空气微生物的培养介质，按常规的自然沉降采样法选取市内不同城市功能区划、方位计 12 个测点/时曝皿 5 min，培养于 37℃左右 2～3d 后计数各皿中的菌落形成数(CFU，下同)，换算为单位体积空气(m^3)中的空气细菌(B)、空气真菌(F)、空气总微生物含量(T)及真菌占总菌量的%(F/T%)，比较分析各测点/时空气微生物含量的时空分布变化，判断微生物引起的空气污染程度以指示所测地区的空气质量[2～3]。

2　测点/时描述

在银川所设 12 个空气微生物测点/时状况列于表 1 之中，结合银川地图，可见所有测点/时可覆盖和代表该市不同环境、位置和时段的空气微生物状态。

表 1　银川空气微生物测点/时环境状况表

编号	测点	一般状况	测定时间*
1	1 路汽车工商银行站	小毛雨，阵风 5 级	2004-08-19. 15:30
2	1 路汽车满城南街站	阵风 4 级，阴云	2004-08-19. 15:45
3	银川火车站花园	3～4 级风，阴，树多	2004-08-19. 16:15
4	兴顺农贸市场内	2 级风，阴云	2004-08-19. 17:21

* 原文刊于《国土与自然资源研究》，2009，2:54-55，本人为第二作者。

（续表）

编号	测点	一般状况	测定时间*
5	4路汽车富康酒店站	2级风，阴云	2004-08-19. 18:27
6	玉皇阁	雨后，1级风，凉	2004-08.19. 20:30
7	宁园	水池边，雨后，1级风	2004-08-19. 23:30
8	宁园	1级风、无人	2004-08-20. 06:00
9	西夏王陵景区内	晴间多云、1级风、树旁	2004-08-20. 08:55
10	钟楼步行街	无风、人来往走动	2004-08-20. 10:15
11	宁园	1级风、有游人休憩	2004-08-20. 10:45
12	汽车站南门车场	多云，人不多、尾气较浓	2004-08-20. 12:40

* 最高气温约24～26℃；测时持续5 min。

3 结果和讨论

3.1 银川空气微生物含量的一般状况

表2列出了12个测点/时所取得的空气微生物含量之一般状况。

由表2可见，银川空气微生物的检出率为100%，极差达91 688.0 CFU·m^{-3}，最高空气微生物含量是最低者的147.0倍，含量之大小排序为No.4＞No.2＞No.8＞No.5＞No.12＞No.1＞No.3＞No.10＞No.6＝No.7＞No.11＞No.9。空气微生物的平均含量为23 306.8 CFU·m^{-3}，它表明银川空气有中度微生物污染，虽然这一数值低于太原、乌鲁木齐等地，但却高于嘉兴、北海和宜昌等类似规模城市[4~8]。个别测点的微生物含量表明其空气污染十分严重，不容乐观。

表2 银川空气微生物含量状况 CFU·m^{-3}

测点/编号	细菌(B)	真菌(F)	总菌(T)	F/T%
1	14 758.0	8 007.0	22 765.0	35.2
2	36 424.0	3 925.0	40 349.0	9.4
3	1 147.0	1 413.0	12 560.0	11.3
4	90 432.0	1 884.0	92 316.0	20.4
5	28 260.0	471.0	28 731.0	1.6
6	4 239.0	0	4 239.0	0
7	1 727.0	2 512.0	4 239.0	59.3
8	38 622.0	785.0	39 407.0	14.9
9	157.0	471.0	628.0	75.0
10	7 379.0	785.0	8 164.0	9.6
11	3 925.0	157.0	4 082.0	3.8
12	22 294.0	628.0	22 922.0	27.4

由表 2 可看出，银川空气细菌的检出率为 100%，极差达 90 275.0 CFU·m^{-3}，最高细菌含量是最低者的 576.0 倍，含量大小排序为 No. 4>No. 8>No. 2>No. 5>No. 12>No. 1>No. 3>No. 10>No. 6>No. 11>No. 7>No. 9，空气细菌的平均含量为21 613.7 CFU·m^{-3}。

表 2 还表明，银川市空气真菌的检出率达 91.1%，极差达 8 007.0 CFU·m^{-3}。最高空气真菌含量是次低者的 51.0 倍，低于细菌的这一比率。其含量大小排序为 No. 1>No. 2>No. 7>No. 4>No. 3>No. 8＝No. 10>No. 12>No. 5＝No. 9>No. 11>No. 6＝0，空气真菌的平均含量为 1 751.2 CFU·m^{-3}，占平均空气微生物总含量的 7.5%。平均的细菌含量是平均的真菌含量的 12.1 倍，亦即真菌量只及细菌量的 8.1%。表明银川空气微生物主要由细菌组成，真菌处于次要地位，因而空气微生物污染主要由细菌造成，这符合银川的气候特征。

3.2　银川空气微生物含量在城市不同功能区划间的差异

按城市功能差异可划分相关测点/时为以下 4 类，即景点/区（No. 9），绿地、休憩处（No. 3、No. 6～8、No. 11）、交通要道等（No. 1、No. 2、No. 5、No. 12）、市场（No. 4、No. 10）。银川空气微生物含量在不同功能区划中的状态如表 3 所示。

由表 2 可见，银川的所测景区空气微生物含量很少，但真菌量却占了很大比例（75.0%），表明那儿绿化较好。树木的蒸腾作用散发较多水分，湿度较大，致具耐湿孢子的真菌增多。空气微生物少意味着那儿空气洁净，污染很轻。

不同功能区域空气细菌含量间的大小排序为市场（D）>要道（C）>绿地（B）>景区（A）。真菌的大小排序为 C>D>B>A。F/T%的大小排序却变为 A>C>B>D。此时的交通要道出现较大 F/T%，与测时的毛雨效应相关。空气微生物总量之差异相同于细菌的。总的看，景区和绿地等环境的空气质量好于闹市交通要道和商业场所。但前者的真菌污染较重些。

3.3　银川空气微生物含量的时间差异

由表 1 中的部分数据统计可得出表 4。

表 4 显示闹区取静处 3 测点 5 时段即 No. 3，No. 6～8 和 No. 11 的空气微生物含量状况。这 5 组数据反映出一昼夜中晨间、上午、下午、晚上和午夜 5 个时段的状况。该表表明，空气细菌含量的大小排序是：晨间>下午>晚上>上午>午夜；空气真菌含量大小排序是午夜>下午>晨间>上午>晚上；空气微生物总含量的大小排序则为晨间>下午>晚上>午夜>上午，即细菌有晨峰夜谷、真菌显夜峰晚谷，总菌似细菌有晨峰，但很快上午则降至谷底之势。由于细菌占总量的比率绝对高（92.5%），因此该类测点的空气微生物含量主要由细菌引起，其峰值也由细菌决定。但总菌量的低谷提前至上午意味着那时经早晨的交通流量高峰后及阳光照射的杀菌作用所带来的总量降低势态。该类测点空气微生物含量的时段变化是否有一定规律可循，尚待验证，但多少反映出银川空气微生物含量不同于一些城市的昼夜变化过程[5,7,8]。

表 3 银川不同城市功能区划测点间空气微生物含量的差异

测区编号	测区类别	空气微生物含量(CFU·m^{-3})等				测点
		细菌(B)	真菌(F)	细菌(B)	真菌(F)	
A	景点/区	157.0	471.0	628.0	75.0	No. 9
B	绿地、花园	11 932.0	973.4	12 905.4	7.4	No. 3、6-8、11
C	交通要道	28 691.8	3 757.8	31 949.6	10.2	No. 1、2、5、12
D	市场	48 905.0	1 334.5	50 240.0	2.7	No. 4、10

表 4 银川闹市取静处空气微生物含量(CFU·m^{-3})的昼夜的变化

类别	时段					均值
	晨间(No. 8)	上午(No. 11)	上午(No. 3)	晚上(No. 6)	午夜(No. 7)	
细菌	38 622.0	3 925.0	11 147.0	4 239.0	1 727.0	11 932.0
真菌	785.0	157.0	1 413.0	0	2 512.0	785.0
总菌	39 407.0	4 082.0	12 560.0	4 239.0	4 239.0	12 905.4

4 结语

夏末的空气温度仍较高，又有一定湿度。人们户外活动的时空扩大，微生物同时乘虚而入，因而空气受微生物、生活工商交通的污染有增涨之势。空气中不乏胃肠炎、呼吸道感染等疾病的微生物，虽然本文未对所得微生物作种类鉴别分析，但调研结果从一个侧面表达出银川空气遭中度微生物污染，不可避免会有病菌污染。此污染的状况在时空的分布是不均衡的。

提醒人们提高环境忧患意识，减少废气废水等废物的排放，合理处置垃圾，讲究个人及环境卫生，科学调控交通秩序。据地理、气象条件布局工矿企业及生产和绿化美化环境，当是银川控制空气乃至整个环境污染，提高人们生活质量，使之健康良性持续发展的要务之一，由此也将改善投资环境。

参考文献 8 篇(略)

(合作者:孙丕喜 于明哲)

ANALYSIS ON CONDITION OF AIRBORNE MICROBIAL CONTENT ABOVE YINCHUAN

(ABSTRACT)

Abstract The content of airborne microbes at 12 sampling sites/moments was determined above Yinchuan city using the method of gravity plate. The result shows that the outdoor air of Yinchuan had been lightly polluted middle with microbes, and to a serious level at individual sampling site. The state of content spatiotemporal distribution was analyzed above Yinchuan. This paper indicates that it is a matter of great importance on monitoring air pollution and improving the quality of air to Yinchuan city.

Key words Yinchuan; Airborne Microbes; Air Quality

瑞士四城市空气微生物考察*

摘　要　用自然沉降法在 Bacto agar 平板上于 2005 年 6 月份对瑞士巴塞尔、伯尔尼、日内瓦和洛桑的 45 个测点/次做了空气微生物考察。结果表明暮春/初夏时节的空气微生物检出率高，其中真菌占了一定份额。四市空气微生物平均含量约 1250 CFU·m^3。洛桑空气微生物含量最低，表明其空气洁净度高。总体上表达出瑞士空气质量良好，少数测点/次表达出较多空气微生物，尚待深究和改进。空气微生物含量表达出一定的时空分布势态，总体看人少和静谧处空气微生物少，晨夜时段空气微生物较少。

关键词　瑞士　空气　微生物　时空分布　污染　空气质量

引言

美丽瑞士是中欧一小国，领土以阿尔卑斯山山区为主，湖泊 1 484 个，以日内瓦湖为最大。莱茵河流贯 375 km，南部暖热，越往北趋温凉。人口七百万左右，生活富足，城市化水平高。主要城市偏西—中部，经济以服务业、纺织、钟表、化学、食品、发电和机械等工业为主，因而其自然环境总体上印象良好，但难免有所污染，包括空气。

空气微生物是空气环境和空气生态系统的一个重要组分，它们与空气中各种气溶胶颗粒物等污染物一起沉浮、扩散、传播，其中包括病原菌[1]。空气微生物状况及其变化深刻影响空气环境质量，与人类动植物息息相关，许多国家对此都有数量、种群等的指标，对空气微生物的研究牵涉到大气、环境、医学、卫生等学科领域，备受相关专家学者和人们的关注[2,3]。

巴塞尔(Basel)、伯尔尼(Berne)、日内瓦(Geneva)和洛桑(Lausanne)是瑞士的重要城市，它们的空气及其微生物状况鲜为人知，也许是质量较好，因而并不被人们所重视。可是空气及其微生物是不断变化的，此一时空代替不了彼一时空。因此必须对有关地区/城市的空气微生物及相关因子有所了解和研究，人们才可能做到心中有数，以此估测现状和掌握未来[4]。为此于 2005 年 6 月对上述四市空气微生物状况作了粗浅考察，目的是从一个侧面了解它的空气状况，并与瑞士等国家的学者作一探讨。

1. 四城市考察概况

巴塞尔位于北部、伯尔尼则中偏西部，洛桑和日内瓦则分别位于日内瓦北部和西南。四城市基本可代表全瑞士政治、经济、环境状况，考察时间以次序分别为 2005 年 6 月 1、2、8～10、13、16～17 日(巴塞尔)、6 月 3 日(伯尔尼)、6 月 10 日(日内瓦)、6 月 15 日(洛桑)。每个城市的考察(测)点(次)分别为 19(巴塞尔)、伯尔尼和日内瓦分别为 10 及洛桑的 6 个。

---

* 原文刊于《科技创新导报》*Science and Technology Innovation Herald*，2010，No. 15:148 转 150，本人为第四作者

2. 材料和方法

取 Bacto agar (Dicklison and company)制成平板培养基，按常规即自然沉降法去设定测点采某空气样本[3]。曝皿时间一般为5～10分钟。将采过空气样的平皿置25/37℃恒温中培养48h。分别计数各皿中的细菌(B)、霉菌(M)、酵母菌(Y)的菌落形成数(CFU)，再换算为单位体积空气中的总微生物菌数为总含量(T)，即(CFU·m^{-3})，得出微生物检出率(%)，结合当时所得的天气和人文活动，分析空气微生物含量的时空分布状况，评价各市空气污染程度和空气质量。相关城市及其测点和环境状况等见文献[5-8]。

3. 结果和讨论

3.1 空气微生物检出率

根据对四市检测结果[5-8]，可知洛桑(L)、巴塞尔(Ba)、日内瓦(G)、伯尔尼(B)的空气微生物检出率分别为83.3、94.7、100.0和100.0%，总体看检出率较高，说明瑞士四市空气适合相应微生物生存。

3.2 空气真菌状况

由于测时正值瑞士的暮春/初夏时节，一些空气有较高湿度，因而出现真菌的机率较大。四市比较[5-8]可见，洛桑空气以霉菌为主，伯尔尼空气真菌检出率为42.3(25℃)－80.8%(37℃)，日内瓦则为75.6(25℃)－66.2%(37℃)，而以巴塞尔空气真菌检出率最低，仅为46.0%[5]。

3.3 空气微生物含量

各市空气微生物平均含量(CFU·m^{-3})分别为：L-471.6，Ba-513.0、B-1500.0和G-2515.2，即洛桑最低，日内瓦的最高，如果微生物含量的高低能反映出某特定区域空气质量的话，那么该结果则说明洛桑空气最洁净[7]，上述四市状况若和国内外许多城市相比，则也显出上乘水平[9-11]，但平均状况并不代表所有状况，比如日内瓦湖中的一测点，空气微生物含量分别达到16 977.6(25℃)和14 462.4(37℃)[6]，这说明当时空气中有许多具耐湿孢子的真菌，尽管如此，它仍比许多城市的低[4,9]。

3.4 温度的影响

比较两种温度培养的结果可知，各市空气微生物反应并不完全一致。25℃培养时，微生物含量有时偏高，但总体看，培养/自然温度提高，大体可促使微生物繁殖。温度低时，微生物检出率则差异较大。四城市微生物参数均归入表1之中[5-8]。

表1 瑞士四城市空气微生物参数比较

Table 1 Comparison on air-borne microbial parameters among four cities, Switzerland

城市名	检测点/次数	总检出率(%)	总菌量均值(CFU·m^{-3})	总菌量波幅(CFU·m^{-3})	真菌检出率%	真菌量(CFU·m^{-3})均值	真菌量波幅(CFU·m^{-3})
巴塞尔	19	94.7	513.0	0～2 436.6	46.0	235.8	0～943.2
伯尔尼	10	100.0 (100.0)	1 524.8 (1 556.3)*	157.2～4 558.2 (314.4～2 829.6)	42.3 (80.8)	644.5 (1 257.6)	157.3～1 100.4 (157.2～2 672.4)

(续表)

城市名	检测点/次数	总检出率(%)	总菌量均值(CFU·m^{-3})	总菌量波幅(CFU·m^{-3})	真菌检出率%	真菌量(CFU·m^{-3})均值	真菌量波幅(CFU·m^{-3})
日内瓦	10	90 (100.0)	2 515.2 (2 719.5)	0～16 977.6 (314.4～1 446.26)	75.6 (66.2)	1 902.1 (1 556.2)	0～16 820.4 (0～13 519.2)
洛桑	6	83.3 (83.3)	471.6 (655.0)	0～1 257.6 (0～1 108.4)	66.7 (83.3)	288.2 (393.0)	0～1 257.6 (0～943.2)

* 括弧()中的数值均系37℃培养所得,余为25℃时的。

3.5 测点间/功能区划间空气微生物状况分析

四市合计45个测点/次所有微生物含量详细状况见文献[5-8],测点间的差异数反映了测点空气状况。对四市测点按区域分类大致可分为10种,可以发现各市各区划间的差异。以空气微生物含量大小排序,即洛桑:湖滨街道铁道,巴塞尔:静谧区河畔桥上街道市场;日内瓦:树下湖内湖滨街道室内;伯尔尼:要道区静谧区广场。从中看出人少、静谧区的空气微生物少些,而人多、繁华之处微生物趋多的机会大。表明空气微生物与空气成分组成的环境、人群活动状况密切相关[12]。

3.6 空气微生物的时间分布状况

以巴塞尔为例,空气微生物含量在昼夜内的不同时段中分布是不平衡的,大体上表达为晨=夜=傍晚<上午=中午<下午,即下午到达顶峰,而晨和夜分别为低谷时光,这与人群作息、户外活动时段相吻合。其他3市不同时段的测定结果也表达了空气微生物状况的时间分布[5-8,12],反映了各市空气在不同时段内经受着复杂多变因子的制约[13-14]。

4. 结语

通过以上对瑞士四市空气微生物状况的测定和分析,可以看出瑞士空气微生物检出率不低,说明瑞士所处空气地理环境颇适于微生物生存,而其中真菌检出率和含量则占相当额份,表明湿润的初夏空气适于许多真菌繁衍。空气微生物含量排序为日内瓦>伯尔尼>巴塞尔>洛桑,四市的均值约为1 250 CFU·m^{-3}(25℃)[5-8]。表明洛桑空气质量上佳,总体看此四市空气微生物含量不算高,如果这能代表瑞士的话,说明瑞士空气良好,有的处于优异状态,但少数测点空气质量并不很好,尚待研究和改进。空气微生物含量呈现出一定的时空分布态势,各市间存在一些差异,反映出各市的空气和环境特征[13],其昼夜变化势基本上与气温、人群活动状况相吻合。

参考文献14篇(略)

(合作者:陈颖稚 李为民 黄秀兰)

INVESTIGATION ON AIR-BORNE MICROBES ABOVE FOUR CITIES, SWITZERLAND

(ABSTRACT)

Abstract The airborne microbial investigation above Basel, Berne, Geneva and Lausanne, Switzerland was finished by method of gravity using Bacto agar culture plate in June, 2005.

The result obtained showed that the examine percentages were higher during later spring/early summer. Among them the fungus occupied some shares. Their average content above the four cities was about 1 250 CFU · m^{-3} (25℃). The airborne microbial content was lowest above lausanne among them, it indicated that the air of it was clean. The result showed that the air quality of the Switzerland was better. The microbiological investigated result of a few sampling sites/times showed that there were more air-borne microbes there, indicated that they would have been to researched deep and improved. The air-borne microbial contents expressed that certain statuses of time-space distribution. General speaking, if there were not a lot of persons or peace and quiet places, there were not a lot of microbes in their air. The time period of early morning or night, a less microbes in the air.

Key words Switzerland, Air-Borne Microbes, Cities, Air Pollution, Air Quality

E_2 海洋(SEAS AND OCEANS)

珠江三角洲及珠江口海区的空气微生物含量*

摘　要　1994 年 6,7 月,调查了广州、深圳、珠海及珠江口海区的空气微生物粒子沉降量状况。发现空气微生物粒子总量、真菌粒子量、真菌粒子数/总菌粒子数百分比在三市中不一致。这三种数据一般是陆上的高于海上。随着距陆距离的增加,都市影响的减弱和海面的开阔,空气微生物含量呈减少之势,显示出海洋新鲜空气对陆上污染空气的调节、净化作用。文中还分析了空气微生物有关参数与气温、风力或相对湿度间的相关关系。

关键词　珠江三角洲　珠江口海区　空气微生物粒子含量　空气环境质量

经济活动和人们生活的迅速扩展,给空气环境带来一定的负荷,监测和提高空气环境质量越来越受到重视。空气微生物是空气中各种媒体,包括污染物上的重要成分,其中许多又属于有害生物。它们能使人、动植物致病,使产品腐败变质。近 20 年来,空气微生物的研究在国际间取得了可喜的进展[1,2]。我国于 20 世纪 80 年代起,才见报道,但研究区域主要在北方[3,4]。

珠江三角洲是我国经济发达地区之一。经济的腾飞和物质生活的丰富,要有高质量的环境和高水准的精神生活相匹配,因而需要对所处空间及毗邻的珠江口海区的空气环境质量包括空气微生物粒子沉降量(即空气微生物含量,下同)及其变动有所了解,以便必要时采取一定措施。本文就珠江三角洲的广州、深圳、珠海及珠江口海区的空气微生物含量及其变动状况作初步报道。

1. 材料和方法

1994 年 6～7 月,在广州、深圳、珠海三市市区及广州新洲码头至珠江口海区共布设 10 个测点(图 1),作空气微生物含量测定。

陆地上空气微生物样品采自地面以上 1.5 m 高度的人群呼吸带的空气。珠江口海区等水面上空气微生物样品是在中国科学院南海海洋研究所“实验 2”号调查船甲板上空取得。采用平皿沉降法取样。采样介质分别是普通培养基和真菌培养基[5]。每个测点重复

---

* 原文刊于《热带海洋》,1996,15(3):76-80.

中国科学院南海海洋研究所“实验 2”号船提供出海机会并得到船员帮助,一并致谢。

3 份样品。盛有样品的培养皿置于合适条件下培养后，计算出单位体积空气中的菌落形成单位数，即 CFU · m^{-3}。采样时，同时取得相应气象参数，然后对所有数据作统计分析[6]。

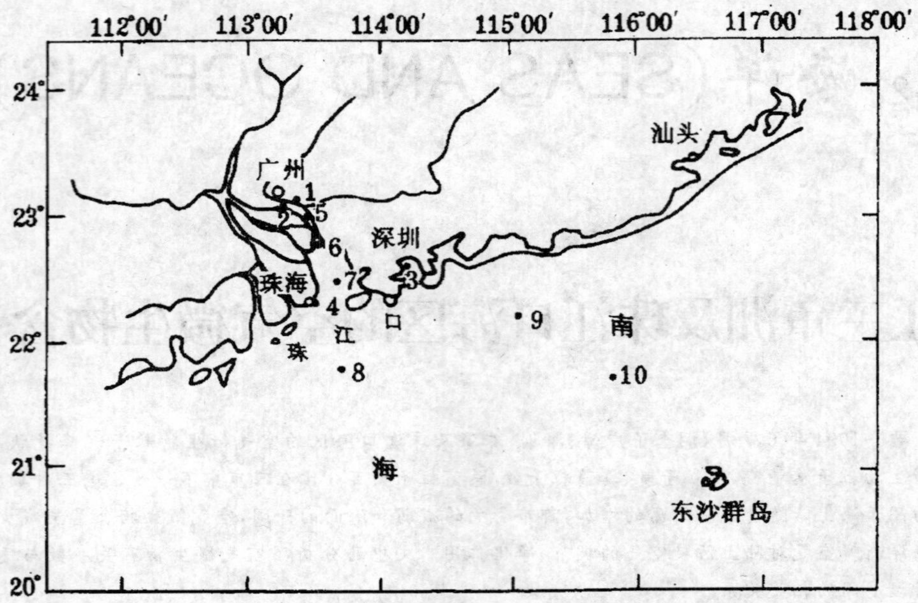

图 1　珠江三角洲及珠江口海区空气微生物测点示意图

Fig. 1　The map of sampling stations for air borne microbes over the delta and estuarine area of Zhujiang River

2. 结果和讨论

2.1　空气微生物含量

表 1 列出每次空气微生物含量的测定结果和相应的有关参数。由表 1 可看出，广州、深圳、珠海三市的空气微生物总量（即总菌数）按大小排列，其次序是深圳＞珠海＞广州。深圳的空气微生物含量高，其主要原因是因为当时正值暴雨来临之前，携带许多尘埃的强阵风经过曝气平皿。在珠海测点，虽值街上人员稀少的晨间，空气微生物在许多情况下应处于平缓或少量状态，但测时湿度大，风小，有些车辆来往。同时这两市也难免各受毗邻城市——中国香港或澳门空气状况的影响。广州的结果表明测定时的空气较洁净。尽管如此，本结果与天津、青岛类似区域相比，仍表明这三个城市的空气微生物含量与之不同[3,6,7]。

从广州新洲码头至珠江口的 10 号站，共设 7 个测点。其空气微生物粒子沉降量的多寡排列依次是 2 号＞5 号＞8 号＞6 号＝7 号＞9 号＞10 号测点。一个明显的趋势是离广州及珠江三角洲城市越近，则空气微生物含量越高。不难看出，人口的稠密，工商业及交通的发达，对空气微生物和空气质量都有重大影响。表明繁杂的人群及其活动促使空气微生物增多。2 号和 5 号两个测点离广州城区最近，空气微生物粒子沉降量平均达到 2 703.1CFU · cm^{-3}。8 号、9 号和 10 号 3 个测点在珠江口外缘，离岸远，其空气生物粒子沉降量平均为 420.7 CFU · cm^{-3}。居中间的两测点，即 6 号和 7 号测点其空气微生物粒子沉降量平均为 903.9 CFU·cm^{-3}。2 号和 5 号两测点空气微生物粒子沉降量分别为 6 号和 7 号测点及 8 号、8 号、9 号和 10 号测点的近 3 倍和 6.4 倍多。表明了大都市和城市群的影响随着离陆距离的增长而减弱，海洋新鲜空气的调节、净化作用增强。这种势头越到开阔海面就越为显著。如开阔的南海海面（主要指广东沿海）又比珠江口海区的空气更为洁净[2]。

表 1　珠江三角洲及珠江口海区空气微生物粒子沉降量及相关参数

Tab.1　Precipitation of airborne microbial particles and relevant parameters over the delta and estuarine area of Zhujiang River

测点编号	所在城市或位置	微生物粒子总沉降量/CFU·cm^{-3}	真菌粒子沉降量/CFU·cm^{-3}	真菌数/总菌数%	气温/℃	相对湿度/%	风力等情况	日期
1	广州新港西路	9 235.5	903.9	9.8	29	80	静风，少云	1994.7.10
2	广州新洲码头	4 244.4	1 886.4	44.4	30	80	南风 2～3 级，晴	1994.6.27
3	深圳	61 622.4	1 729.2	2.8	29	70	西南风 6 级，雨前	1994.7.12
4	珠海	19 807.2	8 331.6	42.4	27	90	南风 1 级，阴	1994.7.13
5	坭沼头锚地	1 161.8	0	0	29	90	东南风 4 级，阴	1994.7.08
6	113°27′25″E，22°44′30″N	903.9	235.8	26.1	29	80	东南风 3 级，晴	1994.7.7
7	113°42′04″E，22°30′28″N	903.9	296.5	21.7	30	85	南风 3 级，晴	1994.7.1
8	113°44′45″E，21°45′07″N	967.4	0	0	26	80	东南风 5 级，阴	1994.6.30
9	115°07′38″E，22°11′33″N	255.5	59.0	23.1	26	85	南风 2 级，多云	1994.6.29
10	115°50′17″E，21°42′54″N	39.3	19.7	50.1	27	75	南风 2～3 级，少云	1994.6.28

2.2　空气真菌粒子沉降量、空气真菌粒子数/总菌数的百分比

不少真菌是致癌因子。它们的含量（即真菌粒子沉降量，下同）及其与总菌数（即微生物粒子总沉降量，下同）百分比的高低也是衡量空气微生物含量组成及环境质量的重要指标。将表 1 的数据分为陆上和水上来看。就 3 城市相比，真菌粒子数大小排列依次是珠海＞深圳＞广州。真菌数/总菌数的百分比依次是珠海＞广州＞深圳。

水面以上空气的真菌粒子沉降量大小排列次序是.2 号＞6 号＞7 号＞9 号＞10 号＞5 号＝8 号测点。平均值为 342.5 CFU·m^{-3}。真菌数/总菌数的百分比的次序是 10 号＞2 号＞6 号＞9 号＞7 号＞5 号＝8 号测点。5 号、6 号、7 号 3 个测点在珠江口河道上，离岸近，其真菌粒子沉降量平均为 144.1 CFU·m^{-3}。真菌数/总菌数的百分比的平均值为 14.6。8 号，9 号和 10 号 3 测点处于开阔海区，其真菌粒子沉降量、真菌数/总菌数的百分比各自的平均值分别为 26.2 CFU·m^{-3}和 6.2 CFU·m^{-3}。而 2 号测点正在码头上，其真菌含量、真菌数/总菌数的百分比明显比其他测点高。10 号测点的真菌含量虽不大，但真菌数/总菌数的百分比很大，可能暗示小风速和适宜的湿润空气有利于一些真菌孢子的萌发（下文详述）。

以上结果大致表明，都市和陆地是真菌的大本营。大河河口、海区离城市越近，真菌粒子沉降量及真菌数/总菌数的百分比越大。海洋空气的真菌比陆上空气的少得多。海洋空气对陆源真菌同样具有稀释、净化作用。

2.3 空气微生物粒子沉降量、真菌数/总菌数的百分比与相应气温-风力或相对湿度间的相关关系分析

表2列出空气微生物粒子沉降量等3种数据,分别与气温等环境参数间的相关系数。

表2 珠江三角洲及珠江口海区空气微生物粒子沉降量与相应参数间的相关系数

Tab. 2 Correlation coefficients between precipitation of air borne microbial particles and relevant parameters over the delta and the estuarine area of Zhujiang River

相关因子	相关系数(r值)	备 注
总菌数一气温	0.156	$n=10$
总菌数一风力	0.443	$n=10$
总菌数一相对湿度	−0.515	$n=10$
真菌数一气温	−0.219	$n=8$
真菌数一风力	−0.274	$n=8$
真菌数一相对湿度	0.483	$n=8$
真菌数/总菌数的百分比一气温	−0.281	$n=8$
真菌数/总菌数的百分比一风力	−0.380	$n=8$
真菌数/总菌数的百分比一相对湿度	0.313	$n=8$

由表2可见,在本监测条件下,总菌粒子沉降量与气温、真菌数/总菌数的百分比与相对湿度间分别有正相关关系,尽管相关性不大。总菌数与相对湿度,真菌数与气温、风力,真菌数/总菌数的百分比与气温、风力间分别为负相关。相关性虽不甚明显,但多少可以看出,总菌数受相对湿度的负面影响较大($R=-0.515, >R_{0.10}$),而真菌数受相对湿度的正面影响稍大($R=0.483, >R_{0.10}$)。这意味着空气微生物中,有一些成员不适应过高的相对湿度而被淘汰。多数真菌,尤其是那些具耐潮湿孢子的真菌,可能喜欢并适应较湿润的空气环境[7,8]。

空气微生物深受环境复杂因子的制约,常与化学污染物发生协同、有害等等的作用,从而使并不洁净的空气环境潜伏危机,殃及人们健康。有关与化学污染及环境保护方面的问题尚待进一步探讨。

3. 结语

迄今,已有几种器皿(方法)用于测定空气微生物。它们各有优缺点。其中的平皿沉降法得出的结果似有粗糙之虞,但具有成本低廉、省时省力等优点,故仍被国内外学者普遍采用。本文用平皿沉降法对广州、深圳、珠海三市及珠江口海区的空气微生物测定的结果只能是初步的。但从这一初步结果可以看出,测区中三城市空气微生物粒子沉降量间的差异,意味着空气质量有待进一步提高。陆上空气微生物含量常比海上的高。随着离陆距离的扩大、城市影响的减弱和海面的宽阔,珠江口海面上空气微生物粒子沉降量呈现出减少之势,显示出海洋空气有强大的调节、净化功能。海、陆空气的交换,海一气的相互作用,对改善陆上和沿海城市环境污染状况大有裨益。

参考文献9篇(略)

THE CONTENT OF AIR BORNE MICROBIAL PARTICLES OVER THE DELTA AND ESTUARINE AREA OF ZHUJIANG RIVER

(ABSTRACT)

Abstract During June—July, 1994, investigations were conducted on the state of precipitation of air borne microbial particles over Guangzhou、Shenzhen and Zhuhai in the delta and estuarine area of Zhujiang River. It is discovered that of the 3 cities, the highest precipi-tation of air borne microbial particles, fungi particles and fungi particles/total microbial precipitation % are in Shenzhen and Zhuhai respectively. Generally. the three kinds of data which are obtained from overland are higher than those from over sea water. With the increase in distance from land, reduction in city influences and extension of the sea surface, the precipitation of air borne microbial particles shows a decreasing tendency. This study shows that fresh sea air has regulative and purificatory effects towards polluted air over the land. The paper analyses the correlative relationships between some parameters of air borne microbes and relevant temperature, wind force or relative humidity respectively.

Key words Zhujiang River Delta, Estuarine Area of Zhujiang River, Content of Air Borne Microbial Particles, Quality of Air Environment

东海、黄海测区的空气微生物研究*

摘　要　对东海、黄海的近岸、远岸海区空气中陆源性和海洋性微生物含量进行了测定。并比较了不同测区间空气微生物含量的差别，分析了它们的昼夜变化状态，以及气温对这些变化的影响。认为测区空气微生物状态指示海洋与陆地、人类活动及环境污染与海洋空气间的相互作用。

关键词　东海　黄海　空气微生物　人类活动　海陆相互作用　环境污染

引言

海—气的相互作用及人为活动的影响，将反映到空气微生物的种群组成、含量波动和生态效应上。反之空气微生物的状态即是环境—生态变化的一个综合性指标。它能指示特定区域的大气污染状态、评价空气质量和推测疾病发生、流行的发展过程，反映所在海区海水微生物状态及被干扰的程度。国际上空气微生物学的研究开展较早[1,2]。在我国相应的研究只是近 20 年的事情[3,4]，研究的区域主要在陆地，集中在城市等人类活动频繁之处。随着经济开发向海洋的倾斜，人类活动的影响已经越来越强烈地影响到近海、大洋甚至南、北极海域等。因此进行包括海洋空气微生物学在内的海洋环境科学研究，将越来越被人们所认识和重视[4]。本文就东海浙江海区及邻近的黄海海域的空气微生物含量研究结果作一报道，以期促进海洋和空气的微生物学、应用与环境微生物学及环境科学的全面发展。

1　材料和方法

于 1995 年 11 月，在“向阳红 9”号船上采用平皿沉降法采样。培养基有海洋细菌培养基和真菌培养基，以在此上面长出的细菌和真菌代表该海区空气中的海洋性细菌和海洋性真菌，两者之和为空气中海洋性微生物总量；以普通培养基和真菌培养基捕捉陆源性细菌和真菌，两者之和为空气中陆源性微生物总量。各类微生物含量计数单位均为每立方米空气中的菌落形成单位数，即 CFU/m^3。并以真菌占总微生物量的百分数来衡量真菌含量变化。采样、培养、计数、分析和评价等均按常规法进行[3、4]。采样站位详见图 1。

2　结果和讨论

* 原文刊于《东海海洋》，1998，16(3)：33-39。

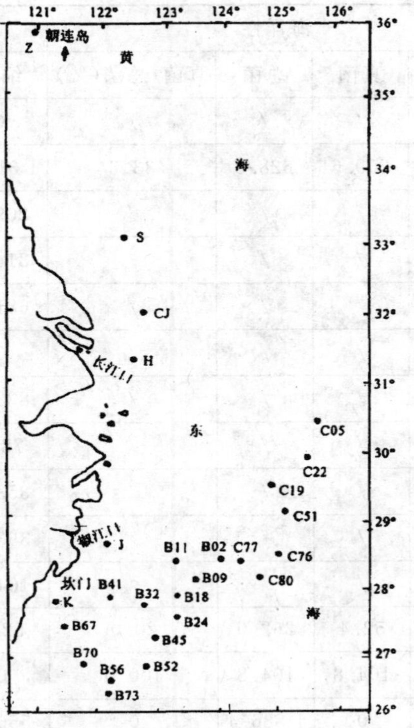

图 1　东海、黄海测区空气微生物测点(站)示意图

为叙述的方便起见,将采样区域分作两部分(见图 1),一是近岸区,它包括浙江的椒江口(J)和坎门港(K)、上海浙江交界处的花鸟山(H)、长江口外北区(CJ)和青岛的朝连岛(Z)等近岸测点,南北跨度在 28°05′～35°54′N 之间。另一区为远岸区,它包括 26°20′～28°25′N、121°30′～124°04′E 的 B 区(设 13 个站)、28°07′～30°30′N、124°15′～125°40′E 的 C 区(设 7 个站)和苏北海区 S(1 个站)。

所有测点空气微生物含量见表 1。

表 1　东海、黄海测区空气微生物含量

测区、站(点)		海洋性				陆源性			
		细菌	真菌	总菌	真菌/总菌(%)	细菌	真菌	总菌	真菌/总菌(%)
B 区	02	104.8	78.6	183.4	42.9	262.0	0	262.0	0
	09	366.8	52.4	419.2	12.5	104.8	0	104.8	0
	18	52.4	104.8	157.2	66.7	104.8	52.4	157.2	33.3
	11	/	/	/	/	1 545.8	0	1 545.8	0
	24	366.8	366.8	733.6	50.0	0	52.4	52.4	100.0
	32	314.4	733.6	1 048.0	70.0	681.2	157.2	838.4	18.8

(续表)

测区、站(点)	海洋性				陆源性			
	细菌	真菌	总菌	真菌/总菌(%)	细菌	真菌	总菌	真菌/总菌(%)
41	/	/	/	/	0	262.0	262.0	100.0
45	419.2	209.6	628.8	33.3	1 310.0	419.2	1 729.2	24.2
52	/	/	/	/	209.6	262.0	471.6	55.6
56	/	/	/	/	314.4	366.8	681.2	53.8
67	/	/	/	/	681.2	157.2	838.4	18.8
70	/	/	/	/	0	52.4	52.4	100.0
73	/	/	/	/	838.4	471.6	1310.0	36.0
C区 05	/	/	/	/	78.7	314.4	393.1	80.0
22	/	/	/	/	353.7	78.6	432.3	18.2
19	/	/	/	/	393.1	104.8	497.9	21.0
51	/	/	/	/	104.8	78.6	183.4	42.9
76	209.6	52.4	262.0	20.0	52.4	0	52.4	0
77	0	104.8	104.8	100.0	0	0	0	0
80	1 886.4	0	1 886.4	0	550.2	0	550.2	0
CJ区	/	/	/	/	366.8	0	366.8	0
H区 H_1	104.8	104.8	209.6	50.0	0	0	0	0
H_2	/	/	/	/	681.2	209.6	890.8	23.5
H_3	/	/	/	/	157.2	52.4	209.6	25.0
J区 J_1	550.2	628.8	1 179.0	53.3	471.6	1 257.6	1 729.2	72.7
J_2	471.6	1 414.8	1 886.4	75.0	1 650.6	1 375.5	3 026.1	45.5
K区 K_1	/	/	/	/	707.4	2 829.6	3 537.0	80.0
K_2	/	/	/	/	78.6	943.2	1 021.8	92.3
K_3	/	/	/	/	1 336.2	786.0	2 122.2	37.0
S区	/	/	/	/	157.2	52.4	209.6	25.0
Z区	/	/	/	/	3 248.8	366.8	3 615.6	10.1

注:各类微生物含量的计量单位为 CFU/m³,0 为未测出,/为未测定,下同。

在各自的同一点上 H 站共测 3 次,J 站测 2 次,K 站测 3 次。

2.1 近岸区的空气微生物含量

若以 CJ 和 H 测区为中部,K 和 J 区为南部,Z 区为北部。经统计(数据见表 1),可以得出近岸区各采样点的空气微生物含量(表 2)。由表 2 可见陆源性空气细菌和总菌含量是北部高于南部,其余两项指标均是南部高于北部。海洋性微生物的 4 个指标均是南部高于中部。南北海区空气微生物的差异显著。如以细菌含量为空气污染指标[5]，似北部海区的污染较南部重。另一方面，测定时南部海区的温度和湿度较高，可能使真菌及其所占百分率提高，而且深秋初冬时节的南方气温比北方较寒冷的气温更合适海洋性空气微生物的生长。中部和南部测区的陆源性空气微生物含量比海洋性的高，说明测区陆地影响的增强和海洋性影响的减弱。

表 2 东海、黄海近岸区空气微生物含量

测区	海洋性				陆源性			
	细菌	真菌	总菌	真菌/总菌(%)	细菌	真菌	总菌	真菌/总菌(%)
J和K	510.9	1 021.8	1 532.7	66.7	848.9	1 438.4	2 287.3	62.3
CJ和H	104.8	104.8	209.6	50.0	301.3	65.5	366.8	17.9
Z	/	/	/	/	3 248.8	366.8	3 615.6	10.1

2.2 远岸区的空气微生物含量

由表1数据所统计的远岸区三个区域空气微生物含量状况如表3所示。由表3可见，B、C两区陆源性微生物4个指标均比苏北海区(S)高；相对近岸的B区的陆源性细菌、真菌和总菌量均比相对远岸的C区高。这意味着近岸一些海区受陆地空气及人类活动的影响和干扰更明显些。C区有较多的海洋性细菌和总菌量，意味着离岸更远的海区其空气的海洋性趋势更明显，陆地影响减弱[6]。但B、C两区空气中的海洋性真菌量却相反，即C区低于B区，这与空气温度有关(下述)，因为真菌(包括海洋性的)一般适于较高温度生长繁殖。

表 3 东海、黄海测区远岸区空气微生物含量

测区、站(点)	海洋性				陆源性			
	细菌	真菌	总菌	真菌/总菌(%)	细菌	真菌	总菌	真菌/总菌(%)
S	/	/	/	/	157.2	52.4	209.6	25.0
C	698.7	52.4	751.1	7.0	219.0	82.3	301.3	27.3
B	270.7	257.6	528.3	48.8	465.6	173.3	638.9	27.1

2.3 近岸区与远岸区空气微生物含量的比较

据表1所列的数据统计，近岸区和远岸区的空气微生物含量是不同的。表4表明，除海洋性细菌在远岸空气中所占百分率较高外，近岸陆源性空气中真菌所占百分率比细菌的高，也比远岸区真菌的高。说明近陆空气可能更适合许多微生物存留。

表 4 东海、黄海测区近岸和远岸两区空气微生物含量的比较

海区	海洋性				陆源性			
	细菌	真菌	总菌	真菌/总菌(%)	细菌	真菌	总菌	真菌/总菌(%)
近岸	869.8	782.1	1 651.9	47.3	375.5	716.1	1 091.6	65.6
远岸	368.7	157.2	505.9	27.1	413.4	189.2	802.6	31.4

2.4 空气微生物含量的昼夜变化

东海、黄海测区空气微生物含量的昼夜变化状态如图2所示。图2a表明,近岸区空气中海洋性细菌的含量呈由晨间向夜间逐降之势;真菌量和总菌量均为中午最高而夜间降低。由此说明海洋真菌在空气微生物含量变动中起主导作用,这与南极麦克斯韦尔湾空气微生物含量的谷值在夜间出现相类似[3]。近岸区空气的陆源性微生物含量有上午高于下午的变化趋势,这与南极的不一致,但与庐山的有类似之处,即中午较高[7]。

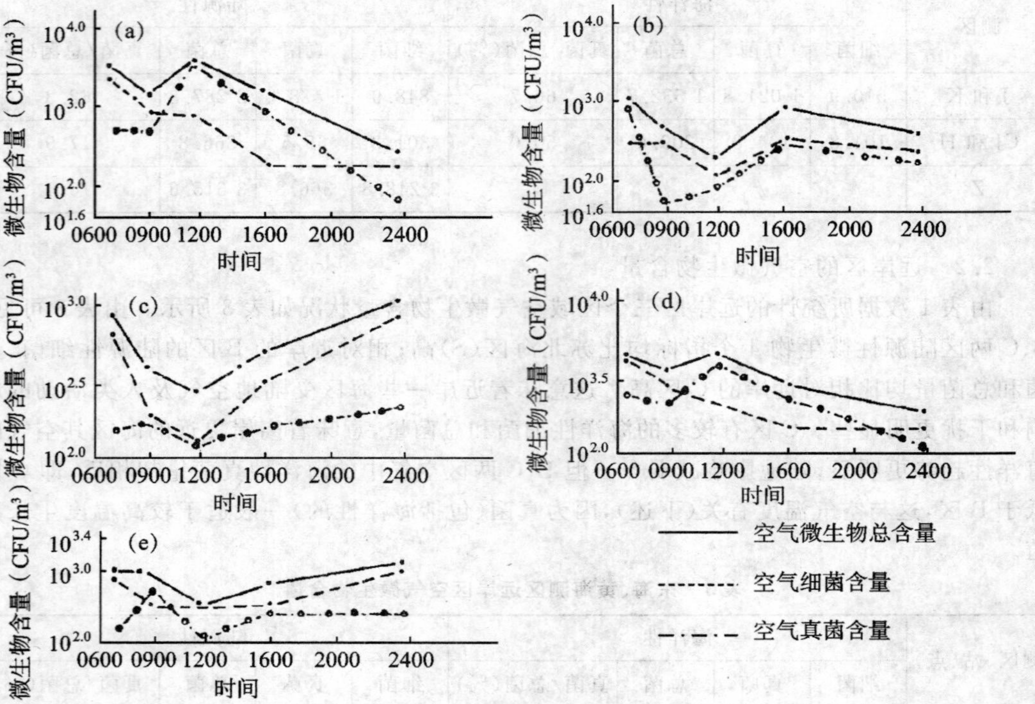

图2 东海、黄海测区空气微生物含量的昼夜变化

a. 近岸区海洋性空气微生物. b. 远岸区海洋性空气微生物;c. 远岸区陆源性空气微生物;d. 全测区海洋性空气微生物;e. 全测区陆源性空气微生物

远岸区空气微生物含量昼夜变化情况见图2b、c。海洋性真菌与总菌量变化趋势基本一致,总趋势是晨间出现峰值,但真菌量在峰后很快跌为上午谷。总菌量下降稍平缓,中午见低谷。细菌与总菌量的谷值同时出现在中午时段,但细菌量峰值出现时间迟于总菌量。下午,真菌和细菌都促使总菌量抬升。远岸区陆源性空气微生物量稍低于海洋性的。其细菌、真菌和总菌数在中午跌至谷底。午后至夜间并维持至翌日清晨均处于上升趋势,之后又不断下跌,其中真菌量上午一度抬升后与细菌、总菌量一起降入上午谷。

图2d显示了整个测区空气中海洋性微生物的昼夜变化状态。真菌、细菌及总菌量均在夜间出现谷值,但峰值出现时间不全一致。细菌、总菌峰值出现在晨间,而真菌峰值在中午前后。

图2e显示整个测区陆源性空气微生物含量的变化势态。细菌、真菌及总菌量谷值出

现在中午，但真菌、总菌量的峰值在上午，细菌的峰值则在夜间出现。

2.5　气温对空气微生物含量的影响

将采样时有气温记录的测点与表1中相应的空气微生物含量作一统计，可得到表5。由表5可见，B区气温高于C区，B区空气中陆源菌各指标均高于C区。C区较低的气温与空气中较多的海洋性细菌相对应，这表明海洋性微生物（尤其是海洋细菌），较适宜于低温，陆源性微生物较适于较高气温生长繁殖。

表5　气温的变化对东海、黄海空气微生物含量的影响比较

测区*	海洋性空气微生物			陆源性空气微生物			气温(℃)
	细菌	真菌	总菌	细菌	真菌	总菌	
B	270.7	257.6	528.4	382.5	183.4	565.9	19.1
C	698.7	52.4	751.4	200.9	0	200.9	17.3

* B区由10个测点平均得出，C区由3个测点平均得出。

将空气微生物各指标与气温之间作相关分析，结果表明陆源性真菌量与气温间有较显著的正相关关系（$n=8,\gamma=0.746,\alpha=0.05$）。这进一步证实了陆源菌，尤其是真菌在一定条件下受气温的影响较大。

近岸区空气微生物含量比远岸区高。远岸区的一些测点空气中存在较多海洋性微生物。空气微生物含量在纬度间有差异，较低纬度的近岸区空气微生物含量常高些。这与工业化城市散发的污浊空气对其毗邻海区空气中微生物（如陆源细菌）含量所起的推波助澜作用有关。也与海水微生物含量在纬度间、海区间的差异有吻合之处[8]。空气微生物含量的昼夜变化在近岸区和远岸区、陆源性和海洋性间显示出不一致的势态。近岸区空气微生物与陆上多数人的日常作息规律有一致之处，如微生物高含量常在中午，而远岸区的峰值避开了中午甚至在午后即锐减至低谷。

海洋的近岸区域如码头、港区、抛锚区或海岛的岸区受陆地的影响大，其空气中含有较多陆源微生物。随着离陆距离的增大和人类活动影响的减弱，空气中所含陆源菌减少，而海洋菌增多[4,6]。虽然大洋上空气微生物本应稀少[4]，但海洋中人为设施的增多，也是陆源菌的一个来源。本研究中相对远岸测点的微生物含量不低，尽管采样方法不尽相同，但与10年前相邻测区相比增加了许多[4]。这可能与陆源菌入侵有关，意味着陆源污染的加重。相对而言的近岸区和远岸区空气中有时也有较多的海洋菌和陆源菌，似乎表明陆源菌与海洋菌在东海、黄海此类边缘海区的缓冲、交叉或消涨，时机一到，如风向、气温、湿度和颗粒污染物等因素都将影响微生物类群和数量组成。因此特定区域空气微生物的状态是相应区域空气成分及其来源、污染类型和空气质量的重要指标。它能印证相应海水微生物状态。人类开发活动的类别和强度、海—气、海—陆间的交换作用都可从空气微生物的监测成果中获得有效信息和启示。因此加强海洋空气微生物的研究是很有必要的。

参考文献8篇（略）

（合作者：王　波）

RESEARCH ON AIRBORNE MICROBES OVER SOME AREAS OF THE EAST CHINA SEA AND THE YELLOW SEA

(ABSTRACT)

Abstract This paper studied the airborne microbes over some areas of the East China Sea and the Yellow Sea. The counts of marine and terrigenous airborne microbes were monitored over the inshore and outshore sea areas. The differences in the counts among the various investigated areas were compared. The diurnal changes of their status and the air temperatures effect on them were analysed. It is considered that the airborne microbial status over the sea areas indicates the interactions between sea and land, as well as those between human activity and environmental pollution and air over seas.

Key words The East China Sea and The Yellow Sea, Airborne Microbes, Human Activity, Interaction Between Sea and Land, Environmental Pollution

海南岛及其毗邻海区的空气微生物浓度*

摘　要　用平皿沉降法测定了环海南岛和粤西两海区及海南岛两城市的空气微生物浓度，比较了测区空气中海洋和陆源性微生物的浓度变化，分析了空气微生物浓度的时空分布和变化势态。反映了海洋与陆地、室内与室外空气微生物的状态以及人类活动对它的影响。

关键词　空气微生物　空气质量　环境污染　海气交换

在空气微生物的研究方面，国内外已取得了许多成果[1-2]，但关于海洋及其沿岸的空气微生物的研究则不多[3-4]，海面上 15 m 左右高度的作业带和地面上 1.5 m 左右的人群呼吸带中的空气微生物与人们的健康有着密切关系，其含量（浓度）和种群的变化能指示特定区域的经济发展和环境污染状况，表达和反映所在区域的空气质量。

海南岛及其环岛海区是改革开放的一片热土，海域内具有石油、天然气的开发前景，这些与热带季风气候相结合，使海上和陆地空气中所含的各类微生物经受着海—气和越来越多的人—自然相互作用的强烈影响，因而研究该岛及其毗邻海区和室内外空气微生物的状况能反映它们的空气污染程度和空气质量水准，促使人们在进行经济开发的同时更科学地关注环境和健康问题，使包括空气微生物在内的环境—生态系统更好地良性循环，持续发展。

本文就 1995 年 5 月和 1996 年 12 月对海南岛及其毗邻海区的空气微生物浓度的测定情况作一报道。

1　材料和方法

1.1　测点概况

于 1995 年 5 月和 1996 年 12 月分别从广州乘船去海南岛近海作业，并逗留于海南八所、临高和海口三地，在广州新洲码头和深圳蛇口海区各设 1 个点，粤西海区设 2 个点，环海南岛海区设 9 个点，八所、海口陆上室内外各设 2 个点，共计 17 个点，19 次测定空气微生物浓度，各测点位置如图 1 中阿拉伯数字所示，其中 17～19 为一个测点 3 个测次，每次采样时的概况见表 1。

1.2　采样和分析

用平皿沉降法取样.采样介质包括海洋细菌培养基和海洋真菌培养基，以其上长出的菌落形成单位数（CFU）作为单位体积（m^3）空气中的海洋性细菌或真菌浓度（记作 CFU/m^3，下同）；同时，用普通培养基和真菌培养基捕捉空气中陆源性细菌和真菌，相应的细菌与真菌之和分别代表海洋性或陆源性空气微生物总（菌量）浓度，采样、培养、计数和分析均按常规方法进行[2,5]。

* 原文刊于《海南大学学报自然科学版》，1998，16(4)：345－350。

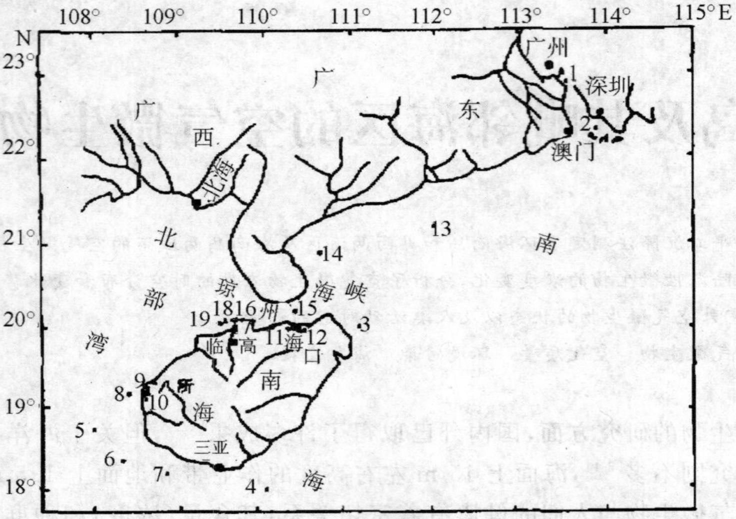

图 1　海南岛及其毗邻海区的空气微生物采样点示意图

（图中阿拉伯数字为采样点号，1995.5，1996.12）

表 1　海南岛及其毗邻海区空气微生物测点概况

测点所在海区	概　况
环海南岛海区	
3. 110°06′E，20°00′N	1995-05-23 T07:00～07:10，阴天，2～3 级风
4. 110°00′E，18°00′N	1995-05-23 T20:00～20:30，阴天，小风，轻雾
5. 107°55′E，18°50′N	1995-05-24 T12:55～13:10，晴，>5 级风
6. 108°27′E，18°23′N	1995-05-24 T16:30～17:00，晴，>5 级风，30.0℃气温
7. 108°48′E，18°17′N	1995-05-25 T08:55～09:15，3～4 级风，28.0℃气温
8. 108°30′E，19°14′N	1995-05-26 T12:30～12:40，晴，3 级陆风，30.0℃气温
15. 110°23′E，20°12′N	1995-05-26 T17:05～17:20，3 级风，相对湿度 70%
16. 109°52′E，20°05′N	1996-12-26 T19:05～19:35，晴，小风转 4 级，相对湿度 85%，21.0℃气温
17. 109°40′E，20°01′N	1996-12-30 T07:40～08:10，零星雨，3 级风，相对湿度 95%，21.3℃气温
18. 109°40′E，20°01′N	1996-12-30 T16:50～17:20，多云，薄雾，2～3 级风，相对湿度 85%，22.5℃气温
19. 109°40′E，20°01′N	1996-12-30 T22:45～23:15，3 级风，21.5℃气温
粤西海区	
1. 广州新洲码头	1995-05-22 T12:00～12:05，阴天，小风
2. 深圳蛇口外 114°08′E，22°20′N	1995-05-22 T16:00～16:10，阴天，小风

(续表)

测点所在海区	概况
13. 112°00′E,21°10′N	1996-12-23 T08:00～08:30,晴,5级风
14. 110°40′E,20°56′N	1996-12-24 T11:45～12:00,晴,3级风,24.0℃,相对湿度90%
海南岛陆上区	
9. 八所某酒店客房	1995-5-27 T06:30～06:35,3人,空调气温25.0℃
10. 八所市街道	1995-05-27 T07:15～07:20,晴,静风,28.0℃气温,车辆来往多,行人不多
11. 海口市街道	1995-05-27 T12:40～12:45,晴,小风,较干燥,36.0℃气温,车、人不多,安宁
12. 海口市室内	1995-05-27 T13:00～13:05,空调气温27.0℃,旅客来往多

2 结果和讨论

2.1 空气微生物浓度的一般状况

海南岛及其毗邻海区空气微生物浓度的测定结果见表2。

表2 海南岛及其毗邻海区空气微生物浓度状况 CFU/m³

测点	海洋性				陆源性			
	细菌	真菌	总菌	真菌/总菌(%)	细菌	真菌	总菌	真菌/总菌(%)
3	52.4	52.4	104.8	50.0	78.6	131.0	209.6	62.5
4	235.8	0	235.8	0	26.2	104.8	131.0	80.0
5	209.7	52.4	262.1	20.0	576.5	471.7	1 048.2	45.0
6	26.2	78.6	104.8	75.0	52.4	0	52.4	0
7	39.2	1 140.0	1 179.2	96.7	117.9	275.2	393.1	70.0
8	314.5	864.7	1 179.2	73.3	78.6	707.6	786.2	90.0
15	—**	—	—	—	1 205.2	314.4	1 519.6	20.7
16	—	—	—	—	52.4	157.2	209.6	75.0
17	—	—	—	—	628.8	314.4	943.2	33.3
18	—	—	—	—	104.8	104.8	209.6	50.0
19	—	—	—	—	120.4	4 087.2	4 207.6	97.1
1	3 615.6	1 336.2	4 951.8	27.0	—	—	—	—
2	393.1	0	393.1	0	393.1	393.1	786.2	50.0
13	—	—	—	—	576.4	1 545.8	2 122.2	72.8
14	—	—	—	—	1 152.8	157.2	1 310.0	12.0
9*	—	—	—	—	5 187.6	0	5 187.6	0
10	3 301.2	471.6	3 772.8	12.5	—	—	—	—

(续表)

测点	海洋性				陆源性			
	细菌	真菌	总菌	真菌/总菌(%)	细菌	真菌	总菌	真菌/总菌(%)
11	—	—	—	—	3 458.4	157.2	3 615.6	4.3
12*	—	—	—	—	786.0	157.2	943.2	16.7

* 为室内采样 ＊＊为未测者,下同。

将表 2 所列的海上各测点空气微生物浓度的数据作一统计,便可得出所测海区空气微生物浓度的一般状况:海洋性空气中细菌、真菌、总菌量浓度及真菌占总菌量的百分率分别为 610.8 CFU/m^3、440.5 CFU/m^3、1 051.3 CFU/m^3 及 41.9%;陆源性空气中细菌、真菌、总菌量浓度及真菌占总菌量的百分率分别为 368.9 CFU/m^3、626.0 CFU/m^3、994.9 CFU/m^3 及 62.9%,这表明全区空气中陆源性微生物浓度比海洋性的小,暗示了所测海区的空气具有较重的海洋性质,而陆地和人为的影响较小

陆上室内空气中细菌、真菌、总菌平均浓度分别为 2 986.8 CFU/m^3、78.6 CFU/m^3 和 3 065.4 CFU/m^3 室外的海洋性空气中细菌、真菌、总菌浓度则分别是 3 301.2 CFU/m^3、471.6 CFU/m^3、3 772.8 CFU/m^3,陆源性细菌、真菌、总菌浓度各为 3 458.4 CFU/m^3、157.2CFU/m^3 和 3 615.6 CFU/m^3,这表明室内空气好于室外,两个室内点之间存在明显差异,No.1 测点之所以微生物含量低,可能与其空间开阔、空气流畅有关,室外空气微生物浓度仅达轻度污染程度[5],空气质量还属上乘。

2.2　各区空气微生物浓度分述

将图 1 所示海区分作两小区,一为珠江口、蛇口以西至海南岛以东,称作粤西海区,二是环海南岛海区,前者包括 No.1、2、13 和 14 四个测点,后者有 9 个测点,它们是海南岛东部的 No.3、南部的 No.4～7(其中 No.3 与粤西海区共有)、西部的 No.8 及北部的 No.15～19(其中 No.17～19 是一个测点 3 个测次)。

由表 2 统计得出粤西海区空气微生物浓度,以近陆的 No 1、2 测点平均,海洋空气中细菌、真菌、总菌浓度分别是 2 004.4 CFU/m^3、668.1 CFU/m^3、2 672.5 CFU/m^3,真菌/总菌%为 25.0%。陆源性细菌、真菌、总菌平均浓度及真菌/总菌%分别为 707.4 CFU/m^3、698.7 CFU/m^3、1 406.1 CFU/m^3,及 49.7%,表明这一带空气中陆源性真菌量不小。

环海南岛海区空气中的微生物浓度仍由表 2 统计得出平均值.即海洋性细菌、真菌、总菌浓度及真菌/总菌%分别是 146.3 CFU/m^3、364.7 CFU/m^3、511.0 CFU/m^3 及 71.4%,陆源性细菌、真菌、总菌浓度及真菌/总菌%分别为 276.5 CFU/m^3、606.2 CFU/m^3、882.7 CFU/m^3 及 68.7%。从总体看,环海南岛海区的空气微生物含量小于粤西海区,但真菌所占的份额比粤西海区多。

2.3　环海南岛海区各测点空气微生物浓度在方位间的差异

如上所述,环海南岛海区的测点可分属东、南、西、北 4 个方位,下面列出这 4 个方位上空气微生物浓度的比较。

表 3 环海南岛海区空气微生物浓度在 4 个方位上的比较 CFU/m³

方位	海洋性				陆源性			
	细菌	真菌	总菌	真菌/总菌(%)	细菌	真菌	总菌	真菌/总菌(%)
东	52.4	52.4	104.8	50.0	78.6	131.0	209.6	62.5
南	127.7	317.8	445.5	71.3	193.3	245.7	439.0	56.0
西	314.5	864.7	1 179.2	73.3	78.6	707.6	786.2	90.0
北	—	—	—	—	422.3	995.6	1417.9	70.2

以上数据表明，海洋性细菌、真菌和总菌浓度大小顺序均是西＞南＞东，真菌/总菌%仍是西＞南＞东；陆源性细菌浓度大小排序是北＞南＞西＝东，真菌和总菌均是北＞西＞南＞东，真菌/总菌%是西＞北＞东＞南。总的来看，环海南岛海区的空气微生物含量是西部多，东部少，北部大于南部。此种状况可能与采样的时节和离岸的远近有关。如西部的采样于湿热的 5 月和盛行的陆风时进行，东部的采样既在较凉的 12 月份，又离岸稍远些，北部的采样点在近岸的抛锚区，而南部的采样点离岸更远些。

将粤西海区和环海南岛海区的空气微生物浓度与以往所测的珠江口海区相比[6]，可以发现珠江口的陆源性细菌浓度居该两者之间，真菌甚少，总菌量也低。

2.4 空气微生物浓度的昼夜变化

对所测海区的空气微生物含量(13 个测点 15 次测定)，按其在昼夜中的上午、中午、下午和夜间来分，再以此含量的常用对数值作纵坐标，以不同的时间段作横坐标. 绘制成图 2。可见海洋性微生物中的细菌、真菌及总菌量浓度均在中午出现峰值，细菌浓度值的低谷却出现在早上，而真菌和总菌的则在夜间出现. 陆源性细菌浓度峰值在中午出现，其低谷在夜间出现，真菌和总菌浓度的峰值均在夜间出现，低谷均在下午出现，总体看海洋性细菌和微生物总量峰值均高于陆源性的，仅就海南岛海区 6 个测点(即 No. 3～8)的情况看，海洋性细菌、真菌、总菌浓度和陆源性真菌、总菌浓度及真菌/总菌%的峰值均在中午出现；海洋性真菌、总菌浓度及真菌/总菌%和陆源性细菌、真菌、总菌浓度的谷值均在夜间出现或浓度较小，这表达了陆上空气微生物浓度的一般规律[5,7]。

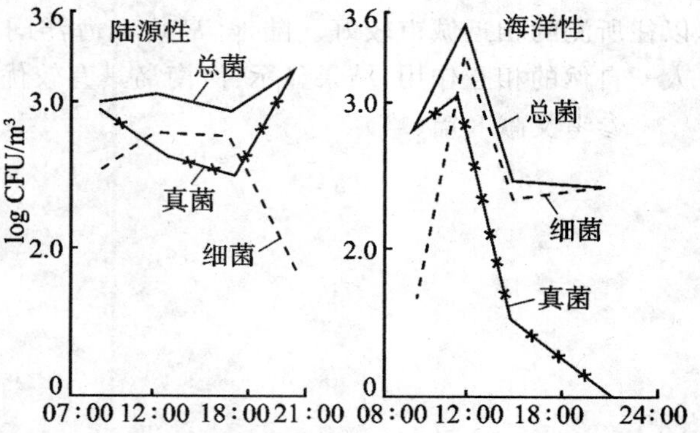

图 2 海南岛及粤西海区空气微生物浓度的昼夜变化

2.5 陆上空气微生物浓度的评价

八所港是海南岛西部的重要港口，宽阔的近岸海面与低平的台地间有较通畅的海陆风交流路径. 测定时，陆风将市内的空气微生物送入锚泊区，致使空气中既有较多海洋性微生物，又有许多陆源性真菌(见表 2 的 No. 8)。八所市内与锚泊区的室外空气微生物浓度差异十分明显，具体表现在市内空气中主要由细菌所引起的微生物总量大幅度增加，致使晨间室外空气中的细菌浓度超过了轻度污染程度[5]，这与来往车辆和行人扰动的不洁空气有关，反映了较密集人口和工交事业对空气微生物的重大影响，也体现了海滨市镇受海洋性空气明显的调节作用[8]。

海口市位于该岛的东北端，新型、整洁及不过度密集的街道上行驶着不多的轿车和走动着较少的行人，所测空气环境显得相当洁净，即使是中午高温时分，空气微生物浓度仍未达到轻度污染上限，表明海口室外空气质量良好[5]。

两市室外空气微生物浓度一般均高于海上的。这仍然表达了陆、岛和人类活动在促进空气微生物密集中的作用[3]。两市的监测结果反映了它们的室外空气的差异，海口大大好于八所，将两市所测的结果平均，其空气微生物浓度水平仍较低。与广州相比，该两市室内空气真菌浓度较低，但真菌的百分率较广州的高，这可能与此时该两市室内湿度较高有关。室外则有些不同，两市各自的真菌比率比广州的小[3]。

3　结语

本文论述了海南岛及其毗邻海区的海洋性和陆源性空气微生物浓度状况及变化。比较了海洋、陆地、室外、室内以及不同海区的空气微生物。认为从陆地到沿岸、远岸海区，空气微生物浓度有不断稀释之势，但海洋空气中仍有不少陆源菌。测区空气微生物性状表达了较浓的海洋气息，与以往邻近海区的测定结果相比较，本海区空气微生物浓度显出上升之势[2]。海南岛海区不同方位中的空气微生物状态有一定差异。

海南岛沿岸城市的室内外空气微生物状态暗示出八所市室外有轻度空气污染，但比以往所测的相近城市较好。陆地、岛屿和海洋间空气微生物状态及变化表达了海—陆和人—自然的相互作用，结果显示：海南岛具有较优越的空气环境。

参考文献8篇（略）

AIR-BORNE MICROBIAL CONCENTRATION OVER HAINAN ISLAND AND ITS ADJACENT SEA AREAS (ABSTRACT)

Abstract The air-borne microbial concentrations over the sea areas around Hainan Island and Western Guangdong. and over two cities of Hainan Island were measured by using the method of gravity plate. The concentrations and their changes of marine and terrigenous air-borne microbial counts in the areas monitored were compared. The spatial and temporal distributions and the changing tendency of the air-borne microbial concentrations were analysed. The conditions of air-borne microbes over the sea and land areas, the indoor and outdoor status of it, and the effects of human activity and environmental pollution on them were reflected.

Key words Air-Borne Microbes; Air Quality; Environmental Pollution; Sea-Air Exchange

渤海湾大气微生物粒子沉降量的研究*

摘　要　对渤海湾大气微生物粒子沉降量作了测定。结果表明：测区大气中海洋性细菌、真菌、总菌粒子沉降量及真菌/总菌百分比分别为1774.5、389.1、2 163.6 CFU/m³ 和 18.0。陆源性细菌、真菌、总菌粒子沉降量及真菌/总菌百分比为1 377.9、440.6、1 818.5CFU/m³，和 24.2。海上大气中所含的海洋性微生物总量较大，仍有不少陆源性微生物。这意味着陆源污染和人群活动对近岸海区大气的影响。陆上和近岸的大气微生物状况暗示了海洋空气对陆地天气的调节作用。渤海湾大气微生物含量受气温的负面影响较小，受风力的正面影响较大。陆源菌和海洋菌昼夜变化的幅度不大，中午常低，但峰值出现时间各不相同。

关键词　渤海湾　大气微生物粒子沉降量　陆源污染 海洋空气

近海海面上10 m左右高度的大气边界层内气生微生物属大气微生物。它们的主要来源有二：一是海洋微生物，由海水及底质的微生物与散发出来的颗粒物共同构成了海洋上空的许多颗粒物[1]。另一个是陆源性的，它包括工农业污染及人群活动带来的含微生物的气溶胶及海洋作业产生的悬浮于空气中的污染物所挟持的微生物。因此海洋，尤其是像渤海湾这样的半封闭海湾，其大气微生物群体(atmospheric microbial populations)既与海洋密切相关，又受陆区重大影响。近些年来日益频繁的渤海湾海洋经济开发活动，对海洋空气环境状况势必产生影响。海域大气微生物反映特定空气受污染状况，与相邻陆区相比可反映海陆间空气的相互作用。已有一些海洋大气微生物的研究[2,3]，但就全球范围而言，它迄今仍是个薄弱环节。

本文报道1996年1月对渤海湾的大气微生物粒子沉降量的初步研究结果。

1　研究概况及所用方法

于1996年1月15～21日在渤海湾118°E以西、38°37′N以北(即塘沽与岐口间)海区内布设站位10个，另设1个滨海陆上站和1个码头站共计12个站以研究大气微生物。此时的气温约在−7～9℃，盛行西北风，最大时为6级(蒲福风级)，无雨。一般无雾。有关采样站位详见图1。采样时间及天气状况等参见表1。

用平皿沉降法采样。采样介质包括海洋细菌培养基、海洋真菌培养基，用以收集海洋性细菌和真菌。普通细菌培养基和真菌培养基则分别采集陆源性细菌和真菌。海洋性细菌和真菌之和、陆源性细菌和真菌之和则分别代表海洋性微生物或陆源性微生物粒子总沉降量。它们的单位均以CFU/m³ 计算，采样、培养、分析和评价等均按常规进行[3,4]。

* 原文刊于《海洋环境科学》，1998，17(4)：27-31

表 1　渤海湾大气微生物采样记录(1996-01)*

Tab. 1　Sampling record from atmospheric microbes above Bohai Bay

站号	采样时间	天气状况	具体方位
1	1996-01-16,16:00～16:10	晴,4 级风,气温:－2.6℃	117°56′17″E,38°39′17″N
2	1996-01-16,20:00～20:10	晴,3 级风,气温:－5.2℃	117°55′30″E,38°39′10″N
3	1996-01-17,07:30～07:45	晴,2 级风,气温:2.5℃	117°54′55″E,38°39′04″N
4	1996-01-17,16:04～16:14	晴,间少云,3 级风,气温:－3.0℃	117°53′48″E,38°37′41″N
5	1996-01-17,21:00～21:15	4 级风,气温:－2.0℃	117°52′23″E,38°38′32″N
6	1996 01-18,02:35～02:50	3 级风,气温:－5.0℃	117°51′15″E,38°38′11″N
7	1996-01-18,21:25～21:40	5 级风,气温:－2.4℃	117°50′00″E,38°39′00″N
8	1996-01-19,06:55～07:25	5 级风,气温:5.0℃	117°50′00″E,38°44′13″N
9	1996.01.19,12:∞～12:20	5 级风,气温:9.0℃	117°48′12″E,38°49′27″N
10	1996.01-19,14:∞～14:10	6 级风,气温:7.0℃	117°47′14″E,38°55′18″N
11	1996-01.15,02:10～02:15	1 级风,轻雾,地上有积雪,气温:－10℃	117°41′17″E,38°00′30″N
12	1996.01.21.07:55～08:05	4 级风,气温:－7.0℃	117°43′15″E,38°58′33″N

* 注:采样时。各站的相对湿度除 11 号站外,其余均小

2　结果和讨论

2.1　渤海湾大气微生物粒子沉降量及其各区比较

表 2 列出了测区内每次获得的大气微生物粒子沉降量。由该表可计算出测定海区大气陆源性微生物粒子沉降量的一般状况,即大气细菌、真菌及总微生物沉降量和真菌/总菌百分比分别为 1 321.8、292.1、1 613.9 CFU/m³ 及 18.1。海洋性的大气细菌、真菌、总菌沉降量及真菌/总菌% 则分别为 1 509.4、486.2、1 995.6 CFU/m³ 及 24.4。将此与海南岛、北部湾和东—黄海所测海区比较,渤海湾的大气中细菌量较大,总菌量也大。重要的差别还在于真菌量及其所占(在总菌量中)百分比一般小于上述有关海区的值。

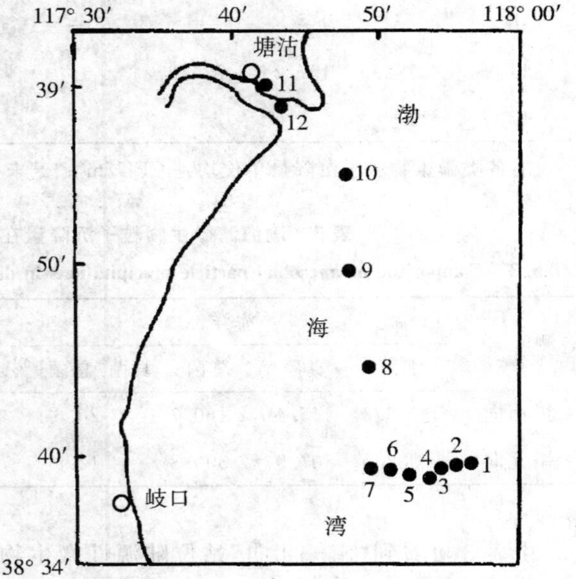

图 1　渤海湾大气微生物采样站位图

Fig. 1　the stations of sampling atmosphericmicrobes above Bohai Bay

以 11 号站代表陆上测区,12 号站

代表近陆岸区，其余站为海上站，比较这3区的大气微生物状况，可以看到的大体趋势是：陆区大气中有较多的陆源细菌，近陆岸区大气中的陆源性细菌大为减少，而海洋大气中则有较多的海洋性细菌和真菌。其中真菌所占百分比较陆上测区的高，而且陆源性微生物粒子沉降量也不低，这表明所测海区受陆源污染影响非常明显。

将海上部分的10个测站分作两类，一类是歧口附近东西向测点，它由1～7号站组成。另一类由8～10号站组成，它们以南北向排列。这两部分大气微生物状况如表3所示。

表2　渤海湾大气微生物粒子沉降量

Tab.2　Atmospheric microbial particle precipitations above Bohai Bay

站号	海洋微生物				陆源微生物			
	细菌	真菌	总菌	真菌/总菌(%)	细菌	真菌	总菌	真菌/总菌(%)
1	707.4	0	707.4	0	314.4	78.6	393.0	20.0
2	471.6	0	471.6	0	157.2	0	157.2	0
3	340.6	104.8	445.4	23.5	209.6	314.4	524.0	60.0
4	2 200.8	1 021.8	3 222.6	31.7	3 458.4	1 100.4	4 558.8	24.1
5	2 627.4	890.8	3 518.2	25.3	2 986.8	995.6	3 982.4	25.0
6	4 296.8	1 100.4	5 397.2	20.4	2 200.8	262.0	2 462.8	10.6
7	1 362.4	524.0	1 886.4	27.8	995.6	576.4	1 572.0	36.7
8	3 615.6	340.6	3 956.2	8.6	2 462.8	576.4	3 039.2	19.0
9	550.2	275.1	825.3	33.3	275.1	117.9	393.0	30.0
10	—	—	—	—	3 065.4	235.8	3 301.2	7.1
11	1 572.0	157.2	1 729.2	9.1	1 100.4	157.2	1 257.6	12.5
12	—	—	—	—	314.4	235.8	550.2	42.9

注：各类微生物粒子沉降量单位均是 CFU/m³；"0"未检出；"—"未检测

表3　渤海湾微生物粒子沉降量在不同方向大气中的比较

Tab.3　Comparison of microbial particle precipitations in different directions in atmosphere above Bohai Bay

测区名	海洋性				陆源性			
	细菌	真菌	总菌	真菌/总菌(%)	细菌	真菌	总菌	真菌/总菌(%)
东-西向	1 715.3	445.4	2 160.7	20.6	1 474.7	475.3	1 950.0	24.4
南-北向	2 082.9	307.9	2 390.8	12.9	1 631.0	524.0	2 155.0	24.3

由表3可看到"南—北向"站位陆源性微生物含量比"东—西向"站位的多，这可能与调查时西北风向带来较多陆源污浊空气有关。整个测区大气中平均的陆源性细菌、真菌、总菌粒子沉降量和真菌/总菌百分比分别为1 377.9、440.6、1 818.5 CFU/m³ 和24.2。海洋性细菌、真菌、总菌粒子沉降量和真菌/总菌百分比值则分别为1 774.5、389.1、2 163.6 CFU/m³ 和18.0。

2.2 大气微生物粒子沉降量的昼夜变化

将所有测点按一日之中的上午、中午、下午和晚间4个时段分，可以得知渤海湾大气微生物粒子沉降量昼夜变化的大致走势，结果以图2示之。由图2可知，海洋性和陆源性大气微生物粒子沉降量昼夜变化趋势基本相似，昼夜变幅不大，而且均显出下午偏低，陆源菌甚至显示出谷值在中午。两者相似之处是真菌/总菌百分比均显出峰值出现在中午，海洋性大气细菌、真菌及总菌沉降量峰值均于夜间出现，而陆源性的峰值则出现在下午。渤海湾大气微生物昼夜变化状态不同于海南岛及邻近海域，而比较相似于东-黄海的。如东-黄海大气陆源真菌峰值、细菌、真菌及总菌的低谷，海洋性细菌及总菌的低谷出现时间基本上与渤海湾的一致。这与两者地理位置和采样季节相对接近有关。

2.3 大气微生物粒子沉降量与一些环境生态因子关系分析

大气微生物受环境—生态因子的制约。这些因子包括气温、风力、湿度等。以表2各项数据之常用对数值与表1相应的气温、风力各作相关分析获得表4中的各相关系数值。

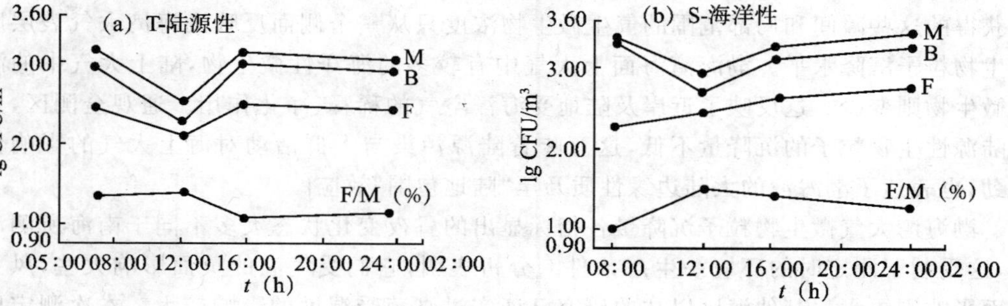

图2 渤海湾大气微生物粒子沉降量的昼夜变化状况

Fig. 2 State of diurnal changes of the atmospheric microbial particle precipitations above Bohai Bay

注：B—细菌 F—真菌 M—总菌量 F/M(%)—真菌/总菌(%)

表4 渤海湾大气微生物粒子沉降量与气温、风力的相关系数

Tab. 4 Corréation coefficients of atmospheric microbial particle precipitations with air temperature, wind strength respectively above Bohai Bay

海洋性		陆源性	
对子	R	对子	R
细菌-气温	$-0.207(n=10;r_{0.25})$	细菌-气温	$-0.252(n=10;r_{0.25})$
真菌-气温	$-0.289(n=8;r_{0.25})$	真菌-气温	$-0.248(n=9;r_{0.25})$
总菌-气温	$-0.269(n=10;r_{0.25})$	总菌-气温	$0.753(n=10;r_{0.25})$
细菌-风力	$0.628(n=10;r_{0.25})$	细菌-风力	$0.352(n=12;r_{0.25})$
真菌-风力	$0.475(n=8;r_{0.25})$	真菌-风力	$0.333(n=10;r_{0.25})$
总菌-风力	$0.178(n=10;r_{0.25})$	总菌-风力	$0.423(n=12;r_{0.25})$
真菌/总菌(%)-风力	$0.276(n=8;r_{0.25})$	真菌/总菌(%)-风力	$-0.333(n=10;r_{0.25})$

由表 4 可见，在本调查时间内，渤海湾大气中海洋性微生物（总菌量）、陆源性细菌和真菌粒子沉降量与气温间各呈现出不显著的负相关关系。陆源性总菌与气温间似存在着较明显的正相关关系。结果意味着在所测条件下海洋性微生物似受气温变化一定的负面影响，陆源性微生物总量与气温的变化似平行不悖。而各项指标与风力间大多存在着正相关关系。此外海洋性细菌、陆源性微生物总量与风力间的正相关性各达显著水平，表明这两者受风力的正面影响大。渤海湾的这一情况大多类似于珠江口空气中总陆源菌、真菌与相应气温、风力间的关系[5]。风向对大气微生物群体有影响[6]，对渤海湾大气微生物的影响主要表现在西北污浊的陆风携来较多陆源菌（前有所述）。有关研究尚待深入。真菌粒子沉降量及其在总菌量中所占百分比常暗示当时风速和湿度的影响力。渤海湾大气真菌量及其所占百分比一般不大于我国南方有关海域的，这与测定时较干燥的天气有关[5]。

3　结语

渤海湾的地理条件及其周边的环境、生态、经济条件决定了它的大气微生物状况。本文获得的这些瞬间和局部范围的气生微生物浓度只从一个侧面反映渤海湾大气表层底部微生物粒子沉降水平。渤海湾海面上大气中有较多的海洋性微生物，陆上大气中的海洋性微生物则变少。这反映了近岸及陆地受海洋空气的稀释、净化作用。通观全测区，大气中陆源性生物粒子的沉降量不低，这意味着陆源污染与人群活动对海上大气的影响仍然强劲，也表达了渤海湾的大陆边缘性质乃至“陆地包围”效应。

渤海湾大气微生物粒子沉降量各指标显出的昼夜变化状态大多不同于南海所测海区的，这受制于测定时的环境一生态条件。分析表明，它们受气温的负面影响大些，风力的正面影响则更大。其他海区以往的研究已证实过真菌受湿度的影响较大。本次测定所获真菌粒子沉降量及其在总菌中的百分比较小，推测此时较小的相对湿度是一个原因。

参考文献 6 篇（略）

STUDY ON ATMOSPHERIC MICROBIAL PARTICLES PRECIPITATION ABOVE BOHAI BAY

(ABSTRACT)

Abstract The monitoring results of atmospheric microbial particles above Bohai Bay indicated that the precipitations of the marine bacterial, fungus, total microbial particles and fungi/total microbial content(%) were 1,774.5, 389.1, 2,163.6 CFU/m^3 and 18.0 respectively. Those of the terrigenous microbes were 1,377.9, 440.6, 1,818.5 CFU/m^3 and 24.2 respectively. In the atmosphere above Bohai Bay there were more of the marine microbes, but a few terrigenous ones. It means that there was the influence of land pollution and human activity on the nearshore waters. The atmospheric microbial conditions in the land and nearshore area showed the regulating effect of the sea air on the land weather. The atmospheric microbial counts above Bohai Bay were less by the effect of the air temperature, but more by the effect of wind strength. The ranges of diurnal changes of terrigenous and marine microbial counts were not wide. The counts were often lower at noon, but their peak values appeared at different times.

Key words Bohai Bay; Atmospheric Microbial Particle Precipitation; Terrigenous Pollution; Air Above Sea

北部湾北海—临高海区的空气微生物*

摘　要　对北部湾的北海—临高海区作了空气微生物含量考察。结果表明,测区的空气平均细菌、真菌/总菌量及真菌,总菌的百分比分别为 632.3,1 324.0,1 956.3 CFU·m^{-3}及 67.7%。数据分析显示:测区空气微生物含量呈北高、南低、中间最少的势态;近岸海区的比相邻陆区的低,海上部分的平均比环海南岛上空的稍大,比粤西海区的稍少。其昼夜变化显示出空气细菌量峰值出现在晨间,谷值在中午。真菌量的峰值出现在黄昏,谷值在上午。其状态基本上不同于北海市区的,但近似于相邻海区的。空气微生物含量与气温间无显著相关关系。结果意味着空气污染近岸区重于海上区,人为活动的影响也较海上区的大。

关键词　北部湾　北海—临高海区　空气微生物　空气污染　人群活动

地球大气圈的空气微生物学研究虽有了较长的历史,但引人注目的发展还是近 20～30 年间取得的。就海洋上空的空气微生物而言,Certes[1] 于 1984 年分别在加勒比海和大西洋上空作过研究。笔者等[2] 在 1994 年发表了环球的空气微生物研究,范围包括世界大洋的空气微生物,并扩展到极地空气。空气微生物学的近代发展与探讨疾病传播和行星生命有关,直接关系到空气污染水平测定、微生物灾害控制和生物武器制剂的研制等问题,因此,包括海洋上空的空气微生物的研究自然就具有现实意义。本文就广西北海市与海南省临高县之间的陆—海区(简称北海—临高海区,下同)陆地上/海面上 15 m 高度内的空气微生物作一初步研究,从微生物学的角度来了解该区近岸和海上的空气污染状况,为该地区的经济开发活动提供有关的基础资料。

1　材料和方法

北海—临高海区位于北部湾近雷州半岛侧,全长 168 km,宽约 0.6 km,面积约 100 km^2,属热带季风气候。考察期间(即 1996 年 12 月底)的风况为静风至北风 5 级。气温波动于 18～25℃之间。相对湿度为 80%～95%。考察区共设采样点 15 个(图 1)。

采用自然沉降法取样,培养介质包括普通细菌培养基②和真菌培养基③两种。以此两种培养基上生长出的菌落数(CFU)之和代表微生物总菌量(单位:CFU·m^{-3}),并计算出真菌占总菌量的百分数。采样、细菌和真菌培养、计数及分析步骤按常规进行[3,4]。

* 原文刊于《热带海洋》,1999,18(3):100-104.

② 细菌培养基:牛肉膏 3g,蛋白胨 10g,葡萄糖 10g,NaCl 5g,琼脂 18g,蒸馏水 1 000 mL,pH 7.0～7.4。

③ 真菌培养基:蛋白胨 5g,葡萄糖 10g,KH_2PO_4 1g,$MgSO_4$ 0.5g,琼脂 18g,蒸馏水 1 000 mL,pH 7.2～7.4。

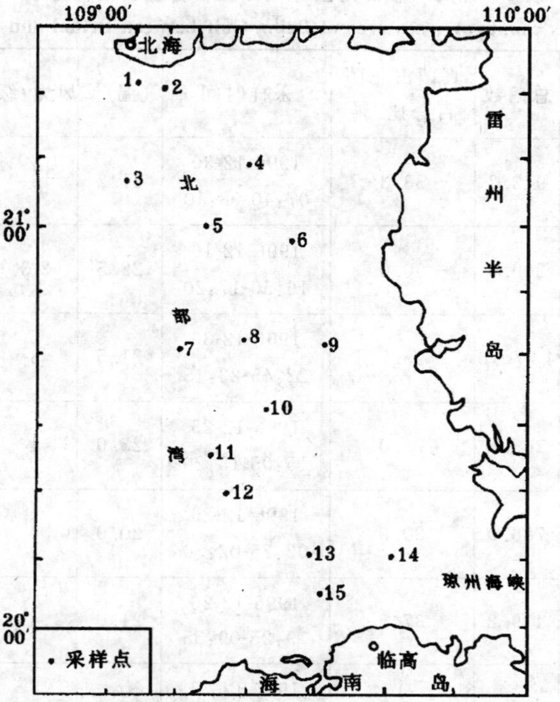

图1　北海—临高海区空气微生物采样点示意

Fig. 1　Sampling stations of airborne microbes over sea area of Beibu Gulf between Beihai and Lingao

2　结果

表1列出各采样点的测定结果。由表1可计算出测区空气微生物有关的平均数据，即空气细菌量、空气真菌量、空气微生物总量及真菌/总菌百分比分别为632.3，1 324.0，1 956.3CFU·m^{-3}及67.7%。表1还表明，测区空气中细菌数量从晨间峰值降为午间谷值，后变化平缓，直至夜间。真菌数量峰值在黄昏前后出现，谷值在上午。总菌数量的变化势态与真菌数量的类似。

将测区分为3部分，即近北海市的1，2站列为北区，近海南省临高县的13～15站为南区，其余各站(3～12站)列为海上区。3区空气微生物的4个参数平均值见表2。表2表明，3区的北区空气中细菌、真菌和总菌量均最高，海上区的最低；反之，北区空气的真菌百分比却最低。

对各类微生物含量与相应的气温作了相关分析，结果未发现它们与气温间有显著的相关关系。

表1 北海—临高海区空气微生物含量及采样记录*

Tab. 1 Airborne microbes counts over sea area of Beibu Gulf between Beihai and lingao and sampling record

采样点	细菌	真菌数	总菌数	真菌占总菌百分比/%	采样时间	气温/℃	风力/级	相对湿度/%	其他
1	628.8	314.4	943.2	33.3	1996-12-30 07:40-08:10	21.3	3	90	零星雨滴
2	104.8	104.8	209.6	50.0	1996-12-10 16:50-17:20	22.5	2-3	85	多云薄雾
3	183.4	4 087.2	4 270.6	95.7	1996-12-30 22:45-23:15	21.5	3	85	阴
4	52.4	157.2	209.6	75.0	1996-12-25 15:05-15:20	23.0	4	—	晴
5	628.8	157.2	786.0	20.0	1996-12-26 02:25-02:30	20.0	4	85	晴
6	262.0	157.2	419.2	37.5	1996-12-26 19:25-09:55	21.0	4	85	晴
7	314.4	183.4	497.8	36.8	1996-12-27 16:40-07:10	18.0	5	90	阴间多云,雾
8	78.6	262.0	340.6	76.9	1996-12-27 11:48-12:18	24.0	3	90	多云雾
9	52.4	7 702.8	7 755.2	99.3	1996-12-27 18:00-18:30	20.0	4	90	多云
10	131.0	0	131.0	0	1996-12-28 0:00-0:30	18.0	5	80	多云
11	943.2	2 200.8	3 144.0	70.0	1996-12-28 05:00-05:32	18.0	5	80	多云
12	104.8	0	104.8	0	1996-12-29 08:30-09:00	19.7	5	85	晴间多云
13	104.8	366.8	471.6	77.8	1996-12-29 14:30-15:00	23.5	静风	90	—
14	3 772.8	183.4	3 956.2	4.6	1996-12-31 08:30-09:00	22.0	2	95	阴,雾
15	2 122.2	3 982.4	6 104.6	65.2	1996-12-31 11:35-12:05	24.5	2	90	阴,雾

注:* 各微生物含量单位:CFU·m^{-3};"—"表示未测。

表 2　北海—临高海区 3 部分测区空气微生物含量比较

Tab. 2　Comparison of airborne microbes counts among three parts of sampling area of Beibu Gulf between Beihai and Lingao

测区名	空气微生物含量指标*			
	细菌	真菌	总菌	真菌/总菌百分数/%
南区	305.7	1 504.1	1 809.8	83.1
海上区	267.2	1 098.8	1 366.0	80.4
北区	2 947.5	2 082.9	5 030.4	41.4

注：* 各微生物含量单位：$CFU \cdot m^{-3}$。

3　讨论

本测区位于北部湾东部，三面为陆地所包围，经琼州海峡与粤西海区相通。因此其空气必然受陆地和岛屿的影响较大。本文论述的该区空气微生物分布状况，其含量一般高于环海南岛上空的，而小于粤西海区的，但是真菌百分数却高些[5]。这与测定时该海区空气的相对湿度高有关。测定结果表明，空气微生物含量在测区中显出北高、南低、中间最少的变化状况。其中，“北高”反映了较高工业化程度造成空气污染的影响，而且人群活动的影响也大。“南低”反映了那里人烟较稀少，工业交通事业还未形成规模，因而空气污染不严重。一般而言，海区上空的空气微生物含量最低，这意味着那里的海洋空气较洁净，尽管此时顺行的是风力不大的北风，但对空气中的陆源菌具明显的稀释作用。

测区空气微生物含量的昼夜变化状况基本不同于相邻区海、陆空气微生物的，仅真菌含量出现的黄昏峰稍类似于环海南岛和粤西海区的[5,6]。测区空气微生物含量在中午未出现峰值，这可能与太阳辐射致使一些微生物死亡有关[7]。虽然陆源菌来到海洋空气中与海洋菌有一定的相互选择和适应的过程，但本测定中空气微生物含量与气温间的相关分析结果意味着测时不大的气温变幅对当时的空气微生物变化影响不大。

本结果初步揭示和分析了北海—临高海区空气微生物含量及变动势态，由此可窥见该区空气污染和人群活动之影响。有关研究应该深入进行。

参考文献 7 篇（略）

AIRBORNE MICROBES OVER SEA AREA OF BEIBU GULF BETWEEN BEIHAI AND LINGAO

(ABSTRACT)

Abstract Investigation on the airborne microbes count over the sea area of the Beibu Gulf between Beihai and Lingao indicates that the averages of airborne bacteria, fungi, total microbes and the percentage of fungi count to the total count are 632. 3, 1 324. 0, 1, 956. 3 CFU · m^{-3} and 67. 7, respectively. Analyses on the relative data show that the airborne microbes over the investigated area display the status of higher count in the north and lower count in the South of the sea area. The microbial counts over the sea area are slightly bigger than that around the Hainan Island, but smaller than that over the western Guangdong sea area, that in the nearshore zone lower than that over its neighbour land area. The diurnal change of airborne microbial counts shows that the peak of bacterial count appears early in the morning, the lower value at noon; the peak of fungi count appears in the evening, and their lowest value in the morning, which is different from the diurnal change of airborne microbes counts over the area of the Beihai city, but similar to those over the adjacent sea areas. No significant correlation is found between the airborne microbial counts and air temperatures there. The obtained results mean that air pollution in the nearshore zone is heavier than that over the sea area, and the same is the influence of human activity.

Key words Beibu Gulf, Sea Area Between Beihai and Lingao, Airborne Microbes, Air Pollution, Human Activity

西部北太平洋测区的空气微生物*

海洋上空气微生物包括海洋性和陆源性两类微生物。海洋性微生物借本身的质量和体积在海洋波涛的上托、海水(包括各种粒子、水雾、冰晶雾等气溶胶)的发散和风力的携带中散布、悬浮、生繁于空气中。同样陆源性微生物包括陆地源(大陆和岛屿)、淡水等自然和人为作用带入(包括各种设施排放物)的微生物借助陆风及其挟持的气溶胶、颗粒物飘移至海洋上空,受重力作用一起沉浮于大气下边界中。因此海洋上空气微生物是海洋、陆源和人类活动的综合的延伸和映证。特定测区的空气微生物反映所在海域的海洋状况、大气、陆源和海洋开发的影响。

现代空气微生物学发迹于 20 世纪 60 年代,研究集中在陆地,尤其是城市室内外空气[1]。对海洋上空气微生物学的研究仅是上世纪 80 年代才兴起[2]。随着海—气、海—陆相互作用研究和海洋开发的拓展,关注海洋环境,包括其空气状况已成必然,海洋空气微生物的研究也就成了一个重要课题[3]。

奄美群岛、吐噶喇群岛一带的西部北太平洋水域处中国台湾和日本九州之间,受东中国海影响大,离"西太暖池"又不远,海—气、海—陆(岛)交换频繁、人类活动较多,但其空气微生物状况不明,这正是本文探讨的课题。

一、测区、测点和测时

本测区位于 27°00′～36°20′N 和 120°20′～134°32′E 之间,西部离岛较近,中东部离岛远、开阔,平均气温 25.8℃,盛行西风或南风,风力不大。共设 16 个测点,具体测点如图 1 所示。此外,为了作对照,选取青岛港为 No. 17 测点。

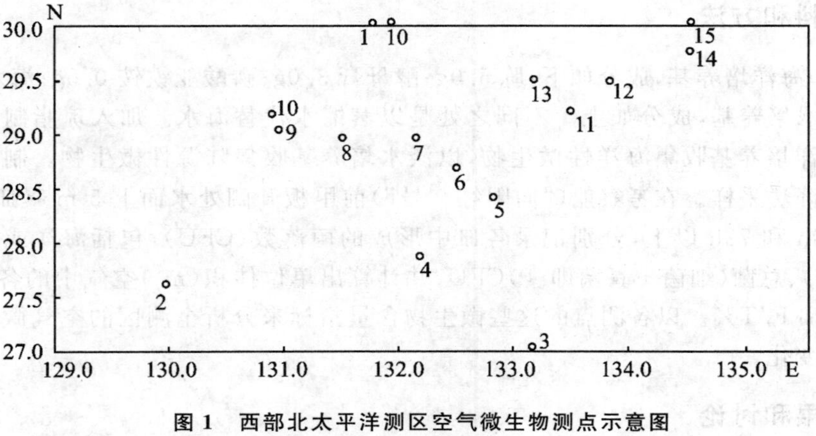

图 1　西部北太平洋测区空气微生物测点示意图

* 原文载于《中国当代思想宝库(七),科学学》,2002,644-647

表1列出的是各测点/时的概况。

表1 西部北太平洋测区空气微生物测点位置及天气状况

测点编号	各点经纬度	天气状况					起始测时
		气温(℃)	风向	风力	能见度	天气	
1	131°33′87E、30°00′62N	24.0	NW		7	云天	2002.6.15 07:25
2	126°55′09E、27°34′57N	24.0	SW	4级	6	阴间天	2002.6.17 05:30
3	131°10′23E、27°02′24N	28.0	SW		7	晴	2002.6.19 05:30
4	132°11′04E、27°50′55N	27.0	SW	2~3级	7	阴	2002.6.21 12:45
5	132°51′23E、28°22′87N	25.5	ESE	2~3级	6	毛毛雨	2002.6.23 14:00
6	132°29′04E、28°39′24N	27.0	SW	2级	7	云天	2002.6.25 05:15
7	132°10′03E、28°55′23N	27.2	ESE	2级	7	晴	2002.6.26 10:20
8	131°30′01E、28°55′26N	25.0	NW	3级	7	半晴	2002.6.26 18:25
9	130°50′79E、28°04′03N	24.6	W	4级	7	晴	2002.6.27 00:10
10	130°53′32E、29°11′21N	24.0	NW	3级	7	半晴	2002.6.27 04:00
11	133°29′74E、29°11′25N	25.0	SW	3~4级	7	晴	2002.6.27 19:30
12	133°50′06E、29°27′85N	26.0	SW	2级	7	半晴	2002.6.28 07:00
13	133°09′85E、29°27′04N	26.1	NW	2级	7	半晴	2002.6.28 12:00
14	134°30′06E、29°43′91N	27.0	SE	3级	7	晴	2002.6.29 11:25
15	134°31′68E、29°59′64N	26.0	SE	2级	7	半晴	2002.6.29 16:40
16	131°55′66E、30°00′30N	27.0	SW	3级	7	阴	2002.6.30 06:25
17	120°20′68E、36°10′58N	24.4	S	4级	7	半晴	2002.7.2 13:30

二、材料和方法

采用(1)海洋培养基,成分如下:胨5.0g;酵母膏3.0g;磷酸亚铁铵0.5g;海水1 000.0 mL。(2)淡水培养基,成分如上。不同之处是以蒸馏水代替海水。加入琼脂制成固体培养基。以海洋培养基收集海洋性微生物,以淡水培养基收集陆源性微生物。制成的平皿用于自然沉降法采样。在考察船("向阳红9号")前甲板开阔处水面上5 m曝皿30 min。室温培养48h和72h以上,分别记录各皿中形成的菌落数(CFU),包括海洋或陆源细菌(B)、真菌(F)、总菌(细菌+真菌即T)CFU,并计算出单位体积(m^3)空气中的各CFU,即CFU·m^{-3}和F/T%。以各测点的这些微生物含量指标来分析全测区的空气微生物含量状况及时空变化。

三、结果和讨论

(一)测区空气微生物含量概况

表2列出的是全测区所有测点空气中海洋性、陆源性微生物含量指标,由该表计算出

海洋性和陆源性细菌、真菌的检出率分别为 93.7、37.5 和 86.7 及 33.3%。表明测区空气中海洋性细菌、真菌出现机率大于陆源性的，但两者差距不很大。细菌机率大大高于真菌。

表 2　西部北太平洋测区空气微生物含量状况 *

测点编号	海洋性				陆源性			
	细菌	真菌	总菌	F/T(%)	细菌	真菌	总菌	F/T(%)
1	0	104.8	104.8	100	262.0	78.6	340.6	23.1
2	183.4	0	183.4	0	209.6	0	209.6	0
3	183.4	0	183.4	0	157.2	0	157.2	0
4	1 414.8	0	1 414.8	0	1 441.0	0	1 441.0	0
5	52.4	0	52.4	0	26.2	0	26.2	0
6	262.0	0	262.0	0	2 953.8	0	2 953.8	0
7	26.2	524.0	550.2	95.2	0	52.4	52.4	100.0
8	26.2	104.8	131.0	80.0	78.6	235.8	314.4	75
9	655.0	0	655.0	0	183.4	0	183.4	0
10	471.6	0	471.6	0	969.4	0	969.4	0
11	0	13.1	13.1	100	0	0	0	0
12	26.2	0	26.2	0	628.8	0	628.8	0
13	131.0	39.3	170.3	23.1	104.8	78.6	183.4	42.9
14	52.4	0	52.4	0	157.2	52.4	209.6	25.0
15	131.0	0	131.0	0	/	/	/	/
16	458.5	0	458.5	0	131.0	0	131.0	0
17	707.4	340.6	1048.0	32.5	969.4	262.0	1 231.4	21.3

* 微生物含量单位：CFU·m^{-3}，下同；* * No.17 为对照测点。

由该表得出的平均海洋性细菌、真菌、总菌量及 F/T%分别为 254.6、49.1、303.7 CFU·m^{-3} 及 16.2。细菌量是真菌的 5.2 倍，占总菌量的 83.8%，平均的陆源性细菌、真菌量及 F/T%则分别为 486.9、33.2、520.1 CFU·m^{-3} 及 6.4。细菌量是真菌的 14.7 倍，占总菌量的 93.6%。海陆两者相比，前者的细菌、总菌量均小于陆源的，但真菌量及 F/T%均大于陆源的。陆源细菌、真菌、总菌量及 F/T%分别是海洋的 1.9、0.68、1.71 及 0.40倍，说明测区空气中陆源性细菌及总菌量均大于海洋性的，即来自大气沉降来的与陆地和岛屿飘来的细菌等和人及人为设施带来的陆源影响较大。这与亚洲东部近海的趋势相类似，但不同于中国沿岸和台湾海峡[4]，也不同于并小于浙江近岸区[5]。与对照点 No. 17(青岛)比，微生物含量低了许多，陆源对西部北太平洋测区的影响也小了许多。测区空气海洋性微生物含量也大大低于表层海水微生物含量(未公布资料)。

(二)空气微生物含量的时空分布

1. 空气微生物含量的空间分布。

自岛屿较多、离陆较近的西部到开阔的东部，将 No. 2、9、10 看作近岛测点，将 No. 3、12、14、15 看作离岛开阔测点。其各自平均的空气微生物含量指标状况如表 3 示。

表 3　近岛测点与离岛开阔测点空气微生物含量比较

测点类型	空气微生物							
	海洋性				陆源性			
	细菌	真菌	总菌	F/T(%)	细菌	真菌	总菌	F/T(%)
近岛测点	436.7	0	436.7	0	454.4	0	454.4	0
离岛开阔测点	87.3	0	87.3	0	314.4	17.5	331.9	5.3

表 3 说明近岛测点平均的海洋性细菌、总菌，陆源性细菌与总菌量均比离岛开阔测点的高，无论是海洋性的，还是陆源性的细菌。在此两类测点中真菌都少甚而无。说明陆、岛细菌对海洋空气微生物有较大贡献，且随离陆、岛距离增而减。

自北至南以 28°～30°N 为界，将测区分为两部，得出北测区(计 11 个测点)和南测区(计 5 个测点)，其各自的空气微生物含量指标状况如表 4 所示。

表 4　南北测区空气微生物含量比较

测区名	空气微生物含量							
	海洋性				陆源性			
	细菌	真菌	总菌	F/T(%)	细菌	真菌	总菌	F/T(%)
北测区	144.1	71.5	215.6	33.2	646.5	49.8	696.3	7.2
南测区	497.8	0	497.8	0	403.5	0	403.5	0

由该表可见，北测区空气中只是海洋性细菌及总菌量小于南测区，其余各项指标均是北测区大于南测区，说明南测区海洋性细菌对其空气微生物的贡献大，北测区空气中真菌量增大，陆源影响大于海洋性的且大于南测区。这反映了两测区所在的海洋(及陆源影响)环境状况。

2. 空气微生物含量的时间分布。

将测时划分为早晨、上午、中午、下午、傍晚和夜间六个时段，它给出了测区空气微生物含量指标的昼夜变化状态。由该图可见，海洋性细菌量在傍晚低至谷后于夜间起高峰。真菌的谷值从夜同持续至早晨，上午才抵峰值。总菌状态与细菌一致。陆源性细菌的峰值在下午，谷值在晚上，真菌峰同细菌，而谷不同于细菌的，出现于早晨、晚上和夜间。总菌峰、谷似细菌。本测区空气微生物昼夜变化状基本不同于东、黄海测区，显示它自身的特点[5]。

四、结语

本文论述了西部北太平洋测区空气微生物的检出率、空气中海洋性、陆源性细菌、真菌、总菌量及F/T%的时空分布状态。

分析了这些状态及差异的原因。测区此间的气温适于海洋性微生物生繁，风力有利于空气微生物稳定，风向适于海陆微生物散布、消涨。比较了以往邻近海区空气微生物状况。结果反映了海洋—大气、海洋—陆地、人—自然在测区的相互作用。

参考文献5篇（略）

AIR-BORNE MICROBES OVER INVESTIGATED AREA OF WESTERN NORTH PACIFIC OCEAN

(ABSTRACT)

Abstract Detection percentages, contents of bacteria, fungi, total and F/T% of marine and terrigenous microbes are determined by using the method of gravity plate during June-July, 2002, this paper analyses their spatial and temporal distributions and variations.

The result obtained discovers that: the occurrence possibilities of marine bacteria and fungi in the air are more than terrigenous ones. Occurrence possibility of bacteria is more than fungous one. Marine microbial contents of two kinds in the air are lower than terrigenous. It illustrates that the stronger sensibility of air-borne microbes in the investigated area on atmospheric precipitation, influence of land and islands and interference in human being and some equipments made by hunan.

There is a comparison of the air-borne microbial conditions between the southern and northern areas investigated. The result shows the character of diurnal dynamics of air-borne microbial contents over this area.

Key words Western North Pacific Ocean; Air-Borne Microbes; Interactions Among Sea-Air, Sea-Land and Human Being-Nature; Marine; Terrigenous

黄海测区空气微生物含量估计*

摘　要　本文对黄海测区空气微生物作了一次测定，结果表明空气中海洋性细菌、真菌的检出率分别为52.4和47.6%。陆源性细菌、真菌的检出率分别为80.0%和60.0%。陆源性微生物出现机率大于海洋性的。空气微生物以细菌为主。测区平均的海洋性空气细菌、真菌、总菌量及真菌/总菌%分别为592.6、329.1、921.7 CFU·m^{-3}及35.4。平均的陆源性空气细菌、真菌、总菌量及真菌/总菌%分别为689.1、377.9、1 067.0 CFU·m^{-3}及35.7。文章分析了空气微生物的时空分布状态。指出测区空气微生物状态反映了海—气、海—陆、人和自然的相互作用。

关键词　黄海　空气微生物、海-气、海-陆和人与自然相互作用

海洋与大气下边界间有各种物质的交换，包括微生物的交换。海洋状况及分布与变化直接影响到海洋上空气微生物的结构与功能，同时海洋上空气微生物又始终经受着不同距离陆源的影响。随着海洋开发的兴起，海洋遭遇日益增强的人类活动干预，因此海洋空气微生物状况是海—气、海—陆、人与自然相互作用的综合作用产物。它反映海洋空气污染程度，是海洋空气质量的重要指标。它直接关系到海上作业人员的健康，也相关着陆上空气质量。

海洋空气微生物学的研究落后于现代空气微生物学的发展，只是20世纪80年代才开始注意[1-2]。作者曾对黄海测区空气微生物作过一些研究，但不深入[3-4]。本文就黄海测区再作一次较广泛的空气微生物调研，以加深对海洋及其空气微生物的认识。

一、测区与测点概况

于2002年4月间乘“中国海监8”作空气微生物测定。测区范围是：27°02′～30°10′N和129°50′～134°32′E，共设12个测点分22次测定。测区与测点详见图1。测点测时及其天气状况如表1所示。

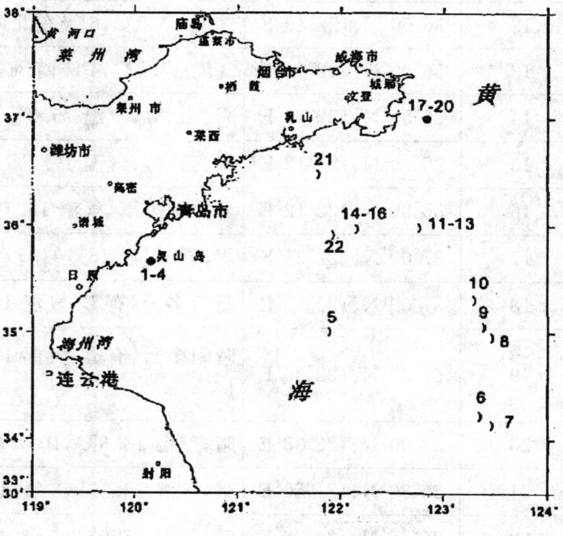

图1　黄海测区空气微生物测点示意图

Fig. 1　Sketch map of sampling stations for air-borne microbes over the investigated area of the Yellow sea

*　原文刊于《海洋湖沼通报》，2003，No. 4：55-60.

表 1　黄海测区空气微生物测时及天气状况

Table 1　Time for air borne microbes over the investigated area of the Yellow sea

测次编号（No.）	站位	天气状况	测时
1	35°41′N、119°21′E	阴间小雨，相对湿度 90%，气温 18.3℃，7～8 级北风	2002.4.15 15:55～16:10
2	35°41′N、119°21′E	阴间小雨，湿度 90%，气温 17.3℃	2002.4.15 20:00～20:17
3	35°41′N、119°21′E	小雨点，气温 17.8℃，6～7 级北风	2002.4.16 03:15～03:30
4	35°41′N、119°21′E	阴间晴，气温 17.0℃	2002.4.16 09:45～10:00
5	35°02′N、121°56′E	气温 15.7℃，7 级以上北风	2002.4.17 18:45～19:15
6	34°15′N、123°23′E	晴朗，但仍灰蒙蒙，气温 15.7℃，4～5 级南风	2002.4.18 18:15～18:45
7	34°10′N、123°30′E	晴，日出半小时，气温 16.4℃，5 级南风	2002.4.19 05:50～06:20
8	34°59′N、123°30′E	晴间多云，气温 17.3℃，5 级以上东北风	2002.4.19 18:00～18:30
9	35°05′N、123°25′E	晴，气温 16.3℃，<7 级东北风	2002.4.20 07:00～07:30
10	35°20′N、123°20′E	晴间多云，气温 17.8℃，5 级南风	2002.4.20 10:00～10:30
11	36°00′N、122°48′E	晴间多云，气温 17.5℃，5 级南风	2002.4.20 15:45～16:15
12	36°00′N、122°48′E	气温 17.5℃，6 级东北风	2002.4.20 20:45～21:15
13	36°00′N、122°48′E	气温 17.4℃，4 级南风	2002.4.21 01:45～02:15
14	35°59′N、122°12′E	阴、云、雾，气温 15.7℃，7～8 级南风	2002.4.21 11:45～12:15
15	35°59′N、122°12′E	阴、云、雾，气温 17.7℃，7～8 级南风	2002.4.21 15:50～16:20
16	35°59′N、122°12′E	雾如细雨下，气温 16.7℃，4 级北风	2002.4.22 08:45～09:15
17	37°00′N、122°35′E	阴，雾，气温 16.7℃，7 级南风	2002.4.22 17:45～18:15
18	37°00′N、122°35′E	晴间多云，有雾，气温 18.0℃，4～5 级北风	2002.4.22 22:15～22:45
19	37°00′N、122°35′E	晴间多云，有雾，气温 17.6℃，8 级以上东北风	2002.4.23 08:00～08:30
20	37°00′N、122°35′E	晴，气温 13.6℃，10 级东北风	2002.4.24 09:20～09:50
21	36°30′N、121°50′E	晴，气温 15.1℃，3 级风	2002.4.25 09:00～09:20
22	35°56′N、121°58′E	晴，气温 16.9℃，3 级风	2002.4.25 13:20～13.40

二、材料与方法

以 Zobell 2216E 固体培养基和营养琼脂分别作为海洋性微生物和陆源性微生物的培养介质。采用平血自然沉降法。在上甲板开阔处（离海面 5 m 以上）曝皿 30 min。室温培养 48h 和 72h 以上，分别计数平皿中长出的菌落即 CFU。并计算出单位体积(m^3)空气中海洋性和陆源性细菌(B)、真菌(F)及总菌(细菌＋真菌，即 T)量(即 CFU・m^{-3})及真菌/总菌量(F/T)%。分析空气微生物含量的时空分布，以评价海洋空气质量[5]。

三、结果和讨论

1. 空气微生物含量概况

表 2 列出了所有测点/时的空气微生物含量指标。

表 2　黄海测区空气微生物含量指标*

Table 2　Indexes of air-borne microbial contents of the investigated area of the Yellow Sea

No.	陆源性微生物				海洋性微生物			
	B	F	F/T%	T	B	F	F/T%	T
1	157.2	0	0	157.2	0	2148.4	100	2 148.4
2	92.4	785.4	89.5	877.8	46.2	0	0	46.2
3	524.0	0	0	524.0	471.6	471.6	50.0	943.2
4	366.8	2 305.6	86.3	2 672.4	262.0	0	0	262.0
5	497.8	2331.8	82.4	2 829.6	262.0	1 048.0	80.0	1 310.0
6	131.0	524.0	80.0	655	209.6	314.4	60.0	524.0
7	1 493.4	550.2	26.9	2 043.6	235.8	1 257.6	84.2	1 493.4
8	0	366.8	100.0	366.8	0	471.6	100.0	471.6
9	0	157.2	100.0	157.2	0	0	0	0
10	52.4	52.4	50.0	104.8	0	78.6	100.0	78.6
11	78.6	183.4	70.0	262.0	0	0	0	0
12	104.8	0	0	104.8	0	0	0	0
13	0	0	0	0	0	0	0	0
14	/	/	/	/	13.1	0	0	13.1
15	0	0	0	0	0	0	0	0
16	/	/	/	/	2 122.2	0	0	2 122.2
17	4 637.4	0	0	4 637.4	0	0	0	0
18	52.4	157.2	75.0	209.6	0	235.8	100.0	235.8
19	707.4	26.2	3.6	733.6	/	/	/	/
20	2 331.8	0	0	2 331.8	2 751.0	0	0	2 751.0
21	2 397.3	0	0	2 397.3	5 895.0	294.8	4.8	6 189.8
22	157.2	117.9	42.9	275.1	176.9	589.5	76.9	766.4

* 各微生物含量单位：CFU・m^{-3}，下同。

由该表可计算出测区海洋性空气细菌、真菌的检出率分别为52.4(11/21)和47.6(10/21)%。陆源性空气细菌、真菌检出率分别为80.0%和60.0%。说明测区空气中陆源性微生物出现机率大于海洋性微生物,细菌出现机会多于真菌。

测区平均的海洋性空气细菌、真菌、总菌及真菌%分别为592.6、329.1、921.7 CFU·m^{-3}及35.7,细菌是真菌的1.8倍。平均的陆源性细菌、真菌、总菌及真菌%则分别为689.1、377.9、1 067.0 CFU·m^{-3}及35.4,细菌是真菌的1.8倍多,说明空气中微生物组成以细菌为主。海洋性与陆源性微生物两相比较,前者的细菌、真菌、总菌量均比后者的低,仅真菌%稍高于后者,如果说某一来源的微生物可以指示某一来源影响的话,那么可以说测区空气的陆源性影响大。本测区离岸最远点(No.7)仅约180海里,而黄海本身又是大陆边缘海,加上测时正值初春,大陆来的北风,甚至沙尘暴挟带来的陆源菌的机会尚多,因而陆地影响必然是明显的。本结果与16年前的测定,即海、陆空气微生物差异相类似[5],比粤西海区空气中海洋性微生物少,但比环海南岛海区空气微生物多了不少[6]。

2.空气微生物含量测点间差异

由表2可见,海洋性空气细菌、真菌、总菌量最高测点分别是No.21、No.1和No.21,有10个测点未测出细菌,11个测点未测出真菌。陆源性空气细菌、真菌、总菌量最高测点分别是No.21、No.5、No.17。未测出细菌和真菌的点分别为4和8个。最高与最低空气微生物含量测点间差距大。

将所有测点按离陆远近分作两区,即近岸区和远岸区。近岸区包括以下测点/时:No.1-4和17-22,远岸区包括测点/时:No.5-16,两区各自空气微生物指标的平均值列入表3。

表3 黄海近岸区、远岸区空气微生物含量指标比较

Table 3 Comparison of indexes of air-borne microbes between investigated inshore and outshore areas of the Yellow sea

测区名	海洋性						陆源性					
	B		F		T	F/T%	B		F		T	F/T%
	检出率(%)	含量	检出率(%)	含量			检出率(%)	含量	检出率(%)	含量		
近岸区	66.7	1 066.9	55.6	415.6	1 482.5	28.0	100.0	1 142.4	50.0	339.2	1 481.6	22.9
远岸区	41.7	236.9	41.7	265.3	502.2	52.8	60.0	262.0	70.0	416.6	678.6	61.4

含量单位:CFU·m^{-3}

由表3可见,近岸区海洋性和陆源性空气细菌、海洋性真菌的检出率均高于远岸区,仅陆源性真菌检出率显出近岸区小于远岸区。近岸区空气海洋性和陆源性细菌、总菌量均大于远岸区的,近岸区空气海洋性真菌量也大于远岸区的,但陆源性真菌量及其%、海洋性真菌%却是近岸区小于远岸区。这一状况与东海不一致[4]。

3.空气微生物含量指标的昼夜变化

将所有测时按早晨、上午、中午、下午、傍晚和夜间六个时段划分,其微生物含量指标的常用对数值为纵坐标、以时段为横坐标,得出图2。由图2①可见全区空气中海洋性细菌和总菌含量峰值均出现在上午,中午见细菌量的低谷,而总菌量的低谷推迟至夜间。海洋性真菌量的峰与谷分别在早晨和随之而来的上午，真菌％的峰和谷则分别出现于上午和下午。

陆源性细菌量的峰和谷分别在上午和夜间，而真菌量的峰和谷却在晚上和下午，总菌波动吻合于细菌，真菌％波动则吻合于真菌量(图 2②)。

由此可见空气细菌和总菌量的峰，无论是海洋性、还是陆源性的，均分别出现于上午。总菌量的谷也无论是海洋性抑或陆源性的均出现于夜间。陆源性细菌量的谷也在夜间，此点类似于东、黄海测区以往结果[4]。

No. 1-4 测点/时的状况则表达了一个测点的空气微生物含量之昼夜变化，它反映了其不同于全测区的变化状态，而显示出海洋性与陆源性微生物昼夜变化在该点上的相似性(见图 2③和④)。

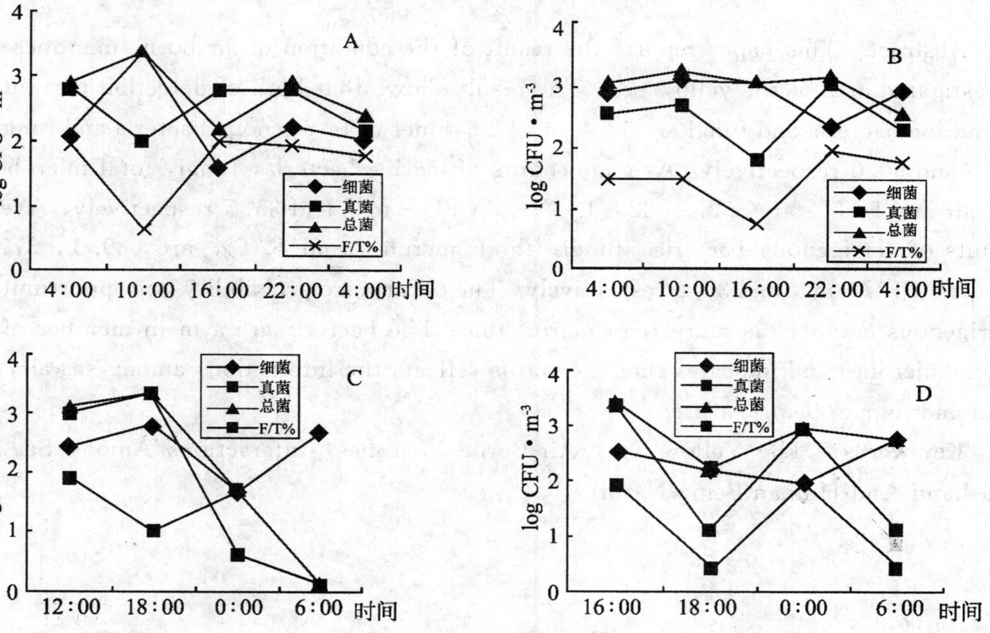

图 2　黄海海区空气微生物昼夜变化状态

Fig. 2　Status of diurnal variation of air-borne microbes over investigated area of the Yellow sea

四、结语

对黄海测区作了一次空气微生物测定，结果表达了空气中海洋性和陆源性细菌、真菌、总菌及真菌％的时空分布状况。陆源菌在测区的广布性和对海洋的扩展及海洋菌的变化意味着海陆两类微生物在测区空气中的交叉、消涨，本空气微生物测定结果反映了测区的边缘海特征。海洋性和陆源性微生物之昼夜变化状态显出异同点。按陆上标准看测区空气仍十分洁净，但比起以往结果来，此间空气微生物含量有了增加[3-4]，在一些近岸测点更为明显。这可能暗示陆源污染的进一步入侵。海洋性微生物的消涨意味着海水运动对空气的影响程度。

参考文献 6 篇(略)

ESTIMATION ON AIRBORNE MICROBIAL CONTENT OVER INVESTIGATED AREA OF THE YELLOW SEA

(ABSTRACT)

Abstract This paper reports the result of the condition of air-borne microbes over investigated area of the yellow sea. The result shows that both of detection percentages of marine bacteria and fungi are 52.4 and 47.6, ones of terrigenous bacteria and fungi are 80.0 and 60.0 respectively. Average counts of marine bacteria, fungi, total microbes in the air and F/T% are 592.6. 329.1. 921.7 CFU · m^{-3} and 35.7 respectively, Average counts of terrigenous bacteria, fungi, total microbes and F/T% are 689.1, 377.9, 1 067.0 CFU · m^{-3} and 35.4 respectively. The occurrence probability (or opportunity) of terrigenous microbes is more than marine one. The bacteria are a main member of air-borne microbes and indicates that the status reflects the interactions among sea-air, sea-land and human being-nature.

Key words The Yellow Sea, Air-Borne Microbes, Interactions Among Sea-Air, Sea-Land And Human Being-Nature

E_3 极地(POLAR REGIONS)

南极菲尔德斯半岛空气微生物含量初报*

摘 要 南极菲尔德斯半岛室外空气微生物的采样站分作3种类型:类型Ⅰ.科考站及附近区域;类型Ⅱ,大型动物出没区;类型Ⅲ,大型动物少,且保持荒凉状态区。结果显示,Ⅰ类型区室外空气微生物含量明显高于Ⅲ类型区。平均Ⅰ区有4 994 CFU·m^{-3},Ⅲ区仅16 CFU·m^{-3}。结果提示,菲尔德斯半岛空气微生物已受到人类活动的干扰。

关键词 空气微生物 南极菲尔德斯半岛 人类活动的干扰

菲尔德斯半岛是南极半岛中位于乔治王岛上最北端的一片陆地。其北部和柯林斯冰盖相接,西北与东南两面分别被德雷克海峡和麦克斯韦尔湾所包围,南端隔菲尔德斯海峡与纳尔逊冰盖相望。因受绕极气旋影响,此处中等以上风力的天气较多。南极夏季降水较丰,温度升高,气温上升。1993.12～1994.02期间的极端最高气温可达7℃左右,最低气温为－3.9℃,一般在0℃左右。

人类介入南极这一带,最早可上溯到19世纪。近20余年来,各国在南极半岛,主要是在菲尔德斯半岛一带相继建起了7个科学考察站。人类活动必将干扰和影响其固有的生态平衡。空气微生物为南极生态系中重要而敏感的成员,因而了解空气微生物的状况,并比较分析不同地区微生物的变化,在研究人类活动或其他外来者对南极地区生态的影响方面具有十分重要的意义[1～3]。10年前一些国际间合作研究组织开始南极空气微生物的研究工作,并取得了一些成果[4]。本文作者于1986年和1987年南极之夏,也曾对南极设得兰群岛地区的空气微生物作过研究[3]。本文则初步报导1993.12～1994.02在南极工作期间对菲尔德斯半岛的一些地区空气微生物含量的考察结果。

1.材料和方法

采用平皿自然沉降法,培养基用普通营养琼脂培养基[3,5]。将制备好的培养基平皿置于较空旷的地方;曝皿时间依实际情况而定,一般为1h,曝皿后的培养皿置12℃左右的培养箱中培养1～2 wk即可计算出单位体积(cm^3),空气中的CFU数即n(cfu)[3],并对之进

* 原文刊于《应用与环境生物学报》,1997,3(2):180-183

“八·五”国家重点攻关项目《南极重点地区生态系统研究》资助

行统计、分析和比较，采样的同时记录必要的天气参数等有关资料。共设测点计 21 个，共测 63 次(详见图 1)。

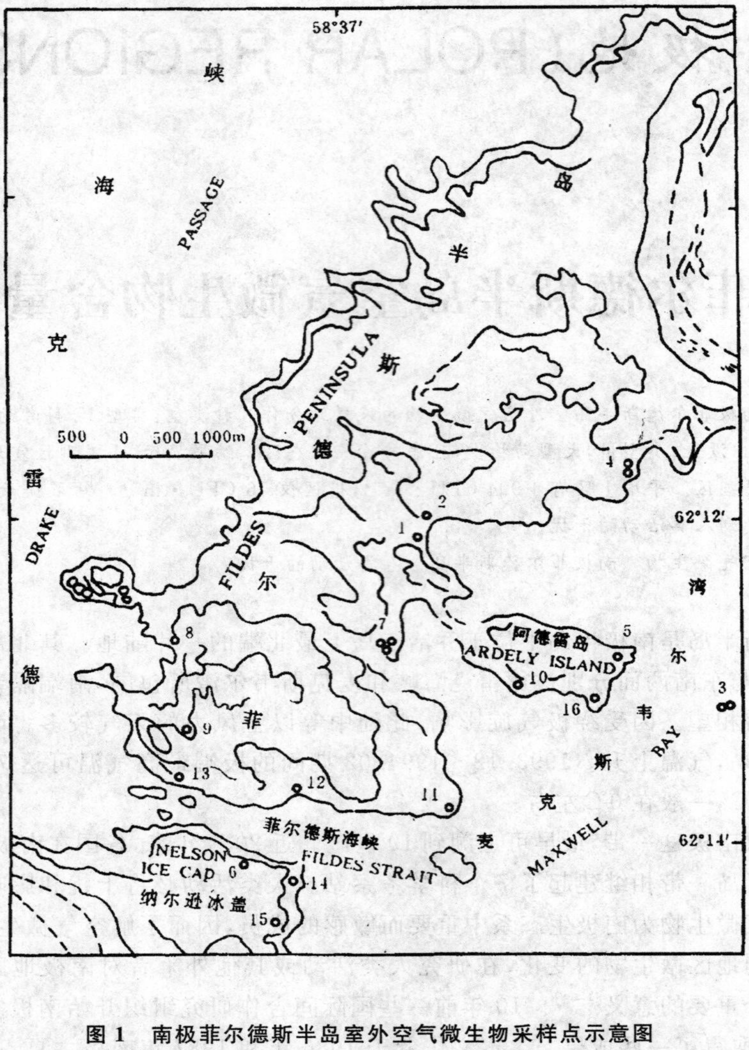

图 1　南极菲尔德斯半岛室外空气微生物采样点示意图

图例：1. 马尔什基地(CL)，2(2-1，2-2)，别林斯高晋站(SU)，3. 金世宗站(KR)，4. 阿蒂加斯站(Uru)，5. 德国观察所(DE)，6. 捷克站(CZ)，7. 长城站(CN)，8. 地理湾，9. 西海岸，l0. 阿德雷岛，11. 半三角，12. 碧玉滩，13. 三门峡口，14. 纳尔逊冰盖，15. 纳尔逊冰盖，离岸稍远处，16. 阿德雷岛南海滩

Fig. 1　Map of the sampling sites for outdoor air-borne microbes over the Fildes Peninsula, Antarctica

Legend: 1. Marsh Base(CL); 2(2-1, 2-2). Bellingshhausen Station(SU); 3. King Se Jong Station(KR); 4. Altigas Station(Uru); 5. Germany Observatory(DE); 6. Czechic Station(CZ); 7. Great Wall Station(CN); 8. Geographer's Bay; 9. West Coast; 10. Ardely Island; 11. Half Three Point; 12. Biyu Beach; 13. The Mouth of Sanmen Strait; 14. Nelson Ice Cap; 15. Nelson Ice Cap. slightly away from the seashore; 16. Southern beach of Ardely Island

2. 结果和讨论

表1列出所有测点上不同时间测得的空气微生物含量状况。

表1　南极菲尔德斯半岛空气微生物含量

Table 1　Counts of the outdoor air-borne microbes over the Fildes Peninsula, Antarctic

测点类型 Type of sites	测点号* No. of sites	n(cfu)/ m^{-3}	测定日期 Surveying dates	备注 Remarks
Ⅰ	1	79	1994-01-20	
	2-1	118	1994-01-20	
	2-2	603	1994-01-30	
	3-1	25 603	1994-02-05	
	3-2	10 847	1994-02-05	
	4-1	2 038	1994-02-11	
	4-2	10 063	1994-02-11	
	5	35	1994-02-23	在阿德雷岛上 On the Ardely Island
	6	393	1994-02-19	纳尔逊冰盖，气象湾海边 Nelson Ice Cap. Seaside of Meterological Bay
	7**	162	1993-12	～1994.12
Ⅱ	8	1 874	1994-01-22	海豹群集处 Seale gathering place
	9	5 616	1994-01-30	死鲸处 The site where a dead whale was found
	10	46	1994-02-23	-山上 On a hill
Ⅲ	11	13	1994-02-05	
	12	28	1994-02-05	
	13	21	1994-02-05	
	14	12	1994-02-19	海岸边 Nearby the coast
	15	21	1994-02-19	离岸稍远处 A little far from the coast

*见图1(See Fig.1)

**是该站3个测点上45次测试结果之平均值 The average value from the results of 45 determinations over three sites.

表1中将测点分作3个类型：Ⅰ类，为科学考察站及附近。这里有较多的人类用固定设施，人类活动较频繁，大型野生动物极少。Ⅱ类，为大型生物较多的野外区域，时有考察队员活动。Ⅲ类，大型生物较少的荒凉区域，人迹甚少。由表1可见，Ⅰ区中包括7个科考站的12个测点。按空气微生物含量大小排列，各站次序如下：3、6、2、7、1、5。总平均 n(CFU)＝4 994/m^3・3测点的数据可能比实际略高，因为在测定时常有挟持站区地面尘埃的风雪干扰，会使微生物落入培养基的机会增多。但是3号测点的空气微生物含量总的来说远高于其他测点。1号测点微生物含量较少，可能与该站相当整洁的环境有关。7号

测点则处于甚少水平。6号和5号测点在Ⅰ区中是人为影响较少的地方。6号测点有时仅有1～2人偶然作短暂的工作停留，其空气微生物理应不多。但本次测试时，上风有5～6人在操作，这可能是微生物含量高于5号测点的原因。5号测点设施极为简陋，一年中极少有人去，成为Ⅰ区中空气微生物最少之处。

Ⅱ类区设有3个测点：8号测点海豹群集、栖息地。虽处海边，但四周被山丘和海岛包围，环境相当稳定。人们可以看到和感到大群海豹身上散发和口中呼出的暖气形成的气雾和排泄物的浓烈腥臭气味。此处空气微生物数之高当与此相关。9号测点有一搁浅死鲸，其腐尸体招致大批港湾鸽啄食；此处还有不少大型海藻残体，加之考察队员的活动等，必然会对该测点的空气微生物数量产生影响. 与上述两测点相反，10号测点的地势较高，虽有些企鹅活动，但空气流动好，空气也较干燥，空气中微生物少了许多[6]。Ⅱ区空气微生物含量范围为46～5 616 CFU/m^3，平均2 512 CFU/m^3。

Ⅲ类区离开人类活动区最远，大型生物少，空气清新度高，因而空气微生物量很低，平均16 CFU/m^3，其变幅为0～28 CFU/m^3。

本结果与毗邻海区以往的资料相比，空气微生物含量均高[3]，表明陆上空气比海洋空气常集聚较多微生物，陆上空气也易受人类影响。

3地区空气微生物含量间有明显差距。其含量由高到低的基本次序是Ⅰ区、Ⅱ区、Ⅲ区。这一结果似可看出人类活动及其他生物对南极菲尔德斯半岛空气微生物的影响。

参考文献6篇（略）

OUTDOOR AIR-BORNE MICROBIAL COUNTS OVER THE FILDES PENINSULA, ANTARCTICA

(ABSTRACT)

Abstract The sampling sites for investigating outdoor air-borne microbes over the Fildes Peninsula, Antarctica were divided into 3 types: 1. the scientific research stations and their adjacent areas; 2. the haunts of large animals; 3. the desolate areas with few large animals. The microbial counts of type 1 were obviously higher than those of type 3. Averagely speaking, the microbial counts of types 1, 2, 3 were 4,994, 2,512 and 16 CFU · m^{-3}, respectively. The results indicated that the air-borne microbes over the Fildes Peninsula were already affected or disturbed by human activities.

Key words Outdoor Air-Borne Microbes; Fildes Peninsula, Antarctica; Disturbance of Human Activities

ESTIMATION OF CONCENTRATION OF HETEROTROPHIC BACTERIA IN SURFACE WATER OF GREAT WALL BAY AND ITS ADJACENT WATERS, ANTARCTICA AND PRELIMINARY ANALYSIS OF SOME RELEVANT FACTORS*

Abstract A survey was made of the number of the heterotrophic bacteria in surface seawater of Great Wall Bay and its adjacent waters, Antarctica during the period of austral summers from Dec. 1993 to Feb. 1994. The result shows that the number of the heterotrophic bacteria in the surface water of the surveyed area is about 798 CFU • cm^{-3}, ranging from 53 to 4,250 CFU • cm^{-3}. Comparison of the numbers for three months shows a declining tendency month by month. Comparing with the previous results, this one shows that the heterotrophic bacteria number changes somewhat, but slightly. It is lower than that in nearshore waters but higher than that in the open area. This shows that there is a positive correlative relationship between CFU and NH_4-N, which means that heterotrophic bacteria should be an important agent in NH_4-N reproduction in the environment surveyed. There is an significant negative correlation both between CFU and NO_2-N, and between CFU and PO_4-P. This means that the NO_2-N, PO_4-P, which were decomposed and transformed by bacteria, were used rapidly by phytoplankton.

Key words Heterotrophic Bacteria, Great Wall Bay In Antarctica, Ecological Factors

1 INTRODUCTION

Being an important member of a marine ecosystem and depending on their two major functions brought into playing, i. e. mineralization and production of organic matter, marine bacteria play a role that can not be replaced any other living bodies in the circulation of material and energy flow, i. e. the role of microbial cycle. Marine microbiologists have attached great importance to the research subject on the content of marine bacteria and

---

* The original text was published in *J. Glaci. Geocr.*, 1998, 20(4): 316-319.

other microbes, and regular patterns of distribution for a long time. Furthermore, their concentration, population, biochemistry and metabolic activities are closely related to marine hydrological factors nutrient salts, primary producers and microphagoes animals. So, marine ecololgists and environmental scientists are also paying more attention to marine microbiological research.

As soon as the Antarctic expedition has stepped into the period large scale comprehensive scientific activity, scientists including Chinese Antarctic research expeditionists have paid close attention to marine microbiological research and some research results have been continuously published (Chen et al., 1990, 1993).

Similarly, marine microbiological expedition of the Chinese Great Wall Bay and its adjacent sea area has been put into the program of marine ecosystem. This paper studied and analyzed the heterotrophic bacterial concentration in the area during the summer from Dec. 1993 to Feb. 1994.

2 MATERIALS AND METHODS

2.1 Sampling

The samples were obtained from surface water of Great Wall Bay and its adjacent area. The locations of the sampling stations were shown in Fig. 1. The area investigated was divided into 3 zones, i. e. zone Ⅰ, zone Ⅱ and zone Ⅲ. Determination was carried out these times, i. e. in Dec. 1993, Jan. 1994 and Feb. 1994 respectively. The samples were collected sterilely according to the method cited in reference (State Technological Superintendency Administration, 1991), and immediately brought back to the laboratory to make a counting experiment on heterotrophie bacteria in petrie dishes using plating smear method (State technological Superintendency Administration, 1991).

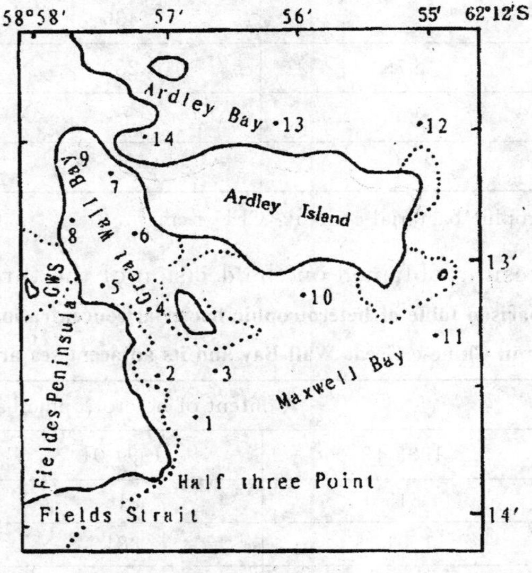

Fig. 1 Sampling stations for Heterotrophie Bacteria in Surface Water in Great Wall Bay and its adjacent aea area, Antarctica during the summer of 1993/1994

2.2 Experimenting

The smearing culture medium was made according to the routine method (Chen et al., 1994). The smeared plates were cultured for about one week at a room temperature below 4℃. The growing condition of the microbes during the culturing period was observed frequently and CFU recorded. Some colonies were picked up and brought back home for further research. At the meantime, some parameters such as water temperature, nutrient salts, chlorophyll-a and primary production were recorded for use in analyzing the results.

3 RESULTS AND DISCUSSION

Table 1 listed all the CFU data collected from the area surveyed total 28.

Table 1 Heterotrophic bacterial concentration in surface layer water in the Great Wall Bay and its adjacent sea area, Antarctica

Station	1993-12-05	1994-01-07	1994-02-03
1	85	155	185
2	53	485	396
3	2,950	607	245
4	353	635	345
5	840	1,050	720
6	653	503	895
7	—	—	895
8	—	1,635	735
9	—	757	1,075
10	—	430	—
11	975	230	—
12	200	—	—
14	4,250	—	—

Note: Unit of heterotrophic bacterial content: CFU · cm^{-3}.

Table 2 gave the results obtained on the decision of the three zones surveyed.

Table 2 Comparison table of heterotrophic bacterial concentration among different surface layer in Chinese Great Wall Bay and its adjacent sea area, Antarctica

Surveyed Zones	Content of heterotrophic bacteria		
	1993-12	1994-01	1994-02
Ⅰ	615	916	778
Ⅱ	1,017	381	—
Ⅲ	2,230	—	—

Note: Unit of heterotrophic bacterial content: CFU · cm^{-3}.

From Table 2, it can be seen that the bacterial concentration in the three zones may be arranged in the descending of its magnitude: in Dee., zone Ⅲ > zone Ⅱ > zone Ⅰ, in Jan., zone Ⅰ > zone Ⅱ. In arranging the bacterial concentration in the respective months in zone Ⅰ. order of the concentration in Jan. > that in Feb. > that in Dec., and in zone Ⅰ, that in Dec. > that in Jan.

By looking on the 3 months together, is was found that the average concentration was 792 CFU · cm^{-3} in zone Ⅰ. In zone Ⅱ, the average was 664 CFU · cm^{-3}. In zone Ⅲ, the value for December only was 2,230 CFU · cm^{-3}. The whole investigated area viewed overall, the descending order of the three months was arranged according to the magnitude of the bacterial concentration in each of them as Dec. > Jan. > Feb.. The total number of the heterotrophic bacteria in the whole area surveyed was 798 CFU · cm^{-3}. The bacterial number fluctuated between 53 CFU · cm^{-3} and 4,250 CFU · cm^{-3}. The minimum bacterial content per cm^3 of water was observed of station 1 (December, 1993). The maximum observed at station 14 (December, 1993) was probably associated with the higher amount of organic material and higher petroleum pollution there, because the station is near by Marsh Base (Chile) and Bellingshausen Station (Russia). To some extent the discharge of dirty, oil and chemicals might be helpful to the increase of heterotrophic microorganism. Except the bacterial concentration monitored respectively only once at station 7, 10, 12 and 14, the descending order of the stations according to the bacterial concentration contained there in was arranged as station 3 > station 8 > station 9 > station 6 > station 11 > station14 > station1. The maximum value was 9 times higher than the minimum. Judging from the regional distribution of water bacterial concentration in the three zones, it was inferred that the increase in the bacterial content within the Bay and in the nearshore waters might be counteracted by the active praying activity of the microphagous organisms and the clean water from outside the Bay and outshore waters (Wheeler and Kirohaman, 1986). As viewed from the tendency of the change in bacterial concentration in the three months, there appeared a tenderly toward a decrease month by month.

A comparison of this result with those from the references cited showed that the bacterial concentration in the surface water of Great Wall Bay was higher than that of the offshore waters of the Bay, though the former was of the same order of magnitude as the later and lower than that from the nearshore water and the beach (Chen et al., 1990, 1993, 1994)

Table 3 listed the statistical results from correlation analysis of the bacterial CFU data measured and the relevant environmental ecological parameters. It was found that there were positive correlation between the CFU and T, NH_4-N, N/P, Chl-a, respectively. Among them, only on positive correlation coefficient shown between CFU and NH_4-N reached a significant level (>R, 0.351), which suggests that the increase in heterotrophic bacteria might accelerate the decomposition of nitrogenous organic matter into

NH_4-N, there by increasing the concentration of NH_4-N and which seems to hint that heterotrophic bacteria is one of the active agents in the regeneration of NH_4-N in the environment surveyed.

Table 3 Result of correlation analysis among concentration of heterotrophic bacteria in surface layer water and relevant environmental-ecological parameters in Chinese Great Wall Bay and its adjacent sea area, Antarctica

Pairs	*R*	*n*
CFU-T	0.262	27
CFU-(NH_4-N)	0.407	23
CFU-(NO_3-N)	−0.146	23
CFU-(NO_2-N)	−0.377	22
CFU-(PO_4-P)	−0.372	23
CFU-ΣN/P	0.139	26
CFU-(Chl-a)	0.244	22

Note: Unit of heterotrophic bacterial content: CFU • cm^{-3}; T-Water temperature, concentration of nutrient salt; g • dm^{-3}.

Table 3 also illustrated that there were negative correlation between CFU and NO_3-N, NO_2-N or PO_4-P respectively, among them, the negative correlation coefficients of CFU-[NO_2-N], CFU-[PO_4-P] were significant respectively, (i. e. $R>0.360$, and $R>0.352$). These data hint a contrary relationship between increase/decrease of bacteria and low/high of concentration of NO_2-N and/or PO_4-P, show a phenomenon of the coexistence of high-abundance of bacteria and low concentration of some nutrient salts in the sea area investigated. The NO_2-N and PO_4-P regenerated through bacterial decomposition could be absorbed rapidly by phytoplankton, so that the concentration of these salts might decrease in the environment. This phenomenon happened to be confirmed by the positive correlation between [Chl-a] and [NH_4-N].

The increasing NO_2-N (nitrous salt) provides the nutrient for fast propagation of phytoplankton, thus increasing the Chl-a content and in this process heterotrophic bacteria perhaps participate in the mineralization which is involved in the transformation.

4 CONCLUSIONS

To sum up, the results obtained from this in vestigation indicated that the concentration of marine aerobic heterotrophic bacteria in the surface water of the Chinese Great Wall Bay and its adjacent sea area during the summer is about 798 CFU • cm^{-3}. The concentration is higher within the Bay than outside, and shows a tendency to decrease progressively month by month. As compared with the previous results the concentration is still in normal condition, i. e. it is lower than that in inshore water, and higher than that from the open waters. The variation of heterotrophic bacterial concentration is restricted by environmental-ecological factors. It might be associated with sea water pollution, so there was higher heterotrophic bacteria concentration at some stations. The negative correlation between heterotrophic bacteria and concentration of NO_2-N or PO_4-P were more significant. The propagation of heterotrophic bacteria might accelerate mineralization in the environment and consequently, it promotes photosynthesis of phytoplankton in the sea area surveyed.

References5(abr.)

(Cooperators: Zhu Mingyuan Yuan Junfeng)

MONITORING THE CONTENT OF AIRBORNE MICROBES OVER THE CHINESE GREAT WALL STATION, ANTARCTICA*

Abstract The outdoor air borne microbial content over Great Wall Station, Antarctica has been monitored, including its diurnal variations and states in fair or foul weather conditions.

The results obtained show: the concentration averaged 161.9 CFU·m^{-3}. When the weather condition was fair, its range of variation is 0～1,336.2 CFU·m^{-3}, the average value was 1,488.3 CFU·m^{-3}, when the weather was foul, the range of variation was 471.4～4,296.8 CFU·m^{-3}. The average value of air borne microbial number in either fair or foul weather was 383.0 CFU·m^{-3}. This value is over than 21 times that obtained during 1986/1987.

The results seem to show that the influence of human activities has been increasing on Antarctic ecosystem. The diurnal variation of the outdoor air borne microbes shows that the peak of content appeared at about 01:00, and the trough at about 13:00 in a whole day. Analyses were made of the relationships between the microbial concentration with its related indexes, i.e. relevant air temperature, relative humidity or wind force.

The results showed some particularities as compared with those from the areas outside Antarctica.

Key words Outdoor Air Borne Microbes, Great Wall Station In Antarctica, Diurnal Variations

1 INTRODUCTION

People often pay attention to chemical and physical pollution in the surrounding environment, but fail to understand the pollution due to biotic factors. In fact, there are often pollution problems due to some biotic factors in the environment, i.e. sea, land and air (atmosphere) upon which human beings and living organisms depend for existence. An important biotic factor is abnormal air borne microbial condition. We have been al-

* The original text was published in *J. Glaci. Geocr.*, 1998, 20(4): 320-325.

ways surrounded by air borne microbes. In the recent 20～30 years, rapid progress has been made in researches on aerobiology in some countries, but these researches focused mainly on some areas where there is human activity and economic livelihood is developed. There are basically links in survey of air borne microbes in untraversed regions (e. g. Antarctic region), even though there have been a few reports on the survey published since the 1970's (Cameron et al., 1972; Chen et al., 1994a). In fact microbial activity exists there more or less any time, so, indeed, air borne microbes are always important member in the specific Antarctic ecosystem. They participate actively all the time in circulation of materials and energy flow (metabolism) there.

The research on airborne microbes over Antarctica will promote man to understand the migration and distribution activity of microbes at global level and to explore the life phenomenon in the outer space. Attention of scholars of various countries have been aroused by the invasion and settlement of the microbes accompanying human beings and other living organisms on the clean Antarctic continent and its islands. ecological changes and environmental pollution problems brought about by them (Chen et al., 1992).

The Chinese Great Wall Station (GWS, the same fellow), is located on King George Island, a part of the Antarctic peninsula nearby the Antarctic circle. The polar and subpolar natural conditions there provide an environment richly endowed by nature for making research on air borne microbes.

We monitored the air borne microbial content over GWS, Antarctica (Chen et al., 1994a). Further, during the austral summer of 1993～1994, on the basis of the former research with the aim of analyzing and comparing the variations in the quantity of outdoor air borne microbes over the station area and as a consequence, formed a judgment of the effort of their relationship with human activity on indoor air borne microbes so as to deepen our understanding of the Antarctic ecosystem.

2 MATERIALS AND METHODS

2.1 Brief description of station area, monitoring sites and sampling

The Chinese GWS is situated at the southeastern part of the Fildes Peninsula (62°12′59″S, 58°58′52″W). Bordering on Maxwell Bay on the east and with its inshore sea area as Great Wall Bay, it faces the Fildes Strait on the south. It is at a distance of about 2.5 km from the west coast of King George Island. The southern-western and northern area of the station is one of the warmest and dampest areas in the Antarctic continent. The austral summer season is between the end of November and next mid-March. During the summer, the air temperature is about －3.9～5.7℃. There are frequent rain, wind, snow and fog. Some birds fly or have a rest or look for foods some times. On the ground, mosses and lichens are growing luxuriantly, full of life. Microbes and their spores diffuse in the air. Outdoor human activities including scientific research and field work are frequent. Guests of various countries coming and going are increasing in summer (day by day). All the conditions stated about would make disturbance to and impact on the original quantity of air borne microbes.

The sampling sites were set up in two ways or divided into two categories. The first way was setting up sites for sampling at a fixed time. These sites were set up and arranged in the east, southwest and north of the station. Samples were taken 5 times per day for 3 inconsecutive days at each site (i. e. 3×5×3 times); The purpose is to understand the regular patterns of dial variations of outdoor air borne microbial concentration. The other way was setting up sampling sites according to bad weather conditions. Altogether 6 sites were set up and samples were taken 9 times. The aim was to supplement information obtained from the time sites mentioned above so as to understand more completely the outdoor airborne microbial content in various weather conditions.

The locations of the sampling sites were shown in Fig. 1.

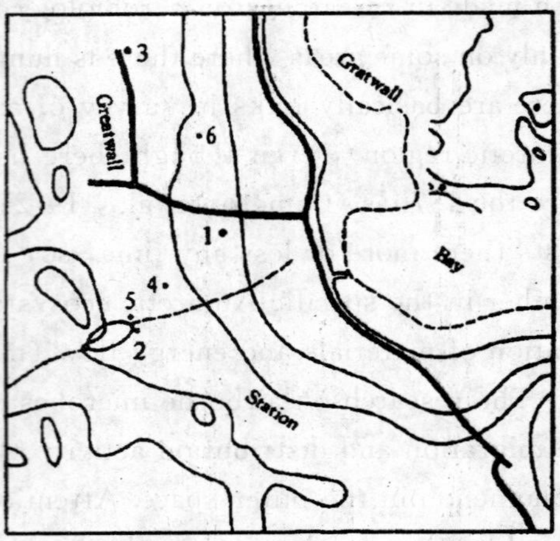

Fig. 1 Outdoor sampling sites for airborne microbes in Great Wall Station. Antarctica
1. Stele marker of the station; 2. Little hill behind the scientific research building; 3. In the weather observation station; 4. Outdoor stairs of the scientific research building; 5. Outside west of the scientific research building; 6. Outdoor side stairs of recreation and sports building

2.2 Methods

Petri dish natural settling method and two kinds of culture media were used. The common nutrient agar medium (produced by Shanghai Medical Testing Laboratory) was used for collecting total air borne microbes and Martin culture medium for collecting air borne fungi. The petri dishes with the prepared culture media were placed at the established positions during the monitoring time periods for collecting microbial samples and taken back at regular times. The time interval is generally 1 hour and meanwhile some meteorological parameters were recorded. After sampling, the dishes were cultured at 12℃ for 1～2 weeks. The outdoor airborne microbial concentration was counted in CFU $\cdot$ m^{-3}, compared and analyzed in combination with the relevant parameters (Yu et al., 1994).

3 RESULTS AND DISCUSSION

The air borne microbial concentrations listed in Table 1 were determined at the three timed sites during three inconsecutive days. They represented the general situation of outdoor airborne microbial content during good weather time interval. From this table, it can be seen that the mean of these outdoor air borne microbial concentrations is 161.9 CFU $\cdot$ m^{-3}, which is 9 times higher than that for the summer of 1986/1987 and nearly 5 times higher than that obtained at Zhong Shan station, East Antarctica. The variation range is 0～1,336.2 CFU・m^{-3}, even though some of them are lower than those in 1986/1987 (Chen et al., 1992). The descending order of the concentration mentioned above the three sites is site No. 1, No. 3, No. 2, but there are not much difference in the respective average for the three sites.

The results of correlation analysis among the outdoor air borne microbial concentration above the stations and the related parameters listed in Table 2, Showing that there was a negative correlative relationship between the concentration and the air temperature ($R=-0.178$, $n=45$). The descending order of the average air temperature at each site is site No. 1, No. 3, No. 2, which showed that the order of the airborne microbes seemed to be suited to lower temperature there. The correlation between the microbial concentration and the relative humidity was better (taking site 3 for example), i. e. the higher the relative humidity, the higher the content of the air borne microbes. This suggests that the air borne microbes above the station located on the Antarctica.

Table 1 The statistics for the outdoor airborne microbial concentration in 15 times intervals above the Chinese Great Wall Station, Antarctica

Sampling date	Sampling site	Microbial concentration/CFU・m^{-3}	Sampling site	Microbial concentration/CFU・m^{-3}	Sampling site	Microbial concentration/CFU・m^{-3}
1993-12-10						
13:00～14:00		13.1		32.8		13.1
19:00～20:00		45.9		39.3		6.6
1993-12-11	1		2		3	
01:00～02:00		26.2		13.1		13.1
07:00～08:00		32.8		13.1		13.1
13:00～14:00		131.0		26.2		32.8
1994-01-10						
13:00～14:00		39.3		32.8		6.6
19:00～20:00		65.5		65.5		26.2
1994-01-11	1		2		3	
01:00～02:00		1,336.2		936.7		930.1
07:00～08:00		30.5		183.4		903.9
13:00～14:00		0		0		0
1994-02-06						
07:00～08:00		255.5		19.7		19.7
13:00～14:00		0		19.7		19.7
19:00～20:00		39.3		0		19.7

(续表)

Sampling date	Sampling site	Microbial concentration/ CFU · m^{-3}	Sampling site	Microbial concentration/ CFU · m^{-3}	Sampling site	Microbial concentration/ CFU · m^{-3}
1994-02-07	1		2		3	
01:00～02:00		641.9		576.4		347.2
07:00～08:00		104.8		163.8		19.7

Note: 1. Stele marker of the station; 2. Little hill behind the scientific research buiding; 3. In the weather observation station.

Peninsula were suited to higher humidity there. At the same time, though the correlation between the microbial concentration and wind strength was positive, yet the correlation was not notable.

Table 2 Correlation coefficients between outdoor airborne microbial concentration (CFU · m^{-3}) and some relevant parameters above the Chinese Great Wall Station, Antarctica

Pair of parameters	Correlation coefficients
CFU-air temperature (℃)	−0.178 (n=45)
CFU-relative humidity	0.657 (n=15)
CFU-wind strength	0.490 (n=16)
Fungous CFU/Total microbial	−0.211 (n=10)
CFU%-air temperature	
Detected rate of fungi reative humidity	−0.662 (n=10)

Note: Unit, counts m^{-3}, same below.

Fig. 2 was constructed from the data in Table1. This figure shows the variation of the concentration of the outdoor air borne microbes over the GWS. At the three sampling sites all the peak concentration appeared at 01:00 (night), whereas the trough concentrations often at 13:00 (soon afternoon) during the three inconsecutive 24 hours, which indicated further that there was a relationship of mutual opposition and yet also mutual complementary between the concentration of airborne microbes and the air temperature to some extent. This variation is not only different from that over some areas in the main land of China (Xie, 1987), but also not completely similar to that over the adjacent sea area to the station, Antarctica (Chen et al., 1994a).

The results of correlation analysis of detection rate of outdoor air borne fungus count against relative humidity were listed also in Table 2, indicating a more significant negative correlation between the two parameters. This seems to show that the higher relative humidity did not become a limiting factor for the air borne fungi, which was not constant with the situation in some areas (Chen et al., 1994b). The negative correlation between fungus content/total microbial content percent and air temperature was not very significant nor with some results. The results obtained indicated that the outdoor air borne fungi seemed to tolerate lower temperatures than those indoor fungi.

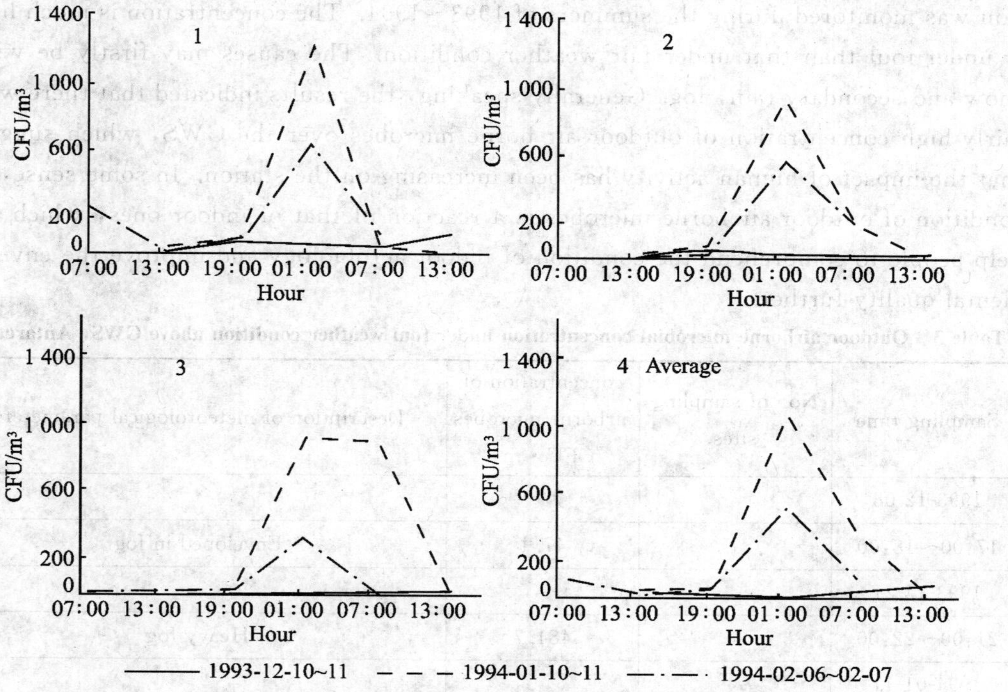

Fig. 2　Diurnal variation of outdoor airborne microbial contents over the GWS, Antarctica

No. 1, 2, 3 represent respectively three sampling sites; No. 4 represents the diel variation situation of out door airborne microbial content over the whole station. The data were averaged basically on the values of three sites

The data obtained in monitoring the outdoor airborne microbial concentration of 9 times under bad weather condition were listed in Table 3.

During the determination, there were fog, strong wind, snow or light rain.

Under those conditions, the outdoor air borne microbial concentration ranged from 471.7 to 4,296.8 CFU · m^{-3}, with an average value of about 1,488.3 CFU · m^{-3} and was 9 times that under good weather condition, which was mainly due to the fact that those particles carrying microbes were blown up by wind from the station area and settled on the media in culture dishes (Cameron et al., 1972, 1977).

According to the statistics from Table 1 to 3, it can be seen that under various weather conditions in GWS, the outdoor air borne microbial concentration range from 0 to 4,296.8 CFU · m^{-3}, with a mean of about 383.0 CFU · m^{-3}. Although this is still lower than the results from some stations in the Fildes Peninsula, but is 21 times as high as the results of the determination of the samples collected in 1986/1987 using the same method (Chen et al. , 1992). This seems to be an indication of an increasing impact from human activity on the Antarctic peninsula and other areas (Cameron et al. , 1972, 1977).

4 CONCLUSIONS

Outdoor air borne above microbial concentration above the Chinese Great Wall Station was monitored during the summers of 1993～1994. The concentration is much higher under foul than that under fair weather condition. The causes may firstly be wind, snow and secondary rain, fog. Generally speaking, the results indicated that there was a fairly high concentration of outdoor air borne microbes over the GWS, which suggests that the impact of human activity has been increasing on the station. In some sense, the condition of outdoor air borne microbes is a reaction of that of indoor ones, which may help people to comprehend the condition of indoor aerobiology and improve the environmental quality further.

Table 3 Outdoor airborne microbial concentration under foul weather condition above GWS, Antarctica

Sampling time	No. of sampling sites	concentration of airborne microbes /CFU・m^{-3}	Description of meteorological parameters
1993-12-08			
17:00～18:00	1	47.1	Enveloped in fog
1993-12-21			
21:00～22:00	1	484.7	Heavy fog
1994-01-11			
7:00～8:00	6	4,296.8	Faced towards force 7 wind fog like drizzle
7:00～8:00	4	1,585.1	Faced sidewards force 7 wind fog like drizzle
1994-01-17			
21:00～22:00	5	8,90.8	Heavy fog like drizzle
22:00～22:00	1	641.9	Heavy fog like drizzle
1994-01-22			
11:45～12:45	1	3,288.1	Sea wind accompanied by snow
18:00～19:00	3	1,087.3	Rain&fog
1994-01-26			
18:00～19:00	1	648.5	Rain&fog

Note: 1. Stele marker of the station; 2. Little hill behind the scientific research building; 3. In the weather observation station; 4. Outdoor stairs of the scientific research building; 5. Outside west of the scientific research building; 6. Outdoor side stairs of recreation and sports building.

The dial variation of outdoor air borne microbial content is effected deeply by complicated environmental factors. The results obtained show that it is higher at night time and lower at day time, which is not completely consistent with the results from near by sea areas and there is some discrepancy between the results obtained in this paper and those in some related papers.

Further research should be made, if possible, on the related problems.

References7(abr.)

中国南极长城站室内空气微生物含量七年前后的比较*

摘 要 中国南极长城站室内空气微生物含量的考察结果表明，七年后，虽然一些室内空气微生物含量有所下降，即空气质量有所改善，但总体来看，略呈上升势。与友邻科考站相比，长城站的多少有点偏高。也比非南极的较高纬度的某些室内空气微生物含量高。结果意味着长城站尚有提高空气质量、改善环境状况的余地。

关键词 中国南极长城站 室内空气微生物含量 七年前后的比较

前 言

尽管1.5亿年前的南极大陆仍是冈瓦纳大陆的一个重要部分，但从那以后，南极便自成体系了。这个体系的明显特征之一是具有独特结构和功能的生态系统。随着科技的发展，人类的足迹越来越多地接近并深入南极大陆。外来生物（首先是包括人类）的入侵，局部地干扰影响着它原有的生态系统的本来面貌，打破着原有的生态格局而不得不建立新的平衡。

南极半岛伸展出南极圈外的部分，便是外来入侵南极半岛纵深和进入南极大陆的重要跳板。本世纪四十年代，人们便最早从它的欺骗岛而入。迄今，在这块狭窄的地盘即该岛的可分部分——乔治王岛上，各国已相继建立了十六个以上的科学考察站。坐落于其西部，临麦克斯韦尔湾的中国南极长城站便是其中之一。长城站建于1984年，建成于1985年初，十五年来，一批批考察队员进出长城站，并以其不断完善的生活设施和科研条件，东方人特有的好客吸引着各国考察队员，迎送过不少旅游者、观光客，这些活动必然无疑地、不可避免地碰撞着站区内外及其周围的生存环境和生态系统。从而向人类自身提出了一个严肃的课题，即如何从生态学的角度来认识它的过去、了解它的现状和保持好它的生态平衡。此课题的一个重要内容便是测定它的立体环境，即海、陆、空及其衔接、交叉或边缘部位中的微生物状况，包括空气微生物。

由于作为整个生态系统中的空气微生物又是环境洁净程度的一个重要参数[1]，它与人们的健康息息相关，因而越来越多地为微生物学家、流行病学家、环境科学家和医疗卫生部门所重视。这项工作，于1986/1987年间曾作过初次研究[2]。时隔七年之后（即1993/1994），长城站室内的空气微生物状况发生了什么变化？与友邻及地球上其他地区的状况有何差异？本文就此作一报道，以利进一步的南极微生物资源、环境保护和生态学研究。

---

* 原文刊于《海洋湖沼通报》，1999，No. 4：48-52.

1　材料和方法

长城站上的测点共19个，共测了20次。它们被归成4类功能房间；Ⅰ类—人群活动频繁的公共间，如No.1、2、3、7、9号；Ⅱ类—人们少去的共用间，如No.11；Ⅲ类—实验工作间，如No.12；Ⅳ类—寝室，如：No.20。其中有8个测点同于1986/1987年测定时的。设点如前文所述[2]。被测的友邻站还包括韩国金世宗站、智利马尔什中尉站、俄国别林斯高晋站和乌拉圭阿蒂加斯站。共测5次。测定时期：1993/1994年南极之夏，所采用的培养基包括普通细菌培养基和真菌培养基，具体的采样和数据统计方法同以往所述[3]。

2　结果和讨论

表1列出了长城站八对室内空气微生物含量七年前后的比较。由此可见，时隔七年，这八对数据中有62.5%(5/8)呈现为减少趋势。余下的三对数据取自餐厅、冷库和一个寝室。前两者虽有增加，但不明显。只是No.20测点的空气微生物看出明显地增多。表明该测处环境状况欠佳。总的讲，七年后的长城站，相对应的室内空气微生物含量相当稳定，且稳中有降，这应该是个好兆头。

表1　中国南极长城站相同室内空气微生物含量七年前后的比较*

Table 1　Comparison of indoor airborne microbial numbers in the same rooms seven years ago and after, in Chinese Great Wall Station, Antarctica

测点编号	测点日期	
	1987.2	1994.2
1	416.6	419.2
2	501.1b)	393.0
3	212.2c)	101.8
7	1,063.4d)	52.1
9	110.1e)	52.4
11	275.1	288.2
12	338.0	235.8
20	393.0f)	1781.6

*空气微生物含量：$CFU \cdot m^{-3}$

a)编号均对应于1994年2月测的　d)狭小，空气污浊　b)1986.12与1987.3两次测定平均值
e)1986.12与1987.3两次测定平均值　c)临时房，小而挤　f)相应于1993年12月的No.214室

表2是对时隔七年的空气微生物含量测定的七个指标作比较的结果。结果表明，表示七年后总体水平的七个指标中有五个呈下降势。这种势头与表1的结果是一致的。但全站的室内空气微生物平均含量呈上升势，虽然势头不大。人群活动相对频繁的Ⅰ类功能室上升势较明显些。这似乎暗示公共场所的空气环境质量有待改善。若仅以1987年2月的指标与1994年2月的比，Ⅲ、Ⅳ类功能房间空气微生物含量降低表明那儿比以前注重了环境质量。Ⅳ类房居住人数的减少也可能与降低其空气微生物浓度有关。但5/7的指标都显现出增高。因而长城站室内空气微生物含量状况仍是值得重视的问题，即使所得结果可能带有偶然性。

表 2 中国南极长城站室内空气微生物含量的几个指标七年前后比较[a)]

Table 2 Comparison of some indexes of indoor airborne microbial contents seven years ago and after, in Chinese Great Wall Station, Antarctica

指标名称		测定时间		
		1986/1987[b)]	1987.2	1994.2
采样次数		29	10	20
微生物含量平均值		628.8	605.5	660.2
霉菌检出率%		65.5	50.0	60.0
霉菌数/总菌数%		26.5	3.0	24.3
不同功能房间数微生物含量平均值	Ⅰ	512.3	0	1,339.9
	Ⅱ	252.3	0	150.7
	Ⅲ	538.1	1,821.6	144.1
	Ⅳ	1510.3	600.2	529.2

a)空气微生物含量 CFU·m^{-3} 参见文献[3]　b)指 1986.12,1987.2 和 1987.3 测定之平均值

表 3 南极菲尔德斯半岛 5 个科考站室内空气微生物含量的比较(1994.2)

Table 3 Comparison of indoor airborne microbial contents among five scientific expeditionary stations in the Fildes Peninsula, Antarctica in 1994.2

考察站名	空气微生物含量 CFU·m^{-3}	测定日期	室温	备注
金世宗站(韩)	419.2	1994.2.5	15.5	会议室
马尔什中尉站(智)	235.8	1994.2.24	18.0	邮局
别林斯高晋站(俄)	78.6	1994.2.24	16.0	休息娱乐厅
阿蒂加斯站(乌)	209.6	1994.2.27	18.5	厨房窗口
阿蒂加斯站(乌)	681.2	1994.2.27	17.0	酒吧
长城站(中)	660.2	1994.2	15.7	20 次平均

表 3 列出了长城站与附近的四个考察站室内空气微生物含量之比较。该表说明了五个站室内空气微生物含量按高低排,其次序是:长城站>阿蒂加斯站>金世宗站>马尔什中尉站>别林斯高晋站。由此可见,长城站的空气微生物含量处于偏高水平。分析其原因,大致可能与房舍的老化、生活习惯、个人/环境卫生状况等有关。其室内较持续的相对不洁可为空气微生物的孳生提供有利条件。

长城站室内空气微生物含量比圣地亚哥机场新候机大楼的高,也比青岛市较干净的居民家庭室内的还高。这同样是上述原因:一是房舍的老化,二是讲究卫生的程度,但长城站室内空气微生物含量比起北京机场的却低了许多。这既反映了南北两半球之差异,又表达了室内微生物与人群活动的密切关系。以上结果说明长城站室内空气微生物状况尚有改善余地。

3　结语

中国南极长城站室内空气微生物含量指标七年前后的比较结果初步呈现出，七年后，总体略呈上升势，虽然有些室内的空气微生物含量有所下降，即空气质量有所改进。与友邻站相比，基本上偏高，也比非南极的较高纬度的一些室内空气微生物含量高。结果意味着长城站空气质量应该提高，环境状态可以改进。

参考文献 4 篇（略）

COMPARISON OF INDOOR AIRBORNE MICROBIAL NUMBERS OF GREAT WALL STATION, ANTARCTICA SEVEN YEARS AGO AND AFTER

(ABSTRACT)

Abstract The results of the overall investigations for indoor airborne microbes of the Great Wall Station, Antarctica show that, in general, their quantity showed a tendency of going up slightly, even if the numbers slightly decreased in some rooms, i. e. the air quality had some what improved seven years later. The numbers of indoor airborne microbes in the Great Wall Station were more or less on the high side, as compared with the amount of some other scientific research stations, Antarctica, and higher than those at some higher latitudes outside Antarctica.

The results suggest that there are still some room for raising air quality and improving the environmental conditions in the Great Wall Station, Antarctica.

Key words Indoor Air Borne Microbial Numbers; Great Wall Station; Antarctica

中国南极长城站室内空气微生物状况*

关键词 南极;长城站;空气微生物;霉菌;环境质量;大气卫生

空气微生物是空气洁净程度的重要指标是环境污染的真实反映[1]。十年来,位于南极半岛的中国长城站常年有考察队员居住和工作,尤其在南极之夏,不时有各国考察、旅游、观光者出入,这些都给长城站空气微生物的变化带入了外来因素[2]。我们于1994年2月,随机对站上19个室内测点作空气微生物考察,以监测和比较室内微生物含量状况,为进一步了解空气微生物的卫生学意义,并服务于提高南极考察队员的身体素质,加深对南极生态系统的理解而提供一些资料。

1. 材料与方法

1.1 测点

1994年2月份6天对19个室内空间作了20次测定,这20次测定被分作四类,即1类,是人群活动相对频繁的公用房间,其测点编号是1~7,Ⅱ类是人少去的公用房间,编号:8~11.室内气温甚低,Ⅲ类是实验、工作用房,编号:12~15.Ⅳ类是寝室,一般是一人居住,室温较高,常在16~20℃之间,编号是16~20。各测点(次)的具体情况详见表1。

表1 中国南极长城站室内空气微生物测点一般状况描述表

Table 1 General conditions of the sampling sites for indoor airborne microbes in Chinese Great wall Station, Antarctica

测点类别 Category of monitoring sites	测点(次)编号 Serial number of monitoring sites(time)	测点 Monitoring sites	环境描述 Enviromental description		
			环境状况 Enviromental conditions	A/m²	B/℃
I	1	餐厅	常有20人用餐,视听设备,小卖部	100	16.0
	2	厨房	炊具,食品,杂物,油烟,有些食物发霉	60	12.0
	3	医务室	医药器材,常有人出入	14	13.6
	4	会议室	少有人,有时开会,会客	20	18.0
	5	文体室	文体用品,时有体育活动	100	36
	6	浴室	温度高,天花板上有冷凝水下滴	50	14.0
	7	厕所	尿屎气味不重,较干燥	10	11.2

---

* 原文刊于《应用与环境生物学报》,2000,6(1):90-92.

(续表)

测点类别 Category of monitoring sites	测点(次)编号 Serial number of monitoring sites(time)	测点 Monitoring sites	环境描述 Enviromental description 环境状况 Enviromental conditions	A/m²	B/℃
Ⅱ	8	车库	车辆、工具、器材,机油味重,时有个别人出入	250	37
	9	干食品库	新,干冷,隔成几室	250	4.9
	10	集装箱	出海器材,杂乱,阴冷	12	4.5
	11	冷冻食品库	常年存放食品,阴冷	20	−27.0
Ⅲ	12	气象工作室	仪器,1～2人工作	14	18.9
	13	生态医学实验室	器材、药品、标本,2人工作	18	12.0
	14	地球物理实验室	微机、器材,2～3人工作	18	18.6
	15	发电栋	机房、发电机运转,震动,噪音大	80	13.0
Ⅳ	16	寝室 204	1人住,东向房	10	18.2
	17	寝室 204	1人住,东向房	10	20.4
	18	寝室 205	1人住,西向房	10	16.0
	19	寝室 207	1人住,西向房	10	19.5
	20	寝室 214	1人住,东向房,杂物,标本	10	16.5

1.2 采样和数据处理

空气微生物的采集用平皿自然沉降法在普通营养琼脂和马丁培养基上进行,<50 m² 间,一式2份,≥50 m² 间,一式3份,分别同时布设平皿,暴皿时间一般为15 min,经1～2 wk12℃的培养,微生物浓度以单位空气体积(m^3)计算平皿中出现的CFU数,表达为n(CFU)/m^{-3}[2],并算出霉菌检出率及霉菌占总菌数%,霉菌检出率指出现霉菌的测定次数与各点测定次数的百分率,分析并挑取单菌落,带回国内作进一步研究。

2.结果和讨论

2.1 总的空气微生物含量状况

表2列出了所有测定的空气含量状况,由该表可以计算出这20次测定数据的平均值,即n_1(CFU)=660.3 m^{-3},这比其相近时期飞行于国际航线上的一些客机内空气含量高,达2倍以上[3]。如果剔除特别潮湿的No.6测点和空气微生物CFU较多的No.20两测点,平均值降为233.0 m^{-3},这一数值比一些候机室内的低,比中山站室内空气微生物含量的平均值高[4],说明西南极比东南极有更优越的条件适于空气微生物的生长,与七年前相比呈现出上升势头,也比多数邻近科考站的高,这意味着站区的老化和改善环境卫生状况的重要。

由表2可计算出Ⅰ类空间的空气微生物含量平均为1,339.9 m^{-3},Ⅱ、Ⅲ、Ⅳ类空间的则分别为 150.7、144.1 和 529.2 m^{-3}，按其大小排列，次序为 I＞IV＞Ⅲ＞Ⅱ，四者间存在明显地与人群活动状况，即人数和活动内容等关系密切(见表 1)，I 类空间，如餐厅、厨房、文体室是人们常聚之处，IV 类空间，即寝室，是人们居住、活动的稳定舒适之处，有些场合中堆放的易腐烂变质物品多或不注意卫生等，致使微生物繁殖和增加，不难想象，此类环境也会滋生适于 37℃温度生长的微生物，包括一些病原菌，相反，Ⅱ类空间的室温低，也很少有人出入，医务室即 No. 3 测点的空气微生物含量状况也处于良好水平。

室内空气微生物含量与相应室温的相关分析表明，两者间仅有很小的负相关性存在($r=-0.05, N=20$)。

霉菌数/总菌数比的平均值约 24.3%，各类测点大小次序排列是Ⅱ＞IV＞I＞Ⅲ。表明长期存放不洁物品之处可能孳生较多霉菌，不良的生活习惯使环境质量变差，IV 类空间的霉菌两参数较高，应引起注意，由于一些霉菌产生致癌的真菌毒素[5]，应对霉菌的状况作经常的监测，以便为控制真菌蔓延，防病治病提供基本资料。

2.2 空气霉菌含量状况

表 3 给出了所测定的空气霉菌的两个有关参数，该表说明了霉菌检出率为 60.0%，各类霉菌检出率大小排列次序为 IV＞I＞Ⅱ＞Ⅲ，这一次序次与上述总的空气微生物含量大小次序有类似之处，即公用房、寝室中的高于人为活动少的房间和实验室的。

表 2 中国南极长城站室内微生物含量状况表

Table 2 Indoor airborne microbial number in Chinese Great Wall Station, Antarctica

测点类别 Category of monitoring sites	测点(次)编号 Senal number of monitoring sites(time)	N(CFU)/m^{-3}
I	1	419.2
	2	393.0
	3	104.8
	4	209.6
	5	969.4
	6	7 231.2
	7	52.4
Ⅱ	8	104.8
	9	52.4
	10	157.2
	11	288.2
Ⅲ	12	235.8
	13	78.6
	14	26.2
	15	235.8

(续表)

测点类别 Category of monitoring sites	测点(次)编号 Senal number of monitoring sites(time)	N(CFU)/m⁻³
IV	16	368.8
	17	157.2
	18	209.6
	19	131.0
	20	1 781.6

表 3 中国南极长城站室内空气霉菌含量状况表

Table 3 Indoor airborne mildew number in Chinese Great Wall Station, Antarctica

测点类别 Category of monitoring sites	测点(次)编号 Senal number of monitoring sites(time)	霉菌检出率% Detection rates of milder	霉菌/总菌数 Muldew/bxal (%)
I	1	100.0	37.5
	2	100.0	66.7
	3	0.0	0.0
	4	100.0	25.0
	5	100.0	13.5
	6	100.0	19.2
	7	0.0	0.0
Ⅱ	8	0.0	0.0
	9	0.0	0.0
	10	100.0	83.3
	11	100.0	72.7
Ⅲ	12	0.0	0.0
	13	100.0	33.3
	14	0.0	0.0
	15	0.0	0.0
IV	16	100.0	25.0
	17	0.0	0.0
	18	100.0	50.5
	19	100.0	20.0
	20	100.0	40.0

3. 结语

从卫生学角度来看中国南极长城站室内空气微生物含量检测，其结果表明该站室内空气质量尚可[6]，但有的研究表明，南极考察站一些室内能生存于 37℃的微生物数量将增多令人担心，尽管不一定是病原菌[7]，它们经适应和扩散将会与土著者产生竞争[8]，不同功能的室内空间，其微生物含量间的差异原因主要是人群活动和卫生状况上的，包括设施的老化。一些测点霉菌含量较高，与不良卫生习惯，过高的湿度或食品变质等有关。空气微生物含量较高的一些测点可能存在着食品处理的不恰当问题。因此必须提高考察队员的素质水准，养成良好的工作、生活习惯。提倡勤俭节约，加强环境生态保护意识，以保证和提高长城站室内外乃至南极上好的环境质量水平。

参考文献 8 篇(略)

(合作者：洪旭光)

SITUATION ON INDOOR AIR BORNE MICROBES OF CHINESE GREAT WALL STATION, ANTARCTICA

(ABSTRACT)

Abstract The total average value (CFU/m^{-3}) of indoor air borne microbial number of Chinese Great Wall Station Antarctica investigated in Feb. 1994 was 660.2 m^{-3}, with variation range <26.2～7,231.2 m^{-3}. Generally the number of indoor air borne microbes was 233.0 m^{-3} on the average except under particular circumstances. There were higher mildew detection rales and mildew number/total microbial number in some sampling sites. The variation in the indoor air borne microbial number was related closely with human activities and environmental hygienical condition. Generally speaking, the indoor air quality of the Great Wall Station was fairly good. But it still needs good hygienical habits and more attention to further improve the indoor air conditions and the whole circumstance condiouns of the great Wall Station.

Key words Antarctica; Indoor Air Borne Microbes; Great Wall Station; Mildew; Circumstance Quality

拉萨、日喀则空气微生物含量状况*

摘 要 文章记述了用平皿沉降法于2001年5月间分别测定拉萨和日喀则两市的空气微生物含量状况。结果表明,拉萨和日喀则的空气微生物平均含量分别为15 161.8CFU/m^3 和 12 062.0 CFU/m^3,说明其空气已达中度污染,拉萨的空气污染重于日喀则。同时分析了空气微生物含量等指标的时空变化及其原因,指出进一步开展两市空气污染监测和空气质量评价的必要性。

关键词 拉萨 日喀则 空气微生物 空气污染

1 引言

空气中飘浮着不同粒径的颗粒物等气溶胶。这些气溶胶来自不同污染源并因空气的流动而散布四面八方。微生物(包括病原菌)以此为载体、媒介或营养源而生长、繁殖、扩散,与人类、动植物的健康密切相关。空气微生物含量及其变化反映了空气污染状况和程度,是空气质量的一个重要指标,日益为环境科学家、生态学家和流行病学家所重视。世界先进国家已有一些关于空气微生物状况的研究报告[1]。本文就西藏自治区拉萨和日喀则两市的空气微生物含量状况作初步报道,以供西部环境保护工作参考。

2 测区环境和经济活动概况

海拔3 500 m的拉萨市位于西藏自治区中部的拉萨河畔,无雾,阳光充足。有电力、煤炭、机械、纺织、化轻工业等,城市设施日益完善,有众多名胜古迹,是西藏的政治、经济、文化和交通中心,流动人口较多。

距拉萨400多km的日喀则市位于自治区南部喜马拉雅山脉北麓,雅鲁藏布江和年楚河在此交汇,空气较温暖湿润。有电力、机械、食品、皮革等工业,手工业发达。有扎什伦布寺等名胜古迹,为西藏第二大城市。

3 测定空气微生物时的天气、环境状况概述

2001年5月间,对拉萨和日喀则两市分别设定了7个空气微生物测点,以代表两市1d中不同时段和不同功能区状况,其具体的天气、环境概况见表1。

4 材料和方法

所用培养基是细菌和真菌培养基,其成分和制作参见文献[2]。所用方法是平皿沉降法,在所设定测点的人群呼吸带内(离地面1.5 m左右),定时开启培养皿收集降落至平皿培养基上的菌落(colony)。采样时间长度一般为5 min,之后培养于25℃室温中3~5d,分别计算各皿中的细菌或真菌CFU(colony-forming units),并换算为CFU/m^3,以细菌和真菌合计为总菌含量,再计算出真菌的检出率及真菌含量占总菌含量的%(F/T%),以此比较分析各测点的空气微生物状况,结合环境、生态因子一起估计两市空气质量。

---

* 原文刊于《青海环境》,2003,13(3):97-99

5　结果和讨论

表 2 列出了拉萨、日喀则两市空气微生物含量的所有测定值。

5.1　拉萨市区空气微生物含量分析。

由表 2 可知，拉萨 7 个测点的空气细菌含量之最高值出现于下午的 No. 4 测点，最低值则出现于下午的 No. 1 测点，两者相差 12. 2 倍，这是由测时两处的特定条件决定的。No. 4 测点虽然人少，但位于风动的树下，这可促使空气微生物降落至平皿中。而 No. 1 测点测时风很小，无人，空间开阔，空气微生物数量自然少了许多。各测点空气细菌 CFU 按大小排序为 No. 4＞No. 5＞No. 2＞No. 6＞No. 7＞N0. 3＞No. 1。7 点空气细菌的平均含量为 14690. 1 CFU/m^3。按照有关推荐标准，空气细菌含量达到 $5\times10^3\sim5\times10^4/m^3$ 时的空气质量为中度污染[2]，拉萨空气已处于此等污染状态。其中少数测点污染更重，比相当规模的南方海滨城市北海及内陆城市天水都高了许多[3]。估计人群集中和工贸交通繁忙之处的空气中细菌更多。

表 1　拉萨、日喀则空气微生物测定时的天气环境概况

测区	测点编号（No.）	测点位置	测点天气、环境概况	测 时
拉萨	1	西藏博物馆院内	软风，无人，太阳直射处气温 23℃	2001. 5. 22　14:15
	2	布达拉宫正门外	阵风，轻风-微风，20℃，游人少，但有摊贩	2001. 5. 22　16:00
	3	大昭寺内二楼院	软风，20℃，游人少	2001. 5. 22　17:10
	4	小昭寺内二楼口	轻风—微风，游人少，树荫下 20℃	2001. 5. 22　17:50
	5	金珠西路武警交通二支队院	软风，18℃，无人，地处较僻	2001. 5. 22　19:30
	6	夺林底路（近北郊）	和风，12℃，碎石路面，车人不多，有些尘土飞扬	2001. 5. 22　07:00
	7	长途汽车站附近街头	轻风，15℃	2001. 5. 22　09:40
日喀则	1	上海路石化招待所对面人行道上	轻风，18℃，有少量车人来往	2001. 5. 22　16:45
	2	市委院内	轻风，18℃，无人	2001. 5. 22　17:05
	3	武警支队门口，扎什伦布寺附近	和风，17. 5℃，有些飞尘，有寺的影响	2001. 5. 22　17:30
	4	德虔格桑普彰门口碑上	无风，安静，17. 5℃，有农田影响	2001. 5. 22　17:45
	5	建设银行中心支行门口（山东路）	软风，17℃	2001. 5. 22　18:00
	6	客运站附近店面外	软风，有少量人走动，16℃	2001. 5. 22　22:00
	7	汽车客运站上海中路街面上	软风，有些人走动，12℃	2001. 5. 22　07:30

* 测时指测定开始时间，一般持续 5 min。

表 2 拉萨、日喀则空气微生物含量的所有测定值

测区	测点编号	微生物含量指标				
		细菌	真菌	总菌	真菌/总菌%	平均真菌检出率
拉萨	1	2 358.5	0	2 358.5	0	
	2	17 924.6	1 415.1	19 339.7	7.3	
	3	6 603.8	471.7	7 075.7	6.7	
	4	28 773.7	943.4	29 717.1	3.2	
	5	22 641.6	0	22 641.6	0	
	6	14 622.7	471.7	15 094.4	3.1	
	7	9 905.7	0	9 905.7	0	
	平均	14 690.1	471.7	15 161.8	3.1	57.1
日喀则	1	21 698.2	0	21 698.2	0	
	2	6 132.1	471.7	6 603.8	7.1	
	3	26 415.2	0	26 415.2	0	
	4	6 132.1	0	6 132.1	0	
	5	4 245.3	0	4 245.3	0	
	6	16 509.5	471.7	16 981.2	2.8	
	7	1 886.8	471.7	2 358.5	20.0	
	平均	11 859.9	202.2	12 062.0	1.7	42.9

按 1d 中的时段分此 7 个测点为上午、下午和晚上，则可见其空气微生物含量排序为晚上＞下午＞上午，表明傍晚时分空气中含有的大量微生物沉降下来。

空气真菌含量最高值在 No. 2 测点，有 3 个测点未检出真菌，总的看真菌的检出率为 57.1%(4/7)。真菌/总菌%(F/T%)最高值在 No. 2 测点，平均的 F/T%为 3.1，平均真菌含量为 471.7 CFU/m^3。这比温带季风气候区渭河平原的宝鸡市及温带半干旱气候区西北高原的兰州都高[5,6]，表明测时的拉萨空气湿度大于上述两市。各真菌含量大小排序为 No. 2＞No. 4＞No. 3＝No. 6＞No. 1＝No. 5＝No. 7＝0。

拉萨各测点的平均空气微生物含量为 15 161.8 CFU/m^3，这意味着细菌加上真菌使得相应的空气微生物含量更高，空气污染加重。

5.2 日喀则市区空气微生物状况。

表 2 中所列数据表明了日喀则市区空气微生物含量状况，其最高细菌含量出现于下午的 No. 3 测点处，较适宜的气温及和风可能促使附近飘来的细菌生长繁殖。最低值出现于晨间的 No. 7 测点处，因为较低的气温和较小的风力使空气微生物少了许多。最高值是最低值的 14 倍，空气细菌含量各测点按大小排序为 No. 3＞No. 1＞No. 6＞No. 2＝No. 4＞No. 5＞No. 7，表明商贸活动、交通要道之处的空气细菌量大于较僻静和开阔、人少之处。平均的细菌含量为 11 859.9 CFU/m^3，该值同样说明日喀则市区空气已处于中度污染。空气细菌含量在一天各时段上的变化是晚上＞下午＞上午。

空气真菌的检出率平均为 42.9%，平均的真菌含量为 202.2 CFU/m^3，F/T%仅为 1.7。这一比值比同为内陆的敦煌和天水高[4,7]，但比南方沿海城市北海和海口低得

多[3,8]，说明日喀则测时的空气中含有一定的湿气。该市空气真菌含量1d中时段上的差异是晚上＝上午＞下午。空气微生物含量的平均值为12 062.0 CFU/m³。真菌加重了日喀则市的空气污染。

5.3　拉萨、日喀则两市空气微生物含量比较。

根据表1和表2所列数据，拉萨的空气微生物含量的5个指标均大于日喀则，说明拉萨的空气污染重于日喀则。拉萨的细菌含量极差：26 415.2 CFU/m³(28 773.7-2 358.5)、真菌含量极差1 415.1 CFU/m³(1 415.1～0)和总菌含量极差27 358.6 CFU/m³(29 717.1～2 358.5)均比日喀则的相应值即分别是24 528.4、471.7和24 056.7 CFU/m³ 为大。这表明拉萨空气微生物含量指标变化大，意味着其空气环境比日喀则的复杂，变数多。但三个共同之点是：①空气细菌含量在1d中时段上的变化趋势一致，即均显出晚上＞下午＞上午；②小风和飞尘促使空气微生物增多。这些反映了测时西藏高原空气微生物的一些特点。③空气真菌检出率小于沿海、南方城市。

6　结语

地处西藏高原的拉萨和日喀则可代表“世界屋脊”之都，测定结果表明，其自然的、充分日照而稀薄的空气微生物状况已受到人为和经济活动的影响。拉萨空气微生物平均为15 161.8 CFU/m³，日喀则为12 062.0 CFU/m³，说明拉萨空气污染重于日喀则。平均地看，两市空气均为中度污染，与国内一些城市相比，污染水平不算很高。但两市个别测点污染较重，这是值得注意的信号。包括拉萨、日喀则在内的城市空气污染是多元性的。本文所述数据仅能表达一点一时的局部情况。鉴于两市所处地域的特殊性和未来的可持续发展，应在可能的条件下逐步开展包括微生物在内的空气生物污染状况及其与环境关系的评价和卫生评价。

参考文献8篇(略)

E_4 环球(ROUND THE EARTH)

环球空气微生物考察*

摘　要　在1986年10月31日～1987年5月17日间,中国"极地"船作环球暨南极的航行中对空气微生物进行了考察。测点分布在其行经的30921海里的空气中,样品计432份。作了对微生物的含量分析及分类研究。结合有关参数,比较各类型环境间微生物含量的差别,指出本环球航行所经各区空气微生物的状况。结果表明,西半球、大西洋、温带、台湾海峡、南美洲、中国、坎帕纳(阿根廷)、黄海等区域的空气微生物含量一般分别比同类环境的高。南大洋空气较洁净,所考察的环球空气至少有15属细菌和5属酵母,葡萄球菌、微球菌较多,长城站空气中有芽孢杆菌。

关键词　环球　空气微生物　微生物含量　微生物属群

中国"极地"号环球航行,属世界第三次。于1986年10月31日～1987年5月17日,进行了赴南极和环球的科学考察,环球空气微生物监测是考察的内容之一,其航行共199天(包括锚泊日63天,靠泊日16天,于南极半岛的77天中,在长城站19天,在其周围海域58天)。"极地"号自中国青岛出发,航程共30 921海里,横贯太平洋,顺智利沿岸南下,越过德莱克海峡至麦克斯威尔湾长城站。回程经阿根廷,穿越大西洋、印度洋,小憩新加坡,沿中国海北上终止于青岛港。采集样品432份。

20世纪60年代以来,空气微生物学的研究成果较多[1~3],但未见关于全球性范围空气微生物的报告,而像"极地"船如此长距离的考察,世间尚属首次。监测并研究全球性空气中微生物状况,是环境科学的重要组分,也有助于微生物学科的发展,本文报道此次考察结果。

1　材料和方法

1.1　培养基

1.1.1　成分

淡水培养基(培养基Ⅰ):牛肉膏3～5g,蛋白胨10g;NaCl 5g;琼脂18g;蒸馏水1 000 mL;pH值7.2～7.4。以能在该培养基上生长的微生物视作陆源性微生物。海水培养基(培养基Ⅱ):酵母膏1g;蛋白胨2g;$FePO_4$ 0.01g;琼脂18g;陈南大洋水1 000 mL;pH值

* 原文刊于《中国环境科学》,1994,14(2):112-118。

8。以能在此培养基上生长的微生物视作海洋性微生物。

1.1.2 制作

高压消毒后，凉至45℃的上述培养基分别注入4.5 mL塑料皿中（直径2.5cm），使皿中培养基保持水平、凝固、备用。

1.2 采样

培养：用国产JWL-4型空气微生物监测仪测。取样：近岸空气暴露5～10 min；远洋空气（包括南极及南大洋的）暴露30 min。空气流量：加采样装置后达$37dm^3/min$。少量用自然沉降法。平均每日采样一次以上。经采样的培养皿置10℃、25℃或37℃中培养5天以上。

1.3 计数、实验和分析

对培养好的培养皿，计算CFU（Colony Forming Units）*，选取单菌落，纯化，保存于冰箱中，回国内作微生物鉴定、生化试验等[4,6]。

2 结果和讨论

2.1 环球空气微生物含量估计

本次环球考察包括东、西半球，南、北半球。贯穿太平洋、大西洋、印度洋及南大洋，行经亚洲、南美洲、非洲和南极洲近海，包括著名的德莱克海峡、麦哲伦海峡、莫桑比克海峡、马六甲海峡和台湾海峡。所经主要国家包括中国、智利、阿根廷、南非、莫桑比克、新加坡等。停靠或锚泊青岛、瓦尔帕莱索（智）、彭塔阿雷纳斯（智）、坎帕纳（阿）、新加坡、中国南极长城湾及邻近海域。沿比格尔水道（智）南下。经拉普拉塔河出入钢城坎帕纳，顺南海、东海、黄海至青岛。

陆源性微生物和海洋性微生物取样分别于每天的同一时间内进行，共438个（219次，每次2个重复）。除冰库内的各两次，另有各一次未计入外，实际所用数据为432个（216次×2），资料分析表明**，陆源性微生物共测216次（不包括室内），其中28次未检出，故其出现的阳性率，即检出率为87.10%。海洋性微生物共测216次（不包括室内），其中80次未检出，故其检出率为63.13%。上述两大类微生物含量（剔除“0”项）各自的中位数值，陆源性微生物是7.50 CFU/m^3，海洋性微生物是4.00 CFU/m^3。结果表明，环球空气中陆源性微生物出现机会比海洋性的高，分布广，含量高。海洋空气原较洁净，由于陆地等因素的影响，其空气已受到污染，出现一定数量的陆源性微生物。

* 军事医科院五所空气微生物学研究组：《空气微生物监测仪使用说明书》

** 有关考察航线、空气微生物含量及取样地理位置、气候状况、微生物属群等资料可参见《中国“极地”船环球暨南极考察航线和空气微生物资料》（未发表）。

表 1　世界 4 大洋空气微生物含量

洋名	微生物含量(CFU/m³)		采样日期	采样次数
	陆源性	海洋性		
太平洋	35.36	26.26	1986.10.31～1986.12.27 1987.5.1～1987.5.17	63 17
大西洋	86.38	122.90	1987.3.17～1987.4.16	34
印度洋	21.91	6.76	1987.4.17～1987.4.30	15
南大洋	6.45	1.53	1986.12.28～1987.3.15	85

表 2　世界各气候带空气微生物含量

气候带名	微生物含量(CFU/m³)		采样日期	采样次数
	陆源性	海洋性		
温带	70.82	5.30	1986.12.6～1986.12.18 1987.3.18～1987.4.12	13 28
亚热带	59.21	6.70	1986.11.17～1986.12.5 1987.4.13～1987.4.22	19 10
赤道、热带	27.63	12.04	1.86.11.10～1986.11.13 1987.4.25～1987.4.30 1987.5.3～1987.5.8	4 6 6
赤道	16.68	2.67	1986.11.11～1986.11.13 1987.4.25～1987.4.29	3 5
南温带南缘极地气候	2.70	0.67	1986.12.19～1987.2.19 1987.2.20～1987.3.17	68 29

表 3　世界 4 个半球空气微生物含量

半球名	微生物含量(CFU/m³)		采样日期	采样次数
	陆源性	海洋性		
东半球	4.00	1.30	1986.10.31～11.14 1987.4.13～5.17	16 35
北半球	12.70	1.30	1986.10.31～11.17 1987.4.27～5.17	19 21
西半球	5.30	1.30	1986.11.14～1987.4.12	164
南半球	7.30	1.30	1986.11.13～1987.4.26	178

表4　世界4大洲部分近海空气微生物含量

洲名称	微生物含量(CFU/m³)		采样日期	采样次数
	陆源性	海洋性		
亚洲、东、南近海	13.30	1.30	1986.10.31～1986.11.4 1987.4.28～1987.5.17	5 20
南美洲、南部近海	20.00	6.00	1986.11.30～1986.12.19 1987.3.18～1987.3.21	21 4
南极洲、南极半岛近海	2.70	0.67	1986.12.28～1987.3.15	84
非洲、东南近海	7.67	0.99	1987.4.13～1987.4.25	14

将这次考察结果按不同洋区划分，得出表1，可见，两类空气微生物含量均是大西洋＞太平洋＞印度洋＞南大洋。若微生物含量多寡可衡量空气洁净程度，则南大洋相对洁净得多。

环球空气微生物含量若按气候带分，则得出表2，表2说明了陆源性微生物含量以温带最多，南温带南缘的寒冷区域最少；海洋性微生物以热带的居首，南温带南缘区域最少。

按经纬度将全球分作4个半球，其空气微生物含量比较情况可从表3看出。

由表3可见，陆源性微生物含量大小排列是北半球＞南半球，西半球＞东半球，这类似于非海盐源硫酸盐浓度在半球间的变化特征*。这是否反映出与北半球、西半球国家的工业污染状况相关，值得探讨。4个半球的空气中海洋性微生物含量则是一致的。

依洲来分的情况如表4示，各洲相比，区域空气中陆源性微生物与海洋性微生物含量均以南美南部的最高，南极洲的南极半岛最低。非洲东南部的海洋性微生物含量也低。南极半岛近海空气微生物的出现，与取样地点，即考察站及其邻近区域有关，这可能意味着这一带正在受到人为因素的影响，尚待进一步查证。

太平洋面积居各大洋之首，本考察在该区域经历的时间和行程均较长。由表5可见，位于太平洋各气候带空气微生物状况，表6列出航行经历六国沿岸所得空气微生物含量，说明空气中陆源性微生物含量以莫桑比克沿海最高；海洋性微生物以中国沿海的最高。两者在阿根廷均显得低。智利沿岸的含量不低，这与其所处地理环境有关。

表5　太平洋所处不同气候带空气微生物含量

气候带名	微生物含量(CFU/m³)		采样日期	采样次数
	陆源性	海洋性		
北温带	24.00	0.00	1987.5.13～1987.5.17	5
南温带	13.30	8.65	1986.11.22～1986.12.11	20
南北温带合	13.30	6.10		

* 陈立奇等，1989，环球海洋大气气溶胶的特征(未发表，下同)

（续表）

气候带名	微生物含量（CFU/m³）		采样日期	采样次数
	陆源性	海洋性		
北热带	2.00	2.00	1986.11.3～1986.11.12	10
南热带	13.30	6.00	1986.11.13～1986.11.21	10
南北热带合	6.60	3.75		
南温带南缘	22.00	8.00	1986.12.12～1986.12.19	9

表 6　世界 6 国沿岸空气微生物含量（CFU/m³）

国名	微生物含量		采样日期	采样次数
	陆源性	海洋性		
中国	12.70	53.00	1986.10.31～1986.11.1 1987.5.9～1987.5.17	2 6
智利	22.00	8.80	1986.11.30～1986.12.19	21
阿根廷	0.00	1.49	1987.3.18～1987.3.21	4
南非	3.30	1.73	1987.4.15～1987.4.19	5
莫桑比克	24.00	0.67	1987.4.20～1987.4.22	3
新加坡	18.70	20.65	1987.5.3～1987.5.8	6

表 7 是所经历的 5 个海峡的空气微生物含量状况，表明莫桑比克海峡空气陆源性微生物含量最高，德莱克海峡的最低；海洋性微生物则以台湾海峡空气中的最高，莫桑比克海峡的最低。总的看，台湾海峡空气微生物含量较高。

对国际间 5 城市*即青岛（中）、瓦尔帕莱索和彭塔阿雷纳斯（智）、坎帕纳（阿）及新加坡的港口空气微生物含量监测结果表明，这两类微生物以坎帕纳的居首，彭塔阿雷纳斯的最低，说明前者空气相当污浊，后者相当洁净。

“极地”船航行过一些沿海河道和内航道，表 8 列举了它们的空气微生物含量。源于内陆的拉普拉塔河空气的微生物含量较高。

自 1986 年 12 月 27 日至 1987 年 3 月 15 日，对南极半岛 3 个测区（南极长城站、麦克斯威尔湾、南设得兰群岛海域）作空气微生物监测**的结果说明长城站的空气微生物含量高于其他两测区的，内海（此处指麦克斯威尔湾）空气的微生物含量高于开阔性海域的（指南设得兰群岛海域）。与其他区域比，南极半岛空气仍较洁净，但长城站的空气有待改善。

当船只进入中国海域时，均测定了空气微生物。结果列于表 9，表中说明黄海的空气微生物较多，该海域空气微生物含量的基本趋势是由北向南递减。

---

* 陈皓文等，国际五城市空气微生物概况。

** 陈皓文等，中国南极长城站微生物考察。南设得兰群岛区域的微生物。

表 7　世界 5 个海峡空气微生物含量

海峡名	微生物含量(CFU/m^3)		采样日期	采样次数
	陆源性	海洋性		
麦哲伦海峡（开阔处）	10.00	12.00	1986.12.19	1
德莱克海峡	2.70	1.55	1986.12.20～1986.12.22 1987.3.16～1987.3.17	3 2
莫桑比克海峡	14.34	0.34	1987.4.19～1987.4.22	4
马六甲海峡	8.00	1.10	1987.5.1～1987.5.3	3
台湾海峡	7.30	24.65	1987.5.12～1987.5.13	2

表 8　南美一些近海河流、水道空气微生物含量

河道、海峡名称	微生物含量(CFU/m^3)		采样日期	采样次数
	陆源性	海洋性		
比格尔水道	4.29	3.99	1986.12.19;12.23～12.25	7
麦哲伦海峡	18.70	0.00	1986.12.12	1
拉普拉塔河	21.91	6.76	1987.3.22～1987.3.28	7

表 9　中国海空气微生物含量(CFU/m^3)

海区名称	微生物含量		采样日期	采样次数
	陆源性	海洋性		
黄海	71.48	117.34	1986.10.31 1987.5.13～1987.5.17	1 5
东海	12.00	58.67	1986.11.1 1987.5.13～1987.5.14	1 2
台湾海峡	7.30	24.65	1987.5.10～1987.5.11	2
南海	0.79	0.26	1987.5.10～1987.5.12	3

2.2　环球空气微生物属群概况

各取得来自培养基Ⅰ上的菌 88 株、培养基Ⅱ上的菌 85 株，它们分布在 1986 年 11 月 2 日～1987 年 5 月 11 日间的 51 天中，即 7 天在长城站，7 天在南大洋、南极半岛海域，37 天在环球航行中。地理范围是 1°15′N～26°05′N 和 30°39′E～171°00′E、1°04′S～62°18′S，2°09′W_1～174°45′W 间。所鉴定出的菌株包括 15 属细菌，即不动杆菌 *Acinetobacter*、芽孢杆菌 *Bacillus*、噬纤维菌 *Cytophaga*、欧文氏菌 *Erwinia*、黄杆菌 *Flavobacterium*、库特氏菌 *Kurthia*、明串珠菌 *Leuconostoc*、微球菌 *Micrococcus*、莫拉氏菌 *Moraxella*、足球菌 *Pe-*

diococcus、假单胞菌 *Pseudomonas*、八叠球菌 *Satcina*、葡萄球菌 *Staphylococcus*、链球菌 *Streptococcus*、动胶菌 *Zoogloea*。5 属酵母，即假丝酵母 *Candida*、隐球酵母 *Cryptococcus*、德氏酵母 *Debaryomyces*、红酵母 *Rhodotorula*、酵母 *Saccharomyces* 和 1 个细菌科即肠杆菌科 *Enterobacteriaceae*。

表 10 各类细菌在所鉴定细菌总数中的百分比值

细菌类别	所占百分率(%)
革兰氏阳性菌	11.76
革兰氏阴性菌	47.06
球菌	23.53
微 球 菌	5.89
多形杆菌	11.76

根据细菌染色特性及形态差别[7]，将上列细菌占细菌总数的百分率计算后得出表 10。上述细菌大部分与人体、动植物腐烂、土壤等有较密切的关系，它们若在空气中大量繁殖并广泛传播，可能诱发一些疾病流行。作为污染指示菌之一的不动杆菌的出现，可能意味所测空气含相当数量污染物，葡萄球菌和微球菌占有相当大的比例，与一些报道相吻合[3]。

尚有少数动胶菌和芽孢杆菌从长城站上被检出，说明长城站及其邻近区域有一定程度人为影响。从麦克斯威尔湾空气中发现肠杆菌科细菌，可能表明南大洋及其空气有人粪便污染。

2.3 温度试验

将分离自培养基Ⅰ的 88 株细菌作了不同温度时的生长试验，结果列于表 11。由该表可见，不少菌株能在 4～14℃中生长，这表明这些菌株可能是耐低温菌，其中有些也许是嗜冷菌，它们与低温的海洋空气密切相关，仍有一定数量的菌株能在 37℃中生长，反映人类影响的程度。

表 11 不同温度时能生长的细菌株数百分比值*

培养温度(℃)	能生长的细菌株数占参试细菌总株数的%
4	25.00(22/88)
10	17.05(15/88)
14	22.73(20/88)
25	17.05(15/88)
37	18.18(16/88)

* 随机抽取在淡水培养基上长出的 88 株细菌参试

3 结论

空气微生物状况受气象、工农业生产、人类活动等复杂环境因素的严重制约，海洋空气微生物与海气的盐浓度、海水的微生物浓度有关。特定测点上的空气微生物状况是环境综合因素的集中表现，是环境质量的重要参数。本文所述环球各测点的微生物数据一般取自海(地)面以上 10 m 内，离人群呼吸带不远，海(地)面因素的作用必然反映到微生物上来，因此这些数据与人类环境状况密切相关。空气微生物含量的一般状况，就环球航行一圈而言是：西半球在 4 个半球中居首，大西洋在各大洋中居第一，温带在各气候带中占首位，5 大海峡以台湾海峡举冠，4 大洲以南美为最，6 国中以中国为第一，5 城市以坎帕纳最高，中国海以黄海最甚，陆地、港口、沿海的一般均分别比海洋、离岸、远洋的高。

环球考察所获空气样品的微生物含量，其中位数值并不很高，但一些城市空气中微生物含量较高。南极，这片“净土”上的空气已有人类影响。所鉴定菌株大部与人体、动植物腐败有关，有可能是潜在的致病菌，有些是污染指示菌，生存能力强的微生物(包括一些致病菌)可被长距离传输。因而加强全球(包括南极)，特别是工业化城市空气微生物监测和空气净化是当今世界一项重要任务。

参考文献 7 篇(略)

(合作者：孙修勤　张进兴)

EXPLORATION ON AIR-BORNE MICROBES ROUND THE WORLD

(ABSTRACT)

Abstract From Oct. 31st, 1986 to May 17th, 1987, an exploration of air-borne microbes on Antarctica and round the world was performed aboard the Chinese polar vessel "JiDi" which took a voyage of 30,921 nautical miles. 432 samples were collected from different climate zones, hemispheres, Oceans, continents, cities, straits, rivers and China seas, each of which was counted and analysed microbiologicaly. The differences of microbial content of every type from air environment were compared in combination with relevant physicochemical parameters. The results show the condition of air-borne microbes round the world, i. e. terrigenous microbes and marine ones generally 7.50 CFU/m^3 and 4.00 CFU/m^3, their detection rates being 87.10% and 63.13% respectively.

The air over the Southern Ocean is still quite clean generally. The content of air-borne microbes over China seas fundamentally shows a trend of higher in the north than in the south. Some strains obtained were studied taxonomically. There were at least fifteen genera of bacteria and five genera of yeasts in the air examined round the world. *Staphylococcus* and *Micrococcus* occupied considerable percentage in generic composition of the air-borne microbes. *Bacillus* was occasionally found in the air over the Great wall station, Antarctica. It is still very important to monitor global air-borne microbes, to improve the condition of global air and keep it clean continuously.

Key words Round The World; Air-Borne Microbes; Microbial Content; Microbial Genera And Groups.

E_5 旅游区
(TRAVELING REGIONS)

庐山旅游区空气微生物污染调查*

摘　要　庐山风景旅游区空气微生物含量的初步调查结果表明，该区空气微生物含量平均值为21976.6 CFU/m^3，已达我国都市城区中度污染程度，个别测点已达严重污染程度。通过分析庐山空气细菌、真菌及总的微生物含量的空间分布及昼夜变化特征，认为空气微生物含量与人类活动有密切关系。

关键词　空气微生物　含量　空气环境质量　庐山

空气微生物与化学污染物、颗粒物并列为空气的3大污染物。它包括细菌、真菌、病毒、立克次氏体等，其中一些种类可引起动植物乃至人类病害，使大宗工农业产品及商品腐败变质[1]。

空气微生物污染已引起广泛重视[2]，其重要内容之一便是监测空气微生物的含量[3]。一些先进国家已相继制订出相应标准，以空气微生物含量作为重要指标评价空气洁净程度和环境质量[4]。

位于长江中游、鄱阳湖畔的庐山是驰名中外的避暑、疗养及旅游胜地。但随着旅游业迅猛发展，庐山生态环境，包括空气生态环境受到了巨大的影响[5]。

1　方法

在庐山主要景区、疗养场所等处设12个监测点，同时在九江市区设1个参比点。

用平皿自然沉降法采集空气微生物样品。平皿直径9cm，曝皿时间5 min。

样品分别采集至细菌培养基和真菌培养基[4]培养48h。培养温度细菌为25℃，真菌为37℃[6]。计数每皿细菌或真菌的菌落形成单位数(Colony Forming Unit 即 CFU)，也可将CFU/皿换算为CFU/m^3，以表示每立方米空气中的菌含量，计算公式如下：

$$CFU/m^3 = 5\times10^4 N/(A\cdot t)$$

其中：A—所用平皿面积(cm^2)；

t—平皿曝气时间(min)；

N—经培养后各皿中的CFU。

* 原文刊于《环境监测管理与技术》，1996，8(2)：21-23.

以空气细菌和真菌含量之和代表空气微生物总量，由此评价空气微生物含量状况[4]。

2 结果和讨论

2.1 庐山空气微生物含量的空间分布

监测调查结果列于表1。整个旅游区空气细菌含量平均为64.2 CFU/皿，极差为197.0 CFU/皿。空气真菌含量平均为75.7 CFU/皿，极差为768.0 CFU/皿，空气微生物总量的平均值为139.8 CFU/皿(21 976.6 CFU/m^3)，按我国的推荐标准[4]，在这12监测点次中，属正常或轻污染的有5个点次；属中度污染的有6个点次；属重污染的有1个点次。这表明，庐山空气的微生物污染已达中度污染水平。值得指出的是，这是用城区标准来评价的，旅游区理应有严于城区的评价标准。

表1 庐山旅游区空气微生物含量一览表

测点编号	细菌		真菌		空气微生物总量	测定时间	备注
	含量	百分比	含量	百分比			
1	8.0	80.0	2.0	20.0	10.0	6:30	凉爽
2	119.0	97.5	3.0	2.5	122.0	12:00	晴，温热
3	13.0	12.1	94.0	87.9	107.0	18:00	干热
4	0.0	0.0	768.0	100.0	768.0	24:00	雨后阵风
5	13.0	86.7	2.0	13.3	15.0	8:50	浓雾、人多
6	135.0	97.1	4.0	2.9	139.0	11:00	雾
7	25.0	80.6	6.0	19.4	31.0	13:55	瀑布
8	61.0	100.0	0.0	0.0	61.0	15:00	餐馆外有游客
9	146.0	94.2	9.0	5.8	155.0	8:30	小草丛边
10	21.0	87.5	3.0	12.5	24.0	11:10	平台上、荫凉
11	32.0	86.5	5.0	13.5	37.0	16:00	雾风飘
12	197.0	94.3	12.0	5.7	209.0	17:30	树林内
参比点	278.0	97.2	8.0	2.8	286.0	8:50	燥热、人多

注：含量单位CFU/皿，百分比指该项占空气微生物总量的百分比(%)。

按人群密集程度及人群活动频度，将全部测点分为3类，即人群密集、人群活动频繁区(1～4,9,10号测点)；人群活动中等频度区(11,12号测点)人群稀少区(5～8号测点)。统计分析结果表明3区的空气细菌平均含量分别为51.2、114.5和58.5 CFU/皿，空气真菌平均含量分别为146.5、8.5和3.0 CFU/皿；空气微生物总量平均分别为197.7、123.0和61.5 CFU/皿，由此可见，空气微生物数量与所在区的人群密集及人群活动频度有关。

2.2 庐山空气微生物含量的昼夜变化

用两个培训中心(距离较近)的4次测定结果(见表1，测点编号1～4)分析空气微生物数量的昼夜变化状态。

将表1中的有关数据换算成相应的常用对数值并绘成图1。由图1可见，空气细菌含量的最低值出现于午夜，峰值出现在中午；空气真菌含量从清晨至中午稳中有升，中午至午夜直线上升。空气微生物总量从清晨到中午及傍晚至午夜均呈直线上升，只是中午至傍晚之间有段平稳稍降。可见，该点的空气微生物含量在下午降低主要由细菌造成，而在午夜提高由真菌引起。

因此，空气细菌含量的昼夜变化与人们昼出夜归的活动规律一致。空气真菌含量是导致庐山区空气微生物含量在时间上突变的主因。

3　结论

庐山旅游区空气微生物平均含量相当于城区空气中度污染水平，个别场所已达到重污染程度。与以往资料相比[5]，庐山旅游空气环境质量呈恶化趋势。空气微生物含量的空间分布受人群密度、活动频度及地形等因素影响，其昼夜变化趋势为中午较高、午夜出现峰值，空气真菌含量对此影响较大。

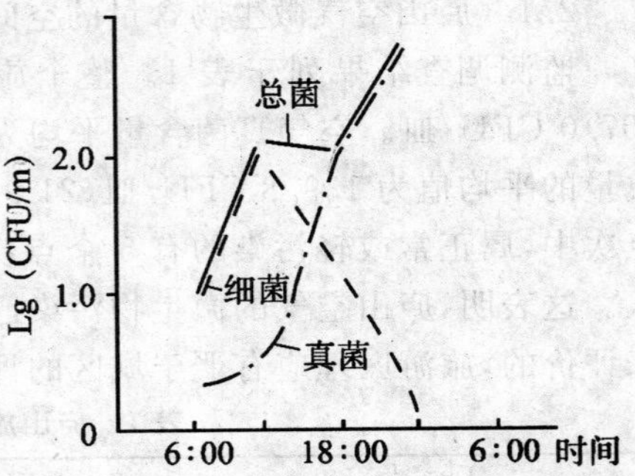

图 1　庐山某培训中心外空气微生物含量的昼夜变化

参考文献 7 篇（略）

INVESTIGATION OF AIRBORNE MICROBIAL CONTENT IN TOURIST ZONE, LUSHAN MOUNTAIN

(ABSTRACT)

Abstract Results are presented from the preliminary investigation of airborne microbial content, above Lushan mountain which are used as biomonitors of air pollution. The average content of airborne microbes is about 21,976.6 CFU/m^3. The data are analyzed in relation to spatial and temporal factors. The airborne microbial content is in relation to human activites.

Key words Airborne Microbe Content Air Environmental Quality Lushan Mountain

空气微生物含量对敦煌空气质量的意义*

一、概述

空气微生物不仅对铁、石、木质文物有破坏作用，对纸、壁画破坏更严重。

处于暖温带干旱气候区的中国河西走廊西端的历史文化名城——敦煌，虽饱经风雨沧桑，却仍以其丝路重镇之典雅风情，世界佛教之艺术精华令世人惊叹，人们期望其艺术生命万古不衰，采用多种技术保存它，使之“健康”延续。但实际上当今的敦煌如同世上其他一些著名艺术遗产一样，正在经受着日益严重的环境污染。这儿除了每年有沙尘暴天气外，还有一定的人为污染。可是人们对它所在的空气环境状况了解得并不多。

一些硫酸盐还原菌的生化产物能腐蚀金属石头及相关物件。许多细菌会引起人和生物患病，物品变质。一些真菌毒素既可致病也可破坏艺术品。敦煌位于沙漠之中，风沙和大气中漂浮的污染物及其挟持的微生物不断侵蚀、腐蚀、破坏它。已有证据表明，敦煌壁画的色变过程有微生物参与。而生存于艺术品中的微生物又与空气环境中的微生物经常交换，并互为依存。因此，掌握当今敦煌空气微生物状况和它对于敦煌环境及艺术的影响，对国宝的保护是十分必要的。本文报道敦煌空气微生物含量及初步调研结果。

二、材料和方法

用平皿沉降法采集空气微生物粒子。培养介质：普通营养琼脂—用以采集细菌。沙氏培养基—用以采集真菌。采样时间：5 min，经三天培养后计算出单位体积空气中的微生物含量（CFU/m^3）以及相应的微生物含量参数。采样点计三处，详见表1。它们分别代表城市功能生态、干旱区天然沙漠生态和古艺术宝库遗产景区生态的空气环境。

表1　敦煌空气微生物采样记录

测点编号	测点名称	环境概况	采样起始时间
1a	飞天宾馆二楼顶	晴天，轻风，15℃，安静	1996.4.20. 22:30
1b	飞天宾馆二楼顶	阴天，轻风，10℃，周围车人少	1996.4.20. 07:45
2	莫高窟正门口	晴天，和风，20℃，树旁有些游客	1996.4.20. 12:00
3	鸣沙山西南沙峰	晴天，阵风为劲风，18℃，脸部感到有细风沙刮过，安静	1996.4.20. 16:35

* 原文载于《现代空气微生物学》，2002，P393-395

三、结果和讨论

表2列出了敦煌空气微生物含量的统计结果。

由该表可见三个测点的空气微生物含量指标间存在一定差异,依大小排序分别是:空气细菌—3,2,1;空气真菌—2,1,3;总菌(为空气细菌和真菌含量之和,下同)—3,2,1;空气真菌/总菌百分比—2,1,3。这表明在鸣泉山山顶测定时,面对敦煌市区方向刮来已污染的风和尘埃,使测点空气挟持有较多的微生物,加上测时的较高气温,均有利于细菌生长繁殖,指示空气污染较重。反观1号点所测的情况,它表明测定时此处空气微生物含量不大,空气较为清洁。2号测点的空气微生物含量居于中偏高状态,尤其真菌量所占百分比较高。这与测定时的环境状况有关(见表2)。也表明此时的空气湿度较高。与色变壁画中的细菌、真菌量比,虽然这儿此时的微生物含量低得多,但也足以令人关注。室外空气微生物含量在许多情况下比室内的低,因此推测在游客不时出入,导致滞留窟内的空气微生物之含量将比室外高。这些空气微生物为附着、繁殖于艺术品内外提供了生物来源,这表明除对艺术品本身做微生物学的研究外,对石窟内的空气微生物做较深的监测分析仍是十分必要的。所幸,此方面的工作已经开展。

表2　敦煌空气微生物含量的统计

测点编号	空气微生物含量			
	细菌①	真菌①	总菌量①	真菌/总菌(%)
1②	4 794.6	78.6	4 873.2	1.6
2②	6 916.8	157.2	7 074.0	2.2
3	12 733.2	0	12 733.2	0

注:①空气细菌、真菌、总菌量,单位均是 CFU/m³　②No.1,No.2 均是两次测定之均值

将表2的数据作相应的计算可得出以下四个平均值:空气细菌量为 8 148.2 CFU/m³,空气真菌量为 78.6 CFU/m³,空气总菌量为 8 226.8 CFU/m³,空气真菌占总菌量的百分率为 0.012 7。如果这能大体上表达出敦煌空气微生物概况的话,那么可以说敦煌空气已处于中度污染范围。不过,将此与国内一些城市、景区相比,此间的敦煌空气质量尚可。

空气中的真菌含量及其所占比率总体看并不大。与国内湿润的沿海区和一些南方城市相比,这儿真菌量应算是很低的。这可能决定于敦煌空气的干燥性。干燥空气有利于文物的保护。在一定条件下,营造树木花草,会起到绿化美化环境的作用,也可改善干旱环境生态。但浓密的植被反过来将增大和保留空气湿度。湿度恰恰是壁画铅丹变色主因之一,而且较大湿度将促使具耐湿孢子的真菌及喜湿真菌的繁殖。过高湿度、过多真菌对人体、文物及其所在环境都有不利的一面。本结果中如莫高窟景区(2号测点为代表)的绿化较好,但空气真菌量也比其他测点增加不少(如表2)。这提示我们要处理好敦煌这类景区绿化——潮湿——微生物繁殖——文物保护这系列矛盾,有关研究应当深入。

四、结束语

对敦煌空气微生物的这一测定工作虽然所涉及的测点不多,次数也有限,但所得结果

已经表明现今的敦煌空气微生物含量总体看并不很大，所反映出的空气质量尚可。但鉴于敦煌这一艺术宝库的价值和作用，仅以通用的标准来要求，看来是不够的。敦煌应有更严格的环境质量，应以更完善的手段来保护它。当今各国面对的环境问题是全球性的，而不仅仅是局部地区性的。这涉及沙漠化扩大、脆弱的大气圈表层底部面临的人为错误干扰，环境良性持续发展问题日益紧迫等等。旅游热点的敦煌避免不了这些问题。为此，密切关注敦煌空气微生物及其环境状况的变化，并采取行之有效的措施来保护好这一瑰宝乃是世人面临的严肃任务。今后，除加强室外空气微生物监测外，还应深化窟内空气、各类艺术品的微生物及其影响的研究，为确定先进科学的保护方法奠定基础。

参考文献 10 篇（略）

张家界、韶山和衡山空气微生物粒子沉降量分析*

摘　要　对张家界、韶山和衡山空气微生物粒子沉降量用自然沉降法作了测定，三地的空气微生物污染处于轻—中度状态，以轻污染为主，衡山的空气污染最轻；三地空气多以细菌污染为主，张家界空气真菌含量较大；分析了三地空气微生物粒子沉降量的差异及原因，讨论了空气微生物粒子沉降量的昼夜变化。

关键词　张家界　韶山　衡山　空气微生物污染　空气质量

张家界、韶山和衡山均位于湖南省。三地均设市，均为湖南风景游览区。三地气候类型属北亚热带季风区，空气温湿。因地理和经济状况的差异导致三地环境生态状况的不同。

随着国民经济，包括旅游业的发展，三地都存在着人为的环境污染，大气污染即是其中最明显、最直接的环境问题。人群呼吸带则属与人息息相关的空气圈。空气中的微生物是环境状况的综合指标，其中的致病菌在一定条件下是疾病感染率增高的要素，因而危害人们健康。调研特定地区空气微生物状况是一项基本的环境监测项目，也是改善环境状况的一项前提条件[1]。

对张家界、韶山和衡山 3 地的空气微生物状况迄今未见研究报道。笔者于 2002 年 3 月间作了随机的空气微生物粒子沉降量调研，旨在了解它们的空气微生物污染状况，评价其空气质量并为改善其环境—生态状况提供初步资料。

1　三地环境概述

三市的基本地理、环境状况如表 1 所示。

2　三地空气微生物粒子沉降量采样点概述

在三地各取 6 个测点，进行空气微生物粒子沉降量的测定。

3　材料和方法

在三地上述的采样点用平皿自然沉降法获取空气微生物菌落粒子即 CFU，以普通营养琼脂测细菌，沙氏培养基测真菌。曝皿 5 min。方法详见参考文献[2]。整理所得单位体积(m^3)空气中的 CFU，分析空气细菌(B)、真菌(F)、总菌(T)(细菌十真菌)的粒子沉降量、真菌粒子占总菌粒子量(即 F/T，下同)的百分率，估计和评价空气环境质量。

4　结果和讨论

4.1　三地空气微生物含量概况

表 2 列出了三地所有测点空气微生物粒子沉降量的上述 4 个指标，由该表可见三地空气微生物污染概况。

---

*　原文刊于《国土与自然资源研究》，2003，No. 2：54-56

国家自然科学基金(40086001)资助

表 1　张家界、韶山、衡山的环境和市政情况

地名	所属气候区	自然环境及测时天气状况	
张家界	北亚热带季风区	湘西山丘 离市区 34 km，森林公园原始大自然风格 阴雨转晴，湿，9～15℃	地级市，奇峰连绵，怪石高耸，沟壑幽深，溪水潺流，奇花异草，珍禽异兽。旅游经济，常住人口不多
韶山	北亚热带季风区	湘中偏东 湘潭西北 40 km，南岳 72 峰之一 阴间晴，小风 10～18℃	群山环抱、松柏葱茏、风景秀丽。清溪镇为市区，县级市
衡山	北亚热带季风区	湘东南大小峰 72 座，祝融峰海拔1 290 km。山势雄伟，盘延百里。风景名胜区 晴，小风，10～24℃	风景绚丽多彩，古木参天，终年常绿，奇花异草，四时郁香，市区在南岳镇

注：各类微生物含量单位为 CFU・m^{-3}

表 2　张家界、韶山、衡山空气微生物粒子沉降

	编号	细菌(B)	真菌(F)	总菌(T)	真菌/总菌(F/T)%
张家界	1	757.6	378.8	1 136.4	33.3
	2	378.8	2 272.8	2 651.6	85.7
	3	1 894.0	378.8	2 272.8	16.7
	4	0	378.8	378.8	100.0
	5	757.6	2 651.6	3 409.2	77.8
	6	8 333.6	1 136.4	9 470.0	12.0
韶山	编号	细菌(B)	真菌(F)	总菌(T)	真菌/总菌(F/T)%
	1	2 651.6	378.8	3 030.4	12.5
	2	3 030.4	378.8	3 409.2	11.1
	3	0	0	0	/
	4	10 227.6	1 894.0	12 121.6	15.6
	5	1 136.4	757.6	1 894.0	40.0
	6	90 912.0	1 136.4	92 048.4	1.2
衡山	编号	细菌(B)	真菌(F)	总菌(T)	真菌/总菌(F/T)%
	1	4 545.6	378.8	4 924.4	7.7
	2	1 136.4	0	1 136.4	0
	3	1 136.4	0	1 136.4	0
	4	0	0	0	/
	5	1 515.2	0	1 515.2	0
	6	6 060.8	0	6 060.8	0

注：各类微生物含量单位为 CFU・m^3。

据表 2 数据计算可得知，张家界空气中平均的细菌、真菌、总菌粒子及真菌粒子(F/T)百分比含量分别为 2 020.3，1 199.5，3 219.8 CFU·m^3 及 37.3。按文献[2]的推荐，本结果表明张家界空气处于轻度微生物污染状态。各测点间存在一定差距。细菌量的极差是 8 333.6m^{-3}。真菌量的极差是 2 272.8 m^{-3}。总菌量极差是 9 091.2 m^{-3}。F/T 百分比极差是 88.0。最高细菌量出现于测点 6。最低细菌量出现于测点 4。最高真菌量在测点 5，最低真菌量出现在 3 处，即测点 1、3、4。最高最低总菌量分别在测点 6 和测点 4。表明张家界空气微生物以细菌为多，但其中有一站未测出细菌。真菌所占份额不少，有 3 站的真菌百分比含量高于细菌的，说明张家界空气湿度大，森林覆盖率高。市区(测点 6 为代表)空气细菌、真菌、总菌及 F/T 百分比含量分别是景区的 11.0，0.94，0.8 及 0.19 倍，表明市区空气污染重于景区，且以细菌污染为主，但景区真菌污染重于市区。与敦煌比，这儿空气细菌、总菌量均低，但真菌量及 F/T 百分比含量均大大高于敦煌，这显示两地空气及生态环境的差异[3]。

统计三市数据可知韶山空气平均的细菌、真菌、总菌粒子沉降量及真菌百分比含量分别为 17 993.0，757.6，18 750.6 CFU·m^{-3} 及 4.0，其极差分别为 90 912.0，1 894.0，92 048.4CFU·m^{-3} 及 40.0 以上，最大与最小间比值差距大。平均细菌量是真菌的 48.0 倍。污染主要由细菌引起。景区内差异也大，细菌、真菌的极差分别为 10 227.6 及 1 894.0CFU·m^{-3}。令人称奇的是测点 3 测点竟未测出任何细菌和真菌，说明那儿十分洁净。市区与景区比，前者的细菌、真菌、总菌量分别是后者的 26.7，1.7 和 22.5 倍，表明市区污染重于景区，且以细菌污染为主。真菌较少。与张家界比，韶山景区真菌量及所占百分比均小。说明韶山真菌污染轻于张家界。景区空气细菌污染最大处在测点 4，已达中度污染水平，但景区由平均细菌水平说明它污染仍轻(3 409.2 CFU·m^3)。市区空气污染已超过中度水平。看韶山所有空气测点各自的细菌、真菌、总菌污染的平均水平，表明它的空气微生物粒子沉降量已处于中度污染，且重于张家界的。与泰山相比，这儿的空气细菌、总菌及 F/T 百分比含量均高，但真菌量只是泰山的 0.73 倍[4]。

据表 2 可计算出衡山空气中平均的细菌、真菌和总菌的粒子沉降量及真菌所占的百分比含量分别为 2 399.1，63.1，2 462.2 CFU·m^{-3} 及 2.6。同理，可以认为衡山空气质量上佳，仅达轻微污染程度，也比庐山的好得多[5]。最大、最小空气细菌量分别出现在测点 6 和测点 4，两者相差 6 060.8 CFU·m^{-3}。5/6 的测点未测出真菌，因而真菌百分比含量很小，最大、最小总菌量分别出现于测点 6 和测点 4，相同于空气细菌。从测点 1 到测点 5，山势越来越高，空气微生物含量也基本上越来越小。测点 5 较多的细菌，可能与当时烧香造成的污染有关。与张家界、韶山相比，除空气细菌粒子沉降量大于张家界外，真菌、总菌量、F/T 百分比均是最低的。衡山市区(测点 6)的空气微生物污染与上述两地的一样，均重于景区。但在三地市区中仍为最好者。市区空气细菌、总菌粒子沉降量分别是景区的 3.6 和 1.5 倍。

4.2　三地空气微生物粒子沉降量的昼夜变化

将张家界空气微生物采集时间分作上午、中午、下午和夜间 4 个时段，各时段的微生物粒子沉降量作成图 1。由该图可大致据时间变化看出其沉降量的变化状况。它表明空气细菌量从上午开始不断攀升经中午至下午的高峰，在傍晚前后下降，总菌量和真菌量大致也呈此态。似乎表明空气微生物经一个白天的生长繁殖和积累，至傍晚前后衰退，这与人们日常起居、生产活动及气温变化的规律相一致。白天热闹的户外旅游活动和频繁的工商业、交通活动使空气中悬浮颗粒增多,微生物依附的条件和机会即多。

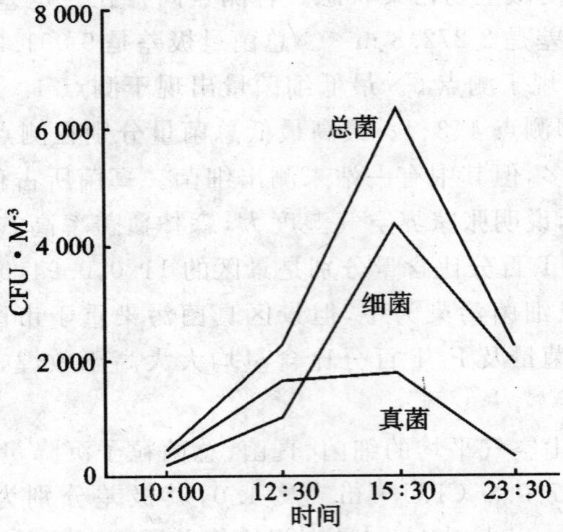

图1 张家界空气微生物粒子沉降量的变化状况

韶山空气微生物测时可分作上、下午两个时段,微生物粒子沉降量的变化态势呈现为上午>下午。但仅就真菌而言,其F/T%则显示为下午>上午。与韶山相反,衡山空气微生物粒子沉降量的4个指标显出下午>上午之势,而且上午未测出真菌。

以上结果显示出三地空气微生物之间存在着一定差异。比较而言,张家界与衡山的空气微生物粒子沉降量日变化较相似,虽然未测衡山夜晚的空气微生物。但张家界空气真菌量及其百分率(即F/T%,下同)均大大高于韶山和衡山的,这可能与测时的张家界空中有微雨和高的空气湿度及很大的森林覆盖率相关。

5　结语

本文对张家界、韶山和衡山三地空气微生物污染状态作了随机的检测,结果表达了三地空气微生物4指标间的差别,空气细菌和真菌污染以韶山为最重,张家界的空气细菌污染最轻。空气真菌污染以衡山最轻,张家界却最重,这是由各地的地理、气象条件和经济发展状况决定的,同时也反映了空气质量上的差异。

结果还表明三地市区空气微生物污染重于景区,说明市区经济、贸易、交通的繁忙和人口的密集导致空气质量下降,也使其相对干燥的空气中真菌及其百分率少于湿润的景区。

与其他地方比,总体看,三地空气微生物污染处于轻—中度状态,以轻污染为主。空气质量,尤其是衡山为上佳水平。

由空气微生物粒子沉降量表达的空气质量反映了特定地域整体环境状况。发展经济,必须考虑生态的可持续良性发展,以此造福后代。

参考文献5篇(略)

ANALYSIS OF AIR-BORNE MICROBIAL PARTICLE PRECIPITATION ABOVE ZHANGJIAJIE, SHAOSHAN AND HENGSHAN

(ABSTRACT)

Abstract Air-borne microbial particle precipitation above Zhang jiajie, Shaoshan and Hengshan was determined by using the method of gravity plate. The result obtained showed that: the air-borne microbial pollution above the three cities above mentioned was at light-middle state and more of them was at light one. The air pollution of Hengshan was lightest and the air quality was fine. The pollution in urban district was more serious than landscape area. The air-borne microbial content determined was mostly consisted of bacteria, but the percentage of air-borne fungus content occupied was higher in Zhang Jiajie, The difference and its causes of air-borne microbial particle precipitation above the three cities were analysed and their diurnal variation was discussed.

Key words Zhang Jiajie; Shaoshan; Hengshan; Air-Borne Microbial Pollution; Air Quality

天山天池与喀纳斯湖测区空气微生物含量的测定与比较*

摘　要　通过测定和比较天池与喀纳斯湖区空气微生物含量,其结果表明两区空气均处于较轻的中度微生物污染水平,质量尚佳,并指出处于干旱少雨大环境中的两区所具有的空气微生物地域性特色。

关键词　天山天池　喀纳斯湖　空气微生物　空气质量

一、引言

天池位于新疆天山博格达雪峰半山腰,系由高山冰川冰蚀冰碛而成。湖面海拔1 928 m,四周群山环抱,云杉密布,绿草如茵,风景秀丽绝伦,堪与瑞士雪山大湖媲美的秀丽景色使之成为我国首批重点保护风景区之一,测时游客甚多。

喀纳斯湖位于阿尔泰山森林带中部,系我国唯一的西伯利亚动植物分布区。周围层峦叠嶂,青山绿水与白雪峻岭交相辉映,湖光山色,美不胜收。湖面海拔1 374 m,面积大天池的8倍,是我国著名的自然保护区之一,但测时游客并不多。天池在新疆中部,喀纳斯湖在新疆北端,两点直线距离约520 km。测时气温虽然均不高于20℃,但喀纳斯湖区的天气及测点环境的变化大于天池。

风景游览区空气中的微生物检测工作,关系着环境的好坏、生态的优劣和人们的健康。我国因相关的研究起步较晚,成果较少,尤其是旅游区的空气微生物状况常常不被人们重视。本文论述新疆两个水体景区的空气微生物状况,并作了两者比较,以了解它们的空气污染/空气质量现状,促进我国景区空气及环境生态状况的改善,使之良性地永续发展。

二、采样测定

(一)材料和方法

采用自然沉降法。将消毒好的营养琼脂和沙氏培养基平皿放在离地/水面约1.5 m的人群呼吸带内曝气5 min,取回在25℃的室温中培养2天以上后,分别计数平皿培养基上生长出的细菌(B)和真菌(F)菌落数(CFU),换算为单位体积空气中的CFU·m^{-3}。

细菌和真菌之和为总菌含量(T),并算出真菌占总菌的比率即F/T(%)。以此评价空气的微生物污染状况和空气质量。方法详见文献[4]。

(二)测点概况

表1列出了天池和喀纳斯湖两景区空气微生物10个测点的环境状况和测定时间。

---

* 原文刊于《国土资源科技管理》,2003,20(3):52-54.

由表1可见,天池测点主要分布于湖岸一侧,测时共持续3小时,气温变幅小。喀纳斯湖测区的测点分布范围广,有在水上、有在山顶。测定持续时间跨度较大,达2天以上,气温偏低。

表1 测区空气微生物测点状况描述

测点编号	测点名称	环境状况	采样起始时间
1	东小天池附近	晴好天气,4～5级风,有些人,车,18℃气温	2002-08-18 13:30
2	北码头附近	海拔1928 m,晴好天气,2级风,人少,11℃气温	2002-08-18 13:55
3	天池东码头与北码头间盘山路边	晴好天气,2级风,人少,11℃气温	2002-08-18 14:35
4	天池中段停车场	晴好天气,微风,车多,有人来往,20℃气温	2002-08-18 15:10
5	天池下段停车场	晴好天气,微风,停车多,少人来往,19℃气温	2002-08-18 16:30
6	喀纳斯边防检查站	晴天,2级风,车人往来,有浮尘,16℃气温	2002-08-20 12:00
7	喀纳斯湖三道湾口	海拔1 374 m,晴天,3级风,水面上气温13℃	2002-08-20 15:40
8	喀纳斯湖回家山庄草坪	晴天,2级风,有些食客,15℃气温	2002-08-20 19:15
9	喀纳斯湖回家山庄草坪	薄雾,轻风,少量游人走动,畜粪气浓,太阳未出,10℃气温	2002-08-23 07:00
10	喀纳斯观鱼亭	海拔2 030 m,阵雨后,5～6级风,8℃气温	2002-08-23 09:15

采样时间持续5 min,采用北京时间,当地时间比北京时间约晚2h。

三、结果和讨论

表2列出了天池和喀纳斯湖两景区测点的空气微生物含量状况。

表2 测区空气微生物含量

测 区	测点编号	空气微生物含量(CFU·m^{-3})			
		细菌	真菌	总菌	F/T(%)
天山天池	1	5 344.8	1 572.0	6 916.8	22.7
	2	2 829.6	1 100.4	3 930.0	28.0
	3	3 301.2	314.4	3 615.6	8.7
	4	9 746.4	2 200.8	11 947.2	18.4
	5	2 672.4	1 100.4	3 772.8	29.2

(续表)

测 区	测点编号	空气微生物含量(CFU·m^{-3})			
		细菌	真菌	总菌	F/T(%)
喀纳斯湖	6	9 117.6	1 257.6	10 375.2	12.1
	7	628.8	471.6	1 100.4	42.9
	8	1 886.4	3 458.4	5 344.8	64.7
	9	1 624.0	1 218.0	2 842.0	42.9
	10	6 090.0	812.0	6 902.0	11.8

由表2数据可得出,天池平均的空气细菌、真菌、总菌及F/T(%)分别为4 778.9、1 257.6、6 036.5 CFU·m^{-3}及20.8。根据文献[4]的推荐,可知天池的空气中细菌污染处于轻度污染状态,而总菌含量则处于较轻的中度污染水平。真菌占总菌的比率与天水相比,高了许多,反映了天池的湿气大于天水。喀纳斯湖平均的空气细菌、真菌、总菌及F/T(%),分别为3 869.4、1 443.5、5 312.9 CFU·m^{-3}及27.2。显示出喀纳斯湖空气微生物含量处于较轻的中度污染状态,其中真菌的贡献率大于天池。

两景区相比,喀纳斯湖的空气中、真菌污染虽重于天池,但细菌污染、总菌污染却轻于天池,这映证出喀纳斯湖水体对空气的调节作用似大于天池。一些真菌毒素是致病源,对空气真菌的种类和含量状况应引起关注。两景区与泰山和庐山相比,均显出空气质量较好。

上述空气微生物四个指标之变幅及最高最低均值经计算列入表3。从表3可见,喀纳斯湖的四个指标均大于天池,表明喀纳斯湖空气环境的参数变动大于天池,空气中的微生物状况也随之而变。

表3 测区空气微生物含量变幅及最高/最低比值的比较

测 区	变幅比值	空气微生物含量(CFU·m^{-3})			
		细菌	真菌	总菌	F/T(%)
天山天池	变幅	7 074.0(9 746.4-2 672.4)	1 886.4(2 200.8-314.4)	8 331.6(11 947.2-3 615.6)	20.5(29.2-8.7)
	比值	3.6(9 746.4/2 672.4)	7.0(2 200.8/314.4)	3.3(11 947.2/3 615.6)	3.4(29.2/8.7)
喀纳斯湖	变幅	8 488.8(9 117.6-628.8)	2 986.8(3 458.4-471.6)	9 274.8(10 335.2-1 100.4)	52.9(64.7-11.8)
	比值	14.5(9 117.6/628.8)	7.3(3 458.4/471.6)	9.4(10 375.2/1 100.4)	5.5(64.7/11.8)

注:变幅=最高值-最低值;比值=最高值/最低值

将测时按时段划分,得出的空气微生物含量状况如表4所示。

表 4 空气微生物含量在时段上的比较

测区	时段	空气微生物含量(CFU·m⁻³)			
		细菌	真菌	总菌	F/T(%)
天山天池	上午	4 087.2	1 336.2	5 423.4	24.6
	下午	5 240.0	1 205.2	6 445.2	18.7
喀纳斯湖	清晨	1 624.0	1 218.0	2 842.0	42.9
	上午	7 603.8	1 034.8	8 638.6	12.0
	下午	1 257.6	1 965.0	3 222.6	61.0

表 4 说明,天池空气细菌、总菌含量均呈下午大,上午小,而真菌含量及 F/T(%)均为下午小上午大。喀纳斯湖空气微生物的这四个指标正好与天池相反,它清晨空气的细菌、真菌及其 F/T(%)均显示为处于上、下午之间,而总菌量则显出清晨最低。微生物含量大的空气说明其湿度、温度与载体,包括污染物气溶胶更适于微生物生存及积聚。

在相似情况下,空气微生物含量常显出海拔高空间的小于海拔低空间的(见表 5),但空气微生物含量易受种种因子的干扰与影响而显出不那么规则。表 5 说明天山天池的空气微生物含量的四个指标总体上显出海拔低(山下)的测点(No. 1～3 测点)高于海拔高(山上)的测点(No. 4～5 测点)。喀纳斯湖的空气真菌含量及 F/T(%)也表达出这一点。这是因为海拔高处空气比海拔低处稀薄,气温、湿度也较低,大小污染物颗粒也少于海拔低处,微生物依附的载体就少。但此种情况又受制于其他条件,如降水、污染和风力等。所以喀纳斯湖海拔高处(No. 10 测点)的空气细菌、总菌含量大于海拔低处(No. 6～9 测点)。

表 5 测区空气微生物含量在地势间的变化

测区	地势	空气微生物含量(CFU·m⁻³)			
		细菌	真菌	总菌	F/T(%)
天山天池	山上	3 825.2	995.6	4 820.8	19.8
	山下	6 209.4	1 650.6	7 860.0	47.6
喀纳斯湖	山上	6 090.0	812.0	6 902.0	11.8
	山下	3 314.2	1 101.4	4 915.6	40.7

四、结语

本文初步论述天山天池和喀纳斯湖两景区测点的空气微生物含量,并作了两者的比较。结果表明,两景区的空气微生物含量指示其空气质量尚佳,污染不严重,景区内与景区间所显出的空气微生物含量的差异,反映了测区的地域特色。总体看,喀纳斯湖空气污染轻于天山天池,但后者真菌污染重于前者。两者的真菌污染表示出虽同处于干旱少雨的大陆性气候大环境之中,但小区域性气候环境又决定了两湖区的湿度较大,有较多的真菌相适应。这显示出它们不同于一般干旱环境的特点。文章分析了这两个景区空气微生物含量的时空分布状况及其原因,同时说明对真菌污染必须引起重视。

参考文献 8 篇(略)

DETERMINATION AND COMPARISON BETWEEN AIR-BORNE MICROBIAL CONTENTS OVER HEAVENLY POND AND KANAS LAKE IN XINJIANG

(ABSTRACT)

Abstract This paper reports the state of air-borne microbial contents over 10 sampling regions of Heavenly Pond, Tianshan and Kanas Lake and makes a comparison. The result obtained shows that the air over the two regions is at the level of slight moderate microbial pollution, its quality being pretty good. The air-borne bacterial pollution over Kanas Lake is lighter than that over Heavenly Pond, but the pollution of air borne fungus over Kanas is more serious than that over the latter. The paper also describes the regional characters of air-borne microbes over the two regions situated in the environment deficient in rain.

Key words Heavenly Pond in Tianshan; Kanas Lake; Air-Borne Microbes; Quality of Air

吐鲁番市空气微生物浓度状况*

摘 要 对吐鲁番市空气微生物进行了初步测定。结果表明,室外平均空气细菌(B)、真菌(F)、总菌(T)的浓度及F/T分别为33 106.3、0、33 106.3 CFU/m^3 及0,说明室外空气处于中度微生物污染状态,一些室外测点的空气已受到微生物的严重污染。室内平均空气细菌浓度为9 982.2 CFU/m^3,说明室内空气污染较重。室内外均未测到空气真菌。本文对吐鲁番市空气微生物浓度指标的时空分布状态作了分析。

关键词 吐鲁番市 空气微生物 空气污染 空气质量

位于新疆东天山南麓的吐鲁番市,面积6.97万km^2,人口20万。有少量的皮毛、葡萄加工工业和经贸市场。该市具有地理和气候方面的3个“中国之最”,即最热、最早、最低,因而给微生物提供了独特的空气条件。

空气是一切生物赖以生存的先决条件,其中的微生物则关系到空气质量、人类健康和生态状况。空气中微生物大多依附灰尘等气溶胶粒子而构成微生物气溶胶。由于微生物气溶胶来源、种类活动、传播、沉积和感染的多样性、复杂性而使空气微生物与人类、动植物、工农业、商业发生种种关系,包括环境污染及某些传染病等疾病的爆发流行,因而对空气微生物的研究越来越受到人们的关注[1-3]。

作为特定地域的吐鲁番市,据知其空气微生物状况的信息迄今是极少可资引用的。随着西部开发热潮的涌起。吐鲁番市的环境,包括空气质量将越来越受到人们的关注,为此本文对吐鲁番市空气微生物状况作了初步探讨。

1 吐鲁番市空气微生物测点情况

2002年8月16～17日,共设10个室外测点,测定10次。另设1个室内测点为对照点(测定2次)。测点情况见表1。

表1 吐鲁番市空气微生物测点情况

测点编号	测点名称	环境状况		采样时间*
1#	旅游文化广场	有些休闲娱乐者,干、晒	气温约42℃	2002-08-16 16:30
2#	高昌路北端一商店外		气温40℃	2002-08-16 17:00
3#	西环路农贸市场路边	人不多,尘土多,5级风	气温39℃	2002-08-16 18:40
4#	大河沿邮局门口	天将暗,3～4级风	气温33℃	2002-08-16 21:00
5#	吐鲁番火车站广场	10余人纳凉,5～6级干凉风	气温27℃	2002-08-16 23:00
6#	大河沿汽车站门口路旁	有些乘客,4级风	气温28℃	2002-08-17 08:50

* 原文刊于《干旱环境监测》,2003,17(4):211-214.

（续表）

测点编号	测点名称	环境状况		采样时间*
7#	交河古城碑前	游人刚过，静，1 级风	气温 37℃	2002-08-17　09:40
8#	坎儿井民俗园内	游人少，0 级风	气温 30℃	2002-08-17　12:25
9#	苏公塔	无游人，3 级风	气温 41℃	2002-08-17　13:35
10#	老城路一商店外	车、人来往不多，2 级风	气温 42℃	2002-08-17　14:20
11(1)**	某旅馆内	有风刮入	气温 26℃	2002-08-17　08:15
11(2)**	某旅馆内	无风	气温 29℃	2002-08-17　16:30

*采样起始时间(北京时间)，持续 5 min。**室内测点。

2　材料和方法

采用自然沉降法。所用固体培养基包括营养琼脂及真菌沙氏培养基。以无菌操作法制作培养基，消毒好的培养基倾碟后备用。方法见文献[4]。在测定现场于人群呼吸带(离地高 1.5 m 处)，开启平皿盖，曝气 5 min。在 27℃左右至少培养 2d 后，分别计数皿中的细菌(B)和真菌(F)，计算出总菌落数(T＝B＋F)，并以相关公式换算为所测单位体积(m^3)空气中的各类及总菌数。以此评价相应空气微生物污染状况[4]。

3　结果与讨论

3.1　室外空气微生物浓度状况

3.1.1　室外空气微生物的一般特点

吐鲁番市室外各测点的空气微生物浓度状况见表 2。由表 2 可知，吐鲁番市室外空气微生物有如下特点。

表 2　吐鲁番市室外空气微生物浓度　CFU/m³

测点编号	空气微生物浓度			
	细菌(B)	真菌(F)	总菌(B＋F)	$\frac{真菌(F)}{总菌(T)}$/%
1#	15 562.8	0	15 562.8	0
2#	12 104.4	0	12 104.4	0
3#	97 149.6	0	97 149.6	0
4#	18 706.8	0	18 706.8	0
5#	35 212.8	0	35 212.8	0
6#	69 325.2	0	69 325.2	0
7#	19 335.6	0	19 335.6	0
8#	57 090.0	0	57 090.0	0
9#	4 558.8	0	4 558.8	0
10#	8 017.2	0	8 017.2	0

(续表)

测点编号	空气微生物浓度			
	细菌(B)	真菌(F)	总菌(B+F)	真菌(F)/总菌(T)/%
11(1)	14 619.6	0	14 619.6	0
11(2)	5 344.8	0	5 344.8	0

1)所测空气中未见真菌，只有细菌，这反映出吐鲁番市空气极其干燥，极少的降水被高气温“蒸发”了，空气湿度常为零[5]。此空气在一定条件下不适于真菌生存，尤其不适于具耐湿孢子真菌的生存。而内陆城市与沿海城市的情况则不同[6-8]，因此吐鲁番市空气微生物污染实质上是细菌污染。

2)吐鲁番市空气细菌，即空气微生物浓度的平均值为 33 106.3 CFU/m^3，据有关资料显示[4]，吐鲁番市空气处于中度微生物污染状态(国标范围 5 000-50 000 CFU/m^3)，大部分测点均处于此水平上下。

3.1.2 吐鲁番市室外空气微生物浓度的空间分布

吐鲁番市空气微生物浓度各测点之间存在差异，其极差为 92 590.8 CFU/m^3。从大到小排序为:3$^\#$>6$^\#$>8$^\#$>5$^\#$>7$^\#$>4$^\#$>1$^\#$>2$^\#$>10$^\#$>9$^\#$。最高浓度是最低浓度的 21.3 倍，3$^\#$、6$^\#$和 8$^\#$点的空气微生物浓度已属重度污染。

根据上述空气微生物浓度状况，可粗略地将测点空气质量状态分为三类。Ⅰ类为 9$^\#$，即空气质量最好之处，空气污染仅为轻度水平；Ⅱ类为 1$^\#$、2$^\#$、4$^\#$、7$^\#$、5$^\#$、10$^\#$，其空气污染处于中等水平，平均空气微生物浓度达 18 156.6 CFU/m^3；Ⅲ类为 3$^\#$、6$^\#$、8$^\#$，空气已处于重度污染水平，空气微生物浓度为 72 521.6 CFU/m^3。

按城市不同功能区划分类，区划 1 为市中心区，区划 2 为交通集散地，区划 3 为市场，区划 4 为旅游景点。见表 3。

表 3 吐鲁番市各功能区划室外空气微生物浓度比较(CFU/m^3)

各功能区划	空气微生物浓度	
	平均值	变幅
区划 1	15 562.8	15 562.8
区划 2	52 269.0	34 112.4(69 325.2-35 212.8)
区划 3	33 994.5	89 132.4(97 149.6-8 017.2)
区划 4	24 994.8	46 531.2(57 090.0-4 558.8)

由表 3 可知，不同城市功能区划间的空气微生物浓度有差异，其大小排序为:区划 2>区划 3>区划 4>区划 1。即区划 2 人群密度大，人气旺，空气尘埃浓度高，空气流动慢，则空气微生物浓度高。相反，区划 l 空间宽阔，有一定绿化，人群密度稀，空气流动好，空气微生物浓度则小。但区划 3 所测的空气微生物虽不算最多，但变幅大，估计与人群活动、所在物质及流动状态关系密切。总体看吐鲁番市的空气多处于中度污染状态，而交通集散地(区划 2)的空气质量已达重污染水平。

3.1.3　吐鲁番市室外空气微生物浓度的时间分布

将所有室外测点按时间分为3个时段(上午、下午、夜间)进行分析。处于上午时段的测点包括6#、7#、8#、9#,下午是1#、2#、3#、10#,夜间测点为4#、5#。空气微生物浓度的昼夜变化状态见表4。

由表4可知,吐鲁番市室外空气微生物浓度昼夜变化状态为上午>下午>夜间。但各时段中的空气微生物浓度变幅并非如此,变幅大小排序为下午>上午>夜间,夜间的变幅为最小。上午的高浓度反映了气温和人群活动的渐升状态。下午的浓度反映出经中午高温之后的逐降气温及趋于平稳之势的人群活动,但其间却有较重的空气污染发生。夜晚的浓度则意味着最低气温的到来及人群活动的消退状态。严格地讲,应在同一测点作不同时段数据的比较,然而影响数据的因子很多。限于测定条件,仅将此测定得出本初步趋向,也大致可知吐鲁番市室外空气微生物浓度状态受制于当地的环境、天气变化及人群活动的重大影响。

表4　吐鲁番市室外空气微生物浓度时间分布状态(CFU/m³)

监测时段	空气微生物浓度	
	均值	变幅
上午	36 077.4	64 766.4(69 325.2-4 558.8)
下午	33 208.5	89 132.4(97 149.6-8 017.2)
夜间	26 959.8	16 506.0(35 212.8-18 706.8)

3.2　吐鲁番市室内空气微生物浓度状况

吐鲁番市室内空气微生物平均浓度为9 982.2 CFU/m³,处于较重空气污染状态。室内空气中也未见真菌。初步结果表明,在所测的普通旅馆中空气污染较重,已超过旅店业卫生标准(GB9663-88)(细菌总数≤4 000个/m³)[3]。

将室内外作比较可发现,室内平均空气微生物浓度比室外平均空气微生物浓度低了一个数量级,前者只及后者的32%。

4　结语

吐鲁番市初次空气微生物测点跨度较大,涵盖了吐鲁番市部分主要地区,因而其测定结果可以粗略地代表吐鲁番市(地区)的空气微生物状况。结果显示,吐鲁番市空气中少有真菌存在,其室外空气微生物浓度大致处于中度污染水平,有些地区空气微生物污染较重。空气微生物浓度存在明显的时空分布变化及差异。室内空气微生物平均浓度低于室外,但已处于较重的污染状态。测定结果既反映了吐鲁番市空气微生物特定地理环境、气候状况特征,也揭示了它与人类活动的密切关系。

参考文献7篇(略)

THE STATUS OF COCENTRATION OF AIR MICROORGANISM IN TULUPAN CITY

(ABSTRACT)

Abstract After primarily determined on air microorganisms in Tulupan city, the results show that concentrations of average bacteria (B), fungi (F), total bacteria (T) in outdoors and F/T% are 33,106.3, 0, 33,106.3 CFU/m^3, 0 respectively. It is suggested that outdoor air is in a state of general microorganisms pollution, air of some monitoring spots is polluted seriously. Cocentration of average bacteria in indoors is 9,982.2 CFU/m^3, which shows air pollution seriously indoors. Then space-time distributing status of air microorganisms, concentration indexes are analysed in this paper.

Key words Tulupan City; Air Microorganism; Air Pollution; Air Quality

灵山岛空气微生物含量的测定*

摘　要　用平皿自然沉降法测定灵山岛空气微生物及肠道致病菌。结果表明该岛空气微生物处于中-轻度污染状态，以轻污染为主，其中细菌为主成分，也含少量肠病菌。空气微生物显出某种时空变化势态，且与气温、人群活动多相关。估计该岛脆弱的生态环境易被外界干扰，应引起关注。

关键词　灵山岛　空气　微生物　污染　环境

引　言

灵山岛位于青岛灵山湾外，离胶南陆地仅 5.7 公里，面积约 7.66 km^2，系中国北方海区第一高岛，主峰歪头顶海拔 513.6 m。该岛 56 峰，峰峦景秀，远近看均似一幅幅山海作品。全岛没什么工业，居民不多，民房散布沿海山崖。有少量驻军。碎石山土路沿山而筑。金秋十月是该岛气候最佳之时，空气质量上乘。多少年来它虽水仙岛灵，但并不为外界多知。近些年来，青岛旅游的影响，越来越波及该岛，它已被辟为风景旅游区，其生态环境质量将会发生变化。空气微生物为空气一重要成员，其种类和数量的多寡及变化与人们健康、生态、环境息息相关，并越来越为大众所熟知[1-2]。本文就灵山岛空气微生物状况作一初步调查，以便社会各界对其状况及变化做到心中有数。

1. 材料与方法

将营养琼脂和沙门氏志贺菌属琼脂分别按要求制成平板，用以测定空气中细菌等微生物总数与肠道致病菌数。制好的平板按灵山岛大体的环境和功能状况曝露于空气中后，分别置于 25℃、37℃和 44℃温度中培养 2 天以上后计数各皿 CFU，并以菌落形态大体分类微生物。一式三份。按文献 2-3 换算为单位空气容积(m^3)中的 CFU 数，即个 CFU · m^{-3}，并以此评价该岛空气质量。测点环境概况列于表 1 之中。

表 1　灵山岛空气微生物测定时的天气环境概况

测点编号	测样号（NO.）	环境、天气概况			采样起始时间	备注
A	1	晴，太阳直射	近海	21℃气温，无风	2007.10.20，14:00	灵山韻海旅馆院内
B	2	晴，背阳	海滨	14.8℃气温，阵海风 4 级	2007.10.20，16:00	背来石公厕墙上，无人，臭味轻
A	3	晴，明月	小风	15.5℃气温	2007.10.20，18:00	灵山韻海旅馆院内
A	4	晴，明月	小风	16.0℃气温	2007.10.20，20:00	灵山韻海旅馆院内

* 原文刊于《海岸工程》，2008，27(4)：37-41，本人为第二作者。

（续表）

测点编号	测样号（NO.）	环境、天气概况			采样起始时间	备注
A	5	晴，明月	小风	14.9℃气温	2007.10.20，22:00	灵山韻海旅馆院内
A	6	晴，月下山，满天星		16.0℃气温，2级风	2007.10.20—10.21，24:00	灵山韻海旅馆院内
A	7	晴，月下山，满天星		14.8℃气温，1～4级风	2007.10.21，02:00	灵山韻海旅馆院内
A	8	晴，月下山，满天星		15.2℃气温，1～4级风	2007.10.21，04:00	灵山韻海旅馆院内
A	9	晴，月下山，满天星		15.0℃气温，1～4级风	2007.10.21，06:00	灵山韻海旅馆院内
C	10	晴，月下山，满天星		16.0℃气温，3～4级海风	2007.10.21，08:00	土路边平房顶上，尘土不多，人少走动
D	11	晴间少云		22.5℃气温，4级阵风	2007.10.21，10:00	海边蓝球场，灌木旁，杨树叶动，游人、汽车不多
E	12	晴，厚云轻雾		23.0℃气温，5～6级阵风	2007.10.21，15:00	游船与码头平，游客来去

*采样持续时间：5 min

2. 结果与讨论

2.1 总空气微生物状况

2.1.1 空气微生物检出率与含量

于2007年10月20日～21日在灵山岛选取5个测点12个时段测定的空气微生物含量状况列于表2之中。

表2 灵山岛空气微生物含量及大类组成状况

测点编号	测样号	空气微生物状况*														
		25℃					37℃					44℃				
		总量	细菌	霉菌	酵母	放线菌	总量	细菌	霉菌	酵母	放线菌	总量	细菌	霉菌	酵母	放线菌
A	1	10049.1	8815.0	176.3	1057.8	0	13751.4	9343.9	705.2	3526.0	176.3	14280.3	7404.6	0	6523.1	352.6
B	2	1763.0	1057.8	705.2	0	0	6875.7	4583.8	1057.8	1234.1	0	7404.6	4760.1	881.5	1763.0	0
A	3	2291.9	1763.0	352.6	176.3	0	6170.5	4583.8	352.6	1234.1	0	6346.8	4054.9	881.5	1410.4	0

(续表)

测点编号	测样号	空气微生物状况*														
		25℃					37℃					44℃				
		总量	细菌	霉菌	酵母	放线菌	总量	细菌	霉菌	酵母	放线菌	总量	细菌	霉菌	酵母	放线菌
A	4	2644.5	2291.9	352.6	0	0	8991.3	5641.6	1410.4	1939.3	0	8462.4	4760.1	1057.8	2644.5	0
A	5	528.9	528.9	0	0	0	4231.2	2291.9	881.5	881.5	176.3	5289.0	2291.9	881.5	1939.3	176.3
A	6	0	0	0	0	0	2291.9	1410.4	528.9	176.3	176.3	2115.6	1057.8	352.6	528.9	176.3
A	7	176.3	176.3	0	0	0	1939.3	0	1234.1	352.6	352.6	2468.2	0	1586.7	528.9	352.6
A	8	352.6	352.6	0	0	0	1763.0	1234.1	352.6	176.3	0	2644.5	1234.1	528.9	705.2	176.3
A	9	528.9	528.9	0	0	0	4583.8	2644.5	881.5	1057.8	0	4231.2	1939.3	528.9	1586.7	176.3
C	10	5817.9	5817.9	0	0	0	13575.1	9343.9	528.9	3702.3	0	9167.6	2997.1	0	6170.5	0
D	11	11106.9	10930.6	176.3	0	0	42488.3	39138.6	176.3	3173.4	0	30676.2	20098.2	0	9520.2	1057.8
E	12	1763.0	1763.0	0	0	0	20450.8	17982.6	528.9	1939.3	0	17101.1	9343.9	705.2	6346.8	705.2

* 单位:CFU · m^{-3}

由该表计算可见,该岛空气中微生物检出率高,起码是91.7%。25℃、37℃和44℃培养得出的空气微生物平均含量分别为3 085.3、10 592.7和9 182.3个 · m^{-3},表明37℃得出的空气微生物含量最高。就空气微生物总量看,该岛空气微生物污染大多处于轻度范围,少量处于中度状况,因而总体看并不严重[4],比庐山、青岛市区空气的低了许多,但重于衡山的[5-7]。

2.1.2　空气微生物类别结构组成

从表2也可得出,空气微生物大类别的结构组成,3种温度中培养得出的微生物均以细菌为主成员,至少占54.4%(44℃),最高占91.9%(25℃),其次是酵母,所占比例从3.3%(25℃)到36.0%(44℃),霉菌的份额变化不大,也不多,放线菌只在37℃或44℃中出现。虽然培养基成分在一定程度上筛选微生物,而且测时的空气湿度不大。本结果表明该岛空气微生物测时以细菌为主,细菌是空气微生物污染的主要成分。

2.1.3　空气微生物含量的时空差异

由表2可看到空气微生物含量在测点间存在的差异,按各测点空气微生物含量大小排序如下:D＞C＞A＞E＝B(25℃)、D＞E＞C＞B＞A(37℃)、D＞E＞C＞B＞A(44℃),这表明景点洁净的旅馆院内空气质量较好,而运动场所、交通要道等处空气质量较差。

表2中也表达出测时空气微生物含量的差异,比如,25℃时,测时差异如下:

10:00＞14:00＞8:00＞20:00＞18:00＞15:00＝16:00＞22:00＝6:00＞4:00＞2:00＞24:00＝0;

37℃时为:10:00＞15:00＞14:00＞8:00＞20:00＞16:00＞18:00＞6:00＞22:00＞24:00＞2:00＞4:00;

44℃时则为:10:00>15:00>14:00>8:00>20:00>16:00>18:00>22:00>6:00>4:00>2:00>24:00。这表明灵山岛空气微生物含量大多以中午前后为最高,午夜前后出现最低值。结果意味着气温是控制空气微生物的主因之一,且与白天人们的室外活动多相关,而微生物的组成是其内因。

2.2 肠道致病菌含量

表3 灵山岛空气肠道病原菌含量

测点编号	测样号(NO.)	测定起始时间	肠道病原菌含量*		
			25℃	37℃	44℃
A	1	2007.10.20,14:00	0	176.3	176.3
B	2	2007.10.20,16:00	0	0	0
A	3	2007.10.20,18:00	0	176.3	176.3
A	4	2007.10.20,20:00	0	0	0
A	5	2007.10.20,22:00	0	176.3	176.3
A	6	2007.10.20-10.21,24:00	0	0	0
A	7	2007.10.21,02:00	0	0	0
A	8	2007.10.21,04:00	0	0	0
A	9	2007.10.21,06:00	0	0	0
C	10	2007.10.21,08:00	0	0	0
D	11	2007.10.21,10:00	0	528.9	528.9
E	12	2007.10.21,15:00	0	176.3	176.3

*单位:CFU·m^{-3}

表3列出了各测时/点空气中肠道菌含量的检测结果。该表说明25℃时测不到空气中的肠道致病菌,意味着温度低的空气中少有可培养成活的肠病菌。37℃或44℃培养所得结果一致,其测时检出率为41.7%(5/12),肠病菌出现机率多以气温较高和人们室外活动多的白天为主,含量平均为102.8 CFU·m^{-3},其变幅为0~528.9。5个测点中有3个测得肠病菌,检出率为60.0%(3/5)。含量在测点间的大小排序为D>E>A>B=C=0,这意味着人走、车动多处的扰动空气将有机会蕴含多量肠道菌,与人体常规温度相近的空气培养条件将是容易得出肠道菌的一个条件,而高温比如44℃将是普通肠道菌不易忍耐、少数肠道菌可以繁衍的条件之一。

2.3 肠道致病菌与总微生物在含量上的比较

根据以上有关数据,可知肠病菌与总菌的检出率分别为41.7和>91.7%,两者的平均含量分别为102.8和10 592.7 CFU·m^{-3},肠病菌与总菌数量比率为1:102.9,说明灵山岛空气中有一定的肠道病原菌存在,但数量不大,所占份额也小。

3. 结语

本文对具有旅游前景/潜力的灵山岛空气用平皿沉降法做了微生物含量测定。初步结果表明,该岛空气微生物检出率较高,其含量一般在3 055.3~10 592.7 CFU·m^{-3}间,以细菌为主。证实其空气中微生物污染不重，空气质量尚佳。空气中含有为数不多的肠道致病菌，在空气总微生物中所占比率、份额小。温度试验表明，气温升高，空气微生物，包括肠道菌含量可能增大。该岛空气微生物显示出一定的时空分布差异及变化。迄今，气温、人群、环境及空气扰动状况是影响该岛空气微生物变动的四大主因。该岛不大，易受外界影响，生态环境脆弱，对其空气污染动向尤应关注。有关研究尚待深入。

参考文献 7 篇(略)

（合作者：孙丕喜　黄秀兰）

E6 交通(TRAFFIC MEANS)

中国“极地”号船的空气微生物*

摘　要　在“极地”号船于1986年10月～1987年5月赴南极暨环球航行的199d中，分7次对该船作了空气微生物监测。结果表明，该船舱内空气微生物含量的大体范围是97.5～4 323.0 CFU/m³，平均在1 000 CFU/m³左右。其分布状况大体上是上层较中、下层的少，霉菌的检出率约45.7%。霉菌占微生物总数量的17.4%。舱内外空气微生物数量变化受制于船只所处地理位置和季节，舱内空气微生物的变化有其特异性。这些结果基本上客观地反映了该船的空气状况。同时，环境洁净程度与人们的主观努力很有关系。“极地”号船尚有改善环境的余地。

关键词　“极地”船空气微生物　空气质量

1　引言

“极地”号船是我国唯一的万吨级极地科学考察船，1986年10月31日～1987年5月16日，它完成了赴南极考察暨环球航行的任务。在这艰苦而多变的长距离航海活动中，保持空气的洁净是保障所有人员健康正常地活动的重要条件之一。其中空气微生物含量是重要指标[5,8]。为此，监测“极地”号船舱内外空气的微生物状况显得有意义，这方面的工作，国内尚未见诸报道。本文论述这次考察中，“极地”号船的空气微生物测定和分析结果。

2　“极地”号船及其航行概况

“极地”号船总长152 m，型宽20 m，吃水7.6 m，型深10 m，续航力20 000 mile。实验室4间。该船船体分老区和新区两部。老区主要有航行和驱动设施，住舱和工作间设施较齐全，空气调节设备较好。新区包括新改建的住舱、实验室、文体活动厅、仓库等。空调设备系新安装的。

1986年10月底至1987年5月间，该船环球航行一周，总航程30 912n mile。总航时2 286h16 min(120d)。锚泊日63d。靠泊日16d。其中在南极佐治亚岛、南设得兰群岛海域计77d。

将该船分22个测区作了空气微生物监测。根据测区位置，将船体从上至下分做五

* 原文刊于《黄渤海海洋》，1994，12(1)：52～59。

层。详见表1。

表1　"极地"船空气微生物测定部位一览表

编号	测定部位	所在层次	所在舱区	一般描述
1	驾驶室	顶层	老区	空间较大,有值班人员,空气流通
2	大餐厅	一层	老区	空间较大,用餐,有时人多
16	16室	一层	老区	空间较大,一人住兼办公
44	44室	一层	老区	空间较大,一人住兼办公
56	56室	一层	老区	空间较大,一人住兼办公
3	医生室	二层	老区	二人住兼诊室
4	厨房	二层	老区	空间大,通风好,杂物多
5	船员A室	二层	老区	一人住
120	船员120室	底层	老区	四人住,通风较差
6	机舱	最底层	老区	空间大,气温较高
7	队长室	二层	新区	一人住兼办公
8	队员01室	二层	新区	一人住兼办公
9	手术室	二层	新区	极少有人进入
220	队员220室	三层	新区	四人住,空间较小
226	队员226室	三层	新区	四人住,空间较小
228	队员228室	三层	新区	四人住,空间较小
229	队员229室	三层	新区	四人住,空间较小
235	队员235室	底层	新区	四人住,空间较小
10	微生物实验室	底层	新区	一人,空间较大,通风差
11	文体活动厅	底层	新区	公共场所,空间大,杂物堆放
12	冰库	底层	新区	冰藏食品等,空间大
13	舱外	中甲板	新区	直接受天气影响
14	空白	对照	/	/

对"极地"号船共作了七次空气微生物测定,取得84个数据。每次测定时,该船所处地理位置及天气状况如表2所示。

3　材料和方法

表 2 测定“极地”号船空气微生物时该船所处地理位置等情况表

测时	所处地理位置	天气状况	备注
1986.11.31 13:00～14:00	南温带北部太平洋上,26°18′S,124°33′W	晴间阴,气温 22.7℃,风 NE6	春末夏初之交
1986.11.12.7 10:00～11:00	智利中部外航道内,36°04′S,73°40′W	晴,气温 16.5℃,风 WS7	船上较脏
1986.11.12.21 9:00～10:00	德来克海峡中南部,60°26′S,58°27′W	阴雾,小雨,气温 3.5℃,风 NE3	/
1987.1.26 8:00～10:00	南极半岛乔治王岛海区,62°18′S,58°23′W	阴,多云,气温 2.0 ℃,风 SW1	船上人少
1987.2.7 11:00～12:00	南极半岛企鹅岛海区,61°19′S,58°53′W	阴,气温 0.5℃,风 SE5	船上人少
1987.4.27 8:00～11:00	赤道 00°36′N,68°01′E	晴,气温 32℃,风 W4	船上人多
1987.5.13 8:00～10:00	福州以北海区 27°07′N,121°28′E	雨,气温 17.5℃,风 EN4	船上人多

3.1 培养基

1.细菌培养基:蛋白胨 10g;葡萄糖 10g;牛肉膏 3g;氯化钠 5g;琼脂 18g;蒸馏水 1 000cm^3。pH 7.2～7.4。

2.霉菌培养基(马丁培养基):蛋白胨 5g;葡萄糖 10g;KH_2PO_4 1g;$MgSO_4$ 0.5g;琼脂 18g;蒸馏水 1 000cm^3。pH 7.2～7.4。

3.2 采样和计数

将以上培养基分别倒入平皿。按不同测区作自然沉降试验[3,5],经一定时间(除特别说明外,一般不长于 10 min)暴露后的平皿置 37℃(霉菌)或 25℃(细菌)温箱培养 3～5d,计算每皿的菌落形成数(colony forming Unit,简写为 CFU,下同)。计算公式如下:

$$\text{CFU/m}^3(\text{每立方米空气中微生物菌落形成数})=1\,000\left(\frac{A}{100\div}\times t\times\frac{10}{5}\right)\times N$$

式中 A 为所用平皿面积(cm^3);t 为平皿暴露于空气中的时间(min);N 为经培养后各皿中的 CFU。

根据以上所得数据,比较分析不同测区空气中的微生物含量状况。

4 结果和讨论

表 3 列出各平皿培养基经自然沉降后培养得出的 CFU/m^3。

表 3 "极地"号船空气微生物含量表

测定部位编号 \ 测定时间	空气微生物含量(CFU/m³)							平均[a)]
	1986.11.23	12.7	12.21	1987.1.26	2.7	4.27	5.13	
1	448.0	652.4	/	275.1	/	283	652.4	462.2
2	172.9	102.2	/	180.8	/	39.3	157.2	130.5
16	/	/	/	/	/	/	1 729.2	1 729.2
44	/	/	/	196.5	/	/	2 475.9	1 336.2
56	/	11 790	/	67.6	/	1 037.5	98.3	3 248.4
3	408.7	1 257.6	/	314.4	/	338	157.2	495.2
4	110	2 774.6	/	707.4	/	/	314.4	937.3
5	3 104.7	2 515.2	/	510.9	/	817.4	1 611.3	1 452.5
120	1 344.1	1 469.8	/	1 265.5		416.6	746.7	1 221.5
6	393	180.8	/	0	/	62.9	982.5	276.7
7	/	/	/	165.1	/	338	628.8	495.2
8	448	683.8	/	70.7	/	3 222.6	78.6	1 010.8
9	133.6	157.2	/	0	/	117.9	/	97.5
220	/	/	4 323.0	/	/	/	668.1	4 323.0
226	1 359.8	841	/	55	/	1 218.3	/	828.4
228	/	/	78.6	817.4	/	/	/	448.0
235	/	1 807.8	/	353.7	/	1 218.3	1 179	1 139.7
10	495.2	1 674.2	/	0	/	550.2	628.8	669.7
11	943.2	707.4	/	117.9	/	157.2	78.6	400.9
12	/	/	/	/	/	/	0	0
13	/	/	212.2	330.1	275.1	0	39.3	171.3
14	/	/	/	/	/	/	0	0
平均[b)]	780.1	1 901.0	1 710.9	299.9		701.2	726.1	1 025.1

a)按测定部位得出的 CFU/m³ 平均值;b)除测定部位 12、13、14 按测定日期得出的 CFU/m³ 平均值。

表 3 反映出这 23 个测区内空气的微生物数量状况,按含量(CFU/m³)多寡排列(12,13,14 三个测区另作讨论),可见其次序是 220>56>16>5>44>120>235>8>4>226>229>10>3=7>1>228>11>6>2>9。这 20 个测区,空气微生物平均为1 071.7 CFU/m³。有七个测区(即 16,44,56,5,120,220,235),其空气微生物数均大于这一平均值。其余测区的则均小于这一平均值。其中,2,6,9 号测区的空气微生物数较少。2 和 9 测区的比甲板部的还少,这是因为 2 号测区空间大,只是用餐时才有些人去,风浪大或人

员少时就不启用了，所以空气较洁净；6 号测区人员和生活污物少，通风较好；9 号，难得有人进出，干净，但有时 9 号测区的 CFU/m³ 比 2 号的还多。作为手术室，必须严格保持高度洁净状态，才能应付不测事故，这是工作性质所需的[1]。

由表 3 可看出舱内六次测定数的比较，从时间上以 CFU/m³ 多寡排，则是 1986.12.7＞1986.12.21＞1986.11.23＞1987. 5.13＞1987.4.27＞1987.1.26＞，表明舱内以 1987.1.26 的最少，以 1986.12.7 的最多。总的看，除 220 号超标，56 号接近标准外，“极地”号船内空气微生物含量不算大，因而空气质量称得上处于良好水平[5]。

舱内空气比陆地空气污染轻得多[2,3]。

表 3 还列出舱外甲板上空气微生物含量的五次测定结果。以 1987.4.27 的最少，CFU/m³ 的平均值为 171.3。与中国一些海滨的空气比，如天津塘沽海滨空气的细菌含量为 1 380 CFU/m³[5] 比，则“极地”号船所测的海洋空气洁净得多。

表 3 列出的舱内外空气微生物数量差按船只所处的地理位置或季节差异分，可得出表 4。由该表可见，“极地”号船舱内外空气微生物数量在不同地理位置上的状况是：舱外，高纬度＞中纬度＞低纬度；舱内，中纬度＞高纬度＞低纬度。内外相比，舱内微生物数量普遍提高，数量差距拉大。舱内外空气微生物数量在不同季节时的变化势态则较为一致，均是夏（南极半岛附近。1986.11.23～1987.1.26）＞春（北半球，1987.4.27～1987.5.13）。由此可见，舱内外空气微生物数量变化势态大体一致：舱外空气微生物多，舱内一般也较多。综合不同纬度上的数量看，舱内是舱外的 9 倍多；综合不同季节的情况看，则舱内是舱外的 6 倍多。表明“极地”号船舱内外空气微生物均受纬度和季节的制约。其不尽一致外，则表达了“极地”号船舱内的特异性：即舱内空气质量主要还受人群活动和食品杂货的影响。如前所述，海洋空气相对来说总洁净得多。

表 4　不同地理位置或季节的“极地”船舱内外空气微生物数量比较*

“极地”船所处地理位置或季节	空气微生物数量(CFU/m³)	
	舱外	舱内
高纬度，约 60°S，58°W	272.5(三次平均)	1 005.4(二次平均)
中纬度，约 26°～36°S，73° ～124°W；27°N，121°E	39.3(一次)	1 135.7(三次平均)
低纬度约 0°	0(一次)	701.2(一次)
夏(1986.11.23～1987.1.26)南极半岛一带	271.2(二次平均)	1 173.0(四次平均)
春(1987.4.27～1987.5.13)北半球	19.7(二次平均)	713.7(二次平均)

* 据表 3 数据统计得出

表 5 列出了将“极地”船测得的空气微生物数，按舱位的不同层次各自平均所得的数据比较。由该表可见，顶层空气微生物数最小，只 462.2 CFU/m³；第三层的最多，达 1 039.1CFU/m³，即顶层的少于二、三层的，新区的比老区的少。这与各层次舱室的通风条件、卫生状况有关。在同一层次的新老区间的差别则主要由清洁程度包括新老状况决定。

表 6 反映出住舱、工作间空气微生物含量的差别。同时也表达了各层间的不同。一般言，住舱的空气微生物数在层次间差别明显，即上层较下层、底层的少。就工作间言，也

有同样趋势。住舱与工作间比，前者一般比后者的空气微生物数多(如第二层，住舱的是工作间的1.9倍多；在底层，住舱是工作间的1.8倍左右等等)。这是因为住舱内食杂品多，人们日常生活以此为据点，停留时间长，加上一些不讲卫生的习惯所致[7]。

表5 "极地"号船各层舱室空气微生物数比较

层别 \ 采样时间 CFU/m³	1986.11.23	12.7	12.21	1987.1.26	4.27	5.13	平均
顶层	448.0	652.4	/	275.1	283.0	652.4	462.2
二层	841.0	1 477.7	/	294.8	966.8	598.1	827.7
老(区)	1 207.8	2 182.5	/	510.9	577.7	209.6	937.7
新(区)	290.8	420.5	/	117.9	1 226.2	563.3	523.7
合	841.0	1 477.7	/	294.8	966.8	386.5	793.4
三层	1 359.8	841.0	1 710.9	436.2	1 218.3	668.1	1 039.1

表6 "极地"号船各层住舱和工作间空气微生物含量比较*

	层别	CFU/m³		层别	测定部位编号	CFU/m³
住舱	一	934.2	工作间	顶	1	462.2
	二	954.2		二	3	495.2
	三	1039.1		三	7	495.2
	底	1201.0		底	10	669.7

* 数据是根据表3算出，一层住舱的CFU/m³由1987.1.26、4.27及5.13的数据平均。

综上所述，船舱空气微生物含量在层次间差别明显，但这只是问题的主要一面；还要看到不可忽视的另一面，即空气微生物含量大的舱室，有的也出现在一、二层中。相反，有几个微生物含量小的舱室，能在底层中出现。这表明空气微生物状况除了受层次差别这个客观因素影响外，保持环境卫生，改善空气状况，改正不良生活、工作习惯是十分重要的主观因素。有了较好的客观条件，加上主观努力，就一定能创造和保持良好的空气环境。客观条件较差的舱室，经过人们的改进，同样能得到良好的空气质量。

再回到表3可以见到，第12号测区用本法一般测不出空气微生物，表明冷库中的空气洁净。

一些霉菌是导致疾病的因素，严重者可以致癌[4]。因而，必须注意"极地"号船上的霉菌状况。表7列出了各次测定中霉菌的检测情况。由表7可见，霉菌的检出率不低，达45.7%。以1986年12月7日的检出率为最高。这与该天空气中微生物数为最高相一致。因为当时船上进了大批水果、蔬菜等新鲜食品难以及时处理，船内外较脏，加上气温偏高，导致霉菌等微生物的检出率增高。霉菌的出现似与气温间有一定关系。表6还列举了当检出霉菌时，其含量占微生物数量的百分率概况。表明霉菌占有不小的比重，平均为17.4%，应引起注意。同时，这一百分率与室温间的关系也较为密切。经计算，其相关系数超过0.687($>r_{0.10}$)。因而，室温较高时，更要注意控制霉菌等的蔓延。

表 7 “极地”号船空气中霉菌检测统计

日期 / 项目	1986		1987			平均
	11.23	12.7	1.26	4.27	5.13	
测定次数	12	18	19	14	15	16.2
检出次数	7	16	3	7	4	7.4
	58.3	88.9	15.8	50.6	22.2	45.7
$\frac{\text{霉菌 CFU} \cdot m^{-3}}{\text{总 CFU} \cdot m^{-3}}\%$	35.7	5.4	4.4	32.5	9.0	17.4
室内气温(℃)	25.4	20.0	11.7	27.0	22.2	21.2

5 结语

通过以上的论述,可以看出南极考察暨环球航行中的“极地”号船舱内空气微生物含量,即一般在1 000CFU/m^3 左右,处于良好状态。上层舱室的较少;中、下层的较多;工作间的比住舱内的少。不时出现一些霉菌,其检出率约45.7%,霉菌占总菌数的17.4%。不洁的物品与较高的室温,促使霉菌增多。这是“极地”号船空气质量的一个客观反映。

结果还表明,舱内空气微生物数一般比舱外的多;舱内外空气微生物含量变化均受所处的地理位置和季节的制约。舱内有其特异性,较好的客观环境尚需人的主观努力,才能得以保持;客观条件较差的环境经过人们的不懈努力,也能得到较大的改善。就整个“极地”号船的空气微生物状况言,改进环境卫生状况,尚有很大余地。有了清新、洁净的空气环境,使全船人员身心健康,这将是持久工作的重要保证。

参考文献 8 篇(略)

(合作者:李振富　陈国纲)

MONITORING OF AIRBORNE MICROBES OF THE CHINESE POLAR RESEARCH VESSEL "JI DI"

(ABSTRACT)

Abstract During the 199 days of the voyage of R/V "Ji Di" to Antarctica and round the earth from Oct., 1986 to May, 1987, monitoring of air-borne microbes in the vessel was performed seven times. The results show that the content of air-borne microbes in the vessel ranged roughly from 97.5-4,323.0 CFU/m^3, averaging about 1,000 CFU/m^3. As to the quantitative microbial distribution, there were fewer microbes in the cabins on the upper deck than those on the middle and lower decks, and fewer in the working rooms than in the cabins. The detection rates of mildew amounted to 45.7% and the content of mildew to 17.4% of the total quantity of the microbes. The numeral change of air-borne microbes inside and outside of the vessel was conditioned by is geographical location and seasons. There was a particularity of change about the air-borne microbes in the cabins. Meanwhile, the purity of the environment was related to management and people's subjective efforts. So there is still room for "Ji Di" to improve its enviroment.

Key words Polar Research Vessel "Ji Di"　Air-Borne Microbes　Air Quality

客机内空气微生物对人体的影响初探*

乘坐飞机的人往往只注意客舱空气中的香水气味，而对空气中的微生物状况大都忽略。殊不知空气中的微生物却恰恰是客机空气质量的重要指标之一，它与旅客的健康、安全至关重要，数量多，必然使空气质量下降、使旅客感到不适乃至出现病症并随之传播开来。因为空气微生物中往往包含着使人生病的因素，如各种病原菌、病毒、立克次体等。它们导致呼吸道传染病，如肺结核、流感、隐性连球菌病等，引起过敏性疾病，如干草热、农夫肺、哮喘等。还有几十种动植物病患及物品霉烂变质也由它们造成。其中一些能经空气传播到几千公里以外。因而，监测客机的空气微生物状况十分重要。人们由此便可找到改善空气质量切实可行的方法。

空气微生物数量以每立方米空气中含多少个微生物为单位来表示，简写作 CFU/m^3。迄今，我国飞行的客机内空气微生物数量，仍沿用旧的标准，而国际间尚无统一标准。这里以国际长途旅行中监测到的客机空气微生物数量为例，说明这一问题。(表 1)。

表 1　飞行于国际航线上八架次客机内空气微生物数量表

客机代号	飞行区间	空气微生物数量 CFU/m^3	飞行时间	客机类型	备注
A	南极智利马尔什基地—智利彭塔·阿雷纳斯	733.8	1994.3.2	大力神	飞越德雷克海峡上空，客满
B	彭塔·阿雷纳斯—圣地亚哥	629.0	1993.3.4	波音 737	智利南部太平洋沿岸，乘客较多
C	圣地亚哥—圣保罗	52.4	1994.3.14	波音 767	南美大陆上空，乘客不足
D	圣保罗—多伦多	340.7	1994.3.14～3.15	波音 767	中美洲，洲际，大西洋沿岸上空，饭后
E	多伦多—温哥华	104.8	1994.3.15	波音 767	横穿加拿大上空，一半以上乘客
F	温哥华—东京	＜52.4	1994.3.15～3.16	波音 767	经阿拉斯加湾上空
G	东京—上海	524.1	1994.3.18	波音 767	飞经北太平洋上空
H	上海—北京	52.4	1994.3.18	波音 767	中国大陆，冬末

---

* 原文刊于《环境与发展》，1995，10(2)40 转 39

由该表计算出当今客机中空气微生物数量的大体范围,即平均为 311.2 CFU/m^3。不同客机内空气微生物数量不一。以其多寡排列,次序为 A>B>G>D>E>C=F。可将这八个航段上的客机空气微生物数大体归为三类,即最少的,它们是 F<C=H;中等的是,E<D<G;最多的,是 B<A。

造成这些差异是有原因的,如在 A、B 中,A 机是军用飞机,主要运输物资。客人多坐的是简易舱位,卫生条件较差。B 机较小,空间小、客人多,他们来自夏末秋初的不同地域。会携入较多的细菌。F、C 和 H 机则空间较大,乘客较少或处于冬令时节的北半球,空气微生物相对较少。

由此可见,国际旅行中客机内空气微生物含量常与机种、型号及其空间大小,客舱的通风、卫生状况,乘客来源和数量及其所带物品,飞行时节及区域等有较大关系,也与采样时间有关。各架客机空气微生物数量的差异反映了它们在卫生状况上的差距,同时也表明它们有改善卫生条件的余地,客人们也要洁身自爱,遵纪守法,不带入有助于空气污染的物品。

上述一些客机中的空气微生物数量虽然在旧标准以内,但随着现代化程度的加深和旅客们卫生要求的提高,严格的空气质量标准将会提出,届时,世人将会过上更舒适、洁净的空中旅行生活。

长江客轮空气微生物含量的测定*

我国船只每年输送的旅客量正在逐年增长。轮船的卫生状况不仅直接关系着船上的每个旅客,而且同时与每个船员也息息相关。船上的卫生状况之一的空气微生物既反映环境污染程度,又与人体健康密切相连。因而它是空气质量的重要参数。公共场所的空气微生物状况,一般人知之甚少。对众多船只中的空气微生物长期以来只做了很少的工作[1~3],而实际上它恰恰是航行船只的重要生态指标,也反映了公民的卫生意识。本文就申汉线"江汉二号"客轮空气微生物含量的测定结果作一报道,以期有助于改善水上客运的卫生状况。

以常规平皿自然沉降法选用细菌和真菌两种培养基[4],测定具代表性的舱室中空气细菌和空气真菌含量。以空气细菌、真菌含量之和代表所测部位空气微生物总量。单位是 CFU/皿,必要时或可换算为 CFU/m^3[5]。共测 5 个部位,即船员住室、二、三、四和五等舱室。最后对该船室内空气质量作出评价。

据我国的公共卫生标准,当客轮舱内空气细菌含量大于 4000.0 CFU/m^3 时,该空气为污浊空气。由附表换算(上同)得出的五等舱空气细菌含量是 4 558.8 CFU/m^3,已超标约 14.0%。其余舱室尚可。以其含量多寡排列,次序是五等>四等>三等>船员室>二等。室内空气真菌含量一般应比细菌含量低不少。本轮四等舱中的空气真菌含量为 6 288.0CFU/m^3。虽然卫生标准未涉及真菌,但估计已大大超出正常限度。这结果意味着四等舱舱内比较湿热和污浊。其余四个部位的空气真菌含量不高。各测点空气真菌含量排列次序为四等>五等>船员室>二等>三等。

表 "江汉二号"长江客轮舱内空气微生物含量表*

所测舱室	细菌		真菌		微生物总量(CFU/皿)	备注
	含量(CFU/皿)	占微生物总量的%	含量(CFU/皿)	占微生物总量的%		
船员室	15.0	75.0	5.0	25.0	20.0	两人
二等	12.0	80.0	3.0	20.0	15.0	两人
三等	19.0	95.0	1.0	5.0	20.0	满员
四等	22.0	35.5	40.0	64.5	62.0	满员
五等	29.0	80.6	7.0	19.4	36.0	十多人,不满员
平均	19.4	63.4	11.2	36.6	30.6	相对湿度 80%~90% 室温范围:28~31℃

*表中的 CFU 值乘以 157.2,即可换算成相应的 CFU/m^3;所用培养皿的直径约 9cm

* 原刊于《中国公共卫生学报 1996 年增刊》,19-20

附表列出的空气细菌、空气真菌含量占各次测定的空气微生物总量的百分数表明其高低次序分别如下，空气细菌：三等＞五等＞二等＞船员室＞四等；空气真菌：四等＞船员室＞二等＞五等＞三等。若空气细菌含量百分数大于空气真菌含量百分数越高，其空气质量越好的话，则三等舱空气质量较好。反之，四等舱的较差。总体上看所测结果，本轮空气中所含的真菌比率似过高。

附表还列出所测空气微生物总的含量状况。各部位的排列次序如下：四等＞五等＞三等＝船员室＞二等。因此，二等舱内所含微生物最少，其真菌比率也最低，空气质量最佳，四等舱空气质量最差。

平均地说，本船舱内空气细菌含量为 3 049.7 CFU/m^3，空气真菌含量为 1 760.6 CFU/m^3，空气微生物总量为 4 810.3 CFU/m^3。虽然细菌含量未超标，但空气微生物总量似过高。本船尚需改善空气微生物状况。

分析各部位空气微生物含量问题的差距，可见二等舱的空间虽小，杂物不多，相当洁净。三等舱与船员室空气微生物含量相同，表明两处空气质量类似。但船员室空间大，人员少，理应空气质量更好些。但由于通气较差，环境稳定，可能为一些微生物的滋生提供了机会，相比而言，三等舱环境尚可，五等舱位于船底，设施最差。人们不得已才住，乘客多，杂乱。所幸的是各室与通道间流通尚好，若设法经常注入新鲜空气，人人注意卫生，空气质量有望进一步提高。房间狭小又多的四等舱是全船重点部位。旅客多而乱，物品散杂，拥挤。不及时清理的脏物在较高温高湿中易腐变质，加上多量 CO_2、CO 浊气，使空气微生物含量较高，空气质量大为下降[2]。故须着力加强四等舱环境管理，以彻底改善空气的生态学环境。

由上可见，本客轮空气微生物平均含量偏高，其中四、五等舱尤为明显，大力改进四、五等舱乃至全船环境卫生状况是旅客和船员的共同责任。只有齐管共抓，才有可能使大家过上舒适的长途航行生活。

参考文献 5 篇(略)

“向阳红 09 号”调查船的空气微生物*

摘　要　本文对“向阳红 09 号”调查船做了空气微生物的监测，结果表明：平均全船舱内空气微生物总量、空气真菌含量、空气真菌量占空气微生物总量的百分数分别是 3 730.9CFU·m^{-3}，168.1 CFU·m^{-3}和 15.3×10^{-2}。尚未超过国家对客船规定的卫生标准，但有 21×10^{-2}的测定结果已超标。两次测定间的差异主要由空气扰动状况所致；层间及舱室间的差异是因人员工作生活习惯等造成的。结果还表明，较好条件的上层乃至全船舱室大有改善环境卫生条件的必要。

关键词　“向阳红 09 号”海洋调查船　空气微生物状况　环境质量

国家海洋局北海分局“向阳红 09 号”海洋科学调查船（简称“向九”，下同）是我国的一艘大型科学考察船，排水量 4 300 t，总长 113 m；舱室由上至下分为五层：一层为驾驶室，二层为船、队员住室和办公室，三层为实验室、餐厅和厨房以及部分宿舍，四层为仓库及少量住室，底层为水仓和油仓等。船中心从底层至三层为机舱部位，该船建于 1978 年，在大浪中经历了 18 个年头，迄今对其舱内的空气微生物状况一无所知，而空气微生物状况恰恰关系着该船所有人员的健康，进而影响着科学考察任务的执行。对我国海洋调查船、客货船空气微生物状况的调查是近几年才引人注意，并取得了一定的成果[1,2]。本文就该船的空气微生物含量作一报道，以促进我国类似的人为环境和空气质量的监测研究。

1. 材料和方法

用常规的平皿自然沉降法收集空气微生物菌落沉降量。培养基有两种：细菌培养基和真菌培养基。曝皿时间 10 min。细菌和真菌分别按要求培养达 48h，然后对各皿中的 CFU 计数，再换算为单位体积空气中的细菌数和真菌数（CFU·m^{-3}），并以两者之和代表总的空气微生物菌落沉降量，由此得出全船空气微生物含量状况并评价其空气环境质量[1]。

2. 结果和讨论

主要测定工作进行了两次，另外做了一次补充测定和船外空气微生物含量对照测定。表 1 列出了对该船的监测结果。舱内的空气微生物监测第一次共做了 24 个样品，其中一层 1 个，二层 11 个，三层 9 个，四层 3 个；第二次共做了 28 个样品，其中一层 1 个，二层 13 个，三层 9 个，四层 5 个。另外在二层补充监测 2 个样，在船外采对照空气样 3 个。

由表 1 可知，全船舱内空气微生物总量的平均值为 3 730.9 CFU·m^{-3}，空气真菌量平均值为 168.1 CFU·m^{-3}，空气真菌占空气微生物总量的 15.3×10^{-2}，与 1989 年前我国南极考察船“极地”号相比高出许多，达 3.7 倍以上[1]，真菌含量百分比有所下降。按国家卫生标准规定，客船舱内空气细菌数不得超过 4 000 CFU·m^{-3}，本船的细菌平均含量仍未超标[3]，但若加上真菌数则接近国标，而且所测舱室已有 21×10^{-2}(11/53)超标，必须指出的是“向九”船并非客船，舱内空气微生物状况理应要好很多。

* 原文刊于《黄渤海海洋》1997，15(2)：53-58。

表 1　"向阳红 09 号"船空气微生物含量一览表*

Table 1　Airborne microbial contents in the air of the R/V "Xiangyang hong No. 9"

舱室号	测次编号	空气微生物总量（CFU·m^{-3}）	空气真菌含量（CFU·m^{-3}）	真菌量/微生物总量（$\times10^{-2}$）	t℃	备 注
106	1	7 860.6	78.6	1.00	18.8	驾驶室
	2	3 458.4	471.6	13.6	20.0	
201	1	1 572.0	78.6	5.0	18.2	
	2	2 200.8	0.0	0.0	13.0	
204	2	235.8	157.2	66.7	9.0	
206	1	2 200.8	0.0	0.0	18.0	
	2	1 807.8	157.2	8.7	12.8	
209	1	2 279.4	78.6	3.4	19.4	
	2	31 597.2	157.2	0.5	12.4	
211	1	5 030.4	78.6	1.6	17.5	
	2	63 430.2	943.2	1.5	12.6	
217	1	3 301.2	235.8	7.1	18.0	
	2	16 584.6	78.6	0.5	12.8	
219	1	550.2	235.8	42.9	16.5	厕所
	2	4 794.6	314.4	6.6	5.3	
220	1	786.0	0.0	0.0	19.0	
	2	943.2	157.2	16.7	12.0	
221	1	5 816.4	235.8	4.1	17.0	
	2	157.2	0.0	0.0	13.0	
224	1	1 021.8	0.0	0.0	18.6	
	2	3 128.3	366.8	11.8		
225	1	550.2	235.8	42.9	19.2	
	2	1 021.8	0.0	0.0	12.0	
227	1	0.0	0.0	0.0	17.7	
	2	4 71.6	0.0	0.0	13.0	
236	2	1 807.8	78.6	4.3	5.0	
300	1	78.6	0.0	0.0	18.0	餐厅
	2	2 751.0	235.8	8.6	11.8	
301	1	628.8	78.6	12.5	17.5	厨房
	2	1 336.2	78.6	5.9	11.8	

(续表)

舱室号	测次编号	空气微生物总量 (CFU·m^{-3})	空气真菌含量 (CFU·m^{-3})	真菌量/微生物总量 (×10^{-2})	t℃	备 注
304	1	786.0	0.0	0.0	21.0	
	2	3 615.2	0.0	0.0		
306	1	2 436.6	235.8	9.7	18.5	
	2	550.2	0.0	0.0	17.6	
310	1	157.2	0.0	0.0	16.8	实验室
	2	786.0	235.8	30.0		
313	1	2 043.6	157.2	7.7	18.0	
	2				13.5	
317	1	471.6	78.6	16.7	18.0	厕所
	2	3 065.4	78.6	26		
327	1	157.2	157.2	100.0	16.0	实验室
	2	3 144.0	314.4	10.0		
333	1	157.2	157.2	100.0	24.0	机房
	2	628.8	78.6	12.5	21.5	
405	1	1 257.6	78.6	6.3	16.0	
	2	4 637.4	235.8	5.1		
407	2	235.8	157.2	66.7	9.0	
408	1	471.6	157.2	33.3	17.0	
	2	1 179.0	157.2	13.3	9.5	
409	2	6 523.8	0.0	0.0	13.0	
410	1	471.6	157.2	33.3	17.0	客厅
	2	2 829.6	235.8	8.3	12.1	
219	3	4 794.6	1 414.8	29.5	5.3	厕所
221	3	31 833.0	0.0	0.0	13.0	
01	3	7 860.0	1 179.0	15.0	8.5	驾驶室外
02	3	6 838.2	550.2	8.0	8.0	三层舷上
			4 087.2	1 021.8	25.0	
03	3	9 353.4	1 650.6	17.6	5.5	顶层外

* 曝皿时间 10 min；微生物含量均由 78.6N 算出；N 为各平皿长出的 CFU 平均值；备注中除说明外，大都是住室。

将表 1 的结果按舱层计算得出表 2。由表 2 可见，空气微生物总量及真菌含量第一次测定结果均小于第二次。主要原因是第一次测定时舱内空气平稳，离岸及城市较远，污脏颗粒少，第二次测定时许多舱室清扫卫生，空气中污脏颗粒较多，离岸近并靠近青岛，故沉降至平皿培养基上的微生物量多；第一次测定中的真菌含量百分率大于第二次，是由于第一次监测在东海海域，空气湿度较大，真菌相对含量较高。

空气微生物总量测定结果由高到低的次序是：第一次，一层＞二层＞三层＞四层；第二次，二层＞一层＞四层＞三层。空气真菌含量测定结果是：第一次，四层＞二层＞三层＞一层；第二次，一层＞二层＞四层＞三层。空气中真菌含量所占百分率是：第一次，三层＞四层＞二层＞一层；第二次，四层＞一层＞二层＞三层；从全船各层平均看：空气微生物总量是二层＞一层＞四层＞三层。真菌含量是一层＞二层＞四层＞三层；真菌含量占空气微生物总量的百分率是四层＞三层＞二层＞一层。与“极地”号船比，明显的差别是空气微生物总量在舱室层次上的混乱。在一般条件下，上层空气微生物含量应小于下层，而“向九”船则大体上是上层高于下层，这意味着那里的空气质量未必很好。相反三、四层空气的真菌状况表明它受到空气湿度的影响较多。

表 2　“向阳红 09 号”船空气微生物含量在舱层间的比较表

Table 2　Comparison of microbial contents among the airs over the various decks of R/V“Xiangyanghong No. 9”

舱层	测次编号	空气微生物总量		真菌含量		真菌量/空气微生物总量		t	
		CFU·m^{-3}	平均值	CFU·m^{-3}	平均值	（$\times10^{-2}$）	平均值	（℃）	平均值
1	1	7 860.0	5 659.2	78.6	275.1	1.0	7.3	18.8	19.4
	2	3 458.4		471.6		13.6		20.0	
2	1	2 100.8	5 980.5	107.2	146.3	19.5	14.2	18.1	14.9
	2	9 860.1		185.4		9.0		11.8	
3	1	768.5	1 376.6	96.1	107.0	27.4	18.1	18.6	18.8
	2	1 984.7		117.9		8.7		19.1	
4	1	733.6	1 907.4	131.0	144.1	24.3	21.5	16.7	13.8
	2	3 081.1		157.2		18.7		10.9	
整船	Σ_1	2 865.7		103.2		18.1		18.1	
	Σ_2	4 596.1		233.0		12.5		15.4	
	Σ	3 730.9		168.1		15.3		16.7	

将空气微生物含量以各室平均，其大小次序为：211＞209＞221＞217＞409＞106＞219＞405＞304＞224＞313＞206＞201＞236＞317＞410＞327＞306＞300＞301＞220＞408＞225＞310＞333＞407＝227＝204 室。因此空气微生物总量的大体差异是小舱室高于大舱室，住室高于实验室，厕所高于餐厅及厨房。全船空气微生物总量最高室是最低室的 145.2 倍（211 室/204 室），且均出现在二层上；第三层最高室是最低室的 5.6 倍（304 室/333 室）；第四层最高室是最低室的 125 倍（405/408 室）。经考察空气微生物含量高的舱室往往是人多或有吸烟者，关门堵窗空气不畅，或有不洁陈旧地毯等杂物，又不常打扫，甚至是恒温室等。因此值得人们注意从工作、生活习惯等方面加以改进。

对空气微生物总量等3个参数与相应的舱室气温作了相关分析1),结果表明虽以正相关为主,但相关系数不大,说明所测3参数受气温影响较小,即各测室气温变化不大。因此,本船空气微生物状况的变化主要受空气扰动状况的影响。驾驶室空气微生物含量较高可能与底面铺有地毯,人员多,流动性大有关;二层空气微生物较多主要因为住室人员和杂物多造成的;三层尤其是四层空气真菌量较高,是由于那里空气流动差,相对湿度大。按理一、二层的环境条件较好,空气微生物含量应较低,但监测结果并非完全如此,说明一、二层大有改善环境状况的必要。另外全船空气微生物状况还与人群状况(除空气流动程度外)有关,这里的人群状况主要指工作、生活习惯及卫生素质等。

舱内、船外空气微生物含量的比较可从表1得知。我国迄今建议的城市室外轻度污染空气细菌量为<5 000 $CFU \cdot m^{-3}$[4],若将船外的结果与第二次舱内的结果比较,可知舱内空气微生物总量(4 596.1 $CFU \cdot m^{-3}$)是国标的1.15倍,船外空气真菌含量(11 00.4 $CFU \cdot m^{-3}$)是舱内(167.6 $CFU \cdot m^{-3}$)的6.57倍,空气真菌含量所占百分率也是舱外大于舱内,船外空气细菌量达5 934·3 $CFU \cdot m^{-3}$,为轻度污染量的1.19倍,若加上真菌量则是1.41倍,这说明室外空气质量更差,这与船泊青岛港及工业区、风向有关。

3. 结语

本文监测了“向九”船的空气微生物状况,结果表明,全船舱内平均空气微生物总量,空气真菌含量及真菌含量占空气微生物总量的百分率分别为3 730.9 $CFU \cdot m^{-3}$,168.1 $CFU \cdot m^{-3}$和15.3×10^{-2},空气微生物含量尚未超过国家对客船规定的标准。空气微生物状况的3个参数在两次测定间差异较大,主要是由于测定时空气的扰动状况影响明显,而受气温影响较小。在舱室层次间,上层有些舱室虽有较好的空气流通和较低的湿度,空气真菌含量及其百分率稍低,但空气微生物总量较高。室间差异显示了室内环境的差异,空气微生物较多的舱室,主要与那里居住人员较多以及不良的工作、生活习惯等有关。微生物较少的舱室表明其环境较好,如207室空间较大,层次好,只住一人,室内无杂物,卫生条件很好,有一次未检出微生物,说明空气相当洁净。舱内、外空气微生物状况是互为影响,互为印证的。结果认为“向九”船大有改善其空气质量的必要。

参考文献4篇(略)

(合作者:王 波)

AIRBORNE MICROBES IN A MARINE SCIENCE RESEARCH VESSEL "XIANG YANG HONG NO. 9"

(ABSTRACT)

Abstract In this paper, the air in a marine science research vessel "Xiang Yang Hong No. 9" was monitored for airborne microbe content.

The results obtained show:

The average total counts of airborne microbes, airborne fungi and airborne fungi count/total count of airborne microbes % in the cabins were 3,730.9 CFU · m^{-3}, 168.1 CFU · m^{-3} and 15.3 $\times 10^{-2}$, respectively. The total count of airborne microbes has not exceeded the Chinese national standards of environmental hygiene stipulated for passenger ships yet, except about 21 $\times 10^{-2}$ of the data obtained. The differences between the two sets of the result of monitoring were mainly caused by disturbance of the air. The differences in the contents of airborne microbe in the air over different decks and cabins suggest that it was due to the environmental changes caused by the health state of the personal, their work and life habits, etc.

The results also indicate that there are great potentialities for improving the hygienic conditions of the environment of the upper cabins with better condition or even those of all the cabins in the vessel.

Key words "Xiang Yang Hong No. 9" Marine Science Research Vessel Airborne Microbes Environmental Quality

乌鲁木齐—布尔津间 216-217 国道沿线空气微生物含量*

摘　要　本文论述乌鲁木齐—布尔津间 216、217 国道沿线的空气微生物含量状况。采用平皿沉降法测定空气微生物含量。结果表明，平均空气细菌、真菌、总菌含量及 F/T%分别为 30 141.1、1 214.1、31 355.2 CFU·m^{-3}及 3.9。说明该公路测线空气已处于中度的微生物污染状态。文章比较分析了所测空气微生物含量指标的时空分布状态及差异，分析了造成此分布状态的相关原因。

关键词　乌鲁木齐　布尔津　国道　空气微生物　空气污染　空气质量

乌鲁木齐往返新疆北端布尔津之间的 216、217 国道所经地理环境多变，包括古尔班通古特沙漠、准噶尔盆地，其间不乏怪异地质地貌景观，如雅丹地貌、乌尔禾魔鬼城、火烧山等。测时天气由凉至热、干燥少雨。最高气温一般不超过 40℃。公路在广阔的沙漠地带、盆地中蜿蜒伸展。汽车在黑色带沙的路面上颠簸前行。到处是大漠荒沙，河流不多不大，绿色植物不多(只是到了城镇、交通集散地才具成荫绿树)，人烟稀少。

干旱少雨的环境，其空气因风沙而易散布各种污染颗粒物，虽然仅少量工业生产源。它们是气生微生物(airborne microbes)赖以生存的载体、传播的媒介。由于空气微生物与人体健康、环境、生态、军事等密切相关，世界先进各国就不同类型的空气的微生物已进行了大量研究[1-2]，但干旱、沙漠等环境的空气微生物研究较缺。本文就乌鲁木齐—布尔津间 216-217 国道沿线及相关城市的空气微生物状况作一初步监测，以引起人们对此类环境的空气状况的关注。

1. 测点描述

本公路沿线空气微生物监测在 2002 年 8 月间进行。行经路线如下：乌鲁木齐—昌吉—火烧山—北屯—富蕴—布尔津—合丰—乌尔禾魔鬼城—克拉玛依—呼图壁—乌鲁木齐。行程约 1 500 km。其间共设 14 个测点作了 16 次测定。测点、测时及环境状况详见表 1。

表 1　216、217 国道沿线空气微生物采样点状况

编号	采样点名称	环境状况	采样时间*
1	昌吉加油站	小风、有少量车来往	2002 年 8 月 19 日 12:00
2	火烧山	无人，气温>40℃	12.25
3	北屯市农十师加油站	车来往，灰尘大	20:30

* 原文载于《当代杰出管理专家人才名典Ⅱ》，P. 180-183，长征出版社，北京，2004.

（续表）

编号	采样点名称	环境状况	采样时间*
4	富蕴县卡库尔图镇(哈拉通克附近)	无风,附近在建房,32℃	16:30
5	离布尔津县城 20 km,饭店	风小,人不多,无车来往	23:00
6	去喀纳斯途中,离布尔津县城 100 km	空气清静,有些树木	8.20.09:45
7	离喀纳斯途中,返程约 200 km	树旁、三岔路口,3～4 级风	16:40
8	合丰县境,离布尔津 150 km	风小,有些车来往	8.22 09:30
9	魔鬼城,离乌尔禾不远	微风,干旱,很少人	11:00
10	克拉玛依—修车停车点饭店	车来往,有些食客,有喷水	13:00
11	离乌市 280 km 处	车来往,但飞扬尘土不多	15:45
12	离乌市 100 km 处,果蔬市场	车来往,尘土飞扬	18:00
13	布尔津县城边	车很少,有些食客,11℃	20.0:15
14	同上	车、人很少,10℃	20.6:30
15	同上	阴天,风沙较大,夹小雨,18℃	21.19:00
16	布尔津县城内闹市区	车、人很少,有点微雨,15℃	21.22:30

* 指采样起始时间,采样持续 5 min,用北京时间,当地时比北京时间推迟 2h。

2. 材料和方法

采用平皿自然沉降法。将消毒好了的营养琼脂平板和沙氏培养基平板分别在选定的上述测点地上 1.5 m 处曝皿 5 min。于室温中培养 72h 以上。计数平皿培养基上长出的细菌(B)、真菌(F)的菌落形成数(CFU),以 B+F 代表总菌落数(T),并换算为单位体积空气中的 CFU,即 CFU·m^{-3}。求出 F/T%。依上述空气微生物含量的四个指标评价测样所代表的空气环境质量,方法详见文献[3]。

3. 结果和讨论

3.1　216-217 国道沿线空气微生物含量的一般状况

表 2 列出了本次空气微生物监测获得的所有数据。由该表可计算出空气细菌、真菌、总菌及 F/T%各自的平均值分别为 30 141.1、1 214.1、31 355.2 CFU·m^{-3}及 3.9。据文献[3]所述,即细菌含量处于 5 000～50 000 CFU·m^{-3}的空气为中度污染,则可知所测沿线空气细菌含量表明测区空气正处于此状态中。空气真菌含量只及细菌的 0.04,虽然真菌加重了空气污染。说明空气微生物污染主要由细菌引起。这反映了测区空气干燥这一特点,但其平均的 F/T%比敦煌、太原、兰州的略高,而小于宝鸡的[4-7]。

表 2　216、217 国道沿线空气微生物含量指标统计 *

测点编号(No.)	空气微生物含量指标			
	细菌	真菌	总菌	F/T%
1	19 335.6	157.2	19 492.8	0.8
2	10 060.8	157.2	10 218.0	1.5

(续表)

测点编号(No.)	空气微生物含量指标			
	细菌	真菌	总菌	F/T%
3	23 737.2	314.4	24 051.6	1.3
4	36 627.6	1257.6	37 885.2	3.3
5	7 388.4	5187.6	12 576.0	41.3
6	5 959.2	2200.8	7 860.0	28.0
7	9 589.2	1257.6	10 846.8	11.6
8	19 335.6	1886.4	21 222.0	8.9
9	15 022.0	406.0	15 428.0	2.6
10	14 616.0	812.0	15 428.0	5.3
11	4 060.0	812.0	4 872.0	16.7
12	39 382.0	1624.0	41 006.0	4.0
13	14 462.4	1729.2	16 191.6	10.7
14	68 208.0	1218.0	69 426.0	1.8
15	188 384.0	0	188 384.0	0
16	6 090.0	406.0	6 496.0	6.3

* 微生物含量单位:CFU·m^{-3},下同

3.2 空气微生物含量指标的测点差距

由表 2 可知,测点空气细菌、真菌、总菌及 F/T% 大小排序。空气细菌含量的极差在 No.15 与 No.6 测点之间,达 182 424.8 CFU·m^{-3},最高含量是最低含量的 31.6 倍。空气真菌含量的极差在 No.5 和 No.15 之间,达 5 187.6 CFU·m^{-3}。空气总菌含量的极差在 No.15 和 No.11 之间,达 183 512.0 CFU·m^{-3},最高是最低的 38.7 倍。F/T% 的极差在 No.5 与 No.15 之间,达 41.3。

高于平均细菌含量 30 141.1 CFU·m^{-3} 的测点有 4 个,即 No.15,14,12,4,占全测点数的 25%。高于平均真菌含量 1 214.1 CFU·m^{-3} 的测点有 8 个,即 No.5、6、8、13、12、4、7、14,占全测点数的 50%。高于平均总菌含量 31 355.2 CFU·m^{-3} 的测点计 4 个,即 No.15、14、12 和 4,占全测点数的 25%。高于平均 F/T% 的测点有 9 个,即 No.5、6、11、7、13、8、16、10、12,占全测点数的 56.3%。

依空气细菌含量高低,可分测点为三类,Ⅰ类为 No.1,它的空气细菌量在 5 000 CFU·m^{-3} 以下,加上真菌,其空气仅处于轻度空气污染状态,此类测点只占全测点数的6.3%。Ⅲ类包括 No.15、14,其空气含量已超过 50 000 CFU·m^{-3},指示其空气达重污染水平,其测点数占全测点数的 12.5%。其余 13 个测点空气细菌含量处于 5 000~50 000 CFU·m^{-3} 之间,为中度空气污染状态,即Ⅱ类。若将空气真菌含量在 0~1 000 CFU·m^{-3} 之间的测点空气作为轻度真菌污染,1 000~11 000 CFU·m^{-3} 为中度真菌污染,>10 000 CFU

$\cdot m^{-3}$为重真菌污染的话，则可知本测区轻度真菌污染空气测点有 No. 1、2、3、9、10、11、15、16 八个，占 50.0%。中污染有 No. 4、5、6、7、8、12、13、14 八个测点，也占 50.0%。表明测区空气受到轻度—中度真菌污染。证实真菌含量与天气状况相关这一看法[8]。总菌污染级别分布如同细菌，仍然表明沿线空气微生物污染以细菌为主。

测定中有一现象十分明显，即 No. 15 的空气细菌含量最高，却未测到真菌。真菌含量高处，细菌含量相对较低，如 No. 5，6，3 测点。似乎这些测点与离城市近、绿化率高、湿度大，使真菌量所占%提高有关。

216 国道在东，南北走向。217 国道在西，北南走向。216 国道沿线有 No. 1～No. 6 六个测点，217 国道沿线有 No. 7～12 六个测点，它们的平均的空气微生物含量四指标比较结果如表 3 所示。

表 3　216 与 217 国道沿线空气微生物含量的比较*

国道名	空气微生物含量指标			
	细菌	真菌	总菌	F/T%
216	17 184.8	1 545.8	18 730.6	8.3
217	17 000.8	1 133.0	18 133.8	6.2

* 指两道各测点各指标的平均值。微生物含量单位：$CFU \cdot m^{-3}$，下同。

表 3 的这一比较表明，两条国道空气微生物污染状态相似，均处于中度污染水平，而以 217 国道稍轻。这是两条国道所处环境类似、测时天气状况相似所致。

No. 6～7 和 5～16 四测点(时)为布尔津县城空气微生物测点，其细菌、真菌、总菌及 F/T%的各自平均值分别为 52 505.6、966.1、53 471.7 $CFU \cdot m^{-3}$及 1.8，表明该县城的空气微生物含量已达严重污染程度，且主要由细菌引起。平均的真菌污染仅达轻度水平。No. 15 测点(时)过高的细菌污染造成了它的空气污染加重，这是由测定时的大风沙挟小雨所造成的，No. 15 空气细菌含量突高。可以想象沙尘暴天气带来多大的空气污染！细菌少的空气同样显出真菌及其所占%的提高。

表 4　乌鲁木齐、布尔津空气微生物含量的比较

城市名	空气微生物含量指标							
	细菌	极差	真菌	极差	总菌	极差	F/T%	极差
乌鲁木齐	28 583.8	80 643.6	780.5	1 572.0	29 364.3	80 486.4	2.7	14.1
布尔津	52 505.6	182 424.8	966.1	2 200.8	53 471.7	181 888.0	1.8	28.0

布尔津县位于新疆北端，离乌鲁木齐跨 4.5 个纬度，直线距离约 270 km，将两地各自平均的空气微生物含量指标作出的比较列于表 4 之中，可以看出两地空气微生物含量指标间存在一定差异。总体看布尔津的空气处于严重污染状态，除 F/T%较小外，其余各项均是布尔津大于乌鲁木齐，乌鲁木齐空气则属于中度污染水平，布尔津空气微生物指标的大极差说明它的空气环境参数变化大于乌鲁木齐的，而且不稳定性也高。

3.3　空气微生物含量指标的时间差异

表 5 216、217 国道沿线空气微生物含量的时间差异

测样时段	空气微生物含量特征			
	细菌	真菌	总菌	F/T%
清晨	68 208.0	1 218.0	69 426.0	1.8
上午	14 054.9	936.6	14 991.5	6.2
下午	55 608.6	990.2	56 598.8	1.7
夜晚	12 919.5	1 909.3	14 828.8	12.9

表 5 列出的是 216、217 国道沿线测点空气微生物含量指标在时间上的变化。这是将测时按清晨、上午、下午和夜晚四个时段划分微生物指标而得出的结果。由该表可见各指标按大小排序分别为细菌:清晨＞下午＞上午＞夜晚，真菌:夜晚＞清晨＞下午＞上午，F/T%:夜晚＞上午＞清晨＞下午。总菌相同于细菌。测时的早晨空气较静，易于细菌粒子沉降，湿度和温度不高，不利于真菌生长繁殖，其含量及所占%降低。反之，测时的夜晚空气温度虽逐降，但较平稳，且有一定湿度，利于真菌生长繁殖，平均地看，此间的真菌含量增大。上午逐升的气温，较小的湿度及阳光的照射不利于真菌生长繁殖，平均地看此间真菌含量变小。

4. 结语

对 216、217 国道沿线空气微生物含量作了初步测定，结果表明，空气细菌等微生物含量指示该测区空气处于中度污染水平，表明其空气质量大多不算很好。比较分析了公路沿线及两端城市空气微生物含量各指标的时空分布变化及其原因，估计和讨论了空气微生物与空气质量的关系。

参考文献 8 篇(略)

AIRBORNE MICROBIAL CONTENT ABOVE ALONG NO. 216、217 STATE HIGHWAYS BETWEEN URUMQI AND BUERJIN

(ABSTRACT)

Abstract This paper expounds the content condition of air-borne microbes along No. 216, 217 state highways between Urumqi and Buerjin. The air-borne microbial content was determined using the method of gravity plate. The result obtained indicates that the average air-borne bacterial, fungus, total microbial counts and F/T% were 30,141.1, 1,214.1, 31,355.2 CFU · m^{-3} and 3.9 respectively. It illustrated that the air environment determined had been suffered middle pollution from microbes. The paper compared and analysed the condition of spatial and temporal distributions of air-borne microbial content. It also analysed the relative causes for the condition.

Key words Urumqi Buerjin State Highways Air-Borne Microbes Air Pollution Air Quality

第六篇

海洋地微生物学

MARINE GEO-MICROBIOLOGY

F_1 普通微生物
(GENERAL MICROBES)

GEOGRAPHICAL DISTRIBUTION OF GENERAL AEROBIC HETEROTROPHIC BACTERIA IN SURFICIAL SEDIMENTS FROM THE CHUKCHI SEA AND CANADIAN BASIN*

Abstract This paper determined the abundance of general aerobic heterotrophic bacteria (GAB) in surficial sediment from the Chukchi Sea and the Canadian basin by using MPN and discussed their geographical distribution. The result shows that the determination percentages of the GAB were high, even till 100 percentage. The abundance range and averages of GAB for 4℃ and 25℃ were from 4.00×10^2 to 2.40×10^6, 1.71×10^6 ind $\cdot g^{-1}$(wet sample) and from 2.40×10^5 to 2.40×10^7, 1.10×10^7 ind $\cdot g^{-1}$(wet sample) respectively. Not only the abundance range but also the averages of GAB in 25℃ were higher than that in 4℃. The abundance of GAB in sediments shows a tendency that it is roughly greater in the lower latitudinal area than in the higher latitudinal area. The abundance of GAB increased from east to west as for the longitudinal distribution. With the water depth increasing, the abundance of GAB at 4℃ decreased, but GAB at 25℃ is not changed obviously with water depth. It seems that warmer circumstantial temperature is more suitable for some GAB.

Key words The Arctic Ocean, The Chukchi Sea, The Canadian Basin, General Aerobic Heterotrophic Bacteria (GAB), Geographic Distribution.

1. Introduction

Although the Arctic Sea area is permanently cold, there are prominent general aero-

---

* The original text was published in《Chinese Journal of Polar Science》, 2007, 18(2): 147-154.
原中文稿刊于《极地研究》,2007,19(3):231-238.

bic heterotrophic bacteria (GAB) distribution in the Arctic seawater and sediment similarly to other oceans. The cosmopolitan of GAB, its large magnitude and biodiversity make GAB take an important role in characteristic forming of environment including the Arctic area. The importance of GAB in environment includes material transportation, energy transition, cycle of ecosystem, rehabilitation of destroyed-envirorment, supply for human as resource of sustainable utilization and so on. Marine surficial sediment, as the interface between bottom water and sediment, has a twofold request form marine bacteria, i. e. it requests marine bacteria not only adapt to bottom water, but also to oxygen-limited sedimental environment, With respect to the Arctic Ocean, it would be adapted permanently cold environment. The Arctic sea nourishes many adaptive bacteria species, including lager amount of GAB, and is not a "desert" as described before[1].

The studies on bacterial content in the Arctic Ocean can be traced back to early or late of last century, such as Kaneko et al., who researched bacteria in the Arctic, subaretic and Beaufort Sea[2,3,4]. During the last decade, such as Sahm et al, Rysgaard, Hagstrom et al., Soltwedel et al., Ravenschlag et al., Junga et al., Ye De-Zan et al., Chen et al. studied marine bacteria in different part of the Polar ocean[5-12]. The improvement of research methods has been made by many researchers[8,10,13,14]. Researchers analyzed the distribution characteristics and behaviors of GAB in marine sediments in different aspects, which included abundance, geographic distribution, adaption to temperature, activity, biomass, structure of community, diversity, productivity, connection with environmental organic matter, mineralization etc. Research areas were distributed all over the Arctic area, which brought in some production of understanding more about the condition of the Arctic seabed microbe. But correlative research in China is very limited[12].

In this paper, the occurrence percentage and abundance of GAB in surface sediment samples of Chukchi Sea and Canadian basin are analyzed by MPN for a geographic research of bacteria, and also discuss the ecologically geographic distribution of GAB.

2. General in formation of research area

Research area is located in Chukchi Sea and Canadian basin (66°—80°N, 148°—170°W) and near the side of the Asia-America continent, which is more extensive than that of first Chinese Arctic Research Expedition. Because of the influence of Bering Strait, sediments in research area contain rich nutrition matter. Selected 24 samples from 43 samples for bacteriological analysis, and location fo sampling stations can be seen in figure 1.

3. Sampling

Samples were collected by Xuelong expedition ship during the second Chinese Arctic Research Expedition from July to September, 2003. After being taken on board the ship, surface 0～1 cm samples were collected immediately and stored in icebox for microbiological analysis in the domestic laboratory.

4. Analysis in the domestic laboratory

Microbiological analysis was started in January 2004. The liquid culture medium plates used were ZoBell 2216E plates, the specific components and preparing method were as those as in referenced paper[15]. Selection of culture temperature followed the study result of first Chinese Arctic Research Expedition[12]. Based on MPN, 1g from each wet sample was taken quantificationally and diluted to specified concentrations series, respectively[16]. The samples were then cultured in plates for over 3 weeks at 4℃ and 25℃, respectively. During culture period, samples were observed at intervals to ensure turbidity or precipitation/bacteria film appeared in tubes as positive. Content calculating was based on GB/T4789. 3-2003[17], and the content as well as the occurrence percentage of GAB were calculated, their distributions were discussed.

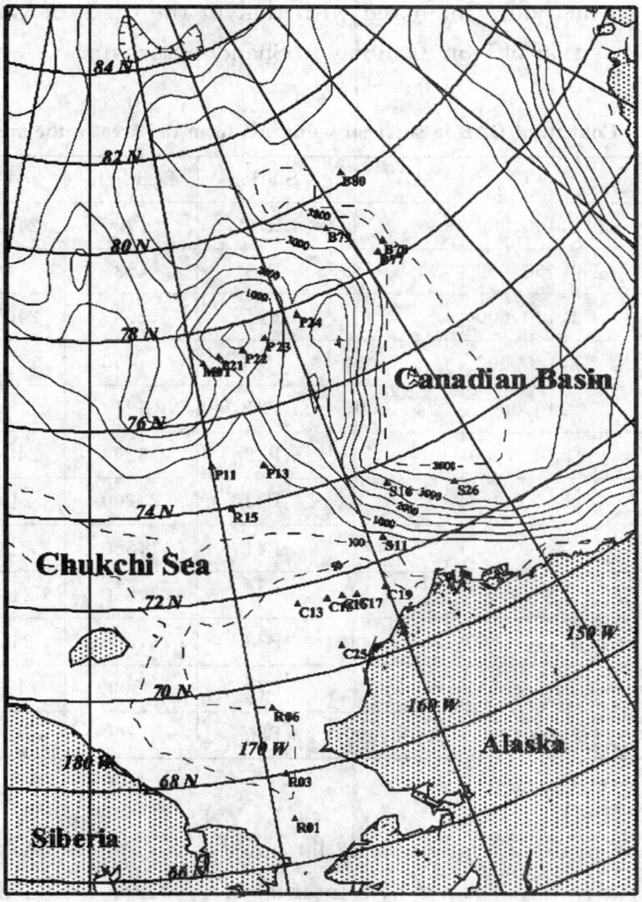

Fig 1 Sketch of sampling stations for GAB in surficial sediments from the area surveyed in the Arctic Ocean

5. Results and Discussion

5.1 Content and Occurrence Percentage of GAB

GAB content at 4℃ and 25℃ are listed in Table 1. It can be calculated by using the data in Table 1 that the overall occurrence percentage of GAB in the sediment samples at all survey sites is 100%, which implies GAB are fairly ubiquitous within our study area and maybe in the whole Arctic ocean. The statistical results show that the overall average abundance of GAB at 4℃ is 71×10^6 ind · g^{-1}(wet) with a range from 400×10^2 ind · g^{-1} to 40×10^6 ind · g^{-1}, and the overall average abundance of GAB at 25℃ is 1.10×10^7 ind · g^{-1}(wet) with a range from 2.40×10^5 ind · g^{-1} to 240×10^7 ind · g^{-1}. Not only content range, but also average abundance at 25℃ are greater than that 4℃. GAB content of this study is greater than that of first Chinese Arctic Research Expedition[12], but is lower than that of Ravenschlag[9] which was carried out along Svalbard island using molecular biological method. Compared with that in the cases of China seas, the only GAB data of which is that of Sang-Gou bay in Shandong province[18], the results are lower.

Table 1 Content of GAB in surficial sediments from the area in the arctic Ocean

K Station	Depth/m	4℃	25℃	Station	Depth/m	4℃	25℃
R01	50	2,400,000		R15	175	240,000	
R03	55	2,400,000		P13	453	440	
R06	53	2,400,000		P11	263	240,000	24,000,000
C25	41	1,100,000		M	1,456	240,000	240,000
C19	50	240,000	240,000	P21	561	400	
C17	46	24,000,000		P22	326	240,000	
C16	43	240,000		P23	2,200	24,000	
C15	42	2,400		P24	1,880	2,100,000	2,400,000
C3	44	240,000		B77	3,850	24,000	
S11	50	2,400,000	2,400,000	B78	3,800	24,000	
S26	3,000	13,000		B79	3800	110,000	
S16	3,800	2,400,000	2,400,000	B80	3,750	70,000	24,000,000

* unit ind · g^{-1}(wet sample)

5.2 Variation of GAB content along latitude

Divided study area into 4 zones by a interval of 4, which included 66°～70°N (3 stations), 70°～74°N(10 stations), 74°～78°N(7 stations) and north of 78°N (4 stations). Then the figures of GAB content distribution along latitudes were drawn (fig 2). Fig 2 shows that at 4℃ GAB content decreases with latitude increase, which is consistent with that of First Chinese Arctic Research Expedition[12]. The data at 25℃ is few, with 3 stations between 70°—74°N, 3 stations between 74°～78°N and 1 station farther north than 78°N. The distribution of GAB content at 25℃ with in 70°～74°N and 74°～78°N are similar, and maximum of 3 latitude interval are fitly same, which is 240×10^7.

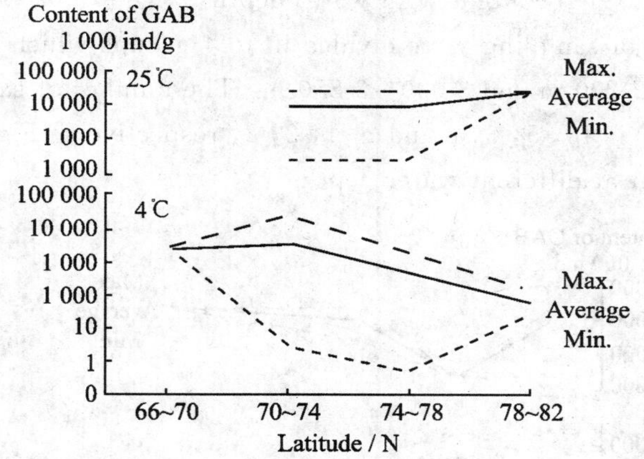

Fig 2　Comparison on content of GAB in surficial sediments among latitudes surveyed in the Arctic Ocean

5.3　Variation of GAB content along longitude

As for longitudinal distribution, the research area was divided into 5 zones by a interval of 5. The first two zones 145°～150°W and 150°～155°W with in Canadian Basin, including 5 stations. 155°～160°W zone with in the continental slope of Canadian basin, distributes 3 stations • 160°～165°W with in eastern part of Chukchi sea has 7 stations, and 165°～170°W with in the middle part of Chukchi sea accounts for 9 stations. Variations of GAB in each longitude zones is show that at 4℃ GAB content decrease from west to east The number of samples at 25℃ in 5 longitude zones from west to east is 2, 1, 3, 0, 1. Nevertheless, GAB content in 5 longitude zones is vary lesser, except station C19 (240×10^5 ind • g^{-1}) with in the east part of Chukchi sea where the GAB content is relatively low.

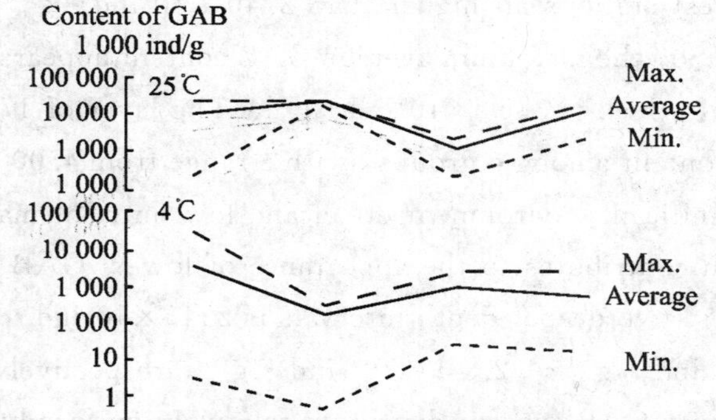

Fig 3　Comparison on content of GAB in surficial sediments among longitudes surveyed in the Arctic a Ocean

5.4 Variation of GAB content with water depth

All water depths of sampling were divided in to 4 groups which include 41～55 m, 175～561 m, 1 456～2 200 m and 3 000～3 850 m. The number of samples with in each group at 4℃ and 25℃ is 10, 5, 3, 6 and 2, 1, 2, 2, respectively. Figure 4 shows distribution of GAB content at different water depth.

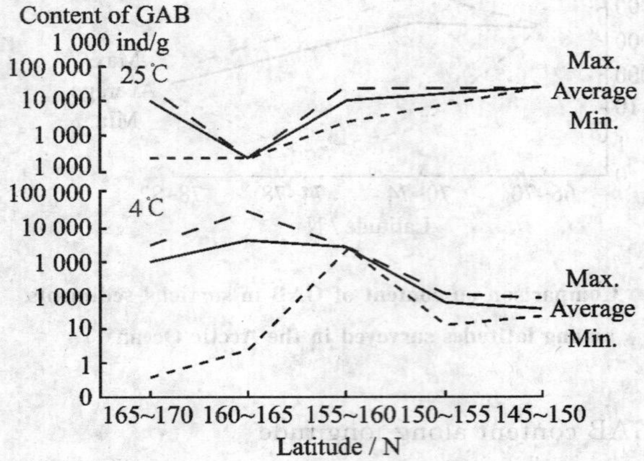

Fig 4 Comparison on contents of GAB in surficial sediments among different water depths in the Arctic Ocean.

As showed in figure 4, at 4℃ average content of GAB descend jumpily with water depth, which present as maximum-low-high-lowest Minima of GAB content change as low lowest-highest-high, and maxima of GAB content vary as highest-lowest-high-higher. How-ever, all conditions at 25℃ run in opposite sequence.

As a result of larger effect from environment and nutrient matter with in continental shelf, at 4℃ GAB content vary observably, with maximum (240×10^7 ind · g^{-1}) appears at station C17. The rest stations all higher than 2.40×10^5 ind · g^{-1} except station C15 (240×10^3 ind · g^{-1}), So, the maximum average GAB content appears with in the shallowest continental shelf, which is 3.54×10^6 ind · g^{-1}. The layer of 175 m～561 m depth has the lowest GAB content among 4 groups, with a range from 4.00×10^2 ind · g^{-1} to 2.40×10^5 ind · g^{-1}. Significant environment effect and low nutrient matter supply may be the main reason which contributes to the appearance of lowest GAB content. The GAB content with in the rest two deeper depth intervals is 2.40×10^4 ind · g^{-1}～2.10×10^7 ind · g^{-1} and 1.30×10^4 ind · g^{-1}～2.40×10^6 ind · g^{-1}, respectively. Because of little effect from environment, GAB content distribute relatively uniformly.

At 25℃, only one data located with in the depth interval 175 m～561 m, and its value is 2.40×10^7 ind · g^{-1} which is equal to the maxinum appears in interval 141～55 m and interval 3 000～3 850 m, but is one-magnitude-order larger than the maximum of interval 1 456～2 200 m. The rest 3 depth intervals have 2 data, and the minimum of the 2 data is

2.40×10^5 ind • g^{-1} which appears both in interval 141～55 m and in 1,456－2,200m. So, it is concluded that 25℃ GAB content has weak or no relation with water depth, and the GAB content in interval 1 456～2 200 m (continental slope) is relatively low.

In conclusion, control factors on GAB content of surface sediments within the Arctic Ocean are complicated, because GAB content does not consequentially vary with water depth.

5.5　temperature effect on GAB content

Samples were cultured at 4℃, 15℃, 25℃ and 28℃ separately during the research in First Chinese Arctic Research Expedition[12]。Experiment indicates that most of GAB are favorable for living at lower temperature and about one third of GAB are suitable for living at higher temperature (25℃ to 28℃). In the study, only 4℃ and 25℃ were selected, then not only content range, but also average abundance at 25℃ are greater than that at 4℃, and the results are greater than that of first Chinese Arctic Research Expedition. More strict sampling and preservation method were adopted in this study. In addition, the results also show the interaction between bacteria and temperature, namely, temperature choose bacteria and bacteria select the adaptable temperature. It is shown in the experiment results that the bacteria in the Arctic sea area have considerable adaptability to the environmental temperature. Surface sediments in our study area contain large amount of psychrophilic bacteria, and also include some mesophilic bacteria.

Unlimitedly, global climate change and ocean flow swarming caused by excessive carbon emission also influence Arctic Ocean, although it has a huge self-adjusted ability. This study has a clear elevation on temperature of bottom water and salinity compared with that of first Chinese Arctic Research Expedition, and maximum, minimum get a rise, too[19]. These changes of environment certainly will impelmicrobe to adjust themselves to adapt new environment. Water temperature affects life activity and metabolic rate of bacteria directly, and then affects the biogeochemical processes. It may imply that the environment changes of the Arctic Ocean must be thought about much more.

Based on above analysis, it may be concluded that GAB is mainly controlled by material sources and ecological environment. Organic matter brought in through Bering Strait, by the east Siberia coast current and Mackenzie River is the main factor in restricting GAB distribution. Water depth, water temperature and genus of microbe also influence GAB distribution. Relative researches remain to be further studied.

Acknowledgements: This study was sponsored by the nationality from finance. And as a part of the second Arctic research expedition (CHINARE-2003) which is organized by Polar Expedition Office of State Oceanic Administration of China, we would like to thank the crew of RV Xue long for their support and assistance. Research funds are supplied by National Natural Science Foundation of China (No. 40576060, 40376017 and

40176017). We thank all these assistance faithfully.

References19(abr.)

Cooperator: Gao Aiguo

漫谈海洋地微生物学*

摘　要　文章由地微生物学谈起，论述海洋地微生物学的研究历史及难度等。讨论海洋地微生物学内容及其环境的特点、研究现状及方向。文章寄希望于海洋环境、地质等地球科学家与微生物学家等的共同努力、密切配合，大力开拓我国的海洋地微生物学研究。

关键词　微型生物　海洋沉积物　海洋地微生物学

1　海洋地微生物学的提出、简史与研究难度

地微生物学（Geomicrobiology）是一门边缘科学，首先它是地球科学与微型生命科学之间的交叉，是探讨不同地质年代沉积物、岩石等地质环境中的微型生物及其在地球化学活动和重要元素循环中作用的科学。于20世纪末才形成为地球科学和生命科学间相互融汇的一支崭新学科。它不同于土壤微生物学的研究领域。

地表是大气圈、水圈、生物圈、岩石圈相互渗透、结合最为复杂的区域。鉴于技术的限制，当今地微生物学的研究范围应是这一地质表层包括风化和沉积层及以下的岩石圈和邻近水圈内地球环境与微生物间的交叉/边缘学科。地表的71%为海水所覆盖，因而，地微生物学研究的主要对象首先应该是海洋沉积物（海底地表均厚6 km）及其下岩石中的微生物，这便是海洋地微生物学（marine geomicrobiology）。这一课题最早可上溯到19世纪30年代[1,2]。可是由于海洋的上覆水最厚之处已超万米，人们要从陆地进入海洋，再从海洋表面向深水之下的沉积物、岩石作研究是相当困难的事情，更不用说要探索底层沉积物及其下岩石中微生物的动态。迄今为止，世界向沉积物的钻探活动所达的最深之处也就12 262 m[3]，这些活动对于地球这一庞然大物而言是微不足道的。而当今海洋地微生物学的研究人员数量及研究力度比陆地的要薄弱多了，然而，人们对地球/海底的探索却是从没停止过。

对海洋地微生物进行研究在20世纪40年代前后一直存在争议，争议之点主要有二：一是已从几千米深处沉积物中发现并培养出细菌等微生物，一些人们怀疑它们是钻探器材（具）带入的或是污染进去的，还是原有的；二是深部沉积物一般处于高压、缺氧甚至无氧环境，其微生物处于什么生存状态，是处于长期休眠状态，还是一直在不断繁殖之中，是一类或几类什么生理类型的微生物，即耐压、厌氧等或是人们未知的生存/生理状态[4]。当然此类问题有的已找到了答案[5]。如今的现代人已经不怀疑深海沉积物有细菌了，但未解之谜太多，尚待探讨。问题的解决首先涉及的是研究手段，人们希望借助先进手段，比较直观地获取相应的可信信息、数据、资料，可以想象最好的方法是能直接深潜入海并深入沉积物、岩石中较快地发现/探明到相关微生物各种活动信息，但这又谈何容易！更

* 基金项目：国家自然科学基金项目（40576060、40376017、40176017）

原文刊于《海洋地质动态》，2008，24(3)：14-17.

何况仪器/器材经一定过程后到达预定深度再返回陆上实验室而进行的探测所得信息未必能保证完全准确无误，有可能形成假象[5,6]。

2 海洋地微生物及其环境的特点

沉积物除与上覆水交界面接触/交流部分及其相邻表、浅层的沉积物外，其下的沉积物之各种环境、生态因子应该是相对稳定的。已有研究证实表层以下沉积物的细菌数量较大，但变动也大。再往下则细菌数量大多下降，但也并不能排除高菌量，这决定于当地的各种因子甚至是微宇宙/环境（microcosmos microenvironment）即微生境（microhabitat）[7,8]的状态，因子还包括分布不匀的有机质等，总体上呈现出不连续、非匀质的斑块或补丁分布状[4,8]；还决定于所用钻探器具、器材、方法，但往下的菌数趋向稳定（主指需氧菌）应该是情理中事。因为较深之处的许多条件应趋于稳定、变化小，但越往下，沉积物所承受的上覆压力将越大。在压力的不断增加中，沉积物环境将会发生变化，因而越往下，其中的细菌等生物必将承受高压，它们所需的不同类型、状态的氧、营养基质及环境、生态条件将随之变化，它们所进行的生理/生化活动也将发生一系列变化[8]。单就这些因素而言，就将产生有别于表层、浅层沉积物、岩石乃至海水微型生物的不同生存、结构、功能与代谢机制，经过亿万年的进化将可能产生不同于我们常规所认知的微生物类群[10-13]。

决定沉积物微生物生长繁衍的首要条件是其生长所需要的营养源，此来源有三，①沉积物本身、上覆水和底部岩石圈；②沉积物本身的矿物质及其盐类；③以及其中的生物遗体降解/分解物等是无机/有机营养源之一[12]。上覆水中各种生物产生的大小有机质颗粒及水中的无机元素，在各种不同的水流、潮汐作用下，部分将沉入海底并被沉积物所吸附[1]。再者便是岩石圈的各种地质构造和活动，包括热液或冷泉带来的含有各种化学元素和溶解气体在全球性大洋水体和沉积物中进行着迁移、流动和沉降，部分将再降入沉积物成微生物营养[12]。营养物将构成微生物的结构、能量以维持正常生命活动。当然上覆水中沉下来的营养物多数将是水中不易被降解者，于是沉积物微生物将担负起对其的降解作用。它们必须产生相应的酶，其有关营养和降解动力学值得人们关注，它们的新型酶类更具吸引力[5-10]。

对沉积物微生物产生决定性影响的另一重大因子是温度[14]。港湾、浅水、养殖设施等的沉积物，因其上覆水薄，其温度与水温差距不是很大，微生物状态与水中的差异并不十分明显，而对于大洋、深海、极地的沉积物，其温度则比表层海水的低得多，通常低于5℃，大多数则是2～3℃，甚或更低。低温造就了其部分微生物活性的低下，甚而休眠下来，因此，大多数深海沉积物的有机、无机物被降解、分解的速度就慢，而温度的升高必然迫使或唤醒低迷中的部分微生物活动（当然也不排除有机质此时的裂解），从而促进和加速其代谢率等，反过来也影响其他环境、水文、生态因子的变化。它们相互促进，相互制约，打破原有平衡，建立新关系。由此持久深刻地改观全球面貌。可以想象全球变暖所造成的后果多么严重！有关的主要影响因子还有溶氧、pH值、氧化还原状态等。

3 海洋地微生物学研究现状及方向

当今人们之所以对探索地微生物学的热情一浪高过一浪，其原因之一是对新、奇、特的生物活性物质的研究兴趣；其二是探索地微生物在地球化学物质迁移、转化和积累等过程中的作用和地位甚至生物起源[3,5]，因为，早在20世纪40年代就孕育出了石油微生物学和海洋底栖细菌学这两个分支学科[1,4]，当今的海洋学家、地质学家和微生物学家的结合将会使海洋地微生物学这朵奇葩开得更艳丽、更久远[15,16]。

3.1 研究现状

①随着钻探的加深和探测技术的进步，世界上已有14个国家实施了大陆科学钻探。大洋钻探考察包括大洋钻探计划(ODP、IODP、DSDP)，并已进行了多次综合性测试，取得了一系列重大成果，对深部地层中气体和强大水流的测定同时，已在寻找微生物等生物成因，并取得了一些重要的成果和认识，推测远古时代就已有微生物存在并活动了[3]。

②对金属矿物及其成矿作用的不同深度钻探的岩性分析，发现各种类型矿物和岩石的形成及矿化分布是按沉积作用进行的，它们各有其自身良好的成矿条件，微生物在其中发挥了作用，主要是使不同类型矿物和岩石成分发生改变，人们期望将相关原理及成果广泛应用于国民经济许多部门中去[17]。

③探索极端条件包括高压、厌氧、高低温、冷泉、热液、碱性、贫营养等状况中的微生物，发现了一些新细菌并探索其培养条件，不断揭示原始生命之谜[18,19]。

3.2 我国海洋地微生物学研究成果

(1)对极地及南海海洋沉积物的微生物，包括硫酸盐还原菌、好气异养菌、厌氧菌、锰、铁细菌作出含量及其分布的分析，对分离所得的一些菌株作了序列分析和系统发育上的研究等等[20-22]。

(2)在我国大洋矿产资源勘探中发现了锰结核中的螺旋状超微生物化石[23]；确认微生物是锰等多金属结核的建造者，其中不乏新的微生物类群[17]。

(3)对取自深海沉积物的微生物活性物质包括酶的活动条件作了研究[22]。

3.3 海洋地微生物学今后的研究方向

随着海洋科学家对海洋地微生物多样性和重要性的关注提高，基因组革命逐渐拓展至与生物地球化学相关的微生物上，海洋地微生物学当今展现出方兴未艾之势，其未来的研究方向应该包括：

(1)大力研发更有效的适于海底地微生物的探测手段和设施，以与钻探相适应、配合。不断更新的分子地微生物学(molecular geomicrobiology)技术肯定大有用武之地，人们有望用其分子尺度来研究地球化学和地质过程的内在机理、微生物进化机理和速率，探索生命起源和进化机制[24-26]，但同时还得开发出选择性更精确、条件更完善的培养系列；

(2)地球上除了光合食物链，是否还有别的类型的食物链？在深海底的极端环境中，一些微生物将利用硫化物、乙酸盐等参与到“黑色大洋”的生命大家庭中去，对洋底下可能存在的另一个生物圈的认识必须不断充实、完善、提高，包括微生物在其中所起的作用[3,13]；

(3)探讨微生物的生存机制，与地质、环境、化学/生态要素间的关系，获取和研究新、奇、特的海洋地微生物产物的模式和路线，以更好地服务于人类。

参考文献26篇(略)

(合作者：高爱国)

DISTRIBUTION OF GENERAL AEROBIC HETEROTROPHIC BACTERIA IN SEDIMENT CORE TAKEN FROM THE CANADIAN BASIN AND THE CHUKCHI SEA*

Abstract The occurrence percentage and abundance of general aerobic heterotrophic bacteria (GAB) were determined by using the method of MPN for 182 subsamples from 10 sediment cores taken from the Canadian basin and the Chukchi Sea at two different culturing temperatures. The results showed that the general occurrence percentage of GAB was quite high, average abundances of GAB at cultured temperatures of 4℃ and 25℃ were 4.46×10^7 and 5.47×10^7 cells · g^{-1} (wt), respectively. The highest abundance of GAB occurred at 20～22 cm section in the sediment. GAB abundances changed among the sections of sediments, but there is a trend: the abundances at the middle or lower sections were lower than those at upper section. Cultivation at 25℃ could improve the occurrence percentage and abundances of GAB, which suggests that the increasing of temperature may change the living circumstances of GAB. The differences of GAB among the latitude areas indicated that occurrence percentage and abundances of GAB in middle latitude areas were higher than those in the higher or lower latitude areas, and were more obvious at 4℃ than those at 25℃. The GAB abundances in sediment under the shallower water seemed to be lower than those in sediments under the deeper water and this status was more obvious at 25℃ than that at 4℃.

Key words The Canadian Basin, The Chukechi Sea, General Aerobic Heterotrophic Bacteria, Spatial Distribution, Sediment Core.

1 Introduction

General aerobic heterotrophic bacteria(GAB)is a kind of bacteria that widely distributes in the Arctic Ocean. They play an important role in the substance transformation,

---

* The original text was published in *Chinese Journal of Polar Science*, 2008, 19(1): 14-22, I was the Second writer.

原中文刊于《极地研究》,2008,20(1):40-47.

circulation of marine sediments, and in marine sediments geochemical activities.

As early as in 1884, Certes first discovered bacteria in the sediment at the water depth of 5,100 m from Talisman; until the beginning of this century, there are still many scholars tirelessly researching the deep-sea sediment microbiology in different habitats, but with some more exciting goals or in a wider geographical area, through accurate observation or using more advanced instrument. There are two reasons: One is that bacteria in marine sediment are involved in various geochemistry and ecological activities: another is that many bacteria (including new types) may have unique metabolism, and may produce some kind of new bioactive substances.

Today's Arctic is no longer difficult for people to reach as in the past. During past recent decades, many reports about the bacteria in the Arctic Ocean sediment have been brought out[1-7]; but there still has been a large number of problems which needed to solve. Based on the study of the GAB distribution in the surface sediments of the research area[7], the 10 sediment cores under the 41～3,850 m water depth in the area of the Canada Basin and the Chukchi Sea were examined by MPN method to determine the occurrence percentage and the abundance of general aerobic heterotrophic bacteria, and geographical distribution of the bacteria are discussed.

2　Materials and Methods

2.1　Sampling

The samples were collected under the depth of 41 m to 3,580 m in the Canada Basin and the Chukchi Sea(69°29′43″N～80°13′25″N, 146°44′16″W～169°59′37″W) during the second Chinese Arctic Research Expedition from July to September, 2003. The 10 cores basically represent each sedimentary unit and the ecological environment of the research area. Figure 1 shows the specific distribution of the sampling locations.

The sediment sampling was taken on deck of RV Xuelong using sediment box-cores and multi-tube cores. After being taken on the board, the 10 high resolution sampling sediments cores were processed according to microbiological standards aseptic condition, that is, removing the floating surface soil of the samples in the sampler, and taking the sample slice in 1 cm intervals in 0～10cm depth, 2cm intervals below 10cm. The total number of the subsamples is 182. All samples were preserved at 4℃ in clean, capped plastic bottles, which had been high temperature sterilized, until analysis in the domestic laboratory.

2.2　Analysis in domestic laboratory

The sediment samples were diluted to ZoBell 2216E liquid culture medium[8] according to MPN. The specific culture temperature was set referring to the research result of the first Chinese Arctic Research Expedition[6]. Based on MPN, 1g from each wet sample was taken quantitatively and diluted to specified concentrations series[9]. After the inoculation, The samples were then cultured in the test tubes for over 3 weeks at 4℃ and 25℃, respectively. During the cultivation, samples were observed at intervals to ensure lasting turbidity, the precipitation or bacteria film in tubes as positive for GAB.

According to the number of positive tubes and the rate of sample dilution, GAB abundance calculating was based on GB/T4789. 3-2003[10], and the abundance as well as the occurrence percentage of GAB were in each core or each sediment layer was statistical analyzed (cells per gram of wet sediment weight, hereafter referred as cell · g^{-1}). The distribution of GAB was discussed referring to the information of sediment lithology and geochem.

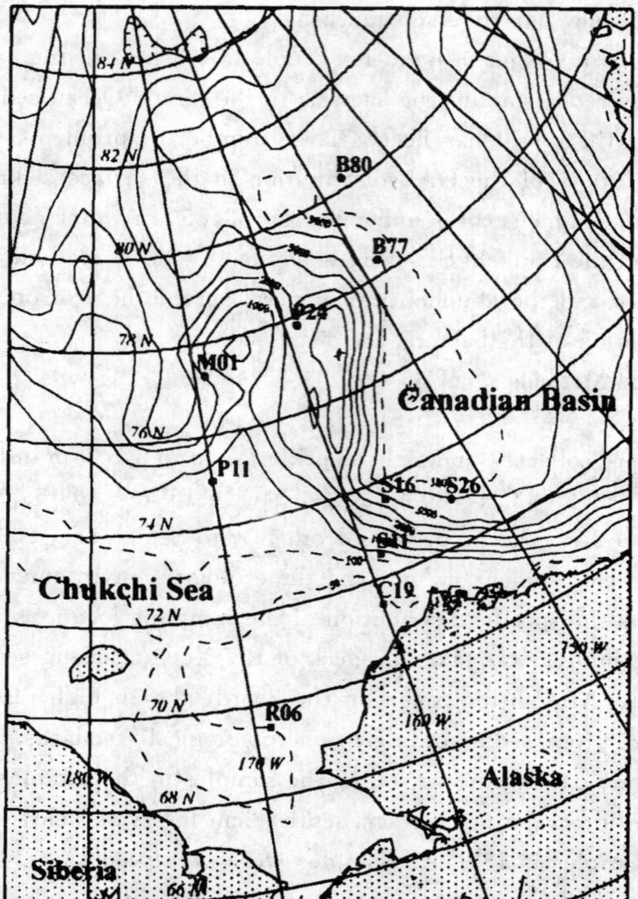

Fig. 1 Sampling stations of GAB from the Chukchi Sea and Canadian basin, the Arctic ocean

3 Result

The length range of the 10 sediment cores is 12～38 cm. The results of GAB Occurrence percentage and abundance in each sediment cores or layer are listed in Table 1 and 2.

Table 1 shows that R06 has the lowest GAB average abundance of 1.44×10^6 cell · g^{-1} while S16 has the hishest abundance of 2.18×10^8 cell. g^{-1} at 4℃, the GAB average abundance ranging from 1.34×10^6 for R06 to 2.19×10^8 cell · g^{-1} for S16 at 25℃. According to table 2, overall occurrence percentage of GAB is 91.3% at 4℃. There are 11 out of 24 layers, occurrence percentage of which all is 100%; the lowest level occurred at 16~18 cm layer which is only 66.7%. The GAB abundance range from 0 to 2.40×10^9 cell · g^{-1}. average is 4.46×10^7 cell · g^{-1}. Overall GAB occurrence percentage is 96.7% at 25℃, which of 17 layers is 100%. The lowest occurrence percentage appears in the 22~24 cm layer, which is 83.3%. The GAB abundance range stay alike with that at 4℃, which is 0~2.40×10^9 cell · g^{-1}, with an average of 5.47×10^7 cell · g^{-1}. From above, the conclusion, which can be easily come to that the GAB occurrence percentage and the abundance are generally higher at 25℃ than those at 4℃.

Table 1　occurrence percentages and contents of GAB in the sediment cores

core	long (cm)	4℃					25℃				
		Num.	Oc. Per.	Min.	Max.	Ave.	Num.	Oc. Per.	Min.	Max.	Ave.
R06	12	11	90.9	0	7.50×10^6	1.44×10^6	10	100.0	1.90×10^4	7.50×10^6	1.34×10^6
C19	22	16	100.0	9.30×10^3	2.40×10^7	7.82×10^6	16	100.0	1.50×10^4	2.40×10^8	5.49×10^7
S11	28	19	79.0	0	2.40×10^8	1.53×10^7	19	94.7	0	2.40×10^8	2.70×10^7
S16	28	19	94.7	0	2.40×10^9	2.18×10^8	19	100.0	2.10×10^5	2.40×10^9	2.19×10^8
S26	26	18	88.9	0	2.40×10^8	4.16×10^7	18	100.0	1.90×10^4	2.40×10^8	7.87×10^7
P11	38	16	81.3	0	2.40×10^7	3.29×10^6	16	87.5	0	2.40×10^7	6.65×10^6
M1	36	24	100.0	2.40×10^3	2.40×10^8	2.30×10^7	24	95.8	0	2.40×10^8	1.10×10^7
P24	28	19	94.7	0	2.40×10^8	1.61×10^7	19	94.7	0	2.40×10^8	1.68×10^7
B77	22	16	87.5	0	1.50×10^7	1.70×10^6	15	86.7	0	2.40×10^7	4.01×10^6
B80	38	24	79.2	0	2.40×10^8	2.09×10^7	24	100.0	3.00×10^4	2.90×10^8	5.94×10^7

Note: Num.: Sample number; Oc. Per.: Occurrence Percentages (%); Min.; abundance minimum; Max.: abundance maximum; Ave. abundance average; unit: cell · g^{-1} wet sediment weight; there are not data in R06-1 and B77-1 at 25℃. Hereafter the same.

Table 2　Occurrence percentages and contents of GAB in different depths in the sediment cores

layer (cm)	4℃					25℃				
	Num.	Oc. Per.	Min.	Max.	Ave.	Num.	Oc. Per.	Min.	Max.	Ave.
0~1	10	90.0	0	1.10×10^7	1.71×10^6	8	100.0	1.90×10^4	2.40×10^7	1.24×10^7
1~2	10	90.0	0	2.40×10^7	6.04×10^6	10	90.0	0	2.40×10^8	3.60×10^7
2~3	10	80.0	0	2.40×10^7	5.35×10^6	10	100.0	2.40×10^5	2.40×10^8	3.00×10^7

(continue)

layer	4℃					25℃				
(cm)	Num.	Oc. Per.	Min.	Max.	Ave.	Num.	Oc. Per.	Min.	Max.	Ave.
3~4	10	80.0	0	2.40×10^{8}	2.68×10^{7}	10	90.0	0	2.40×10^{8}	7.37×10^{7}
4~5	10	90.0	0	2.40×10^{7}	6.51×10^{6}	10	100.0	3.00×10^{4}	2.40×10^{8}	3.19×10^{7}
5~6	10	90.0	0	2.40×10^{7}	3.06×10^{6}	10	100.0	3.00×10^{4}	2.40×10^{8}	2.83×10^{7}
6~7	10	100.0	1.90×10^{4}	2.40×10^{8}	2.84×10^{7}	10	1400.0	1.90×10^{4}	2.40×10^{8}	2.98×10^{7}
7~8	10	100.0	1.60×10^{4}	2.40×10^{8}	5.40×10^{7}	10	100.0	9.30×10^{4}	2.40×10^{8}	6.41×10^{7}
8~9	10	90.0	0	2.40×10^{7}	6.37×10^{6}	10	100.0	4.30×10^{4}	2.40×10^{8}	2.79×10^{7}
9~10	10	100.0	2.40×10^{3}	2.40×10^{8}	5.12×10^{7}	10	100.0	2.40×10^{3}	2.40×10^{8}	5.36×10^{7}
10~12	10	80.0	0	2.40×10^{7}	6.14×10^{6}	10	0.0	0	2.40×10^{7}	7.56×10^{6}
12~14	9	88.9	0	2.40×10^{8}	3.00×10^{7}	9	100.0	9.00×10^{2}	2.40×10^{8}	5.66×10^{7}
14~16	9	77.8	0	2.40×10^{7}	3.42×10^{6}	9	88.9	0	2.40×10^{7}	6.39×10^{6}
165~1	9	66.7	0	2.40×10^{7}	5.87×10^{6}	9	88.9	0	2.40×10^{7}	6.13×10^{6}
18~20	9	100.0	2.40×10^{5}	2.40×10^{7}	5.05×10^{6}	9	100.0	2.40×10^{5}	1.10×10^{8}	1.99×10^{7}
20~22	9	88.9	0	2.40×10^{9}	2.75×10^{8}	9	88.9	0	2.40×10^{9}	2.76×10^{8}
22~24	6	100.0	1.50×10^{6}	2.40×10^{7}	9.45×10^{6}	6	83.3	0	2.40×10^{7}	1.28×10^{7}
24~26	6	100.0	2.40×10^{6}	1.10×10^{9}	1.95×10^{8}	6	100.0	2.40×10^{5}	1.10×10^{9}	1.93×10^{8}
26~28	5	80.0	0	2.40×10^{6}	1.63×10^{6}	5	100.0	1.60×10^{6}	2.40×10^{7}	6.56×10^{6}
28~30	2	100.0	2.40×10^{6}	2.40×10^{8}	1.21×10^{8}	2	100.0	2.40×10^{6}	3.60×10^{6}	3.00×10^{6}
30~32	2	100.0	3.40×10^{6}	2.40×10^{7}	1.37×10^{7}	2	100.0	1.60×10^{4}	4.20×10^{6}	2.11×10^{6}
32~34	2	100.0	2.40×10^{7}	2.90×10^{7}	2.62×10^{7}	2	100.0	2.40×10^{6}	2.90×10^{8}	1.46×10^{8}
34~36	2	100.0	2.40×10^{6}	2.40×10^{7}	1.32×10^{7}	2	100.0	2.40×10^{6}	2.40×10^{7}	1.32×10^{7}
36~38	2	100.0	1.10×10^{8}	2.40×10^{8}	1.75×10^{8}	2	100.0	1.10×10^{8}	2.40×10^{8}	1.75×10^{8}

4 Discussion

4.1 comparison of GAB in sediment cores

Table 1 shows that core C19 and M1 with 100% occurrence percentage of GAB account for 20% of all cores when cultivated at 4℃. The lowest occurrence percentage (79.0%) appears in core S11. When cultivated at 25℃, samples from core R6, C19, S16, S26 and B80 present 100% occurrence percentage of GAB, and account for 50% of all cores. The sample of the lowest occurrence percentage, which is 86.7% locates in B77. It is suggested that not only the proportion of cores with 100% occurrence percentage of GAB, but also occurrence percentages are higher at 25℃ than those at 4℃, meanwhile, which shown that GAB distributes widely in sediment cores.

When cultivated at 4℃, the layers at 6～7cm、7～8cm and 9～10cm contain GAB in all cores, whereas at 25℃ the layers containing GAB in all cores increases, at 2～3cm、4～5cm、5～6cm、6～7cm、7～8cm、8～9cm and 9～10cm. Table 2 also shows that the depth of the lowest occurrence percentage is deeper when cultivated at 25℃ than that at 4℃, which indicates that the occurrence percentage and abundance of GAB increase along with temperature increase.

Divided all cores into 3 segments, 0～10cm, 10～24cm and 24～38cm. The occurrence percentages and average abundances of GAB of 3 segments at 4℃ are 91.0%, 86.0%, 97.1% and 1.89×10^{8} cell·g^{-1}, 4.79×10^{7} cell·g^{-1}, 7.81×10^{8} cell·g^{-1}, respectively. At 25℃ the data are 98.0%, 91.4%, 100.0% and 3.88×10^{8} cell·g^{-1}, 5.50×10^{7} cell·g^{-1}, 7.70×10^{7} cell·g^{-1}, respectively.

Those data show that occurrence percentage of GAB present higher within top and bottom segment, and lower in middle segment. Occurrence percentage of GAB at 25℃ are higher than that at 4℃. Occurrence percentage of GAB in short cores (<26cm) is higher than those in longer cores, if cultivated at the same temperature. Occurrence percentage of GAB at 4℃ and 25℃ in short cores are 90.4% and 98.8%, in longer cores are 88.9% and 93.9%, respectively.

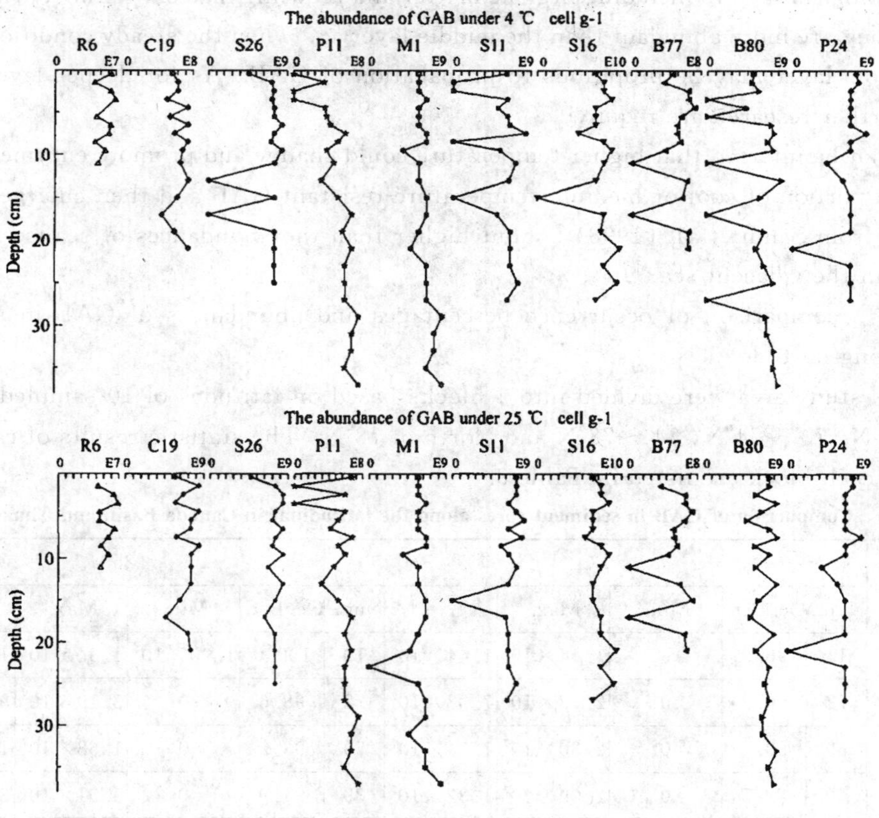

Fig 2 Distribution of GAB among the layers of sediments in the Canada Basin and the Chukchi sea

Figure 2 shows distribution of GAB abundances of 10 cores at 4℃ and 25℃. Based on statistic of Table 2, it is known that fluctuating ranges of GAB abundance are identical at 4℃ and at 25℃. But average abundance of GAB is higher at 25℃, which is 1.2 times of that at 4℃. By statistic analysis based on Table 1 and Table 2, the variation characteristics of GAB abundance in each layers of cores can be found. Highest average abundance of GAB appear within 20～22cm layer at both 4℃ and 25℃, and the value is 2.75×10^8 cell·g^{-1} and 2.76×10^8 cell·g^{-1}, respectively. Lowest average abundance of GAB at 4℃ appears within 26～28cm layer, and is 1.63×10^6 cell·g^{-1}, which is just equal to 6‰ of highest average abundance. Lowest average abundance at 25℃ presents within 30～32cm layer, and is 2.11×10^6 cell·g^{-1}, not enough to 8‰ of highest average abundance. It is suggested that bacterial abundance increasing in deeper layer may be caused by raising temperature. Figure 2 also indicates that high GAB abundance does not necessarily appear at surface or near surface sediment, which is controlled by environmental factors.

Table 2 and Figure 2 show that GAB abundance of cores do not distribute regularly. GAB abundance of cores present high value within surface and bottom layer, but the middle layer has low abundance. It may be the main reason that biogeochemical condition, ecological factors including organic matter and reducing matter, within surface sediments they are more abundant than the middle layer[11]. Then the steady condition within the bottom layer may be propitious to propagation of GAB. As for deeper layer condition, further research are required.

It can be inferred that higher temperature could change and promote enzyme production in a portion of cool or medium temperature-resistant GAB and then enlarge its survival PI from Sahm et a1. (1998)[2], but higher than the abundances of sea-ice microorganism in the adjacent sea[5].

4.2 comparison of occurrence percentages and abundances of GAB in sediment cores along latitude

The study area were divided into 4 blocks based on latitudes of 10 sampled cores:. 66°～70°N, 70°～74°N, 74°～78°N and North of 78°N. The statistic results of calculated GAB in each block are listed in Table 3.

Table 3 Comparison of GAB in sediment cores along the latitudinal in Canada Basin and Chukchi Seas

latitudinal (N)	4℃					25℃				
	Num.	Oc. Per.	Min.	Max.	Ave.	Num.	Oc. Per.	Min.	Max.	Ave.
66～70	11	90.9	0	1.44×10^6	1.44×10^6	10	100.0	1.90×10^4	1.34×10^6	1.34×10^6
70～74	72	90.3	0	2.18×10^8	7.38×10^7	72	98.6	0	2.19×10^8	9.69×10^7
74～78	59	93.2	0	2.30×10^7	1.54×10^7	59	93.2	0	1.68×10^7	1.17×10^7
>78	40	82.5	0	2.09×10^7	1.32×10^7	39	94.9	0	5.91×10^7	3.79×10^7

The arrangement of occurrence percentages of GAB at 4℃ in 4 blocks is 74°～78°N >66°～70°N>70°～74°N>North of 78°N, and those of the maximum or average abundance of GAB in 4 blocks is 70°～74°N>74°～78°N>North of 78℃N>66°～70°N., respectively.

The arrangement of occurrence percentages of GAB at 25℃ in 4 blocks is 66°～70°N >70°～74°N>North from 78°N>74°～78°N, and those of the maximum or average abundances of GAB in 4 blocks is 70°～74°N>North from 78°N>74°～78°N>66°～70°N, respectively.

The occurrence percentages and abundances of GAB are higher for medium latitude area than those for the southern part and northern part of study area. This may be relative with the stability of sediment environment. According to the First and Second Chinese National Arctic Research Expedition, GAB in surface sediments shows the tendency of abundances decreasing with latitude[6,7]. But the GAB abundances in sediment cores are different.

4.3　comparison of GAB in sediment cores from different water depth

GAB of the sediment cores were statistic analyrzed after dividing into 4 water depth division: 41～55 m, 175～561 m, 1,456～2,200 m and 3,000～3,850 m. Comparative results are listed in Table 4.

Table 4　Comparison of GAB in sediment samples under different water depths

Depth (N)	4℃					25℃				
	Num.	Oc. Per.	Min.	Max.	Ave.	Num.	Oc. Per.	Min.	Max.	Ave.
41～55	46	89.1	0	2.40×10^8	9.37×10^6	45	97.8	0	2.40×10^8	3.12×10^7
175～561	16	81.3	0	2.40×10^7	3.29×10^6	16	87.5	0	2.40×10^7	6.65×10^6
1 456～2 200	43	97.7	0	2.40×10^8	2.00×10^7	43	95.4	0	2.40×10^8	1.36×10^7
3 000～3 850	77	87.0	0	2.40×10^9	7.05×10^7	76	97.4	0	2.40×10^9	9.29×10^7

The occurrence percentage maximum of GAB at 4℃ appears in the division 1,456～2,200 m, and the minimum is in division 175～561 m. The abundance maximum and minimum of GAB locates in division 3,000～3,850 m and 175～561 m, respectively. Moreover, average abundance of GAB generally increases with increase of water depth except for division 175～561 m, which locates in the continental slope region with especial sediment environment.

Cultivated at 25℃ the occurrence percentage maximum of GAB presents in division 41～55 m, and the minimum is in division 175～561 m like that at 4℃. Moreover, the abundance maximum and minimum of GAB at 25℃ are consistent with that at 4℃, in the division of 3,000～3,850 m and 175～561 m, respectively. However, average abundance of GAB at 25℃ presents fluctuating variation along with water depth, with maximum in shallow water area, and minimum in the division 175～561 m. These distribution characteristics differ from that in surface sediment[7]. Both minimum of occurrence percentage and abundance appears in the division 175～561 m, it indicates the division may locate in transition region of geological, geographical and ecological variation. The condition like this may not suitable to the living of GAB, but the actual reason needs further study.

As for all cores, the high abundances of GAB largely exists in the middle-lower layers, and do not appear at the surface. It may be related with the possible that the decomposition of organic matter and enhancement of oxidation in middle-lower layers[11], they are propitious to propagation of GAB. Medium latitude area exists higher occurrence percentage and abundance of GAB. Although northern high latitude area holds relative steady environment, but the nutritional matter required by microbes is lacking. Southern low latitude shallow water area really has abundant nutritional matter, but variation of environment is acute, such as icing and higher salinity water in winter. warm and low salinity water in summer et a1.. Such a condition will go against propagation of some microbe. However, medium latitude area exists some advantage on source of nutritional matter and ecological environment, so higher occurrence percentage and abundance of GAB appear. Growth of GAB is affected more distinctly by water depth at 4℃ than at 25℃. Moreover the thicker the overlying water, the lower the abundance of GAB, but except for the continental slope area.

5 Conclusion

GAB in 10 sediment cores from the Canada Basin and Chukchi Sea were analyzed by the methods of MPN. The occurrence percentages, range of abundance and averages of GAB cultivated at 4℃ and 25℃ are 91.3%, 0～2.40×10^9 cell · g^{-1}, 4.46×10^7 cell · g^{-1} and 96.7%, 0～2.40×10^9 cell · g^{-1}, 5.47×10^7 cell · g^{-1}, respectively. Comparisons of GAB in 10 cores indicate that GAB distribute extensively in the study area, and the occurrence percentages of GAB in shallow layer are higher than that in middle layer.

The experiments, cultivating at two temperature, show occurrence percentage and abundance of GAB increase when temperature rises. Although the cultivation temperature can't reflect the in situ environmental temperature correctly, yet the results imply that environmental temperature could be a significant factor for microbes in sediments in Arctic Ocean to some extent. Temperature change will bring the variety of physiological and biochemical state of microbes. Global climate warming is consequentially influencing the surviral and evolution of microbes in sediments in Arctic Ocean, and further rearches are expected to carry out.

Acknowledgements This study was sponsored by the national ministry of finance. And as a part ofthe second Arctic research expedition (CHINARE-2003) which is organized by Polar Expedition office of State Oceanic Administration of China, we would like to thank the crew on RV Xuelong for their support and assistance. Research funds are supplied by National Natural Science Foundation of China (No. 40576060, 40376017 and

40176017). We thank all these assistance faithfully.

References12(abr.)

Cooperator: Gao Aiguo

我国深海地微生物学研究状况分析*

摘　要　本文从引用国际相关趋势入手论述我国深海地微生物学的研究状况。该命题论及我国此方面相关的研究进展，包括介绍国际间相关研究的历史状况、动态，我国研究的主要课题/内容/成果。文章最后展望了我国本领域的前景。

关键词　深海　大洋　地微生物学　中国

1. 国际深海微生物学研究概况

19 世纪 30 年代，微生物学家将眼光扩大至海洋。于 1838 年，德国学者 Ehrenberg 在世界上首次分离鉴定出了第 1 种海洋细菌——叠折螺旋体（*Spirochaeta plicatilis*）。之后，Fischer、Certes、Russell、Issatchenvo、Butkevich、Kusnetzow、Rubentschik、克里斯等对海洋微生物学的发展作出过巨大的贡献。美国学者佐贝尔则于 20 世纪 30 年代始专注海洋微生物，并于 1946 年发表专著——海洋微生物学，由此奠定了现代海洋微生物学的基石[1]。在迄今的 170 余年间，世界海洋微生物学经历了翻天覆地的变化，它借助不断进步的生物医学技术和日益完善的取样手段，已从传统进入分子生物学阶段，并从单一学科向综合交叉边缘学科相结合发展，海洋微生物学的研究深度和广度等都获得了巨大的扩展。如果说以往的海洋微生物学研究在海洋科学中仅起配角作用，并不起眼，而且主要局限于海水、近海、沿岸、表层沉积物，那么自 20 世纪 70 年代以来，人们将目光还进一步伸向深海一高或深于千米的广阔大洋的空气、深水和海底及沉积物/岩石中/下的地壳圈，并成为海洋生命科学和环境科学的焦点之一。对海底的探测自 1968 年以来，国际间以开展 DSDP（1968～1983）、ODP（1985～2003）和 IODP（2003～2013）最为著称，并取得和将取得一系列的成就[2~3]。虽然 ODP201 航次（2000 年）前的 ODP 项目没有一次研究将海底下生命作为主要目标，但仍有一些学者参与了其他领域（如“深海之星”）的航次，并获得了结果[4]。近 30 年来，一系列成果的取得使人们认识到海底广阔领域—由开初表层沉积物几米到表面下 800 m 的沉积物中均有生物存在[5]。对这一范围的探索及对一些迹象的推断，使人们大胆地提出了海底深部生物圈（subseafloor deep biosphere 缩写为 SFDB）这一概念，并使该生物圈年龄终于可上溯到 1500 万年前[6]。SFDB 是深海微生物学研究的主要内容和领域之一。它的重中之重是由深海（洋）地学和深海（洋）微生物学等生物学科相交叉而形成的边缘科学，即深海（洋）地微生物学（deep sea geomicrobiology）。迄今相关研究及其前景吸引并推动多学科学者的热烈参与。

该学科最先并不是对洋底生物圈直接进行微生物学研究，而是通过对沉积物样品、孔隙水、地化和有机地化的研究，了解到沉积物序列中固相和液相有机/无机质总量及分布中的变化，再推断出相关的微生物活动。这些活动的状况包括岩芯和孔隙水中的 SO_4^{2-} 损

* 原文刊于《海洋地质动态》，2009，25（7）：20-25 转 41。

耗、CH_4 生成、碱度浓度、Mn 还原及 NH_4^+ 消涨等[7]。许多研究借助大陆钻探经验，最大程度地克服和排除微生物污染的可能和机率，最终获得了可信的资料（讯），证实了全世界大洋底广布微生物[5]，而且证实了其群落（包括原生物的）生物种群大小、新陈代谢、生存战略，并进行一些培养试验等[8~9]。以上成果进一步推动人们更大胆地将海底生物圈向远远超过 800 m 深度（如几千米）处进军，因为那儿因热催化反应生成的 H_2 等还原性气体有可能支撑微生物代谢、生长和 CO_2 固定。这些迷津可能正在被日本“地球”号钻探船及地心探索计划等开端的科研活动所指点、解析[10~12]。

2. 中国深海地微生物学研究成果

中国海洋微生物学研究起步于 20 世纪 50 年代，之后于“文革”有所停滞。而中国学者对深海微生物的研究始于南大洋、南极及环球考察[13]，关注/从事深海地微生物学的研究之于中国学者主要是 20 世纪末叶以来的事。这些年来主要的课题来自国家重点基础研究发展规划、包含《深海微生物抗癌等活性成分及药学研究》的 863 计划、国家自然科技资源平台、自然科学基金等。如“地球圈层相互作用中的深海过程和深海记录”、“深海生物圈在循环中的作用”、含“深海热液区微生物筛选、基因克隆及其应用”的“大洋专项”、包括“深远海极端生物及其基因资源开发关键技术研究”在内的“海洋 863 计划”、“国家 973 项目”、“中国科学院知识创新工程重要方向项目”等等。中国学者历年来所取得的深海地微生物学研究成果主要表现在以下几个方面[32]：

2.1 关注国际动态：这些年来，相当多的中国学者关注国际间相关动态，包括介绍研究概念思路、方法、仪器、设备、成果和展望，所涉及的论文约 20 篇[13~33]。

2.2 主要研究课题/内容

表 1 中国深海地微生物学研究成果一览表

海区	主要研究内容/成果	引证文献号
太平洋	环己酮降解菌的筛选及功能	34
太平洋	抗肿瘤微生物，筛选活性及冷致代谢产物	35
东太平洋	微生物多样性、热液区沉积物微生物、细菌调查	33、83
东太平洋	铁锰结核区微生物丰度及成矿	36
东太平洋	海隆深海热液区沉积物微生物多样性	78
西北太平洋	沉积物、锰结核中微生物丰度、锰细菌	39
太平洋暖池	微生物生态	37
西太平洋暖池	细菌类群及其与环境的关系	38
西太平洋暖池	古菌、嗜泉古菌与沉积物深度关系等	38
热带太平洋	微生物生理生态特征，抗肿瘤活性微生物	35
冲绳海槽	硫化物中微生物矿化	40
冲绳海槽	黑烟囱中微生物化石的发现	41
南沙	细菌、古菌 16*S* rDNA 多样性	42
南沙	细菌 16*S* rDNA 多样性、可培养微生物	43、92

（续表）

海区	主要研究内容/成果	引证文献号
南海中部	异养细菌生态分布	82
南海北坡深水区	沉积物细菌多样性	90
南海西沙海槽	沉积物微生物/细菌多样性	44、45
南海深海	沉积物烷烃降解菌	46、89
南海南坡表层沉积物	细菌、古菌多样性	91
墨西哥湾 GC238 区冷泉	碳酸盐岩微结构与石化微生物特征	47
大西洋	酵母菌 MAR-MC(3)的鉴定及生理特征	48
印度洋	克隆食烷菌(*Alcanivorax* sp. p40)烷烃羟化酶基因、抗砷菌	49、79
北冰洋	硫酸盐还原菌生物地球化学要素间相关分析	50
白令海和楚科奇海	好气异养细菌丰度及对温度的适应性	51
白令海和楚科奇海	硫酸盐还原菌	52
北冰洋	铁细菌分布、锰细菌分布	53
加拿大海盆与楚科奇海	好气异养菌分布	54
深海	抗菌抗肿瘤活性真菌	55
深海	抗锰细菌的分离与鉴定	56
极地深层冰川	微生物	29
深海	铁锰结核微生物成因、淀粉酶细菌	57、85
	极端嗜盐古菌的嗜盐菌素	58
	冷适应微生物的冷活性酶	19、59
	细菌的 Fe(Ⅱ)还原、深部生物圈微生物	2、60
深海	微生物活性成分及药学	61
深海	微生物培养系统流动体系的建筑与仿真	62
深海	细菌的分子鉴定分类、微量 DNA 提取与应用	63、64
深海	适冷菌 *Pseudomoncs* sp. SM9913 的低温蛋白酶，适冷蛋白酶 MCP-01 基因结构与功能、细菌基因组文库	65、66、93
深海	微生物培养模拟平台温度控制技术	67
深海	微生物研究平台技术	68
深海	取样器保压过程、保真采样	70、84
深海	锰结核中螺旋状微生物化石的发现	71
深海	高压技术在兼性嗜压菌的筛选与鉴定中的应用	74
深海	地球化学新指标、微生物多样性形成的机制、地微生物学	31、32、77

（续表）

海区	主要研究内容/成果	引证文献号
深海	低温碱性淀粉酶菌	72
南大洋	蓝细菌和微微型光合真核生物丰度与分布	13
深海	古菌 16SrDNA 分析、萘降解菌	73、80
深海	具抗肿瘤活性微生物及其产物	76
深海热液喷口、热泉	生物群落	75、86
深海极端环境深部生物圈	微生物及微生物学综述	2
冷泉极端环境深部生物圈	碳酸盐岩微结构，石化微生物	47
南冲绳海槽	沉积物多环芳烃降解菌	81
南极	深海底泥中度嗜盐单胞菌、细菌生产力	69、87
南极	底泥中度嗜盐菌	88

表 1 列出了相关课题的主要研究内容和成果，由该表可见，所取得的研究成果主要包括：

2.2.1　深海地微生物学的基础研究，如方法、仪器、器材、设备、过程、分类、生态、生理特征、与相关地球化学因子间关系分析、矿化作用、生命起源探讨、微生物化石、低温嗜压菌、铁锰结核成因及环境基因等等[18,41,54,55,58,64,69,76]。

2.2.2　应用基础研究，如冷活性酶、Fe（Ⅱ）/Mn（Ⅱ）还原与氧化、活性成分如抗肿瘤物的确认，一些物质的降解菌、DNA 的提取及应用等[2,36,52,63,66,76]。

2.2.3　微生物资源的开发性利用研究，如嗜盐菌素提取与应用、活性成分的药学研究等等[18,37,61]。

相关研究涉足全球大洋一些海区，包括东西太平洋、热带太平洋、冲绳海槽、印度洋、大西洋、北冰洋、南大洋、南海及其他深海区。

参与本学科研究的中国学者大约有 150 个人次，他们发表的论文所刊登的刊物约计 38 种，包括 SCI 刊物，其中以《微生物学报》、《厦门大学学报（自然科学版）》、《地球科学进展》、《台湾海峡》、《海洋科学》、《微生物学通报》等为多，以上 6 刊，均有至少 4 篇论文发表。此外尚有在网上发表的文章。

表 2　中国深海地微生物学论文数历年统计

年份	1989	1990	1991	1992	1993	1994	1995	1996	1997	1998	1999	2000	2001	2002	2003	2004	2005	2006	2007	2008
论文（篇）数量	2	0	0	0	0	0	1	1	0	6	3	3	0	5	7	7	12	19	6	17

据不完全统计，表明发表论文数量在年度间的变化如表 2 所示。由表 2 可见，自 1989 年迄今的约 20 年间，总数约 89 篇的论文数在年度间有起伏，其中有 7 年没有论文发表，且主要在 20 世纪末。进入 21 世纪，论文数量呈上升趋势，以 2006 年为峰，2008 年论文数仍

在上升，表达出中国深海地微生物学研究正在攀登，并接近/跻身国际水平。

3. 中国深海地微生物学研究展望

综观国内外深海地微生物学研究，可以看出，世界先进海洋国家的深海地微生物学家纷纷与其他学科/领域的学者、工程技术人员紧密结合，争取以更先进实用的手段，有更多的机会实现参与现场取样，涉足尽可能多的各种场合如热液喷口、冷泉、海底深处等极端环境，包括《深海之星》、《嗜极端微生物作为细胞工厂》、《极端环境生命》、《洋中脊多学科全球试验》等，向极限进军。认识和开发更多更新的生物种类及其结构功能，以造福人类。以微生物为武器/手段/工具，探索古今海洋环境、地球化学、地质环境的演变历史、现状，并预测地球变化。

在这一国际大背景中，中国深海地微生物学成果迄今与国际间存在明显的差距，体现在深度和广度上，力度不够，其原因是起步晚，缺乏统筹谋略与周全考虑，因而显得步子新、碎、轻。所做的课题项目表现出分散、零乱局面，迄今全国相关学者尚捏不起一个有力的拳头来[34]*。但在我国面前展现的前景十分远大，空间很大。中国学者必须瞄准新动态，急起直追，必须长期落实相关人选，真切关注国内外动态，形成高瞻远瞩的领导层和智囊团，以提出协调和安排好相应项目规划和实施。安排好自身的计划，引进、研制先进而实用的方法和仪器设备，有所侧重地做好做深工作。将有限的经费发挥出极致水平。力争多参与国际学术交流/活动和实地调研，多拿第一手材料。同时加紧自身海域深海地微生物学的研发。不该全面铺开，而是量力而行，有所侧重，做到有所突破，跻身先进。强调加强基础性研究，切记勿急功近利。同时大力培养和引进相关的学术人才，鼓励陆地学者参与。提供多学科交叉/渗透机遇，相互促进，共同提高。

深海底部生物圈既有微生物学本身的广深未知，与其他学科又有巨大的合作、交叉领域，因而是任重道远，我国一定能在相关领域取得骄人业绩。

参考文献 93 篇(略)

(合作者:陈颖稚　麦迪逊技术学院　麦迪逊　威斯康星州　美国)

* 陈皓文等，微生物学向深海进军(待发表)。

ANALYSIS ON SITUATION OF DEEP SEA GEOMICROBIOLOGICAL RESEARCH IN CHINA

(ABSTRACT)

Abstract This paper relates situation of deep sea geomicrobiological research in China. The thesis quotes from the international trend of this field, discusses the research progression including introduction of relative international development history/state, main subjects/contents researched in our country. Lastly, it looks forward to the future of this domain in china.

Key words Deep Seas/Oceans, Geomicrobiology, China

西北冰洋海域表层沉积物中厌氧菌分布特征*

摘　要　用逐步稀释法在4℃和25℃培养条件下测定了西北冰洋海域24个表层沉积物样品中厌氧菌(anaerobic bacteria, AAB)的检出率和含量。同时,分析了这两项指标的水平分布(纬度间、经度间)差异,以及在不同水深的变化特征。结果表明,在4℃和25℃培养条件下厌氧菌检出率高达100%,AAB含量范围分别为$9.00\times10^2\sim2.40\times10^7$ cell·g^{-1}和$2.90\times10^4\sim2.40\times10^7$ cell·g^{-1},平均含量分别为4.54×10^6 cell·g^{-1}和3.99×10^6 cell·g^{-1}。AAB含量存在水平分布差异,随着纬度升高,或经度自西向东,或水深的加大,AAB的含量均呈现逐渐降低的趋势。

关键词　西北冰洋　加拿大海盆　楚科奇海　厌氧菌　表层沉积物

1　引言(Introduction)

海洋沉积物中的有机质受到微生物的降解,将使环境中的氧被逐渐消耗,并产生CH_4、H_2S、CO_2、H_2等代谢产物,从而形成少氧或缺氧的厌氧微环境,厌氧菌(AAB)随之增多(Parkes et al., 2000)。厌氧菌与好氧菌并没有明确的界限,为了讨论方便,通常将生长在没有氧气或培养基$E_h<0V$的细菌称为厌氧菌。厌氧菌存在于所有的海洋沉积物或大部分海水样品中,它们许多是兼性厌氧(Facultative)或兼性需氧(Microaerophic)的,有一些则是专性厌氧的(薛庭耀,1962)。厌氧菌与好氧菌共存于沉积物中,构成了沉积物中的微生物生态系统(Robador et al., 2007; Parkes et al., 1994; D'-Hondt et al., 2002),在适当条件下它们之间甚至会发生相互转变。

厌氧菌在元素的地球化学循环中起着重要作用,它们与其他微生物一起参与有机碳、硫、氮、磷等的矿化过程,对沉积物的组成和性质变化、早期成岩过程等发挥着重要的作用(Schink et al., 2000; Guldberg et al., 2002; Rysgaard et al., 2004; Salminen et al., 2004; Arnosti et al., 2005)。目前,研究较多的有硫酸盐还原菌(Knoblauch et al., 1999a; 1999b;陈皓文等,2007;高爱国等,2003;2008a)、铁细菌(高爱国等,2008b)、锰细菌(高爱国等,2008c)、甲烷菌(Thomsen et al., 2001; Singh et al., 2005)、反硝化菌(Hulth et al., 2005)等。

在微生物的生物地球化学研究中,不仅要了解微生物的种类、特定微生物的生物地球化学行为等(Rivkin et al., 1996; Rossello-Mora et al., 1999; Zhou et al., 2004),还需要了解微生物的丰度变化特征(Knoblauch et al., 1999a; Parkes et al., 2000)。对于进行大量样品的分析而言,经典的逐步稀释法虽然工作量较大,但仍是简便易行的方法之一。虽然厌氧菌的发现已有330年,由于要计算海洋厌氧菌面临很多困难,如受制于实验条件

*　基金项目:国家自然科学基金(No. 40576060, 40376017)
Supported bytheNationalNaturalScience Foundation of China (No. 40576060, 40376017)
原文刊于《环境科学学报》,2009,29(10):2209-2214,本人为第二作者。

和认识水平，且难以保证所用方法中各步骤的条件均达到厌氧状态等。因此，这方面的资料较少，而在西北冰洋海域海洋沉积物中的研究更是少见。考虑到某些功能性细菌同时具有多种功能，如同时兼具硫酸盐还原菌与铁细菌的功能等，因此，厌氧菌的数量将小于各功能细菌数量之和。

近年来，对于适冷菌的研究受到越来越多的关注(Shcherbakova et al.，2005)，为了有效地评价厌氧菌在西北冰洋海域表层沉积物地球化学中的作用，从总体上认识厌氧菌的分布及变化特征，更全面地了解研究区的生物地球化学过程，本文借鉴前人的方法，用逐步稀释法在4℃和25℃培养条件下测定西北冰洋海域24个表层沉积物样品中厌氧菌的检出率和含量，所涉及的厌氧菌是相对于特定培养基及培养条件(薛庭耀，1962)而言的。同时，分析这两项指标的水平分布(纬度间、经度间)差异，以及在不同水深的变化特征。

2　材料与方法(Materials and methods)

2.1　研究区简介

研究区位于北冰洋及其邻近的楚科奇海与波弗特海(66°～80°N，148°～170°W，水深41～3 850 m)。其中，加拿大海盆是西北冰洋的主体，而楚科奇海与波弗特海是北冰洋周围的边缘海，为叙述方便，本文将其统称为西北冰洋海域。相比较而言，在靠近亚美大陆一侧的沉积物含有较多营养物质(薛斌等，2006)，推测可能分布有较多的厌氧菌，并在生物地球化学中发挥作用。

2.2　样品采集

2003年7～9月的中国第二次北极考察期间，在“雪龙”号考察船上进行表层沉积物采样。所用采样器分别为箱式采样器和多管取样器。将沉积物采至甲板后，在现场按微生物学采样要求去掉上覆水及浮泥，采集深度0～1cm的表层沉积物样品作为微生物分析所需样品，并迅速将其置于无菌塑料瓶中，使其尽可能不/少暴露于空气中，封盖保存于4℃以下的冰箱中，带回国内实验室分析。研究区采样站位参见图1，共计24个站位。

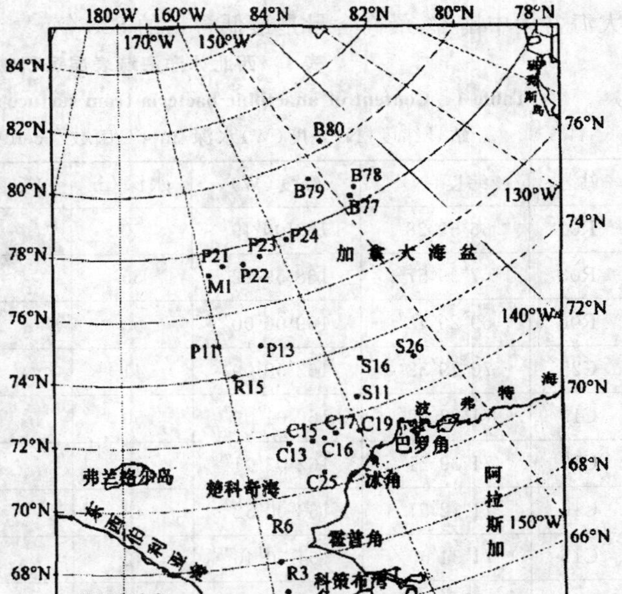

图1　北极沉积物细菌分析样品采样站位图

Fig. 1　Map of sampling stations for bacteriological analysis from surficial sediments in the Western Arctic sea

2.3　厌氧菌分析

样品运回国内后立即开展实验室研究，运用逐步稀释法进行厌氧菌分析。首先将样品转移到试管中，所用培养基(薛庭耀，1962)为：陈海水1 000 mL、消化蛋白

5g、磷酸高铁 1g、琼脂 15.0g、质量分数 1.0%葡萄糖、质量分数 0.1%硫化甘醇酸钠、质量分数 0.000 2%美蓝。在 120℃消毒 20 min，pH 值为 7.5～7.6。

用逐步稀释法组成稀释序列，橡皮塞塞紧后置于 4℃或 25℃的恒温中分别培养 3 周以上，期间不断观察培养情况，并与空白作对照，以试管中出现混浊/菌膜/沉淀者为厌氧菌阳性。参照文献（薛庭耀，1962；陈皓文等，2007）要求，计算厌氧菌的检出率和含量（cell · g^{-1}，以湿重计，下同）。

3 结果（Results）

对研究区表层沉积物厌氧菌分析表明，样品中厌氧菌的检出率为 100%（4℃时 24 个样品，25℃时 8 个样品），结果见表 1。由表 1 可知，4℃和 25℃两种培养温度中 AAB 的检出率均为 100%，说明在表层沉积物中广泛分布着 AAB。在 4℃时，样品中 AAB 含量范围为 9.00×10^{2}～2.40×10^{7} cell · g^{-1}，平均为 4.54×10^{6} cell · g^{-1}。4℃培养时，AAB 最高含量同时出现于 4 个站位，即 R01、R03、R06 和 C17 站，均为 2.40×10^{7} cell · g^{-1}，它们处于研究区南部。最小 AAB 值则出现于 P21 站，它位于研究区北部。最高与最低含量相差悬殊，达 2.7×10^{4} 倍，而 AAB 含量均值为最低含量的 5000 余倍，AAB 含量均值与最高含量差为 50 余倍。这表明 AAB 低含量站位相对较少，各站位间 AAB 含量有显著差异。25℃时，AAB 含量范围为 2.90×10^{4}～2.40×10^{7} cell · g^{-1}，平均为 3.99×10^{6} cell · g^{-1}。25℃培养时，AAB 最高值和最低 AAB 值分别出现于 S26 和 S16 站，它们均出现于研究区加拿大海盆偏中南部，最高含量是最低含量的 828 倍。

表 1　西北冰洋海域表层沉积物中厌氧菌含量

Table 1　Content of anaerobic bacteria from surface sediments in the Western Arctic Sea

站号纬度（N）经度（W）水深/m 4℃菌数/（cell · g^{-1}） 25℃菌数/（cell · g^{-1}）

站号	纬度（N）	经度（W）	水深/m	4℃菌数/（cell · g^{-1}）	25℃菌数/（cell · g^{-1}）
R01	66°59′28″	169°00′49″	50	2.40×10^{7}	
R03	67°59′57″	168°59′22″	55	2.40×10^{7}	
R06	69°29′43″	169°00′00″	53	2.40×10^{7}	
C25	70°29′38″	163°58′09″	41	1.10×10^{6}	
C19	71°27′49″	160°01′09″	50	1.10×10^{6}	2.40×10^{5}
C17	71°29′21″	162°02′01″	46	2.40×10^{7}	
C16	71°32′51″	163°00′52″	43	2.40×10^{5}	
C15	71°34′45″	164°00′46″	42	2.40×10^{6}	
C13	71°36′51″	165°59′51″	44	2.40×10^{6}	
S11	72°29′24″	159°00′00″	50	9.30×10^{4}	2.40×10^{6}
S26	73°00′00″	152°40′00″	3000	7.50×10^{5}	2.40×10^{7}
S16	73°35′28″	157°09′50″	3800	1.90×10^{4}	2.90×10^{4}
R15	73°58′58″	169°04′23″	175	1.10×10^{6}	
P13	74°48′02″	165°48′24″	453	2.40×10^{5}	

(续表)

站号	纬度(N)	经度(W)	水深/m	4℃菌数/(cell·g^{-1})	25℃菌数/(cell·g^{-1})
P11	75°00′24″	169°59′37″	263	1.10×10^5	1.10×10^6
M1	77°17′56″	169°00′46″	1456	2.40×10^5	2.40×10^5
P21	77°22′44″	167°21′38″	561	9.00×10^2	
P22	77°23′43″	164°55′59″	326	1.50×10^4	
P23	77°31′40″	162°31′05″	2200	2.40×10^4	
P24	77°48′38″	158°43′16″	1880	2.40×10^6	2.40×10^6
B77	78°18′55″	148°06′57″	3850	2.40×10^4	
B78	78°28′43″	147°01′41″	3800	2.40×10^5	
B79	79°18′52″	151°47′09″	3800	1.10×10^5	
B80	80°13′25″	146°44′16″	3750	4.30×10^5	1.50×10^6

4　讨论(Discussion)

4℃和25℃两种温度培养中,AAB的平均含量分别为 4.54×10^6 cell·g^{-1} 和 3.99×10^6 cell·g^{-1}。在数量上与同研究区表层沉积物和岩芯中的好氧菌相比,含量差距不大(Gao et al., 2007; 2008),这表明相当数量的厌氧菌在研究区沉积物中同样起着重要的地球化学和生态学作用。

相关研究表明,控制沉积物中细菌生存的主要因素是输入海底的有机碳通量及其可利用性(Deming et al., 1992),表层沉积物中微生物量与有机碳显著相关、与沉积速率也成正相关(Boetius et al., 2000)。Kroencke 等(2000)对北冰洋欧亚海盆从巴伦支海陆坡到罗蒙诺索夫海岭的细菌分布研究发现,随着水深增加,纬度升高,细菌生物量减少,呈现细菌的地理分布差异特点。温度对细菌的影响也不容忽略。以往的研究表明,该研究区沉积物中细菌的适宜生长温度为4～15℃,大部分细菌为嗜冷菌或适冷菌,有少量的中温菌存在(Chen et al., 2004)。然而,也有关于适宜细菌生长更高的温度报道,如太平洋沉积物细菌的适宜生长温度为25℃(Parkes et al., 1994),北冰洋沉积物中硫酸盐还原菌的适宜生长温度为20℃(Knoblaucht et al., 1999a)。为了揭示研究区表层沉积物中厌氧菌在不同地理区域内的分布差异,讨论如下。

4.1　厌氧菌分布的纬度差异

将表层沉积物的AAB含量按纬度区间进行统计,结果列于表2。由表2可知,4℃时,AAB含量均值在4个纬度区间中呈现出随纬度升高而降低的趋势;25℃时,AAB含量均值也是随纬度升高而降低。尽管培养温度不同,AAB高含量的站位均出现于较低纬度的海区中,这表明靠近亚美大陆一侧的海区可能更适于厌氧菌,尤其是嗜中温厌氧菌的生长繁殖。推测该区域可能有更适于厌氧菌生存的条件,如较多的生源物质供应(薛斌等,2006)。低含量AAB的站位大多位于中、高纬度的海区中,并以4℃培养结果更为明显。这可能是受沉积物的化学成分、海底水温等因子的制约(Rivkin et al., 1996; Rossello-Mora et al., 1999; Robador et al., 2007)。本研究结果表现出测站间、海区间表层沉积物中 AAB 含量分布的不均匀性,类似的现象也见于北冰洋欧亚海盆(Kroencke et al.,2000)。

表 2 西北冰洋海域表层沉积物厌氧菌含量纬度区间的比较

Table 2 Comparison of anaerobic bacteria numbers in surface sediments by latitude in the Western Arctic Sea

纬度(N)	站位	4℃含量/(cell·g^{-1})			
		样品数	最小值	最大值	平均值
66°~70°	R1、R3、R6	3	2.40×10^7	2.40×10^7	2.40×10^7
70°~74°	C25、C19、C17、C16、C15、C13、S11、S26、S16、R15	10	1.90×10^4	2.40×10^7	3.32×10^6
74°~78°	P13、P11、M1、P21、P22、P23、P24	7	9.00×10^2	2.40×10^6	4.33×10^5
>78°	B77、B78、B79、B80	4	2.40×10^4	4.30×10^5	2.01×10^5
纬度(N)	站位	25℃含量/(cell·g^{-1})			
		样品数	最小值	最大值	平均值
66°~70°	R1、R3、R6				
70°~74°	C25、C19、C17、C16、C15、C13、S11、S26、S16、R15	4	2.90×10^4	2.40×10^7	6.67×10^6
74°~78°	P13、P11、M1、P21、P22、P23、P24	3	2.40×10^5	2.40×10^6	1.25×10^6
>78°	B77、B78、B79、B80	1			1.50×10^6

4.2 厌氧菌的经度分布差异

将表层沉积物的所在站位按经度区间划为 5 个分区后进行 AAB 指标比较分析,结果列于表 3。从表 3 可以看出,经度由东向西,4℃时的 AAB 平均值呈明显的增高趋势,而 25℃时这种增高趋势不明显。4℃和 25℃时,西部(165°W ~170°W)的 AAB 值分别是东部(145°W~150°W(4℃)、150°W~155°W(25℃))的 36.6 和 5.8 倍,这表明在所测站位由东部、中部向西部,生态环境逐渐有利于表层沉积物中 AAB 的生长繁衍。这与北太平洋的营养物质通过白令海峡进入西北冰洋,进而对研究区的生物地球化学活动产生影响,并呈由西向东减弱的趋势一致。

表 3 西北冰洋海域表层沉积物厌氧菌含量经度区间的比较

Table 3 Comparison of anaerobic bacteria numbers in surface sediments by longitude in the Western Arctic Sea

经度(W)	站位	4℃含量/(cell·g^{-1})			
		样品数	最小值	最大值	平均值
145°~150°	B80、B78、B77	3	2.40×10^4	4.30×10^5	2.31×10^5
150°~155°	B79、S26	2	1.10×10^5	7.50×10^5	4.30×10^5
155°~160°	S16、P24、S11	3	1.90×10^4	2.40×10^6	8.37×10^5
160°~165°	C19、C17、P23、C16、C25、C15、P22	7	1.50×10^4	2.40×10^7	4.13×10^6

（续表）

经度(W)	站位	4℃含量/(cell·g^{-1})			
		样品数	最小值	最大值	平均值
165°～170°	P13、C13、P21、R3、R6、M1、R1、R15、P11	9	9.00×10^{2}	2.40×10^{7}	8.45×10^{6}

经度(W)	站位	25℃含量/(cell·g^{-1})			
		样品数	最小值	最大值	平均值
145°～150°	B80、B78、B77				
150°～155°	B79、S26	1			2.40×10^{5}
155°～160°	S16、P24、S11				
160°～165°	C19、C17、P23、C16、C25、C15、P22	4	2.90×10^{4}	2.40×10^{7}	6.88×10^{6}
165°～170°	P13、C13、P21、R3、R6、M1、R1、R15、P11	3	2.40×10^{5}	2.40×10^{6}	1.38×10^{6}

4.3　不同水深沉积物中的厌氧菌分布

统计不同水深沉积物样品中厌氧菌含量，结果如表4所示。由表4可知，AAB含量在不同水深范围间均有差异。4℃时，AAB均值随水深的增加呈现出先略减小后增大，然后又减低到最小的态势，最大值和最小值分别出现在41～55 m和3 000 ～3 850 m。25℃时，AAB均值随水深增加呈现先增大而后减小的趋势，最大值和最小值分别出现在175～561 m和1456～2200 m。总体来看，两种培养温度条件下的AAB含量随水深增加而减小，这表明水深的增大不利于部分AAB的生存。这是水压增高、水温降低和来自海水的生源物质减少等生态环境因子综合影响的结果，但这些因素却可能趋向有利于表层沉积物中部分兼性好氧、兼性厌氧菌，包括一些适压(barophilic)和耐压(barotolerant)细菌的生长，而不利于严格厌氧菌的生存，具体原因尚待探讨。

表4　西北冰洋海域表层沉积物厌氧菌含量变化对上覆水厚度的响应

Table 4　Response of sediment anaerobic bacteria content to overlying water thickness in the Western Arctic Sea

水深/m	站位	4℃含量(cell·g^{-1})			
		样品数	最小值	最大值	平均值
41～55	C25、C15、C16、C13、C17、S11、C19、R1、R6、R3	10	9.30×10^{4}	2.40×10^{7}	1.03×10^{7}
175～561	R15、P11、P22、P13、P21	5	9.00×10^{2}	1.10×10^{6}	2.93×10^{5}
1 456～2 200	M1、P24、P23	3	2.40×10^{4}	2.40×10^{6}	8.88×10^{5}
3 000～3 850	S26、B80、B78、B79、S16、B77	6	1.90×10^{4}	7.50×10^{5}	2.62×10^{5}

(续表)

水深/m	站位	25℃含量($cell \cdot g^{-1}$)			
		样品数	最小值	最大值	平均值
41～55	C25、C15、C16、C13、C17、S11、C19、R1、R6、R3	2	2.40×10^5	2.40×10^6	1.32×10^6
175～561	R15、P11、P22、P13、P21	3	2.90×10^4	2.40×10^7	8.38×10^6
1 456～2 200	M1、P24、P23	1	2.40×10^5	2.40×10^5	2.40×10^5
3 000～3 850	S26、B80、B78、B79、S16、B77	2	1.50×10^6	2.40×10^6	1.95×10^6

5 结论(Conclusions)

(1)利用逐步稀释法对西北冰洋海域 24 个表层沉积物作了厌氧菌检出率和含量测定,厌氧菌的检出率高达 100%。在 4℃和 25℃培养条件下,厌氧菌的含量范围为 9.00×10^2～2.40×10^7 $cell \cdot g^{-1}$ 和 2.90×10^4～2.40×10^7 $cell \cdot g^{-1}$,平均含量则分别为 4.54×10^6 $cell \cdot g^{-1}$ 和 3.99×10^6 $cell \cdot g^{-1}$,证实研究区沉积物中广泛分布着大量的厌氧菌。

(2)研究区表层沉积物中厌氧菌含量均值大致呈现随纬度升高而降低,随经度由东向西升高,随水深增加而减小的趋势。厌氧菌的分布与沉积环境、沉积物中有机质含量、营养物质来源、水深等理化和生态因子有关,其相互作用关系还有待深入研究。

参考文献 35 篇(略)

(合作者:高爱国　赵冬梅)

DISTRIBUTION OF ANAEROBIC BACTERIA IN SURFACE SEDIMENTS FROM THE WESTERN ARCTIC SEA

(ABSTRACT)

Abstract The occurrence and content of anaerobic bacteria in 24 surface sediments in the Western Arctic Ocean were measured via progressive dilution under laboratory incubation at 4℃ and 25℃, and the spatial patterns of anaerobic bacteria distribution were further examined. All the samples have anaerobic bacteria. The contents of anaerobic bacteria cultivated ranged from 9.00×10^2 to 2.40×10^7 cell · g^{-1} wet sample (the same hereafter) and 2.90×10^4 to 2.40×10^7 cell · g^{-1} with an average of 4.54×10^6 cell · g^{-1} and 3.99×10^6 cell · g^{-1} in the 4℃ and 25℃ samples, respectively. The distributions of anaerobic bacteria varied by both latitude and longitude, and also in sediment samples with different water depths. The anaerobic bacteria content decreased from low latitude to high latitude, from the western area to the eastern area, and from shallow water to deep basin.

Key words Western Arctic Sea; Canada Basin; Chukchi Sea; Anaerobic Bacteria; Surface Sediment

北冰洋沉积物岩芯中厌氧细菌分布*

摘　要　用逐步稀释法对北冰洋研究区10个沉积物岩芯(计189份样品)的厌氧细菌(AAB)进行分析。用改进的ZoBell 2216号培养基,在4℃和25℃温度中分别培养3周以上,根据培养结果统计样品中厌氧细菌的检出率与含量,分析厌氧细菌含量的分布情况。结果表明,4℃时AAB检出率为85.71%,含量范围0～2.40×10^9个·g^{-1},平均4.42×10^7个·g^{-1};25℃时检出率为93.05%,含量范围0～1.10×10^9个·g^{-1},平均5.31×10^7个·g^{-1},略高于4℃时的平均值。研究表明研究区岩芯中广布AAB;温度的提高,可能提高AAB的相关指标;沉积物深度增加可能有利于AAB生长,但太深也不利于AAB生长,呈中层增高现象,但并不规则;在一定水深范围内,水深增加也有利于AAB的生长。同时给出了研究区内不同海域AAB分布差异。

关键词　北冰洋;沉积物岩芯;厌氧细菌;空间分布

微生物在自然界中的分布受制于所在的环境及其相关的生态因子,其中地理环境是重要的条件之一[1]。海底沉积环境不同,细菌生态学指标也将不同。同时,沉积物细菌的生态学指标也反映相应的地理学状况[2]。海洋厌氧细菌(anaerobic bacteria,简写为AAB,下同)大多生存于底层缺氧环境,包括不同深度的沉积物中,在海底生物圈中有着独特的地位和作用[3]。由于受采样条件、培养与分析方法限制等原因,对它们的专门研究,迄今并不是很多。以至于对海洋厌氧细菌的研究远滞后于海洋普通好氧菌或陆地与淡水环境细菌的研究[4,5]。

海洋沉积物中的厌氧细菌对游离氧有着不同的要求,它们中包括有比较严格意义上的厌氧细菌(专性厌氧细菌)、普通厌氧细菌、兼性厌氧细菌、甚或微好氧细菌。这表明所谓厌氧,只是相对好氧而言。一些条件下得出的研究报告大多包含了上述各类群的厌氧细菌,通称AAB,这在以往的研究中是司空见惯的[1]。在漫长的地质年代中,海底沉积物中厌氧细菌逐渐进化形成了有别于同一海区其他生境微生物的性状,表现出异样的生态及分布状态[2],这在北冰洋低温缺氧的海底环境中更为明显。在无溶解氧的条件下,专性厌氧细菌和兼性厌氧细菌通过厌氧分解将复杂的有机物分解成简单的有机物和无机物(如有机酸、醇、二氧化碳、氨、硫化氢以及其他硫化物等),再被甲烷菌进一步转化为甲烷和二氧化碳等,在这一代谢过程中,pH先下降(酸性发酵阶段)后上升(碱性发酵阶段)。它们改变沉积物的物理化学环境,对元素迁移、沉积物的成岩作用产生影响,并对全球的气候变化作出自身的响应[3,4]。

对厌氧细菌的研究已有了较久的历史,也有关于海洋厌氧细菌作用的阐述,但近年来对海洋沉积物厌氧细菌的研究并不多见[3,4],尤其对北冰洋沉积物厌氧细菌的专门研究成

---

*　原文刊于《海洋科学进展》,2009,27(4):469-476,本人为第二作者。

资助项目:国家自然科学基金——楚科奇海及其邻近海域沉积物中细菌生物地球化学(40576060),楚科奇海沉积物中古环境变化的地球化学响应(40376017)

果迄今仍十分少见[5-9]。本文对取自北冰洋楚科奇海和加拿大海盆的沉积物岩芯中厌氧细菌进行研究，分析沉积物岩芯厌氧细菌检出率和含量指标等生物地理学状况，以期有助于学科的发展及加深对北极地区生物地球化学研究。

1　研究区概况

研究区位于北冰洋楚科奇海与加拿大海盆（66°～80°N，148°～170°W，水深 41～3 850 m）。由于它靠近亚美大陆一侧，推测其沉积物应含较多营养物质[10]，研究区大致可分作楚科奇海台、南楚科奇海陆架、北楚科奇海陆架、北加拿大海盆、南加拿大海盆及西波弗特海六部分，在研究区共设 10 个站位（图 1）。

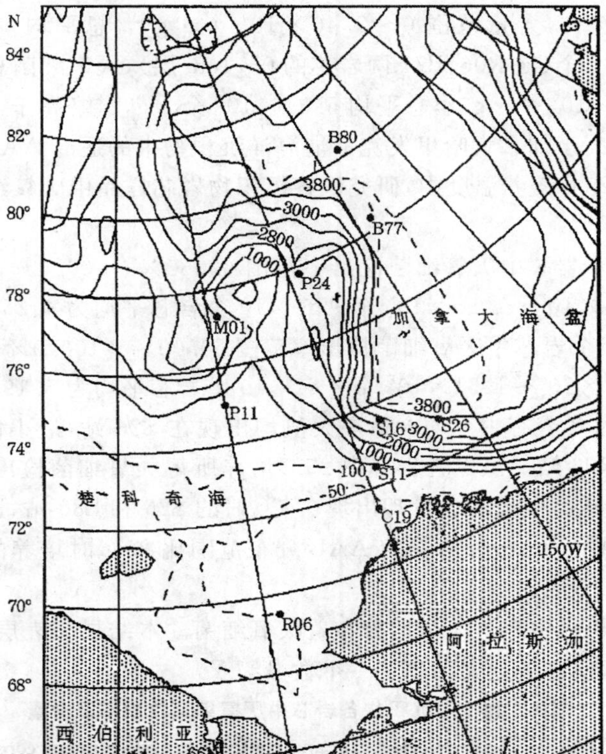

图 1　加拿大海盆与楚科奇海柱状样站位图

Fig. 1　Sampling stations of AAB from the Chukchi Sea and Canadian basin, the Arctic ocean

2　样品采集

2003 年 7～9 月的中国第 2 次北冰洋考察期间，在考察船"雪龙"号上，用箱式采样器或多管取样器采得沉积物岩芯样，岩芯长 12～38cm。沉积物岩芯采至甲板后，按微生物学采样要求进行样品采集，0～10cm 内按 1cm 间隔、10cm 以下按 2.0cm 间隔取样，装入无菌塑料瓶中，使其尽可能少暴露于空气中，封盖保存于 4℃ 以下的冰箱中，直到实验室分析。

3　厌氧细菌分析

在国内实验室对沉积物岩芯样进行厌氧细菌分析，所用培养基为改进的 ZoBell 2216 号培养基[1]。经逐步稀释的样品用橡皮塞塞紧后置于 4℃或 25℃的恒温中分别培养 3 周以上，期间不断观察培养情况，并与空白对照比较，以试管中出现混浊/菌膜/沉淀者为厌氧细菌阳性，参照文献[1]要求，计算样品厌氧细菌检出率(%)、含量(个 g-1 湿重，下同)等，以此统计所有岩芯中的厌氧细菌，分析、推断研究区厌氧细菌在岩芯中的垂直分布及其他变化状况。

4　结果

对 2 种培养温度时厌氧细菌的检出率和含量统计表明：4℃时样品数为 189 个，厌氧细菌总检出率为 85.71%，含量范围 0～2.40×10^9 个・g^{-1}，总平均 4.42×10^7 个・g^{-1}。25℃时样品数为 187 个(缺 R06、B77 两岩芯的 0～1cm 层)，厌氧细菌总检出率为93.05%，含量范围 0～1.10×10^9 个・g^{-1}，总平均 5.31×10^7 个・g^{-1}，略高于 4℃时的平均值。研究区厌氧细菌含量高于加拿大哈里发克斯港海洋沉积物中的适冷 AAB 含量[11]。

表 1 为 4℃和 25℃培养温度中，研究区各沉积物岩芯样品中厌氧细菌检出率和含量的统计结果。

4℃培养时 10 个岩芯中厌氧细菌检出率范围 62.50%～100.0%，平均为 85.71%，含量平均值范围 7.61×10^5～1.90×10^8 个・g^{-1}，10 个岩芯平均为 4.25×10^7 个・g^{-1}。

25℃培养时 10 个岩芯中厌氧细菌检出率范围 60.00%～100.0%，平均为 91.87%；含量平均值范围 2.43×10^6～1.51×10^8 个・g^{-1}，10 个岩芯平均为 5.19×10^7 个・g^{-1}。

2 个培养温度时各岩芯中平均含量最大值均出现在 S26 站，最小值 4℃时出现于 P11 站，25℃时出现于 R06 站。虽然各岩芯中 25℃培养所得厌氧细菌检出率与平均含量均大于 4℃时的，但单一样品中最高含量却出现在 4℃时的 S26 和 B80 站，达到 2.40×10^9 个・g^{-1}。这表明培养温度为 4℃时培养的 AAB 分布范围比 25℃时培养的 AAB 分布范围要宽。

这说明所测沉积物中广泛大量地分布着厌氧细菌。本结果与表层沉积物 AAB 相比，其检出率有所降低，但含量则大约高了一个数量级[12]。

表 1　北冰洋沉积物各岩芯中厌氧细菌检出率及含量

Table 1　AAB-detected rate and contents in the sediment cores

站位	岩芯长/cm	样品数/个	4℃(个・g^{-1})				25℃(个・g^{-1})			
			最小值	最大值	平均值	检出率%	最小值	最大值	平均值	检出率%
R06	12	11*	—	2.40×10^7	4.40×10^6	63.64	—	1.10×10^7	2.43×10^6	60.00
C19	22	16	9.00×10^3	2.40×10^8	1.71×10^7	100.00	2.60×10^4	2.40×10^8	6.25×10^7	100.00
S11	28	19	—	2.40×10^7	3.59×10^6	68.42	—	2.40×10^8	3.15×10^7	94.74
S16	28	19	1.90×10^4	2.40×10^8	4.58×10^7	100.00	1.20×10^4	2.40×10^8	6.93×10^7	100.00
S26	26	18	—	2.40×10^9	1.90×10^8	94.44	4.20×10^5	1.10×10^9	1.51×10^8	100.00
P11	38	24	—	5.30×10^6	7.61×10^5	79.17	—	2.40×10^8	2.29×10^7	91.67

（续表）

站位	岩芯长 /cm	样品数 /个	4℃（个·g^{-1}）				25℃（个·g^{-1}）			
			最小值	最大值	平均值	检出率 %	最小值	最大值	平均值	检出率 %
M1	36	23	—	2.40×10^8	1.22×10^7	95.65	—	2.40×10^8	1.22×10^7	95.65
P24	28	19	1.50×10^3	2.90×10^8	4.57×10^7	100.00	1.50×10^3	1.10×10^9	7.98×10^7	100.00
B77	22	16 *	—	2.40×10^7	2.47×10^6	93.75	—	2.40×10^8	2.00×10^7	93.33
B80	38	24	—	2.40×10^9	1.03×10^8	62.50	—	1.10×10^9	6.76×10^7	83.33
平均值					4.25×10^7	85.71			5.19×10^7	91.87

注：“*”表示 25℃时 R06 为 10 个样品，B77 为 15 个样品；“—”表示未检出

5 讨论

5.1 厌氧细菌的垂直分布

对比各岩芯内不同层位 4℃和 25℃培养时厌氧细菌检出率大小及平均含量（表 2）。4℃培养时全检出的层位占 33.3%，检出率最低值位于 16～18cm 层，为 55.56%；各层平均含量范围为 1.08×10^5～2.95×10^8 个·g^{-1}，平均 3.94×10^7 个·g^{-1}，最小值位于 36～38cm 层，最大值位于 12～14cm 层。25℃培养时全检出的层位占 58.3%，检出率最低值位于 14～16cm 层，为 77.78%；各层平均含量范围为 9.57×10^5～1.77×10^8 个·g^{-1}，平均 5.35×10^7 个·g^{-1}，最小值位于 28～30cm 层，最大值位于 12～14cm 层。这说明沉积物中的 AAB 检出率或含量在不同层位间的垂直分布是不均匀的。

表 2 4℃、25℃培养时沉积物岩芯中厌氧细菌指标的垂直分布

Table 2 Vertical distribution of the AAB indices of the sediment cores respectively with the 4℃ culture temperature and the 25℃

层位 /m	样品数 /个	4℃（个·g^{-1}）				25℃（个·g^{-1}）			
		最小值	最大值	平均值	检出率 %	最小值	最大值	平均值	检出率 %
0～1	10*	—	2.40×10^7	2.82×10^6	90.00	2.90×10^4	2.40×10^7	3.99×10^6	100.00
1～2	10	—	1.10×10^7	1.96×10^6	80.00	—	2.40×10^8	2.76×10^7	90.00
2～3	10	—	2.40×10^8	5.17×10^7	80.00	—	2.40×10^8	5.03×10^7	90.00
3～4	10	—	2.90×10^8	3.06×10^7	80.00	—	1.10×10^9	1.37×10^8	90.00
4～5	10	—	2.40×10^9	2.44×10^8	80.00	—	4.20×10^7	1.45×10^7	90.00
5～6	10	—	2.90×10^7	5.85×10^6	80.00	—	2.40×10^7	5.38×10^6	80.00
6～7	10	6.00×10^3	2.40×10^8	2.92×10^7	100.00	5.30×10^4	2.40×10^8	4.60×10^7	100.00
7～8	10	—	2.40×10^7	8.81×10^6	90.00	1.50×10^3	2.40×10^8	3.01×10^7	100.00
8～9	10	—	2.40×10^8	4.84×10^7	90.00	4.00×10^4	2.40×10^8	4.87×10^7	100.00
9～10	10	—	2.40×10^8	2.94×10^7	80.00	—	2.40×10^8	5.33×10^7	90.00

（续表）

层位/m	样品数/个	4℃（个·g^{-1}）				25℃（个·g^{-1}）			
		最小值	最大值	平均值	检出率%	最小值	最大值	平均值	检出率%
10～12	10	—	1.50×10^{8}	1.98×10^{7}	80.00	—	2.40×10^{8}	2.81×10^{7}	80.00
12～14	9	1.10×10^{4}	2.40×10^{9}	2.95×10^{8}	100.00	4.20×10^{3}	1.10×10^{9}	1.77×10^{8}	100.00
14～16	9	—	1.50×10^{7}	2.62×10^{6}	77.78	—	2.40×10^{7}	6.39×10^{6}	77.78
16～18	9	—	2.90×10^{7}	6.42×10^{6}	55.56	—	2.40×10^{8}	3.01×10^{7}	88.89
18～20	9	2.40×10^{5}	1.50×10^{7}	4.14×10^{6}	100.00	2.40×10^{5}	2.40×10^{8}	6.19×10^{7}	100.00
20～22	9	3.00×10^{4}	2.40×10^{8}	2.83×10^{7}	100.00	3.00×10^{4}	2.40×10^{8}	5.61×10^{7}	100.00
22～24	7	—	4.20×10^{7}	8.59×10^{6}	85.71	—	1.10×10^{9}	1.62×10^{8}	85.71
24～26	7	2.40×10^{5}	2.40×10^{8}	3.83×10^{7}	100.00	4.20×10^{5}	2.40×10^{8}	7.26×10^{7}	100.00
26～28	6	—	2.40×10^{7}	4.87×10^{6}	83.33	5.30×10^{3}	2.40×10^{7}	1.28×10^{7}	100.00
28～30	3	—	2.40×10^{5}	1.17×10^{5}	66.67	2.40×10^{5}	2.10×10^{6}	9.57×10^{5}	100.00
30～32	3	—	5.30×10^{6}	1.78×10^{6}	66.67	2.90×10^{4}	2.40×10^{8}	8.80×10^{7}	100.00
32～34	3	2.30×10^{3}	4.40×10^{6}	1.55×10^{6}	100.00	6.40×10^{3}	2.40×10^{8}	8.01×10^{7}	100.00
34～36	3	2.00×10^{6}	2.40×10^{8}	8.16×10^{7}	100.00	4.20×10^{6}	2.40×10^{8}	8.94×10^{7}	100.00
36～38	2	6.00×10^{3}	2.10×10^{5}	1.08×10^{5}	100.00	2.10×10^{5}	2.40×10^{6}	1.31×10^{6}	100.00
平均				3.94×10^{7}	88.07			5.35×10^{7}	94.27

注："*"表示25℃时0～1cm层位为8个样品，"—"表示未检出

若将样品分3段，计算两指标在0～8、8～22和22～38cm三段中的统计值，结果如表3所示。在2个培养温度时，检出率均呈下段大于中段、上段，表明较深部可能有利于厌氧细菌的生存。从厌氧细菌平均含量看，4℃培养时为中段大于上段、下段，表明上段是氧化还原环境的制约、下段可能是物质与能量来源的制约使厌氧细菌含量低于中段；但对于25℃时，厌氧细菌平均含量为下段大于中段、上段。这也表明厌氧细菌在沉积物中分布的复杂性，受多种因子制约，包括沉积速率、细菌移动速度、各种生物/非生物因子，包括营养物和氧气等的迁移和扰动[5,13,14]。

表3　不同深度沉积物中厌氧细菌检出率与含量

Table 3　AAB-detected rates and contents in the sediments at the different depths

深度/m	样品数/个	特征值	24℃/（个·g^{-1}）		25℃/（个·g^{-1}）	
			含量	检出率%	含量	检出率%
0～8	80*	最小值	—	85.00	—	92.31
		最大值	2.40×10^{9}		1.10×10^{9}	
		平均值	4.68×10^{7}		4.03×10^{7}	

(续表)

深度/m	样品数/个	特征值	24℃/(个·g^{-1})		25℃/(个·g^{-1})	
			含量	检出率%	含量	检出率%
8～22	75	最小值	—	85.33	—	92.00
		最大值	2.40×10^9		1.10×10^9	
		平均值	5.34×10^7		5.72×10^7	
22～38	34	最小值	—	88.24	—	97.06
		最大值	2.40×10^8		1.10×10^9	
		平均值	1.80×10^7		7.34×10^7	

5.2　厌氧细菌在不同水深下沉积物中的分布

表4为各岩芯按不同水深统计AAB分析结果。4℃与25℃时，AAB检出率均随水深增加而呈现为低—高—最高—次高的趋势，含量则呈现为随水深增加而呈高—低—更高—最高的态势，最高含量均出现于深水区，而最低含量则出现在175～561 m水深区。基本上表现为随水深增加而增加的分布格局。

表4　北冰洋不同水深下沉积物岩芯中厌氧细菌指标比较

Table 4　Comparison among the AAB indices of sedimentary cores from various water depths in the Arctic Ocean

水深/m	站号	样品数/个	4℃/个·g^{-1}			25℃/个·g^{-1}		
			最大含量	平均含量	检出率/%	最大含量	平均含量	检出率/%
41～55	R06,C19,S11	46*	2.40×10^8	8.47×10^6	78.26	2.40×10^8	3.61×10^7	88.89
175～561	P11	24	5.30×10^6	7.61×10^5	79.17	2.40×10^8	2.29×10^7	91.67
1 456～2 200	M1,P24	42	2.90×10^8	2.74×10^7	97.62	1.10×10^9	4.28×10^7	97.62
3 000～3 850	S16,S26,B80,B77	77*	2.40×10^9	8.84×10^7	85.71	1.10×10^9	7.83×10^7	93.42

注:"*"表示25℃时比4℃时少1个样品

总体看25℃培养得出的AAB两指标均高于4℃时的。因此厌氧细菌在深水沉积物中不但不减反而有增加的可能，这与文献所载的厌氧细菌与所在沉积物pH，Eh变化相适应是一致的[15,16]。

5.3　厌氧细菌在不同区域内分布

如将研究区按地理位置分成6个区域，各区域沉积物岩芯AAB统计结果见表5。

表5　北冰洋沉积物岩芯厌氧细菌指标在分海区中的比较

Table 5　Comparison among the AAB indices of the sedimentary cores from various regions in the Arctic Ocean

区域及编号	站号	样品数/个	4℃/(个·g^{-1})			25℃/(个·g^{-1})		
			最大值	平均值	检出率/%	最大值	平均值	检出率/%
①楚科奇海陆架北部	P11	24	5.30×10^6	7.61×10^5	79.17	2.40×10^8	2.29×10^7	91.67

(续表)

区域及编号	站 号	样品数/个	4℃/(个·g^{-1})			25℃/(个·g^{-1})		
			最大值	平均值	检出率/%	最大值	平均值	检出率/%
②楚科奇海台	M1、P24	42	2.90×10^{8}	2.74×10^{7}	97.62	1.10×10^{8}	4.28×10^{7}	97.62
③楚科奇海陆架南部	R06	11*	2.40×10^{7}	4.40×10^{6}	63.64	1.10×10^{7}	2.43×10^{6}	60.00
④波弗特海西部	C19、S11	35	2.40×10^{8}	9.75×10^{6}	82.86	2.40×10^{8}	4.57×10^{7}	97.14
⑤加拿大海盆北部	B77、B80	40*	2.40×10^{9}	6.29×10^{7}	75.00	1.10×10^{9}	4.93×10^{7}	87.18
⑥加拿大海盆南部	S16、S26	37	2.40×10^{9}	1.16×10^{8}	97.30	1.10×10^{9}	1.09×10^{8}	100.0

注:“*”表示25℃时比4℃时少1个样品

4℃培养得出的各区域AAB检出率范围为63.64%～97.62%,在各区域间大小排列为:②>⑥>④>①>⑤>③;在楚科奇海为②>①>③,即由研究区北部向中部渐增,然后再往南减少;在加拿大海盆为⑥>⑤,即由北向南增加。

25℃时AAB检出率范围为60.0%～100.00%,在各区域间表现为:⑥>②>④>①>⑤>③;在楚科奇海为②>①>③,在加拿大海盆的2个区域中为⑥>⑤,增减趋势,两温度时一致。这表明培养温度虽然不同,但AAB检出率的变化基本一致,即随着纬度的升高,检出率呈渐升之势。

4℃培养时各区域AAB平均值范围为7.61×10^{5}～1.16×10^{8}个·g^{-1},各海区大小排序为:⑥>⑤>②>④>③>①。25℃培养时AAB平均值范围为2.43×10^{6}～1.09×10^{8}个·g^{-1},各海区大小排序为:⑥>⑤>④>②>①>③。两温度时AAB的分布基本一致:在楚科奇海为由北向中递增而再向南减,加拿大海盆为由北向南增,分布格局与检出率相似。在楚科奇海、加拿大海盆和波弗特海西部三个海区中,4℃时AAB含量:加拿大海盆>楚科奇海>波弗特海;25℃时含量:加拿大海盆>波弗特海>楚科奇海,与4℃时基本一致,由西向东逐增之势更明显些。

AAB的这种分布格局与沉积物类型、沉积物中有机质含量、微生物的分解代谢、合成代谢、早期成岩作用等密切相关,并与离物源区的距离、微生物生境处的水深、温度、沉积物的孔隙度、氧化还原条件等有关[7-9,17],其分布机理与控制因素还有待进一步的研究。

5.4 温度的影响

由上述各表可知:沉积物样品中2种温度时的培养结果表明适当增高培养温度,会提升厌氧细菌的检出率和含量。温度的适当提升,也可能部分改善那些适应力强的嗜冷/适冷厌氧细菌或者原本受压抑的中温厌氧细菌酶活[18-19],这也许表明未来北极厌氧细菌资源的开发利用价值。结合其他研究结果[20],在全球气温变高的背景下,海水温度的升高对深海沉积物中微生物产生的影响也许不是立竿见影,甚而在某个时间段中也不一定产生明显的作用,但是气候变暖、水温升高,乃至沉积物温度的轻微提升将对海洋沉积物中的微生物生态系,包括所处的物理、化学、水文、地质环境形成潜在的影响,如被称为“死亡区”的低氧区或无氧区的扩大,进而对沉积物中微生物产生直接或间接的影响。届时包括北冰洋厌氧细菌在内的海洋地微生物的生理生化反应、生态系统的结构和功能有可能发生不可逆转的变化,产生新的平衡,祸福孰多孰少,难以预料,应引起足够的重视。

6　结语

本文对取自北冰洋楚科奇海与加拿大海盆的沉积物岩芯样，用改进的 ZoBell 2216 培养基进行了厌氧细菌检出率和含量的测定。结果表明：4℃时厌氧细菌总检出率为 85.71%，含量范围 0～2.40×10^9 个·g^{-1}，总平均 4.42×10^7 个·g^{-1}。25℃时总检出率为 93.05%，含量范围 0～1.10×10^9 个·g^{-1}，总平均 5.31×10^7 个·g^{-1}，略高于 4℃时的平均值。

在垂直方向上，以 0～8、8～22 和 22～38cm 三段中的统计值分析，检出率均呈下段大于中段、上段，表明深部有利于某些厌氧细菌的生存。厌氧细菌平均含量变化趋势为：4℃培养时为中段大于上段、下段，25℃时为下段大于中段、上段。

AAB 在不同水深中分布表现为随水深增加而增加的趋势，最低含量出现在 175～561 m 水深区，而最高含量均出现于深水区。

在 3 个海区中基本呈现出由南向北渐增、由西向东渐增的趋势。波弗特海区的分布特征处于楚科奇海与加拿大海盆之间，显示出介于两者之间的过渡中间类型。

培养温度的适当增高会增加厌氧细菌的检出率和含量。在全球气温变暖的大背景下，温度升高对深海沉积物中微生物产生的影响也许会导致低氧区或无氧区的扩大，其利弊应引起足够的重视。

参考文献 20 篇(略)

(合作者：高爱国　赵冬梅)

DISTRIBUTION OF ANAEROBIC BACTERIA IN SEDIMENTARY CORES FROM THE ARCTIC OCEAN

(ABSTRACT)

Abstract 189 samples divided from 10 sedimentary cores collected in the studied area of the Arctic Ocean are analyzed for anaerobic bacteria (abr. AAB) detection by means of the step-by-step dilution method. The samples respectively with the 4℃ temperature and the 25℃ have been cultured for 3 weeks to determine the AAB-detected rate and the content in the samples and to analyze the AAB distribution. As it is shown from the results, the AAB-detected rate of the 4℃ samples is 85.71% and the AAB contents range between 0 and 2.40×10^9 ind./g with an average of 4.42×10^7 ind./g, and the rate of the 25℃ is 93.05% and the contents range between 0 and 1.10×10^9 ind./g with their average of 5.31×10^7 ind./g, slightly higher than the 4℃ average. Therefore AAB distributed widely in the cores from the studied area. It seems that the temperature increment is helpful for the increments of those indices. In addition, the sedimentarv depths within the cerntain limit might be favourable to AAB growth. In general, the AAB contents become higher in the middle layers although it is not definitely regular. Water depth within a certain limit in the ocean might be also advantageous to AAB growth. The difference among the AAB regional distributions in the studied area is shown in the paper.

Key words Arctic Ocean; Sedimentary Cores; Anaerobic Bacteria; Distribution

F_2 硫酸盐还原菌

（SULFATE REDUCING BACTERIA）

海洋硫酸盐还原菌及其活动的经济重要性*

摘　要　文章论述了海洋硫酸盐还原菌的分类学、生态学、生化学和方法学等方面的一些问题。讨论了硫酸盐还原作用、微生物腐蚀作用及防腐措施。阐述了它们在海洋工程建设、水产养殖事业中的作用和地位。分析了我国这一领域的研究现状。指出了进一步开展这一工作的关键问题及深入研究海洋硫酸盐还原菌的必要性将越来越被人们所认识。

关键词　海洋硫酸盐还原菌　微生物腐蚀　控制腐蚀　SRB活动的经济重要性　海洋工程　水产养殖

1　引言

硫酸盐还原菌(sulfate reducing-bacteria,缩写为 SRB,下同),顾名思义是利用硫酸盐等取代好氧生物所用的氧为呼吸的氧化剂的一类微生物,因而也称之为硫酸盐还原微生物或硫酸盐还原者。在其代谢活动中,除 CO_2 和水外,还产生高浓度的 H_2S 为呼吸终产物,并以此与在腐败过程中形成少量 H_2S 的其他许多细菌相区别。经典意义中的 SRB,是专性厌氧者,但一般不为空气所致死。一旦环境变得厌氧,它们就繁殖起来。H_2S 是有害气体,在适宜条件下,它的大量积累造成钢铁、金属等材料的腐蚀,使工程设施受损;它败坏水环境等,使水产养殖业蒙受灾难。Gaines 于 1910 年首次发现了这一微生物腐蚀(microbial corrosion)现象之后,人们开始注意腐蚀问题的严重性。据近几年的不完全统计,由 SRB 等微生物腐蚀造成的年损失。美国高达 US$ 20×10^{10},日本为 US$ 4.6×10^{8},英国则达 10^{9} 英镑。海陆底质/地下管线腐蚀的 $(50\sim80)\times10^{-2}$ 由微生物引起。因腐蚀造成的损失约占国民生产总值的 $(3\sim5)\times10^{-2}$ 以上。SRB 成了腐蚀的主要原因。许多先进国家不得不将工程总费用的 $(15\sim20)\times10^{-2}$ 用于防蚀。连年来国内外水产养殖业滑坡虽然最终表现在病害致死上,但其重要原因之一则是养殖水体环境的恶化。养殖环境,尤其是底质的严重受污给大量 SRB 创造了繁殖的机会。污染环境与生物间的恶性循环使养殖业难以风光再现。SRB 虽在 19 世纪末已被分离到,20 世纪 30 年代提出了 SRB 参与金属腐

* 原文刊于《黄渤海海洋》,1998,16(4):64-74.

蚀中阴极去极化理论，但对它们参与的经济活动引起普遍关注则是近代的事情。当今国外行家及国内部分专家在工程开发活动和水产养殖事业中都重视调查研究、运用 SRB 的有关参数，SRB 的深入研究被提到议事日程上来[1]。SRB 作为自然界中的一类生物，是硫化物转化、物质循环和能量流动中不可缺少的参与者、发动者，但限于篇幅，本文仅就海洋 SRB 的分类、生态、生化学、研究方法和腐蚀及防护等经济活动作一不成熟的论述。

2　SRB 的分类学

SRB 的分类研究史已逾一个世纪，当今对它的认识仍在扩展、加深之中。对 SRB 的种属归类犹如其他微生物分类一样，仍在变动着。1965 年，Campbell 等提出将 SRB 以产生芽孢与否而分为 2 类[2]。产芽孢的有机体归属为脱硫肠状菌属(*D. sulfotomaculurm*)，其中的致黑脱硫肠状菌(*D. nigrificans*)为其模式种。不产孢子者归于脱硫弧菌属(*Desulfovibrio*)。当时涉及的 SRB，只有 5 种。它们分别是脱硫肠状菌属的 2 种，即 *Desulfovibrio desulfuricans*(脱硫弧菌)、*D. rubents chikii*(鲁氏脱硫弧菌)、*D. alstuarii*(河口脱硫弧菌)等。

在这之前，SRB 并非完全依此归类，被定为致黑芽孢梭菌(*Clostridum nigrificans*)的脱硫脱硫弧菌，最初却被定为脱硫芽孢弧菌(*Sporovibrio desulfuricans*)。以前的一些 *Sporovibrio* 被译作为有孢弧菌属，而现在它的一些 SRB 则都归入以前被译作为去磺弧菌 *Desulfovibrio* 之中了。最初发现螺菌属(*Spirillium*)的 3 个种即 *S. desulfuricans*(脱硫螺菌)、*S. aestuarii*(海水螺菌)和 *S. themodesulfurican*(高温脱硫螺菌)均为 SRB。在 1930 年，Bars 将它们看作为一个种——*Vibrio desulfuricans*(脱硫弧菌)的 3 个不同品系[3]。我国学者研究过微球菌(*Micrococcus*)产生 H_2S 的现象。SRB 在营养上和形态上的多样性，以基因 DNA 的 G+C 含量、色素等来分则有 11 个属。Skyring 等对采集的新 SRB 菌株作了数值分类研究后归纳出 8 个类群[6]。这些表明现代测定和分类技术的出现和完善，使 SRB 的种群范围比以前有了很大的扩增[2]。不能仅以成孢与否简单区分归类相关的真细菌(Eubacteria)。而且在古细菌(Archaeobacteria)中，也发现了 SRB。因此，1982 年由科学出版社出版的"细菌名称"一书，不可能包括以后被发现和定名的 SRB 种群。下面罗列散见于各时期相关文献中的 SRB 等名录(表 1)。

表 1　常见的硫酸盐还原菌种类(类别)

Table 1　Common species (Categories) of sulfate-reducing bacteria

Archaeobacteria 古生(细)菌纲	*D. hydrogenophilus* 嗜氢脱硫杆菌(嗜水脱硫菌)
Archoeoglobus 古生球菌属	*D. latus* 侧脱硫杆菌(阔脱硫菌)
A. fulgidus 火焰色古球菌(闪烁古生球菌)	*D. postgatei* 波氏脱硫(杆)菌
Bacillus 芽孢杆菌属	*D. propionicus* (丙酸脱硫杆菌)
Bacteriodaceae 拟杆菌科	*Desulfobacter* sp. *Strain* 3ac10 脱硫杆菌 3ac10 系
Bacteriun 杆菌属	*Desulfobacter* sp. *Strain* 4ac11 脱硫杆菌 4ac11 系
Butyuibrio 丁酸弧菌属	*Desulfobacter* sp. *Strain* B5401 脱硫杆菌 B5401 系
Clostridium 梭菌属	*Desulfobacter* sp. *Strain* DSM2035 脱硫杆菌 DSM2035 系

（续表）

C. granularum 颗粒梭菌※	*Desulfobacter* sp. *Strain* DSM2057 脱硫细菌 DSM2057 系
C. nigrificans 致黑梭菌	*Desulfobacterium* 脱硫杆菌属
Desulfoarculus 脱硫弓形菌属	*D. catechalicum* 儿茶酚脱硫杆菌
D. baarsii 巴氏脱硫弓形菌	*D. dautotrophicum* 自营脱硫杆菌
Desulfobacter 脱硫细菌属	*D. indolicum* 吲哚脱硫杆菌
D. curvatus 弯曲脱硫细菌	*D. niacini* 烟酸脱硫杆菌
D. phenolicum 酚脱硫杆菌	*D. desulfuricans* 脱硫脱硫弧菌
D. vacuolatum 具泡脱硫杆菌	*D. desulfuricans subs p* 脱硫脱硫弧菌亚种
Desulfobulbus 猎肠脱硫杆菌属	*Desulfouibrio desulfuricans subs p. desulfaricams Essex*6. 脱硫脱硫弧菌亚种脱硫 *Essex*6.
D. sapowrans	*Desulfouibrio desulfuricans subs p. desulfaricams El Agheila*. 脱硫脱硫弧菌亚种脱硫亚种 *EL Agheila*.
Desulfobulbus 脱硫球状菌属	*Desulfouibrio desulfuricans subs p. desulfaricams sylts*. 脱硫脱硫弧菌亚种脱硫亚种 *sylt*3.
D. elengatus 优雅脱硫球状菌	*Desulfovibrio gigas* 巨大脱硫弧菌
D. Propiraiuss 丙酸脱硫球状菌	*D. longus* 长脱硫弧菌
D. sp. strain DSM2058 脱硫球状菌 DSM2058 系	*D. salexigens* 需盐脱硫弧菌
*Sp. Strain M*161 脱硫球状菌 161	*D. sapowrans* 食皂脱硫弧菌
D. sp. strain 3*pr*10 脱硫球状菌 3*pr*10 系	*Desulfovibrio* sp. *Strain* DSM2056 脱硫弧菌 DSM2056 系
Desulfococcus 脱硫球菌属	*Desulfovibrio* sp. *Strain* 3210 脱硫弧菌 3210 系
D. multivorans 杂食脱硫球菌	*Desulfovibrio* sp. *Strain* 4207 脱硫弧菌 4207 系
D. niacini 烟酸脱硫球菌	*Desulfovibrio* sp. *Strain* 4303 脱硫弧菌 4207 系
Desulfomicrobium 脱硫微菌属	*Desulfovibrio* sp. *Strain* 9301 脱硫弧菌 9301 系
D. apshercnum	*Desulfovibrio* sp. *Strain* G4701 脱硫弧菌 G4701 系
D. baculatum 棒形脱硫微菌	*Desulfovibrio vulgaris* 普通脱硫弧菌
D. baculatum DSM1743c 棒形脱硫微菌 DSM1743C.	*Eubacterium* 真杆菌属
*D. baculatum Noruny*4 棒形脱硫微菌挪威 4	*Holococcus* 盐球菌※
Desulfomonas 脱硫脱硫单胞菌	*H. morrhuoe* 鳕盐球菌※
D. pigra 惰性单胞菌属	*Micrococcus* 微球菌属※
Desulfonema 脱硫丝状菌属	*M. cinnabareus strain D* 朱红微球菌 *D* 系※
D. limicola 泥生脱硫丝状菌属	*M. cinnabareus strain E* 朱红微球菌 *E* 系※

（续表）

D. magnum 马格南脱硫丝状菌	*M. citreus Migula var. marinus n. var* 柠檬黄微球菌海洋变种※
Desulfrhabdus※ 脱硫杆状菌属	*M. candidus corbert—strain A* 李氏微球菌海洋变种A系※
D. amnigenus※	*M. candidus* 亮白微球菌※
Desulforbopalus 脱硫感棍菌属	*Pasteurella* 巴斯德氏菌属※
D. vacuolatus 具泡脱硫感棍菌	*P. tularensis PRTOL* 1 土拉热巴斯德氏菌 1(*Francisella tularensis* 土拉热弗朗西丝氏菌)(野兔热弗朗西丝氏菌)
Desulfosarcina 脱硫八叠球菌属	*Spirillum* 螺菌属
D. variabilis 可变脱硫八叠球菌	*S. aestuarii* 河口螺菌
Desulfotomaculum 脱硫肠状菌属	*S. desulfuricans* 脱硫螺菌
D. acetoxidans 醋酸氧化脱硫肠状菌	*S. themodesulfurican* 高温脱硫螺菌
D. antarcticum 南极脱硫肠状菌	*Sporovibrio* 芽孢弧菌属
D. guttoideum 滴状脱硫菌	*S. desulfuricans* 脱硫芽孢弧菌属
D. kuznetsovii 库氏脱硫肠状菌	*Syntrophobacter S. funaroxidans* 合胞细菌属(互营杆菌属、共养杆菌属)
D. nigrificans 致黑脱硫肠状菌	*Thermodesulfobacterium* 高温脱硫杆菌属
D. orientis 东方脱硫肠状菌	*T. commune* 普通高温脱硫杆菌属
D. ruminis 瘤胃脱硫肠状菌	*T. mobile* 能动高温脱硫杆菌
D. sapomasndens	*Vibrio* 弧菌属
Desulfotomaculam strain T90A 脱硫肠状菌 T90A 系	*V. desulfuricans* 脱硫弧菌
Desulfotomaculam strain T93A 脱硫肠状菌 T93A 系	
Desulfotomaculam thermoacetoxidans 热氧化醋酸脱硫肠状菌	
Desulfovibrio 脱硫弧菌属	
D. aestuarii 河口脱硫弧菌	
D. africanus 非洲脱硫弧菌	
D. baarsii 巴氏脱硫弧菌	
D. carbinolicus	

※与 SRB 相近，产生硫化物或 H_2S。《细菌名称》，科学出版社 1982 年版中未有载入者。若干菌名的汉译可能不确切。

从染有(柴)油的 Suvsurlace 土壤中分离出一株新奇的细菌，暂定为 PRTOL1 的土拉热巴斯德菌(*Pasteurella tularensis*)。它生于淡水环境，是胞内含新型的 16S rRNA 基因序列的，能降解甲苯的 SRB[2]。

此外与 SRB 相关、参与硫循环和转换的微生物种类还有不少，在此不赘述。

3　SRB 的若干生态学问题

最早发现海洋 SRB 的是 Zelinski。他于 1893 年从黑海海泥中分离出 2 种 SRB。它们能将硫酸盐、亚硫酸盐及硫代硫酸盐还原为 H_2S。此后人们对 SRB 生存的历史、地域等的生态学知识有了很大的突破。

3.1　温度

迄今发现的大部分陆生 SRB 是中温(mesophilic)菌，最适生长温度为 30～40℃，也能在 10～15℃的温度中存活。

分布最广的 SRB 是海洋 SRB。这是由海洋面积在地球上占的份额所决定的。它们大多属低温性。其中一些是专性嗜冷菌(obligate psychrophiles)，其最适温度为 15～18℃，≥20℃时即可死亡。其他的是兼性嗜冷菌(facultative psychrophiles)，其最适生长条件与中温菌相似，但也可在≤20℃的温度中缓慢生长，所以象南极和北极那样的冷境中也存在有 SRB。

嗜热的(thermophilic)SRB 一般适于 55～75℃的温度中生长，最低生长温度为 35～40℃。有些嗜热 SRB 适应范围更宽。如一种耐热的 SRB，其最高生长温度为 70～85℃。而 Stettle 等(1987)发现的古细菌 *Archaeoglobus fulgidus*(闪烁古生球菌)的最适温度为 83℃，最高温升至 92℃时仍可存活[3]。少数可形成芽孢的 SRB 可以在 131℃的高温中存活 20 min。耐热 SRB 大都分离自热火山区、深油井、海底热泉或热液系统之中。

3.2　盐度

不同生存环境中的 SRB 适应不同的盐度条件。陆生或淡水环境中的 SRB 可不要求盐分或所需盐度很低。海洋 SRB 或咸水 SRB 所需盐度则因区域而异。盐度可以从 0.076×10^{-2} 到饱和浓度。河口脱硫弧菌需要海水或 3×10^{-2} 的 NaCl 才能生长。而脱硫螺菌在 3×10^{-2} NaCl 的培养基中即可活动了。含浓度为 24×10^{-2} NaCl 的泻湖水覆盖的泥淖中也可分离到脱硫脱硫弧菌。少数成孢(spore-forming)SRB 从高盐环境中可分离得到。

3.3　氧气

狭义的 SRB 生存于缺乏游离氧的环境中，它们对厌氧条件有苛刻的要求。所以一般的实验室培养不易成功。

自 1977 年 Jφrgensen 在氧化的海洋沉积物的还原微生境(microniches)中发现了细菌性的硫酸盐还原作用后，传统的 SRB 观念已有变化。1991 年 Canfield 等进一步在微生物垫(丛)中找到了好气的硫酸盐还原作用证据。该垫发现于墨西哥下加利福尼亚。此垫中除了土质体微鞘蓝细菌(原型体微鞘蓝藻 *Microcoleus chthonoplastes*)等成员外，还有高盐细菌。这表明在该垫中，硫酸盐还原作用可能有新的生化途径和复杂的微生物相互作用。这些发现大大突破了 SRB 认识的传统观念，SRB 的研究必将取得新型进展。

这样的特定好氧环境还包括海洋沉积物的有氧表层、废水处理厂滴滤器光合作用生物膜、甚而还有被硫化物污染的某些空气。

3.4　压力

大多数陆生 SRB 生长于常压下。海底 SRB 生存环境的静水压常常较高。它们是耐压(baroduric)菌或嗜压菌(barophiles)。如北海某处海平面下 2～4 km 的生油库区其压力为 200～500bar(20～50 mPa),仍有 SRB(Rosnaes 等,1991)。

3.5　深度

在海洋或油田的不同深度发现有 SRB,Rittenberg(1941)发现,加利福尼亚海岸和海湾的几百个不同钻探深度的沉积物样品中均有 SRB。我国东海某海洋开发工程项目区不同水深下的 39 个沉积物样品中,SRB 的检出率为 70×10^{-2}。沉积物中的 SRB 一般随钻探深度的增大而减少。在紧接表层沉积物与底水交界处下的有氧带(oxic zone)内,SRB 最多。但是海水,尤其是较清洁的大洋水,SRB 一般不易检测到,即使检测到,也主要出现于海底。在河口底或养殖水体污浊的底层海水中可以检出该菌。

3.6　SRB 生存的广泛性

由上可见,SRB 在地球上的生存环境很广,从地球的两极到较深的海洋,从较冷的区域到高温的热泉。与陆上地下管线和海洋工程设施相关的 SRB 则严重关系到经济效益的好坏。SRB 还活动于生物体内,因而又关系到生物的健康与生存。如反刍动物胃中有一种瘤胃球菌(*Ruminococcus*),牛瘤胃含物中有脱硫脱硫弧菌,羊胃中有东方脱硫弧菌等等。它们参与相关动物的营养、消化和转换活动。SRB 还可生存于并抵抗远远超过多数食物可保留风味和品质以外的高温之中,因而引起罐头工厂的严重污染。以上这些表明,SRB 可生存于各种相异生境之中,包括土壤、岩石、水体、污泥、空气以及一些人为的大小环境中,从这一角度看,SRB 是无处不在的(ubiquity)。

SRB 不仅有较年轻的物种,也有更古老的物种。如耐热的古生细菌 *Archaeoglobus fulgidus*(闪烁古生球菌)。还有属于古生细菌纲 Archaeobacteria 的一种极端嗜热的硫酸盐还原者。它们分离自意大利海底热液系统的深油井中,使热的硫酸盐还原为 H_2S[4]。

3.7　SRB 的含量

SRB 在不同生境中的含量是不同的。自本世纪初就有学者记录过 SRB 的含量状况。如 Issatchenk。在 1914 年就注意北极的某区海泥中广泛存在着河口脱硫弧菌。Eillis(1932)曾对克莱德海海泥中的 SRB 含量作过估计。Zobell 等在 20 世纪 30 年代分析过加利福尼亚州外太平洋底海泥中的 SRB。我国学者对我国海区一些沉积物的 SRB 作了一些估测,如长江口、山东乳山湾及其邻近养虾场等等[5,6]。

在许多情况下,沉积物表层中的 SRB 数量都多于表层以下的,但也不尽然。这在我国所测得的结果及德国的一些成果中都可反映出来。不同深度的沉积物可能有不同的 SRB 种类和含量。Lilleback 等 1995 年曾指出过这一点[7]。

Ramsing 等(1993)曾测定过废水处理厂滴滤器生物膜不同厚度处的 SRB。结果证实在不同性质和成分的沉积物介质中,因条件不同(如光照或黑暗)可以有不同的 SRB 含量[8]。

表 2 列出了一些海区沉积物或介质中的 SRB 含量。

表 2　一些海区沉积物、介质中的 SRB 含量※

Table 2　Sulfate-Reducing Bacterial content of sediment from some sea areas and media

沉积物介质所在区域	SRB 含量(CFU/g)※	样品所取 层次	备注	引证文献号
克莱德海	2.7×10^{7}	表层		
加利福尼亚外太平洋	$1.0\times10^{4}\sim1.0\times10^{6}$	表层		
32°51′02″N,117°28′03″W	1.0×10^{3}	780 米水深下表层		
33°25′09″N,118°06′05″W	1.0×10^{3}	505 米水深下表层		
32°44′02″N,118°46′01″W	1.0×10^{4}	1 322 米水深下表层		
中国长江口	1.0×10^{3}	表层		[5]
中国山东乳山湾	$4.0\times10^{3}\sim1.1\times10^{6}$	表层		[6]
乳山湾畔养虾厂	$7.0\times10^{3}\sim5.4\times10^{5}$	表层		[6]
胶州湾潮间带和沿岸区	4.1×10^{7}	表层		
54°29′26″N,09°59′15″W	$1.5\times10^{5(a)}$	表层下 3.0～6.0cm		[7]
54°29′26″N,09°59′15″W	$6.0\times10^{7(a)}$	表层下 18.0cm	脱硫脱硫弧菌	[7]
54°29′26″N,09°59′15″W	$7.0\times10^{5(a)}$	表层下 18.0cm	*Desulfovibrio baculatus*	[7]
丹麦 SKΦdstrup 废水	$1.0\times10^{9(b)}$	2.0～3.0 mm 厚度内	有光培养中	[8]
处理用滴滤器生物膜	未测出或很少$^{(b)}$	0.5～1.0 mm 厚度内	有光培养	[8]
	$1.0\times10^{9(b)}$	5.0 mm 厚度外	黑暗培养	[8][12]
荷兰某污水处理厂	$1.0\times10^{4}\sim1.0\times10^{10(c)}$	厌氧颗粒污泥		[9]

※一般用 MPN 法测得,(a)间接荧光免疫法,(b)低核苷酸探针法,(c)dot blot 杂交和 PCR 扩增法。

SRB 含量呈现一定的季节变化趋势。如乳山湾 5～6 月份的少,而 8～9 月份达峰值[15]。

对海洋沉积物 SRB 的估算,受诸多因素的影响,但重要的 2 点是肯定的:①尽可能取得不扰动的样本;②选取合适的培养基和培养条件。

由此可见,掌握特定环境中 SRB 的时空分布及其变化关系着工程项目的设计和防蚀技术的制订与实施。对 SRB 含量状况,相关的工程技术部门和人员必须做到心中有数。

4　SRB 的研究法

在尽可能不扰动地取得样品后,重要的工作是尽早地采用合适方法培养、进行硫酸盐还原速率的测定和对金属等材料的腐蚀作用的研究。

4.1　SRB 所用培养基举例

中温性 SRB 的培养基[6]

乳酸钠(60×10^{-2}),6.0cm^3;氯化铵,1.0g;磷酸二氢钾,0.5g;硫酸镁,(含 7 分子水)。2.0g;硫酸钠,0.5g;陈海水,1.0dm^3,pH7.0～7.5。煮沸后滤去絮凝物,以 15cm^3 分装于试管中,高压灭菌后即用。使用前每试管滴加过滤除菌的新配制的硫酸亚铁铵(含 6 个分子水)溶液(10×10^{-2})0.2cm^3 及抗坏血酸液(1.0×10^{-2}),0.1cm^3。用 MPN 法接种倍比稀释的样品,于 37℃中,培养 7d 后观察结果。

用于脱硫弧菌的培养基(即改进的 42# 培养基)[10]

蛋白胨,5.0g;牛肉膏,3.0g;酵母膏,0.2g;葡萄糖,5.0g;半胱氨酸水溶液,0.5g,刃天青(0.025×10^{-2}溶液),4.0cm^3;蒸馏水,1.0dm^3。pH7.0±0.2(对海洋菌,可以陈海水代替蒸馏水)。

三糖铁琼脂培养基

牛肉膏,3.0g;酵母膏,3.0g;细菌蛋白胨,15.0g;胨,5.0g;乳糖,10.0g;蔗糖,10.0g;葡萄糖,1.0g;硫酸亚铁,0.2g;氯化钠,5.0g;硫代硫酸钠,0.3g;琼脂,12.0g;酚红,0.024g;蒸馏水,1.0dm^3。

以上培养基分装消毒后,制成底柱为 1 寸、斜长 1.5 寸的试管斜面,用作穿刺培养。

4.2 硫化氢的测定

4.2.1 硫化氢是 SRB 生化活动的重要指示性产物

H_2S 的定性测定一般用下列 2 法测。①醋酸铅纸条法:如果有 H_2S 产生,即可以使醋酸铅条变黑。②次甲基蓝:在酸性溶液里,对位氨基一二甲基苯胺和 H_2S 在有氯化铁存在时形成次甲基蓝。

4.2.2 H_2S 的定量测定.

用示踪元素术测硫酸盐的还原率。以 35S 作标记,用 $CuSO_4$-HCl 试验,并辅以微电极测定,得出 H_2S 的产率[8]。次甲基蓝或硫酸铜法可测出 H_2S 的浓度[11]。如 Rosene 等(1991)以此测定了分离自油田水的耐热 SRB——一种脱硫肠状菌的 H_2S 产生速率[4]。而甲基蓝法可查出 0.01×10^{-6} 的硫化物浓度。Kuhl 等(1992)则用微传感器(microsemor)法测定厌氧生化膜中密集的微生物群落之硫酸盐还原作用和硫化物的氧化作用[11,12]。

4.3 SRB 腐蚀钢铁速率的测定

SRB 对钢铁腐蚀的作用常靠失重法测得。将已知的 SRB 菌株接种入相应的培养基中培养,插入一定规格精确处理和称重过的钢、铁或其他金属试片。待适当培养一段时间后取出,测定经 SRB 的腐蚀作用和腐蚀率[13]。所用菌株可以是单一菌种,也可是混合菌种。试验设计可以根据要求进行各种有控模拟测定。

野外测定可做现场埋片试验,但必须首先了解现场的腐蚀环境状况、选择有代表性和相对稳定性的条件中进行。为此,诸如沉积物的类型及沉积物扰动状况、温度、电阻率及氧化—还原状态、重金属和硫化物含量及 SRB 等生物的含量、有机物类型及含量等都必须预先测得。海底试验还必须了解底层海水的温度、各营养盐成分及含量、动力学条件及其他因素等。经过一定时期的现场试验后,将埋片取回观察分析并计算材料被腐蚀的情况。

5 SRB 进行的硫酸盐还原作用和腐蚀活动

脱硫脱硫弧菌等 SRB 细胞含有氢化酶、重亚硫酸盐(bisulfite)还原酶-desulfoviridin。杂食脱硫球菌-*Desulfococcus multivorans* 和泥生脱硫丝状菌 *Desulfonema limicola* 也含有 desulfoviridin 等生化活动催化剂。SRB 摄氢使腐蚀反应的阴极去极化。

在有机质不多的环境中,多数 SRB 对无机硫酸盐的还原作用过程如下:在含水环境中的铁可能经受下列初始反应:$8H_2O\rightarrow8H+8OH^-$,$4Fe+8H\rightarrow4Fe^{2+}+8[H]$,之后在金属表面形成氢的保护膜。阻止该金属被进一步水解。SRB 用氢还原硫酸盐:$H_2SO_4+8[H]$

$\rightarrow H_2S+4H_2O$。这一步即是阴极去极化作用。由此铁受蚀：$Fe^{2+}+H_2S\rightarrow FeS+2H^+$。这里，阳极也发生去极化作用：$3Fe^{2+}+6H_2O\rightarrow 3Fe(OH)_2$，铁受蚀加重。因此，SRB 对金属材料还起着加速腐蚀的作用。SRB 产生的腐蚀产物包括 3 克分子的铁氧化物和 1 克分子的硫化物等。就是说并非环境中所有的硫化铁均被 H_2S 所攻击，其他含硫化物如亚硫酸盐、硫代硫酸钠等也可被 SRB 所还原。因此，阴极氢的分解是 SRB 等厌氧菌致腐的主要过程，又加速了铁的腐蚀作用。此间有机质的存在更助长了 SRB 的繁盛。

在含过量残饵、水循环差的养殖池塘或有机污染严重的近海等水环境中，在其他微生物配合下，有机物经 SRB 等的作用产生了大量 H_2S 等。其化学反应大致如下[6]：

$C_6H_{12}O_6+6H_2O\rightarrow 6CO_2+24[H]$，

$SO_4-2+8[H]\rightarrow H_2S+2H_2O+2OH^-$

$SO_4-2+8[H]+ATP\rightarrow S^{-2}+4H_2O+AMP+PP$

$$\underset{\displaystyle NH_2}{HSCH_2\underset{|}{C}HCOOH}\rightarrow H_2S+NH_3+CH_3COOH$$

此时的 H_2S，一部分对工程设施产生腐蚀作用，大部分则毒害所养生物。

这些有机硫化物大多以半胱氨酸、胱氨酸、蛋氨酸、甲酸盐、乙酸盐、延胡索盐、苯甲酸盐、丁酸盐及其他硫氢基氨基酸形式存在，少数有机硫化物则是硫酸酯、高碳链脂肪酸、有机硫酸盐、磺酸盐（sulfonetes）、含硫葡萄糖苷等形式存在。还原和利用它们的 SRB 以及相关微生物广泛分布于相应环境之中。

许多 SRB 所引起的腐蚀作用需要一个厌氧环境，其产物一硫化物也要在厌氧环境中才得以不被氧化，而且与 SRB 培养物相连的样本之电极势必须是低的——这有利于腐蚀作用的进行。但若此处有二甲基吖啶或有铬酸盐存在，这些 SRB 的生长将受抑[10]。研究表明。当中性土壤的 Eh 低于 +200 mV 时，腐蚀过程受阻。>400 mV 时，腐蚀作用全无。由此，启示人们创造合适条件以抑制腐蚀过程。

SRB 除引起铁的腐蚀外，还使 Cu、Pb、Zn 等金属材料腐蚀，尤其是当铜与少量碳之类的阴电材料相接触时腐蚀更为严重。

除了脱硫脱硫弧菌进行的此种典型的腐蚀作用外，东方脱硫弧菌和致黑梭菌还引起不同的腐蚀活动。它们均不含氢化酶。前者引起的是非阴极去极化作用的腐蚀。后者使热的变压罐和糖厂的热糖蜜罐腐蚀。

一些 SRB 的细胞内含有 Desulfofuscidin、desulforubidin 或 desulfoviridin 之类的脱硫酶类物质。如一种嗜热脱硫杆菌（*Thermodesulfobacterium*）有 desulfofuscidin。*Desulfomicrobium* baculatum Norway 4 和 DSM1743C 胞内有 desulforubidin。据报，这些酶类均参与硫酸盐的还原活动[14]。

不同生境中的 SRB 在适当条件中因其生化活动而引起的生态学变化基本上包括以下 7 点：

(1)形成硫化物。在 pH 为 7、Eh 为-300 mV 的区域中，硫酸盐变为硫化氢之类的硫化物，腐蚀电池形成。这种环境于好气菌不利，使高等生物中毒。在油田区，使油井堵塞、油品变质并影响有关操作人员的健康。

(2)pH 值改变为碱性。

(3)形成硫化铁。重金属发生迁移活动,促进阳极化过程。

(4)氢被移去,使环境中缺氢,管线、设施、贮油器或养殖场构件的金属等材料受腐。

(5)因硫酸盐被消耗而停止了相关的还原活动。

(6)共栖微生物区系发生改变,好气生物趋于消失,甲烷菌失活。对无色和有色硫细菌及兰细菌的生存有利。硫黄氧化菌受到鼓舞。

(7)同位素硫发生分部分离。SRB一般优先利用较轻的同位素S32。这样就为相关生物提供了初始的好气和光合生命进程条件,从而使SRB在地质学年代历程中起作用[13]。

6 SRB腐蚀作用的防护措施

可以从物理、化学和生物学3个方面采取防护措施[15]。

6.1 物理学的

(1)减少沉积物扰动和运移机会,减少设施受机械磨损的机会。

(2)对可能造成SRB适宜环境之处加强通气、调节温度。

(3)用紫外线、超声波或新生态氧等作为消毒手段。

6.2 化学的

(1)阴极保护。用锌、镁或铝—镁合金等比铁更负的金属阳极材料作牺牲电极以提供足够的保护电流或外加电流使相应材料保持足够的负电位以阻止金属被溶。

(2)创造条件提高 Fe^{3+}/Fe^{2+} 值,使SRB繁殖活动处于不利状态。

(3)提高pH值。当 $pH>9$ 时,可以抑制一些SRB活动。

(4)选用合适的药剂达到抑/杀菌、防蚀和去垢的作用。

6.3 生物学的

(1)控制环境中的有机质、铵盐、磷酸盐、铁/亚铁离子的浓度和含量。

(2)调整和处理好大小生境中的生态结构。利用生物间的各种关系,达到除害防蚀效果。

(3)掌握SRB的生态习性。使它们附着在有关金属等材料上以形成腐蚀电池。

据此选用耐蚀性更好的金属或合金、非金属材料来代替钢铁。在必要的钢铁设施、构件内外涂复防护层或隔离层以阻止它们与环境间SRB、离子、电子的侵蚀和渗透。总之无论采取何种防腐措施,都必须综合考虑、取长补短、经济实惠、简便易行[11]。

7 我国海洋SRB研究展望

我国海洋SRB的研究史,可上溯到20世纪50年代。之后一部分工作集中在海洋SRB对钢铁等金属材料的腐蚀作用上[13,17]。在蓬勃开展的海洋工程建设和水产养殖事业中,人们体会到底质环境调查、管线敷设、路由勘定及水产病害的防治等经济活动都离不开有关SRB的知识,因而对SRB的生态学[5,6]、腐蚀行为[18]、机理及防护[19,20]、污染防治[6]等的研究日益受到重视。过去几年间,我国在从北至南的海区及沿岸地带开展了大范围较广泛的调查研究工作,为有关项目的论证、规划和实施等曾提供了重要的数据和信息。但总体水平仍落后于国内陆土SRB的研究。回顾这一历程可以发现,我国海洋SRB的研究仍缺乏统一规划,采用的方法仍较原始和传统。与相关学科间的渗透也不够深入。因而成果难免显得粗浅。与国际先进水平相比,差距拉大[1,7]。SRB的研究是基础性和应用性都很强的工作,提高我国SRB研究领域的水平,必须采用先进的方法。要使SRB资料及时准确地应用于经济活动中,必须广泛紧密结合相关的调研项目。随着海洋SRB研究力度的加大,SRB在自然界和国民经济中的作用最终会被众多学者和决策人所深刻认识和发挥出来。

参考文献20篇(略)

(合作者:战　闰)

MARINE SULFATE-REDUCING BACTERIA AND THE ECONOMIC SIGNIFICANCES OF THEIR ACTIVITY

(ABSTRACT)

Abstract This paper expounds some aspects of taxonomy, ecology, microbial chemistry, methodology of marine sulfate-reducing bacteria (SRB). The sulfate reduction, microbial corrosion and some measures of anti-corrosion were discussed together with an explanation of their function and position in the marine engineering construction and the undertaking of aquaculture. In addition, analysis is made of the research situation of marine SRB in China and key is pointed out to the question of further research of marine SRB in China. The necessity of deep research into marine SRB will be recognized more and more by men.

Key words Marine Sulfate-Reducing Bacteria Microbial Corrosion Corrosion Control Economic Significances Of SRB Activity Marine Engineering Aquaculture

北部湾东侧沉积物硫酸盐还原菌含量及其意义*

摘　要　用 MPN(the Most Probable Number)法测定了北部湾东侧沉积物硫酸盐还原菌(SRB)含量,表层沉积物 SRB 的检出率约 76.3%,浅层(表层下 0.5 m～3.7 m)中是 31.3%。表层和浅层中的 SRB 平均含量分别是 1 760.0 个每克(湿重,下同)和 344.4 个每克。分析了测区不同区域的 SRB 含量分布。讨论了 SRB 含量与硫化物(W_S^{2-})、Eh、T(℃)、pH 值间的关系和对腐蚀环境评价的意义。

关键词　硫酸盐还原菌　北部湾　沉积物　腐蚀　环境

大多数硫酸盐还原菌(SRB)生存于厌氧环境,其生化活动的一个重要产物—硫化物腐蚀工程、水产养殖设施的金属等构件,使石油变质,气体堆积[1]。SRB 成了环境腐蚀的主因[2]。环境监测、海洋工程等均要将防 SRB 引起的腐蚀问题列为重要项目之一,因此,就特定海洋环境中 SRB 含量变化的研究既是寻找其自身生态规律的必需,也是海洋经济建设所必要。

北部湾位于热带,其东侧区沉积物 SRB 处于较高的温度、较丰富的营养及渐增的海洋开发活动影响之下。本文就该海区沉积物 SRB 含量及对环境腐蚀的意义作一分析研究。

1　环境概况

研究区域位于广西北海市和海南临高县连线间约 158 km² 的范围中(E109°00′～109°40′,N20°10′～21°15′)。共设 21 个测站,其 15 个测站位于 N21°30′,E109°06′和 N19°50′,E109°36′两点的中轴线上(它们分别为 No. 2、3、4、5、6、7、8、9、10、11、12、13、14、17、18、19 站)。该线两侧各平行扩展 0.4 km 成东西两线,每线设定 3 个测站。东线有 No. 7、20、21 等 3 站。西线有 No. 1、15、16 等 3 站,或将 No. 1～7 的 7 个站划为北区,No. 15～21 的 7 个站划为南区,其余 7 个站划为中区,详见图 1。

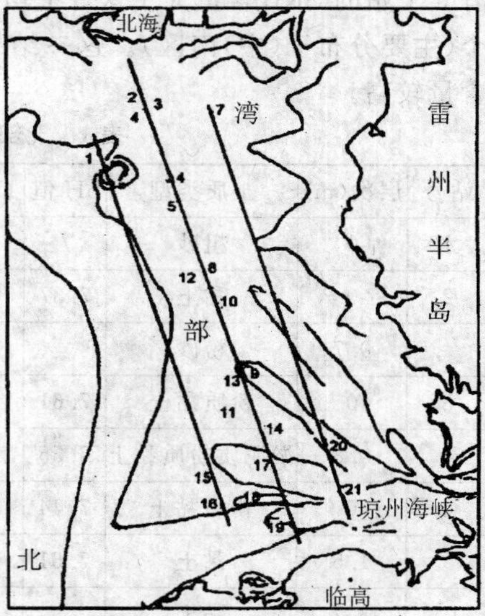

图 1　硫酸盐还原菌采样站点示意图

测区表层沉积物类型以黏质粉土为多见,约占 47.1%,其次为粉砂、黏土、粗砂质各占 17.6%,浅层沉积物类型以淤泥质较多,约占 47.1%,次为粉砂、淤泥质粉质黏土,各占 23.5%,次砂质沉积物占 5.9%。详见表 1。

* 原文刊于《广西科学院学报》,1999,15(3):103-107

测时所得表层样品的温度（T℃），酸碱度（pH 值），氧化还原电位（Eh），硫化物活度（W_S^{2-}（10^{-6}））分别为 20.3（变幅：19.2～22.0）、7.56（变幅：7.37～7.89）、90.63（变幅：56.00～173.00）、24.79（变幅：9.76～44.96）。相应地，浅层这 4 个参数分别为 23.0（变幅：22.5～24.5）、7.70（变幅：7.37～8.40）、99.3（变幅：55.00～272.00）、177.13（变幅：101.68～238.40）。这表明表层的 4 个参数各自的均值均比浅层的低，各自的变幅除温度外，也均比浅层的小。说明表层的这 3 个因子变化较小，而浅层温度较高且稳定。

2 材料和方法

2.1 取样

表层沉积物用青岛产 HNM 型蚌式采泥器取样，浅层沉积物（表层下 0.5 m～4.0 m 间）用青岛产重力活塞柱状取样器（带塑料衬管）取得。

2.2 测定

2.2.1 SRB 在船上实验室按需挑取样品入采样瓶中，取少量沉积物依文献[3]要求用 MPN 法稀释、培养后，计算样品中的细胞数。为了对照，取表层水样作同样处理。各个步骤均须无菌操作。

2.2.2 理化参数 样品温度：当样品来到船甲板上时即刻测定其温度（T℃）。pH 值：用 CP5943-05 型 pH/mV 计测。Eh：用铂—甘汞电极对测。HS^-：所需样品先封存于聚乙烯塑料袋中（－20℃），再用碘量法测。

3 结果和讨论

3.1 表层沉积物 SRB 含量

经计算所得表层和浅层的 SRB 含量均列入表 2 之中，由该表可得出，表层沉积物的 SRB 检出率为 76.3 %，SRB 含量变幅是 0～11 000.0 个每克，平均值为 1 760.0 个每克，有 6 个站的 SRB 含量大于这一平均值。最高值出现于北区近岸的 No.1 站，未检出站计 5 个（主要分布于（中）南区）。这一结果与东海测区比，可以发现 SRB 检出率较低，但 SRB 含量较高。

表 1 北部湾东侧沉积物 SRB 的环境参数

站号	层次(m)	土质类型	pH 值	Eh(mv)	W_S^{2-}（10^{-6}）	T(℃)	经纬度
1	0	粗砂	/	/	/	/	E109°04′18.43″，N21°07′13.48″
2	0	黏土	7.37	66	27.52	20.0	E109°10′24.37″，N21°13′32.02″
	0.7	粉砂	/	/	/	/	
3	0	粉质黏土	7.61	75	28.12	19.2	E109°10′49.06″，N21°13′41.18″
	0.8	淤泥质粉质黏土	7.55	109	184.60	23.0	
4	0	粉质黏土	7.44	84	44.09	20.0	E109°14′09.89″，N21°01′13.49″
	1.5	黏土	7.61	78	197.08	22.5	
	3.7	黏土	/	/	/	/	
5	0	粉质黏土	7.50	99	44.96	20.2	E109°15′48.0″，N20°57′05.80″
6	1.5	淤泥	7.75	109	184.60	23.0	E109°14′58.00″，N21°01′22.29″

(续表)

站号	层次(m)	土质类型	pH 值	Eh(mv)	W_S^{2-} (10^{-6})	T(℃)	经纬度
7	0	粗砂	7.70	65	20.53	19.3	E109°20′04.18″,N21°11′11.35″
8	0	粉质黏土	7.46	109	15.06	20.2	E109°19′13.08″,N20°45′09.82″
	1.5	淤泥	7.58	160	141.59	23.0	
9	0	粉质黏土	7.56	173	28.50	20.0	E109°24′27.76″,N20°28′13.82″
	1.2	淤泥	/	/	/	/	
10	0	粉质黏土	7.45	100	15.99	21.0	E109°20′58.50″,N20°40′38.20″
	1.5	淤泥	7.39	101	185.28	23.2	
11	0	粉砂	7.58	105	23.07	21.4	E109°26′06.02″,N20°24′10.13″
	1.5	淤泥质粉质黏土	7.81	107	169.63	23.5	
12	0	黏土	7.52	86	18.00	20.4	E109°19′54.03″,N20°44′57.00″
	1.5	淤泥	7.61	104	210.99	24.5	
13	0	黏土	7.54	84	27.78	19.8	E109°24′46.51″,N20°28′41.69″
	1.5	淤泥	7.37	87	108.69	22.5	
14	0	粉质黏土	7.50	100	23.71	19.5	E109°28′54.01″,N20°16′00.55″
	1.06	淤泥质粉质黏土	8.40	272	101.68	/	
15	0	/	/	/	/	/	E109°20′08.09″,N20°12′09.52″
16	0	/	/	/	/	/	0E109°21′37.13″,N21°07′19.81″
17	0	粉质黏土	7.48	56	53.41	20.7	E109°28′22.07″,N20°15′50.69″
	1.5	淤泥	7.45	−55	238.40	23.0	
18	0	粉砂	7.89	83	9.76	20.5	E109°30′53.55″,N20°07′35.25″
	1.04	淤泥质粉质黏土	7.66	112	177.01	22.6	
19	0	/	/	/	/	/	E109°32′42.06″,N20°03′25.66″
	1.5	淤泥	7.82	39	272.07	/	
20	0	粉砂	7.63	63	16.19	22.0	E109°29′42.07″,N20°11′44.44″
	0.8	粉砂	8.06	68	131.09	22.5	
21	0	粗砂	7.73	102	11.45	20.0	E109°40′11.72″,N20°09′33.45″

* pH 值、Eh、W_S^{2-}、T 分别为氢离子浓度、氧化—还原电位、硫离子活度、温度，下同。

以 3 条测线分，可知东线 SRB 的平均含量是 2 533.3 个每克、中线为 1 224.5 个每克、西线为 3 690.0 个每克，即显出西＞东＞中线之势。以南北分，则北、中、南 3 区的 SRB 平均含量分别为 2834.4 个每克、1 873.1 个每克和 583.6 个每克，即北＞中＞南。因此，测区的西北侧 SRB 含量较高，这可能与近岸、近岛及较多的海洋开发活动有关。

研究表明，在严格的实验条件下，SRB 的活动使硫酸盐被还原，硫化物含量增加，氧化还原电位(Eh)降低。同时SRB也与其他微生物类似,即环境中的温度适当增高,将促进它们的生长繁殖。SRB活动的结果之一可能使 H^+ 浓度增高,从而降低 pH 值。但 SRB 含量在一定条件下又可能显出与 pH 值呈相逆之势[4,5]。

表3列出的是测区东、中、西3线和北、中、南3区 SRB 含量与 W_S^{2-}、pH 值、Eh、T 等之间的比较结果。

由表3可知,在东、中、西3线上,SRB 含量差异与 Eh 差异、与 W_S^{2-} 一差异相逆,而与 T 差异、pH 值差异同向地增减,其中只是 SRB 含量与 Eh、T、pH 值间的关系似合乎正常。在北、中、南3区,SRB 含量与 Eh 均显出中>北>南;相反 T 和 pH 值均为南>中。W_S^{2-} 则是北>中>南,其中只有 SRB 含量与 W_S^{2-} 的变量方向间有近似理想状态的关系。

表2 北部湾东侧沉积物 SRB 含量

站号	层次	SRB(个每克)	站号	层次	SRB(个每克)
1	S	11 000.0		1.5 m	0
2	S	4 600.0	12	S	0.4
	0.7 m	9.0		1.5 m	0
3	S	11.0	13	S	11.0
	0.5 m	0		1.5 m	0
4	S	360.0	14	S	0.4
	1.5 m	0		1.0 m	0.4
	3.7 m	280.0	15	S	0
5	S	160.0	16	S	70.0
6	S	110.0	17	S	0
	1.5 m	0		1.5 m	15.0
7	S	3 600.0	18	S	15.0
8	S	1 500.0		1.5 m	0
	1.5 m	0	19	S	0
9	S	7 000.0		1.5 m	0
	1.2 m	40.0	20	S	0
10	S	4 600.0		0.5 m	0
	1.5 m	0	21	S	4 000.0
11	S	0			

S 表层沉积物。

3.2 浅层沉积物 SRB 含量

表2中也列出了浅层沉积物 SRB 含量状况,综观 0.5 m~3.7 m 这段浅层沉积物,其 SRB 检出率约 31.3%,平均 SRB 含量为 21.5 个每克。与表层比,无论是检出率和平均含量,浅层 SRB 均显示为较低。无论是表层,还是浅层,本测区 SRB 有关参数也比东海测区

的低。

该测区浅层 SRB 含量似显出随样品所在深度的下降而呈跳跃式地变化(增高)之势,这表明本区与东海测区的 SRB 生存环境有所差别[6]。

水样中未测出 SRB,联系到以往的研究,可以推测本区水中一般不易测出 SRB,除非有 SRB 合适的生存条件。

3.3 SRB 与一些环境参数间的相关分析

据表 1 和表 2 有关数据,对 SRB 含量与 W_S^{2-} 等 4 个参数分别作相关分析(表 4)。由表 4 可见,表层 SRB 与 pH 值、W_S^{2-} 间各有不显著的负相关。与 Eh、T 间各有一定的正相关性。由于浅层 SRB 的检出率较低,因而未单独进行此层 SRB 与有关参数间的相关分析。将浅层与表层样品统算,SRB 与这 4 个参数间各自的相关系数与表层的相比时,就发生了变化。

将前述按测线和测区分析 SRB 含量及其与环境参数间关系分析的结果比较,大体可以看出 SRB 含量的增加可能促使环境中硫化物(包括硫化氢)的增多之端倪[4]。

3.4 SRB 含量状况对环境腐蚀的意义

已知 SRB 确是环境的重要腐蚀因子[2],其含量的高低意味着腐蚀性的强弱。据此,可以分析本测区的腐蚀环境。如果以 SRB 含量达 1 000 个每克～10 000 个每克的环境属强腐蚀区,＜1 000 为腐蚀中区,＜100 为腐蚀弱区,＜1 的为基本无腐蚀区的话,那么可以看到测区 33.33％的表层样品代表所在环境为强腐蚀区,另有 33.33％样品为中度腐蚀区,而仅有 9.52％为腐蚀弱区。总体看,尚有 23.82％为无腐蚀区。表层样品的 66.67％有中度以上的腐蚀可能发生。浅层样品的腐蚀状况轻于表层,有 6.31％显中等腐蚀,但从 W_S^{2-} 含量看,浅层比表层的高,仍表明那儿有一定的腐蚀作用。25.00％为弱腐蚀区,尚有 68.75％为无腐蚀区。

表 3 北部湾东侧表层沉积物 SRB 含量与相关 4 个参数在不同地理方向间的比较

地理方向	SRB(个每克)	W_S^{2-} (10^{-6})	Eh(mv)	pH 值	T(℃)
东(n=3)	2 533.3	16.06	76.67	7.69	20.43
中(n=14)	1 312.6	26.52	95.64	7.53	20.21
西(n=3)	3 630.0	/	/	/	/
南(n=4)	1 003.9	17.70	76.00	7.68	20.80
中(n=7)	187.31	21.73	7.52	20.33	/
北(n=6)	1 473.5	32.75	84.67	7.52	17.78

括号中的数字均为样品数。

将此按东、中、西 3 线和北、中、南 3 区分(表 4),腐蚀可能依次是中＞北＞南区和西＞东＞中线,即北区和中区处于中等腐蚀区,南部为弱区。3 线虽均为中等腐蚀区,但其中以中线的腐蚀强度较弱些,即中线腐蚀较轻。线、区重叠起来看北(西)区的腐蚀可能较其他区的强。

结果还表明 W_S^{2-}、Eh 等参数与 SRB 含量间有一些不确定的(模糊)关系存在。这表明 SRB 的生存和发展尚有其他重要因子在影响甚而支配着。

表 4　北部湾东侧沉积物 SRB 含量与相关参数间的相关分析结果

层次名称	$-W_S^{2-}$	$-$Eh	$-$pH 值	$-T$
表	-0.01 ($<\alpha_{0.5}, n=14$)	0.242 ($<\alpha_{0.5}, n=14$)	-0.10 ($<\alpha_{0.5}, n=14$)	0.175 2 ($<\alpha_{0.5}, n=14$)
表＋浅	0.325 6 ($>\alpha_{0.5}, n=16$)	0.583 6 ($<\alpha_{0.02}, n=16$)	-0.779 3 ($>\alpha_{0.001}, n=16$)	-0.066 7 ($<\alpha_{0.5}, n=16$)

4　结语

用 MPN 法测定了北部湾东侧表层和浅层沉积物的 SRB 含量。发现全测区表层和浅层沉积物 SRB 的检出率、平均含量分别为 76.3%、31.3%和 1 760.0 个每克，344.4 个每克(湿重)。SRB 含量的分布状况是东＞西＞中和北＞中＞南。表层沉积物中的 SRB 含量大多高于浅层的。浅层 SRB 含量呈现出随深度而变化，偶显增高之势。结合分析样品的 SRB、硫化物含量、Eh、pH 值和 T(℃)等环境参数间的相关关系，发现 SRB 含量与 W_S^{2-} 间有相一致变化现象，表明 SRB 应是环境硫化物的转换者，是环境腐蚀的主因之一。由此试图分析测区的腐蚀范围，即浅层大多为无腐蚀环境，表层沉积物发生腐蚀的可能性大于浅层。SRB 含量与一些环境参数间同时出现某种不确定关系，意味着 SRB 的生存与发展受着其他重要因素的制约。

致谢：本文理化参数由国家海洋局第一海洋研究所有关人员测定，特此致谢。

参考文献 6 篇(略)

(合作者：徐家声)

CONTENT OF SULPHATE-REDUCING BACTERIA IN THE SEDIMENT ON THE EAST SIDE OF BEIBU GULF AND ITS SIGNIFICANCE

(ABSTRACT)

Abstract The content of sulphate reducing bacteria (SRB) in the sediment on the east side of Beibu Gulf was determined by using the most probable number(MPN)method. The percentage detection of SRB in the surface sediment was about 76.3%, that in the shallow layer of the sediment (0.5 m～3.7 m under the surface sediment) was 31.3%. Average content of the SRB were 1,760.0 cells · g^{-1}. (wet weight), (the same below)and 344.4 cells · g^{-1} in the surface sediment and the shallow one respectively. The content distribution of the SRB was analysed in the different area determined. The relationships between SRB content and sulphide(W_S^{2-})、Eh、T(℃)、pH value were discussed respectively, and the significance of SRB to evaluate the environmental corrosion was also explained.

Key words Sulphate Reducing Bacteria, Beibu Gulf, Sediment, Corrosion, Environment

胶州湾潮间带和沿岸区硫酸盐还原菌含量分布*

摘　要　用 MPN 法对胶州湾潮间带和沿岸区的硫酸盐还原菌(缩写为 SRB)含量做了研究。结果表明表层沉积物的 SRB 含量大约是 4.1×10^7 个细胞/g,波动范围较大。水中的 SRB 含量约 2.5×10^2/mL。分析了 SRB 含量的分布状况,发现调查区内的 SRB 含量基本上呈现出北高南低、西高东低之势,即工业区、港口区的高于海水浴场、海滨旅游区的。证实了生活废弃物的排放致其环境 SRB 含量高。SRB 含量分布状况大体上与以往所测环境污染状况一致。对虾养殖池的 SRB 含量意味着水产养殖必须考虑这一因子的作用。

关键词　硫酸盐还原菌　胶州湾潮间带和沿岸区　环境污染

海洋环境,尤其是受城市化和工业化影响严重的潮间带和沿岸区,广泛分布着较多的硫酸盐还原菌(sulphate-reducing bacteria,缩写为 SRB,下同)。SRB 虽然有其在环境中存在的必然性,但如果大量出现,必定是环境质量恶化的结果和标志之一。已有文章指出 SRB 是钢铁等金属材料的腐蚀主因[1]。SRB 对经济产生的负面作用已引起各国重视,并采取了一些措施。SRB 的生态学是阐明硫元素在环境中循环和材料腐蚀活动的内在依据之一。国际间已有一些关于 SRB 生态学的研究[2,3],而国内公开报道的研究成果则不多[4,5]。本文就青岛胶州湾潮间带和沿岸区的 SRB 含量分布作一初步报道。

1　采样区域和测点概况

图 1 描绘出本调查范围内 SRB 的测点位置。测点按由北向南、自西至东、从小而大编号共计 35 个。

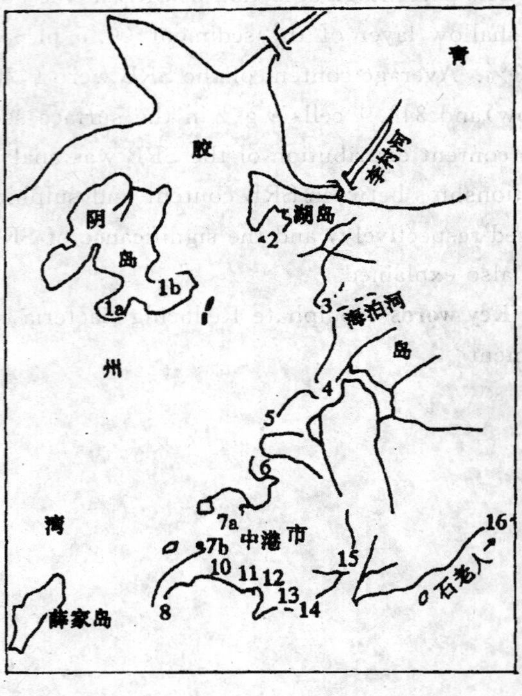

图 1　胶州湾潮间带和沿岸区硫酸盐还原菌测点示意图

Fig. 1　Sketch map of sampling sites of Jiaozhou Bay for sulphate-reducing bacteria in the intertidal and coastal areas

5 包括 5a、5b、5c、5d、5e、5f;9 包括 9a、9b;10 包括 10a、10b、10c、10d、10e、10f;13 包括 13a、13b、13c、13d、13e;

为分析不同功能的城市区划对测

* 原文刊于《海洋环境科学》,1999,18(2):37-40.

区 SRB 的影响，将测点归为 6 区。A 区为水产养殖区，养殖品种主要是对虾、贝类，包括 1a、1b 和 16 号测点。B 区为工业区，包括 2、3、4、5a、5b、5c、5d、5e、5f 及 6 号测点，位于海泊河至石化厂间。C 区为港口区，有 7a、7b 两点。D 区为海滨旅游区，指 8、10a、10b、10c、10d、10e、10f 及 12 号测点。E 区为海水浴场区，有 9a、9b、13a、13b、13c、13d、13e、14a、14b 及 14c。分布在 3 个浴场，F 区为生活污水排放区，包括 11、15 两测点。

2 材料和方法

无菌采得的各表层沉积物和表层水样，迅即送入预先消毒好的小广口瓶中。运输途中，样品保存于冰瓶。在最长不超过 3h 内将样品按 MPN 法稀释入 SRB 培养液中。培养基如文献[5]中所述制作。将各试管稀释好了的样品于 30℃ 培养 7d 后作检查。以培养基呈现黑色甚而见黑色沉淀为 SRB 阳性管，查表判断 SRB 含量，分析所调查区域内 SRB 含量的分布状况。SRB 含量最后以个细胞/g(湿样)或个细胞/mL(水)表示。

3 结果和讨论

表 1 列出了各测点的 SRB 含量。该表左侧是夏季测区 SRB 含量状况。

夏季沉积物中的 SRB 含量平均为 8.0×10^{7} 个/g(中位数为 1.1×10^{3} 个/g)。最高值出现在 11 号站，最低值出现在 8 和 16 号测站，均未测出(记作 0)，两者差距很大。离 11 号测站很近的测点 9 和 10 号，其 SRB 含量降得很快，表明相对洁净的涨潮水的稀释作用和表层沉积物曝气对 SRB 有抑制作用。各点 SRB 含量按大小排序如下：11，6，4，7b，3，5b，16，10a，9a，1a，2，13e(14a)，5c 及 8(16)号测站。

表 1 右侧是冬季状况。可以统计出沉积物中总的 SRB 含量平均值为 5.3×10^{-1}/g(其中位数为 9.0×10^{-1}g)，最高值出现在 No. 5e 测点，最低值在 No. 13b。极差 $<4.3\times10^{1}$/g。冬季所有测点沉积物 SRB 按其含量大小，排序为 No. 5e、5f、9b(10b，10c)、10d(10f、12、13a、13c)、10e、13d(14b，14c)及 13b。

比较表 1 左右两部分情况可见，表层沉积物中平均 SRB 含量的夏冬之差是很明显的。夏季 SRB 含量比冬季的高出 8 个量级。这主要是由夏季 6 和 11 号两测点与冬季 10e、13d、14b、14c、13b 这 5 点间的差异造成。夏季雨水多，陆上排污多，使 SRB 含量增加，因潮汐关系，使沉积物样的夏冬测点不完全一致。但据相近测点状况综合看，其夏冬差异也很明显。如夏天的 5b、5c、9a、10a、13e 和 14a 号测站与冬天的 5e、5f、9b、10f、13a 和 14c 号测站，其各自的 SRB 含量平均值各为 1.7×10^{4} 和 1.2×10^{1}，夏冬两季样品温度差明显。对 10 个沉积物温度与 SRB 含量作的相关分析表明，两者间有一定的正相关关系存在，($R=0.452, >R_{0.20}, n=10$)。指示了一定范围温度的升降与 SRB 含量的增减并行不悖。

表 1　胶州湾潮间带和沿岸区硫酸盐还原菌含量(个/g)

Tab. 1　Contents of SRB in the intertidal and coastal zones of Jiaozhou Bay

区域	测点	SRB 含量	区域	测点	SRB 含量
A	1a	8.1×10^{2}	B	5d*	2.1×10^{2}
	1b	2.2×10^{4}		5e	4.3×10^{1}
	16	0		5f	2.3×10^{1}
	2	9.3×10^{1}	C	12	9.0×10^{-1}
	3	1.1×10^{5}	D	10b	2.3×10^{0}
	4	9.0×10^{5}		10c	2.3×10^{0}
B	5a*	4.3×10^{1}		10d	9.0×10^{-1}
	5b	9.0×10^{4}		10e	$>4.0\times10^{-1}$
	5c	2.1×10^{0}		10f	9.0×10^{-1}
	6	$>1.1\times10^{0}$	E	9b	2.3×10^{0}
C	7a*	6.0×10^{2}		13a	9.0×10^{-1}
	7b	3.5×10^{5}		13b	4.0×10^{-1}
D	8	0		13c	9.0×10^{-1}
	10a	$>1.1\times10^{4}$		13d	$<4.0\times10^{-1}$
E	9a	1.1×10^{3}		14b	$<4.0\times10^{-1}$
	13e	1.5×10^{1}		14c	$<4.0\times10^{-1}$
	14a	1.5×10^{1}			
F	11	$>1.1\times10^{9}$			
	15*	1.6×10^{2}			

* 系水样。余均为沉积物样，左部为夏季样，右部为冬季样

水中的 SRB 只测了 4 个点的，即 3 个夏季样，1 个冬季样。夏季的平均值约为 2.7×10^{2} 个/mL，比沉积物中的明显低，并呈现出测点间有差距，但波动范围小于沉积物中的。冬季水样的 SRB 含量为 2.1×10^{2} 个/mL。夏冬季比较，波动不大。通算表 1 数据，可知测区沉积物和水中平均的 SRB 含量分别为 4.1×10^{7} 个/g 和 2.5×10^{2} 个/mL。

表 2 列出的是各测点按受城市不同功能区划影响归类比较其 SRB 含量的结果。

由表 2 可见，夏季各类测点沉积物中平均的 SRB 含量大小排序为 F、B、C、A、D 和 E。水中 SRB 含量是 C>D>B。冬季，沉积物中 SRB 含量排序为 B>D>C>E。沉积物和水中的 SRB 显示出基本一致的含量变化势态。结果意味着受工业区排污影响大的潮间带和沿岸区测点类，其 SRB 含量一般较大。而海水浴场等地的海滨沙滩之 SRB 含量一般较低。生活污物排放口(如 11 号测点)，由于历年来的积淀，滋生特多的 SRB，并扩散至毗邻地区，应引起注意。在市区东部的 16 号测点，随着经济和生活的发展，若不注重监控排污和保护环境，其 SRB 含量也可能呈上升趋势。

两个对虾养殖池(1a 和 1b 号站)沉积物 SRB 含量显示出一定差异。这意味着 SRB 的大量出现,其还原作用产生的 H_2S 恶化环境,毒害养殖对象[8]。因此水产病害的防治工作应考虑到 SRB 这一因子的作用。

由上结果可推测污染较重区的样品所在的沉积物之颗粒在经受各种因子干扰中不断以粒径大小分级积淀,逐步形成相对稳定的环境。在丰富的有机质条件中,于缺氧环境中 SRB 大量繁殖。相对洁净区,砂粒较粗,氧气尚可,SRB 较少。但是一些研究表明,造成 SRB 含量测点差异,其原因复杂,现象各异,尚待进一步探讨。

表 2 胶州湾潮间带和沿岸区受城市不同功能区划影响的测点类的硫酸盐还原菌含量比较

Tab. 2 Comparison of SRB contents in classification of sampling sites in the intertidal and coastal zones of Jiaozhou Bay according to the influence degree by the different functional regions of Qingdao

测点类别	SRB 含量(个/g)			
	夏		冬	
	沉积物	水	沉积物	水
A	7.6×10^3			
B	1.9×10^7	4.3×10^1	3.3×10^1	2.1×10^2
C	3.5×10^5	6.0×10^2	9.0×10^{-1}	
D	$>5.5\times10^3$		1.4×10^0	
E	3.8×10^2		8.1×10^{-1}	
F	$>1.1\times10^9$	1.6×10^2		

4 结语

本文对测区的 SRB 含量作了初步的时空分布分析。结果表明测区表层沉积物中 SRB 含量的大体范围波动于测不出与 1.1×10^9 个/g 之间,平均值约为 4.1×10^7 个/g。表层水中的 SRB 含量波动于 4.3×10^1 与 6.0×10^2 个/mL 之间,平均约为 2.5×10^2 个/mL。波动范围小于沉积物中的。这意味着测区的环境有一定的 SRB 污染,有些测点污染严重。

就沉积物 SRB 含量显出的测点(及其测点类别)间的差异而言,工业、港口影响区的 SRB 含量高,旅游、浴场区的低,即大体呈现出北高南低和西高东低之势。SRB 含量还显出夏季高冬季低,差异很大,表明测区沉积物及水深受工业废弃物和生活排污的明显影响。样品的温度与 SRB 间有正相关性关系。

SRB 的繁盛使环境中积累较多的 H_2S 等硫化物,造成腐蚀环境,损害构件和设施,毒害动植物。推测 SRB 受沉积物所含颗粒大小、有机质多寡和含氧量高低(氧化还原电位高低)等的较多影响。

参考文献 8 篇(略)

DISTRIBUTION OF SULPHATE-REDUCING BACTERIA IN THE INTERTIDAL AND COASTAL ZONES OF JIAOZHOU BAY

(ABSTRACT)

Abstract The content of sulphate-reducing bacterial (SRB) in the intertidal and coastal zones of Jiaozhou Bay was researched by MPN method. The results showed that content of SRB in the surface layer sediments averaged about 4.1×10^7 cells/g, and its fluctuation was rather big. The concentration of SRB in the water wasabout 2.5×10^2 cells/mL. The distribution of SRB was analysed and found out that the tendancy of the contents of SRB was high in the north and west areas, and low in the south and east areas, i. e. concentration in the industrial or harbour areas were higher than in the bathing beaches and tourist areas. The specially high content of SRB in the environment was induced by the discharge of domestic waste. The distribution of the SRB content was roughly consistent with that of the environmental pollution confirmed early.

The SRB contents in the ponds for culturing shrimp were similar to those reported previously.

Key words Sulphate-Reducing Bacteria, Intertidal and Coastal Zones, Environmental Pollution

海南岛西南侧海域底质硫酸盐还原菌分析*

摘　要　论述了用MPN法测定海南岛西南侧海域底质硫酸盐还原菌含量的结果。全测区SRB平均检出率、含量及其变幅分别为60.0%,5 680.3细胞·$100g^{-1}$(湿)和0～110 000细胞·$100g^{-1}$(湿),其含量一般呈现出表层大于浅、深层,近岸区大于中部和远岸区。由此判断测区大部分范围底质环境的表层和近岸区发生腐蚀的强度可能大于浅、深层和中部至远岸区,少数特殊的深层底质也不排除有较强腐蚀性的可能。SRB含量与温度、pH、Eh、Fe^{3+}/Fe^{2+}、电阻率、含水率及有机质含量等因子分别做的相关分析结果表明,它们的相关性都不具显著意义,说明在该低硫化物含量的环境中,由SRB等产生的腐蚀作用不严重。但随着海洋开发活动的扩展,与SRB相关的腐蚀活动将可能增强,这是必须引起注意的。

关键词　海南岛　底质　硫酸盐还原菌　环境腐蚀性　环境因子　海域

海南岛西南侧海域海底可能蕴含丰富油气资源,纷呈的海洋开发活动要求对环境腐蚀性作科学调查。一些研究已指出硫酸盐还原菌(sulphate reducing bacteria,缩写为SRB,下同),既与油气生成有关,又与其他因子共同对环境、设旅及材料产生不可低估的腐蚀作用[1]。对该类菌数量分布及有关因子做出分析对海洋工程的实施和防腐手段的采用至关重要[2]。本文论述该区底质硫酸盐还原菌的分布状态,分析与相关因子的关系并推断环境腐蚀性。

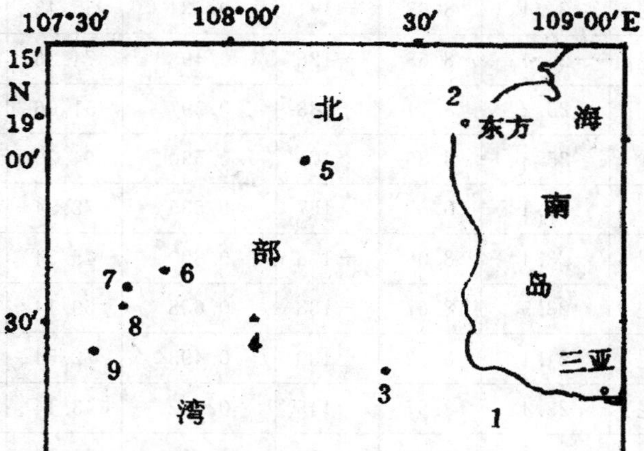

图1　海南岛西南侧海域底质硫酸盐还原菌测站示意图

* 原文刊于《海岸工程》,2000,19(2):10-15

调查时间在1995年5月。调查海域位于海南岛西偏南(107°35′～108°50′E,18°17′～19°03′N),水深18.5～76.0 m,设置站位9个,其地理位置和站位分布如图1所示。

1　样品采集和分析

1.1　工程地质取样

用DDC1-4型采泥器、DDC-7型箱式采样器和DDC4-2型重力活塞取样管,分别将底质样按表层(水底下表层沉积物)、浅层(表层下至0.4 m内)及深层(表层下1.9～3.4m)三个层次取得。以海南岛西南岸为准,将1、2号两个站划为近岸区,3～5号三个站划为中部区,6～9号四个站划为远岸区,记录并分析沉积物类型及性质。

表1　海南岛西南侧海域底质工程地质理化参数一般状况

站号	层次	工程地质、理化参数							
		土质类型	T(℃)	pH	Eh	Fe^{3+}/Fe^{2+}	电阻率	含水率	有机质
1	表	含砾贝岩砂	27.1	8.43	−185	0.857	88.74	26.58	0.340
2	表	泥质粉砂	24.3	8.22	−225	0.658	49.46	51.92	1.102
	浅	砂泥砾	25.1	8.55	−219	0.606	49.00	45.92	1.105
	深	淤泥含贝	25.0	8.63	−157	0.568	53.66	40.00	1.203
3	表	泥质贝壳砂	24.6	8.39	−201	0.478	63.20	32.03	0.622
	浅	泥质贝壳砂	26.7	8.86	−194	0.387	90.53	26.79	0.390
	深	黏土	26.1	8.06	−212	0.700	88.94	23.16	0.274
4	表	粉砂淤泥	23.3	8.51	−146	0.768	53.86	45.63	0.959
	浅	粉砂淤泥	24.4	8.63	−171	0.634	55.43	28.98	0.964
	深	粉砂淤泥	24.4	8.68	−126	0.493	74.26	33.333	0.874
5	表	砂质泥	25.2	8.71	−183	0.597	51.06	34.78	0.845
	浅	泥质粉砂	25.3	8.86	−175	0.590	64.20	31.45	0.605
	深	/	26.4	8.62	−167	0.535	72.34	28.44	0.809
6	表	粉砂淤泥	23.4	8.08	−178	0.800	45.54	47.99	1.021
	浅	粉砂淤泥	23.9	8.67	−138	0.658	69.14	41.70	1.071
	深	粉砂淤泥	24.1	8.52	−130	0.493	72.54	21.95	0.862
7	表	粉砂淤泥	23.4	8.57	−145	0.882	43.14	36.83	0.878
	浅	粉砂淤泥	24.0	8.63	−153	0.629	51.66	42.00	1.165
	深	粉砂淤泥	24.9	8.32	−124	0.562	59.80	40.53	1.048
8	表	粉砂淤泥	22.8	8.2	−170	0.743	44.46	47.39	0.886
	浅	粉砂淤泥	23.4	8.65	−167	0.589	40.66	47.17	1.015
	深	粉砂淤泥	23.9	8.79	−156	0.493	55.34	37.29	1.045

（续表）

站号	层次	工程地质、理化参数							
		土质类型	T(℃)	pH	Eh	Fe^{3+}/Fe^{2+}	电阻率	含水率	有机质
9	表	粉砂淤泥	23.4	8.23	−161	0.904	44.26	49.69	0.938
	浅	粉砂淤泥	24.0	8.69	−137	0.683	57.94	43.10	1.119
	深	粉砂淤泥	23.9	8.44	−120	0.524	66.26	38.34	1.060

* Eh(mV)相对于饱和甘汞电极，电阻率：(·cm)、含水率(10^{-2})、有机质($\times 10^{-2}$)

1.2 环境理化因子分析

用水银温度计测底质物温度(℃)。PⅡS-2C 型酸度计测 pH。PⅡS-2C 型酸度计、铂电极、饱和甘汞电极、非参比电极测 Eh(mV)。EDTA 容量法测 Fe^{3+}/Fe^{2+}，四电极法等测电阻率(·cm)，重铬酸钾氧化还原容量法测有机质(or$\times 10^{-2}$)，重量法测含水率(10^{-2})，硫离子选择电极法测硫化物(检出限：0.14×10^{-6})。表 1 列出了上述 8 个参数在各站各层次中的类型或含量。

1.3 SRB 含量测定

当底质样来到船甲板上后，立即取足样品进实验室，用 MPN 法在 SRB 培养基中稀释样品和培养。一切以无菌操作进行，具体方法和步骤如文献[3]所述。

1.4 数据处理

将所得 SRB 检出率、含量及其变幅等按样品不同的层次、离岸远近等进行归类分析，结合有关工程地质、理化分析 SRB 的三个参数之变化和分布特征，指出 SRB 对腐蚀性评价的意义。

2 结果和讨论

2.1 SRB 的三个参数及其变化

表 2 列出了所有样品(25 份)中 SRB 含量状况。由该表可计算出所有样品中 SRB 的检出率、平均含量及其变幅分别为 60%、5 680.3、0～110 000 个细胞·$100g^{-1}$(湿)。由表 2 可以得出表 3，它比较了 SRB 三参数在样品层次间、离岸远近间的状况。从中可看出以下趋势，SRB 检出率是表大于浅大于深层，及近大于远大于中部。平均含量是表大于深大于浅层，及近大于中大于远部。变幅是表大于深层大于中大于远部。与邻近的北部湾东侧比，SRB 检出率较高，但含量比其低[2]。近岸和表层的 SRB 三个参数值大，深层样品有时含较多 SRB，但其变化也大。远岸样品的 SRB 有时检出率也不低。表明这些样品所在环境有时也比较适于 SRB 生存。

表 2　海南岛西南侧底质硫酸盐还原菌含量状况

站号	层次	硫酸盐还原菌含量 (个细胞/100g)(湿)	站号	层次	硫酸盐还原菌含量 (个细胞/100g)(湿)
1	表	15	6	表	240
2	表	110 000		浅	9
	浅	0		深	11
	深	15			
3	表	0	7	表	43
	浅	0		浅	0
	深	20000		深	0
4	表	11 000	8	表	420
	浅	0		浅	1.5
	深	0		深	0
5	表	210	9	表	20
	浅	1.5		浅	21
	深	0		深	0

表 3　海南岛西南侧海域底质 SRB 三个参数在不同层次或离岸远近间的比较

层次	检出率(%)	平均含量	含量变幅	站区	检出率(%)	平均含量	含量变幅
表	88.9	13 549.8	0～110 000	近岸	75.0	27 507.5	0～110 000
浅	50.0	4.1	0～21	中部	44.4	3 467.9	0～20 000
深	37.5	2503.3	0～20 000	远岸	66.7	63.8	0～420

* SRB平均含量及其量变幅的单位均是个细胞·100^{-1}g(湿重)。

2.2　SRB 与环境理化参数的分析

将 SRB 含量以 0.15～420 和 11 000～110 000 个细胞·100^{-1}g(湿)分测样为三类，计算出各自的 SRB 含量平均值，以比较 pH 等几个参数值与它们的关系，可得出表 4。

表 4　海南岛西南侧海域底质各类 SRB 含量与相应 pH 等参数的比较

样品数	SRB 含量	pH	Eh	含水率(10^{-2})	有机质含量($\times 10^{-2}$)
10	0	8.60	-157	35.80	0.899
12	84	8.56	-161	39.05	0.898
3	47 000	8.26	-194	40.24	0.778

由表 4 说明在本测区中，SRB 含量以中等居多，占 48%(12/25)的样品数，未检出者占 40%(10/25)，高 SRB 含量的样品少，只占 12%(3/25)。

SRB含量与pH、Eh间各呈现相悖趋势,与含水率间则有一致趋势。这与表5所列数据基本一致。

表5 海南岛西南侧海域底质SRB含量与有关因子的相关系数*

对子	相关系数	n
SRB-T	0.016	15
SEB-pH	−0.129	15
SRB-Eh	−0.206	15
SRB-Fe^{3+}/Fe^{2+}	−0.024	15
SRB-电阻率	−0.018	15
SRB-含水率	0.093	15
SRB-有机质	0.041	15

*所有出现SRB的样品数中各SRB含量与相应因子间的相关分析

表5列出的是测区所有SRB含量与相应温度等7个参数值之间进行的相关分析结果(所有样品中的硫化物含量均低于本检出限0.14×10^{-6},故未将它作为影响因子讨论)。由该表可见,SRB与这些参数间均无很有意义的相关性存在。其稍好的正相关性仅分别在与Eh或pH之间。这与北部湾东侧相比有一定差异,主要表现在SRB含量与Eh、T的相关性上,原因可能是地质环境和采样时间的差异[2],表明了在本特定条件下,SRB的环境与它间未形成严格的制约关系,即SRB没有产生大面积范围严重的腐蚀性可能,尽管少数样品有较多SRB存在。

分析以上情况,可以看出,含水率和有机质含量较高、Eh和电阻率偏低的泥质粉砂(如2号表层样)或不大的碱性、较少有机质、较大电阻及Fe^{3+}/Fe^{2+}的黏土可能含较多SRB(如3号深层样)。有适中含水率和有机质的粉砂淤泥也可能利于SRB生存(如4号表层)。反之,pH、Eh较高、含水率较低之处,其SRB含量较低甚而测不出来。这意味着颗粒细小的底质物,可吸附较多水分和有机质,利于SRB生长繁殖。SRB繁殖可促使硫酸盐还原,有机质降解及含量降低,pH、Eh也降低,电阻率升高,最终表现出环境腐蚀性增强。

3 SRB含量对环境腐蚀性的意义

SRB是环境腐蚀重要因子,其含量的高低意味着腐蚀性的强弱,将它结合相应地质性状和理化参数分析,可以指示出特定环境的腐蚀性。假设样品中SRB含量(个细胞·$100g^{-1}$湿样,下同)达到或超过10^5时,推测该环境有强的腐蚀性,10^4时的环境有中等腐蚀性,小于10^4的环境为腐蚀弱区或无腐蚀区的话,那么本测区表层有11.11%(1/9)为腐蚀性强区,另有11.11%区域有中等强度的腐蚀性,有12.5%(1/8)的深层区域可能有中等的腐蚀性发生迹象,所代表的测区多数层次及环境似无发生腐蚀的可能。前述2号、4号站的表层样品有适宜的含水率、电阻率和Eh,3号站深层样的电阻率较高,而Eh和含水率却较低,但均有较大的SRB含量,意味着那些地方很有较强腐蚀发生的可能。

结果还表明,不同特定环境中的SRB不是仅仅受某个因子的制约,而是许多因子共同作用的结果。这些复杂因子间本身也可能产生对SRB控制不力的作用。

根据以上分析，可以认为在硫化物含量甚低的本测区，大部分底质环境腐蚀性弱。腐蚀性强度以层次分，一般是表层的强于浅层和深层，但深层也偶有较强的腐蚀性迹象。以站区划分，近岸站位的腐蚀有条件强于中部和远岸站的。

4 结语

经对海南岛西南侧海域底质 SRB 含量等及其相关的地质理化因子初步分析，认为虽然测区大部分环境由 SRB 产生的腐蚀性不强，但就全测区而言，底质多粉砂淤泥，它易产生宏观电偶腐蚀。粉砂有使去氧去极化腐蚀增大的趋势。表层底质环境的腐蚀性可能强于浅、深层，近岸强于中部、远岸。这意味着近岸区及水—底质界面的还原环境可能有较强的腐蚀性诱因。随着海洋开发活动的加快，环境的腐蚀性可能会增强。在海底的管道外保护、阴极保护和管道的铺设过程中，要充分认识 SRB 及其所在的不同底质环境因子可能引起的腐蚀问题的复杂性。

致谢：取样、地质和理化参数等由国家海洋局第一海洋研究所相关人员提供，在此一并致谢。

参考文献 3 篇（略）

（合作者：徐家声　战　闰）

ANALYSIS ON THE SULPHATE REDUCING BACTERIA IN THE SEDIMENT OF THE SEA AREA ON THE SOUTH-WEST SIDE BY HAINAN ISLAND (ABSTRACT)

Abstract The content of sulphate-reducing bacteria (SRB) in the sediment of the sea area on the south-west side by Hainan Island was determined by using MPN method. The result showed that the detection ratio, average content and its variation range of SRB count were 60%, 5680.3 and 0～110 000 cells · $100g^{-1}$ (wet) respectively. The content appeared generally that in surface layer is larger than shallow layer and deeper one, inshore area is larger than the central and off shore ones. Then the judgement was obtained that the corrosiveness is stronger in the surface layer and inshore area than shallow and deeper ones of the main area investigated, and there might be have stronger corrosiveness in a few particular samples of deeper layer.

The result of correlation analysis between the SRB content and temperature, Eh, electronic resistance ratio or content of organic material etc factors respectively showed that there were not statistically significant correlations. It indicated that the corrosive action produced from the SRB etc, was not serious in this sedimental environment with low sulphate content, but the following problem would be noticeable that the complexity of environmental corrosion including SRB is increased accompanied with the expansion of marine engineering.

Key words Hainan Lsland; Sulphate-Reducing Bacteria; Corrosion; Sediment Evironmental Factors; Sea Area

浙江—闽北陆架沉积物硫酸盐还原菌及与生物地球化学因子关系的分析*

摘　要　采用 MPN 法测定浙江—闽北陆架沉积物的硫酸盐还原菌(SRB)。结果表明,在表层其出现率和平均含量各为 84.2%和 436.8 个(细胞/g(湿重)),浅层则各为42.4%和 164 个/g。近岸的南区表层和浅层沉积物 SRB 出现率、平均含量分别为 93.0%、576 个/g 和 33.3%、267 个/g;远岸的北区则分别为 75.0%、302 个/g 和 53.3%、41 个/g。分析了 SRB 含量与相应样品中硫化物、有机质含量、Fe^{3+}/Fe^{2+}、pH、Eh、电阻率等生物地球化学因子间的相关关系,发现 SRB 含量与它们间以正相关关系为主,其中 SRB 与 Fe^{3+}/Fe^{2+}、Eh 间的正相关性最为明确,与电阻率间以负相关关系多见。讨论了 SRB 含量及其分布状况对腐蚀性的意义。

关键词　硫酸盐还原菌　沉积物　生物地球化学因子　腐蚀性评价　浙江—闽北陆架区

随着东海油气资源开发和海底工程建设的迅速发展,研究海底沉积物对钢铁等结构物腐蚀问题已成为工程界关注的课题之一。海洋沉积物既是硫酸盐还原菌(sulphate-reducing bacteria,简写为 SRB,下同)等微生物重要的存在场所,也是 SRB 作为硫酸盐还原作用的催化剂参与地球化学活动的一种产物。SRB 生化活动后果之重要方面便是对金属等材料的腐蚀作用。一些相关研究已有公开发表的研究成果[1-5],但总体看来仍是很缺乏的。本文初步报道东海浙江—闽北陆架沉积物的 SRB 含量,分析它与若干生物地球化学因子之间的关系,逐步掌握海洋 SRB 的活动规律,以服务于海洋经济良性的持续发展。

1　测区和测点概况

调查范围位于浙江—闽北海区(图 1),大体以 28°N 为界,分为相对近岸的南区和远岸的北区。南区位于 26°20′－28°48′N 和 121°10′－124°00′E 间。北区位于 27°50′－30°50′N 和 124°02′－126°15′E 间。该调查区的底质属全新世浅海相松散沉积物。表层和浅层底质以黏土质粉砂、粉砂或细砂为主。南区表层(0～0.4 m,下同)沉积物岩性为细砂、黏土质粉砂或中细砂,浅层(0.4 m～4.0 m,下同)岩性为细砂或黏土质粉砂。北区表层底质岩性主要为黏土质粉砂或细砂,浅层岩性为细砂、粉砂或黏土质粉砂。两区沉积物的硫化物、有机质含量、Fe^{3+}/Fe^{2+}、pH、Eh 及电阻率参数的一般状况列于表 1 之中。

在南区设定的表层测点有 71 个,浅层测点 18 个,北区设表层点 68 个,浅层点 15 个。

* 原文刊于《环境科学学报》,2000,20(4):478-482.

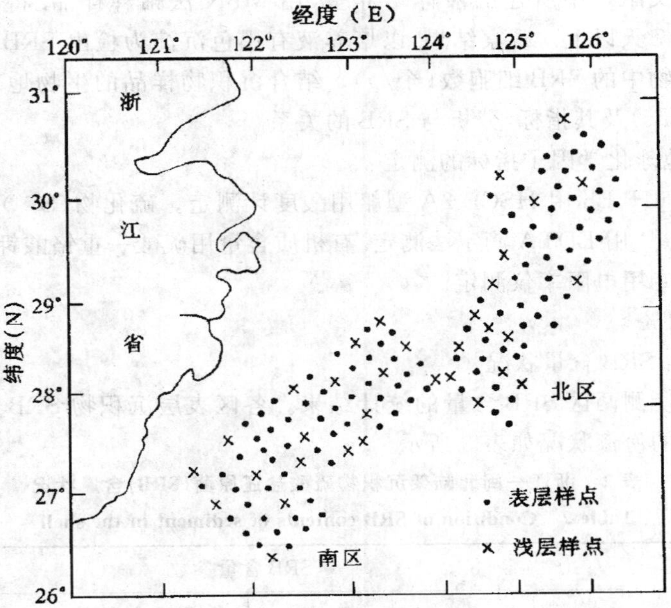

图 1　浙江—闽北陆架硫酸盐还原菌采样点示意图

Fig. 1　Sketch map of sampling for sulphate reducing bacteria in the sediment of Zhejiang-North Fujian continental shelf

表 1　浙江—闽北陆架沉积物与 SRB 相关的一些生物地球化学参数值

Table 1　Some values of Biogeochemical Parameters in relation to SRB of Zhejiang-North Fujian shelf

测区及层次	参数					
	S^{2-}，10^{-6}	Or，%（W/W）	Fe^{3+}/Fe^{2+}	pH	Eh，mV	电阻率，·cm
南，表层	0.43 （0.14～1.91）	0.55 （0.13～1.41）	1.00 （0.34～5.05）	7.73 （7.33～8.18）	233 （95～437）	51.62 （33.88～64.37）
南，浅层	0.29 （<0.14～0.83）	0.72 （0.38～1.60）	0.31	7.84	142	61.72 （45.29—75.97）
北，表层	0.21 （<0.14～0.85）	0.43 （0.21～1.18）	0.74 （0.32～1.73）	7.72 （7.48～8.06）	124 （−107～375）	54.13 （33.85～68.66）
北，浅层	0.19 （<0.14～0.44）	0.53 （0.21～1.00）	0.37 （0.1～0.76）	7.90	86	70.10 （50.23～81.62）

2　材料和方法

2.1　SRB 的测定

表层沉积物用 DDC1-4 重力活塞取样器取得，浅层沉积物岩芯用 DDC4-2 重力活塞取样器获取之后，用刚消毒的刮铲取出沉积物表膜下的样品，即刻将其投入无菌试管中，塞紧。回实验室以文献[6]中所述的液体培养基,用 MPN 法稀释样品. 将稀释好的样品置25℃室温中培养 7 天以上。观察结果,以培养液有黑色沉淀为检出 SRB 的阳性者。计算出单位湿重沉积物中的 SRB 细胞数(个/g)。结合沉积物样品的生物地球化学因子(biogeochemical agents)及其指标,分析与 SRB 的关系。

2.2　生物地球化学因子指标的测定

沉积物样的 pH、Eh 用 HSCL-2A 型船用酸度计测定。硫化物(S^{2-})用 PXJ-1 型离子计测定。Fe^{3+}/Fe^{2+} 用 EDTA 滴定法测定. 有机质含量用硫酸—重铬酸钾氧化还原滴定法测定。电阻率用专用电阻率仪测定。

3　结果和讨论

3.1　测区内 SRB 含量状况

表 2 列出了所测两区 SRB 含量的统计结果。各区表层沉积物 SRB 含量及其所在样品数占样品总数的份额状况如表 3 所示。

表 2　浙江—闽北陆架沉积物硫酸盐还原菌(SRB)含量状况

Table 2　Condition of SRB contents of sediment in the shelf

测区及层次	SRB 含量参数					
	测定样品数	平均值	波动范围	阳性率,%	出现频率最高的样品数及其 SRB 含量*(%)	
南,表层	71	576.8	0～24 000	93.0	36.7	230
南,浅层	18	267.0	0～4 600	33.3	66.7	0
北,表层	68	302.8	0～4 600	75.0	25.0	0
北,浅层	15	41.0	0～230	53.3	46.7	0

* SRB 含量单位:个(细胞)/g(湿沉积物),下同。

由此可见,表层 SRB 检出率、含量波动范围、浅层 SRB 含量及其波动范围均是南区>北区,仅浅层 SRB 检出率是南<北。表、浅两层比较,SRB 检出率、含量及波动范围均是表层>浅层。这些差异可能反映出纬度较低、离岸较近的陆架近岸区有利于 SRB 的生长繁殖。在特定条件下,也意味着表层比浅层有更多的营养及硫酸盐可供 SRB 利用(见表 1),而生存环境就某种意义言则是浅层比表层稳定。

表 3　浙江—闽北陆架表层沉积物不同含量硫酸盐还原菌的样品数及其占样品总数的百分比值

Table 3　Numbers of Samples with different content of SRB in the surface layer of sediment investigated and percentages occupied in total numbers of Zhejiang-North Fujian shelf

SRB 含量,个/g	所在区域			
	南		北	
	样品数	样品数/样品总数,%	样品数	样品数/样品总数,%
0	5	7.0	17	25.0
40	10	14.1	11	16.2
70	/	/	3	4.4

（续表）

SRB含量，个/g	所在区域			
	南		北	
	样品数	样品数/样品总数，%	样品数	样品数/样品总数，%
90	23	32.4	7	10.3
150	1	1.4	4	5.9
200	1	1.4	2	2.9
230	26	36.7	16	23.5
430	1	1.4	3	4.4
750	1	1.4	1	1.5
1 500	/	/	1	1.5
2 400	2	2.8	/	/
2 800	/	/	1	1.5
4 600	/	/	2	2.9
24 000	1	1.4	/	/

南、北区样品数各为71和68个

就全区言，表层SRB检出率、平均含量及其波动范围分别是84.2%、436个/g及0～24 000个/g。浅层SRB检出率、平均含量及其波动范围则分别是42.4%，164个/g及0～4 600个/g。这一结果与其他区域相比，可以发现SRB的检出率低于胶州湾、高于北部湾北海—临高海区的，而SRB含量则是低于上述两区的[4-5]。表层SRB含量的平面分布呈现出不均匀状态，也即"补丁式的(patchy)"。南区以230个/g和90个/g这两块"补丁"最大。最高SRB含量(24 000个/g)仅出现在No.33测点。北区以0和230个/g这两块最大。最高SRB含量(4 600个/g)则出现于No.1和54两测点，出现较多SRB含量处可能受海洋开发活动影响较多(见表3)。浅层SRB也基本如此。这反映出测区间、测点间生物地球化学参数间的差异。它为划分调查区腐蚀性的级别提供了信息。

3.2 SRB含量与若干生物地球化学因子间关系分析

对南、北两区表、浅层SRB含量与相应的硫化物、有机质含量、Fe^{3+}/Fe^{2+}、pH、Eh、电阻率等参数间作的主要相关分析结果列入表4之中。由该表可见，南区表层、表层＋浅层出现高值SRB处，SRB含量与Eh间的关系最为一致，其一致程度达99.9%，尤其在南区表层更为明显。另一类较好的相关性是在SRB含量与S^{2-}含量间。这一正相关关系主要表现在南区出现高量SRB处。在北区表层，该正相关关系尚可。SRB含量与Fe^{3+}/Fe^{2+}间均有正相关性存在，最具显著意义的正相关关系主要体现在有高量SRB处(即出现高量SRB的表层、浅层样品)。就浅层言，这一正相关性是南区好于北区。SRB含量与有机质含量间的相关关系较为复杂，在南区表层呈现为一定意义的负相关，在北区浅层却是正的，在高含量SRB处，也是明确的正相关关系。SRB-pH的相关系数都为正值，南区表层好于北区浅层。SRB-电阻率间的关系是负相关的，它主要存在于南区浅层、北区表层和高量 SRB 处。从这 17 个相关系数中，可以发现其正负比值分配中，正值的几率(76.5%，13/17)大于负值的(23.5%，4/17)。说明 SRB 含量与这 6 个参数间的变化以一致为主，其中尤以与 Eh、与 Fe^{3+}/Fe^{2+} 的正相关性为多。与电阻率的相关系数是负的。表明当所测样品中的 Eh、Fe^{3+}/Fe^{2+} 增减和电阻率降升时，相应的 SRB 含量增减的可能性也大。pH、S^{2-} 含量的增减一般也伴随着 SRB 的增减。与有机质含量的变化有多种因果关系存在，当有机质对 SRB 的生物有效性(biological effectiveness)提高时，SRB 含量可能增多，反之有机质含量的增加不一定意味着 SRB 的增多。

表 4　浙江—闽北陆架沉积物硫酸盐还原菌含量与有关生物地球化学因子间的相关分析结果

Table 4　The Result of correlation analysis of SRB contents and their relative biogeochemical agents in Zhejiang-North Fujian shelf

测区及层次	相关因子	相关系数(r)	显著性水平(a)	样本数(N)
南，表层	SRB-Or	−0.125	0.20	63
	SRB-pH	0.113 7	0.20	66
	SRB-Eh	0.414 3	0.001	60
南，浅层	SRB-Fe^{3+}/Fe^{2+}	0.507 7	0.20	8
	SRB-电阻率	−0.721 3	0.20	5
南，表*＋浅层	SRB-S^{2-}	0.883 1	0.02	6
	SRB-Or	0.7401	0.10	6
	SRB-Fe^{3+}/Fe^{2+}	0.897 3	0.02	6
	SRB-Eh	0.919 0	0.001	6
北，表层	SRB-S^{2-}	0.220 4	0.20	50
	SRB-电阻率	−0.173 8	0.20	51
北，浅层	SRB-Or	0.407 7	0.50	6
	SRB-Fe^{3+}/Fe^{2+}	0.607 7	0.20	6
	SRB-pH	0.498 6	0.50	6
北，浅＋表层*	SRB-Fe^{3+}/Fe^{2+}	0.623 4	0.05	10
	SRB-Eh	0.317 7	0.50	10
	SRB-电阻率	−0.650 8	0.10	7

* 表＋浅层，指表、浅层沉积物中出现 SRB 高含量时，与相应因子间作分析用的样品

SRB 含量与上述 6 个参数间的差异意味着南、北两区，表、浅两层间的差异，因而形成了不同的腐蚀环境。

3.3　SRB 含量等参数与沉积物岩性类型的关系

将 SRB 含量等 7 个参数及一些相关情况按沉积物的岩性类型比较，可以发现它们受不同岩性的影响，并表达出不同的分布势态。在南区，淤泥质粉质粘土表层中的 SRB 含量、Fe^{3+}/Fe^{2+} 分别平均为 860 个/g 和 0.74，而中砂质表层中的 SRB 含量及 Fe^{3+}/Fe^{2+} 则

平均仅为 40 个/g 和 0.54[7]，粉砂质浅层中的 SRB 含量及 Eh 分别为 20 个/g 和202.8 mV，一些淤泥样品中甚而未测得 SRB，Eh 则降为 122.0 mV。SRB 含量与 Eh 值在北区各类岩性的表层沉积物中显示出一致的差异趋势，如 SRB 的平均含量的高低排序是，淤泥：240 个/g；淤泥质粉质粘土：46 个/g；细砂中是 135 个/g；粉砂中是 40 个/g。相应地，Eh 的平均值大小排序也是：淤泥，134.0 mV；淤泥质粉质粘土，121.2 mV；细砂，34.5 mV；粉砂，－69.0 mV。SRB 含量与电阻率间的负相关性在北区表层的淤泥质粉质粘土和细砂中体现得较好，如，淤泥质粉质粘土沉积物中的 SRB 含量为 416 个/g，细砂中仅为 135 个/g。相反，电阻率在细砂中达 60.6(Ω·cm)，而淤泥质粉质粘土中便降为 52.9(Ω·cm)。一些研究还表明，在特定条件下，同类岩性的沉积物对 SRB 的生存和腐蚀参数可能产生不一致的作用[8]，比如粒径大的沉积物可有利于以氧去极化的腐蚀过程，却不利于 SRB 的生存。反之亦然. 因此不同岩性的沉积物类型与其相应的生物地球化学因子共同作用于海底，塑造出 SRB 等生物的大、小、微生境，它们间的关系无疑是错综复杂的，因而不能一言以蔽之[9]。

4 结语

本文分析了全测区表、浅层沉积物 SRB 检出率，一般是表层高于浅层，SRB 含量一般是近岸高于远岸。比较了两参数在南、北区间的差异. 主要发现南区表层 SRB-Eh 间、浅层及高 SRB 含量处 SRB-Fe^{3+}/Fe^{2+} 间有最显著正相关，南区浅层见 SRB-电阻率有最显著负相关。SRB 是腐蚀过程主要因素[10]。本 SRB 检出率表明测区大部分沉积物都有强化腐蚀过程的条件，北区重于南区。表、浅两层 SRB 含量虽同处一量级，但表层高于浅层，说明浅层强化腐蚀范围小。以 SRB 等参数对环境作粗略的腐蚀性分类[5]，表明测区大部分范围腐蚀性为中强度，少数表层区可能有小范围特强腐蚀性，进行施工尤要加强防腐措施. 对本区腐蚀性作出高度概括和精确评价，尚需对 SRB、生物地球化学参数、沉积物状况等作更深研究。

沉积物样品及地球化学参数由国家海洋局第一海洋研究所有关人员取得分析，在此一并致谢。

参考文献 10 篇(略)

(合作者：李培英　王　波)

SULPHATE-REDUCING BACTERIA RELATIONSHIP BETWEEN THEM AND BIOGEOCHEMICAL AGENTS IN ZHEJIANG-NORTH FUJIAN SHELF

(ABSTRACT)

Abstract The SRB content in the sediment of the Zhejiang-North Fujian continental shelf was determined using the method of MPN. The results indicated that: the percentage of occurrence and average content of SRB in the surface sediment surveyed were 84.2% and 436.8 cell/g wet sediment (the same below) respectively, and in the shallow layer of the sediment were 42.4% and 164 cell/g respectively. As to the distribution state of SRB contents, it was discovered that the percentage of occurrence and average content of SRB in the surface sediment surveyed in the inshore area were 75.0% and 302 cell/g, and in the offshore area were 53.3% and 41 cell/g respectively. Enquiry was made as to the significance of SRB contents and their distribution condition to the extent of corrosiveness of the environment surveyed. An analysis of the correlation between the SRB content and the contents of sulphide, organic matter, Fe^{3+}/Fe^{2+}, pH, Eh and electric-resistivity respectively revealed that: there were mainly some positive correlations among them, especially there were very significant positive correlations between SRB content and Eh, Fe^{3+}/Fe^{2+} respectively. A negative correlation was found between SRB content and electric resistivity. The seven biogeochemical parameters including SRB condition etc. have provided important information for correct assessing the phenomenon of environmental corrosiveness in the area surveyed.

Key words Sulphate-Reducing Bacteria; Sediment; Biogeochemical Agents; Assessment on Corrosiveness; Zhejiang-North Fujian Continental Shelf

白令海和楚科奇海测区沉积物硫酸盐还原菌初步研究*

摘　要　用MPN法测定了取自白令海和楚科奇海35个沉积物样品中的硫酸盐还原菌(SRB)含量。所有样品的SRB平均含量约1.68×10^5个·g^{-1},变幅为$0\sim228\times10^5$个·g^{-1}。分析并比较了SRB含量在表层、次表层或混合层样中及不同站位、测区间的差异。大体看来,SRB含量表层少于次表层。若以白令海峡为界,在海峡两侧SRB含量的增加呈相反趋势,即在白令海呈向南增加、在楚科奇海呈向北增加;经向分布也显示出相反的趋势:在白令海呈向东增加、在楚科奇海呈向西增加的趋势。结果还表明SRB在不同培养温度中有各自的出现率,反映测区SRB以适冷菌为主。温度是其主要控制因子。相当多的中温菌也有出现。

关键词　白令海　楚科奇海　硫酸盐还原菌　温度效应　适冷菌

1　前言

硫酸盐还原菌(Sulphate-Reducing Bacteria,缩写为SRB,下同)广泛分布于世界不同的生态环境之中,即使是寒冷的极地海洋或深海之中也有其踪迹可寻。它们发挥着独特的生物地球化学作用(Smith,1993)。凭借主要进行的硫酸盐还原活动,它们常常对环境和各种工程设施产生腐蚀作用,造成经济损失。针对SRB含量及生态分布的研究正愈益受人重视(陈皓文,1998)。Vainshtein et al.(1994)曾分析过俄国Kdvlna低地永冻层(<−5℃)的SRB,但北极及附近海区的SRB的研究迄今较少(Arnosti et al.,1998;Isaksen and Gensen 1996)。根据我国首次北极科学考察的野外观察,推测考察区海底仍会有SRB的活动。本文就白令海峡两侧的白令海(Bering Sea)和楚科奇海(Chukchi Sea)两测区沉积物SRB含量状况等作一报道.为研究SRB在极地环境中的作用打下基础

2　测区和站位概况

白令海和楚科奇海测区范围在58°33′～73°30′N和158°56′～179°33′W之间,水深范围30～840 m,其环境状况如表1。测站所在位置如图1、图2所示。

---

*　原文刊于《极地研究》,2000,12(3):211-218.

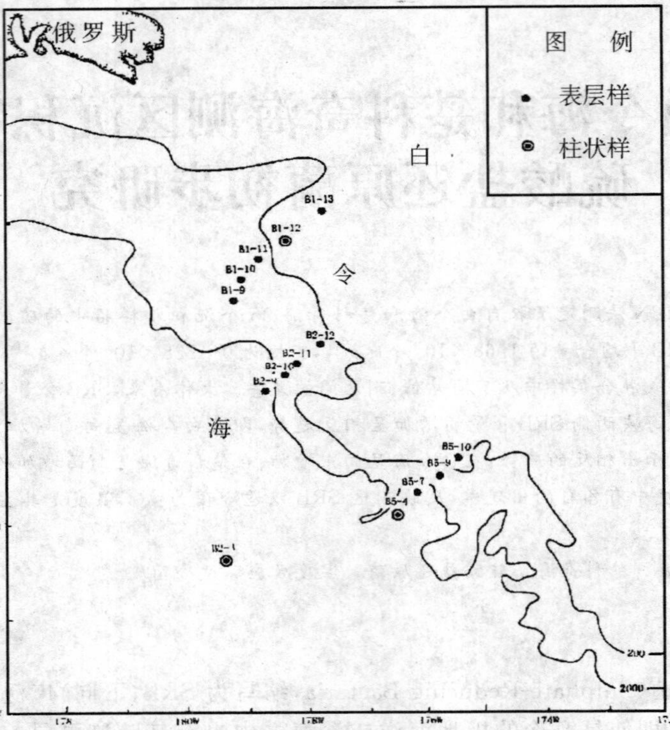

图 1　白令海测区沉积物硫酸盐还原菌站位示意图

Fig. 1　sketch map of sampling stations for sulphate-reducing bacteria in sediments of the survey areas of Bering Sea.

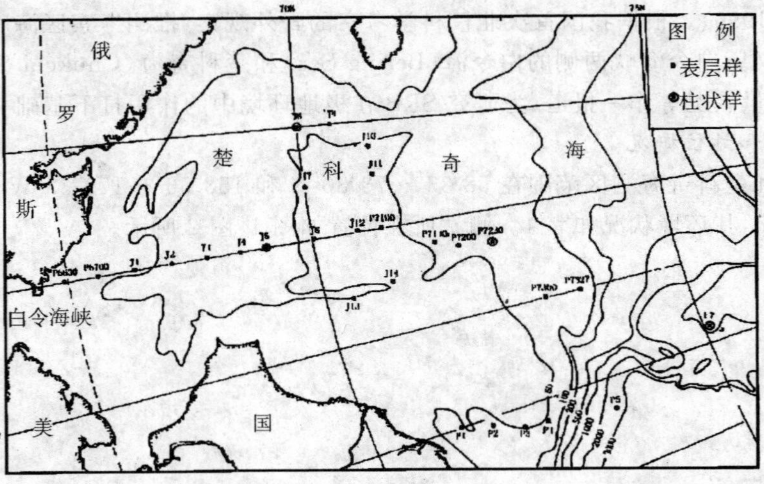

图 2　楚科奇海测区沉积物硫酸盐还原菌站位示意图

Fig. 2　sketch map of sampling stations for sulphate-reducing bacteria in sediments of the survey areas of Chukchi Sea.

表 1　白令海、楚科奇海测区环境简况*

Table 1　brief environmental conditions of the survey areas in Bering Sea and Chukchi Sea

测区	范围	测点数	水深(m)	盐度(‰)	底水温(℃)	沉积物描述
白令海	58°33′～60°56′N 175°33′～179°33′W	10	288 (139～840)	33.60 (32.70～34.60)	1.6 (1.0～2.0)	白令海陆架—陆坡。浅水陆架区沉积物以细砂或粉砂为主。生物沉积较多。深水沉积物以黏土或软泥为主。
楚科奇海	66°30′～73°30′N 158°56′～175°02′W	24	51 (30～92)	32.60 (31.00～32.20)	－1.5 (－1.75～－0.65)	多数样品呈浅、淡灰色，少数灰黑色。以粉砂质黏土或黏土质粉砂为主。少数为软泥。多数样品含泥质团块。有机质多。大部分样品有浓 H_2S 气味。

* 括号前为算术平均值，括号中为变化范围。

3　样品采集和分析

3.1　采样

两测区沉积物样品均由中国首次北极考察队提供。本文所用表层沉积物样品全由箱式取样器采集。当样品采到船甲板后即刻将它们封存于干净塑料袋中带回国内实验室。根据实验设计，按照表层(上 0～5cm)、次表层(上 5～10cm)或不分层次进行取样。样品采集按微生物学要求操作。

3.2　室内分析

将所得样品定量地按最大可能数量法(Most Probable Number，缩写为 MPN)进行系列稀释至 SRB 液体培养基中(陈世阳等，1987；陈皓文，1998)，在 4℃、15℃、25℃和 28℃温度下分别培养 2 周以上，进行观察记录。以试管中见黑色沉淀者为 SRB 阳性。统计实验结果，分析测区 SRB 含量状况和不同培养温度中 SRB 的出现率。

4　结果和讨论

4.1　测区 SRB 含量状况分析

表 2　白令海、楚科奇海测区沉积物 SRB 含量状况表

Table 2　SRB content state of sediments in survey areas of Bering Sea and Chukchi Sea.

样品编号	站位号	SRB 细胞数·[10^2·g^{-1}(湿)]		
		表层(上 0～5cm)	次表层(上 5～10cm)	混合层(不分层)
1	J-1	33.30	/	/
2	J-2	/	0.02	/
3	J-3	1.20	/	/
4	J-4	/	/	3.00

(续表)

样品编号	站位号	SRB 细胞数·[10^2·g^{-1}(湿)]		
		表层(上 0~5cm)	次表层(上 5~10cm)	混合层(不分层)
5	J-5	0.02	/	/
6	J-7	/	/	0
7	J-8	/	0.70	/
8	J-9	2 800.00	/	/
9	J-10	/	0.10	/
10	J-11	5.45	/	/
11	J-12	/	15.00	/
12	J-14	/	/	2.50
平均		567.99	3.96	11.83
13	P6630	/	/	79.75
14	P6700	/	/	11.08
15	P7100	/	/	2.85
16	P7130	/	/	0.03
17	P7200	/	360.00	/
18	P7230	0.02	/	/
19	P7300	0.20	/	/
20	P700	/	1.00	/
21	P7327	6.00	/	/
22	P1	/	/	0.5
23	P2	/	0.02	/
24	P3	0.23	/	/
25	P4	/	580.00	/
平均		1.61	235.26	18.84
26	B1-9	/	/	201.00
27	B1-10	/	/	22 750.00
28	B1-11	/	/	3 750.38
29	B1-12	0.05	/	/
30	B1-13	/	16.70	/
31	B2-10	3.78	/	/
32	B2-11	/	7 375.00	/

(续表)

样品编号	站位号	SRB 细胞数·[10^2·g^{-1}(湿)]		
		表层(上 0～5cm)	次表层(上 5～10cm)	混合层(不分层)
33	B2-12	33.30	/	/
34	B5-9	/	/	20 500.00
35	B5-10	/	/	381.00
平均		12.38	3 695.85	9 516.48
总平均		240.30	834.85	3 667.85

* "/"表示未测定

表 2 列出了所有样品的 SRB 含量状况. 由表 2 可见,除了 J-7 站 SRB 未检出到外,其余各站均有检出,两测区沉积物的 SRB 检出率高达 97.1%以上,尤其是白令海测区检出率达 100%,说明 SRB 在该类区域中的广泛存在。测区 SRB 平均含量约为 168×10^5 个细胞·g^{-1}(湿重,下同)。与以往其他海区 SRB 测定结果比,本测区 SRB 检出率和含量均较高(陈皓文,徐家声,1999;陈皓文等,2000;Hines et al.,1999)。总体看表层 SRB 含量小于次表层,这意味着表层以下的许多环境适于 SRB 生存活动(Arnosti et al.,1998;Hines et al.,1999)。

表 3 是白令海与楚科奇海两测区 SRB 含量状况的总体比较。它说明楚科奇海测区表层沉积物 SRB 平均含量高于白令海区,变化范围也大。次表层及混合层的情况却与此相反,即白令海测区的 SRB 含量高、楚科奇海区的 SRB 含量低。这可能与此时白令海相对较高的温盐度有关。

表 3 白令海、楚科奇海沉积物 SRB 含量比较

Table 3 comparison of SRB content in the sediments between bering Sea and Chukchi Sea

海区	SRB 含量(个·10^2·g^{-1})					
	表层		次表层		混合层	
	平均值	范围	平均值	范围	平均值	范围
白令海	12.38	0.05～33.30	3 695.85	16.70～7 375.0	9 516.48	0.75～20 500.0
楚科奇海	316.27	0.02～2 800.0	119.61	0.02～580.00	12.46	0～79.75

表 4、5 分别代表了两测区各自的 SRB 含量的空间分布差异。表 4 说明了在白令海测区,SRB 含量在纬向(南、北区间)的差异是低纬度区的高于高纬度区的。经向的差异则是东部高于西部。表 5 则说明了楚科奇海测区 SRB 含量在纬向的上述差异只体现在混合层样中. 表层和次表层则是相反的。东、西区间在表层和混合层样中却显出西部 SRB 含量高于东部,次表层中又似乎反了过来. 即东部 SRB 含量高于西部。

表 4 白令海测区纬向或经向区间 SRB 含量比较

Table 4 comparison of SRB content along the latitudinal or longitudinal areas in Bering Sea

区间	SRB 含量(个·10^2·g^{-1})		
纬度(N)	表层	次表层	混合层
61-60	0.05	16.70	7 775.46
60-59	18.54	7 375.00	/
59-58	/	/	10 440.50
平均	9.30	3 695.85	9 107.98
经度(W)			
180-179	3.78	/	7775.46
179-175	16.68	3 695.85	/
177-175	/	/	10 440.50
平均	10.23	3 695.85	9 107.98

*"/"表示未测定

表 5 楚科奇海测区纬向或经向 SRB 含量比较

Table 5 comparison of SRB content along the latitudinal or longitudinal areas in Chukchi Sea

区间	SRB 含量(个·10^2·g^{-1})		
纬度(N)	表层	次表层	混合层
75-70	468.65	134.90	1.47
70-66.30	11.51	0.02	2 663.03
平均	240.08	67.46	1 332.25
经度(W)			
175	2 800.0	0.7	/
170	11.51	1.26	1 997.98
165	3.1	1.0	/
160	0.23	290.01	0.5
平均	703.71	73.24	999.24

*"/"表示未测定

两测区相比,似乎可将白令海峡作为界线,SRB 含量由该海峡向南、北两侧呈反向增加趋势。两测区的东西之间也因此海峡而呈反向增多之势。

从以上结果可以大致看出,两测区沉积物 SRB 含量除在样品所处层次间有一定差异外,在地理分布上也有一定差异。影响此差异的因子推测除了地理位置不同外,沉积物的来源、沉积环境、生物地球化学因子的影响也是十分重要的,对此有必要作进一步的探讨。

4.2 SRB 对不同温度的感应性

长期处于一定环境条件中的微生物对稳定的生态因子应有较好的适应性,SRB 的情况如何,为此测试了 SRB 对不同培养温度的感应性状。

表 6　白令海、楚科奇海测区沉积物中 SRB 在不同培养温度中的出现率*

Table 6　occurrence percentages of SRB from the sediments in Bering Sea and Chukchi Sea

培养温度(℃)	SRB 出现率(%)
4	25.70
15	40.00
25	25.70
28	2.90

* 参试样本数 35,培养时间 2 周以上

表 6 列出的是经一定稀释的沉积物在 4 种培养温度时 SRB 出现率(同一培养温度时 SRB 阳性反应数/总样品数)的比较,由该表可见,以 15℃培养出的 SRB 出现率最高,在 28℃培养时的 SRB 出现率最低,而 4℃和 15℃的 SRB 出现率共达 65.7%,根据 Isakem and Gensen(1996)的观点,这一结果表明测区沉积物的 SRB 大都是适冷菌(cold-adapted bacteria),即包括嗜冷菌(psychrophiles bacteria)和适冷菌(psychrotrophic bacteria)。25℃和 28℃中出现的 SRB 表明它们是中温菌(mesophilic bacteria)(Isaksen and Gensen,1996)。而且 25℃时的数量也不少,其中包括部分适冷菌,由此推测沉积物中适于 SRB 的基质经历着融化期之后带来的环境升温效应有利于它们的生存。一旦条件合适,尤其是环境温度的升高,适冷菌即可借助这温度利用合适的有机质等来迅速发展壮大自己。表明在北极、亚北极此类长期寒冷的环境条件中,低温是对 SRB 起制约作用的重要的生态因子,这些 SRB 仍可保留强劲的硫酸盐还原作用等活力(Arnosti et al.,1998)。

5　结语

本文对白令海、楚科奇海测区的 35 个沉积物样品作了 SRB 含量的测定,比较并分析了它们在不同海区、站位、层次间及不同培养温度中的差异。不同培养温度中这些样本的 SRB 出现率研究表明,测区适冷菌较多,表明低温是起制约作用的重要的生态因子。适冷菌对研究区域温度的适度升高有良好的适应性,尚有相当比率的中温菌存在,因此在测区此类冷环境中各类 SRB 仍有强大的生存能力和生化活动能力。

致谢:感谢国家海洋局极地考察办公室及“雪龙”号极地考察船为本次科考所作的协调及条件保障工作.使我们能取得研究区珍贵的沉积物样品。程振波、陈荣华、李秀珠、李亮、郑风武参加了采样工作,在此一并致谢。

参考文献 8 篇(略)

(合作者:高爱国)

PRELIMINARY STUDY ON SEDIMENT SULPHATE REDUCING BACTERIA IN THE SURVEY AREAS OF BERING SEA AND CHUKCHI SEA

(ABSTRACT)

Abstract The content of sulphate-reducing bacteria (SRB) of the 35 sediment samples was determined by using MPN technique in the survey areas of Bering Sea and Chukchi Sea. The average content of SRB in all samples tested is about 1.68×10^5 cells · g^{-1} (wet.), its variation is $0\sim2.28\times10^6$ cells · g^{-1}. The differences of SRB content were analyzed and compared among the surficial layers, subsurficial or mixed ones in varied areas and stations surveyed. Generally speaking, the SRB content in the surficial layers was lower than that in the subsurface and mixed ones. The determinable SRB content appears presumedly opposite trend, i. e. it seems that the content of SRB in creases south or east in Bering Sea and north or west in Chukchi Sea respectively if drawing a line on Bering Strait. The result indicates the occurrence percentages of SRB at 4 different cultivation temperature and suggests that there are a lot of cold-adapted bacteria there. The low environmental temperature is still mainly a controlling factor. There are quite a few mesophiolic bacteria, and it can be deduced that an effect of suitable organic matter supply to the SRB from raising temperature after melting frozen period.

Key words Bering Sea, Chukchi Sea, Sulphate Reducing Bacteria, Temperature Effect, Cold-Adapted Bacteria.

北极沉积物中硫酸盐还原菌与生物地球化学要素的相关分析*

摘　要　对北极楚科奇海、白令海测区 34 个沉积物样品的硫酸盐还原菌(SRB)含量与 $CaCO_3$、Si、S、P、Cd、Fe、Ti、Pb、Hg 等地球化学要素作了相关分析。同时分析了 SRB 与其所在沉积物粒度与底层水温间的关系。结果表明：SRB 与 Si、砂及底层水温为正相关；SRB 与 $CaCO_3$、Fe、Ti、P、Pb、Hg、粉砂、粘土为负相关；而 SRB 与 Cd、S 的相关性因海区不同而不同。既有负相关，也有正相关。分析了这些相关状况及其原因。

关键词　硫酸盐还原菌　地球化学要素　相关分析　楚科奇海　白令海　北极

处于分解系统末端的硫酸盐还原菌(Sulphate Reducing Bacteria，缩写为 SRB，下同)，利用其他细菌的终产物作为自己的能源而生长，其本身又是还原活动的参与者，在极地海洋环境中进行硫酸盐还原活动. 在北极海洋沉积物缺氧条件下，由于大量适冷、甚至中温 SRB 等微生物的存在，使一些无机物、有机物(包括含硫有机物)的分解、营养盐再生以及元素赋存状态的变化等一系列地球化学活动得以发生，从而使元素在沉积物、间隙水、海水间循环迁移。同时 SRB 复杂的生态学和生物地球化学活动又受环境中众多因子的影响和左右[1]。前人及我们的研究曾经表明楚科奇海及白令海沉积物中 SRB 在不同的培养温度中有不同的出现率，在不同测区、站位间也存在地理差异[2,3]。本文试图论述楚科奇海和白令海测区沉积物 SRB 与 S、Fe 等一些生态环境因子、地球化学因子间的关系，有助于了解极地环境下 SRB 在生物地球化学过程中的作用和意义。

1　材料和方法

1.1　样品的采集

1999 年 7～9 月中国首次北极考察期间，用箱式取样器在楚科奇海及白令海采集了沉积物样品。所有样品在采上船后，在现场进行分层取样，将 0～5cm 层作为表层上部；5～10cm 层作为表层下部；有的站位样品较少，不再分层(相当于表层上部样)作为表层混合样。将样品封入干净的塑料袋中保存，运回国内实验室启封分析。在实验室根据研究工作的实际需要，分别从楚科奇海沉积物样品中取 24 站，白令海取 10 站样品，并于同年 10～11 月进行了 SRB 及相关因子的测定。研究区范围和站位布设见图 1 所示。

1.2　微生物测定和粒度、化学分析

启封的沉积物样品按需进行取样，在无菌条件下进行 SRB 分析. 样品接入预先消毒的试管中，按 MPN 法定量逐步稀释样品至 SRB 液体培养基中恒温(10℃)培养 2 周以上(样

*　基金项目：国家自然科学基金(40176017、40086001、49873015)

原文刊于《环境科学学报》，2003，23(5)：619-624，本人为第二作者。

品稀释范围：10^{-1}～10^{-5})，检查、记录和分析细菌生长状况，以培养液见黑色沉淀为阳性者获得单位重量(1.0g，湿重)沉积物中SRB含量(细胞个数·g^{-1})，培养基成分和SRB含量测定参见文献[2]。

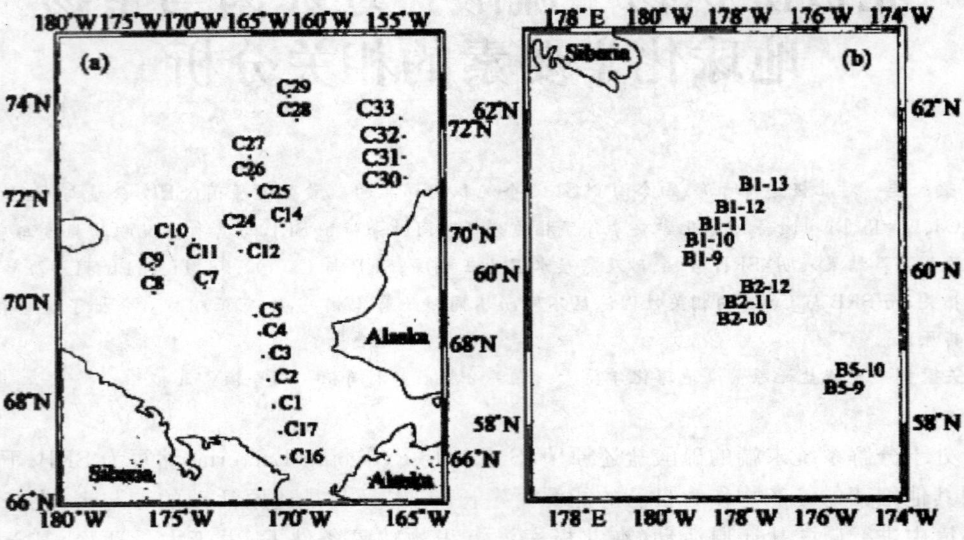

图1 楚科奇海(a)与白令海(b)沉积物采样站位图

Fig. 1 Map of sampling stations for SRB in the Chukchi(a) and Bering Seas(b)

样品的化学分析包括X射线荧光光谱法测定Si、Fe、Ti、S、P；石墨炉原子吸收法测定Cd；冷原子荧光法测定Hg；容量法测定$CaCO_3$。样品的化学分析采用外控及实验室内部质量控制二级质量监控，采用的监控样为国家一级标准物质GBW07039、GBW07408。各组分的检出限分别为Si、Fe：0.1%；Ti、$CaCO_3$：0.01%；S：50×10^{-6}；P：20×10^{-6}；Cd、Hg：10×10^{-9}。粒度分析采用激光粒经谱仪与筛析法相结合的分析方法，误差小于5%。样品测试在国家海洋局第一海洋研究所与国土资源部物探与化探研究所进行。底层水温用CTD进行现场测定。

2 结果及讨论

2.1 沉积物岩性特征

楚科奇海沉积物为浅灰色至深灰色粉砂质泥或泥质粉砂，局部站位沉积物表层为浅黄色。往下颜色变深，大部分样品有H_2S气味。波弗特海一侧沉积物的颜色主要为浅黄～浅灰色软泥或粘土，略有H_2S气味。白令海沉积物为浅黄至浅灰色粉砂或砂质沉积物，沉积物颗粒较粗，H_2S气味很轻或无H_2S气味。从沉积物岩性看，两海区在接近白令海峡的区域，有较多的有机质，向两侧有机质含量呈降低趋势，沉积物的孔隙度为白令海大于楚科奇海，而后者则是近岸大于远岸。两区沉积物的岩性差异有可能影响SRB的地域差异，沉积物样品的岩性特征见文献[4]。

2.2 沉积物SRB的含量

沉积物样品中的SRB含量统计结果见表1，由表可看出楚科奇海SRB的表层上部平均含量高于白令海，在表层下部及表层混合样的含量却均表现为楚科奇海低于白令海。研究区内、站位间、经度或纬度间SRB含量的差异和分布状况见文献[2]。

表1　楚科奇海、白令海测区沉积物SRB含量统计表

Table1　SRB contents of the sediments in survey areas of the Chukchi sea and Bering Sea

海区	含量特征	SRB，细胞数·10^2·g^{-1}（湿）		
		表层上部（上0～5cm）	表层下部（上5～10cm）	不分层
楚科奇海	含量范围	0.02～2 800	0.02－580	0－79.8
	平均值	316	120	12.5
白令海	含量范围	0.05～33.3	16.7－7 375	201－22 750
	平均值	12.4	3 696	95.7

2.3　元素地球化学要素概况

研究区沉积物中与SRB分析相对应的各样品中元素的含量范围及平均含量见表2，由表中可知，除了Si与Hg外，其余各要素的含量均是楚科奇海大于白令海。

表2　楚科奇海及白令海沉积物中元素的含量

Table 2 Contents of elements in the sediments in the Chukchi and Bering Seas

含量	楚科奇海		白令海	
	范围	平均值	范围	平均值
Si，%	26.23～40.11	31.56	32.58～35.58	34.38
Fe，%	1.23～4.06	2.96	2.05～2.61	2.28
$CaCO_3$，%	0.799～3.516	2.409	1.553～2.577	1.921
Ti，%	0.178～0.421	0.357	0.249～0.360	0.319
S，10^{-6}	753～3 568	2 063	726～1 894	1 085
P，10^{-6}	468～1 499	1 055	692～900	809
Pb，10^{-6}	7.9～26.7	15.8	5.2～15.6	10.4
Cd，10^{-9}	89～481	252	131～330	225
Hg，10^{-9}	17～36	24	17～46	25

2.4　SRB含量与生物地球化学要素间的相关性

生活于海底沉积物中的SRB与其生存环境之间有着相当密切的关系，其活动的强弱受多种因素制约。根据SRB在自然界的分布特征，即具有对数分布的特点，为此将SRB取对数后，计算其与各要素之间的相关系数，结果显示在80%置信水平上呈现相关的相关系数有80个，在95%和99.9%置信水平上相关的分别有46个和6个（表3）。

由表3列出的相关系数可看出，SRB与Si、砂及底层水温为正相关。SRB与$CaCO_3$、Fe、Ti、P、Pb、Hg、粉砂、粘土为负相关，SRB与Cd、S既有负相关，也有正相关。

2.5　SRB与地球化学要素间相关性分析讨论

2.5.1 SRB 与 Si 的正相关性

SRB 作为环境中的一类微生物，必然与所在环境密切相关。Si 作为地壳中的主要元素，其含量既与物源区母岩的风化产物有关，又与海区硅质生物的发育程度有关。在楚科奇海，SRB-Si 的正相关性说明特定沉积物中活性 Si 为 SRB 提供了有利的生存环境，沉积物中硅质生物增加，可为微生物提供更充足的物质来源，有利于 SRB 的生存。在白令海硅质生物含量相对较少，因此 SRB 与 Si 之间的相关性减弱。当沉积物中生源硅含量相近时，硅的多少又可反映沉积物中石英、长石等矿物的多少以及沉积物的通透性的差异，一定的厌氧环境有利于嫌气 SRB 生存。反过来，SRB 参与 Si 地球化学过程，还利用和腐蚀硅酸盐类[5]。因此在楚科奇海区硅的增加对 SRB 影响表现在 SRB 的营养上，而白令海 Si 的增加对 SRB 的影响则表现在环境通透性增加上。Si 与 SRB 在两海区的粗粒沉积物中相关性较好，在表层混合样或表层浅部样品中均在 95％置信水平上呈显著相关（表 3）。

表 3　楚科奇海及白令海沉积物 SRB 含量与生物地球化学要素含量间的相关性

Table3　Correlation between the contents of SRB and biogeochemical factors in the sediments in the Chukchi and Bering seas

要素	楚科奇海				白令海				全海区		
	表上*	表混	表浅	全样	表混	表浅	全样	表上	表混	表浅	全样
Si		0.5647					0.5301		0.5554	0.4284	0.3274
$CaCO_3$	−0.5247	−0.7460	−0.5561	−0.5065		−0.7523	−0.6122	−0.4284	−0.6637	−0.6238	−0.5934
P		−0.9081	−0.4180	−0.3622		−0.8684	−0.7251		−0.8038	−0.5880	−0.5182
Ti	−0.5877	−0.7193	−0.4514			−0.8515	−0.7153	−0.4643	−0.5791	−0.6140	−0.3973
Pb		−0.6616	−0.3440						−0.6292	−0.4118	−0.2844
Fe		−0.6972							−0.5930	−0.4118	−0.2844
Hg	−0.6536	−0.6636	−0.6389					−0.5483			
S	0.6115				−0.6921			0.4618	−0.8306	−0.3027	−0.3674
Cd			0.3518		−0.7792	−0.6296	−0.6243	0.3942			
粉砂		−0.7575	−0.4138			−0.7648	−0.6142		−0.6144	−0.5109	−0.3368
粘土					−0.7312	−0.7756	−0.5634		−0.6884	−0.4615	−0.3501
砂		0.5200				0.7834	0.6244		0.6065	0.4857	0.3401
底层水温		0.7820				0.5722			0.9191	0.6219	0.5277
样品数	9	8	17	24	5	8	10	12	13	25	34

*：表上为表层上部，表混为表层混合样，表浅为表层上部与表层混合样之和，全样为各种表层样之和。

2.5.2 SRB 与 $CaCO_3$、P、Fe、Ti、Pb、Hg 之间的负相关关系

如表 3 所示，SRB 与 $CaCO_3$、P、Fe、Ti、Pb、Hg 之间均为负相关关系，负相关程度因采样区域及深度、沉积物岩性等不同而异。

海洋沉积物中的碳酸钙与生物活动关系密切，常作为生源物质的指标。北极地区由于其纬度高、温度低，不仅钙质生物活动相对较少，而且溶蚀也较严重，因此沉积物中 $CaCO_3$ 含量总体较低。当沉积物中生源物质含量较低时，或生源物质的补充小于 SRB 对生源物质的消耗时，SRB 含量越多，其所消耗的生源物质也就较多，沉积物中的 $CaCO_3$ 便

较少，呈现一种此消彼长的互补关系。在研究海区 $CaCO_3$ 含量较低，与 SRB 均呈较好的负相关关系（表 3）。在所研究的 34 个样品中，或在表层浅部的 25 个样品中在 99.9%水平上呈负相关（图 2）。

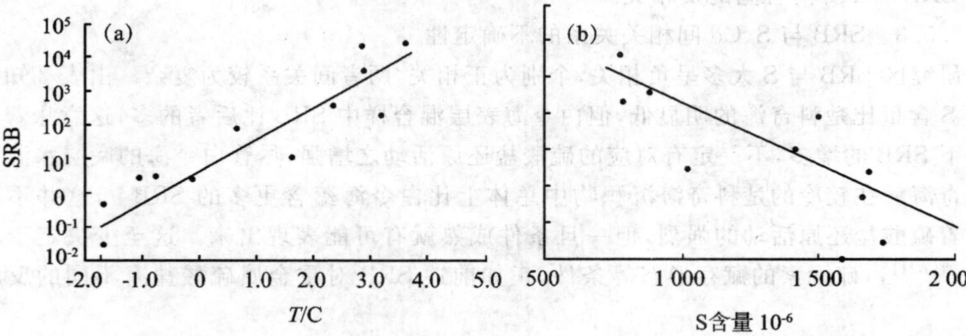

图 2　研究区 SRB 与底层水温(a)及硫(b)之间的相关图($n=13$)

Fig. 2　Correlations between SRB and bottom water temperature(a) or sulfur(b) in the sediment in the Chukchi and Bering Seas($n=13$)

适当的 P 是维持 SRB 正常生存环境和营养代谢之需[6]，PO_4^{3-}-P 与有机物间呈正相关关系，当有机物较少时，PO_4^{3-}-P 也少。P 与 SRB 的关系类似于 $CaCO_3$ 与 SRB 的关系，呈此消彼长的负相关。研究区沉积物中有较多的磷分布，其中楚科奇海沉积物中磷含量是中国浅海沉积物中磷含量的两倍多[7]。在一些站位，过多的 P 可能不利于 SRB 的生存，反之在另一些站位，P 的生物学有效性发挥又受制于它在沉积物中的存在形式，从而表现为 SRB-P 的负相关性。

铁对 SRB 的作用主要取决于 Fe 的生物学有效性，在研究区这种 pH 较高、Eh 较低的还原性沉积环境中，Fe^{2+} 较为稳定。在 SRB 进行的硫酸盐还原活动中，键产生的 H_2S 与 Fe^{2+} 在 pH≈7 时可生成非品质硫化铁。在研究区可能生成一种包含 PO_4^{3-} 和 OH^- 的亚稳态铁化物[8]。SRB 与 Fe 的负相关可能意味着本研究区沉积物中铁氧化菌较缺乏，铁被氧化得少了，而 SRB 的还原却占了主导地位，因而 SRB 的增多，使环境中的 Fe 减少[9]。在楚科奇海较粗的沉积物中，SRB 与铁在 90%显著水平上呈现负相关，而在整个研究海区则在 95%显著水平上相关（表 3）。

钛是自然界中较稳定的亲氧元素，也是铁族元素中被人们认为唯一的非人体必需元素。在生物地球化学研究中很少涉及该元素，但是在研究区 SRB 与 Ti 相关性好于其他铁族元素的相关性。在楚科奇海表层混合样沉积物中，Ti 与 SRB 在 95%水平上相关。在白令海表层上部与混合样中相关系数的置信度大于 99%，在两海区甚至高达 99.9% 以上（表 3）。Ti 与 SRB 之间这种较为密切的相关关系可能暗示存在某种共同因子，将两者维系在一起。

铅是亲硫元素，在还原环境下与硫结合形成方铅矿（PbS），在氧化环境下方铅矿易被氧化生成铅矾（$PbSO_4$），SRB 还原所产生的 S 可为铅的硫化物提供物质来源，同时影响铅的存在形态。另一方面当铅的含量过高时，它又是一种有毒元素，对生物活动将产生抑制作用（喜铅生物除外）。研究区铅与 SRB 在 95%显著性水平上呈负相关（表 3）。

汞也是一种亲硫元素，它与 SRB 为负相关。SRB 活动使环境中硫酸盐被还原，这其中可包括含汞的硫酸盐。无机汞在合适条件下(包括微生物生化活动)可以转化为一甲基汞或二甲基汞，当这些有机汞积累到一定浓度时，其毒性将影响 SRB 的生存状态，结果表现出 SRB 与 Hg 两者间的负相关。

2.5.3　SRB 与 S、Cd 间相关关系的不确定性

研究区 SRB 与 S 大多呈负相关，个别为正相关，两者间关系较为复杂。由表 2 知白令海的 S 含量比楚科奇海的明显低，但白令海表层混合样中 SRB 比后者的多，这意味着特定条件下 SRB 的增多，不一定有对应的硫酸盐还原活动之增强，尽管白令海的底层水温高于楚科奇海。在较冷的楚科奇海沉积物中总体上比白令海蕴含更多的 SRB[2]，这并不直接意味着硫酸盐还原活动的强烈，但一旦条件成熟就有可能表现出来。这至少决定于环境温度[3,10,11]。硫元素的赋存形态等条件，正如前述 SRB 对重金属硫酸盐有不同的反应那样。

在中低纬度的许多海区 SRB 与 S 常常为正相关关系[12,5]，这说明 SRB 的增多可能导致硫酸盐还原活动的强化。但 SO_4^{2-} 与 SRB 的关系还取决于 SO_4^{2-} 的浓度。正如 Amosti 等所证明的那样：只有当 SO_4^{2-} 浓度≤10 mmol/L 时，纯培养的 SRB 数量的增长与 SO_4^{2-} 两者才有直接关系。当 SO_4^{2-} 浓度为 10～100 mmol/L 时，SRB 增长速率却与此无关[3]。本研究也表明，样品中一些 SO_4^{2-} 或 H_2S 浓度甚高，其起因可能包括 SRB 自身活动产物，它对 SRB 可起抑制作用。

SRB 与 Cd 的负相关表明 Cd 的积累不利于 SRB 的生存。表明包括 Cd^{2+} 的含硫化物对 SRB 可能产生的毒害作用大于其他金属的硫酸盐类。SRB 与 Cd 的负相关性也表现为研究区的 SRB 可能不属于耐 Cd 菌，至少在本次测定中是如此。它们对 Cd 的耐性(tolerance)甚或抗性(resistance)还很小[13]。两海区镉的含量较高，均高于中国浅海沉积物中 Cd 的丰度[7]。而 Cd 与 SRB 的相关性则表现在当 SRB 多时，其所还原生成的 S 多，在楚科奇海还原环境中 S 易保存，并为 Cd 的硫化物形成提供了足够的硫，使 Cd 形成硫化物，从环境中移去，以致 SRB 少受伤害。而在白令海为氧化环境所生成的 S 不易保存，Cd 的毒性对 SRB 影响较大。

2.6　SRB 与沉积物性质及沉积物环境温度间的关系

沉积物性质对 SRB 的影响表现在样品的粒度和通透性等对 SRB 含量的影响上，在粗粒沉积物中 SRB 较多；而细粒沉积物中 SRB 相对较少。这与粗粒沉积物含有较丰富的有机质有关，SRB 将选择性地繁殖和聚集其中，促进有机质分解，从而降低 Eh，随之而来的是硫酸盐还原活动强化[8]。SRB 与砂、粉砂、粘土之间的相关分析表明：SRB 与砂之间有较好的正相关关系，而 SRB 与粉砂、粘土之间则有较显著的负相关关系，显著性水平甚至达 99.9%(表 3)，表明在研究区粗粒沉积物有助于 SRB 的生存，细粒沉积物不利于 SRB 的繁衍。

研究表明，研究区沉积物中的 SRB 为适冷菌，温度的升高有利于 SRB 的繁殖。SRB 与底层水温之间的相关系数显示两者之间呈良好的正相关，其中全海区表层混合样中的这对相关性是所有 80 对系数中相关程度最高的(表 3、图 2)。

3　结语

简要介绍了楚科奇海和白令海研究区沉积物的岩性特征，论述了沉积物中 SRB 含量

状况、沉积物中地球化学要素的含量以及两者之间的相关关系。在此基础上列举并分析了 SRB 与 Si、$CaCO_3$、P、Ti、Fe、Pb、Hg、S、Cd 及沉积物粒度、底层水温之间的相关关系。结果表明，研究区沉积物中的 SRB 与 Si、砂及底层水温为正相关，SRB 与 $CaCO_3$、Fe、Ti、P、Pb、Hg、粉砂、粘土为负相关，SRB 与 S、Cd 的相关性因海区不同而不同，既有负相关，也有正相关。在所有的相关系数中以 SRB 与底层水温的相关性最好，在元素中以 SRB 与 $CaCO_3$、P、Ti 及 S 间的相关程度为高。极地寒冷环境下化学风化作用及各种化学过程相对较弱，元素与 SRB 之间的相关性受沉积物性质、沉积环境，尤其是底层水温的控制。粗粒富有机质的沉积物、较高的底层水温有利于 SRB 的活动。另一方面元素的地球化学性质对 SRB 的生存状态有着重要的制约作用。

致谢：底层水温资料由矫玉田同志提供，在此表示感谢。

参考文献 13 篇（略）

（合作者：高爱国　孙海青）

ANALYSIS ON CORRELATION BETWEEN SULPHATE-REDUCING BACTERIA AND BIO-GEOCHEMICAL FACTORS OF SEDIMENT IN THE CHUKCHI SEA AND BERING SEA, ARCTIC

(ABSTRACT)

Abstract This paper analyses and discusses the correlations between content of sulfate-reducing bacteria(SRB) and $CaCO_3$, Si, S, P, Cd, Fe, Ti, Pb, Hg in 34 sediments samples from the Chukchi sea and Bering sea, Arctic. The relationships between SRB and bottom water temperature as well as granularity of the sediment are also analyzed. The results show that there are positive correlations between SRB and Si, sand, as well as bottom water temperature. The negative correlations present between SRB and $CaCO_3$、Fe, Ti、P、Pb、Hg、silt、clay. The correlations between SRB and Cd、S are different in two seas, some are negative ones, others are positive ones. The paper analyses and discusses the relative statuses and their condition.

Key words Sulphate-Reducing Bacteria; Geochemical Factors; Correlation Analysis; Chukchi Sea; Bering Sea; Arctic

加拿大海盆与楚科奇海表层沉积物中硫酸盐还原菌丰度分布状况*

摘　要　对中国第2次北极科学考察海洋表层沉积物(0～1cm)中硫酸盐还原菌进行4℃、25℃的培养实验，结合首次北极科学考察海洋沉积物硫酸盐还原菌研究成果，研究了研究区硫酸盐还原菌丰度的分布。结果表明，4℃培养得出硫酸盐还原菌丰度为0～2.4×10^4个/g(湿样)，平均3 433个/g(湿样)，并呈现有规律的分布趋势：低纬区的高于高纬区的，浅水区的高于深水区的。本区硫酸盐还原菌丰度高于中国东海、南海和黄海部分海区沉积物中的，低于胶州湾以及西北冰洋某些海区中的。25℃培养时硫酸盐还原菌丰度为0～2.4×10^4个/g(湿样)，平均4 062个/g(湿样)，在1 880 m以浅水域其丰度与4℃培养时的一致，而在1 880 m以深水域其丰度高于4℃培养时的丰度。

关键词　加拿大海盆　楚科奇海　表层沉积物　硫酸盐还原菌

硫酸盐还原菌(Sulphate-Reducing Bacteria，SRB)，是生存于地球不同环境中的广谱性微生物。它们在包括硫、碳等元素在内的矿化、再矿化及其他元素的地球化学循环中进行不可替代的活动，并在人类的经济生活中发挥着不可忽视的作用，因而长期广受人们重视。对北极一些环境的SRB状况已有所研究[1-6]，但迄今仍留有大量疑问，亟待人们去探索。根据对中国首次北极科学考察沉积物样品中SRB的研究，在中国第2次北极科学考察中我们又重新设计和采集了大量样品，对SRB进行更深入细致的研究。本文报道北极研究区不同水深条件下，表层沉积物中SRB的检出率与丰度状况，并比较其区域分布差异。

1　研究区概况

研究区位于北冰洋靠亚美大陆一侧的加拿大海盆和楚科奇海的广阔海域(66°～80°N，148°～170°W)。从白令海峡流入的北太平洋海水带来了大量的营养物质和有机质，使研究区具有较高的生产率[7]，为海区沉积物中的SRB生长提供了充足的养分。为了解不同环境中SRB的分布状况，在中国第2次北极科学考察期间根据研究区的自然环境状况，在测区设定了41个测点进行样品的采集，并从中选取24份表层沉积物样进行了SRB的室内分析。各取样站的地理位置如图1所示。

* 基金项目：国家自然科学基金项目——楚科奇海及其邻近海域沉积物中细菌生物地球化学(40576060)；楚科奇海沉积物中古环境变化的地球化学响应(40376017)

原文刊于《海洋科学进展》，2007，，25(3)：P295-301.

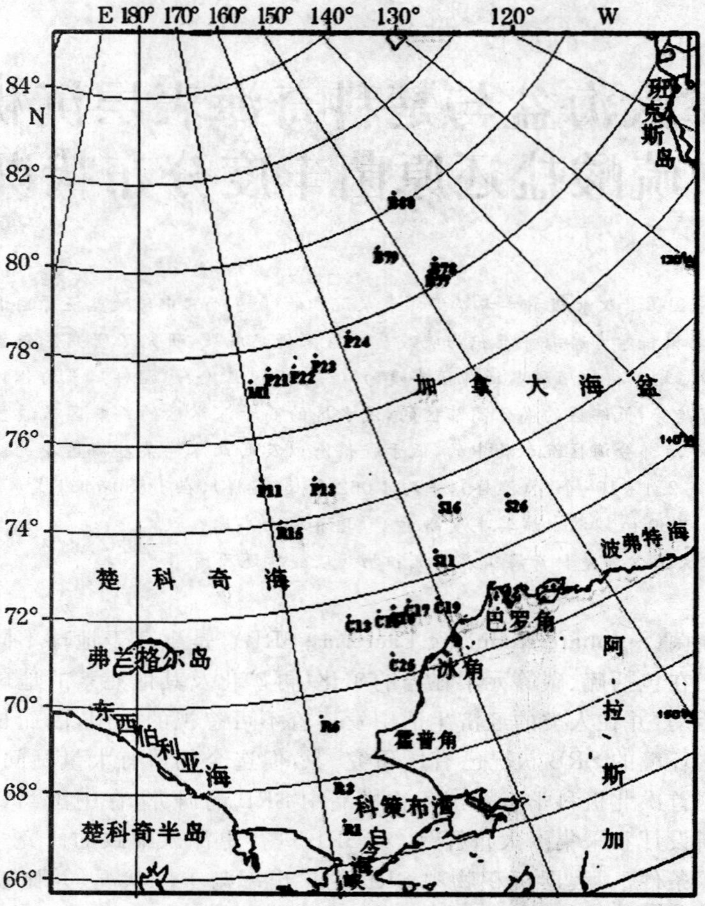

图 1　北极沉积物细菌分析样品站位图

Fig. 1　Location of sampling stations for analyzing bacteria from surficial sediments in the Arctic Sea

2　样品的采集和分析

2.1　采样

样品的采集是中国第 2 次北极科学考察期间在"雪龙"号考察船上完成的，采样时间为 2003 年 7 月 30 日—9 月 7 日。分别用小箱式取样器、多管取样器和抓斗采集，采样器到达甲板后，先按微生物学采样要求采集表层 0～1cm 样作为表层样，然后再采集柱状沉积物样，并按要求采集不同深度的沉积物样，样品存放于事先经高温灭菌的干净塑料瓶中，于 4℃的冷柜中保存。

样品的采集范围为 66°59′28″～80°13′25″N，146°44′16″～169°59′37″W，水深 41～3 850 m，其中水深大于 1 000 m 的站位有 9 个，覆盖区域比首次北极的调查范围(66°30′09″～ 73°26′58″N，158°56′30″～175°01′42″W，水深 30～92 m)明显增大(图 1)。采样站点

设置上既有深海盆和陆架浅海，也有陆坡和海台，并关注测线的延伸及控制区域，使样品的代表性更强。在样品的采集层位上更精细，所选的 24 个表层样品，全为 0～1cm 的样品，而区别于首次北极样品少和分层较粗的不足(0～5cm 9 站，5～10cm 8 站，不分层 8 站)，为了探讨垂直变化，又采集了短岩芯进行研究，使样品分层更为精细。而外业条件的改善及前期更充足的准备工作，使采集后样品的保存与分析条件也明显优于首次北极科学考察，因此，与首次北极科学考察 SRB 调查相比，中国第 2 次北极科学考察 SRB 调查样品具有覆盖面广、代表性强、分层精细、保存良好等特点。

2.2　国内陆上实验室分析

样品送回陆地实验室后，按微生物学要求保存于 4℃的冷柜中，并从 2004 年 1 月 17 日起，开始对所选取的 24 份表层沉积物样进行微生物学分析。将样品按 MPN 法进行系列稀释入 SRB 液体培养基中[8]，分别将接种了样品的试管培养于 4℃和 25℃条件下 3 周以上，其间每 3 天观察培养情况，以试管见黑色(颗粒)为 SRB 阳性，统计分析比较 SRB 的检出率，根据 MPN 检验表[9]确定 SRB 的丰度状况。

3　结果与讨论

3.1　表层沉积物中的 SRB 分布

表 1 为中国第 2 次北极科学考察表层沉积物样品中 SRB 测定结果，其中 4℃培养测定的样品 24 个，检出率为 50%，丰度为 0～2.4×10^4 个/g(湿样)，平均 3 442.7 个/g(湿样)。本次 SRB 检出率小于首次北极科学考察表层样(含不分层的 8 个，共计 17 个)中 SRB 检出率(94.1%)，且 SRB 丰度上也与首次调查结果存在着明显差异[6,10]。25℃培养测定的 7 个样品中检出 5 个，检出率为 71.4%，丰度也为 0～2.4×10^4 个/ g (湿样)，平均 4 062.0 个/g(湿样)。对比 4℃与 25℃分析结果时发现，就同一样品而言，在水深 1 456 m 以浅的 4 个站位样品中，4℃与 25℃分析结果是一致的，而在 1 880 m 以深的 3 个样品中 25℃培养的 SRB 丰度均高于 4℃培养的 SRB 丰度。陈皓文等[6]研究表明，所研究 SRB 以适冷菌为主，也有相当多的中温菌出现，25℃时 2 种菌均得到不同程度的生长。这可能是因为深海环境底层水温较为稳定，而浅海底层水温易变且较低，对某些 SRB 的生长产生不利影响，或许还有其他原因，有待进一步研究。

表 1　研究区表层沉积物硫酸盐还原菌检出率与丰度

Table 1　Detection rate and abundance of SRB in surficial sediment from the study area

个 · g^{-1}(湿样)

统计参数	水深/m	4℃	25℃
最小值	41	0	0
最大值	2 200	24 000	24 000
平均值	432.7	3 442.7	4 062
均方差	680.5	7 075.9	8 844.7
样品数/个		24	7
检出率/%		50	71.4

3.2 SRB 的纬度变化

在研究区按 4°为间隔，统计各纬度区间 SRB 分析结果，从低纬度到高纬度 SRB 的检出率呈递减趋势（表 2），这种趋势在 169°W 线剖面非常明显，沿 169°W 线，从 67°N 附近的 R01 站向北到 77°18′N 的 M1 站 SRB 丰度快速下降（图 2），推测本测区低纬度更适合于 SRB 的生存与繁衍，但纬度只是一种地理因素，还应受其他因子的制约。

表 2 不同培养温度时不同纬度表层沉积物中硫酸盐还原菌检出率与丰度

Table 2 Detection rate and abundance of SRB at two culture temperatures in surficial sediments from different latitudinal zones

个 · g^{-1}（湿样）

纬度	4℃					25℃				
	样品数 /个	检出率 /%	最小值	最大值	平均值	样品数 /个	检出率 /%	最小值	最大值	平均值
66°～70°N	3	100	1 500	4 300	2 433.3					
70°～74°N	10	60	0	24 000	6 423	3	66.7	0	2 100	776.7
74°～78°N	7	42.9	0	11 000	1 584.9	3	66.7	0	24 000	8 001.4
>78°N	4	0	0	0	0	1	100	2 100	2 100	2 100

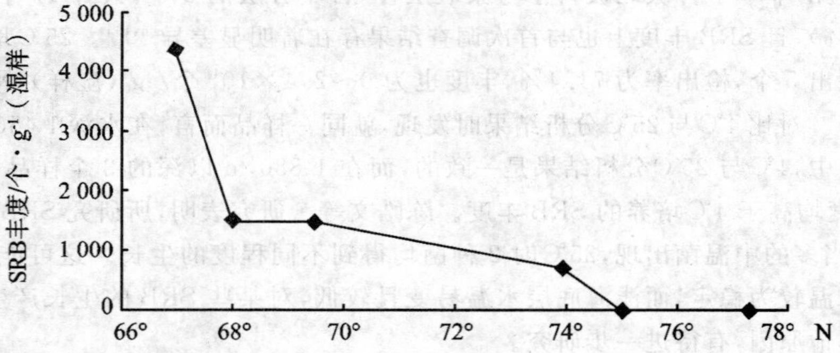

图 2 SRB 丰度沿 169°W 的变化

Fig. 2 Variations in SRB abundance Along the 169°W longitude

3.3 不同经度区间内的 SRB 丰度

根据采样点分布情况将研究区分为 3 个经度区间。145°～158°W 为加拿大海盆区，包括 B77，B78，B79，B80，S16 和 S26 共 6 个。4℃培养的 SRB 均未检出；25℃培养的 2 个样品均检出，且结果都是 2.1×10^3 个/g（湿样）。158°～165°W 为楚科奇海的东部，包括 P22，P23，P24，C15，C16，C17，C19，C25 和 S11 共 9 个站，4℃培养 SRB 检出率为 66.7%，丰度为 $0\sim2.4\times10^4$ 个/g（湿样），平均 5 624.4 个/g（湿样）；25℃培养的 3 个样品检出率也为 66.7%，丰度为 $0\sim2.4\times10^4$ 个/g（湿样），平均 8 076.7 个/g（湿样）。165°～170°W 楚科奇海中部的 9 个样品中，4℃培养样品的检出率为 66.7%；25℃培养的 2 个样品检出率为 50%，丰度为 0～4.3 个/g（湿样），平均 2.2 个/g（湿样）（表 3）。从丰度上来看，在 78°

N 附近 MPB 线，即 M1，P21，P22，P23，P24，B77 和 B78，因该线所在纬度较高，大部分水深较深，丰度仍低至未检出，但水深 1 880 m 的 P24 站丰度最高。其次 72°N 附近的 C 线，由 C13，C15，C16，C17 和 C19 组成，变化呈弧形，东侧的 C17 丰度相对前 3 站约低 60%，而 C19 站则未检出。

表 3 不同培养温度下不同经度表层沉积物中硫酸盐还原菌检出率与丰度

Table 3 Detection rate and abundance of SRB at two culture temperatures in surficial sediments from different longitudinal zones

个 · g^{-1}（湿样）

经度	4℃					25℃				
	样品数/个	检出率/%	最小值	最大值	平均值	样品数/个	检出率/%	最小值	最大值	平均值
145°～158°W	6	0	0	0	0	2	100	2 100	2 100	2 100
158°～165°W	9	66.7	0	24 000	5 624.4	3	66.7	0	24 000	8 076.7
165°～170°W	9	66.7	0	24 000	3 556	2	50.0	0	4.3	2.2

3.4 不同水深范围样品中的 SRB

不同水深范围样品中 SRB 丰度的统计结果表明，从浅水到深水，SRB 的检出率逐渐降低，而且前述 169°W 剖面线和 C 线均可较直观地反映不同水深的影响（表 4），类似的还有 RCS 线，该线 R06，C25，C19，S26 四站中除 R06 的丰度为 1.5×10^3 个/g（湿样）外，其余均未检出。

表 4 不同培养温度下不同水深表层沉积物中硫酸盐还原菌检出率与丰度

Table 4 Detection rate and abundance of SRB at two culture temperatures in surficial sediments at different water depths

个 · g^{-1}（湿样）

水深/m	4℃					25℃				
	样品数/个	检出率/%	最小值	最大值	平均值	样品数/个	检出率/%	最小值	最大值	平均值
41～55	10	80	0	24 000	7 083	2	50	0	230	115
175～561	5	60	0	700	158.9	1	100	4.3	4.3	4.3
1 456～2 200	3	33.3	0	11 000	36 966.7	2	50	0	24 000	12 000
3 000～3 850	6	0	0	0	0	2	100	2 100	2 100	2 100

3.5 与首次北极 SRB 调查结果差异分析

第 2 次北极科学考察 SRB 调查的经纬度与水深范围远较首次调查范围大，发现 SRB 检出率低于首次调查的结果，在 SRB 丰度上也存在一定的差异，原因可能有 3 个：

（1）首次北极科学考察 SRB 分析的样品是在 1999 年 7 月 14 日—8 月 9 日采集的。采样点大多水深较浅，底水温度相对较低，为－1.753～1.642 1℃，平均－1.026℃。第 2 次北极科学考察的样品采于 2003 年 7 月 30 日—9 月 7 日（季节上偏后），水深相对较深，底水温度为－1.703～4.270℃，平均 0.031℃。对比同一纬度区间（67°～73°N）2 次北极科学考察时的底水温度值，则首次北极科考时的底水温度平均值为－1.099℃，第2次北极科学考察时的底水温度平均值为0.068℃（表5，表6）。显然较高的底层海水温度和相对较长的生长期可能更有利于部分SRB的生长繁衍。

(2)近几十年来北极地区正在发生着一系列变化：气候变暖，海冰面积缩小，生物生产率提高等，这些变化有可能使得生活在研究区的部分SRB获得更多有机质，在更为适宜的环境温度中生存。

(3)样品保存明显优于首次北极科学考察时的SRB样品保存，从而保证分析结果更准确可靠。

表5　两次北极科学考察表层沉积物中硫酸盐还原菌采样站环境参数对比

Table 5　Comparison between environmental parameters at SRB sampling stations during two Antarctic scientific expeditions

项目	首次北极科学考察					第2次北极科学考察				
	Z/m	$t_{表}$/℃	$S_{表}$	$t_{底}$/℃	$S_{底}$	Z/m	$t_{表}$/℃	$S_{表}$	$t_{底}$/℃	$S_{底}$
样品数		24	24	24	24		24	24	24	24
最小值	30	－1.397 4	20.719 8	－1.752 9	31.836 4	41	－1.535 1	27.756 2	－1.7038	31.853 9
最大值	92	7.841 3	32.510 1	1.642 1	33.502 2	3 800	7.171 8	31.948 4	4.2704	34.956 8
平均值		1.944 4	29.337 8	－1.026 0	32.822 0		0.053 6	29.500 0	0.031 3	33.876 6
均方差		3.001 4	3.039 3	0.897 8	0.461 4		2.624 6	1.036 1	1.444 2	1.193 7

表6　两次北极科学考察67°～73°30′N内硫酸盐还原菌采样站环境参数对比

Table 6　Comparison between environmental parameters at SRB sampling stations from 67°N to 73°30′N during two Antarctic scientific expeditions

项目	首次北极科学考察					第2次北极科学考察				
	Z/m	$t_{表}$/℃	$S_{表}$	$t_{底}$/℃	$S_{底}$	Z/m	$t_{表}$/℃	$S_{表}$	$t_{底}$/℃	$S_{底}$
最小值	30	－1.397 4	20.719 8	－1.752 9	31.836 4	41	－1.024	27.756 2	－1.703 8	31.853 9
最大值	92	7.841 3	32.5101	1.6421	33.502 2	3 800	7.171 8	31.948 4	4.270 4	34.954 3
平均值	50.5	1.974 6	29.207 0	－1.098 5	32.831 8	47.4	1.568 5	29.936 8	0.068 2	32.904 1
均方差	11.6	3.062 4	3.037 8	0.845 5	0.468 8	4.6	3.030 1	1.267 5	1.983 2	0.965 8

3.6　与其他海区的SRB比较

将研究区表层沉积物中SRB与我国学者采用相同方法在中国近海的研究结果进行对比，研究区表层沉积物中SRB的丰度及其范围大于海南岛西南侧、北部湾东侧、浙江—闽北陆架沉积物及乳山湾沉积物中的SRB丰度，低于胶州湾潮间带和沿岸区表层沉积物中的SRB丰度(4.1×10^{7} 个/g)[11-13]。

4　结　语

与首次北极科学考察沉积物中SRB调查研究相比，第2次北极科学考察采集了覆盖范围更为广泛、分层更为精细的表层沉积物样品，对SRB进行了更为细致的研究，发现海区 SRB 丰度明显高于首次调查的结果，而且表现为低纬度区的高于高纬度区的；SRB 的检出率表现为西部与中部一致，高于东部，SRB 丰度以 158°～165°W 海区的较高；就水深而言，浅水区 SRB 的检出率与丰度也高于深水区的。虽然研究区水温较低，但沉积物中 SRB 仍较为丰富，高于我国学者在中国沿海所获得的大部分调查结果，但低于北冰洋大西洋侧（西北冰洋）表层沉积物中 SRB 的调查结果[4]。

参考文献 13 篇（略）

（合作者：高爱国　林学政）

DISTRIBUTIONS OF SULPHATE-REDUCING BACTERIA ABUNDANCE IN SURFICIAL SEDIMENTS FROM THE CANADA BASIN AND CHUKCHI SEA

(ABSTRACT)

Abstract The surficial sediment samples from the Canada Basin and Chukchi Sea were collected during the second Chinese Arctic scientific expedition, the culture experiments on sulphate-reducing bacteria (SRB) in surficial sediments at 4℃ and 25℃ temperatures were made, and the distributions of SRB abundance in surficial sediments from the study area were studied in combination with the SRB study results for the first Chinese Arctic scientific expedition. It is shown from the study results that the SRB abundance cultured at 4℃ is in the range 0 to 2.4×10^4 cells/g (wet sample) with an verage of 3,433 cells/g, and shows a distribution trend, that is, the SRB abundance is higher in low latitude area than in high latitude area, and higher in shallow water area than in deep water area. The SRB abundance in surficial sediments from the sea area is higher than those from the East China Sea, South China Sea and part of the Yellow Sea and lower than those from the Jiaozhou Bay and some sea areas of the western Arctic Ocean. The SRB abundance cultured at 25℃ is in the range 0 to 2.4×10^4 cells/g (wet sample) with an average of 4,062 cells/g, the SRB abundance cultured at 25℃ from water depth less than 1,880 m is consistent with that cultured at 4℃, but the SRB abundance cultured at 25℃ from water depth greater than 1,880 m is higher than that cultured at 4℃.

Key words Canada Basin; Chukchi Sea; Surficial Sediment; Sulphate-Reducing Bacteria (SRB)

加拿大海盆与楚科奇海柱状沉积物中硫酸盐还原菌的分布状况*

摘 要 对取自北极楚科奇海及加拿大海盆的10个沉积物岩芯分别在4℃,25℃培养进行硫酸盐还原菌(SRB)分析,结合首次北极科考海洋沉积物SRB的研究成果,探讨了研究区SRB的分布特点。研究结果表明,4℃与25℃温度培养的SRB含量均为0~2.4×10^6个·g^{-1}(湿样);4℃时SRB的检出率与平均含量分别为45.5%和2.06×10^4个·g^{-1}(湿样),25℃培养条件下分别为73.7%和4.70×10^4个·g^{-1}(湿样);柱状沉积物中SRB的检出率、含量范围、平均含量都明显高于表层沉积物中SRB的相关指标;岩芯中SRB含量分布与采样点的纬度、深度有一定关系,但这种关系不如表层沉积物中SRB分布表现的那么明显;4℃培养时,各层位SRB含量的平均值范围为51~1.2×10^6个·g^{-1}(湿样),25℃时为2.04×10^2~2.47×10^5个·g^{-1}(湿样);在所研究的深度范围内,4℃时培养SRB的垂直变化较为明显,而25℃时SRB的垂直变化相对缓和;根据4℃、25℃2个不同培养温度时SRB的检出率、含量对比看,似乎25℃时更有利于某些SRB的繁衍。

关键词 加拿大海盆 楚科奇海 柱状沉积物 硫酸盐还原菌

1 引言(Introduction)

硫酸盐还原菌(Sulphate—Reducing Bacteria,SRB)是海洋沉积物中常见的微生物,广泛分布于从低纬度到高纬度的各个海域。作为一类既包含厌氧性、又包含兼性厌氧性的腐生微生物,SRB以有机物为能源,以海水硫酸盐为硫酸供体,以硫酸盐作为有机物氧化时的最终电子受体,最终产物以富含H_2S和S^{2-}为特征。由于SRB独特的生命过程,在海洋沉积物早期成岩过程中起着独特的作用,从而驱动着有机质的矿化过程,促进了S、C、N、P等元素的迁移,并因其对环境氧化还原条件的影响驱使变价元素Fe、Mn的迁移,因此,SRB成为早期成岩过程的积极参与者(Glud, 1998; Bruechert et al., 2001; Knoblauch et al., 1999)。

沉积物中的SRB与环境中有机质含量,环境的温度、盐度、压力、氧气含量、埋藏深度以及氧化还原电位等有密切的关系。早在1914年Issatchenk就从北极海洋沉积物中检测到大量SRB,而后西方学者在东北冰洋进行了一系列研究(Glud, 1998; Sagemann, 1998; Sahm et al., 1998, 1999; Ravenschlag et al., 2000, 2001)。但是,在西北冰洋对SRB研究较少,Steuard等(1996)在西北冰洋的楚科奇海等海区进行了水体中的细菌与病毒研究。1999年中国首次北极科学考察对白令海和楚科奇海沉积物中SRB进行了初步研究,在楚科奇海分别采集了9份0~5cm、8份5~10cm、8份不分层的样品,并进行了

* 原文刊于《环境科学学报》,2008,28(5):1014-1020,本人为第二作者。

基金项目:国家自然科学基金(No. 40576060,40376017,40176017)

Supported by the National Natural Science Foundation of China(No. 40576060,40376017,40176017)

SRB 研究。结果表明，表层 SRB 含量少于次表层含量（陈皓文等，2000）。高爱国等(2003)则进行了 SRB 与生物地球化学要素的相关分析，探讨了北极沉积物中 SRB 与环境的关系。结果表明，由于沉积后的环境变化，SRB 在垂向上变化受多种因素控制。为了对研究区沉积物中 SRB 的垂向分布增进了解，开展本项研究，进行 SRB 含量测定和更深入细致的研究，以便比较其水平和垂直分布差异。

2　研究区概况(The study area)

研究区位于北冰洋靠亚美大陆一侧的加拿大海盆和楚科奇海。从白令海峡流入的北太平洋海水带来了大量的营养物质和有机质，使研究区具有较高的生产率（陈敏等，2002)，为海区沉积物中的 SRB 生长提供了充足的养分。为了解不同环境中 SRB 的分布状况，在对中国首次北极科考沉积物样品中 SRB 研究的基础上，我们在中国第二次北极科考期间根据研究区的自然环境状况，又重新设计采样站位，从中选出 10 个代表性岩芯进行了 SRB 分析。10 个岩芯的采样区范围覆盖 69°29′43″N～80°13′25″N，146°44′16″W～169°59′37″W，水深范围为 50～3 850 m，基本上可代表研究区各沉积单元和生态环境状况（见图 1)。

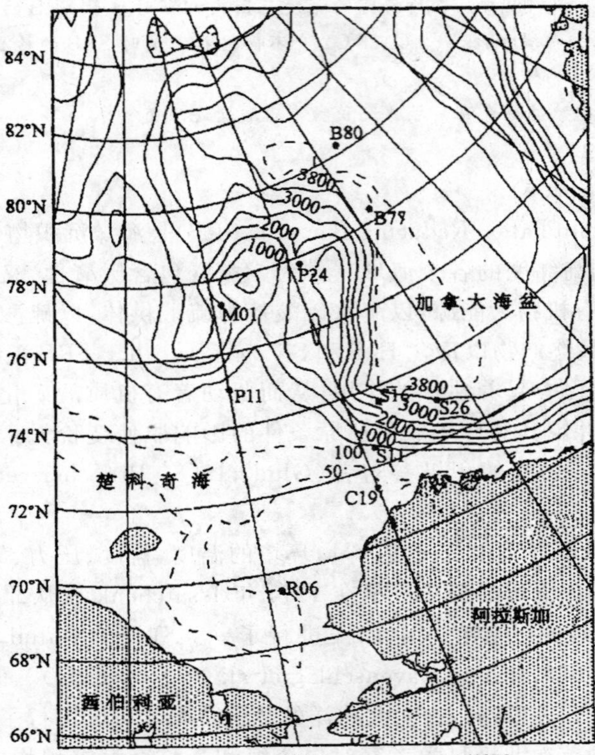

图 1　北极沉积物 SRB 分析岩芯采样站位图

Fig. 1　Locations of sediment cores for SRB analysis

3　样品的采集和分析(Sampling and Analysis)

3.1　采样

样品的采集是 2003 年 7 月～9 月中国第 2 次北极科考期间在“雪龙”号船上完成的。分别用小箱式取样器、多管取样器采集，样品采到甲板上后，先按微生物学采样要求，对采样工具进行消毒，去掉最上层的泥水，然后采集表层样，再采集柱状沉积物样. 在现场按要求对柱状沉积物进行精细分层取样，即：0～10cm 以 1cm 间隔分样，10cm 以下以 2cm 间隔分样。采样过程尽可能干净利索，减少污染和化学变化，样品存放于事先经高温灭菌的干净塑料瓶中，于 4℃环境中保存。

3.2　实验室分析

对所采集的 10 个柱状沉积物样进行微生物学分析。根据陈皓文等(2000)对该海域的 SRB 研究，此次选用 4℃和 25℃两个培养温度进行分析。将样品按 MPN 法进行系列稀释入 SRB 液体培养基中(陈世阳等，1987)，分别于 4℃和 25℃中培养 3 周以上，其间按常规的微生物培养要求经常观察培养情况，以试管见黑色(颗粒)为 SRB 阳性，统计分析比较 SRB 的检出率，查 GB/T 4789.3-2003 中所附的 MPN 检验表，确定 SRB 的含量状况。

4　结果(Result)

研究区柱状沉积物样品中不同培养温度下 SRB 分布的统计结果见表 1。将柱状沉积物样从表层到 38cm 深分为 24 个层次。培养温度为 4℃时，分析了 189 份样品的 SRB，检出 86 份，检出率为 45.5%，含量范围 0～2.4×10^6 个・g^{-1}(湿样)，平均 2.07×10^4 个・g^{-1}(湿样)。培养温度为 25℃时，分析了 186 份样品的 SRB，检出 137 份，检出率为 73.7%，含量范围为 0～2.4×10^6 个・g^{-1}(湿样)，平均值 4.70×10^4 个・g^{-1}(湿样)。这表明，较高的培养温度可使样品中 SRB 的检出率和含量都有一定提高，但仍比 Sahm 等(1999)所报道的结果低，这可能与研究海区、方法等方面的差异有关。

由表1可知，虽然培养温度不同，但 SRB 的含量范围基本一致，均为 0～2.4×10^6 个・g^{-1}(湿样)，但在检出率与 SRB 平均含量上以 25℃培养条件时为高。需要指出的是，与同航次所采集的表层沉积物中 SRB 研究结果(陈皓文等，2007)相比，岩芯 SRB 的检出率、含量范围、平均含量都明显高于同航次表层沉积物中，这与我们首次北极 SRB 研究结果一致(陈皓文等，2000)，也与东北冰洋沉积物中 SRB 研究结果一致(Knoblauch et al.，1999)，与陈世阳等(1987)及陈皓文等(1999)的 SRB 分析结果也具可比性。这充分表明，SRB 具厌气性或兼性厌气性细菌的特征。从岩芯中 SRB 含量范围看，本结果比 Sahm 等(1999)在东北冰洋用分子生物学手段测得的 SRB 含量低 2 个数量级，比 Ravenschlag 等(2001)在东北冰洋用分子生物学手段测得的细菌数量低 3 个数量级。一般来说，分子生物学方法测定结果均比培养法测定结果高 2～3 个数量级，本研究结果与此一致，表明用 MPN 法进行大量样品的 SRB 分析，虽然较为粗放，但仍不失为一种经典可靠的方法。

表 1　各柱状沉积物中硫酸盐还原菌的检出率与含量

Table 1　Occurrence percentages and contents of SRB from the sediment cores

站位	岩芯深度/cm	4℃				25℃			
		样品数	检出率	最大值/(个・g^{-1})	平均值/(个・g^{-1})	样品数	检出率	最大值/(个・g^{-1})	平均值/(个・g^{-1})
R06	12	11	72.7%	3 000	736.4	10	100.0%	2 400	957.0

（续表）

站位	岩芯深度/cm	4℃				25℃			
		样品数	检出率	最大值/(个·g^{-1})	平均值/(个·g^{-1})	样品数	检出率	最大值/(个·g^{-1})	平均值/(个·g^{-1})
C19	22	16	50.0%	4 360	80.9	15	80.0%	11 000	1 487.6
S11	28	19	68.4%	11 000	882.1	19	89.5%	240 000	18 023.7
S16	28	19	10.5%	1 500	80.5	19	78.9%	2 400 000	127 828.4
S26	26	18	66.7%	460 000	4 6220.0	18	94.4%	460 000	76 605.0
P11	38	24	79.2%	2 400 000	103 309.2	24	66.7%	15 000	1 374.6
M1	36	23	8.7%	400	21.3	23	8.7%	400	21.3
P24	28	19	68.4%	240 000	17 857.4	19	94.7%	240 000	33 062.1
B77	22	16	43.8%	110 000	10 013.8	15	80.0%	2 400 000	215 966.7
B80	38	24	8.3%	46 000	2 541.7	24	75.0%	240 000	27 120.1
小计		189	45.5%	2 400 000	20 635.3	186	73.7%	2 400 000	4 6957

除了25℃时R06站SRB的最小值为90个·g^{-1}外，其余所有的情况下最小值都为0个·g^{-1}

5　讨论(Discussion)

5.1　SRB在柱状沉积物中的垂直分布

SRB在沉积物中垂直分布受多种因素控制。统计4℃、25℃培养条件下各层位SRB的样品数、检出率、含量范围与平均值(见表2)。由表2可知，4℃时各层位SRB的平均值范围为51～1.2×10^6个·g^{-1}(湿样)，就大部分层位而言，介于2.0×10^2～4.6×10^4个·g^{-1}(湿样)之间。25℃时各层位的平均值范围为2.04×10^2～2.47×10^5个·g^{-1}(湿样)，大部分层位介于4.0×10^3～6.0×10^4个·g^{-1}(湿样)之间。与4℃培养所获得的SRB相比，25℃时层位间差异较小。而从各层位中SRB平均含量的变化看，至少在该研究区域，在所研究深度内SRB含量总体变化并不明显，由此推测SRB的生存深度范围较大。

表2　各采样层位沉积物中硫酸盐还原菌检出率与含量

Table 2　Occurrence percentages and contents of SRB from different depths in the sediment cores

层位/cm	4℃				25℃			
	样品数	检出率	最大值/(个·g^{-1})	平均值/(个·g^{-1})	样品数	检出率	最大值/(个·g^{-1})	平均值/(个·g^{-1})
0～1	10	40.0%	11 000	1 273.4	7	71.4%	24 000	4 062.0
1～2	10	60.0%	1 200	188.2	10	80.0%	11 000	1 295.2
2～3	10	30.0%	420	57.0	10	70.0%	35 000	7 086.0
3～4	10	60.0%	24 000	3 353.0	10	80.0%	24 000	9 445.0
4～5	10	50.0%	110 000	11 726.0	10	70.0%	2 400 000	241 903.0

(续表)

层位/cm	4℃				25℃			
	样品数	检出率	最大值/(个·g^{-1})	平均值/(个·g^{-1})	样品数	检出率	最大值/(个·g^{-1})	平均值/(个·g^{-1})
5～6	10	50.0%	110 000	11 856.0	10	80.0%	240 000	59 806.0
6～7	10	60.0%	46 000	4 718.0	10	90.0%	240 000	24 841.0
7～8	10	50.0%	460 000	46 268.4	10	80.0%	460 000	49 064.4
8～9	10	60.0%	110 000	11 148.0	10	80.0%	240 000	26 188.0
9～10	10	30.0%	240	51.0	10	80.0%	290 000	31 538.3
10～12	10	60.0%	15 000	3 041.9	10	80.0%	2 400 000	246 642.0
12～14	9	44.4%	1 100	203.5	9	88.9%	24 000	3 959.0
14～16	9	44.4%	29 000	6 236.7	9	88.9%	240 000	31 214.4
16～18	9	11.1%	24 000	2 666.7	9	55.6%	240 000	30 063.3
18～20	9	55.6%	110 000	12 580.2	9	55.6%	240 000	29 403.3
20～22	9	55.6%	240 000	29 470.0	9	55.6%	240 000	31 200.0
22～24	7	42.9%	4 000	914.3	7	57.1%	210 000	30 301.4
24～26	7	28.6%	27 000	3 862.9	7	57.1%	240 000	39 352.9
26～28	6	33.3%	24 000	4 000.4	6	83.3%	24 000	5 318.4
28～30	3	33.3%	300	100.0	3	66.7%	35 000	11 800.0
30～32	3	33.3%	28 000	9 333.3	3	66.7%	3 000	1 140.0
32～34	3	33.3%	15 000	5 000.0	3	66.7%	19 000	7 000.0
34～36	3	33.3%	2 100	700.0	3	66.7%	600	204.3
36～38	2	50.0%	2 400 000	1 200 000.0	2	50.0%	15 000	7 500.0

注：所有层位的最小值都为 0 个·g^{-1}

4℃时 SRB 最高检出率出现在 1～2cm、3～4cm、6～7cm、8～9cm、10～12cm 5 个层位，达到 60.0%；最低则发生在 16～18cm 层，仅为 11.1%。25℃下培养得出的最高 SRB 检出率出现于 6～7cm 层中，高达 90%；最低检出率(50.0%)则出现于 36～38cm 层。检出率大于 60.0%(4℃时的 SRB 最高检出率)的层位有 18 层，占总层位的 75.0%。2 种培养温度相比，4℃时的最高、最低 SRB 检出率和检出率变幅均低于 25℃时的。

4℃培养得出的 SRB 最高含量与平均含量最大值均发生在最底层中，即 36～38cm 层；最低 SRB 含量与平均含量最小值出现于 9～10cm 层。最高、最低 SRB 含量相差 4 个数量级。SRB 含量在所有柱状沉积物 24 个层次中的分布呈现出不规则、不均匀状态。25℃培养得出的 SRB 最高含量与平均含量最大值均发生在 10～12cm 层中，最低 SRB 含量与平均含量最小值则出现于 34～36cm 层中，最高、最低 SRB 含量相差 3～4 个数量级。

以上情况说明，SRB 的检出率和含量在所测柱状沉积物样中的垂直分布是相当随机的，这意味着 SRB 的这 2 项指标受制于沉积环境的各种理化、地质、生物因子和样品的保存状况等。Sahm 等(1998)在研究斯瓦尔巴德附近海域沉积物中 SRB 分布时，发现 SRB 呈表层富集趋势。RayeYischlag 等(2000)发现，斯瓦尔巴德附近海域中 SRB 在表层 5 mm 沉积物中最高，然而后来的研究又发现在表层 19cm 深度内，细胞总数并不显示随深度增加而减少的趋势，似 SRB 受多种因素影响，而且显示出某种随机性(Ravenschlag et al.，2001)。

图 2 为各岩芯在 4℃、25℃培养温度时所获得的 SRB 含量的垂直分布，横坐标为指数坐标，分别对应 0、1、10、10^2、10^3、……个 · g^{-1}(湿样)。比较各岩芯中 SRB 含量的垂直分布可知，4℃时 SRB 含量的变化较为明显，而 25℃时 SRB 的含量变化相对缓和。对于大部分样品而言：2 个温度时的培养结果在含量变化上均较为相似，在 M1 岩芯中 2 种培养温度时 SRB 分布完全一致；在 P11 岩芯中 SRB 的分布趋势很吻合，但在 S16 与 B80 岩芯中 2 种培养温度时的 SRB 含量的垂直变化差异较大。

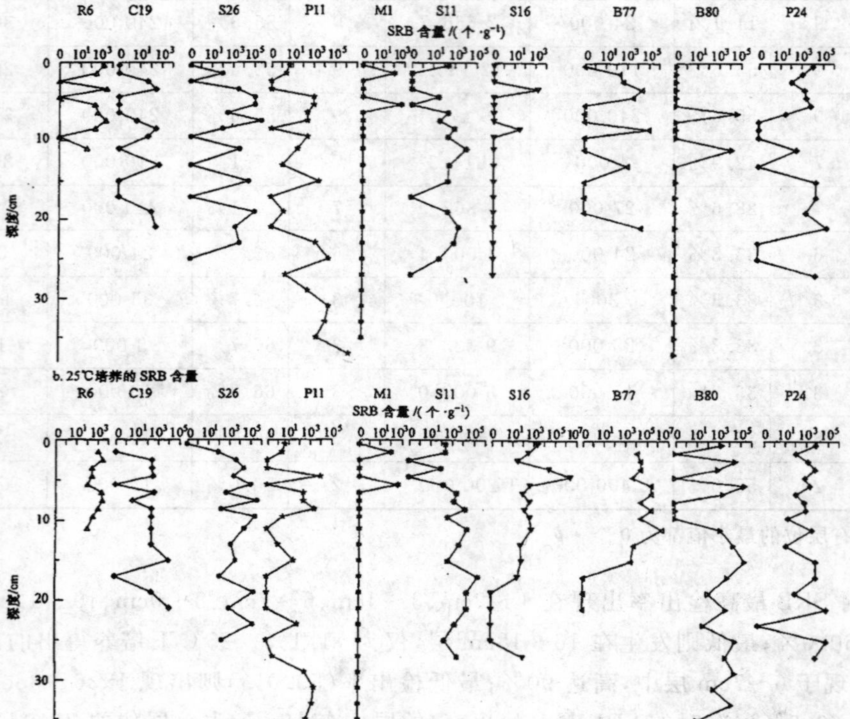

图 2　4℃、25℃培养的 SRB 的垂直分布图

Fig. 2　Depth profiles of SRB cultivated at 4℃ and 25℃ in sediment cores

5.2　沉积物 SRB 指标在海区内的比较分析

表 1 给出了 10 根柱状沉积物中 SRB 指标(检出率与含量)在海区间的分布状况:4℃时的 SRB 检出率大约呈现为西、东两侧较低,中部偏高之势,而含量呈高低起伏状,最高检出率出现于中偏西的柱状沉积物中,最低出现在东部的 B80 柱,两者相差 9.5 倍。最高含量出现于西部 P11 柱,最低的则在 M1 柱,尽管两者毗邻,却相差 6 000 倍。25℃时的检出率和含量均显出由西往东渐高之势。总体来看,25℃的两指标高于 4℃条件下的指标的。最高检出率在白令海峡附近的 R6 柱,含量最高的站位是 B77 站;最低检出率与最低含量均出现在西部的 M1 站。最高、最低检出率相差 11.5 倍,最高、最低含量相差 10 139 倍。

Kroencke 等(2000)在北冰洋欧亚海盆进行细菌学调查时发现,随纬度升高,细菌丰度呈减少趋势。在研究区表层沉积物中这种趋势也很明显(陈皓文等,2007),但从各柱状样中的 SRB 分布看,虽然仍有随纬度升高,检出率降低的趋势,但已不像表层中那样明显;而其平均含量变化无这种规律。这表明,随着沉积物的埋藏深度增加,SRB 受沉积物—水界面过程各种物理化学条件影响逐渐变小,而主要受控于沉积物中的有机质含量、O_2、SO_4^{2-} 供应量及环境的氧化还原状态。4℃、25℃培养温度时 SRB 各指标的纬度分布特征见表 3。

表 3　不同纬度区间柱状沉积物中 SRB 含量指标

Table 3　Comparison of SRB contents along the latitude in the Chukchi Sea and the Canada Basin

纬度区间	4℃					25℃				
	样品数	检出率	最小值/(个·g^{-1})	最大值/(个·g^{-1})	平均值/(个·g^{-1})	样品数	检出率	最小值/(个·g^{-1})	最大值/(个·g^{-1})	平均值/(个·g^{-1})
66°N～70°N	11	72.7%	0	3 000	736.4	10	100%	90	2 400	957
70°N～74°N	72	48.6%	0	460 000	1 586.4	71	85.9%	0	2 400 000	10 454
74°N～78°N	66	51.5%	0	2 400 000	8 240	66	54.5%	0	240 000	1 168.6
>78°N	40	22.5%	0	110 000	936.7	39	76.9%	0	2 400 000	17 390.1

5.3　水深对柱状沉积物中 SRB 含量的影响

将测样按所在水深归成 4 类,即 41～55、175～561、1 456～2 200 和 3 000～3 850 m,分析不同水深沉积物中 SRB 指标的结果列于表 4 之中。

由表 4 可见,4℃培养的结果,SRB 平均含量大小的水深排列为 175～561 m>3 000～3 850m>1 456～2 200 m>41～55 m;25℃培养时,其排序为 3 000～3 850 m>1 456～2 200m、41～55 m>175～561 m。这表明,深水处的沉积物 SRB 含量不一定很低,浅水处沉积物 SRB 含量不一定很高。

从 SRB 的检出率可知,4℃时沉积物中的 SRB 检出率小于 25℃培养的 SRB,4℃时 SRB 检出率有随水深增加而变低之趋势;而 25℃时 SRB 含量检出率有随水深加大先降低后增高的趋势。从平均含量上看,25℃培养的 SRB 的变化较有规律,从浅水向深海盆呈增加趋势。

表 4　不同水深环境下沉积物中 SRB 含量指标

Table 4　Comparison of SRB content under at different water depths in the Chukchi Sea and the Canada Basin

水深范围/m	4℃					25℃				
	样品数	检出率	最小值/(个·g^{-1})	最大值/(个·g^{-1})	平均值/(个·g^{-1})	样品数	检出率	最小值/(个·g^{-1})	最大值/(个·g^{-1})	平均值/(个·g^{-1})
41～55	46	63.0%	0	11 000	80.7	44	88.6%	0	24 0000	1 339
175～561	24	79.2%	0	24 000 000	103 309	24	66.7%	0	15 000	1 374.6
1456～2200	42	35.7%	0	240 000	1 299.6	42	47.6%	0	240 000	1 753.5
3000～3850	77	29.9%	0	460 000	1 935.6	76	81.6%	0	2 400 000	17 927.5

6　结论(Conclusions)

1)通过对研究区不同地理单元的沉积岩芯中 SRB 的研究表明,4℃与25℃温度时培养的 SRB 含量范围基本一致,多为 $0\sim2.4\times10^6$ 个·g^{-1}(湿样),但在 SRB 检出率与平均含量上以 25℃培养条件时的为高,25℃时检出率为 73.7%,平均值 4.70×10^4 个·g^{-1}(湿样),而4℃时的检出率 45.5%,平均 2.06×10^4 个·g^{-1}(湿样)。研究区沉积物不低的 SRB 检出率和含量,表明在北极海这种冷酷环境中 SRB 仍保留活性,并在适宜时空中进行着微生物的生物地球化学活动。

2)研究区柱状沉积物中 SRB 的检出率、含量范围、平均含量都明显高于表层沉积物中 SRB 含量,这与首次北极科考获得的 SRB 研究结果一致,也与东北冰洋沉积物中 SRB 研究结果一致。这是由 SRB 的厌气性特征、沉积物一水界面附近含氧量及氧化还原电位相对较高等特征决定的. 虽然岩芯中 SRB 分布与采样点的纬度、深度有一定关系,但这种关系并不如表层沉积物中 SRB 分布表现得那么明显。

3)在垂直方向上,SRB 的分布受多种因素控制,变化较为复杂,似无明显的规律。环境温度应是其主导因子之一,如:4℃时各层位的平均值范围为 $51\sim1.2\times10$ 个·g^{-1}(湿样),25℃时各层位的平均值范围为 $2.04\times10^2\sim2.47\times10^5$ 个·g^{-1}(湿样)。在所研究的深度范围内,研究区各层位 SRB 平均含量变化不明显,表明 SRB 的生存深度范围较大。至于 4℃时 SRB 的垂直变化较为明显,而 25℃时 SRB 的垂直变化相对缓和,层位间差异较小,这还需要从生物地球化学角度作进一步的研究。

4)根据 4℃、25℃ 2 个培养温度中 SRB 的检出率、含量状况看,似乎 25℃条件时更有利于研究区某些 SRB 的繁衍。这对高纬度寒冷的北极海域海底沉积物而言,是个值得深思的问题。由此推测,环境温度的提高将改变、加速或增强某些 SRB 的生理生化活性和代谢能力,进而影响现场地球化学和地质环境状况。这启示我们随着全球气候变暖和环境因子的变化,北极微生物群落在经历一段时间后有可能产生结构性的变化,其深远影响值得人们关注。

参考文献 19 篇(略)

(合作者:高爱国　林学政)

DISTRIBUTIONAL CONDITION OF SULPHATE REDUCING BACTERIA IN CORE SEDIMENTS FROM THE CANADA BASIN AND THE CHUKCHI SEA

(ABSTRACT)

Abstract Ten sediment core samples (water depth from 50～3 850 m) were collected in the the Canada Basin and the Chukchi Sea during the Second Chinese National Arctic Research Expedition. Each core was sliced on board at 1 cm intervals from 0～10 cm and at 2 cm intervals below 10 cm as subsamples, the content of sulfate-reducing bacteria (SRB) of these subsamples were analyzed by MPN (Most Probable Number) methed. The contents of SRB cultivated at both 4℃ and 25℃ ranged from 0 to 2.4×10^6 cell · g^{-1} wet sample and the average SRB cultivated at 4℃and 25℃were 2.06×10^4 cell · g^{-1} wet sample, and 4.70×10^4 cell · g^{-1} wet sample, respectively. Unlike surface sediments in the study area, the latter shows a tendency to increase from low latitude to high latitude, or from shallow water to deep basin. From surface to depths, the content of SRB in sediment changes irregularly, depending on the sedimentary environment. The average contents of SRB cultured at 4℃ and 25℃ in the same layer ranged-from 51 to 1.2×10^6 cell · g^{-1} wet samples and 2.04×10^2 to 2.47×10^5 cell · g^{-1} wet samples. Comparing the contents, content range, and occurrence rate of SRB cultivated at 4℃ and 25℃, it seems that 25℃ is more suitable than 4℃ for survival and propagation of some SRB.

Key words The Canada Basin, The Chukchi Sea, Sulphate Reducing Bacteria, Sediment Core

F_3 铁细菌和锰细菌
(IRON-OXIDIZING BACTERIA AND MANGANESE BACTERIA)

铁细菌在北极特定海区沉积物中的分布*

摘 要 用平板法对取自北极加拿大海盆和楚科奇海10个沉积物岩芯中184个样品进行铁细菌检出率和含量分析。结果表明,4℃培养时铁细菌的检出率为10.33%,含量范围为未检出～8.40×10^6 个/g(湿样,下同),平均含量为 7.76×10^4 个/g。25℃培养时铁细菌的检出率为40.22%,含量范围为未检出～1.00×10^8 个/g,平均含量为 1.40×10^6 个/g。研究区沉积物中铁细菌含量高于太平洋深海沉积物中铁细菌含量;与前人在淡水沉积物中所作的调查相比,研究区的铁细菌含量落在前人调查结果范围内,但比细菌簇中铁细菌含量要低,这表明研究区海底沉积物的微生境较适合铁细菌的繁衍。对铁细菌在不同水深、经纬度以及沉积物深度中的分布进行了讨论,结果表明铁细菌检出率随沉积物深度增加而增大,含量变化呈现出浅层少、下层多之势。铁细菌的纬度分布范围较大,并显出有扩大的趋势。铁细菌含量与检出率随水深加大也呈跳跃式变化,水较深时铁细菌有较高的检出率和数量,但离陆距离的增加又会在物质供应上对其生长产生制约。培养温度对铁细菌分布有一定影响,温度提高不仅提高了铁细菌检出率,也增加了铁细菌的含量。这也许与部分铁细菌的最适宜生长温度较高有关,温度升高有利于某些铁细菌的生长发育。

关键词 铁细菌 北极 加拿大海盆 楚科奇海 沉积物岩芯

铁是地球上重要的元素之一,它不仅作为常量元素广泛存在于地壳中,参与各种地球化学过程;而且铁也是生物体中一种重要的微量元素,是生物酶中辅因子的一部分,是生物体细胞中电子转移反应的理想辅助因子,因此在生物过程和生物化学领域也有重要意义。在自然界中二价铁常呈可溶态存在,而三价铁则趋于形成羟基络合物,其在形成氢氧化物胶体时会捕集许多重金属离子,对微量元素起到清扫作用[1,2]。影响铁元素在自然界循环的因素很多,其中细菌对铁元素的迁移转化有着不可忽视的作用。

与铁相关的细菌有铁氧化菌和铁还原菌,人们一般将前者简称为铁细菌(Iron-Oxidi-

---

* 原文刊于《海洋科学进展》,2008,26(3):326-333.本人为第二作者。

资助项目:国家自然科学基金项目——楚科奇海及其邻近海域沉积物中细菌生物地球化学(40576060);楚科奇海沉积物中古环境变化的地球化学响应(40376017)

zing Bacteria,简称 FeOB,下同)。铁细菌是指能将环境中二价铁氧化为三价铁的细菌,迄今已发现了一些铁细菌,研究得较多的有:*Gallionella*,*Leptothrix*,*Sphaerotilus*,*Metallognium*,*Thiobafillus* 和 *Siderocapsa*[1,3-5]。铁细菌参与环境中铁元素的提取、转化、迁移、富集和沉淀等元素循环过程。铁细菌对铁的作用可导致铁制品(如管道)的锈蚀与堵塞[4],在矿山废水中,铁氧化细菌的作用使黄铁矿被氧化,导致体系的氧化还原电位升高,酸度增加,促使微量重金属的溶解,使水体污染加剧[6]。

早在 1836 年 Ehrenberg 就已发现了铁细菌[7],早期的铁细菌研究主要与矿山酸性废水处理、城市供水系统及管道系统[8-11]、前寒武纪条状铁燧岩的成矿机理研究等相关[12],而后拓展到海底铁锰结核的成矿机理[13],海底热水区附近铁细菌[14],细菌作用下深海含铁矿物溶解机理[15],湖相沉积物中铁细菌[16,17],湿地生态系统中铁细菌研究等领域[18],并注重探讨铁细菌在铁的地球化学循环中的作用,以及在生源要素的生物地球化学循环中的作用[19,20]。已有学者对海洋中包括南极海域的铁细菌作过相关研究,指出那里有铁细菌活动,但北极海域铁细菌的相关研究很少且不深入[21,22]。对北极地区海洋沉积物中铁细菌的研究,有助于更全面地了解北极生物地球化学过程。对铁细菌等微生物的研究将有助于加深人们对海底沉积中生物地球化学循环的认识。本文就北极加拿大海盆与楚科奇海水深 41～3 850 m 处的沉积物中铁细菌检出率与含量的地理分布加以讨论。

1 材料与方法

1.1 样品的采集与保存

样品的采集是 2003 年 7 月～9 月中国第 2 次北极科学考察期间在“雪龙”号考察船上完成的。研究区概况、采样站位图、采样方法等请参见文献[23]。

1.2 实验室分析

将第 2 次北极科学考察中所得沉积物的微生物样品转至国内陆上实验室后,于 2004 年 1 月起开始铁细菌分析。首先定量称取样品,不断稀释至一定浓度,取 0.1 mL 合适浓度的稀释样品至铁细菌固体培养基平板上均匀涂布,然后分别置 4℃或 25℃的温度中恒温培养 3 周以上,其间经常观察平板上菌落等状况,最后计数平板上长出的菌落,以此为铁细菌 CFU,计算出样品中铁细菌的检出率和单位重量样品中的铁细菌含量(单位:个/g,湿重样品,下同),分析铁细菌的地理分布状况。铁细菌培养基成分及分析方法参见文献[24]所述。

2 结果

对 184 个样品的分析结果表明,4℃与 25℃温度培养的铁细菌未检出样品数分别为 165 个和 110 个,检出率分别为 10.33%和 40.22%;含量范围分别为未检出～8.40×10^6 个/g 与未检出～1.00×10^8 个/g,平均含量分别为 7.76×10^4 个/g 和 1.40×10^6 个/g。表 1 为 4℃和 25℃培养时各岩芯铁细菌检出率和含量状况,岩芯各层位中铁细菌检出率和含量状况见表 2。

表 1 加拿大海盆与楚科奇海沉积物铁细菌测定结果

Table 1 Occurrence percentages and contents of FeOB in the sediment cores from the Canada Basin and the Chukchi Sea

个/g(湿重)

站位	岩芯深度/cm	样品数	4℃			25℃		
			最大值	平均值	检出率/%	最大值	平均值	检出率/%
R06	12	10	—	—	0	1.00×10^7	1.00×10^6	20.00
C19	22	16	—	—	0	1.00×10^4	1.33×10^3	25.00
S11	28	19	8.00×10^3	4.21×10^2	5.26	1.20×10^5	9.23×10^3	26.32
16	28	19	1.30×10^4	6.84×10^2	5.26	4.00×10^4	2.21×10^3	15.79
S26	26	18	—	—	0	9.00×10^3	5.07×10^2	22.22
P11	38	21	8.40×10^6	4.63×10^5	42.86	8.40×10^6	6.51×10^5	76.19
M1	36	23	—	—	0	3.00×10^5	2.61×10^4	47.83
P24	28	19	1.00×10^4	5.32×10^2	10.53	1.60×10^6	8.86×10^4	36.84
B77	22	15	2.00×10^5	1.41×10^4	20.00	4.00×10^6	2.89×10^5	20.00
B80	38	24	3.30×10^6	1.79×10^5	12.50	1.00×10^8	9.43×10^6	79.17
小计		184	8.40×10^6	7.76×10^4	10.33	1.00×10^8	1.40×10^6	40.22

注:"—"为未检查,下同;表 1 中最小值均为未检出。

表 2 加拿大海盆与楚科奇海各层沉积物铁细菌检出率与含量比较

Table 2 Occurrence percentages and contents of FeOB at different depths of the sediment cores from the Canada Basin and the Chukchi Sea

个/g(湿重)

层位/m	样品数	4℃				25℃			
		最小值	最大值	平均值	检出率/%	最小值	最大值	平均值	检出率/%
0~1	8	—	—	—	0	—	3.00×10^4	6.00×10^3	50.00
1~2	10	—	1.00×10^4	1.01×10^3	20.00	—	1.13×10^6	1.16×10^5	50.00
2~3	10	—	—	—	0	—	1.00×10^8	1.10×10^7	30.00
3~4	10	—	—	—	0	—	8.00×10^4	9.90×10^3	30.00
4~5	10	—	—	—	0	—	2.10×10^5	2.20×10^4	30.00
5~6	10	—	—	—	0	—	2.00×10^4	4.04×10^3	40.00
6~7	10	—	1.30×10^4	1.40×10^3	20.00	—	5.60×10^6	5.96×10^5	40.00
7~8	10	—	1.00×10^6	1.30×10^5	20.00	—	3.10×10^7	3.13×10^6	20.00
8~9	10	—	1.00×10^4	1.00×10^3	10.00	—	8.00×10^4	9.05×10^3	50.00
9~10	10	—	1.00×10^4	1.98×10^3	30.00	—	1.50×10^4	2.29×10^3	30.00
10~12	10	—	1.30×10^4	1.30×10^3	10.00	—	2.40×10^7	2.44×10^6	50.00

(续表)

层位/m	样品数	4℃				25℃			
		最小值	最大值	平均值	检出率/%	最小值	最大值	平均值	检出率/%
12～14	9	—	3.30×10^6	3.67×10^5	11.11	—	2.60×10^6	2.93×10^5	22.22
14～16	9	—	—	—	0	—	4.00×10^6	4.75×10^5	44.44
16～18	9	—	1.00×10^6	1.11×10^5	11.11	—	2.00×10^6	4.00×10^5	44.44
18～20	8	—	2.00×10^5	2.50×10^4	12.50	—	4.00×10^6	5.38×10^5	37.50
20～22	9	—	2.00×10^3	2.22×10^2	11.11	—	2.20×10^7	2.45×10^6	33.33
22～24	6	—	—	—	0	—	4.00×10^5	6.85×10^4	50.00
24～26	7	—	2.00×10^2	2.90×10	14.29	—	2.00×10^5	4.19×10^4	42.86
26～28	6	—	—	—	0	—	5.00×10^6	8.39×10^5	50.00
28～30	2	—	—	—	0	—	—	—	0
30～32	3	—	2.00×10^2	6.70×10	33.33	—	3.30×10^3	2.10×10^3	66.67
32～34	3	—	—	—	0	—	3.00×10^7	1.00×10^7	66.67
34～36	3	—	1.00×10^3	3.33×10^2	33.33	—	2.40×10^6	8.02×10^5	66.67
36～38	2	—	8.40×10^6	4.20×10^6	50.00	6.00×10^4	8.40×10^6	4.23×10^6	100.00
小计	184	—	8.40×10^6	2.02×10^5	12.00	—	1.00×10^8	1.56×10^6	43.53

3　讨论

3.1　与其他地区的比较

据史君贤等对太平洋铁锰结核区 39 个站位海洋沉积物中铁细菌调查资料，在远离大陆的太平洋深海底部沉积物中铁细菌含量为未检出～2 143 个/g，检出率为 43.59%。比该处(17 个站位)沉积物上覆水中铁细菌含量(3～53 个/mL)高 40 余倍，比该处(17 个站位)铁锰结核中铁细菌含量(7～138 个/mL)也高[24]，但均比本研究区低。

Straub 等对德国不来梅附近水沟沉积物中铁氧化硝酸盐还原菌进行研究，其含量为 $1\times10^5\sim5\times10^8$ 个/g，其中兼养型铁细菌达到 $9\times10^6\sim5\times10^8$ 个/g，无机型铁细菌含量大约为 $5\times10^3\sim2\times10^4$ 个/g[25]。HaLucmk 等调查了 Constance 湖相沉积物中铁细菌含量，兼养型铁细菌达到 $1.0\times10^4\sim5.8\times10^5$ 个/mL，无机型铁细菌含量比兼养型铁细菌大约低 1 个数量级[17]。而 Weiss 等分析了湿地土壤及植物根围中铁氧化菌含量，分别为 3.7×10^6 个/g 和 5.9×10^5 个/g[26]。Emerson 和 Revsbech 在丹麦阿尔路斯港市(Aarhus)野外发现细菌簇中铁细菌含量高达 $1\times10^8\sim1\times10^9$ 个/cm^3[27]。

研究区铁细菌的检出率低于太平洋铁锰结核区，但铁细菌含量相对较高，前者可能与样品分析方法不一有关，后者与两区的物质供应与水深环境条件不同有关。与淡水沉积物中的分析结果相比，研究区的铁细菌含量落在前人调查的范围内，这表明研究区海底沉积物的微生境也较适合铁细菌的繁衍。

3.2　沉积物不同层次间铁细菌检出率与含量比较

表 2 和表 3 列出了各岩芯不同层次或深度沉积物中铁细菌经 2 种温度培养所得检出率和含量状况，从中可看出 2 种培养温度中得出的铁细菌含量的垂直分布趋势。

4℃培养时检出率范围为 0～50.00%，各层平均为 12.00%，有 10 层检出率为 0，底层检出率最高，为 50.00%。各层铁细菌平均含量范围为未检出～4.20×10^{6} 个/g，层间平均含量为 2.02×10^{5} 个/g，最高值为 4.20×10^{6} 个/g，出现于 36～38cm 层。若将所有层次以 8 层为一段分为上、中、下三段，可得出 0～8cm、8～22cm、22～38cm 三段的铁细菌平均检出率分别为 7.50%，11.98%和 16.37%，这表明尽管一些次表层的检出率很低，但总的趋势是铁细菌的检出率随沉积物深度增加呈增高之势。在上、中、下三个层段中铁细菌平均含量分别为 1.66×10^{4}，6.34×10^{4} 和 5.25×10^{5} 个/g，也显出随沉积物深度增加而增高之势。

表 3　沉积物中铁细菌检出率与含量统计

Table 3　Statistics of occurrence percentages and contents of FeOB at different depths of the sediment cores from the Canada Basin and the Chukchi Sea

个/g(湿重)

深度/cm	层位	4℃			25℃		
		含量	平均值	检出率/%	含量	平均值	检出率/%
0～8	最小值	—	—	0	—	4.04×10^{3}	20.00
	最大值	1.00×10^{6}	1.30×10^{5}	20.00	1.00×10^{8}	1.10×10^{7}	50.00
	平均值		1.66×10^{4}	7.50		1.86×10^{6}	36.25
8～22	最小值	—	—	0	—	2.29×10^{3}	22.22
	最大值	3.30×10^{6}	3.67×10^{5}	30.00	2.40×10^{7}	2.45×10^{6}	50.00
	平均值		6.34×10^{4}	11.98		8.25×10^{5}	38.99
22～38	最小值	—	—	0	—	—	0
	最大值	8.40×10^{6}	4.20×10^{6}	50.00	3.00×10^{7}	1.00×10^{7}	100.00
	平均值		5.25×10^{5}	16.37		2.00×10^{6}	55.36

25℃培养各层位检出率范围为 0～100%，各层平均为 43.53%，在 24 个层次中只有 28～30cm 层为未检出，底层检出率最高，为全检出。各层铁细菌平均范围为未检出～1.10×10^{7} 个/g，最高值和次高值分别出现于 2～3cm 和 32～34cm 层，层间平均含量为 1.56×10^{6} 个/g。将所有层次分作上、中、下三段(同上)，三段铁细菌检出率分别为 36.25%，38.99%和 55.36%，显出随深度增加而增长的趋势。在 3 个层段中铁细菌平均含量分别为 1.86×10^{6}，8.25×10^{5} 和 2.00×10^{6} 个/g，中层段的铁细菌含量最少，分布趋势与检出率比稍有变化。

综上所述，铁细菌检出率随沉积物深度增加而提高。4℃时铁细菌含量在层次间的分布大体上呈随沉积物深度增加而提高，而 25℃时为上层高而中、下层稍低之势。培养温度的差异对其含量的层次分布趋势有一定影响，温度提高不仅提高了铁细菌检出率，也增加了铁细菌的含量。前人研究表明，铁细菌的生长温度为 6～37℃，不同种类的最适宜生长温度区间不同，都落在 15～30℃内[10]。显然较高的温度有利于铁细菌的生长发育。

3.3 沉积物所在经度和纬度区对铁细菌检出率及含量的影响

将研究区按纬度划为4个区，即66°～70°N，70°～74°N，74°～78°N和＞78°N，比较铁细菌两指标状况，表4列出了测区铁细菌两指标在纬度上的比较。由此可见，在2种培养温度，即4℃和25℃时的铁细菌检出率和含量基本上呈现为随纬度增高而增高后趋缓的变化。这表明铁细菌的纬度分布区间可能相当广泛，而温度的提高则可能有助于它们分布范围的缓慢扩大化，这暗示气温的变暖对当地原有遗传保护性强的物种的影响，有一个渐进而深刻的过程。

表4 北极加拿大海盆与楚科奇海沉积物铁细菌检出率和含量在纬度区间的比较

Table 4 Comparison of occurrence percentages and contents of FeOB in the sediment cores with latitudes in the Canada Basin and the Chukchi Sea

个/g(湿重)

纬度	站号	样品数	4℃			25℃		
			含量	平均值	检出率/%	含量	平均值	检出率/%
66°～70°N	R06	10	—	—	0	1.00×10^7	1.00×10^6	20.00
70°～74°N	C19,S11,S16,S26	72	1.30×10^4	2.76×10^2	2.78	1.20×10^5	3.32×10^3	22.22
74°～78°N	M1,P11,P24	63	8.40×10^6	1.55×10^5	17.46	8.40×10^6	2.55×10^5	53.97
＞78°N	B77,B80	39	3.30×10^6	9.66×10^4	15.38	1.00×10^8	4.86×10^6	56.41

将研究区按经度划为5个区间，各经度区间铁细菌的分布见表5。由此可见，在2种培养温度时铁细菌的检出率和含量基本上随经度呈不规则变化，东西差异不明显。

表5 不同经度区间柱状沉积物中铁细菌含量指标

Table 5 Comparing the occurrence percentages and contents of FeOB in the sediment cores with longitudes in the Canada Basin and the Chukchi Sea

个/g(湿重)

经度	站号	样品数	4℃			25℃		
			最大值	平均值	检出率/%	最大值	平均值	检出率/%
145°～150°W	B77,B80	39	3.30×10^6	9.66×10^4	15.38	1.00×10^8	4.86×10^6	56.41
150°～155°W	S26	18	—	—	0	9.00×10^3	5.07×10^2	22.22
155°～160°W	P24,S11,S16	57	1.30×10^4	5.46×10^2	7.02	1.60×10^6	3.33×10^4	26.32
160°～165°W	C19	16	—	—	0	1.00×10^4	1.33×10^3	25.00
165°～170°W	M1,P11,R06	54	8.40×10^6	1.54×10^5	16.67	1.00×10^7	5.60×10^5	53.70

3.4 水深对沉积物铁细菌检出率及含量的影响

按取样站位的不同水深范围比较铁细菌分布(表6)，结果表明，2种培养温度得出的检出率均为随着水深的增大而呈最低→最高→低的趋势。而含量则随水深加大也呈跳跃式变化。水较深时，沉积物的铁细菌有较高的检出率和数量，也许是较深的水下环境，使沉积物中铁细菌有一定的适应性，但离陆距离的增加又会在物质供应上对其生长产生制约，因此这也许与沉积环境的适宜性和细菌所需物质的供应等因素的协同制约有关，相关研究尚待深入。

表6 北极加拿大海盆与楚科奇海沉积物铁细菌在不同水深区间的比较

Table 6 Comparing the occurrence percentages and contents of FeOB in the sediment cores with water depth in the Canada Basin and the Chukchi Sea

个/g(湿重)

水深/m	站号	样品数	4℃			25℃		
			最大值	平均值	检出率/%	最大值	平均值	检出率/%
41～55	C19，R06，S11	45	8.00×10^3	1.40×10^2	2.22	1.00×10^7	3.37×10^5	24.44
175～561	P11	21	8.40×10^6	4.63×10^5	42.86	8.40×10^6	6.51×10^5	76.19
1 456～2 200	M1，P24	42	1.00×10^4	2.66×10^2	4.76	1.60×10^6	5.73×10^4	42.86
3 000～3 850	S16，S26，B77，B80	76	3.30×10^6	4.85×10^4	9.21	1.00×10^8	2.43×10^6	38.16

4 结语

铁在生物体内外的存在状态与活动是多态的、多样性丰富的，海洋中的铁含量与全球大气 CO_2 增减和温室效应相关。铁参与地球生态圈循环。铁细菌是地球铁活动的一类活跃成员，其活动作用方式有些类似于锰细菌，并在大洋多金属结核中共同发挥重要作用。本文的研究结果只是初步分析了这一海区的铁细菌状况，由此获知铁细菌的存在机率及含量在不同站位及沉积的层次、沉积物所在经纬度、水深间的差异，此差异有明显的温度效应，它们在这些方面都多少显示出某些规律，在一定程度上表达并反映了该特定海区铁细菌的生存状态。本结果还表明，虽然测区处于寒冷水底，但依然广泛分布着铁细菌，意味着铁细菌在全球范围的广布性。温度试验的结果意味着环境的改变必将使其原有的铁细菌生态随之发生缓慢、持续的变化。包括铁细菌与含铁物质及其它们之间互动关系在内的课题尚待深入研究。

参考文献27篇（略）

（合作者：高爱国）

DISTRIBUTIONS OF IRON-OXIDIZING BACTERIA IN SEDIMENT CORES FROM THE SPECIAL AREAS IN THE ARCTIC OCEAN

(ABSTRACT)

Abstract Analyses of iron-oxidizing bacteria (FeOB) were made of 184 samples of 10 sediment cores from the Canada Basin and the Chukchi Sea by means of plate culture, and their occurrence percentages and contents were studied. The results show that the occurrence percentages of FeOB in the samples cultivated at 4℃ and 25℃ are 10.33% and 40.22% respectively and the contents of FeOB (in wet samples, the same hereinafter) range from non-detected (ND) to 8.40×10^6 CFU/g, with an average of 7.76×10^4 CFU/g, in the samples cultivated at 4℃ and from ND to 1.00×10^6 CFU/g, with an average of 1.40×10^6 CFU/g, in the samples cultivated at 25℃. The contents of FeOB in the sediment cores of the study area are higher than those in the abyssal sediments of the Pacific Ocean. Compared with previous studies carried out in fresh water, the FeOB contents in the study area are all dropped in the range of the previous results, but are lower than those in the bacteria mat. This indicates that the micro-habitat-environment in the sediments of the study area is favorable for the regeneration of FeOB. The distributions of FeOB in different water depth, different longitude and latitude and different sedimentary depth have been discussed. It has been shown that the occurrence percentages of FeOB tend to increase with sedimentary depth and the contents of FeOB tend to be lower in the shallow layers and higher in the deeper layers. The latitudinal distributions of FeOB show a wide range and an expanding tendency. The occurrence percentages and contents of FeOB increase with water depth and vary in a jumping way. Higher occurrence percentages and contents of FeOB occur in relatively deeper water. With increasing in distance away from the coast, however, the growth of FeOB has been restricted due to the lack of nutrient subsfances supply. The cultivation temperature can also exert influences, to certain extent, to the distributions of FeOB. With the increasing in temperature not only the occurrence percentage but also the content of FeOB is increased. This may probably indicate that relatively higher temperature is more favorable for the growth of some of the FeOB.

Key words Iron-Oxidizing Bacteria; Arctic Ocean; Canada Basin; Chukchi Sea; Sediment Cores

锰细菌在加拿大海盆、楚科奇海沉积物中的分布*

摘 要 用平板法分别在4℃和25℃对北极加拿大海盆和楚科奇海10个沉积物岩芯中的锰细菌进行培养，并测定了检出率和含量。分析结果表明，4℃与25℃温度时培养的锰细菌含量范围均为(一)～3.3×10^8 个·g^{-1}，锰细菌平均检出率分别为77.78%、86.03%。平均含量分别为5.60×10^6、7.91×10^6 个·g^{-1}。该结果高于太平洋某深海沉积物中的锰细菌含量，比北极阿拉斯加淡水湖湖水中的锰细菌含量高1个数量级，比印度洋海岭Carlsberg Ridge区热水区海水样品中的锰细菌含量高3个数量级。证实寒冷北极海同样广布锰细菌。对锰细菌在不同水深、经纬度以及沉积物深度中的分布进行了讨论；结果表明，锰细菌检出率随沉积物深度的增加而增大，含量变化呈现浅层少、下层多之势；纬度分布范围较大，并显出有扩大的趋势。水深对沉积物中锰细菌也有一定影响。尽管锰细菌对温度有较强的适应能力，环境温度的升降对本研究区锰细菌具有双重作用。

关键词 锰细菌 北极 加拿大海盆 楚科奇海 沉积物岩芯

1 引言(Introduction)

锰是元素地球化学循环的一个活跃组分，它通过氧化还原状态的改变，可在溶液—固体界面发生迁移，同时通过对环境物理化学条件的影响以及吸附—解吸作用而影响其他元素的地球化学循环，是元素地球化学中研究较多的一个元素。在海洋中，锰的地球化学行为包括锰离子的氧化、还原、迁移、富集、沉淀，乃至形成结核或结壳等成矿作用；它不仅与物理、化学过程有关，且与生物过程也有关，包括一些微生物参与的过程(Chapnick, 1982; Emerson et al. 1982; Sunda et al., 1990; 阎葆瑞等, 1992; Moffett et al., 1996; Francis et al 2001; 2002)。这些能够参与锰地球化学行为的细菌被统称为锰细菌，其大多指能氧化 Mn^{2+} 为 Mn^{4+} 的细菌；锰细菌并不是分类学上的概念，而是具有能氧化锰这一属性的细菌，隶属于不同的科、属(翁酥颖等, 1985)。人们对大洋底部富含Mn、Fe、Co等的多金属结核的不倦探索，以及利用自大洋分离出来的锰细菌去开发富集锰矿资源和治理环境锰污染的愿望，进一步激发着对这类细菌的研究(Mandernack et al., 1995; 胡文宣等, 1999; Villalobos et al., 2003; Loy et al., 2005; roner et al., 2005; 田美娟等, 2006)。目前，人们已认识到锰细菌是环境中介导Mn(Ⅱ)氧化作用的最重要因子(Johnston et al., 1988, Francis et al., 2001)。早在1913年Beijerinck就报道了对锰细菌的研究。自

* 原文刊于《环境科学学报》, 2008, 28(11): 2369-2374, 本人为第二作者。

基金项目：国家自然科学基金(No. 40576060, 40376017, 40176017)

Supported by the National Natural Science Foundation of China(No. 40576060, 40376017, 40176017)

1966 年 Ehrlich 从海洋中分离出锰细菌以来，相关研究不断出现，包括对太平洋、大西洋、波罗的海、北极湖泊等环境中的锰细菌丰度、作用、与生态关系、环境等方面(Rosson et al.，1982;史君贤等，1996；1998)。对东地中海的研究表明，气候变化对深海细菌及底栖群落已产生影响(Danovaro et al.，2001)。深海生态系统对气候变化响应较敏感，但北冰洋锰细菌的研究尚未见诸报道。北冰洋虽处地球严寒区域，但据现有微生物资料可知，那里的微生物活动仍较为活跃，甚至不亚于中低纬度海区。在全球变暖过程中，作为对全球变化敏感的北极地区，它正发生着明显的环境变化。因而对北极海洋沉积物中的锰细菌含量及分布状况进行估测、探讨锰细菌在元素地球化学循环及全球变化中的作用就具有一定意义。本研究就北极加拿大海盆与楚科奇海沉积物的锰细菌作初步探索，旨在促进相关研究的深入开展。

2　材料与方法(Materials and methods)

2.1　样品的采集与保存

2003 年 7 月～9 月中国第 2 次北极科考期间在“雪龙”号船上完成样品采集，共采集了 10 个岩芯的沉积物样品。研究区概况、采样站位图、采样方法等请参见文献(高爱国等，2008)。

2.2　实验室分析

2004 年 1 月对所采集沉积物中的微生物样品，先按培养要求逐步稀释至 ZoBell 2216E 液体培养基中，以确保锰细菌的生长，再选择合适的稀释度以 0.1 mL，涂布到锰细菌培养基平板上。该培养基成分见史君贤等(1998)所述。接种好的平板分别于 4℃和 25℃时培养 3 周以上，期间定期观察菌落生成及其色泽情况，最终以褐色菌落为锰细菌菌落，计算其 CFU，并按稀释度换算为单位沉积物(湿重)的 CFU，即：个 CFU · g^{-1}(湿重)(简写为个 · g^{-1}，下同)；并以此与所测沉积物的层位、水深、经纬度等地理位置资料进行分析。

3．结果(Results)

3.1　各岩芯中锰细菌检出率和含量状况

对 180 个样品的分析结果表明，4℃与 25℃温度时培养的锰细菌含量范围均为(—)～3.3×10^8 个 · g^{-1}。平均含量分别为 5.61×10^6 个 · g^{-1} 和 7.91×10^6 个 g^{-1}，样本总体的标准偏差分别为 2.93×10^7、3.24×10^7，检出率为 77.78%和 86.03%。

表 1　加拿大海盆与楚科奇海沉积物岩芯中锰细菌测定结果

Table 1　Occurrence percentages and contents of MnB from the sediment cores

站位	岩芯深/cm	4℃				25℃			
		样品数	最大值/(个 · g^{-1})	平均值/(个 · g^{-1})	检出率	样品数	最大值/(个 · g^{-1})	平均值/(个 · g^{-1})	检出率
R6	12	10	1.00×10^6	1.06×10^5	50.00%	10	2.00×10^6	2.96×10^5	60.00%
C19	22	16	3.00×10^7	2.12×10^6	62.50%	16	1.60×10^8	1.09×10^7	75.00%
S11	28	19	1.00×10^6	7.47×10^4	73.68%	18	1.00×10^6	8.04×10^4	83.33%
S16	28	19	1.00×10^8	5.78×10^6	94.74%	19	1.30×10^8	8.11×10^6	100.0%

(续表)

站位	岩芯深/cm	4℃				25℃			
		样品数	最大值/(个·g^{-1})	平均值/(个·g^{-1})	检出率	样品数	最大值/(个·g^{-1})	平均值/(个·g^{-1})	检出率
S26	28	18	1.50×10^7	1.57×10^6	88.89%	18	4.00×10^7	4.89×10^6	100.0%
P11	26	21	3.30×10^8	2.70×10^7	90.48%	21	3.30×10^8	2.70×10^7	90.48%
M1	38	20	1.15×10^7	1.40×10^6	90.00%	20	1.15×10^7	2.32×10^6	95.00%
P24	36	19	1.61×10^7	1.48×10^6	94.74%	19	1.61×10^7	1.52×10^6	89.47%
B77	28	14	1.71×10^5	3.01×10^4	35.71%	14	1.60×10^7	1.17×10^6	42.86%
B80	22	24	9.00×10^7	8.82×10^6	70.83%	24	9.20×10^7	1.41×10^7	95.83%
小计	38	180		4.83×10^6	75.16%	179		7.03×10^6	83.20%

注:25℃条件时 S11 站 3~4cm 层样,下同;4℃时最小值均为(—)(未检出,下同),25℃时 S16、S26 最小值分别为 1.58×10^4 个·g^{-1}和 2.60×10^2 个 g^{-1},其余均为(—)。

表 1 为 4℃和 25℃培养时各岩芯锰细菌检出率和含量状况。4℃培养条件中各岩芯锰细菌的检出率范围为 35.71%~94.74%,各岩芯平均检出率为 75.16%;平均含量范围为 3.01×10^4~2.70×10^7 个 g^{-1},最低值出现在 B77 站(9 层未检出),次低为 B80 站(7 层),最高值见于 P11 站。25℃培养条件中各岩芯锰细菌的检出率范围为 42.86%~100%,各岩芯平均检出率为 83.20%;平均含量范围为 8.04×10^4~2.70×10^7 个 g^{-1},最低值出现在 S11 站,最高值出现于 P11 站,与 4 ℃培养时同。

3.2　各层位中锰细菌检出结果

岩芯各层位中锰细菌检出率和含量状况见表 2,各层位中锰细菌的出现机率和含量呈不均匀分布,对温度的响应也不一致。结果表明,温度的适当提高可促使锰细菌含量提升,但在部分测点上温度提高可减少锰细菌的检出率;这也许意味着尽管从总体看 2 种温度培养时检出率基本一致,但一些锰细菌,尤其是嗜冷菌会因不适应温度的提高而死去。

表 2　加拿大海盆与楚科奇海沉积物层位间锰细菌检出率与含量

Table 2　Occurrence percentages and contents of MnB from different depths in the sediment cores

层位/cm	样品数	4℃				25℃			
		最小值/(个·g^{-1})	最大值/(个·g^{-1})	平均值/(个·g^{-1})	检出率	最小值/(个·g^{-1})	最大值/(个·g^{-1})	平均值/(个·g^{-1})	检出率
0~1	8	3.00×10^2	2.00×10^6	2.84×10^5	100.0%	8.80×10^3	1.40×10^7	1.80×10^6	100.0%
1~2	10	(—)	1.00×10^6	1.45×10^5	80.00%	(—)	1.08×10^6	2.73×10^5	90.00%
2~3	10	(—)	3.00×10^6	4.46×10^5	60.00%	(—)	3.00×10^7	3.50×10^6	80.00%
3~4	10*	(—)	1.00×10^6	1.33×10^5	60.00%	(—)	8.00×10^6	1.14×10^6	66.67%
4~5	10	(—)	1.50×10^7	1.67×10^7	60.00%	(—)	3.40×10^7	4.58×10^6	70.00%
5~6	10	(—)	3.00×10^7	4.07×10^6	70.00%	(—)	1.60×10^8	2.01×10^7	80.00%

（续表）

层位/cm	样品数	4℃				25℃			
		最小值/(个·g^{-1})	最大值/(个·g^{-1})	平均值/(个·g^{-1})	检出率	最小值/(个·g^{-1})	最大值/(个·g^{-1})	平均值/(个·g^{-1})	检出率
6～7	10	(—)	9.40×10^6	1.04×10^6	70.00%	(—)	9.40×10^6	1.23×10^6	80.00%
7～8	10	(—)	3.30×10^8	3.35×10^7	90.00%	(—)	3.30×10^8	3.68×10^7	90.00%
8～9	10	(—)	3.20×10^6	4.20×10^5	80.0%	(—)	3.20×10^6	4.33×10^5	90.00%
9～10	10	(—)	1.00×10^7	1.03×10^6	80.00%	(—)	1.00×10^7	1.06×10^6	90.00%
10～12	10	(—)	1.61×10^7	2.13×10^6	70.00%	(—)	3.00×10^7	5.13×10^6	90.00%
12～14	9	(—)	7.80×10^7	9.41×10^6	77.78%	(—)	7.80×10^7	9.43×10^6	77.78%
14～16	9	(—)	1.15×10^7	1.44×10^6	77.78%	(—)	1.60×10^7	3.30×10^6	88.89%
16～18	9	(—)	5.70×10^6	8.18×10^5	66.67%	(—)	5.70×10^6	9.65×10^5	77.78%
18～20	9	1.00×10^3	2.00×10^6	5.04×10^5	100.00%	4.00×10^3	2.00×10^6	5.43×10^6	100.0%
20～22	7	(—)	1.00×10^8	1.49×10^7	85.71%	(—)	1.30×10^8	2.03×10^7	85.71%
22～24	7	(—)	9.00×10^7	1.30×10^7	85.71%	4.60×10^3	9.20×10^7	1.47×10^7	100.0%
24～26	6	(—)	3.70×10^7	6.29×10^6	66.67%	(—)	3.20×10^7	5.62×10^6	66.67%
26～28	5	(—)	1.32×10^5	7.30×10^4	80.00%	2.90×10^4	9.10×10^6	1.90×10^6	100.0%
28～30	3	1.72×10^3	1.67×10^6	8.91×10^5	100.0%	1.72×10^3	7.00×10^6	2.89×10^6	100.0%
30～32	2	3.00×10^3	1.40×10^8	7.00×10^7	100.0%	8.30×10^4	1.40×10^8	7.00×10^7	100.0%
32～34	2	1.20×10^6	4.00×10^6	2.60×10^6	100.0%	1.20×10^6	2.40×10^7	1.26×10^7	100.0%
34～36	2	3.10×10^3	6.00×10^7	3.00×10^7	100.0%	6.60×10^3	6.00×10^7	3.00×10^7	100.0%
36～38	2	3.20×10^4	1.04×10^7	5.22×10^6	100.0%	5.80×10^4	1.04×10^7	5.23×10^6	100.0%
层平均				8.33×10^6	81.68%			1.08×10^7	88.48%

*25℃时样品数为9个

3.3 沉积物不同层位间锰细菌的检出率与含量比较

从表2中可看出2种培养温度中得出的锰细菌含量的垂直分布趋势。4℃培养时未检出样品40个，各层的检出率范围为60.00%～100%，各层平均为81.68%。未检出样均不出现于表层或底层，表、底层检出率均为100%，有7个层位的检出率为100%，占所有层位数的29.2%。锰细菌含量最高值(3.30×10^8 个·$^{-1}$)、次高值(1.40×10^8 个·g^{-1})分别出现于P11站7～8cm层和30～33cm层。各层锰细菌平均含量范围为7.30×10^4～7.00×10^7 个·g^{-1}。若将所有层位以8层为一段分为上、中、下3段，可得出0～8cm、8～22cm、22～38cm 3段的锰细菌平均检出率分别为73.08%、79.45%和86.21%，平均含量分别为5.28×10^6、3.42×10^6和1.20×10^7 个·g^{-1}。这表明，尽管一些次表层的检出率很低，但总的趋势是锰细菌的检出率随沉积物深度增加呈增高之势，而锰细菌含量则显上段高、中段偏低，下段最高之势。

25℃培养时未检出样品 25 个，各层位检出率范围为 66.67%～100%，各层平均为 88.48%，在 24 个层位中有 15 个层位未检出。锰细菌检出率最低的层位是 3～4cm，全检出的层位出现于 0～1 cm、18～20cm、22～24cm 以及 26～38cm 中。锰细菌含量最高值和次高值出现的站号、层位与 4℃时相同，平均含量范围为 2.73×10^5～7.00×10^7 个·g^{-1}；以 30～32cm 层为最高，最低值则出现于 1～2cm 层，即次底层最高，次表层最低，两者相差 2 个数量级。按上、中、下 3 段（同上）统计，锰细菌检出率分别为 81.82%、87.67% 和 93.10%，显出随深度增加而增大的趋势；而锰细菌平均含量分别为 8.95×10^6、4.61×10^6 和 1.35×10^7 个·g^{-1}，与 4℃时一致，呈上段高、中段偏低，下段最高之势，这与深部二价锰较多、中层段的锰细菌含量最少相一致。

综上所述，锰细菌检出率随沉积物深度的增加而提高，而锰细菌含量在层位间的分布大体上呈现出下层高、上层次高、中层偏低之势。这可能受早期成岩过程中沉积物氧化还原体系的垂直分带制约，取决于体系中有机质及各种电子供体与受体的含量、氧化还原电位的变化等因素，其机理有待进一步的研究。培养温度的差异并未对其含量的层位分布趋势有明显作用，只是温度的提高，增加了锰细菌含量。

4 讨论(Discussion)

4.1 与其他海区比较

史君贤等(1998)在太平洋铁锰结核区海洋沉积物中进行的锰细菌调查结果表明，在远离大陆的太平洋深海底部沉积物中锰细菌含量范围为未检出～3 900 个·g^{-1}，比该处沉积物上覆水中的未检出～160 个·mL^{-1} 高 25 倍，但比本研究中所获得的结果明显要低。Johnston(1988)对北极阿拉斯加 Toolik 湖，及 Chapnick 等(1982)对 Oneida 湖水中的锰细菌调查结果表明，2 个湖湖水中锰氧化菌在 10^4～10^5 CFU·mL^{-1}。而 Femandes 等(2006)对印度洋海岭 Carlsberg Ridge 区热水区附近采得的海水样品分析，结果表明，锰细菌含量范围为未检出～3.21×10^3 CFU·mL^{-1}，略低于 Johnston(1998)等的观察值；造成这种差异的原因可能主要与物质供给的影响和较大的静水压力有关。由史君贤等(1998)的研究结果以及微生物生态研究经验可知：在沉积物中锰细菌的含量一般要比水体中的高。相比之下，考虑水深及物质供应条件等差异，本研究所获得的锰细菌范围为(一)～3.30×10^7 个·g^{-1}应是较为客观的。以上结果也表明，研究区沉积物中锰细菌含量不低于中、低纬度海区。

4.2 沉积物所在纬度、经度对锰细菌检出率及含量的影响

将研究区按纬度划为 4 个纬度区，即 66°～70°N、70°～74°N、74°～78°N 和>78°N 4 个区，分别比较锰细菌 2 指标状况（表 3）。在 4℃和 25℃条件培养时，锰细菌的检出率和含量基本上呈随纬度增高而增高、然后变化趋缓. 纬度高于 78°N 后又降低，这可能与水深增加有关。显然纬度增高、温度降低这 2 个因素由于受巨厚水体的缓冲，没有对锰细菌分布产生多少负面影响。在这里主要表现为生存环境的稳定性与物质供应的可利用性及巨大的静水压力的影响。这表明，锰细菌在纬度上的分布区间可能相当广泛，而温度的提高可能有助于它们中部分成员分布范围的缓慢扩大化，这暗示气温的变暖对当地原有遗传保守性强的物种的影响是一个渐进而深刻的过程。

表 3　北极加拿大海盆与楚科奇海沉积物锰细菌检出率和含量在纬度区间的比较*

Table 3　Comparison of MnB content along the latitudinal gradient in the Chukchi Sea and the Canada Basin

纬度(N)	站号	样品数	4℃			25℃		
			最大值/(个·g^{-1})	平均值/(个·g^{-1})	检出率	最大值/(个·g^{-1})	平均值/(个·g^{-1})	检出率
66°～70°N	R06	10	1.00×10^6	1.06×10^5	50.00%	2.00×10^6	2.96×10^5	60.00%
70°～74°N	C19、S11、S16、S26	72**	1.00×10^8	2.41×10^6	80.56%	1.60×10^8	5.89×10^6	90.14%
74°～78°N	M1、P11、P24	60	3.30×10^8	1.04×10^7	91.67%	3.30×10^8	1.07×10^7	91.67%
＞78°N	B77、B80	38	9.00×10^7	5.58×10^6	57.89%	9.20×10^7	9.31×10^6	76.32%

*最小值均为(一)；**25℃时样品数为 71 个

将研究区按经度划为 5 个区间,各经度区间锰细菌的分布见表 4。在 2 种培养温度时锰细菌的检出率和含量基本上随经度呈不规则变化,东西差异不明显。

表 4　楚科奇海和加拿大海盆柱状沉积物中锰细菌含量指标*

Table 4　Comparison of MnB content along the longitudinal gradient in the Chukchi Sea and the Canada Basin

经度(W)	站号	样品数	4℃			25℃		
			最大值/(个·g^{-1})	平均值/(个·g^{-1})	检出率	最大值/(个·g^{-1})	平均值/(个·g^{-1})	检出率
145°～150°	B77、B80	38	9.00×10^7	5.58×10^6	57.89%	9.20×10^7	9.31×10^6	76.32%
150°～155°	S26	18	1.50×10^7	1.57×10^6	88.89%	4.00×10^7	4.89×10^6	100.00%
155°～160°	P24、S11、S16	57**	1.00×10^8	2.45×10^6	87.72%	1.30×10^8	3.29×10^6	89.29%
160°～165°	C19	16	3.00×10^7	2.12×10^6	62.50%	1.60×10^8	1.09×10^7	75.00%
165°～170°	M1、P11、R06	51	3.30×10^8	1.17×10^7	82.35%	3.30×10^8	1.21×10^7	86.27%

注:*除 25℃时 150～155W 最小值为 260 个·g^{-1}外,其余均为(一);**25℃时样品数为 56 个。

4.3　水深对沉积物锰细菌检出率及含量的影响

按取样站位的不同水深范围比较锰细菌分布(表 5),2 种培养温度时检出率均显示随着水下深度的增加而呈低高更高趋低的变化,而含量则随水深加大呈跳跃式变化。较深水下环境,沉积物中锰细菌能具有较高的检出率和数量。这可能是由于较深的水下环境使沉积物中锰细菌有了一定的适应性,但离陆距离的增加又会在物质供应上对其生存产生制约,相关研究尚待深入。

表 5 北极加拿大海盆与楚科奇海沉积物锰细菌在不同水深区间的比较*

Table 5 Comparison of MnB content among different water depth in the Chukchi Sea and the Canada Basin

水深/m	站号	样品数	4℃			25℃		
			最大值/(个·g^{-1})	平均值/(个·g^{-1})	检出率	最大值/(个·g^{-1})	平均值/(个·g^{-1})	检出率
41～55	C19、R06、S11	45**	3.00×10^7	8.09×10^5	64.44%	1.60×10^8	4.07×10^6	75.00%
175～561	P11	21	3.30×10^8	2.70×10^7	90.48%	3.30×10^8	2.70×10^7	90.48%
1456～2200	M1、P24	39	1.61×10^7	1.44×10^6	92.31%	1.61×10^7	1.93×10^6	92.31%
3000～3850	S16、S26、B77、B80	75	1.00×10^8	4.67×10^6	74.67%	1.30×10^8	7.94×10^6	86.67%

注：* 最小值均为(—)；* * 25℃时样品数为 44 个。

4.4 环境温度对锰细菌检出率及含量的影响

初步的温度试验结果显示，锰细菌对温度有一定的适应能力，不管其出现机率或者是含量均如此。本研究及其他研究结果显示，在锰细菌的生长温度范围内，温度的提高大多可提升锰细菌的检出率和含量，这意味着环境温度是影响锰细菌在北冰洋沉积物中分布的重要因子。温度的持续提高可能激发原本受压抑的部分中温锰细菌的复苏且增强其活力，也可能提高部分嗜冷锰细菌对增温的逐步适应，从而提高或加速锰的氧化迁移速率；但是，同时也会加速对仅适于低温的锰细菌的死亡。

本研究启示我们，温度对北冰洋锰细菌的影响是正反两方面的，暗示气候的全球变暖无疑已对土著的锰细菌产生了一定的影响，尽管它们有较强的适应能力，但温度变化产生的生态影响是长期的、严重的，而且这种后果对环境及人类社会的影响也是难以预料的，十分令人担忧。

5 结论(Conclusion)

1)在北冰洋海底沉积物中不仅有锰细菌的存在，而且还具有较高的丰度，表明它们能生存于寒冷深水之下的沉积物中，进一步证实了锰细菌较宽泛的适应性。

2)在研究区所测定的沉积物样品中，在 2 种培养温度时锰细菌检出率为 81.91%，含量范围为(—)～3.30×10^8 个·g^{-1}，平均含量在 6.76×10^6 个·g^{-1}左右。

3)在锰细菌的生长温度范围内，温度的提高大多可提升锰细菌的检出率和含量，这意味着环境温度是影响锰细菌在北冰洋沉积物中分布的重要因子。

4)在沉积物中，锰细菌检出率有随深度增加而提高的趋势，这可能与早期成岩过程中沉积物氧化还原体系的垂直分带有关。

5)在一定水深范围下，随着水深的增大，锰细菌的检出率和含量均有相应的增加，表现出一定的适应力；但若上覆水体太厚，又会影响到营养物质的供应及细菌耐高压特性。

6)锰细菌在纬度上的变化表明，其分布范围较大，而且有扩大的趋势。

参考文献 22 篇(略)

(合作者：高爱国)

DISTRIBUTION OF MANGANESE BACTERIA IN THE SEDIMENT CORES FROM THE CANADA BASIN AND THE CHUKCHI SEA (ABSTRACT)

Abstract The distribution of manganese bacteria (MnB) in 10 sediment cores from the Canada Basin and the Chukchi Sea were analyzed by the methods of MPN and plate culture. The cores were sliced in 1 cm intervals from 0～10 cm and at 2 cm intervals below 10 cm on board as subsamples. Retrievability of MnB cultivated in both 4℃ and 25℃ ranged from non-detectable levels (ND) to 3.3×10^8 CFU · g^{-1} of wet sample. The occurrence percentages and average numbers of MnB cultivated at 4℃ and 25℃ were 77.78%, 5.60×10^6 CFU · g^{-1} wet samples, and 86.03%, 7.91×10^6 CFU · g wet samples, respectively. The contents of MnB in sediment cores were higher than those in sediment from the deep Pacific Ocean or in water from Alaska Lake, similar to the results reported in water from the Carlsberg Ridge in the Indian Ocean. The result showed a tendency of the MnB content increasing from low latitude to high latitude, or from the shallower continental shelf to deeper basin in the south of 78°N. From the surface to deeper sediment, the content of MnB in sediment changed irregularly, depending on the sedimentary environment. It seems that the variability of MnB content was larger at 25℃ than at 4℃.

Key words Manganese Bacteria; The Arctic Sea; The Canada Basin; The Chukchi Sea; Sediment Core

北极海洋沉积物中锰细菌的分离与系统发育*

摘　要　对中国第二次北极科学考察采集的北极海洋沉积物中的锰细菌进行了筛选、分离和系统发育分析。根据其在筛选平板上菌落的形态学特征，分别从站位 P11 和 S11 采集的沉积物中分离到了 21 株和 19 株锰细菌。系统发育分析表明，两个站位的锰细菌群落组成有着明显的差别。站位 P11 分离的可培养锰细菌主要由细菌域(Bacteria)中变形杆菌门的-变形杆菌纲(γ-Proteobacteria)和放线菌纲(Aetinobacteria)组成，二者分别占 86%和 14%；γ-变形杆菌纲主要包括嗜冷杆菌属(*Psychrobacter*)、希瓦氏菌属(*Shewanella*)、假交替单胞菌属(*Pseudoalteromonas*)、不动杆菌属(*Acinetobacter*)、海杆菌属(*Marinobacter*)，其中以嗜冷杆菌属为主，其比例可达 67%。从站位 S11 分离到的可培养锰细菌主要包括细菌域中变形杆菌门的 γ-变形杆菌纲(α-Proteobacteria)和 γ-变形杆菌纲以及拟杆菌门(Bacteroides)中的黄杆菌纲(Flavobacteria)；γ-变形杆菌纲主要包括希瓦氏菌属、海单胞菌属(*Marinomonas*)和交替单胞菌属(*Ateromonas*)，γ-变形杆菌纲主要为鞘氨醇单胞菌属(*Sphingomonas*)。实验菌株均对 Mn^{2+} 有着较强的抗性，其中以菌株 *Marinomonas* sp. S11-S-4 耐受性最高。

关键词　北极　海洋沉积物　锰细菌　微生物区系　系统发育分析

海洋由于其特殊的环境特征，存在着一个适应高压、高盐和低温等的微生物生态系统，同时，随着深海"热液"和"冷泉"生物群的发现，越来越多的研究表明，在超过地球表面 50%的深海区域中存在着强烈而独特的生物地球化学过程，而海洋微生物是这些过程中的关键环节。

在海洋沉积物中生存着大量的微生物，可促进海洋中许多化学变化与成岩作用。陆源可溶态和悬浮态的锰进入海洋后，被海洋生物结合入体内，当这些有机体的残骸下沉时也将这些元素带到海底，经过生理生化功能各异的细菌的矿化作用，会引起环境中 Eh、pH、Mn^{2+}/Mn^{4+} 和 Fe^{2+}/Fe^{3+} 等生物地球化学参数变化。尽管 Mn^{2+} 在海水中含量较少，但是其氧化还原过程在海洋的地球化学循环中可能起着关键的作用。目前已对海洋沉积物中细菌在铁锰结核形成过程中所起的作用做了一些研究工作，但大多是在实验室内研究某一种细菌对金属离子的转化作用[1,2]。在绝大部分环境中，Mn^{2+} 的氧化被认为是细菌介导(bacterial mediation)的。锰氧化细菌不仅种类数量繁多，而且分布广泛，已从各种不同的环境，如海水、淡水、土壤沉积物、热液口等分离到了锰氧化细菌。从系统发育角度看，锰氧化细菌的种类看起来非常广泛，其主要归类为低 G+C 含量 G^{+} 菌、放线菌、变形杆菌门(Proteobacteria)的 α、β 和 γ 变形杆菌纲。目前研究的最多的锰氧化细菌是枯草杆菌(*Bacillus* sp.)SG-1、生盘纤发菌(*Leptothrix discophora*)SS-1、恶臭假单胞菌(*Pseudomonas putida*)MnB1 和 GB-1[3-5]。国内对深海铁锰结核区微生物的丰度、成矿作用和系统发育分析也已进行了初步研究[1,6,7]，但对极地海洋沉积物中锰细菌的多样性和系统发育

* 基金项目：国家自然科学基金资助项目(40576060)，本人为第三作者。

原文刊于《生态学报》，2008，28(12)：6364-6370。

分析尚未见报道，国外也较少见。

本研究将通过对中国第二次北极科考采自 P11 和 S11 两个站位沉积物的锰细菌的菌群组成和系统发育的比较分析，了解锰细菌的多样性，为研究微生物在极地海洋沉积物元素生物化学循环中的作用打下基础；同时还可获得锰去除率高、适应性强、生命旺盛的细菌，以服务于污染环境的生物修复。

1　材料与方法

1.1　样品的采集

北极海洋沉积物样品在中国第二次北极科学考察中采得，站位：P11：169°59′37″W，75°00′24″N；水深 263 m；采样时间，2003.8.10；青灰色软泥。S11：159°00′00″W，72°29′24″N；水深，50 m；采样时间，2003-8-17；灰色硬泥。各站分别按上下两层取样：表层（S）0～10cm，按 1cm 间隔取样；下层（B）11～底层，按 2cm 间隔取样，于 4℃无菌保存。

1.2　锰细菌的筛选与分离

培养基：$MnSO_4 \cdot 4H_2O$ 0.2g，$FeSO_4 \cdot 7H_2O$ 0.001g，蛋白胨 2g，酵母膏 0.5g，琼脂 15g，过滤陈海水 1 000 mL。

取适量沉积物经无菌海水系列稀释后，将其涂布于筛选培养基平板上，于 4℃倒置培养 30d，期间不断观察，分上下两层按其菌落形态学特征挑取褐色单菌落，并进一步纯化后获得纯培养。

1.3　锰细菌的分子鉴定与系统发育分析

可培养锰细菌的分子鉴定与系统发育分析中 DNA 模板的制备、16*S* rDNA 的 PCR 扩增与序列测定以及系统进化关系的分析均采用文献[8]方法。

1.4　锰抗性能力的测定

将各待测菌株置含不同 Mn^{2+} 浓度的 Zobell 2216E 培养基中，接种量为 1%，于 5℃，150r·min^{-1} 培养 3d 后用分光光度计在 550nm 处测定其光密度值。

2　结果与分析

2.1　锰细菌的分离与筛选

根据锰细菌筛选平板上菌落出现的时间、大小、形态和颜色等，分别从站位 P11 和 S11 的表层和下层沉积物样品中分离得到 8、13 株和 11、8 株细菌，分别编号为 P11-S-1～P11-S-8、P11-B-1～P11-B-13 和 S11-S-1～S11-S-11 和 S11-B-1～S11-B-8。

2.2　锰细菌的分子鉴定

将锰细菌的 16*S* rDNA 的阳性克隆子进行测序，获得长度约为 1.5 kb 的 16*S* rDNA 序列，并将其提交 GenBank 数据库，获得序列注册号分别为 EU016143～EU016182（表 1）。将这些序列与 GenBank 数据库进行比对分析，与所分离的锰细菌同源性最高的菌株如表 1 所示。

从表 1 可见，本研究分离的锰细菌大多与庞大的 GenBank 数据库中已有菌株的 16S rDNA 序列均存在着较高的相似性，大多相似性在 99%以上，其中 P11-B-2、P11-B-8、P11-B-11、P11-B-12 分别与 *Psychrobacter* sp. BJ-3（DQ298407）、*Rhodococus* sp. SGB1168-118（AB010908）、*Shewanella* sp. BSw20461（EF639386）、*Pseudoalteromonas* sp. BSw20513（EF551376）的相似性达到 100%。S11-S-2 与 Genbank 数据库中已有的 16*S* rDNA 序列相似性最低，只有 96.88%。其次是 P11-B-6，与已有的 16*S* rDNA 序列相似性为98.40%，

其余的多介于99.26%～100%之间。

2.3　系统发育分析

将从站位P11和S11海洋沉积物中分离到的锰细菌的16*S* rDNA序列分别用BioEdit软件进行序列比对，采用Mega3.0[9]软件的邻接法（neighbor-joining method）构建系统发育树，如图1和图2所示。

从站位P11分离到的锰细菌主要包括细菌域（Bacteria）变形杆菌门（Proteobacteria）的γ-变形杆菌纲（γ-Proteobaeteria）和放线菌纲（Aetinobaeteria），二者分别占86%和14%。γ-变形杆菌纲包括嗜冷杆菌属（*Psychrobacter*）、不动杆菌属（*Acinetobacter*）、海杆菌属（*Marinobacter*）、假交替单胞菌属（*Pseudoalteromonas*）和希瓦氏菌属（*Shewanella*），其中嗜冷杆菌属细菌最多，其比例可达整个γ-变形杆菌纲的67%。放线菌纲包括短杆菌属（*Brevibacterium*）和红球菌属（*Rhodococcus*）。

从站位S11分离到的锰细菌主要包括细菌域变形杆菌门的γ-变形杆菌纲和α-变形杆菌纲（-Proteobacteria）以及拟杆菌门（Bacteroides）的黄杆菌纲（Flavobacteria），其中γ-变形杆菌纲和α-变形杆菌纲分别占63%和31%。γ-变形杆菌纲包括希瓦氏菌属（*Shewanella*）、海杆菌属（*Marinobacter*）、海单胞菌属（*Marinomonas*）、交替单胞菌属（*Alteromonas*）、科尔韦尔氏菌属（*Colwellia*）和食烷菌属（*Alcanivorax*）；α-变形杆菌纲包括亚硫酸盐杆菌（*Sulfitobacter*）和鞘氨醇单胞菌属（*sphingomonas*），其中交替单胞菌属、海单胞菌属和希瓦氏菌属菌株较多。

2.4　锰耐受性分析

将各待测菌株接种入含不同Mn^{2+}浓度的Zobell 2216E培养基中，5℃、150r·min^{-1}培养3d后，测得的培养液OD值如表3所示。可以看出，实验菌株均对Mn^{2+}有着较强的抗性，其中以*Marinomonas* sp. S11-S-4的耐受性最强。Mn^{2+}浓度为10 mmol·L^{-1}时，与对照（Mn^{2+}浓度为0）相比对菌株的生长影响很少。当Mn^{2+}浓度为50 mmol·L^{-1}时，可较显著抑制菌株的生长，其中对P11-B-4的影响最大；当Mn^{2+}浓度为100 mmol·L^{-1}时，除S11-S-4和P11-S-1的生长明显，其余只显微弱生长。

表1　北极海洋沉积物锰细菌16*S* rDNA序列同源性比较结果

Table 1　Homology comparison of 16*S* rDNA sequences of the manganese bacteria isolated from sediment in the Arctic ocean

菌株 Strain	注册号 Accession number	相似性最高菌株（%） Closet match（% similarity）	注册号 Accession number	门 Phylum
S11-S-1	EU016172	*Sphingomonas* sp. GCl4	AY690679	α-Proteobacteria
S11-S-2	EU016173	*Marinobacter* sp. BSa20186（96.88）	DQ514302	γ-Pmteobactcria
S11-S-3	EU016174	*Marinomonas* sp. BSi 20591（99.80）	En82678	γ-Pmteobacteria
S11-S-4	EU016175	*Marinomonas* sp. BSi 20591（99.86）	EF382678	γ-Pmteobactena
S11-S-5	EU016176	*Marinomonas* sp. BSi 20591（99.20）	VEF382678	γ-Pmteobacteria

（续表）

菌株 Strain	注册号 Accession number	相似性最高菌株(%) Closet match(% similarity)	注册号 Accession number	门 Phylum
S11-S-6	EU016177	uncultured bacterium clone WLBl3-118 (99.52)	DQ015843	α-Proteobaeteria
S11-S-7	EU016178	Arctic seawater bacterium Bsw20352 (99.93)	DQ064636	γ-Pmteobacteria
S11-S-8	EU016179	uncultured bacterium clone WLBl3-118 (99.45)	DQ015843	α-Proteobacteria
S11-S-9	EU016180	uncultured bacterium clone WLBl3-118 (99.45)	DQ015843	α-Proteobaeteria
S11-S-10	EU016181	*Shewanella* sp. BSw20461(99.73)	EF639386	γ-Proteobacteria
S11-S-11	EU016182	uncultured bacterium clone W1213-118 (99.45)	DQ015843	α-Proteobacteria
S11-B-1	EIJ016164	*Alcanivorax* sp. BSs20190(99.93)	DQ514306	γ-Proteobacteria
S11-B-2	EU016165	*Shewanella* sp. BSw20461(99.88)	EF639386	γ-Proteobaeteria
11-B-3	EUOl6166	*Flavobacterium* sp. BSs20191(99.26)	DQ514307	Flavobacteria
S11-B-4	EU016167	*Sulfitobacter* sp. ARCTIC—P49(99.86)	AY573043	α-Prateobactcria
S11-B-5	EIJ016168	*Colwellia* sp. P28(99.80)	EF528008	γ-Proteobacteria
S11-B-6	EU016169	Alteromonas sp. KC98716-13(99.26)	AB072388	γ-Proteobacteria
S11-B-7	EU016170	Alteromonas sp. KC98716-13(99.26)	AB072388	γ-Proteobacteria
S11-B-8	EU016171	Aleromonas sp. KC98716-13(99.53)	AB072388	γ-Proteobaotcria
P11-S-1	EU016156	*Psychrobacter* sp. EP06(99.60)	AM398215	γ-Proteobacteria
P11-S-2	EU016157	*Shewanella* sp. BSw20461(99.53)	EM39386	γ-Proteobacteria
P11-S-3	EU016158	*Psychrobacter* sp. BBTRl010(99.80)	EFl471232	γ-Proteobaeteria
P11-S-4	EU016159	*Psychrobacter* sp. BJ-3(99.3)	DQ298407	γ-Proteobacteria
P11-S-5	EU016160	*Psychrobacter* sp. BBTRl010(99.60)	EF471232	γ-Pmteobacteria
P11-S-6	EU016161	*Rhodococus* sp. SGB1168-118(99.93)	AB010908	Actinobaeteria
P11-S-7	EU016162	*Psychrobacter* sp. EP06(99.66)	AM398215	γ-Proteobacteria
P11-S-8	EU016163	Arctic seawater bacterium BSW20352 (99.80)	DQ064612	γ-Proteobacterla
P11-B-1	EU016143	*Psychrobacter* sp. BJ-3(99.80)	DQ298407	γ-Proteobacteria

(续表)

菌株 Strain	注册号 Accession number	相似性最高菌株(%) Closet match(% similarity)	注册号 Accession number	门 Phylum
P11-B-2	EU016144	*Psychrobacter* sp. BJ-3(100)	DQ298407	γ-Proteobaeteria
P11-B-3	EU016145	*Psychrobacter* sp. BBTRl010(99.73)	EF471232	γ-Proteobacteria
PU-B-4	EU016146	uncultured *Acinetobacter* sp. Clone MRT87(99.87)	EF371492	γ-Proteobacteria
P11-B-5	EU016147	*Brevibacterium* casei strain 3Tg(99.33)	AY468375	Actinobacteria
P11-B-6	EU016148	*Psychrobctaer* sp. BBTRl010(98.40)	EF471232	γ-Proteobacteria
P11-B-7	EU016149	*Psychrobacter* sp. BBTRl010(99.67)	EF471232	γ-Proteobaeteria
P11-B-8	EU016150	*Rhodococus* sp. SGB1168-118(100)	AB010908	Actinobaetena
P11-B-9	EU016151	*Psychrobacter* sp. BBTRl010(99.73)	EF471232	γ-Proteobaeteria
P11-B-10	EU016152	*Marinobacter* sp. BSa20186(99.93)	DQ514302	γ-Proteobaeteria
P11-B-11	EU016153	*Shewanella* sp. BSw20461(too)	EF639386	γ-Proteobacteria
P11-B-12	EU016154	*Pseudoalteromonas* sp. BSw20513(100)	EF551376	γ-Proteobacteria
P11-B-13	EU016155	*Shewanella* sp. BSw20461(99.73)	EF639386	γ-Proteobaeteria

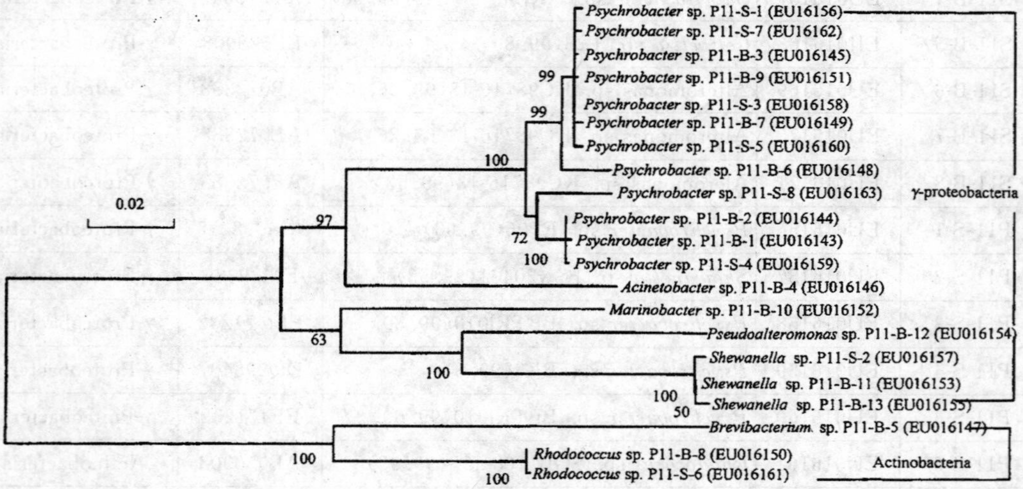

图 1 站位 P11 可培养锰细菌的系统发育分析

Fig 1 Phylogenetic analysis of cultivable manganese bacteria isolated from station P11

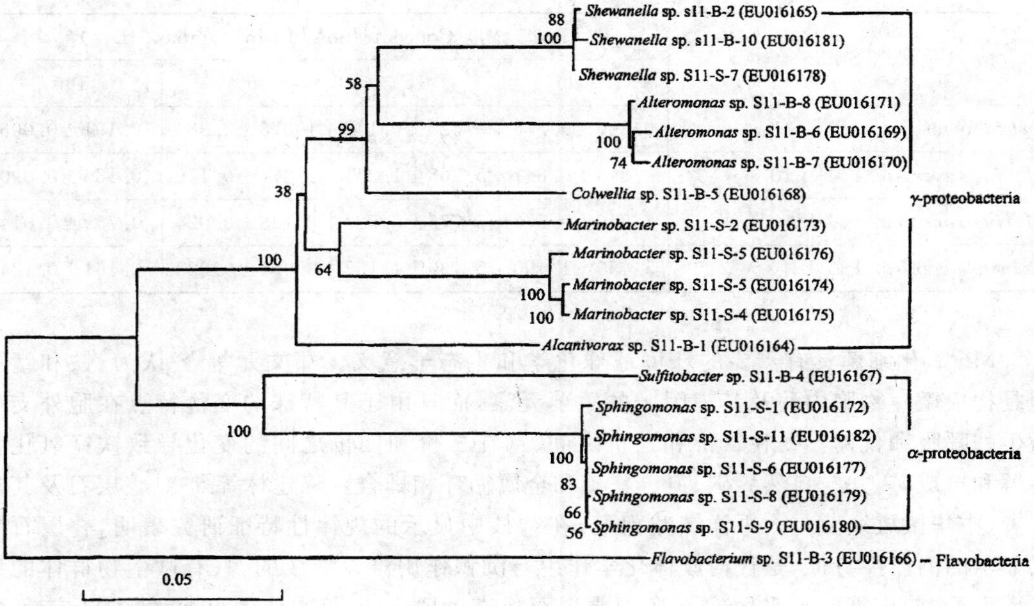

图 2　站位 S11 可培养锰细菌的系统发育分析

Fig 2　Phylogenetic analysis of cultivable manganese bacteria isolated from station S11

3　讨论

地(质)微生物学是在 20 世纪末发展起来的新的地学研究分支,主要研究地质环境中的微生物活动过程及其形成的各种地质地球化学记录。微生物与矿物的相互作用、极端环境中的微生物和生态以及分子地质微生物学是当前地质微生物学研究的重要方向[10]。Mn 的生物地球化学循环被认为是重要的环境过程之一,这不仅是因为 Mn 是所有生物的必需元素之一,而且 Mn^{2+} 的氧化与还原同其他必需元素的循环密切相关[11,12]。

表 3　12 株测试锰细菌的 Mn^{2+} 耐受性

Table 3　Tolerant ability on Mn^{2+} of 12 species of manganese bacteria tested

菌株 Strain	Mn^{2+} 浓度 Concentration of Mn^{2+} (mmol · L^{-1})			
	0	10	50	100
Alteromonas sp. S11-B-8	5.658±0.965	5.493±0.965	1.116±0.724	0.081±0.017
Cowellia sp. S11-B-5	4.890±1.520	6.061±0.069	1.317±0.581	0.174±0.018
Shewanella sp. P11-B-13	6.425±0.629	4.943±1.601	1.412±0.260	0.097±0.016
Shewanella sp. S11-B-2	5.750±0.181	5.770±0.297	1.731±0.131	0.242±0.072
Sulfitobacter sp. S11-B-4	5.465±0.491	4.105±0.308	1.492±0.292	0.244±0.068
Pseudoalteromonas sp. P11-B-12	4.580±0.481	5.593±0.441	1.473±0.022	0.121±0.024
Rhodococcus sp. P11-S-6	6.250±0.849	5.925±0.728	1.173±0.968	0.159±0.097
Psychrobacter sp. P11-S-1	4.868±0.067	5.703±1.128	1.381±0.343	0.330±0.025

（续表）

菌株 Strain	Mn^{2+}浓度 Concentration of Mn^{2+} ($mmol \cdot L^{-1}$)			
	0	10	50	100
Sphingomonas sp. S11-S-1	6.260±0.276	5.520±0.148	1.078±0.981	0.191±0.006
Marinobacter sp. P11-B-10	5.893±0.647	5.368±1.241	1.317±0.116	0.155±0.090
Marinomonas sp. S11-S-4	5.840±0.573	5.775±0.523	2.188±0.024	0.759±0.014
Acinetobacter sp. P11-B-4	5.450±1.400	5.190±1.167	0.904±0.023	0.161±0.124

Mn 氧化细菌具有重要的环境地球化学和生态学意义。郑文元等[12]认为，铁和锰不但是构成微生物胞内酶及其辅因子的重要元素，而且由于其特殊的变价特点在胞外支配微生物呼吸和代谢。铁在二价和三价之间、锰在二价和四价之间的变化导致铁锰氧化物溶解和再成矿，并与自然界碳、氮、磷、硫和金属循环相耦合。陈建林等[13]对叠层石及其建造者（中华微放线菌与太平洋螺球孢菌）在结核中显示的规律性特征研究表明，叠层石纹层显示的韵律性特征，是任何胶体化学作用与沉积作用难以形成的，只有微生物群体的世代繁衍，才能构筑成几乎固定不变的叠层石形态。Francis 等[3]认为可溶性的 Mn^{2+} 氧化成不溶性 Mn^{3+}/Mn^{4+} 氧化物是一个非常重要的环境过程，由于其结构和氧化还原特征，生物源的 Mn 氧化物能氧化许多有机和无机化合物，如腐殖质、Cr^{2+}、Fe^{2+} 等，对许多金属离子如 Cu、Co、Cd、Zn、Ni、Pb 等具有较强的吸附能力，而且可以作为厌氧呼吸的电子载体[3,14]。陈祥军等[15]的研究则认为，细菌对锰具有明显的氧化作用，结合矿床的其他方面特征，认为自然界中微生物在锰的氧化富集过程中起着积极的作用，作用表现在两个方面：一方面微生物可以直接对锰进行氧化，使原生碳酸锰矿转变为氧化锰矿；另一方面微生物又可以改变环境的 pH 值，促进锰的化学氧化。

已有的研究表明，各种各样的细菌、真菌和藻类均能催化 Mn^{2+} 的氧化。本研究进一步表明，锰细菌不仅是广泛存在的，而且不同地点锰细菌群落的组成也并不相同。本研究的两个站点的锰细菌中 G^- 菌比例可占 90%以上，以变形杆菌门的 γ-变形杆菌纲最多。二者的锰细菌微生物区系也存在着较大的差异，如站位 P11 分离到的锰细菌主要包括 γ-变形杆菌纲和放线菌纲，二者分别占 86%和 14%；站位 S11 分离到的可培养锰细菌属于 γ-变形杆菌纲和 α-变形杆菌纲以及黄杆菌纲，其中 γ-变形杆菌纲和 α-变形杆菌纲分别占 63%和 31%。两个站位的 γ-变形杆菌纲的组成也不一样，站位 P11 的 γ-变形杆菌纲包括嗜冷杆菌属、不动杆菌属、海杆菌属、假交替单胞菌属和希瓦氏菌属，其中嗜冷杆菌属细菌最多，其比例可达整个 γ-变形杆菌纲的 67%。而站位 S11 的 γ-变形杆菌纲包括希瓦氏菌属、海杆菌属、海单胞菌属、交替单胞菌属、科尔韦尔氏菌属和食烷菌属。而田美娟等[7]对 2003 年“大洋一号”采集的太平洋 4 个站位的沉积物样品的 50 株锰细菌的分子鉴定和系统发育分析表明，50 株菌中 G^+ 菌占 90%；芽孢杆菌（*Bacillus*）最多（40 株），可占 80%，其余为少量的短杆菌（3 株）、副球菌（4 株）等，而且芽孢杆菌在 4 个站点细菌中分别占 70%、93.34%、68.42%、100%，都占了最大比例。史君贤等[6]对东太平洋铁锰结核区微生物丰度的研究表明，锰细菌可达 3929 $CFU \cdot g^{-1}$ 沉积物。Johnston 等[5]对北极贫营养湖泊

(Toolik Lake)锰细菌介导的锰氧化速率的研究表明,与其他水体相比,尽管其锰氧化速率较低,但仍能在3个月内全部氧化湖中的Mn^{2+},湖水中锰细菌浓度最高可达9×10^{4} CFU·mL^{-1}。Mn^{2+}的氧化大多是微生物介导的,因为叠氮化钠可抑制其氧化,而叠氮化钠已经被证明是研究海水中Mn^{2+}氧化合适的微生物杀灭剂。以上研究结果均说明锰细菌存在的广泛性和多样性,这些可能都受制于锰细菌生存的环境条件。

本研究表明,锰细菌筛选培养基分离的锰细菌均对Mn^{2+}有着较强的抗性,这与已有的研究相一致。田美娟等[7]的研究表明,锰细菌对Mn^{2+}有着较强抗性,而锰抗性菌株对锰具有高去除率的能力,因此可将之应用于环境污染治理领域,如城市饮用水净化、污水处理等。从微生物学角度讲,铁锰氧化细菌的应用潜力是巨大的[14,16]。

参考文献16篇(略)

(合作者:林学政　高爱国)

ISOLATION AND PHYLOGENETIC ANALYSIS OF CULTIVABLE MANGANESE BACTERIA IN SEDIMENTS FROM THE ARCTIC OCEAN

(ABSTRACT)

Abstract The biogeochemical cycling of manganese is recognized as an important environmental process because manganese is not only an essential nutrient for all organisms but also its oxidation and reduction are intimately coupled with the cycling of other essential elements. Studies have demonstrated that Mn^{2+} oxidizing bacteria are abundant and distributed widely. A diverse array of bacteria, fungi, and microalgae hase been shown to have the ability to catalyze oxidation or reduction of manganese. The oxidation of soluble Mn^{2+} to insoluble Mn^{3+}/Mn^{4+} oxides and oxy-hydroxides is an environmentally important process because the solid-phase products oxidize a variety of organic and inorganic compounds, scavenge many metals, and serve as electron acceptors for anaerobic respiration. In most environments, Mn^{2+} oxidation is believed to be bacterially mediated. Over the years, Mn^{2+}-oxidizing bacteria have been isolated from wide variety of environments, including marine and freshwaters, soils, sediments, water pipes, Mn nodules, and hydrothermal vents. Phylogenetically, Mn^{2+}-oxidizing bacteria appear to be quite diverse, with all isolates analyzed to date falling within either the low G+C gram-positive bacteria, the Actinobacteria, or the α, β and γ subgroups of the Proteobacteria branch of the domain Bacteria.

In order to investigate the biodiversity of manganese bacteria in polar region, isolation, molecular identification and phylogenetic analysis of manganese bacteria were carried out in the sediments which were collected from Arctic ocean during 2nd Chinese Arctic Scientific Expedition. Twenty one and nineteen species of cultivable strain were isolated from the sediments of station P11 and S11 respectively according to their distinct morphology on screening plate of manganese medium. Molecular identification and phylogenetic analysis showed that the cultivable manganese bacteria from station P11 were basically composed of 1 subgroup of the Proteobacteria branch of the domain Bacteria (γ-Proteobacteria) and Actinobacteria, which accounted for 86% and 14% respectively. The γ-Proteobacteria mainly included *Psychrobacter*, *Shewanella*, *Acinetobacter* and *Marinobacter*, of which *Psychrobacter* was the major genus, it accounted for 67% of γ-Proteobaeteria in sediments of station P11. The cultivable manganese bacteria from station S11 included α-proteobaeteria, γ-Proteobacteria and Flavobacteria of Bacteroides. γ-Proteobacteria included *Shewanella*, *Marinomonas* and *Alteromonas*. The majority of α-Proteobacteria was Sphingomonas. The phylogenetic analysis indicated that bacteria from the sediments of station P11 and S11 had different cultivable manganese bacteria flora. All tested strains had higher resistance to Mn^{2+}, of which *Marinomonas* sp. S11-Sboth −4 had highest resistance.

Key words Arctic; Ocean Sediments; Manganese Bacteria; Microflora; Phylogenetic Analysis

ISOLATION AND PHYLOGENETIC ANALYSIS OF CULTIVABLE MANGANESE BACTERIA IN SEDIMENTS FROM THE ARCTIC OCEAN*

Abstract Isolation, molecular identification and phylogenetic analysis were carried out to investigate the biodiversity of manganese bacteria in sediments which were collected from the Arctic Ocean during the 2nd Chinese Arctic Scientific Expedition. 21 and 19 species of cultivable strains were isolated from sediments at Stations P11 and S11, respectively, according to their distinct morphological character on the screening plate of manganese medium. Molecular identification and phylogenetic analysis showed that the cultivable manganese bacteria from Station P11 were basically composed of γ-Proteobacteria (γ subgroup of the Proteobacteria branch of the domain Bacteria) and Actinobacteria, which accounted for 86% and 14%, respectively. The isolates of γ-Proteobacteria mainly included *Psychrobacter*, *Shewanella*, *Acinetobacter* and *Marinobacter*, of which *Psychrobacter* was the major genus, which accounted for 67% of the γ-Proteobacteria. The cultivable manganese bacteria from Station S11 included α-Proteobacteria, γ-Proteobacteria and Flavobacteria of Bacteroides. The γ-Proteobacteria mainly included *Shewanella*, *Marinomonas* and *Alteromonas*. The majority of α-Proteobacteria was *Sphingomonas*. The phylogenetic analysis indicated that bacteria from sediments at Stations P11 and S11 had different cultivable manganese microbial communities. All tested strains had higher resistance to Mn^{2+}, of which *Marinomonas* sp. S11-S-4 had the highest resistant ability.

Key words Arctic; Ocean Sediments; Manganese Bacteria; Microbial Community; Phylogenetic Analysis

Specific microbial ecosystems exist in ocean and adapt their peculiar environmental stress such as high pressure, high salinity and low temperature. With the discovery of biological communities in deep-sea hydrothermal vents, more and more evidences show

* The original text was published in *Acta Ecologica Sinica*, 2008, 28(12):6364 - 6370, I was the third writer.

原中文稿刊于《生态学报》,2008,28(12):6364-6370.

that there are distinct biogeochemical processes, in which microorganisms play a pivotal role.

Although manganese is a minor chemical constituent of marine water and freshwater, its oxides may play an important role in aquatic geochemical cycles because their remarkable surface area and charge distribution make it a potentially rich reservoir of adsorbed metals[1,2].

The biogeochemical cycling of manganese is recognized as an important environmental process because manganese is not only an essential nutrient for all organisms, but its oxidation and reduction are also intimately coupled with the cycling of other essential elements. Studies have demonstrated that Mn^{2+}- oxidizing bacteria are both abundant and distributed widely[3]. A diverse array of bacteria, fungi and microalgae has shown an ability to catalyze oxidation and reduction of manganese. The oxidation of soluble Mn^{2+} to insoluble $Mn^{3+/4+}$ oxides and oxyhydroxides is an environmentally important process because the solid-phase products oxidize a variety of organic and inorganic compounds, scavenge many metals, and serve as electron acceptors for anaerobic respiration[4]. In most environments, Mn^{2+} oxidation is believed to be bacterially mediated. Over the years, Mn^{2+}-oxidizing bacteria have been isolated from a wide variety of environments, including marine and freshwater, soil, sediments, water pipes, Mn nodules and hydrothermal vents[4,5]. Phylogenetically, Mn^{2+}-oxidizing bacteria appear to be quite diverse, with all isolates analyzed to date falling within either the low G+C gram-positive bacteria (Actinobacteria), or the α, β and γ subgroups of the Proteobacteria branch of the domain Bacteria. Former studies focused on *Bacillus* sp. strain SG-1, *Leptothrix discophora* strain SS-1, *Pseudomonas putida* strain Mn B1 and GB-1[4]. Preliminary study has been performed on bacterial abundance, mineralization mechanism and phylogenetic analysis in ferromanganese nodule area[1,6,7]. There have been fewer reports on diversity and phylogenetic analysis of manganese bacteria in marine sediments in polar regions[2,5].

Arctic marine sediments were obtained to screen manganese bacteria in order to learn their diversity and phylogenetic status. This would provide fundamental information to evaluate the role of bacteria in element biochemical cycles in polar regions. Bacteria with higher manganese removal ability might also be obtained, which could be used in bioremediation of polluted environments at low temperature.

1 Materials and methods

1.1 Samples

The marine sediments were collected during the 2nd Chinese Arctic Scientific Expedition. The sampling sites and their character are described in Table 1. Sediments were shipped at 4℃ and studied upon their arrival at the lab. The sediment samples were divided into 2 groups of surface (0 - 10 cm) and bottom (11 cm to the bottom).

1.2 Screening of manganese bacteria

The medium used to screen manganese bacteria was as follows (per liter of filtered seawater): $MnSO_4 \cdot 4H_2O$, 0.2g; $FeSO_4 \cdot 7H_2O$, 0.001g; peptone, 2g; yeast extract, 0.5g; agar, 15g. Manganese bacteria were obtained according to their distinct morphological characters on the screening plate of manganese medium such as size, shape and color of bacterial colonies.

1.3 Molecular identification and phylogenetic analysis

DNA template, amplification of 16 S rDNA, sequencing and phylogenetic analysis of cultivable manganese bacteria were performed with methods described previously[8].

1.4 Determination of manganese-resisting ability

The strains tested were inoculated in medium Zobell 2216E with different Mn^{2+} concentrations. The OD culture values were measured with a colorimeter at wave length of 550 nm after incubation at 5℃ with continuous shaking at 150 $r \cdot min^{-1}$ for 3 d.

2 Results and analysis

2.1 Screening and isolation of manganese bacteria

According to the distinct morphology of bacterial colonies on the screening plate of manganese medium, 8, 13, 11 and 8 species of the cultivable strains were isolated from surface and bottom sediments at Stations P11 and S11, respectively. These strains were named as P11-S-1 to P11-S-8, P11-B-1 to P11- B-13, S11-S-1 - S11-S-11 and S11-B-1 - S11-B-8 (Table 2).

Table 1 Arctic marine sediments used to isolate manganese bacteria

Station	Date	Site	Depth (m)
P11	2003-08-10	169°59′37′W, 75°00′24″N	263
S11	2003-08-17	159°00′00″W, 72°29′24″N	50

2.2 Molecular identification of manganese bacteria

Amplified 16 S rDNAs of the 40 strains of cultivable manganese bacteria were sequenced. These sequences, with average 1.5 kb in length, were aligned and compared with the 16 S rDNA sequences from GenBank by BLAST search. The 16 S rDNA sequences of 40 strains were deposited to GenBank nucleotide sequence database with the accession numbers of EU016143 - EU016182 (Table 2). The strains with closest match with the cultivable manganese bacteria were shown in Table 2.

The similarity analysis showed that most of the isolates had higher similarity to the 16 S rDNA sequences deposited in enormous GenBank database, which had the similarity more than 99%. The strains P11-B-2, P11-B-8, P11-B-11 and P11-B-12 were 100% similar to *Psychrobacter* sp. BJ-3 (DQ298407), *Rhodococus* sp. SGB1168-118 (AB010908), *Shewanella* sp. BSw20461 (EF639386) and *Pseudoalteromonas* sp. BSw-20513 (EF551376), respectively. The strain S11-S-2 showed the lowest similarity (96.88%) to the 16S rDNA sequences in GenBank database. The next lowest similarity strain was P11-B-6, which was 98.40% similar to *Psychrobacter* sp. BBTR-1010 (EF471232). All the rest showed similarity between 99.26% and 100%.

2.3 Phylogenetic analysis

The 16 S rDNA sequences of cultivable manganese bacteria isolated from sediments at Stations P11 and S11 were compared and analyzed using software BioEdit, and then phylogenetic trees were constructed by the neighbor-joining method of software Mega3.0[9] (Figs. 1 and 2).

The phylogenetic analysis indicated that bacteria from the sediments at Stations P11 and S11 had different cultivable manganese bacterial communities. The cultivable manganese bacteria from Station P11 were basically composed with γ-Proteobacteria (γ subgroup of the Proteobacteria branch of the domain Bacteria) and Actinobacteria, which accounted for 86% and 14%, respectively. The γ-Proteobacteria mainly included *Psychrobacter*, *Shewanella*, *Acinetobacter*, *Pseudoalteromonas and Marinobacter*, of which *Psychrobacter* was the major genus, which accounted for 67% of the γ-Proteobacteria in sediments at Station P11. Actinobacteria included 2 genera, *Brevibacterium* and *Rhodococcus*. The cultivable manganese bacteria from Station S11 included γ-Proteobacteria, α-Proteobacteria and Flavobacteria of Bacteroides, of which γ-Proteobacteria and α-Proteobacteria accounted for 63% and 31%, respectively. γ-Proteobacteria included *Shewanella*, *Marinomonas*, *Alteromonas*, *Marinobacter*, *Colwellia* and *Alcanivorax*. The majority of α-Proteobacteria was Sphingomonas.

2.4 Analysis of manganese resistant ability

The cultivable manganese bacteria tested for manganese resistant ability were inoculated in the medium Zobell 2216E with different Mn^{2+} concentrations. After incubation at 5℃ with continuous shaking (150 r · min^{-1}) for 3 d, their growth rate was calculated with a colorimeter at wave length of 550nm (Table 3). The results showed that all the strains tested had higher resistant ability for Mn^{2+}, of which the strain S11-S-4 showed the highest resistant ability. There were little influences on the growth when Mn^{2+} concentration was 10 mmol · L^{-1} compared with the control (no Mn^{2+} had been added to the medium Zobell 2216E). The growth of the strains tested was inhibited markedly when Mn^{2+} concentration was 50 mmol · L^{-1}, and the growth of strain P11-B-4 was most inhibited. When Mn^{2+} concentration was up to 100 mmol · L^{-1}, all of the strains grew weakly except that strains S11-S-4 and P11- S-1 showed an obvious growth rate.

Table 2 Homology comparison of 16*S* rDNA sequences of the manganese bacteria isolated from sediments in the Arctic Ocean

Strain	Accession number	Closest match (% similarity)	Accession number	Phylum
S11-S-1	EU016172	*Sphingomonas* sp. GC14	AY690679	α-Proteobacteria
S11-S-2	EU016173	*Marinobacter* sp. BSs20186 (96.88)	DQ514302	γ-Proteobacteria

(continue)

Strain	Accession number	Closest match (% similarity)	Accession number	Phylum
S11-S-3	EU016174	*Marinomonas* sp. BSi 20591 (99.80)	EF382678	γ-Proteobacteria
S11-S-4	EU016175	*Marinomonas* sp. BSi 20591 (99.86)	EF382678	γ-Proteobacteria
S11-S-5	EU016176	*Marinomonas* sp. BSi 20591 (99.20)	EF382678	γ-Proteobacteria
S11-S-6	EU016177	Uncultured bacterium clone WLB13-118 (99.52)	DQ015843	α-Proteobacteria
S11-S-7	EU016178	Arctic seawater bacterium Bsw20352 (99.93)	DQ064636	γ-Proteobacteria
S11-S-8	EU016179	Uncultured bacterium clone WLB13-118 (99.45)	DQ015843	α-Proteobacteria
S11-S-9	EU016180	Uncultured bacterium clone WLB13-118 (99.45)	DQ015843	α-Proteobacteria
S11-S-10	EU016181	*Shewanella* sp. BSw20461 (99.73)	EF639386	γ-Proteobacteria
S11-S-11	EU016182	Uncultured bacterium clone WLB13-118 (99.45)	DQ015843	α-Proteobacteria
S11-B-1	EU016164	*Alcanivorax* sp. BSs20190 (99.93)	DQ514306	γ-Proteobacteria
S11-B-2	EU016165	*Shewanella* sp. BSw20461	EF639386	γ-Proteobacteria
S11-B-3	EU016166	*Flavobacterium* sp. BSs20191 (99.26)	DQ514307	Flavobacteria
S11-B-4	EU016167	*Sulfitobacter* sp. ARCTIC-P49 (99.86)	AY573043	α-Proteobacteria
S11-B-5	EU016168	*Colwellia* sp. P28 (99.80)	EF628008	γ-Proteobacteria
S11-B-6	EU016169	*Alteromonas* sp. KC98716-13 (99.26)	AB072388	γ-Proteobacteria
S11-B-7	EU016170	*Alteromonas* sp. KC98716-13 (99.26)	AB072388	γ-Proteobacteria
S11-B-8	EU016171	*Alteromonas* sp. KC98716-13 (99.53)	AB072388	γ-Proteobacteria
P11-S-1	EU016156	*Psychrobacter* sp. EP06 (99.60)	AM398215	γ-Proteobacteria
P11-S-2	EU016157	*Shewanella* sp. BSw20461 (99.53)	EF639386	γ-Proteobacteria
P11-S-3	EU016158	*Psychrobacter* sp. BBTR1010 (99.80)	EF471232	γ-Proteobacteria
P11-S-4	EU016159	*Psychrobacter* sp. BJ-3 (99.3)	DQ298407	γ-Proteobacteria
P11-S-5	EU016160	*Psychrobacter* sp. BBTR1010 (99.60)	EF471232	γ-Proteobacteria
P11-S-6	EU016161	*Rhodococus* sp. SGB1168-118 (99.93)	AB010908	Actinobacteria
P11-S-7	EU016162	*Psychrobacter* sp. EP06 (99.66)	AM398215	γ-Proteobacteria

（续表）

Strain	Accession number	Closest match (% similarity)	Accession number	Phylum
P11-S-8	EU016163	Arctic seawater bacterium BSW20352 (99.80)	DQ064612	γ-Proteobacteria
P11-B-1	EU016143	*Psychrobacter* sp. BJ-3 (99.80)	DQ298407	γ-Proteobacteria
P11-B-2	EU016144	*Psychrobacter* sp. BJ-3 (100)	DQ298407	γ-Proteobacteria
P11-B-3	EU016145	*Psychrobacter* sp. BBTR1010 (99.73)	EF471232	γ-Proteobacteria
P11-B-4	EU016146	Uncultured Acinetobacter sp. Clone MRT87 (99.87)	EF371492	γ-Proteobacteria
P11-B-5	EU016147	*Brevibacterium casei* strain 3Tg (99.33)	AY468375	Actinobacteria
P11-B-6	EU016148	*Psychrobacter* sp. BBTR1010 (98.40)	EF471232	γ-Proteobacteria
P11-B-7	EU016149	*Psychrobacter* sp. BBTR1010 (99.67)	EF471232	γ-Proteobacteria
P11-B-8	EU016150	*Rhodococus* sp. SGB1168-118 (100)	AB010908	Actinobacteria
P11-B-9	EU016151	*Psychrobacter* sp. BBTR1010 (99.73)	EF471232	γ-Proteobacteria
P11-B-10	EU016152	*Marinobacter* sp. BSs20186 (99.93)	DQ514302	γ-Proteobacteria
P11-B-11	EU016153	*Shewanella* sp. BSw20461 (100)	EF639386	γ-Proteobacteria
P11-B-12	EU016154	*Pseudoalteromonas* sp. BSw20513 (100)	EF551376	γ-Proteobacteria
P11-B-13	EU016155	*Shewanella* sp. BSw20461 (99.73)	EF639386	γ-Proteobacteria

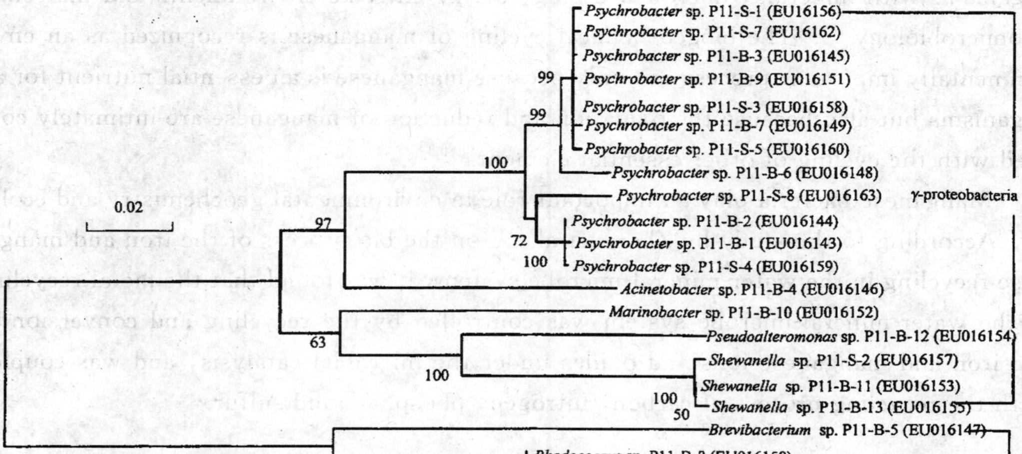

Fig. 1 Phylogenetic analysis of cultivable manganese bacteria isolated from Station P11

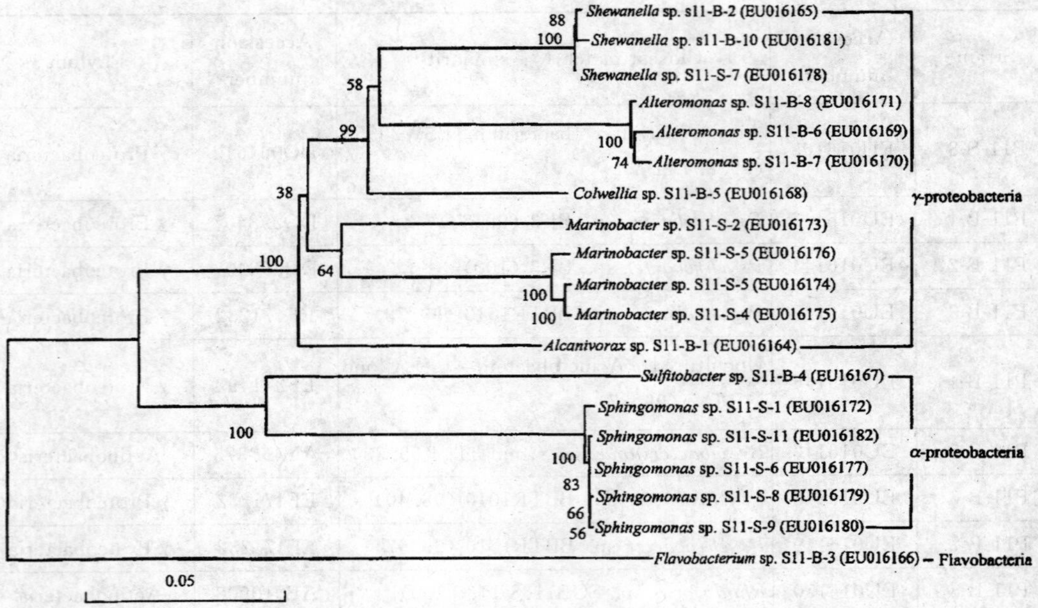

Fig. 2 Phylogenetic analysis of cultivable manganese bacteria isolated from Station S11 with a colorimeter at wave length of 550 nm

3 Discussion

Geomicrobiology is a newly-developed branch of earth sciences since the end of the twentieth century. It mainly studies the microbial processes in geologic environment and all kinds of geochemical records generated in these processes, i. e. , interaction of microorganisms with minerals, microbial ecology at the extreme environments and molecular geomicrobiology[10]. The biogeochemical cycling of manganese is recognized as an environmentally important process not only because manganese is an essential nutrient for all organisms but also because the oxidation and reduction of manganese are intimately coupled with the cycling of other essential elements[11,12].

Manganese bacteria play an important role in environmental geochemistry and ecology. According to the study by Zheng et al. [12] on the bio-process of the iron and manganese recycling in the water-mineral-microbe system, it was found that the metal recycling in the water-mineral-microbe system was controlled by the recycling and conversion of the iron and manganese ions and oxides under the microbial catalysis, and was coupled with the recycling process of carbon, nitrogen, phosphate and sulfur.

Table 3　Tolerant ability on Mn^{2+} for 12 species of manganese bacteria tested

Strain	Concentration of Mn^{2+} (mmol · L^{-1})			
	0	10	50	100
Alteromonas sp. S11-B-8	5.658±0.965	5.493±0.965	1.116±0.724	0.081±0.017
Cowellia sp. S11-B-5	4.890±1.520	6.061±0.069	1.317±0.581	0.174±0.018
Shewanella sp. P11-B-13	6.425±0.629	4.943±1.601	1.412±0.260	0.097±0.016
Shewanella sp. S11-B-2	5.750±0.181	5.770±0.297	1.731±0.131	0.242±0.072
Silfitobacter sp. S11-B-4	5.465±0.491	4.105±0.308	1.492±0.292	0.244±0.068
Pseudoalteromonas sp. P11-B-12	4.580±0.481	5.593±0.441	1.473±0.022	0.121±0.024
Rhodococcus sp. P11-S-6	6.250±0.849	5.925±0.728	1.173±0.968	0.159±0.097
Psychrobacter sp. P11-S-1	4.868±0.067	5.703±1.128	1.381±0.343	0.330±0.025
Sphingomonas sp. S11-S-1	6.260±0.276	5.520±0.148	1.078±0.981	0.191±0.006
Marinobacter sp. P11-B-10	5.893±0.647	5.368±1.241	1.317±0.116	0.155±0.090
Marinomonas sp. S11-S-4	5.840±0.573	5.775±0.523	2.188±0.024	0.759±0.014
Acinetobacter sp. P11-B-4	5.450±1.400	5.190±1.167	0.904±0.023	0.161±0.124

The study by Chen et al.[13] on stromatolites (*Minima* and *Admirabilis*) and their constructors (*Miniactinomyces chinesis* and *Spirisosphaerospora pacifica*) indicated that the rhythmic features shown by the laminae of stromatolites were hard to result from any colloid chemical action or sedimentary action, and that only the reproduction of microorganisms for generations could produce the almost unchanged pattern of stromatolite columns.

Francis et al.[4] thought that the oxidation of soluble Mn(Ⅱ) to insoluble Mn(Ⅲ, Ⅳ) oxides and oxyhydroxides was an environmentally important process because the solid-phase products oxidize a variety of organic and inorganic compounds (e. g., humic substances, Cr(Ⅲ) and Fe(Ⅱ)), scavenge many metals (e. g., Cu, Co, Cd, Zn, Ni and Pb), and serve as electron acceptors for anaerobic respiration[4,14]. The study by Chen et al.[15] indicated that microorganisms had manganese(Ⅱ)-oxidizing activity. Integrated with other features of the deposit, it was believed that microorganisms played an important role in the natural oxidization and enrichment process of manganese. The effects might be shown in 2 aspects. On the one hand, microorganisms can oxidize manganese; on the other hand, microorganisms can change pH character of environments, which leads to accelerate the chemical oxidation of manganese.

Previous studies have demonstrated that Mn^{2+}-oxidizing bacteria are abundant and distributed widely. Our study also indicated that manganese bacteria were not only distributed widely, but also displayed a distinction in their community. Gram-negative bacteria accounted for more than 90% of the total cultivable manganese bacteria isolated from 2 sample sites, and most of them belonged to γ-Proteobacteria. There were also differences in the community composition of manganese bacteria. For example, the cultivable manganese bacteria isolated from Station P11 were mainly composed of γ-Proteobacteria and Actinobacteria, which accounted for 86% and 14%, respectively. Manganese bacteria isolated from Station S11 belonged to γ-Proteobacteria, α-Proteobacteria and Flavobacteria, of which γ-Proteobacteria and α-Proteobacteria accounted for 63% and 31%, respectively. There was also a discrepancy in the composition of γ-Proteobacteria from the sample sites. γ-Proteobacteria isolated from Station P11 included *Psychrobacter*, *Acinetobacter*, *Marinobacter*, *Pseudoalteromonas* and *Shewanella*, of which *Psychrobacter* was the major genus, accounting for 67% of the γ-Proteobacteria. γ-Proteobacteria isolated from Station S11 included *Shewanella*, *Marinobacter*, *Marinomonas*, *Alteromonas*, *Colwellia* and *Sphingomonas*. The phylogenetic analysis by Tian et al.[7] on 50 isolates of manganese resistant bacteria obtained from 4 sediment samples in deep Pacific sea showed that the strains belonged to *Bacillus* (40 isolates), *Brevibacterium* (3 isolates), *Brachybacterium* (1 isolate), *Kocuria* (1 isolate), *Paracoccus* (4 isolates) and *Alcanivorax* (1 isolate). 80% of the 50 isolates belonged to genus *Bacillus* which contains 6 species and 90% of the 50 isolates were gram-positive. The study by Shi et al.[6] on bacterial abundance in iron-manganese nodule area of the eastern Pacific Ocean showed that manganese bacteria in the sediments were up to 3,929 CFU · g^{-1}. Johnston[3] investigated the oxidation rate of Mn(Ⅱ), mediated by microorganisms in the water column of Toolik Lake, Alaska. They revealed that the measured rates were lower than those observed in other aquatic systems, but were sufficient to oxidize all Mn(Ⅱ) in the lake within a 3-month period. Mn(Ⅱ) oxidation was largely microbially mediated, as indicated by inhibition of oxidation rates by sodium azide. Azide had been previously demonstrated to be a suitable microbial poison for studying Mn(Ⅱ) oxidation in seawater. These studies indicated the universality and diversity of manganese bacteria that existed, probably depending on the environments where they existed.

Our study also showed that the isolates of manganese bacteria had higher manganese resistant ability to Mn^{2+}, which was consistent with previous study. The study by Tian et al.ˢ [7] indicated that microorganism with higher manganese resistance had higher manganese removal rate. Biogenic Mn oxides have a high absorption capacity for metal cations and an ability to oxidize numerous inorganic and organic compounds, owing to their structural and redox features. Thus, the microbial process is of significance in both biogeochemical and biotechnological contexts[14,16].

Acknowledgements

The project was financially supported by the National Natural Science Foundation of China (No. 40576060)

References16(abr.)

(Cooperators: Lin Xuezheng   Gao Aiguo)

第七篇

海洋微生物学等的研究述评

REVIEW ON RESEARCH OF MARINE MICROBIOLOGY ETC.

G_1 海洋微生物学
(MARINE MICROBIOLOGY)

回顾二十世纪70年代世界海洋微生物学研究*

摘　要　本文从21世纪之初的角度回顾20世纪70年代前后世界海洋微生物学研究概况，以温故而知新，并促进中国相关学科的健康发展。全文共分十部分，即：①海洋微生物学方法和普通海洋微生物学的研究；②海洋微生物分类学的进展；③海洋微生物生态学进展；④入海陆地微生物的变化及归宿；⑤弧菌研究；⑥海洋微生物在物质循环中的作用；⑦海洋环境微生物学的研究进展；⑧海洋微生物与其他生物的关系；⑨其他；⑩我国海洋微生物学研究现状。

关键词　世界　海洋微生物　海洋环境　生态学　物质循环

海洋微生物学作为微生物学的一个重要分支，其研究始于1884年A. Certes的著作的发表。在这约120年间，海洋微生物学经历了发生、发展和成熟的阶段。如果说20世纪70年代前是海洋微生物学蓬勃发展阶段，那么70年代以来，就是其不断成熟的阶段。在1970年代前后，世界海洋微生物学取得了巨大的进展，本文就此作一简单综述，一起温故知新。

一、海洋微生物方法学和普通海洋微生物学的研究

海洋微生物学的正确取样是一项十分复杂而重要的问题。长期以来，海洋微生物学家就取样技术、测定方法和微生物分析方法做了大量研究。如Collins就底质微生物，Morita、Colwell等、Litchfield等就河口和海洋微生物，Jannasch就深海微生物，Rodidna、冈见吉郎就水生微生物方法取得了很大进展，推动了采样这个任务的不断完善。

培养法和镜检法对海洋微生物的形态和分类研究作出了重要贡献。佐培尔等着重于培养法，克里斯等偏重于镜检法。这两方面如今仍在不断改进，如对培养和培养方法的改进[1-7]。随着各种电镜和过滤技术的应用[6-7]，人们可直接发现不同形态的微生物。Waton(1977)叙述了3种测定海洋环境细菌数量的技术。Evans(1978)用鲎试剂快速测定水中细菌。Kogure和Patrick(1978)还比较了这两种方法。研究的目标是如何尽可能更接近客观地正确反映各种环境微生物数量、种群和生理生化活动。由于许多技术的应用，如示

* 原文载于《共和国建设者论文成果选编》，2004，660-664.

踪元素[7-8]、透析培养[9]、荧光抗生一膜滤术[10]、自动放射照相术[11]、电泳、气相液相色谱、红外分光、细胞成分与结构的精细分析[12]、荧光显微[13]以及生理生化如类脂生物合成相对率测活性[14]、ATP、DNA、RNA、蛋白质、叶绿素 a 及细菌数、生物量参数等的应用及数理分析统计，使人们对海洋环境从个体到群体的微生物活动规律有了更为合乎逻辑的估计[15-18]。

二、海洋微生物分类学的进展

由于化学分类、数量分类的引进并与传统的形态、生理分类相结合，使海洋微生物分类学有可能不断从细胞水平提高到分子水平[19]，对大量的微生物作较快而精确的分类。如 Corrado(1969)以免疫荧光法简化鉴定过程，制定分类索引，对海洋细菌以编号法作分类。Hendrie 等(1970)对发光菌的鉴定和分类。Boumann 等(1972)对非发酵海洋细菌的研究，Rüger(1974)对海洋细菌分类、Boeye 等(1976)对北海沉积物芽孢杆菌数量分类、Bahnweg 等(1974)对南北极水中动孢子真菌种类和 Roth 等(1962)对海洋酵母的分类。Austin 等(1977)对石油降解菌、耐重金属菌的分类，Kasama 等(1978)对日本沿海好气胶原细菌的分类等等。发现了一些新微生物种或对一些原来的分类地位作更正，如海洋螺菌的一些种、贝氏发光杆菌[19]、火神发光杆菌[20]、藤黄紫交替单胞菌[21]、埃氏交替单胞菌、水蛹交替单胞菌[22]、海圆杆菌[23]、海泥黄杆菌[24]、多形挠杆菌[25]、冷红科尔韦尔氏菌[26]、细菌科(chlorokiaceae)、着色菌科[27]等。进入 80 年代以来，在对容易分离和鉴定的海洋微生物有了相当认识的基础上，采用先进技术和方法，包括分子生物学技术，从海洋这个资源宝库发现更多的微生物种，包括趋磁细菌、古菌等。

三、海洋微生物生态学进展

现代海洋微生物生态学是建立在广阔而宏厚的分类生理生化基础和大量理化参数统计分析之上的。生态研究已在世界各大海洋中取得了成果，著名的有前苏联克里斯等所作的大规模调查研究，他在 1959 年和 1964 年出版专著后，1976 年又编录了 *Нникробиолтическая* 0 *кеанографuя*、Olitchfied 的 *Marine microbiology*、Rehamamer 的 *Microbial Ecology of a Brackish water Environment* 及 Kushnr 的 *Microbial life in extreme environments* 等。这四本书对我们开展不同生态环境微生物学研究提供了很大便利。而 Kuzenetsov(1977)[2]又向我们提示了生态微生物学的新动向，即各类微生物生态小生境及在物质流中各类微生物过程的强度、硫酸盐还原和化学合成强度、化学合成过程的强度、分子固氮强度和有机质降解等等。

从海区看，除克里斯等的调查外，还有亚速海水层细菌分布和季节动态，中部太平洋异养微生物分布的地理规律和活动，西南巴伦支海浮游细菌数量和生产，印度西及东南海洋沉积物异养细菌分布，黑海水和底质微生物分布、丰度和代谢状态，Mikawa 和 Akkeski 湾异养菌生态，东京湾微生物[59]及该湾与切萨皮克湾好气异养菌区系的比较，香港 Tolo 港异养菌分布，珊瑚礁生态，深海沉积物细菌的性质等等。就特定微生物言，有对含叶绿素 a 的好气异养菌分布，近岸海水发光菌种类组成和季节变化，日本中北部淡水和海水区中寡、富营养细菌分布，切萨皮克湾弧菌生态，纽约湾底质粪便菌和肠细菌分布，入海口附近海域大肠菌分布，产气荚膜杆菌分布及特性，维生素 B_{12} 及其产生菌的分布和生态，美国东北海域海洋动植物区系的高等真菌、子囊菌、半知菌及担子菌类研究。对两极海域微生物学研究也得到迅速开展。

关于环境因子与微生物相互关系调查研究已进展得相当深入，就温度、盐度、压力、营养、污染物对特定海区微生物种群数量、生理生化等活动的影响在 Colwell 等(1972)编的 *Effect of the Ocean Environment Microbial activities*、多贺信夫(1974)编的《海洋微生物》及 Jannasch(1978)的 *Experirments in deep—sea Microbiology* 中得到了新的体现。

在不同环境因子作用下，微生物细胞结构和功能发生变化与适应，相应地发展出一系列研究、分析方法和内容，如用^{14}C葡萄糖法测细菌异养活动，寻找不同水层与营养带的关系。分析营养、氧含量、水温及细菌耗氧量关系。某些放线菌分离频率、季节变化、耐盐力和抗生素产生的关系分析，适盐细菌长列电子态相互作用、阳离子特异性和静水压效应，适冷细菌呼吸、生活能力、可透性及大分子合成等。

由此可见，当年海洋微生物生态研究就既有范围较广调查，又有针对特定海区或特异微生物的生态和它们与环境参数间关系的深入仔细研究。

四、入海陆地微生物的变化和归宿

陆地微生物入海后的行为和归宿是微生物海洋学的一个大课题，其中对肠系细菌、致病菌和病毒等的研究较多。大量陆地细菌入海后就迅速地与海洋环境和海洋微生物发生直接反应，其中有许多就归于死亡。杀伤原因可能是海水的[28-33]、重金属离子的[34-37]、温度的[37]和低营养浓度的[34]作用。存活者主要集中在河口、近岸和浅海环境中，并随排水距离延长而减少。因它们与海洋环境因子的关系，它仍有特性生态分布[38-40]，且对海洋中悬浮有机物等的形成有重要作用。Baumann 等(1977)专门论述海洋肠细菌—贝内克氏菌和发光杆菌的生物学性状，Vind 等、Jenes 等就污水细菌，主要是大肠杆菌在大洋不同深度处存活问题作深入探讨。将入海陆地微生物种群组成和数量分布与环境因子结合起来考虑，可寻找陆地微生物与海洋微生物的差异及亲缘关系，也可看做污染水平和公共卫生重要指标[30]。据其分布情况作出运动模式以检测和预报特定海区污染程度和趋势[38、41-43]。

五、弧菌研究

在许多海区和不同基质上已分离到各种弧菌，对其生态或时空分布特征做了大量研究[44-48]，有关生理生化活动也有大量报导[53]。据其温度反应[26]、形态学和行为、繁殖及与其他生物的关系[45-53]，分离并鉴定出新弧菌。由于弧菌能引起海鲜变质、食物中毒、养殖病害[44-54]等在经济上的重要问题，对它的流行病学、血清学、与环境的关系及与初级生产力关系和在食物中作用方面的研究取得了一定进展[55-56]。

六、海洋微生物在物质循环中的作用

海洋微生物积极参与海洋环境的 C、N、S、Fe、Ca、P 等物质循环，现就其中三方面的研究进展作些介绍。

1. 在硫酸盐活动中的作用

海洋是多种形态的硫贮藏处。据小山(1966)估计，海洋中约含 3.62×10^5 吨的硫。各步循环活动，都有微生物参与，其中硫酸盐还原活动规模较大，研究得也较透彻。Zobell

(1958)、Wood(1965)都对硫酸盐还原的微生物学作过研究。后几年对沿岸沉积中的硫酸盐活动作了深入研究,如 Jrgenson(1977)用放射性示踪术测沿岸海洋沉积物表面氧化层的细菌硫酸盐还原作用,Nakanara 等(1977)调查了东京湾沿岸海水污染和硫酸盐还原菌关系,指出大量硫酸盐还原菌活动与高数量有机质相关,而该菌可能引起金属腐蚀。Ornemland 等(1978)研究了热带海草底沉积物中硫酸盐还原和甲烷来源的关系,Bohlood 等(1978)则发现盐沼沉积中硫酸盐还原速度与硫酸盐原菌数量不相关,盐沼底栖藻丛阻碍硫酸盐杆菌消散,使硫酸浓度保持较高水平。Inohebc 等(1979)研究了初级生产与硫酸盐还原的关系。Dlanczuk-Nogreen 等(1977)则测定了脱硫弧菌在海洋沉积的呼吸-转运系统中的活动。Jorgenson 还指出无色细菌分布和 H_2S 产生速率与沉积物物理结构、氧化还原状态相关,并认为对 S、C 循环有主要作用。Hashwa 等(1978)对黑海底的光营硫细菌存在问题作过探讨。Naziona 等(1978)深入研究了油层嗜热硫细菌,发现一个新亚种:*Desulfotomesalum nigrificans eupapsalinus*,并对它作了温、盐度试验。相比之下,中国的有关研究直至 20 世纪 90 年代仍开展得少且不系统化。

2. 海洋氮循环中的作用。

海洋中存在着大约 1×10^{12} 吨的氮素,许多氮素以生物活动为基础,其过程包括:①氮的亚硝化和硝化及其它们的还原,②固氮与脱氮。

在 20 世纪 60 年代,木俣正夫、丸山芳治、吉田阳一等曾对硝化细菌培养、分布和作用系统地研究过,对海洋来源、硝酸盐还原和硝化过程及有关的细菌群分布有所阐明。后来几年,Herbland 等(1977)和 Voittrriez 等(1977)对热带大西洋初级生产力和硝酸盐分布作研究,Tan 等(1979)调查了威悉河和德国湾沉积中的硝酸盐和亚硝酸盐异化细菌的密度,Rajendran 等(1977)对与营养再生的微生物学如 VeIlar 河口硝化细菌分布的调查,指出沉积中的硝化过程比水中的更大,盐度流向和基质浓度控制海洋和湖沼的细菌硝化。Rainte(1977)主要从细胞水平论述了无机氮的微生物转化,就营养、生理,通路和动力作研究,其中特别考虑溶氧影响,还谈及一些实际用途。Castellvi 等(1977)从西北非海域上升流区分离到一株异养细菌,并研究了它的硝化活动。

据 Werher 等(1974)主要根据兰绿藻活动,估计全球性生物固氮约为 175×10^6 吨/年,其 20%发生在海洋,海洋中细菌固氮的第一次证据是在 60 年代前就发现了,Shewan 等(1960)对海洋固氮菌的性质和分类曾做过研究。Brooke 等(1971)和 Sugabars 等(1970)研究了河口环境和水境中的微生物固氮。Maruyama 等(1970)和丸山(1973)对太平洋固氮菌的分布,生态调查及对相模湾、浚河湾、东京湾海水固氮菌固氮活性、Hevbert(1975)对浅海河口沉积异养性固氮等都作过研究。Eppiev(1977)指出过太平洋中陆水非光合微生物对上真光层内浮游植物的氮同化率有所贡献。此外,Pshenin 在 1959 年就发现梭菌和固氮细菌的固氮活动。Werner 等和 Sisler 及 Zobell 也研究过硫酸盐还原菌的固氮。Herbett 等对海洋环境氮同化的限制因子研究中得出有效碳和水的盐度是主因,同化依赖谷氨酸合酶进行。Bohlool(1978)对新西兰 Waimea 入口处污染的潮间带沉积物作了固氮活动调查。Green(1974)对佛罗利达彭萨科拉海湾盐沼中的固氮作用调查后指出,盐沼极适固氮。Hanson 等(1976)和对珊瑚礁生态系统固氮的研究揭示出夏威夷 Dahu kaekke 湾的固氮细菌有 50%依赖光固氮。具有重要意义的是在潮间带生态系中的固氮活动,群集着固氮菌、藻类植物、动物的该沉积物尽管只占表面积的 1%,但在固氮能力上却贡献了 50%。其中浒苔—颤藻—眼虫藻—光合菌群落起重要作用。由于人口的激增，伴随以日益加重的食粮问题，人们对海洋环境固氮问题越来越重视。

3. 有机质循环中的微生物活动

海洋环境有机质由生物合成，而分解则主要由微生物完成。微生物分解的有机质又参与各级食物链活动。人们早就对纤维素、淀粉、多糖类、藻酸、洋菜、几丁质、天然高分子化合物、腐殖质、脂质、尿素和蛋白质等的微生物分解进行了研究。Miyoshi(1978)从海洋中分离出三群厌氧纤维素分解菌，并定了一新种。Wolter 等(1977)研究基尔湾拍岸浪区分解洋菜、淀粉和纤维素的腐生菌与藻类的密切关系。Chandramokan 等(1972)指出海洋纤维素分解菌——*Streptomycetes* 在破坏含纤维素的材料上起重要作用。1977 年，Nitlsowski 对纽约湾顶和双迪霍克及新泽西湾海底沉积物的解脂、解朊菌作比较，发现污染区解脂或解朊菌都比未污染区的高，而且主要菌都是弧菌和假单胞菌。Bruni 等(1976)指出海水化学组分对某些海洋细菌解脂活动及正常代谢是基本的。Tsyban 等(1977)还就黑海、里海、波罗的海、北太平洋的解脂菌生态问题作调查。Kado 等(1978)于 1973-1977 年间对日本沿岸好气胶原降解菌作分离和计数研究，Meshol 等(1975)对解胶原的分解菌活动作过研究。Dartevelle 等(1977)发现潮间带沉积物好气异养细菌主要是解朊的和溶甘氨酸的(Glucidolytic)。对微生物在海洋环境物质循环中作用的研究，可加深理解海洋的地球化学和生物化学。

七、海洋环境微生物学的研究进展

自 20 世纪 60 年代以来，特别是 70 年代以来，由于经济活动和物质生活水平的提高以及对废弃物危害性的逐渐认识，一门以生态学等为研究手段，以污染物等为研究对象的边缘学科—环境科学蓬勃发展起来了，环境微生物便是其中的一个重要分支。海洋环境的污染问题越来越被人们所重视，包括石油及其炼制品、重金属、农药及许多人工合成物、放射性物质等对微生物的影响和微生物降解作用研究正在大力开展，它涉及分类、生态等基础课题，又涉及污染物的微生物降解活动及其后的变化，还引进了一些新技术和新方法，密切结合实际，因此大大推动了新兴学科如分子生物学等在海洋微生物学中的发展。较著名的科学家有 Colwell、Walker、Atlas、Anon、Chakrabarty、Austin、Seki、Letit 和 дьI бань 等[49-52]。他们对石油的海洋微生物降解问题做了大量而深入研究，有关动态可从《海洋环境中石油的微生物降解》、《海洋微生物降解油污物的若干生态学问题》及由 Sharpley-kaplan 主编的 *Proceedings of the Third international Biodegradation symposium*、Walters 等主编的Ⅰ、Ⅱ卷 *Biodeterioration of Materials*、Kenneth Mellaflby 编的 *Environmental pollution* 中得到了解。

重金属、农药、卤族烃等的微生物降解方面也做了大量研究，人们对一些海区特别是海湾、河口或沿岸区污染物分布变化状况作调查，如纽约湾、拉里坦湾、马赛福斯湾、新泽西沿岸、阿苏海和 Hiachi 海等的重金属或类金属，如 Cu、Cd、Cr、Ni、Pb、Zn、Ag、As 等的分布及变化。由于微生物与环境因子息息相关，所以从各种特殊微生物数量和种群变化及相关参数的调查分析，可反映污染水平以及生物自净能力，对环境质量作出评价。Melvin(1972)研究水中污染物如重金属等无机或有机物的测定法，环境卫生细菌学测定法。Приела(1977)介绍了美国长岛与纽约间海区的生态平衡问题。Jones 作了关于重金属离子毒性研究[34-35,48]。Waldhauez(1978)调查拉里坦湾和下纽约湾水中Pb、Cu对海洋生物的影响，Mills等(1977)研究切萨皮克湾水和沉积物重金属离子微生物效应，仓田(1978)对燧滩海域细菌的耐重金属谱研究后指出底泥菌比海水菌耐性强而广，Timaney等(1978)就纽约湾沉积物细菌区系的耐重金属和抗生素问题做了研究。Thermann等(1959)对咸水和海洋沉积细菌被Ca和Pb等的抑制效应。Tatsuyam等(1975)研究了抗Cd真菌对其他重金属的耐性问题及Cd与微生物数量的关系。Carlcucc(1974)研究Cu对海洋细菌的影响，Wallcer等(1976)指出Hg的抑菌性。Mearns(1976)和Goulder等(1979)描述了Cr对生物的影响。Siman—Puzol等(1978、1979)研究了一株G^-抗Cr菌的耐药性问题。Kurata(1977)调查了阿苏海水和沉积物中耐Ni菌分布。Austin等(1977)对230株河口耐重金属菌作系统的数量分类研究，1978年伊琦和夫作决定于质膜的细菌耐重金属性综述，对细菌耐性和耐性因子遗传学问题提高到分子水平上来认识。为进一步探求重金属离子的细菌效应，Vaccaro(1977)和Azam(1977)设计并进行受控生态系实验，国际DSSO论丛(1978)*Discharge of sweage from sea outfalls*一书引用了大量有关海洋环境金属与微生物的文献。有关方面的工作还推动人们对微生物在河口和海洋环境中矿物成因和转化过程中的作用研究[58]。

至于有机物污染问题，如1976年以波罗的海浮游细菌数量分布评价有机污染程度，对污染环境和未污环境可溶性有机物被细菌消耗的利用率比较研究，Islida(1977)对相对分子质量小于10 000的DOC被细菌分解的利用率测定。此外，Bourquin(1975)研究了PcBs对河口细菌的影响，Blokemore(1978)研究PcBs对异养菌大分子合成、生长和呼吸的影响，Ituraiaga等(1972)、Happe(1979)关于解酚细菌的研究，以及有关农药的微生物降解活动等等。

八、海洋微生物与其他生物的关系

微生物与其他生物的关系，包括鱼虾贝藻养殖中出现的饵料、病害和在各级生产力、食物链中的作用问题。

1. 饵料

小川数也(1977)通过对海洋动物饵料试验，发现用海洋细菌喂养，能使其细胞分裂加快、体长增加、寿命延长。粪便中大量细菌形成悬浮物，是有机质的分解者，能作为动物的直接食料，因而成为海洋生物食物链中的重要一环。Rieper(1978)以各种海洋和半咸水的细菌作食物，对桡足类的猛水蚤类作摄食试验，并测定了它仍以此作食物的消耗量。另外在食料中掺以维生素、氨基酸等生物活性物产生菌、蛋白质含量较高的酵母等，对海洋动物养殖业大有好处。

2. 鱼病

Rodsaether等(1977)对鱼病原——鳗弧菌(*Vibrio anguillaum*)的生化性质作了研究，Larsem等(1977)对沿岸区鳄鱼溃疡病综合征微生物学的研究。Hasteim(1977)对挪威海鱼和野鱼培养中*V. anguillaum*引起的弧菌病做了研究。Spceekens就某些鱼虾腐败细菌特征的描述，指出三甲基昆布氨酸由腐败极毛杆菌、某种发光杆菌及似莫拉氏菌的某些细菌所产生。Ewing等(1978)就耶尔森氏菌*Y. Xrwkerri*. sp. nov与虹鳟及某些鱼类的红口病和有关杀菌剂作过研究。Shieh(1978)观察了美洲红点鲑被杀鲑气单胞菌感染后血液中的酶发生的变化，Zachang 等(1977)探讨了切萨皮克条纹鲈的皮囊肿病(*Epithe liocyetis*)病因，LiM. F(1978)研究了底鳉被食爬虫假单胞菌感染后发生的组织病理学和血清学变化等等。随着海鱼养殖业的发展，鱼病研究必将深化。

3. 从养殖生态角度，探讨了作为环境因素的微生物区系与养殖对象的关系，如 Karthiayani 等(1975)就 Cochin 水中与环境相关的某些鱼、虾、贝等的细菌区系、Morton 等(1976)对香港喂食和消化周期与大肠菌的关系、Kess(1977)就某些鱼和发光菌的共生关系及 Joseph 等(1972)及 Ruby 等(1979)关于鱼消化道微生物相的研究等、Sleeter (1978)对与海洋钻木等足类钻孔活动中相关微生物的电镜观察。

4. 与海藻相关的微生物学，有藤田雄二等(1977)对紫菜红腐病原菌腐霉菌的性状研究及其生育环境主要因子和营养问题的探讨，Ragha(1978)对海洋硅藻楔形藻被真菌孔外弧菌感染后的生理学研究，Kong 等(1979)对海藻的细菌区系调查表明不同门类的藻体有不同属的细菌相关连，而这些细菌种群与海水中的不同。Kenneth 等(1971)研究了海生双鞭毛藻的共生细菌。人们对大型海藻和微观海藻的培养都注意到细菌等微生物的作用，如 Kogure(1979)就中肋骨条藻与海洋细菌的关系。Wooleng1977)、Kogure(1978)、Bonneau(1977)、Groereueg(1978)等都注意到海藻的无菌培养技术。Bahnweg(1978)还就与落叶腐解有关的霜霉目生理学进行了研究。

由上看到，这些研究都涉及到一个基本问题，即海洋微生物在生产力和食物链中的地位和作用，反映出微生物与各级生物量的密切相关性，因此，海洋环境中细菌等微生物与浮游动物、底栖动物、高等动植物间的关系正受到人们的更大重视[57]。

中国海洋水产微生物学的研究则是在 20 世纪 90 年代达到了顶峰[60]。

九、其他

海洋微生物与其他环境中的微生物一样，其活动方面是包罗万象的，很难用一篇文章把它全部阐明，但许多方面都又是相当重要的。比如，海洋嗜菌体、病毒、海气界面、海洋空气的微生物活动，高压、低温、低营养、嗜盐微生物的活动，海洋地质的微生物活动以及海洋微生物资源的开发利用等[10]，国外在这些方面已有所开展，中国在 20 世纪 70 年代可说尚未引起重视。

十、我国海洋微生物学研究现状

由于我国相关研究基本上是在新中国成立后从无到有地建立起来的，所以起步较晚。但是我们也取得了一些成果，如自 20 世纪 60 年代中期以来，在海洋微生物调查、有关纤维素、木质素、藻朊酸、几丁质等降解菌的应用研究，维生素 B_{12} 产生菌的调查和培养、金属的微生物腐蚀、溶菌酶的分离和测定、经济海藻病原菌分离和防治，石油降解菌的分离、培养和生态研究，发光菌的分离培养和耐药性问题等等。改革开放以来，我国普通海洋微生物学，水产微生物学的研究取得了前所未有的成就，与国外的水平差距进一步缩短。我国学者关注国际新动向，采用新方法，注重原创性。比之 20 世纪 90 年代以前，思路和成果更广泛和深刻。此处不赘述，将另文发表(即英文《中国海洋微生物学 1950—1990 年间研究进展》一文)。

综上所述，我们可以看到国际海洋微生物学在若干年前的研究动向。美国学者研究得比较全面。基础理论比较扎实，对推动学科的发展起了决定性的带动作用。前苏联学者则看重于生态方面。日本学者的研究重心主要在应用方面，后来，对基础理论也日益重视。欧美其他国家的研究则各有特色，对某些问题研究得相当透彻，如德国有相当一部分人对真菌的研究等等。相比我国，则显得研究课题较为零乱，缺乏长远设想和统一安排，这与我国辽阔海域、丰富资源、世人期望是不相称的。进入 20 世纪 90 年代，新思路、新方法的引入，国际间的交流，渗透更为深入、拓宽，所得成果更为本质和深刻。我国同国际间的接轨加快加宽，研究成果不断涌现[60]。为此，①在这片大好形势下，回顾过去，主要是为了立足当今。展望未来，我们应该登高望远，胸怀祖国、放眼世界，与时俱进，脚踏实地干好工作。②加强方法学的研究，把先进思路技术和设备引入本学科领域。③大力开展基础理论和分类、生态、生理生化和遗传学（包括分子遗传）的研究。把基础打好，向深度和广度扎实发展。④适当开展应用基础和应用课题的研究。⑤大力培养、造就和稳定海洋微生物学的研究人才，充分调动发掘各种积极因素，把有限资金和力量投入最富意义的、具原创性的研究课题上。⑥加强国内国际间的学术交流，资源共享、共同提高。

SCIENTIFIC RESEARCH ACTIVITIES IN MARINE MICROBIOLOGY*

Marine Microbiological Resource Investigation

Marine microbiology is a pretty young discipline in China. The category of microbes, their quantitative distribution and so on in the sea areas of China are not clear to us—the marine research workers. During 1970's, we investigated the microbes in sea、water and sediments from the East China Sea, the Huang hai Sea, the Bo hai Sea and their shelves by means of the research vessel "Venus". Several thousands of microbial strains were isolated from samples collected. They were indentified to genus and their ecological distribution was researched preliminarily. Some biochemical substances, i. e. vit. B_{12} were examined[1]. Some strains that have stronger ability to produce vit. B_{12} were selected, and a lot of experiments about using them in production were done.

Before that, extensive investigations were made into the corrosion of jute hemp outlayer of submarine cable and the mechanism of anti-corrosion. On this basis, we selected some anticorrosive agents with chinese characteristics, and used them in submarine cable production and submarine communication.

The researches mentioned above have extended research area and opened up a new page in large-scale investigation and surveying in marine microbiology in China.

Research on Marine Environmental Microbiology

With the global environmental pollution becoming more and more serious, marine environment has unavoidably suffered pollution[2-5]. Since 1970s, marine environmental microbiological research which emerged as the times require has been started in China. Bays of coastal areas and intertidal zones of China have undergone impacts of environmental pollution in different degrees. Pollutants include petrolum and its refined products with some of their degraded components, heavy metal ions and their salts of Cr, Cd, Hg, Pb, waste material and waste water from industrial and agricultural production and

* 原文于 1992 年在美国 TA&MU、西弗罗理达大学、威廉玛丽大学海洋科学研究所学术交流会上宣读过。

daily domestic life, etc. Relevant microbial quantity and group were investigated. The relationships between microbes and some pollutants were analysed. The abilities to tolerants or degrade pollutants and conditions for the occurrence of degradation activities were researched[6-10]. All of those researches are occupied leading position and attained international levels at that time in some aspects.

Research on Antarctic Microbes

The start of the research on antarctic microbiology in China is rather late, But the antarctic ecosystem as the largest extreme environment on the earth had aroused close attention of our scholars long age. Since 1983, research work on antarctic microbes in China has been started. During the period from the autumn of 1986 to the spring of 1987, the microbes of the Great Wall Station and its adjacent area, some sea areas of Euphacie antarctic and its shoaling were investigated. Those researches include regime analysis of the microbes of the stereo structure of land, air and sea[11]. As a result, a preliminary general role of microbes in the areas explored has been obtained. Those data provided a scientific basis for realizing, exploring, utillzing and protecting the antarctic microbial resources.

A global exploration of air-borne microbes together with their situation over the route of "JiDi" vessel has been performed during the round trip to Antarctica from Qingdao, China. That exploration was made round the earth, the range of the voyage was 30921 miles and stay the Marxwell Bay lasted 77 days. The results of exploration represent the condition of air-borne microbial quantitative distribution from different geographic longitudes, latitudes, climate zones, administrative areas (continents, countries) and types of cities. These results are achievements seldom seen internationally.

Research on Prevention and Treatment of Diseases of Maricultured Prawns

The work includes two aspects:

1. Prevention and treatment of diseases of *Penaeus chinensis* during the overwintering, hatching, culturing periods.

There were chances of going to parent shrimps overwintering ponds, rearing rooms and culturing farms to investigate shrimp diseases, and their growth and decline of quantity and quality of those shrimps. Some achievements were obtained together with an increase of economic benefits*.

---

* Chen hao-wen et al. Microbiological research on parent Penaeid shrimp, *Penaeus chinesis* and its water environment overwintering period. (send to press).

2. Microbial investigation on the internal and external environments of culturing ponds.

Long-term and systematical investigations on microbes combined with relevent sea water chemistry, geochemical and geological factors of sediments from inside and outside shrimp culturing ponds have been periormed. We hope that can analyse and forecast the relationships between these factors and shrimps cultivation and offer suggestions or measures for improvement.

That kind of work has both theoretical significance and practical and guiding effects.

Till now there have been only a few papers dealings with the comprehensive data on this aspects of the research work both in china and abroad.

References11(abr.)

发展和应用分子生物学技术，开发未知海洋细菌*

摘　要　回顾国内外海洋细菌学研究史，以往海洋细菌及其多样性研究发展缓慢之状况及原因。应用和发展分子生物学技术于海洋细菌学研究具历史必然性。引用微生物研究中的一些分子生物学技术成果，展望国际间海洋细菌分子生物学技术的应用前景，论述海洋细菌的多样性。指出此研究对保护和开发海洋细菌资源的重要意义。

关键词　海洋细菌　生物多样性　分子生物学技术　海洋微生物

一、认知丰富的海洋细菌世界是当代生命科学中最富挑战性的课题之一

当今虽然科技高速高质发展，甚至冲出地球奔向太空，但全球人口及其增长的压力、陆地资源的短缺都迫使人们比以往更多地将眼光注视海洋。因为唯有海洋才是解决人类自身生存和发展问题的一个比较现实且能有所获的巨大潜力实体。海洋蕴含包括细菌等微生物在内的丰富资源，如海洋微生物基因组中的遗传序列。一些先进国家着手实施的"海洋生物基因组计划"等项目，目的在于通过获得相应的遗传序列，全面寻找有价值的遗传信息，更有效地开发和应用海洋微生物等资源。同时对海洋中新型的活性物质的认知、开发和利用也越来越受重视。但据马克平等(1993)的推测，全球微生物的多样性，迄今被描述的细菌种数不足总数量的5%，已知细菌总种数也只占应知种数的12%。另一个估计是对海洋细菌的，迄今最多只认知了其总量(100万～2亿种)的1%[1]。这99%未知的物种是巨大的资源。认知和开发它是一项艰巨的任务、是重要的使命，为生命科学中最重要和最富挑战性的区域之一。它有非常诱人的前景，充满生机，需要有强劲实力、不断创新、善于进取并持续发展的研究群体，以挑战之势和实战能力去潜心艰苦工作[2]。必须对传统方法进行创新革命，应用和发展高新技术以扩展和加深对多种多样海洋细菌的认识。

二、分子生物学技术应用于海洋细菌研究的必然性

细菌作为一类生命实体，由于其中一些能使人致病，在300年前就开始被人们所意识，并开始注意到它在生物圈中(或许还有地球之外)无处没有的广泛存在性。150年前人们开始用丰富培养法(enrichment cultivation method)，第一次分离并描述了一种海洋细菌，即折叠螺旋体(*Spirochaeta plicatilis*)。60年前的《勃吉氏鉴定手册》中记载有86种海洋细菌，它们仅占当时已确认的所有细菌种数的6.4%(86/1335)。40年前，人们推测

* 国家自然科学基金资助项目(40086001)

原文刊于《自然杂志》，2002，24(3)：129-134.

凡是陆上发现的细菌所有属目，海洋中均有其代表存在，它们分布于至少 67 个属中。我国虽然海域辽阔，但直至 20 世纪 90 年代前期，记录在案的海洋细菌种（包括变种、生物型、系）仅约 33 个属中的 79 个种。它们只占中国海记录到的所有物种总数的 0.39%（79/30278）。其他的微生物尚有 18 种放线菌、131 种蓝细菌、1 种原绿菌（*Chloroxy bacteria*）、61 种酵母、127 种其他菌物及菌藻类的一种[3-4]。如果说中国海洋生物占全球海洋生物 1/10 的话，那么我国海洋细菌等微生物迄今的记录显然就太少了。这些年来，世界（包括我国在内）陆续发现了一些新的细菌[5]，但总的来讲，尤其是与浩大的未知菌数比仍是很少很少的，其原因是：①海洋，尤其是远洋、深海及极端环境，对大多数人来讲是难以到达的，更要有先进的设备和合适方法。②局限于历代海洋细菌学的发展状况，即基本上没有脱离过丰富培养法的研究程序。人们主要借助非常有限的特定培养基在设定的培育条件中培养出一定类型的细菌菌落、显微镜检别形态、生理生化性状测定等方法确认分类特征，得出另一估计是迄今世界充其量能培养出的海洋微生物不多于 10%[6]。虽然这无疑是以往海洋细菌学坚实发展的关键，但人们为其制备的培养条件、培养基类型比起浩瀚海洋、复杂生态和缤纷类别等状况，这些手段毕竟是非常有限的，即使是单一的高压或热泉环境，一般实验室也都难以创造，更何况许多因子结合成的复杂多样生境！因而仍无力在许多实验室中创造出同自然界相一致的微生境（小宇宙），难以去界定和廓清庞杂类别，包括一些处于非可培养状态中的原核生物，而迄今这些非可培养状态的种类占自然环境中的绝大多数[7-8]！试图解决这一难题始终萦绕于科学家心间。这是一条漫漫探索长路，只是到了近来（近 10 年），才理清了这一头绪，那就是运用分子生物学方法鉴定自然样品中细菌的各种类型。总体言，这是基于遍及所有细菌的 DNA 序列方法。基因序列在漫长的进化岁月中发生着缓慢的变化，可将其分出高度保守区和相对的可变区。后者的序列往往专一（特化）而成为各个细菌类型差别的指示器。凭借分子探针，识别可变区中合适的信号部位，结合其他工具，直接鉴定从简单到复杂的自然或人为环境中的各类型细菌，也可研究细菌的多样性及各种生态学问题。

三、分子生物学技术在海洋细菌学中的应用

分子生物学技术在微生物学研究中得到了迅猛发展，可是国内直接用于海洋细菌等微生物学上的例子并不多见[2,5,9,10]，所以我们应该借鉴和引用非海洋微生物的分子生物学技术，并发展出一套行之有效快速精确的测试海洋及其环境的微生物学方法。

迄今对海洋细菌研究可借鉴的分子生物学技术主要有下列五大类。

1. 核酸杂交技术

核酸杂交技术是包括整个基因组的重组，可测核苷酸序列相似程度。用 DNA-DNA 杂交法判同一种或两种间关系。用 DNA-rRNA 杂交法测同源性。此法特异性强、灵敏度高、快速、简便[11]。借助测定核酸序列，并与光密度测量法直接比较，由核酸杂交所得阳性条带或斑点，可得定量结果。

该技术包括四种测定法：

(1)质粒 DNA 印迹杂交(blot hybridization)。Ogunseiton 等在 1991 年用此法从含水杨酸盐的土壤中分离出降解萘的假单胞菌。

(2)狭线印迹杂交法(slot blotting hybridization)[12]。此法已用于测量被汞污染的新

鲜水中微生物群落的汞还原酶基因。

(3)斑点杂交(dot-blot hybrfidization)[1,13]。用PCB分解基因的斑点杂交测土壤微生物群落中降解PCB者之状况。

(4)核酸杂交技术与MPN(最可能细胞数)法结合,已用于对芳香族化合物有脱氧作用的微生物研究。此法可估计序列相似度,划分特定分类单元,以可能相近的微生物属为参照作出种水平上的分类。

2. PCR及相关技术

PCR之原理是体外扩增核酸序列获得多个核酸拷贝,包括对双链DNA的直接测序或克隆后测序。用于鉴定分枝杆菌、立克次体及蛋白分解菌(无形体属类)等。琼脂糖凝胶电泳纯化,检验全细胞提取液等复杂混合物中低量的核酸并扩增核酸序列,可达到定性、定量、克隆:转化或测序之目的.它包括以下三种方法:

(1)嵌套式(nested)PCR[14]。已用此法测定过实验淤泥生态系统中3-氯苯脱氢细菌外源基因种类。

(2)PCR扩增。已用于分批式反应器中含酚废水的假单胞菌降解作用研究。可检测混合菌群中原本含量很低的DNA序列。PCR扩增从菌落中分离到的DNA,对相应待测物种作分子鉴定。已鉴定出新的嗜压深海菌。我国学者戴欣等也从南海沉积物中用分子鉴定法检出一株适于海洋环境的枯草芽孢杆菌(*Bacillus subtilis*)[9]。

(3)竞争性(competitive)PCR。测量多环芳烃污染沉降物中编码儿茶酚2,3-过氧化物酶的dmpB基因浓度。

rDNA序列分析常与基因克隆、PCR扩增术等交替使用。

3. 基于多态性的技术

将PCR技术与限制酶酶切技术合用以测自然沉降物中nahAC基因等情况[6]。

包含下列两种方法:

(1)随机扩增DNA多态(random-amplified polymorphic DNA,即RAPD)分析[14]。用标记rRNA探针作核酸类型分析,即DNA指数多态性分析术。此法分析过混合微生物群系中(如各种生物反应器中)微生物的多样性。PCR-RFLP(下述)法研究不同菌株间的亲缘关系。

(2)变性梯度凝胶电泳(denaturing-gradient gel, electrophoresis, DGGE)[14]。它可分离PCR扩增产物,测定扩增物中单个碱基的替代,也用于多样性测定。

凝胶电泳可从土中直接分离DNA。在菌落水平上用基因序列及PCR引物扩增基因后再测,传统的限制性酶切消化分析用于那些缺乏PCR引物的基因分析。Ferris等(1998)已用DGGE分析了16*S* rRNA基因的PCR扩增产物,绘制一组微生物种群16*S* rRNA基因图谱。

4. 核酸探针检测技术

借助能与特定核苷酸序列发生特异性互补的已知核酸片段作探针为有效方式标记,即探针灵敏检测杂交结果,分辨样品中特定DNA序列。

这一技术包括:

(1)稳定同位素探针法(stable isotope probe)[15]。已鉴定了土壤中的甲醇利用菌等[2,15]。

(2)寡聚核苷酸探针法。这是据 rRNA 基因的多拷贝且高度保守的 DNA 片段设计的。检测环境细菌大致种类,并鉴别特定种,已测出大肠杆菌、利斯特菌、耶尔森氏菌、鸟分枝杆菌等。该法包括:

①有基于 DNA 序列的探针、以 DNA 制备的 cDNA 探针、双链 cDNA 探针或单链 cDNA 和 RNA 探针等法。如小核糖体 RNA 的 DNA 序列(16*S* rDNA)分析法。经 N-端氨基酸来推测某段 DNA 序列,由此与基因库中相关基因的探针作对比。可同时鉴定某样品中多种类型细胞,如海胆(*Meoma ventricosa*)的一新病原菌 VL-菌即如此鉴定出来[16]。还有南极某砂岩中新发现的弧菌和古菌(*Archaea*)[11]。

②16 S rRNA 扩增和测序。该法采用 Sanger 的双脱氧末端终止法。已用于检测活性淤泥中微生物种群结构,判断相应有机体。Vaainio 等(1997)已从 516 种孤立的菌落中抽提并测序 16 S rDNA[9]。

③16 S rDNA 直接寡核苷酸探针。这是一种 rRNA 和 rRNA 基因序列分析,作原位杂交鉴定,来调查群落结构、功能间关系、鉴定物种间的亲缘关系。三界系统理论的出现,即是建立在对原核生物等的系统发育新认识基础上的[17,18]。与此相似的是 5 S rRNA,已用于物种鉴定,包括一些难培养者。

Axnosti 等(1998)用此新探针获得的资料显示嗜冷菌在北极 Svalbard 沉积物中的重要性[19]。

基于 DNA 序列的探针可迅速地改进探索和界定细菌代谢和生理学方面的多样性。这些细菌以前尚未在分子水平上被广泛研究过。这一探针包括两个主要方面:一是克隆专事特定细菌代谢或生理特性的基因,将这些基因组经重组而返回去构筑限定的基因突变株的转移并描述那些基因的突变。二是将传递转座子(delivering transposons)作为诱变工具去帮助鉴定特定性质的基因。已用此技术检测一些新的致病菌[20]。还可测知某样品中细菌的大致种类[21]。

rRNA 和 rRNA 基因序列分析,可确定某分离物为某一分类单元或证明为一新分类单元,指出分类单元间系统进化关系。检测存在专一性探针靶位点的新序列,为系统发育株用微生物方法检测未知菌。

5. 分子流行病学技术

分子流行病学技术包括:

(1)DNA 的限制性内切酶消化。该酶将细胞基因组 DNA 切割,在琼脂糖凝胶上电泳分离,示出不同种群基因组 DNA 限制性片段长度多态性。

(2)质粒指纹图谱分析。经"指纹图谱"检测、分析基因组中内切酶限制性切点的变化以比较不同基因组间的核苷酸差别。用标记 rRNA 探针作核糖类型分析。

凭借数量相对稳定的质粒,区分同一物种不同菌株。用于病原菌感染而致的流行病学调查、海洋水产病害研究等。

用 DNA 指纹术研究南极东方湖(Lake Vostok)冰下冰的地质微生物。16*S* 核糖体 DNA 基因表明,那儿含有高达 $2.8\times10^3\sim3.6\times10^4$ 个细胞/cm^3 浓度的微生物,但多样性不高[15,20,21]。Hines 等[21]从北卡(美)附近的 Eel 河盆地的海洋沉积物中发现了一半以上(30/56)的消耗甲烷细菌是新种,且大多是古菌。Lsaksen 等[22](1996)借助分子系统发育和生物地球化学方法,分离出以往难以鉴定的新硫酸盐还原菌菌种。

(3)低相对分子质量 RNA 分布图。

低相对分子质量 RNA 分布图是 tRNA(75～90 个核苷酸)和 5*S* rRNA(105～120 个核苷酸)的双相电泳分离,是一种有效的鉴定工具。

(4)限制性内切酶酶切片段长度多态性分析(restriction endonuclease fragment length polymolphism,RFLP)。该法是对 PCR 扩增后的 rRNA 基因,用限制性内切酶片段的电泳图谱分析,即借助标记 rRNA 探针进行核糖类型分析,鉴定细菌[18]。

PCR-RFLP 法,用于细菌种间及种内株间的分型鉴定。用来研究不同菌株间的亲缘系统。为 DNA 指纹图谱中最有效分型方法[2]。

RAPD 和 AP-PCR(arbitmy primed PCR)两法是建立在对未知 DNA 区段扩增基础上的,可用于多态性分析,也研究菌株间亲缘关系。

对海洋细菌作精确的系统分类,各项分子生物学技术无疑都有其自身的优良之处,都将发挥各自的用途。但实际解决还是需要各项技术的优化组合。如对嗜盐古菌新种的系统分类工作即是如此完成的。它包括极性脂组分分析、G+C mol% 含量测定,16*S* rRNA 扩增和测序,系统发育树、DNA-DNA 杂交等。Fiala 和 Stetter 于 1986 年发表的异养性海洋超高温古菌——强烈炽热球菌(*Pyoococcus furiosus*)也是用类似此法鉴定出来的。

任何技术都是时代的产物,都有其特定的历史局限性。比如,美国 1982 年在太平洋某水深 2600 m 下的一高压热隘口发现的詹氏甲烷球菌(*Methanococcus jarlnaschii*),只是到 14 年后的 1996 年,经美国基因组研究所对其全部(167 万对)碱基测序后,才将其确认为是另一种生命形式即古菌——生命第三域(以前称为 *Archaebacteria*,古细菌)。这说明了新技术的神奇用途,即更能揭示海洋生命的本质[1,12,17,18]。目前的分子技术虽促进甚或部分地代替了纯培养鉴定,但仍需不断完善和发展。这包括它在阐明自然种群结构上的许多工作。正在发展的标记寡核苷酸探针与胞内特定靶序列杂交技术正是此方面的代表。该探针穿透经甲醛固定的微生物胞壁后鉴定该细胞的基因型,这是基于每个细胞均含许多个含大量分类特异性信息的 rRNA 拷贝,常以 16*S* rDNA 探针为主同时鉴定某样品中的多种类型细胞。另一办法是从自然样品中抽提细胞,再从细胞提取核酸。其原理是直接进行菌落杂交以鉴定环境样品,借此可直接探测某细菌群落的时空变化。已从美国长岛南部、加勒比海和马尾藻海分离到不同浮游细菌的集合物。发现其总 DNA 杂交表现出巨大的相似性,相反百慕大附近珊瑚泻湖群落的浮游细菌集合物总 DNA 相似性都甚低[8]。第三个办法是直接从自然样品中抽提核酸而不必先分离细菌。就是用凝胶电泳从土中直接分离 DNA,使可用的基因序列增多以在菌落水平上靠 PCR 引物扩增基因后再检测,将被检测到的专一性探针靶位点新序列作为系统发育株,流式细胞计富集未知物细胞,用包括化学分类、生化、形态及超微结构等行之有效的微生物法等检测手段即可探清其归属。这些工作相当于将现代分子生物学技术与经久不衰的经典技术相结合,这种结合显得更为有力。伍兹霍尔海洋研究所的 Jannasch 教授即靠如此技术结合而发现了深海热液孔(deep sea hydrothennal vents)的一些新种细菌[3]。Sahm 等[24]借此分析北极海沉积物嗜冷硫酸盐还原菌的种系发生关系,并作了量化分析,鉴定出属于 Protebacteria 亚纲中的 13 个新种。类似工作也基于基因序列构架的识别。帮助了确立特定海洋细菌在系统发育(进化水平)上的多样性[3]。新技术的互相渗透、借用,可达到优势互补之效果,如彭宣宪等采用通用引物 PCR(universal primer PCR,即 UPCR)与单链构象多态性(SSCP)

和 RFIP 术相结合的方法，不但快速确诊检测到一些鱼病原菌，还检出了大肠杆菌、双歧杆菌等非鱼病病原菌[10]。

四、认识海洋细菌多样性，保护并开发这一宝贵信息和资源

海洋细菌多样性可从其基因、物种和生态系统 3 个层次来理解[25,26]。它表现为海洋中细菌内在和形态的生命多样性。

从宏观看，海洋细菌多样性即是海洋细菌细胞与其一定的环境发生相互作用的结果。这些环境就物理、化学和生物学等条件对某细菌而言都是特定的。细菌细胞于是借助这些条件在漫长的进化世代过程中统合并具备了生存条件和代谢通路，形成了多姿多彩的细菌机体多样性，包括体态、生理特征等反映出的亲缘关系等等。同时细菌等生物也在改变和塑造环境，因而该多样性又表现为各种生态过程的总和[27]。

从微观看，环境与细菌间的平衡及统一性建立在后者的分子生物学基础即基因组上。内在的基因组及其变异借助其序列主动编码蛋白质，进而作各色代谢活动和生理适应，细菌便应付自如地统一于环境之中，不适者即被淘汰[28]。从此意义言海洋细菌多样性既基于其自身基因多样性，也依赖其生存环境(包括其他生命体)的多样性。两者形成了复合体，即细菌生态系统[7]，这是各种生态过程矛盾的统一体。它是不断推陈出新筛选来的可良性持续发展的生态平衡过程[1]。基因对海洋细菌来说，其环境，即使是几毫米范围，其复杂性和差异性就可能有明显的不同，诸如产生的抗冻、嗜冷、嗜盐碱、嗜高压、抗辐射等极端微生物的多样性。

亿兆细菌间不得不产生巨大的代谢和生理生化方面的多样性、变异性和生态复杂性，包括代谢类型和潜力及生理属性的多样性。它们在核酸、蛋白质序列间的差别更本质地反映出在分类和进化上的关系。Dunlap 等[3]对费氏弧菌(*Vibrio fischeri*)已有效地用分子生物学手段作过此方面有意义的探索。动植物与之相比，它们的多样性就惊人地小了。

不可忽视的环境因子还有来自生物学本身间的。这包括对营养和所需资源的生物竞争、捕食者和寄生物活动等等，如许多对特定类型细菌专一的病毒在海洋细菌物种(类别)及群系组成的形式和发展过程中也是关键作用之一。代谢和生理生化类型的多样性决定了细胞形态和生长类型的多样性。各类海洋环境要求并规定耐受、适应、嗜好海洋的细菌以适当的生理代谢类型和生长方式发展，形成富有自身特色的生长和物种类型，而异彩纷呈的细胞形态则是海洋环境与细胞内在因子共同塑造出的完善、和谐外在表现之一。

分子生态学的方法在海洋细菌多样性探索中已有并将仍有广阔的用武之地。如 Beja 等在 RNA 的反转录作用后用 PCR 特定的分子技术，鉴定未知菌活跃的光合作用活动，从海洋中发现了以往从未认识的光合作用基因簇，发现了以往实验室从未培养出来的 11 类新菌种，即好气、厌氧光合生物(aerobic，anoxygenic phototrophs，缩写为 AAnP 或 AAP)。这类菌估计占大洋表面光合电子转运的 5%、总微生物群系的 11%[29]。分子技术的应用和发展将大大丰富海洋细菌的多样性，并改变以往人们对海洋生态系的传统观念和估计。一门微生物海洋学(microbiological oceanography)学科正在兴起、扩展。

五、结语

过去对海洋细菌的研究只给出了分子水平上的极少关注。可喜的是我国已有学者关

注和实践分子生物学技术在海洋细菌研究上的应用[2-3,12-13]。崭新的分子生物学技术对传统海洋细菌分类学、生态学、海洋环境微生物学及进化研究的渗透不仅为它注入活力，获得新生，其途径为探索海洋细菌多样性包括其环境和生态的多样性，更提供了一系列强大武器，尤其在包括对受威胁物种的保护上将发挥生力军作用[7,31,32]。新方法、新途径在未来岁月的海洋细菌研究中增长着的用途当将产生难以预测的巨大动力，并将为许多新的生物技术产物的开发和应用形成新的基础[33]。在刚刚开始的新世纪中，海洋细菌学家在分子生物学家的通力协作中将有关方法和技术的研究转变成应用的研究，透过进一步打开的海洋细菌世界新窗口，能不断扩展新视野，提高认知和开发海洋细菌的效率。中国加入 WTO，无疑为中国学者展开了与国际接轨千载难逢的大好时机，缩短这一领域与世界先进水平不大的差距，可能提升中国海洋微生物学的基础和应用研究水平，我们应以实际行动迎接之。

参考文献 33 篇（略）

（合作者：孙修勤）

MARINE MICROBIOLOGICAL RESEARCH IN CHINA (1950—1990)*

Abstract The forty years under consideration of marine microbiological research in China can be divided into three successive periods: (i) 1950-1965, after the creation of the People's Republic of China up to the year before the start of the Cultural Revolution; (ii) 1966-1976, a dormant period in marine microbiological research, during which few national organizations or other bodies participated in research and no publications exist on what, if anything, was studied; and (iii) 1977-1990 (beginning one year before China adopted its policy of reform and opening up), marked by a very noticeable increase of information as a result of a larger body of researchers, who was prolific in the production of publications. The eight main research areas covered include: pollution, environmental (sanitary) microbiology, prevention and control of agricultural diseases, photobacteria, mineral resources and metal corrosion, marine yeasts (molds), as well as Antarctic and general marine microbiology. Academic activities and exchanges are also described.

Key words Marine Microbiology, Environmental Microbiology, Diseases Of Aquatic Organisms, Marine Yeasts, Microbial Corrosion.

Overview

Research in marine microbiology in PR China has been accompanied by the emergence and development of marine scientific research in general in the country. Prior to 1950, these research areas were not yet developed. Only after 1950 has development in these areas occurred. The research periods in China are divided into three stages as following.

1950-1965

During this period research efforts in this area were scarce and the work that was performed was very basic. For example, only a few microbial strains were obtained from the sea, identified and tested to determine physiological and biochemical characteristics. Only 13 papers were published in this period (Hsysen et al., 1960; Chen and Li, 1961).

1966-1976

During this period research work in marine microbiology remained, for the most

* The Original text was published in *Ocean sciences bridging the millennia-A spectrum of historical accounts*, 2004, P. 399-405.

part, in a dormant state with little progress. The corrosion of the yellow hemp (jute fibre) that formed the protective outer layer of submarine cables was surveyed and studied by young scientists from the Institute of Oceanology, Academia Sinica, Qingdao, China. They investigated the causes leading to the destruction of the outer layer and developed replacement outer layers with improved anti-corrosive protection. Later microbial investigations in China's sea areas, including the Huanghai, Bohai and East China Seas were conducted. The investigations included: collection of the microbial strains with the ability to produce specific physiological and biochemical substances-for example, vitamin B12-and their ecology; selection of certain superior strains for producing vitamin B12; and exploring the prospect of launching industrial production, and distribution, of certain microbial strains, as well as studying their taxonomies.

The staff at the Shangdong College of Oceanography [*renamed later Ocean University of China* (*OUC*)], Qingdao, studied the corrosion of iron and steel in the marine environment. However, there were no papers published on this subject.

1977-1990

Intensive work was carried out in marine microbiology during this period. Young researchers, recruited into the discipline, continuously worked together conducting a considerable amount of scientific research. Research areas expanded, so as to include both basic and applied research. Significant papers were published with an output of more than 158 papers recorded and a noticeable improvement in the quality of research. Up to 1991, marine scientific research and educational units often engaged in marine microbiological research in coastal cities, such as Qingdao, Shanghai, Hangzhou, Xiamen, Guangzhou, Tianjin and Dalian. These units were attached to various government departments, such as the State Oceanic Administration (SOA), Chinese Academy of Sciences, State Committee of Education, State Committee of Agriculture, Ministry of Forestry and Fishery as well as coastal provinces, cities or regions. Senior research scientists created the necessary conditions and basis for the development of marine microbiology in the country.

Marine microbiological research in China

Pollution microbiology

A great deal of work was conducted on pollution microbiology during the third period. Major research efforts were focused on petroleum refinement, heavy metals and waste materials in the marine environment (Chen, 1981, 1983, 1984, 1989). Research areas extended from the South China Sea to the northern part of China's sea areas, including Daya Bay (Li and Wang, 1987), Xiamen Harbour (Ni et al., 1984), the mouth of the Yangtze River, Qingdao and the adjacent Jiaozhou Bay, Tianjin and the adjoining Bohai Bay up to Dalian Bay etc. (Sun et al., 1988). Research areas included: the classification of microbial groups, quantities, ecology and relationships, tolerance, degradation, modeling experiments on pollutants including those in enclosed ecosystems, application in purification of pollutants, environmental protection, and effects on aquaculture (Chen et al., 1987).

Environmental and sanitary microbiology

Studies were conducted on a group of fecal coliform bacilli, the distribution, status and its quantity in the sea areas of Xiamen Harbour, the mouth of the Yangtze River and Jiaozhou Bay etc. (Lin et al., 1984; Qian and Zhang, 1985; Hu and Cao, 1990).

The life history of *Vibrio cholerae* bacterium had not been clearly observed owing to its non-culturable state. Xu and Colwell (1989) developed monitoring methodology and studied the wintering-over stage of this bacterium. From this work, a new effective means were provided for further research on the bacterium. In addition, an investigation was conducted in the field of environmental sanitation to monitor the status of a variety of marine recreational and sports areas (Li, 1985).

Prevention and treatment of aquatic organism diseases

Much has been achieved in this research, but the understanding of diseases and pests of aquatic organisms is still very superficial. In the development of mariculture, proper methods are necessary for the prevention and control of certain mariculture-associated pathogenic organisms. If prevention and control does not keep pace with the development of breeding areas, certain adverse effects will occur. Therefore careful attention has been given to this aspect by the Chinese government.

Although considerable achievement had been attained in related research, as discussed in other documentation, the knowledge of aquatic organism diseases and pest control techniques was still at an early stage. An insufficient understanding of, and lack of proficiency in monitoring methods can put lives at risk in the event of a large-scale epidemic outbreak. The problem is compounded by a reticence to accept the argument that additional funds are needed to increase the stability of qualified staff and to ensure that comprehensive research will be conducted in the domain of prevention and cure of diseases and pest control.

Photobacteria

Investigations in China of marine luminescence began in the 1960s. Initially, some microbiologists studied the biochemical luminescence phenomena of marine luminescent organisms; however, this work stopped for about 20 years. Since the 1980s, a number of papers on luminescent bacteria have been published via academic for a (Cao and Hu, 1982). Research included taxonomy, ecological distribution, population constitution, ultramicroscopic structure and application of photobacteria. The sea areas investigated were the East China Sea, the mouth of theYangtze River and the Huanghai Sea (Wu and Cao, 1982). Several strains were first recorded in the areas concerned (Shen andYang, 1987; Yang et al., 1984).

Mineral resources and metal corrosion

Microbes identified in sediments in the North Pacific Ocean were investigated and linked to the existence of manganese nodules. The abundance in manganese nodules of such microbes and their relationship with manganese bacteria were analysed (Liang, 1986).

Research on microbial corrosion led to an understanding of its effect on steel, copper and aluminum (Yi and Liang, 1985). Research showed that the sulphate-reducing bacteria (three species of *Thiobacillus*) in the mold, when in contact with the three metals mentioned above as well as with various types of steel, demonstrated a corrosive effect. The corrosive rate of microbes on metal materials and various factors leading to corrosion were analysed. Further, the related investigative work also included applied research towards the utilization of fixed *Thiobacillus* as an antifouling organism. The effect of bacteria on fishing nets was also investigated (Ji et al., 1989).

Marine yeasts and molds

Research on marine yeasts began in the 1970s. Since this period, preliminary investigations were conducted on groups of marine yeasts along the coastline. Applications included the ability to feed sea cucumber larvae. Research leading to the use of yeasts to degrade petroleum was demonstrated in the laboratory (Wang et al., 1980).

Zhou (1989) led a research group from Nankai University in a basic, thorough investigation of the Bohai Sea and along the nearby shore. About 50 species in five genera, such as *Candida*, *Rhodotorula and Cryptococcus* etc., were identified. Among these, ten new species were recorded in China.

Several investigators studied the effect of salinity on the growth and reproduction of certain species of yeast. Another study group concentrated on specific *streptomycetes* from a selected coastline in China (Zhou et al., 1989).

Antarctic microbiology

Since the 1980s, research on Antarctic microbes has been conducted. Since 1981 Chinese scientists have made investigations in the Antarctic, e. g. in: Victoria Valley, Rose Island, the Great Wall Station, the South Shetland Island, King George Island (Ni et al., 1987) and the Southern Ocean. The taxonomic groups, quantity and ecological distribution of microbes from the ocean, land and air in the above locations were explored. In particular, the relationships between shoaling *Euphausia superba* and environmental characters (i. e. chemical and biological) were analysed and relevant papers were published (Chen et al., 1990). Those papers provided the basis for recognizing, exploring, understanding and protecting Antarctic microbial resources, all of which has led to solid progress in this field. The global status of airborne microbes along the route of the polar scientific research vessel *Ji Di* was surveyed during the vessel's 122-day voyage from Qingdao to Antarctica and the return voyage to China.

General marine microbiology

The ability to conduct practical marine microbiological research in China was developed early on. Initial investments were made in order to conduct basic research in several areas. Published papers dealt with: bacterial distribution in the continental shelf areas and in bays; the isolation, culture, and identification of *Azotobacter*, *Thiobacillus*, *Micrococcus*, the *Thiorhodaceae*, etc.; research on halophilic pathogenic bacteria, for example halophilic vibrio; the quantities of anaerobic sulphate-reducing bacteria, applications of cyanobacteria, bacteriostasis of soft coral, endo-toxin, marine medicine and biomass determination, etc (Zheng, 1988; Zhang et al., 1990). Scientists plan to investigate microbial ecology on a large scale and flora and relevant physical, chemical and biological parameters and plan to increase their understanding of the roles of marine microbes and their role and functions in the vast seas and oceans.

Academic activities and exchanges

Academic activities

Since the 1980s, academic activities in marine microbiology have begun to develop. Marine microbiologists have had opportunities to join local or national academic associations and to attend academic meetings. These opportunities have led to important academic exchanges; opportunities to attend international academic conferences and travel abroad for further training. International scholars from countries such the USA, Japan, Germany, France, United Kingdom, the former Soviet Union and Finland plus other countries visited China to exchange academic views with Chinese scholars.

Academic reviews and comments

In summary, marine microbiological research in China is still in a young stage and in many respects needs to be further developed. A refined methodological approach to develop this field in line with the assigned tasks and orientation of the scientific community has not yet been established. A considerable number of problems await investigation and discussion. A proposal to enhance the orientation of research and inspire the scientific community was published by eminent academics with the ability to lead people both in discussion and in the laboratory (Sun and Chen, 1987). All this is needed to broaden the Chinese field of vision, enliven the spirit of academic cooperation to promote scientific exchanges as well as to improve the quality of research.

Acknowledgement

I am grateful to Professor Qianwen LAN for his skillful help.

References30(abr.)

中国普通海洋微生物学的研究进展*

摘　要　本文论述了1992～2003年中国普通海洋微生物学的研究状况，全文由13部分组成，分别是：①海洋微生物学状况基本参数的测定；②分子生物学技术的兴起；③微生物分类学研究；④海洋微生物学研究方法的进步；⑤环境与生态微生物学研究；⑥微生物活性物质的研究；⑦硫酸盐还原菌；⑧石油降解微生物研究；⑨空气微生物；⑩极地微生物；⑪藻菌关系研究；⑫极端环境微生物；⑬拓宽思路，不断探索。

关键词　中国　海洋微生物学　研究状况　探索与追求

从1992年至2003年，中国海洋微生物学的研究在1949～1991年42年的研究基础上跨入了新的历程，各类研究取得了丰硕的成果。除了病毒和海产养殖生物病原微生物的研究外，中国普通海洋微生物学研究所发表的论文起码有200篇以上，包括普通海洋微生物、分子微生物学、微生物分类学、环境与生态微生物学、微生物活性物质研究以及拓展思路、发现新途径等展望性的探讨等。

本文从13个方面论述中国普通海洋微生物学这十余年来的研究进展，并期望得到更好的发展。

1. 海洋微生物生物学状况基本参数测定

各种海洋环境中微生物的含量、生物量、生产力、生态特征等的估计仍是当前海洋微生物学研究的基本内容之一。所进行的研究有细菌计数、生物量、生产力及培养方法研究及改进，细菌ATP、POC、呼吸作用：增长及被摄入速率、多样性，不同介质中各类细菌数量、时空分布及变化与作用等[1-12]。

2. 分子生物学技术的兴起

分子生物学技术是近十几年来兴起的学科，传入我国也很快，如今它已经在病原体检测、疾病诊断、分类、新物种的发现上发挥重要作用。在我国陆源微生物研究上分子生物学技术应用得较早，在海洋微生物研究上稍后一步，但相关研究也已开展，如戴欣等用分子生物学技术对海洋细菌作的分子鉴定分类、新物种的探讨以及环境微生物和病原菌的检测[13-15,197]。

3. 微生物分类学研究

应用数值分类法于酵母分类上，如孙修勤等做的酵母支序分类，将隐球酵母等7属归为与子囊菌酵母平行发展，且同属一个发育系统。还有发光细菌分类、产芽孢细菌、河流弧菌、深海锰结核微生物的分类等[16-20]。

4. 海洋微生物学研究方法的改进

有关方法的研究包括水生细菌计数、微生物技术、分子生物学技术、生物量及多样性

---

* 原文刊于《中国水产》2004年专刊，青岛·全国水产养殖研讨会论文集：253-257.

研究法的改进、病原微生物检测技术、PCR 快速测定副溶血弧菌、酵母的培养、二维电泳术测嗜水气单胞菌及其蛋白质表达的影响因子、热休克应答中蛋白质表达度、药物资源微生物分类技术、气质联用技术，使得对相关微生物的生态学、影响因子和认知水平在本质上有了更深的理解，也提高了对它们开发利用的水平[21～27]。

5. 环境与生态微生物学的研究

我国海洋环境与生态的微生物学研究得到了广泛的开展，所涉区域包括我国四大海及邻近海区、极地及大洋。表 1 列出了这一研究所在海区及主要研究内容。

表 1　我国已开展的海洋微生物生态调查成果(1991～2003)

海区		主要内容	引文号
南海	南海	*Fusarium* sp(2849)代谢产物	28
	大亚湾	*Pseudomonas* sp 发酵产物成分	29
	珠江口亚热带红树林	真菌、子囊菌、半子囊菌、担子菌	30
	闽南-台湾浅滩上升流区	水、沉积物中拮抗菌	31
	厦门同安西柯虾场	总菌、弧菌、发光菌数量变化与环境生态关系	32
	厦门西海域	细菌数量与活动、与赤潮关系	33
	厦门西海域水体	细菌耗氧、动力学、环境容量、BOD	34
	厦门海区潮间带动植物	共附生微生物抗菌活性	35
	大亚湾、厦门港水	嗜盐弧菌数量及季节变化	36
	红树林叶凋落区	土壤细菌、真菌的作用及变化	37
	厦门西海区	细菌丰度、生物量、生产力、总大肠菌时空分布	38
	台湾东北角沿岸	氯化钠对酵母生长及蛋白合成的影响	39
	海洋水、泥、牡蛎、红树林落叶	河流弧菌生物Ⅰ型	16
	海南岛西南侧底质	SRB、环境生态因子	40
	北部湾东侧沉积物	SRB 含量分布	41
	气单胞菌广东海南株	气溶素类毒素基因分布	42
东海	南麂列岛海域	氨化细菌、解油菌、大肠菌	43
	长江口水	异养细菌、粪大肠菌、与环境-生态因子关系	4
	长江口及附近水	细菌数量、ATP、时空分布	44
	东海	异养细菌生产力	6
	浙江海岛海域	石油降解细菌生态分布	45
	长江口及冲淡水区	叶绿素 a、细菌 ATP、POC、呼吸作用速率	3
	东海	秋季异养菌分布，春秋蓝细菌生态分布及作用	8,46-47
	浙江-闽北陆架沉积物	SRB、生物地球化学因子	48
	黄海	秋季异养菌分布，春秋蓝细菌生态分布	8,46

(续表)

	海区	主要内容	引文号
北海	胶州湾潮间带沿岸区	硫酸盐还原菌(SRB)分布	49
	乳山湾东流区	细菌数量分布、环境相关	2
	辽东湾海冰	异养细菌、真菌、弧菌、解烃菌、粪大肠菌	50
	渤海	发光细菌分离测定	18
	海洋沉积物	SRB与钢铁腐蚀正因子	51
外海	桑沟湾水及底质	各类细菌、环境因子	52-53
	北海油田	高温 SRB(NM2)和古菌、H_2S	54
	白令海、楚科奇海沉积物	SRB	55
	南极菲德尔斯半岛环境	微生物含量	56
	英吉利海峡	蓝细菌增长速率、被摄入速率	
	南极长城湾及邻近海域	烃氧化菌、分布、变化	57
	太平洋、南极普里兹湾及邻近海域	微生物含量、分布	58
	南极菲德斯半岛南部及近岸水	粪大肠菌群	59
	环球、南极长城站、中山站及邻近区域	空气微生物	60
	白令海	蓝细菌分布	11

由表1可知,相关的生态、环境微生物学研究方面是相当广的,包括各种类群的微生物数量及时空分布,生态学及与相应因子的关系分析,具特定生理生化功能活性物质的微生物分离、测定及分布等。极地微生物及硫酸盐还原菌的研究也较深广。

6. 生物活性物质的研究

众所周知,海洋微生物是人们梦寐以求的巨大资源宝库。由于历史的局限性和技术条件的不足,限制了人们的追求。随着经济实力的雄厚,直到20世纪90年代,人们探求的眼光才越来越多地移向海洋,中国海洋微生物学家也不例外。世界上迄今为止已从海洋微生物中分离出800多个新的代谢产物。表2列出了这些年来微生物活性物质的研究状况。

由该表可见,我们所追求或进行的活性物质研究主要是维生素、酶、肽或蛋白质、甙等化合物,色素和抗生素等抗菌抗癌药物等。由此可见,海洋微生物展现出诱人的应用前景,具有可持续发展和易于产业化的独特优点,将会有更高更大的实用价值。

其中值得一提的是,以厦门学者为代表的福建学者所进行的新抗等筛选研究工作,取得了综合而丰硕的成果[81,84,86,94];另一个群体是青岛学者,他们对一株具有高产低温碱性蛋白酶活性菌株做了从筛选、分离直到活性成分分析和实用的系列研究[76~79];第三个群体是以林永成等为首的中山大学海洋天然产物化学研究室,他们从20世纪90年代初以来,不仅应用新方法、新技术进行海洋微生物区系、分类及性状研究,还从近百种海洋菌株中分离出一大批结构独特的新化合物[87~196]。这些研究表明了海洋微生物产物的多样性和优越性。

表 2　海洋微生物产物及其用途

微生物所属类别	所研究的微生物	主要产物及用途	引文号
细菌	北极 G-杆菌	胞外碱性蛋白酶	61
	乳酸链球菌 L-318	抑海洋弧菌活性物	62
	深海适冷菌 SM. 9913	适冷蛋白酶	63
	假单胞菌 SM. 9915	冷活性酶/低温蛋白酶	64
	海洋弧菌(*V. pacini* 4B-7)	碱性蛋白酶,用于核酸提取	65
	海洋假单胞菌 7～11	纤溶酶及蛋白酶	66-67
	埃氏交替单胞菌	褐藻酸裂解酸	68
	爱德华氏菌	菌膜蛋白	69
	芽孢杆菌 B-9987	酚类化合物,用于抑植病真菌的农用抗生素	70
	芽孢杆菌及假单胞菌	灵菌红素	71
	海洋弧菌 Z010	几丁质酶	81
	塔式弧菌	琼脂酶	73-74
	海洋细菌	Vit. B_{12}	75
	黄海黄杆菌 *Flavobacterium yellow sea* sp. Nov. ys-9412130	低温碱性蛋白酶	76-79
	芽孢杆菌	胞外抗菌蛋白	80
	海洋细菌	抑菌物	90
放线菌	海洋放线菌 2B	丁酰苷菌素 A	81
	植物上游动放线菌(链孢囊菌)	抗菌活性	82
	海洋放线菌	体外细胞毒,抑癌	83
	海洋放线菌	胞外免疫调节活性多糖	84
	海洋放线菌	壳聚糖酶,抗菌	85
	海洋动植物共附生放线菌	抗真菌活性物	86
	链霉菌 146	几丁质酶	87
	链霉菌 S-128	几丁质酶	88
	链霉菌 *S. xiahai ensis zhou*	碱性物、抗生素 s-111-9	94
	底栖放线菌	抗菌物质	83
	放线菌	胞外多糖	89

(续表)

微生物所属类别	所研究的微生物	主要产物及用途	引文号
菌物	海泥真菌 *Penicillium* sp. FS010441	干曲酶,低温淀粉酶	91
	鲁特格斯链霉菌鼓浪屿亚种	新抗生素,醇胺霉素,春日霉素	92
	海洋真菌	活性物	93
	海葵上真菌	抑菌物	95
	黄海葵附生真菌 *Penicillium* sp	4 种化合物,其中(E)N-[2-(4-羟基苯基)乙烯基]甲酰胺为首次分离的天然物	96
	产蛋白酶菌	蛋白酶	97
	鲣节霉的 5 株散囊菌	酸性蛋白酶,酸性羧肽酶	98
	Fusarium sp(# 2849)	10 种化合物,有鞘胺醇甙等	29
	海洋丝状真菌	活性物	100
	红树林内真菌 1356、细菌	代谢产物、4 个环二肽	101
	炭团菌 *Hypoxylon* sp	沸点代谢物计 18 种	102
	微生物	DHA	103
	红树林内真菌 2524	肽	104
	鹿角菌	含丙二烯醚的 N-肉桂环肽	
	N350 菌株	对 P388 和 KB 肿瘤细胞强烈抑制物	104

7. 硫酸盐还原菌(SRB)

我国海洋 SRB 的研究起始于 20 世纪 70 年代[105]。由于 SRB 广泛参与地球化学硫循环、石油形成与金属腐蚀,随着海洋石油等开发工程的逐步拓展而越来越受到人们关注。此方面的成果一部分,是在不同生态环境中的腐蚀活动和防腐工程的应用项目中以技术资料出现;一部分涉及环境与生态调查;第三部分则是探讨它们在地球生物化学活动中的作用[40,41,48,49,54,54～55,105～107]。

8. 石油降解活动

参与石油降解活动的海洋微生物类群不少。已进行的研究包括不同类群细菌、真菌的解油活动,相应的生态调查及模拟实验,降解条件等。研究表明微生物在防止石油污染等环境保护方面起着重要作用[108～118]。

9. 空气微生物

海洋空气微生物的研究始于 20 世纪 80 年代,进入 90 年代以来已出成果,包括测定和分析海洋空气中海、陆源微生物含量及其时空分布和变化。所研究地域范围是我国四大海、近海、海滨直至极地和环球,由此海洋空气微生物成为全球空气微生物状况的一部分,与海洋微生物学结成了一个整体[119～137]。

10. 极地微生物

我国极地微生物学的研究始于20世纪80年代初。在我国自行组织的极地考察之前，就已有学者进行大洋海水微生物学研究。自1984年以后，南极考察常有微生物学研究项目，内容涉及微生物分类、生态、污染、环境和应用及活性物。自20世纪90年代后期，涉及北极考察；进入21世纪之初，涉及第三极——西藏高原。发表论文数十篇之多，有关方面将另著文论述[55~61,138~142]。

11. 藻菌关系研究

所有海洋微生物在自然状态中均与周围环境的生物、非生物接触，因而其生存状态与作用已受人们关注。除了病原菌、养殖海藻病菌外，其中之一便是藻菌关系。这包括赤潮中的微生物类群与作用、有害有毒藻类对微生物的影响、饵料藻类培养中微生物的作用、藻类对某些菌类的促进作用、微藻的抗菌活动、藻菌的跨界融合等。这些研究为处理藻菌关系、海藻养殖、赤潮治理、环境和生态保护、趋利避害、生物工程等开辟了新思路[143~149]。另一些研究则是关于与污损相关的微生物[150~152]。

12. 极端环境微生物

极端环境主要指极端海洋环境如高盐、高/低温及高压、高碱、高酸及热液口等环境，对其中的微生物研究，近年来也十分引人注目。何若丸等介绍过嗜盐细菌的研究进展；方金瑞等做过嗜盐(杆)菌的色素、耐/嗜碱放线菌的抗生素研究，介绍过嗜碱耐冷深海微生物的分离法及其活性物。极地嗜冷、耐冷微生物也引起了专家的关注[153~158]。海洋超高温古菌、高压热溢口区微生物及趋磁细菌的研究在国外相当活跃[159~162]。

13. 开拓思路，不断追求

我国学者密切关注国际间海洋微生物学发展新动向，纷纷推荐和采纳新思路、新技术和新方法，这包括药用资源微生物技术及开发利用[23,170,177,182]、脂多糖活性物[163]、lectins[163]、极端环境微生物[156~158,176,180,183]、非可培养状态活细菌[165~166]、海洋细菌BAC活性物文库[167]、分子生物学技术等[113~115,168,197]。分子生物学技术与传统有效方法的结合，正在或将要使人们对海洋微生物有更新、更全面的认识和理解。海洋中存在着的、至今仍未被认知的微生物以及有神奇活性作用物质的微生物之开发利用等都已在人们探索、思考之列[169,181,184,186]。海洋微生物学科正在拓展和增长自身的分支学科，它们与其他领域的结合使得海洋微生物学充满生机。知识经济时代的到来大大唤醒经济观念，因而开发和应用海洋微生物资源比以往任何时期更为迫切和热门，海洋微生物菌库、代谢产物库、基因库TE在催生，但彻底认识它需要在以时代为计的时间中孜孜不倦、默默无闻、持之以恒、不断创新地奋斗不息。中国海洋微生物学的巨大突破和成功腾飞必将建立在对广阔海域不同生境微生物广泛而深入认知的基础上，即需要自身广泛而坚实的基础、系统并有原创性的见地，才能达到和完成。到那时，中华民族就能在海洋微生物学科上缩短与国际先进水平的差距，为全人类作出更大贡献。

参考文献197篇(略)

（合作者：孙丕喜）

PROGRESS OF GENERAL MARINE MICROBIOLOGY IN CHINA DURING 1992-2003

(ABSTRACT)

Abstract This paper discusses the situation of general marine microbiology in China during 1992～2003. The paper is composed of thirteen parts. ①Determination of elemental parameters of biological conditions of marine microbiology. ②Rise of molecular biology; ③Research on microbial taxonomy; ④An improvement on research methods of marine microbiology; ⑤Research on environmental and eco-microbiology; ⑥Research on microbial bioactive substances; ⑦Sulphate-reducing bacteria; ⑧Research on petroleum-degrading microbes; ⑨Air-borne microbiology; ⑩Polar microbiology; ⑪Research on relationship between algae and microbes; ⑫Extreme environmental microbiology; ⑬Open up train of though and explore continually.

Key words China; Marine Microbiology; Situation Of Research; Explore & Seek

中国海洋病毒研究进展*

摘　要　论述了近10余年来(1990～2003)我国海洋病毒研究状况,包括对虾和其他无脊椎动物、鱼的病毒病及其诊断检测技术和防治,并对我国海洋病毒学今后研究提出一些看法。

关键词　中国海洋　病毒　病毒病防治

自20世纪60年代以来,世界各国对海洋病毒的研究取得巨大进展。迄今已知海洋甲壳动物病毒已达70余种,海洋哺乳动物病毒3种。我国养殖水产动物的病毒病计60余种。对海洋病毒研究应包括病毒类在海洋中的分布特征、生物学、生态学性状作用与地位及应用、在寄(宿)主体内外的发生发展过程、归宿及相应诊断检测防治技术措施等。

本文论述我国对虾、其他无脊椎动物及鱼等的海洋病毒研究现状,并展望未来的研发方向。

1　对虾病毒

由于对虾的可食用性和经济价值,使得沿海各国都十分重视对虾养殖。大面积、高密度的对虾养殖终致暴发流行性病毒病等病害,人们开始重视病毒病的发生发展规律研究。由于病毒分类的难度,真正被国际病毒分类委员会认可的病毒种迄今只有BP[1,2]。下面就我国对虾主要病毒作一阐述。

1.1　对虾白斑综合征病毒(WSSV)

WSSV是近年来研究的热门课题之一。它于20世纪90年代初首先发现于中国。WSSV因对虾种类(原始宿主)、地域、形态发生及病理症状等差别而在国内外有不同命名(至少有7种叫法,4种分类单位),由于研究的广泛和深度,它在病毒学研究中是屈指可数的,故成为迄今第一个被获得基因组序列的海洋动物病毒,且是目前已知最大动物病毒。人们就WSSV大小、形态结构、化学组成和理化特性、基因组及其编码的多肽、致病机理、宿主和传播途径及流行病学等作了广泛研究[3,4]。已经推断出对虾养殖池WSSV同时存在口感染和浸泡感染,WSSV对上皮组织和造血组织有较强嗜性。WSSV囊膜中可能有增效蛋白或类增效蛋白。WSSV除感染中国对虾等绝大多数种类对虾外,还感染非对虾种类,如端足类、介形类、蟹、龙虾、桡足类、水蝇等甲壳动物和少数淡水虾。它们可能成为WSSV传播的中间宿主。已证实WSSV的传播,而垂直传播尚未肯定。对WSSV检测包括如PCR生物1447bp核酸片段,可用于早期检测。PRDV(RVPJ)基因组*EcoRI*部分酶切片段为灵敏套式PCR术。中国对虾WSSV部分基因组酶切片段经光敏生物素或地高辛标记的探针用于检测。海洋局三所已在世界上率先破译和分析了对虾白斑杆状病毒(WSBV,同WSSV)基因组全部密码,它将对海陆生物分子进化关系研究产生重大影响。

---

* 国家海洋局生物活性物质重点实验室资助。

原文刊于《齐鲁渔业》,2005,22(4):11-14

1.2　对虾的其他病毒

从中国对虾身上发现的病毒除 WSSV 外，还有中国对虾杆状病毒（*Penaeus chinesis baculovirus*，PCBV），血细胞感染的非包涵体杆状病毒（*Hemocyte-infecting nonoccluded baculovirus*，HB），类淋巴器官细小样病毒（*Lymphoidal parvo-like virus*，LOPV）；中国对虾细小样病毒（*Penaeus Chineses parvo-like virus*，PCPV）；肝胰腺细小（样）病毒[5]、肝胰腺坏死病毒[6]、幼虾肠组织中似痘状病毒[7]、肝胰腺上皮细胞质中杆状病毒[8]、卵细胞一种杆状病毒和一种球形病毒[9]、呼肠孤病毒Ⅳ型（REO-Ⅳ）、囊膜状病毒科 C 亚群病毒、C 亚群杆状病毒及淋巴组织培养中发现的病毒[10]等等。

从长毛对虾发现无包涵体具囊膜 C 亚群杆状病毒[11]、东方对虾的杆状病毒[12]、刀额新对虾的无包涵体杆状病毒（WSDV）和南美白对虾的桃拉病毒等等。

1.3　斑节对虾病毒

斑节对虾病毒病是中国学者研究的另一重点，但 Lighter 等 1981 年最先从台湾地区报导斑节对虾杆状病毒（MBV）[13]。其后就该病毒检测检查法、不同时期虾体病毒感染[14]，在虾体不同部位及细胞内的分布[15]及发病理化因子、防治方法、流行病等作了广泛研究。

斑节对虾尚有 C 型杆状病毒（TCBV），也有 WSSV、球形病毒及肝胰腺细小样病毒[16]、呼肠孤病毒—Ⅲ型（REO-Ⅲ）。

黄头症杆状病毒（YHV）除感染斑节对虾外，也感染甲壳纲其他非对虾种类而构成中间宿主。相关病理、电镜观察和 RT-PCR 检测技术可检测出来。日本对虾有呼肠孤病毒—Ⅲ型（REO-Ⅲ）。

2　*其他无脊椎动物病毒*

由于病毒存在的广泛性、普遍性，因而在许多宿主上都可找到病毒的踪影，迄今世界上已发现的贝类病毒有 20 余种，重要一点是发病不明显情况下病毒与带毒宿主关系常被人忽视，一旦有较重大病情或威胁经济价值，才重视它的存在。此时检测诊断和防治显得重要。随着病害蔓延和人们的重视，已从一些其他无脊椎动物发现了病毒，它们是：杂色鲍裂壳病球形病毒[17]、九孔鲍病原之一的球形病毒、皱纹盘鲍裂壳病球形病毒、牡蛎肝胰腺病毒、缘膜病毒（属虹彩病毒）病（OVVD）、面盘病毒、疱疹型病毒、食用牡蛎 Birnavirus、卵巢囊肿病肿疡样病毒、三种虹色病毒病、Papovaviridae 科 papova 病毒样和 Herpes 病毒样感染、呼肠孤病毒、三角帆蚌瘟病病原似砂粒病毒、海湾扇贝疱疹病毒、中华绒螯蟹幼蟹球状病毒及蟹的呼肠孤病毒（REO-Ⅰ和 RED-Ⅱ型）、杂色花蛤 1 种杆状病毒和 2 种球状病毒、栉孔扇贝囊膜病毒（球形病毒）[18]及急性病毒性坏死症（AVND）病毒。

3　*鱼病毒*

对于鱼病毒研究，也引人注目，这得益于鱼病毒诊断技术不断改进和完善。这方面研究包括 4 条检测技术的出现，即电镜观察、细胞培养、免疫学和分子生物学技术[19]。目前就世界范围而言，发现的鱼病毒不少，仅大菱鲆就有 11 种病毒，而国内迄今尚未有相关研究成果。淋巴囊肿病毒（LCDV）感染牙鲆等 30 个科的海水鱼。迄今鱼病毒病研究内容主要是病原、症状、诊断、形态观察（包括超微结构）、病理学、预防等。病毒病防治方面已发明病毒疫苗，但仍缺乏有效治疗手段[20,21]。

发现的鱼病毒包括有欧鳗狂游症圆形病毒[22]、真鲷球形病毒、杆状病毒[23]、牙鲆淋巴囊肿病病毒[24]、条斑星鲽的病毒性神经坏死症(VNN)、云纹石鱼淋巴囊肿病病毒、青鱼和草鱼出血症病毒、草鱼呼肠孤轮状病毒、传染性胰脏坏死病病毒、传染性造血器官坏死病病毒(IHNV)、鲤痘症病毒(引进鲤痘疮病)、台湾石斑鱼彩虹病毒、红鳍东方鲀弹形病毒。此外,从条石鲷、尖吻鲈上也检出过病毒。

4　*病毒病的诊断检测技术*

病毒学研究成果建立在相关方法上,迄今比较成功的早期快速诊断法有 9 种,即:直接压片光镜术、暗视野光镜术、组织病理学测定、透射电镜、扫描电镜、荧光抗体技术、酶联免疫技术(ELISA)、核酸探针和 PCR 方法[25]。其中电镜术可用于探测病毒有无及病毒类型初鉴、病毒引起的组织细胞病理变化、病毒在宿主细胞中的形态、结构及繁殖动态。其所用方法是病原体负染色、真空喷镀、对病灶组织超薄切片后观察[26]。细胞培养技术用于分离和培养病毒,据其特征作初步鉴定,以终点稀释法或空斑技术定量研究。荧光抗体术是将免疫化学和血清学高度特异性、敏感性与显微术高度精确性相结合方法。鱼酶联免疫吸附试验为抗原——抗体免疫反应和酶的催化反应相结合技术(即 ELISA),用于虾,ELISA 则包括抗 IHNV 单克隆抗体、抗 HNBV 单克隆抗体间接 ELISA 检测法。多克隆抗血清以建立病毒间接 ELISA 检测法及双抗体夹心 ELISA 检测法。

鱼病毒核酸探针是针对病毒特异 DNA 序列的基因探针诊断法,它为检测早期感染、研究特定病毒在宿主中传播途径提供有效方法。PCR 法则借助特定引物引导,靠 DNA 聚合酶催化,经高温变性、低温退火、适温延伸为一周期,循环进行,使目的 DNA 迅速扩增。组织病理学测定靠 Giemsa 染色涂片,已成功用于 HPV 检测。FE 染色法可现场测定。至今用于虾病毒诊断以组织病理学测定。透射电镜、核酸探针和 PCR 法较为成熟,诊断的敏感性最高[27-29],如组织病理学测定透射电镜对黄头症杆状病毒(YHV)的诊断,核酸探针、PCR 对传染性皮下及造血器官坏死病毒(IHNV)的检测。尽管已研究出上述一些快速精确诊断检测技术,但并不十全十美,只有少数技术具有商业性诊断试剂盒水平,如已有 IH-HNV 基因探针作为诊断试剂盒。

5　*病毒病的防治*

病毒的发生、生存、发展都有其自身的规律,这个规律又受内外环境因子的制约。连年来病毒病爆发流行引起人们重视,已发表了一些成果,这些成果包括病毒在宿主细胞内的装配表达、超微结构、侵染途径,在各种类型细胞中的分布[30],传播途径及与其他生物的生态学关系,包括与其他微生物间相互作用及生物学特征,如细菌并发症等[31,32]。

防治措施应该是综合性的。早期诊断、预警系统的建立、养殖池塘的科学设计及预防措施的采用,包括多效 S 型净化池、围栏封闭预防、科学用药等等[33-35]。

虽然对病毒基本特征和检测技术的研究有较大发展,但病毒病预防和治疗发展仍相当缓慢,也无很有效药物治疗法。

总之病毒病防治主要应抓三大环节:控制和消灭病原体、提高养殖对象机体抗病力和改善养殖环境[2],以有效避免病毒病原最为有效[36]。

6　*对海洋病毒今后研究的几点看法*

我国海洋病毒研究总的来看,起步较晚,相关基础研究较薄弱。提出以下 4 点建议供大家讨论:

6.1　我国现有病毒学研究人才应向海洋方面倾斜并增加海洋病毒研究人才的培养。

6.2　科学规划我国海洋病毒研究总体设想，统筹安排、协调好研究区域（领域）、研究人力资源和研究方向。

6.3　海洋动植物病毒，包括浮游病毒[37]，在中国广大海域及特定区域中的生物学、生态学状态及在各环节中的作用和地位的基础研究应大大加强[38]。

6.4　大力加强重点养殖对象病毒病的发病机理、发展过程诊断、检测技术和病害防治方法的实用研究，进一步减轻病害损失。

参考文献 38 篇（略）

现代海洋微生物学之父
——佐贝尔(Zobell C E)的早期成就*

纪念佐贝尔诞辰102周年

摘　要　记述国际著名海洋微生物学家,亦即被许多人推崇的现代海洋微生物学之父、"著名老人"佐贝尔从出生(1904年)到1946年,他出版专著《海洋微生物学》这段时期中的履历和主要研究内容、工作成绩和世人评论,目的是为了激励年轻学者早立志献身科学,抓住机遇,勤奋努力,为人类多作贡献!

关键词　佐贝尔　海洋微生物学　专性海洋微生物

2006年是著名海洋微生物学家——佐贝尔(1904—1989)诞辰102周年。他毕其一生致力于海洋微生物学研究,对海洋微生物学的开拓、发展起到了极为重要而关键的作用,迄今仍为世人所怀念和尊敬。我国海洋微生物学起始于20世纪50年代末[1],这比国际间晚了几十年。时差与国情拉大了我国与国际先进水平的距离,但国人已意识到并奋起努力。本文回顾佐贝尔早期工作成就,旨在激励人们,尤其是年轻学者奋发努力,赶超国际海洋微生物学领先水平。

1　佐贝尔早期简历

被国际海洋学界许多人称之为"现代海洋微生物学之父"的Zobell C E于1904年生于美国犹他州Provo的8口之家,1905—1922年在爱达荷蛇河畔的农场别墅中度过。1927年从犹他州农学院获得细菌学学位;两年后获得临床外科硕士学位;1931年取得贝克利加州大学博士学位。此时他撰写了长达175页的学位论文。此年前后一段时间,他的主要研究目标为非海洋哺乳动物病原——布鲁氏菌。

美国斯克里浦海洋研究所(Scripps Institution of Oceanography,缩写为SIO,下同)的这一名称是1925年才采用的,其前身是斯克里浦生物学研究所。1932年佐贝尔进入SIO,起初,他延续了原来的工作,其主要兴趣之一是细菌生理学,后来便逐步转向海洋细菌学/海洋微生物学,直至毕其一生。从此一系列桂冠直至"现代海洋微生物学之父"等便逐步当之无愧地戴到他头上。

2　国际海洋微生物学的早期研究

对海洋微生物学的研究,国际间可追溯到1875年,Pflugere在此年确认了海鱼有发光细菌;1884年Certes在Talisman探险活动中,从100个海水样品中发现了96个有好气细菌;Fischer在1884—1904年间连续发表论文9篇,在以浮游植物为主要对象的探险研究中,也涉足海洋细菌等微生物,包括发光细菌、固氮细菌、芽孢杆菌等;Russell于1892—

* 原文刊于《海洋地质动态》,2007,23(5):38-41.

1936年间发表论文4篇，论述伍兹霍尔附近大西洋海区的细菌区系、海发光问题；沙俄/前苏联学者 Issatchenvo 于1912—1940年发表论文10篇/部，其中以1914年的关于北冰洋的细菌调查最为著名；其他前苏联早期学者如 Butkevich、Kusnetzow 和 Rubentschik 等对海洋微生物学研究也作出了不可磨灭的贡献。这些学者的贡献主要在于用传统的、初步的方法确认海洋中广布活细菌，并开启了海洋微生物学研究方向，使之逐步孕育成一个独立的分支学科的雏形[2]。

但是以上的研究大部分中断于1914年，也即国际海洋微生物学的开拓时期因第一次世界大战而中止了十年。其后于20世纪20年代末、30年代初才开始了应用及定量水生细菌学研究，海洋微生物学研究得以恢复，才有佐贝尔等为现代海洋微生物学的创建和发展时期[3]。

3　佐贝尔早期研究的主要内容和主要成就

3.1　佐贝尔早期研究的先决和重要条件

佐贝尔在 SIO 的研究得益于其早期时代的各种合适条件，当时正逢国际间人类医学的研究开始扩展至非人类医学上，人们开始探索研究环境中细菌在医学上的作用。他在 SIO 的前任 Gee(1901—1961)和 Bigelow 1931年的重要文章、所长(前后两届)Vaughn T W 和 Sverdrup H U 的高瞻远瞩及 Shor E N 的“对探索海洋”的支持以及1936年 SIO 向广泛海洋科学研究的转变都是佐贝尔研究的助推器和强大后盾。佐贝尔1934年的宣言性论文为研究海洋细菌打下了基础，在这篇论文中，他宣布自己最迫切的问题是研究海洋中细菌及相关微生物是如何在生物学、化学和地质学过程中发挥作用，该作用的程度和范围是什么[4]。

3.2　佐贝尔明确专性海洋微生物的土著性

当初佐贝尔创建现代海洋微生物学仍需冲破许多陈规陋俗。首先人们普遍怀疑海洋微生物不是土著的(autoch thonous)，而是陆源污染来的。佐贝尔与同事(Feltham)，一起进行了严肃的研究[4,5]。他们发现淡水或非海洋的细菌在海水培养基上生长少，而海洋来源的细菌要多得多。他的人工海水培养基满足了70%～75%的海洋种之需，再加10%的真正海洋水进去；将使80%～85%的海洋种生长。后来的研究进一步明确，远岸海区，尤其是深海大洋，更多的是土著者。相关论文发表于1933年[5]。这也是专门论述专性海洋细菌的第1篇，尽管此年他已发表了海水光化学硝化作用和海洋细菌对浸没载片附着的研究[6]。这段时间，他和他的同事发明并运用了一系列海上采样装置和培养基，迄今仍是行之有效的，如J Z海水采样瓶和 Zobell 2216 系列培养基，并在以后的年代中不断改进完善，经得起考验。

3.3　浸水表面和贮水容器的微生物

对浸水表面(船身)生物沾污及附生(Periphytic)细菌的研究，佐贝尔作了相当的工作[7]。在其所研究的试样培养物中发现32.9%是趋触性(thigmotactic)细菌。对表面沾污现象作了较深研究，发现此机理以微生物首先在表面形成细菌膜为先导，其后大生物以各种方法和营养需要而吸引上去，致污沾的生物如藤壶、贻贝等软体动物，海鞘等被囊动物及其他附着生物和定居生物及相关植物顺序延续全部形成。这一机理为海洋防污损等的材料、油漆工业提供了理论根据。贮存海水中的细菌群落的变动，与贮水体表面/体积比动力的大小、氧的有效性、贮存时间长短和营养物的浓度相关[7]。这些研究即是佐贝尔举起的另一面海洋微生物学大旗。

3.4　盐水的杀菌性

美国大盐湖(Great Salt Lake,缩写为 GSL,下同)中的土著微生物是佐贝尔 20 世纪 30 年代的一个工作内容。包括用浸没玻璃技术研究菌落生长、盐水杀菌效果的持续性及嗜盐微生物包括耐盐淡水种对盐度的适应等等。此时佐贝尔工作的 4 个重点即表面沾污、土著微生物、细菌生理和盐水杀菌性融为一体了[8,9]。

3.5　海洋环境因子与海洋微生物

作为佐贝尔海洋微生物学研究早期关注的另一个问题是温度等环境因子。4/5 以上的海底温度低。海洋细菌(微生物)是如何在生理功能上适应的？研究发现在环境温度相当均匀的海洋,都有其相应的微生物存在。温度的突然升降对它们的代谢与生殖起暂时作用。但温度与其他因子如有机质等浓度状况的共同作用,将使微生物类群与代谢率产生重大变化。挥之不去的另一个问题是高渗高压因子。初步研究表明海洋菌嗜好海水,但对渗透压改变的抗力却没淡水菌强,也不喜好淡水或与海水等渗的氯化钠溶液。而深海静压不抑制那里细菌的生命活动,即使是深海者来到表层水也不受压力降低的影响。对此等环境,年轻的佐贝尔意识到海洋微生物在其中必定起着重要的作用,心中又升腾起高压生物学,启动高渗环境中的细菌研究并发现了海洋细菌的可驯化性[10]。

佐贝尔进入 SIO 的第 1 篇关于海洋环境中的氮循环的论文作于 1932 年,并于 1933 年发表[6]。他创造了适于海洋环境硝化细菌的研究方法,发现了硝化作用的一部分是由光化学激活的,其中专性海洋尿素细菌也起作用。它们与氮循环相关。佐贝尔继续研究海中细菌的垂直分布现象,初步发现阳光的杀菌作用以及静水压、溶解氧、浮游生物和季节变化现象。由此将研究扩展至海洋沉积物,创立并发展了海洋的地微生物学(Geomicrobiology)。所涉及内容有沉积物细菌等的种群、数量、分布、活动范围、影响因子等。这一工作还包括海底微生物作为底栖无脊椎动物的食物,它们本身的厌氧生活、物质的分解作用和计数技术等等的研究[11,12]。

关于海洋细菌在石油形成中的作用的研究也令人注目,并得到著名地质学家的赞许[11],由此对海底环境因子包括氧化还原状态(条件)、有机质成分及状态对底栖微生物的影响作了一些研究,冲破了 19 世纪以来的"无生物区 azoic zone"这一概念禁区,证明高压和低温于那里的细菌之存在及其活动不仅无害,而且其生化活动在地质学上改变着环境,包括石油等的生成。

3.6　应用和基础海洋微生物学

佐贝尔的兴趣是广泛的,他还关注海洋水产的微生物病原机理研究,对具有较高经济价值的品种特别感兴趣,包括太平洋鳝鱼和加州贻贝等。他们不仅找到了病原菌,而且使基础海洋微生物学得到了发展。1934 年发现了新种微生物——鱼皮无色杆菌即鱼鳞假单胞菌[13]。在直至 1944 年的约 10 年中,他们一共发现了 60 个海洋新种[14]。这一成就是难能可贵的。他鉴定的物种(包括以后鉴定的)迄今仍大多为国际命名委员会所认可。佐贝尔从应用和基础两方面对海洋微生物学产生了浓厚的兴趣,并一直贯彻于其整个生涯。此外佐贝尔对海水中细菌的有机质氧化速率、好气细菌的培养要求(技术)、采样方法改进、大洋中肠系细菌的生存问题、海洋空气中的微生物、石油及烃的降解、几丁质降解等广泛而普遍的课题也有广泛的研究。

佐贝尔在1933—1944年间发表论著约80部/篇。这在他同时代同专业的学者中是首屈一指的。这些成果为他于1946年出版专著即《海洋微生物学》奠定了基础。在开创和引导国际海洋微生物学的多方位研究上,他的早期成就为他成为"现代海洋微生物学之父"埋好了基石。他毕生创作了约250部/篇论著(包括与他人合作)。"现代海洋微生物学之父"这一桂冠是他于1989年逝世之后获得的[16]。他被公认为该领域的杰出老人"Grand old man",因而根据学识、成就和对学科发展的划时代贡献,他获得如此桂冠是当之无愧的,是经得起时间和空间的考验的。他成了全球此学科领域的开拓者和纯真老手,因而被世人所推崇和尊敬。

4 结语

回顾佐贝尔成为现代海洋微生物学之父的早期成功,其主要成就大体可归结为以下5方面:①明确了海洋微生物这一特殊群体的普遍意义;②基本确立海洋微生物学的研究方法;③认识并加深钻研海洋微生物与环境—生态因子的关系;④开创和发展海洋微生物在物质转化、能量流动中的作用和地位的研究;⑤关注海洋微生物的经济重要性,等等。当然他的早期成就同样也是时代的产物,也是建立在其专业及相关领域学者成果的基础上的。由此不难看出一个人的成功在很大程度上取决于他早期的状态。佐贝尔有了他早期发展的先决条件:恰好的时势及和谐的周围、人事环境加上他自己的审时度势,勤奋努力和不倦进取,终于用不长的时间造就和奠定了自己未来发展的基础。1946年以来的60年中,国际海洋微生物学发生了革命性变化,从传统方法走向分子生物学水平,其资源的开发利用越来越受世人关注,时势创造了英雄用武之地。当今的中国,就一般而言,已为年轻学者提供了较好的外部条件和必需的工作环境,那么余下的部分就是要靠自己的眼光和先进方法夯实功底,抓住机遇。将来之不易的经费尽量用于工作,而不是散掉。不好高骛远,而持之以恒求真务实。真正与时俱进,奋斗创新。

参考文献16篇(略)

中国海洋微生物学 2006 年研究足迹

摘 要 居新世纪第一个年代中间年份的 2006 年，数百名中国学者继往开来研究基础和应用海洋微生物学，发表文章至少 363 篇，其中环绕海产养殖病害及防治、机理的研究占了 63.9%。基础性海洋微生物学研究，包括先进的方法学、室内外所用仪器的研发及向深广海洋的拓展尚需加强。

关键词 2006 年 海洋 微生物学 基础和应用性研究

进入 21 世纪，中国海洋微生物学研究驰入较宽阔的快车道。作为 21 世纪第一个年代中的中间年份，2006 年，中国海洋微生物学研究取得了丰硕的成果。中国学者既关注相关动态，尤其是跟踪国际先进，又重视并实践自身的课题研究。这表现在对基础学科课题的重视，但更多的精力则用于海产养殖动植物病害的防治、机理研究和应用中。本文归纳 2006 年全年中国这方面研究的主要论文，向世人展示成就和去向，以期利于中国海洋微生物学研究更健康发展。

1. 关注国内外动态

1.1 综合：海洋微生物研究的回顾与展望，发展海洋微生物学及微生物技术的思考与实践[1]及核糖体工程技术二次开发微生物资源、化学生态研究，治理污染中所起作用，海洋生物药、生物资源与可持续利用对策、近海微生物源示踪监测技术、多样性、抗菌活性物质、毒素。深海极端环境深部生物圈微生物[2]、深海热液生物群落。病原微生物快速检测、在水产养殖环境生物修复中的作用机制。鳗弧菌铁吸收系统毒力作用、动植物共附生微生物的分离和抗菌活性、基因资源的研究与利用。

1.2 病毒等：对虾细胞培养与对虾病毒体外培养、鱼淋巴囊肿病毒、噬菌体治疗及在水产养殖中的用途、应用噬菌体随机肽——研究抗溶藻弧菌及其外膜蛋白单抗识别的特异性。

1.3 古菌：多样性。

1.4 细菌：活性代谢产物[3]、胞外多糖、细菌素及在水产养殖中的应用。养殖环境中细菌耐药性的危害、嗜水气单胞菌毒力与毒力基因分布相关性、一株病原假单胞菌对氯霉素中间体的专性转化。

1.5 蓝细菌等：微藻利用与开发，赤潮生物图像实时采集系统、微藻产特种天然类胡萝卜素，氢代谢相关酶及其分子生物学。

1.6 真菌：抗肿瘤活性物、红树林内生镰刀菌 332 代谢产物 Ennictin 发酵[4]。

2. 基础及应用基础

2.1 方法学

自主式深海浮游生物浓缩保真取样器控制系统设计、T-RELP 分析测定胶州湾细菌和蓝细菌多样性、海中蛭弧菌分离与纯化、微藻活体及乙醇固定时基因组 DNA 微卫星提取法。

2.2 病毒：研究病毒受体的生物技术途径[5]。

2.3 细菌：WPZ深海细菌基因组文库构建和克隆子测序对比分析，南极深海沉积物宏基因组DNA中低温脂肪酶基因克隆表达及序列[6]、东太平洋结核区细菌调查及鉴定、深海抗锰细菌分类鉴定、热带太平洋活性微生物菌株筛选及鉴定、西北太平洋沉积物细菌多样性[7]、热带西太平洋暖池深层沉积物SRB的分子系统分析、北极太平洋扇区沉积物细菌多样性系统发育分析及北极高纬度海域海冰嗜冷菌系统发育多样性及其低温水解酶，南极假交替单胞菌S-15B的分子鉴定及胞外多糖促小鼠脾淋巴细胞转化的作用、南极微生物分离及抗肿瘤活性、南极海洋假交替单胞菌胞外多糖生产与特征。产DHA南极菌筛选及培养条件、低温蛋白酶产生菌YS-9412-130原生质体制备与再生[8]，产冷适应复合酶的南极假单胞菌NJ.197的选育、发酶及酶学特性。一种水产益生菌的鉴定。

最小弧菌热不稳定型溶血素基因克隆与序列分析，新鞘氨醇菌PAHS降解基因和降解特性、溶藻弧菌菌体及其外膜蛋白多克隆抗体、假单胞菌cn4902磷酸甘油酸脂酶基因克隆与鉴定。红树林海洋细菌分类鉴定及活性物。枯草芽孢杆菌液体发酵条件。

嗜热菌分离、筛选、分子分类、聚磷菌YSR和YSC-1的分离与鉴定及YSC-1的生物学特性、光合细菌的鉴定、生长特性及应用、光合细菌PS培养参数及光合菌对活性染料的脱色作用、好气不产氧光合细菌。长赤菌生长的环境控制和Bchla的表达。黄海冷水团浮游细菌分布[9]和该水团附近沉积物趋磁细菌及其磁小体特性、细菌对海水中各形态氮的影响、环境因子对乳酸杆菌LH生长的影响。

沉积物中反硝化细菌的分离及硝酸盐、氮作用、一株海洋菌的鉴定及活性物、哈维氏弧菌TS-28菌株抗原性，植物生长激素对细菌生长及净化N、P能力的影响、枯草芽孢杆菌BS-1拮抗溶藻弧菌、解油菌群构建和解油过程。假单胞杆菌QD80低温碱性蛋白酶的化学修饰、粘质沙雷氏菌发酵条件。生物共栖细菌抑藻活性。小单胞菌与大洋螺菌发酵液化学成分，胶州湾链霉菌鉴定及抑菌活性，抑真菌活性的全霉素链霉菌。

2.4 放线菌

深圳福山湾红树林放线菌筛选及其抗菌活性，产杀虫活性放线菌筛选及鉴定，菌株M326活性代谢产物，活性组分AM105-Ⅱ的分离与结构鉴定[10]，M313菌株的鉴定及抗肿瘤活性物，海绵上抗肿瘤放线菌分离鉴定及菌株A0105发酵条件。

2.5 真菌

Aspergillus sp. MF13Y的抗菌特性及BH4次级代谢物。深海沉积物抗菌抗肿瘤活性真菌的筛选[11]，舟山近海底栖真菌及其抗生活性，红树林内生真菌种群动态及活性，M095菌株遗传转化体系的建立、抗MRSA真菌的分离鉴定及W0707菌株的鉴定，南极酵母*Rhodtorula* sp. NJ-298产虾青素[12]，海绵源黄灰青霉SP-19抗肿瘤活性物。

2.6 微藻、蓝细菌

微藻脂肪酸去饱和酶，南海三亚湾及邻近海域蓝藻类群组成、分布及演替特征，聚球蓝细菌的16*S* rDNA系统性及生理生态特征[13]，两株蓝细菌的固氮生理。

2.7 海绵：抗MRSA的海绵真菌分离及W0707株的鉴定、可培养与原位微生物组成的DGGE及PCR-DGGE指纹分析、共附生微生物活性生物与化学防御、宏基因组文库构建及抗菌肽功能基因初选[14]。

3. 应用

3.1　养殖海产病害防治、机理

水生动物溶菌酶及病原微生物快速检测研究进展、病毒受体研究的生物技术途径。细菌素及在水产养殖中的应用前景。发光细菌测生物毒性、育苗时“荧光病”防治、荧光抗体术测体内河流弧菌、弧菌病及致病机理、斑点 ELISA 术检测副溶血弧菌。杀鲑气单胞菌一新亚种，嗜水气单胞菌病的免疫技术及该菌温敏胞外蛋白酶基因 eprj 的克隆与检测[15]，中草药杀嗜水气单胞菌，蛭弧菌 BDSG1 发酵，防疫、检疫与病害测报的重要性、WSSV 提取物中 43×10^3 蛋白来源的 MALDI-TOF 分析。

溶藻弧菌外膜蛋白（va-OMP）的免疫原性及免疫保护性。淋巴囊肿病毒结构蛋白及其抗原性。乳酸杆菌 LH 对几种病原细菌的抑制。噬菌体的治疗及应用、养殖场沉积环境微生物总 DNA 提取法、鳗弧菌 W-1 外膜蛋白 ompu 基因克隆及在大肠杆菌中的表达，哈维氏弧菌外膜蛋白 Ompk 基因的克隆及原核表达、病原弧菌的 16*S*-23*S* rDNA 区间（IGSs）酶基因的克隆、测序、鉴定与分析。乳酸菌及其代谢物对鳗弧菌生长的抑制。育苗中发光弧菌病防治、芽孢杆菌菌剂的应用、微生物肥料的研发及应用，白斑综合症病毒提取物中 43Kda 蛋白来源的 MALD-TOF3 分析[16]。药敏试验在防治病害中的应用、益生菌鉴定、褐藻酸菌生长条件及生理效应。

3.2　养殖环境

养殖环境生物修复研究进展，乳山对虾养殖场白斑综合征调研。乳酸菌对牙鲆稚鱼养殖水体及肠道菌群的影响，乳酸杆菌 LH 对病原细菌的抑制。虾池好氧反硝化菌新菌株分离鉴定，虾池沉积环境功能菌及弧菌时空变化。微生物及其制剂和 EM 产品在养殖环境生物修复及养殖中的作用机制、应用和前景。用芽孢杆菌调控虾池微生物、养殖环境细菌耐药性的危害。凡纳滨对虾池细菌数量变化受环境因子影响及用 PCR-DGGE 分析咸淡水养殖系统内细菌群落组成[17]。大亚湾网箱养殖水体弧菌种类组成及变化。

4. 养殖对象患病各论

4.1　鱼

病毒性疾病——淋巴肉瘤及淋巴囊肿病、诺达病毒及其所致疾病。靶器管组织病理、表皮增生症、白点病防治、苗的红圈病防治、荧光病防治、鱼实验动物在鱼病防治上的应用、近海网箱鱼病害防治。

大黄鱼：溃疡病病原菌及其防治药物、不同条件对溶藻弧菌粘附大黄鱼肠粘液的特性及影响，冷藏中细菌相组成和优势腐败菌鉴定及品质变化特征，诺卡氏菌病，结节病病原——诺卡氏菌鉴定及系统发育[18]。

卵形鲳鲹：幼鱼死因。

牙鲆：淋巴囊肿病毒——抗原蛋白确认与定位，核酸疫苗安全性、衣壳蛋白基因片段真核表达载体构建、表达与免疫效果[19]，乳酸菌对稚鱼养殖水体和肠道菌群的影响。抗菌肽 hepcidin 基因在成鱼组织和不同发育期胚胎中的表达、达氟沙星对健康和鳗弧菌感染的牙鲆体内药物代谢动力学比较，杀鱼美人鱼发光杆菌病、肠道弧菌感染症及病原特征、秦皇岛弧菌感染症及病原菌、网箱养殖病害、漠斑牙鲆亲鱼病害，腹水病及病原、疾病诊断。

19 株致病性弧菌 OMPK 基因序列及其抗原性[20]。

比目鱼：爱德华氏菌和病原菌。

石鲽：病原琼氏不动杆菌形态型Ⅰ的鉴定。

大菱鲆：红体病彩虹病毒体外增殖条件、常见病毒病及防治术、淋巴病腹水病病原及药敏试验、SRI的分离鉴定，出血性败血症病原菌致病性及鉴定，复合益生菌对育苗水净化的效果，病原鳗利斯顿氏菌鉴定[21]。（南方）常见病、溶藻弧菌、鳗弧菌二联免疫的免疫效果、烂鳍病鳗弧菌流行及人工感染、溶藻胶弧菌脂多糖对其白细胞吞噬活性的影响。

真鲷：鳃组织cDNA文库构建与hepcidin抗菌肽基因序列扩增[22]。黄鳍鲷白细胞介素1B基因克隆及在大肠杆菌中的表达。

虹鳟：壳寡糖对肠道菌群的影响。

鲀：暗纹东方鲀毛霉菌的鉴定、致病性和病理。非01霍乱弧菌的鉴定及毒力基因。多糖和益生菌对免疫功能的调节。红鳍东方鲀病害。

石斑鱼：溃疡病病原哈维氏弧菌胞外产物、溶藻弧菌胞外产物注射鱼体后组织病理观察。

海马：病原菌分离鉴定及抗生素敏感。

军曹鱼：淋巴囊肿病毒研究[23]、病原弧菌鉴定及系统发育。

鳗：嗜水气单胞菌主要外膜蛋白基因克隆及表达、肠道益生菌。欧鳗：肠炎。日本鳗：夏秋肠炎。

黄姑鱼：5种常见病害防治。

鲟：常见病防治。

梭鱼：疾病防治、鱼苗弹状病毒病。

罗非鱼：病害及链球菌病[24]。

4.2 虾

对虾：细胞培养及病毒体外培养研究进展。多重PCR检测白斑综合征病毒（WSSV，下同）和传染性皮下组织坏死病毒。WSSV结构蛋白质VP41、VP24定位、P9蛋白转录水平的微阵列分析及其免疫荧光标记、基因组同源重复区结合蛋白的筛选。基因组细菌双杂交系统的构建[25]、WSSV和传染性皮下组织坏死病毒用多重PCR法测[26]。cDNA芯片结合抑制性差减杂交建对虾WSSV表达基因。疾病要因。池养时的综合防治。复方新诺明的毒性。育苗期异养菌。光合细菌的应用。芽孢杆菌合生素在养殖中的应用。人工饲料对肠道菌及水体细菌区系影响。噬菌蛭弧菌的应用。中国明对虾蜕皮抑制激素在大肠杆菌中的表达、分离、纯化。

凡纳滨对虾受WSSV感染后的血液病理。防治WSSV措施。抗菌短肽的分离与功能。WSSV与免疫、病毒检测术在商业化应用中的限制。抗WSSV单链抗体P1D3基因在毕赤酵母中的分泌表达。WSSV结构蛋白VP.28和VP26的功能[27]、肠道菌群与WSSV的关系。汉虾王控制病毒。噬酸小球菌对体液免疫因子的影响及对血清氧化氮合酶的影响、副溶血弧菌、脂多糖和坚强芽孢杆菌对幼体变态的影响。过氧化氢杀虾仁菌的效果。虾白蓝蛋白活性。高位池细菌数量变化与水环境因子关系。益生菌在育苗中的应用、水温骤降后虾血清No、NOS水平及虾溶血弧菌的敏感性。

南美白对虾：病毒病防治。红体病彩虹病毒体外增殖条件。WSSV病爆发与克氏原螯虾的关系。中草药防WSSV病、感染WSSV的雄性虾生殖系统PCR检测与电镜观察[28]。游塘现象及对策。大规格仔虾红体病防治、细菌性白斑病防治、北方养殖疾病、肠道菌群分析及抗菌活性、溶藻弧菌疫苗注射对免疫功能的影响。

日本对虾:仔虾红线病。

斑节对虾:血淋巴细胞清除鳗弧菌。凡纳滨对虾、斑节对虾对 WSSV 敏感性比较。病毒与弧菌共同引发虾病。苗期"红圈病"防治。

克氏螯虾:WSSV 浙江株在组织中的分布。

4.3　蟹:

青蟹:常见病及防治。

梭子蟹:常见病、育苗期与成蟹病害原因及防治肌肉乳化病原及其致病性、春季亲蟹死因及对策。

缢蛏:死因分析、对策。

4.4　贝类:立克次体感染引起的病患[29]。滩涂养殖贝发病、死亡原因及预防措施

鲍:微生物性疾病研究进展、病害的 ELISA 法分析、工厂化养殖病害防治、杂色鲍幼苗急性死亡脱落症病原菌。健、病苗弧菌胞外产物。异养菌致病力。消除九孔鲍细菌病原的微生物法、死因、病原菌分离及致病性。九孔鲍苗掉板病病原菌、苗及药敏试验。肠道及养殖水体异养细菌产酶力。

扇贝:微生物病研究进展。常见病病因及防治。病害的 ELISA 分析[30]。常温贮藏烤扇贝品质及潜在病原。栉孔扇贝:急性病毒性坏死症病毒 cDNA 文库构建及部分序列。大量死亡的流行病学。盐度突降对抗病力指标的作用。

牡蛎:常见敌害防治。

文蛤:病害防治。

毛蚶:对单胞藻的滤除率、选择性和消化状况。

4.5　棘皮动物

海参:病因研讨及可持续发展之路[31]。微生态制剂。蛭弧菌的生物学特性及应用。海参病因及夏病死亡原因、预防措施、病毒。海参液态发酵制备枯草杆菌蛋白酶。

刺参:有益微生物养刺参。保苗期病害及腐皮综合症病原分离鉴定及感染源。苗越冬及养成期病害防治。有益微生物制剂养刺参。溃疡病病原。一种球状病毒。仿刺参幼体烂胃病病原灿烂弧菌及致病性。幼参口围肿胀症细菌性病原[32]。幼刺参溃烂病细菌性病原、蛭弧菌。

海胆:虾夷马粪海胆红斑病病原弧菌特性及致病。

4.6　海蜇:筛选抗真菌的细菌。

4.7　海藻:海带:烟台沿海腐烂病病因及对策。

条斑紫菜:叶状体色落症。

4.8　海产病害防治通报:

冀、浙、粤、鲁等省及全国范围水产病害监测月报或年报、中国北方海鱼、虾蟹病防治。

5. 论文数量分析

以上对 2006 年的分析,是基于从 71 份中国大陆、台湾、香港刊物中 52 份刊有相关论文得出,共引用文献约 363 篇。一年中刊登有≥10 篇相关文章的刊物计 15 份。达≥20 篇相关文章的刊物计 5 份。有 30 篇相关文章的刊物仅 1 份,即《中国水产》。因此,≥20 篇

相关文章的5份刊物排序为《中国水产》30篇，《齐鲁渔业》22篇=《中国水产科学》22篇>《厦门大学学报(自然科学版)》21篇=《水产学报》21篇。以上5份刊物相关文章计116篇，占全年文章总数约32％(116/363)。52份刊物中有14份是纯水产学科的，占刊用刊物数的26.9％(14/52)。

全年论文数中至少约有239篇是专门论述或与病害相关的文章，即占总数的63.9％(239/363)。如上所述，相关文章较多(≥20篇)的5份刊物中，有4份是专业水产学科的，均说明此年中国海洋微生物学研究的着力点在海产养殖对象病害防治上，其中有一部分文章已经注意到跟踪先进，在思路和方法上有所创新、水平趋高。

参考文献32篇(略)

(合作者：孙丕喜)

RESEARCH FOOTPRINT OF MARINE MICROBIOLOGY IN CHINA DURING 2006

(ABSTRACT)

Abstract In 2006, a median year during the first decade years of the new century, i. e. 21st century, about hundreds of Chinese scholars had been researched basic and applied marine microbiology continuously, 363 papers /articles were published at least.

Research articles on principles and prevention/treatment of diseases and pasts of marine cultivations occupied about 63.9% among them. It showed stronger characteristic of application. It needs to be strengthened on basic research including exploiting advancing methodology and studying/manufacturing indoor/outdoor instruments for using to research/develop on and exploration to deep/wide seas and oceans.

Key words 2006, China, Marine Microbiology, Basis/Application

2007 年的中国海洋微生物学研究状况浅析

摘　要　本文就 2007 年中国大陆海洋微生物学研究所发表的文章作统计分析。认为其成就喜人。全年基础/普通海洋微生物学研究论文约 132 篇，涉及病毒、古菌、新细菌、放线菌、南北极和深海大洋等约 18 个方面。约有 183 篇文章是关于养殖海产品病害防治的，包括鱼、虾、海蜇、海参等。一些论文有了相当的深度，方法较新。总体看，全年文章表达了较强的实用性。应加强基础、深海微生物学和方法学的研究。

关键词　中国　海洋微生物　基础及应用研究

虽然海洋微生物学作为海洋科学的成员发展至今已逾 170 年，但其重要性和地位的提升只是近几十年，特别是 20 世纪 80 年代以来的事。中国海洋微生物学在国际学术界算是年轻的，但发展可谓日新月异。一年，在时间长河中只是一瞬间，但在过去了的 2007 年中，中国海洋微生物学研究取得了可喜的成就：基础、应用基础海洋微生物学研究在 18 个方面展开。海产品的微生物学研究则涉及 13 个方面，即 14 种鱼、8 种甲壳动物和腔肠、棘皮动物及紫菜等养殖对象病害防治等。本文就此作了粗浅剖析和论述，并就其发展趋势作了初步探讨。

Ⅰ. 基础海洋微生物学研究

1. 病毒：诺如病毒基因原位表达。

2. 古菌：古菌研究及展望[1]。张晓华发表了《海洋微生物学》一书，其中的《海洋古菌》为专门一章列入。超嗜热古菌硫化叶菌（*Sulfolobus*）*S. takodaii* Rad A 蛋白的克隆表达及其辐射可诱导性。

3. 新细菌（珠植氏菌）：*Zooshikella* sp. syo 培养条件研究。

4. 弧菌：对虾弧菌病病原种类、致病机制、症状、组织病变及防御机制研究进展[2]。弧菌、病毒共同引起虾病的预防及治疗。弧菌 QTH-2 发酵产琼脂酶条件优化。

4.1　副溶血弧菌：在天然海水中的饥饿耐受。以 16*S* 和 16*S*-23*S* rDNA 间区为靶区建立该菌 PCR 快速检测术[3]。一株该菌的分离鉴定、海产品和临床分离株的表型。乳酸菌胞外产物对它的抑制作用。该菌噬菌体衣壳蛋白组成分析及主组分的分离纯化。

4.2　鳗弧菌：有关其毒力相关基因研究进展、该菌致病相关因子及分子生物学。牙鲆抗该菌力、该菌的（灭活）菌苗及对牙鲆外周血细胞吞噬活性的影响。MVAV6203 基因缺失株的生物反应器培养及应用[4]。间接 ELISA 术在该菌 SMP1 快速检测中的应用。

4.3　河流弧菌：对牙鲆黏液的趋化和黏附。

4.4　溶藻弧菌：大黄鱼受该菌人工感染后对免疫功能的影响。用 RD-PCR 术构建 cDNA 文库[5]。该菌对 *Pseudosciaena Crocea* 鳃黏液的黏附。负载型纳米 Tioz 光催化的杀灭效果。

4.5　哈维弧菌：生物学特性、流行病学及检测术。它的溶血素基因 WhhA 在大肠杆菌中的表达及活性、其外膜蛋白与溶藻弧菌脂多糖偶联疫苗的效果。

4.6　蛭弧菌：研究进展。噬菌蛭弧菌特性、作用、机理和应用及在水产养殖中的应用(后述)。

4.7　海洋弧菌：几种该菌密度感应系统及其信号干扰、不同海洋弧菌 5 类溶血素基因的分布及其与溶血活性和磷脂酶活性的相关性[6]。

4.8　其他细菌：一株产黑色素细菌的鉴定及相关新基因(簇)的分离[7]。侧孢短芽孢杆菌 Lh-1 抗菌物[8]、E18 的生物学特性及其蓝紫色稳定性。

5. 放线菌：新活性代谢物重要来源及次级代谢产物研究进展。生物多样性。H2003 代谢产物中环二肽成分及其抗肿瘤活性。放线菌 XS904 的初步鉴定。嗜盐放线菌分离法。具抗肿瘤活性的链霉菌 WBF 鉴定、发酵条件及抗肿瘤物。JMC06001 发酵条件、培养基优化及抑菌物性质。链霉菌 GB-2 发酵产物抗细菌活性及性质、另一株放线菌发酵液的抗肿瘤活性、又一株放线菌分类鉴定及抗菌活性。

6. 嗜盐菌：在环境修复中作用及中度嗜盐菌相溶性溶质机制研究进展。极端嗜盐古菌 *Natrinema* sp. R6-5 胞外嗜盐蛋白酶纯化与性质。

7. 嗜热菌：嗜热嗜碱菌超氧化物歧化酶克隆。

8. 真菌：海洋破囊壶菌 *Throus tothgtrium* sp. FjN-10DHA 生物合成。裂殖壶菌发酵产二十二碳六烯酸(流加培养法)[9]。真菌分离及抗肿瘤活性筛选。粘帚霉对几种水生生物有害菌的抑制作用。烟曲霉 H1-04 发酵液中硫代二酮哌嗪类产物及抗肿瘤活性、产淀粉酶海洋酵母筛选及鉴定。一株产纤维素海洋酵母筛选鉴定及发酵条件优化。皮质丝胞酵母用大米草水解液发酵产微生物油脂。海藻酸裂解酶基因在毕赤酵母中的表达及重组酶性质、水溶性海参皂苷的分离纯化及抗真菌。酵母在水产养殖中的应用。真菌 K38 号代谢产物。两株真菌的三种鞘脂类结构测定。两种海洋寡糖对植物病原真菌的抑制。活性物质多样性。

9. 微藻(菌)：研究进展。束氏藻及微藻间竞争。微藻多糖微波提取。营养条件对铜绿微囊藻胞外多糖产生的影响。微藻体外抗肿瘤活性筛选。蓝细菌基因组学[10]。南海蓝细菌化学成分。

10. 红树林：未经分离的微生物群体培养物生物活性。内生细菌 CⅢ-1 菌株胞外代谢产物 GC/MS 分析[11]。南海红树林内生真菌ⅡF79 中吡喃酮类代谢产物。No. 1403 真菌产木黄霉素发酵条件优化。真菌 SBE-14 次级代谢产物及其活性。真菌 E33 代谢产物的分离及衍生物制备。共生真菌 *Paecilomycos* Sp. Tree1-7 代谢产物。

11. 海绵等：共附生微生物的 PKS 基因筛选及次级代谢产物研究进展。与湛江硇洲岛海葵相关可培养细菌系统发育多样性。Scleractinian 珊瑚 *Symphyllia gigantea* 黏液的抗细菌物质。

12. 生物降解：菲降解菌的分离鉴定及降解性。石油降解菌解烃力及影响因素、一株 AHL 降解菌株的高效筛选。

13. 南北极考察

13.1　南极：假交替单胞菌 sp. 5-1513 胞外多糖低温保护作用及其 EPS-Ⅱ对小鼠 S180 肉瘤的抑制。产低温超氧化物歧化酶菌株的筛选、鉴定及其部分酶学性质。深海底泥色盐杆菌属 Njs-2ect ABC 基因克隆及二氨基丁酸转氨酶 ectA 基因的表达[12]。深海单胞菌 NJ223ectC 基因的克隆表达及 ectoine 合成酶性质分析。深海科尔韦尔氏菌低温适应性。石油烃降解嗜冷菌的筛选及其降解特性。适冷红酵母 *Rhodotorula* sp. NJ298 在低温中质膜流动性及温度对其抗氧化系统及质膜流动性的影响。

13.2　北极:海冰可培养细菌的系统发育分析。楚科奇海和加拿大海盆表层沉积物普通好气细菌地理分布、硫酸盐还原菌丰度[13]。海冰低温微生物生长条件及其胞外水解酶、极区大洋可培养浮游细菌的系统发育多样性和表型特征、培养温度对耐冷假交替单胞菌胞外蛋白酶合成的影响、极地微藻研究进展。

14.深海大洋:深海嗜冷细菌假交替单胞菌 sp. SM9913 适冷 *Chaperones* Dnak 和 DnaJ 的基因克隆和序列分析[14]。低温碱性淀粉酶菌 *Halomonas*(盐单胞菌)sp. W7 的筛选及发酵条件。南海深海沉积物烷烃降解菌的富集分离与多样性、南海北部及珠江口细菌生产力、南海蓝细菌化学成分。一株太平洋深海环己酮降解菌的筛选与功能。深海沉积物微生物间位基因组 DNA 和 RNA 的共浸提。中印度洋洋中脊 Edmand 热液场 Fe、Si 沉淀与微生物关系。高压技术在深海沉积物兼性嗜压菌筛选和鉴定中的应用。

15.生态学:嗜盐微生物在环境修复中作用的研究。硝化细菌分子生态学研究。气单胞菌生态学分布研究概况。好氧氨氧化菌种群生态学。分子世界及其生态学研究进展、趋势[15]等都有报道。

石油污染海域微生物群落及其对烃的降解。石油降解菌解油力及影响因素。渤海湾近岸海域细菌数量。大肠杆菌和粪大肠杆菌分布数量。罗源湾海水中粪大肠菌群分布和年际变化。长江口邻近海域浮游细菌分布与环境因子关系。海洋发光细菌生长和发光条件。南海珠江口 TCBS 和 EMB 类群细菌动态监测(1992～2002)。湛江硇洲岛海葵相关可培养细菌系统发育多样性。围隔中异养细菌与环境中氨氮、磷关系。河北沿海异养菌。微生物多样性在赤潮调控中的作用、微藻间竞争。

16.方法:宏基因组工程进展与展望。分子世界里的海洋微生物生态学[15]。

17.应用及相关研究进展:微生物抗肿瘤天然产物、胞外多糖结构与生物活性。多聚酮的生物合成及其多样性和国外抗菌活性物研究进展。海水中 DHA 产生菌筛选及一株高产菌鉴定。聚壳多糖高产菌株筛选及发酵条件。青岛近海琼胶降解菌筛选和多样性。褐藻胶裂解酶产生菌 *Vibro* sp. Oy102 发酵条件的优化。几丁质酶产生菌筛选及产酶条件。乳酸菌胞外产物对副溶血弧菌的抑制。海洋细菌 201721 抗菌物质的发酵优化与性质。链霉菌 GB-2 发酵产物抗细菌活性及性质、细菌 S-12-86 的产溶菌酶条件。

18.综合:铁载体。趋磁细菌基因组和磁小体成链机制研究进展。趋磁细菌磁小体结构相关蛋白研究进展[16]。脉冲磁场对超磁螺细菌 AMB-1 磁小体形成及相关基因表达的影响。嗜酸氧化亚铁硫杆菌生理特征比较。嗜冷枯草芽孢杆菌低温脂肪酶纯化与酶学性质。碳样小单胞菌产大豆黄素和染料木素。蜡样芽孢杆菌对凡纳滨对虾幼体变态的影响。

Ⅱ.海洋水产的微生物学研究

1.方法学:分子微生物学技术测细菌病原的应用。分子生物学技术在水产养殖动物细菌性病原检测中的应用。LAMP 法在快速测病中的应用及对取自渔场水细菌群落基于

16 S rDNA 基因的DGGE 分析的 DNA 提取和万能引物的效果。间接 ELISA 技术在病原性鳗弧菌 SMP1 快速检测中的应用[17]。抗菌渔药筛选中最佳参数选择与论证。鳗弧菌 MVAV6203 基因缺失株的生物反应器培养及检测。鱼类虹彩病毒 PCR 快速检测试剂盒的准确度评价及其应用[18]。用 RD-PCR 术构建溶藻弧菌 cDNA 文库。白斑综合征中国对虾肝胰腺蛋白质组学研究技术探索。噬菌蛭弧菌——养殖刺参疾病生物防治新工具。对虾弧菌病病原种类、致病机理、条件、症状与组织病变及对虾的防御机制[2]。

2. 环境质量:大鹏澳网箱养殖区沉积物硫酸盐还原菌。粤东增养殖区拓林湾浮游细菌时空分布。大连海区海藻和无脊椎动物附生细菌对养殖网笼污损细菌的拮抗性。

水质等:EM 菌、高效菌、底净剂调节水质[19]。养殖区水浮游细菌分布。沉积物硫酸盐还原菌。

3. 流行病:气单胞菌及其对水产的危害。

4. 白斑综合症(WSSV)等:WSSV 宿主和抗 WSSV 感染途经研究进展。WSSV 全基因组展示库构建[20]。WSSV 传递至浮游动物的新途径、雄性虾生殖系统感染 WSSV 在垂直传播中的作用。抗神经坏死病毒全序列测定。病毒—浮游植物的粘附。

5. 弧菌病等:*Phaeobaeter*(褐杆菌)DL2 的鉴定及对致病弧菌的抑制作用。迟钝爱德华氏菌及其病理控制。

6. 气单胞菌:生态分布研究概述。气单胞菌及其危害。嗜水气单胞菌研究进展。致病性嗜水气单胞菌多重 PCR 检测法的建立。嗜水气单胞菌重组-HemA-ISCOMs 对鳗鲡的浸泡免疫效果。中草药对嗜水气单胞菌体外抑菌作用。豚鼠气单胞菌 CB101 中几丁质酶的纯化及其基因克隆。

7. 鱼:正常肠道菌及菌群结构研究进展。常见细菌性病、春夏常见鱼病防治术。以鱼养鱼防病增效。几种特殊鱼种高温季节病。北方暴发性鱼病成因、对策。网箱养殖的病理、分析及措施。4 种海鱼淋巴组织病理特征比较。鱼用哈维氏弧菌亚单位缓释微球口服疫苗免疫效果。链球菌分离、鉴定及其 16*S* rDNA 分析[21]。生产鱼松废水培养苏云芽孢杆菌的可行性。酵素菌生物有机肥的应用。

7.1　鲷:红笛鲷弧菌病血清流行病学研究方法。

7.2　军曹鱼:养殖水体及其肠道弧菌耐药性。淋巴囊肿病毒系统关系及基因型。

7.3　鲟:苗种阶段气泡病危害及防治。俄罗斯鲟肝病及肝性综合征防治。

7.4　罗非鱼:越冬期病害及防治。

7.5　石斑鱼:青石斑鱼肠道菌群。

7.6　剑尾鱼:Hsp70 基因表达受溶藻弧菌感染后的影响。

7.7　卵形鲳鲹:用美人鱼弧菌与创伤弧菌人工感染后组织病理学研究。

7.8　鲀:红鳍东方鲀病原杀对虾弧菌的生物学特性。暗纹东方鲀病害防治。

7.9　半滑舌鳎:病原发光杆菌杀鱼亚种分离鉴定[22]。

7.10　比目鱼:败血病病原。

7.11　鳗:鳃霉病防治。疾病无公害防治。

7.12　黄鱼:副溶血性弧菌单克隆抗体制备及应用[23]。三联灭活细菌疫苗诱肾组织 SMART cDNA 文库构建。网箱技术及病害防治。中草药抑病原弧菌显效果。贮藏中细菌菌相。

7.13　鲈鱼:868菌发酵物作饵料添加剂。

7.14　鲆:流行病学调查。淋巴囊肿病毒核酸疫苗表达及免疫效果评价、日本比目鱼中的中国淋巴囊肿病病毒(LCDV-cn)监测中定量PCR法的应用[24]、腹水病病原及防治。工厂化养殖中病的防治。肌细胞生成素(Myogenin蛋白)在大肠杆菌中重组表达。病原秦皇岛弧菌的血清型及荧光抗体试验。抗鳗弧菌病的AFLP分子标记筛选。迟钝爱德华氏菌对牙鲆的半致死浓度。大菱鲆养殖中水质净化菌筛选与系统发育分析。牙鲆抗菌肽(hepciain)基因的克隆及表达和它在大肠杆菌中的融合表达。不同组织中抗菌肽提取及部分性质检测。

8.甲壳动物

8.1　中国明对虾等:患病基本征兆、防病毒12项措施。虾蟹苗期病害检索表。弧菌病研究进展。虾池桡足类及其休眠体携带WSSV的调查。中国对虾杂交优势对自然感染WSSV的抗病力。控制环境养殖中近交虾早期体重与抗WSSV性状影响。病毒、弧菌共同诱发疾病的预防与治疗。虾池弧菌对抗菌药物的耐药性与水质评价。虾池中降解硝酸盐、亚硝酸盐的细菌的多样性。有益微生物对虾池细菌动态变化的影响。明对虾T7噬菌体展示文库构建与免疫相关基因淘选[25]。虫草菌粉对日本沼虾免疫功能的影响、3种生物制剂调控养殖、卤虫对日本对虾育苗防病中的作用。

8.2　斑节对虾:北方沿海暴发性疾病防治。芽孢杆菌对饵料表观消化率的影响。

8.3　白对虾等:WSSV的控制。凡纳滨对虾桃拉综合征病毒主要结构蛋白基因克隆及原位表达[26]。肠道微生物群落的分子分析。光合细菌、噬菌体的应用、副溶血弧菌对南美白对虾生理生化指标的影响。病害的控制与全面防治及常见异常现象的防治。健康养殖疾病预防。枯草杆菌制剂调节水质。芽孢杆菌胞外产物对凡纳滨对虾蛋白酶活性影响的体外试验。铜绿微囊藻对低盐度养殖的危害。溶藻弧菌在血清中NO及自由基的影响。用中草药治“死底症”。黑鳃烂鳃的防治。红线病和仔虾红体病防治。

8.4　红螯螯虾:WSSV的感染。

8.5　扇贝等:常见病害及流行趋势防控。大规模死因及防治。栉孔扇贝血细胞类型、功能受病毒感染的影响分析。崆峒岛扇贝死因及预防。海湾扇贝亲贝死因及预防。贮存扇贝品质及细菌相变化。翡翠贻贝多糖抑制流感病毒在鸡胚中增殖的研究。

8.6　鲍:光合细菌育鲍苗的技术。杂色鲍暴病超微病理学。微生物黏膜对杂色鲍幼虫附着和变态的影响。九孔鲍消化道及水体中弧菌胞外毒力因子。健康/患病黑鲍微生物产胞外酶分析。

8.7　蛤:文蛤暴病机理及综合防治对策。浅海滩涂文蛤病害防治术。菲律宾蛤仔冬季育苗系统中几类细菌数量变化。泥蛤血细胞耐饥饿及抗菌特性。

8.8　牡蛎:寄生虫病和微生物病。

8.9　蟹:梭子蟹WSSV感染试验及组织病理学观察。牙膏病病原溶藻弧菌病研究和该菌鉴定及其系统发育分析。苗种的粘爪病防治。中华绒螯蟹:维生素E对酚氧化酶。抗菌力和溶菌酶活性的影响。幼体的丝状细菌扫描电镜观察。锯缘青蟹:呼肠孤病毒的分离、纯化及鉴定。

9.棘皮动物:噬菌蛭弧菌用于养殖刺参疾病防治。刺参病因及防治、刺参溃疡病。杀鲑气单胞菌分离、致病性及胞外产物特性。刺参腐皮综合征两种致病菌Dot ELISA快速检测。蛭弧菌特性及其用于养殖海参。EM 菌育苗。乳山湾海参病。海参海胆养殖水体中多抗性细菌抗性基因筛选。

10. 腔肠动物(海蜇):死因及净水复合菌在养殖中的应用研究。

11. 褐藻酸降解:褐藻酸降解菌的筛选、产酸条件优化及 *Vibrio* sp. Oy102 的发酵条件、该菌感染海带过程中几种酶作用的比较。

12. 紫菜:条斑紫菜丝状体病害防治及拟油壶菌病。坛紫菜常见病害防治。外生细菌抑菌活性及其多聚酮合酶(PKSI)基因筛选。

13. 藻—菌关系:塔玛亚历山大藻藻际细菌溶藻过程。

14. 有益微生物:用于影响对虾池中细菌动态变化研究。微生物制剂的应用。硝化细菌的应用及实践、分子生态学和好氧反硝化菌研究进展。蛭弧菌对西施舌幼虫培养水质和生长的影响。噬菌蛭弧菌和粉红酵母对中国对虾生长及非特异性免疫因子的影响。

15. 综合:鱼肠道正常菌群研究进展。河北海水养殖主要流行病害调查、春、夏季鱼病防治。迟钝爱德华氏菌及其致病机理。鳗弧菌致病相关因子及分子生物学。乳酸菌胞外产物对副溶血弧菌的抑制。EM 菌、高效菌、底净剂调节水质、病原体形成耐药性原因、后果及对策。大型海藻的细菌降解。渔用微生物絮凝剂产生菌筛选及培养条件。细菌性肠炎防治法。红酵母和光合细菌及其应用、几种有益微生态制剂。无公害水产品生产中病害防治。细菌耐药性根源及对策。3 种微生物制剂调控工厂化对虾养殖。

Ⅲ. 学术论文数量与趋势的分析

2007 年中国的海洋微生物学研究取得了颇丰的成就。翻阅其大多数文章,可将其分为两大支:第 1 支是基础及应用基础的,第 2 支是海产品的微生物。全年发表的文章数超过 315 篇。第 1 支论文 132 篇,第 2 支文章 183 篇。第 2 支的文章数是第 1 支的 1.39 倍。说明研究偏向海产品养殖及其病虫害防治。文章所刊刊物数超过 47 份。文章所载刊物最多的是《齐鲁渔业》,有 34 篇之多,以下依次为《微生物学报》的 23 篇,《河北水产》和《中国水产》均为 22 篇及《海洋科学》的 21 篇。此外尚有 8 刊刊登 10 篇(含 10)以上的。说明学术水平有所提高,基础和应用基础研究数量增加、质量提高,但仍尚待大力加强。全年有 14 篇文章以英文发表,说明中国学者信心增强、水平提高,敢于并重视同国际同行交流、接轨,相关速度和广度正在加强。

参考文献 26 篇(略)

(合作者:孙丕喜　王宗兴)

PRELIMINARY ANALYSIS ON STATUS OF MARINE MICROBIOLOGICAL RESEARCH IN CHINA DURING 2007

(ABSTRACT)

Abstract This paper statistic and analysed status of research article on marine microbiology in the main land of China during 2007. It considered that the achivements obtained were heartening. Its basic/general research papers were roughly 132 in a whole that year. It was related to 18 divisions including virus, archaea, new bacteria, actinomyces, Arctic/Antarctic, and deep seas and oceans etc. There were about 183 articles related to prevention and treatment of diseases and evils on cultured marine organisms such as fishes, shrimps, jelly-fish, and sea cucumber etc. 15 divisions. The level of some papers was rather high and deep, and the research methods were rather news. General speaking, the articles in the whole year expressed more obvious applicational characterization. It ought to augment on researchs of basic marine, deep sea and Oceanic microbiology, and relative methodology.

Key words Marine Microbes, Basic/General, Applied Microbiology, China, 2007

微生物学向深海进军*

摘　要　论述全球的深海概念，包括深于1000 m的各种海域，如大洋、远海、一些海湾、海峡、海沟、海槽、深渊、超深渊等。认为深海环境包括3个单元：深海海平面上的空气、海洋表层至海底间的深水体和表层沉积物/岩石及以下部分。在此环境繁衍了深海上空气微生物、深水微生物及深海地微生物，对这三者的研究构成了深海微生物学，而其中的嗜极微生物则十分重要。尤其关注深海地微生物学的研究现状并展望其研究前景。

关键词　深海环境　深海空气微生物　深水微生物　深海地微生物　嗜极微生物

20世纪70年代以来，对海洋、尤其深海的探索不断升温，因为人们对资源的渴望日甚，认识和技能也日隆。本文刍议深海及其环境，尤其关注深海微生物研究。

1　深海

深海顾名思义即深深的海洋，它与"大洋"基本上一致。海洋占全球面积的71%，≥1 000m水深的海洋为深海，它占全球海洋总面积的3/4。全球海洋水均深3 795 m。4 000～6 000 m为深渊，>6 000 m处为超深渊。>6 000 m的深海只占全球海洋的0.1%。

(1)太平洋占全球海洋面积的50%以上，其最深处斐查兹(即查林杰海渊的马里亚纳海沟)达11 034 m，即地表最深处。太平洋、大西洋、印度洋和北冰洋的平均水深分别为4 282m、3 825 m、3 963 m和1 300 m，还有远离南极洲周围的南大洋深处，它们除了有些近陆海区较浅外，均为深水海域。

(2)有些水体虽以"洋"命名，但并不是真正的深海所在，如中国南方近海的大戢洋、伶仃洋、七洲洋等(共约13个)。

(3)除了大洋外，不少海是真正的深海，如珊瑚海，水均深为2 394 m，最深达9 140 m。再如白令海，水均深为1 598 m，最深达4 151 m。全球水深千米以上的海计18个。

(4)另外，尚有些以"海"为名的水域，只不过是咸/盐湖，却并不浅，如黑海，均深达1 200m以上，南部最深达2 212 m，盐度达17～22，为一深水内海水体。而咸海(位于中亚)是咸水湖，盐度达10～14，但水深最多只68 m，并非深水水域。美国的大盐湖(Great salt lake)盐度是海水的4～8倍，达151～288，是残迹盐湖。以青海命名的内陆青海湖，虽湖水属咸，但水深不超过32.8 m，不属"深海"范围。

(5)有些不以"海"或"洋"称呼的海域，如"湾"，也为真正的深海，如墨西哥湾，水均深达1 500 m。再如孟加拉湾，其水深范围为2 000～4 000 m。几内亚湾均深2 996 m，最深6 318m。世界水深超千米的海湾计5个。

(6)一些并非以"海"或"洋"命名的水域，实际上却也是真正的深海海域，如"峡"等等。莫桑比克海峡，水体大，均深大于2 000 m，最深3 533 m，为真正的深海。再如德雷克海

* 原文刊于《海洋地质动态》，2009，25(2)：14-20。

峡，它连接大西洋和太平洋，水均深 3 400 m，最深处 5 248 m，为一大深水道深海海域。

(7)还有一些海沟、海槽，深度分别达万米以上（计 14 个）和 2 300～7 440 m 间（26 个）[1]。一些大陆边缘水，有的也很深。一些洋盆、岛屿水深也大。中国东、黄、渤海是较浅的边缘海或内海，唯南海水均深约 1 212 m，最深达 5 559 m，也属深海系列。所以我们不能仅望“深”或“海”之文来生“深海”之义，也不能仅以“洋”即认定它为大洋。所谓“深海”即是远离大量人类活动、难以直接到达的深水海域，尤其是海底及其下的沉积物或岩石地带，即为大洋海域。

2　深海环境

深海环境可以分为 3 个单元，其一是表层水面上的空气，主指水表之上、大气下边界 10 m 内的空气，可以上至百米左右甚或更高的空气环境。由于受海水蒸发作用和天体、阳光、气候等的多重影响而形成海气相互作用系统，深海空气比陆上或近海地面空气的湿度大，氧气负离子多，太阳（紫外线）辐射较强、平均风况和海况面积较大、气温较均衡和低一些。向上气体密度和气压渐稀/小，空气质量较好。其二是表层水至海底表层沉积物或岩石之间的水体。在该范围内向下，各级各类生物经受着不断加大的静水压，水温变低，及至底部，水温处于 4℃左右且处较稳定状态中。水的流体力学和海水化学因地因时而变，同时深部水又受到沉积物/岩石及生物的向上作用，使得其中的生态环境和生态系统一直处于既稳又动的过程中。其三是海底表层沉积物（占全球表层面积的 50%以上）至不同深度处的沉积物/岩石间的地质/地理环境。这里有来自水体的更大的压力，虽有相对稳定的沉积环境，但要经受不时的流体、地质、地球物理、地球化学等异常事件的干扰乃至破坏。另外两个重要状况是缺氧甚至无氧以及持续低温，偶有高温或冷泉[2]。因而构成了众多的极端环境条件，对生物造成了不同的“补丁”式沉积环境，进而形成千变万化的小环境，以至“微生境”[3]。

虽然这 3 个单元垂直相距千米以上，但相互间即气、水、固体间还是有一定联系的。在如此的深海环境中仍可孕育常规的海洋生物，同时也必然地筛选和繁衍了更多、更强、更耐极端环境的生物，即嗜极微生物（extremophiles）等，包括原核生物，它们具有特定的生理、生态、生化多样性[4,5]。

3　深海微生物

由上所述，深海海面上的环境造就了空气微生物，深水中孕育了深水微生物，沉积物/岩石中繁衍出众多的地微生物（geomicrobes），这些微生物包括如病毒、亚病毒和噬菌体、古菌、细菌原核生物等微生物和真核生物，如真菌、微藻，及一些原生动物等，其中的嗜极微生物尤其引人注目。

3.1　空气微生物

受海气的交互作用，海面上的空气微生物一般适于较湿较凉的空气环境，因而其中含较多产耐湿孢子的、嗜中/低温度的微生物。空气中较多的气溶胶和较少的颗粒物成为微生物的附着媒体和介质，较多的氧气供给和较好的空气质量为普通真菌、细菌等提供了优越的生存条件，好气/嗜气的微生物增多，总的看，深海上空气微生物种群和含量不如城市、港口、陆地[6]。

3.2　深水微生物

水深深于 1 000 m 的海水域其水体众多的生物构成了复杂的生态系统，其中的（超）微

(型)生物是该海洋环境中最重要的组分之一。这些微生物包括病毒、噬菌体、细菌、古菌、真菌、微型藻类和原生动物等。真光层下降来的颗粒物是其下水层物质的重要来源。而海洋浮游古菌则自 20 世纪 90 年代以来成为海洋生态系统功能研究方向新观点、新认识的一个生长点。深水微生物的研究成果构成了几十年来海洋微生物学知识的支柱之一。深水微生物在生态系统内外均发挥不可或缺的作用,对营养物质转化、能量流动以及海水化学循环和生态环境、保护修复等一直起着难以替代的作用[3]。

3.2.1　深水静压的影响

深水环境对微生物的一个重要影响因子即是静水压,最深海水处的静水压约1 000 atm,但许多微生物对此压力应付自如。一些研究证实海水静压力对它们的垂直分布没显著影响,即使压力高至 2 900atm 时也如此,对于产孢子的细菌尤其如此。研究认为,只要微生物能经常生存在 1 个大气压中,它就不怕增加的压力,只要压力的增加不是太突然,对微生物的危害就不会很严重,这就是为什么从深海仍能检出许多活细菌的原因之一。深海压力是深水微生物与生俱来的活动环境条件,同时又经受低温和低营养(当然除了那些区域性的富有机质)的持续作用,形成了专性/极端嗜压菌及对各种压力的适应菌。分子生物遗传学的研究结果表明,深海发光杆菌(*Photobacterium profundum*)和一种希瓦氏菌(*Shewallella* sp.)对压力有其调控机制,后者的一些压力反应基因可被分成几种压力调控操纵子,压力反应使它们更易吸收寡营养成分[7]。总之,压力的影响主要体现在对微生物结构功能与代谢等方面。

3.2.2　深水水体运动的影响

虽然深水的化学成分和性质相当稳定,但是由于水团的不断运动,微生物的成堆出现并与其他物质相结合,使得它们的种群和数量分布出现每时、每日、每季的、垂直的和地理(水平)的变化,即它们的时空分布有明显的不均匀性,除非检样很多,方法一致,长期坚持才可得到一个相对平均的数据来。潮水的涨落(即潮汐)对深水微生物波动影响不大,但不同洋流(潮流)的影响则十分明显。这包括大洋风生环流、大洋热盐环流、北部环流、南部环流等,虽然它们生成或运动深度不一,但对其流经的周边和上下水体产生深远影响。比如黑潮,虽活动于水深 400 m 左右,但对其流经之处的深层会产生巨大作用。洋流引起的水体运动及携带的生物/非生物都对深海微生物有重大效应。而深水底流包括等深流、内波和内潮汐,影响底质底层水元素及化合物分布,乃至微生物生存状态[8]。

3.2.3　深水水温的影响

阳光对浅层(光照层)海水微生物有一定影响,某些紫外线有一定的杀菌作用,随着水深的加大,光线越来越少而至黑暗世界,但在此处仍有各级各类生物活动,因此必然有相应的微生物生存。

深水中水温常年偏低,相当稳定,一些冷水团(如日本远洋滩)从上到下水温降得很低,在 2 500～4 000 m 深处仅为 1.5～2.0℃。适应此种低温的微生物多为嗜冷(psychrophilic cold-living)原核生物——细菌和古菌及适冷菌。一种深海的泉生古菌——嗜冷产甲烷菌(*Methanogens* sp.)有大量的延伸因子Ⅱ(elongation factor Ⅱ)极有利低温中的蛋白质合成。又发现了古菌域与细菌域的一些成员中均有冷休克蛋白的同源基因,证实它们在低温中在两域某些成员间发生了基因转移。深海微生物中发现的多种不饱和脂肪酸类(PUFAs)为生物工程上的应用提供了巨大潜力[9]。

与低温相反，深海还存在地热区域，来自于海底热液系统，如黑烟囱和活火山。这在太平洋深水区尤为显著，其地震和活火山分别占全球的85%和80%，在西缘的岛弧更为剧烈。大西洋和西太平洋的暖池产生的暖水库作用可一直深入水下3 000 m，高(有的可达350℃)、低水温间形成的温度梯度造就了不同的嗜热或嗜冷原核生物。它们在此或营好氧/厌氧生活，或进行化能自养/异养生活。环境迫使相应微生物产生高温条件中照常活跃而稳定的酶和结构蛋白，并发挥持续功能和正常的运转[4]。

3.2.4　深水微生物对盐的忍耐

海洋微生物与非海洋微生物的一个明显区别就是对盐的忍耐力，但前者对渗透压改变的抵抗力却没有淡水微生物强。随着水深加大，海水盐度也呈加大之势，越往下微生物对高盐度海水的承受、适应能力也逐渐增强。全球海水盐度大体上呈现赤道附近低，南北回归线附近最高，中纬度区随纬度升高而下降之势。盐度最大的红海(达42)有较多的嗜盐菌。深海中也生存着对盐有不同适应、嗜好能力的原核生物。极端嗜盐菌(extreme halophiles)主要是古菌，而极端嗜盐细菌迄今仅有红色盐杆菌(*Salinibacter rubrum*)，它们靠积累非抑制性物质(相容性溶质)使细胞液保持高浓度而不致脱水。有些则将外界K^+泵到胞内，使K^+的胞内浓度与胞外Na^+浓度平衡，或者通过高比例酸性氨基酸的酶和结构蛋白保护细胞构象不受高浓度盐的破坏。在不断加浓的盐环境中，微生物多样性降低。但当NaCl浓度超过15时，有些古菌可成为优势种，迄今已发现67余属海洋古菌[4,7]。

3.2.5　深水微生物的营养供给

深水微生物得以在水中生存，还决定于营养物质的供给状况。营养来源主要有上层水体和沉积物。上层水体主要提供下沉的各种颗粒物——对微生物丰度和活性影响最大的是颗粒有机物(POM)。在沉降过程中，POM遭遇破坏降解直至到沉积物表层间，POM等营养源越来越少，越来越难以降解吸收利用，从而在沉降过程中产生了对营养源及状态利用的各种微生物。但反过来，作为营养第2来源即沉积物，借助各种生物/非生物方式向水体微生物提供营养，这包括沉淀在沉积物上的少部分返回的POM(来自碎屑、粪便、尸体等)。第3来源，即地层及岩隙热液或冷泉液体扩散来的无机营养，越往上则越少越稀。总体看，深水微生物生存于趋向均匀与不均匀稀释扩散的寡营养中，但水与顶层沉积物(1 m)界面间(SWI)的深水微生物丰度估计为$(1\sim5)\times10^9$ mL^{-1}沉积物，底栖边界层(BBL)中数十米水柱中则约为$(3\sim60)\times10^4$ mL^{-1}，其个体大，分化频率也高[3]。

3.3　深海地微生物

地微生物(geomicrobes)即地质微生物但又不局限于与地质学相关的微生物，对其的研究目前已形成了地微生物学[10]，此处指的是深海底部表层沉积物及以下若干深度岩石圈间的微生物。表层及浅层沉积物是地微生物最不稳定的生存环境。它们要承受来自水柱的压力、底流的波动和沉积物、地层运动的干扰。但这里有较多的经过上部水筛漏下来的各种颗粒、碎屑所提供的营养，因而大多地微生物是嗜压/适压的，并与其他底栖生物关系较密切的原核生物。由于渗透下去的氧气越来越少，以至于完全缺氧，因而微生物多为厌氧或兼性厌氧者，但仍有相应的好氧生物[2]。

深海75%的底质为深海黏土及淤泥，其中大多是砂质、硅质软泥或黏土，形成了远洋沉积物。这在印度洋和大西洋尤甚，主要由原生生物的碳酸钙外壳沉积来。大西洋和太平洋中大部区域的沉积物约500～1 000 m厚。2007年，“地球”号开始在日本附近的太平洋水深 2 500 m 水域钻探，计划用 1 年时间，入钻 5 000～7 000 m 深度，在地壳和上地幔间，寻找细菌等微生物，以探索生命起源。虽然海洋微生物学先驱佐贝尔教授于 20 世纪 30 年代就研究过水深 1 322 m 下的沉积物细菌，但那时国际对表层以下的探测不超过 5 182 m(17002ft)，随着技术的进步，钻探才逐渐加深[1]。

沉积物在海底经历漫长的年代，逐渐下沉并变得坚实，为岩石的形成提供了一种长远的基础和动力。海底岩石圈已成为人们探索"地下生物圈"和"深部生物圈"的重要部位。但这一生物圈大体上仍属于地球表层(10 km 左右)范围。陆地地下 6 000 m 花岗岩岩心微生物的发现(瑞典锡利延)更加鼓舞人们探索深海岩心微生物的热情，估计地球上 2/3 的微生物可能藏于洋底沉积物和地壳中，但是迄今人们掌握的海洋微生物知识大多来源于海水微生物，包括深水微生物，对于海洋地微生物学(marine geomicrobiology)的认识以往大都来自表层沉积物。包含大量谜团的深海地微生物越来越成为人们探索追求的热点[10]。

4　*深海地微生物学研究概况*

海洋地微生物学创立于 20 世纪 30 年代，1977 年"阿尔文"潜水器于 2 500 m 深水下首次发现热泉口有众多微生物，直到上世纪末才得到快速发展。地微生物学主要研究目标为不同时期地质环境中微生物的生存与适应机制、活动过程及其对地质地球化学和生态学的作用，探讨微生物在全球或局部尺度上的生命重要元素循环中的地位、作用和对全球气候影响的理解，并提供开发应用前景。深海地微生物学内容基本相同于此[10]。

深海沉积物中存在超乎人们想象的微生物种类和数量，甚至超过了海水中的。即使在 10 898 m 深水下的沉积物中仍可发现放线菌、真菌及对碱、酸、热、压或冷嗜好的原核生物[11]，已证实沉积物中有嗜盐真核生物，嗜盐真细菌、嗜盐甲烷菌、嗜盐古菌、嗜热酸古菌、产甲烷古菌等。沉积和岩石环境温度可造就出有不同适应能力的物种，因而形成了嗜冷、嗜温、嗜热甚至嗜超热的微生物，其营养方式有自养、异养、化能/光能营养，对气体有好氧/厌氧等十分新奇的生存方式，产甲烷或产氢的等等。有的生存于海洋温泉，有的极端嗜冷嗜压[12]。还有的喜好火山口(高达 250℃～400℃)或热液处，如硫氧化菌等 6 类微生物。受海底压力的影响，沉积物或岩石滋生出耐压、嗜压和极端嗜压的微生物，如詹氏甲烷球菌(*Methanococcus jannaschii*)，既嗜热又嗜压[13]。沉积物微生物还积极参与地球化学循环，如铁锰等金属的氧化还原等[2]。以上这些微生物，构成了深海嗜极微生物。而病毒、噬菌体在深海底则不均衡分布。由于海底有大量原生和后生原生动物，它们引来与之对应的共生/寄生菌，目前已分离培养出与海绵共生的少数菌[7]。

深海地微生物学作为海洋微生物学的一部分，近 30 余年来得到了较大的发展，这得益于人们对海洋知识的探索及对海洋微生物应用的追求，迄今全球已进行了几十个热液区研究。全球深海大洋钻探活动(DSDP、ODP 和 IODP)使对深海微生物的探索到达地壳和上地幔间，一系列的成果使人们认识到海洋微生物资源的举足轻重。

迄今的深海沉积物微生物调研大多取水底表层沉积物样品，范围遍及全球各大洋及主要海区，如中大西洋热液孔[14]、中印度洋某处烟囱[15]、北大西洋[16]、西北太平洋铁锰结核区[17]、黑海某海区颗粒和沉积物[18]、南北极海区[19,20]，其中对大西洋区的调查较为充分，所调查的内容涉及微生物的许多方面，如在热液孔 Chinney 发现新菌种[22]、中白垩纪海洋无氧事件期古菌的巨大扩展[20]、深水下次表层沉积物中微生物过程[23]、甲烷微生物[13]、微生物群落的定量分子分析[19]、病毒及其生物地化、生态作用[24]、深海细菌压力调控操纵子分布[25]等等。对马里亚纳湾微生物、黑烟囱热液孔(系统)、铁锰结核等的微生物、微生物在地球化学中的作用等等有了较深广的研究[22,26-29]，对深层沉积物微生物的研究并不多见[30-33]。大洋钻探提供了向深部探索微生物的机会和可能[34-36]。近些年来，中国学者从关注方法学研究到参与世界不同海区深海微生物的调研，取得了令人鼓舞的成果[10,37-44]。

5　深海微生物学的研究前景

当今的深海、尤其深海地微生物学的知识都是借助十分有限的探测器获取。没有续航能力强大的船只，没有精良的保真的采样设备和后续精准的实验仪器、材料等等，要做好深海微生物学研究等于天方夜谭。因此，近些年来，我国除了引进国外先进设施外，也开始试制相关高水平的仪器，努力赶超国际先进。分子生物学知识和技术应用于深海微生物研究是必然趋势，海洋学界要大力借鉴陆上的先进经验，同时更要发展特别适合深海的微生物学技术，还要注意与传统方法相结合，创新分离培养技术与策略，开发更多的新菌株新用途。深海地微生物学研究是深海、大洋、极地综合科考的重要组成部分，随着人们对深海微生物学认识的扩展和加强，有可能让微生物学在地质、矿物、海水下的古今生态、环境，包括极端环境、深海及其下部沉积物和岩石生物圈、元素及物质的循环、生物地球化学活动，甚至气候变化等学科方面发挥作用。

迄今为止，人们认知的海洋细菌大约只占所有海洋细菌资源总量的1%，而细菌仅是原核生物的一部分。未知生物绝大部分蕴藏于深水、海底沉积物及岩石层之中，这是个巨大的资源库。人们对深海微生物寄予厚望，它们可能具有大量新奇特的次级代谢产物、生物活性物、各种活性酶，尤其是极端酶等，对它们结构和功能的研究将大大拓展人类关于生命科学的认识。

参考文献44篇(略)

(合作者：孙丕喜　高爱国)

MICROBIOLOGY ADVANCES TOWARDS DEEP SEA

(ABSTRACT)

Abstract　This paper discusses the concept of the deep sea in the world including the seas deeper than 1 000 m, such as Oceanic provinces, offlying seas, some gulfs, straits, sea trenchs, troughs, ocean deeps and super abyses etc.

It considers that the environment comprises three divisions, i. e. air-above deep sea level, deep waters between surface water and surficial sea floor, and surficial sediment/rock, and parts under them. It propagates airborne microbes above deep seas, microbes in deep sea water and geomicrobes under water in this type of the environment. Three dicisions and research on them organize microbiology in deep seas, meanwhile, the extremophiles are very importants. The paper pays close attention to the research situation on geomicrobiology of abyssal regions especially and looks forward to its future.

Key words　Deep Sea, Environment of Deep Sea, Airborne Microbes Above Deep-Sea, Deep-Sea Water Microbes, Deep-Sea Geomicrobes, Extremophiles.

海洋的古菌*

摘　要　笔者概述古菌的研究方法，论述海洋中古菌的特征，包括它们的广布性、多样性和适应性等，最后文章展示了研究海洋古菌的前景。

关键词　古菌　海洋　广布性　多样性　适应性　前景

1　引言

古菌(Archaea)于1977年被首次发现，当时它们只是被看作为细菌中一个独特的子分支(subset)，因而被称作为古细菌(Archaebacteria)。直到进入20世纪90年代，古菌才被较多的发现。古菌与细菌虽均属于原核生物，但它们之间有明确的区别，后来两者均独立为各自的世系——域(Domain)——古菌域和细菌域，但对所有生物划域的学说仍在争论之中[1]。古菌大多为超微型原核生物，其有别于细菌的差异主要表现在进化史和细胞特征上。古菌细胞的DNA中有组蛋白、延伸因子，对白喉毒素具有敏感性，有多种RNA聚合酶，需转录因子，而细菌则没有或不需要，或有不一样的脂膜结构连接方式[2,3]。

古菌的发现有赖于分子生物学技术的发展，所以其发现时间比细菌晚了许多年。迄今所认可的古菌只不过99属，而且种数也不多，这是因为它们难以用传统细菌学方法分离培养，并且古菌常生存在人迹罕至之处，直到20世纪90年代发现海洋浮游古菌之后，人们才对古菌的生境、分布及作用刮目相看。

2　海洋古菌研究法简述

海洋是古菌的发祥地、摇篮，首先发现的古菌来自有限的极端环境，随着技术条件的逐渐先进化，海洋古菌不断得到发掘、认识并展示出其开发价值，这些方法大体上分为以下几种。

2.1　收集

以一定孔径的核孔滤膜加压抽滤水样获得含古菌的浮游生物样本。普通浅层沉积物生物体的古菌样本经抽提或高速离心沉淀获得。深海样品则经历低温(≤20℃)和高压等保真系统细心耐心地取得。

2.2　分析

从样本中提取总DNA，扩增片断，构建基因库、进行限制性酶片多态片断长度性分析(RFLP)、测定其序列，分析序列间相似性、比较国际基因数据库信息，绘制系统进化树，设计出引物和探针。

2.3　含量和多样性测定

比较已知浓度靶序列和待测序列在同条件中的扩增效率以推算原始浓度及比例数；

---

*　国家自然科学基金项目(40576060)资助

原文刊于《自然杂志》，2009，31(2)：94-99。本人为第二作者。

或用携有同位素标记的探针与样品杂交法；或非放射法灵敏计数；或将古菌的特异探针挂上适当的荧光素，借核酸分子杂交术将荧光探针与样品中靶序列结合，在荧光显微镜下或流式细胞仪上凭荧光底物检测法更为常用。

总之主要应用分子生物学技术来研究古菌，它主要包括核酸杂交技术，PCR 技术等。借助相应的方法，可以揭示出古菌的分布频度，绘制出它们的系统进化树，设计某些探针、识别和发现新种；纯化菌种，进一步研究其生物生化性状并着眼于应用，包括用于环境污染检测和治理。

3　海洋古菌的特征

3.1　广布性

海洋极端环境大体上指高压、高温、低温、非常规的氧气状态、紫外线辐射、高盐、高碱、低营养等等，其中以浅海或深海的缺氧沉积物、深海火山口和高盐海区为人们较为熟知。在以上不同生境中，不同状态的极端因子以某些方式结合造成了特定的环境状况，比如深海，可以随深度而产生不同压力、氧气状态。在不同沉积物深度中，都有其独特的氧气和营养条件，或者受制于地质状况，形成火山热液口等特异环境。在这些极端环境，都发现过古菌。比如缺氧沉积物的产甲烷古菌（*Methanogenic archaea*），火山口的产甲烷古菌、硫酸盐还原古菌、极端嗜热菌（*extremely thermoph*），高盐区的极端嗜盐菌（*extremely halophilic archeao bacteria*）等等。这些极端环境实际上是指我们人类所认知的非通常状态，但是绝大部分环境中都会有其相应的生物存在。在这种环境，那些人们所理解的极端生物，对其所在的环境并非觉得太过分（假设它们也有感知的话），必须而且应该能适应下去。人们一旦在那些不寻常环境中发现它们就顺理成章将其称为极端生物。事实上许多古菌既可在极端生境中被发现，也可在常态环境中被发现，这便是古菌的广布性。

海洋古菌最早从极端环境中分离出来，迄今这一局面并未得到根本转变。随着分子生物学技术对海洋微生物研究的渗透和扩展，对广布的海洋古菌生境的多样性认识有了很大的突破，这一认识起始于 20 世纪 90 年代海洋浮游古菌的发现[4-6]。海洋浮游古菌的确认大大开启和拓展了人们的视野。迄今的十余年间，人们认识到海洋古菌还生存于不同的地域、不同层次的海水、不同深度的沉积物中，或以不同的生存/营养方式与其他生物发生的某些关系中，包括附着、寄/共生等等，因而海洋古菌在不同的海洋生态系统中的广泛分布是十分明确的了[7,8]。Karnet 等根据对太平洋某水域中层海水古菌丰度的研究，推测古菌细胞体个数与细菌处于同一量级[9]，Vetriani 等研究大西洋 1 500 m 深海沉积物古菌（丰度），认为它们占原核生物的 2.5%～8.0%[10]。此外人们还从海绵体内发现了一些新种古菌，其与当前基因库收录的古菌序号 652DNA 相似性不超过 90%[11]。

分布极不平衡的海洋古菌在分布上受制于海洋的多种因子，包括海水、沉积物的盐浓度、营养成分的含量与分布状况，因不同深度水层和沉积物而引起的光强、压力、气体状况和温度等的变化，还有海洋生态系统各类/个成员与古菌的关系等等。但这些因子并非杂乱无章或平均地影响古菌，其中必有主次之分。比如常态极地海洋，低温可以是主因子；在深海，压力可能起重要作用。但是无论外界因子如何变异，必然地有相应古菌应运而生，从而构成海洋微生物类群的主组分之一。

3.2　多样性

迄今全世界发现的古菌绝大多数来自极端环境，它们被分成 2 或 3 界计 10 纲，约 99

属。每个属的种数不一,但普遍不多,合计约有195种。已发现海洋古菌67属184种。常见海洋古菌约6属16种,如盐杆菌属的盐沼盐杆菌(*Halobacterium solinarum*)、脱硫球菌属(*Desulfurococcus*)、火源甲烷球菌(*Methanococcus igneus*)等。由此可见海洋古菌的种类多样性至今肯定无法和海洋细菌比,其原因主要决定于研究方法对它们生存环境的难以仿真以至于(海洋)古菌的成功分离和纯培养少之又少。至今仅成功分离出少数产甲烷古菌、硫酸盐还原型古菌。但从已有资料可推测海洋古菌生境的多样性,其丰富程度是无可争议的。它们在不同水层中不仅数量有变化而且主要种群也不一致[7,12]。比如,研究表明,表层水域海水的古菌分布广泛,南极海域古菌可占到浮游微生物的10%~30%[13]。平和海域与极端海域、不同深度的海洋沉积物等环境中,古菌的系统发生多样性实际上比人们预先想到的多得多[14]。我国学者也从太平洋近赤道区水深5 774 m下和西沙海槽的沉积物中发现了不同类型古菌[15,16]。

3.3　适应性

古菌对温度有强大的适应能力。广域古菌中有一类是甲烷产生菌,它们包括对温度的要求可以相当宽泛而成嗜冷、嗜温、嗜热甚或超嗜热者,其生境还可以有动物消化道、缺氧沉积物乃至腐烂物品、表面微生物垫及海水中。它们形态和生理上的多样性形成了对各种生境的适应性。亿万年来它们之中有的还以产生甲烷而影响气候变化[2]。甲烷火菌属(*Methanopyrus*)成员很嗜热,它们能在海底热液喷口的黑烟囱壁上快速繁殖。比它更嗜热的是延胡索酸火叶菌(*Pyrolobus furmurii*),其最高生长温度可达113℃。与此相反,在寒冷的常规极地或深海中,温度低至2℃左右,16*S* rRNA引物分析得出那儿广布有泉生古菌,甚而其数量占了微型浮游原核生物的很大一部分,其中的共生泉生古菌(*Crenarchaeum symbiosium*)与冷水中海绵等的共生关系可能反映了它们适应冷境的机制。而不少泉生古菌还能生存于高温中,包括海底热液喷口处。与多数常规极地环境相反,生活科学网(美)2008年8月报道,格陵兰—挪威间中大西洋脊上却有海底黑烟囱,该黑烟囱泉位于洋底2 000 m处,水温达400℃,周围却生有古菌等微生物,适于80~120℃温度中生存。相反相成的是冷泉,那里同样有古菌。

古菌对盐度也有广泛的适应性,已发现的20属极端嗜盐古菌,不仅适应高盐度,而且适应于各类不同的以高盐为主的环境,这决定于其特大质粒紫膜和化能异养方式等。与此相反许多古菌也生存于普通盐度的海水和沉积物中。

海洋中包括氧气在内的气体,其分布也不呈平均状态。10~20 m水层的氧气有最高的溶解度,越往下,氧气浓度越低。深至200~1 000 m的水层,氧气浓度达低谷,甚而有可能呈无氧区。深于此水层后,因低温又可使氧溶度增加,因此不同水层、不同海区氧梯度变化大,在从有充分氧气的水层至无氧水层的变化中,不同的古菌对氧有其相应的适应性。比如表层水,浮游古菌广泛存在;中层水的原核生物中浮游古菌虽然还占优势[7],但比表层水低了一个量级[17]。不同水层古菌类群成员组成也发生变化。太平洋某区海水,其透光层下(150 m左右),浮游的*Crenarchaeota*为浮游生物主成员之一,水深增大,该类古菌渐变为只占可测浮游生物的39%[8]。但是在无氧区或者缺氧区,包括不同深度的沉积物,乃至海底热/冷泉,仍然可以发现活跃着的古菌,这些表达出古菌对不同氧浓度有广泛的适应性。但在缺氧或无氧环境,包括产甲烷古菌在内的厌氧型古菌,必定是重要的原核生物成员。表层海水的浮游古菌,无疑生存在轻松的大气压力下,但同时不同大气压的海洋环境催生了相应的古菌。迄今已分离出的海洋古菌基本上能耐高压(barotolerant)，但其适压性宽泛，从 $1\times10^5\sim400\times10^7$ Pa 不等。高压对它们来说只是降低了生长速度和代谢活性。必须在 $\geqslant4\times10^7$ Pa 下生长的古菌构成了嗜压菌或专性嗜压古菌(obligate or extreme barophilies, or piezophilies)，这包括热液喷口处的专性嗜压化能自养古菌，其承受压力更大。多数生存于海岸/浅海环境的古菌，可承受的最大压力仅达 2×10^7 Pa。一些古菌对压力的适应可以说是随遇而安，且有巨大的应变能力。高压环境对许多古菌形成了不可抗拒的力量，而其他因子又给它们创造了利弊不一的影响，这些都促使它们顺应地产生相适的细胞结构、生理生化功能和互补的分子、基因组[2]。随着深探和仿真系统技术的发展和完善，人们有可能在更深的大洋底层及下部发现耐受/嗜好更大压力的古菌。

4　*海洋古菌研究前景*

分子生物学技术对海洋原核生物研究的扩展和渗透向人们展示了海洋古菌研究的美好前景。借助设计不同的探针，估测不同水域中的含量，由此确认古菌在特定海区总原核生物和浮游生物中的份额和地位。从自然环境提取 DNA，对比已知数据库相关的古菌 16*S* rRNA 部分片段序列，绘制古菌的系统进化树，进一步认识生物的进化学说。由此分析比较归类资料，设计新探针优选条件，来纯化培养样品，找出与其他属种古菌的亲缘关系，有可能找到新菌种，包括骑行纳古菌、初生古菌，对生命起源学说有价值。如前所述古菌的特性，对纯化古菌作生化、代谢和调控机制及酶学等的研究，有可能发现和利用人们所需的资源，包括像对无氧环境古菌的认知和利用，以探讨一些污染物的代谢机制[18,19]。

对深海热液喷口处古菌的研究，既可发现它们的特殊细胞结构功能，也可探索它们与其他生物关系中的作用和地位，又可将其对环境的适应能力用于生物工程研究。

极端嗜盐古菌对外界/胞外高浓度 Na^+ 等离子的调控活动和能力必然与其细胞的结构功能息息相关，对它们的研究将可用于仿生学，并可使人们进一步认识海洋生态系统的物质转化和能量循环活动。

对泉生古菌生境多样性的研究使人们对海洋原核生物多样性、生态学和海洋生物地球化学循环的认识有了重大改变。对共生古菌及古菌与其他生物间的各种生存关系的研究则展示了一些共生生物(如冷水海绵)的许多天然产物可能是其体内聚集的古菌等微生物所为，它们将给人们带来新的产物，有些使人们获得更多更好的代谢活性物[20]。海洋古菌与其他微生物发生的互养共栖(Syntrophis)关系已在沉积物(甲烷储层)中得到了证实，甲烷从海底的释放过程包括古菌的厌氧活动。古菌由此也合成了脂类，而这些菌又被硫酸盐还原菌(SRB)包围，再与其他生物在营养上构成了化能合成共生关系，由此可进一步去推测远古时代以来生物的新奇营养关系和生存历程。

认识和评估古菌在海洋及其生态系统中的地位和作用。海洋浮游古菌的发现和确认是海洋生态学发展的一大里程碑。它们在海洋中的普遍存在具有物种和代谢类型、功能基因组成和生态功能上的多样性和适应性，这些是有重要理论和应用价值的。对沉积物中普遍存在的古菌的研究必然地拓展对微生物在地质学、生物地球化学、物质转化和能量流动、循环中的作用的认识。对新的古菌物种的探索，一方面可以更新和发展系统进化树，推测远古时代以来的生物进程和发展，推动生命起源学说发展。另一方面对古菌的纯化培养研究则可以寻找新型的生化代谢方式和途径，获取新的生物工程学概念、理论和实践，满足人类日益扩大和提高的物质需求、改进环境污染治理手段，使海洋生态系建立在更加良性和可持续发展的基础上。

对古菌的研究，尤其是对海洋古菌的研究，尤其在中国，是星星之火、方兴未艾的事业。

参考文献 20 篇（略）

（合作者：高爱国　孙丕喜　王　波）

ARCHAEA IN SEAS AND OCEANS

(ABSTRACT)

Abstract The research ways of Archaea were briefly outlined and the characterizations of marine Archaea were discussed including their universal distribution, diversity and adaptability. The research prospect of marine Archaea was also presented.

Key words Archaea, Ocean And Sea, Distribution, Diversity, Adapt-Ability, Foreground

原核生物概况及海洋原核生物组成*

摘　要　概述原核生物及其分类，细菌域和古菌域主要成员构成状况，海洋原核生物的主要类别、属群组成及海洋细菌和海洋古菌的组成。

关键词　原核生物　古菌　细菌　海洋古菌　海洋细菌

在亿万年的空间变化中，各种生物经历自身的遗传、变异与选择，最终形成了当今多样性的生物世界。原核生物尽管相对低等、简单，但始终是生物世界中的重要成员。人们对它们的认识，包括其地位，属群和分类等并不容易，在历经几百年研究后才形成了今日的格局。海洋作为地球空间的最大组成部分，人们迄今对其中的原核生物知之甚少。原核生物的研究，不仅对认识生物历史和发展有重要的作用，而且对相关知识的开发与利用也具有广阔前景。本文就原核生物分类及海洋原核生物组成作一概述。

1　原核生物概况及其分类

1951年多尔蒂提出原核(细胞)生物的概念，它是指没有真正形态细胞核但是有分布于细胞内的核质，其周围没有与胞质隔开的核膜的原始微型生物。原核生物在微生物中占据重要地位，其个体小，常以单细胞为独立生存单位。迄今发现的原核生物主要指古菌和细菌。古菌是一类发现较晚，异军突起的生物。细菌包括真细菌、蓝细菌、原绿菌、放线菌等[1]。它们分别归入古菌域和细菌域。

地球是个巨大完整的生态系统，其原核生物十分丰富，孕育和繁衍着众多的微生物。人们对微生物的了解古来有之[2]，真正的微生物学研究迄今已经历了5个时期。如今人们应用遗传分子生物学理论对微生物，包括细菌等的识别——分类学研究，已经进入最新阶段。《伯捷氏系统细菌学手册》第2版[3]已被广为运用。

1.1　古菌域

1977年Woese等发现古菌，最初被看成细菌成员，后来人们根据其与细菌的渊源关系而判定它为古菌。它的地位与细菌一样都各自成为一个域。20世纪70年代至90年代初，人们一直认为古菌只是生存在极端环境中，包括陆域极端环境中的一类原核生物，1992年才发现古菌也存在于一般状态的环境中，包括海洋[4]。古菌具有对广泛生境的适应性、营养/生存方式的多样性以及分子遗传的变异性，这使得他们基本上广布地球各处。

1994之前原核生物中的古菌被称作为古细菌，一直被认为是细菌中的一个纲——古生菌纲，即被归为一个独特的子分支而已[4]。随着科学技术的不断提升，古菌有别于细菌的特征、性状日益明朗化，原核生物中的古菌不同于细菌的学说逐步确立起来，古菌域已

*　国家重大基础研究前期研究专项项目——我国近海浮游植物多样性演变机理及其对生态系统功能影响(2002CCA4900)资助。

原文刊于《广西科学院学报》，2010，26(2)：140～142转145，本人为第四作者。

成为生物三大域之一。

按任立成等[5]引证的观点，古菌域被分作三界，即广古菌界、嗜温泉古菌界和初生古菌界。按潘晓驹等[6]引证的观点，古菌域只有两界，即广域古菌界和嗜温泉古菌界，其下共有 5 个类：产甲烷型古菌、硫酸盐还原型古菌、极端嗜盐型古菌、无细胞壁型古菌和极端嗜热型硫代谢菌，其下还有 8 目 12 科。而据陶天申等人[1]的看法，古菌分为两界——泉古菌和广古菌。泉古菌包括 4 目，广古菌至少有 6 目，有 5 个生理类群，分属 7 纲。张晓华和陈皓文[7]根据系统发生树的分法，把古菌分为 4 个主分支，即 4 门，此门又相当于界，这是为了对应于细菌域的分类方式而已。目前有明确性状特征的古菌属有 99 个，即广域古菌门 8 纲共计 68 属，泉生古菌门 6 纲 30 属、纳古菌门 1 纲 1 属，而初生古菌门尚未建属，约 195 种，常见的古菌有甲烷杆菌、甲烷球菌、盐球菌、热原体、热球菌、热变形菌、纳古菌等。

1.2 细菌域

由于微生物分类系列、方法等的变迁，包括重新认识、变位、拆并、取消、新建等，使得相应的微生物归属、名称也处在不断变化之中。

1965 年出版的《拉汉微生物名称》[8]所载细菌、放线菌名称约计 2 800 条。1977 年出版《细菌分类基础》记载有 138 属 155 种细菌。1980 年出版的《细菌名称(第 1 版)》收录 5 000条细菌名称。1982 年的《细菌名称》载有 5 000 条名称。1994 年第 9 版 *Bergey's Manual of Determinative Bacteriology* 将世界细菌划分为 35 群，它们被归入 4 大类目，即具胞壁的 G^- 真细菌、具胞壁的 G^+ 真细菌，无胞壁的真细菌及古细菌。它们被分作 574 属，在属之前已被归入不同纲、目、科等类别之中。

1999 年第二版第 4 次印刷的《细菌名称》中已载入细菌名称 15 000 余条，其中包含蓝细菌、古细菌和放线菌。2000 年出版的《细菌名称英解汉译经典》也作过相关估计。陶天申等[1]在《原核生物系统学》一书中将细菌归成 23 门 9 纲。而截至 2007 年，*Taxonomic Outline of the Bacteria and Archaea*[3]中收集的资料经统计后得出世界细菌域成员被分成 26 门(一说为 23 门)36 纲，1 未定纲 5 亚纲，88 目 4 未定目 15 亚目，241 科 13 未定科，1 684属。截至 2008 年 4 月张晓华、陈皓文[7]统计的有关资料，得出世界原核生物约 8 500 余种，被归入 1 799 余属中，其中细菌 1 700 属 8 305 种。赵乃昕等统计原核生物至少高达 9 380 种，而细菌种数达 9 185 个。

各原核生物的科所含属数不一，细菌域中达到和超过 10 个属的科约有 48 个。达到和超过 24 个属的科有 10 个，以黄杆菌科所含的属最多，有 72 个。各原核生物属所含种的数量也不一，其中有 562 个属分别只包含 1 个种，尚有 11 个属未列出任何种名。有 18 个属的细菌域成员，其种数各达到或超过 50 个，其中有 7 个属的种数至少有 106 个，它们分别是分枝杆菌属(种数达 106 个，下同)、支原体属(119 个)、乳杆菌属(126 个)、芽孢杆菌属(147 个)、假单胞菌属(155 个)、梭菌属(176 个)和链霉菌属(533 个)，链霉菌属的种数最多。这 18 个属的菌种数合计 2 040 个，占细菌域总种数(8 305 个)的 24.6%，与蔡妙英等[8]的估计比，现在的细菌数只占 1995 年细菌数的 55.4%左右(8 305/15 000)。至今无种数记录的 11 个属有 10 个在蓝细菌中。而据以往的记录，除吉特勒氏蓝细菌和 *Leptolyngbya* 属外，均是属于有种的属，种数波动于 1～38 之间。除了上述大属和无种属外，余下的菌种数共计 4 406 个，它们分布于 639 个属中，其种数≥2，平均每属含种数约 6.9 个(4 406/639)，而细菌域所有菌属平均菌种数则为 4.89 个(8 305/1 700)。

2 海洋原核生物组成

海洋是原核生物的发祥地，其面积覆盖全球的70%左右。文献[5]认为全球迄今发现认定的细菌等原核生物目均可在海洋中找到其代表。但是世界海洋，尤其大洋从来都是互通的，因而其原核生物生境，除极端环境外，在各海洋间具有许多共性，亦可以认为海水原核生物的物种多样性在某种程度上，比陆地和淡水的要简单一些，尤其是海洋浮游古菌种的多样性是相对简单的。目前地球上发现的原核生物中，有1/4～1/3是已经生活或者可以生活在海洋中。与海洋有点类似的环境还包括盐、碱水/地/滩等及极地环境也是原核生物包括古菌的滋生地。

2.1 海洋古菌

海洋古菌的最初发现地实际上是海洋极端环境，包括高压、高盐/咸、缺氧、高/低温、低营养等。20世纪90年代人们发现了浮游海洋古菌，由此改变了古菌只存在于极端环境的这一传统观念，从此人们对海洋生态学的认识发生了质的变化[6]。海洋古菌如同其他生物一样，其种群数因环境/生存条件变化而变化。学者们推断世界海洋的古菌细胞体约有 1.3×10^{28}，与细菌细胞体个数处于同一个量级(后者为 3.1×10^{28})[6,8,9]。同样，海洋古菌的多样性在海洋沉积物、在与其他生物发生的不同关系状态中均有显示[9,10]。

世界上陆地生境的古菌，大部分都能在海洋中发现，但是初生古菌尚未在海洋中被发现。不过在海水和沉积物中却发现了其特有的古菌类群——类群Ⅰ、Ⅱ、Ⅲ和Ⅳ等新属种，如其中的产甲烷古菌(*Methanogenic archaea*)和硫酸盐还原古菌(*Archaeasulfate reducers*)。据记载现在共有海洋古菌67属184种。海洋古菌属种分别占已发现的古菌属种的67.7%和94.4%。常见海洋古菌约4属45种，4属占所有属数的6.0%，45种占所有种数的24.5%。每属均超过10种，如盐盒菌、甲烷球菌、盐杆菌等属及共生泉生古菌(*Arcfheoglobus fulgidus*)、火源甲烷球菌(*Methanococcus igneus*)、速生热球菌(*Thermococcus celer*)等种。

古菌域家属虽然古老，但是发现较晚，所以并不庞大，原因是大多数古菌难以用传统方法分离纯化培养出来。人们对它们的认识虽然日趋全面，但新兴分子生物学技术包括16*S* rDNA方法、核酸杂交技术、PCR技术等大都源于对细菌域成员的研究，对古菌的研究技术须进一步改进和完善[11]。

古菌在中国的研究并不多，20世纪90年代已报道过一些陆境古菌的研究，也纯化出少数嗜盐、嗜碱的新杆菌[12]。有学者[13]从海洋盐场分离出嗜盐小盒菌。对海洋古菌的研究已经开始引起国内学者关注[14,15]，但成果还不多。

2.2 海洋细菌

目前已发现的细菌门，无疑在海洋中均有，其中变形菌门，即Proteobacteria成员最多，生理状态最多样。它分为5纲，即α-、β-、γ-、δ-和ε-变形菌纲，而海水中则以浮游的α-和γ-变形菌最为重要，但α-变形纲成员优势最大。这是根据16 S rDNA法得出的结论，非培养法所证实。不过γ-变形纲菌在培养法中最常见。海水中最占优势者在α-变形纲中。

据*Taxonomic Outline of the procaryotes*记载，以海洋、盐碱、极地等拉丁字开头的属数多达126个。其中以ha-(盐、海-)开头的细菌属有25个，若将Natro[(-c)、喜-、嗜-盐]加入，则更多(约4个)。以Mar-(海洋-)开头的有24个、以Ocea-(大洋-)开头的有15个，以S(al-)(盐-)开头的约有13个、以Pol-(极地-)开头的有5个等，这些细菌是纯海洋的或主要(首先)发现于海洋环境。

迄今,海洋细菌属 798 个,种数约 2 345 个(包括亚种、变种),占地球细菌种数的 25.5%(2 345/9 185),种数在 10 个以上的属有 29 个,占海洋细菌属总数的 36.3%,29 属菌种数达 618,占所有海洋细菌总种数的 26.4%,其中以弧菌种数最多(80 个),其次为芽孢杆菌(52 个),假单胞菌(32 个)等等。常见海洋细菌约 64 属,约占地球细菌属数的 3.8%(64/1 700)。假交替单胞菌(24 个)、盐单胞菌、黄杆菌(同为 27 种)也较多,大洋螺菌 18 种(及亚种)和海杆状菌(21 种)。

海洋,包括海水或咸水中还存在活的不可培养的细菌,即处于不良环境条件中仍存活而常规条件培养不出的细菌,迄今报道有 29 种。

3　结束语

目前,人们对非海洋细菌的认识比较丰富、全面,但有关海洋细菌的知识并不丰富,缺乏深入了解[15]。不过近来有些学者对各种生境的海洋细菌产生了浓厚的兴趣[1,5]。古菌,尤其海洋古菌展现给人们许多美好的前景和应用价值,其中包括生命起源、生活特性的探索和资源的开发利用等等[1,6]。研究原核生物尤其是海洋原核生物的任务十分艰巨但很有意义。世界原核生物有多少,迄今还是个谜[15],它尚待有志者的进一步探索。

参考文献 15 篇(略)

(合作者:孙丕喜　陈颖稚　王宗兴)

GENERAL SITUATION OF PROKARYOTES AND THE COMPOSITION OF MARINE PROKARYOTES

(ABSTRACT)

Abstract Prokaryotes and their classification, outline of constructional states of most members of bacterial and archaeal domains, compositions of marine bacteria and marine archaea, main and common species, genera of marine prokaryotes are briefly introduced.

Key words Prokaryote, Archaea, Bacteria, Marine Archaea, Marine Bacteria

2008 年中国海洋微生物学研究概况*

摘　要　本文参阅中国 2008 年全年多数海洋微生物学研究论文，可将其分为两个方面：即基础海洋微生物学和海洋水产微生物学。文章梳理其主题、内容，从中把脉该年中国此方面研究的主流、成果和进展。认为中国海洋微生物学研究正处于成长期，其研究内容较为宽泛，尤其表现在养殖海产品的病害机理、防治等方面。基础研究仍显得单薄，这是值得注意并需强化的。

关键词　中国　海洋　微生物　基础　应用

中国海洋微生物学发端于 20 世纪 50 年代末。在迄今不足 60 年的历程中，它由渺小逐步走向壮大，这体现在其研究的广度和深度上。如今海洋微生物学的重要性越来越成为世人的共识。当今中国海洋微生物学研究正在蓬勃发展。2008 年，对中国海洋微生物学研究来说是平凡的一年。翻阅此年我国相关论文，可以把脉和梳理其发展内涵和趋势，以期与同行探讨我国本学科的发展前景。

1. 基础海洋微生物学研究

1.1　病毒：近岸海洋环境人类致病病毒研究进展。青岛近海水域夏冬季、北黄海夏季浮游病毒丰度、分布。大连金石滩海水浴场人肠道四种病毒的分布。重组蓝藻抗病毒蛋白 N 的纯化、复壮及抗单纯疱疹病毒Ⅰ型活性等等。

1.2　原核生物

1.2.1　古菌：极端嗜热菌的热休克蛋白、超嗜热古菌硫化叶菌(*Sulfolobus*)S. tokodaii RadA 蛋白的克隆表达及其辐射可诱导性、南海古菌多样性。

1.2.2　细菌：关注海单胞菌研究[1]。氨氧化细菌系统发育。耐盐细菌：16S rRNA 基因序列分析、HBCC-2 菌株的基因测序、南极海冰耐盐菌 NJ82 的分子鉴定及耐盐性、南极海冰嗜冷细菌培养条件和胞外水解酶、产四氢嘧啶的中度嗜盐菌 TA-4 的分离、特性。

克雷伯氏菌：胞外多糖制备及组分。硝化细菌：富集、驯化和筛选。弧菌：哈维氏弧菌：TS-628 菌株胞外产物(ECP)蛋白酶活性、抗该菌的脂多糖单克隆抗体的制备及鉴定。副溶血弧菌：致病因子与耐热直接溶血素研究进展。河流弧菌：在天然海水和人工海水中的饥饿效应。二氟沙星对 3 种弧菌的抗菌后效应。5 种致病弧菌对 3 种中草药敏感性测定。烷烃降解菌：深海多环芳烃降解菌新鞘氨醇杆菌 H25 的降解特性及基因、相关菌株的筛选与特性[2]；T_1 和 T_2 菌株的分子鉴定分类、深海嗜低温萘降解菌 Nah-1 的分离及降解基因。铁、锰细菌分布。聚磷菌：核苷二磷酸激酶基因的克隆及序列。病原菌：近海水域中该类菌的研究进展。放线菌：具抗菌活性的菌株 JMC06001 的分离和鉴定、XS904 菌株的分类鉴定及发酵液抑菌活性、海绵源放线菌 5h6004 发酵得活性生物碱[3]、北极放线菌 AFN2001 的培养和发酵。发光细菌：生长和发光条件。

* 原文刊于《管理观察·学术观察(学术论坛版)》，2010，11 月：143-146.

蓝细菌:聚球蓝细菌引起的赤潮及生态、固 N 蓝细菌的基因。

1.3 益生菌:多数是围绕养殖进行的研究,如一些光合细菌的分离及培养条件、对虾肠道益生菌筛选与鉴定、中国明对虾 1 株益生菌的筛选与鉴定、鱼消化道微生物分布与调控、对凡纳滨对虾生长和虾体营养组分的影响、植物乳杆菌发酵对牡蛎成分的影响、养殖中的应用[4]。明对虾 Rel/NF-KB 家族基因在弧菌刺激下的表达。外来病因所致的机体内包括抗力、免疫力等的变化如溶菌酶、抗菌肽的产生和作用发挥、相关蛋白的增多、白蓝蛋白与病原菌起凝集作用或可致抑菌/抗病力提高。对日本囊对虾 Kunitz 型蛋白酶抑制剂在毕赤酵母中的表达等作了探讨。对虾相关疾病检测与诊断方法在不断提高和完善中。

1.4 真核生物

1.4.1 耐(嗜)盐真菌:耐盐 *Aspergillus variecolor* B-12 产烷基苯甲醛衍生物及细胞毒活性、台湾海峡耐(嗜)盐真菌及其生物活性。

1.4.2 酵母/真菌:破囊壶菌(*Throustochytrium* sp. F)N-10 DHA 生物合成途径相关延长酶克隆与表达、产虾青素酵母 YS-185 的原生质体制备与紫外诱变、黏红酵母表面黏附蛋白及肠黏液受体、海洋酵母胞外酶及其基因研究进展、海洋动植物区系真菌的分离及抑菌、产淀粉酶的海洋曲霉的分离及酶学特性、近海潮间带植物及其生境中微生物特性和药用前景、产蛋白酶的海洋酵母的筛选。南极假丝酵母脂肪酶 B 的酿酒酵母表面展示及其催化已酸乙酯的合成[5]、

1.5 沉积物微生物:南海南部陆坡表层沉积物细菌、古菌多样性、南沙海域沉积物可培养微生物及其多样性、浙江舟山海岸带渗漏水沉淀铁泥中微生物矿化作用、舟山潮间带底泥微生物分离及抗菌活性、海河-渤海湾沿线沉积物微生物群落代谢特征、氟苯尼考对沉积物微生物活动影响。

1.6 南北极微生物:极地微生物的工业应用及与天体生物学研究进展、南极嗜冷假单胞菌 7197 的分离和无机焦磷酸酯酶(PPASE)基因 PPa 的克隆、降解石油烃的嗜冷菌筛选及降解特性、北极海沉积物锰细菌分离与系统发育、北极海沉积物铁细菌[6]、北极微生物分离及抗菌抗肿瘤活性筛选、北极放线菌培养和发酵、加拿大海盆、楚科奇海沉积物岩芯普通异养菌分布[7]、极地环境污染物降解、修复。

1.7 近海微生物:病原菌多样性分析进展[8]、大连湾、辽东湾夏季海水细菌多样性 16SrDNA 分析、盐度对粪大肠菌群数影响、自然水体中溶藻弧菌的饥饿效应。

1.8 深海(地)微生物:2008 年大约有 17 篇论文对取自不同深海海域(沉积物)微生物作了广泛的调研和论述,包括南沙海域沉积物可培养微生物,有关成果将另文报道*[9]。

1.9 天然气水合物:微生物成因及识别[10]。

1.10 抗肿瘤活性物及应用微生物:活性物等相关研究进展[11]、共附生微生物天然产物生物合成基因研究进展[12]、深海来源微生物抗肿瘤活性筛选及次级代谢、放线菌抗肿瘤物发酵条件优化、纤溶活性化合物筛选及分离、多菌灵的光化学降解、富含古罗糖醛酸的褐藻多糖生产菌株构建。

---

* 我国深海地微生物学研究状况分析

1.11　藻菌关系：溶藻菌溶藻效果、一株 *Pseudoalteromonas* sp. 的分离、鉴定及溶藻活性、(微)藻(细)菌关系及应用、红树林黄杆菌抑藻物及对亚历山大藻的抑制、溶藻菌对棕囊藻溶藻的电镜观察、塔玛亚历山大藻的溶藻菌筛选法、抗生素抑菌及对中肋骨条藻生长的影响。

1.12　普通微生物学及研究方法：分子印迹术用于生物碱活性成分研究、DGGE 技术及在微生物多样性中的应用[13]、低温微生物的冷适应机理及应用、从产 Macrolactin 细菌 X-2 中筛选Ⅰ型 PKS 基因簇、鉴定与功能分析、1 株细菌 MZ0306A1 的 16SrDNA 扩增及序列分析、细菌逃避补体识别和攻击机制、鱼源乳酸菌抑菌特性。

2. 海洋水产微生物学研究

本课题发表的文章占全年文章数的近半，表达出我国海产工作者对此的关注。所涉及的海产养殖生物多达约 30 种。下面着重介绍涉及较多的海产品之微生物学研究。

2.1　甲壳动物

2.1.1　对虾：以中国明对虾疾病为主要目标，其中有白斑综合征及其病毒(WSSV)，包括 WSSV 病毒蛋白质组研究进展介绍、WSSV 膜蛋白[14]和囊膜蛋白、线性抗原表位单链抗体的研究。日本对虾、凡纳滨对虾也得此病，对后者身上出现的新症状作了调研。经受 WSSV 感染的日本对虾子三代抗病力的研究。壳聚糖硫酸酯抗 WSSV 感染的能力、温度对 WSSV 感染的影响。传染性皮下造血组织坏死病毒、桃拉综合征病毒 RT-PCR 检测法改进、应用及受其感染后体内同工酶酶谱变化。弧菌病、鳗弧菌病、“偷死病”、红体病等均有调研。

2.1.2　扇贝：大规模死亡原因及防治、常见病毒种类及流行趋势和防控、富集文库—菌落原位杂交法筛选栉孔扇贝的微卫星[15]；虾夷扇贝面盘幼虫细菌病病原及该贝死因探讨、控制对策；预防海湾扇贝亲贝死亡措施。

2.1.3　牡蛎：诺如病毒的分子流行病学研究进展、太平洋牡蛎诺沃克样病毒检测法及植物乳杆菌对牡蛎体成分的影响。

2.1.4　蛤：菲律宾蛤仔育苗用水系统中细菌状况；南通市文蛤的污染现状及大肠菌群的净化、文蛤副溶血弧菌间接 ELISA 检测技术。

2.1.5　青蟹：电镜观察呼肠孤样病毒粒子、常见病及其防治技术。

2.1.6　鲍：受副溶血弧菌感染后的九孔鲍细菌均一化全长 cDNA 构建。

2.1.7　缢蛏：养殖期病因及防治。

2.1.8　螯虾：借助单克隆抗体对抗螯虾白斑综合征做活体内中和作用测定[16]。

2.2　棘皮动物：海(刺)参常见疾病预防/诊治术、蛭弧菌的生物学特性及在刺参养殖中的应用前景；海胆养殖水体中多抗性细菌的抗性基因；海参、海胆卵内共生菌的分离及其活性探讨。

2.3　鱼类：三重 PCR 检测鱼类致病性嗜水气单胞菌

2.3.1　牙鲆：淋巴囊肿病毒灭活疫苗制备、应用及药物防治试验；褐牙鲆腹水病防治试验、癸甲溴铵对褐牙鲆急性毒性及抑菌试验；牙鲆弧菌(新种)主要生物学特性；河流弧菌对牙鲆表皮的黏附作用[17]。

2.3.2　大菱鲆：虹彩病毒(TRBIV)PCR 检测方法的建立及应用[18]；两种 ELISA 法快速检测细菌性出血性败血症病原菌、体表出血症病原菌分离鉴定、腹水病防治、鳗利斯顿氏菌的药物敏感性测定分析、副溶血弧菌的致病性、溶藻胶弧菌脂多糖对血清溶菌酶活性影响。

2.3.3　大黄鱼:副溶血弧菌及抗副溶血弧菌菌株筛选、主要病害诊治、溶藻弧菌和哈维氏弧菌的分子鉴定及其系统发育学分析、假单胞菌病病原分离鉴定;Piscidin 样抗菌肽基因部分 cDNA 克隆与序列分析[19]。

2.3.4　鳗:水霉病防治技术、新病及控制。

2.3.5　红鱼:血清溶菌酶性质及用途。

2.3.6　石斑鱼:原位杂交技术检测神经坏死病毒及其病情流行调查、河流弧菌对鱼体表黏液黏附作用影响。

2.3.7　军曹鱼:淋巴囊肿病毒系统关系及基因型研究、养殖水体和肠道异养细菌耐药性及产酶能力。

2.3.8　鲀:红鳍、暗纹东方鲀皮肤溃疡病病原菌分离与鉴定[20]、东方鲀几种组织中可培养细菌组成分析。

2.3.9　花鲈:鳗弧菌 W-1 对鱼苗的致病性。

2.3.10　罗非鱼:海豚链球菌疫苗对免疫功能的影响。

2.3.11　鲑鳟鱼:弧菌病主要病因及防治。

2.4　海绵:共附生放线菌抗肿瘤物。

2.5　植物

2.5.1　紫菜:壇紫菜叶状体细菌性红烂病;条斑紫菜外生细菌遗传多样性。

2.5.2　红树林:黄杆菌抑藻活性物鉴定及对亚历山大藻的抑制活动;海莲内生真菌 *Penicillium* sp. 091402 化学成分[21];盐度对该林土壤微生物数量影响、盐度和 pH 值对根际土壤源曲霉 F3 生长及对其分泌活性物质的影响。

2.6　海产养殖:对益生菌的积极作用已开始调动,包括有益微生物菌群的应用、越冬棚养虾中微生物的应用等。围绕病害问题,对病毒感染中 WSSV 被膜蛋白 VP124、鱼病毒性神经坏死病毒衣壳蛋白在昆虫细胞中的表达及特性、病原检测基因芯片应用及前景、几种弧菌外膜蛋白结构比较分析、ELISA 技术在病害诊断和生物毒素检测中的应用[22]做了研究。固定化微生物的应用、滩涂养贝环境中细菌分布;海产品副溶血性弧菌的分离鉴定及与临床分离株的生化性状比较、溶菌酶和抗菌肽的应用;豚链球菌和停乳链球类 M 蛋白及其抗原性、豚鼠气单胞菌 CB101 的 G1cNAC 利用系统及调控蛋白 NagC 对 Chil 转录的调控;小球藻与芽孢杆菌对对虾养殖水质的调控;疾病防控策略[23]。

3.论文数和所载刊物分析

据不完全统计,2008 年中国大陆发表的海洋微生物学论文数约 220 篇(这比 2007 年的论文数少了 30.2%),它们分布在约 45 种刊物上。此年发表有 10 篇及以上相关文章的刊物约有 7 份,它们分别是《中国海洋大学学报》、《中国水产科学》、《微生物学通报》、《海洋环境科学》、《海洋水产研究》、《海洋科学》、《水产科学》。在上述 3 份纯粹水产学科杂志上发表的海洋微生物学论文数计 45 篇,占全年海洋微生物学论文数的 21.0%左右。与海产养殖等直接相关的文章约 103 篇,占全年相关论文数的 47.2%,加上相关的其他文章,海产微生物学研究论文数已约占海洋微生物学研究论文数的半壁江山。纵观深海微生物学的研究论文约 17 篇,占全年本学科论文数的 7.9%。统观全年状况,可知我国海洋微生物学研究的面较宽泛，其中应用及其基础研究较多。1篇文章的参与者数＞3人的论文约有118篇。全年参与论文的写作者总数约在900人以上。

参考文献23篇(略)

(合作者：孙丕喜)

GENERAL SITUATION OF MARINE MICROBIOLOGICAL RESEARCH IN CHINA DURING 2008

(ABSTRACT)

Abstract This paper consults nearly all papers of research on marine microbiology in China during a whole year of 2008, might divided them into two fields, i. e. basic marine microbiological and marine fishering microbiological research. The paper combs and puts in order of their themes/subjects and contents etc, and tries to find the main currents, results and progresses of the research in China from those papers/articles. It considers that the research in China is in growing period, its contents are rather spacious and in deep going, specially in the principles, controls and treatments of the diseases and evils etc. of cultured marine fisherings. Its basic research is still thin and weak. It is noticeable and has to be strengthenning.

Key words China Seas/Oceans Microbes Basis Application 2008

中国海洋微生物学 2009 年研究进展*

摘　要　本文归类剖析 2009 年中国海洋微生物学文章，全年有 429 篇之多，攀上了新高峰，其中海产养殖方面的微生物学文章占半数以上，基础海洋微生物学文章仍少。学者们关注和跟踪国际先进新理念/新方法，包括分子生物学方法和技术的研究和应用得到较大的发展。与前几年相比，本年度该领域的研究与国际间的交往正在加强，研究势头正在强劲之中。

关键词　中国　海洋　基础/养殖微生物学　2009 年

引言

时光荏苒，2009 年离我们而去已有一整年多了。中国海洋微生物学在以往大约 60 年的基础上迈上了新台阶，创造出新成就，以此为新世纪第一个年代画上了完美的句号，令人难忘。据我们不完全统计，此年海洋微生物学文章数量超过了历年水平达 429 篇以上。不少论文质量上佳，并有新书出版，使我国的相关研究离跻身世界先进水平越来越近。

本文就 2009 年我国海洋微生物学研究发表的文章进行归类剖析，将其分为基础和海产养殖两大分支阐述。

1　基础海洋微生物学

1.1　关注国内进展，追踪国际动向

在浮游病毒遗传多样性，鱼虾贝等水产品病毒及病原菌分离纯化检控技术，PCR 技术在对虾病毒检测上的应用[1]，微生物培养新技术[3]，古菌，包括以硫化叶菌为代表的超嗜热古菌遗传操作系统[3]，氨氧化古菌（古核生物）[4]；深海及其地微生物[5]、生物圈酵母[6]、极端酶；极端微生物、极端环境嗜热/酸甲烷营养细菌；超磁/磁敏细菌、趋磁菌生态[7]；对虾肝胰腺坏死性细菌、乳酸菌在水产业上的应用、鱼表皮黏液抗菌蛋白胶、贻贝抗菌肽等上的动态都有报道。

1.2　研究方法

近几十年来，分子生物学理论与技术风起云涌，传遍世界。无论从病毒到真核生物，如今都广泛运用上了，且还在不断发展。中国海洋微生物学研究方法的改进和完善也在不断进行中。这包括：对虾白斑综合征病毒（WSSV，下同）的提纯及应用[8]。WSSV 单克隆抗体制备及 BA-ELDA 法的建立、对虾病毒的 RNA 干扰术、栉孔扇贝急性坏死病毒荧光定量 PCR 检测，巢式逆转录 PCR 检测鱼病毒性出血性败血症病毒（VHSV），间接 ELISA 术检测近江牡蛎类立克次体和迟钝爱德华氏菌。微微型真核生物分子多样性。热泉嗜热细菌依赖培养的检测，高磁性、无磁性、趋磁细菌的获得和磁小体释放术。PCR-变

* “十一五”国家科技支撑计划“滩涂养殖污染控制与清洁生产技术研究与开发”(2006BAD09A01)
原文刊于《广西科学院学报》，2011，27(4)：341-347，本人为第二作者。

性高效液相色谱检测水产品溶藻弧菌和超声波平板法测该菌 Hy9901 生物被膜、牛津杯法测五倍子对大黄鱼病原弧菌的体外抗菌力。RAPD 法分析花鲈皮肤溃疡病抗性和敏感群体。毕赤酵母重组子 PCR 模板制备法比较。发光细菌法测水产品中氯霉素、微生物显色法测水产品中恩诺沙星含量。

1.3　普通微生物

这一类成果主要体现在各种海洋环境中一些类别微生物在不同生态因子中发生的生物学多样性[9]、生物学或化学作用等。有古菌、浮游细菌生物量、异养细菌、多样性、氮再生、厌氧氨氧化、生物量与环境因子关系，难培养微生物的寡营养培养等[10]。

分类鉴定和系统发育分析：借助分子生物学技术对特定菌株作了较深研究，使人们对其分类地位和系统发育有了更深更广的理解。这些微生物包括嗜水气单胞菌、短小芽孢杆菌 ZH1-6、蛭弧菌、硝化细菌、反硝化细菌、溶血性葡萄球菌[11]、巨大芽孢杆菌、芽孢杆菌 HN07、石油烃降解菌[12]、解烷烃的戈登氏菌、耐盐原油降解菌、高效降解有机物和促藻生长的菌株、甲胺磷高效降解菌(巴氏葡萄球菌)、中等嗜热嗜酸菌、海南红树等植物根系/根际解磷菌、肺炎克雷伯氏菌、拮抗香蕉枯萎病的细菌、抗肿瘤放线菌、产纤维酶细菌、舟山抗菌活性菌、条石鲷出血病病原菌、大菱鲆迟钝爱德华氏菌、海南、北极根际放线菌，冷藏罗非鱼特定腐败菌、分离自鲍环境的细菌、文蛤菌、牙鲆肠弧菌等。

相关研究除了表达在菌株分类鉴定和系统发育关系上，还在遗传基因克隆、序列分析、菌种选育等上，比如印度洋深海热液区可培养细菌研究[13]，几丁质酶产生菌筛选及几丁质酶编码基因，沉积物宏基因组文库中产蛋白酶克隆的筛选及性质[14]、细菌铁吸收调节蛋白(Fur)基因的克隆表达、链球菌疫苗免疫罗非鱼前后差异表达基因的鉴定与分析、破囊壶菌 57 延长酶和 4 脱饱和酶在酿酒酵母(尿嘧啶缺陷型菌株 INVSc1)中的共表达等等。此外分子生物学知识和技术也在海产养殖上得到应用[15]。

海绵、海葵等的微生物：温度对厚指海绵体内真细菌的影响、相关微生物的分离培养及产物、海葵(产自浙江硇洲岛)的可培养细菌系统发育多样性及海藻表面细菌分离鉴定[16]。

1.4　生态学

海洋微生物生态学研究以往注重异养细菌本身的生态，这些年来研究的眼光更大、更远、更细，研究扩及病毒、不同类别微生物内部及之间、微生物与动植物间、微生物与环境、生态/微生态因子关系等等[17]。2009 年的相关成果有：一些海区病毒生态分布、浮游细菌、异养细菌分布与环境因子关系、细菌群落分布、发光细菌生态分布与种类组成、蛭弧菌生长特性、异养菌对小球藻生长和无机营养盐吸收的影响、聚球藻和 Picoeuraryates 时空变化、微藻与其病毒间微生态关系，米草对底泥微生物群落的影响等等。

1.5　环境微生物学

海洋环境的微生物已从狭义的海洋环境保护意义中扩展出来，它包括海洋及其邻近环境评价中的微生物及其在不同生态环境的长久良性发展中发挥的作用[18]。相关的研究如下：沿岸海水甲肝病毒、造纸废水灌溉对芦苇湿地微生物影响、海水及其贝类生物中粪大肠菌群[19]、环境因子对某些弧菌生长和胞外蛋白酶表达的影响、海滨区海水评价、发光细菌测沉积物毒性、河口水菌群脱氧、大西洋戈登菌降解柴油、$C_{10}CC_{36}$ 直链烷烃[20]、低温解烃的希瓦氏菌生长和解油、氮源对解油菌解柴油影响、植物根际解磷菌、解磷菌剂对盐土有效磷和油料作物生长的影响、巨大芽孢杆菌和巴氏葡萄球菌的高效甲胺解磷、近岸污染与半知菌群体关系、浅海养殖环境复合生态净化菌群及其净化力、盐沼底栖微藻分布特征及有机质产出的贡献等,由此可见此年的相关研究已覆盖了不少生境微生物作用上。

深海微生物:2009 年中国学者比以往任何年份都关注相关的研究动向,这表明中国学者的研究触角正在伸向深邃的大洋。他们既紧盯国际动态,又开展实际的研究活动,如,深海及其地微生物等研究状况分析[5],酵母、适冷菌、假交替单胞菌 SM9913 胞外多糖对 Pb^{2+} 和 Cu^{2+} 的吸附、沉积物宏基因组库中产蛋白酶克隆、劳盆地热液喷口沉积物细菌多样性、冲绳海槽黑烟囱微生物矿化作用[21]、印度洋热液区可培养细菌[13]及盐单胞菌 YD-7、大西洋洋中脊多环芳烃降解菌群[18]、西太平洋"暖池"沉积物异化型硫酸盐还原菌基因多样性、中等嗜热嗜酸菌、南冲绳海槽沉积物中度嗜盐菌及极端酶等的研究都令人兴奋。

极地微生物:多年来中国学者一直关注并从事着极地微生物学研究工作[12],是年的工作包括产低温淀粉酶的微球菌、特定细菌胞外多糖溶液冻结特性的差示扫描量热[22]、超微细菌——极地之星交替单胞菌外膜蛋白和脂多糖提取物对黑鲷的免疫活性、放线菌 NJ-F2 萃取物的抗菌性、低温降解石油烃的希瓦氏菌 NJ49 生长和降解活动、白令海北部表层沉积物细菌多样性、北极海洋微生物石油降解菌筛选及系统发育[12]、北极黄河站植物根际土壤放线菌、西北冰洋表层沉积物厌氧菌[24]及岩芯中厌氧菌分布[25]和极地微生物菌种数据库。

红树林:相关研究已持续了一些年数。南方学者在以往研究基础上,正在扩展相关研究深度,如,南海红树林内生真菌 ZZF13、B2、次级代谢产物及筛选重组人 DNA Topojsomerase(hTopo Ⅰ)拓扑异构酶Ⅰ抑制剂[24]、溶磷菌鉴定及溶磷力测定、优化培养、根际解磷细菌筛选及其鉴定、混合发酵的产氢菌群中梭菌 Fe-氢酶基因(hydA)的克隆及序列分析、根际土壤分离小双孢菌[24]。

1.6 病原微生物学

病毒:作为生态系统结构与功能的重要调控者,人们对它比以前引起了更大注意。除了分析浮游病毒生态分布特征和遗传多样性外,2009 年中国学者对病毒的关注则主要放在养殖水产品的相关研究上,如就对虾白斑综合征病毒(WSSV)发表了不少论著,包括提纯方法、VP28 基因表达及与血细胞的结合、基因组同源重复区结合蛋白 WSSV021 的鉴定、囊膜蛋白 VP19 的单克隆抗体制备及定位、早期启动子的新型杆状病毒表达载体的构建与分析[21]、卵黄抗体对凡纳滨对虾免疫相关酶活力和抗病毒能力的影响、抗 WSSV 囊膜蛋白(VP28、VP19)单抗的体内中和效果[15]、WSSV 对锯缘青蟹的致病性及血清酶指标的影响。对淋巴囊肿病毒也作了一些研究,如它的实时定量 PCR 检测方法、锌指蛋白基因 Kn140 的原核表达与结构、牙鲆的该病毒纯化及不同来源病毒免疫特性、双抗体夹心 ELISA 检测法。虹彩病毒也为学者们所关注,这包括大菱鲆被它感染后的病理学、主要衣壳蛋白基因在毕赤酵母中的重组分泌表达。鱼被它感染,但无症状时用套式 PCQ 分析。对于急性病毒性坏死病毒(AVNV)在栉孔扇贝中的存在用荧光定量 PCR 检测,对在卵内的垂直传播途径也作了研究。对贝类产品中诸如病毒的感染和流行作了探讨。对鱼的病毒性出血性败血症病毒(VHSV)已用巢式逆转录 PCR 检测。此外,对球石藻的病毒之杆状菌主要外壳蛋白基因作了克隆与序列分析。

致病菌:6 种弧菌 36×10^3 外膜蛋白特性、副溶血弧菌耐热直接溶血毒素基因克隆及原核表达、副溶血弧菌温一盐度双因素预测模型、副溶血弧菌Ⅲ型分泌系统、鳗弧菌 rpos 基因克隆及在大肠杆菌中表达和溶血毒素基因 Vah4 的基因克隆及表达[25]、PCR 法检测溶藻弧菌、及用 PCR 微阵列法检测沿岸海水创伤弧菌,并对其危害作评估。迟钝爱德华氏菌致病相关因子。杀鲑气单胞菌杀日本鲑亚种胞外产物毒性及免疫原性。由此可见相关研究均比过往更深更细。

1.7　有益菌

芽孢杆菌:在水产养殖中的作用机制、发酵代谢产物。海洋芽孢杆菌 B-9987 菌株代谢产物 BMME1 抑制病原真菌茄链格孢菌的致病机理、B09 菌株的筛选及其发酵条件。饲料中加该菌和中草药制剂对凡纳滨对虾免疫力的影响、对泥蚶和养殖场底质中菌群的影响。

蛭弧菌:该菌制剂用于检测并应用于水产养殖、Bdh5221 株理化特性和 16 S rDNA 序列分析、噬菌蛭弧菌防对虾的弧菌感染、鳗肠道蛭弧菌的防病作用。

光合细菌:展望它在水产养殖中的应用前景,研究了对罗非鱼鱼苗养殖水质及抗病力影响、对凡纳滨对虾养殖池水质及抗病力影响及液态浓缩制剂保藏。

放线菌:对两株抗肿瘤病的放线菌作了分离鉴定、云南红豆杉 4 株内生放线菌对半滑舌鳎病原弧菌的拮抗作用、分离并纯化出产抗真菌物质的链霉菌 GB-2,对其稳定性也作了研究。

真菌(菌物):对真菌代谢产物研究作了展望。除了对红树林的内生真菌有所研究外,青岛农大学者就山东渤黄海海岸的木生海洋真菌作了研究,连续发表了 7 篇论文[26,27]。发现了 *Monosporascces cannonbullus*。鉴定出抑制烟草花叶病毒及抗肿瘤的两株真菌,对其次级代谢产物作了研究。研究胶州湾近岸污染与半知菌群体的关系。就产生碱性蛋白酶的 yeast fi 的生物活性肽生产作了评估。对德巴利酵母、克鲁维酵母、虫道酵母的生化营养成分作了分析、南极地衣提取物抗氧化能力。酵母不同处理后分散度和对轮虫培养的影响。

1.8　活性物质及微生物制剂·酶

一些海洋微生物体内有活性物质自古以来就引人注意,现代人从更深广的角度更本质地去认识和开发利用它,这一势头正在加强之中[16]。是年,人们关注其国内外动态(包括真菌代谢产物),进行的具体研究包括筛选具抗弧菌的细菌、分析玫瑰弧菌发酵液活性成分、从小单胞菌 FIM03-1149 获得抗真菌抗生素。从酵母 f'_i 获取碱性蛋白酶及活性肽、海绵的真菌等微生物的抗肿瘤等活性物筛选。

微生物絮凝剂对微颗粒高岭土悬浊液的絮凝特性,用于养殖环境调控的微生物制剂评价法[28]、含嗜水气单胞菌气溶素的免疫制剂刺激复合物用于欧鳗免疫、微生物制剂对桉树人工林土质的影响、复合剂的应用、植物源杀菌剂、制剂养欧鳗、光合细菌液态浓缩制剂保藏、食品添加剂、乳酸菌在水产养殖上的用途。红树林内生真菌的重组 LDNA 拓扑异物酶Ⅰ抑制剂、地衣芽孢杆菌 MP-2 酯酶性质、细菌低温溶菌酶的抗 anqiogenic 活性。假交替单胞菌 LJ1 菌株产褐藻胶裂解酶。产琼脂酶细菌 DF-7 培养基优化、琼胶酶(分离自白色食琼脂菌)纯化、酶学性质及降解产物分析。噬纤维菌(*Cytophaga fucicola*)和芽孢杆菌 HN07、产碱性纤维素酶、沙蚕消化道 D2 菌株胞外蛋白酶抗肿瘤。

2　海水养殖的微生物学

2.1　海水养殖动物疾病及养殖环境总论

溶藻弧菌检测法[29]、柱状黄杆菌研究动态、硝化/反硝化细菌的应用、阿维菌素杀线虫、中草药对致病菌的抑制作用、疾病流行特点及预防措施、病害(包括山东的)测报[30]。洞头海区鱼病预报模式、温州鱼病害调查、底泥芽孢杆菌多样性[31]、溃疡病预报模型[32]。高位池 WSSV 病症病毒变化。对虾养殖池沉积物细菌遗传多样性[33]、抗生素对水体细菌区系的影响、光催化降解虾池废水中有机物及灭菌、有效微生物菌群对养殖池水体细菌生态和水质的影响、益生菌调控水中浮游生物生态、细菌与水质关系、养殖模式与清洁养殖关键技术、池内细菌群落的 PCR-DGGE 分析[34]、两种菌剂对对虾存活及消化酶活力的影响、对虾池套养白鲳鱼防虾病、浮游微藻、乳酸杆菌不同生物量对池内优势微藻的影响、浙江海产品环境副溶血弧菌微毒力基因。

2.2　养殖品种

2.2.1　无脊椎动物

南美白对虾(凡纳滨对虾):生态养殖与防病、综合防治 WSSV 及常见病防治、乳酸菌发酵饲料养殖、梅雨季时防病、爆发性传染病病原。WSSV 卵黄抗体对对虾免疫相关酶及抗病毒力的作用、芽孢杆菌和中草药制剂对免疫功能的影响、副溶血弧菌的表型及分子特征、弧菌拮抗菌对对虾的抑菌防病作用、虾溶菌酶基因在大肠杆菌中的表达和活性、黑鳃及红体病诊治。

中国明对虾:蚯蚓、蝇蛆对对虾生长及抗 WSSV 感染、WSSV 非结构蛋白 ICP11、WSSV 囊膜蛋白 VP19 的单克隆抗体制备定位、对 WSSV 的抗性遗传标记筛选和遗传多样性[35]、WSSV 的 VP2f 基因表达及其与血细胞结合、传染性肌肉坏死病毒、新型杆状病毒[21]、肝胰腺坏死细菌研究现状。

螯虾:WSSV 单抗的制备及检测、克氏原螯虾病因及防治。

蟹:WSSV 对锯缘青蟹的致病性及血清酶指示作用、溶藻弧菌对溶菌酶和磷酸酶活性的影响、微藻强化酵母、致病性及血清酶指示作用、三种易发病、早期病因、常见病。三疣梭子蟹苗种病害、抗特定病原(SPR)三疣梭子蟹养殖性能与免疫力、轮虫对中华绒螯蟹溞状幼体的影响。

蛤及蚶:文蛤:疾病研究进展、内脏产气夹膜毒素基因的 PCR、需钠弧菌的鉴定和生物学特性、哈维弧菌引起暴死;泥蛤:芽孢杆菌对蛤及底质的影响;蚶:生蚶中细菌生理生化特性、枯草芽孢杆菌对泥蚶及底泥中细菌、弧菌总数的影响。

扇贝:栉孔扇贝急性病毒坏死病毒(AVNV)荧光定量 PCR 检测及卵内的垂直传播途径[40]、海湾扇贝弧菌病超微病理微生物学[36]。

方斑东风螺:吻管水肿病病原菌。

缢蛏:养殖中的病害、大面积死因。

海参:刺参及其育苗中常见病病因及防控、腐皮综合征病原灿烂弧菌检测探讨[37]及中草药对病原菌体外抑菌;灭活气单胞菌菌苗对部分免疫因子的影响、附着期“化板症”病原菌(弧菌)分离鉴定及来源。

鲍:九孔鲍附着基质上细菌膜系统发育多样性[38]、患病苗中异养菌产胞外酶、脱板期养殖环境细菌分离及分子鉴定。耳鲍血淋巴抗菌性。福建鲍病因。

海胆：硇洲岛马粪海胆可培养细菌多样性。

海绵：相关微生物分离培养，获得一活性物。

2.2.2 脊椎动物

鱼病防诊控治术及春冬季病防治、虹彩病毒无症状感染的套式PCR分析、类立克次氏体肝胰腺坏死细菌（NHP）的a蛋白菌、弧菌的耐药性及主要外膜蛋白OmPK基因检测[34]、弧菌病免疫技术、嗜水气单胞菌对三种鱼血红蛋白的影响、链球菌病及诊治、乳酸菌分离及其抑菌机制、细菌性败血症疫苗制备检测、出血病及爆发性出血病防治、肝胆综合征、水霉病防治。

大菱鲆：杀鲑气单胞菌无色亚种鉴定及组织病理学。虹彩病毒感染后的病理学，它可引起红体病，研究了该病毒主要衣壳蛋白基因在毕赤酵母中的重组特征表达[39,40]。带细菌的一种颗粒饲料小杂鱼与发病相关性、迟钝爱德华氏菌致红体病、病理、疫苗及其免疫应答。红嘴病病原及防治。

牙鲆肠弧菌分离鉴定及组织病理、蒜素治肠炎。

牙鲆：淋巴囊肿病毒靶器官上病毒结合蛋白的鉴定该病毒的纯化及不同来源病毒免疫特性比较、双抗体夹心ELISA检测、海水循环养殖系统生物膜上3种细菌数量与代谢活性、鱼苗病害防治、河流弧菌对表皮黏液的趋化性、南方鲆（斑漠牙鲆）病毒、细菌、寄生虫防治[41]。

鲈鱼：地衣芽孢杆菌对尖吻鲈鱼血液生化生理指标的影响、引起内脏白点病的哈维氏弧菌[42]。加州鲈鱼病害防治。澳洲宝石鲈鱼细菌病防治。花鲈由哈氏弧菌引起的皮肤溃疡敏感或抗性群体RAPO分析及常见病防治。

鳗：中草药对致病菌的抗菌效果、肺炎克雷伯氏菌的分离鉴定[43]、夏季病害的防治。鳗肠道蛭弧菌的作用。日本鳗体表溃疡病病原嗜水气单胞菌分离鉴定及抗克隆制备[18]及水霉科真菌所致的腐皮病。

石斑鱼：斜带石斑鱼幼鱼消化道与养殖水体中可培养菌群。龙胆石斑鱼源美人鱼发光杆菌生物学特性及系统发育分析。

鲷、鲀：黑鲷抗菌肽hepcidin在毕赤酵母中的表达及抗菌性。条石鲷出血病病原分离鉴定。暗纹东方鲀常见病、育苗及越冬期病害防治。红鳍东方鲀鱼肠道弧菌生物学特性。

条纹黑鲽：病毒和机械损伤后的细菌感染。

黄颡鱼：红头病病原、疾病防治术。

鲇鱼：骨蛋白酶酶解物抗菌性酶解工艺优化。

罗非鱼：链球菌疫苗免疫前后差异表达基因的鉴定分析、枯草芽孢杆菌对鱼苗养殖水体水质及抗病力影响、常见细菌性病及防治、鱼及养殖环境中食源性致病菌菌谱[44]、冷藏时腐败菌鉴定和生长动力学、微生物发酵对下脚料蛋白酶解液脱腥去苦效果。

倒刺鲃：养殖中常见病防治。

舌鳎：土霉素对半滑舌鳎肠道细菌耐药性影响与药物残留分析、鳗利斯顿氏菌表型及分子特征。

大黄鱼：哈氏弧菌、溶藻弧菌对该鱼6种酶活性的影响、该弧菌单克隆抗体制备及应用、弧菌多重PCR检测[45]、拮抗菌对溶藻弧菌黏附大黄鱼表皮黏液的影响、网箱养殖病与环境因子关系[46]及溃疡病预报。

鳕鱼：加工过程中的磷酸盐抑制大肠菌群的作用。

比目鱼:接种海豚链球菌后免疫球蛋白水平变化。*Pseudosciaena crocea*:注射溶藻弧菌的免疫感应。

鲨:长鳍真鲨源哈氏弧菌生物学性状。

2.2.3　海藻养殖及其他

条斑紫菜:丝状体细菌附着症、外生细菌及其多聚酮合酶(PKSI)基因筛选[47]。坛紫菜:病烂及烂苗对策。麒麟菜死因对策。除菌处理对强壮前沟藻和青岛大扁藻生长的影响。不同浓度碳源中肋骨条藻和海洋异养细菌对磷酸盐吸收的竞争、藻叶面附着细菌生理特性及荧光杂交分析[42]、羊栖草提取物抗动植物致病菌活性及化学成分[48]。龙须菜附生细菌对铜离子的吸附性能比较。几种藻表面细菌分离及鉴定[49]。Pb^{2+}、Cd^{2+}对单胞兰细菌聚胞藻PCC6803生长和色素的效应。藻蓝蛋白类重组大肠杆菌[50]、蓝细菌附着水生生物[51]。

3　结语

本文是在翻阅89份中国大陆、港澳台当年出版中确认出69份有相关文献计429篇文章及有关书籍的基础上写就的。由此的粗线条罗列,可见2009年中国海洋微生物学发展的历程。可将这年刊载有本学科文章的刊物大体分为水产学科、海洋学科、微生物学、环境科学、综合类等5种。发表在水产学科刊物上的文章为190篇,以篇数多寡排序为《中国水产》、《渔业科学进展》、《中国水产科学》、《科学养鱼》。由此可见2009年中国海产学科文章占总数的44.3%(190/429),加上在其他学科刊物上发表的海产养殖微生物学文章,可知海洋微生物学研究成果仍与以往一样,与海产养殖密切相关,占了很大份额。在海洋学科等刊物上发表的论文计147篇,以篇数多寡排序分别为《中国海洋大学学报》、《海洋与湖沼》(包括英文版)、《海洋环境科学》、《海洋科学》。海洋学科文章数占文章总数的34.3%(147/429)。全年以英文发表的论文计19篇,这表明中国学者同国际交流的愿望和能力越显增强。跟踪先进,新方法的研究已开始崭露头角。以往对病毒的研究,清一色成果基本都来自养殖病害。2009年有了变化,学者们关注起海洋中这一广泛分布、作用和机制并不十分明晰的病毒。海洋古菌、微微型真核生物已引人注目。另外全球海洋以深海大洋为主体,随着学者们眼光的深广化和国力的增强,有关深海及其他微生物学研究正方兴未艾。参与文章发表的作者已逾千人。2009年由科学出版社出版了一部重要的工具用书,即《海洋原核生物名称》*Names of Marine prokaryotes*,该书是我国首部包括海洋古菌、海洋细菌中文/拉丁文名称的专述著作,它必将大大方便相关科学工作者的研究[52]。可以预见,随着我国从事海洋微生物学及其相关学科的学者数量的增多和质量的提高,研究水平将会提高。总体看,我国学者跟踪先进较为积极,"拿来主义"较为盛行,但创新不多,亮点太少。要做到原创有自身特色不易,占领国际制高点更难。多年来,形成了自身优势团队,但强有力的、令国际同行刮目相看的领军人物尚无。多年来,中国海洋微生物专业学会/委员会似没开过大会,即便在国内同行间的规模交流也不多,沟通不多,信息不甚畅通。中国海洋微生物学的发展任重道远,应减轻乃至抵御急功近利思想及其影响。静下心来,从长计议,挑起振兴祖国的大梁。随着我国科研资金投入的实际增多,研究质量和水平的大幅提高将是不太遥远的事。

参考文献52篇(略)

(合作者:郑风荣)

RESEARCH PROGRESS OF MARINE MICROBIOLOGY IN CHINA DURING 2009

(ABSTRACT)

Abstract This paper focuses on adoances in China's marine microbiology article during 2009. These articles throughout the year about 429 and one book were published, climbed to new heights. Among the 429 papers including microbiological aspects of aquaculture accounts for more than half of the article, but the papers on basic marine microbiology were still less. Scholars of international attention and tracking new and advanced concepts / new methods, including methods and techniques of molecular biology research and applications have been developed greatly. Compared with previous years, this year the field research and international exchanges is intensifying, the momentum is strong among the study.

Key words China, Marine Microbiology, Basis/ Maricultivation, 2009

2010年中国海洋微生物学研究成就*

摘　要　本文参阅2010年中国海洋微生物学研究成果文章，发表约350篇之多。文章分析和评述成果的主要内容，指出其发展趋势。本半年关于普通微生物学的文章数已超过了海水养殖方面的。中国海洋微生物学研究进入成熟期，涵盖了许多方面，预示着它正在世界上广泛地提升。

关键词　中国　海洋　微生物学　2010年

历经60年(1950～2010)的中国海洋微生物学研究已经进入全面成长接近成熟期，相关的各类研究正在蓬勃开展，并取得了丰硕而令人振奋的成果。通观2010年的成果，可见本学科的研究已得到较广泛的开展，更涵盖了基础的和应用的两个方面，有些是这两方面研究的结合产物，包括传统的和分子的。可以预见中国海洋微生物学研究将会在更加通盘考虑和更加深广化的努力中再取辉煌。本文解析本年度本学科的研究状况，以引起相关学者关注，共同促进学科发展。

1.关注国内外研究动向

中国学者关注本学科研究方向的势头，今年以来有增无减，涉及方面进一步扩大，这包括海洋生态、古菌、环境、多样性及代谢产物、水产养殖生物病害、防控治技术等20个方面。对国内外的研究前沿引起了更多关注[1-2]。相关论题见表1。

表1　2010年中国海洋微生物学家关注的研究动向

作者	论题
李洪波等[1]	浮游病毒研究法
王新等	细菌生态学研究前沿
韩荣乔	环境微生物
贾仲君等	氨氧化古菌生态
沈萍等	微生物学复兴、挑战、趋势
李巧连等，王海鹰等，吕静等	放线菌多样性及产物；细菌活性蛋白，活性肽
张润等	固氮
袁于军等，陈雪红，等，任虹，等	以细胞骨架为靶点的抗肿瘤大环内酯化合物；抗肿瘤活性物
韩雷，等	抗肿瘤药物的临床研究
吕静，等	海洋细菌活性蛋白、活性肽研究

*　原文刊于《广西科学院学报》，2012，28(4)：277-286转297。
本人为第三作者。

(续表)

作者	论题
张文燕,等	趋磁细菌地域分布特征
韩峰,等	细胞生物膜感染的防治
王瑞旋,等	细菌耐药性
冀世奇,等	微包埋培养及应用
汪岷,等	噬藻体遗传多样性
吴淑勤,等,李清	病害防控技术、疾病免疫、药物防治、趋势
沈伟,等	原位 PCR、检测细菌
陈爱平,等	WSSV、传染性脾肾坏死病(ISKV)、病毒性神经坏死病(VNN)、流行性造血器官坏死病(EHN)、彩虹病毒
陈皓文,等[2]	中国海洋微生物学研究概况
黄玲英,等	珊瑚疾病的主要类型、生态危害及其与环境的关系

2. 方法学的探求

这些探求分别表达在病毒、致病菌等的检测方法、免疫分析、培养方法和芯片构建等中,如 WSSV 环介导等温扩增检测,极早期基因启动子筛选文库构建,WSSV 和血卵涡鞭虫多重 PCR 检测、WSSV 的 VP37 基因诱饵重组载体构建、转化与自激活作用检测。大口黑鲈溃疡病综合征病毒 MCP 基因序列分析及 PCR 快速检测[3]。用于现场检测动物病毒的免疫芯片构建等等。细菌方面有希瓦氏菌单抗 介导间接 ELISA 检测。副溶血弧菌双重 PCR 检测,创伤弧菌检测非培养法,溶藻弧菌外膜蛋白 OmPK 基因表达和间接 ELISA 检测。鲍爱德华氏菌间接酶联免疫检测。刺参腐皮综合征 2 种致病菌间接荧光抗体检测。3 种病原菌多重 PCR 检测。6 种鱼病原菌免疫反应分析及其检测免疫芯片构建,贝类食用安全微生物检测及原位 PCR 检测细菌等[4],这说明方法学越来越追求精准快速并向着检测对象的多样性发展。

3. 分子生物学的研究与应用

世界上用分子生物学理念和方法来研究海洋微生物比中国要早了不少年,中国相关的研究只是进入 21 世纪后才有了实质性的开展和进步,2010 年则取得了不少成绩,具体而言,表达在以下几个方面。

3.1 海洋病毒

2010 年中国海洋病毒的研究与此前不同之处在于题目不局限于养殖生物疾患,研究范围更为扩大(表 2),因而取得的成果是多方位的,其中包含了基础性研究,可以预示中国海洋病毒学的研究将跨上更全面发展之路。

除表 2 所示外,还有 WSSV 相关基因表达特征 DDRT-PCR 分析,WSSV 的 VP37 基因诱饵重组载体构建转化自激活,酵母双杂交系统用于 WSSV 结构蛋白 VP39 相互作用宿主蛋白及基因[5],鲈溃疡病综合征病毒 MCP 基因序列分析、石斑鱼虹彩病毒 ORF086 蛋白的原核表达、纯化及抗体制备。

表2　中国海洋病毒研究2010年成果

作者	主要成果
徐晓明	水产动物病毒现场检测免疫芯片
韦秀梅,等	兔抗对虾 WSSV 抗体制备、特性
王轶南,等	WSSV 对克氏原螯虾血细胞的感染
陈爱平,等	WSSV、ISKN、VNN、EHN、彩虹病毒、海水浴场肠道病毒
赵丽丽,等	传染性胰腺坏死病毒 VP3 蛋白原核 表达及抗原性分析
许海东,等[6]	神经坏死病毒对卵形鲳鲹的致病性、外表蛋白基因序列
夏立群,等	新加坡石斑鱼彩虹病毒 ORF162 蛋白表达、纯化、抗体制备
赵花,等	胶州湾浮游病毒分布
陈豪,等	南极细菌包胞外多糖体外抗单纯疱疹病毒Ⅰ型
施慧,等	WSSV 和血卵涡鞭虫多重 PCR 检测
侯捷,等	红藻共附生微生物及抗植物病毒活性
张彦峰,等	球石藻病毒硫氧化还原蛋白基因克隆、生物信息学
张婧宇,等	微藻带 AVNV(急性坏死症病毒)
Liu et al.[7]	嗜热细菌嗜菌体 βV1 完整基因组序列和 proteomic 分析
蔡俊鹏,等	赤潮霍乱弧菌噬菌体分离筛选、特性
董浩,等	噬菌体 EJ(3)9P1 ORF172 基因克隆、表达
Zheng et al.[8]	淋巴囊肿病毒 DNA 疫苗在细胞及牙鲆体内的分布与表达

3.2　弧菌

牙鲆弧菌 HQ010712-1 外膜蛋白,DDR-PCR 分离鳗弧菌刺激牙鲆差异 cRNA 及表达,坎氏弧菌热不稳定溶血素基因克隆表达、蛋白纯化及其特性,副溶血弧菌质粒图谱,溶藻弧菌质粒 pVAE259 全序列与分子生物学[9]。

3.3　弧菌以外的细菌

地中海富盐菌 phαβ 基因鉴定及 pHBV 前体供应、红灯食烷菌黄素结合单加氧酶(AlmA)基因克隆及其烷烃诱导表达、黄海希瓦氏菌单柱介质间接 ELISA 检测、放线菌次级代谢力的核糖体工程改造、沼泽红假单胞菌核酮糖-1S-三磷酸羧化酶/氧化酶基因克隆及表达、抗鲇爱德华氏菌多克隆抗体制备及特性、南极假交替单胞菌多糖合成酶 UGD 基因表达[10]、病原菌多重 PCR 法、假单胞菌 CI4 抗菌蛋白纯化。

3.4　石油等环境微生物

石油降解菌 16S rDNA 序列、富集文库—菌落原位杂交筛选仿刺参微卫星标记、浮游藻与巨型细菌 EhV 互换基因、PCR-DGGE 解析阴离子交换膜生物反应器反硝化过程中微生物群落结构、深海沉积物微生物基因组文库源的酯酶基因克隆、表达及酶学性质[11]、离子色谱法同测嗜热厌氧菌发酵液有机磷液有机磷及无机阴离子。

3.5　抗菌肽

厚壳贻贝两抗菌肽cDNA基因克隆及序列、三疣梭子蟹Ⅰ型两抗菌肽基因克隆力基因表达、家蝇抗菌肽对凡纳滨对虾生长性能及免疫、红树林内生真菌CⅢ-1菌抗菌蛋白[12]。

3.6 其他

侧孢短芽孢杆菌Lh-1株多肽R-1性质及作用。海洋金藻16S rRNA基因序列分类、白色琼脂胶菌QM38琼脂酶基因agaDoz的克隆表达。

4. 鉴定、分类、系统发育

对气单胞菌、产电希瓦氏菌sp. S2、丝状蓝藻等约34个微生物类别作了相关研究，详情见表3。由该表可见，相关研究涵盖了以细菌为主微生物的分离、纯化、生理生化性状，乃至分子生物学测得的序列、系统发育分析等，这表明研究步入了新阶段，主要体现在新方法、广泛生境和研究深度上。

表3　中国海洋微生物分离、鉴定或系统发育分析等2010年成果表

微生物名称/类别	所用方法、主要成果	作者
气单胞菌	缢蛏、致病、分离、鉴定、毒力	王光强等
迟钝爱德华氏菌	二/三聚吲哚类生物碱；黏附、侵袭特性	王斌等，杨春丽等
鲇爱德华氏菌	间接酶联免疫检测；多克隆抗体	李重实等，李华等
鳗利斯顿氏菌	基于16S rRNA和recA进行的分离鉴定	王成义等
CⅢ-1菌等	红树内生细菌、内生真菌无孢类群（桐花树内），促生菌遗传分析、鉴定、促生力	张兴锋等，邓祖军
链球菌/无乳链球菌	致罗非鱼病、分离鉴定及药敏试验；主养区菌分离、鉴定及致病性	陆俊锟等，祝璟琳等，柯剑等
zy-23(系沙雷氏菌)	河肠毒素产生菌	池珍等
产电希瓦氏菌sp. 32	*Shewanella* sp. s2筛选、产电分析、具电催化活性	王彪等，陈雷等
美人鱼发光杆菌杀鱼亚种	卵形鲳鲹	王瑞旋等
恶臭假单胞菌	条石鲷尾白浊病、分离、鉴定及系统发育；参腐皮病；黑鲷肠炎	毛芝娟等[13]，黄华伟等，闫茂仓
芽孢杆菌	沼虾养殖	张云
whw5菌（与盐芽孢杆菌似）	产自盐场、中度嗜盐、分离、鉴定及系统发育	冯二梅等
鲁氏耶尔森氏菌	斑点叉尾，分离鉴定系统发育	范芳玲等
弧菌		
溶藻弧菌	可溶性蛋白二维图谱，部分蛋白分子	庞欢瑛等
创伤弧菌	非培养法	Li et al.
产过氧化氢酶菌株CE-1	筛选、鉴定	张增祥等

（续表）

微生物名称/类别	所用方法、主要成果	作者
副溶血弧菌	质粒图谱及抗药性、三疣梭子蟹病原、分离鉴定	王安纲，阎斌伦等[14]
坎氏弧菌，热不稳溶血素基因克隆、表达	蛋白纯化及特性	张婧等
哈维氏弧菌	鳗弧菌病病原、分离鉴定	张凤萍等
霍乱弧菌噬菌体	深圳赤潮中，分离筛选、抗肿瘤活性	蔡俊鹏等
黄鱼腐败菌	腐败能力、菌种鉴别；假单胞菌、嗜冷菌、产 H_2S 细菌	李学英等，郭全发等
黄鱼弧菌拮抗菌	取自肠道、筛选、鉴定	王娟等，徐晓军等
(β)-半乳糖苷酶产生菌	低温，筛选、鉴定，生长特性，酶学性质	周春雷等，徐明星等
沼泽红假单胞菌	快速批量培养	王妹等
好氧反硝化细菌	YX-6 株，特性、鉴定、分析；好氧反硝化特性	高春燕等，宋增福等
亚硝酸盐降解菌	筛选、鉴定及降解条件、效果	熊焰军
罗非鱼耐冷腐败菌	CO 发色罗非鱼细菌分离鉴定及生长特性	张韵思等
锯缘青蟹混合感染菌	分离鉴定、感染治疗	夏小安
珊瑚礁—海草床固氮菌	分离鉴定	凌娟等
光营细菌	系统发育分析	Feng et al.
欧鳗肝肾病原菌	分离、鉴定	吴斌等
与海绵相关细菌	在南海中，海绵为 *Agelas robusta*，系统发育多样性	Sun et al.[15]
防污细菌	筛选、鉴定	段东霞等
丝状蓝藻	分子鉴定、对原油的耐受力	唐霞等

另一个可喜之处是，基础研究得到重视，是年发现了一些新菌种，如 *Aestuariibacter aggregatus*，*Oceanbicola nitratireducens*，*Vibrio atypicus*，*V. marisflavi* 等并被国际权威机构确认[16]。

5. 生态和环境微生物学研究

与生态学研究关系密切的一类课题便是环境微生物学研究。是年相关研究课题不少于 27 项。内容涉及不同类别微生物在不同生境中的活动，与环境、生态因子的关系等，其中一个突出的内容便是微生物对原油/石油的降解活动及其影响因子[17]。另一个方面则是养殖环境中微生物活动的研究，涉及有害微生物防治，有益微生物（包括固定化微生物）的利用。研究的地域比较广，说明中国学者相关研究的眼光、角度、广度和深度都有了更新、更大的变化，相关研究成果可从表 4 中得知。

表 4 海洋环境微生物学研究 2010 年成果

作者	成果主要内容
刘蕾等	侧扁轮珊瑚已醇提取物抗氧化、抑菌
高孟春	PCR-DGGE 解析阴离子交换膜生物反应器及硝化过程中微生物群落结构
熊焰	1 株亚硝酸盐降解菌的筛选、鉴定及降解力
王亚丽等	伯克霍尔德氏菌 DAZ 降解邻苯二甲酸、二甲酯
郑伟等	均匀设计优化新鞘氨醇菌 US6-1 降解高相对分子质量多环芳烃
曹煜成等	地衣芽孢杆菌 De 株降解凡纳滨对虾粪便
包木太等	解油微生物筛选及降解条件
陈亮等	厦门金海水多环芳烃降解菌原位富集与其多样性
吴鹏等[18]	红树林湿地希瓦氏菌 W3 还原腐殖质
崔志松等	食烷菌、海杆菌降解石油的协同效应
白树猛等	*Penicillium griseo fulvum* 对重金属的反应
张明磊等	光合细菌对重盐碱地养殖质池水质的作用
梁静儿等	原油降解微生物的 16*S* rRNA 序列
刘海燕等	产生物表面活性剂石油降解菌的筛选及降解特性
徐士庆等	深海沉积物微生物元基因组文库来源
王秋璐等	海洋异养细菌对无机氮的吸收
刘芳明等	南极低温降解菌、筛鉴及对低温降解的适应
宋学章等	菌藻系统处理养殖用海水
梁利国等	中草药抗菌
唐霞等	1 株丝状蓝藻分离、鉴定及对原油的耐受性
吴常亮等[19]	印度洋表水石油降解菌多样性
王丽平等	1 株高效降解多环芳烃芘的筛、鉴、特性
宋兴良等	多环芳烃蒽的高效降解菌筛选及中间降解产物分析
迟明磊等	南极假交替单胞菌净化模拟海水养殖水
肖慧等[20]	天津渤海湾近岸表层沉积物细菌丰度与环境因子关系
张嘉昌等	繁茂膜海绵、抗生素和加大水交换量对大菱鲆水境病原细菌的影响
周鑫等	海水养殖含氮废水复合微生态制剂处理

相关研究涉及地域，包括深海、近岸、海湾、河口等微生物本身或与其他生物、环境生态因子间关系分析。如南海北部陆坡某海域某岩芯表层沉积物古菌多样性、深海耐/嗜压菌细胞毒活性生化代谢[21]、黄海西北近岸沉积物细菌群落空间分布，超磁细菌地域分布特征、趋磁球菌 QH-3 鞭毛特征；胶州湾表层沉积物细菌多样性；红树林区微生物群落结构等

等，尚有浙江南麂岛海洋沉积物抗弧菌的放线菌。东海赤潮高发区浮游细菌分布与活性。长江口粪大肠菌等夏秋季变化。罗源湾水粪大肠菌分布年际变化。深圳海区弧菌生态、大鹏湾异养细菌群落结构、丰度及与环境因子培养力关系、强降雨对虾池微生物的影响和细菌群落结构分析，海水浴场肠道病毒、南黄海夏季微微型浮游植物丰度。

6. 海洋真菌的研究

如果说以往海洋真菌的研究比较零星，甚至边缘化的话，那么2010年的研究可说是成果不少，甚至到了集体亮相的程度。是年涉及的海洋真菌地域之广、种类之多、手段之新等，都是历年之最。对红树林微生物的研究已持续多年，今年对其共内生真菌等微生物的研究尤为突出，研究方法/手段更新、目标多样化更明确（详见表5）。

表5　海洋真菌研究成果

作者	成果主要内容
白树猛等[22]	灰黄青霉分子检测、蛋白融合基因
Li et al.	寄生海藻的卵菌和真菌
唐天乐等	毕赤酵母高效电转化
杜涛等	酵母、轮虫、小球藻、螺旋藻强化的轮虫培育3种仔鱼
徐新霞等	100株真菌次生代谢产物的生物活性
白树猛等	半知菌JYX-Z-1菌株（*Penicillium grisco falvum*）对重金属的反应
陈萌等	海绵内生真菌胞外多糖理化性质及清除自由基活性
郭守东等	南极树粉孢鼠真菌胞外多糖分纯、结构
陈立等	海绵源真菌具抗肿瘤活性的代谢产物
夏雪奎等	用HPLC/ESI技术分析南海红树林内生真菌1403#代谢产物
王芳芳等、邓祖军等	桐花树内生真菌GT_{2002}1545发酵液化学成分，无孢类群真菌的分子鉴定
王鸿等	毛霉 *Mucor* sp. MNP801转化香豆素得发挥性成分
周雅林等	环境胁迫对珊瑚共附生真菌 *Aspergillus ochraceus* LCTJ11-102次生代谢的影响
常敏等	红海榄根际土壤泡盛曲霉F12及其代谢产物
滕宪存等[21]	花刺柳珊瑚共生真菌 *Penicillium* sp. gx wz406的次生代谢产物
许鹏等	水霉病治疗法
霍娟等	酵母双杂交系统用于WSSV结构蛋白VP39相互作用的宿主蛋白基因
赵永刚等	副溶血弧菌tdh基因在毕赤酵母中表达及溶血性
梁昌聪等	海南某热带雨林丛枝菌根真菌

7. 养殖生物病患的微生物学研究成果

表6　与海产养殖相关的微生物学研究成果一览表

养殖对象名称		病原微生物研究成果概述	作者
贝类生物			
青蛤		鳗弧菌刺激后溶菌酶基因表达、磷酸酶活性	潘宝平等，辛欣等
菲律宾蛤仔		育苗中光合菌 RPD-1 的作用	王芳等
牡蛎		抗菌蛋白	王春波等
方斑东风螺		肿吻症病原菌分鉴、药敏	张新中等
大扇贝		单孢藻育幼体	苏浩等
厚壳贻贝		抗菌肽、克隆、序列	廖智等[24]
缢蛏		气单胞菌分离鉴定及毒性	王兴强
贝		食用安全微生物检测	兰睿赜等
棘皮动物			
海参		常见病因及防治；腐皮病、表皮溃烂和肿嘴病病原；肠道及养殖塘菌群组成 PCR-DGGE 分析；肠道益生菌；越冬池浮游菌；越冬期病因；育苗期病因；灿烂弧菌胞外生物毒性，内生真菌胞外多糖；养殖池浮游病毒与环境因子	王印庚等[25]；徐广远等；王轶南等；黄伟华等；高菲等；李彬等；张豫等；周慧慧等；Liu et al.[26] 王斌等；许乐乐等；卢超等；姜北等
海绵		真菌产物	陈萌等
甲壳动物			
蟹	三疣梭子蟹	副溶血弧菌分离鉴定；底泥异养细菌群落；抗菌肽基因克隆、重组、表达	阎斌伦等[27]；孙苏燕等 申望等
	锯缘青蟹	弧菌；病害防治术；致病菌分离、鉴定、治疗，围垦区爆病	邱庆连、张水波、夏小安等 吴松、吴清洋等
鲍		弧菌病，罗源湾，爆发性死亡原因	张凤萍等；吴松等
缢蛏		气单胞菌	王兴强等
虾	对虾	对虾养殖前期水细胞；WSSV 病毒病；WSSV 早期基因启动子、蓝藻	林梵宇等；周凯等；侯传宝等；王凤敏等；张莉等；陈爱平等
	中国对虾	WSSV 基因表达；5-溴嘧啶处理受精卵；河北；兔抗 WSSV 抗体；自然与人工感染 WSSV；益生菌 2D0Z 分离及在精养中作用	赖晓芳等；张天时等[28]；丰秀梅等；张天时；丁贤等
	日本对虾	芽孢杆菌用于育苗	常传刚等
	南美白对虾	褐斑病防治；红体病（桃拉病毒）；常见病；冰温下菌相	李永仁等；赵润朝等；张文革等；郭红等

（续表）

养殖对象名称		病原微生物研究成果概述	作者
虾	凡纳滨对虾	美人鱼发光杆菌免疫、抗病；养殖中；家蝇抗菌肽及抗菌肽应用；抗 WSSV 选育、抗病免疫；芽孢杆菌解粪；枯草芽孢杆菌改水质作饲料；固定化微生物在养成中应用；强降雨影响菌群落菌	兰萍等；煜成等；陈冰等；黄永春等；宋理平等；于明起等；陆家昌等；胡晓娟等
	沼虾	芽孢杆菌作为益生菌	张云
	克氏原螯虾	斑症病毒、WSSV 对血细胞的感染	王轶南等
付军曹鱼		神经坏死病毒致病性	闫云锋等
软体动物			
海蜇		腐烂等防治	杨辉等
鱼			
牙鲆		苗种培养中微藻；前肠黏液	曾马
		淋巴囊肿病毒；河流弧菌感染后血清抗菌；抗鳗弧菌病家系筛选；鳗弧菌刺激诱导，cRNA；秦皇岛弧菌外膜蛋白等；鲇爱德华氏菌	邢明青等；邹文政等；徐田军等[29]；范玉顶等；Zheng et al[30]；程顺峰等
大菱鲆		腐败菌生长动力、货架期；中草药治肠炎；工厂化中的病害防治；病诊断、药及投药；海绵等对环境病原菌调控	崔正翠等；王丰等；丁美林等；殷禄阁等[31]；张喜昌等
鲈		大口黑鲈溃疡病毒	马冬梅等
卵形鲳鲹		神经坏死病毒致病、外壳蛋白基因序列；发光菌分离鉴定	许海东等；王瑞旋等[32]
红鳍东方鲀		哈维氏弧菌胞外产物致病	王斌等
石斑鱼		虹彩病毒 ORF162 蛋白表达、纯化、抗体；乳酸菌；RNA 技术等	夏立群等；孙云章等
斜带石斑鱼		乳酸菌在模拟胃肠环境中存活	杨红玲等，孙云章等
点带石斑鱼		溶藻弧菌脂多糖对其毒性、免疫的影响	徐先栋等
中华倒刺鲃		灭活菌免疫后血免疫指标	罗芬等
鲷	黑鲷	恶臭假单胞菌分离、鉴定	毛芝娟等
	条石鲷	尾白浊症病原恶臭分离、鉴定	闫茂仓等
罗非鱼		益生菌；分鉴链球菌病及诊治；无乳链球菌分离、鉴定、致病性；耐冷腐败菌分离、鉴定及生长；海藻酸钠涂膜保鲜；芽孢杆菌制剂	王艺红，李南青等；祝璟琳等[33]；柯剑等；张韵思等；张杰等；袁翠霞等
条纹斑竹鲨		哈氏弧菌对酶活影响	马赛红等
斑点叉尾鱼		鲁氏耶尔森氏菌分、鉴及系统发育；虹彩病毒	范芳龄等；夏立群等

(续表)

养殖对象名称	病原微生物研究成果概述	作者
黄鱼	冷藏中腐败菌鉴别及能力;抗弧菌的放线菌;染哈氏弧菌后血液及组织病理;溃疡病组织病理;放线菌;货架	李学英等、王娟等;徐晓津等;戴瑜来等[34];赵淑红等;郭全友等
尖吻鲈	凝结芽孢杆菌对消化酶及非特异性免疫酶的影响	袁丰华等
鲟	烂鳃、胃充气,病害调研防治	陈弟伟等;蒋晓红等
大泷六线鱼	链球菌(与海肠链球菌似的)防治	冯春明等[35]
鳗	利斯顿氏菌分离鉴定、胺肾病病原分离、鉴定	王成义等;吴斌等

病患从来都与生物相伴。海洋的养殖生物也不例外,因此,有养殖对象,必有相应的病患发生。与此相伴而行的便是对病患微生物的研究。多年来,中国海洋微生物学的研究,围绕养殖生物病患的课题始终是重头戏。2010 年在以往的基础上,对病患等的微生物研究则更上一层楼,其研究规模、研究手段和研究成果更大、更新和更多。

是年研究对象已涉及约 40 种各级各类动植物。方法包罗传统的和现代分子生物学的,因而成果就显出丰富多样和方法多样化。从表 6 所罗列的内容可见一斑。表 6 说明主要对象生物中以鱼类和甲壳动物为多(尚可见诸表 3 和表 4)。病原生物以病毒和弧菌为多,所用研究手段和阐明问题以分子生物学见长。反映了中国海洋微生物学研究手段正在不断更新中,因而其研究成果也正在赶超世界先进水平之中。对养殖生物病患的研究正在从"头痛医头,脚痛医脚"的穷于应付状态中走出。人们比以往更多地重视养殖环境和生态状态的综合研究治理,从更高更新更深的层次上阐明问题。

8. 2010 年论文(文章)初步统计

就相关刊物上发表文章进行数量分析,可知本年文章数约 350 篇,比上年的少了一些*。将文章大体分为基础和水产两类看,水产学科的文章约 150 篇,余为基础(包括应用基础)性质的,后者比前者多了,这似乎说明学者们对基础性课题比以往有所重视。所载文章在相关刊物间的比较,按数量多寡排序,前 8 份刊物分别是《中国海洋大学学报》51 篇(下同)>《海洋科学》(26)>《海洋与湖沼》(25)>《微生物学报》=《水学学报》(23)>《海洋学报》(20)>《中国海洋药物》(19)>《中国水产科学》(16),这 8 份刊物所载文章数占文章总数的 58%,若加上《渔业科学进展》、《齐鲁渔业》两刊,则该比率升至 67.7%。学者们在学术性较强刊物上发表的论文比以往增多,少量国外刊物上也有中国学者的文章[16,26,30],这意味着学术水平有所提高。这些刊物应该说学术性、权威性都比较高,说明了主要刊物性质由以往以水产学科类为主向包括广泛的微生物学在内的综合性刊物转移,即表明人们对海洋微生物学的主要眼光正在从水产方面扩展,即从侧重病害研究向全面研究转移。全年英文论文至少发表了 20 篇,作者来自中国大陆、香港或在国外工作/学习的华人。一些相关的国际会议在中国举行。证明中国学者在世界范围内论述观点、展现成果的信心和愿望有所增强,能力和水平均有提高,世人正刮目相看。以上各点表明中

* 中国海洋微生物学 2009 年研究进展(待发表)

国海洋微生物学研究正日臻成熟。

9.结语

作者自20世纪末以来,关注海洋微生物学研究的国内外动态,连续发表了一些论述。就所掌握材料看,中国海洋微生物学历经60余年的研究,整体正在进入成熟期。中国学者数量、质量和成果的水平都得到了很大的提高,有些已跻身世界水平。当前我们不仅要追求论文数量、质量,更要高瞻远瞩、关注外界动态[Marine microbiology: eligibility, colleges, companies related to marine microbiology (Christo Benz, 2010); Marine microbiology: Ecology & Applications (Colin Munn)],加强加大经费投入,而且应全面考虑、统筹安排、集中优势,有大的气魄,建立和巩固自身特色的研究领域和路线。不仅要搞好自己领海的,更要有深海大洋意识,做大做强,为世界范围的海洋微生物学研究做出我海洋大国乃至强国的应有贡献。加强薄弱环节,比如深海水、空、地微生物学及微生物在生态系统中的作用和地位等等的研究。这包括研发方法、思路,仪器设备手段等的研制、更新和创新,以形成自身特色并行之有效、可持续发展的群体,而不仅仅是跟踪先进。只有这样,方能在世界范围内屹立于不败之地,方能成为真正的海洋微生物学研究强国。

参考文献176篇(略)

(合作者:孙丕喜 郑风荣)

ACHIEVEMENT OF MARINE MICROBIOLOGICAL RESEARCH IN CHINA DURING 2010

(ABSTRACT)

Abstract This paper read up the literature of research results of marine microbiology in China during 2010. It was found that the 350 articles at least were published.

The paper analysed and dissected the achievement of main contents and pointed out the expanding trend.

In that year (2010), the articles numbers on general marine microbiology had been exceeded ones of fishery (maricultured) subjects. The results are considered that marine microbiology in China has been entering the mature period. The subjects it involved have been covering many fields. It betokens that the marine microbiology in China is comprehensively rising in the world.

Key words China, Seas, Oceans, Microbiology, the Year of 2010

G2 空气微生物学

(AIR MICROBIOLOGY)

中国空气微生物学的研究进展*

摘　要　本文就近 40 多来中国空气微生物学研究的方法与技术、范围与领域、成果的分析与统计及以后的发展趋势作一述评,并期盼有更丰硕的成果跻身于世界前列。

关键词　空气微生物学　综述文献

空气微生物学,即研究空气环境中微生物的科学。1861 年巴斯德首次获取空气微生物,1881 年柯赫用固体法(平皿暴露)首采空气微生物。迄今的 143 年间空气微生物学发生了巨大变化,除了与人类健康息息相关外,还与国防及国民经济密切相关,须运用各种手段、仪器设施进行操作、调研,因而现代空气微生物学是科技发展的产物和成果。

1　空气微生物学方法学/技术学的研究

1.1　采样介质

空气微生物的采样介质主要是指能捕获并让空气微生物生存且可作进一步研究的物质。严格地讲,此介质必须是性能稳定、尽可能不干扰原生态、在不同环境中具有相当黏性、无毒、水溶且不易干燥的物质。不同类群的微生物对介质的要求不尽相同,因而有通用和特定使用之分。比如采集空气病毒的介质原则上不同于其他微生物。目前广泛应用的采样介质有硅酮类、糖类和蛋白类等 8 大类,病毒介质则更专一,解放军 302 医院发明了采病毒的首选介质[1]。

1.2　培养基

空气微生物的培养基名目繁多,是可以收集空气微生物并经培养就可直接在其上显出菌落的介质。迄今相关的培养基包括 5 种普通细菌培养基,26 种致病细菌培养基,2 种放线菌培养基,51 种真菌培养基,3 种病毒、立克次体保存培养基/保存液和 6 种支原体培养基等。国内学者为此做出大量工作,并筛选出一些行之有效的、具有中国特色的培养基[1]。

1.3　采样方法

---

* 原文《空气微生物学的国内研究进展》刊于《疾病控制杂志》,2005,9,(6):627-629.

与采样介质同步发展的采样方法得到了长足进步，但空气微生物采样法主要是传统的自然沉降法。后经奥梅梁斯基的修正，该法对获得空气中自然沉降入固体培养基上的较大微生物粒子，可粗略估计其含量和进行生物学观察研究，对推动空气微生物学的普及和发展起了很大作用。二战以后，各种原理、制式的采样器相继问世，主要趋势是机器采样。归纳起来有：①17种固体撞击式采样器，其中以Anderson采样器最为有名；②离心式采样器；③气旋式采样器；④3种液体冲击式采样器；⑤6种滤器的过滤式采样器；⑥大容量静电沉降采样器。国产仿制/自行研制的采样器有LWC-Ⅰ型、Ⅱ型离心式采样器。固体撞击式采样器有JWL的计4型及JWL-ⅡB型等。THK-201型采样器是JWL型的改进，DCQ-Ⅰ型采样器则是LWC-1型的改进。此外，还有温差迫降类和生物类[2~4]。

不可否认，各种采样器的研制和使用大大推动了国际/国内空气微生物学的发展，但一个不可辩驳的事实是采样器价格不菲，均有某种难以排除的缺陷（限度）。使用有限定的范围，互相之间的换算也繁杂，因而推广有难度，特别是对中国这样的发展中国家推广更难。目前许多人都主张联合采样，即不同性能、规格的采样器联合应用，但会造成人力物力的巨大代价，因而采样方法的改进仍是必须大力加强的课题[1]。

1.4　分子生物学技术的应用

生物和质粒DNA大分子气溶胶，PCR及相关技术、索氏杂交等分子生物学的概念及技术已不断改变着传统的分类、检测及鉴定过程。分子生物学技术与行之有效的传统方法相结合，将使空气微生物学测试、鉴定和研究更快捷、更精确和更具创新性[1,5]。

2　广泛开展的空气微生物学研究

几十年来，我国的空气微生物学家已经在医院感染、临床医学、预防医学、公共卫生、军事医学、生物工程和消毒学等方面做了大量工作，大大提高了相应学科的理论和应用水平。空气微生物学观念得到普及，医疗水平得到提高，预防医学及其理念深入人心。

2.1　空气微生物的定性/定量及其变化的研究

空气微生物学的一个重要分支即是对公共卫生学的参与和渗透。自20世纪80年代以来，我国就不同环境/地理及其景观/景点的室外空气，不同功能、性质、类型的室内/人工空气环境的微生物作了调研，这包括不同状态、不同时段的高山、大海、江河、湖泊、内陆、沙漠、极地、高空和大中小型城市不同功能、类型和场合[6~15]；对不同规模、性质的医疗单位、交通设施、生产现场，不同性能的实验室、军事设施、工厂的空气微生物作了定性、定量研究[16~24]；探讨促使它们变动的内外因子及相互关系，因而具有重要的公共卫生、环境和生态学意义[25]。同时研究了不少消毒设备、器具和药剂等[26,27]，使我国消毒水平和洁净技术大大提高[28]，生物安全的概念和应用得到规范和标准并逐步与国际接轨[1]。

2.2　空气微生物学知识的实际研究应用

空气微生物学的成果广泛应用于临床医学、预防医学和生物医学中，这催生了洁净技术和消毒学的大力发展[29~31]。虽然我国相应研究及成果出现较晚，但已获得了长足进步。开发研制了一些消毒仪器、设施、药剂并广泛应用，进行MKJ型空气消毒洁净器的研制和应用、生物安全实验室的设计和应用，DX-OIB臭氧消毒机的应用，及蒲公英合剂的研究，对空调机的广泛应用也提出合理建议[16,32]。

3　中国空气微生物学研究成果的初步统计

据不完全统计，从1961～2003年的43年间，我国空气微生物学研究共发表了超过

230篇的论文(图1),相关的专著达十多部,举行过3次全国性会议。

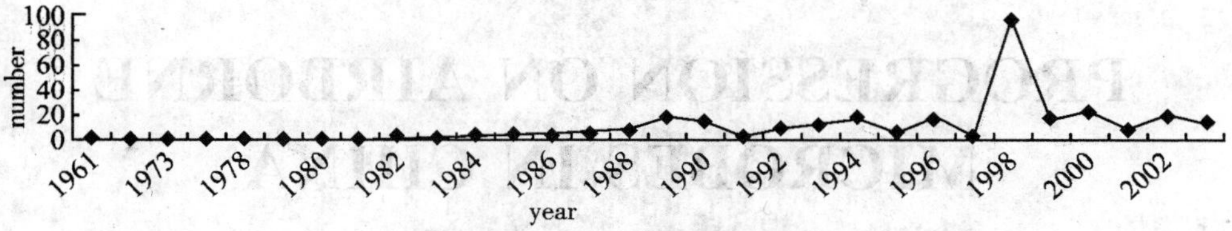

图1　中国空气微生物研究论文数历年统计(1961～2003年)

Figure1　Statistics of the number of papers about research on airborne microbes in China, 1961～2003

这些论文分别发表于约98种刊物上,涉及微生物、医学、环境生态、内陆、海洋和国土资源管理,发表文章最多的刊物(≥5篇)依次是《中国卫生检验杂志》(78篇)、《中国公共卫生学报》(23篇)、《环境与健康》(18篇)、《中国环境科学》(9篇)、《消毒与灭菌》、《微生物学通报》(各6篇)、《卫生研究》、《中国公共卫生》与《中国医院感染学杂志》(各5篇)。

中国在20世纪80年代前,相关工作做得极少,1960～1979年只发表论文4篇,20世纪80年代后期开始进入增长期,至1989年的10年间共发表论著42篇/部。20世纪90年代是高产期,10年间共发表论著183余篇/部,其中以1998年的95篇达最高峰。进入新世纪相关论著数进入平缓期,此时的中国空气微生物学研究接近成熟。据1991～2003年不完全统计的论文内容组成分析,大致可知有关方法/技术学的研究占21.3%,室外空气微生物调研占25.0%,室内包括医疗单位、学校托幼单位、旅店、居民房、殡仪馆等的空气微生物研究占3 4.0%,工厂畜舍空气微生物研究占8.7%,综述/评论等占7.4%,其他3.6%。研究水平提高,其中以于玺华等62位学者编著的《现代空气微生物学》[1]及1998年的《第三届全国空气微生物暨医院感染学术研究特刊》为代表。

4　中国空气微生物学研究的未来

中国辽阔的国土和疆域及不同特性的室内空间要求我们对不同生态类型、不同功能性质的大气和室内空气微生物状况做到心中有数,相关研究必须向深度发展,因而必须安排好人力、财力、物力及必要的涵盖面广、代表性强的调研,摸清特定区域和环境的空气微生物状况及变动规律。对不同功能、类型、性质和部位的医疗单位室内外空气微生物状况作全面深入研究,制订出我国自己的标准,并与国际接轨。

对未来可能进行的战争中空气微生物威胁(微生物气溶胶制剂等)的防范手段、措施要早作准备和演练。增强和加大空气微生物采、检、鉴新方法/新技术的研制和使用,尽可能向廉价、实用、方便、精准方向发展,使我国空气微生物学研究水准更上一层楼。大力培养和造就空气微生物学的研究通才和专才。空气微生物学是交叉学科,而且其调研已向多兵种、多功能联合协作方向发展,需要多专业的复合型工作者同心协力、持之以恒,才能使科研成果更丰硕。

参考文献32篇(略)

(合作者:陈　阳)

PROGRESSION ON AIRBORNE MICROBES IN CHINA

(ABSTRACT)

Abstract The research history of air microbiology in the world has longer than 140 years. In China, it has been getting a great advance from 1980s. This paper reviews its methodology and technology, scope and field, analysis and statistics of achievement, and developmental trend in future.

Key words Air Microbiology; Review Literature

G3 极地微生物学
（POLAR MICROBIOLOGY）

中国极地微生物学调查研究进展*

摘　要　中国自行组建的极地微生物学调研起始于 20 世纪 80 年代初，迄今 20 余年来，中国极地微生物学研究取得了丰硕的成果，调查范围涉及南北极，包括南极长城站、中山站及其附近区域土壤、空气、水体和南大洋水域，及从中取得的生命和非生命材料、白令海和楚科奇海、波弗特海、加拿大海盆等北冰洋海区，及青藏高原等等。对极地微生物的种群、含量、多样性、生长特征、生理生化特性、系统发育等作了广泛而有深度的研究分析．并就其中的一些菌株的活性物质开发利用进行了有意义的探索。最后对今后中国极地微生物学的研究提出了一些建议。

关键词　中国　极地　微生物学　生态学　生物多样性　生物活性物

1　前言

极地包括寒冷海域、陆地等的微生物在元素的生物地球化学循环、食物链和生态系统的物质、能量转化中起着十分重要的、不可代替的作用，长期以来引人注目。中国极地科学考察与研究，在 20 余年间经历了从无到有、并逐步壮大的过程，其间极地微生物学调查研究也取得了长足的发展。中国极地微生物学家对不同时段、不同地域、不同介质、不同来源的微生物进行了生态、分类、区系、与环境相互作用、低温适应机制和生物活性物质等的探索性研究，取得了丰硕的成果。新世纪展现在我们面前的极地微生物学调查研究事业任重道远，要在现有基础上，加强合作，全面规划，高起点、多学科地开展研究，争取进一步提高中国的极地微生物学调研水平。本文主要阐述相关的研究进展，并对未来提出一些期盼。

2　中国极地微生物学考察研究简史

中国极地微生物调查研究起始于 20 世纪 80 年代。从 1981 年起，中国微生物学家即涉足南极，并于 1987 年发表了第一篇论文（Nichuzhi et al.，1987），即 *The hetertrophic microbes in dry Victoria land and Ross Island Antarctica*。在 1984 年之前，中国学者曾参与南大洋水体的细菌和生物量考察，并发表过数篇论文（Chen Haowen，2004）。从 1984 年起，中国开始独立组队进行南极考察，中国海洋微生物学家参加了相关的考察活

*　原文刊于《极地研究》，2005，17(4)：299-307.

动。据不完全统计，进行南极微生物考察活动的年份有1984/1985、1986/1987、1988～1994、2001/2002年等等，积累了不少样本，并进行了研究。“八五”期间，国家攻关课题之一的“南极菲尔德斯半岛及其附近地区生态系统研究”，其中微生物学考察研究取得了较丰硕的成果。20世纪末，我们进行了中国首次北极考察，微生物研究涉足北极地区（50年代吴宝铃先生曾随前苏联学者赴北极作过考察），并发表过论文（陈皓文等，2000b；曾胤新等，2000；肖天等，2001；高爱国等，2003）。中国南北极微生物考察研究的成果为掌握极地微生物的基本状况及研究全球变化对该特定生态环境系统的影响奠定了初步基础。21世纪初，中国学者又开始关注世界第三极青藏高原的微生物及其环境状况（陈皓文，2002，2003）。

3 中国极地微生物考察研究主要地域及样本来源

表1列出了中国极地微生物考察研究主要地域及样本来源。由该表可见，考察研究的地域范围主要集中在长城站、菲尔德斯半岛、南设得兰群岛等邻近地区及南大洋海域和北极楚科奇海、波弗特海、加拿大海盆、白令海等。样本来自生物体的和非生物体的、室内的和室外的，海、陆、空等不同类别的生存环境和介质。

4 中国极地微生物调查研究主要成果

4.1 南极微生物主要类别

目前中国所取得的南极微生物主要类别见表2。由该表可见，研究的微生物以细菌为主，次为菌物（即真菌）。不完全统计表明所研究的细菌大约有38属（种），丝状菌物6属，酵母8属（种），此外尚有肠系细菌。不同功能类型的微生物包括固氮细菌、硝化细菌、石油烃降解菌、耐重金属菌、耐LPS菌、耐酸菌及卫生状况指示菌等等（陈皓文等，1993；肖昌松等，1994；袁峻峰等，1998）。

4.2 极地微生物含量特征

表3是已发表的不同类群、不同生存状态、生态环境中微生物含量估计。结果表明，虽然极地环境总体上温度较低，但不同环境、时空间差异较大，包括采样季节等，时间以及人为环境温度差异。不同生存状态、生态环境中，尤其是小/微生境造成的千差万别变化，使得环境中既有大量嗜冷/耐冷者，也有相当数量的中温甚至耐高温菌类的极端微生物，它们在极地的物质迁移和流动、转化和循环中多起着不可忽视的作用。

4.3 极地生态环境与环境微生物学的研究

这方面的研究主要将微生物置于整个生态系统中分析，如南极磷虾集群过程、集群区生态系统、食物链中的作用，菲尔德斯半岛及附近区域生态系统中微生物的多样性、生态结构特征、食物链及与生物/非生物因子间关系分析（吴宝铃等，1998）。

表 1 中国极地微生物学调查的地理范围及样本类别

Table 1 Geographical areas and kinds of samples Chinese polar microbiological investigation and research

样本类别	主要地域	引证文献
空气包括雾、雨、飞雪	长城站及附近室内、室内室外衔接部、室外、野外、马尔什基地	
	别林斯高晋站	陈皓文等(1992a,b,1997,1998,1999,2000)
	金世宗站	
	阿蒂加斯站	
	德国观察所	
	捷克站	
	中山站	于永中等(1998)
	《极地》船	陈皓文等(1994)
菲尔德斯半岛	地理湾	陈皓文等(1997)
	西海岸	
	阿德雷岛及海滩	
	半三角	
	碧玉滩	
	三门峡口	
	纳尔逊冰盖及海滨	
	麦克斯威尔湾	
	大型生物出没区	
	荒凉状态区	
冰、积雪、海水、湖水、饮用水、陆土、砂、碎石、海滨砂、浅海沉积物	南大洋(南设得兰三群岛与南极半岛间)	张建中等(1989),陈皓文等(1992,1998)
	菲尔德斯半岛	肖昌松等(1994),陈皓文等(2001)
	阿德雷岛	陈皓文等(2001)
	长城湾远岸表层海水(外湾)	
	长城湾海滨海水(内湾)	
	欺骗岛	
	智利 Skur 站室外	
	阿根廷 Marina De Gena 站室外	
	长城站及海滨和附近地区	肖昌松等(1995)刘大力等(1994)

（续表）

样本类别	主要地域	引证文献
	西湖	陈皓文(1997)
	阿德雷湾	
	生物湾	
	乔治王岛	孙华忠等(1991)，胡继兰等(1993)
	南大洋磷虾集群区(W52°－62°，S54°30′－63°50′)	陈皓文等(1993a)
	南大洋(普里兹湾、长城站—中山站间环南极水域)	宁修仁(1996)
	南大洋普里兹湾、长城站—中山站间环南极水域	宁修仁(1996)，邱雨生等(2003)
	北极白令海、楚科奇海、波弗特海、加拿大海盆测区	陈皓文等（2000b)，高爱国等(2003)，肖天等(2001)，陈波等(2003)，He Jianfen et al.（2004)，Chen Bo(2005)
	植物：镰刀藓、卷毛藓、裂叶苔、石萝、倒卵银杏藻，小腺串藻	陈皓文等(1998)
	动物：沙蚕、小红蛤、帽贝、南极鳕鱼、阿德雷企鹅、逆戟鲸、鸟粪	
青藏高原	冰雪、拉萨、日喀则	沈亮*、陈皓文(2003)

*参阅沈亮，2011，No.4 的《大自然杂志》《青藏高原里的微生物》一文

表 2　极地微生物主要类别

Table 2　Main kinds of polar microbes researched by Chinese microbiologists

样本所在地区及来源	主要微生物属群	文　献
菲尔德斯半岛土壤	单胞菌、芽孢杆菌、微球菌、葡萄球菌、产碱杆菌、节杆菌、黄杆菌、硝化杆菌、硝化球菌、小单孢苗、诺卡氏菌，以假单胞菌和芽孢杆菌为优势。青霉菌、枝孢霉、金孢霉、曲霉、散子囊菌其中以青霉菌和枝孢霉为多。	肖昌松等(1994)
长城湾水	纤维黏细菌、薰瑟氏菌、产碱杆菌、八叠球菌、节杆菌、乳酸杆菌、肠杆菌、曲挠杆菌、假单胞菌、不动细菌、黄杆菌、黄单胞菌	Chen Haowen et al.（1998)陈皓文等(2001)
长城站	黄杆菌，纤维黏细菌、莫拉氏菌、不动细菌、库特氏菌、假单胞菌、芽孢杆菌、动胶菌、葡萄球菌、足球菌	肖昌松等(1995)
中山站空气	木糖葡萄球菌、柠檬色葡萄球菌、白色葡萄球菌、甲型链球菌、假单孢菌、不动细菌、微球菌、曲挠杆菌、黄单胞菌等	于永中等(1998)

(续表)

样本所在地区及来源	主要微生物属群	文　献
南极土壤	耐冷细菌、放线菌、金孢霉等菌物	于永年等(1996)、王泽君等(1999)、胡继兰等(2000)、Tamatsu Hoshino et al. (2001)、曾胤新等(2001)、鲁敏等(2002)
南极乔治王岛	金孢霉等丝状菌物链霉菌、诺卡氏菌、类诺卡氏菌等	孙华忠等(1991) 胡继兰等(1993)
环南极水域	蓝细菌	宁修仁(1996)
南极淡水	菌物	于永年等(1996)
南大洋磷虾集群区水	细菌、噬胞苗、葡萄球菌、芽孢杆菌、肠杆菌科细菌;菌物:青霉菌、曲霉、拟青霉、枝孢霉	Chen Haowen et al. (1990),陈皓文等(1993a)
磷虾	细菌:假单胞菌、不动细菌、莫拉氏菌、黄杆菌、弧菌、葡萄球菌、微球菌;菌物:枝孢霉、红酵母	陈皓文等(1993b)
南大洋(南极半岛与南设得兰群岛间)	细菌:气单胞菌、色杆菌、弯曲杆菌、纤维黏细菌、可变单胞菌、土壤杆菌、棒状杆菌、发光杆菌、黄单胞菌、肠杆菌、假单胞菌、黄杆菌、不动杆菌、产碱杆菌、弧菌、莫拉氏菌、微球菌;菌物:假丝酵母、隐球酵母、红酵母、球拟酵母、红色冬孢酵母;丝状菌物:头孢霉、青霉、曲霉、枝孢霉	张建中等(1989),Ye dezan et al. (1989),林燕顺等(1998)
北极楚科奇海和白令海	属细菌、硫酸盐还原菌、菌物、聚球菌	陈皓文等(2000),肖天(2001),陈波等(2003)

对北极白令海和楚科奇海水细菌、菌物置于环境与生态学的高度,考虑其在群落结构中的特征,并提高到分子生物学的水平上来认识(陈波等,2003),也关注细菌多样性对环境变迁的响应(张锐等,2003)。环境微生物学包括室内及野外空气微生物监测(陈皓文等,1994,1999;于永中等,1998),生活污水入海后大肠菌群监测(黄凤鹏等,1992b),石油烃降解菌含量分析(袁峻峰等,1998),细菌与营养盐关系分析(Chen Haowen, et al, 1998)等。这些成果为研究更大尺度生态系统及人类影响中的极地微生物打下了基础(赵烨等,1998)。

4.4　极地微生物的开发利用研究

对极地这一极端环境中的微生物调查研究,一开始就带有“开发和利用”的目的,中国相关研究开始时基本处于推介和引进国外的研究动态上(曾胤新等,1999;任修海,1993)。近年来,人们对此更注重实际(张晓君等 2000,曾胤新等,2000;俞勇等,2003),研究的重点在低温酶上,其中以低温适应性分子机制研究最为详尽(胡继兰等,2000;陈熹兮等,2001;吴虹等,2001)。人们期望极地微生物将成为研究低温生物学的良好试验材料及新到活件物质的潜在来源,开发出有价值的生理活性物质和产品(曾胤新等,1999;鲁敏等,2002),

相关研究进展见表4。

表3　极地微生物含量估计

Table 3　Estimation on polar microbial contents

样本类别	主要地域及微生物含量	文　献
空气	菲尔德斯半岛,科考站及附近:4 994 CFU・m^{-3}(下同) 大型动物出没区:2 512 大型动物少,荒凉区:16 室内:381 室外:303	陈皓文等 (1992,1997,1999,2000a,2001)
	南设得兰群岛:海洋性微生物:0.776 陆源性微生物:5.56	Chen Haowen et al (1994) 陈皓文等(1992a)
	麦克斯韦尔湾:海洋性微生物:1.53 陆源性微生物:6.45	陈皓文等(1992a)
	长城站:野外海洋性微生物:148.7 野外陆源性微生物:17.9 室外陆源性微生物:211.1 室内—室外衔接部:陆源性细菌:425.6 室内衔接部:陆源性细菌:645.9	陈皓文等(1992b) Chen Haowen (1998)
	中山站:室内细菌:137.34~516.09,无溶血性链球菌 室外细菌:16.8	于永中等(1998)
	青藏高原:西藏——13611.9 青海——20691.3	陈皓文(2002,2003)
海水和沉积物生物体	菲尔德斯半岛南部近海表层水粪大肠菌:20个・dm^{-3}左右;还有石油烃降解菌 南大洋(南极半岛与南设得兰群岛间) 表水:—HB(异养细菌,下同)310个/L(范围:20~10000)— —yeast 8.5个/L(范围:0.3~79) —丝状菌物:1.5 CFU・m^{-3},范围:0~5.3 底质:表层:—HB1.6×10^3个・g^{-1}, —yeast 0.85×10^2个・g^{-1}, —丝状菌物:2.1×10^2个・g^{-1}, 表层下:各类菌均少得多	黄凤鹏等(1992a,1992b) 袁俊峰等(1998) Ye Dezan et al. (1987,1989) 张建中等(1989) 林燕顺等(1998)
	生物体微生物含量排序:苔藓>海藻>地衣 鲸>企鹅>蛤>沙蚕>贝>鱼	陈皓文等(1998)

(续表)

样本类别	主要地域及微生物含量	文　献
北极	沉积物硫酸盐还原菌：1.68×10^5 ge·g^{-1}（$0\sim2.28\times10^5$ 个·g^{-1}） 好气异养菌：水——2.22×10^2·cm^{-1} 沉积物——$3\ 166.3\times10^2$·g^{-1} 白令海和楚科奇海水：细菌 $10-10^5$ cells·cm^{-3} 等，菌物 $0-10^4$ cells·cm^{-3} 白令海蓝细菌：水——$0\sim7.93\times10^6$·dm^{-3}	陈皓文等(2000b) 陈皓文等(2003) 陈波等(2003)，He Jianfeng et al. (2004) 肖天等(2001)
海　滨	土壤：固氮细菌 10^5-10^7 个·g^{-1} 耐 Cr^{6+}、Cd^{2+}、Hg^{3+}、Fe^{3+}、LAS 的菌中以耐 Cr^{6+} 菌最多，达 8.93×10^3 个·g^{-1}	肖昌松等(1994) 陈皓文等(2001) 陈皓文等(1998)

表4　极地微生物生物活性物的开发研究一览表

Table 4　Exploration results of bioactive substances of microbes from polar regions

微生物类别	生物活性产物	文献
真菌	新核苷转运抑制剂	粟俭等(1994)
南极土壤放线菌	抗肿瘤抗生素 C_{3905} A	王泽君等(1999)，胡继兰等(2000)
南极假丝酵母	生物表面活性剂	华兆哲等(2000)
南极中山站 G^- 耐冷细菌	胞外酸性蛋白酶	曾胤新等(2001)
南极土壤嗜冷金孢霉 *Chrysosporium* sp. C_{3438}	C_{3438} A，抗枯草杆菌	鲁敏等(2002)
北极海水细菌 Ar/W/b/75°N/10/5	碱性蛋白酶	曾胤新等(2000)，Zeng Yinxin et al(2001)
北极海冰细菌	胞外蛋白酶、脂质水解酶、抗肿瘤、抗精原细胞	曾胤新等(2003)
南极乔治王岛丝状菌物、链霉菌、诺卡氏菌等	纤维素酶	胡继兰等(1993)
低温假单胞菌	低温脂肪酶	俞勇等(2003)
极地微生物	海藻糖合成酶	徐骥等(2004)
北极浮游细菌	Protests	He Jianfeng et al (2004)
极区海水低温细菌	蛋白酶、淀粉酶、琼脂酶、纤维素酶	曾胤新等(2002)
北极细菌	抗肿瘤的活性物	曾润颖等(2002)

5　中国极地微生物考察研究的未来

国际南极微生物考察始于1901年，直至20世纪50年代后才有了蓬勃发展。极地微生物学是研究极地这一特定时空环境中微生物状况并使之服务于人类的科学。回顾中国20余年来的极地微生物学考察历程，虽然已取得了相当丰富的成果。但成果仍缺乏系统性、完整性，同时在深度和广度上也相当有限。这是由各种条件的限制所造成的。国际极地微生物考察研究在经历探险、观察、分类描述、获取和保存样品、起源、系统和生理学之后，已开始转入起源、系统、生理及生物学适应机制及其他生态系统、能流、物流中的作用、地位研究。一些成果涉及微生物适应性机理的分子水平的探讨，以及活性物质的应用前景研究，并趋向作为一个分支而参与多学科、综合性的跨地区、跨学科的考察研究。因此摆在中国学者面前的任务也必须高起点、高标准。就是要高瞻远瞩，有一个系统化的全面综合考虑。投入足量的经费、配备先进的仪器设施，吸引高素质人才，并积极参与国际项目的调查研究，如冰盖、干谷、内陆山脉、极地荒漠、冻土带、深海、环境等领域。尽力争取更多机会参与实地考察，为采集各种样品提供保障，以取得符合微生物学研究要求的第一手样品，为后续的精心研究提供支持。扩大研究内涵，并不断深化。从微生物的鉴定、生态环境、微生物与环境的相互作用，如在元素迁移循环、食物链/网/环、生态系统中的作用，治疗人体疾病及生物活性物质的开发利用等等，持之以恒地研究下去。打好坚实的基础，注重原创性研究.尽可能高水准地科学认知、开发和利用好极地微生物这一宝贵资源，真正造福于全人类，并使之永久持续健康发展。

参考文献63篇(略)

(合作者:高爱国　战　闰)

PROGRESS ON INVESTIGATION AND RESEARCH OF POLAR MICROBIOLOGY IN CHINA

(ABSTRACT)

Abstract Investigation and research organized and constructed on polar microbiology of China own accord began at the early of the eighties of the last century. During about twenty years till now, the plentiful and substantial achievement of the research on polar microbiology was obtained by Chinese scientists. The areas surveyed included the locations of great wall station and Zhongshan station, and their surrounding countries and air, water, the waters of the southern ocean, Antarctic region, Bering Sea and Chukchi Sea etc, the Arctic Ocean, the living and nonliving materials and Qinghai, Tibet plateau. The samplings were studied and analyzed extensively and more deeply. On the kinds and groups, content, biodiversity, characteristics of growth, specific physiological and biochemical properties, and phylogeny development. The significant discovery was progressed on some of polar microbial bioactive substances. Lastly, this paper looks forward to the future on Chinese polar microbiological research.

Key words Polar Regions, China, Microbes, Biodiversity, Bioactive Substances.

附录 1

论文编年史*

1978 年

1. 海洋环境中石油的微生物降解(译),应用微生物,1978,No. 4:70-75。

1979 年

2. 海洋微生物降解油污染的若干生态学问题,环境科学,1979,1(2):71-76。

1980 年

3. 海洋中维生素 B12 供需变化及其生态意义,海洋科学,1980,4(3):51-54。

1981 年

4. 胶州湾潮间带和沿岸区的耐铬菌,环境科学学报,1981,1(4):313-323。

5. 促进石油的生物降解(译),海洋环境保护,1981,No. 4:32-46。

1982 年

6. 渤海湾表层海水和沉积物中石油利用菌的数量分布,中国科学院环科委,环境科学学术报告会论文摘要,1982,P142～143(第二作者,合作者另有 5 人:张景镛、李士荣、边嘉敏、李士安、谭步云)。

1983 年

7. 胶州湾潮间带和沿岸区的耐汞菌,Ⅰ. 耐汞菌的数量分布,环境科学,1983,4(1):15-20。

1984 年

8. 微生物富集铀,工业微生物,1984,No. 5:44-47。

9. 胶州湾潮间带和沿岸区的耐汞菌,Ⅱ. 耐汞菌的耐汞力,海洋通报,1984,3(5):70-74。

1985 年

10. 海洋细菌计数和生物量测定方法述评,1985 全国第一届微生物生态学术会议论文集,昆明。

11. 鉴定海洋细菌的分类图解(译),海洋译丛,1985,No. 1:177-179。

12. 渤海湾耐重金属菌数量分布的初步分析,海洋通报,1985,4(6):27-33。

1986 年

13. 一些海藻对黄曲霉毒素累积的抑制活性(译),应用微生物,1986,No. 3:57-58。

14. 潮间带耐铅菌时空变化,中国环境科学,1986,6(6):30-35。

---

* 除注明者外,均为第一作者或个人著述。

15. 铅污染与微生物,环境科学,1986,7(6):87-89。

1989 年

16. 潮间带耐铅菌与铅等重金属、有机质关系初析,生态学报,1989,9(3):280-282。

1990 年

17. Microbes in the surface water of the shoaling areas of the Antarctic krill and their ecological analysis, Collected Oceanic works, 1990, 13(2): 85-99(与孙修勤、宋庆云、张进兴、卢颖合作)。

1991 年

18. Scientific Research Activities in Marine Microbiology, 1991。

1992 年

19. 南设得兰群岛区域的微生物,南极研究,1992,4(4):1-6。

20. 南极长城站微生物考察研究,南极研究,1992,4(4):7-13。

1993 年

21. 国际五城市空气微生物概况,黄渤海海洋,1993,11(1):50-57(与宋庆云合作)。

22. 越冬亲虾及其环境的细菌学研究,黄渤海海洋,1993,11(2):38-46(与袁春艳、王伟合作)。

23. 南设得兰群岛海域磷虾集群区表层水微生物及其生态分析,中国第三次南大洋考察暨环球重力调查报告,1993,海洋出版社,北京,P93~102(与孙修勤,宋庆云合作)。

24. 南极磷虾的微生物,中国第三次南大洋考察暨环球重力调查报告,1993,海洋出版社,北京,103-107(与孙修勤,宋庆云合作)。

25. Summary Brief on Cytobiological structure and classification of shrimp, 1993.

1994 年

26. 中国"极地"号船的空气微生物,黄渤海海洋,1994,12(1):52-59(与李振富、陈国纲合作)。

27. Microbes in the Area of the South Shetland Islands, Antarc. Res. 1994, 5(2): 1-9 (与宋庆云合作)。

28. 环球空气微生物考察,中国环境科学,1994,14(2):112-118(与孙修勤、张进兴合作)。

29.《再赴南极之行(纪实文学连载 1~15)》,青岛科技报,青岛,1994-1995。

1995 年

30. 外源凝集素——水产动物御敌的有力兵器,黄渤海海洋,1995,13(1):50-58(与孙丕喜、宋庆云合作)。

31. 客机内空气微生物对人体的影响初探,环境与发展,1995,10(2):39-40。

32. 青岛城区室外空气微生物数量的测定,上海环境科学,1995,14(2):31-32。

33. 十三城市空气微生物含量研究,环境与发展,1995,10(4):5-9 转 18。

34.《青岛人在南极——写在我国南极长城站建站十周年之前》,青岛科技报,1995.2.23,第 8 期 4 版。

1996 年

35. 庐山旅游区空气微生物污染调查,环境监测管理与技术,1996,8(2):21-23。

36. 珠江三角洲及珠江口海区的空气微生物含量，热带海洋，1996，15(3)：76-80。

37. 空气微生物对铜陵市环境质量的指示作用，长江流域资源与环境，1996，5(3)：259-261。

38. 南极特定区域污染及其影响，环境与发展，1996，11(4)：35-37，(与战闰合作)。

39. 中国南极长城湾及附及海区 1993/1994 夏季表层水样无机氮、磷酸盐的浓度及其分布状况，海洋湖沼通报，1996，No. 4：13-19(与朱明远、洪旭光、袁峻峰、陸华合作)。

40. 长江客轮空气微生物含量的测定，中国公共卫生学报，1996，15(增刊)：19-20。

1997 年

41. "向阳红 09 号"调查船的空气微生物，黄渤海海洋，1997，15(2)：53-58(与王波合作)。

42. 南极菲尔德斯半岛空气微生物含量初报，应用与环境生物学报，1997，3(2)：180-183。

43. 中国南极长城湾 1994 年 1～2 月间海水初级生产初析，植物生态学报，1997，21(3)：285-289(与朱明远、洪旭光、袁峻峰合作)。

44. 中国的来克丁研究，广西科学，1997，4(3)：161-169(与孙丕喜合作)。

1998 年

45. 从来克丁物质与微生物相互作用看水产无脊椎动物对疾病的防御活动，黄渤海海洋，1998，16(1)：55-72(与孙丕喜合作)。

46. 北海市空气微生物含量的时空分布，广西科学，1998，5(2)：83-86。

47. 空气微生物粒子沉降量指示兰州空气质量，内蒙古环境保护，1998，10(2)：11-13。

48. 南极麦克斯韦尔海湾阴离子洗涤剂污染研究，中国环境科学，1998，18(2)：151-153(第二作者，与袁峻峰、吴宝铃、李永祺合作)。

49. 天水市空气微生物含量分析，干旱环境监测，1998，12(3)：152-154。

50. 南极麦克斯尔湾及邻近海域高锰酸钾指数研究，极地研究，1998，10(3)：212-216(第二作者，与袁峻峰，李永祺、吴宝铃合作)。

51. 东海、黄海测区的空气微生物研究，东海海洋，1998，16(3)：33-39(与王波合作)。

52. 长城湾夏季初级生产力及其产量与产值，黄渤海海洋，1995，16(3)：52-59(第四作者与吕瑞华，朱明远，夏滨合作)。

53. 广州市区的空气微生物含量，环境监测管理与技术，1998，10(4)：14-16。

54. 南极长城湾及邻近海域的烃类氧化菌的分布及变化研究，极地研究，1998，10(4)：26-30(第三作者，与袁峻峰，李永祺、吴宝铃合作)。

55. The amount of hydrocarbon bacteria in the Great Wall Bay and its adjacent area, Chin. J. Pol. Sci., 1998，9(1)：66-70，(第三作者，与袁俊峰，李永祺，吴宝铃合作)。

56. 南极菲尔德斯半岛生物体的微生物含量分析，极地研究，1998，10(4)：288-293(与袁峻峰、张树荣合作)。

57. 海洋硫酸盐还原菌及其活动的经济重要性，黄渤海海洋，1998，16(4)：64-74(与战闰合作)。

58. Concentration and distribution of Inorganic Nutrient salts of Nitrogen and phosphorus in surface water of Chinese Great Wall Bay and its adjacent sea area, Antarctica,

during Austral summer of 1993/1994, J. Glaciol. Geocryol., 1998, 20(4):306-311(与朱明远、洪旭光、陸华、袁峻峰合作)。

59. A primary analysis on concentration and distribution of chlorophyll-a in surface water of Great Wall Bay and its adjacent area during Austral summer 1993/1994, J. Glaciol. Geocryol., 1998, 20(4):312-315(与朱明远、洪旭光、吴宝铃合作)。

60. Estimation of concentration of heterotrophic bacteria in surface water of Great Wall Bay and its adjacent waters, Antarctica and preliminary analysis of some relevant factors. J. Glaciol. Geocryol., 1998, 20(4): 316-319(与朱明远、袁峻峰合作)。

61. Monitoring the content of airborne microbes over the Chinese Great Wall station, Antarctica, J. Glaciol. Geocryol., 1998, 20(4): 320-325。

62. 海南岛及其毗邻海区的空气微生物浓度，海南大学学报(自然科学版)，1998，16(4):345-350。

63. 渤海湾大气微生物粒子沉降量的研究，海洋环境科学，1998，17(4):27-31。

1999 年

64. 空气微生物含量对敦煌空气质量的意义，《中国卫生检验学杂志》，第三届全国空气微生物暨医院感染学术研讨特刊，1999，46-48。

65. 北部湾北海——临高海区的空气微生物，热带海洋，1998，18(3)100-10416(4):345-350。

66. 中国南极长城站室内空气微生物含量七年前后的比较，海洋湖沼通报，1999，No. 4:48-52。

67. 济南、泰安、曲阜空气微生物监测，环境与开发，1999，14(4):43-45。

68. 毛蚶(*Arca subcremate*)体液来克丁的凝集作用，黄渤海海洋，1999，17(4):60-65(与孙丕喜合作)。

69. 南极长城湾夏季叶绿素 a 变化的研究，南极研究，1999，1(2):113-120(第四作者，与朱明远、李葆华、黄凤鹏合作)。

70. 空气微生物粒子沉降量指示太原空气污染状况，内蒙古环境保护，1999，11(4):9-10。

71. 北部湾东侧沉积物硫酸盐还原菌含量及其意义，广西科学院学报，1999，15(3):103-107(与徐家声合作)。

72. 胶州湾潮间带和沿岸区硫酸盐还原菌含量分布，海洋环境科学，1999，18(2):27-30。

2000 年

73. 中国南极长城站室内空气微生物状况，应用与环境生物学报，2000，6(1):90-92(与洪旭光合作)。

74. 微生物凝集素的研究，《自然杂志》，2000，22(2):95-100(与牟敦彩、李光友合作)。

75. 微生物来克丁分析方法，青岛科学，2000，2(3):32-38(与李光友合作)。

76. 海南岛西南侧海域底质硫酸盐还原菌分析，海岸工程，2000，19(2):10-15(与徐家声、战闰合作)。

77. 浙江—闽北陆架沉积物硫酸盐还原菌及与生物地球化学因子关系的分析，环境科学学报,2000,20(4):478-482(与李培英、王波合作)。

78.南海沉积物石油开发的微生物调查。

79.白令海和楚科奇海测区沉积物硫酸盐还原菌初步研究,极地研究,2000,12(3):211-218(与高爱国合作)。

80.宝鸡市空气微生物粒子沉降量及与铜陵市的比较,宝鸡文理学院学报,2000,20(4):292-294。

81.海洋维生素B12产生菌培养条件研究,中国海洋药物,2000,19(6):9-13(与刘秀云、高月华合作)。

2001年

82.大力开展海洋无脊椎动物防卫素研究,2001,中国生物化学与分子生物学会第八届会员大会暨全国学术会议论文摘要集。

83.桑沟湾细菌的研究,海岸工程,2001,20(1):72-80。

84.南极石油烃污染的自然消除过程研究,首都师范大学学报(自然科学版),2001,22(3):44-47(与袁峻峰、李永祺合作,本人为第三作者)。

85.南极菲尔德斯半岛环境微生物含量估计,黄渤海海洋,2001,19(1):49-54(与袁峻峰、曹俊杰、张树荣合作)。

86.桑沟湾表层水细菌与生态环境因子的关系,海洋环境科学,2001,20(3):29-33。

87.西宁、青海湖、格尔木的空气微生物含量,国土资源科技管理,2002,19(1):62-66。

2002年

88.发展和应用分子生物学技术,开发未知海洋细菌,自然杂志,2002,24(3):129-134(与孙修勤合作)。

89.西部北太平洋测区的空气微生物,中国当代思想宝库,马骏如主编,2002,中国经济出版社,北京,P644-647。

90.四种海产双壳贝类血淋巴中抗菌物质的诱导及其活性测定,海洋科学,2002,26(8):5-8(第四作者,与魏玉西、郭道森、李丽合作)。

2003年

91.青岛空气微生物状况的测定,山东科学,2003,16(1):9-13。

92.乌鲁木齐空气微生物含量的测定,新疆环境保护,2003,25(1):17-20。

93.张家界、韶山和衡山空气微生物粒子沉降量分析,国土与自然资源研究,2003,No.2:54-56。

94.贻贝防卫素的研究进展,广西科学,2003,10(2):129-134(与魏玉西、郭道森合作)。

95.菲律宾蛤仔(*Ruditapes philippinarum*)血淋巴中抗菌活性物质的活性影响因素,应用与环境生物学报,2003,9(3):271-272(第四作者,与魏玉西、郭道森、李丽合作)。

96.拉萨、日喀则空气微生物含量状况,青海环境,2003,13(3):97-99。

97.白令海和楚科奇海总好气异养细菌丰度和对温度的适应性,极地研究,2003,15(3):171-176(与高爱国、孙海青、矫玉田合作)。

98.防卫素——先天免疫的一种中枢组分,自然杂志,2003,25(3):129-135(与洪旭光、张进兴合作)。

99.天山天池与喀纳斯湖测区空气微生物含量的测定与比较,国土资源科技管理,

2003,20(3):52-54。

100. 吐鲁番市空气微生物浓度状况,干旱环境监测,2003,17(4):211-214。

101. 黄海测区空气微生物含量估计,海洋湖沼通报,2003,No. 4:55-60。

102. 北极沉积物中硫酸盐还原菌与生物地球化学要素的相关分析,环境科学学报,2003,23(5):619-624(与高爱国、孙海青合作)。

103. 鲎肽素(tachyplesins)、tachystatins 和 polyphemusins,生命科学,2003,15(5):304-306。

104. 襄樊、荆门、宜昌空气微生物含量状况,卫生软科学,2003,17(6B):149-152。

105. 菲律宾蛤仔(*Ruditapes philippinesis*)血浆中防卫素的纯化及其抑菌功能,生物化学与生物物理学报,2003,35(12):1145-1148(第四作者,与魏玉西、郭道森、李荣贵、陈培勋合作)。

106. 中国海产养殖生物病原微生物研究进展,中国水产,2003,专刊:60-66 转 45。

2004 年

107. Abundance of general aerobic heterotrophic bacteria in the Bering sea and Chukchi sea and their adaptation to temperature. Chinese Journal of Polar science, 2004, 15(1):39-46(与高爱国、孙海青、矫玉田合作)。

108. 对虾一帖的研究进展,海洋科学,2004,28(2):75-76。

109. 中国的防卫素研究进展状况分析,自然杂志,2004,26(5):264-268(与孙丕喜、战闰合作)。

110. Marine microbiological research in China (1950-1990), Ocean sciences Bridging the millennia Aspectrum of historical accounts, Edi. S. Morcos et. al Pub. spons SOA, IOC, FIO, IOI, 2004, P399-405。

111. 中国普通海洋微生物学的研究进展,中国水产,2004,专刊:253-257(与孙丕喜合作)。

112. 养殖鲍的疾病与敌害,中国水产,2004. 专刊:221-226(与王波合作)。

113. 海洋微生物,人类取之难竭的资源宝库,2004.

114. 南京空气微生物含量状况的研究,中华名人文论大全Ⅰ,中国文联出版社,北京,2004,P1393-1395。

115. 乌鲁木齐—布尔津间 216-217 国道沿线空气微生物含量,刊于《当代杰出管理专家人才名典Ⅱ》,P180-183,长征出版社,北京,2004。

116. 东海、黄海及渤海测区维生素 B12 产生菌生态学研究,刊于《中国改革发展理论文集》,P416-420,中国文艺出版社,北京,2004(与刘秀云、高月华合作)。

117. 回顾二十世纪 70 年代世界海洋微生物学研究,刊于《共和国建设者》,P660-664,中国文联出版社,北京,2004。

2005 年

118. 珠母贝的疾害,水产养殖,2005,26(1):34-37。

119. 蓝老,随风飘去,中国海洋报,2005. 3. 11 第四版(与徐家声合作)。

120. 中国海洋病毒研究进展,齐鲁渔业,2005,22(4):11-14。

121. 扇贝的疾病,载于《学习和实践"三个代表"的重要思想》,P325-326,亚洲传媒出版社,北京,2005。

122. 中国空气微生物学的研究进展,疾病控制杂志,2005,9(6):627-629(与陈阳合作)。

123. 无脊椎动物高血糖肽激素和速激肽相关肽,海洋科学,2005,29(9):83-87,(与刘洪展合作)。

124. 生命的三界学说以及生命的第三形式,生物学通报,2005,40(9):19-20(第三作者,与刘洪展、郑凤荣合作)。

125. 中国极地微生物学调查研究进展,极地研究,2005,17(4):299-307(与高爱国、战闰合作)。

126. 青岛十大山头空气大肠菌含量分析,创新与发展,2005,总 No. 67,10-12(与陈阳合作)。

127. 一位老者,静静地逝去了,南大校友通讯(冬季号 No. 31):2005,51-52(与徐家声合作)

2006 年

128. 巴塞尔(Basel)空气微生物含量状况,中国现代化理论成果汇编,中国文史出版社,北京,P724-728,2006。

129. 青岛夏冬空气副溶血性弧菌含量分析,世界优秀学术论文(成果)文献,中文版(下),世界文献出版社,北京,P64-66,2006。

130. 大同市空气中肠炎致病菌含量调查,刊于中国延安文艺学会编的《永葆共产党人先进本色论文集》,人民日报出版社,北京,P730-731,2006。

131. 呼和浩特空气胃肠炎病原菌含量的初步分析,《世界重大学术成果精华》,华人卷Ⅱ(2006)(与孙丕喜合作)。

132. 胶州湾表层海水石油烃含量状况与评价(与孙丕喜等合作)。

133. 延安空气微生物含量状况并与韶山北海比较,创新与发展,2006,总 No. 30:86-89。

134. 青岛空气中胃肠炎病原菌含量的分析,国土自然资源研究,2006,No. 1:51-52。

135. 日内瓦空气微生物含量,刊于《中国学术大百科全书》学术卷Ⅱ,中国文化传媒出版社,北京,P954-956,2006。

136. 古都洛阳与中国内陆三城市空气微生物的含量比较,《创新中国探索集》,中国文史出版社,北京,2006,P410-413。

137. 瑞士一月,2006.

2007 年

138. 加拿大海盆与楚科奇海表层沉积物中硫酸还原菌丰度分布状况,海洋科学进展,2007,25(3):295-301(与高爱国、林学政合作)。

139. 楚科奇海和加拿大海盆表层沉积物中好气异养细菌的地理分布,极地研究,2007,19(3):231-238(与高爱国合作)。

140. 现代海洋微生物学之父——佐贝尔(ZoBell C. E.)的早期成就——纪念佐贝尔诞辰 102 周年,海洋地质动态,2007,23(5):38-41。

141. 牡蛎的微生物疾病,水产科学,2007,26(9):531-534(与陈阳合作)。

142. 中国海洋微生物学2006年研究足迹(与孙丕喜合作),2007。

143. 牡蛎的寄生虫病,河北渔业,2007,No.10:6-11转22(与孙丕喜合作)。

144. 洛桑空气微生物检测,刊于《世界重大学术成果精选——华人卷Ⅲ珍藏版》,世界文献出版社,2007,北京,P73-75(与李为民、陈颖稚合作)。

145. 我以研究海洋为荣,海洋研究以我为荣,2007.

146. Geographical distribution of general aerobic heterotrophic bacteria in surficial sediments from the Chukchi sea and Canadian Basin. Chinese Journal of Polar Science, 2007, 18(2): 147-154.(合作者:高爱国)

147. 异域边境赏别样风光,半岛都市报,2007.3.7,B51版。

2008年

148. 养殖牡蛎的敌害,河北渔业,2008,No.1:31-32转42(与王波合作),《广西水产科技》转载,2008,No.2:42-45.

149. 加拿大海盆与楚科奇海柱状沉积物中硫酸盐还原菌的分布状况,环境科学学报,2008,28(5):1014-1020(第二作者,与高爱国,林学政合作)。

150. 2007年的中国海洋微生物学研究状况浅析(与孙丕喜,王宗兴合作)。2008。

151. 加拿大海盆与楚科奇海沉积物岩芯中好气异养细菌的分布,极地研究,2008,20(1):40-47(与高爱国合作)。

152. 铁细菌在北极特定海区沉积物中的分布,海洋科学进展,2008,26(3):326-333(第二作者,与高爱国合作)。

153. 锰细菌在加拿大海盆、楚科奇海沉积物中的分布,环境科学学报,2008,28(11):2369-2374(第二作者,与高爱国合作)。

154. 漫谈海洋地微生物学,海洋地质动态,2008,24(3):14-17(高爱国为第二作者)。

155. 伯尔尼(Berne)空气微生物含量状况,刊于《中国党政干部理论文选》,P.461-465。中共中央党校出版社,北京,2008。

156. 北极海洋沉积物中锰细菌的分离与系统发育,生态学报,2008,28(12):6364-6370(第三作者,与林学政、高爱国合作)。

157. Isolation and phylogenetic analysis of cultivable manganese bacteria in sediments from the Arctic ocean(第三作者,与林学政、高爱国合作).

158. 灵山岛空气微生物含量的测定,海岸工程,2008,27(4):37-41(第二作者,与孙丕喜、黄秀兰合作)。

159. Distribution of general aerobic heterotrophic bacteria in sediment core taken from the Canadian basin and the Chukchi sea. Chin. J. Pol. Sci., 2008, 19(1): 14-22(第二作者,与高爱国合作)。

2009年

160. 银川空气微生物含量状况分析,国土与自然资源研究,2009,No.2:54-55(第二作者,与孙丕喜、于明哲合作)。

161. 微生物学向深海进军,海洋地质动态,2009,25(2):14-20(孙丕喜,高爱国为第二、三作者)。

162. 海洋的古菌,自然杂志,2009,31(2):96-99(第二作者,与孙丕喜、王波合作)。

163. 我国深海地微生物学研究状况分析，海洋地质动态，2009，25(7)：20-25 转 41(与陈颖稚合作)。《海洋地质前沿》转载。

164. 青岛的发展靠人才，载于《中国科教兴国战略研究文库》，中共中央党校出版社，2009，北京，P. 147-149(与陈颖稚合作)，《中国科协年会论文集(四)转载》。

165. 西北冰洋海域表层沉积物中厌氧菌分布特征，环境科学学报，2009，29(10)：2209-2214(第二作者，与高爱国、赵冬梅合作)。

166. 北冰洋沉积物岩芯中厌氧细菌分布，海洋科学进展，2009，27(4)：469-476(第二作者，与高爱国、赵冬梅合作)。

167. 美利坚三季，2009

168. 胶州湾表层海水石油烃含量状况与评价。(第二作者，与孙丕喜、战闰合作)。2009。

2010 年

169. 瑞士四城市空气微生物考察，科技创新导报，2010，No. 15，148 转 150(第四作者，与陈颖稚、李为民、黄秀兰合作)。

170. 原核生物概况及海洋原核生物组成，广西科学院学报，2010，26(2)：140-142 转 145(第四作者，与孙丕喜、陈颖稚、王宗兴合作)。

171. 2008 年中国海洋微生物学研究概况，管理观察(学术论坛版)，2010 年 11 月：143-146(与孙丕喜合作)。

2011 年

172. 中国海洋微生物学 2009 年研究进展，广西科学院学报，2011，27(4)：341-347(第二作者，与郑凤荣合作)。

173. 包围地球的大气圈及室内外空气微生物，2011。

2012 年

174. 南大精神，永在心间，长驻青岛，发扬光大。2012。

175. 桑沟湾大肠菌和粪大肠菌群状况，2012(第二作者，与孙丕喜等合作)。

176. 2010 年中国海洋微生物学研究成就，广西科学院学报，2012，28(4)：277-287 转 297(第三作者，与孙丕喜，郑凤荣合作)。

177. 南大—青岛友情源远流长，2012。

178. 旅迹，2013。

179. 2012 年中国海洋微生物学成果初析。

附录 2

参与的专著和技术报告

1.《南大洋考察报告》中的《南大洋磷虾集群区表层中微生物及其生态分析》,1987,海洋出版社,北京。

2.国家南极考察委员会编《中国第三次南大洋考察暨环球重力调查报告》中第 4 章《测区的微生物与浮游植物》微生物部分的写作,海洋出版社,北京,1992,P93-107。

3.朱明远等主编的《石油开发废水对对虾食用价值的影响》,海洋出版社,北京,1993。

4.武衡等主编《当代中国的南极考察事业》第 16 章《南大洋生物学》微生物部分的写作,当代中国出版社,北京,1994,P375-377。

5.程波主编《养殖环境中的生态因子》中第 3 章《养殖池环境中的生物因子》微生物部分的写作,海洋出版社,北京,1994,P145-157。

6.《中国当代地球科学家大辞典》,气象出版社,北京,1994,P29。

7.徐家声主编《北海—海口光缆通信干线工程.北海—临高段海洋路由勘察报告》,1997。

8.参与国家海洋局极地考察办公室编,《南极菲尔德斯半岛及其附近地区生态系统的研究》·《中国南极考察科学研究》·《成果与进展》,海洋出版社,北京,1998,P63-139。

9.龙江平主编《芦洋跨海大桥海洋环境防腐研究,第一阶段认可阶段研究报告》,1999。

10.《吴宝铃文选》,编辑出版组第二成员,1999。

11.于玺华主编《现代空气微生物学》第 21 章《空气微生物含量对敦煌空气质量的意义》的写作,人民军医出版社,北京,2002,P393-395。

12.参与陈立奇等编著《北极海洋环境与海气相互作用研究》的《北冰洋沉积物组成与古环境》一章的写作,海洋出版社,北京,2003,P181-225。

13.参与中国海洋学会编的《中国海洋学会学术期刊优秀论文精品集》中《铁细菌在北极特定海区沉积物中的分布》的写作,海洋出版社,北京,2009,P.613-619。

14.《海洋原核生物名称》,主编之一,科学出版社,北京,2009,P248。

15.《来克丁学导论》,第一作者,海洋出版社,北京,2011,P.285.